CW00557366

BIOLOGY

BIOLOGY

EDITORS

BRUCE KNOX

PAULINE LADIGES

BARBARA EVANS

McGRAW-HILL BOOK COMPANY Sydney

New York   San Francisco   Auckland   Bogotá
Caracas   Lisbon   London   Madrid   Mexico City
Milan   Montreal   New Delhi   San Juan
Singapore   Tokyo   Toronto

Cover photo This Pacific anemone fish, *Amphiprion perideraion*, is immune to the stinging hydra of the anemone's tentacles, and so can live within the tentacles unharmed. The fish benefits from the shelter afforded by the anemone, and the anemone benefits from food it receives from the fish. Anemone fish are also known to pick off micro-organisms and slime from the tentacles, and sometimes to eat the wastes of certain anemones.

National Library of Australia Cataloguing-in-Publication data:
   Biology.

   Includes index.
   ISBN 0 07 452757 6.

   1. Biology. I. Knox, R. Bruce. II. Ladiges, Pauline Y. III. Evans, Barbara K.

574

Published in Australia by
McGraw-Hill Book Company Australia Pty Limited
4 Barcoo Street, Roseville NSW 2069, Australia
Typeset in Australia by Midland Typesetters, Victoria
Printed in Singapore by Toppan Printing Company Pty Ltd

Publisher: Jane Mackarell
Sponsoring Editor: Rebecca Browning
Production Editor: Carolyn Pike
Designer: Wing Ping Tong
Computer Illustrators: Diane Booth, Lorenzo Lucia
Illustrators: Mike Gorman, Terryn Hough, Christine Turnbull
Permissions Editors: Kerstin Broden, Karin Riederer
Indexer: Michael Wyatt

FOREWORD

It is a great privilege to introduce this truly Australian textbook of biology for tertiary level students. It is a curious anomaly that Australian universities, until now, have used textbooks written in a Northern Hemisphere context for teaching advanced biology. These texts have usually been supplemented with material from the literature to provide an understanding of the unique Australian landscape, our unique biology and, of course, our unique problems. Researchers in Australia have made enormous efforts to understand our own biology. As a result, we contribute about 4% of the world's new knowledge in the field of biology, whereas our contribution overall in all fields of knowledge is about 2%. This level of 2% is about right for our size. The fact that our contribution in biology is twice that figure reflects the need for us to understand our own situation in which *our biology* plays a key role. It also provides us with a strong base on which Australia's emerging biotechnology-based industries are built. Indeed, we are now entering a golden age of biology. The rapid advances in knowledge brought about by the techniques of molecular genetics, coupled with the pressing ecological problems, global climate change, biodiversity and the need to feed and care for a rapidly increasing global population, bring the critical importance of biology as a discipline into sharp focus.

Our researchers who create new knowledge in these fields are experts acknowledged worldwide for their contributions. Now, for the first time, these researchers have worked together to produce a truly Australian advanced biology text. It provides Australian students with a current view of biology built around knowledge of Australian organisms.

We are very fortunate in Australia to have some of the last great wilderness areas on earth. These are places of wild beauty in which we can refresh our spirits, as the original inhabitants of this continent did for thousands of years. Many young Australians enjoy recreations such as bushwalking, birdwatching, surfing and fishing, which bring them into direct contact with our biology in action. Many are inspired by the experience to study biology in the Australian context. This book, written by experts from around Australia, will give them a modern text for guidance and inspiration.

Professor Adrienne E. Clarke, AO, FTS, FAA

BRIEF CONTENTS

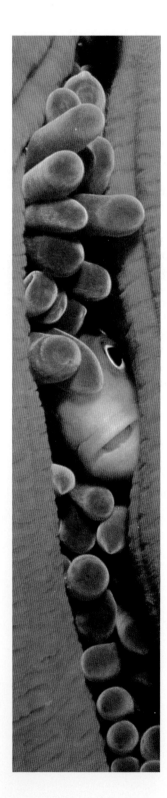

DETAILED CONTENTS

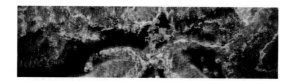

PART 4   REGULATION OF THE INTERNAL ENVIRONMENT ▪ 351

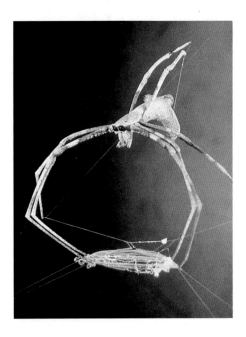

PART 6   EVOLUTION AND BIODIVERSITY ▪ 659

PART 7 ECOLOGY ▪ 935

Chapter 42 Population ecology ▪ 937

Chapter 43 Living in communities ▪ 954

Chapter 44 Ecosystems ▪ 978

ABOUT THE AUTHORS

EDITORS

Bruce Knox, FAA

Bruce Knox is a Professor in the School of Botany, The University of Melbourne. He is a distinguished researcher and Fellow of the Australian Academy of Science, specialising in the reproductive biology of flowering plants, particularly at the molecular and cellular level; he has contributed to our knowledge of how pollen affects asthma and hay fever. He is an author of scientific books and numerous papers.

Pauline Ladiges

Pauline Ladiges is a Professor and Head of the School of Botany, The University of Melbourne. Her research is in the field of plant systematics and biogeography, and she has a particular interest in the diversity, evolution and ecology of the Australian flora. She is an expert on the genus *Eucalyptus*. She has contributed to the design and teaching of university biology courses, including the development of multimedia teaching materials, and has co-authored two biology texts for secondary schools.

Barbara Evans

Barbara Evans is a Reader and Associate Professor in the Department of Zoology, The University of Melbourne. She is an expert in comparative animal physiology, specialising in cardiovascular and respiratory regulation. Her studies also include the reproductive biology of Australian animals such as the platypus and echidna. She is a renowned teacher, contributing to the design of both tertiary and secondary level courses, and is author of two successful biology texts for secondary schools in Victoria.

CONTRIBUTING AUTHORS

The editors would like to thank the following people for their contributions to part of or complete chapters of the text.

	CHAPTER
Professor Craig Atkins, The University of Western Australia	2, 5
Dr Russell Baudinette, Flinders University of South Australia	29
Dr Michael Bennett, The University of Queensland	27
Dr Prem Bhalla, The University of Melbourne	1, 2
Dr Mark Burgman, The University of Melbourne	43, 45
Dr Christa Critchley, The University of Queensland	5
Dr Rodney Devenish, Monash University	11
Dr Andrew Drinnan, The University of Melbourne	6, 15, 17, 37
Dr Mark Elgar, The University of Melbourne	28
Dr P. Finnegan, Monash University	11
Dr Paul Fisher, La Trobe University	7
Dr Bill Foley, James Cook University	19
Dr Peter Frappell, La Trobe University	20
Professor John Furness, The University of Melbourne	26
Professor Adrian Gibbs, Australian National University	34
Dr David Guest, The University of Melbourne	36
Dr Michael Guppy, The University of Western Australia	2, 5

Dr Adrienne Hardham, Australian National University — 3, 6, 8

Professor Barrie Jamieson, The University of Queensland — 38, 39, 40

Dr Peter Kershaw, Monash University — 41

Dr Jill Landsberg, CSIRO Division of Wildlife and Ecology — 44

Dr Barry Lee, The University of Melbourne — 16

Dr Ian McDonald, The University of Melbourne — 25

Dr Geoff McFadden, The University of Melbourne — 35

Dr David Macmillan, The University of Melbourne — 26

Professor Jennifer C. Marshall-Graves, La Trobe University — 12

Dr Chris Moran, The University of Sydney — 32

Dr David Morris, University of Cambridge — 4

Professor Phillip Nagley, Monash University — 11

Dr Tom Neales, The University of Melbourne — 18

Dr Gareth Nelson, American Museum of Natural History, New York — 30, 31

Tim Offer, The University of Melbourne — 45

Professor Jeremy Pickett-Heaps, The University of Melbourne — 8

Dr Hugh Possingham, University of South Australia — 42

Professor Jim Reid, The University of Tasmania — 24

Dr John Ross, The University of Tasmania — 24

Dr Gordon Sanson, Monash University — 19

Dr Roland Scollay, Centenary Institute of Cancer Medicine and Cell Biology, Sydney — 23

Dr Roger Seymour, The University of Adelaide — 21

Dr Geoffrey Shaw, The University of Melbourne — 13

Professor Andrew Smith, The University of Adelaide — 4

Dr Peter Stewart, Australian National University — 33

Dr Mark Tester, University of Cambridge — 4

Dr Steve Tyerman, Flinders University of South Australia — 29

Dr Graeme Watson, The University of Melbourne — 41

Dr Anthony Weiss, The University of Sydney — 10

Dr Paul Whitington, The University of New England — 14, 16

Dr Phil Withers, The University of Western Australia — 22

OTHER CONTRIBUTORS

Dr Marilyn Ball, Australian National University

Professor Andrew Beattie, Macquarie University

Dr Mike Borowitzka, Murdoch University

Dr Peter Crane, Field Museum of Chicago

Dr Gustav Hallegraeff, The University of Tasmania

Professor Don Metcalfe, Walter and Eliza Hall Institute of Medical Research

Dr Ebbe Nielsen, CSIRO Division of Entomology

Dr Pat Rich, Monash University

Dr George Scott, The University of Melbourne

Dr Lynne Selwood, La Trobe University

Dr David Smith, *imaginACTION films*, Melbourne

Dr Rick Wetherbee, The University of Melbourne

Dr Paul Willis, Macleay Museum, The University of Sydney

PREFACE

Biology has undergone a revolution during the past two decades following exciting developments in molecular biology, such as the isolation and sequencing of genes from humans, other animals, plants and micro-organisms. These developments have led to applications throughout biology. At the level of DNA, for example, comparison of organisms is allowing discovery of the evolutionary history of life; and genetic engineering is providing new therapies and changing some characteristics of our domestic animals and plants.

This book applies these new developments, introducing new concepts that have changed the contemporary face of biology as we approach the twenty-first century. It presents to us a view of the evolution and diversity of life at all levels of organisation, from the molecular and cellular levels to the functioning of whole organisms and ecosystems. The approach is integrative, emphasising the biological principles that relate to all organisms but, at the same time, clearly focusing on either plants or animals. This provides a flexible structure for different ways of teaching biology.

Work on the text began in 1989, when the three editors met to consider the challenge of producing an authoritative and contemporary Australasian biology text in all its facets and fascinations. Teaching of first year biology in Australian universities and tertiary institutions has always made use of Australian examples, yet the overseas textbooks available to students at this level seldom take account of Australia's unique plants and animals. Also, Australian science has advanced so that our scientists produce research discoveries of international standing and acclaim. Thus, the book has been designed to present contemporary biology in the context of Australasian examples. Each chapter relates to Australasian organisms in Australasian environments.

To ensure that the textbook is as up-to-date as possible in all aspects of biology, the editors initially invited more than 30 contributors—experts in their fields—to join them in writing the 45 chapters. These contributors came from many different universities in all states of Australia. Subsequently, more than 30 additional contributors provided further material.

Australasian biology presents many colourful subjects, as depicted throughout the book. The beauty of the Australian bush and its landscapes, animals and plants, are presented in over 750 photographs, and over 1200 illustrations represent state-of-the-art technology, depicting exciting new concepts.

The book is divided into seven parts, which encompass the diversity of life processes and forms.

Part 1: Cell biology and energetics

Part 1 introduces cell biology and energetics. Firstly, biologically important molecules and their chemistry are addressed, as an understanding of the structure and function of these molecules is an essential introduction to living systems. This is followed by chapters describing how molecules are organised into cellular membranes and organelles in both prokaryotes and eukaryotes; how molecules cross biological membranes, forming the basis of transport systems; how energy is harvested by organisms including glycolysis, the utilisation of fuel molecules, photosynthesis and the importance of light energy; and how cells form tissues, communicating with and recognising one another, and dividing and multiplying.

Part 2: Genetics and molecular biology

Part 2 begins with the principles of classical genetics, many of which were originally demonstrated experimentally by Mendel in the nineteenth century. Subsequent chapters explain the structure and organisation of genes, chromosomes and DNA, comparing the genomes of prokaryotes and eukaryotes; how genes encode RNA and proteins in cells and tissues at the molecular level; and how recombinant DNA technology has made possible the genetic engineering of animals and plants.

Part 3: Reproduction and development

Part 3 provides a comparative overview of the strategies and mechanisms of asexual and sexual reproduction and development. Using Australian animals and plants as examples, we show how reproduction enables the survival of species and how it is often dependent on environmental cues. The details of animal and plant development are considered separately in terms of how gametes and germ cells form, fertilisation and embryonic development, and the regulation of development.

Part 4: Regulation of the internal environment

How the internal environment of animals and plants is regulated is presented in Part 4. The structure and function of plants and animals are approached in a comparative and integrative way, including, for example, different patterns of heterotrophic nutrition, exchange of respiratory gases, circulatory and excretory systems, and the immune system in animals. Australian flora and fauna, and the particular environments in which they live, are emphasised.

Part 5: Responsiveness and co-ordination

Using Australian examples in Part 5, the bases of hormonal control of plants and animals is examined, followed by chapters on nervous systems, animal movement and animal behaviour. Finally, these threads are drawn together in an integrated and comparative overview of how different organisms cope with the demands of particular Australian environments.

Part 6: Evolution and biodiversity

The theme in Part 6 deals with how life and earth evolve together, presenting modern approaches to discovering the history of organisms (phylogeny) and mechanisms of evolution. The most up-to-date view of the tree of life and classification of organisms (based on molecular and morphological data) form the structural basis of chapters dealing with bacteria, viruses, protists, fungi, animals and plants. Finally, the evolution of Southern Hemisphere biotas and Australia in particular is addressed.

Part 7: Ecology

In Part 7, modern ecological theory is presented at the population, community and ecosystem levels, and the impact of humans on the biosphere is debated. The uniqueness of Australasian environments and habitats are examined, unlike any other first year biology textbook.

An innovative support package for the lecturer accompanies this book, including both a printed and computerised testbank of multiple choice questions, and the art program, along with other textual material, is available on CD-ROM for lecturer presentations.

Acknowledgments

The book has undergone an exhaustive and careful process of review. Each chapter has been reviewed by experts throughout Australia, revised and reviewed again. We would like to thank the following reviewers for all their constructive comments and suggestions.

Dr Jim Akers, University of Southern Queensland

Dr Elizabeth Alexander, Curtin University

Dr Michael Augee, University of New South Wales

Dr Robert Ballantyne, Charles Sturt University

Dr Ian Bennett, Edith Cowan University

Dr Stuart Bradley, Murdoch University

Professor Don Bradshaw, University of Western Australia

Dr Neil Brink, Flinders University of South Australia

Dr Max Cake, Murdoch University

Dr Jim Campbell, Wollongong University

Dr Geraldine Chapman, The University of Sydney

Dr Andrew Collins, University of New South Wales

Dr Ron Crowden, University of Tasmania

Dr John Dearn, Canberra University

Dr Janet Gorst, University of Tasmania

Dr David Happold, Australian National University

Dr A. Chris Hayward, University of Queensland

Dr Chris Hill, James Cook University

Dr Sidney James, University of Western Australia

Dr Jacob John, Curtin University

Dr Michael Johnson, University of Western Australia

Dr Graham Kelly, Queensland University of Technology

Dr Jim Kohen, Macquarie University

Dr William A. Loneragan, University of Western Australia

Dr Hamish McCallum, University of Queensland

Dr Peter McGee, The University of Sydney

Dr David Morrison, University of Technology, Sydney

Dr Ray Murdoch, Newcastle University

Dr Mary Peat, The University of Sydney

Dr Hugh Possingham, University of South Australia

Associate Professor Nallamilli Prakash, University of New England

Dr Rob Rippingale, Curtin University

Dr Liz Smith, Macquarie University

Associate Professor Ted Steele, Wollongong University

Dr Robert Vickery, University of New South Wales

Dr Helen Wood, Charles Sturt University

We would also like to thank the following biology departments for assisting in reviewing the art program at various stages of development: Macquarie University, Newcastle University, The University of Sydney and Wollongong University.

The ancillary package could not have been completed without the efforts of Wollongong University. Ian Tait, in conjunction with the Interactive Multimedia Unit, has worked tirelessly on the testbank and the lecturer CD-ROM.

Finally, the three editors would like to thank their families for their patience and understanding during the many weekends spent writing and editing this book, and the staff of McGraw-Hill Book Company, Sydney, for their help, especially the Publisher, Jane Mackarell. Jane has been very supportive of this project and we could not have asked for a better publisher. We are very appreciative of the help of the Production Editor, Carolyn Pike, who not only has a good sense of humour but a keen eye for detail, and of former staff Elizabeth McMahon, Rosemary Gibbs and Rebecca Browning, for their help in earlier stages.

PART 1

CELL BIOLOGY

AND

ENERGETICS

MOLECULES OF LIFE

The variety of life forms that exists on earth is immense (Fig. 1.1). Variety in size can be seen with mammals, which range from the Etruscan shrew, the size of your little fingernail, to the blue whale, more than 30 metres in length. Plants range from tiny mosses to mountain ash eucalypts standing almost 100 metres tall. Some animals fly, some swim, some burrow, some climb. The adaptations associated with such diverse activities are reflected in an almost bewildering array of body forms.

This diversity, however, masks fundamental underlying similarities, which are especially evident at the molecular level. DNA is DNA whether you are studying slime moulds, slugs or seaweeds; when a cane

(b)

(a)

(c)

Fig. 1.1 Variety of life forms (a) Mass of water lilies against a backdrop of paperbark at Jim Jim billabong, Kakadu National Park. (b) Several beetles feeding on the flowers of *Angophora* (family Myrtaceae). (c) Long-nose hawkfish feeding on young fish among coral

toad is stressed, it releases similar hormones to those released in your body when you get a fright. The chemistry of cellular processes is, likewise, strikingly similar in the vast majority of organisms. Because outwardly diverse life forms are so similar at the biochemical level, genes can be transferred from one type of organism to another. This is why genetic engineering has such widespread applications in modern biology. Indeed, it is our recognition of the commonality of life at the molecular level that has earned for biology the importance and popularity it currently enjoys.

> There is an immense diversity of living organisms but their underlying chemical processes are strikingly similar.

THE CHEMISTRY OF ORGANISMS

Atoms

The materials of the earth's crust are made up of about 90 naturally occurring elements, the most abundant of which is oxygen (O), constituting about 50% of the crust, followed by silicon (Si), constituting nearly 28%. An **element** is a substance made up of one type of atom only and an **atom** is the smallest part of an element that can exist and retain the properties of that element. An atom is made from still smaller subatomic particles—electrons, neutrons and protons. Protons are positively charged particles and, together with neutrons, which are neutral particles, form the nucleus (core) of each atom. Electrons are negatively charged particles, which are much smaller, fast-moving and orbit the atomic nucleus. In most atoms, the number of negatively charged electrons equals the number of positively charged protons, so that an atom is effectively neutral in electrical charge.

The number of protons in a nucleus is the **atomic number**, which is characteristic for each element. The **mass number** is the combined number of protons and neutrons in each nucleus. For example, in the common form of the element carbon (C), which has six protons and six neutrons, the atomic number is 6, and the mass number is 12. This atomic form (**isotope**) of carbon is known as carbon-12 ($^{12}_{6}C$), where 12 is the mass number, and 6 is the atomic number. Other isotopes of carbon, which have the same chemical properties as carbon-12, have different numbers of neutrons and thus different mass numbers: carbon-13 ($^{13}_{6}C$) has seven neutrons, and carbon-14 ($^{14}_{6}C$), rare in nature, has eight neutrons.

> Atoms consist of a central nucleus of protons (positively charged) and neutrons (neutral charge), and orbiting electrons (negatively charged).

Energy levels and excitation of electrons

Electrons in an atom are attracted to the positively charged protons in the nucleus but repelled by other electrons. They move around the nucleus in orbitals, which are zones of space in which electrons exist at any one moment. Only one or two electrons can occupy the same orbital. In an atom of hydrogen, the single electron is in the 1s (spherical) orbital, the innermost orbital to the nucleus (Fig. 1.2). This electron is at the *lowest* energy level. In atoms with more than two electrons, the electrons exist in further orbitals at greater distances from the nucleus. For example, sodium has 11 electrons orbiting the nucleus: two electrons orbit in the first energy level, the second energy level contains eight electrons (two in each of the four available orbitals) and the third energy level contains only one electron (the third level contains up to eight electrons in four orbitals; Fig. 1.3).

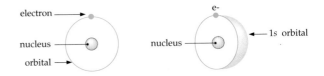

Fig. 1.2 Hydrogen atom. Model showing central nucleus of 1 proton and orbital of 1 single electron

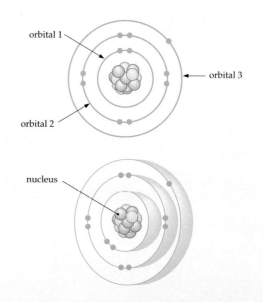

Fig. 1.3 Sodium atom. Model showing two electrons in orbital 1, eight in orbital 2 and one in orbital 3

Electrons can move from one orbital to another by gaining or losing energy in discrete amounts or quanta. Electrons can move to a higher energy level when a particular amount of energy is absorbed by an atom and to a lower energy level when the additional energy is released (Fig. 1.4). For example,

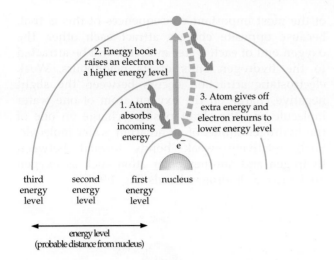

Fig. 1.4 An electron can move to a higher energy level when a particular amount of energy is absorbed by the atom or molecule (positions 1 to 2). An electron can move to a lower energy level when the additional energy undergoes controlled release from the atom (positions 2 to 3)

during photosynthesis by leaf cells in sunlight, electrons in pigments absorb quanta of light energy, which shifts the electrons to higher energy levels. Electrons do not remain in this excited state for very long but fall back to their original orbital, with the release of energy in the form of heat or a flash of light (fluorescence), or excitation of a neighbouring atom, or driving a chemical reaction such as occurs in cells.

Electrons orbit a nucleus in zones of space (orbitals). Electrons can move between orbitals of different energy levels, with the gain or loss of energy.

Molecules

If the outer level of an atom is not filled completely with electrons, two or more atoms may combine to form a stable association, a **molecule**. The chemical forces binding the atoms together in a molecule result from the pairing of electrons, the attraction of negative and positive charges, or the tendency for energy levels of the orbitals to be completely full. In gaseous oxygen (O_2), the two oxygen atoms each share two electrons with one another, thereby filling the outer orbital of each atom (Fig. 1.5). Atoms of elements with the maximum number of electrons possible in outer orbitals (e.g. helium) are stable and tend not to combine with other atoms.

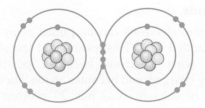

Fig. 1.5 Gaseous oxygen is composed of two atoms of oxygen, and each oxygen atom shares two electrons with its partner so that there are eight electrons in the outermost energy level

A **compound** is a molecule composed of more than one type of atom. For example, water is composed of two atoms of hydrogen bound to one atom of oxygen (H_2O). The oxygen atom shares one electron with each hydrogen atom, again filling the outer orbitals (Fig. 1.6).

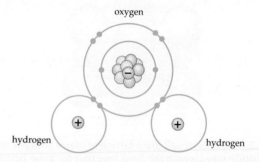

Fig. 1.6 A molecule of water is composed of two atoms of hydrogen and one atom of oxygen. The oxygen atom shares one electron with each hydrogen atom

Chemical forces in atoms can lead to the combining of two or more atoms into a stable association, a molecule.

Chemical bonds

Chemical bonds involve attraction between two atoms or molecules. There are three types of chemical bonds—covalent, ionic and hydrogen.

Covalent bonds

A **covalent bond** forms when an atom neither loses nor gains an electron but shares some of the electrons in its outermost orbital with another atom, as in a water molecule (Fig. 1.6). Single covalent bonds readily form between two atoms, for example, in a hydrogen molecule (H—H). In some cases, two atoms can share two electrons to form double covalent bonds, for example, an oxygen molecule (O=O), or three electrons to form triple covalent bonds, for example, a nitrogen molecule (N≡N).

Ionic bonds

One or more electrons can be pulled away from an atom, or added to it. This means that the atom that loses or wins electrons becomes positively or negatively charged, and becomes an **ion**. For example, an atom of calcium (Ca) has 20 protons and 20 electrons. When the atom loses two electrons, it becomes a positively charged calcium ion, Ca^{2+}.

For an ion to form, there must be another atom or molecule to act as an acceptor or donor of electrons. Both atoms become ionised as a result of this process and may remain together or separate depending on the environment. An **ionic bond** is formed when ions of opposite charge are attracted to each other. For example, common table salt, sodium chloride (NaCl), is a lattice of ions of Na^+ and Cl^-, which remain in association as a result of ionic bonding (Fig. 1.7).

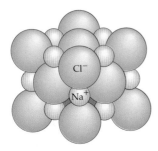

Fig. 1.7 Crystals of salt are a lattice of many sodium and chloride ions, which are held together by ionic bonds

Hydrogen bonds

A water molecule has an overall neutral charge. However, the oxygen atom strongly attracts electrons, including those of hydrogen. Thus, the oxygen end of a water molecule tends to become slightly negative with respect to the hydrogen end and the covalent bond is, in effect, polarised. This results in a dipolar V-shaped molecule (Fig. 1.8). One

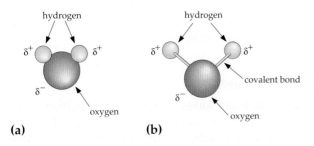

(a)   **(b)**

Fig. 1.8 In a water molecule, each hydrogen atom is linked to an oxygen atom by a covalent bond. **(a)** The space-filling model shows the physical relationships between oxygen and hydrogen atoms. **(b)** Ball-and-stick model shows the V-shape that develops due to the strong attraction of the oxygen atom for electrons, causing the oxygen atom to be slightly negative (δ^-) and the hydrogen atoms slightly positive (δ^+)

of the most important consequences of this is that, because opposite charges attract each other, the oxygen end of each water molecule will be attracted to the hydrogen ends of its neighbours. Weak electrostatic attraction occurs between the slight negative charge on the oxygen atom of one water molecule and the slight positive charge on one of the hydrogen atoms of an adjacent water molecule. Such relatively weak bonds formed between hydrogen and another polar atom such as oxygen are known as **hydrogen bonds** (Fig. 1.9).

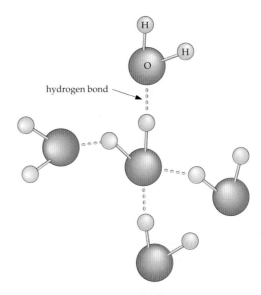

Fig. 1.9 Hydrogen bonding between water molecules. In liquid water, each molecule apparently forms transient hydrogen bonds with other molecules, forming a fluid network

In water, hydrogen bonds are broken and formed readily. They have a short lifetime of 10^{-11} seconds and, as one is broken, another is formed. The cumulative effect of large numbers of these bonds generates considerable strength, causing water molecules to cling together as a liquid under normal temperature and pressure. In many large molecules, hydrogen bonds help maintain structural stability.

> Chemical bonds involve the attraction between two atoms or molecules. Covalent bonds involve the sharing of electrons, while hydrogen bonds and ionic bonds involve attraction between opposite charges.

WATER

Water is the most abundant molecule found in living organisms, which contain 70–90% of water by weight. Water is usually a liquid under the relatively cool conditions of the earth, and provides a medium in which molecules can interact. Because of its special properties, water is an ideal medium for living organisms (Fig. 1.10), for example, as a carrier of

(a)

(b)

Fig. 1.10 Diversity of life in pond water. (a) Egg mass and newly hatched tadpoles of the spotted grass frog hanging below froth, with filamentous green algae in the water and water reeds in the background. (b) Red snow algae at Paradise Bay, Antarctica

nutrients, oxygen and waste products to and from cells. Hydrogen bonding between water molecules is responsible for many of the unusual and important physical properties of water, such as its solvent properties, high boiling point, high specific heat, high latent heats of fusion and vaporisation, cohesion and surface tension (see Box 1.1).

> The properties of water, especially hydrogen bonding, make it an ideal medium for living organisms.

Water as a solvent

Water, being polar, is an excellent solvent for many substances, particularly those such as sodium chloride that dissociate to form ions (**ionisation**).

When sodium chloride is added to water, the negative oxygen ends of the water molecules are electro-statically attracted to the positive sodium ions, and the positive hydrogen ends are attracted to the negative chloride ions, forming weak bonds (Fig. 1.11a). The combined strength of the new bonds is sufficient to overcome the ionic bonds between sodium and chloride ions and the sodium chloride molecule dissociates.

In aqueous solution, an ion such as Na^+, K^+, Mg^{2+}, Ca^{2+} or Cl^- is surrounded by a loosely bound array of water molecules forming a hydration sphere (Fig. 1.11b). Hydration increases the effective size of an ion. The orientation of water molecules in the sphere will reflect the charge on a particular ion. Na^+ and

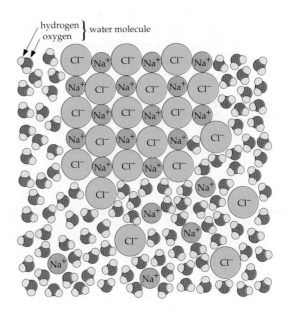

(a)

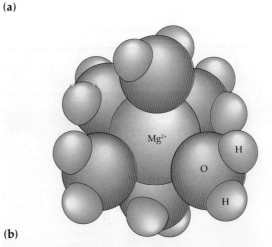

(b)

Fig. 1.11 Dissociation of sodium chloride in water. (a) When a salt crystal is placed in water, the water molecules cluster around the individual sodium (Na^+) and chloride (Cl^-) ions, forming hydrogen bonds with them, resulting in separation of the NaCl molecule. (b) Hydration sphere around an Mg^{2+} ion results from electrostatic attraction with six water molecules

Mg^{2+} attract the oxygen ends of water molecules, and Cl^- attracts the hydrogen ends.

Polar molecules such as ethanol, ammonia and sugars are very soluble in water because they can readily form hydrogen bonds with water molecules. Substances that dissolve readily in water are **hydrophilic**.

In contrast, non-polar molecules, such as oils and fats, are very insoluble in water. This is because the hydrogen bonds between water molecules tend to exclude non-polar molecules, which often form globules. These molecules are **hydrophobic**. Both repulsion between polar and non-polar molecules (hydrophobic forces) and hydrogen bonding play crucial roles in maintaining the structure and function of large biologically important molecules, particularly those that make up membranes (see pp. 42–3).

> Water is an excellent solvent for polar molecules, which dissociate in solution to form ions (ionisation). Water repels non-polar molecules.

BOX 1.1   PROPERTIES OF WATER

The chemical processes of life evolved within a set of constraints imposed by the physicochemical properties of water. These properties include high specific heat, high heat of vaporisation, high heat of fusion, surface tension and cohesion, and density.

Specific heat

Specific heat is the amount of heat required to raise the temperature of 1 g of a substance by 1°C. Compared with other solvents, the specific heat of water is unusually high, which means it can absorb considerable amounts of heat with little change in temperature. In a living organism, the heat generated by the vast array of chemical reactions occurring within cells, together with the heat absorbed from the external environment, could damage cells were it not for the 'heat-buffering' effect of water.

Vaporisation

The change from a liquid to gaseous state, such as water to steam, is **vaporisation**. Water has a high heat of vaporisation (the energy absorbed per gram of liquid vaporised). Water evaporating from a surface will draw the heat from the surface, thereby cooling it. This property allows overheated organisms to lose heat to the environment (see Fig. a).

Heat of fusion

A property of water that helps make life possible in near-freezing conditions is its high latent heat of fusion. A gram of water at 0°C loses eight times as much heat in turning to ice as it does in cooling from 1°C to 0°C. This property of water slows down the reduction in temperature of water within organisms during winter.

Cohesion and surface tension

Cohesion is the attraction between similar polar molecules, such as hydrogen bonding between water molecules. At the interface between water and air, the result is a special force, **surface tension**, which is so powerful that water striders and other insects can walk on water almost as though it were solid (see Fig. b).

(a) Because of its high heat of vaporisation, evaporating water draws heat from surfaces. Kangaroos lick their forearms to lose heat in this way during hot weather

(b) Cohesion between water molecules at a surface allows these water striders to mate on the pond surface

Water has the highest surface tension of any molecule except mercury (atoms of mercury are so strongly attracted to each other that they tend not to adhere to anything else). Attraction between *different* polar molecules is **adhesion**. Water adheres to any charged substance with which it can form hydrogen bonds. This is why some things get 'wet' when dipped in water and why others, composed of non-polar molecules such as waxy substances, are water-repellent and stay dry (see Fig. c).

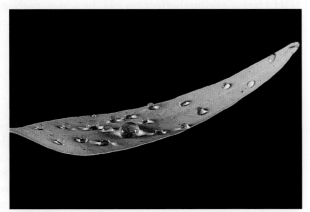

(c) Cohesion causes water to form droplets on the waxy surface of this eucalypt leaf

Capillary action is the result of the combined effects of cohesion and adhesion. Because of capillary action, water can creep up a piece of blotting paper or rise above the surrounding fluid in a very fine glass tube. Capillarity causes water to move through minute spaces between soil particles and to rise to considerable heights in tall trees (Chapter 18).

Density

With decreasing temperature, solutions increase in density as individual molecules move more slowly and the spaces between them decrease. Water has a maximum density at 4°C, at which temperature the water molecules are so close and are moving so slowly that each one of them can form hydrogen bonds simultaneously with four other water molecules. At higher temperatures this would be impossible. As the temperature drops below 4°C, a stable structure is formed in which each molecule is positioned at a small distance from its neighbours. At 0°C water freezes, creating an open latticework that forms the most stable structure for an ice crystal (see Fig. d).

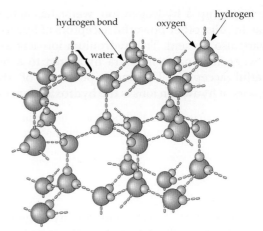

(d) When water freezes to form ice, each water molecule has formed hydrogen bonds with four other water molecules in a three-dimensional open lattice. The water molecules are further apart in ice than in liquid water

Water as a solid takes up more volume than water as a liquid. Therefore, ice is less dense than water and floats on it (see Fig. e). Oceans, lakes and streams often do not freeze solid but retain a liquid layer beneath an insulating cap of ice, protecting aquatic organisms living in the water beneath. Some aquatic organisms have chemicals in their body fluids that effectively lower their freezing point, thus protecting them from the damaging effects of ice crystal formation. These chemicals function in the same way as does antifreeze, which is used in car radiators.

(e) Iceberg floating in the ocean

Acids and bases

Many chemical reactions in cells involve a hydrogen atom being donated by one molecule and accepted by another. When a hydrogen atom dissociates from a molecule, it leaves its electron behind and becomes for an instant a hydrogen ion (proton, H^+). Substances that release hydrogen ions into solution are **acids**, while those that accept hydrogen ions are **bases**.

Hydrochloric acid, HCl, which dissociates in water into the ions H^+ and Cl^-, is a typical acid. The hydrogen ion (proton) from HCl then becomes attracted to an adjacent water molecule forming a hydronium ion, H_3O^+.

$$HCl \longrightarrow H^+ + Cl^-$$
$$H^+ + H_2O \longrightarrow H_3O^+$$

By taking up a hydrogen ion, water has acted as a base. In cells, other powerful acceptors of hydrogen ions are also present. The hydronium ion acts as an acid when it donates its extra hydrogen to a more powerful acceptor. The most powerful of these acceptors of hydrogen ions is the hydroxide ion, OH^-.

> Substances that release hydrogen ions into solution are acids, while those that accept hydrogen ions are bases.

pH

Cells are subject to changes in hydrogen ion concentrations resulting from the release or binding of hydrogen ions during continuing chemical reactions. In pure water, the concentration of H^+ is 10^{-7} moles per litre and is exactly equal to the concentration of OH^-. (For simplicity, we refer to H^+, although in solution, most H^+ are actually hydronium ions, H_3O^+). An acidic solution has a higher concentration of H^+ while a basic (alkaline) solution has a lower concentration. The concentration of H^+ in solution is represented by the pH scale, in which H^+ concentrations are expressed as the negative logarithm to the base 10 within a possible range of 0 to 14 (pH = $-\log[H^+]$). Because the pH scale is logarithmic each unit represents a ten-fold change in H^+ concentration. The midpoint of the pH scale is 7, which is the pH value for pure water. Acid solutions have a pH lower than 7, and basic solutions have a pH greater than 7 (Fig. 1.12).

The internal pH of cells is near neutral (pH 7.3–7.4) but the external environment may be quite different. For example, for plants and animals living in a creek, the pH of the surrounding water may vary from 6.5 to 8.5. Sphagnum moss grows in acidic peat bogs of low pH (3–4) while the shrub wirilda (*Acacia retinodes*) grows in coastal limestone soils (pH 8.5) of southern Australia.

> pH is the concentration of H^+ in solution. The pH scale is logarithmic and ranges from 0 to 14.

Buffers

Living cells have different ways of maintaining a relatively constant internal pH. One way involves **buffers**, which are substances that act as a reservoir for hydrogen ions, accepting them as pH falls and releasing them as pH rises. Consider a solution of carbonic acid, H_2CO_3, in water. The addition of further acid (H^+) in small amounts has little or no effect upon pH. Similarly, the addition of small amounts of a base (OH^-) has a minimal effect on pH. In water, carbonic acid dissociates to form bicarbonate ions and hydrogen ions:

$$H_2CO_3 \rightleftharpoons HCO_3^- + H^+$$

| carbonic acid | bicarbonate ion | hydrogen ion |

Carbonic acid is acting as a buffer. If more acid (H^+) is added to the solution, it combines with the bicarbonate ion, effectively removing it by acting as a proton acceptor or base. If OH^- is added, it combines with H^+ to form water. This removal of H^+ causes further dissociation of carbonic acid, releasing more H^+ into the solution. In this manner, the pH of the solution is stabilised by equilibrium between the forward and backward reactions.

The carbonic acid–bicarbonate buffer system is the key buffer system in mammalian blood. Buffering blood against changes in pH is needed because many cellular activities are sensitive to changes in pH.

> A buffer solution such as carbonic acid–bicarbonate solution maintains relatively constant pH by acting as a reservoir for hydrogen ions.

BIOLOGICALLY IMPORTANT MOLECULES

Biologically important molecules are often grouped into two major classes, organic and inorganic. **Organic compounds** contain one or more carbon atoms in their

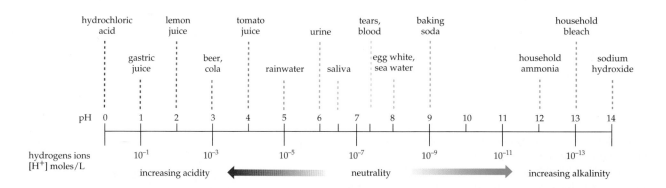

Fig. 1.12 The pH scale, showing the pH of some common substances

molecules and they are mostly produced by living organisms. They are so-called because the first known examples were natural products synthesised by plants or animals. For example, the carbon in coal comes from the remains of long-dead plants and animals. Organic compounds include carbohydrates (e.g. cellulose, starch and sugar), lipids (e.g. fats, waxes and steroids), proteins (e.g. haemoglobin) and nucleic acids (e.g. deoxyribonucleic acid [DNA] and ribonucleic acid [RNA]; Table 1.1).

Table 1.1 Types of biologically important molecules

Macromolecule	Subunit	Composition	Function	Example
Carbohydrates				
Storage polysaccharides	Glucose	Carbon, hydrogen and oxygen; except chitin has nitrogen	Storage of energy	
Starch				Potatoes
Glycogen				Animal tissues
				Bulbs of plants
Insulin				(dahlias, artichokes)
Structural polysaccharides			Cell wall components	
				Paper, cotton
Cellulose	Glucose			
Chitin	Modified glucose (*N*-acetyl-D-glucosamine)			Insect shell, crab shell
Pectin	Arabinose, galactose, galacturonic acid			
Hemicellulose	D-xylose			
Lipids				
Simple lipids		Carbon, hydrogen, oxygen; except phospholipids have phosphorus		
Fats	3 fatty acids and glycerol		Storage of energy	Butter
Waxes	Fatty acids and long chain alcohols		Protection	Coating on skins and fur of animals
Compound lipids				
Glycolipids	2 fatty acids, glycerol and carbohydrate (1–15 monosaccharide monomer)		Cell membranes	Higher plants, neural tissues of vertebrates
Phospholipids	2 fatty acids, glycerol and phosphate			Lecithin (phospho)
Steroids	4 carbon rings		Membranes and hormones	Cholesterol
Proteins				
Structural	Amino acids	Carbon, hydrogen, oxygen, nitrogen and sulfur	Support and structural material	Hair, silk, nails
Functional	Amino acids		Catalysis, transport defence hormones	Proteinases (enzymes) Transport of Cl⁻ ions Antibodies, snake venom Insulin
Nucleic acids				
DNA RNA	Nucleotides	Carbon, hydrogen, oxygen, nitrogen and phosphorus	Storage and dictation of genetic information Transcription and translation of genetic material	Nucleus or chromosome Viruses

All other compounds, that is, those that do not contain carbon atoms, are **inorganic compounds**. Inorganic compounds are usually associated with non-living sources but many are essential requirements of living organisms. These include minerals and their salts as well as, for example, oxygen and water. Phosphorus, nitrogen and trace elements are essential for plant nutrition; iron prevents anaemia; calcium is needed for bones; and, without water, there would be no life.

Why carbon?

Carbon is the principal structural element of living matter and forms the backbone of organic molecules. What are the chemical properties of carbon that resulted in carbon compounds becoming fundamental to the evolution of life?

A carbon atom has six electrons, with two electrons in the first level and four electrons in the second level. This means that it has no overall tendency to gain or lose electrons and instead can form four covalent bonds by sharing electrons with other atoms (e.g. $O=C=O$). Carbon also forms strong carbon–carbon bonds, which remain strong when carbon atoms are bonded to other elements. They can form chains of nearly any length, with linear, branched or circular shapes. Theoretically, an infinite number of different carbon compounds exist.

The carbon of organic compounds is cycled from the atmosphere (Fig. 1.13), which contains approximately 0.033% by volume of carbon dioxide (CO_2). Photosynthetic organisms trap light and convert CO_2 to sugars, some of which are eaten by animals. Carbon dioxide is released back into the atmosphere as a result of the decay of organic material and as an end product of respiration in all living organisms.

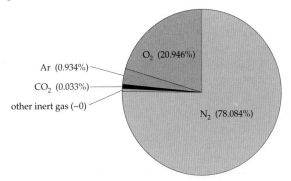

Fig. 1.13 Major constituents of atmospheric air

Carbon is the principle structural element of organic compounds. It can form strong $C=C$ chains and an almost infinite number of different compounds. Photosynthetic organisms produce organic compounds using atmospheric carbon dioxide.

Oxygen

Oxygen in the earth's crust exists in the forms of silicates, oxides and water. Gaseous oxygen (O_2) is consumed during respiration in living cells and combustion (burning) processes. It is liberated during photosynthesis by plants and other photosynthetic organisms, so the concentration in the atmosphere, like CO_2, remains nearly constant at 21% by volume.

One important property of O_2 is its low solubility in water (4.89 parts of oxygen in 100 parts of water at 0°C) yet sufficient must be available in seas, lakes, rivers and ponds to support aquatic organisms. The content of dissolved oxygen is of prime concern in aquatic ecosystems that are subject to pollution. Streams used for waste disposal have, in many instances, been fully depleted of dissolved oxygen, killing aquatic organisms.

Oxygen also exists in a unique triatomic form, **ozone** (O_3). This is a highly reactive molecule and its increase in the lower atmosphere as a result of air pollution is a hazard to living organisms. Paradoxically, the presence of a small amount of ozone (one microlitre per litre) in the upper atmosphere is crucial to the survival of life on earth because of its ability to filter harmful ultraviolet radiation from sunlight (see Chapter 43).

Oxygen is essential for respiration in living cells and is produced during photosynthesis.

CARBOHYDRATES

Carbohydrates are the most abundant organic compounds in nature. They are a source of chemical energy for living organisms, form structural components and combine with other molecules such as proteins (Table 1.1). Carbohydrates are composed of carbon, hydrogen and oxygen with the general formula $(CH_2O)_n$ (where n is the number of carbon atoms); hence the name 'hydrates' of carbon. Useful extractable energy resides chiefly in the carbon–carbon (C—C) and carbon–hydrogen (C—H) bonds of these molecules. The basic unit of carbohydrates is a sugar molecule (saccharide). The three main groups of carbohydrates are monosaccharides, disaccharides and polysaccharides.

Monosaccharides

Monosaccharides are simple sugars composed of a single sugar molecule and are sweet-tasting. The most abundant monosaccharide is glucose, with six carbon atoms (Fig. 1.14). Glucose is the primary product of photosynthesis, and is the monosaccharide from which most other carbohydrates, such as starch and cellulose, are derived. Monosaccharides can be grouped according to the number of carbon atoms they contain. Trioses (e.g. glyceraldehyde) contain three carbon atoms, pentoses (e.g. ribose) five, and hexoses (e.g. glucose, fructose and galactose) six. Fructose and galactose are alternative forms (isomers) of glucose.

Within a monosaccharide, one of the carbons forms a double bond with an oxygen atom, while the other carbons in the chain are bonded to one hydrogen atom and one hydroxyl group. The hydroxyl groups make the molecule polar and readily soluble in water.

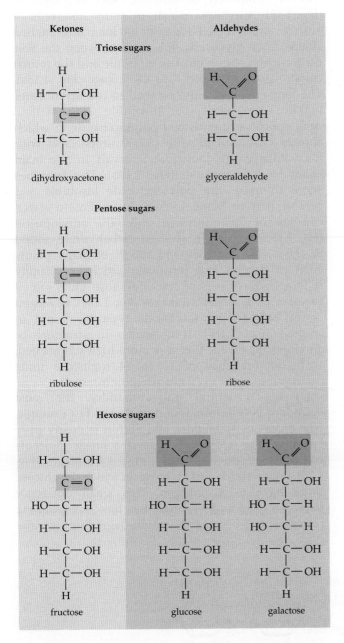

Fig. 1.14 Types of monosaccharides based on the number of carbon atoms they contain and their functional groups

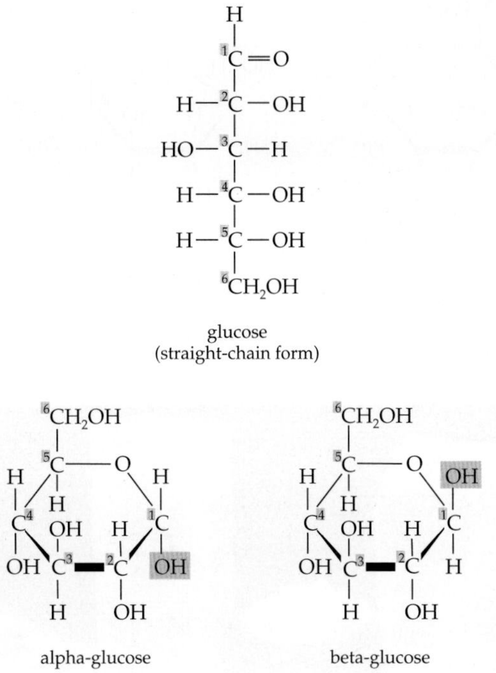

Fig. 1.15 Different structural forms of glucose

Two monosaccharides can have the same chemical formula but differ in the location of the C=O group. When this bond lies in the middle of the molecules, it forms a ketone (—C=O) group. Ketose sugars include dihydroxyacetone, ribulose and fructose. When the C=O bond lies at the end of the carbon chain, it forms an aldehyde (—C$\lessgtr^{O}_{H}$) group. Aldose sugars include glyceraldehyde, ribose and glucose (Fig. 1.14).

Monosaccharides can exist in a straight-chain form but in water they almost always form a ring that can close in one of two ways. In the alpha form, the position of the hydroxyl group attached to carbon atom 1 is below the plane of the ring and in the beta form it is in the plane (Fig. 1.15). This small difference between the alpha and beta forms of glucose can lead to very significant differences in the properties of large molecules formed from glucose. For example, polymers of α-glucose form starch and polymers of β-glucose form cellulose.

Disaccharides

Disaccharides consist of two monosaccharides joined together by a **glycosidic bond**, in which the first carbon atom of one sugar molecule reacts with a hydroxyl group of another sugar molecule, with the loss of a water molecule (Fig. 1.16). Sucrose, maltose and lactose are examples of common disaccharides. Sucrose (cane sugar; Fig. 1.17) is composed of equal amounts of glucose and fructose, and is the form in

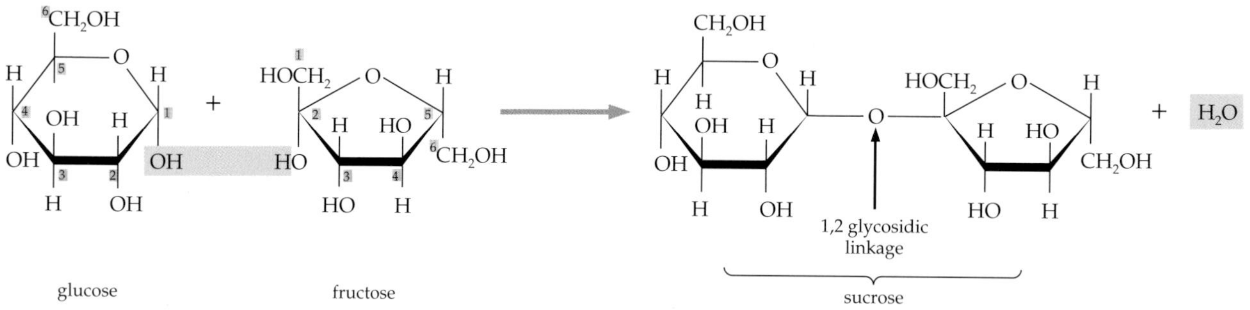

Fig. 1.16 Formation of the disaccharide, sucrose, from two monosaccharides, glucose and fructose

Fig. 1.17 Cane sugar, or sucrose, is composed of equal amounts of glucose and fructose and is the form of glucose transported in plants

which plants transport glucose from the site of photosynthesis in the leaves to other parts of the plant. Lactose is a disaccharide composed of glucose and galactose and is the form of sugar that mammals produce in their milk for suckling their young. In insects, sugar is transported through the haemolymph ('blood') as trehalose, a disaccharide consisting of two glucose molecules.

Polysaccharides

As the name implies, **polysaccharides** consist of many monosaccharides joined in long linear or branched chains (polymers). They can be made from a single type of monosaccharide or two or more types. They are usually tasteless, insoluble compounds with high molecular weight, and serve a storage or structural function.

Starch is the storage polysaccharide of higher plants and is composed of amylose and amylopectin (Fig. 1.18). Amylose consists of many hundreds of glucose units linked together in long unbranched chains. Each glycosidic linkage is between the first carbon atom of one glucose molecule and the fourth carbon atom of another (α-1,4 linkage; Fig. 1.18a). The long chains of amylose tend to coil up in water making the molecule insoluble (Fig. 1.18c). Amylopectin is a branched polysaccharide in which shorter chains of glucose (20–30 units), also with α-1,4 linkage, are joined at intervals to the main chain by α-1,6 linkage (Fig. 1.18b). The branches serve to make the polysaccharide insoluble and protect it from being broken down. Potato starch consists of 20% amylose and 80% amylopectin (Fig. 1.18d). Inulin, a storage carbohydrate found in the roots, rhizomes and tubers of many plants such as dahlias and Jerusalem artichokes (family Asteraceae), consists of fructofuranose units joined together by glycosidic linkages (Fig. 1.18e). Glycogen is the storage polysaccharide typically found in animals (Fig. 1.18f). It is similar in structure to amylopectin except that it is more highly branched, with side-branches of 16 to 24 α-glucose units occurring at eight to 10 glucose unit intervals.

Of the structural polysaccharides, the most abundant is **cellulose**, found in plants. For example, 50% of wood is cellulose and cotton is nearly pure cellulose (Fig. 1.19). Cellulose is a linear molecule composed of glucose units with β-1,4 linkages (Fig. 1.20), unlike starch and glycogen in which the glucose units have α-1,4 linkages. This small difference between α- and β-linkages has far-reaching consequences because the starch-degrading enzymes

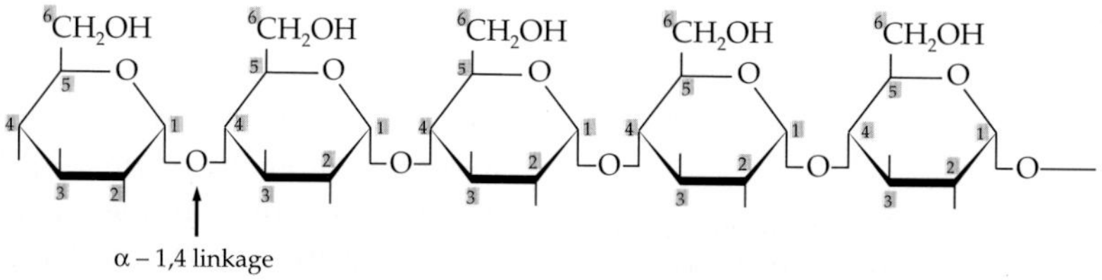

Fig. 1.18 (a)

(b)

α – 1, 6 linkage

(d)

(e)

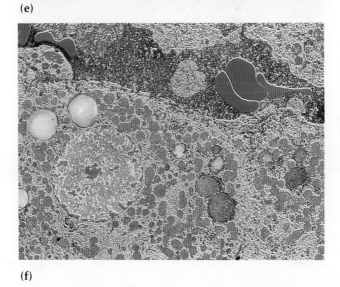

(f)

Fig. 1.18 Starch is composed of **(a)** amylose and **(b)** amylopectin. **(c)** In water, amylose coils to form a helix. **(d)** Potato starch is 20% amylose and 80% amylopectin. **(e)** Inulin is the storage carbohydrate of Jerusalem artichokes. **(f)** Liver cell (with blue nucleus) containing many vesicles of stored glycogen (deep pink)

(c)

Fig. 1.19 Cotton crop ready for harvest

of most organisms are unable to digest cellulose. The reason is not that the glycosidic linkages between glucose subunits in cellulose are stronger, but that their cleavage requires a different enzyme, cellulase, not usually present in the animal kingdom. Certain bacteria, protozoa and fungi are able to hydrolyse cellulose and use it as an energy source. Animals with a diet high in cellulose have symbiotic organisms in their gut to help digestion; for example, kangaroos have bacteria and termites have protozoans.

Plants contain various other structural polysaccharides including pectins (composed of arabinose, galactose and galacturonic acid) and hemicellulose. Chitin is the structural material found

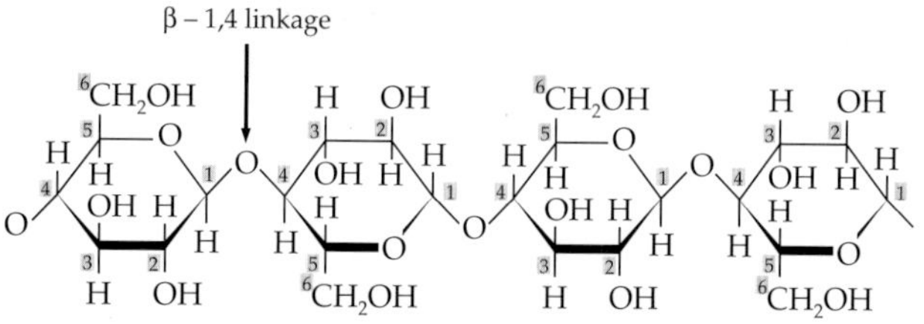

β – 1,4 linkage

Fig. 1.20 In a cellulose molecule, glucose molecules are linked to each other by β-1,4 linkages

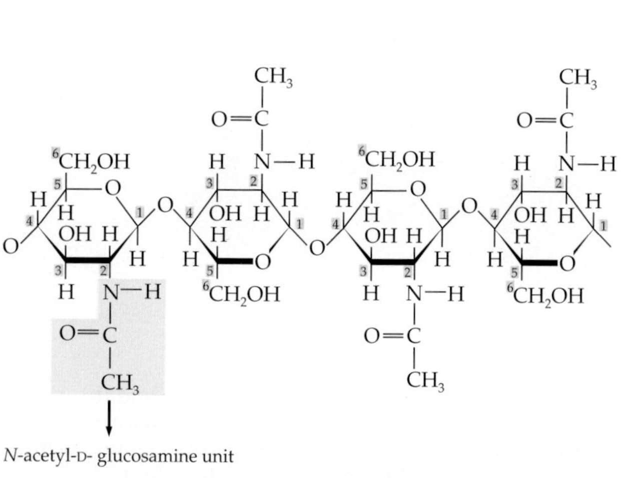

CH₃

N-acetyl-ᴅ- glucosamine unit

Fig. 1.21 Chitin is a polymer of N-acetyl-ᴅ-glucosamine

in the exoskeletons of insects and crustaceans, and the cell walls of fungi. Chitin is a modified form of cellulose in which a nitrogen-containing group has been added to each glucose unit (N-acetyl-ᴅ-glucosamine; Fig. 1.21).

> Carbohydrates, including simple sugars and their polymers, are a source of energy for immediate use and storage, and form structural components of living organisms.

LIPIDS

Lipids perform many biologically important functions (Table 1.1). They are a structural component of membranes and, as fats or oils, they have roles in energy storage and transport, and insulation. They form a protective and water-repellent coating at the surface of many organisms and sometimes act as chemical messengers both within and between cells.

Lipids are insoluble in water as a result of the non-polar (hydrophobic) nature of their numerous C—H bonds, which prevent water molecules from penetrating and cause the lipid molecules to cluster together as undissolved droplets. However, lipids dissolve readily in organic solvents such as chloroform. Lipids are composed principally of carbon, hydrogen and oxygen but differ from carbohydrates in having a much smaller proportion of oxygen. They also contain other elements, particularly phosphorus and nitrogen. The different types of lipids include fats, oils, waxes, phospholipids, glycolipids and steroids.

Lipids for energy storage: fats and oils

Fats and oils contain a higher proportion of energy-rich C—H bonds than do carbohydrates per unit weight and, as a result, can function as efficient stores of chemical energy (Chapter 2). On average, fats yield about two to three times as much energy per gram as do carbohydrates or proteins. Oils in seeds and fats in animals provide weight-efficient energy storage.

All fats have a backbone of glycerol, containing three carbon atoms (Fig. 1.22a), to each of which is attached a fatty acid chain (Fig. 1.22b) forming a **triglyceride**. Fatty acids are hydrocarbon chains of variable length with a carboxyl (—COOH) group at one end. The fatty acid chains are joined to the glycerol molecule at their carboxyl groups. Fatty acids in edible fats and oils contain an even number of carbon atoms, ranging from four to 24. Three of the most common fatty acids are stearic acid (18 carbon atoms), palmitic acid (16 carbon atoms) and oleic acid (18 carbon atoms; Fig. 1.22b).

(a)

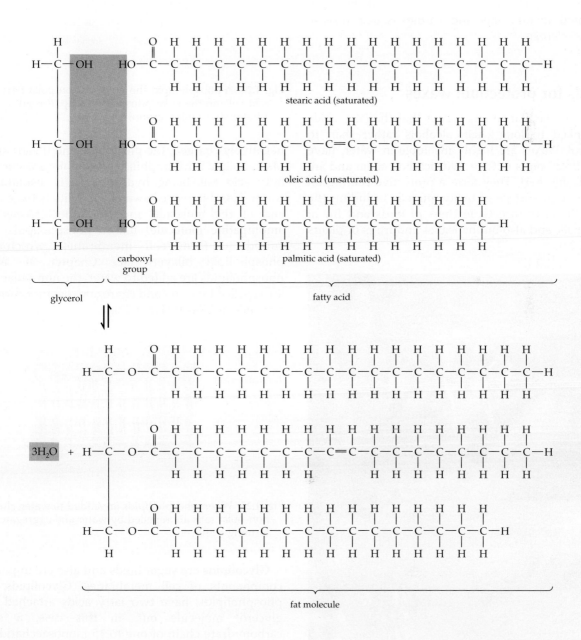

(b)

Fig. 1.22 Structure of lipids. **(a)** Structure of glycerol. **(b)** Structure of saturated and unsaturated fatty acids. A fat molecule forms by a condensation reaction. As the three fatty acids join to glycerol, three water molecules are released

Fatty acids with the maximum possible number of hydrogen atoms attached to each carbon (e.g. stearic acid) are said to be *saturated* because they contain no carbon–carbon double bonds. Examples include butter, lard and animal fat. Fatty acids with double bonds between one or more pairs of successive carbon atoms (e.g. oleic acid) are said to be *unsaturated*; they therefore contain fewer than the maximum possible number of hydrogens.

Unsaturated fats (oils), such as canola, olive, and corn oils, are usually liquid at room temperature. Oils with more than one double bond are *polyunsaturated*. Polyunsaturated margarine contains 70% of polyunsaturated oils and has tended to replace butter in the Western diet because of the link between saturated dietary fats and cardiovascular disease (atherosclerosis).

Lipids for protection: waxes

Waxes are similar to fats except that the fatty acids are linked to long chain alcohols rather than to glycerol. Waxes are highly insoluble in water, have no double bonds in their hydrocarbon chain and are chemically inert. They form a protective coating on the external surface of many animals including the exoskeleton of insects, feathers of birds and fur of mammals, and also on the leaves and fruits of plants (Fig. 1.23).

Fig. 1.23 Waxes form a protective, water-repellent coating on the exoskeleton of this honey bee

Lipids for membranes: phospholipids and glycolipids

Phospholipids are similar to fats and oils except that they have two fatty acids attached to each glycerol molecule and a phosphate group attached to the third carbon (Fig. 1.24). As the phosphate groups are

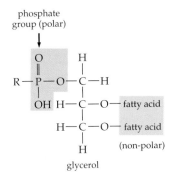

Fig. 1.24 A phospholipid showing the non-polar fatty acid tail and the polar phosphate group (R is an additional chemical group)

negatively charged, the phosphate end (head) of the phospholipid is hydrophilic and soluble in water. The fatty acid tail, being hydrophobic, is insoluble in water. This difference in water solubility between one end of the molecule and the other (termed an **amphipathic molecule**) makes phospholipids well suited to form cell membranes, which are phospholipid bilayers (see Chapter 3). When phospholipids are added to water, the non-polar tails are repelled by water and aggregate together, forming a micelle or bilayer (Fig. 1.25).

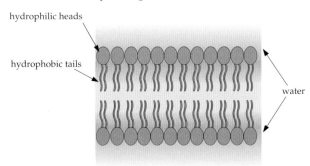

Fig. 1.25 When phospholipids are added to water, the non-polar tails are repelled by water and aggregate, forming a bilayer

Glycolipids are sugar lipids and also are important components of cell membranes. Glycolipids, like phospholipids, have two fatty acids attached to a glycerol molecule but, in this case, a short carbohydrate chain of one to 15 monosaccharides is attached to the third carbon atom. Glycolipids are also amphipathic and form micelles when added to water in a way similar to phospholipids.

Steroids

The molecular structure of steroids is quite different from other lipids. They are classified as lipids primarily because they are insoluble in water and soluble in fat solvents such as chloroform. Steroids are based on a primary structure of four carbon rings, derived from cholesterol, to which may be added a variety of hydrocarbon tails (Fig. 1.26).

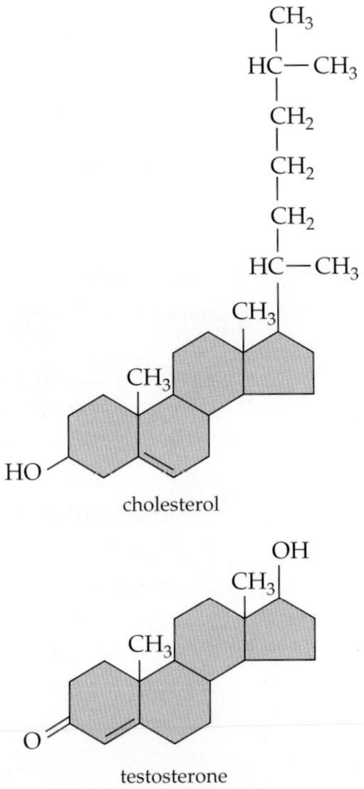

Fig. 1.26 Cholesterol and testosterone are examples of steroids. Testosterone is an example of a hormone with a four ring structure but without the hydrocarbon tail

Steroids are important molecules in vertebrate animals, where they are found primarily in the heart, blood vessels and liver. Some hormones and vitamins are also steroids. One of the most abundant, cholesterol, is a component of membranes in all animals and the starting point for the synthesis for several hormones.

Cholesterol, being insoluble in water, may precipitate from bile in the gall bladder, forming pebble-like gallstones. Its accumulation in the atherosclerotic plaques in arteries contributes to high blood pressure and increases the risk of heart attack and strokes in humans and other animals. The level of cholesterol found in blood is not necessarily related to a person's dietary intake but appears to be more closely related to the rate of synthesis of cholesterol by the body.

> Lipids are water-insoluble carbon compounds and include fats, oils, waxes, phospholipids, glycolipids and steroids. They function as structural components of membranes, sources of energy storage and transport, insulators and chemical messengers.

PROTEINS

Proteins are structurally varied molecules that perform a wide range of functions: enzymes for metabolism; proteins for structure, movement, storage, nutrition and transport; antibodies and toxins for defence and attack; and hormones for regulation of body functions (Table 1.1).

Proteins contain carbon, hydrogen, oxygen, nitrogen and usually sulfur. The building blocks of proteins are **amino acids**. There are only 20 different kinds of amino acids found in proteins, however, the number of possible combinations of amino acids, and hence the variety of proteins, is enormous.

Amino acids

All amino acids have a similar structure, each containing an amino group ($-NH_2$, except for proline which has an imino group, $-NH-$), an acidic carboxyl group ($-COOH$), a hydrogen atom and a unique side chain R-group all bonded to a central carbon atom (Fig. 1.27). The 20 different amino acids

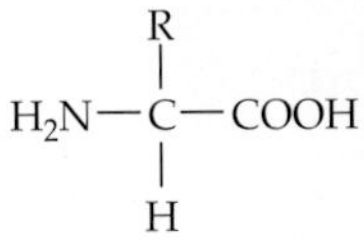

Fig. 1.27 General formula of amino acids found in proteins. The R-group is distinctive but the remainder of the molecule is the same for all amino acids

differ in their side chain R-groups, which may be simply a hydrogen atom, as in glycine, or complex, such as a double-ring structure in tryptophan (Fig. 1.28). The identity and unique chemical properties of each amino acid, and the way in which each amino acid affects the shape of a protein, are determined by its R-group.

Three amino acids have special functions. Methionine initiates protein synthesis, proline causes kinks in chains of amino acids and cysteine links amino acid chains together. Certain R-groups are hydrophilic (polar) while others are hydrophobic (non-polar). Hydrophobic side chains make proteins highly insoluble in water. Because there are many amino acids in a protein, each is referred to by a code, which initially was composed of three letters but now is just a single letter of the alphabet (e.g. alanine, ala, A; Fig. 1.28).

Polypeptides

A chain of amino acids is a **polypeptide**. Amino acids in a polypeptide chain are linked together by **peptide bonds**, which form when the ($-COOH$) group of one

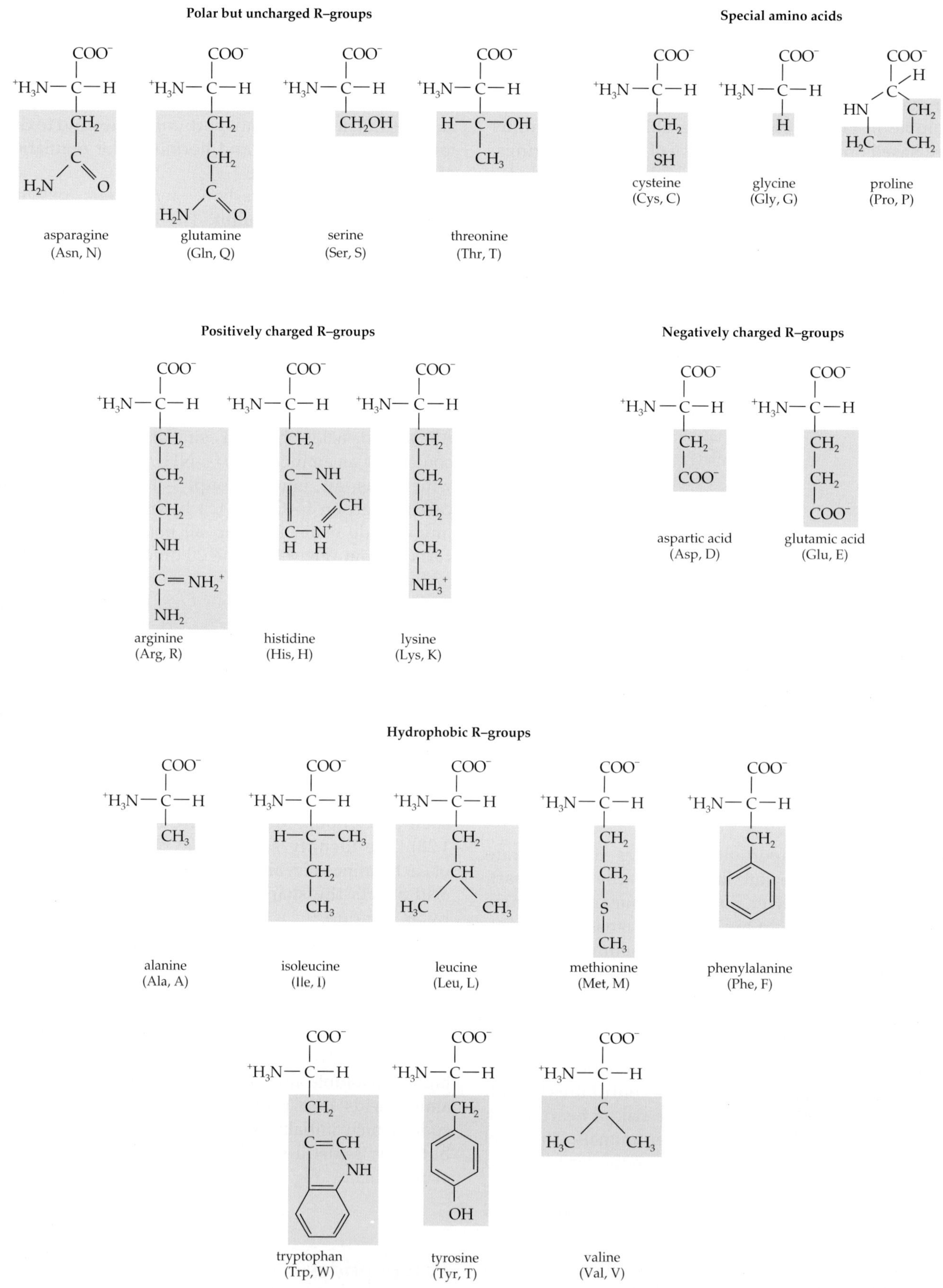

Polar but uncharged R-groups

asparagine
(Asn, N)

glutamine
(Gln, Q)

serine
(Ser, S)

threonine
(Thr, T)

Special amino acids

cysteine
(Cys, C)

glycine
(Gly, G)

proline
(Pro, P)

Positively charged R-groups

arginine
(Arg, R)

histidine
(His, H)

lysine
(Lys, K)

Negatively charged R-groups

aspartic acid
(Asp, D)

glutamic acid
(Glu, E)

Hydrophobic R-groups

alanine
(Ala, A)

isoleucine
(Ile, I)

leucine
(Leu, L)

methionine
(Met, M)

phenylalanine
(Phe, F)

tryptophan
(Trp, W)

tyrosine
(Tyr, T)

valine
(Val, V)

Fig. 1.28 Structures of the 20 amino acids found in proteins, with the R-groups shaded, showing the code by which they are referred

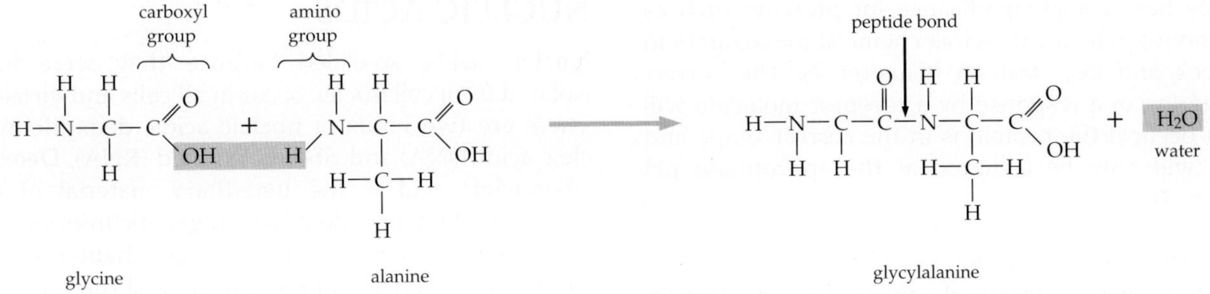

Fig. 1.29 Formation of a peptide bond between two amino acids, glycine and alanine, with the production of glycylalanine and water

amino acid attaches to the ($-NH_2$) group of another, with the release of a molecule of water (Fig. 1.29). When amino acids are joined together by peptide bonds, they are referred to as amino acid residues. There is always a free amino group at one end of the chain, the N-terminus, and a free carboxyl group at the other end, the C-terminus. A protein may comprise one or more polypeptide chains, having up to several hundred amino acids in a specific sequence.

Polypeptide chains can also be cross-linked by covalent bonds, **disulfide bonds**, formed between the sulfur atoms of two cysteine residues. Disulfide bonds may link cysteine residues from within a single polypeptide chain or they may cross-link between different chains. These linkages affect the three-dimensional structure of proteins.

Structure and conformation of proteins

Proteins are divided into two major classes: simple and conjugated. **Simple proteins** consist solely of amino acids. **Conjugated proteins** consist of amino acids and other organic or inorganic components, for example, nucleo-, lipo-, phospho- and glycoproteins. The non-amino acid portion of the conjugated protein is called its prosthetic group.

The **primary structure** of a protein is the linear sequence of amino acids in the polypeptide chain (Fig. 1.30a). Each kind of protein has a distinct primary structure. A protein's **secondary structure** results from hydrogen bonding between amino acid residues lying close to one another within the primary structure (Fig. 1.30b). An example of a secondary structure is a right-handed α-helix which resembles a spiral staircase.

Imagine taking a very tall spiral staircase and folding it in upon itself until it forms a sphere. That mental image represents a protein's **tertiary structure** and is the result of bending or folding of the polypeptide chain in three dimensions to form a compact, tightly folded structure (Fig. 1.30c). Folding normally results from the interaction between amino acid residues relatively far apart in the sequence. The final shape is also affected by the environment in which the protein finds itself, especially pH. This is due to the presence of charged regions on their molecules, which also affect the shape of the molecule.

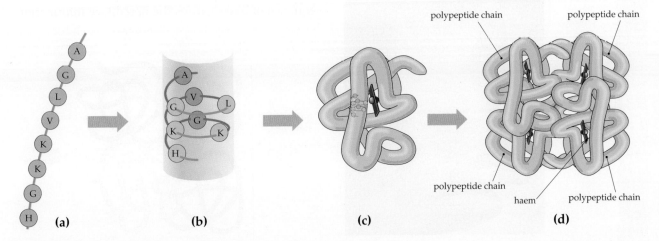

Fig. 1.30 Haemoglobin molecule. **(a)** Primary structure showing the linear sequence of amino acids in the polypeptide chain. **(b)** Secondary structure formed as a result of hydrogen bonding between nearby C—O and —NH groups. **(c)** Tertiary structure formed after folding of the chains in three-dimensional shapes. **(d)** Quaternary structure resulting from the combination of the four polypeptide chains into a single functional molecule

Shape has special significance for proteins such as hormones, which can act as chemical messengers in a 'lock and key' fashion (Chapter 7). The correct triggering of a response by a receptor molecule will only occur if the protein is in the correct shape and that will only be acquired at the appropriate pH (see p. 31).

Some larger proteins, such as haemoglobin, contain more than one polypeptide chain, which allows yet another level of structural organisation, **quaternary structure** (Fig. 1.30d). Most large proteins, particularly enzymes, contain two or more polypeptide chains.

The three-dimensional shape of a protein is referred to as its **conformation**. Proteins can be placed in two major classes based on their conformation: fibrous and globular. In **fibrous proteins**, polypeptide chains are arranged in parallel along a single axis forming long fibres or sheets (e.g. collagen, keratin, elastin). Fibrous proteins are insoluble in water (Fig. 1.31a), often strong and hard, and therefore ideal constituents of hair, nails, horns and claws (Fig. 1.31b). In **globular proteins**, polypeptide chains are folded into a compact spherical or globular shape. Globular proteins are soluble in water and include most proteins functioning as enzymes and hormones (Fig. 1.31c).

> Proteins are composed of one or more polypeptides, which are polymers of up to 20 different types of amino acids. The function of a protein depends on its three-dimensional shape.

NUCLEIC ACIDS

Nucleic acids, so-called because they were first isolated from cell nuclei, occur in all cells and viruses. There are two kinds of nucleic acids, deoxyribonucleic acid (DNA) and ribonucleic acid (RNA). **Deoxyribonucleic acid** is the hereditary material of an organism, the molecule in which genetic information is stored in coded form (genes; see Chapter 10). It can produce precise copies of itself, and these copies are passed from cell to cell, and thus to an organism's descendants during reproduction. **Ribonucleic acids** (RNAs) are involved in transmitting and translating the coded information of DNA during the production of proteins.

Nucleotides

Nucleic acids are chains of repeating nucleotide units; DNA and RNA are thus polynucleotides. Each **nucleotide** is composed of three subunits: a five-carbon sugar (pentose), a phosphate group, and a nitrogenous base (Fig. 1.32). Nucleotides are linked together by phosphodiester bonds in which the phosphate group of one nucleotide is linked to the hydroxy group of another to form a chain (Fig. 1.33). Thus, a nucleic acid is simply a chain of pentose sugars linked together by phosphodiester bonds with a nitrogenous base protruding from each sugar. The bases in each nucleotide are not directly involved in the linkage, so they may occur in a variety of sequences.

There are five nitrogenous bases, which are either pyrimidines or purines. **Pyrimidines** are single-ring molecules: *cytosine, uracil* and *thymine*. **Purines** are double-ring molecules: *adenine* and *guanine* (Fig. 1.34). These nitrogenous bases are the molecular entities that form the alphabet of the genetic code. The sequence of bases carries the hereditary information.

helical coil

fibrous protein

(a)   **(b)**

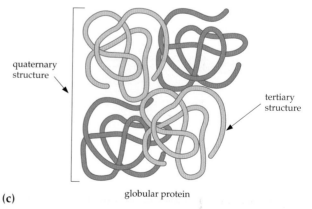

quaternary structure

tertiary structure

globular protein

(c)

Fig. 1.31 Conformation of proteins. **(a)** Fibrous proteins. **(b)** The strength and hardness of the horns of many animals, including the black rhinoceros, are due to fibrous proteins. **(c)** Globular proteins

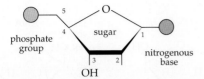

Fig. 1.32 Chemical structure of nucleotides of DNA: both the phosphate group and the nitrogenous base are covalently bound to the sugar

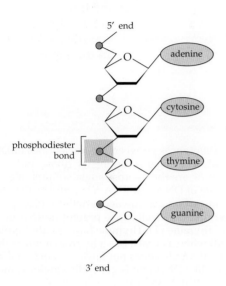

Fig. 1.33 Part of a single chain of DNA showing phosphodiester bonds between the sugar and phosphate groups of nucleotides

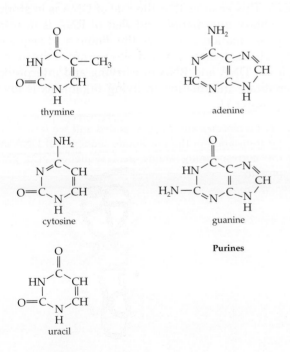

Fig. 1.34 Pyrimidines (thymine, cytosine and uracil) and purines (adenine and guanine) found in nucleic acids

DNA

The type of pentose sugar in DNA is *deoxyribose*, in which the oxygen at carbon 2 in the ribose molecule is absent (hence 'deoxy', Fig. 1.32). The bases in DNA

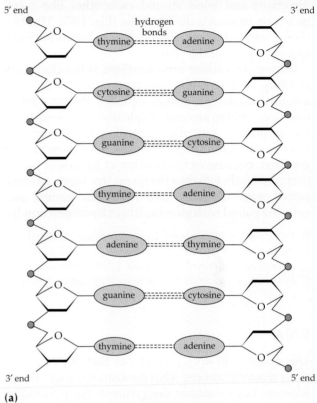

(a)

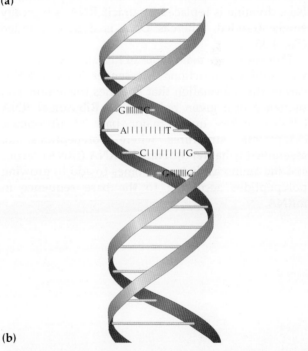

(b)

Fig. 1.35 DNA molecule. (a) Portion of an uncoiled DNA molecule showing ladder-like structure formed by two polynucleotide chains, running in opposite directions. (b) A double helix of DNA consists of two polynucleotide chains held together by hydrogen bonds between bases of opposite strands

are the purines adenine and guanine, and the pyrimidines cytosine and thymine. The pioneering work of Nobel prizewinners James Watson and Francis Crick, in 1953, showed that DNA molecules occur as double chains orientated in opposite directions and wind around each other like a coil in a rope to form a double helix (Fig. 1.35). The two chains are held together by hydrogen bonds between opposite bases of the chains.

During the earliest investigations of the chemistry of DNA, an interesting observation was made. The amount of adenine always equalled the amount of thymine, and the amount of cytosine always equalled the amount of guanine. According to Watson and Crick's model, the simplest explanation of this finding was that, because of the number of hydrogen bonds that could form between the respective bases, adenine paired with thymine (two hydrogen bonds) and cytosine paired with guanine (three hydrogen bonds).

> DNA is a double-helical molecule comprising nucleotides that are made up of a pentose sugar, phosphate and nitrogenous base. DNA stores hereditary information as a coded sequence of nitrogenous bases.

RNA

The structure of RNA is similar to that of DNA, with three main differences: RNA contains the sugar ribose whereas DNA contains deoxyribose; the pyrimidine base thymine is replaced by uracil; RNA is typically single-stranded, whereas DNA is double-stranded (Fig. 1.36).

There are three main types of RNA molecules, each with different functions. **Messenger RNA** (mRNA) carries the information that specifies the amino acid sequence of a given polypeptide. **Ribosomal RNA** (rRNA) makes up a major part of ribosomes, cytoplasmic structures where polypeptides are assembled (Chapter 3). **Transfer RNA** (tRNA) carries specific amino acids to ribosomes to add to growing polypeptides according to the base sequence in mRNA.

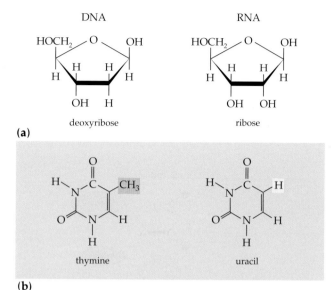

(a)

(b)

Fig. 1.36 Differences between the nucleotides of DNA and RNA. The sugar components: ribose and deoxyribose are the pentose sugar subunit of a nucleotide of DNA and RNA respectively. Oxygen is absent at carbon 2 in the deoxyribose molecule. The pyrimidine, uracil (U), is present in RNA in place of thymine (T). Thymine has a methyl group (shaded) whereas uracil has a hydrogen ion at the same position. RNA also does not have a regular helical structure like DNA and is generally single-stranded

Cells do not possess the enzymes necessary to assemble double strands of RNA as they do with DNA. This ensures that the role of DNA is in storing hereditary information and that of RNA is in using this information to specify the amino acid sequence of proteins. A more detailed description of the unique roles of DNA and RNA in carrying and interpreting hereditary information in living organisms is given in Chapter 11.

> RNA molecules are single-stranded and are responsible for translation of the nucleotide sequence of DNA into the corresponding amino acid sequence in a given protein.

SUMMARY

- There is an immense diversity of living organisms but their underlying chemical processes are strikingly similar. They contain atoms combined together to form organic compounds, for example, carbohydrates, lipids, proteins and nucleic acids. The various properties of biologically important molecules are a function of the atomic structure of individual elements.

- Atoms consist of a central nucleus of protons (positively charged) and neutrons (neutral charge), and orbiting electrons (negatively charged). Electrons can move between orbitals of different energy levels with the gain or loss of energy.

- Chemical forces between atoms can lead to stable associations found in molecules and compounds. Forces between molecules include hydrogen and ionic bonds, which involve attraction between opposite charges, and covalent bonds, which involve the sharing of electrons.

- The special properties of water make it an ideal medium for living organisms. It is an excellent solvent for polar molecules, which dissociate in solution to form ions (ionisation); however, it repels non-polar molecules.

- Substances that give up H^+ into solution are acids, while those that accept H^+ are bases. The concentration of H^+ in solution is pH, which is measured on a logarithmic scale ranging from 0 to 14. A buffer solution is one that maintains a relatively constant pH.

- Carbon is the principal structural element of living matter and forms the backbone of organic molecules. Carbon atoms have no overall tendency to gain or lose electrons, and form strong C—C bonds.

- Carbohydrates, which are simple sugars and their polymers, function for short-term energy supply and storage, and as structural molecules.

- Lipids, which include fats, oils, waxes, phospholipids, glycolipids and steroids, are insoluble in water due to the non-polar nature of their numerous C—H bonds. Lipids contain a higher proportion of energy-rich bonds than do carbohydrates per unit weight.

- Proteins are composed of one or more polypeptides, which are polymers of up to 20 different types of amino acids. The function of a protein depends on its three-dimensional shape, and the nature of the R-groups of its component amino acids.

- Nucleic acids are chains of nucleotides, each nucleotide being composed of a pentose sugar, a phosphate and a nitrogenous base. There are five nitrogenous bases, which are either pyrimidines or purines.

- DNA is a double-helical nucleic acid, which stores hereditary information as a coded sequence of nucleotide bases. RNAs are single-stranded nucleic acids, responsible for translation of the nucleotide sequence of DNA into the amino acid sequence of a protein.

QUESTIONS

1. Draw several water molecules. Indicate on your diagrams the forces holding the atoms together and the forces between the molecules in liquid water. Describe the difference between the two types of forces.

2. Explain how the distribution of electrons in the orbitals of atoms can lead to the formation of a molecule and a compound.

3. Explain how the special properties of water contribute to the functions of living organisms.

4. What is pH and how does a buffer work?

5. Construct a table summarising the properties of the four macromolecules that are the chemical building blocks of all organisms. Include their basic subunits and chemical composition, and their biological functions.

6. Plants store energy as carbohydrates, whereas seeds and animals store energy as lipids. Suggest an explanation for these differences.

7. Compare and contrast the storage carbohydrates used by plants and animals in terms of their structure and linkages.

8. (a) What is the R-group of an amino acid? (b) Draw an example of one amino acid with a simple R-group and one with a complex R-group.

9. (a) How are the subunits of proteins held together to form a functional molecule? (b) Distinguish between fibrous proteins and globular proteins.

10. (a) Name the three components of a nucleotide. (b) Contrast comparisons of the structure and function of DNA and RNA.

THE CHEMISTRY
OF LIFE

In living organisms, cells are continuously using energy to do the work involved in growth and reproduction. Cells must replace damaged proteins, repair DNA, and maintain concentrations of various ions within the cell. Energy has to be expended simply to stay alive. Where does this energy come from? It comes from the breakdown of **fuel molecules**, such as carbohydrates or fats, that have energy-rich chemical bonds. The amount of energy contained in a molecule is its free energy. When fuel molecules are broken down, the energy they contain is converted into a usable form, which drives other reactions, resulting in work being done in a cell. This pattern of energy flow, in which fuels are broken down and energy is released to do work, is typical of all living organisms.

In this chapter we will consider:

- the nature of energy and how it is involved in chemical reactions;
- chemical reactions in cells;
- ATP, the molecule that provides immediate energy in a form usable by cells; and
- fuel molecules, their structure and the chemical energy they contain.

THE NATURE OF ENERGY

In simple terms, **energy** is the capacity to do work. Energy exists in a number of forms, such as heat, sound, electricity and light. **Kinetic energy** is the energy of movement, as in running water (Fig. 2.1). Heat or thermal energy is the kinetic energy of randomly moving molecules. Radiant energy is the kinetic energy of photons of light. **Potential energy** is stored energy and is the energy usually involved in biological systems. Chemical energy is the potential energy stored in the bonds of atoms and molecules.

Fig. 2.1 The potential energy of water in this stream in Litchfield National Park, Northern Territory, is converted to kinetic energy as it tumbles to a lower level. This form of energy is harvested by hydroelectric schemes

Much of the work carried out within organisms involves the transformation of potential energy to kinetic energy and vice versa (Fig. 2.2). The transformation of energy, both in the physical and biological worlds, is governed by the laws of thermodynamics.

(a)

(b)

Fig. 2.2 Examples of the transformation of potential energy to kinetic energy. (**a**) Australian wedge-tailed eagle swooping to catch its prey. (**b**) An emerging pea seedling converts potential chemical energy into the kinetic energy of movement as it pushes through soil towards the light

Laws of thermodynamics

The **first law of thermodynamics** states that *energy can be neither created nor destroyed*. According to this law, energy may be transformed from one form to another but the total energy of the universe remains constant. No energy transformation is 100% efficient. In biological systems, the efficiency of energy conversions from stored to usable energy is never

higher than 30%. When energy changes from one form to another, some is always lost as heat (energy unavailable for work). As all forms of energy can ultimately be converted into heat, energy is usually measured in terms of its equivalent heat (in joules).

Biological systems are highly organised and ordered structures that are formed and maintained as a result of highly organised energy input. Since an input of energy is required to maintain an ordered state, the continual loss of energy as heat in every energy conversion results in increasing disorder or **entropy**. This is the basis of the **second law of thermodynamics**, which states that *the entropy of the universe is increasing.*

> Energy is the capacity to do work and exists in two general forms: potential and kinetic energy. Energy transformations are described by the laws of thermodynamics.

CHEMICAL REACTIONS AND EQUILIBRIA

In the microcosm of a cell, there is a vast array of chemical reactions in which reactants are continuously being changed into products. When a chemical reaction is at **equilibrium**, there is no net change in the concentration of either the reactants or the products (Fig. 2.3). This is because the rates of the forward and reverse reactions are the same. Within the confines of the system in which they occur, chemical reactions that are at equilibrium are in a state of maximum disorder (possessing maximum entropy). Increased order can be achieved by expending energy to push a reaction away from its equilibrium. Energy is released from any reaction as it proceeds towards equilibrium, and energy is required to move a reaction away from equilibrium. Therefore, in a thermodynamic sense, a reaction that is not in equilibrium has potential energy, which can be released if the reaction proceeds spontaneously.

The equilibrium position of a chemical reaction is described by the **thermodynamic equilibrium constant** (K_{eq}):

$$K_{eq} = \frac{\text{concentration of product(s)}}{\text{concentration of reactant(s)}}$$

when the forward and reverse reactions rates are equal.

If the reactants and products contain the *same* chemical energy per molecule, the K_{eq} is 1.0 (Fig. 2.3). For such a reaction to do work, that is, to be far enough away from equilibrium, there must be a high concentration of reactants or a low concentration of products. However, if the reactants or products contain very different amounts of chemical energy

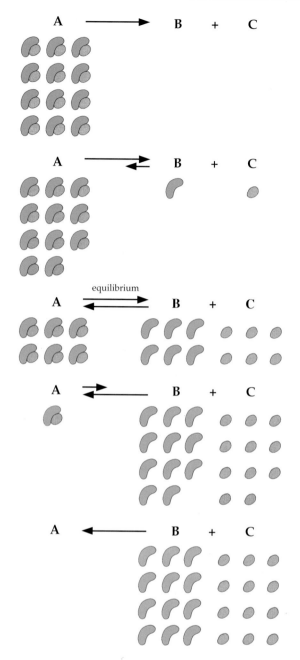

Fig. 2.3 With a high concentration of reactant A, the reaction will proceed strongly in a forward direction. With a high concentration of products B and C, the reaction will proceed strongly in the opposite direction. At equilibrium, the rates of the forward and reverse reactions are the same. (In this example, $K_{eq} = 1.0$)

Reactions must be out of equilibrium in order to do work. The most useful reactions for providing energy are those with equilibrium constants far higher or far lower than 1.0.

Free energy and the equilibrium constant

In effect, the equilibrium constant describes quantitatively how far towards completion a reaction will proceed but gives no indication at all about the *likelihood* of the reaction occurring in the first place. The feasibility of a reaction occurring requires some estimate of change in the orderliness of the system in question. This is measured in terms of the degree of disorder of the system, entropy (S).

Reactions that are not at equilibrium have the capacity to do work, and the extent of this work (or its value in energy terms) depends on the concentration differences between the reactants and products and the intrinsic energy content of each. The relation between the equilibrium constant for a reaction and the energy available to do work as the reaction proceeds is embodied in the concept of free energy.

Free energy (G, after the chemist Josiah Gibbs, the founder of the science of thermodynamics) is defined as the algebraic difference of the change in heat content (ΔH) and change in entropy (ΔS) at temperature T (measured in degrees Celsius above absolute zero) according to the equation:

$$\Delta G \text{ (change in free energy)} = \Delta H - T\Delta S$$

This equation is useful in predicting whether or not, and in which direction, a reaction will proceed. If ΔG is negative, the reaction will proceed spontaneously and give off heat, and is said to be **exothermic**. Conversely, if ΔG is positive, the reaction will not proceed spontaneously but will require absorption of heat. Such a reaction is **endothermic**. If ΔG is zero, the reaction is at equilibrium.

Sometimes these ideas are expressed in terms of energy rather than heat (Fig. 2.4). When ΔG is negative and energy is released in a reaction, the reaction is **exergonic**; that is, the reaction is spontaneous or goes 'downhill' in energy terms. When ΔG is positive and energy is needed for a reaction to proceed, the reaction is **endergonic**; that is, the reaction cannot proceed without the input of extra free energy. In cells, endergonic reactions can be driven by being coupled with exergonic reactions so that one supplies energy to the other. In biological systems, the coupling of such reactions results in the formation of chemical pathways in which non-spontaneous energy-requiring reactions are linked to spontaneous energy-yielding reactions.

per molecule, the K_{eq} will be far higher or far lower than 1.0. The reaction will thus be out of equilibrium when the concentration of reactants and products are equal.

Cells have a limited solvent capacity and yet contain many solutes (several thousand enzymes and chemical reactants) in relatively low concentrations. Therefore, given low and similar concentrations of reactants and products, reactions with a K_{eq} far higher or far lower than 1.0 are the most useful for doing work within the physical constraints of a cell.

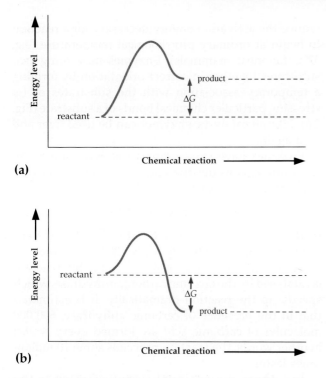

(a)

(b)

Fig. 2.4 (a) In an endergonic reaction, ΔG is positive; that is, energy must be supplied in order for the reaction to proceed. **(b)** In an exergonic reaction, ΔG is negative; that is, energy is released

Change in free energy (ΔG) is the energy that can be drawn upon to do work. It is the energy available above and beyond the energy changes due to the making and breaking of chemical bonds and the overall change in orderliness of the system. As long as ΔG is negative, the reaction will proceed spontaneously.

Since it is difficult to measure the concentrations of reactants and products in cells, change in free energy can be used to express K_{eq}. Change in free energy is related to K_{eq} by the equation:

$$\Delta G^\circ = -RT \log_e K_{eq}$$

where ΔG° is the standard free energy change (the gain or loss of free energy when one mole of reactant is converted to one mole of product under defined conditions in solution), R is the universal gas constant (8.314 J mol^{-1} degrees Kelvin^{-1}), and $\log_e K_{eq}$ is the natural logarithm of K_{eq}.

The actual ΔG relates the concentrations of reactants and products in a cell to the standard free energy by the equation:

$$\Delta G = \Delta G^\circ + RT \log_e \frac{products}{reactants}.$$

Standard free energy change relates the *actual* concentrations of products of a reaction to their concentrations at equilibrium. This allows biochemists to work backwards and to evaluate the potential for work under the conditions that actually occur in cells.

The amount of energy available in a molecule is its free energy. A reaction will proceed spontaneously if its products contain less free energy than the reactants (an exergonic reaction). In spontaneous reactions, ΔG is always negative. A reaction will not proceed without the addition of energy if its products have more energy than the reactants (an endergonic reaction).

Activation energy and catalysts

Many chemical reactions normally proceed far too slowly to be of use to organisms, especially at the temperatures at which most function. The rate at which a chemical reaction will proceed towards equilibrium is independent of the equilibrium constant and depends on the kinetic energy of the molecules involved. A collection of molecules in solution may have a constant average energy at a given temperature but the energy of individual molecules will vary widely. For a chemical reaction to occur between any two molecules, the pair must possess more than some minimal level of energy necessary to break existing bonds at the instant they collide. Those with less than this amount will not react at all. The energy required to initiate a reaction is the **activation energy** of that reaction (Fig. 2.5).

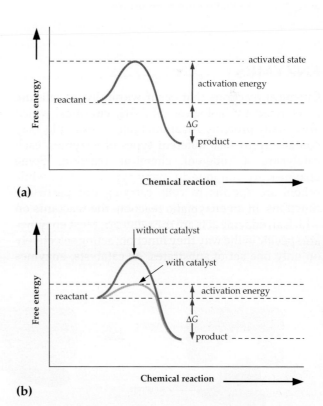

(a)

(b)

Fig. 2.5 Energy required for a chemical reaction. **(a)** For a reaction to occur, energy must be supplied to alter chemical bonds. This energy is the activation energy. **(b)** A catalyst speeds up a reaction by lowering the amount of activation energy required

The simplest way to increase the rate of a chemical reaction is to raise the temperature. Heating increases the kinetic energy of molecules and thus increases the proportion of molecules whose energy exceeds the activation energy; thus the rate of reaction increases. Obviously, living systems could not rely solely upon increasing their temperature to achieve useful reaction rates. Apart from anything else, their proteins would denature (change shape) at temperatures required to achieve useful reaction rates. There is another option and that is catalysis.

Catalysis is the process of lowering the activation energy of a reaction by the use of a catalyst (Fig. 2.5). A catalyst affects only the *rate* of a reaction by reducing its activation energy and allowing it to occur more readily at low temperature. It only affects reactions that are going to occur but at a slower rate; it cannot cause an endergonic reaction to proceed spontaneously. The reduction in activation energy by a catalyst accelerates *both* forward and reverse reactions by exactly the same amount without changing the final equilibrium position. At the end of a reaction, the catalyst itself remains unchanged and so can be used over and over again.

> The amount of activation energy required to break existing bonds determines the speed of a reaction. Catalysis is the process of lowering activation energy so that a reaction occurs more readily.

ENZYMES

Chemical reactions that occur within living cells are accelerated by biological catalysts, **enzymes**, which are usually proteins. An animal cell, for example, may contain up to 4000 different types of enzymes, each catalysing a different chemical reaction. Some enzymes are common to many types of cells, while others are specific to cells carrying out particular functions. In an enzymatic reaction, the reactants on which an enzyme acts are the *substrates*. Most enzymes are specific in the way they function, acting selectively on only one set of substrates. As catalysts, enzymes

reduce the activation energy necessary for a reaction to begin at ordinary physiological temperature (e.g. 37°C for most mammals). Enzymes may bring two substrates together in correct orientation by forming a temporary association with the substrates or by stressing particular chemical bonds of a substrate (Fig. 2.6). Like all catalysts, enzymes can be used over and over again.

Enzymes have extraordinary catalytic power. As an example, consider the reaction of carbon dioxide and water to form carbonic acid:

$$CO_2 + H_2O \rightleftharpoons H_2CO_3$$

In the absence of an enzyme, this reaction is very slow, perhaps producing only 200 molecules of carbonic acid per hour. In living cells, this reaction is catalysed by the enzyme carbonic anhydrase, which speeds up the reaction dramatically. It is estimated that in the presence of carbonic anhydrase, 600 000 molecules of carbonic acid are formed every *second*; in other words, the reaction proceeds some 10 million times faster.

In three-dimensional conformation, the polypeptide chains of an enzyme are folded in such a way as to form one or more pockets or grooves on the surface, forming a specialised region into which the substrate molecules can fit. This region of the enzyme is known as the **active site** (Fig. 2.6). The active site of each enzyme has a specific shape and only one sort of substrate molecule can fit into it. Active sites are not only sites at which substrates may be bound, but they also orientate substrate molecules in a particular direction. This is because an active site has charged and uncharged and hydrophilic and hydrophobic regions that bind precisely with particular parts of the substrate molecule. For example, if a substrate has a positive charge on a portion of its molecule that binds to the active site, then the corresponding part of the enzyme's active site will have a negative charge.

An active site is formed from only a few of an enzyme's total complement of amino acids. These may be adjacent to one another in a polypeptide chain or on different parts of polypeptide chains that are

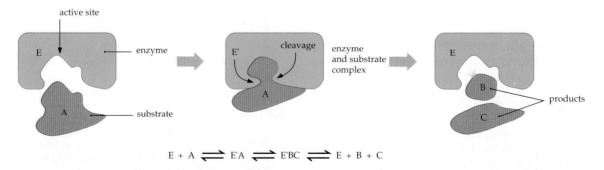

$$E + A \rightleftharpoons E'A \rightleftharpoons E'BC \rightleftharpoons E + B + C$$

Fig. 2.6 When an enzyme binds to a substrate, the interaction places stress on particular bonds in the substrate molecule, in this case causing it to break into two parts

brought together in three dimensions by the tertiary or quaternary structure of the enzyme. The exposed R-groups of the amino acids forming the active site and their detailed arrangement in relation to one another determine the binding of enzyme with substrate (Fig. 2.7).

Fig. 2.7 A three-dimensional model of the enzyme α-chymotrypsin, showing the three amino acid residues of the active site (red)

Enzymes are proteins that are the catalysts of reactions in cells. The specificity of an enzyme is attributed to its active site, which fits only one type of substrate molecule.

Models of enzyme action

The lock and key model

According to this hypothesis, a substrate fits exactly into the active site just as a key fits a specific lock (Fig. 2.8). The substrate and enzyme come into intimate contact to form a transient enzyme–substrate complex. The substrate is altered in a highly specific manner and, subsequently, the product or products are released.

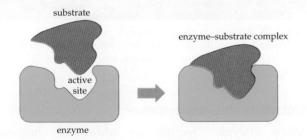

Fig. 2.8 In the lock and key model of enzyme action, the substrate molecule fits directly into the active site of the enzyme

The induced-fit model

Studies on enzyme structure suggest that the active site is considerably more flexible than a keyhole. In some cases, during binding of a substrate, conformational changes may be induced in the shape of the active site of the enzyme, producing a slightly better fit (Fig. 2.9). This induced fit alters the strength of particular bonds of the reacting molecule, thereby facilitating the reaction being catalysed.

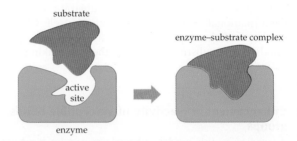

Fig. 2.9 In the induced-fit model of enzyme action, the substrate induces a conformational change in the enzyme that positions the substrate so that a reaction will occur

Factors affecting enzyme activity

In a living cell, chemical processes are usually the result of the action of enzymes. Enzyme activity is regulated so that appropriate amounts of products are made at the required rate. Underproduction would result in a slowing of cellular activity. Overproduction would result in wastage of both energy and raw materials and is often avoided by the end product itself inhibiting enzyme activity (end-product inhibition).

The rate of a reaction is partly a function of the concentration of substrates, products and enzymes. For example, the higher the concentration of substrate molecules, the more frequently the substrate molecules are likely to encounter the active sites of enzyme molecules. However, the reaction rate will eventually reach an upper limit when all active sites of enzymes are occupied. A further increase in rate would require an increase in enzyme concentration. Apart from the concentrations of the enzyme, substrate and product molecules, environmental factors regulate enzyme activity. These include pH, temperature and cofactors.

pH

pH can affect enzyme activity by changing the conformation of the enzyme, particularly the interaction between negatively charged (acidic) and positively charged (basic) amino acid residues (p. 19). With change of pH, the shape of the active site of

---

BOX 2.1   HOW ENZYMES WORK

There are many different hypotheses to explain how enzymes reduce the amount of activation energy needed for a reaction to occur. These are not competing hypotheses as one or more may be relevant to a particular reaction.

Conformational changes

Binding of an enzyme and its substrate results in conformational changes in the enzyme itself. Changes in shape of the enzyme impose stress on susceptible bonds of the substrate molecule, thereby making them easier to break.

Enhancement of reactivity of susceptible bonds or groups

According to this model, substrate molecules bind with the active site of an enzyme by forming weak bonds (sometimes covalent bonds). As a result, redistribution of electrons within the substrate molecules takes place, leading to weakening of susceptible bonds, which makes them more easily broken and thus more reactive.

Maximum collision effect

In an enzyme-substrate complex, binding at the active site ensures that substrate molecules are in a precise orientation. Also, the active site of the enzyme itself is aligned to promote the formation of an enzyme-substrate complex upon collision. In this situation, a great force of collision is no longer necessary for interaction because the enzyme surface is oriented for maximum effect.

Microenvironment

The chemical environment of the active site, at which the reaction takes place, is very different from that of the surrounding medium. The active site of each type of enzyme has an ionic microenvironment specially suited to maximise interaction between the substrate and enzyme. For example, if the active site has amino acids with acidic R-groups, its microenvironment would be a local region of low pH.

---

an enzyme may alter such that binding between the active site and substrate no longer occurs. Most enzymes have an optimum pH in the range 6–8. However, pepsin, which initiates protein digestion in the stomach, acts optimally at a very low pH (pH 2). Pepsin has an amino acid sequence that maintains its ionic and hydrogen bonds even in the presence of strong acid.

pH can also affect reactions in which H⁺ is a component of the chemical changes catalysed.

Temperature

Hydrogen bonding and hydrophobic interaction help maintain the tertiary and quaternary structure of proteins, including enzymes. These bonds are easily disrupted by changes in temperature. Organisms usually have enzymes that function optimally at the temperature at which they live. Animals such as the spangled and sooty grunter fishes in Central Australia have proteins that can function at high temperatures and enable them to live in hot springs (Fig. 2.10). In mammals, the optimum body temperature is 35–40°C. Below this temperature, the bonds that maintain the shape of the protein are not flexible enough to permit the shape changes that are necessary for enzyme action. Hence, their enzymes do not function efficiently at temperatures lower than 35°C. Above 40°C, hydrogen and other weak bonds are unable to hold proteins together in their correct conformation.

An enzyme that has lost its characteristic three-dimensional structure is denatured. Enzymes that are partially denatured by heat have a slightly distorted structure and their polypeptide chains can regain their correct shape on cooling, but complete denaturation, as occurs to albumin protein when an egg is boiled, is irreversible.

Fig. 2.10 Animals such as the sooty grunter fish in Central Australia have enzymes that can function at high temperatures and enable them to live in hot springs

Cofactors and coenzymes

Many enzymes require an additional chemical component, a **cofactor**, in order to function. Cofactors may be a metal ion or an organic molecule. When the cofactor is a non-protein organic molecule, such as a vitamin, it is a coenzyme.

Many enzymes have metal ions at their active site that help to draw electrons away from substrate molecules, thereby helping to fracture bonds. For example, the enzyme carboxypeptidase breaks proteins into amino acids by employing a zinc ion to help draw electrons away from the bonds joining the amino acids. In some cases, ions serve to hold the enzyme protein together.

| pH, temperature and cofactors are important in regulating enzyme activity.

CHEMICAL REACTIONS IN CELLS

All of the chemical processes going on inside the cells of a living organism—the building up, maintenance and breaking down of living tissue—constitute the **metabolism** of cells. These processes occur simultaneously. The energy released as some compounds are broken down is used to build other cellular components.

Of the vast array of chemical reactions essential to life, many fall conveniently into two categories. **Condensation reactions** involve the removal of water molecules in the assembly of complex molecules from simple ones, such as polysaccharides from simple sugars, fats from fatty acids and glycerol, polypeptide chains from amino acids (Fig. 2.11). **Hydrolysis reactions** involve the addition of water in the breakdown of complex molecules to simple ones, such as sugars from carbohydrates, fatty acids from fats, amino acids from polypeptide chains (Fig. 2.12).

In later chapters, we will present the major biochemical pathways associated with energy transformations. These are *glycolysis* and *cellular respiration*, which involve degradative reactions (breakdown of molecules, **catabolism**), and *photosynthesis*, which involves synthetic reactions (building of molecules, **anabolism**). In both pathways, adenosine triphosphate (ATP) is produced by processes that are dependent on electron transport systems. An **electron transport system** is a group of membrane-bound enzymes and cofactors that operate sequentially in a highly organised manner. The basis of electron transport is that electrons are transferred stepwise from one molecule (a donor) to another (an acceptor). When an atom or molecule loses one or more electrons, it is *oxidised*. Conversely, when it accepts one or more electrons, it is *reduced*. Such electron transfer reactions are oxidation–reduction reactions.

When a fuel such as a carbohydrate is broken down, molecular oxygen (O_2) combines with atoms of the fuel to form oxides of carbon and hydrogen. For this reason, the fuel is said to have been oxidised. More accurately, as mentioned above, **oxidation** of a compound involves the removal of electrons while in **reduction**, electrons are added (Fig. 2.13). Thus, when one substance is oxidised another is reduced. The gain or loss of electrons may be accompanied by the gain or loss of protons (H^+) and the removal or addition of oxygen.

Breakdown of fuel molecules to release energy involves the oxidation of a C—H bond, lowering the total number of reduced bonds by one. In terms of potential energy, C—C bonds have as much as C—H

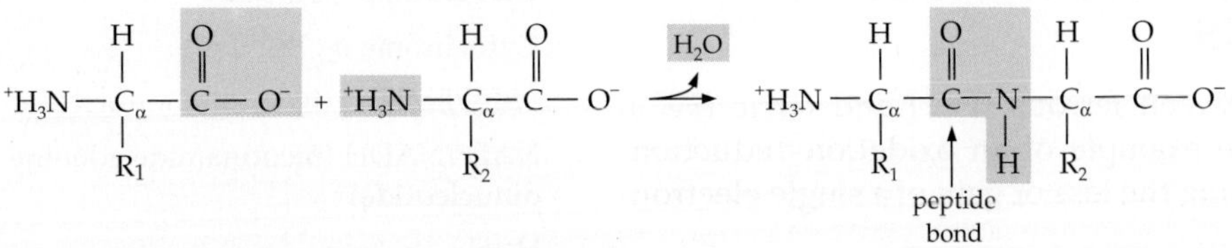

Fig. 2.11 In a condensation reaction, a water molecule is liberated; for example, in the formation of a peptide bond between two amino acids in a polymer

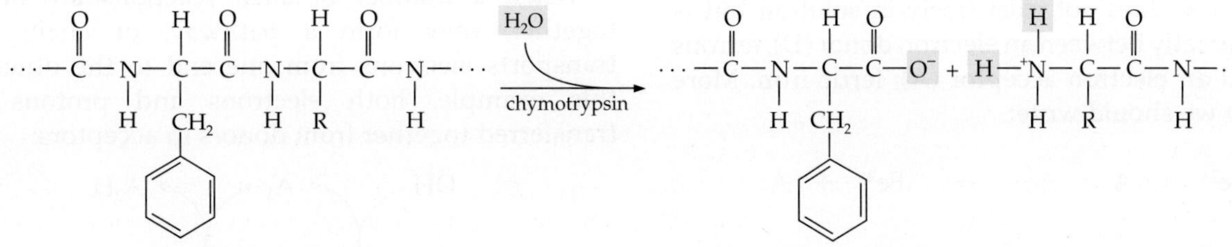

Fig. 2.12 In a hydrolysis reaction, a water molecule is used; for example, the enzyme chymotrypsin catalyses the hydrolysis of peptide bonds in proteins

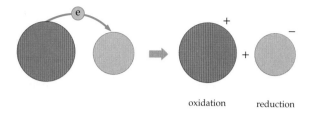

oxidation   reduction

Fig. 2.13 Oxidation is the loss of electrons and reduction is the gain of electrons

bonds but they cannot be oxidised directly. Therefore, some of the reactions in the degradative pathways are preparation reactions that convert C—C bonds to C—H bonds. The C—H bonds can then be oxidised.

Electrons removed from bonds are eventually shared with oxygen but this does not happen immediately. Electrons are first transferred to dedicated electron-carrying organic compounds that in turn transfer the electrons to other molecules and eventually to molecular oxygen. These sorts of oxidation and reduction reactions are important components of energy metabolism.

> Metabolism is the sum of all the chemical reactions occurring in cells. These commonly involve the removal of water molecules (condensation), the addition of water molecules (hydrolysis), removal of electrons (oxidation) and addition of electrons (reduction).

OXIDATION AND REDUCTION REACTIONS

Conversion between ferrous (Fe^{2+}) and ferric (Fe^{3+}) ions is a simple example of an oxidation–reduction reaction, involving the loss or gain of a single electron (e^-)

$$Fe^{2+} \xrightarrow{\text{oxidation}} Fe^{3+} + e^-$$

and

$$Fe^{3+} + e^- \xrightarrow{\text{reduction}} Fe^{2+}$$

The electron does not exist freely in solution but is passed directly between an electron donor (D), ferrous iron, and an electron acceptor (A), ferric iron. More correctly we should write:

$$Fe^{2+} + A \longrightarrow Fe^{3+} + A^-$$

and

$$Fe^{3+} + D^- \longrightarrow Fe^{2+} + D$$

or

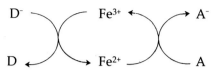

Each donor has a tendency to give up electrons, while each acceptor has a tendency to take up electrons. The tendency to donate or accept electrons can be measured as electrical potential, the oxidation–reduction (or redox) potential, which is expressed in volts or millivolts and indicated by the symbol E_0'. Any redox reaction is thermodynamically favourable (ΔG is negative) if electrons are transferred from a more electronegative potential to a less electronegative (or more electropositive) potential.

Some oxidation–reduction reactions together with their standard redox potentials are shown in Table 2.1. Cytochromes, flavin mononucleotide (**FMN/FMNH₂**) and nicotinamide adenine dinucleotide (**NAD⁺/NADH**) participate in the oxidation reactions involved in cellular respiration (Chapter 5). Oxygen (with a high potential) will accept electrons from NADH (with a low potential; Table 2.1). The products of this exchange are water and NAD⁺.

Table 2.1 Standard reduction potentials for reactions of biological importance (*a*)

Reaction	E_0' (mV) (*b*)
O_2/H_2O	815
Fe^{3+}/Fe^{2+}	770
Cytochrome *a*; Fe^{3+}/Fe^{2+}	290
Cytochrome *c*; Fe^{3+}/Fe^{2+}	220
Cytochrome b_2; Fe^{3+}/Fe^{2+}	120
FMN/FMNH₂ (flavin mononucleotide)	–120
NAD⁺/NADH (nicotinamide adenine dinucleotide)	–320
H⁺/H₂	–420

(*a*) Adapted from H.R. Mahler and E.H. Cordes, *Biological Chemistry*, Harper and Row, New York, 1966, p. 207.
(*b*) E_0' is the standard redox potential relative to that of the H₂ electrode at pH 7 (–420 mV).

When a number of such reactions are linked together, they form a pathway, or chain that transports electrons from one end to the other. In this example, both electrons and protons are transferred together from donors to acceptors:

Or, using specific examples from Table 2.1, a possible electron transport chain, including the voltage, is:

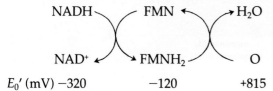

E_0' (mV) -320 $\qquad$ -120 $\qquad$ $+815$

Two electrons and one proton are transferred along the chain from NADH to reduce oxygen and form water. In doing this, the two electrons traverse an electrical potential of 1.135 V (from -320 to 815 mV).

The potential of such a reaction to do work is quantified by converting the electrical potential into the units of free energy, ΔG. The change in potential is related to the change in free energy by the equation:

$$\Delta G^\circ = -nF\Delta E_0$$

where n is the number of electrons transferred, F is a constant relating heat equivalent to electrical potential (the Faraday constant: 9.64 C mol^{-1}) and ΔE_0 is the change in redox potential.

In our example above, a ΔE_0 of 1.135 V corresponds to a ΔG of -218.8 kJ mol^{-1}($-2 \times 96.4 \times 1.135$), which is a strongly exergonic reaction sequence. This energy would all be lost as heat were it not for the fact that biological systems are able to conserve the energy derived from an ordered sequence of oxidation–reduction reactions.

> Oxidation–reduction reactions involve the transfer of electrons from a donor molecule to an acceptor molecule, and are measured in terms of their electrical (redox) potential.

Biological electron carriers

NAD$^+$ and FMN are examples of biological electron carriers, which carry electrons between oxidation and reduction reactions. Biological electron carriers act alternatively as electron acceptors or donors. Some of these compounds only carry electrons and some carry both electrons and protons, but it is the electron-carrying function that is important in terms of their redox potential.

Some electron-carrying compounds utilise a metal atom, frequently Fe, as the electron transporter. The metal atom is bound into the structure of a large organic molecule, for example, haem (Fig. 2.14), which is incorporated into a protein. For example, cytochrome c is a haem-containing protein, normally located in membranes, which is involved in the ferrous/ferric oxidation–reduction reaction.

There are several different cytochromes, all of which have slightly different haem structures. The chemical structure of the haem in which the Fe atom is

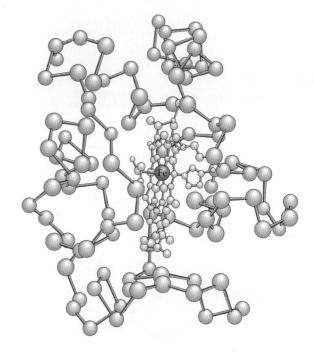

Fig. 2.14 Structure of cytochrome c. An iron-containing group, the haem group, is at the centre of the molecule

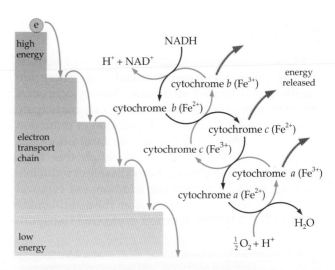

Fig. 2.15 An electron transport chain. Different cytochromes have different redox potentials and determine the unidirectional flow of electrons. As electrons move down the chain, energy that can be used for ATP synthesis is released

embedded alters the microenvironment around the Fe and, as a consequence, the redox potential of its reaction. Thus, different cytochromes, such as a, c and b_2, have different redox potentials (Table 2.1). These differences determine the unidirectional electron flow along a chain; for example, cytochrome b_2 is oxidised by cytochrome c, which is oxidised by cytochrome a (Fig. 2.15). All of these cytochromes can oxidise *free* Fe^{2+}.

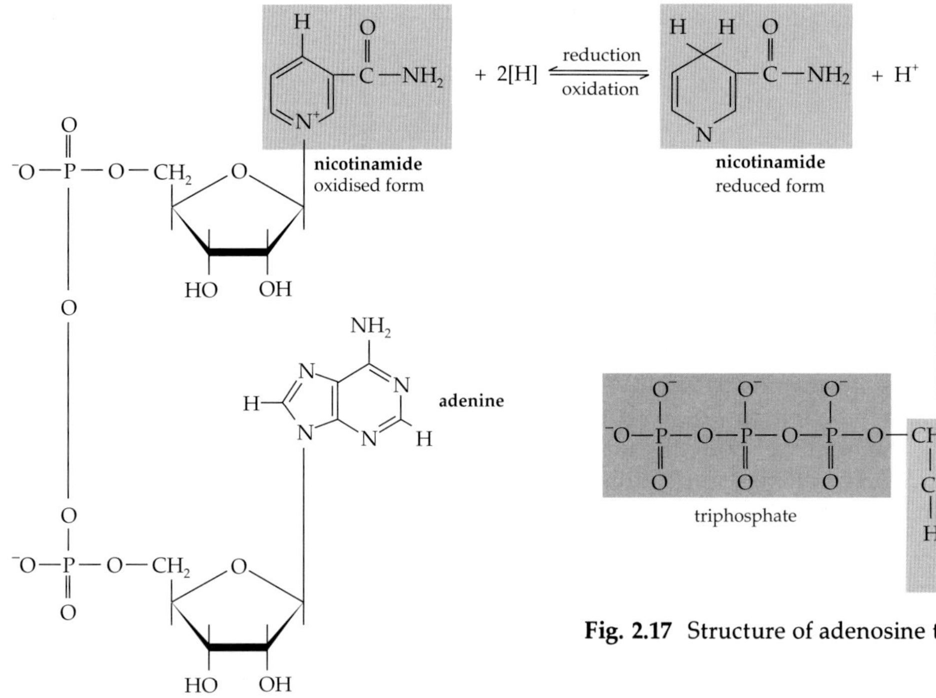

Fig. 2.16 The structure of the electron acceptor nicotinamide adenine dinucleotide (NAD⁺), showing the oxidised and reduced forms that occur in cells

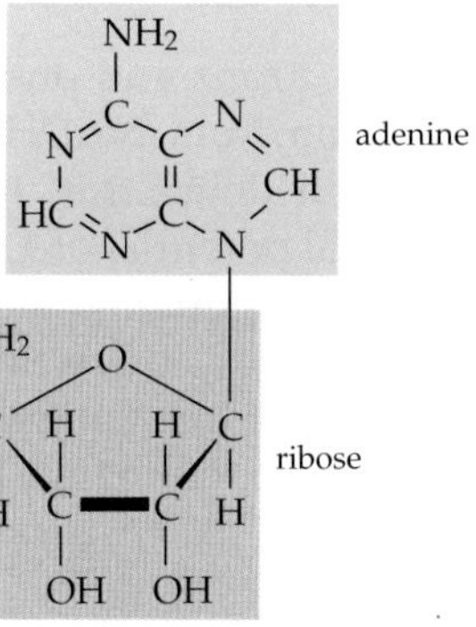

Fig. 2.17 Structure of adenosine triphosphate (ATP)

Many electron carriers accept and donate electrons without the participation of metal ions. These are usually much smaller molecules than cytochromes although still quite complex. They can participate directly in reactions involving oxidation or reduction of C—C, C—H or C—N bonds. The most important of these is NAD⁺ (Fig. 2.16). Although this large molecule has many potentially oxidisable or reducible covalent bonds, it is only the addition and removal of electrons and protons to and from the nicotinamide ring (see Fig. 2.16) that function in electron transport.

> Biological electron carriers (e.g. NAD⁺ and FMN) carry electrons between oxidation and reduction reactions.

ATP as an energy carrier

Reactions that oxidise reduced bonds in fuel molecules, releasing energy, are not coupled directly to reactions that require energy. The released energy is conserved in **adenosine triphosphate** (ATP) and released again when ATP is hydrolysed to ADP. ATP is a high energy compound with a high energy terminal phosphate bond (Fig. 2.17). Hydrolysis of ATP forms adenosine diphosphate (ADP) and inorganic phosphorus (P$_i$) and releases this energy, which can be used to drive non-spontaneous reactions.

ATP is well-suited to its role as an energy intermediate or 'energy currency' for a number of reasons.

1. The equilibrium constant of the ATP hydrolysis reaction is high. Consequently, the reaction is out of equilibrium at the low concentrations of ATP, ADP and P$_i$ that occur in cells and can provide energy as it moves towards equilibrium. The $\Delta G°$ for the hydrolysis of ATP to ADP is large and negative.
2. Since ATP levels can be low and the reaction is out of equilibrium, a small change in the concentration of ATP represents a large percentage change.
3. The energy conserved as ATP is released in a single step, which is both efficient and rapid.
4. Energy is released in small amounts, therefore it becomes available in small packets. Work that requires the equivalent of only half a reduced bond can occur without wasting much energy.
5. ATP is a common intermediate between degradative and synthetic pathways. Thus, the concentration of ATP is a regulator of metabolic pathways.

There are several phosphate compounds of biological importance, of which ATP is only one. For example, the major phosphate compound in vertebrate striated muscle is phosphocreatine, which is a higher energy compound than ATP (Table 2.2).

In the scale of $\Delta G°$ for phosphate, and therefore energy transfer, ATP is intermediate. In this way ATP can be formed from even higher energy compounds (e.g. phosphocreatine or 1,3-diphosphoglycerate) and can transfer this energy in the formation of phosphorylated compounds with a lower $\Delta G°$.

Table 2.2 Free energy of hydrolysis of phosphate compounds of biological importance (a)

Compound	$\Delta G°$ (kJ mol^{-1})
Phosphoenolpyruvic acid	−53.56
1,3-diphosphoglyceric acid	−49.37
Phosphocreatine	−43.93
ATP	−29.29
Glucose 1-phosphate	−20.92
Fructose 6-phosphate	−15.90
3-phosphoglyceric acid	−12.97

(a) Adapted from A.L. Lehninger, *Bioenergetics. The Molecular Basis of Biological Energy Transformation*, W.A. Benjamin Inc., New York, 1965, p. 59.

ATP is a high-energy compound. Energy is released when the terminal phosphate bond is hydrolysed.

Energy in fuel molecules

Carbohydrates, lipids and proteins can be used by cells as fuel molecules. There are three main differences between these fuels: firstly, the C:O ratio; secondly, the number of reduced bonds per carbon atom; and thirdly, the presence of nitrogen (N).

There is a lot of energy in every atom of a molecule, but as far as a cell is concerned, extractable energy is in the C—C, C—H and C—N bonds of the fuel molecules. These bonds are reduced because their electrons are not being shared with oxygen. During hydrolysis of fuel molecules, electrons are removed from the reduced bonds to become part of a bond with oxygen, which has a much greater affinity for electrons than either carbon or hydrogen. Reduced bonds contain vastly more extractable energy per molecule than do oxidised bonds. Oxidation reactions of these carbon bonds have high equilibrium constants and these reactions are out of equilibrium under all situations in the cell. Therefore, work can be done as a reaction moves towards equilibrium. The complete oxidation of glucose is a good example of such a reaction; its K_{eq} is approximately $10^{500\ 000}$.

$$C_6H_{12}O_6 + 6O_2 \longrightarrow 6CO_2 + 6H_2O$$

The other carbon bond found in fuels is the C—O bond but this is worth no energy in a biological sense as the electrons in the bond are already part of a bond with oxygen. Thus, the relative energy value of any fuel molecule to a cell is easily calculated by simply counting the number of C—C and C—H bonds.

The energy available to a cell depends on the type of fuel. For example, lipids represent more energy per carbon atom than do carbohydrates because lipids have a higher proportion of energy-rich C—H bonds. If the energy in lipids is expressed in terms of unit weight, then lipids have an even greater advantage as fuel molecules. Lipid molecules with reduced bonds (C—H bonds), compared with carbohydrates with oxidised ones (C—O bonds), weigh less. Gram for gram, lipids represent a greater store of energy.

Carbohydrates, lipids and proteins are fuel molecules. Extractable energy is contained in their C—C, C—H and C—N bonds. Lipids provide more energy per carbon atom than do carbohydrates.

SUMMARY

- Energy is the capacity to do work and exists in two general forms: potential energy or stored energy and kinetic energy or the energy of movement. Energy transformations are governed by the laws of thermodynamics. The first law states that energy can neither be created nor destroyed. However, energy can be transformed from one form to another. The second law states that the entropy (disorder) of the universe is increasing.

- Chemical reactions that are not at equilibrium have the capacity to do work. The most useful reactions for providing energy are those with equilibrium constants far higher or lower than 1.0.

- The amount of energy available in a molecule is known as free energy. A reaction will proceed spontaneously if its products contain less free energy than the reactants (an exergonic reaction). Other reactions will not proceed without the addition of energy if their products have more energy than the reactants (an endergonic reaction).

- The speed of a reaction is determined by the amount of activation energy required to break existing bonds. Catalysis is the process of lowering activation energy.

- Enzymes are the catalysts of living cells. The specificity of an enzyme is attributed to its active site, which fits only one type of substrate molecule. Temperature, pH and cofactors are important in regulating enzyme activity.

- Metabolism is the sum of all the chemical reactions occurring in cells. These commonly involve the removal of water molecules (condensation) or the addition of water molecules (hydrolysis). Oxidation–reduction reactions involve the transfer of electrons from a donor molecule to an acceptor molecule and are measured in terms of their electrical (redox) potential. Biological electron carriers (e.g. NAD^+ and FMN) carry electrons between oxidation and reduction reactions.

- Cells use ATP, a high-energy compound, to drive endergonic reactions. Energy is released when the terminal phosphate bond is hydrolysed.

- Carbohydrates, lipids and proteins are fuel molecules used in the synthesis of ATP. Extractable energy is contained in their C—C, C—H and C—N bonds. Lipids provide more energy per carbon atom than do carbohydrates.

QUESTIONS

1. State the first and second laws of thermodynamics. How do living organisms maintain a high level of organisation and yet conform to the second law?

2. (a) What does equilibrium mean in a chemical reaction? (b) Why is it that the most useful reactions for providing energy are those with equilibrium constants far higher or lower than 1.0?

3. Define entropy. What is the ΔG of a reaction? On what factors does ΔG depend?

4. Define (a) oxidation, (b) reduction and (c) hydrolysis. Give an example of each.

5. What do you understand by exothermic and endothermic reactions? What is activation energy and why is it important for biological systems?

6. What is an enzyme? What is the basis for the specificity of enzyme action?

7. What are coenzymes and cofactors, and what are their functions?

8. Describe the factors that affect enzyme activity.

9. Explain why ATP is well suited to its function as an energy intermediate in cells.

10. Why do fats provide a more efficient store of energy for cells than do carbohydrates in terms of (a) chemical bonds of the fuel molecules and (b) the relationships of the molecule to water?

FUNCTIONING CELLS

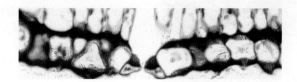

All living organisms are made up of cells and the materials produced by cells (Fig. 3.1). Cells are small membrane-bound compartments in which most biological reactions occur. They contain a diverse range of ions and molecules, both in aqueous solution and organised into complex subcellular structures. These subcellular structures provide compartments

(a)

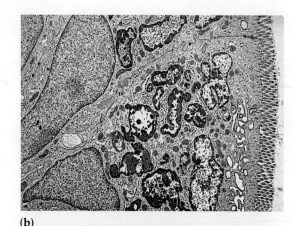

(b)

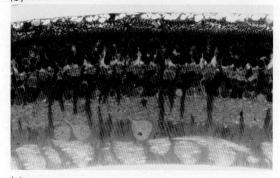

(c)

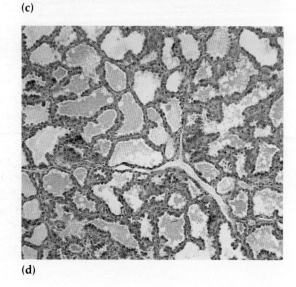

(d)

Fig. 3.1 All living organisms are made of cells.
(a) Multicellular organisms, like this Bennett's wallaby, contain millions of cells of many different types. The inserts show (b) intestinal epithelium (magnification × 4800); (c) retinal cells (magnification × 260) and (d) mammary gland tissue of the mother of *Macropus eugenii*

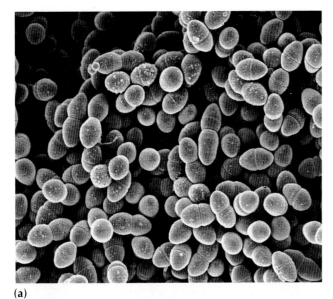

(a)

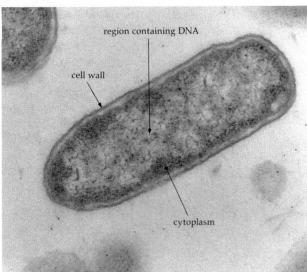

region containing DNA

cell wall

cytoplasm

(b)

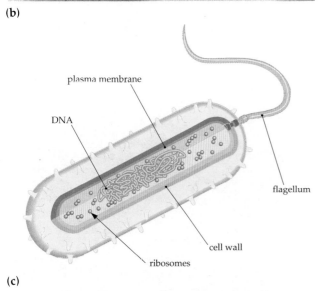

plasma membrane

DNA

flagellum

cell wall

ribosomes

(c)

in which particular chemical reactions can be contained. Localisation of enzymes and reactants within compartments increases the efficiency of reactions.

Some organisms consist of only one cell, while others are multicellular. Within the body of a multicellular organism, cells communicate and co-operate with one another so that their individual activities are integrated. The functions of an organism are the result of the activities of its cells.

During evolution, the internal complexity of cells increased greatly. The first cells to evolve were bacteria (see Chapter 30), which have a simple structure and lack internal compartments (Fig. 3.2). They are **prokaryotic cells** (from the Greek *pro*, before and *karyon*, kernel or nucleus), in which the hereditary material, a double-helical strand of DNA, lies free within the cell. The more complex cells that arose from bacterial ancestors are **eukaryotic cells** (from the Greek *eu*, proper and *karyon*, nucleus) (Fig. 3.3). Eukaryotes include algae, fungi, plants and animals, and they range from single-celled organisms such as amoebae to the largest and most complex plants and animals.

Despite the enormous variety of outward form seen in eukaryotes, their cellular structure is fundamentally the same (Figs 3.4, 3.5). One of the main distinguishing features of eukaryotic cells is their possession of internal, membrane-bound compartments, **organelles**. The most important of these is the **nucleus**, the control centre of the cell and the compartment within which DNA is stored. Other organelles include mitochondria, chloroplasts, vacuoles and endoplasmic reticulum. Eukaryotic cells also contain structures that are not surrounded by a membrane, such as ribosomes and microtubules. Each subcellular component carries out specific functions. The organelles and other components lie in the **cytosol**, an aqueous solution of molecules with a gel-like consistency. The cytosol and subcellular components, excluding the nucleus, constitute the **cytoplasm**. The cytoplasm and the nucleus together constitute the **protoplasm**.

Fig. 3.2 Prokaryotic cells. (**a**) Scanning electron micrograph of prokaryotic cells, *Streptococcus thermophilus*. (**b**) Transmission electron micrograph of a section through a prokaryotic cell, *Escherichia coli*, showing cell wall, dense cytoplasm with numerous ribosomes and less dense regions containing DNA. (**c**) Diagram of a motile prokaryotic cell that has a single flagellum

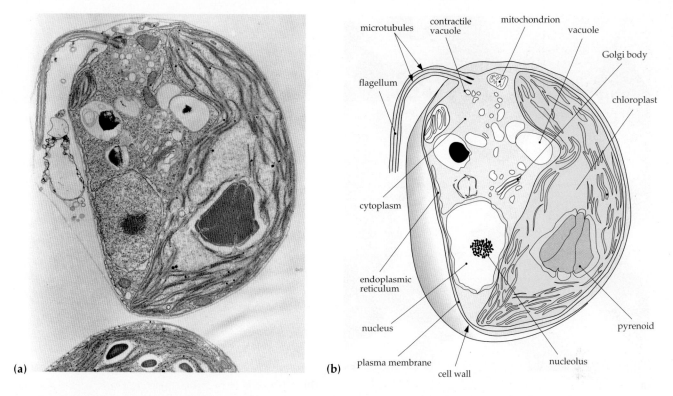

(a)

(b)

Fig. 3.3 Unicellular eukaryotic organism. **(a)** Transmission electron micrograph of the unicellular green alga, *Tetraspora* (magnification × 12 800). **(b)** Diagram showing cell structures. Note the greater number of membrane-bound organelles and compartments, including a nucleus, compared with prokaryotic cells (Fig. 3.2).

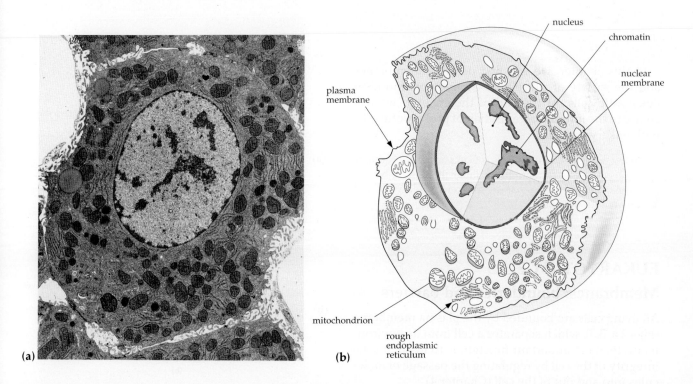

(a)

(b)

Fig. 3.4 **(a)** Transmission electron micrograph of a liver hepatocyte showing the typical features of an animal cell. **(b)** Model of an animal cell

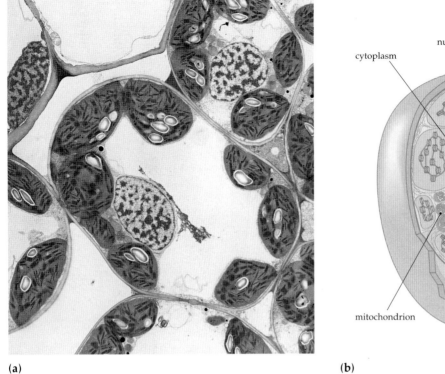

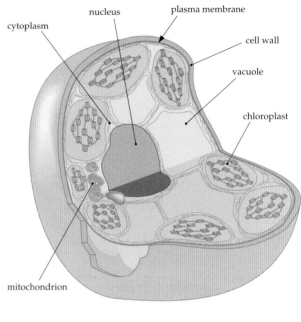

(a)   **(b)**

Fig. 3.5 A plant cell. **(a)** Transmission electron micrograph of a daisy leaf cell showing the characteristic cell wall (magnification × 3000). **(b)** Model of a plant cell

In this chapter, we will first describe membranes and the variety of subcellular components found in eukaryotic cells. At the end of the chapter, we will compare the internal organisation of prokaryotic cells with eukaryotic cells.

> Living organisms are made up of cells and cell products. Eukaryotic cells contain intracellular membrane-bound compartments that separate different molecules and metabolic reactions.

EUKARYOTIC CELLS
Membranes: boundaries and barriers

All living cells are bounded by the **plasma membrane** (Figs 3.4, 3.5), which separates a cell from its environment. Its most important function is to maintain the integrity of the cell by regulating the passage of molecules into and out of the cell (Chapter 4).

The basic structure of all membranes is a lipid bilayer composed mainly of phospholipids (Fig. 3.6). As discussed in Chapter 1, phospholipids are amphipathic molecules having a hydrophilic (water-loving) polar head and a hydrophobic (water-fearing) tail of fatty acids. In water, phospholipid molecules

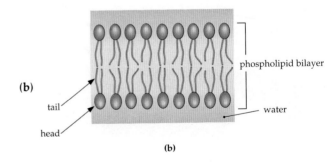

Fig. 3.6 Membrane lipids. **(a)** Phospholipids form micelles spontaneously. **(b)** In a membrane bilayer, the two monolayers of lipid molecules arrange themselves back to back in such a way that the polar heads face outwards to the water and the hydrophobic tails face each other in the centre of the bilayer. Phospholipids will behave this way in a test tube

cytoplasm   ribosomes   plasma membrane

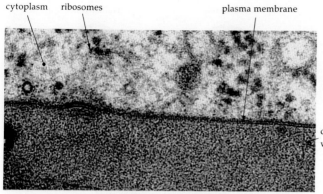

cell
wall

Fig. 3.6 (c) Electron micrograph of a transverse section of a root cell of *Eucalyptus sieberi* through a plasma membrane showing 'tramline' pattern corresponding to polar and hydrophobic portions of the bilayer. The positions of hydrophilic–hydrophobic–hydrophilic regions across a membrane are seen as a dark–light–dark structure after staining for electron microscopy because the protein heads absorb more electrons than the lipid tails (magnification × 113 000)

outside cell

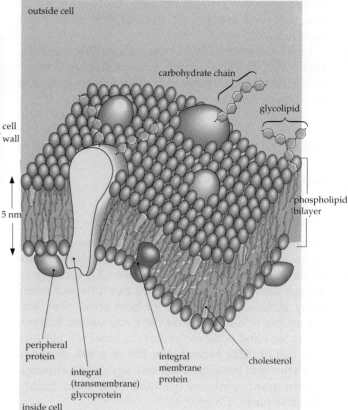

carbohydrate chain

glycolipid

5 nm

phospholipid bilayer

peripheral protein

integral (transmembrane) glycoprotein

integral membrane protein

cholesterol

inside cell

Fig. 3.7 Three-dimensional model of a membrane showing phospholipid bilayer, cholesterol, glycolipids, integral proteins and peripheral proteins. Carbohydrate chains bind to some proteins and lipids, forming a glycocalyx or sugar coating on the outer surface of the membrane. In many membranes, about 50% of membrane mass is made up of proteins, although this proportion can be as low as 25% (in myelin sheath of nerve cells) or as high as 75% (in internal membranes of mitochondria and chloroplasts)

aggregate and align themselves so that the polar heads face outwards and interact with the polar water molecules. The hydrophobic tails cluster together, away from contact with the water molecules. As a result, phospholipids tend to either bunch up into small spheres, **micelles**, or flatten out, forming a **lipid bilayer** (Fig. 3.6).

In addition to lipids, membranes contain two types of proteins (Fig. 3.7). Peripheral membrane proteins are loosely associated with the membrane surface by non-covalent interactions and can be removed with mild treatments such as washing in various salt solutions. Integral membrane proteins interact with the inner hydrophobic regions of the membrane, sometimes through covalent bonding to fatty acid chains or lipid molecules within the membrane. Some integral proteins extend right through the membrane and are known as transmembrane proteins. Integral membrane proteins can be extracted only by disrupting the membrane bilayer by washing with detergents.

The thickness of a membrane depends on its type. The plasma membrane is the thickest membrane in a cell (about 9 nm); membranes of the endoplasmic reticulum are the thinnest (about 6 nm). Membranes are usually, if not always, asymmetric, the properties on one side differing from those on the other. The lipid composition of each monolayer and the proteins exposed on each face are usually different. Carbohydrates, which may constitute 2–10% of the membrane mass, also contribute to membrane asymmetry (Fig. 3.7). Chains of sugar molecules may be attached to membrane lipids, forming **glycolipids**, or proteins, forming **glycoproteins**. The sugar chains occur on the non-cytosolic side of the membrane.

For intracellular organelles, this means that the sugar chains face the inside (lumen) of the organelle. For the plasma membrane, carbohydrate chains on the outer surface of the cell form a **glycocalyx**, which is a cell coat.

Fluidity of membranes

Another important aspect of membranes is their *fluidity*: membranes are a **fluid mosaic**. The fluidity arises because lipid molecules can move laterally (Fig. 3.8); *mosaic* refers to the irregular arrangement of proteins throughout the lipid bilayer. Lipid membrane molecules can travel the length of a bacterial cell (2.5 μm in length) in just one or two seconds. In eukaryote cells, cholesterol, a common membrane lipid, is a major determinant of fluidity. Some proteins move laterally within the membrane, while others are immobilised by attachment to filaments or tubules either inside or outside the cell.

Other proteins are immobilised in aggregates or because they are bound to proteins in adjacent membranes.

Lipid and protein molecules are also able to rotate, that is, flip about their axis perpendicular to the plane of the membrane. Movement of molecules from one monolayer to the other is a rare event. For a phospholipid molecule, movement from one monolayer to the other is so energetically unfavourable that it is unlikely to happen more than once a month! Cholesterol is one of the few molecules that can 'flip-flop' relatively easily.

Permeability of membranes

Membranes are *selectively permeable*. This means that some molecules can pass through the membrane while others cannot. Water, O_2 and CO_2 are able to diffuse freely across membranes but ions and other polar molecules, regardless of how small they are, are unable to diffuse across the hydrophobic bilayer. Many ions and polar molecules do, however, get into and out of cells; indeed, the life of a cell depends on their rapid passage. Their passage occurs through certain transmembrane proteins that form highly selective pores (Chapter 4).

Other membrane proteins are enzymes, which may be arranged in a linear sequence to drive a series of reactions in metabolic pathways, such as cellular respiration and photosynthesis (Chapter 5). Another class of membrane proteins functions as receptors for chemical signalling within and between cells (Chapter 7). Cell surface glycoproteins identify a cell as being of a particular type and are important in recognition processes (Chapter 23).

> Membranes are selectively permeable, fluid mosaic structures that are composed of a phospholipid bilayer containing proteins, glycoproteins and glycolipids.

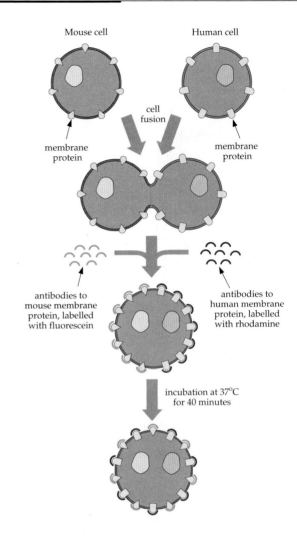

Fig. 3.8 An experiment in which human and mouse cells are fused in culture shows the mobility of cell-surface proteins. At first, the human and mouse marker proteins remain on their respective plasma membranes. After 40 minutes, the two sets of proteins intermingle. The proteins are detected by fluorescence microscopy

BOX 3.1   STUDYING THE STRUCTURE OF CELLS

Cells and subcellular components are usually too small to be seen by the naked eye. Our inability to see very small structures is a function of the limit of resolution of our eyes. *Resolution*, which is the ability to see two small particles as discrete objects, depends on the wavelength of light illuminating them and on the numerical aperture of the lens collecting the image. Numerical aperture is related to the angle subtended by the lens at the object. The human eye's limit of resolution is about 200 μm. Cells are usually only 10–20 μm in diameter and are thus well below our limit of resolution. We can, however, use microscopes to magnify images of cells and thus reveal details of their intracellular organisation. There are two main types of microscopes: light microscopes and electron microscopes (see Figs a–c).

Light microscopes use glass lenses to focus visible light onto an object and collect the light that passes through it. Using green light with a wavelength of about 550 nm, the limit of resolution of the light microscope is 250 nm. This is about 800 times better than can be achieved by eye. Thus, the upper limit for magnifications used in light microscopy should be about 800×. This can be obtained, for example, by using a 100× objective lens and an 8× ocular lens. At this magnification, particles at the limit of resolution of the light microscope will

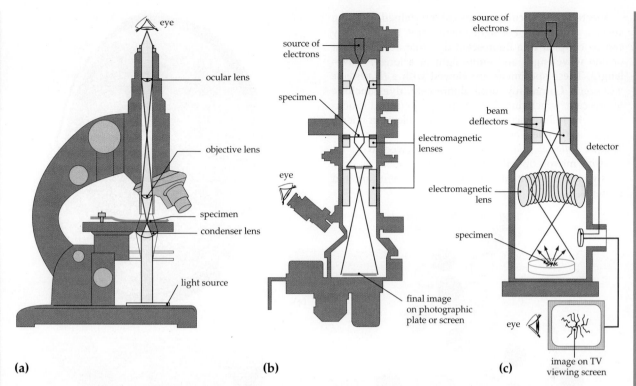

(a)   (b)   (c)

Comparison of different microscopes for viewing cells showing light and electron paths. (**a**) A simple light microscope (LM). (**b**) A transmission electron microscope (TEM). (**c**) A scanning electron microscope (SEM)

be magnified to a size close to the limit of resolution of the eye. Greater total magnifications than this will not reveal any further details of the object.

Another factor that determines whether an object is visible or not is its contrast with its surroundings. Many cells are quite transparent and no details of their internal structure can be seen with ordinary light microscopy. This problem can be solved in two ways. Optical systems that take advantage of a change in the phase of light waves passing through thick or dense parts of the sample can be used to generate contrast. Methods include *phase contrast microscopy* and *differential interference contrast microscopy*, both of which have the important advantage that cells can be observed while they are alive. Another approach is to stain the cell contents so that they contrast with the background. A wide variety of coloured dyes react with different types of organic molecules and these are used in light microscopy to visualise selected structures (see Fig. d).

In many cases where samples are too large to be observed intact, they can be embedded in a supporting resin or wax, which is cut into very thin slices (sections). Before doing this, tissues need to be treated in such a way that the contents of their constituent cells are preserved in a form as close to that in the living cell as possible. The most common method is to 'fix' cells in a chemical that cross-links proteins, such as formaldehyde or glutaraldehyde. Chemical fixatives do not act instantaneously and in some cells considerable disruption of cellular structure can occur during the fixation process. For this reason, it is important to observe samples that have been prepared by a variety

of methods in order to identify fixation artefacts. An alternative to chemical fixation is rapid freezing of the tissue. Frozen samples may be observed either directly or after cutting frozen sections (cryosections). Alternatively, water in a sample can be removed while the tissue is frozen, the process of freeze-substitution. The dehydrated sample is embedded in resin, then sectioned and observed in the same way as chemically fixed samples.

Electron microscopes use magnets to focus a beam of electrons onto a sample. The wavelength of electrons in the beam is of the order of 0.004 nm, which is much lower than the wavelength of visible light. Electron microscopes thus have a theoretical resolution of 0.002 nm. In practice, the resolution normally attained is about 2 nm, which is still 100 times better than that of light microscopes.

In *transmission electron microscopes*, electrons pass through a sample and are projected onto a phosphorus screen. Thin sections of 70–100 nm are routinely used and are usually stained to enhance electron contrast. Electron-opaque dyes, which contain atoms such as lead, osmium or uranium, absorb electrons, leaving a shadow on the phosphorescent screen. The structures that react with the dyes are contrasted against surrounding electron-lucent areas (see Fig. e).

In *scanning electron microscopy*, the beam of electrons is scanned across the sample, and electrons that bounce off the surface are collected and displayed on a video monitor. Scanning electron microscopes have a great depth of focus and reveal three-dimensional details of the surface (see Fig. f).

Selected cellular components can be stained by using specific antibodies. For light microscopy, the antibody can be coupled to a fluorescent dye that absorbs light of one wavelength and emits light at a longer wavelength. These specimens are viewed with a *fluorescence microscope*. Commonly used fluorescent dyes include fluorescein, which fluoresces a green colour, and rhodamine, which fluoresces a red colour. For electron microscopy, antibodies are usually coupled to electron-opaque *colloidal gold particles*. When viewed with an electron microscope, the gold particles appear as dark dots where the antibodies have bound to their targeted proteins.

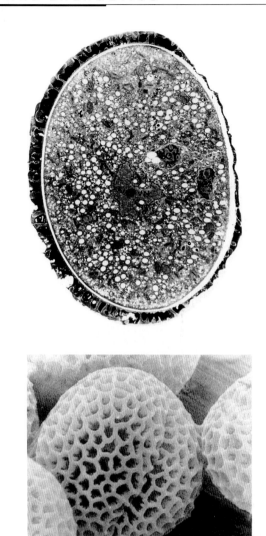

(e)

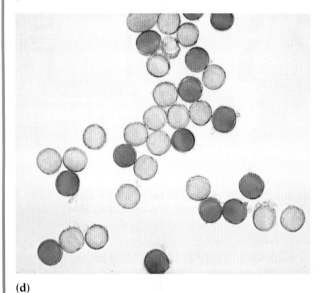

(d)

Examples of the same cell type: pollen grains of oil seed rape (*Brassica*) viewed using (**d**) light microscopy, (**e**) transmission electron microscopy, (**f**) scanning electron microscopy

(f)

The nucleus

The main feature of eukaryotic cells is the **nucleus**, which contains most of a cell's DNA (Figs 3.4, 3.5). The nucleus is surrounded by a double membrane, the **nuclear envelope**, and usually contains one or several **nucleoli**, which are darkly staining regions that contain high concentrations of RNA and protein as well as DNA.

The two membranes of the nuclear envelope are separated by about 50 nm (Fig. 3.9). They are perforated at intervals by **nuclear pores**, channels that allow the movement of certain molecules between the cytoplasm and nucleoplasm. The outer membrane of the nuclear envelope is continuous with the endoplasmic reticulum, a system of membranes that branches throughout the cytoplasm. The inner membrane is continuous with the outer membrane at the nuclear pores. However, the inner and outer membranes have distinct chemical compositions. The

nuclear lamina is a scaffolding meshwork of *lamins*, a class of proteins found only on the inner surface of the nuclear envelope giving it shape. Lamins form specific anchor sites that bind to DNA molecules and may prevent them becoming entangled.

Nuclear pores are not simple holes perforating the double membrane. A nuclear pore complex consists of two rings of globular proteins associated with the cytoskeleton. The diameter of the channel inside the rings is 65–75 nm. Molecules of up to 9 nm in diameter (equivalent to about 40 kD in molecular weight) pass freely through the pores but, in general, the entry of larger molecules is prevented. However, some proteins possess a signal sequence of amino acids. When this signal interacts with a receptor on or near the pore, the pore enlarges, allowing molecules much larger than 9 nm to be actively transported into the nucleus. Similarly, large molecules inside the nucleus, such as mRNA molecules or ribosomal subunits

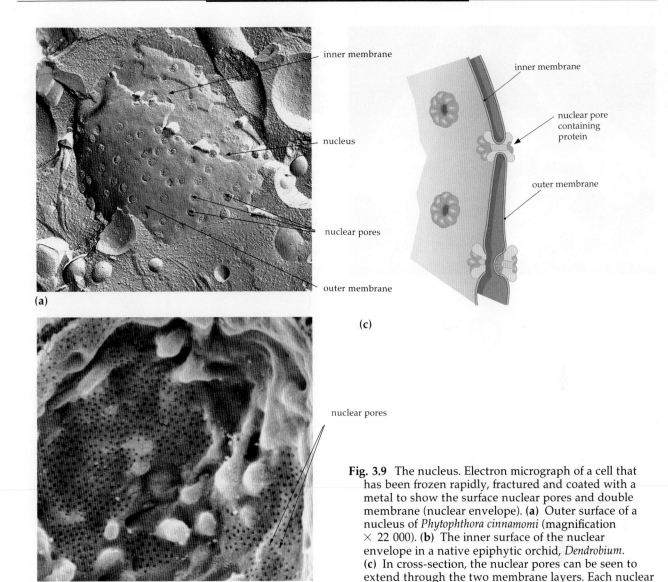

(a)

(b)

(c)

inner membrane

nucleus

nuclear pores

outer membrane

inner membrane

nuclear pore containing protein

outer membrane

nuclear pores

Fig. 3.9 The nucleus. Electron micrograph of a cell that has been frozen rapidly, fractured and coated with a metal to show the surface nuclear pores and double membrane (nuclear envelope). (**a**) Outer surface of a nucleus of *Phytophthora cinnamomi* (magnification × 22 000). (**b**) The inner surface of the nuclear envelope in a native epiphytic orchid, *Dendrobium*. (**c**) In cross-section, the nuclear pores can be seen to extend through the two membrane layers. Each nuclear pore contains protein

(which may be 15 nm in diameter), can interact with the inner face of the pore to be actively exported to the cytoplasm.

> The nucleus contains DNA and is surrounded by a nuclear envelope that is perforated by nuclear pores. Nuclear pores regulate the passage of proteins and RNA into and out of the nucleus.

Chromosomes

In the nucleus the long DNA molecules wind around clusters of **histone** molecules to form **nucleosomes**, resembling beads on a string. This arrangement allows DNA to twist into a helix, forming a chromatin strand (Fig. 3.10).

When a cell is not actively dividing, some chromatin strands aggregate, forming densely staining regions of **heterochromatin**, while others disperse, forming lightly staining regions of **euchromatin**. Euchromatic regions are sites of active gene transcription, where molecules of mRNA are manufactured. When a cell divides, the chromatin strands bunch up and thicken to form **chromosomes**, which stain densely and are large enough to be visible by light microscope (Fig. 3.10).

Nucleolus

The **nucleolus** is a suborganelle of the nucleus, and is a darkly staining spherical region easily recognised under the light microscope (Figs 3.4, 3.5). It is the site of synthesis and maturation of rRNA, and assembly of ribosomal subunits for export to the cytoplasm for protein synthesis. Cells typically make about 10 000 ribosomes per minute and, to do this, the nucleolus contains hundreds of copies of rRNA genes.

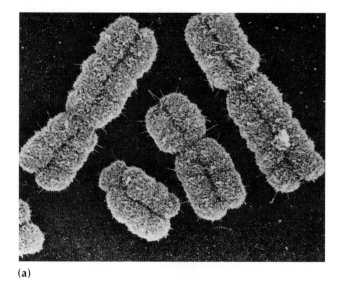

(a)

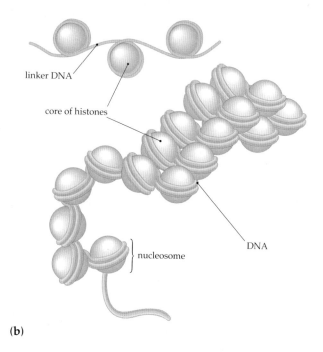

linker DNA

core of histones

DNA

nucleosome

(b)

Fig. 3.10 DNA and chromosomes of eukaryotic cells. **(a)** Scanning electron micrograph of human chromosomes. **(b)** Histones are associated with DNA molecules forming nucleosomes like beads on a string. The DNA molecule winds tightly around the nucleosome. Histones are proteins rich in the basic amino acids, arginine and lysine. This allows them to interact with the acidic phosphate groups in DNA

Ribosomes

The cytoplasm of eukaryotic cells contains several million **ribosomes**, small granular structures which are each 25–30 nm in diameter (Fig. 3.11). Ribosomes are composed of two subunits, each of which is assembled in the nucleolus from RNA and protein molecules. The subunits move through the nuclear pores into the cytosol where they associate with an mRNA molecule, forming a functional ribosome that facilitates protein synthesis (Chapter 11).

As a ribosome moves along an mRNA molecule synthesising a polypeptide chain, sufficient room (about 80 nucleotides) becomes available for another pair of subunits to attach to the mRNA molecule, forming a second ribosome. As ribosomes continue to move along the mRNA strand, more ribosomes become bound, forming a polyribosome or **polysome**. A polysome may remain free in the cytosol or it may become attached to the surface of the endoplasmic reticulum. When cytosolic proteins are made, the polysome remains free in the cytosol.

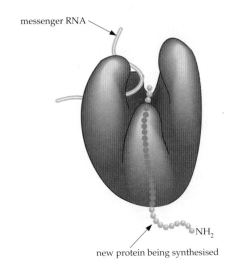

messenger RNA

NH$_2$

new protein being synthesised

Fig. 3.11 A ribosome. Ribosomes bind with mRNA and are the sites of protein synthesis

Ribosome subunits are assembled in the nucleolus and move into the cytosol where they associate with an mRNA molecule, forming a functional ribosome.

Endoplasmic reticulum

Endoplasmic reticulum (ER) is a network of membranous sacs (cisternae), which extends throughout the cytoplasm of eukaryotic cells (Fig. 3.12). It is continuous with the outer membrane of the nuclear envelope. Cisternae are usually flat and sheet-like but are often linked by tubular cisternae. The whole array forms an interconnected system composed of a continuous membrane enclosing a cisternal space. Although the cisternal space occupies only a small fraction of the cell volume, the surface to volume ratio

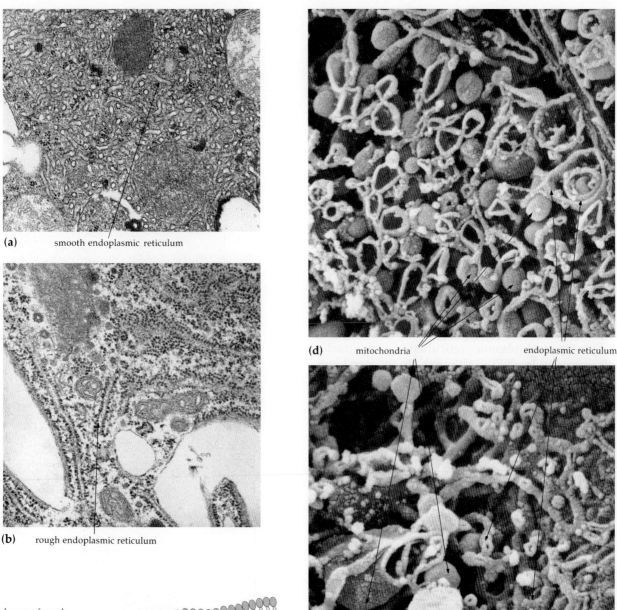

(a)   smooth endoplasmic reticulum

(b)   rough endoplasmic reticulum

(d)   mitochondria   endoplasmic reticulum

(e)   plasma membrane

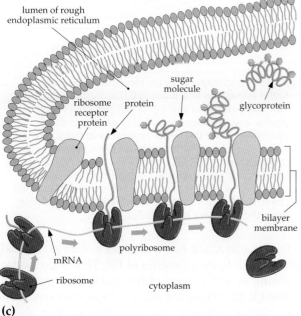

lumen of rough
endoplasmic reticulum

sugar
molecule

ribosome   protein   glycoprotein
receptor
protein

bilayer
membrane

polyribosome

mRNA

ribosome   cytoplasm

(c)

Fig. 3.12 Endoplasmic reticulum (ER). Transmission
electron micrographs of (**a**) hair cell of the alga
Bulbochaete showing smooth ER (magnification
× 41 000) and (**b**) gland cell of *Primula kewensis*
showing rough ER (with ribosomes; magnification
× 25 000). (**c**) Proteins assembled on ribosomes of
rough ER pass immediately into the lumen of the ER.
Sugar chains may be added to the proteins within the
ER to make glycoproteins. Scanning electron
micrographs of developing pollen of the native
epiphytic orchid, *Dendrobium*, show that (**d**) rough ER
forms flattened sacs (cisternae), which form in the
cytoplasm among mitochondria, while (**e**) smooth ER
is more tubular, and is here seen in association with the
plasma membrane

of this compartment is so great that ER often constitutes over half of the total membrane in the cell. In plant cells, ER not only pervades the cell but also forms direct connections between adjacent cells as it passes through the centre of pore-like structures, plasmodesmata (Chapter 6) between cells.

Most ER in a cell is **rough endoplasmic reticulum**, which is ER with ribosomes bound to its surface giving it a 'rough' appearance (Fig. 3.12). These ribosomes are involved in the synthesis of proteins that will be incorporated into membranes, secreted at the cell surface or transported into vacuoles or lysosomes. Newly synthesised polypeptides pass into the lumen of the rough ER. Once inside the ER, the protein folds into its preferred configuration, depending on the particular chemical environment. In addition, carbohydrate chains may be added to those proteins whose sequence contains accessible asparagine residues. When a carbohydrate molecule attaches to the amino group on the side chain of asparagine, an N-linked glycoprotein forms (Fig. 3.12).

Smooth endoplasmic reticulum lacks attached ribosomes (Fig. 3.12). Although both rough and smooth ER can synthesise lipids, in cells that specialise in producing large amounts of particular lipids, such as steroid hormones produced in testicles, smooth ER proliferates and makes the majority of these compounds. Endoplasmic reticulum manufactures nearly all the phospholipids and cholesterol that are needed by the cell for membrane repair and synthesis. Fatty acid precursors occur in the cytosol and the ER enzymes involved in lipid synthesis have their active site facing the cytosol.

Rough and smooth ER thus perform different functions even though their membranes are continuous. Different biosynthetic activities require different assemblages of membrane proteins. This means that the ER must be able to maintain regions that have different molecular compositions.

> Endoplasmic reticulum is a membrane system, composed of cisternae, with a large surface area. It is involved in protein and lipid synthesis.

The Golgi apparatus

The **Golgi apparatus** consists of stacks of four to 10 disc-shaped cisternae, which are 1–2 μm in diameter and positioned close together (Fig. 3.13). The cisternae may be flat or curved, and are dilated at their margins. Some algal cells have just one stack, while animal cells typically contain 10–20 stacks and plant cells may possess hundreds or even thousands of stacks. Each Golgi stack is usually surrounded by a cloud of small vesicles.

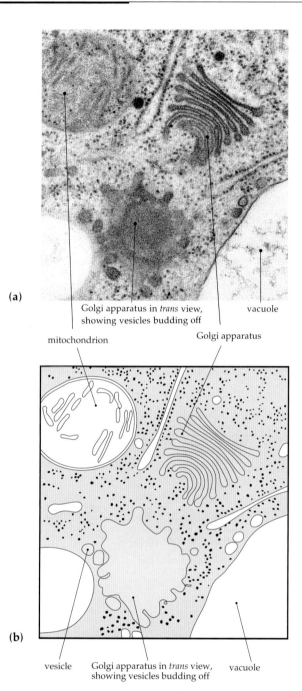

(a)
Golgi apparatus in *trans* view, showing vesicles budding off   vacuole

mitochondrion   Golgi apparatus

(b)
vesicle   Golgi apparatus in *trans* view, showing vesicles budding off   vacuole

Fig. 3.13 Golgi apparatus. **(a)** Electron micrograph of Golgi apparatus in a plant cell (*Eucalyptus sieberi*). The Golgi apparatus at the top of the picture is seen in profile and vesicles are budding off the ends of cisternae, while the Golgi apparatus in *trans* view shows vesicles at right angles to the upper structure (magnification × 43 000). **(b)** Two-dimensional model showing Golgi apparatus in an animal cell. Small vesicles bud off from the edges of stacks of flattened cisternae. Large secretory vesicles form by budding from the *trans* face of the Golgi

Each Golgi stack has a distinct polarity. One side of the stack faces a cisterna of ER. This is the *cis* face (forming or entry face) of the stack. The membrane of the ER that faces the *cis* face lacks ribosomes, and the cytosol between the ER cisterna and the *cis* cisterna of a Golgi stack is filled with small vesicles that are about 50 nm in diameter. The opposite side of the Golgi stack is the *trans* face (maturing or exit face). It is associated with a system of tubular membranes, the *trans*-Golgi network.

Proteins and glycoproteins enter each Golgi stack at the *cis* face. They are transferred from the ER in small transport vesicles that bud off the ER membrane and fuse with the membrane of the *cis* cisterna. As they progress through the stack, proteins and glycoproteins are modified in many ways. Sugar units of *N*-linked glycoproteins may be trimmed and other sugars added to the core of sugars that is left. *O*-linked glycoproteins are produced in the Golgi apparatus from proteins containing serine, threonine or hydroxyproline residues. Sugars are added to hydroxyl groups of side chains (i.e. *O*-linked) on these amino acids. The resulting molecules often have many sugar molecules added to them, thus forming **proteoglycans**, which form surface slimes and mucus. Polysaccharides are also formed in the Golgi apparatus. Polysaccharides and glycoproteins mature as they progress across the stack. Once mature products reach the *trans* face of the stack, they must be sorted and targeted to their correct destinations. Much of this sorting and packaging occurs in the *trans*-Golgi network.

> The Golgi apparatus is involved in the processing and packaging of glycoproteins and polysaccharides.

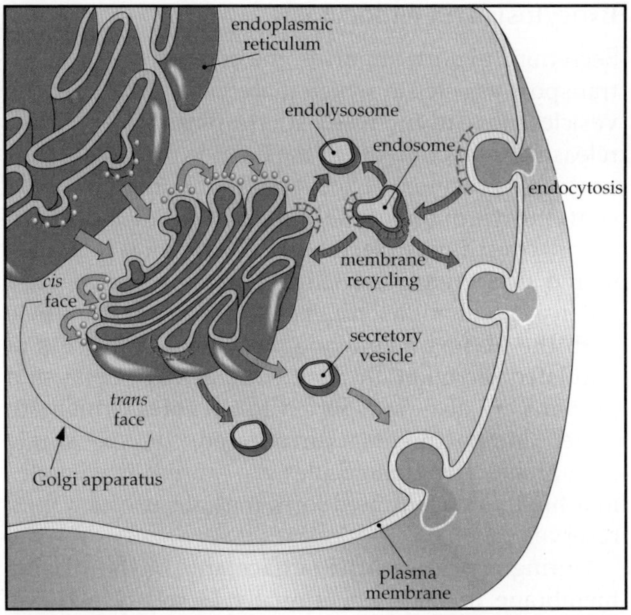

Fig. 3.14 Movement of vesicles (protein transport) from endoplasmic reticulum, through Golgi apparatus to the cell surface. Model shows unselected (constitutive) transport occurring by means of uncoated vesicles (blue arrows). The various kinds of signal-mediated transport are carried out by clathrin-coated vesicles (red arrows) budding from the *trans* face of the Golgi apparatus

Intracellular sorting and transport

The Golgi apparatus processes a large number of different products, which must be packaged and transported to different locations in the cell. Some molecules may be enzymes required for degradative processes within the cell. Some may be glycoproteins or polysaccharides for export across the plasma membrane.

Vesicle formation is the means by which these materials are sorted and distributed throughout the cell. The formation of vesicles is aided by the binding of special scaffolding proteins to patches of membrane, which then round up and pinch off. The bound protein, clathrin, gives the vesicles a coated or spiny appearance. Once the coated vesicles are free in the cytoplasm, the coat of clathrin dissociates and the vesicle is free to fuse with the membrane of its target site (Fig. 3.14).

The sorting of molecules in the *trans*-Golgi network and the mechanisms of targeting vesicles in which molecules are packaged, suggests that routing to the plasma membrane for export does not require any specific signalling mechanism, whereas routing to other destinations requires precise tagging. The best understood pathway is that to lysosomes.

Lysosomes

Lysosomes are membrane-bound organelles of animal cells involved in the breakdown of many types of molecules. They enable a cell to break down worn-out organelles, allowing molecules to be recycled to form new organelles. In this way, animal cells maintain their functions despite wear and tear. Lysosomes contain about 40 different hydrolytic enzymes, including nucleases, proteases, lipases and glycosidases. These enzymes are *N*-linked glycoproteins manufactured in the Golgi apparatus. The enzymes are tagged with mannose-6-phosphate residues attached to their oligosaccharide chains. This tagging, which happens in the *cis* cisternae of the Golgi apparatus, enables the glycoprotein enzymes to be sorted into specific vesicles in the *trans*-Golgi network (Fig. 3.14). Markers on the surface of these vesicles are recognised by receptors on the surface of other organelles, **endolysosomes**. The vesicles fuse with the endolysosome membrane and release their contents into the lumen of the endolysosome. This is accompanied by active uptake of hydrogen ions, which decreases the pH of the endolysosome, converting it to a mature lysosome. Only at a low pH do lysosome enzymes work at full capacity.

Exocytosis and endocytosis

Secretion (export) involves the movement of small transport vesicles in which material is packaged; the vesicles move to and fuse with the plasma membrane, releasing their contents to the outside of the cell. This process is **exocytosis**. Exocytosis of some components continues throughout the life of the cell and is referred to as **constitutive secretion**. Other molecules are stored in secretory vesicles and are exported only when a specific signal triggers the fusion of the vesicle with the plasma membrane, which is the process of **regulated secretion**. In plant cells, glycoproteins and proteins are also transported to **vacuoles**, which are large compartments surrounded by a single membrane. These molecules are stored in vacuoles in a highly concentrated form for later use as a food reserve.

During exocytosis, the surface area of the plasma membrane increases continually as newly arriving vesicle membranes are incorporated. Obviously this could not go on forever. Membrane is retrieved from the plasma membrane through the process of **endocytosis**, in which clathrin molecules associate with small areas of the plasma membrane, folding inwards so that a coated pit forms, which rounds up and pinches off, taking with it a portion of extracellular fluid (Fig. 3.14). In endocytosis, particular molecules that bind to specific receptors in the plasma membrane can also be transported. Once inside a cell, coated vesicles quickly lose their coat and fuse with an intermediate compartment, the **endosome**. Many molecules, including intact receptors, are retrieved from the endosome membrane and recycled. Others pass to endolysosomes and then to lysosomes where they are degraded.

> Small coated vesicles transport molecules and membranes between the ER, Golgi apparatus, cell surface, lysosomes, endosomes and vacuoles. Membrane is retrieved from the plasma membrane and specific molecules are taken into the cell by endocytosis.

MITOCHONDRIA

Mitochondria are very different from the organelles that we have looked at so far. They have their own DNA, make some of their own proteins and are able to grow and divide, although not without help from other components in the cell. Mitochondria are thought to have evolved from bacteria (prokaryotes) that were engulfed by ancestral eukaryotic cells (see Chapter 35).

Nearly all living eukaryotic cells contain mitochondria and it is within mitochondria that cellular respiration, which is the release of energy during the oxidation of sugars and fats, takes place. This process, also known as oxidative phosphorylation, will be examined in detail in Chapter 5. The released energy is stored in molecules of ATP, which are used by cells to drive a vast range of chemical reactions. The need for a continuous supply of ATP by metabolically active cells, such as those of heart muscle, is reflected in the fact that these hard-working cells are richly endowed with well-developed mitochondria. Less-active cells have fewer and less elaborate mitochondria (Fig. 3.15).

Mitochondria are large enough to be seen using a light microscope and can be observed in living cells (Fig. 3.15). Their size, shape and number varies widely, not only in different cell types, but also within a single cell. They also change during the cell cycle and in response to changing metabolic or environmental conditions. Mitochondria may be spherical or elongated, and are able to fuse and fragment. Liver cells each contain 1000–2000 roughly ovoid mitochondria whereas some small algal cells contain only one mitochondrion, which forms a branched reticulum throughout the cell. In some cells, mitochondria are anchored close to sites that require specially large amounts of ATP. In other cells, they are transported along cytoskeletal elements (p. 56) or as part of cytoplasmic flow.

Mitochondria are surrounded by a double membrane. The outer membrane is highly permeable, allowing ions and small molecules with a molecular weight of up to 10 kD to pass freely through abundant protein channels. The structure and chemical composition of the inner membrane contrasts sharply with that of the outer membrane. Whereas the outer membrane is smooth, the inner membrane is thrown into folds, **cristae**, which greatly increase the surface area of the membrane. The inner membrane is also highly *impermeable*, especially to ions. This property allows the generation of an electrochemical gradient due to unequal distribution of ions and uncharged molecules across the membrane. This gradient is a source of potential energy available for use by mitochondria (Chapters 2 and 5). Rapid import and export of molecules required for or produced by oxidative reactions is maintained by specific transport proteins within the inner membrane.

The inner surface of the inner mitochondrial membrane is lined with numerous knob-like structures. These are the enzyme complexes responsible for ATP synthesis (Fig. 3.15). These enzymes use the electrochemical gradient established across the inner membrane to generate ATP from ADP and inorganic phosphate. The number of enzyme complexes that can be accommodated within a mitochondrion depends on the surface area of the inner membrane. Thus, cells that synthesise large

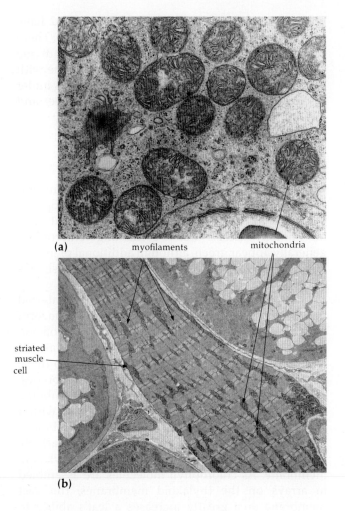

(a)

myofilaments   mitochondria

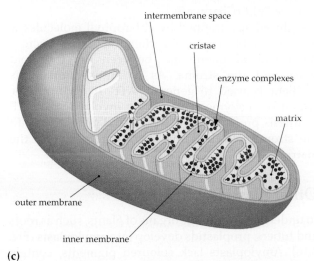

striated
muscle
cell

(b)

intermembrane space

cristae

enzyme complexes

matrix

outer membrane

inner membrane

(c)

Fig. 3.15 Transmission electron micrographs showing mitochondria in **(a)** a plant cell, the broad bean, *Vicia faba* (magnification × 14 000) and **(b)** skeletal muscle fibre showing many mitochondria regularly arranged throughout the myofilaments. **(c)** Diagram of mitochondrion showing double membrane, cristae and enzyme complexes (knob-like structures), which are responsible for ATP synthesis. The outer membrane is separated from the inner membrane by the inter-membrane space

amounts of ATP not only have many mitochondria but also have numerous, extensively folded cristae within them.

The core of the mitochondrion is the matrix space. It contains mitochondrial ribosomes, one or more copies of mitochondrial DNA, and enzymes and structural proteins that are needed for oxidative reactions and DNA replication, transcription and translation. Mitochondrial ribosomes are smaller than those in the cytosol and have many properties similar to those of prokaryotic ribosomes. Mitochondrial DNA is a circular molecule, having features in common with prokaryotic DNA. Mitochondrial DNA codes for rRNA, tRNA and mRNA. Messenger RNA is translated on mitochondrial ribosomes in the matrix.

Although mitochondria possess their own DNA, their growth and function is under cellular control. During evolution, most of the ancestral bacterial genes became relocated to the cell nucleus. As a consequence, mitochondria manufacture few of their own proteins, importing 90% from the cytosol. While a cell is growing, the number (or volume) of mitochondria doubles in readiness for their partitioning to the daughter cells during cell division. In cells with only one mitochondrion, this divides before the cell itself divides. Mitochondria divide by simply pinching in two (fission), but they need cellular proteins to do so. Isolated mitochondria cannot divide but they can synthesise DNA, RNA and proteins for a brief period.

> Mitochondria carry out cellular respiration and generate ATP for use throughout the cell. They are surrounded by two membranes, the inner one being highly impermeable and folded into cristae.

PLASTIDS

Plant and algal cells contain **plastids**, which are a family of organelles, including chloroplasts that, like mitochondria, have evolved from bacteria, which were engulfed by ancestral eukaryotic cells (Fig. 3.16). These bacteria were able to use energy from sunlight to incorporate carbon from CO_2 into organic molecules—the process of photosynthesis (Chapter 5). In this section, we will look at the structure of a chloroplast and other plastids found in photosynthetic eukaryotes.

Plastids have several features in common with mitochondria. They are surrounded by a double membrane envelope, and contain DNA, which is present as multiple copies of circular molecules, RNA and small ribosomes. The outer membrane is highly

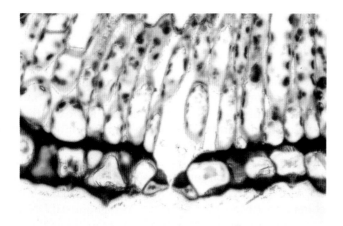

(a)

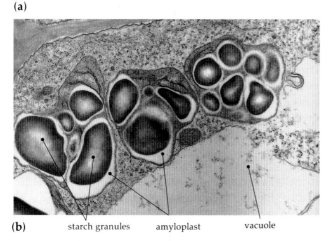

(b)   starch granules   amyloplast   vacuole

Fig. 3.16 Plastids are a family of organelles that occur in algae and plants, characterised by their own DNA and bounded by a double membrane. **(a)** Chloroplasts in leaf mesophyll cells of Geraldton wax plant, *Chamaelaucium*, appear blue-black when stained with iodine. **(b)** Electron micrograph of amyloplasts in the soybean root cap cell. Starch reserves in amyloplasts are present as starch granules (magnification × 18 000)

permeable to ions and small molecules. The inner membrane is more selective, containing carrier proteins responsible for transporting molecules into and out of a plastid. The matrix enclosed within the inner membrane of a plastid is the **stroma**. Within the stroma lies a third membrane system. This internal membrane system possesses structural and biochemical characteristics that distinguish plastids from mitochondria and also different types of plastids from one another.

Many of the proteins that are needed in plastids are encoded by nuclear genes and thus have to be imported from the cytosol into the stroma through contact sites at which the outer and inner membranes come close together, such as occurs in mitochondria. In addition to providing a plant or algal cell with energy and sugars, plastids also synthesise amino acids and most of a cell's fatty acids.

Proplastids are precursors of all types of plastids and occur in actively dividing cells. They are

colourless, contain small starch granules and have only a rudimentary internal membrane system. There are usually 10–20 proplastids per cell. They divide by fission and their rate of division keeps pace with that of the cell. Differentiation of proplastids is under nuclear control and depends on cell type and environmental conditions.

Chloroplasts

Chloroplasts develop from proplastids in plant and algal cells grown in light (Fig. 3.17). Chloroplasts contain **chlorophyll**, a light-absorbing pigment used in photosynthesis. Cells containing chlorophyll generally appear green, although additional pigments in some groups of algae make them various shades of brown or red.

Chloroplasts have a highly developed internal membrane system. Some parts of the membrane form flattened disc-like sacs, **thylakoids**, which lie on top of each other forming stacks, **grana**. The thylakoids within a stack are joined to each other and to those in adjacent stacks by flattened tubular membranes. The whole system forms a continuous membrane network surrounding a single internal space within a thylakoid. This arrangement of internal membranes means that although chloroplasts have a relatively small volume, they contain a large area of thylakoid membrane. As chlorophyll molecules are assembled in arrays on the thylakoid membranes, the vast membrane area greatly increases a leaf's ability to capture light (Chapter 5).

Light energy trapped by chlorophyll molecules is used to establish an electrochemical gradient across the thylakoid membrane, between the thylakoid space and the stroma. As in mitochondria, this gradient is used to generate ATP. In chloroplasts, knob-like enzyme complexes involved in ATP synthesis project from the thylakoid membrane into the stroma, which is the region equivalent to the matrix space in mitochondria.

Other plastids

In underground storage organs of plants, such as roots and tubers, proplastids develop into **amyloplasts** (Fig. 3.16). Amyloplasts lack coloured pigments, contain large reserves of starch and have very few, if any, membranes within the stroma. Proplastids may also develop into non-photosynthetic **chromoplasts** (Fig. 3.16), which contain carotenoid pigments and are responsible for many of the yellow, red and orange hues of fruits, flowers and leaves.

Interconversions can occur between most plastid types. For example, amyloplasts can differentiate into **elaioplasts** (oleoplasts), which contain reserves of oil droplets in many seeds and pollen grains. In shoots

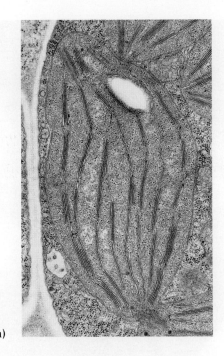

(a)

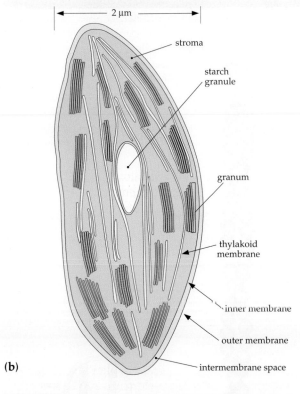

(b)

Fig. 3.17 Transmission electron micrograph showing a chloroplast in bean leaf cell (magnification × 18 200)

grown in the dark, proplastids develop into **etioplasts**, which on exposure to light develop into normal chloroplasts.

> Chloroplasts, the most important type of plastid, are responsible for the capture of energy from light during photosynthesis. Photosynthetic pigments and enzymes are assembled on internal thylakoid membranes.

MICROBODIES

Several organelles are involved in the breakdown of unwanted molecules. Lysosomes hydrolyse material originating both inside and outside a cell. Smooth ER detoxifies drugs and some metabolites. However, **microbodies** are the main organelle involved in the removal of compounds generated within cells.

Microbodies occur in all eukaryotic cells: a liver cell has about 1000. They are spherical to sausage-shaped with homogeneous, granular contents and sometimes crystalline inclusions. As with lysosomes, they are surrounded by a single membrane, contain a variety of degradative enzymes, and lack DNA, ribosomes and the internal membrane elaborations seen in chloroplasts and mitochondria. However, in contrast to the acidic interior of lysosomes, microbodies have a pH close to neutral, and the types of enzymes within the two organelles also differ.

Lysosomes contain hydrolytic enzymes, whereas microbodies house oxidative enzymes that remove hydrogen from molecules to be degraded and couple it to oxygen. This reaction within microbodies generates hydrogen peroxide, H_2O_2, a very reactive molecule which, if not confined by a membrane, would cause considerable cell damage. Microbodies contain large amounts of an enzyme, **catalase**, which breaks down H_2O_2. Catalase constitutes about 40% of a microbody's protein and not only degrades excess H_2O_2 into water and oxygen but also harnesses the reactivity of H_2O_2, exploiting it to oxidise other molecules such as alcohols and phenols. Microbody membranes contain special transport channels for a large number of metabolites that are moved between microbodies and other organelles.

Microbodies can be either peroxisomes or glyoxysomes according to the enzymes they contain and the functions they perform. **Peroxisomes** oxidise amino acids and uric acid. In leaf cells, peroxisomes are involved in the oxidation of glycollate, a by-product of carbon fixation in the chloroplast (Chapter 5). **Glyoxysomes** occur in germinating seeds, whose energy reserves are stored in the form of lipids, and are involved in the conversion of fatty acids to sugars.

The exact complement of enzymes within a microbody depends on the cell type and cell activity. Growth of yeast cells on methanol induces an enlargement of peroxisomes and synthesis of enzymes involved in methanol metabolism. Transfer to a fatty acid source induces production of enzymes that break down fatty acids. As a young, growing plant uses up lipid stores in the seed, the types of enzymes in microbodies in the first leaf cells change and become more typical of leaf peroxisomes.

> Peroxisomes and glyoxysomes are microbodies that contain different complements of enzymes involved in oxidative degradation of molecules.

THE CYTOSKELETON

So far we have looked at different organelles without particular reference to their location in the cell. Organelles are not scattered randomly throughout the cytoplasm; in general, there is considerable order in their arrangement. The nucleus may be suspended in the centre of the cell; Golgi stacks may cluster around the nucleus; and chloroplasts may be oriented with their grana facing towards a light source. Many organelles move within the cell in an orderly way.

Many aspects of cytoplasmic organisation and motility are a result of the interaction between organelles and three different elements of the **cytoskeleton**: microtubules (mostly tubulin), microfilaments (mostly actin) and intermediate filaments (several types, each composed of a distinct protein). Networks of these components interact to generate structure and order, to fix organelles in particular positions or to move individual organelles or the whole cytoplasm from one place to another. They also confer the ability to maintain and remodel cell shape.

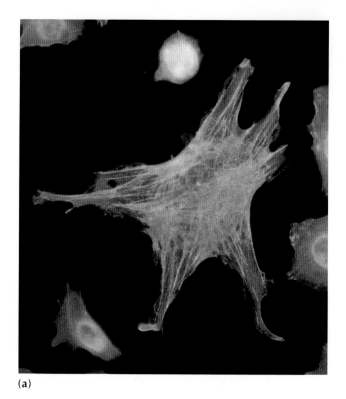

(a)

Microfilaments

Microfilaments are fine fibres, 7–8 nm in diameter, composed of actin, a globular protein with a molecular weight of 42 kD (Fig. 3.18). Free actin molecules in solution are known as G-actin (globular actin). Each molecule of G-actin can interact with other actin molecules to form a chain or filament, F-actin. The two ends of a filament are different because each actin molecule in the filament is oriented in the same direction. Filaments thus have a distinct polarity (+ and −), which affects their biochemistry and function. For example, G-actin molecules add on to the positive end of the filament more quickly than they do to the negative end.

Microfilaments are highly dynamic structures. If there is a net addition of G-actin (polymerisation), microfilaments become longer; with a net loss of G-actin (depolymerisation), microfilaments become shorter. Cells control these two processes by regulating the number of actin molecules that are synthesised and hence the amount of G-actin available. In addition, filament length, stability and organisation are controlled by a wide range of *actin-binding proteins*.

Some actin-binding proteins allow net-like and bundled arrays of actin microfilaments to form, while others break down existing arrays. The actin-binding protein, profilin, for example, binds to G-actin and makes the molecules unavailable for polymerisation. Another actin-binding protein, gelsolin, binds to and severs actin filaments, causing whole actin arrays to

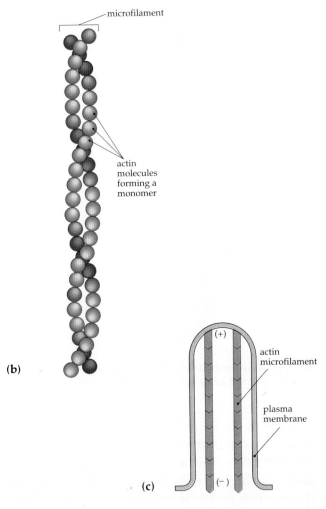

(b)

(c)

Fig. 3.18

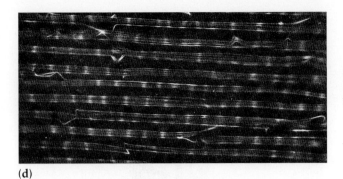

(d)

Fig. 3.18 Actin microfilaments of the cytoskeleton.
(a) Fluorescence micrograph showing actin
microfilaments in bundles (stress fibres) in cultured
human fibroblasts, which form connective tissue. The
green actin fibres provide a supporting network in the
cell (magnification × 460). **(b)** A microfilament is
composed of actin monomers. **(c)** Actin microfilaments
are attached to the plasma membrane in intestinal
microvilli, which are in constant movement.
Attachment occurs at the (+) end of the microfilaments.
(d) Fluorescence micrograph showing actin
microfilaments formed into bundles in a large plant
cell, *Chara corallina*. In giant cells, diffusion cannot
ensure mixing of metabolites. This is achieved by
movement of the cytoplasm (cytoplasmic streaming)
along actin cables, bundles of microfilaments all aligned
with the same polarity. The streaming moves
mitochondria, nuclei, internal membranes and cytosol,
but chloroplasts are fixed in place beneath the plasma
membrane by the microfilaments (magnification × 600)

fragment. The association of other actin-binding
proteins with F-actin helps to build up networks of
microfilaments that hold small organelles in place.
For example, spectrin interacts with actin to form a
network that is cross-linked to the plasma membrane,
giving mechanical stability to the surface of many
animal cells. Other actin-binding proteins link
microfilaments into rigid bundles that support the
fine projections of the plasma membrane (microvilli)
in intestinal epithelial cells (Fig. 3.18).

Interactions of actin microfilaments with myosin
are the basis of many forms of cytoplasmic, organelle
and cell movement. The sliding of actin and myosin
filaments in muscle cells is the basis of muscle
contraction (see Chapter 27). The association of
myosin molecules with organelle membranes enables
organelles to be moved along actin microfilaments.
In large plant cells in which diffusion cannot ensure
adequate mixing of metabolites, cytoplasm is moved
at a rate of several micrometres per second along
bundles of actin microfilaments (Fig. 3.18). This
cytoplasmic streaming may occur in thin strands of
cytoplasm that traverse large vacuoles or it may occur
throughout a thin layer of cytoplasm sandwiched

between the plasma membrane and large central
vacuoles.

> Microfilaments, which are part of the cytoskeleton, are
> composed of the protein actin. They contribute to the
> structural organisation of cells and facilitate intracellular
> movement.

Microtubules

Microtubules are hollow cylinders, 25 nm in diameter.
They are composed of **tubulin**, a globular protein that
exists in two closely related forms, **α-tubulin** and
β-tubulin. Both forms of tubulin have a molecular
weight of 55 kD and in solution they associate into
a dimer composed of one alpha and one beta subunit.
Tubulin dimers bind to each other forming
protofilaments, which are chains of alternating alpha
and beta subunits (Fig. 3.19). Because the dimers all
have the same orientation, microtubules are polar
structures with the properties of one end being
different from those of the other. For example, one
end grows faster than the other.

Microtubules are repeatedly assembled and
disassembled by a number of microtubule-associated
proteins. Microtubule polymerisation usually begins
at specific locations, *microtubule nucleating sites* (MTNS).
The slow-growing end of the microtubule is at the
nucleating site. Dimer addition to the fast-growing
end causes microtubules to grow out from the
nucleating sites, forming radiating, bundled or parallel
arrays. Individual microtubules in a cell may grow
slowly for a while and then rapidly depolymerise;
selected microtubules may become stabilised by
special capping molecules.

Microtubules are more rigid than microfilaments
and can provide greater mechanical support. Groups
of microtubules support specialised appendages that
project from a variety of cells. In nerve cells, for
example, bundles of microtubules extend along the
long thin axons. Within axons, numerous vesicles are
transported rapidly along microtubules by the action
of *microtubule-associated proteins* (MAPs), which are
ATPases. MAPs control the polymerisation of tubulin
and stabilise microtubules by forming cross-bridges
between adjacent microtubules. Vesicles are moved
towards the axon tip by the action of the MAP,
kinesin, and in the opposite direction by the MAP,
dynein.

Within a cell, the arrangement of microtubules
depends not only on cell type but also on the stage
of the cell cycle. In many animal cells when a cell
is not dividing, microtubules emerge from near the
nucleus and radiate out through the cytoplasm
towards the plasma membrane in an array that is
constantly changing (Fig. 3.19). As cells begin to

divide, the arrangement of their microtubule cytoskeleton alters dramatically. Microtubules form spindle fibres, which attach to chromosomes and draw them apart during cell division (Chapter 8).

Microtubules are dynamic scaffolding elements in eukaryotic cells. Microtubule arrangement and stability are determined by microtubule-associated proteins that initiate microtubule formation, cross-bridge microtubules into bundles and networks, and move organelles along microtubules.

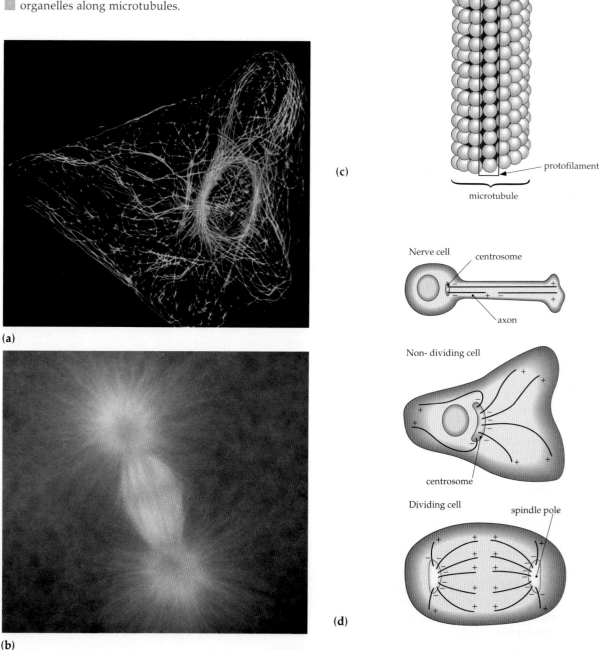

(c)

(a)

(b)

(d)

Fig. 3.19 Microtubules of the cytoskeleton. Microtubule arrangement during two stages of cell division in an animal cell: (a) in a cell before division (interphase) and (b) when the chromosomes are about to separate on the spindle, which comprises ropes of microtubules. (c) Structure of microtubules. Each microtubule, assembled from pairs of tubulin subunits (α- and β-tubulin), consists of 13 protofilaments aligned side by side. (d) Diagram showing microtubule polarity and different nucleating sites in animal cells. The negative ends of microtubules are usually embedded in the microtubule organising centres, and the positive ends are near the plasma membrane

Intermediate filaments

Intermediate filaments are 8–10 nm in diameter, making them intermediate in size between microfilaments and microtubules. The proteins of intermediate filaments vary from about 40 to 130 kD in molecular weight. Intermediate filaments are strong and resist stretching and, in general, do not undergo the rapid and repeated assembly and disassembly that occurs in microfilaments and microtubules. Arrays of intermediate filaments are stable and change little during cell growth. They are fibrous rather than globular and more heterogeneous than actin or tubulin. Pairs of protein molecules (dimers) associate laterally in overlapping arrays to build up long filaments with a high tensile strength.

Intermediate filaments provide mechanical support for the cell and the nucleus. For example, keratins are extremely stable and insoluble proteins in epithelial cells and form the basis of hair and nails (Fig. 3.20); desmin forms intermediate filaments in muscle cells; and neurofilaments occur in nerve cells.

Nuclear lamins are intermediate filaments that are cross-linked into a square lattice, the nuclear lamina. The nuclear lamina is disassembled and reassembled as the nuclear envelope breaks down and reforms during cell division (Fig. 3.20).

Intermediate filaments are strong, stable and resist stretching.

MOTILITY: CILIA AND FLAGELLA

In eukaryotes, cilia and flagella are fine, long cellular projections. They produce movement of a cell or of liquid over a cell surface. Flagella are usually 20–100 μm in length and occur in small numbers (Fig. 3.21). Cilia have the same internal structure but are shorter (2–20 μm) and occur in much greater numbers, up to several thousands on the surface of some cells (Fig. 3.21).

Cilia and flagella are covered by plasma membrane and are supported by a precise array of microtubules, which forms the **axoneme** (Fig. 3.21). The axoneme consists of a cylinder formed by nine pairs or doublets of microtubules enclosing two central microtubules. Adjacent doublets are linked together by fine fibres and are connected to the inner sheath by radial spokes. On one side of each doublet are two short arms formed of the protein dynein.

Movement of cilia or flagella results from the sliding of adjacent microtubule doublets relative to each other. Dynein first attaches to the adjacent doublet, undergoes a conformational change and then dissociates. During this process, dynein hydrolyses

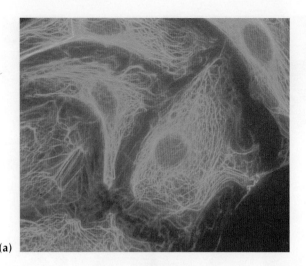

(a)

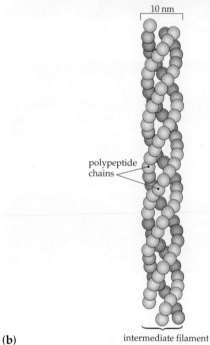

(b)

Fig. 3.20 Intermediate filaments of the cytoskeleton. **(a)** Fluorescence micrograph showing intermediate filaments (vimentin) in kidney epithelial cells of the kangaroo rat. The filaments form three-dimensional networks involved in the movement of cell organelles. The dark spheres in the centre of the cells are the sites of the nucleus (magnification × 270). **(b)** Structure of an intermediate filament

ATP to provide energy. The conformational change in dynein causes sliding of adjacent doublets, resulting in a cilium or flagellum bending (Fig. 3.21).

Axoneme microtubules originate at a basal body, which is a highly structured microtubule nucleating site. The basal body is composed of a ring of nine triplet microtubules and acts as a template for the axoneme. The slow-growing end of the microtubules is attached at the nucleation site and tubulin dimers add onto the distal end at the tip of the growing flagellum.

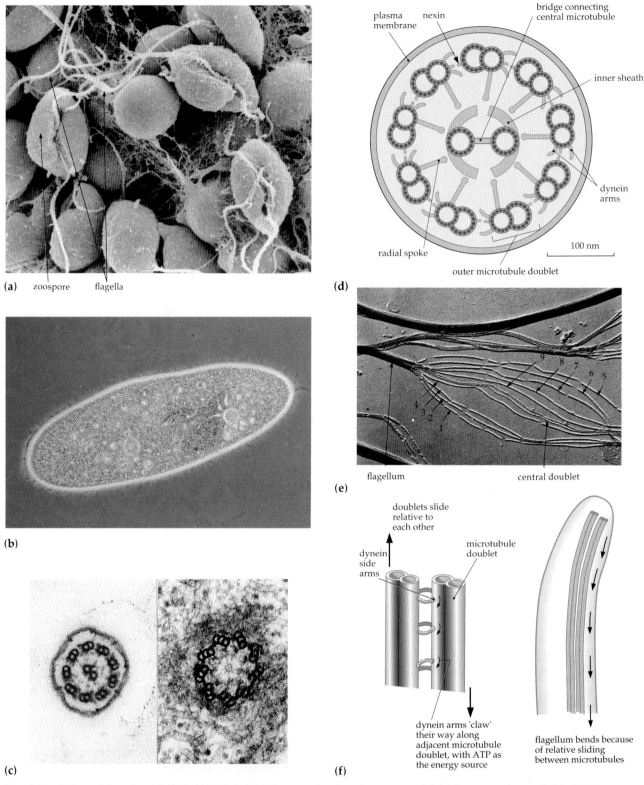

(a)   zoospore   flagella

(b)

(c)

plasma
membrane   nexin   bridge connecting
central microtubule

inner sheath

dynein
arms

radial spoke

100 nm

outer microtubule doublet

(d)

flagellum   central doublet

(e)

doublets slide
relative to
each other

dynein
side
arms

microtubule
doublet

dynein arms 'claw'
their way along
adjacent microtubule
doublet, with ATP as
the energy source

flagellum bends because
of relative sliding
between microtubules

(f)

Fig. 3.21 Cilia and flagella. (**a**) Scanning electron micrograph of motile spores of the 'cinnamon fungus', *Phytophthora cinnamomi*, each with two flagella (magnification × 2400). (**b**) Living ciliated cell of *Paramecium multinucleatum* (magnification × 265). (**c**) Transmission electron micrographs of the single-celled green alga, *Chlamydomonas*, showing transverse section through a flagellum (left) and basal bodies (right). (**d**) Structure of axoneme of flagellum, composed of an outer circle of nine microtubule doublets (each with five dynein arms, a radial spoke, and nexin links), surrounding a central sheath containing a pair of microtubules. (**e**) An Australian first. Grigg and Hodges (1949) observed the nine doublets and central doublet of the sperm flagellum three years before European scientists. This photograph shows the preparation of a fowl sperm. (**f**) Sliding of adjacent microtubule doublets causes bending of a cilium or flagellum, generating movement. This involves a type of 'climbing action' by the contractile dynein arms along the microtubules, using energy provided by the hydrolysis of ATP

Basal bodies also occur as a pair of centrioles in animal cells. Centrioles lie at right angles to each other close to the nucleus. Before cell division, each centriole replicates, ensuring allocation of a pair of centrioles to each daughter cell (Chapter 8).

> Cilia and flagella are composed of nine interconnected microtubule doublets. Cilia and flagella bending is a result of sliding of these adjacent doublets and is powered by dynein hydrolysis of ATP.

COMPARISON OF PROKARYOTIC AND EUKARYOTIC CELLS

The most ancient cells, the prokaryotes (bacteria) are smaller (about 1 μm in diameter) and have a simpler structure than eukaryotic cells. Bacteria have a semirigid cell wall that surrounds a plasma membrane. They have no membrane-bound organelles; that is, they lack a nucleus, endoplasmic reticulum, Golgi apparatus, vacuoles, mitochondria and chloroplasts. They also lack a cytoskeleton of microfilaments and microtubules.

Although some bacteria have flagella, which propel the cells, they have a completely different structure and mechanism of action (Fig. 3.22). Bacterial flagella are composed of flagellin (protein) fibrils that form a stiff coiled filament. The flagellum projects from a special structure in the bacterial membrane that serves as the motor for movement. Movement of the cell is generated by rotation of the curved filament, which thus acts like a propeller. To change the direction of movement, the direction of rotation is temporarily reversed.

The cytosol of bacteria contains a circular DNA molecule and a few hundred to a few thousand ribosomes, about 15 nm in diameter (compared with 25–30 nm in eukaryotes). Because DNA is not segregated into a nucleus, bacterial ribosomes can attach directly to mRNA molecules even while mRNA is being transcribed. In eukaryotes, mRNA molecules form in the nucleus and pass across the nuclear membrane into the cytoplasm before they interact with ribosomes during protein synthesis. In bacteria, the enzymes corresponding to those found in eukaryote mitochondria are located in the plasma membrane, and thus a mitochondrion is similar to a whole bacterial cell. In cyanobacteria, internal membranes contain light-trapping pigments, and the eukaryote chloroplast is similar to a cyanobacterial cell.

The structure and diversity of prokaryotes is considered in greater depth in Chapter 33, and theories of the origin of eukaryotes from prokaryote ancestors are discussed in Chapter 35.

> Bacterial cells have a simple structure and lack membrane-bound organelles.

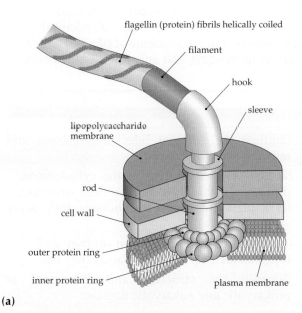

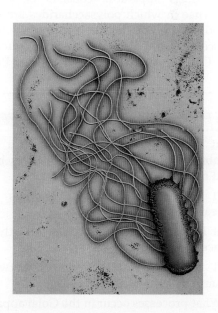

(a)

(b)

Fig. 3.22 (a) A flagellum in a bacterial cell has a different structure (made of flagellin) and mechanism of movement than flagella of eukaryotic cells. Movement of the bacterial cell is generated by rotation of the rigid spiral-shaped flagellum, which acts like a propeller. Changes in direction of movement are achieved by temporarily reversing the direction of rotation. Rotation is achieved by a motor-like structure embedded in the plasma membrane and wall of the bacterial cell. (b) Colour-enhanced scanning electron micrograph of *Salmonella* bacteria

SUMMARY

- All living organisms are made up of cells and cell products. Cells are bounded by the plasma membrane. Eukaryotic cells contain intracellular membrane-bound compartments that segregate different molecules and metabolic reactions. Prokaryotic cells lack internal subdivisions.

- Membranes consist of two layers of phospholipid molecules and contain integral and peripheral proteins. Membranes form selectively permeable barriers, allowing only certain molecules to pass through them.

- In eukaryotic cells, the nucleus is a compartment surrounded by a double membrane and contains DNA in the form of chromosomes. Ribosomal subunits are assembled in the nucleolus and move into the cytosol through pores in the nuclear envelope.

- Proteins are synthesised on ribosomes associated with mRNA molecules. During protein synthesis, some ribosomes remain free in the cytoplasm, while others become associated with the surface of the endoplasmic reticulum.

- Endoplasmic reticulum is an extensive system of membranous cisternae in eukaryotic cells in which lipid synthesis and the initial processing of secretory and membrane proteins and glycoproteins occurs.

- The Golgi apparatus consists of stacks of cisternae in which polysaccharides are synthesised, and polysaccharides and glycoproteins are processed, sorted and packaged. Products leave the Golgi apparatus in coated vesicles. Products

are secreted if the vesicles fuse with the plasma membrane or are stored within the cell if the vesicles are targeted to fuse with specific vacuoles. Membrane is retrieved and specific molecules are taken into cells by endocytosis.

- Mitochondria and plastids are surrounded by a double membrane. Although they contain small amounts of DNA, most of the proteins needed by these organelles are synthesised on cytosolic ribosomes. Mitochondria compartmentalise reactions that oxidise sugars and fats and synthesise ATP to be used as an energy source throughout the cell. Chloroplasts contain pigments that absorb energy from sunlight and enzymes that incorporate carbon from CO_2 into organic molecules.

- Microbodies contain oxidative enzymes involved in the digestion of unwanted molecules. Compartmentalisation of these enzymes and their reactive products protects other cell components from degradation.

- Three types of cytoskeletal elements— microfilaments, microtubules and intermediate filaments—form bundles and extensive networks throughout the cytoplasm of eukaryotic cells. They provide a structural framework that supports extended projections of the plasma membrane and they are responsible for transporting and positioning many subcellular components. Microtubules form the spindle, which separates chromosomes during cell division. Sliding of microtubule doublets within the axoneme of cilia and flagella results in bending and subsequent movement.

QUESTIONS

1. What is the chemical composition of a plasma membrane and why is it described as being a 'fluid mosaic'?

2. What are the distinguishing features of the nuclear membrane?

3. What is the functional significance of the large surface area to volume ratio of the ER?

4. What are the basic differences between a prokaryotic and eukaryotic cell?

5. (a) What processes occur in the Golgi apparatus?
   (b) In which part of this organelle does each process occur?

6. Explain the importance of the limited permeability of the inner membranes of mitochondria and chloroplasts.

7. Contrast the different roles of lysosomes and microbodies within a cell.

8. (a) List the components of the cytoskeleton.
   (b) Describe the differences in their structure.
   (c) Explain the different functions they serve within the cell.

9. (a) What are MAPs and what do they do?
   (b) What are MTNSs and what do they do?

10. Contrast the structure and movement of prokaryotic and eukaryotic flagella.

MOVEMENT ACROSS MEMBRANES

All living cells are surrounded by an aqueous med-ium that comprises their immediate environment. For a unicellular alga, this environment may be pond water; for a cell in a multicellular animal, it is intercellular (interstitial) fluid. Cells are physically separated from this environment by the plasma mem-brane (Figs 3.4, 3.5). This membrane functions not as an impervious barrier but as a selective dynamic boundary to the cell. Without the correct functioning of this membrane, cellular processes would fail and the cell would die.

Invariably, the composition of a cell is different from that of its surrounding environment. The stability of the intracellular environment (homeostasis) must be maintained in the face of both physical and chemical fluctuations in the external environment. In addition, metabolic processes occurring within living cells actively generate changes in the intracellular environ-ment. For example, oxygen and fuel molecules are used and carbon dioxide is produced during cellular respiration; and pH is affected by the production of H^+ and HCO_3^-.

In order for metabolism to proceed in a controlled and sustained manner, the concentrations of solutes within each cell must be kept within well-defined limits. Protein structure, and thus the activity of enzymes, for example, is affected by pH and the concentration of ions such as Ca^{2+}, Na^+ and K^+. Therefore, these ions must be kept at relatively constant levels inside cells.

Generally, the cytoplasm of all cells contains low levels of free Na^+, Cl^- and Ca^{2+} and high levels of K^+ at a pH of between 7 and 8 (Fig. 4.1). The high level of cations is partly balanced by the net negative charge on proteins and nucleic acids inside cells and partly by other anions, both organic and inorganic. Various solutes, again both organic and inorganic, must enter cells so that they can be converted to other compounds. Waste products and compounds specially synthesised for export need to leave the cell.

For these functions to occur, there must be selective movement of solutes across the plasma membrane.

Water is the most abundant component of cells and is the major determinant of cell volume. If too much water enters a cell, components dissolved in the cytosol may become too dilute for efficient metabolism. The reverse may also occur. Therefore, movement of water across the plasma membrane must be controlled. Water moves into and out of cells

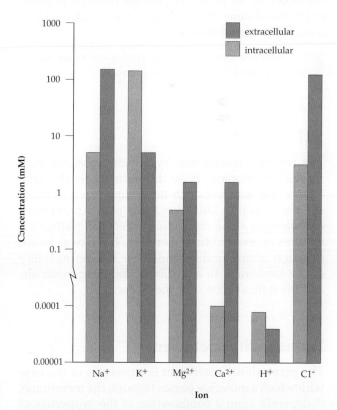

Fig. 4.1 Comparison of concentrations of free ions inside and outside a typical mammalian cell. Ions such as K^+ and Mg^{2+} have higher intracellular (cytosolic) concentrations than those externally, whereas the intracellular concentrations of Na^+ and Cl^- are lower than the extracellular concentrations

along osmotic gradients (p. 66) and cells control osmotic gradients by regulating the internal quantity of solutes. In this way, they control the net movement of water across the plasma membrane.

Within a cell, there are many different membrane-bound compartments (Figs 3.4, 3.5), including mitochondria, endoplasmic reticulum and, especially in the case of plant cells, vacuoles. These compartments have different functions and, as a consequence, they have different concentrations of solutes. Exchange of solutes between these compartments must therefore be controlled. For example, calcium has a central role in controlling metabolism, and its movement between compartments is highly regulated.

Overall, cells are able to control the composition of the cytosol and of their internal compartments because of a special property of biological membranes, **selective permeability**. Selective permeability means that membranes function as a kind of molecular sieve. Because of the hydrophobic nature of lipid membranes, water-soluble components of the cytosol tend to be retained in cells whereas lipid-soluble substances pass freely through the plasma membrane. However, movement across membranes involves more than mere filtration. Some water-soluble substances that pass from one side of a membrane to the other do so by direct passage through proteins in the membrane; other substances are packaged into membrane-bound vesicles.

> The selective permeability of lipid membranes allows cells to control their internal composition.

DIFFUSION

The simplest mechanism by which molecules pass through membranes is by diffusion across the lipid bilayer and, given enough time, most types of molecules will do this. **Diffusion** is the *passive* movement of molecules along their electrochemical gradient; it requires no expenditure of energy. The rate at which diffusion occurs is determined by the permeability of the membrane to a particular diffusing molecule, and the diffusive force on the molecule.

Permeability coefficient

The **permeability coefficient** is a measure of the ease with which a molecule passes through the membrane. It depends upon a combination of the properties of the membrane itself and of the diffusing molecule. The most important factor affecting the permeability coefficient is the lipid solubility of a molecule. This determines how readily the molecule can dissolve in the lipid bilayer and, consequently, how rapidly

it can diffuse across the membrane. The higher the lipid solubility, the greater the ability of a molecule to diffuse across membranes (Fig. 4.2). However, for some molecules other factors also influence permeability. In particular, very small molecules that are relatively insoluble in lipids, such as water and urea, appear to have a higher permeability coefficient than would be expected from their lipid solubility. Evidence suggests that these molecules simply pass through transient pores (discontinuities) in the lipid structure, or through proteins that are embedded within the lipid bilayer. Such an increase in the permeability coefficient with decreasing molecular diameter, including the hydration sphere (p. 7), is shown in Figure 4.3.

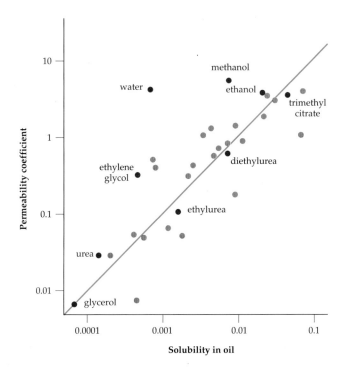

Fig. 4.2 The higher the lipid solubility of a molecule, the greater its ability to permeate a membrane by diffusion (permeability coefficient). Note that codeine, a common constituent of pain-killers, passes easily through the membrane

The phospholipid bilayer of membranes is highly permeable to a wide range of non-polar molecules, such as the dissolved gases, O_2 and CO_2, and very small polar molecules, such as water and urea (Fig. 4.4). The bilayer is not particularly permeable to larger molecules that have polar side groups, such as hydroxyl groups; examples of these molecules include sugars, such as glucose and sucrose. The bilayer is almost completely impermeable to charged compounds (ions) and effectively impermeable to macromolecules, such as proteins and nucleic acids.

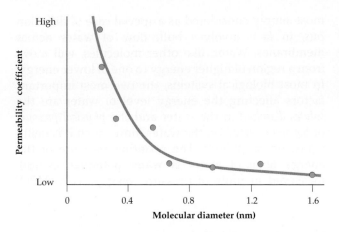

Fig. 4.3 This graph shows the relationship between molecular size and permeability coefficient for polar molecules. Small molecules are more permeable than large molecules. Molecules larger than 0.8 nm diameter do not readily cross the membrane

Diffusive force

The **diffusive force** on molecules across a membrane depends on the difference between their energy levels on either side of the membrane. For uncharged solutes, such as carbon dioxide or glucose, these energy levels are affected only by solute concentration. Diffusion will occur simply from a region of higher concentration to one of lower concentration, that is, from a region of higher energy to one of lower energy, until an equilibrium is reached. When there is no difference in concentration between the two compartments, there will be no *net* movement.

The force leading to diffusion of ions across a membrane is the difference in electrochemical potential across the membrane, which takes into account the difference in both concentration and electrical potential across the membrane (Fig. 4.5). The dual action of concentration and charge on ions means that, if the electrical component exceeds the concentration component, it is possible for ions to diffuse up a concentration gradient, that is, from a region of lower concentration to one of higher concentration. This observation is explained by the fact that the 'push' given to the ions by an electrical potential difference can be greater than the opposing force 'pulling' the ions down their concentration difference.

A difference in electrical potential exists across the plasma membrane surrounding most cells, with the inside of the membrane being negatively charged with respect to the outside. For animal cells, this is often around 50 mV (or one-twentieth of one volt), but in plants it is often much higher, at around 200 mV. A negative charge inside the membrane relative to the outside means that positive ions are attracted to the inside of the cell, even though this may be into a region of higher concentration. The magnitude of this effect is staggering. An apparently small difference of 180 mV, common across a plant plasma membrane, is sufficient to cause a thousand-fold accumulation of K^+ inside the plant cell.

Although diffusion across the lipid phase of membranes is not easily regulated, it is nevertheless the main means of exchange of certain solutes across membranes. For example, in all organisms it is the way that oxygen and carbon dioxide are exchanged with the environment. In humans, diffusion is the means by which volatile anaesthetics are absorbed across the lungs, and the way alcohol is rapidly absorbed into the bloodstream. Chemical pesticides, such as DDT, are highly soluble in lipids and are thus easily able to penetrate cells and accumulate in organisms.

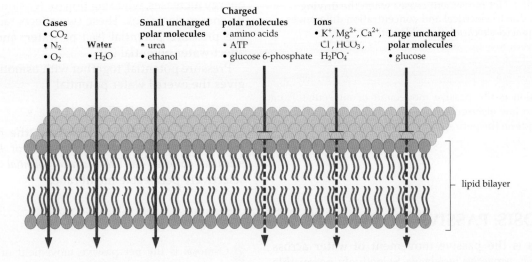

Fig. 4.4 Permeability of a lipid bilayer to different types of molecules. Small molecules, especially those that form few hydrogen bonds with water, diffuse more rapidly across a membrane than do larger and charged molecules

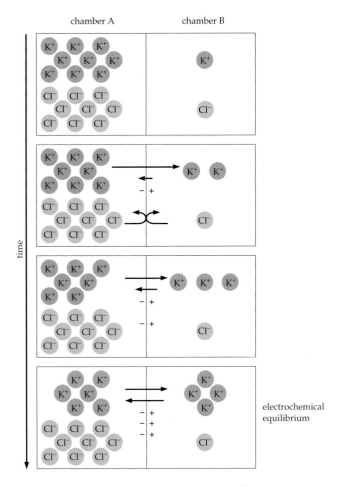

Fig. 4.5 This model illustrates the movement of K⁺ in response to both concentration and electrical gradients across a membrane. Chamber A, containing high concentrations of K⁺ and Cl⁻, is separated from chamber B, containing low concentrations of these ions, by a membrane permeable to K⁺ but not Cl⁻. K⁺ will diffuse from A to B along its concentration gradient. In the model, Cl⁻ is unable to move, so chamber B becomes positively charged. This creates an electrical gradient favouring movement of K⁺ in the opposite direction. Net movement ceases when the driving forces due to electrical and concentration differences are equal (*electrochemical equilibrium*)

> Diffusion is the passive movement of molecules from high to low electrochemical potential. Rate of diffusion depends on the permeability coefficient and the diffusive force.

OSMOSIS: PASSIVE MOVEMENT

Osmosis is the passive movement of water across a *selectively permeable membrane*. Selectively permeable membranes allow water to move at a greater rate than the solutes dissolved in the water. Osmosis is most simply considered as a special case of diffusion but, in fact, involves bulk flow of water across membranes. Water, like other molecules, will move from a region of higher energy to one of lower energy. In most biological systems, the two most important factors affecting the energy level of water are the *solutes dissolved* in the water and the *physical pressure* or *tension* exerted on the water (how much it is being squeezed or pulled). The resulting measure of the energy level of water is **water potential**, ψ (*psi*), measured in units of pressure, megapascals (MPa).

Osmotic potential

The addition of solutes to pure water decreases the energy level of the water, thus lowering the water potential by decreasing its osmotic potential (ψ_π). This occurs for two reasons, both of which cause a reduction in the amount of freely diffusing water molecules in a particular volume (a reduction in the concentration of free water).

1. Solutes increase the volume of a solution without increasing the water content, so the concentration of water in the solution is reduced. In other words, solutes are diluting the water.
2. Solutes interact with water, reducing the ability of water to diffuse freely.

However, some solutes may also interfere with the bonds *between* water molecules, increasing the ability of water molecules to diffuse freely. This occurs when urea is added to water, so that the water potential is raised.

Pressure potential

When hydrostatic pressure is applied to water, its energy increases; when suction (or tension) is applied, its energy decreases. These two effects raise or lower the pressure potential (ψ_P) of water, and therefore affect water potential (ψ).

Pressure potential, together with osmotic potential, gives the overall water potential,

$$\psi = \psi_\pi + \psi_P$$

Using this terminology, osmosis is the net passive movement of water from a region of *higher water potential* to one of *lower water potential* through a selectively permeable membrane.

> Osmosis is the net passive movement of water from a region of higher water potential (ψ) to one of lower water potential through a selectively permeable membrane.

Osmosis and cells

Water will move into or out of cells depending on the difference in water potential across the plasma membrane. If a cell has no cell wall, as in the case of animal cells, there can be very little difference in water potential between the inside of the cell and the surrounding solution because any slight difference will cause the cell to shrink or swell (Fig. 4.6). Thus, any difference in water potential across the plasma membrane will be due only to differences in osmotic potential, that is, in the concentrations of solutes. As water enters or leaves a cell by osmosis, this changes internal solute concentration and thus changes water potential.

In those cells constrained by a relatively rigid cell wall, the situation is different, primarily because the cell wall limits volume expansion. Thus in plant, fungal and bacterial cells, *both* osmotic potential and pressure potential contribute to water potential. When walled cells are placed in a solution with a less negative water potential, water will enter. The cells can only expand by about 10% before pressure (**turgor**) starts to build up inside them. At a particular pressure, the intracellular water potential will come to equal that of the external solution (Fig. 4.7). Turgor in cells is responsible for maintaining the shape of many plant parts. It also causes the crunch you hear when you bite into a piece of celery: the crunch is the explosion of pressurised cells as the walls are broken by your teeth.

The change in the shape of plants as they dry out and wilt is the result of a loss of turgor. Because the atmosphere has a very low water potential, water continually moves from cells to the air. If water is not replaced, the pressure potential (and hence turgor) reduces, allowing gravity to determine the shape. The action of some important antibiotics (e.g. penicillin, vancomycin, cephalosporins and bacitracin) is to weaken the wall structure of growing bacteria to the extent that they may burst under their own turgor.

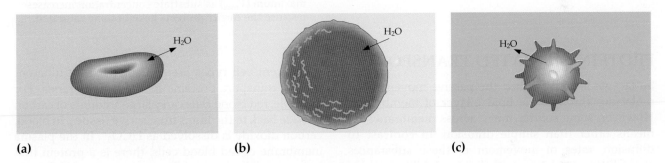

(a) (b) (c)

Fig. 4.6 Osmotic movement of water across the plasma membrane of a red blood cell placed in solutions of different osmotic potential. (**a**) The osmotic potential inside and outside the cell is equal: no net movement of water. (**b**) The osmotic potential inside the cell is lower (a higher concentration of solutes) and so water enters the cell, causing it to swell and possibly burst. (**c**) The osmotic potential inside the cell is higher (a lower concentration of solutes) and water leaves the cell, causing it to shrink

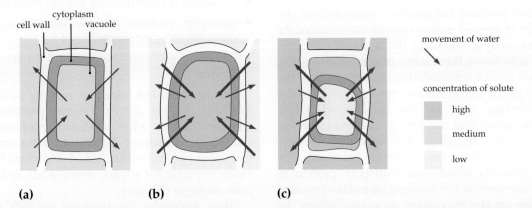

(a) (b) (c)

Fig. 4.7 Due to the presence of a cell wall, diffusion of water across the plasma membrane of a typical plant cell placed in different solutions is dependent on both osmotic potential (concentration of solutes) and pressure potential. (**a**) Water potential inside and outside the cell is equal. (**b**) The osmotic potential inside the cell is lower (a higher concentration of solutes) and so water enters the cell, increasing the hydrostatic pressure until the cell becomes more turgid. (**c**) The osmotic potential inside the cell is higher (a lower concentration of solutes) and so water leaves the cell, in this case causing the plasma membrane to pull away from the cell wall (plasmolysis)

If plant cells are placed into a solution that is more concentrated than their cellular contents, water will leave the cells and the pressure potential will move towards zero. Water may continue to leave until the volume of the cells has decreased to the point at which the internal and external osmotic potentials are equal, just as happens in the case of cells without walls. The shrinkage of the cell usually draws the cell membrane away from the cell wall, leaving a gap between the membrane and wall. This shrinkage is the process of **plasmolysis** (Fig. 4.7c). The point at which pressure potential reaches zero, without causing shrinkage, is the point of *incipient plasmolysis*. At this point, the osmotic potential of the external solution equals that of the cell when it was turgid and so provides a way to measure the osmotic potential of walled cells.

> Water moves into or out of cells depending on the difference in water potential across the plasma membrane. In animal cells, water potential is due only to osmotic potential. Because of the presence of a wall, water potential in plant cells depends on both osmotic and hydrostatic pressures.

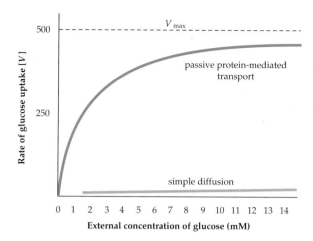

Fig. 4.8 The rate of transport of glucose (in micromolecules per millilitre of cells per hour) into red blood cells in relation to the external concentration of glucose illustrates the difference between simple diffusion and passive protein-mediated transport (facilitated diffusion). The rate of simple diffusion into cells is low and is always proportional to substrate concentration. In passive protein-mediated transport, the rate of transport is much higher, reaching a maximum (V_{max}) as substrate concentration increases because the carrier proteins have become saturated

PROTEIN-MEDIATED TRANSPORT

So far, we have described the passive movement of molecules through the lipid bilayer of membranes. However, some solutes move across membranes by means other than simple diffusion. In contrast to diffusion, rates of movement of these substances show little correlation with lipid solubility and do not increase linearly as the concentration of substrate increases (Fig. 4.8). This form of transport involves intrinsic membrane proteins, called transport proteins, that span the lipid bilayer (Chapter 3) and facilitate the movement of particular molecules across the membrane.

Transport proteins are enzymes that catalyse the movement of specific solutes across membranes. As with other catalytic reactions, the action of transport proteins is to lower the energy of activation (p. 29), in this case for the transmembrane movement of the solute. As solute concentration increases, more of the enzyme-binding sites for the transported solute become occupied and the rate of transport reaches a plateau, so showing the *saturation kinetics* characteristic of all enzyme-catalysed reactions. The reaction between solute and transport protein is usually quite specific. For example, one of the proteins responsible for the movement of sugars across the plasma membrane of red blood cells favours the transport of glucose over similar sugars such as galactose and fructose. In the internal membranes of skeletal muscle cells, there is a transport protein that moves K^+ and almost no other ion.

Different cell types are characterised by different sets of transport proteins in their membranes. For example, red blood cells carry large amounts of carbon dioxide back to the lungs from active tissues. In blood, carbon dioxide is dissolved as HCO_3^-. In the plasma membrane of red blood cells, there is a protein that is responsible for rapidly transporting HCO_3^-, in exchange for Cl^-, into and out of these cells. This protein is more abundant in the plasma membrane of red blood cells than in any other membrane.

The general properties of protein-mediated transport that distinguish it from diffusion are:

■ transport is *faster* than by simple diffusion;

■ transport proteins become *saturated* as substrate concentration increases (Fig. 4.8);

■ transport proteins are *specific* for particular substrates.

Protein-mediated transport may be passive (*facilitated diffusion*) or active. Passive protein-mediated transport occurs where diffusion gradients are favourable but transport is required more rapidly than can occur by simple diffusion. A wide variety of molecules, in particular sugars, amino acids and inorganic ions, move in this way.

Some molecules are required by cells in greater amounts than can be provided by passive protein-mediated transport. Even more importantly, cells often need to transport molecules *against* their prevailing energy gradient. In both these circumstances, energy must be expended to move the required solute across cell membranes at an appropriate rate. The principal features of active transport are that it requires the expenditure of energy and that it can move substances against a concentration or electrochemical gradient.

> Protein-mediated transport is faster than simple diffusion, is specific for particular substrates and shows saturation as substrate concentration increases. It may be passive or active.

Channels and carriers

Two different types of transport proteins catalyse the transmembrane movement of solutes—channels and carriers. *Channels* effectively form hydrophilic pores through the membrane and, when the channels are open, they allow very high rates of movement of certain solutes. Channels are not directly coupled to an energy source, so movement through channels is always passive. *Carriers* tend to move solutes at a slower rate than channels and undergo a more radical conformational change during the passage of the solute (Fig. 4.9). Many carrier proteins only allow passive transport down an energy gradient. However, some carriers are coupled to an energy source and carry out active transport (p. 73). The reversal of all passive ion movements involves active transport to regenerate the original electrochemical gradients.

Channels

Channels are the fastest enzymes known, catalysing the net movement of 10^6 to 10^8 molecules per second.

(By comparison, the fastest water-soluble enzyme known, catalase, can perform 10^5 reactions per second, and most enzymes have much slower turnover rates.) Most channels transport ions and are therefore referred to as ion channels.

Channels often show a high degree of selectivity in the solutes transported. For example, certain channel proteins will catalyse mainly the movement of the monovalent cations K^+ and NH_4^+; other ions such as Na^+ are transported at less than one-tenth the rate that K^+ is moved. Other channels transport Ca^{2+}, and will do so at ten times the rate at which they transport Mg^{2+}. This selectivity suggests that there is an intimate association between the transported ion and the walls of the channel, and that any hydration sphere around the ion (p. 7) is shed during its passage. Ion movement through channels shows saturation kinetics.

Channels may be open or shut in response to certain conditions. Change in the voltage difference across the membrane (the membrane potential) alters the three-dimensional configuration of certain channel proteins, causing the aqueous pore to open or close (Fig. 4.10a). Such channels are *voltage-gated*. Other channels have binding sites for specific signal molecules (ligands), for example, the molecules released from nerve cell endings (neurotransmitters). When bound to such a ligand, the conformation of the channel protein alters, opening or closing the pore. These are *ligand-gated* channels (Fig. 4.10b).

By virtue of the remarkable speed with which channels transport ions, these proteins not only have a role in transport, but also can be used by cells for the rapid transmission of messages. For example, the electrical impulses that move along nerves are caused

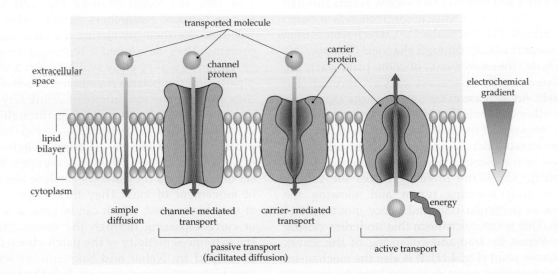

Fig. 4.9 Channels and carriers are two different types of transport proteins. Movement through channels is passive. Carriers undergo a more radical conformational change and may allow passive transport down an energy gradient or be coupled to an energy source for active transport

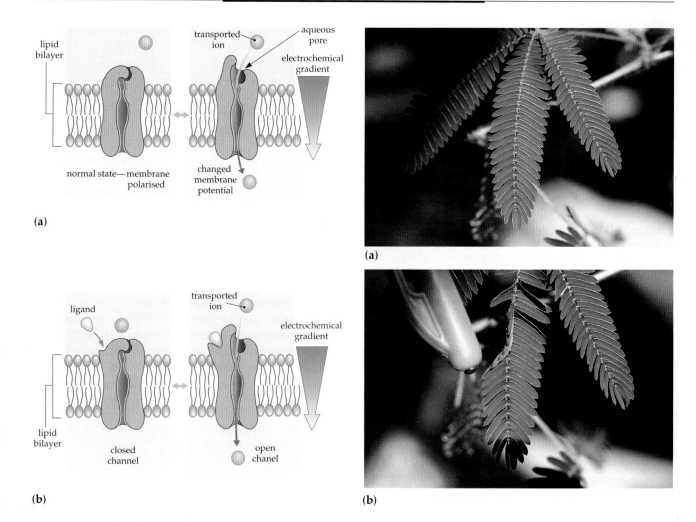

Fig. 4.10 Types of channels. (**a**) Voltage-gated channel. (**b**) Ligand-gated channel

Fig. 4.11 The sensitive plant *Mimosa pudica*. Sudden movements of leaflets (opening and closing) are generated by the movement of ions through channels

by the rapid and selective movement of ions through channels (Chapter 26). Muscle contraction is initiated by an increase in intracellular Ca^{2+}, which results from the transport of Ca^{2+} through channels (Chapter 27).

Similarly, the movements of some plant parts are stimulated by the transport of Ca^{2+} through ion channels. An increase in Ca^{2+} levels in the cytoplasm opens other channels that allow the transport of Cl^- and K^+ ions out of the cell. The consequent decrease in the intracellular KCl concentration causes an increase in the osmotic (and water) potential inside the cell (p. 65). Water therefore leaves the cell by osmosis, thus reducing turgor and allowing the collapse of particular cells and hence movement of tissues. This is the mechanism that underlies closing of the Venus fly trap and collapsing of the leaves of a *Mimosa* plant (Fig. 4.11). It is also the mechanism for the closure of pores in leaves (*stomata*) through which gases diffuse, in particular through which water leaves the plant and carbon dioxide enters to be fixed by photosynthesis.

In 1991, the Nobel Prize for Medicine was given to two German researchers, Erwin Neher and Bert Sakmann, for their pioneering research into channel proteins. They developed a technique enabling the formation of a very close seal between a small glass pipette and the plasma membrane of cells (Fig. 4.12), thus enabling the measurement of any tiny electrical currents carried by ions flowing through channels in the membrane. It has been discovered that channel proteins behave like an electrical switch—they are either open or closed. When they are open, for a given set of conditions they always catalyse the same rate of movement of ions. They flicker open and close at very high speed, which can be seen as a flickering of current passing through the membrane. Due to the extreme sensitivity of the patch-clamp technique developed by Neher and Sakmann, as well as the very high rates of transport catalysed by ion channels, it is possible to observe a single channel protein turning on and off as it bursts into transport activity and then falls quiet (Fig. 4.13).

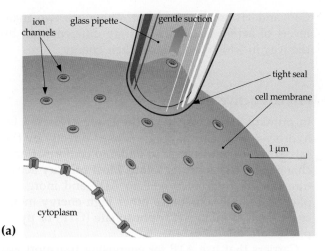

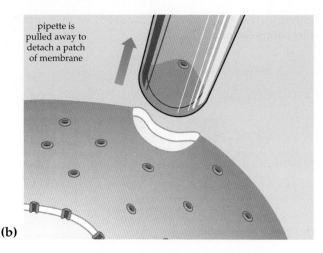

(a)

(b)

Fig. 4.12 In the Neher and Sakmann technique, a tiny patch of membrane is sucked onto the tip of a micropipette, forming a tight seal. Electrical current can only pass across the tip through the ion channel. (**a**) The patch of membrane remains continuous with the plasma membrane. (**b**) The patch of membrane is detached, allowing modification of the solution on either side of the membrane for experimental purposes

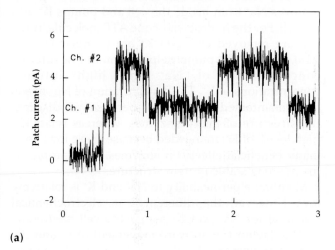

(a)

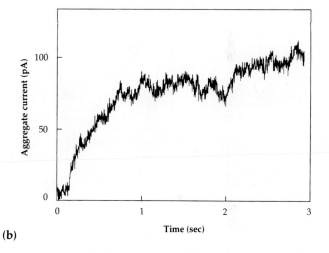

(b)

Fig. 4.13 Recordings of current through individual voltage-gated potassium channels in a patch of plasma membrane detached from a wheat root. (**a**) The patch probably contained just two channels (Ch. #1 and Ch. #2). The membrane was abruptly depolarised from −50 mV to + 50 mV causing the channels to open more frequently. Each upward step in current flow represents the opening of a single channel; downward steps represent closure. When two channels open together in the patch, a current double the magnitude of the single channel current is obtained (i.e. 5 picoamps instead of 2.5 picoamps). The minor fluctuations in the current arise from electrical noise. (**b**) The sum of 40 single channel responses similar to those in (**a**) gives a reconstruction of the current similar to that observed in the whole membrane of a cell. The probability that an individual channel will be in the open state is reflected by the time course of the aggregate current. The potassium channel shown in these experiments is common in plant cells and is thought to be responsible for potassium efflux from stomatal guard cells (causing closure) and in control of the membrane potential. This channel is similar to a potassium channel that contributes to the action potential (delayed outward rectifier) in animal cells but the plant channel is much slower

Ion channels are the fastest enzymes known. They allow passive protein-mediated transport and are highly selective for particular ions. They are opened by change in voltage across the membrane (voltage-gated channels) or by binding with specific signal molecules (ligand-gated channels).

Carriers

The other type of membrane transport proteins that catalyse passive transport are called carriers. These bind to a solute on one side of the membrane, change conformation, and then release the solute on the opposite side of the membrane (Fig. 4.14). On releasing the substrate, the carrier molecule returns to its original orientation within the membrane. Each carrier has a specific binding site for its substrate molecule and the rate of transport reaches a plateau as the concentration of substrate increases (saturation kinetics). Some carriers transport only a single solute across the membrane (*uniport*), while others have two binding sites and co-transport two different solutes, either in the same direction (*symport*) or in opposite directions (*antiport*) (Fig. 4.15).

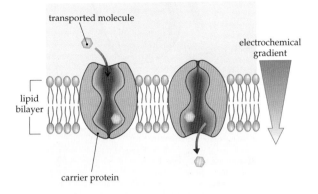

Fig. 4.14 Model of a carrier mediating passive transport as a result of a conformational change after binding to its substrate

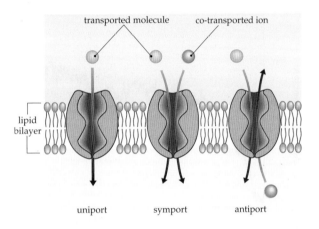

Fig. 4.15 Carriers may be uniporters, symporters or antiporters

Carriers undergo a more radical conformational change than channel proteins during the passage of a solute. Transport may be passive or active.

Active transport

Carriers linked to a source of energy, such as hydrolysis of ATP, are able to move solutes against an energy gradient. For a neutral solute, this means from an area of lower concentration to one of higher concentration; if the solute is charged, this is against an electrochemical potential difference. Metabolic energy is used to drive the solute through the transport protein or 'pump'. The active transport of solutes can result in the accumulation of solutes inside cells; alternatively, solutes may be actively removed from cells. For example, in most plant cells, Cl^-, NO_3^- and $H_2PO_4^-$ are normally pumped in, whereas Na^+,

Ca^{2+} and H^+ are pumped out. There are two main forms of active transport that are differentiated by the way in which they obtain their energy.

Primary active transport

In primary active transport, the energy released by hydrolysis of chemical bonds is used directly by the carrier to provide energy to undergo conformational change and transport its solute. Most primary active transporters hydrolyse ATP to ADP and inorganic phosphate, but sometimes other high-energy molecules, such as guanosyl triphosphate (GTP) or pyrophosphate, are used.

Carriers that use ATP for energising transport are referred to as solute-translocating ATPases. There is a relatively small number of these. One example is the *sodium–potassium pump* (Na^+-K^+ translocating ATPase), which is common in animal cells. This pump removes Na^+ from inside the cell and pumps K^+ into the cell. For the hydrolysis of one ATP molecule, three Na^+ are removed and two K^+ are pumped in (Fig. 4.16). The Na^+-K^+ pump helps maintain low internal concentrations of Na^+ and high internal concentrations of K^+. The concentration of Na^+ needs to be maintained at low levels because Na^+ disrupts the structure of many enzymes, whereas relatively high levels of K^+ are needed because these ions have many beneficial effects on enzymes, for example, in maintaining stable protein structure.

Membrane permeability to Na^+ and K^+ is relatively low. However, the differences in electrochemical potential for Na^+ and K^+ across the cell membrane tend to favour the outward movement of K^+ and the inward movement of Na^+ (Fig. 4.16). If Na^+-K^+ pumps were not operational, internal Na^+ levels would slowly rise and internal K^+ levels would slowly fall. Proteins with a very similar structure to the Na^+-K^+ pump are found in a wide range of eukaryotic cells (animal, plant and fungal) where they catalyse the removal of Na^+, Ca^{2+} or H^+.

The plasma membranes of plant and fungal cells are characterised by an ATPase that removes protons (H^+) from the cell—the *proton pump*. This pump removes protons produced internally (from the synthesis of organic acids, etc.) and protons that move into the cell. In doing so, the proton pump helps to regulate cellular pH, maintaining it just above neutral.

The ATPases described above belong to a group of closely related ATPases called P-type ATPases. They are proteins with a single polypeptide chain with a molecular mass of approximately 100 kD, and they have similar amino acid sequences. It seems that all P-type ATPases evolved from a common ancestral ATPase. There are, however, two other types of ATPases with fundamentally different structures.

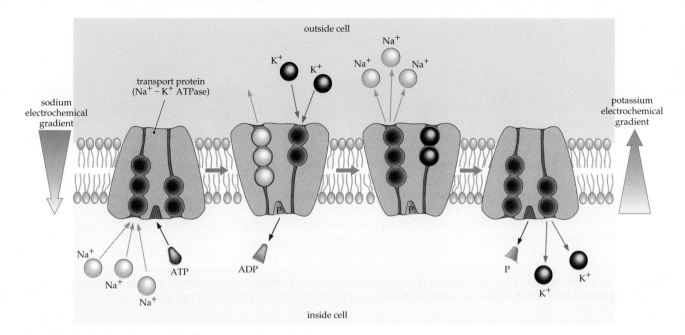

Fig. 4.16 Na⁺–K⁺ ATPase transports three Na⁺ out of the animal cell and two K⁺ into the cell at a cost of hydrolysis of one ATP molecule

V-type ATPases are found on some internal membranes of eukaryotic cells, such as on the tonoplast (the membrane surrounding the vacuole) of plant cells and on the membrane of coated vesicles (as seen in Fig. 3.14). A V-type ATPase consists of many polypeptide chains which together form a molecular mass of about 500 kD.

F-type ATPases are found in the plasma membrane of bacteria, the inner membrane of mitochondria and the thylakoids of chloroplasts. Instead of consuming ATP to pump solutes, these proteins use the passive movement of H⁺ to drive the synthesis of ATP from ADP and inorganic phosphate. This reaction is the reverse of that catalysed by the other types of ATPase. The gradient of H⁺ is initially established by a series of redox reactions (Chapters 2 and 5). The similarities between these reactions, and between the structure of the ATPases of bacteria, chloroplasts and mitochondria, forms part of the evidence to support the theory that mitochondria and chloroplasts originated by the non-destructive engulfment of smaller prokaryotes by larger prokaryotes (Chapter 35).

> In primary active transport, energy is used directly by the carrier protein to pump solutes across a membrane.

Solute-coupled active transport

Both Na⁺–K⁺ and H⁺ pumps 'push' positively charged ions out of eukaryotic cells. They therefore contribute to the difference in electrical potential found across most cell membranes, where the inside of cells is negative with respect to the outside. The pumps thus maintain an electrochemical potential difference for ions across the plasma membrane. The tendency for Na⁺ and H⁺ to move energetically 'downhill' and diffuse back into the cells from which they have been pumped is exploited in *solute-coupled active transport*. In this form of transport, solutes are cotransported through transport proteins against an energy difference using energy derived from the 'downhill' movement of ions. Thus, the energy used initially to create the electrochemical potential differences is used secondarily to power solute transport. (For this reason the term 'secondary active transport' is sometimes used.)

In animals, the driving ion is usually Na⁺ and in plants, H⁺. In fungi and bacteria, some organisms use H⁺ and some use Na⁺. As explained above, in eukaryotes these ions are pumped out of the cell through a primary pump, with the hydrolysis of ATP usually providing the necessary energy. In prokaryotes, the energy comes from a series of redox reactions. In all cells, a regulated number of these ions then moves energetically downhill back into the cell through specialised proteins, the solute-coupled transporters. This movement provides the necessary energy to transport other solutes against their energy differences (Fig. 4.17). Solute-coupled active transport may involve symport or antiport. Important examples of symports include the Na⁺–glucose pump in animal cells and the 2H⁺–Cl⁻ pump in plant cells. Both of these pumps are responsible for the accumulation of the solute (glucose or Cl⁻), which moves in with the driving ion (Na⁺ or H⁺).

The use of energy differences of ions across membranes is a universal process in organisms and occurs across most biological membranes. It was first described in bacteria in the early 1960s by Peter

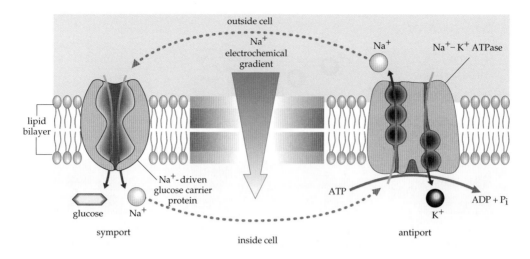

Fig. 4.17 In animal cells, the Na^+-K^+ ATPase (antiport) contributes to a strong electrochemical potential difference for Na^+. Downhill movement of Na^+ through the solute-coupled transporter (symport) brings glucose with it into the cell

Mitchell. He demonstrated the importance of gradients of H^+ for both ATP production and transport of other solutes; for this, he was awarded the Nobel Prize for chemistry in 1978. Differences in energy levels of ions are used not only for the indirect harvesting of the energy of hydrolysis of ATP for the transporting of solutes, but also for the *synthesis* of ATP from ADP and P_i in mitochondria and chloroplasts, where the difference in the energy levels of H^+ built up by an electron transport chain is allowed to dissipate through ATP synthases (Chapter 5).

> In secondary active transport, solutes are dragged through carrier proteins against their energy gradient, using energy derived from the downhill movement of ions (usually Na^+ in animal cells and H^+ in plant cells).

VESICLE-MEDIATED TRANSPORT

The plasma membrane is not a fixed structure; new molecules are synthesised and inserted into the membrane, and existing components are retrieved and recycled or broken down. This dynamic process of insertion and retrieval is due to the pinching in of small portions of membrane to form *vesicles*, the process of endocytosis, or the fusing to the plasma membrane of small vesicles constructed inside the cell, exocytosis (Figs 4.18, 4.19).

During endocytosis, a small area of plasma membrane enfolds or 'invaginates', enclosing substances that are outside the cell (Fig. 4.18). The invagination then pinches off to form a vesicle in which the substance is contained. The vesicle is then transported within the cell. This process brings about the inward movement of solids (*phagocytosis*) or of liquids (*pinocytosis*).

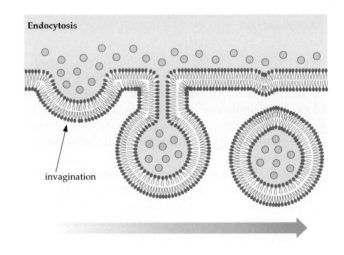

Fig. 4.18 Formation of a transport vesicle during endocytosis

In some cases, endocytosis is stimulated by initial binding of the solute to a receptor molecule on the membrane. An example of *receptor-mediated endocytosis* is the uptake of low-density lipoproteins (LDL) into animal cells to be broken down to form cholesterol (Fig. 4.20). Specific receptor proteins, present in the plasma membranes of animal cells, bind specifically to LDL particles. LDL-receptor complexes then cluster in depressions (pits) of the cell surface. The internal surface of the membrane at the pit is coated with the protein clathrin (Chapter 3). Clathrin-coated vesicles containing LDL particles form by endocytosis.

In the process of exocytosis, intracellular vesicles fuse to the plasma membrane and the contents of the vesicle are deposited on the outside of the cell (Fig. 4.19). Many hormones and enzymes are released in this way. In the germinating seeds of many plants, the enzymes required to break down the storage compounds of the seed are secreted by exocytosis from a special layer of cells upon receiving a hormonal signal from the germinating embryo.

In endocytosis, small portions of the plasma membrane form vesicles, enclosing substances to be imported into a cell. In exocytosis, vesicles fuse with the plasma membrane to export their contents.

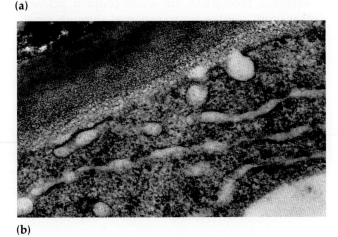

(a)

(b)

Fig. 4.19 **(a)** Fusion of a transport vesicle with the plasma membrane during exocytosis. **(b)** Transmission electron micrograph of transport vesicle during secretion from a sugar-secreting cell in a gland on the foliage of the sunshine wattle, *Acacia terminalis*

Fig. 4.20 Endocytosis of low-density lipid (LDL) particles in animal cells. LDL particles bind tightly and specifically to surface receptors and are then internalised as complexes in a clathrin-coated pit. Clathrin-coated vesicles are pinched off into the cytosol, where clathrin is depolymerised, forming an uncoated vesicle (endosome), which can fuse with special uncoupling vesicles. The pH of these vesicles is low (~5.0), and causes the dissociation of the receptor and the LDL particle. The receptor protein is recycled direct to the plasma membrane within its own vesicle, while the LDL particle enters the lysosome system where the particles are degraded. Apoprotein B is broken down into amino acids, and the cholesterol esters into fatty acids and cholesterol, which is incorporated into cellular membranes

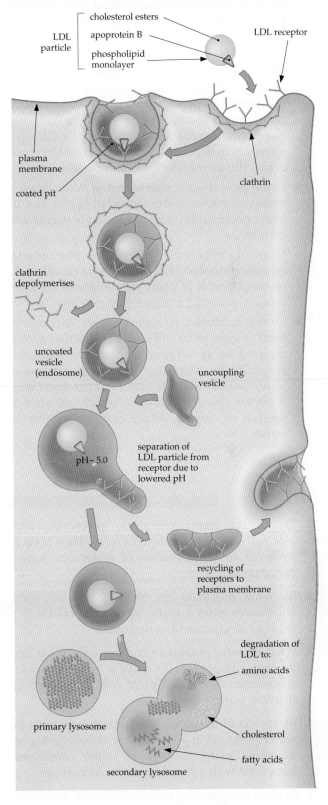

SUMMARY

- All living cells are surrounded by a plasma membrane and an aqueous medium that make up their immediate environment. The selective permeability of the lipid bilayer allows cells to control the composition and stability of their contents.

- Diffusion is the passive movement of molecules along an electrochemical gradient. The rate of diffusion depends on the permeability coefficient, a measure of how easily a molecule passes through the membrane, and the diffusive force, which depends on the size of the difference across the membrane for that molecule.

- Osmosis is the net movement of water from a region of higher water potential (ψ) to one of lower potential through a selectively permeable membrane.

- In animal cells, water potential is due only to osmotic potential (ψ_π). Owing to their cell wall, water potential in plant cells depends on both osmotic potential and pressure potential (ψ_p).

- Protein-mediated transport allows the movement of some polar and ionised molecules across membranes. It is faster than simple diffusion, is specific for particular substrates, and shows saturation as substrate concentration increases. It occurs through channels, which are always passive, or carriers, which may be passive or active.

- Ion channels are the fastest enzymes known. They are highly selective for particular ions and are opened by a change in voltage across the membrane (voltage-gated channels) or by binding with specific signal molecules (ligand-gated channels). The 'patch-clamp' technique enables the operation of a single channel to be studied.

- Carrier proteins undergo more radical conformational change than channel proteins during solute transport. On releasing the solute, the carrier returns to its original orientation. Some carriers transport only a single solute across the membrane (uniport), while others co-transport two different solutes, either in the same direction (symport) or in opposite directions (antiport).

- In primary active transport, energy is used directly by the carrier protein to pump solutes across a membrane. In solute-coupled active transport, solutes are transported through carrier proteins against their energy gradient using energy derived from the downhill movement of ions (usually Na^+ in animal cells and H^+ in plant cells).

- Another way that molecules can cross the plasma membrane is by vesicle-mediated transport. In endocytosis, small portions of the plasma membrane form vesicles, enclosing substances to be imported into a cell. In exocytosis, vesicles fuse with the plasma membrane to export their contents.

QUESTIONS

1. Define the terms selectively permeable and electrochemical potential difference. How is each involved in the diffusion of molecules across membranes?

2. By what processes do each of the following substances enter a white blood cell? (a) O_2 (b) H_2O (c) glucose (d) a bacterium

3. What is water potential? In plant cells, what are its two components? Compare what happens when animal cells and plant cells are placed in distilled water.

4. By means of diagrams, illustrate the changes in water potential in a typical leaf cell occurring in a wilting plant growing in saline conditions compared with a similar plant in normal soil.

5. Explain how passive protein-mediated transport enables glucose to pass through the membranes of red blood cells faster than by simple diffusion.

6. Outline the characteristic features of membrane transport proteins that are channels and carriers.

7. By means of diagrams, distinguish between the mechanism of action of voltage-gated channels and ligand-gated channels.

8. List three key features that distinguish between primary active transport and solute-coupled active transport.

9. Describe the action of the Na^+–K^+ pump in animal cells. Explain how this is used indirectly to transport glucose into a cell. In your answer distinguish between a symport and an antiport.

10. Discuss how the plasma membrane controls the internal stability and overall metabolism of cells.

HARVESTING
ENERGY

The earth formed about 4.5 billion years ago and, as it evolved (see Chapter 31), a crust solidified over the hot mantle and an atmosphere gradually developed. The primitive atmosphere contained water vapour, carbon dioxide, hydrogen, nitrogen, ammonia, methane, sulfur dioxide and hydrogen sulfide, much like volcanic gases do today. Because the atmosphere contained ample hydrogen and *lacked oxygen*, it was a reducing one, a condition that favoured the formation of molecules rich in C—C bonds. In 1953, S.L. Miller performed a classic experiment to show that an electrical discharge (such as lightning) in such an atmosphere could have led to the formation of complex carbon molecules. The waters of oceans or lakes, which formed from the steam of volcanoes, became a 'dilute soup' of carbon-containing molecules—amino acids, carbohydrates and nucleic acids. Simple cell-like structures may have formed from aggregations of these molecules; for example, in water, phospholipid molecules can spontaneously form spheres (Chapter 3) and double-layered membranes. Although the exact way in which cells first formed is unknown, primitive self-replicating structures could have evolved from such aggregations of molecules.

The first organisms to evolve were prokaryotes. They were presumably **heterotrophs**, organisms unable to synthesise their own food. They would have used the complex carbon molecules in the 'soup' for raw materials and energy. However, there are no fossils of these heterotrophs and the oldest evidence of life, dating back 3.5 billion years, is of **autotrophs**, bacteria able to synthesise their own fuel molecules. The first of these (photosynthetic bacteria) had a non-oxygen requiring, or **anaerobic**, metabolism and had evolved the ability to trap the energy of sunlight and use it to synthesise carbohydrate directly from carbon dioxide and water; in the process, they produced oxygen as a by-product. The atmosphere thus gradually became oxygen-rich, destroying the reducing conditions of the earlier atmosphere and, hence, the very conditions that led to the formation of organic molecules and the evolution of life (Fig. 5.1).

Modern photosynthetic organisms include plants, algae (protists) and many bacteria (green, purple and cyanobacteria). Modern heterotrophs include other forms of bacteria, protists, animals and fungi. Heterotrophs are ultimately dependent on the chemical energy produced by autotrophs. However, whether organisms produce organic molecules or feed on ready-made ones, all organisms use chemical energy in the form of ATP to drive metabolic reactions (Chapter 2). Therefore, this chapter deals first with how all organisms convert the chemical energy of fuel molecules to the usable energy of ATP. The second part of the chapter discusses how photosynthetic organisms trap the radiant energy of sunlight and store it in the chemical bonds of glucose.

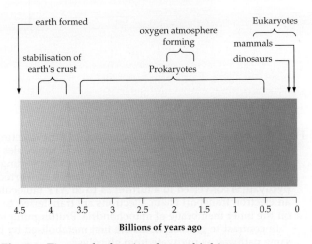

Fig. 5.1 Time scale showing the earth's history

The first organisms were heterotrophic and used molecules that formed spontaneously in the absence of oxygen as a source of raw materials and energy. Photosynthetic organisms evolved later, enriching the atmosphere with oxygen.

HARVESTING CHEMICAL ENERGY

Energy is released from fuel molecules along metabolic pathways that are initially different for carbohydrates (glucose) and lipids. Glucose is processed by glycolysis, while lipids are processed by β-oxidation. The products of both of these pathways can then act as the substrate for cellular respiration, the major process of energy extraction (Fig. 5.2). **Cellular respiration** involves oxidation of fuel molecules, that is, removal of electrons from the C—C and C—H bonds. Electrons extracted from these bonds are accepted by the coenzymes NAD (nicotinamide adenine dinucleotide; see p. 34) and FAD (flavin adenine dinucleotide) and passed down electron transport chains, driving proton pumps that are coupled to the synthesis of ATP.

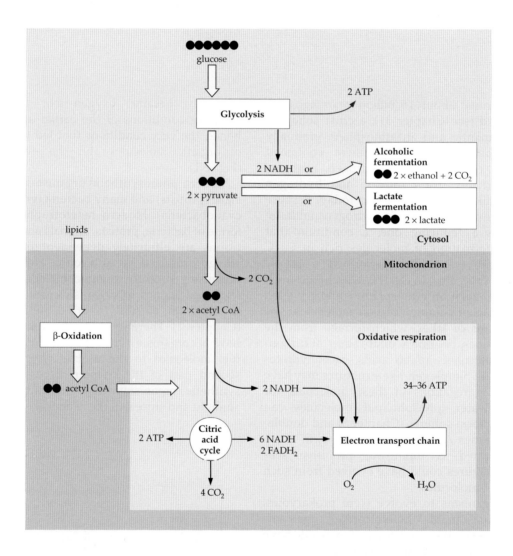

Fig. 5.2 Overview of the metabolic pathways for extracting energy from fuel molecules. Glucose is first metabolised by glycolysis, with a net production of two ATP molecules per glucose molecule. A 3-carbon molecule, pyruvate, from glycolysis can then enter different pathways depending on the organism and whether the environment is anaerobic or aerobic. In the absence of oxygen, pyruvate is converted to lactate or ethanol. In the presence of oxygen, the energy in pyruvate is converted to a further 34 to 36 ATP molecules. The energy extraction pathway is the citric acid cycle and an electron transport chain. Electrons are transferred to coenzymes NAD^+ and FAD and used to drive proton pumps on the inner membrane of mitochondria. Proton pumping is coupled to ATP synthesis.

In contrast to glucose, lipids are first metabolised by β-oxidation, with the product acetyl CoA being oxidised by the same pathway as pyruvate from glycolysis

Glycolysis: the initial processing of glucose

Complex carbohydrates, such as starch (plants) and glycogen (animals), are broken down by catalysed hydrolysis into glucose subunits before entering energy-releasing pathways. In all cells, both prokaryotic and eukaryotic, glucose is processed initially in the cytosol by glycolysis (Fig. 5.3).

Glycolysis involves three stages: rearrangement and splitting of the glucose molecule; energy production; and recycling. In the first stage, two

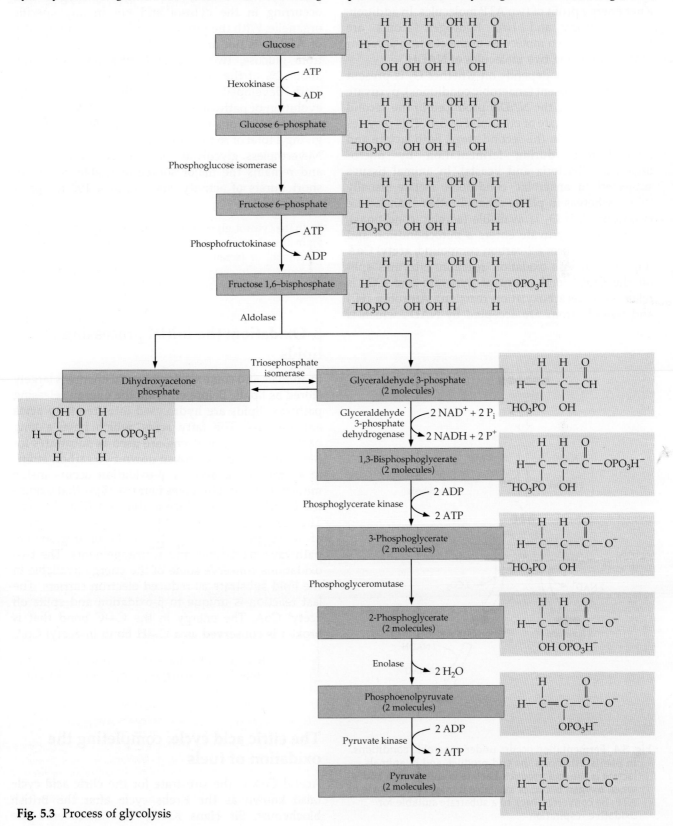

Fig. 5.3 Process of glycolysis

molecules of ATP are used to phosphorylate and change glucose in preparation for splitting it into two 3-carbon molecules (glyceraldehyde 3-phosphate). In the second stage, oxidation of glyceraldehyde 3-phosphate to **pyruvate** ($C_3H_3O_3^-$) is coupled to ATP synthesis: four ATP molecules are produced giving a net energy profit of two ATP molecules. In addition, four electrons and two hydrogen atoms are transferred to two molecules of the electron acceptor NAD^+, producing two molecules of NADH.

The final stage depends on whether O_2 is present or not. If O_2 is absent (anaerobic conditions), pyruvate is reduced using the NADH produced in the second stage to form lactate or ethanol (Fig. 5.4). This recycles NAD^+ for reuse in the second stage. Both lactate and ethanol form during fermentation by micro-organisms (bacteria and yeasts). In animal tissues subjected to anaerobic conditions, lactate usually forms, whereas in plants ethanol rather than lactate is produced. If O_2 is available (aerobic conditions), the 3-carbon pyruvate enters a mitochondrion and is oxidised to form a 2-carbon compound, **acetyl CoA** (Fig. 5.2). In this oxidation reaction, enzymes split off the COO^- end group of pyruvate and CO_2 is released (a decarboxylation reaction). A proton (H^+) and two electrons are accepted by NAD^+, which is reduced to NADH (used later to produce ATP). The 2-carbon molecule remaining is linked to coenzyme A, forming acetyl CoA, which is the substrate for the citric acid cycle.

Glycolysis was one of the earliest biochemical pathways to evolve, requiring no oxygen and occurring in the cytosol and not in any specific organelle. With the production of only two molecules of ATP, it is not an efficient way to extract energy from glucose; two ATP molecules represent only about 2% of the chemical energy available. As the earth developed an oxygen-rich atmosphere, the evolution of pathways for *oxidative respiration* allowed the extraction of a further 34 to 36 molecules of ATP, giving a total of 36 to 38 for every molecule of glucose. Nevertheless, glycolysis generates ATP very rapidly and remains the major source of usable energy in short bursts of activity, such as in a 100 m sprint by an athlete.

In the cytosol, glycolysis produces two molecules of ATP. In the absence of O_2, the end product of glycolysis is ethanol or lactate. In the presence of O_2, pyruvate is converted to acetyl CoA, which is the substrate for the citric acid cycle.

β-Oxidation: the initial processing of lipids

In animals and some seeds, chemical energy is largely stored as lipids. Before entering the energy-releasing pathway, lipids are hydrolysed into free fatty acids and glycerol. The fatty acids, which have a long backbone of carbon atoms, are used as substrates for β-oxidation, and are broken down two carbon atoms at a time. In eukaryotes, β-oxidation occurs inside mitochondria and involves four reactions that oxidise the β-carbon and produce acetyl CoA (Figs 5.2, 5.5). The first three reactions, like glycolysis, are typical of those that make up the bulk of degradative pathways, oxidations and rearrangements. The two oxidations conserve some of the energy available in the lipid substrate as reduced electron carriers. The last reaction is unique to β-oxidation and splits off acetyl CoA. The energy in the C—C bond that is broken is conserved as a C—H bond in acetyl CoA.

β-oxidation degrades long chain fatty acids by two carbon atoms at a time to form acetyl CoA, which enters the citric acid cycle.

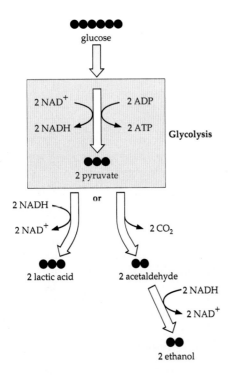

Fig. 5.4 Fermentation occurs under anaerobic conditions. In plants, ethanol is the end product, and in animals, lactate is the end product. Lactate is energy-rich and when oxygen becomes available it may, with the use of ATP, be converted back to a substrate suitable for oxidative respiration

The citric acid cycle: completing the oxidation of fuels

Acetyl CoA is the substrate for the **citric acid cycle** (also known as the Krebs cycle after the British biochemist, Sir Hans Krebs, who discovered the

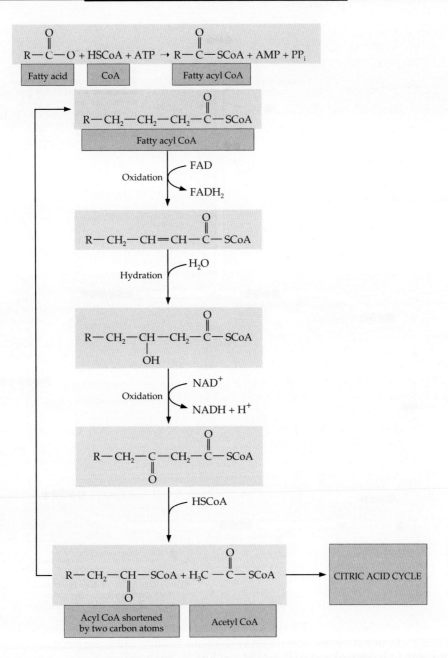

Fig. 5.5 β-oxidation of fatty acids from lipids. In four reactions catalysed by enzymes, a fatty acyl CoA molecule is converted to acetyl CoA and another fatty acyl CoA molecule with two fewer carbons. During oxidation, electrons are transferred to FAD and NAD+ to produce FADH$_2$ and NADH. The cycle is repeated until fatty acids are completely converted to acetyl CoA, which enters the citric acid cycle

pathway in animal tissues). Acetyl CoA enters the citric acid cycle and combines with a 4-carbon molecule, oxaloacetate, releasing coenzyme A and forming a 6-carbon molecule, citrate (Fig. 5.6). Citrate is rearranged into the 6-carbon molecule isocitrate, which is the substrate for a series of oxidation reactions. Isocitrate is stripped of two electrons and one H+, which are transferred to NAD+ to form NADH. One molecule of CO$_2$ is also released. The resulting 5-carbon intermediate is similarly stripped of electrons and H+, forming another NADH and CO$_2$.

The 4-carbon product, succinyl CoA, is converted in four reactions to oxaloacetate to complete the cycle. In these reactions electrons and H+ are transferred to form FADH$_2$ and NADH, and one molecule of ATP is produced.

Oxidation of acetyl CoA yields CO$_2$, ATP molecules and energised electrons that are used to generate reduced electron carriers (three NADH and one FADH$_2$). In eukaryotes, the citric acid cycle occurs in mitochondria.

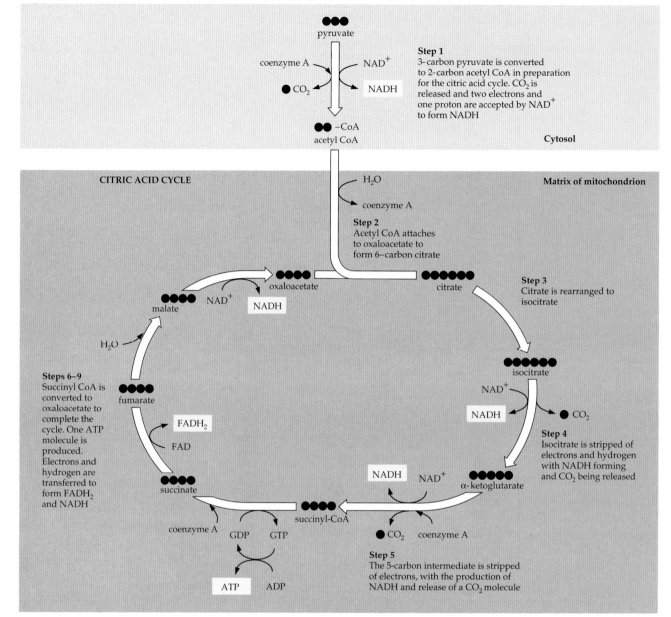

Fig. 5.6 The citric acid cycle. In eukaryotes, acetyl CoA enters mitochondria and is progressively oxidised through a series of reactions with CO_2 released and ATP, NADH and $FADH_2$ produced

Recycling electron carriers

In the citric acid cycle, three of the four oxidation reactions require the electron carrier NAD^+ as a substrate. The quantity of NAD^+ in cells is very small when compared with the continuous flux of carbon through the cycle that is required to maintain metabolic rate. In the human brain, for instance, there is only enough NAD^+ for 3 minutes at resting metabolic rate. The secret is that the NAD^+ used to make NADH, and the FAD used to make $FADH_2$, cycle back after having been oxidised during the next stage, the electron transport system (Fig. 5.2). Recycling is characteristic of a number of biochemical pathways in cells.

The electron transport system: generating ATP

The electrons released in glycolysis and the citric acid cycle are temporarily stored in the reduced electron carriers NADH and $FADH_2$. The energy now conserved in these molecules is then converted into ATP. The electron transport chain is responsible for 85–95% of the ATP that is ultimately produced from fuel molecules. The amount is variable because some of the energy released during the oxidation reactions is used for other purposes in mitochondria.

NADH and $FADH_2$ transfer electrons to special carrier proteins imbedded in the plasma membrane of prokaryote cells or the inner membrane of the

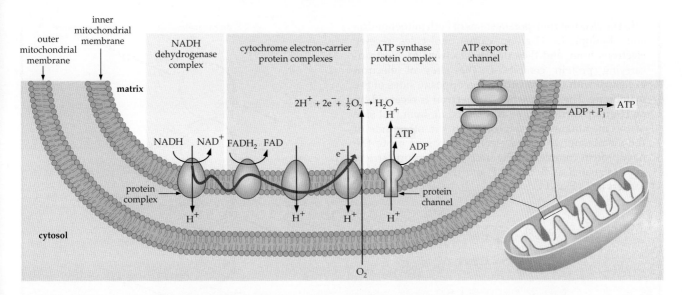

Fig. 5.7 The electron transport system and oxidative phosphorylation in the inner membrane of a mitochondrion. NADH and FADH$_2$ from glycolysis and the citric acid cycle transfer electrons to protein complexes. The transport of electrons is coupled to pumping protons to the outside of the inner membrane. Protons move back across the membrane and drive ATP synthesis. Oxygen acts as the final electron acceptor

mitochondrion of eukaryote cells (Fig. 5.7). Here the energy of NADH and FADH$_2$ is harvested in an electron transport system to produce ATP. (The mechanism of electron transport and phosphorylation is described in Chapter 2.)

Electrons from NADH and FADH$_2$ are removed and passed through three protein complexes. Associated with this transfer of electrons is the translocation of protons from the matrix to the outer side of the inner membrane. The final protein complex uses four electrons and four protons to reduce one molecule of O$_2$ to two of H$_2$O. Oxygen is thus the final electron acceptor with water forming as the end product (Fig.

5.7). In the process, the oxidised forms of the coenzymes NAD$^+$ and FAD are regenerated.

The proton concentration gradient created by proton pumping provides the electrochemical force to drive ATP synthesis (Box 5.1). As protons diffuse down the concentration gradient back across the mitochondrial membrane into the matrix, they pass through special protein channels (Fig. 5.7). The passage of protons drives the phosphorylation of ADP, resulting in the production of ATP.

> In the electron transport chain, electrons donated by NADH and FADH$_2$ from the citric acid cycle drive proton pumps coupled to ATP synthesis.

BOX 5.1   BIOLOGICAL ELECTRON TRANSPORT SYSTEMS

Electron transport systems occur on and in membranes. In prokaryotes they occur on the plasma membrane or endoplasmic membranes, and in eukaryotes the inner membrane of mitochondria and the thylakoid membranes of chloroplasts (Fig. a). Mitochondria are relatively large organelles (Chapter 3), with the cristae (folds) of their inner membrane providing a large surface area for ATP synthesis. It is perhaps not surprising to find that heart and skeletal muscle cells, which have a high requirement for ATP, have mitochondria with particularly large numbers of cristae (see Fig. 3.15).

Electron transport systems are composed of proteins and smaller compounds that have the ability to accept and donate electrons (for example, cytochrome *c*, Fig. 2.15). These components are an integral part of the membrane, and each has a specific affinity for electrons

(redox potential, see Chapter 2). They interact with each other such that electrons move sequentially from high to low potential (Fig. b). This forms a thermodynamically favourable sequence, which achieves a 'flow' or transport of electrons down a gradient in potential, that is, from electronegative towards electropositive.

Electrons from the reduced electron carrier are passed to a component in the chain with a slightly higher affinity for electrons (less negative redox potential) than this carrier. They are then passed to the next component, which has a slightly higher affinity again, and so on until they are passed to the final acceptor, oxygen, which has the highest affinity of all. Water is the product. Each electron transfer can be considered as a reaction coming towards equilibrium because there is more energy in the reduced donor than in the reduced acceptor.

In the electron transport systems of both mitochondria and chloroplasts, H-group carrier molecules receive electrons from, and then donate electrons to, electron-carrying protein complexes. The sites of electron exchange with the proteins are on opposite sides of the membrane. Therefore, when an electron is received by an H-group carrier, a proton is picked up from that side of the membrane and deposited on the other side of the membrane where the electron is passed on to the next protein complex. In Figure b, the H-group carriers are designated as B, D and F and the electron protein carriers as C, E and G.

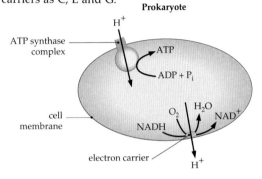

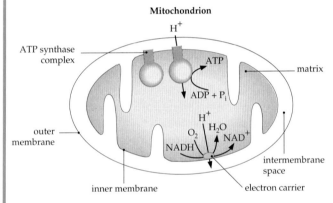

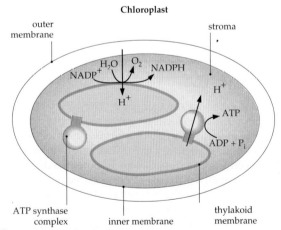

(a) Comparison of a prokaryote, a mitochondrion and a chloroplast showing the direction of proton (H⁺) movement and the position of the ATP synthase complex where ATP is produced. The arrangement of electron carriers and the ATP synthase complex in the cell membrane of bacteria is very similar to the sequence found in the inner membrane of mitochondria

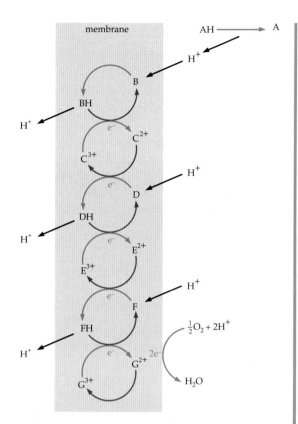

(b) Cytochrome electron transport chain. Electron carrier molecules in membranes interact with each other so that electrons move sequentially from high to low redox potential. This achieves a flow of electrons down a potential gradient

With the transport of protons and electrons, the membrane becomes polarised with a higher concentration of H⁺ (lower pH) on one side than the other. The two sides of the relatively impermeable membrane are in different compartments so that this pH difference (or proton motive force) cannot simply dissipate. In the case of the prokaryotic plasma membrane, the two compartments are the cytoplasm of the cell and the outside environment. In the case of mitochondria and chloroplasts, the inner membranes of the organelles form closed envelopes or vesicles.

As a result of the flow of electrons through the graded series of electron-carrier proteins in the membrane, a proton gradient is generated. In thermodynamic terms, the change in free energy as electrons traverse the series of carriers with decreasing redox potential is conserved as the proton gradient across the membrane.

Synthesis of ATP

Transmembrane proteins are involved in the passage of protons. The protein is attached to an enzyme that can use proton translocation to synthesise ATP (Fig. 5.7). This enzyme, which has been found in a similar form in membranes of many kinds, is ATPase. As its name implies, it normally catalyses the reaction:

$$ATP + H_2O \xrightarrow{\text{ATPase}} ADP + HPO_4 + H^+$$

When ATP is hydrolysed, the reaction proceeds with a $\Delta G'$ of -29.3 kJ mol^{-1}. However, because the enzyme in this case is physically attached to a proton channel, its catalytically active site has an almost unlimited supply of protons. The high concentration of protons effectively allows the reversal of the ATPase reaction to form ATP; the enzyme is thus functioning as an *ATP synthase*.

This 'chemiosmotic' coupling between electron transfer through localised carriers in membranes and the synthesis of ATP through utilisation of a proton gradient, was originally proposed by Peter Mitchell in the 1950s and 1960s. It has proved to be a unifying concept in our understanding of energy conservation and transfer in living systems and, for this, Mitchell received the 1978 Nobel Prize for Chemistry.

HARVESTING LIGHT ENERGY: PHOTOSYNTHESIS

Photosynthesis is the process by which solar energy is harvested and used to convert CO_2 and H_2O into carbohydrates. A considerable amount of energy reaches the earth as radiation from the sun each day. Most is absorbed by the earth as heat and about one-third is reflected back into space. Photosynthetic organisms absorb less than 1% of the solar energy reaching the earth but use it to produce billions of tonnes of carbohydrate each year.

The light-absorbing pigment **chlorophyll** makes it possible for a photosynthetic organism to use light in a way no other organism can. In eukaryotes, chlorophyll is associated with the thylakoid membranes of the chloroplast (Chapter 3). Chloroplasts occur in a wide variety of sizes and shapes and in quite variable numbers per cell; mangroves typically have 20 to 30 chloroplasts per leaf cell whereas the green alga *Chlamydomonas* has only one (Fig. 5.8). A leaf the size of your hand can contain a total of three to five billion chloroplasts (roughly equivalent to the total human population on the earth!).

The overall process of photosynthesis involves a series of chemical reactions, which are summarised by the equation:

$$6CO_2 \;+\; 12H_2O \;\xrightarrow{\text{visible light}}\; C_6H_{12}O_6 \;+\; 6O_2 \;+\; 6H_2O$$

| carbon dioxide from atmosphere | water | sugar | oxygen from original water molecules | water |

In eukaryotes, there are three essential chemical processes in photosynthesis.
1. *Absorption of energy from sunlight by pigments—light-dependent reactions.* The harvesting of light energy for the light reactions of photosynthesis proceeds through two photosystems (photosystem I and photosystem II) operating concurrently. These photosystems consist of packages of pigment and protein molecules embedded in the thylakoid membranes of chloroplasts. Light energy is absorbed by the pigment molecules and

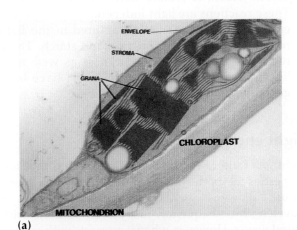

(a)

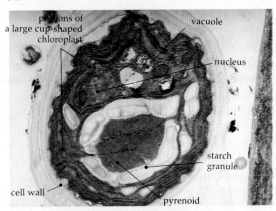

(b)

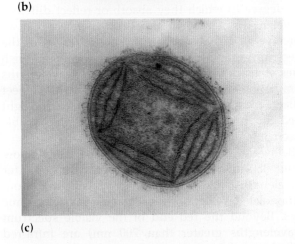

(c)

Fig. 5.8 Electron micrographs of cross-sections through (a) a chloroplast from a plant leaf, *Avicennia marina*, the white mangrove, (b) a unicellular green alga, *Chlamydomonas reinhardii* and (c) a prokaryotic cell, *Prochloron*, a cyanobacterium

transferred to a special part of the photosystems, the reaction centre. Here electrons are energised and removed to be used in a chain of reactions in which the energy is eventually stored in the molecules ATP and NADPH.

2. *Reactivation of pigments.* Pigments are replenished by replacing electrons removed during the light reactions. Water acts as a source of these electrons and oxygen is produced as a by-product.

3. *Carbon fixation to produce sucrose in 'dark reactions'.* These reactions are light independent and use the energy (ATP and NADPH) captured in the light reactions to synthesise sucrose and starch. These reactions can function in darkness as well as in light provided that ATP and NADPH have been generated in the light reactions.

Light energy

Light is a form of electromagnetic radiation that has properties of both waves and particles. Its wave nature allows it to be described in terms of wavelength or frequency, just as can be done with sound waves. However, when it interacts with matter, light can be considered as discrete packages, **photons** (from the Greek, *phos* [gen. *photos*], light). Photons have different amounts of energy, the amount being inversely proportional to the wavelength of a photon. Blue photons, with a wavelength of 400 nm, have more energy than red photons, with a wavelength about 700 nm.

Sunlight is a mixture of photons with a range of energy levels, some of which our eyes perceive and some of which are out of the visible range (Fig. 5.9). Visible light ranges from violet through the colours of the rainbow to red. Our ability to perceive colours relies on the fact that different substances vary in the degree to which they absorb or reflect different wavelengths of incident radiation. An object that we perceive as red absorbs in the green region of the spectrum and either reflects or transmits reddish hues. A black object absorbs all visible wavelengths of radiation and reflects none, whereas a white object reflects most wavelengths and absorbs few. The light used in photosynthesis falls within the visible spectrum (Fig. 5.9).

Below the lower limit of the visible spectrum (below about 380 nm) photons have progressively shorter wavelengths and greater energy, ranging from ultraviolet (UV) light, X-rays, gamma rays to cosmic rays. Beyond the red end of the visible spectrum (wavelengths greater than 700 nm) are infra-red radiation and long-wave radiowaves, which contain less energy than visible light. Infra-red radiation is emitted by warm bodies and is used by organisms such as some snakes to locate the small mammals upon which they prey.

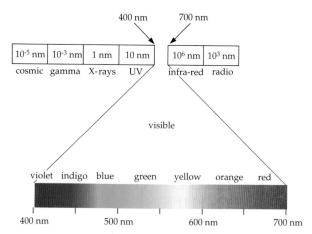

Fig. 5.9 The solar radiation spectrum is composed of extremely short (but high energy) wavelength cosmic, gamma, X-rays, ultraviolet (UV), visible, infra-red radiation and radio waves. Photosynthetically active radiation is the visible spectrum and accounts for only a very narrow waveband, ranging in colour from violet to red (4×10^{-5} cm = 400 nm)

Pigments: molecules that absorb light

Pigments are coloured molecules. We see them as coloured because they absorb photons with particular energy levels and reflect others. Chlorophyll, for example, is a rich green colour because it preferentially absorbs photons of blue and red light, and reflects photons in the green portion of the spectrum.

A pigment is best characterised by its pattern of absorption of photons, **absorption spectrum**. The absorption spectrum of chlorophyll is similar to the wavelengths of light that activate photosynthesis (**action spectrum**) (Fig. 5.10), indicating that chlorophyll is the main pigment involved in photosynthesis. High values on the action spectrum indicate high rates of photosynthesis where photosynthetically active radiation is absorbed.

Chlorophyll absorbs light energy by a process of excitation, involving a central metal atom (magnesium) surrounded by alternating single and double bonds that form a **porphyrin ring** (Fig. 5.11). When photons are absorbed by the pigment molecule, electrons in the magnesium atom are excited and the energy is funnelled off through these bonds. Chlorophyll *a* is the principal photosynthetic pigment in prokaryotes and eukaryotes. Green algae and plants also contain chlorophyll *b*, which has a slightly different absorption spectrum.

Other photosynthetic pigments are *carotenoids*, such as the orange pigment β-carotene (Fig. 5.11), which can absorb light over a wide range of the visible spectrum. Each carotenoid molecule has a carbon ring and a long hydrocarbon chain in which single and double bonds alternate.

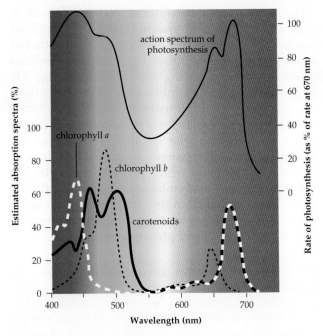

Fig. 5.10 Absorption spectra for chlorophyll *a* and *b* and carotenoids showing the particular wavelengths each pigment absorbs most. The sum of the absorption spectra corresponds to the action spectrum of photosynthesis

Photons of visible light are absorbed by chlorophyll and carotenoid pigments, providing energy for photosynthesis.

LIGHT-DEPENDENT REACTIONS OF PHOTOSYNTHESIS

Chloroplasts trap photons

In eukaryotes, chlorophyll and carotenoid pigments that trap sunlight are located in chloroplasts (Fig. 5.12). You will remember from Chapter 3 that a chloroplast is bounded by a double membrane envelope. The membranes function as both border and barrier, defining the volume and shape of the organelle. They enclose the matrix (stroma) and control the movement of materials between the chloroplast and cytosol (see Chapter 4). Chloroplasts also have a third, innermost membrane system, the **thylakoid membrane system**, which is the site of conversion of light energy into electrical and then into chemical energy.

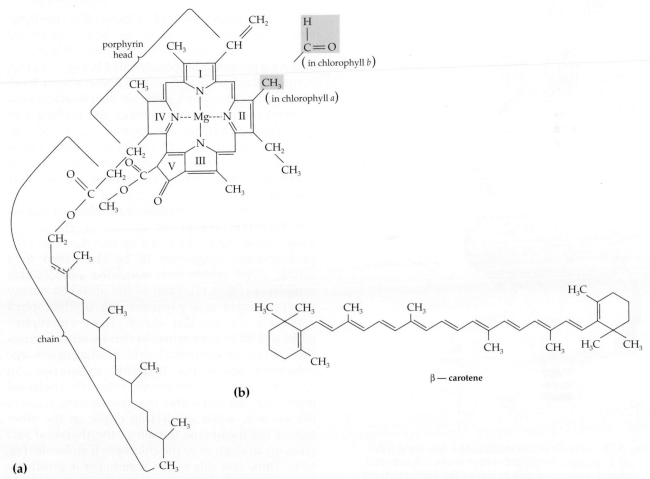

Fig. 5.11 Molecular structure of chlorophyll *a* and β-carotene. Chlorophyll consists of a porphyrin ring to which a phytol chain is attached. Chlorophyll *b* is different from chlorophyll *a* only by a substitution of the methyl group on ring II with an aldehyde group

Thylakoid membranes form an elaborately folded, continuous membrane system (Fig. 5.13). Thylakoid membranes occur in two configurations: as stromal and as granal lamellae. Stromal lamellae are individual cisternae (sacs) traversing the chloroplast (Figs 5.8, 5.12) whereas granal membranes form stacks of from two to more than 50. Integrated into the thylakoid membrane are several types of protein complexes that together make up the photosynthetic electron transport system. The protein complexes include *light-harvesting complexes*, in which pigment molecules are bound to proteins, *electron-transport complexes* and *ATP-synthesising complexes*. Associated with the light-harvesting complexes are two complexes known as photosystem I (PSI) and photosystem II (PSII).

> The chloroplast envelope forms a selective barrier that regulates the transfer of molecules into and out of the chloroplast. Thylakoid membranes contain light-harvesting, electron-transport and ATP-synthesising complexes.

Photosystems I and II

The protein-bound chlorophyll molecules of the light-harvesting complexes serve as light-intercepting 'antennae' (Fig. 5.13). When a chlorophyll molecule absorbs a photon, it becomes excited (Chapter 1), that is, one of its electrons moves to a higher energy level. The excited chlorophyll molecule then interacts with its neighbour, transferring energy and exciting it in turn. Excited electrons are not transferred, only the energy is; the transferred electrons return to their original energy level in the same molecules. This type of energy transfer ('excitation transfer') is extremely efficient.

The orientation and positioning of pigments allows the excitation energy of dozens of chlorophyll molecules in the light-harvesting complexes and the photosystems themselves to be channelled to a certain point within the associated photosystem complexes (Fig. 5.14). Most of the absorbed energy is concentrated at a particular pair of chlorophyll molecules, the **reaction centre**. These chlorophyll molecules differ from others in that excitation causes them to *expel* an electron, thereby producing a charge separation across the thylakoid membrane. An electron acceptor on the stromal side of the thylakoid membrane, but still within the photosystem, receives the electron, while an electron donor on the other side of the membrane (lumen of the thylakoid sac) gives up an electron to the chlorophyll molecule (Fig. 5.15). Thus, one side of the membrane is positively charged and the other is negatively charged. Only one out of every 300 chlorophyll molecules can function in this way.

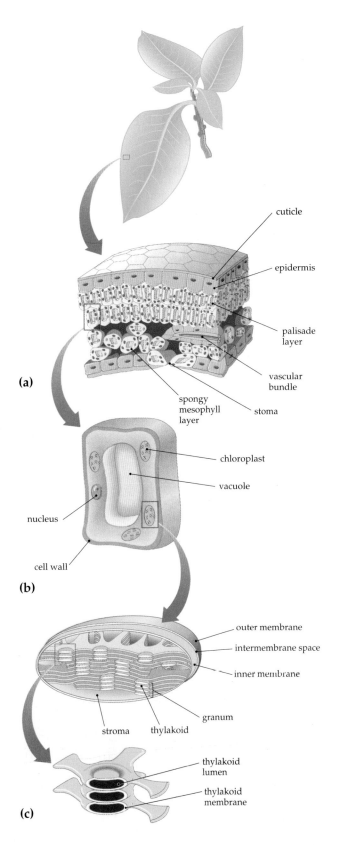

(a)

cuticle
epidermis
palisade layer
vascular bundle
spongy mesophyll layer
stoma

(b)

chloroplast
vacuole
nucleus
cell wall

(c)

outer membrane
intermembrane space
inner membrane
granum
stroma
thylakoid
thylakoid lumen
thylakoid membrane

Fig. 5.12 Location of chloroplasts in a flowering plant. (**a**) In plants, chloroplasts occur in the palisade and spongy mesophyll cells of leaves. (**b**) Mesophyll cell. (**c**) Chloroplast showing the double outer membrane (envelope) and the grana and stroma lamellae embedded in the stroma

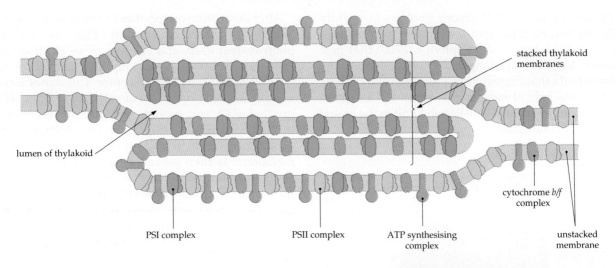

lumen of thylakoid

stacked thylakoid membranes

cytochrome *b/f* complex

PSI complex PSII complex ATP synthesising complex unstacked membrane

Fig. 5.13 A model showing the distribution of protein complexes (light-absorbing photosystem complexes PSI and II, cytochrome *b/f* complex, and ATP-synthesising complex) on the thylakoid membrane system

BOX 5.2 CHLOROPLASTS IN SUN AND SHADE

Australian researchers played a key role in the discovery of the structural and functional complexities of the thylakoid membrane. Dr Keith Boardman, former Chief Executive Officer of the CSIRO, and Dr Jan Anderson, from the CSIRO Division of Plant Industry in Canberra, were among the first in the 1960s and 1970s to describe the structural and functional differences between chloroplasts from leaves growing in the sun and in the shade. The figure shows the different morphological features of chloroplasts in the leaves of *Alocasia macrorrhiza*

(birdsnest fern) grown in a deeply shaded rainforest understorey or in an open, very sunny spot by a roadside. Note the very large granal stacks in the shade chloroplast that seem to pack the entire chloroplast.

Boardman and Anderson also provided the first evidence that all chlorophyll and other pigments involved in photosynthetic light capture and energy conversion are bound to proteins that are arranged in discrete supramolecular complexes, forming functional units in the thylakoid membrane.

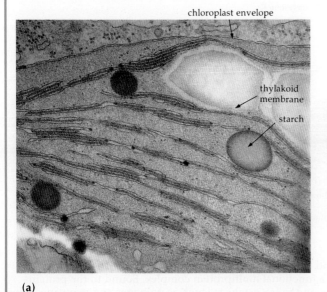

chloroplast envelope

thylakoid membrane

starch

(a)

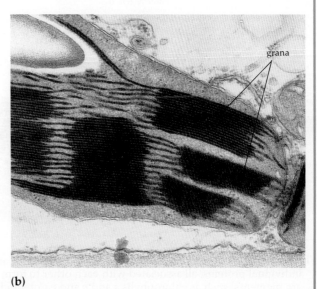

grana

(b)

Chloroplasts in (a) sun and (b) shade leaves of the birdsnest fern, *Alocasia macrorrhiza*. In deep shade under the canopy of rainforest trees, chloroplasts have very large granal stacks

Photosystems I and II absorb photons at slightly different wavelengths and have different functions. PSI includes a type of chlorophyll *a* (P700) that best absorbs light of wavelength 700 nm, whereas the chlorophyll *a* that characterises PSII, chlorophyll P680, best absorbs light of wavelength 680 nm. PSI is located primarily in the stromal lamellae and PSII in the granal lamellae of chloroplasts.

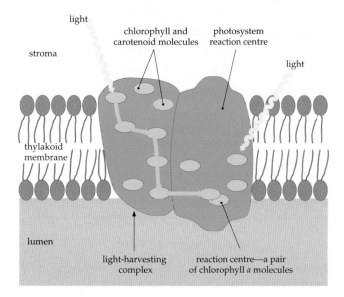

Fig. 5.14 A model of the light-harvesting complexes associated with both photosystems I and II. Pigments in the light-harvesting complex absorb photons, the energy of which is transferred to the reaction centre of a photosystem. Photons can also be absorbed by chlorophyll molecules within the photosystems themselves

After absorption of a photon by the light-harvesting complexes associated with PSII or by the photosystem itself, and the loss of an electron from P680, an electron donor rapidly neutralises the positively charged reaction centre. This donor is then neutralised by removal of electrons from H_2O, producing O_2 and protons. The protons accumulate in the thylakoid sacs (Fig. 5.15). The electron on the acceptor molecule of PSII on the stromal side of the membrane is then passed by an electron carrier to another protein complex, cytochrome *b/f* complex, to the electron donor molecule of PSI. During the process, at least one more proton is pumped into the lumen of the thylakoid sac.

The light-harvesting complex associated with PSI or the photosystem itself absorbs an additional photon, the energy of which allows an electron of chlorophyll P700 to move to an electron acceptor molecule on the stromal surface of the membrane. Here electrons are given to an acceptor, ferredoxin, which passes them on to $NADP^+$. Two electrons and one proton (H^+) reduce $NADP^+$ to NADPH (Fig. 5.15). Thus, electrons ultimately derived from H_2O are passed from PSII to PSI to produce NADPH for the light-independent reactions of photosynthesis. The two photosystems each energise an electron part along the way, and both PSII and PSI are needed for the electron transport.

Photosystems I and II use solar energy to move electrons across thylakoid membranes. Electrons, ultimately derived from H_2O, are passed from PSII to PSI to produce NADPH for the light-independent reactions. Oxygen atoms from H_2O form O_2 as a by-product.

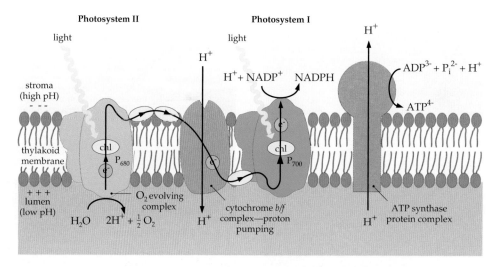

Fig. 5.15 Thylakoid membrane complexes. Photosystem II and the light-harvesting complex consist of at least a dozen individual proteins, all associated with each other to form a functional multiprotein complex. Bound to the proteins are pigments, such as chlorophylls *a* and *b* and carotenoids, and redox catalysts (such as quinones and manganese) for electron transfer. Cytochrome *b* is a protein with unknown function but essential for the reaction centre to operate. The cytochrome *b/f* complex consists of five proteins, which carry two different cytochromes and an iron–sulfur centre. Photosystem I consists of more than 10 proteins that bind pigments and electron transfer catalysts such as iron–sulfur centres and possibly quinones. The ATP synthase (exclusively in stromal lamellae) consists of two parts: a proton channel and a catalytic site where ATP is synthesised

The protons that accumulate in the lumen of the thylakoid sacs create an electrochemical proton (pH) gradient across the membrane. This provides potential energy that is utilised in the synthesis of ATP. Protons moving down the electrochemical gradient cross the membrane through ATP-synthesising protein complexes to the stroma. This is similar to the proton-driven ATP generation in mitochondria. (The process of using an electrochemical gradient to drive the synthesis of ATP is known as chemiosmosis.)

The electron flow from water through PSII and PSI to NADP is one-way (Fig. 5.15) and is referred to as non-cyclic electron transport. The ATP synthesis coupled to this is **non-cyclic photophosphorylation**. Electrons excited by PSI, however, can take an alternative *cyclic* route. They can be transported by ferredoxin and the cytochrome *b/f* complex back to PSI. During this process protons are pumped across the membrane, providing energy for ATP synthesis, but the recycled electron is not used for NADPH production. This process, **cyclic photophosphorylation** (Fig. 5.16), can vary the amount of ATP relative to NADPH produced by the electron transport chain.

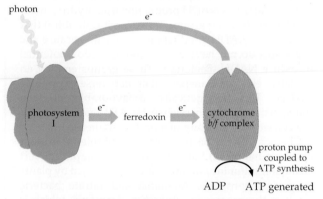

Fig. 5.16 ATP can be produced by cyclic photophosphorylation involving photosystem I. Excitation energy passes to ferredoxin and the cytochrome *b/f* complex, with the electron being returned to PSI

photon

e⁻

e⁻ photosystem I → e⁻ → ferredoxin → e⁻ → cytochrome *b/f* complex

proton pump coupled to ATP synthesis

ADP   ATP generated

Protons, which accumulate in the lumen of the thylakoid sacs, move across the membrane to the stroma, providing energy for the synthesis of ATP.

Prokaryotes and the evolution of photosystems I and II

In contrast to eukaryotes, photosynthetic prokaryotes lack chloroplasts. In primitive photosynthetic bacteria, such as green and purple bacteria, chlorophyll and the protein complexes associated with the light-dependent reactions of photosynthesis are located on the plasma membrane. In cyanobacteria, such as *Prochloron*, which is believed to be related to eukaryotic green algae, chlorophyll is located on thylakoid membranes within the cytoplasm (Fig. 5.17).

Green and purple bacteria lack PSII, do not use H_2O as a source of electrons and therefore do not produce oxygen. Electron flow is usually cyclic, with photosynthesis resulting in the production of energy in the form of ATP rather than being geared to reducing CO_2. When electron transport is non-cyclic in these bacteria, molecules such as hydrogen sulfide (H_2S) or hydrogen gas (H_2) are used as a source of electrons. ATP and NADPH are then produced to drive the reactions for the synthesis of sugar from CO_2. For example, light-dependent oxidation of H_2S produces sulfur (S):

$$12H_2S + 6CO_2 \xrightarrow{\text{light}} C_6H_{12}O_6 + 12S + 6H_2O$$

With the evolution of cyanobacteria came PSII, which provided the mechanism for using H_2O as a source of electrons, and resulted in an atmosphere containing oxygen. The removal of an electron from the P680 chlorophyll *a* molecule of PSII provides enough energy to split H_2O. The process is facilitated by the presence of a cluster of manganese ions bound to the reaction centre proteins. Removal of an electron from the P700 chlorophyll *a* molecule in PSI only yields enough energy to split electrons from H_2S, not H_2O.

Green and purple bacteria lack PSII, do not use H_2O as a source of electrons, and therefore do not produce oxygen. With the evolution of cyanobacteria came PSII, which provided the mechanism for using H_2O as a source of electrons, and provided the earth with an atmosphere of oxygen.

LIGHT-INDEPENDENT REACTIONS OF PHOTOSYNTHESIS

Carbon fixation and the Calvin–Benson cycle

The capture of atmospheric CO_2 and its incorporation into carbohydrates is **carbon fixation**. In eukaryotes it occurs in the stroma of chloroplasts. The stroma contains, among other things, multiple copies of the chloroplast genome, ribosomes and the enzymes needed for converting carbon dioxide to sugar, as well as those engaged in DNA duplication and the synthesis of RNA and protein. Not all of the proteins used in photosynthesis are, however, synthesised in the stroma (or encoded for in the chloroplast genome). Some of the carbon fixed during photosynthesis can be stored temporarily as starch in granules found in the stroma of chloroplasts.

BOX 5.3   CHEMOSYNTHETIC BACTERIA

Autotrophic bacteria have two basic sources of energy available to them for synthesising organic molecules. They can use sunlight in *photosynthesis* or they can use simple inorganic molecules in *chemosynthesis*. All autotrophic bacteria do, however, use carbon dioxide as their source of carbon and, as far as we know, all chemosynthetic bacteria use the Calvin–Benson cycle to fix carbon dioxide.

There are five groups of chemosynthetic bacteria, which use different inorganic substrates for chemosynthesis (see table). The enzyme-catalysed oxidation of these substrates by bacteria releases electrons that can be used by an electron transport chain to produce ATP. This electron transport is similar to mitochondrial oxidative electron transport in plants.

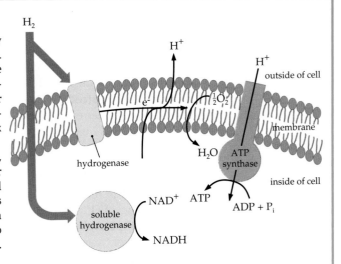

NADH production in a chemosynthetic bacterium, the hydrogen bacterium

Chemosynthetic bacterial group	Half reaction for substrate oxidation
Iron bacteria	$Fe^{2+} \longrightarrow Fe^{3+} + e^-$
Ammonia bacteria	$NH_4^+ + 2H_2O \longrightarrow NO_2^- + 8H^+ + 6e^-$
Nitrate bacteria	$NO_2^- + H_2O \longrightarrow NO_3^- + 2H^+ + 2e^-$
Sulfur bacteria	$H_2S \longrightarrow S + 2H^+ + 2e^-$
	$SO_3^- + H_2O \longrightarrow SO_4^- + 2H^+ + 2e^-$
	$S + 3H_2O \longrightarrow SO_3^- + 6H^+ + 5e^-$
Hydrogen bacteria	$H_2 \longrightarrow 2H^+ + 2e^{3-}$

The production of a pH gradient is coupled to the transport of electrons, and ATP is synthesised by a membrane-spanning enzyme that dissipates the proton gradient. Moreover, like plants, almost all chemosynthetic bacteria use oxygen as their final electron acceptor.

NADH production in chemosynthetic bacteria occurs by two basic mechanisms. In the hydrogen bacteria, where the substrate has sufficient redox potential to reduce NAD^+ to NADH, it is thought that a hydrogenase enzyme catalyses NADH production from hydrogen gas (see figure). In other groups, however, it is doubtful that such direct NADH production can occur because the substrates do not have a high enough redox potential to reduce NADH. Bacteria in these groups employ an oxidative electron transport chain in reverse to produce NADH. This energetically unfavourable electron transport is driven in reverse by ATP hydrolysis and the consequent build-up of a proton gradient.

Chemosynthetic bacteria play a vital role in nutrient cycling (Chapter 42). Sulfur bacteria, for example, convert elemental sulfur into sulfate, which can be used by plants in protein synthesis. Ammonia and nitrate bacteria catalyse the process of *nitrification*. Ammonia, which is in equilibrium with ammonium ions, is converted into nitrate, the main form of nitrogen used by plants.

The first step is carbon fixation: the attachment of CO_2 to the 5-carbon sugar **ribulose biphosphate (RuBP)**. This carboxylation of RuBP is one of a series of reactions known as the Calvin–Benson cycle (Fig. 5.17), after its discoverers Melvin Calvin and Andrew Benson. A short-lived, 6-carbon intermediate is formed in a reaction catalysed by the enzyme **ribulose bisphosphate carboxylase-oxygenase (Rubisco)** (Fig. 5.18.) Owing to it constituting 50% of the protein in chloroplasts, Rubisco is the most abundant protein on earth. The 6-carbon intermediate splits rapidly into two molecules of phosphoglyceric acid (PGA), a 3-carbon molecule (Fig. 5.17). From PGA, two enzymatic steps using NADPH and ATP take place. In the first step, PGA is phosphorylated using ATP, and in the

second step the intermediate compound is reduced and dephosphorylated to form glyceraldehyde 3-phosphate (PGAL) in a reaction requiring NADPH.

PGAL, the primary carbohydrate product of the chloroplast, can follow two paths. Two in every 12 PGAL molecules are exported from the chloroplast into the cytoplasm (Fig. 5.17). These PGAL molecules are combined and rearranged into fructose and glucose phosphates. These two intermediate compounds condense to form sucrose, the major transport form of carbohydrate in plants. Inorganic phosphate is imported into the chloroplast to replace that exported with PGAL. Through another series of reactions in the stroma, the remaining 10 PGAL molecules are used to form six RuBP molecules to

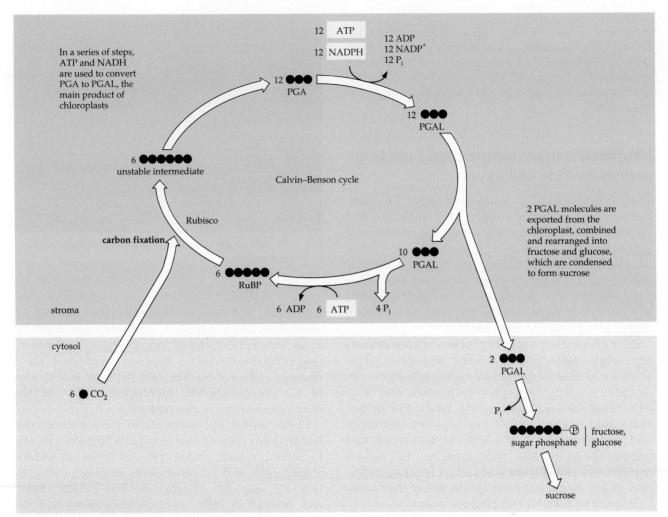

In a series of steps, ATP and NADH are used to convert PGA to PGAL, the main product of chloroplasts

12 ATP
12 NADPH

12 ADP
12 NADP$^+$
12 P$_i$

12 ●●● PGA

12 ●●● PGAL

6 ●●●●●● unstable intermediate

Calvin–Benson cycle

Rubisco

carbon fixation

6 ●●●●● RuBP

stroma

cytosol

6 ● CO$_2$

6 ADP   6 ATP   4 P$_i$

10 ●●● PGAL

2 PGAL molecules are exported from the chloroplast, combined and rearranged into fructose and glucose, which are condensed to form sucrose

2 ●●● PGAL

P$_i$

●●●●●●–Ⓟ | fructose, glucose
sugar phosphate |

sucrose

Fig. 5.17 The Calvin–Benson cycle in the stroma of chloroplasts. At each full turn of the cycle, one molecule of CO$_2$ enters. Six turns are shown here. Six molecules of CO$_2$ combine with six molecules of the 5-carbon compound ribulose 1,5-biphosphate (RuBP). The product, an unstable 6-carbon compound, splits to give the 3-carbon compound, phosphoglyceric acid (PGA). In a series of steps using the energy of ATP and NADPH, PGA is converted to glyceraldehyde 3-phosphate (PGAL). Two PGAL molecules are exported from the chloroplast and used to form sucrose, while 10 remain in the chloroplast to form six molecules of RuBP to continue the cycle

complete the cycle. The regeneration of six RuBP molecules uses a further six ATP molecules.

The Calvin–Benson cycle is an economic way to provide for a continuing supply of the acceptor molecule RuBP. Since the cycle is dependent upon an adequate supply of NADPH and ATP, its operation is closely linked with the electron transport chain. It is not surprising to find that the activity of certain key enzymes, such as fructose bisphosphatase, is light-dependent. These enzymes function only when the electron transport chain is operating. The activity of the key enzyme Rubisco is itself regulated by a specific activating protein, activase.

CO$_2$

CH$_2$OPO(OH)$_2$
|
C=O
|
HCOH
|
HCOH
|
CH$_2$OPO(OH)$_2$

ribulose 1,5 — bisphosphate (RuBP)

Rubisco

CH$_2$OPO(OH)$_2$
|
O   C—COH
‖   /
C
HO
|
C=O
|
HCOH
|
CH$_2$OPO(OH)$_2$

intermediate

+ H$_2$O

CH$_2$OPO(OH)$_2$
|
HCOH
|
HO—C=O

CH$_2$OPO(OH)$_2$
|
HCOH
|
HO—C=O

two ×3 phosphoglycerate (PGA)

Fig. 5.18 Carbon fixation step showing the catalytic function of Rubisco

The stroma is a matrix containing enzymes for carbon fixation as well as DNA, RNA, ribosomes and the machinery for protein synthesis. CO_2 is fixed to RuBP in the first step of the Calvin–Benson cycle in a reaction catalysed by the enzyme Rubisco. Intermediate 3-carbon compounds are rearranged into sugar phosphates and new RuBP molecules.

Photorespiration: competition between carbon dioxide and oxygen

The enzyme that operates in carbon fixation, Rubisco, can also use oxygen as a substrate and *oxygenate* RuBP. This process and the subsequent production of CO_2 from a product of oxygenation is **photorespiration**. Under normal atmospheric conditions, O_2 is much more abundant than CO_2. Consequently, O_2 competes effectively for the binding site on the enzyme despite the enzyme having a higher affinity for CO_2 than for O_2.

Photorespiration was not a problem billions of years ago when early photosynthetic organisms were evolving in an atmosphere lacking oxygen. However, in today's atmosphere, photorespiration results in CO_2 being released, undoing the binding of carbon that would otherwise result in carbohydrate synthesis and production of ATP and NADPH. This means that photorespiration has the capacity to reduce significantly the efficiency of carbon fixation. In this way, many plants lose from 25% to 50% of the carbon that could be fixed during photosynthesis.

Photorespiration occurs because the carbon-fixing enzyme Rubisco interacts with O_2 as well as CO_2. The process leads to less carbon being fixed during photosynthesis.

The C_4 pathway of photosynthesis

Plants have adapted to different environments and there are several variations in the photosynthetic pathway and the structure of chloroplasts. The process of carbon fixation that we have described so far is C_3 **photosynthesis** because the first stable product of carbon fixation is the 3-carbon compound, PGA (Fig. 5.17). However, many flowering plants, including tropical and subtropical grasses such as maize, sugarcane and millet, have a different carbon fixation pathway, C_4 **photosynthesis** (Fig. 5.19). While each C_4 plant group has its own characteristic enzymes, chloroplasts and leaf anatomy, in each case the first stable product of carbon fixation is a 4-carbon compound.

The C_4 photosynthetic process was first discovered independently in the 1960s by American and Russian workers who showed that a C_3 compound was not always the first stable product of carbon fixation in

Fig. 5.19 Sugarcane is an example of a plant that has C_4 photosynthesis

sugarcane and maize. Marshall (Hal) Hatch, an Australian, and Roger Slack, an Englishman, working at the time in the Colonial Sugar Refining Company's research laboratories in Brisbane, confirmed these findings and went on to resolve the basic mechanism of C_4 photosynthesis, identifying many of the enzymes involved in the pathway.

Plants with C_4 photosynthesis have a distinctive leaf anatomy ('Kranz anatomy') in which the vascular bundles are surrounded by a cylinder of bundle sheath cells and an outer layer of mesophyll cells (Fig. 5.20, see also Chapter 18). The bundle sheath and mesophyll cells contain chloroplasts that are different in structure and function.

In C_4 plants, an additional carboxylation enzyme, phosphoenolpyruvate carboxylase (PEP carboxylase), operates in the cytoplasm of the leaf mesophyll cells. The enzyme catalyses the carboxylation of a C_3 compound, phosphoenolpyruvate (PEP). The product of the carboxylation reaction is a C_4 organic acid, oxaloacetate (Fig. 5.21), which is immediately converted into another C_4 compound. In sugarcane, this is malate, which is transported into bundle sheath cells where the Calvin–Benson cycle takes place. The mesophyll chloroplasts provide ATP and NADPH to make PEP and malate but they lack RuBISCO; they have only the enzymes of the Calvin–Benson cycle that are needed to reduce PGA to PGAL.

Once in the chloroplasts of the bundle sheath cells (Fig. 5.22), malate (C_4) is decarboxylated to CO_2 and pyruvate (C_3). Carbon dioxide is then fixed into carbohydrates by Rubisco and the other Calvin–Benson cycle enzymes. The pyruvate is actively transported back into the mesophyll cells and there converted back into PEP to complete the cycle. The C_4 decarboxylation reaction generates NADPH, which is consumed in bundle sheath chloroplasts during PGA reduction. Bundle sheath chloroplasts often have only a few granal stacks and little PSII

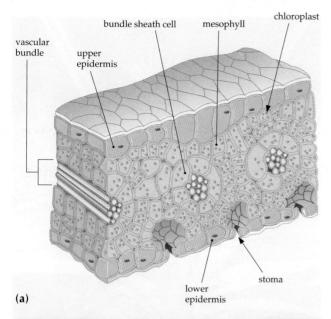

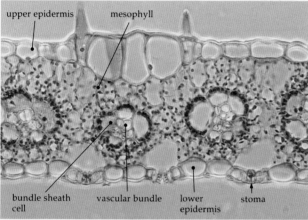

Fig. 5.20 **(a)** Distinctive leaf ('Kranz') anatomy of C_4 plants. The bundle sheath cells form tight wreaths around the vascular bundles. They have very thick cell walls, traversed by many plasmodesmata for metabolite exchange, which do not permit diffusion of CO_2, thus allowing high CO_2 concentrations to be achieved. The Calvin–Benson cycle occurs in the chloroplasts of these cells. Phosphoenolpyruvate carboxylation takes place in the mesophyll cells surrounding the bundle sheath cells. **(b)** Micrograph of maize leaf cell showing bundle sheath cells

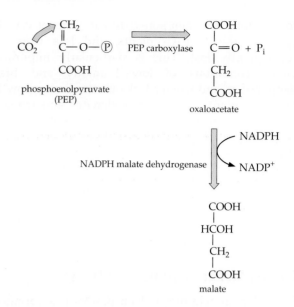

Fig. 5.21 The C_4 pathway of carbon fixation showing the carboxylation reactions

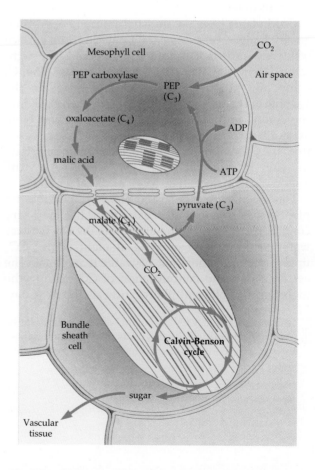

Fig. 5.22 Different cell types and chloroplasts involved in the metabolite transfers and gas exchange reactions

activity, and hence little capacity for light-dependent NADPH formation.

C_4 photosynthesis is a mechanism for concentrating CO_2 in bundle sheath cells, which are relatively impermeable to CO_2 and tend to hold it within them. The relatively high CO_2 concentration within these cells has the added advantage of inhibiting photorespiration. Although the C_4 pathway uses more ATP than the C_3 pathway, the reduction of photorespiration offsets this cost, especially when light is abundant. In addition, because C_4 plants can concentrate CO_2, their pores (stomata) are not as wide open as those of C_3 plants, and less water is consequently lost. This is particularly important under conditions of low humidity and high temperature, and when a lack of water limits growth. Many C_4 plants grow in hot climates and tropical crop plants such as sugarcane have a net rate of photosynthesis two to three times greater than temperate C_3 crop plants such as wheat.

> In C_4 plants, carbon fixation occurs in mesophyll cells and CO_2 is then concentrated in bundle sheath cells, inhibiting photorespiration and increasing carboxylation efficiency.

Crassulacean acid metabolism

Crassulacean acid metabolism (CAM) is a variation on the C_4 pathway of photosynthesis and was first discovered in succulent plants of the flowering plant family Crassulaceae. It has evolved independently in other plants such as pineapple and 'Spanish moss', both members of the family Bromeliaceae (Fig. 5.23). Photosynthesis in CAM plants involves both the C_4 pathway and the Calvin–Benson cycle, but the reactions, which happen in the same cell, occur at different times.

CAM plants have the unusual ability to open their gas-exchange pores (stomata, located on the surface of leaves) and fix CO_2 into C_4 compounds at night rather than the day as in other plants. The C_4 compound produced in darkness is stored in the vacuole of the mesophyll cell and exported back into the cytoplasm of the *same* cell to be decarboxylated as soon as daylight appears. Carbon dioxide released from the C_4 compound is then fixed in the chloroplasts in the normal manner via RuBP and the Calvin–Benson cycle (Fig. 5.24).

This peculiar biochemistry makes it possible for plants to survive under extremely hot and dry conditions. Closure of stomata during the day reduces water loss and their opening at night allows CO_2 uptake when evaporation of water is minimal. The accumulation and concentration of CO_2 during the night ensures efficient photosynthesis the next day.

> CAM plants fix CO_2 at night and convert it to carbohydrate during the day. This allows them to close their gas-exchange pores during the day, minimising water loss.

(a)

(b)

Fig. 5.23 CAM plants occur in many families of flowering plants and include the (**a**) pineapple (*Ananas comosus*, family Bromeliaceae) and (**b**) the tropical epiphytic fern, *Pyrrosia longifolia*. Measurements of net CO_2 exchange show carbon gain occurs via CO_2 dark fixation

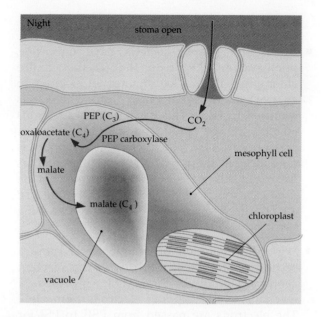
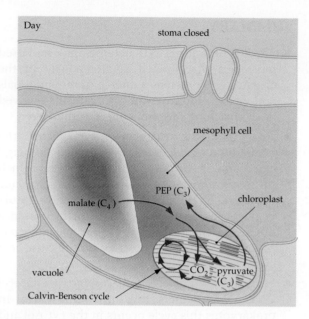

Fig. 5.24 CAM pathway in the cell. Gas exchange takes place at night

SUMMARY

- The first organisms were heterotrophic and used molecules that formed spontaneously in the absence of oxygen as a source of raw materials and energy. Photosynthetic organisms evolved later, enriching the atmosphere with oxygen.

- All organisms extract energy from the breakdown of glucose. The first stage of energy extraction in the cytosol is glycolysis, which produces two molecules of ATP per molecule of glucose. In the absence of O_2, the end product of glycolysis is ethanol or lactate. In the presence of O_2, the end product is a 3-carbon molecule, pyruvate, which is a substrate for the citric acid cycle.

- Pyruvate from glycolysis is converted to acetyl CoA, which enters the citric acid cycle. In prokaryotes this cycle occurs in the cytosol and in eukaryotes it occurs in mitochondria. Oxidation of acetyl CoA yields ATP and energised electrons, which are used to generate reduced electron carriers, NADH and $FADH_2$.

- In an electron transport chain on the plasma membrane (prokaryotes) or inner mitochondrial membrane (eukaryotes), electrons donated by NADH and $FADH_2$ from the citric acid cycle drive proton pumps coupled to ATP synthesis. The net amount of ATP produced from the complete oxidation of glucose to CO_2 is commonly 36 to 38 ATP molecules.

- The bulk of ATP in a eukaryotic cell is produced in mitochondria. The large surface area of the folded inner membrane of these organelles provides a large surface area for ATP synthesis.

- All energy supporting life on earth originates from the sun. Only green land plants, algae and photosynthetic bacteria can make use of this energy directly by the process of photosynthesis.

- All photosynthetic organisms have photosystems in which pigments that absorb specific wavelengths of light are bound to proteins arranged on membranes. In photosynthetic eukaryotes, the protein-pigment complexes are located on the thylakoid membrane system of chloroplasts. A photon captured by a pigment energises an electron in the molecule and the excitation energy is transferred to a pair of specific chlorophyll molecules, the reaction centre. From there, the energy is used to move electrons across the thylakoid membrane for the reduction of NADP to NADPH. Protons also move across the membrane, providing energy for the synthesis of ATP.

- Green and purple bacteria have only one photosystem (PSI) with the reaction centre chlorophyll P700. P700 donates an electron to an electron transport chain from where it is returned to P700. This cyclic flow of electrons results in the production of ATP (cyclic phosphorylation). When electron transport is non-cyclic in these bacteria, molecules such as hydrogen sulfide (H_2S) or hydrogen gas (H_2) are used as a source of electrons. NADPH and ATP are produced to drive the reactions synthesising sugar from CO_2.

- With the evolution of cyanobacteria came a second photosystem (PSII), with the reaction centre chlorophyll P680. PSII provided a mechanism for stripping electrons from H_2O. The electrons are passed from PSII to PSI for the production of NADPH and ATP by non-cyclic photophosphorylation. The residual oxygen atoms from water form O_2 as a by-product.

- ATP and NADPH generated in the light-dependent reactions drive the Calvin–Benson cycle (located in the stroma of eukaryote chloroplasts). Carbon dioxide is fixed to RuBP in the first step of the Calvin–Benson cycle in a reaction catalysed by the enzyme Rubisco. Intermediate 3-carbon compounds are rearranged into sugar phosphates. It takes six turns of the Calvin–Benson cycle to produce one 6-carbon sugar phosphate because only one of every six glyceraldehyde 3-phosphate (PGAL) molecules is funnelled into carbohydrate synthesis. The remainder are used to form new RuBP molecules.

- Because the carbon-fixing enzyme Rubisco can use O_2 as well as CO_2 as a substrate, photorespiration is an inevitable consequence in an oxygen-rich atmosphere. Photorespiration leads to a loss of carbon fixed during photosynthesis.

- Several variants of the carbon-fixation process have evolved in plants. C_4 photosynthesis and crassulacean acid metabolism (CAM) are characterised by an additional carboxylation reaction and carboxylation enzyme, phosphoenolpyruvate (PEP) carboxylase. Leaf anatomy and physiological properties allow spatial (C_4) or temporal (CAM) separation of the two carboxylation (C_3 and C_4) reactions. The advantage of C_4 photosynthesis and CAM is a CO_2-concentrating effect, leading to reduced photorespiration, higher carboxylation and better water use efficiency of photosynthesis. These features are important in arid environments, where combinations of high irradiation, high temperatures and limited water occur.

QUESTIONS

1. What is glycolysis and where does it occur in a cell? How much ATP is produced from glycolysis compared with oxidative respiration?

2. Explain how NADH and $FADH_2$ are used by the inner membrane of mitochondria to drive ATP synthesis. Of what significance are the cristae of mitochondria?

3. Summarise the different biochemical pathways by which energy is extracted from carbohydrates and lipids.

4. What is the significance of fermentation reactions in cells? How does fermentation differ in a yeast cell and a muscle cell?

5. Make a table showing the major differences between energy-harvesting pathways (glycolysis, oxidative respiration and photosynthesis) in prokaryotic and eukaryotic cells. Indicate where the different reactions occur.

6. Explain the relationship between the action spectrum of photosynthesis and the absorption spectra of chlorophylls and carotenoids.

7. How do pigment molecules in the light-harvesting complexes trap and transfer solar energy? What is the role of the reaction centres in the photosystems?

8. What organisms use water in the light-dependent reactions of photosynthesis? What is the role of water?

9. If ATP and NADPH were supplied externally to leaves, leaf cells or isolated intact chloroplasts, would you expect photosynthetic carbon fixation to continue in the dark? Explain your answer.

10. Why is the enzyme ribulose bisphosphate carboxylase-oxygenase so-named?

11. How does C_4 photosynthesis differ from CAM? What are the advantages of these two pathways for plants living in hot or dry environments?

12. What would be the significance for plants of increased carbon dioxide concentration and predicted increased temperature caused by the 'greenhouse effect'?

CELLS AND TISSUES

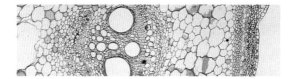

In multicellular organisms, such as animals and plants, cells are specialised for particular purposes and usually linked together into tissues where they function in a co-ordinated way to meet the needs of the whole organism. In earlier chapters, we described the internal structure and functions of cells. This chapter is concerned with the associations between cells that are functioning together as a tissue. **Tissues** are groups of similar, differentiated cells and associated extracellular matrix, which together carry out a particular function (Fig. 6.1): muscle tissue contracts, glands secrete and epidermis covers the surface of an animal or plant. Cells are organised into tissues in all animals except sponges (phylum Porifera, Chapter 38).

Plants are obviously different in appearance from animals. However, as we saw in Chapter 3, plant and animal cells have many similarities. The most obvious difference is that plant cells, like bacteria and algal and fungal cells, are encased in a cell wall external to the plasma membrane. It is the cell wall, more than any other feature, that has influenced the evolution of plants. Cell walls affect how plant cells are organised, interact with each other and function. In fact, the fundamental differences between plants and animals in regard to mechanisms of nutrition, growth, reproduction and defence can be attributed largely to the presence of cell walls in plants.

Animal cells, by contrast, are surrounded by an extracellular matrix that forms a loose and flexible lattice through which many animal cells can move. The matrix is quite different in nature to a cell wall, and confers some of the unique properties of animal cells and their capacity to organise into different tissues. Unlike plant cells, the plasma membranes of adjacent cells may be in direct contact with each other. The surface adhesive properties of plasma membranes are important during developmental processes (Chapter 15). Chemical interactions between the surfaces of adjacent cells influence control of cellular functions, such as growth and division. Within some tissues, such as muscle, many cells are co-ordinated to function as a unit by direct chemical and electrical communication between them. In the absence of a

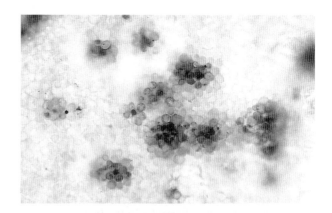

Fig. 6.1 The surface tissue of a rose petal is composed of two cell types: clusters of scent-releasing cells (stained red) and epidermal cells (unstained)

cell wall, support through turgidity is not generally possible; larger animals have internal and external skeletal structures to provide support and protection for their bodies.

In multicellular organisms, cells are specialised for particular purposes and linked together into tissues.

DEVELOPMENT OF ANIMAL TISSUES

At an early stage of development, all cells of animal embryos have a similar appearance. As development continues, the structure and properties of different groups of cells diverge—cells **differentiate**. Under normal conditions, this divergence is irreversible. Even when isolated and grown in tissue culture for many generations, differentiated animal cells usually maintain their distinct characteristics (Fig. 6.2).

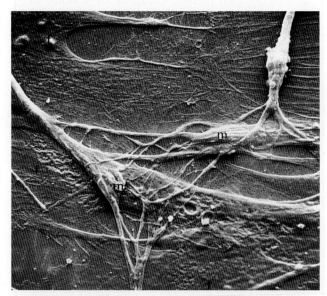

Fig. 6.2 When embryonic nerve and muscle cells are grown together in tissue culture, they develop many of the characteristics of normal differentiated nerve and muscle cells. Muscle cells (m) attach to each other and begin to contract and nerve cells develop long processes, which grow out and innervate the muscle cells

During development, cells with the same properties tend to adhere together to form tissues by the process of selective cohesion. In the same way that cells can aggregate to form tissues, so a number of tissues may combine to form a functional unit, an **organ**. In more complex animals, several organs may combine to form **systems**, which carry out entire processes, such as digestion or gas exchange.

> Animal tissues are groups of similar, differentiated cells and associated extracellular matrix, which together carry out a particular function.

Differentiation of cells often involves considerable **specialisation**, that is, the development of certain cellular functions at the expense of others. To take an extreme example, mature red blood cells in mammals have lost their nucleus and most of their cellular organelles; they are essentially a bag of haemoglobin with a limited life span. As a result of specialisation of function, some differentiated cells depend to a greater extent on the extracellular

environment for the provision of necessary materials and appropriate physical conditions. The maintenance of this environment is therefore of vital importance to survival. In mammals, there is a diverse group of cells involved in maintaining the extracellular environment. These include fibroblasts (cells that secrete the extracellular matrix), macrophages (phagocytic cells that dispose of unwanted materials), and leucocytes (white blood cells that help combat infection).

The extracellular environment is important in modulating the levels of activities of many types of cells. For example, circulating hormones can alter rates of cellular metabolic reactions, a change in diet may alter the rate of secretion of certain digestive enzymes, and increased stimulation by nerves can increase the size of muscle cells. In some tissues, the release of specific substances, such as growth factors, from adjacent tissues is important in the development and maintenance of fully functional, differentiated cells (Fig. 6.2).

> As a result of specialisation of function, some differentiated cells rely to a greater extent on their extracellular environment.

THE EXTRACELLULAR MATRIX

Within animals, the extracellular environment is a fluid containing the **extracellular matrix**, an extensive network of proteins and polysaccharides that fills the spaces between cells. The amount of matrix relative to the cells embedded in it varies in different kinds of tissue. Connective tissues such as cartilage and bone have large amounts of matrix; muscle tissue has very little (Fig. 6.3).

There are two main forms of extracellular matrix—**interstitial matrix**, prominent in connective tissues, and **basement lamina**, which underlies epithelial cell layers. Both consist of strong protein fibres embedded in a highly hydrated polysaccharide gel. In both, large molecules are linked together by covalent and non-covalent bonds.

In the past, the extracellular matrix was regarded simply as a space-filler, but recent research has changed this view dramatically. We now know that the extracellular matrix contains adhesive proteins unique to the tissue from which it originates (see Box 6.1) and that cell surfaces have specific receptors that bind to these matrix components. Strength and elasticity are provided by collagen and elastin. Hydration and porosity are generated by diverse proteoglycans. Cohesion of cells to the matrix is mediated by fibronectin, laminin and other adhesive proteins. The extracellular matrix is a versatile structure that integrates cell activities and guides cell movement.

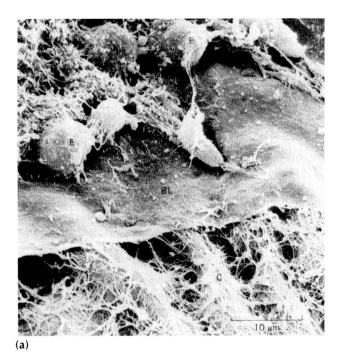

(a)

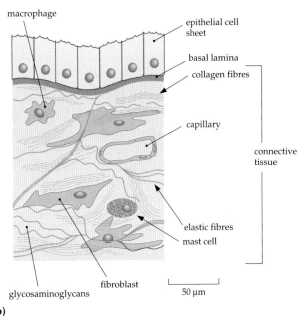

(b)

Fig. 6.3 (a) Scanning electron micrograph showing an epithelial basal lamina and underlying network of connective tissue microfilaments. (b) Extracellular matrix is abundant in connective tissue. The extracellular fluid gel contains scattered cells and an abundance of collagen fibres

The extracellular matrix contains proteins unique to each tissue and integrates cell activities and guides cell movement.

BOX 6.1   COMPONENTS OF THE EXTRACELLULAR MATRIX

There are three main components in the extracellular matrix of animal cells: structural protein fibres, adhesive proteins and glycosaminoglycans (large polysaccharide molecules usually linked to a protein core).

The structural proteins are of two main types. The predominant one is **collagen**, probably the most abundant protein in mammals, constituting about 25% of the total protein. Collagens are a family of proteins composed of three polypeptide chains that entwine into a tight helix. In basement laminae, collagen associates into a strong sheet-like meshwork, whereas in interstitial matrices, collagen associates to form fibrils 3–100 nm in diameter. These, in turn, assemble into fibres of high-tensile strength, several micrometres in diameter.

The second type of structural protein found in some extracellular matrices is **elastin**. Elastin is an unusual protein because it remains in an unfolded, random coil configuration. The elastin forms a highly cross-linked network that can be stretched extensively but, when released, recoils to its original configuration.

The structural proteins, collagen and elastin, are embedded in a highly hydrated gel composed of complex polysaccharides, **glycosaminoglycans**. These are long, unbranched polysaccharides composed of repeating disaccharide units. The polysaccharides have a strong negative charge that attracts cations. This draws water molecules into the gel by osmosis, thus providing a cushioning against compressive forces in some connective tissues. Localised secretion of glycosaminoglycans can cause the extracellular matrix to swell due to gel hydration, and this assists cells in their migration through the matrix.

Adhesive proteins are the third major component of the extracellular matrix. The best characterised examples are **fibronectin**, which occurs in interstitial matrices, has a high molecular weight (about 460 kD) and two polypeptide chains, and **laminin**, which is found in basement lamina, is a large glycoprotein nearly twice the size of fibronectin (about 850 kD) and has three polypeptide chains arranged in the form of a cross.

Fibronectin and laminin molecules mediate attachment of cells to the matrix and, by so doing, affect cell polarity, growth, migration and differentiation. Both molecules contain multiple binding sites that interact with collagen, proteoglycans and specific receptors on the surfaces of cells (see also Box 14.2).

MAINTENANCE OF ANIMAL TISSUES

Tissues that compose various organ systems are maintained for the life of an animal in several ways. Some tissues, such as those that interact with the external environment, face continual wear and tear, and therefore require repair or replacement of cells.

The skin and the lining of the intestine are two examples. Red blood cells also undergo considerable buffeting as they shuttle around the circulatory system. In general, the greater the wear and tear on a tissue, the greater the turnover rate of its cells (and therefore, incidentally, the greater susceptibility of these tissues to anti-cancer drugs that operate by inhibiting cell division).

Some cells, such as red blood cells, have become so highly specialised that they have lost their ability to undergo cell division; they therefore cannot replace themselves. However, the functions that they carry out are essential for continued survival of the animal. These tissues are replaced by division of small populations of relatively unspecialised cells known as **stem cells.**

Tissues are maintained by repair and/or replacement throughout life.

Regenerating tissues

The individual cells of some tissues, for example, liver cells and endothelial cells lining blood vessels, are continually replaced by new cells as a result of simple cell division. The average life span of an endothelial cell is several months. As old cells become damaged or die, surviving differentiated endothelial cells are stimulated to divide, thereby maintaining the population of cells. In addition, when tissues surrounding a blood vessel are deprived of oxygen,

they may release factors that stimulate endothelial cell division, leading to the formation of entirely new capillary networks.

In certain tissues, such as mammalian skin and blood, regeneration is not due to division of differentiated cells but to proliferation from a residual stem cell population. Stem cells are relatively unspecialised groups of cells that are able to divide repeatedly during the life of the animal. Although stem cells may not have a distinctive appearance, they have nevertheless undergone partial differentiation and their potential fate is determined (see Chapter 15). Stem cells give rise to cells that may either remain as stem cells or may continue development to become a fully differentiated cell. Some stem cells, such as skin stem cells, are **unipotent**: they produce only one type of differentiated cell. Blood-forming stem cells are **pluripotent**: they give rise to the whole range of blood cell types (Box 6.2).

Stem cell regeneration is found where there is a continuing need for replacement of cells but where the fully differentiated cell is itself either incapable of or unavailable for cell division. For example, red blood cells and keratinised skin cells have no nuclei, striated muscle cells are almost completely filled with actin and myosin fibres, mature sperm cells are released from the testes, and gut epithelial cells are lost from the tips of villi.

Differentiated skeletal muscle cells are unable to divide. The adult number of muscle fibres is laid down very early in development. Increase in the size of

---

BOX 6.2 BLOOD CELL FORMATION

The formation of new mature blood cells (haemopoiesis) must occur continuously throughout adult life because mature cells of this type have a lifespan of only days or weeks. Most blood-forming tissue is located in scattered deposits within the marrow of various bones. There are eight major types of specialised blood cells (lineages) to be formed red cells, granulocytes, monocyte–macrophages, eosinophils, megakaryocytes, mast cells, and T and B lymphocytes. These are all generated from a small common population of pluripotential stem cells, which form larger numbers of more mature haemopoietic cells in a series of steps (Fig. a).

Stem cells

Only one in every 10^5 haemopoietic cells are stem cells. Stem cells are small to medium-sized mononuclear cells with two distinctive features—an ability to renew themselves by cell division and an ability to produce several hundred daughter cells (committed progenitor cells) that enter a pathway of irreversible differentiation to form different types of blood cells.

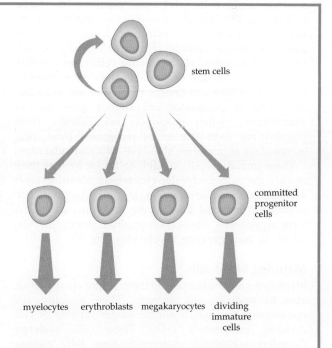

stem cells

committed progenitor cells

myelocytes   erythroblasts   megakaryocytes   dividing immature cells

(a) A schematic representation of the way stem cells generate committed progenitor cells, each of which then generates large numbers of maturing blood cells

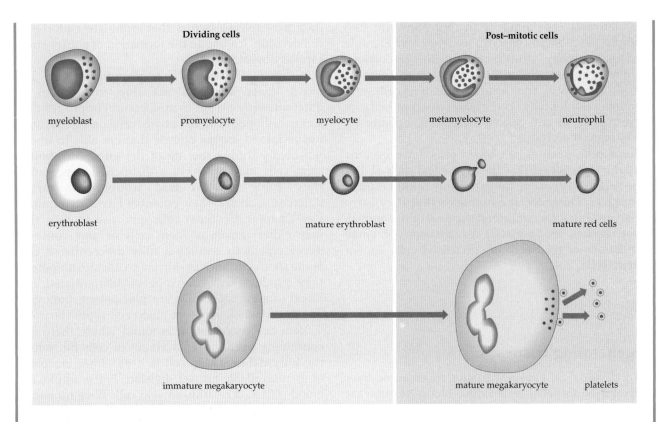

(b) Maturation occurs progressively in dividing haemopoietic cells and continues after the cells lose the capacity for further division. Only fully mature cells are released into the blood circulation

Committed progenitor cells

Committed progenitor cells constitute about 1% of haemopoietic cells and are medium to large undifferentiated *blast cells* that have the capacity to form very large numbers of maturing progeny in their particular lineage. Up to 10^5 cells are able to be produced by each progenitor cell. Progenitor cells cannot revert back to stem cells and cannot alter their differentiation commitment from one lineage to another. For example, an erythroid progenitor cell cannot give rise to granulocytes. When progenitor cells divide, their progeny are soon able to be recognised from their morphology as immature blood cells of a particular type.

These immature cells are able to divide and, as they mature, the progeny develop characteristic features such as nuclear shape or granules in the cytoplasm. For example, committed granulocyte progenitor cells first form myeloblasts. These then generate promyelocytes, which in turn generate myelocytes (Fig. b).

Maturing blood cells

Myelocytes are the last cells in the granulocyte sequence able to undergo cell division. Their progeny are metamyelocytes, which are no longer capable of cell division (post-mitotic cells). These cells undergo extensive maturation changes to form fully mature granulocytes (neutrophils), which enter the peripheral blood where they have a lifespan of, at most, a few days.

This pattern of continuous production of non-dividing mature cells with a limited lifespan also applies to the formation of red cells where maturation changes include extrusion of the nucleus from the cell before it is released to the blood. Platelets represent an even more extreme degree of maturation because they are merely membrane-bound portions of cytoplasm from mature megakaryocytes, which are pinched off and released to the circulation (Fig. b).

In some blood cell types, the apparently mature cells that enter the blood can be reactivated by appropriate stimuli and then undergo further cell divisions. This occurs with some T and B lymphocytes and, less frequently, can occur with some mast cells, eosinophils and macrophages.

Organisation of haemopoietic cells

In blood cell formation, cells at various stages of development in the different lineages are intermingled within the bone marrow. The properties of various ancestral cells and their progeny have only been able to be deduced by the development of cell sorting techniques, which use particular marker molecules expressed on cell surfaces, combined with tissue culture of colonies of maturing blood cells. In this way, pure populations of each developmental stage can be obtained.

Cell division in haemopoietic cells follows stimulation by specific regulatory molecules. For each lineage, multiple regulators (glycoproteins active at pg/mL concentrations) interact to control development. So far, 20 of these have been discovered, the genes encoding them isolated and the regulators mass-produced by genetic engineering. It is suspected that possibly a total of 50 different regulators may exist to control the multiple events of cell division and maturation occurring in the eight lineages of haemopoietic cells. Co-ordination of cell production by haemopoietic tissues scattered throughout the body is achieved by the interactions of these haemopoietic regulators.

When populations of blood cells become depleted, for example, by loss of red blood cells during bleeding, or when additional cells such as granulocytes are needed to respond to an infection, haemopoietic regulators rapidly increase the rate of cell production by haemopoietic cells. In part, this accelerated blood cell production is achieved by shortening the cell cycle times of the dividing cells, but most of the amplification is achieved by increasing the numbers of progenitor cells derived from the stem cells. Where demand is extreme, additional haemopoietic tissue develops in the spleen and liver to allow production of the required number of cells.

This process of accelerated replacement or cell production is known as **hyperplasia**. The hyperplasia ceases when the numbers of cells have reached their required levels. For example, the need of tissues for a minimum level of oxygen requires a certain number of oxygen-carrying cells (red blood cells) to be present and, when that number is reached, blood oxygen levels return to normal and red cell formation returns to basal levels.

muscles, as seen in weight-lifters, occurs largely as a result of hypertrophy (increase in size) of individual cells. There is, however, a small population of muscle stem cells that can, when muscle tissue is damaged, give rise to new cells, which fuse to form mature multinucleate muscle fibres.

> Tissues may be regenerated by division of differentiated cells or by proliferation from a relatively unspecialised stem cell population.

'Permanent' cells

In some tissues there is no turnover of cells. The cells of these tissues are produced during development in sufficient numbers to last the entire life of the animal. These cells are permanent and, if some die, they cannot be replaced. Examples include nerve cells, cells of the retina and lens (Fig. 6.4), oocytes and cardiac muscle cells. In the case of the nervous system, permanency of the developing complex network of nerves is understandably an advantage. Laying down this network during development is an ongoing interactive and sequential process. It is difficult to imagine how the precise relationships between individual nerve cells could be maintained, or how properties such as memory could be retained, if there were a significant turnover of nerve cells throughout the life of an animal. However, in the case of cardiac muscle cells, it is difficult to imagine any advantage in not having the ability to regenerate new cells.

Most permanent cells are able to repair themselves by turnover of cell components. With increased use, cardiac muscle cells become hypertrophied. Damaged nerve fibres may be able to regrow. Retinal cells, rods and cones continually replace the folded stack of photoreceptive membranes that form the sensory receptor that responds to light.

> In some tissues there is no turnover of cells. Most permanent cells are able to repair themselves by turnover of cell components.

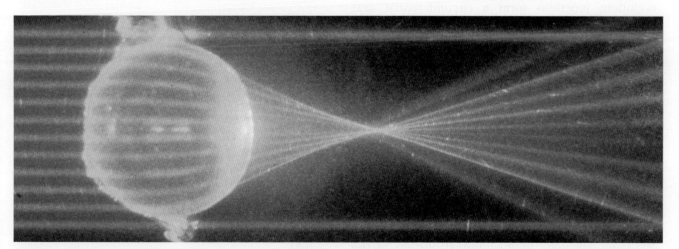

Fig. 6.4 Cells forming the lens of a vertebrate eye are highly specialised to allow unobstructed passage of light. Once the lens has formed, individual cells cannot be replaced. This lens is from a trout eye with laser beams 0.5 mm apart shining through it

INTERCELLULAR CONNECTIONS

Cells in animal tissues are interdependent, bound together to some degree by a non-cellular matrix and in communication with each other for co-ordination and control. Chemical communication occurs by means of signal molecules that pass through the extracellular matrix. Light microscopy has shown that there are also physical connections between the plasma membranes of adjacent cells. The nature and variety of these connections was discovered with the advent of high-resolution electron microscopy.

Junctions between plasma membranes can be grouped according to their function. Examples of each of these types of connections can be seen in the layer of epithelial cells that lines the lumen of the intestine in mammals (Fig. 6.5).

■ **Occluding junctions** are those where the plasma membranes of adjacent cells are fused tightly, preventing the passage of even small molecules through the extracellular space.

■ **Anchoring junctions** provide sites of attachment for the mechanical support of tissues. They link adjacent cells, attach cells to the extracellular matrix, and form attachment sites for fibres of the cytoskeleton.

■ **Communicating junctions** are those that are specialised for chemical and electrical communication between adjacent cells.

> Cells in animal tissues are bound together by junctions between plasma membranes of adjacent cells.

Occluding junctions

These junctions, also known as tight junctions, are found linking epithelial cells, forming a barrier to the free movement of molecules between the cells of the epithelium. In the epithelium lining of the intestine, occluding junctions form a circumferential seal, preventing free movement of molecules between the lumen and interstitial space. Ultrastructural studies show that the plasma membranes are fused together by continuous bands of transmembrane junctional proteins (Fig. 6.6). The permeability of these junctions to small molecules varies in different epithelia. In the intestine, occluding junctions between epithelial cells are 10 000 times more leaky to ions than are those in the epithelium of the urinary bladder.

Interestingly, occluding junctions also serve to maintain the correct location of membrane protein channels involved in the transport of digested food from the gut lumen into the circulatory system. For example, active glucose channels are confined by occluding junctions to the luminal portion of the endothelial plasma membrane, whereas passive glucose carrier proteins are confined to the basolateral regions of the plasma membrane (Fig. 6.7).

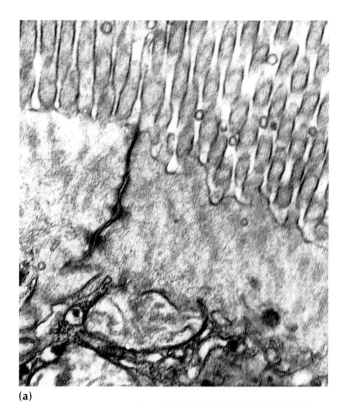

(a)

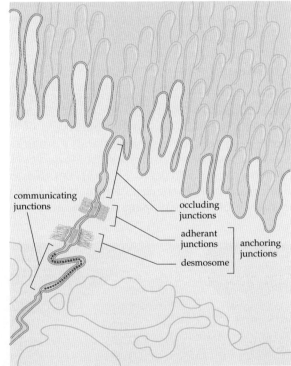

(b)

Fig. 6.5 Cells of (a) the epithelial layer lining the gut are (b) linked by occluding junctions, which form a barrier to the movement of gut contents between the cells; anchoring junctions, which hold the cells together when physical strains are placed on the gut wall; and communicating junctions for ion exchange between cells

> Occluding (tight) junctions form an impenetrable seal between cells and restrict the movement of membrane proteins.

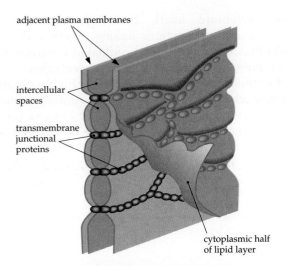

Fig. 6.6 Frozen cells can be fractured in such a way that they split down the middle of the plasma membrane bilayer. Here the fracture has exposed the occluding junctional proteins between two adjacent cells

Anchoring junctions

There are several types of anchoring junctions. Some have localised thickening due to the presence of a dense cytoplasmic layer (plaque) and a wider intercellular space that may show an intermediate line between the plasma membranes (Fig. 6.8).

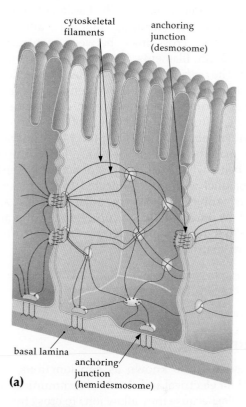

(a)

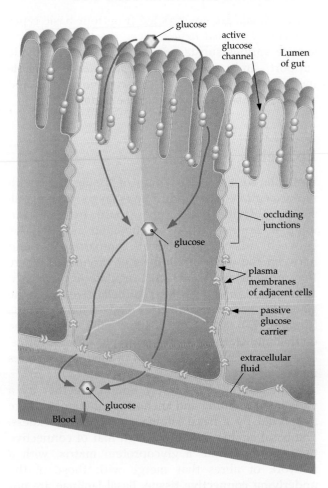

Fig. 6.7 The band of occluding junctions in the gut epithelium prevents active glucose transport channels moving away from the luminal surface. They actively transport glucose from the lumen against a concentration gradient, raising the intracellular glucose concentration. Glucose then passes from these cells into the blood via passive glucose carriers, which are, in this case, confined to the internal walls of the cells by the presence of occluding junctions

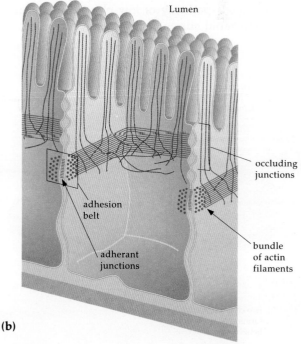

(b)

Fig. 6.8 Several types of (**a**) anchoring junctions (desmosomes, hemidesmosomes) and (**b**) adherant junctions can be seen between gut epithelial cells and linking to cytoskeletal filaments

Intermediate filaments of the cytoskeleton (Chapter 3) attach to the plaque and pass back into the cytoplasm to connect with other structural elements within the cell. These junctions, also known as **desmosomes**, provide structural support to tissues by cross-linking between the cytoskeletal networks of adjacent cells. **Hemidesmosomes** anchor cells to the extracellular matrix, such as the basement lamina.

Another form of anchoring junction, **adherant junctions**, forms cross-links between contractile bundles of actin filaments located in the cortical cytoplasm. The *contractile* nature of adherant junctions makes them important in the movements of cells and tissues that occur during embryonic development, for example, during the formation of the nervous system in vertebrates (see Chapter 14). Adherant junctions may be focal or belt-like, for example, adhesion belts, which are bundles of actin filaments running parallel with and connected to the plasma membrane. Adhesion belts of adjacent cells are connected by intercellular glycoproteins. Adherant junctions, like desmosomes, can also anchor to the extracellular matrix.

Anchoring junctions provide mechanical support and are important in developmental processes.

Communicating junctions

These junctions, also known as gap junctions, are specialised for electrical and chemical communication between cells. Because they allow ions to cross freely, they provide a pathway of low electrical resistance and permit rapid current spread from cell to cell. In the vertebrate heart, it is the presence of communicating junctions linking together all atrial cells that results in the co-ordinated contraction of both atria, expelling blood into the ventricles.

Communicating junctions are highly regular protein channels composed of six protein subunits that span the plasma membrane and link to a similar unit in the adjacent cell (Fig. 6.9). The diameter of the central aqueous channel appears to be about 1.5 nm, but, at least in some tissues, permeability can apparently be regulated.

Communicating (gap) junctions allow ions to pass freely between adjacent cells.

PRINCIPAL ANIMAL TISSUES

Most animals are constructed from four basic types of tissue. *Epithelial tissues* cover most surfaces and secrete various substances, *connective tissues* provide support and strength, *muscular tissues* enable movement, and *nervous tissues* receive signals and conduct messages throughout the body.

Epithelia

Epithelia are a diverse group of tissues that, with rare exceptions, form continuous layers covering all surfaces, cavities and tubes, both external and internal. Epithelia therefore function as interfaces between adjacent biological compartments, and between the organism and its environment, where they provide protection and regulate exchange. In addition to forming barriers, epithelia give rise to the glandular tissue of endocrine and exocrine glands (Chapter 25) and parts of some sense organs. The epithelium of the gut, for example, is absorptive, secretes enzymes for digestion of food and secretes mucus for protection.

Epithelial cells are closely bound together by a variety of occluding and anchoring junctions and are supported by a basal lamina of variable thickness. The basal lamina is a thin, tough mat of connective tissue formed from a glycoprotein matrix, with a network of fibres that merge with those of the underlying connective tissue. Basal laminae are not penetrated by blood vessels, so epithelia are dependent on diffusion through the lamina for exchange of nutrients and wastes with blood.

Epithelia are defined according to the number of layers, and the shape of cells. They may be simple, forming a single layer, or stratified, composed of more than one layer; and, according to the shape of the

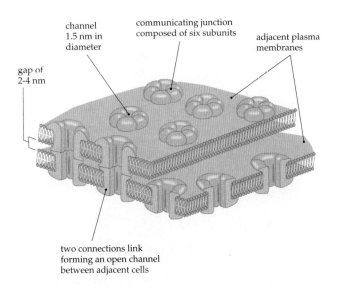

channel 1.5 nm in diameter

communicating junction composed of six subunits

adjacent plasma membranes

gap of 2-4 nm

two connections link forming an open channel between adjacent cells

Fig. 6.9 Six membrane-bound protein subunits in each plasma membrane link across the intercellular space to form the channel of a communicating junction

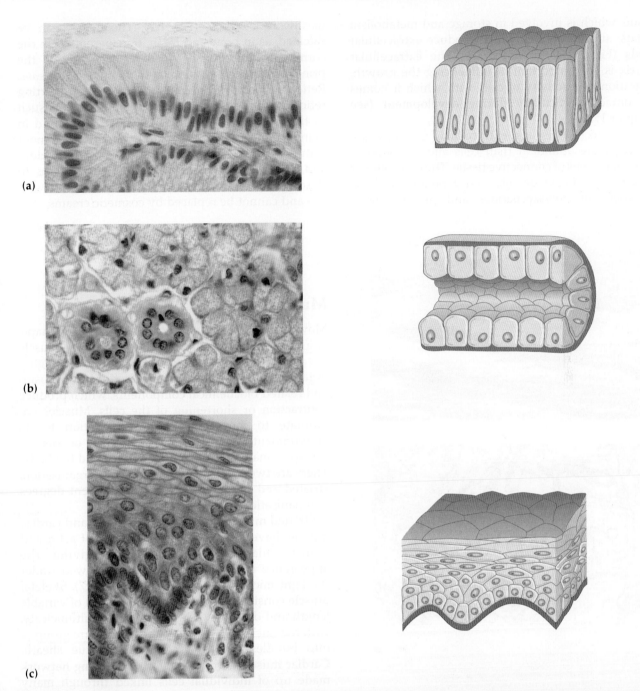

Fig. 6.10 Epithelial cells, as shown in these light micrographs, are usually described in terms of their histological organisation. **(a)** Simple columnar epithelium, composed of tall cells with an oval nucleus at the base, line the human gall bladder. **(b)** Simple cuboidal epithelium forms the ducts of many glands, as shown in these two ducts in the human parotid gland. **(c)** Stratified epithelium, as in the human oesophagus, is a thick, many layered sheet of cells varying from columnar at the base, to cuboidal and squamous (flattened) on the outer surface. Stratified epithelium forms the skin of many vertebrates

cells, may be squamous, cuboidal or columnar (Fig. 6.10). Epithelia may also be categorised according to the presence of specialisations, such as cilia or deposition of keratin.

> Epithelia function as interfaces between biological compartments and with the environment, and in secretion.

Connective tissue

In general, tissues that provide basic structural, metabolic and defensive support for other tissues of the body are **connective tissues**. These include bone and cartilage, which provide physical support; blood, which provides defence and transports important materials throughout an animal's body; adipose tissue, which is involved in storage and metabolism of fats; and fibroblasts, which produce extracellular fluids (Fig. 6.11). In vertebrates, the extracellular matrix is also important in influencing the growth, migration and activity of cells with which it comes in contact, particularly during development (see Chapter 15).

In connective tissues, extracellular materials are usually more abundant than cells and characterise particular types of connective tissue. The extracellular matrix is produced by cells and generally contains a variety of polysaccharides and proteins in a

meshwork of fibres (Box 6.1). The matrix can be calcified, as in bone and teeth, transparent, as in the cornea, or fluid, as in blood. Collagen forms the principal fibre type found in all connective tissues. Reticulin fibres predominate in the supporting reticular, connective tissues that enfold organs such as the kidney and liver. Elastin fibres are found in many connective tissues and they are prominent features of the elastic connective tissues of the bladder and dermis of the skin. Despite advertisements to the contrary, elastin fibres in the skin deteriorate with age and cannot be replaced by cosmetic creams.

> Connective tissues are composed of cells, fibres and fluid matrix. They provide structural, metabolic and defensive support for other tissues.

Muscle

Movement, such as motility and change in cell shape, is an inherent property of all animal cells. Muscle is composed of specialised cells that contain highly organised fibrillar components (actin and myosin) and related biochemical components, which produce contraction or shortening of the cells. Muscles co-ordinate to move an animal in relation to its environment and move its internal organs to transport materials into, out of and around the body. There are two basic types of muscle cell organisation, striated and smooth, which reflect different degrees of organisation of the fibrillar network.

Striated muscle cells, found in skeletal and cardiac muscle, have a highly organised array of actin and myosin filaments (myofilaments) giving the appearance of cross-striations when viewed under the light microscope (Fig. 6.12; Chapter 27). **Skeletal muscle** consists of cylindrical cells (fibres) of variable length and diameter. These fibres are multinucleate, with the nuclei located peripherally, and are grouped into bundles within a connective tissue sheath. **Cardiac muscle** is composed of a branching network made up of individual cells linked through many communicating junctions.

The **smooth muscle** cells of internal (visceral) organs are spindle-shaped cells with a central nucleus. They may be loosely scattered or aligned together to form dense bands or sheets. In the gut wall, individual smooth muscle cells are closely packed with many communicating junctions. In contrast to striated muscle, myofilaments in smooth muscle cells are less regularly arranged and there is no striated appearance (Fig. 6.12). Although smooth muscle myofilaments are not visible under the light microscope, they can be discerned with the electron microscope.

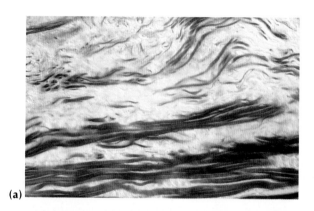

(a)

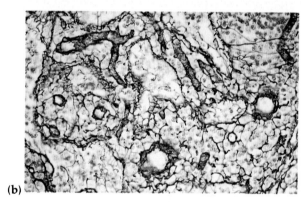

(b)

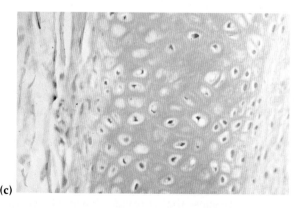

(c)

Fig. 6.11 Different types of connective tissue are characterised by differing types and proportions of fibres and matrix. Light micrographs of
(a) elastic connective tissue from a ligament,
(b) black-stained reticular fibres in a human lymph node, and **(c)** cartilage from the human ear

> Striated muscle cells contain a highly organised array of actin and myosin filaments. Myofilaments in smooth muscle cells are less regularly arranged.

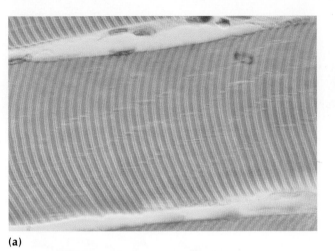 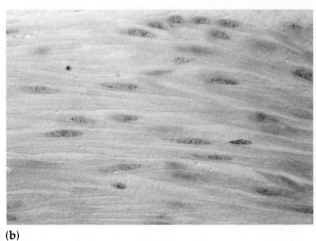

(a)   **(b)**

Fig. 6.12 Light micrographs of **(a)** human skeletal muscle showing the striations due to highly organised bands of fibrils forming the contractile mechanism, and **(b)** smooth muscle, in which the contactile fibres are not regularly arranged and no striations are evident

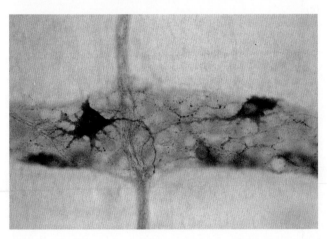 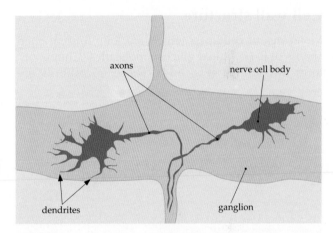

Fig. 6.13 Two nerve cells in a ganglion (cluster of nerve bodies) in the wall of the mammalian intestine. The short processes (dendrites) receive incoming information and outgoing signals pass down the longer single process, the axon

Nervous tissue

Nerve cells, **neurons**, are specialised to carry information rapidly and precisely from one part of an animal to another, often over long distances. Neurons form an interconnecting network through which information passes from sensory structures (receptors) through integrating circuits to effector structures such as muscles or glands (Chapter 26). The basic structure of neurons in all animals is similar. Information is received by relatively short cellular processes, **dendrites**, and passed into the nerve cell body (Fig. 6.13). If the nerve cell responds, a signal is passed down a single process, the **axon**, which in some animals, such as whales and giraffes, may be metres long.

Neurons are highly specialised cells that are unable to undergo cell division. They are only able to carry out their normal functions because of their association with a group of supporting cells—glia. **Glial cells** are derived from the same embryonic tissue as nerve cells

but they are responsible for maintaining the composition of the extracellular environment, which is critical for the proper functioning of neurons. In vertebrates, glial cells also form myelin sheaths, which allow very rapid conduction of signals along the axons of some cells (Chapter 26).

> Nerve cells are specialised to conduct signals rapidly and precisely throughout the body. They are unable to divide and are supported by glial cells.

PLANT CELLS: THE IMPORTANCE OF THE CELL WALL

Unlike animal cells, plant cells are encased in cell walls that form box-like compartments (Fig. 6.14). An important function of cell walls is in limiting the size and shape of cells. A plant cell wall is an extracellular matrix, and is thicker, stronger and more rigid than the extracellular matrix of animal cells. The characteristic component of plant cell walls is cellulose. The strength of the wall allows plant cells to exist surrounded by a hypotonic environment (with a lower concentration of solutes than the cytosol). Without a restraining wall, the inflow of water by osmosis (Chapter 4) would cause plant cells to burst.

Cell walls are also important in metabolic processes: they contain specific enzymes; act as a pathway for transport, absorption and secretion; and act as a defence barrier against pathogens. Recent discoveries indicate that some wall components are similar to plant hormones and thus can play a role in cell to cell communication.

> Plant cells characteristically have walls, which limit cell size and shape.

Cell wall structure

Molecules of cellulose, which are long, unbranched chains of β-1,4 linked glucose molecules (Chapter 1), form **microfibrils** (Fig. 6.14). These microfibrils are

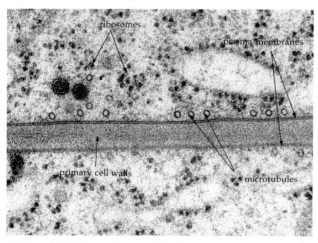

(a)

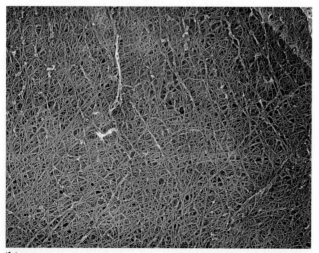

(b)

Fig. 6.14

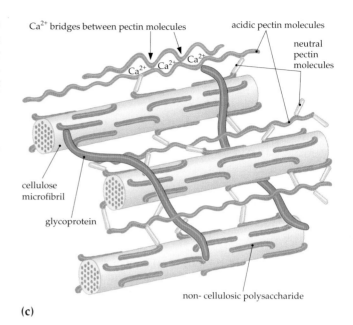

(c)

Fig. 6.14 Plant cell walls. The cellulose forms microfibrils, which are interconnected by pectic polymers and proteins to complete the matrix. (a) Thin section of a primary cell wall of *Eucalyptus sieberi* prepared by rapid freezing in liquid nitrogen to retain lifelike appearance (magnification × 68 200). (b) Surface view of microfibrils of cell wall prepared by fast freeze, deep etch, rotary shadowing. (c) Diagram showing interaction between cellulose microfibrils, hemicelluloses and other cell wall molecules

embedded in and cross-linked to a porous, hydrated gel-like matrix of polysaccharides other than cellulose (*non-cellulosic polysaccharides*), pectins and proteins (Box 6.3). Non-cellulosic polysaccharide molecules form bridges at regular intervals between adjacent cellulose microfibrils, spacing them about 20–30 nm apart. The pectins link with them to form a gel. This network forms a wall that has pores of about 5 nm, large enough to allow water and small molecules to diffuse freely but small enough to inhibit the movement of larger molecules such as proteins.

Cell walls of bacteria, fungi and some groups of algae differ from those of plant cells. In fungi, the main component is not cellulose but chitin, a polymer of glucosamine. Bacterial walls are varied but all are composed of polysaccharides cross-linked by amino acids. In red algae (Chapter 35), cell walls are composed of microfibrils of cellulose (or another polysaccharide) and a mucilaginous matrix of components such as agar and carrageen.

> Plant cell walls comprise cellulose microfibrils embedded in and interconnected to a matrix containing non-cellulosic polysaccharides, pectins and proteins.

BOX 6.3   COMPONENTS OF PLANT CELL WALLS

The cell wall matrix in which cellulose microfibrils are embedded is composed of two main types of branched polysaccharides: pectins and non-cellulosic polysaccharides (Fig. 6.14).

Pectins have a backbone of galacturonic acid residues to which other types of sugars are linked. Pectins have a strong negative charge that attracts cations (especially Ca^{2+}) and associated water molecules, making them highly hydrated molecules. Being divalent, calcium ions can form cross-bridges between two pectin molecules and this greatly strengthens the gel. Many pectins can be removed from cell walls by mild chemical treatments, but others remain tightly bound to cellulose microfibrils. Pectins are common in the pith of citrus fruits and, when making jam, adding lemon skin to fruit will aid in the setting.

Non-cellulosic polysaccharides are removed from walls by harsher chemical treatment, such as with alkali. They have a backbone made of one type of sugar with short side chains of other sugars linked to it, for example, xyloglucans have a backbone of glucose units to which side branches of xylose are linked. The glucose backbone can form hydrogen bonds with cellulose microfibrils.

Many cell walls also contain small amounts of protein. For example, extensin is a glycoprotein which, like collagen, is rich in the unusual amino acid, hydroxyproline. Glycoproteins contribute to the cross-linking of other components in the wall.

By varying the relative proportions of these basic components, by controlling their organisation and by adding more specialised components when appropriate, plant cells are able to generate cell walls with a wide variety of structures, properties and functions. Indeed, most cell types in a plant organ can be recognised by the morphology of their cell walls.

A variety of cell wall components are of industrial and agricultural significance—in the food industry, animal and human nutrition (mostly as dietary fibre), the building industry, paper and pulp industry, and biotechnology. For example, processed foods (such as

Sedimentation of cocoa powders in chocolate drinks is a common defect (right) that is usually alleviated by incorporating a stabilising agent, in this case, a plant cell wall carbohydrate (left). Stabilising agents increase viscosity of the milk and thereby slow sedimentation of particles

ice-cream) have plant or seaweed wall polysaccharides added to modify texture, taste and consistency (see figure). In Australia, plant biotechnologists are investigating the potential to develop industries based on exploiting certain cell wall polysaccharides. When grown in cell-suspension culture, plant cells secrete wall components directly into the surrounding medium. In this way, cell wall components can be harvested for commercial use.

Development of plant cell walls

During the formation of cell walls, glucose precursor molecules are transported across the plasma membrane by specific carrier proteins (Chapter 4). Cellulose chains are assembled on the external face of the plasma membrane by large complexes of membrane-bound enzymes. As the chains are polymerised, they quickly become hydrogen-bonded together into microfibrils (Fig. 6.14).

New cell walls are laid down initially as partitions (cell plates) consisting of sheets of membranes formed after nuclear division in the parent cell (Fig. 6.15). The cell plate grows and fuses with the plasma membrane of the parent cell. The two newly formed cells then add wall material on either side of the cell plate. As the common wall develops, the cell plate material is modified but persists as a thin layer, the **middle lamella**, between the walls of the adjacent cells. The middle lamella is rich in pectin molecules. Walls of growing cells are **primary cell walls**. Primary walls are relatively thin, typically about 100 nm in width, highly hydrated and extensible, allowing the cell to expand in size (Fig. 6.15).

As plant cells grow, the cell wall yields to the internal hydrostatic force of turgor pressure. The direction of expansion, and therefore growth, is

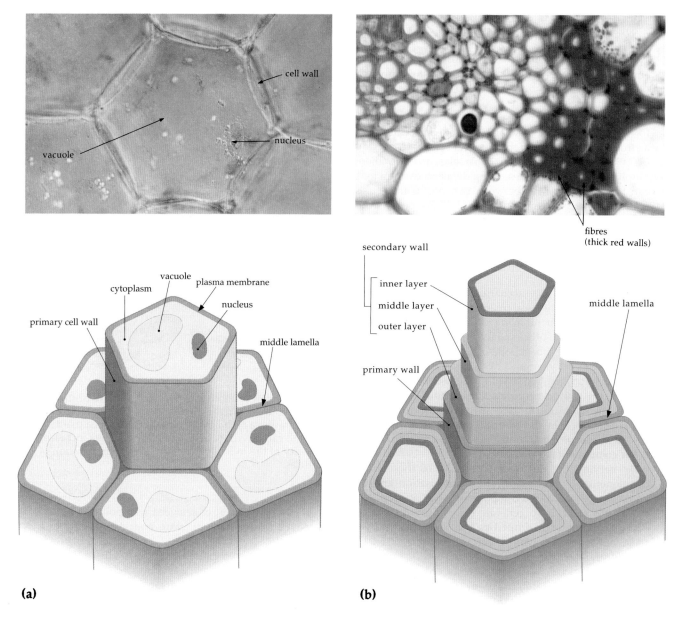

Fig. 6.15 Primary and secondary cell walls. (**a**) Transverse section of parenchyma cells from *Coleus* stem showing thin primary cellulosic cell walls and large vacuole-filled cell with nucleus and peripheral cytoplasm. (**b**) Secondary cell walls of fibres from a transverse section of *Hakea* leaf. The sequential lignin deposits (red stain) are added over the primary cell wall

determined by the arrangement of cellulose microfibrils in the wall. Microfibrils do not extend lengthwise, and any cell expansion occurs by their separation (Fig. 6.14). If, for example, cellulose microfibrils are deposited circumferentially around the side walls of a cell, lateral expansion will be inhibited and almost all growth will extend the cell lengthwise. The extent and direction of growth of cells is controlled so that connections between cells are not disrupted.

As plant cells differentiate and mature, primary cell walls may be added to, producing thicker, more rigid **secondary walls**. The main component of secondary walls is the polyphenolic compound, **lignin**.

Lignification makes walls very strong and rigid by cementing together and anchoring the cellulose microfibrils, thus greatly reducing pore size. Deposition of lignin can occur within the primary wall or new layers may be deposited upon the primary wall, building up a wall that may be several micrometres thick. Secondary walls may completely surround a cell or they may be formed only in localised regions.

Cellulose is a major component of primary walls. Cellulose microfibrils give cell walls strength, and their orientation determines the direction of cell expansion. Lignin is characteristic of many secondary walls.

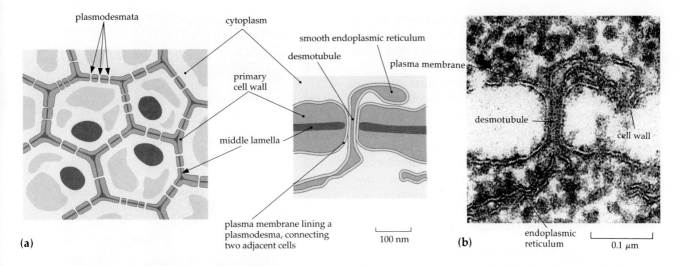

Fig. 6.16 Structure of plasmodesmata. (**a**) Plasmodesmata provide a direct connection between two cells. (**b**) This is achieved by means of a common plasma membrane and internal desmotubule; ER is commonly associated with the plasmodesma

Plasmodesmata: links between plant cells

The rigid wall of plant cells limits contact between adjacent plasma membranes and limits the exchange of molecules between the cell cytoplasm and the environment and between neighbouring cells. Although cell walls allow small molecules and ions to pass between cells, plants also have special channels, **plasmodesmata** (Fig. 6.16), that directly link the plasma membranes and cytosols of adjacent cells. Plasmodesmata are similar to communicating junctions in animal cells.

Each plasmodesma is a fine cytoplasmic channel linking adjacent cells across a cell wall. The channel is bounded by plasma membrane and is cylindrical in shape. A narrower cylindrical structure, the desmotubule, runs through most plasmodesmata and is continuous with the endoplasmic reticulum of each cell.

Evidence that plasmodesmata are involved in transport comes from experiments using dyes that cannot cross the plasma membrane. When such a dye is injected into a plant cell, it passes readily to adjacent cells through plasmodesmata. Furthermore, since the plasma membranes of adjacent cells are continuous across plasmodesmata, they provide no barrier to ion movement and act as conduits for electrical signals that are transferred from one cell to another. In this way, plasmodesmata have a similar function to communicating junctions between animal cells.

> Plant cell walls are penetrated by plasma-membrane-lined channels, plasmodesmata, which provide a means of direct communication and transport between cells.

PLANT TISSUE SYSTEMS

At the tip of a growing shoot or root is a specialised region of cells—the **apical meristem**. These small clusters of cells are continually dividing, each cell producing two daughter cells, one of which remains as part of the meristem and one that differentiates as part of the mature body of the plant. Meristematic cells have a similar function in plants to stem cells in animals (p. 103).

Dividing meristematic cells are typically small, with a dense cytoplasm, small or no vacuole, and a large, active nucleus. The first two apical meristems of a plant develop as part of the embryo while it is still enclosed within the protective covering of a seed. One of these, at the tip of the epicotyl (or stem precursor), is the shoot apical meristem; the other,

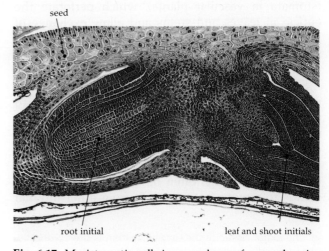

Fig. 6.17 Meristematic cells in an embryo of a cereal grain of wheat, showing the leaf initials around the epicotyl (upper) and the root initials of the radicle (lower)

at the tip of the radicle (or root precursor), is the root apical meristem (Fig. 6.17). Additional apical meristems form at the tips of new shoots and roots as a plant grows and branches.

The daughter cells that differentiate from an apical meristem form many different types of cells varying in size, shape, cell wall construction and organelle composition, features that are linked to the specialised functions of the individual cells. Different cell types become aggregated into tissues, which, in vascular plants (including ferns and seed plants), form three broad functional systems. These tissue systems are the *dermal* (covering) system, the *ground* (supporting) system, and the *vascular* (transporting) system.

> Vascular plants have three major tissue systems: the dermal (covering), ground (supporting) and vascular (transport) systems. All these arise from apical meristem tissue.

Dermal tissue

The dermal system forms the outer covering of a plant and is the interface at which plants interact with the external environment. The major tissue type is the **epidermis** (Fig. 6.18), which consists of a layer (occasionally several layers) of closely packed cells that secrete a cover of water-resistant cuticle. The **cuticle** is composed of cutin, a water-insoluble polymer of fatty acids. A layer of wax platelets is deposited on its surface. The cuticle protects plants against evaporative water loss, physical damage and invasion by pathogens. Because the cuticle is also impermeable to air, the epidermis contains pores (stomata in vascular plants), which perforate the cuticle in leaves and stems and allow exchange of gases with the internal tissues. Stomata have two specialised epidermal cells, guard cells, which control opening and closure (Chapter 18).

Epidermal cells also develop as hairs or trichomes— unicellular or multicellular projections that help to trap moisture and reduce water loss or help to deter insect herbivores (Fig. 6.18). Trichomes are sometimes associated with a gland producing toxic or unpalatable substances. Halophytes (plants tolerant of saline conditions) often have elaborate secretory glands to remove excessive salt from other internal tissues.

> The major tissue of the dermal system is the epidermis, some cells of which are modified as guard cells for opening and closing stomata, hairs, trichomes and glands.

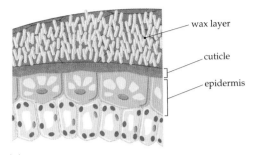

(a)

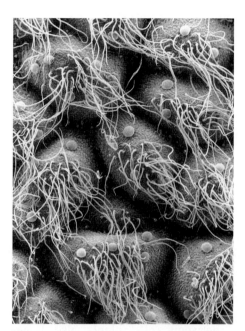

(b)

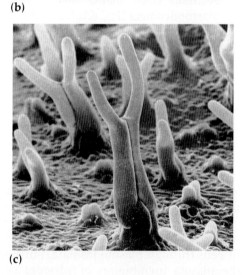

(c)

Fig. 6.18 Epidermis is the major tissue type of the dermal system which forms an outer covering. **(a)** The outermost wall layer is the cuticle, which lies over the primary cellulosic wall and is coated by wax platelets. **(b)** The herb, sage (*Salvia*) has a leaf epidermis covered by long, filamentous hairs and globular oil glands, which give sage its characteristic odour (magnification × 42). **(c)** Trichomes on juvenile leaves of *Eucalyptus wandoo* from Western Australia. A pair of long hairs are associated with an oil gland, (magnification × 18) and probably play a role in deterring insect herbivores

Ground tissue

The ground system consists of storage and structural tissues in which the vascular system is embedded. It is composed largely of three tissue types: parenchyma, collenchyma and sclerenchyma.

Parenchyma

Parenchyma tissue is composed of cells that are typically large, with a thin primary cell wall and well-defined pectin-rich middle lamella, a large vacuole that fills most of the cell, and an active nucleus with dispersed chromatin. Parenchyma cells are commonly polyhedral (many-sided) and loosely packed together, separated by a network of intercellular spaces.

Different types of parenchyma have different functions. Photosynthetic parenchyma, often called **chlorenchyma**, contains chloroplasts and is found in leaves and the outer regions of photosynthetic stems. **Storage parenchyma** can contain either nutrient reserves in the form of starch granules or oil droplets (Fig. 6.19), or various metabolic products that accumulate in the vacuole. Aquatic plants have **aerenchyma**, a spongy parenchyma with a large network of air spaces, which allows aeration of tissues in an environment low in oxygen (Fig. 6.19).

A special type of parenchyma cell, a **transfer cell**, is characterised by ingrowths of the primary cell wall and associated plasma membrane (Fig. 6.19). The increased surface area of membrane allows rapid transfer of molecules to and from adjacent cells of the vascular system.

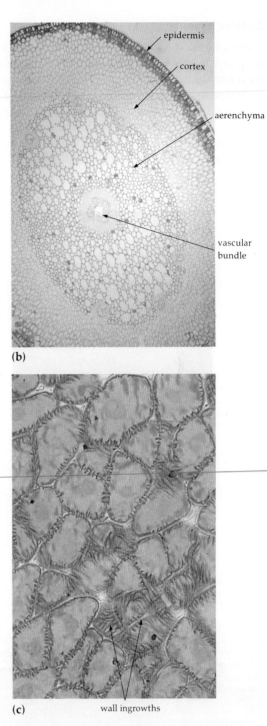

(b)

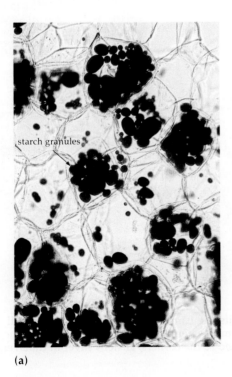

(a)

(c)   wall ingrowths

Fig. 6.19 Parenchyma cells form tissues with defined functions. (**a**) Parenchyma cells of potato tuber store starch granules (here stained blue–black with iodine). (**b**) Aerenchyma cells increase the circulation of gases in underground stems (rhizomes) of the seagrass *Amphibolis*. (**c**) Transfer cells with conspicuous wall ingrowths in an embryo of the seagrass *Amphibolis* increase the surface area of plasma membrane for exchange of substances with parental tissue

Parenchyma cells are living, vacuolated cells with thin primary cell walls. They form the major component of the ground tissue system.

Collenchyma

Collenchyma is a supporting tissue; it gives strength to plant parts where bending and flexibility are required and is commonly found in leaf stalks (petioles) and young stems. It is composed of strands of elongated cells (Fig. 6.20). Collenchyma cells have their cell walls thickened with additional quantities of cellulose. This extra thickening can either be in the corners of the cells, *angular collenchyma*, as in the stringy tissues of celery petioles, or along tangential walls, *lamellar collenchyma*, the latter more common in stems. Although thickened, the cell walls are flexible and can be stretched.

Collenchyma forms strands of living cells with differentially thickened primary walls that are strong but flexible and have a support role.

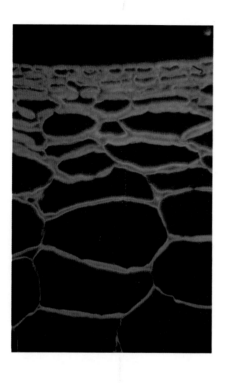

(a)

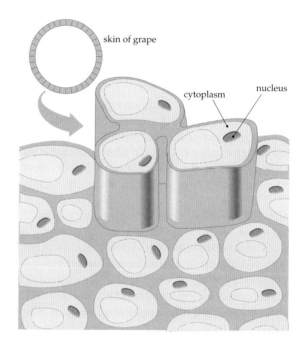

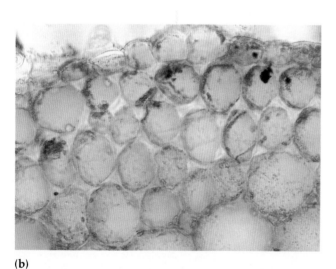

(b)

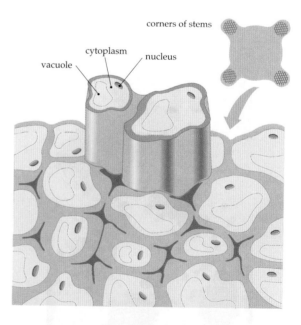

Fig. 6.20 Collenchyma cells form living support tissue in stems and fruits. **(a)** Lamellar collenchyma, with its wall thickening confined to the tangential walls, is responsible for the dermal system (skin) of fruit such as the grape. **(b)** Angular collenchyma, with its wall thickening uneven and confined to the corners, glistens silver in transverse sections of celery stem

Sclerenchyma

Sclerenchyma is also a supporting tissue but it imparts rigidity as well as strength. This rigidity is a property of the cell walls, which are impregnated with lignin. The lignin thickening is deposited in layers, which can be distinguished as concentric rings or lamellae in the cell wall. Because the cell wall is so thick, in mature sclerenchyma cells the protoplasm usually degenerates to leave a gap or lumen in the centre of the cell.

Sclerenchyma includes two cell types—fibres and sclereids (Fig. 6.21). **Fibres** are elongated cells with tapering end walls that overlap adjacent cells. They are often arranged in a continuous layer in stems or in longitudinal bundles associated with vascular tissue. **Sclereids**, or stone cells, are branched or more or less even-shaped, and play a role of protection as much as support. They form the hard tissue of seed coats and give pears and some other fruits their gritty texture.

> Sclerenchyma cells have rigid secondary walls with lignin deposited in concentric rings, forming elongated fibres or short sclereids.

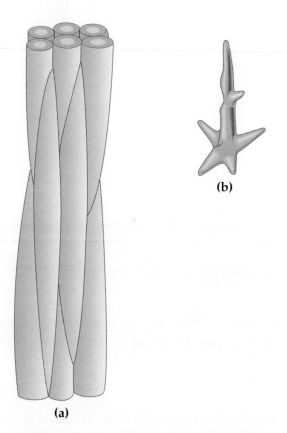

(b)

(a)

Fig. 6.21 Sclerenchyma cells form strengthening tissue. **(a)** Fibres are long tubular cells with secondarily thickened walls (see Fig. 6.15b). **(b)** Sclereids form irregularly shaped cells usually scattered within cortical tissue

Vascular tissue

The vascular system is composed of two tissues—xylem and phloem. **Xylem** transports water and nitrogenous compounds such as nitrate and ammonia ions from the point of uptake in roots to the point of loss in leaves (see Chapter 18). **Phloem** transports photosynthetic products, mostly sugars, from their place of production in leaves to their place of use or storage.

Xylem

Xylem is a complex tissue consisting of several cell types. Most distinctive are the water-conducting elements, vessels and tracheids (Fig. 6.22). In addition, sclerenchyma fibres, with a support role, and parenchyma cells, with a metabolic role, are usually prominent. Tracheids and vessels differ from most other cell types in that at maturity they are dead, consisting only of hollow cell walls without any cell contents. In **vessels**, which are typical of flowering plants, the end walls are broken down to allow unimpeded flow of water from one cell to another. **Tracheids**, which are water-conducting cells of all vascular plants, have intact end walls and water moves from cell to cell through special pores in the cell wall. These allow water movement from one cell to another. They often have an elaborate, overarching lip and are then referred to as **bordered pits** (Fig. 6.22).

Vessels and tracheids are produced by meristem divisions in the same way as other cells but differentiate early. They become elongated and develop a secondary wall heavily impregnated with lignin. The lignin thickening acts as an internal skeleton for the cell and prevents it from collapsing under the internal pressure developed as water is drawn through the plant. After lignin deposition is complete, the cell dies and the wall assumes its role in water transport.

The pattern of lignin thickening in xylem elements varies depending on their position within the plant. At the tips of growing shoots and roots, the first formed primary xylem, protoxylem, has lignin deposited in a series of very closely spaced transverse rings (annular thickening). Protoxylem cells need some flexibility to accommodate elongation of the tissues around them. As the protoxylem is stretched, the rings of lignin separate but can still maintain their support function. Eventually they are stretched too far and they break and cease functioning. Water conduction is then taken over by slightly later maturing primary xylem, the metaxylem. Because metaxylem elements are formed in slightly more mature regions of stem and root, they do not require the same tolerance to stretching as protoxylem. Early metaxylem elements have lignin laid down in a spiral;

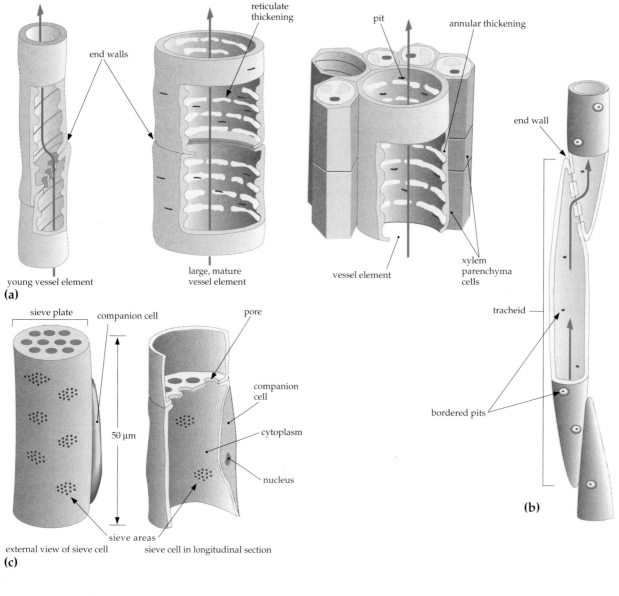

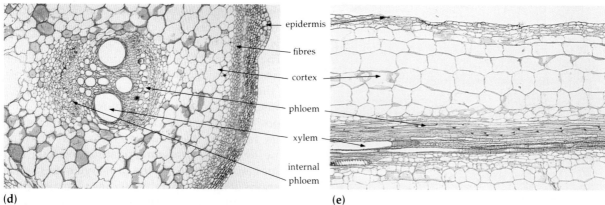

Fig. 6.22 Vascular tissue. **(a)** Xylem vessels have secondary thickening laid down in annular, coiled, or reticular patterns. The mature vessels generally have open ends, so that a series of vessels forms a tube. The walls are perforated by small clusters of pits. **(b)** Tracheids differ from vessels in having rows of bordered pits, which provide the means of water transfer from tracheid to tracheid as they have closed pointed ends. **(c)** Phloem: sieve and companion cells. **(d)** Transverse section of pumpkin stem, showing site of vascular strand central to epidermis and cortex but external to pith. **(e)** Longitudinal section of pumpkin stem, showing the elongated phloem sieve cells (green stained) with conspicuous red-stained sieve plates, and annular and spiral patterns of xylem secondary thickening (red-stained walls)

they look very much like a spring and have a reasonable tolerance to stretching. Later metaxylem elements have a reticulate (network) or mesh-like lignin ornamentation, and have a much more restricted ability to elongate. Xylem elements that mature in fully formed regions of the stem have no need for flexible walls and are provided with a uniform coating of lignin over the entire inner surface of the cell.

> In the vascular system, xylem is a complex tissue composed of tracheids, vessels, parenchyma and sclerenchyma. When mature, tracheids and vessels consist only of cell walls without any cell contents.

Phloem

Like xylem, phloem is also a complex tissue comprising several cell types. The cells through which photosynthetic products are transported are **sieve cells**, and these are usually associated with parenchyma cells and bundles of sclerenchyma fibres that prevent the sieve cells from being crushed. Sieve cells (Fig. 6.22) are quite unlike vessels and tracheids, their functional counterparts in the xylem. They have thin cell walls that consist only of cellulose, with the exception of some callose deposition around pits that connect adjacent cells by their ends. These pits look like the holes in a sieve, hence the name of the cells. Sieve cells are elongated and form a continuous conducting pathway throughout a plant.

Sieve cells are living, retaining a plasma membrane and some cell contents. In flowering plants, each sieve cell is associated with a specialised parenchyma cell—a **companion cell**. Although sieve cells and companion cells are the products of a single mother cell that undergoes longitudinal division, their development and functions are quite different. As a sieve cell matures, the nucleus, vacuole, membrane system and most of the cell organelles degenerate to leave a dense, granular cytoplasm containing energy-producing mitochondria. It is through these cells that active (energy-requiring) transport occurs. Companion cells retain all their cytoplasmic contents and have a well-developed endoplasmic reticulum with ribosomes and a large, dense nucleus. The function of the companion cell is to control the activity of the sieve cell. At sites where products are being loaded into and out of sieve cells, companion cells have extensive ingrowths of the cell wall, which increase the surface area for cell to cell interchange.

> Phloem tissue consists of sieve cells associated with parenchyma cells and bundles of sclerenchyma fibres. In flowering plants, sieve cells lack a nucleus but are associated with companion cells.

SUMMARY

- In multicellular organisms, cells are specialised for particular purposes and linked together into tissues. Tissues are groups of similar, differentiated cells and associated extracellular matrix, which together carry out a particular function.

- As a result of specialisation of function, differentiated cells rely, to a greater extent, on the extracellular environment for the provision of necessary materials, appropriate physical conditions and circulating hormones and growth factors.

- Animal cells are surrounded by an extracellular matrix that forms a loose and flexible lattice. The extracellular matrix contains proteins unique to each tissue and plays a vital role in regulating many aspects of cellular activity.

- Animal tissues are maintained by repair and/or replacement throughout life. Some tissues are regenerated by division of differentiated cells or by proliferation from a relatively unspecialised stem cell population. In other tissues, there is no turnover of cells; most 'permanent' cells repair themselves by turnover of cell components.

- Cells in animal tissues are bound together by junctions between plasma membranes of adjacent cells. Occluding (tight) junctions form an impenetrable seal between cells and restrict lateral movement of membrane proteins. Anchoring junctions provide mechanical support and are important in developmental processes. Communicating (gap) junctions provide direct communication between adjacent cells.

- There are four major types of animal tissues. Epithelia function as interfaces between biological compartments and with the environment, and in secretion. Connective tissues, composed of cells, fibres and fluid matrix, provide structural, metabolic and defensive support for other tissues. Striated muscle cells contain a highly organised array of actin and myosin filaments; myofilaments in smooth muscle cells are less regularly arranged. Nerve cells are specialised to conduct signals rapidly and precisely throughout the body; they are unable to divide and are supported by glial cells.

- Unlike animal cells, plant cells are characterised by a cell wall, which limits cell size and shape and is also important in metabolic processes. Primary cell walls comprise cellulose microfibrils embedded in and interconnected to a matrix containing non-cellulosic polysaccharides, pectins and proteins. Cellulose microfibrils give cell walls strength and their orientation determines the direction of cell expansion. Lignin is characteristic of many secondary walls.

- Plant cell walls are penetrated by plasma-membrane-lined channels, plasmodesmata, which provide a means of direct communication and transport between cells.

- Vascular plants have three major tissue systems: the dermal (covering), ground (supporting) and vascular (transport) systems.

- The major tissue of the dermal system is the epidermis, which lays down cuticle. Some epidermal cells are modified as guard cells for opening and closing stomata, hairs, trichomes and glands.

- The ground tissue system consists of storage and structural tissues. Parenchyma cells are living, vacuolated cells with thin primary walls. Collenchyma forms strands of living cells with differentially thickened primary walls made of cellulose and pectins that are strong but flexible. Sclerenchyma cells have rigid secondary walls with lignin deposited in concentric rings, forming elongated fibres or short sclereids.

- In the vascular system, xylem is a complex tissue composed of tracheids, vessels, parenchyma and sclerenchyma. When mature, tracheids and vessels lack cell contents, and consist only of cell walls. Phloem tissue consists of sieve cells associated with parenchyma cells and bundles of sclerenchyma fibres. In flowering plants, sieve cells lack a nucleus but are associated with companion cells.

QUESTIONS

1. (a) What is a tissue? (b) How do the functions of cells in tissues differ from the functions of individual cells?

2. What are the significant differences between plant and animal tissues?

3. Contrast the properties and functions of the extracellular matrix of animal tissues and plant cell walls.

4. What are stem cells? In what tissues are they found? How are they involved in the maintenance of animal tissues?

5. Describe the types of connections that occur between animal cells. Use examples that illustrate the function of each type.

6. What are the principal features of (a) epithelia, (b) connective tissue, (c) muscle and (d) nervous tissue?

7. What are the distinguishing features of parenchyma, collenchyma and sclerenchyma tissues in plants?

8. What is the dermal tissue system of plants? What is the role of the epidermis? What are trichomes and what are their functions?

9. What types of cells compose xylem tissue? What are the structural and functional features of each?

10. Distinguish between phloem sieve cells and companion cells. Are these cells living or dead when mature?

RESPONDING TO SIGNALS

All cells constantly gather information about their surroundings and use it to control their activities. In slime moulds, for example, chemical signals passing between cells co-ordinate the aggregation of free-living cells into a colony for feeding and reproduction (chemotaxis; Fig. 7.1). In the cellular slime mould *Dictyostelium discoideum*, starving amoebae produce and release cyclic adenosine monophosphate (cAMP), which stimulates neighbouring amoebae to do the same. The extracellular cAMP signals trigger cell aggregation. The amoebae change shape, become more motile and move towards the cAMP source.

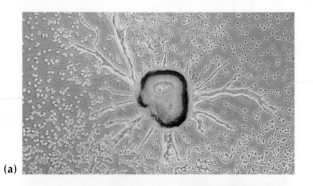

(a)

(b)

Fig. 7.1 (a) Individual cells of a slime mould stream towards one another and aggregate to form a motile 'slug'. (b) The 'slug' forms a stalked fruiting body for spore dispersal

RECEIVING SIGNALS

Stimuli that act as signals for cells may be physical (e.g. light or heat) or chemical (e.g. food or hormones). These stimuli may come from the external environment or from other parts of the body of an organism. In multicellular organisms, certain cells are highly specialised to receive particular external stimuli. Other cells are specialised to send or receive internal signals to co-ordinate functions between different parts of their bodies. Most animals use a variety of internal chemical messengers in an elaborate communications network involving nerves and hormones to control the activities of body cells. Although much is known of the ways that animal cells and micro-organisms receive signals, little is known of these mechanisms in plants.

A **signal** may be a particular chemical or physical stimulus. In order to respond, cells must be able to detect the signal. To do this, cells must contain the right **receptor**, which is a molecule that undergoes some kind of change as a result of interaction with the incoming signal (Fig. 7.2). Examples of receptors are found in the specialised nerve cells that detect stimuli such as light, **photoreceptors**; heat, **thermoreceptors**; and mechanical pressure or stretch, **mechanoreceptors**. Photoreceptors absorb light of a particular wavelength. **Chemoreceptors** bind with specific signal molecules, **ligands**, which have the appropriate molecular structure to interact with the

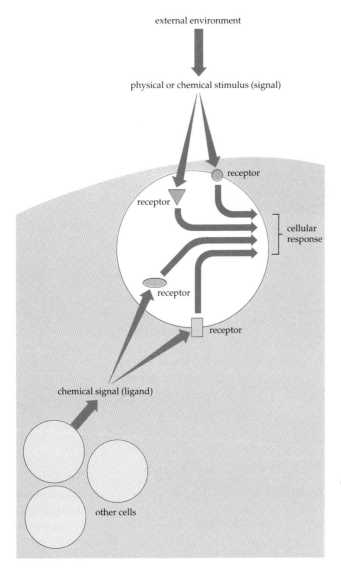

Fig. 7.2 A cell can respond to signals from the external environment and from other cells. Signals are received by specific receptors located on the surface of or inside the cell

receptor. In the case of light or chemical stimuli, receptors are known to be specific proteins, often embedded in the plasma membrane.

Signalling pathways in cells lead ultimately to the regulation of proteins involved in any of a variety of cellular activities including cell division, differentiation and growth, metabolism, secretion and motility. In this chapter, we will consider the different signals that cells receive, and the mechanisms by which they process these signals to produce a response.

> Cells use a range of chemical and physical stimuli to regulate their activities. They have receptors, special proteins often embedded in the plasma membrane, which can receive particular signals.

Chemical stimuli

Cells receive chemical stimuli by direct interaction between the ligand and a specific receptor located in or on the surface of the responding cell. Chemical signals may be lipid-soluble, water-soluble, or bound to a surface, such as the plasma membrane of another cell or the extracellular matrix. Only lipid-soluble molecules are able to pass freely through the cell membrane to interact with receptors located inside the cell. Water-soluble and surface-bound signals (and a few lipid-soluble signals) interact with receptors located at the cell surface.

Lipid-soluble chemical signals

Steroid hormones and thyroid hormones are chemical signals for which the receptors are intracellular; they are lipid-soluble and pass freely into cells (Fig. 7.3a).

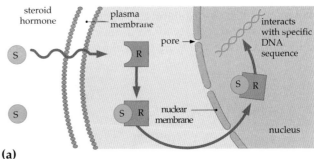

(a)

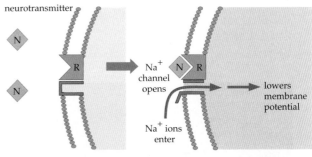

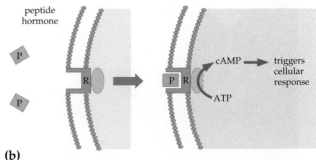

(b)

Fig. 7.3 **(a)** Steroid hormones pass through the plasma membrane, combine with a cytoplasmic receptor, and then pass into the nucleus to interact with DNA.
(b) Peptide hormones and neurotransmitters cannot pass through the plasma membrane; they interact with a receptor on the surface

Receptors for steroids are water-soluble proteins in the cytoplasm or nucleus of responsive cells. When binding occurs in the cytoplasm, the ligand–receptor complex passes into the nucleus through pores in the nuclear membrane. In the process of binding to a steroid molecule, the receptor changes its conformation in such a way that its affinity for specific nucleotide sequences in DNA increases. The ligand–receptor complex binds to DNA, altering the expression of adjacent genes. The subsequent increase (or occasionally decrease) in production of protein initiates a cellular response.

Water-soluble chemical signals

Most extracellular signal molecules are water-soluble and are thus unable to pass freely through the plasma membrane. Water-soluble signals are detected by receptors located on the cell surface (Fig. 7.3b). Like most other membrane proteins, these receptors are usually anchored in the cell membrane by one or more hydrophobic regions (Chapter 3). In most animals, water-soluble signals include **neurotransmitters**, which are released from nerve endings to act on other nerve cells, muscle cells or glands (Chapter 26); some local hormones such as growth factors, which act on nearby responsive cells; and some endocrine hormones, which are distributed through the bloodstream to cells throughout the body (Chapter 25).

Unlike steroids, water-soluble hormones do not regulate protein production by DNA directly. They bind to surface receptors with high affinity. The receptors may be coupled directly to ion channels in the plasma membrane (Fig. 7.3b), linked to a specific type of protein, or may be associated with a specific enzyme, regulating catalytic activity. By these means, the signal is relayed into the cell, where it usually triggers one or more intracellular signals that produce the cellular response.

Surface-bound chemical signals

Some cells respond to molecules located either on the surface of other cells or bound to the extracellular matrix (Chapter 6). As the signal molecule binds to a receptor on the responding cell's surface, there is adhesion between the receptor-bearing cell and the ligand-bearing surface. For this reason, the receptors for such ligands are also known as *cell adhesion molecules*. Surface-bound ligands play important roles in development (Chapter 15) and in immune responses involving killer T cells (Chapter 23).

> Lipid-soluble chemical signals, such as steroid hormones, can enter cells freely and interact with intracellular receptors. Water-soluble and surface-bound ligands interact with receptors at the cell surface.

Physical stimuli

Cells can respond not only to chemical signals but also to a great variety of physical stimuli, including light, temperature, pressure and stretch, and electric and magnetic fields.

Light

Light-sensitive responses involve photoreceptors, which are proteins containing a pigmented chemical group, a **chromophore**. A chromophore absorbs light of a particular wavelength only.

Light regulates many aspects of plant growth and development. Plants have several types of photoreceptors but the most important of these is **phytochrome**, which can exist in two interconvertible forms (Fig. 7.4). Phytochrome is synthesised and accumulates in the inactive form (P_r). Absorption of a quantum of red light converts the inactive form to an active form (P_{fr}). The active form can be reconverted rapidly to the inactive form by the absorption of a quantum of far red light or slowly and spontaneously in the dark. Most physiological responses are produced by the active form of phytochrome (P_{fr}) by a number of different mechanisms. For example, in fern spores, activated phytochrome triggers a transient influx of Ca^{2+}, which increases internal levels of Ca^{2+} and stimulates germination. Activated phytochrome also controls the transcription of many light-activated genes (Chapter 24).

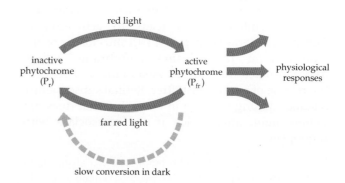

Fig. 7.4 In plants, most light-sensitive responses involve the active form of the pigmented protein, phytochrome. The interconversion of the active and inactive forms depends on absorption of red or far red light

In animals, the photoreceptor whose structure and function is best understood is **rhodopsin**, which is found in rod cells in the retina of the vertebrate eye. Rhodopsin is associated with the chromophore, retinal, which undergoes a chemical rearrangement after absorbing light (Fig. 7.5). The new form of retinal

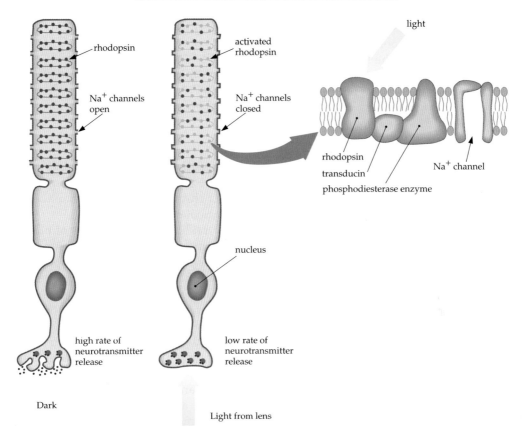

Fig. 7.5 Response to light of a rod photoreceptor cell in the retina of the human eye. Activated rhodopsin leads to closure of Na⁺ channels, increasing the membrane potential. This reduces the release of neurotransmitter from the rod cell

then dissociates itself from rhodopsin, which alters its shape, triggering an interaction with a protein (the G-protein, transducin; p. 129) on the cytoplasmic side of the membrane. This leads to activation of phosphodiesterase enzymes (PDE) and closure of Na⁺ channels in the plasma membrane, which hyperpolarises the rod cell membrane and reduces the release of neurotransmitter. Retinal subsequently resumes its original form in a reaction that does not require light, after which it can reassociate with rhodopsin.

Photoreceptors contain chromophores that absorb light of a particular wavelength and trigger a cellular response.

Mechanical stimuli

Many microbes and certain cells in animals and plants respond rapidly to physical contact or stretching by generating an electrical signal (action potential, Chapter 26). There are many different types of mechanosensory cells, each modified to respond best to a particular type of stimulus. However, the mechanisms involved in mechanosensitivity are not well understood. The simplest model is that distortion of the receptor cell membrane opens or closes ion channels, altering the passage of cations such as Na⁺.

Specialised cells in the leaves of the sensitive plant, *Mimosa pudica* (p. 70), respond to touch by producing an electrical signal that triggers the opening of K⁺ channels in cells at the base of the leaf stalk. Potassium ions leave these cells, dragging water by osmosis, which leads to loss of turgor. This causes the leaves to fold and droop. The closure of the Venus flytrap (Fig. 7.6) when an insect touches sensory hairs, involves a similar mechanism that causes the leaf to snap shut, trapping the insect within.

Fig. 7.6 The Venus fly trap responds to touch by producing an electric signal that leads to loss of turgor and closing of the trap

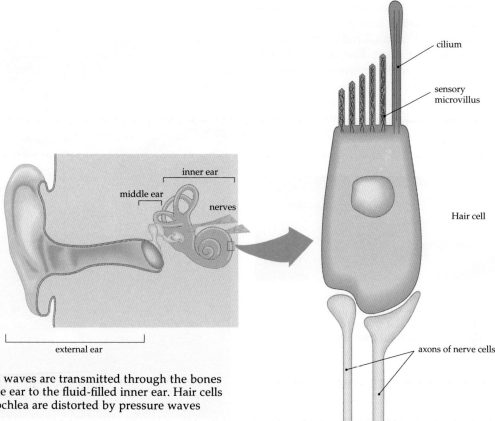

Fig. 7.7 Sound waves are transmitted through the bones of the middle ear to the fluid-filled inner ear. Hair cells lining the cochlea are distorted by pressure waves

In animals, mechanoreceptors range from simple, naked nerve endings that respond to force, as in skin pressure receptors, to the highly specialised non-neuronal hair cells of vertebrates (Figs 7.7, 7.8). In humans, receptors sensitive to stretch are used to monitor everything from blood pressure to bladder distension, from balance to the position of limbs.

The sense of hearing is a specialised form of detection of a mechanical stimulus due to changes in air pressure. In mammals, pressure waves set up resonant vibrations in the eardrum, which are transmitted by a chain of bones in the middle ear

to the fluid within the inner ear (Fig. 7.7). Cilia protruding from hair cells lining the cochlea of the inner ear are distorted by the pressure waves and respond by increasing or decreasing the rate of nerve impulses passing along the auditory nerve. The cilia-bearing hair cells thus translate a mechanical stimulus into electrical signals that can be relayed by the auditory nerves to the brain.

Mechanoreceptors respond to physical contact or stretch, usually by opening ion channels in the plasma membrane.

Temperature

Temperature affects all biological activity in a general way because it determines the rates of chemical reactions and the activities of enzymes. It is therefore not surprising that many animals have specialised heat-sensing capabilities. Mammals, such as ourselves, have cold-sensitive nerve cells, which respond to decreasing temperature, and warm-sensitive nerve cells, which respond to increasing temperature. The response in both cases is an increased frequency of nerve impulses. Thermoreceptors may be temperature-sensitive channel proteins or active carriers (Na^+–K^+ ATPase). In some organisms, a temperature-induced change in local fluidity of the cell membrane may regulate the activity of the membrane proteins.

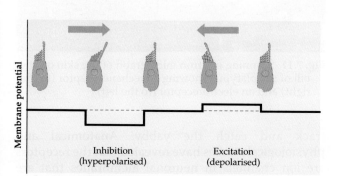

Fig. 7.8 Bending towards the long cilium depolarises the hair cell, increasing the rate of action potentials in the sensory nerve (excitation). Bending in the other direction causes the opposite effect (inhibition)

Blood-sucking insects, ticks, leeches and some snakes, such as rattlesnakes and many Australian pythons, detect infra-red thermal radiation to locate their endothermic host or prey. The snakes do this by means of warm-sensitive thermoreceptors located in pits on their heads (Fig. 7.9). The high sensitivity of these receptors and their precise location on either side of the head, provide directional information and the snakes are able to strike accurately at prey, even in total darkness.

Fig. 7.9 Thermoreceptors located in pits on either side of the lower jaw of the green Australian python, *Chondropython viridis*, allow the snake to strike accurately at prey, even in total darkness

Thermosensitive nerve cells respond to temperature by a change in frequency of nerve impulses.

Electric and magnetic fields

Sharks, skates and rays rely for navigation on electric fields generated by the magnetic field of the earth as they move through the water (Fig. 7.10). Other animals, such as homing pigeons and honey bees, are believed to use tiny particles of magnetic material (magnetite, Fe_3O_4) to navigate using the earth's magnetic field. Certain bacteria also contain magnetite crystals, which physically align the axis of the cell in a north–south direction, like a tiny compass needle. Such *magnetotactic* bacteria in the Northern Hemisphere move towards the north magnetic pole; in the Southern Hemisphere, they move southwards. From most points on the earth's surface, the shortest distance to either pole is *through* the earth; the bacteria therefore move downwards into the sediment, which, for them, is a good place to be.

Recent research has revealed that fishes are not the only animals that can detect electric fields: the Australian monotremes, the platypus and echidna, both have specialised nerve endings that can detect minute electric fields. These electroreceptors, located in the bill of the platypus and snout of the echidna, assist these animals to catch their prey (Fig. 7.11). The freshwater crustaceans (yabbies), which form an important part of the platypus's diet, emit a small electric discharge when their tail muscles contract as they attempt to escape. The platypus, which dives with its eyes closed, uses these electrical signals to

Fig. 7.10 Specialised electroreceptors located in pits on the underside of the head of a shark are sufficiently sensitive to detect electric fields generated by the earth

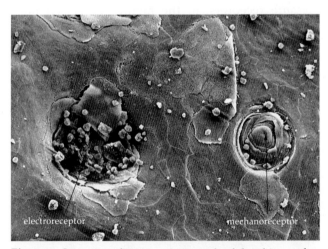

Fig. 7.11 Scanning electron micrograph of the skin on the bill of the platypus showing a mechanoreceptor (to the right) and an electroreceptor (to the left)

track and catch the yabby. Anatomical and physiological studies have revealed that the receptors are ion channels in neuronal membranes that are sensitive to changes in the electric field across the membrane.

Some animals are able to detect electric and magnetic fields by means of modified ion channels.

SIGNAL PROCESSING

Once an extracellular signal has interacted with a receptor, it must be processed into information that produces the appropriate cellular response. Signal processing may be direct or may involve one or more intracellular molecular steps (such as G-proteins and second messengers), which pass the signal from the receptor to specific effector molecules that trigger the cell's response. Effector proteins are regulated by intracellular signalling, which frequently involves phosphorylation by specific kinases or binding of Ca^{2+}.

Direct receptor-mediated responses

The simplest signalling pathways are those in which the activated receptor acts directly to produce a cellular response. This occurs, for example, with steroid hormones, where the hormone–receptor complex binds to a specific region of DNA and alters the rate of synthesis of a particular protein. For example, during egg production in hens, oestrogen stimulates the synthesis of the protein ovalbumin in the oviduct.

Steroid hormones can also produce delayed secondary responses in a target cell. These occur when the protein produced as a result of the primary ligand–receptor interaction binds to a different region of DNA to activate another gene. The same steroid can regulate different genes in different target cells. For example, the male steroid hormone, testosterone, is responsible for the development of the various male secondary sexual characteristics, such as deepening of the voice and beard growth. Testosterone always binds to the same receptor protein but produces different responses in different target cells.

Other forms of direct receptor-mediated response involve regulation of dynamic changes of cytoskeletal proteins on the inner face of the plasma membrane by activation of cell adhesion receptors (Chapter 3), and channel-linked receptors. When activated, channel-linked receptors transiently open or close, regulating the movement of specific ions such as K^+ or Ca^{2+} across the plasma membrane. These are ligand-gated channels (Chapter 4). This alters the balance of ions across the membrane and therefore changes the electrochemical membrane potential (p. 69). In excitable cells, the change in potential may be sufficient to affect other voltage-gated channels in plasma membranes (p. 69) through which ions such as Na^+ or Ca^{2+} pass and cause hyperpolarisation (inhibition) or depolarisation (excitation). For example, acetylcholine is a neurotransmitter released by some nerves of vertebrates which binds to several different receptor proteins. When acetylcholine released from one nerve cell binds to a nicotinic receptor on the membrane of the next nerve cell, a channel of 0.7 nm diameter opens in the post-synaptic membrane, allowing Na^+ to enter the cell and lowering its membrane potential (as in Fig. 7.3b). If the depolarisation is sufficient, it may initiate an action potential in the responding nerve cell.

> The simplest mechanism of intracellular signal processing involves an activated receptor acting directly to produce a cellular response.

G-protein-linked receptors

Many signalling pathways require the action of intermediate proteins, most commonly **G-proteins** (guanosine triphosphate [GTP] binding regulatory proteins). The intracellular pathway leading to the aggregation of slime moulds involves a G-protein that changes cell shape and movement (p. 123).

G-proteins are coupled to receptors that have seven transmembrane domains. G-protein-linked receptors indirectly alter the activity of an ion channel or intracellular enzyme (Figs 7.12, 7.13). They usually do this by altering the concentration of small molecules, such as cAMP or Ca^{2+}, known as second messengers. For example, the photoreceptor rhodopsin is a G-protein-coupled receptor that brings about the closure of Na^+ channels in rod cell membranes by activating the enzyme cGMP phosphodiesterase, thereby decreasing cGMP levels in the cell.

Second messengers

A common feature of G-protein-linked receptors is that their activity changes the concentration of one or more small intracellular signalling molecules (Fig. 7.12). These are known as **second messengers** since they are released in response to extracellular signals (the first messengers). We will consider three of these: the cyclic nucleotides cAMP and cGMP, and Ca^{2+}.

Cyclic AMP was the first of the intracellular second messengers to be identified. Cyclic AMP is produced from ATP by the membrane-associated enzyme, adenylate cyclase, and degraded to AMP by cAMP phosphodiesterase (Fig. 7.12). Cyclic AMP concentration can be increased either by inhibition of phosphodiesterase or, more commonly, by activation of adenylate cyclase, the usual target of regulation by G-proteins. The increase in cAMP concentration leads to phosphorylation of a variety of target proteins by a specific protein kinase enzyme, a process that can regulate their activity.

Cyclic GMP is also a second messenger in some cells (Fig. 7.12). As with cAMP, the levels of cGMP may be regulated by external stimuli and many cells contain a protein kinase whose activity depends on cGMP.

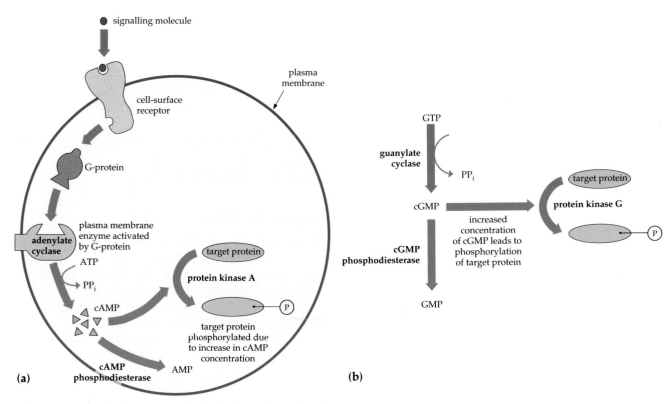

Fig. 7.12 Major pathways by which G-protein-linked cell surface receptors generate intracellular messengers. **(a)** A signalling molecule binds to a cell-surface receptor causing it to bind to a G-protein. The G-protein then activates a plasma membrane enzyme (adenylate cyclase), which produces the second messenger, cAMP. Increase in cAMP concentration leads to phosphorylation of a target protein by a specific enzyme, a protein kinase. cAMP is degraded to AMP by cAMP phosphodiesterase. **(b)** Cyclic GMP is also a second messenger, which can lead to phosphorylation of a target protein. However, the activity of the enzyme guanylate cyclase does not directly involve a G-protein as does adenylate cyclase

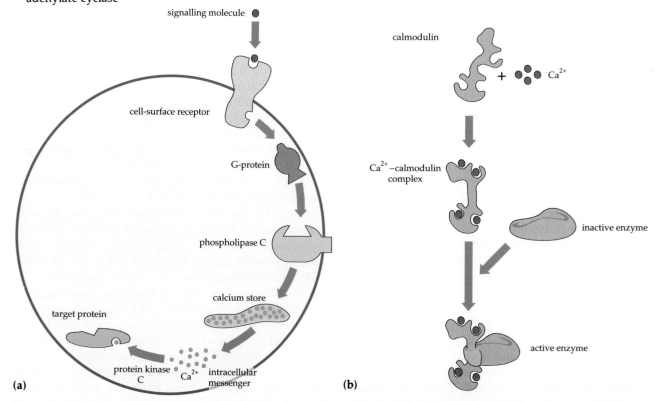

Fig. 7.13 Ca^{2+} acts in cells as a second messenger. Ca^{2+} regulates target proteins either **(a)** directly or **(b)** by means of an intracellular Ca^{2+} receptor, commonly calmodulin. A calmodulin molecule has four Ca^{2+} binding sites. On binding with Ca^{2+}, calmodulin undergoes a conformational change, which allows it to interact with and activate other proteins (enzymes)

Some second messengers are activated directly by ligand binding without the mediation of any G-proteins. The enzyme guanylate cyclase, which synthesises cGMP from GTP, is found in both a soluble and a membrane-associated form. The membrane-associated cyclase appears to be the signalling region of a receptor for some peptide hormones.

Calcium also acts in cells as a second messenger. It may be present in soluble form as free Ca^{2+}, or in bound form associated with specific proteins. In cells, the concentration of free Ca^{2+} is regulated by controlling the opening and closing of membrane channels, which allow Ca^{2+} to enter the cell or be released from intracellular stores.

In a cell, soluble Ca^{2+} is capable of regulating the activity of a variety of proteins, such as protein kinase C, by binding to them (Fig. 7.13). In general, Ca^{2+} regulates target proteins either directly (e.g. protein kinase C), or by means of an intracellular Ca^{2+} receptor, most commonly calmodulin. The Ca^{2+}–calmodulin complex regulates protein activity either directly or through the activation of a specific Ca^{2+}–calmodulin-dependent protein kinase, which has the ability to phosphorylate a variety of target proteins.

Most of the calcium in a cell is either bound to calcium-binding proteins, or stored inside the endoplasmic reticulum or mitochondria. The Ca^{2+} in these stores can be drawn upon in response to a stimulus (Fig. 7.13) and external Ca^{2+} may be drawn into the cell through transmembrane channels. The high binding and storage capacity of the cell for calcium means that concentration increases can be localised to particular regions of the cell. The Ca^{2+} does not diffuse freely to other parts of the cell because it is rapidly removed from the cytoplasm back into the intracellular calcium stores.

Phosphorylating proteins

The mechanism of action of intracellular second messengers in many signalling pathways is to activate protein phosphorylation. This is achieved by one of several protein kinases specific to the particular second messenger. The protein kinases operate by phosphorylating the amino acids serine and threonine in target proteins. The intracellular signalling regions of some receptors possess protein kinase activity, which enables phosphorylation of both the receptor itself and other proteins, adding phosphate groups to the amino acid tyrosine in the polypeptide chain of the protein. In many cases, phosphorylation causes conformational changes in the protein that are accompanied by changes in activity. The active form of the target protein can be either the phosphorylated or dephosphorylated form in different situations.

If protein kinases were continuously active in cells, all target proteins in cells would be permanently phosphorylated and their activities would not change. Cells therefore need enzymes to remove the phosphates from proteins, a process usually carried out by specific phosphatases, the activity of which may also be regulated by extracellular signals. Recent research has shown that the intracellular signalling region of some receptors is a phosphatase that removes phosphate from the amino acid tyrosine in target proteins.

> Intracellular processing of signals may involve activation of G-proteins, leading to changes in the concentrations of second messenger molecules such as cAMP, cGMP and Ca^{2+}. Second messengers act by phosphorylating or dephosphorylating proteins.

Catalytic receptors

Some receptors consist of a single polypeptide having both an external ligand-binding region and an internal signalling region that exhibits tyrosine kinase enzyme activity (Fig. 7.14), a property that enables such receptors to attach phosphate groups to the

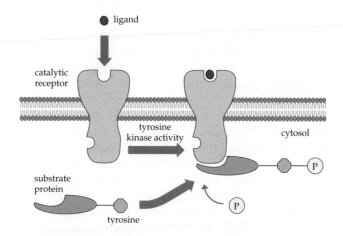

Fig. 7.14 Tyrosine kinase activity is triggered when the ligand binds to a catalytic receptor. The activated receptor phosphorylates the tyrosine side chain of a substrate protein, altering the activity of the protein

amino acid tyrosine where it occurs in proteins. The catalytic activity of the receptor is triggered when the ligand binds to it. Catalytic receptors in mammals include the receptor for epidermal growth factor, which stimulates epidermal and many other cell types to divide, and the insulin receptor, which regulates glucose metabolism. Other kinds of catalytic receptor include those with guanylate cyclase activity (Fig. 7.12) and protein tyrosine phosphatase activity.

> Intracellular processing of a signal can involve a change in the catalytic activity (phosphorylation of amino acids) of the intracellular region of a receptor protein.

PRODUCING A RESPONSE

After the receipt of an input signal, the molecular events that are set in train are used by cells to generate a particular response. The result is a regulated change in cellular activities such as metabolism, growth, phagocytosis, secretion, cell division, movement or differentiation. Some dramatic changes occur as the result of a single stimulus, while other responses require stimulation of several receptors. Also, cells do not always produce the same response to a particular stimulus. A response may be affected by recent stimulation events, or by other input signals for which the cell has appropriate receptors.

Signal amplification

How is it that the male *Bombyx mori* moth can produce a response to a single molecule of the sex pheromone, bombykol? (Sex pheromones are volatile chemical attractants emitted during courtship; Chapter 25.) The molecular events after activation of certain receptors can amplify (increase the strength of) the signal, thereby increasing the sensitivity of the system. Binding of a single molecule of bombykol to a receptor cell on the antennae of this moth causes a response that is amplified by intracellular molecular pathways. Photoreceptor cells in the vertebrate retina that respond to absorption of a single photon of light are another example. In both of these cases, activation of a single receptor leads to the closing or opening of many, perhaps thousands, of membrane ion channels.

How amplification occurs is well illustrated by rhodopsin (Fig. 7.15). Absorption of one photon of light can convert rhodopsin to its excited form; in this form rhodopsin can bind to the G-protein, transducin. After a molecular exchange, transducin dissociates from the receptor and, through a second messenger cGMP system, closes off membrane Na^+ channels. A single excited rhodopsin molecule can activate many transducin molecules before being converted back to its resting form. This causes a relatively large change in concentration of the second messenger cGMP and the closure of not one but many Na^+ channels.

> The sensitivity of some sensory pathways can be increased by signal amplification. For example, activation of a single receptor molecule can cause the opening or closing of many membrane ion channels.

Convergent pathways

Convergent pathways are those in which signals from separate receptors for different stimuli regulate the same intracellular event. This can allow the effects

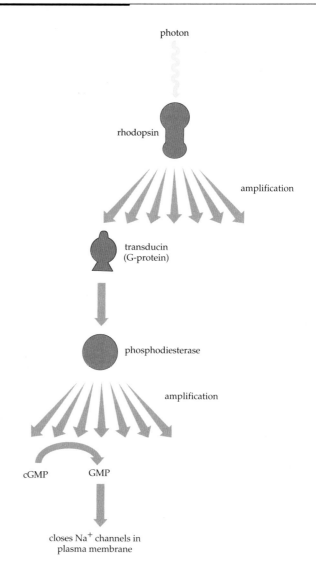

Fig. 7.15 Amplification of an extracellular signal. Absorption of a single photon of light by rhodopsin results in the activation of many transducin molecules, each of which activates the enzyme phosphodiesterase. Each molecule of phosphodiesterase catalyses the production of many GMP molecules, which then close Na^+ channels in the plasma membrane

of two different extracellular signals to be integrated. An example of two convergent stimulatory pathways is seen in liver cells responding to the hormones adrenaline and glucagon (Fig. 7.16). These hormones bind to different receptors, each of which stimulates a G-protein to activate adenylate cyclase. The resulting increase in cAMP levels stimulates glycogen breakdown. At hormone concentrations that elicit less than maximal responses, the effect of both hormones is greater than it is for either hormone alone. However, if either hormone is present at a concentration high enough to evoke the maximal response (in terms of glycogen breakdown), presence of the second hormone has no additional effect.

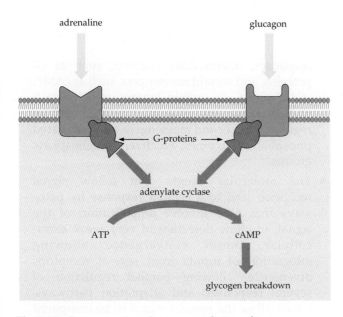

Fig. 7.16 Convergent pathways regulating glycogen metabolism

In other cases where an inhibitory and a stimulatory signalling pathway converge, the two stimuli oppose each other and each can reduce the magnitude of the response to the other. When convergent stimulatory and inhibitory pathways are involved, the response can be different for different signal intensities. For example, many unicellular organisms move towards or away from light depending on light intensity. This can result from convergent pathways starting from different photoreceptors, one of which leads to movement towards light and the other to movement away from light.

Divergent pathways

Divergent signalling pathways allow a number of cellular responses to be regulated in parallel by the same stimulus. This involves intracellular signalling molecules, such as second messengers like cAMP, modulating the activities of a number of different cellular proteins. For example, neutrophils (a type of white blood cell) have receptors for specific protein breakdown products released by bacteria. Ligand-binding to these receptors regulates a variety of responses, including changes in cell adhesion, shape, motility, oxygen uptake and release of toxic oxygen metabolites (e.g. hydrogen peroxide) and release of proteases (enzymes that degrade proteins).

> Convergent signalling pathways in cells involve integration of signals from separate receptors to regulate the same intracellular response pathways. Divergent pathways allow regulation of different cellular responses by the same stimulus.

Excitation and adaptation

For many extracellular signals, the responding cell is sensitive to *relative* differences in the level of the signal rather than to the *absolute* level. In bright light, more light is reflected from the *black* letters on this page than is reflected from the *white* background if the light is very dim. Yet, in both situations, the letters are perceived as black on a white background because the *relative* brightness of the page and the letters on it remains the same. If the book is taken from bright sunlight into a poorly lit room, the brightness of both the print and the paper decreases dramatically relative to what it was, and reading is difficult. It takes a few seconds or minutes for the eyes to adapt to the change in absolute stimulus intensity. This process of adaptation allows the photoreceptor cells in the retina to respond to the intensity of light from the page relative to the overall background intensity.

The process of photoreceptor adaptation has two aspects. Firstly, the cell responds similarly to relative differences in the signal (e.g. two-fold differences) over a wide range of absolute stimulus intensity. In the retina, each photoreceptor cell measures light intensity relative to some average of the light intensity over the whole retina. Relative to that average, the print is dark and the page is bright. Secondly, the cell responds initially to a change in signal intensity, but this response declines over time as the cell adapts to the new absolute stimulus levels. Until the retina adapts relative to that average, the perceived signal intensity is low everywhere when the book is taken from bright sunlight into dim room light. To behave this way, photoreceptor cells must compare, moment by moment, the signal that they are presently receiving with the background level.

The implication from this example is that, in photoreception, extracellular stimuli are usually transduced into two intracellular processes. One is *excitation*, a rapid measure of the intensity of the present stimulus. The second is *adaptation*, a slower estimate of the average background stimulus in the recent past. Both processes are activated by the same receptor so that, at some point, the signalling pathways for excitation and adaptation must diverge. The two pathways must also converge at some point because the present stimulus, measured by the excitation process, is compared with past stimuli, in the adaptation process.

> Excitation pathways activate a cellular response while slower adaptation pathways inhibit excitation. Responding cells are thus sensitive to relative differences in stimulus intensity.

SUMMARY

- The activities of cells are regulated by extracellular signals for which cells have specific receptors. Receptors that interact with light or chemical stimuli are proteins, usually embedded in plasma membranes.

- Extracellular chemical signals may be lipid-soluble, water-soluble, or surface-bound. Lipid-soluble signals such as steroid hormones pass freely across the plasma membrane and bind to intracellular receptors. Water-soluble and surface-bound ligands interact with membrane protein receptors at the cell surface.

- Certain cells also respond to physical stimuli. Photoreceptors contain chromophores, which respond to a particular wavelength of light. Mechanoreceptors respond to pressure and stretch by altering ion flow across the cell membrane. Thermoreceptors respond to temperature change by altered nervous activity. Electric and magnetic fields are detected by modified ion channels.

- In eukaryotic cells, intracellular signal processing may involve direct receptor-mediated responses; intermediate proteins, such as G-proteins, and second messengers, such as cAMP, cGMP and Ca^{2+}; protein kinases, which attach phosphate groups, that may or may not be regulated by second messengers; and associated phosphatase enzymes, which remove phosphate groups.

- The molecular events that follow signal reception lead to a cellular response by pathways that may involve amplification of the signal, to allow detection of very weak extracellular stimuli; convergence, allowing integration of inputs from several receptors; divergence, allowing parallel regulation of several responses; and adaptation pathways, which allow the present signal to be compared with recent signals, so that the cell can respond to a relative change in the stimulus.

- Responses that are regulated by these signalling pathways include all of the major activities of the cell, including metabolism and growth, division, differentiation, secretion and motility.

QUESTIONS

1. Define the terms signal and receptor in relation to cellular responsiveness. List the types of signals to which cells can respond.

2. How do photoreceptors respond to light? What is the function of (a) phytochrome in plants and (b) rhodopsin in the retina of the eye?

3. What types of receptors are involved in hearing? Describe the way that sensory cells of the ear are involved in hearing.

4. What are the three main types of chemical signals to which cells respond? How does receptor interaction occur in each case?

5. Draw a simple diagram illustrating the differences between the signalling pathways of direct receptor-mediated receptors, G-protein-linked receptors and catalytic receptors.

6. What is a second messenger? Use an example to show how they are involved in intracellular signalling pathways.

7. Describe the role of Ca^{2+} in regulating the activity of cellular proteins.

8. What is meant by signal amplification? Give an example.

9. Discuss the possible advantages of divergent signalling pathways in the response of a single-celled organism to an external stimulus.

10. Cells continuously receive many different stimuli from their environment. By what mechanisms do they integrate all this information to produce the most appropriate response?

CELL DIVISION

All cells are derived from other cells by cell division. Single-celled organisms, such as bacteria, protozoa and many algae, simply divide into two similar individuals that grow and divide again. The cells of multicellular organisms arise from a single cell as a result of cell division, followed by differentiation, which creates the various tissues that make up the mature organism (Chapter 6). Cell division occurs in most tissues to replace ageing or damaged cells. The processes of cell division must be strictly controlled for accurate transfer of DNA to daughter cells. The rate of division of cells must be regulated for the correct formation and subsequent maintenance of tissues. Also, the size of a mature organism, be it ant or elephant, is usually a function of cell number.

The mechanisms that control cell division in multicellular organisms are major areas of research in biology and medicine. Many aspects of abnormal development, disease, healing and ageing are related to changes in the abilities of cells to divide.

When a cell divides, its cytoplasm and copies of its DNA are transmitted to each daughter cell. DNA molecules pass to daughter cells in an organised way so that each cell receives a full complement of genetic material. In eukaryotic cells, DNA molecules are organised into chromosomes, and mitosis is the process that ensures that each daughter cell receives one copy of every chromosome.

In most organisms, each cell is **diploid**, that is, it has two copies of each chromosome, one inherited from the egg and one from the sperm following sexual reproduction. The two copies of a chromosome are **homologous**, that is, similar but not identical. Each cell is described as $2n$, where n is the number of homologous pairs of chromosomes (the **haploid** number).

A major advantage of diploidy is that every cell has two copies of each gene. Mutations are often deleterious, altering and perhaps destroying the function of gene products. When a cell has two copies of each gene, a mutation in one of these copies is not necessarily deleterious since the cell can survive by using the other copy of that gene.

During sexual reproduction in eukaryotes, two reproductive cells, usually sperm and egg, fuse to form the first cell (zygote) of a new generation. Reproductive cells are haploid. Haploid cells are the products of meiosis, a special type of cell division that reduces the chromosome number to half that of the parent cell. On fusion of two haploid cells, the zygote will have the diploid number of chromosomes.

> All cells arise from other cells by division. DNA molecules are accurately passed to daughter cells by the processes of mitosis or meiosis.

CELL DIVISION IN PROKARYOTES

Cell division in prokaryotes such as bacteria is simple and rapid. Some bacteria in culture can divide every 10 minutes providing sufficient nutrients are available. Their single, circular molecule of DNA is attached to the plasma membrane at a specific point (Fig. 8.1). Between divisions, the double-stranded DNA molecule is replicated (Chapter 10) by a system of enzymes that produces an identical copy. When the growing cell reaches a certain size and after replication is complete, the new DNA molecule has a separate point of attachment to the plasma membrane. The cell then commences to divide into two. The two DNA molecules separate by growth of the plasma membrane and cell wall between their attachment points. The plasma membrane and wall grow inwards across the middle of the cell. This process of cleavage, **binary fission**, creates two equal-sized cells, each containing one copy of the genetic material and approximately half the cytoplasm. The two cells may separate or remain joined to form a growing filament of cells, depending on the type of bacterium.

In some bacteria, there are additional smaller, unattached circular molecules of DNA, **plasmids**, which also replicate so that several copies may be present in each cell. At fission, plasmids pass as part of the cytoplasm to the daughter cells.

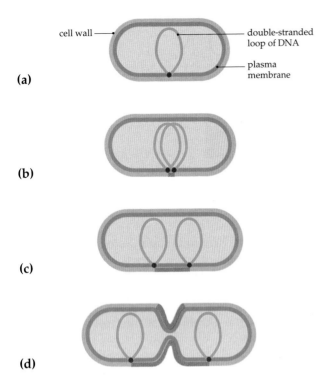

Fig. 8.1 Cell division in a prokaryotic bacterium.
(a) The circular DNA molecule (chromosome) is attached to the cell membrane at a specific point.
(b) DNA replicates as the cell grows. **(c)** The two copies of DNA are separated by expansion of the plasma membrane and cell wall between the attachment points. **(d)** The membrane and wall furrow inwards to divide the cell in two

In a prokaryotic cell, the single circular molecule of DNA is attached to the plasma membrane. After replication, the attachment points of the two DNA copies are separated by growth of the plasma membrane and cell wall, and the cell divides by binary fission.

CELL DIVISION IN EUKARYOTES

Eukaryotic cells are larger and have more internal structures than prokaryotes. Their cell division is correspondingly more complex, particularly since they contain more genetic material. They usually contain many molecules of DNA, each organised into a chromosome contained within the membrane-bound nucleus. In preparation for cell division, there is condensation of the chromosomes from chromatin.

The appearance of chromosomes during cell division attracted the attention of the German microscopist, Walther Flemming, who first described their behaviour in 1882. Early microscopists studying mitosis soon appreciated how genetic information passes from cell to cell, and thus from organism to organism. The discovery of mitosis provided the cellular mechanism for understanding genetics.

The cell cycle

In actively growing eukaryotic cells, division occupies only a small part of the cell cycle (Fig. 8.2). For the rest of the time, cells are in interphase.

During **interphase**, DNA and other molecules are synthesised. DNA is dispersed fairly evenly throughout the nucleus as a finely divided meshwork of chromatin. In interphase cells, one or more nucleoli, producing RNA for export to the cytoplasm for protein synthesis, are usually conspicuous. The first part of interphase, the **G1 phase** (first gap), is often the longest part of the cell cycle. In multicellular organisms, many highly differentiated cells remain permanently in the G1 phase and eventually die. Actively growing cells enter the **S phase** (synthesis), during which DNA is replicated. At the end of the S phase, the nucleus is appreciably larger and contains twice the amount of DNA. The cell then enters the **G2 phase** (second gap), during which the main synthesis of other cellular molecules occurs.

Some time later, the cell prepares for division and enters the **M phase** (mitosis). On the basis of appearance, it is difficult to distinguish between cells that are in G1, S or G2 phases, but the beginning of mitosis is recognisable, even with a light

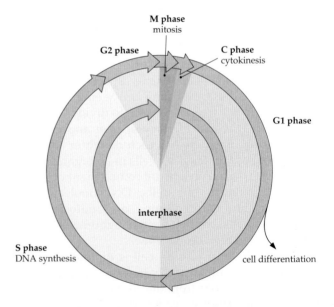

Fig. 8.2 The cell cycle. In dividing cells, interphase alternates with cell division. After cell division, a cell enters G₁ phase, when normal cellular activity, temporarily slowed down during division, resumes. Many of the new cells formed in tissues now differentiate and do not divide further. In cells continuing to divide, DNA synthesis starts in the nucleus after some hours. At the end of this S phase, the DNA content has doubled. The cell enters G₂ phase for a further period before entering mitosis. In cells with about a 24-hour cell cycle, mitosis lasts about 1 hour, G₁ phase about 9–12 hours, S phase 10 hours and G₂ phase about 2 hours

microscope, because of the appearance of chromosomes within the nucleus. As the cell prepares for **mitosis**, the chromatin steadily condenses into chromosomes. Chromosomes are the means by which the cell packages its long strands of DNA into compact manageable bodies. During condensation, DNA is repeatedly coiled and supercoiled around nucleosomes (histone proteins, p. 47). The end result is a set of chromosomes whose number and individual shape are specific for each organism; a few organisms (one kind of ant and some fungi) have only one pair of chromosomes, while some plants (such as ferns)

have over a thousand. (In prokaryotes, the DNA molecule does not condense.)

Mitosis is complete when two daughter nuclei have formed. After nuclear division, division of the whole cell occurs in the **C phase** (cytokinesis).

> In the cell cycle of eukaryotes, the G1, S and G2 phases constitute interphase, during which DNA and other molecules are synthesised. Cells about to divide enter the M (mitosis) phase, during which long strands of DNA condense into chromosomes. Following nuclear division, the whole cell divides by cytokinesis (C phase).

BOX 8.1   CANCER: CELL DIVISION OUT OF CONTROL

When something goes wrong with the control of cell division and differentiation, a mass of cells, a tumour, may form. Benign tumours, such as warts, remain localised and are not usually life-threatening. However, serious problems may arise with malignant tumours, which are highly invasive. In life-threatening forms of cancer, cells grow rapidly and uncontrollably at the expense of their neighbours, depriving them of nutrients and destroying the organisation and function of surrounding tissues. Cells of malignant tumours detach and spread throughout the organism, invading and disrupting other tissues (see figure).

Cancer cells can be identified by microscopy. They have a relatively large nucleus and prominent nucleoli, are often dividing and are not differentiated as are normal cells in surrounding tissue. Cancer cells often have abnormal numbers of chromosomes or chromosomal abnormalities. They also produce new proteins. For example, for malignant cells to penetrate tissues, they must secrete enzymes that digest the basal lamina that normally underlies epithelial surfaces.

The causes of cancer remain enigmatic but many things can stimulate cancerous behaviour, including constant injury or aggravation of tissue, mutagens and viruses. Cancer cells result from changes in the DNA of somatic (body) cells. Many cancers seem to start from a single aberrant cell, probably as a result of several successive changes within it. The changes transform a normally regulated cell to one that does not respond to signals that control the cell cycle, that is, the orderly progression from G1, S, G2 to the M phase.

When cells derived from normal animal tissues are grown in the laboratory in culture, they may proliferate and become independent of the cues that normally control them in tissues, and many such cell lines divide repeatedly without ever reforming any tissue or performing the function they exhibited in the animal from which they were derived. These cells are therefore similar to cancer cells. Cells derived from cancerous tissues are often easier to grow in culture, apparently because they are already less dependent on the external growth signals that regulate normal cell growth.

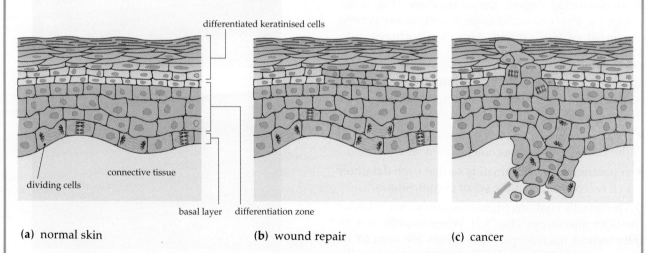

(a) normal skin   **(b)** wound repair   **(c)** cancer

Tissue organisation in the skin. **(a)** Normal cell division in the basal layer generates a supply of cells, which differentiate, become keratinised, die and thereby create the outer resistant layer of the skin. **(b)** During wound repair, there is temporary disruption in tissue organisation and stimulation of cell division to reform the normal tissue structure. **(c)** In cancer, there is uncontrolled proliferation of cells, which do not differentiate properly and disrupt tissues. In dangerous cases, cancerous cells migrate away from where they are formed and invade other tissues (arrows) where they continue to grow

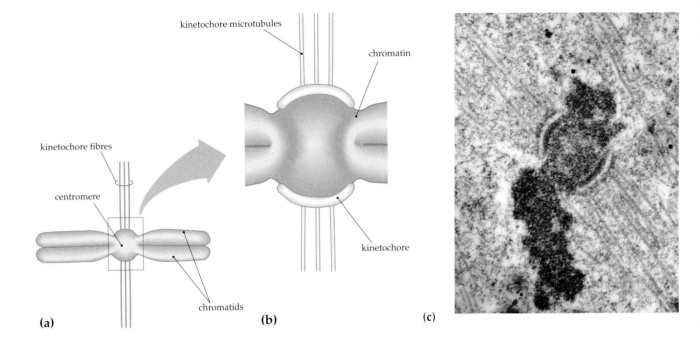

Fig. 8.3 The attachment region of chromosomes to spindle fibres. (**a**) Kinetochore fibres attach to chromosomes, which, after DNA replication, consist of a pair of chromatids at the centromere. (**b**) Under the electron microscope, these kinetochore fibres are seen to consist of a number of microtubules, which terminate in the kinetochore itself. The chromatin in this region often shows no apparent separation into two chromatids. (**c**) Transmission electron micrograph of a chromosome in the green alga *Oedogonium*. This cell has particularly prominent kinetochores. Spindle microtubules are seen here inserted into the outer layers of the paired kinetochores (magnification × 24 000)

Mitosis

After DNA replication in the S phase, and before mitosis, each chromosome consists of two identical **chromatids**, that is, sister copies of the chromosome. The chromatids of a chromosome are held together at a constricted region, the **centromere** (Fig. 8.3a). Under the electron microscope, a centromere is seen to consist of two protein discs, **kinetochores**, into which are inserted microtubules (Fig. 8.3b, c).

The precise behaviour of chromosomes during mitosis is dependent upon the **mitotic spindle**. This is an elaborate cytoskeletal structure that provides the structural basis for:

- movement of chromosomes toward the equator of the cell where they become aligned; and

- separation of the chromatids so that each daughter cell receives a complete set of chromosomes.

The spindle contains numerous fibres visible under the light microscope (Fig. 8.4). When examined with the electron microscope, these fibres are seen to be composed of microtubules. To construct a spindle, cells mobilise proteins by breaking down most microtubules in the interphase cytoskeleton. These protein subunits, tubulin, are reassembled as the spindle fibres. After division, these fibres are broken down in turn, and the proteins recycled into the interphase cytoskeleton once more.

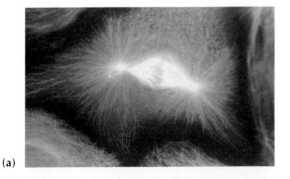

(a)

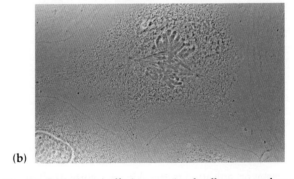

(b)

Fig. 8.4 Mitotic spindle in an animal cell, at metaphase. (**a**) The yellow fluorescent stain has selectively bound to the spindle fibres, consisting of microtubules. The asters are also evident. (**b**) The same cell seen unstained in phase contrast. The chromosomes are now clearly visible, while the fibres can just be distinguished

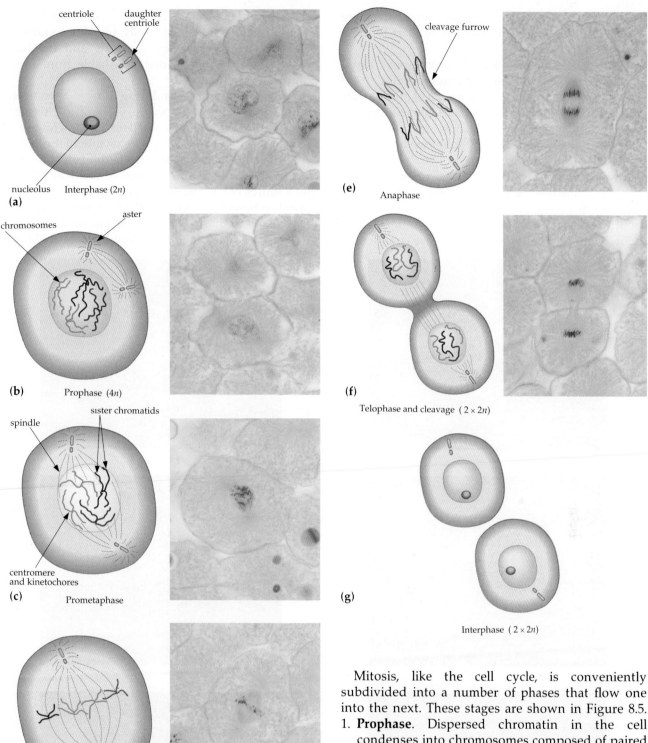

Fig. 8.5 Stages of mitosis in animal (white fish) cells:
(a) interphase; (b) prophase; (c) prometaphase;
(d) metaphase; (e) anaphase; (f) telophase; and
(g) cleavage and cell division complete—interphase

Mitosis, like the cell cycle, is conveniently subdivided into a number of phases that flow one into the next. These stages are shown in Figure 8.5.

1. **Prophase**. Dispersed chromatin in the cell condenses into chromosomes composed of paired chromatids; the mitotic spindle that will organise and move them about, begins to form outside the nucleus.

2. **Prometaphase**. The nuclear envelope enclosing the chromosomes breaks down and allows the growing spindle to interact with and move the chromosomes.

3. **Metaphase**. The spindle arranges precisely all chromosomes so that sister chromatids will later move in opposite directions, into the new daughter cells.

4. **Anaphase**. The two kinetochores of the centromere separate and sister chromatids move apart, forming two identical groups.
5. **Telophase**. New nuclear envelopes form, separating the two groups of chromosomes.

Timing of these phases and duration of mitosis vary widely between different organisms. Some fungi and algae complete mitosis in a few minutes. More typically, mitosis lasts 30 to 60 minutes. In a typical mammalian cell, prophase slowly becomes obvious over about 20 minutes as the chromosomes thicken. Prophase finishes abruptly when the nuclear envelope disperses and chromosomes start moving about. Cells stay in metaphase for some time (often about 20 minutes), after which anaphase commences quite abruptly. Most of anaphase is completed in about 15 minutes. In many cells, telophase and the final stage of cell cleavage are slow and the two progeny cells become indistinguishable from their interphase neighbours.

Changes in the nucleus during mitosis have been known and studied for many years using tissues preserved, sectioned and stained for light microscopy. Less familiar to biologists is the behaviour of the living cell during these stages. It is quite difficult to study live dividing cells within tissues so one has first to find a source of flat, transparent cells (Fig. 8.6). Special optical systems are needed for the microscope in order to see detail in these delicate cells. Mitosis is usually prolonged and movements of chromosomes are quite

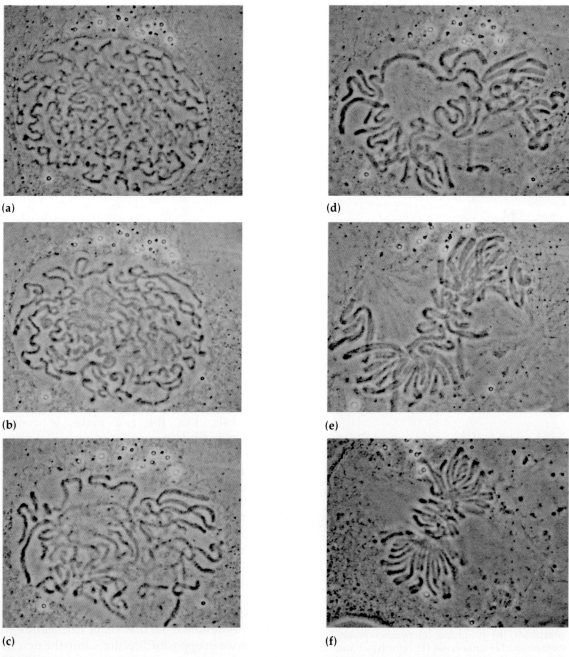

(a)

(b)

(c)

(d)

(e)

(f)

Fig. 8.6

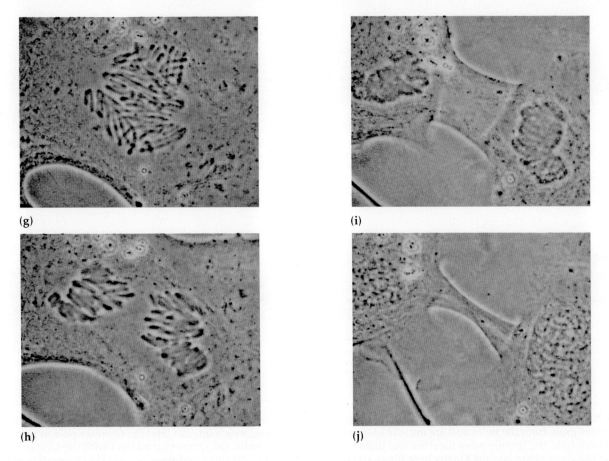

Fig. 8.6 Time lapse sequence of living cultured newt lung cells showing a cell undergoing mitosis: **(a)** prophase, **(b)–(d)** prometaphase; **(e)** early metaphase; **(f)** metaphase; **(g)** anophase; **(h)** late anaphase; **(i)** telophase; and **(j)** cleavage (cytokinesis)

slow. As a result, time-lapse cinematography is needed to visualise clearly what is happening. The following description is based on living cells as well as fixed and stained material.

> For mitosis, tubulin is mobilised and assembled into spindle fibres. The spindle is essential for correct organisation and separation of the chromosomes.

Mitosis in animal cells

The behaviour of chromosomes and spindle fibres in animal cells is summarised in Figure 8.7. At first, spindle fibres are all identical. How they behave subsequently is dependent upon how they interact, either with each other (to form continuous fibres) or with kinetochores (to form kinetochore fibres). In cells, these fibres are all intermingled.

Prophase

During prophase, chromosomes slowly condense inside the nucleus. During the G1 phase, the microtubule organising centre of the interphase cell,

the **centrosome**, contains two cytoplasmic organelles, **centrioles**. During the S phase, the centrioles separate and two new centrioles arise alongside them. Thus, the centrosome has already replicated by the time the cell proceeds into prophase. At this stage, the centrosomes are activated to produce increasing numbers of fibres (bundles of microtubules). The two resultant arrays of radial fibres are **asters**, which slowly move apart, apparently due to interaction between some of their fibres (Fig. 8.7a).

It has recently been discovered that microtubules, particularly in asters, are far more dynamic than expected. Each microtubule grows steadily and then shrinks rapidly in what appears to be a remarkably irregular fashion. This phenomenon has been called 'dynamic instability'. Both growth and shortening are due to tubulin subunits being added and removed from the end of the microtubule distant from the centrosome.

During late prophase, RNA synthesis ceases and the prominent nucleoli shrink and disappear completely. They reappear at telophase.

> During prophase, chromosomes slowly condense within the nucleus. Centrosomes move to opposite poles and form asters.

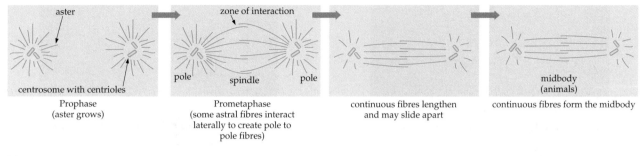

(a)

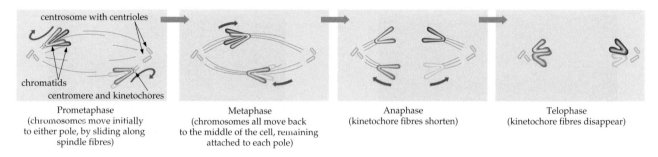

(b)

Fig. 8.7 This diagram shows the behaviour of the two sets of fibres that compose the spindle structure in animal cells. All fibres grow from the poles and are initially identical. (a) Astral fibres, which interact laterally to create pole to pole fibres, hold the spindle poles together and define the axis of the spindle. (b) Fibres that interact with the kinetochores of the centromere become kinetochore fibres and are responsible for attaching and then moving the chromatids to the poles

Prometaphase

Initially, microtubules form a cage-like array over the surface of the nuclear envelope. As the size and activity of asters increase, the nuclear envelope immediately next to them becomes increasingly deformed inwards. Finally, the envelope ruptures and, in a few seconds, the whole membrane breaks down and the microtubules move into the nuclear region. This dramatic moment marks the beginning of prometaphase. As the growing fibres penetrate rapidly and deeply into the nucleus, many chromosomes react by randomly moving close to either centrosome (Fig. 8.7b). Asters also become firmly attached to each other by their fibres to form the biconical (twin cones) structure that gives the spindle its name. The interaction is due to overlapping of a number of fibres. As a result of this overlapping, pole to pole fibres are much more stable than the remainder of the astral fibres. The centrosomes at the centre of each aster become the **poles** of the growing spindle (Fig. 8.7a).

Prometaphase movements of chromosomes are difficult to follow. The function of the pair of kinetochores in the centromere of each chromosome is to interact with spindle fibres and thereby bring about chromosomal movement. Kinetochores commence this interaction by sliding over microtubules towards either pole. At some stage, these microtubules become inserted into (i.e. terminate in) one of the kinetochores, forming a kinetochore fibre. These fibres are stabilised by their insertion into the kinetochore. Thus, two sets of relatively stable fibres are now formed from the large number of microtubules from the aster—those attached to kinetochores, and those that overlap to form the fibres that run from pole to pole.

Having moved to either pole, the chromosomes oscillate gently around that pole. This movement is possibly a result of the population of rapidly shortening and elongating microtubules around them. Very soon each chromosome encounters some microtubules from the other pole and its second kinetochore attaches to these microtubules. Each chromosome then steadily moves to an equilibrium position midway between the poles.

During prometaphase, microtubules become inserted into the kinetochores and the chromosomes begin to move, firstly to the poles, and then steadily to the middle of the spindle.

Metaphase

The central positioning of each chromosome on the spindle indicates that the chromatids (the two copies of each original chromosome) have become correctly attached to opposite poles. As the chromosomes accumulate in the centre of the spindle, the cell reaches the metaphase configuration with kinetochores all in one plane. However, the chromosomes continue to oscillate gently on either side of the central position.

Metaphase occupies a considerable proportion of mitosis. Just why this is so is not clear. Perhaps it is to give all chromosomes time to reach their correct orientation and position.

> By metaphase, the chromosomes have assembled in the centre of the spindle and the chromatids are attached by kinetochore fibres to opposite poles.

Anaphase

The next stage of mitosis, anaphase, is dramatic. Each centromere divides, freeing the attachment of sister chromatids to one another. Sister chromatids separate simultaneously and are steadily drawn to each pole. Careful analysis shows that anaphase has two distinct components: anaphase A, which is movement of chromosomes to the pole; and anaphase B, when the poles move further apart, elongating the spindle. The motile mechanism that generates anaphase A is not understood, although it is clear that kinetochore fibres shorten by disassembly of microtubules at the kinetochores (Fig. 8.8). Anaphase B seems to involve several phenomena. Overlapping spindle fibres elongate and/or slide apart.

> Anaphase involves the rapid separation of sister chromatids to opposite poles of the cell.

Telophase

During telophase, chromosomes clump tightly near the poles of the elongated spindle as a new nuclear envelope appears around them. Then the whole dense mass of chromosomes starts swelling steadily as chromosomes disperse into chromatin. Nucleoli reform and become conspicuous.

> During telophase, chromosomes clump tightly near the poles as a new nuclear envelope appears around them.

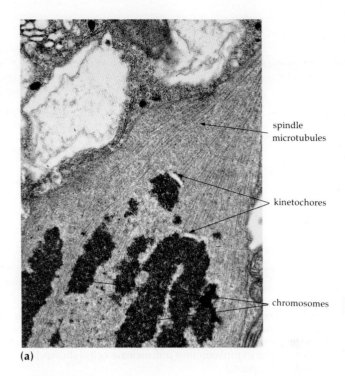

(a)

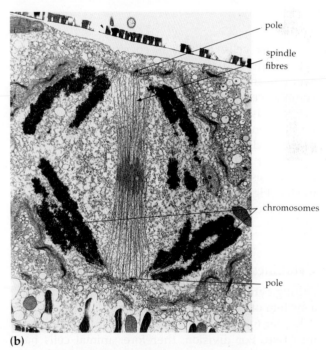

(b)

Fig. 8.8 Anaphase. (a) *Oedogonium*: two single kinetochores, attached to their spindle microtubules, are seen while moving to the pole (magnification × 7700). (b) The spindle of the diatom *Nitzschia* is large and beautifully organised, in contrast with the spindle of most types of cells. Thus, it has been useful in analysing the structural aspects of how the spindle functions (magnification × 3900)

Cytokinesis

Once mitosis is complete, a cell divides by cytokinesis. **Cytokinesis**, the means by which the plasma membrane comes to enclose the cytoplasm of new daughter cells, is different in walled and unwalled cells. In animal cells, it occurs by simply pinching in to form the two daughter cells. In plant cells, division occurs by laying down a cell plate between the two daughter nuclei.

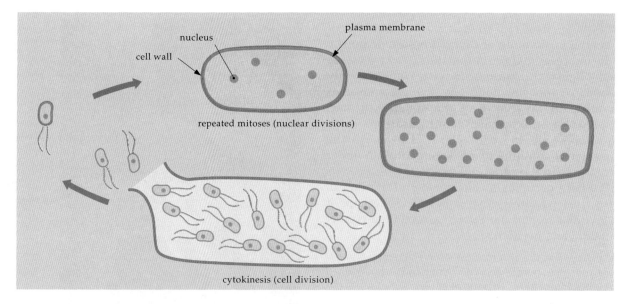

Fig. 8.9 A cell (e.g. certain algae and fungi) in which mitosis and cytokinesis are clearly separate events. A flagellated cell (e.g. a small swimming zoospore) undergoes rounds of mitosis. Accompanying cell growth generates a large cell (sometimes centimetres or more in length) containing hundreds or thousands of nuclei. A single round of cytokinesis cleaves the cytoplasm into uninucleate cells (shown here developing flagella), which are released from the parental wall for dispersal

In most cells, mitosis is so closely followed by cytokinesis that the two appear to be part of one event. In animal cells, for example, cytokinesis usually starts as cells enter anaphase. However, mitosis and cytokinesis are quite separate events, brought about by different cytoplasmic systems. In some cells, for example, during spore formation in algae and fungi, a number of cycles of mitosis can occur without cytokinesis, resulting in one large cell with many nuclei (Fig. 8.9). Cytokinesis in such cells involves the sudden cleavage of the whole cytoplasm into numerous tiny, mononucleate cells.

Cytokinesis in animal cells

During cytokinesis, many animal cells create and use a system of actin filaments, forming a contractile ring (Chapter 6), which constricts the cell by pinching it into two. For division, therefore, animal cells have to establish independently and quite rapidly two cytoskeletal systems—the contractile ring and the spindle. Each system has to be put together in the right place so that cells divide in the correct plane between the two daughter nuclei, and the right time so that cleavage happens after chromosomes are separated.

The commencement of cytokinesis during anaphase causes animal cells to look as if a string is tightening around them, first indenting and then deeply constricting them. The remaining overlapping fibres that run between poles become aggregated into a tight rod (the **midbody**, Fig. 8.10). The midbody

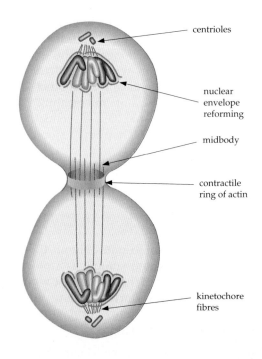

Fig. 8.10 Cytokinesis in animal cells. As anaphase is completed in an animal cell, the contractile ring of actin filaments constricts the cell and eventually divides it. Meanwhile, the continuous fibres remaining in the spindle are compressed into a single rod whose dense central region, the *midbody*, is derived from the region of overlap of the interacting polar fibres. The nuclear envelope reforms as the kinetochore fibres shorten and disappear

is quite stable and often continues to hold the cell together for some time. The ingrowing cleavage furrow is lined with an even layer of actin filaments. Once these filaments start constricting the cell, they will continue to do so even if mitosis is stopped experimentally.

In preparation for cytokinesis, animal cells mobilise actin filaments into a contractile ring that cleaves the cell in two after mitosis.

Division in plant cells

Mitosis in plant cells proceeds much as in animal cells, but plant cells display two major differences (Fig. 8.11). Firstly, plant cells do not have centrioles (although some green algae do). As a result, they also do not have astral spindles, whose fibres are focused upon discrete poles. Instead, the poles are broad with fibres forming a barrel-shaped spindle.

(a)

(b)

(c)

(d)

(e)

(f)

Fig. 8.11 Stages of mitosis in plant cells. These cells were fixed and then released from a wheat root tip by enzymatic digestion of their cell walls. They are unstained and photographed by differential interference optics: (a) interphase; (b) prophase; (c) prometaphase; (d) metaphase and late cytokinesis; (e) anaphase; (f) telophase and early cell plate formation

Secondly, cells are enclosed in a rigid wall and do not undergo cleavage as animal cells do. Instead, as they go through anaphase, fibres remaining between the chromosomes thicken and proliferate. Soon, a densely fibrous **phragmoplast** forms (Fig. 8.12a). This steadily grows out laterally until it reaches the older walls of the cell. Inside the phragmoplast, tiny droplets appear and slide along the fibres, collecting halfway between the reforming telophase nuclei. These droplets coalesce into the **cell plate**, which thickens and separates the two new cells. Under the electron microscope, the cell plate is complex (Fig. 8.12b). It contains many tiny vesicles, probably derived from the Golgi apparatus, along with elements of the endoplasmic reticulum. Once the vesicles fuse together, fibrous wall material appears within them.

Plant cells also show another feature that has no equivalent in animal cells. Before prophase is clearly established, microtubules all group into the **preprophase band of microtubules** close to the wall. With remarkable accuracy, this band predicts exactly where the phragmoplast later will join the growing cell plate to the older walls. This predictive ability is particularly striking in highly asymmetric cell divisions involved in numerous differentiation events in plant tissues, such as those giving rise to root and epidermal hairs, and the guard cells of stomata. Thus, the preprophase band indicates that the plane of cell division has been decided upon well before the cell shows any sign of impending mitosis. The preprophase band disappears completely as the cell

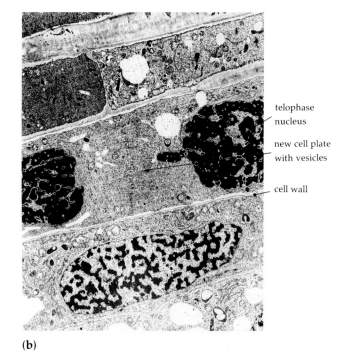

telophase nucleus

new cell plate with vesicles

cell wall

(b)

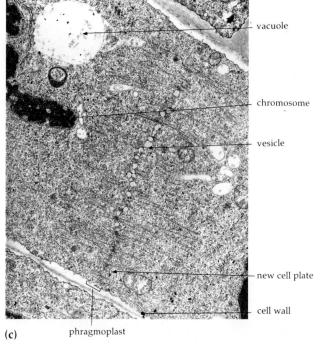

vacuole

chromosome

vesicle

new cell plate

cell wall

(c)   phragmoplast

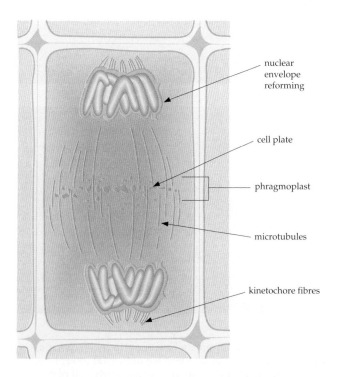

nuclear envelope reforming

cell plate

phragmoplast

microtubules

kinetochore fibres

(a)

Fig. 8.12 (a) Cytokinesis in a typical plant cell. The overlapping spindle fibres remaining between the chromosomes thicken and proliferate, forming the fibrous *phragmoplast*. Much cellular material (tiny vesicles, membranes etc.) accumulate along the midline of the phragmoplast and fuse to form the *cell plate* or new cross-wall that thickens and divides the cells. A **(b)** low (× 3400) and **(c)** high (× 8400) magnification view of the phragmoplast in a wheat cell. Vesicles are collecting among the microtubules of the phragmoplast; these vesicles are guided into the correct position in the cell, whereupon they fuse together and form the cell plate

enters prophase; its complement of microtubules is mobilised into the spindle fibres. The function of the band is not understood. Not all eukaryotic cells that have walls divide in this manner. All the land plants and a few green algae use the phragmoplast for cytokinesis (Chapter 37). However, many algal cells cleave in a similar fashion to animal cells (Fig. 8.13).

> During cytokinesis in plants, the cytoskeletal microtubules form a phragmoplast and a preprophase band that predicts the position of the cell plate.

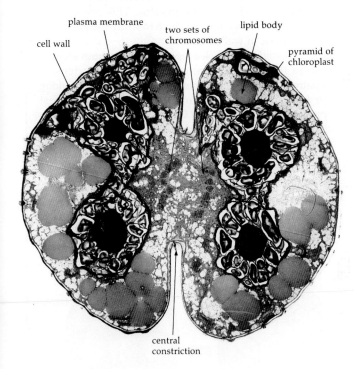

Fig. 8.13 Telophase in the green alga *Cosmarium*. The two sets of chromosomes are separated across the central constricted region of the cell, where cleavage is beginning. Many algae cleave in a similar fashion to animal cells. Only a few green algae (and all the land plants) use the phragmoplast for cytokinesis (magnification × 16 800)

Drugs that block mitosis

Two types of drugs, anti-microtubule and anti-actin drugs, have proved of great use experimentally in investigating mitosis, cytokinesis and many other cellular activities. By using antimitotic drugs to break down the cytoskeleton, cell biologists can investigate the role of the cytoskeleton in the cell.

The anti-microtubule drugs cause disassembly of microtubules of the spindle. The best known of these, **colchicine**, is extracted from certain plants (autumn crocus, *Colchicum*). It is highly poisonous and binds to tubulin. If colchicine is applied before prophase, the cell undergoes chromosome condensation and normal chromosomes appear. The chromosome splits

simultaneously into two chromatids. However, no spindle forms and chromosomes cannot become organised or move about. If colchicine or other antimitotic drugs are applied to a cell during mitosis, the spindle rapidly dissolves and leaves chromosomes stranded. In either case, the cell enters telophase and then interphase without division. The cell now has a double complement of chromosomes and DNA in its nucleus. If it were originally diploid, it is now tetraploid.

The second type of drug that interferes with cell division acts as an anti-actin agent; the best known are **cytochalasins**, derived from certain fungi. These drugs specifically disrupt actin-based cytoskeletal systems and therefore stop cytokinesis. They also interfere with other actin-based phenomena such as cytoplasmic streaming.

These two types of drugs show quite clearly that mitosis and cytokinesis are brought about by different cytoskeletal systems. Cytokinesis can proceed in the presence of colchicine or even, for example, when the spindle is mechanically removed from the cell with a micropipette. Conversely, events of mitosis proceed normally in the presence of cytochalasin.

Meiosis and the formation of haploid cells

Sexual reproduction involves the fusion of two reproductive cells (gametes) resulting in a doubling of the amount of genetic material in the zygote. At some stage in the life cycle before gamete formation, there is a corresponding halving in the amount of genetic material. This occurs by the reduction division, **meiosis**. In meiosis, there is *one* round of DNA replication in a diploid cell ($2n$ to $4n$) as in the S phase of mitosis, followed by *two* separate cell divisions, resulting in four haploid ($1n$) cells (Fig. 8.14). Each haploid cell has one copy of each original homologous chromosome. The two divisions are called **meiosis I** and **meiosis II**.

Some organisms, for example, vertebrates and seed plants, are diploid for most of their life cycle. In animals, meiosis immediately precedes formation of male or female gametes. In seed plants, gamete formation occurs after two or three rounds of mitotic divisions of the haploid cell arising from meiosis. Other organisms, for example, mosses and many algae, are haploid for the larger part of their life cycle and diploid only briefly after sexual reproduction. In these, meiosis occurs well before the formation of gametes. Gametes are formed by mitosis during the haploid stage of the life cycle (Chapter 14).

> In meiosis, there is *one* round of DNA replication in a diploid cell ($2n$ to $4n$) followed by *two* separate cell divisions resulting in four haploid cells (n).

BOX 8.2 CYTOKINESIS IN WALLED CELLS

After mitosis in walled cells, additional plasma membrane is needed for cytokinesis to complete the formation of the new cells. How is new plasma membrane generated during cytokinesis? Two models have been proposed. The first model suggests that networks of vesicles become aligned along the future plane of cleavage and subsequently fuse together to form the new plasma membrane (see figure). The second model suggests that there is a progressive extension and fusion of flat sheets of plasma membrane in the plane of cleavage (see figure).

There is considerable evidence supporting both models from studies of cell division in different eukaryotes. Evidence supporting the second model has come recently from experiments at the Australian National University, where Dr Geoff Hyde and Dr Adrienne Hardham have used rapid freeze fixation techniques rather than traditional chemical fixation before transmission electron microscopy. Rapid freeze fixation is considered superior to chemical fixation in providing more life-like images of cells, their organelles and membrane systems. Chemical fixation can generate artefacts by altering the form of membrane components.

The production of the motile, infective cells (zoospores) of the cinnamon fungus, *Phytophthora*

cinnamomi, have been studied. This pathogen attacks roots and kills many Australian native plants (e.g. dieback disease of Jarrah forest in Western Australia, Chapter 44). The zoospores are produced within a special sac, the sporangium. Initially, as a result of numerous mitoses without cytokinesis, many zoospore nuclei lie free within the cytoplasm of the sporangium. As the sporangium matures, it is subdivided by plasma membranes and extracellular matrix into compartments surrounding each zoospore nucleus. This process has provided an ideal system in which to study the processes of cytokinesis in walled cells.

In *Phytophthora*, cell cleavage results not from aligned vesicles (model 1) but from the progressive extension of paired sheets of membrane which interconnect along the future partition site (model 2). These sheets first appear as vesicles from the Golgi stacks and are initially clustered at one pole of each nucleus. The vesicles then become flattened membranous structures, before travelling along microtubules and coming to rest at the future partition site, where they fuse to form the new cell plate.

This model of cytokinesis in *Phytophthora* may prove to be the mechanism of cell division in a variety of eukaryotes with walled cells.

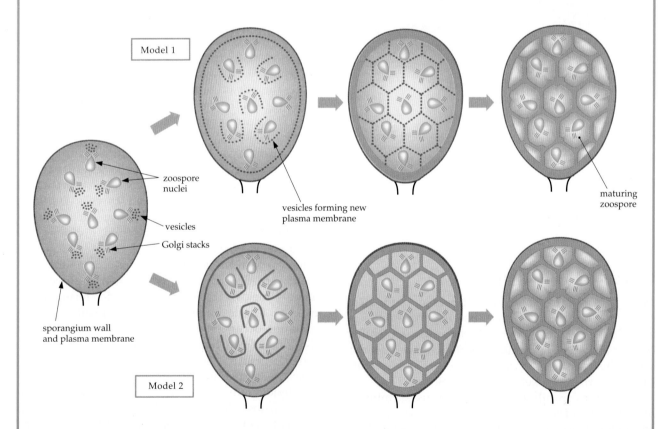

Two models of how new plasma membrane is generated during cytokinesis in the production of zoospores in the sporangium of the cinnamon fungus, *Phytophthora cinnamomi*. In model 1, networks of *vesicles* align and fuse in the plane of cell cleavage. In model 2, *flat sheets* of membrane extend and fuse

Meiosis I

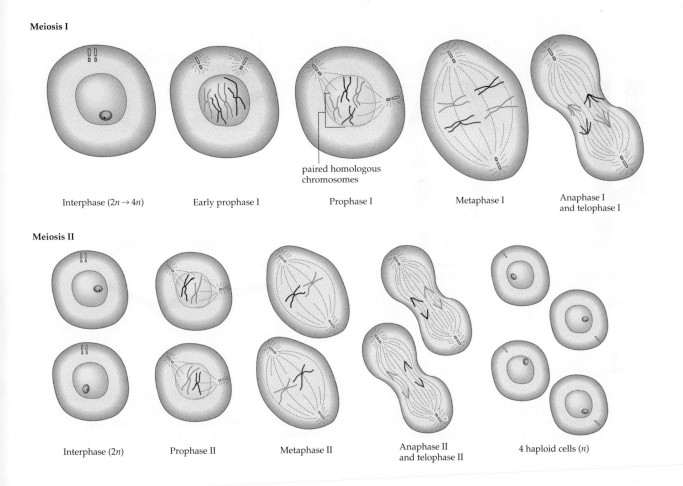

Interphase ($2n \rightarrow 4n$)   Early prophase I   paired homologous chromosomes / Prophase I   Metaphase I   Anaphase I and telophase I

Meiosis II

Interphase ($2n$)   Prophase II   Metaphase II   Anaphase II and telophase II   4 haploid cells (n)

Fig. 8.14 Meiosis in an animal cell. In meiosis I, interphase and early prophase I are similar to mitosis. In prophase I, homologous chromosomes, each consisting of two chromatids, pair and may exchange genetic material by crossing over (see Fig. 8.15). By metaphase I the chromosomes are aligned in the central plane of the spindle. During anaphase I, homologous chromosomes move to opposite poles. The two kinetochores of each centromere act as a unit, ensuring that the two chromatids of each chromosome remain attached and move to the same pole. Meiosis II is similar to a mitotic division except that there is no DNA replication. The end result of the two divisions of meiosis is four haploid cells

Meiosis I: prophase

Prophase of meiosis I is prolonged and the events accompanying chromosome condensation are more complicated than in mitotic prophase. During mitosis, homologous chromosomes are completely independent of one another. However, during prophase of meiosis I, homologous chromosomes pair up precisely along their length (Fig. 8.15a). This pairing is called **synapsis**. As the chromatin condenses, the two sister chromatids, joined at the centromere, become apparent in each homologous chromosome, which often appear attached to the nuclear envelope. At this time, chromatids of homologous chromosomes may exchange portions of their genetic material by the process of **crossing over** (Fig. 8.15a). The DNA strands from a maternal and paternal chromatid are cut at the equivalent point

and reconnected precisely. When crossing over occurs, new combinations of genetic information are formed.

The point where crossing over occurs is a **chiasma** (pl.: chiasmata). Under the electron microscope, chiasmata display a characteristic **synaptonemal complex**, consisting of a central core flanked by several layers between chromatids (Fig. 8.16). This is the molecular scaffold on which precisely controlled crossing over occurs. It consists of a long protein core, which the two homologous chromosomes align to, and has been likened to a zipper, with the DNA of each chromatid unwound and placed in register with the DNA of the homologue. Once crossing over at the DNA level is completed, this scaffold disappears and chromosomes separate from the nuclear envelope.

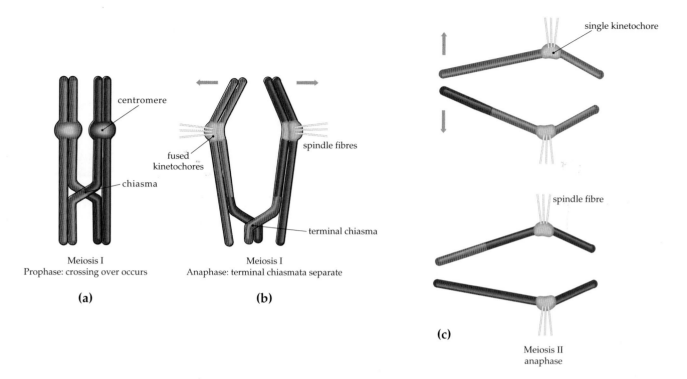

Fig. 8.15 Crossing over. **(a)** During synapsis in meiosis I, homologous chromosomes pair and then undergo crossing over at a chiasma. **(b)** Later, during anaphase I, sister kinetochores on each homologue act as single fused units, attaching to spindle fibres and moving to the poles together. The chiasma has moved along to the ends of the chromatids, becoming a terminal chiasma. **(c)** During meiosis II, these sister kinetochores now act individually and segregate as they would in normal mitosis

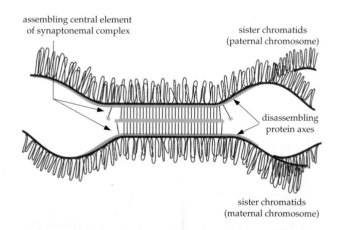

Fig. 8.16 A synaptonemal complex is a ladder-like protein structure, which keeps homologous chromosomes together and closely aligned, and is involved in crossing over and thus recombination of genetic material

> During prophase I of meiosis, crossing over between chromatids of homologous chromosomes may occur, allowing the formation of new combinations of genetic information.

Meiosis I: metaphase and anaphase

As prophase I undergoes transition to metaphase I, the spindle forms as in mitosis (Fig. 8.14). However, again there are important functional differences between meiosis and mitosis at this stage. During mitosis, kinetochores of sister chromatids always end up connected to opposite poles, to ensure accurate segregation of chromatids. In meiosis I, however, kinetochores in sister chromatids appear to be fused and act as a single unit. One homologous chromosome (with its two chromatids) attaches to one pole, while the other attaches to the opposite pole (Fig. 8.15b). The chiasmata continue to affect the behaviour of chromosomes, moving to the tips of the pairs of chromatids and becoming **terminal chiasmata** (Fig. 8.15b). By holding the chromosomes together, they assist in making sure that the two sets of paired kinetochores end up oriented towards opposite poles. During anaphase, the chiasmata eventually separate completely. The orientation and subsequent assortment of maternally and paternally derived homologues to the two poles is random (Fig. 8.15c).

At the end of anaphase I, two nuclei are formed, each containing one set of chromosomes. The cell itself does not necessarily undergo cytokinesis at this stage.

At the end of metaphase I, homologous chromosomes are aligned on the central plane of the spindle. During anaphase I, homologous chromosomes move to opposite poles.

In animals, these cells differentiate directly as gametes, or in plants, fungi and many protists, including algae, they may divide again by mitosis, delaying the formation of gametes to a later stage.

Meiosis II

Meiosis II is similar to mitosis except that there is no S phase of DNA synthesis preceding it. The two kinetochores of the sister chromatids, fused during meiosis I, separate at anaphase II and act as normal single kinetochores, becoming attached to opposite poles of the meiosis II spindle. The end result, therefore, is four haploid sets of chromosomes. In many organisms, nuclear envelopes form around each set and the cytoplasm is divided to form four cells.

Genetic consequences of meiosis

The two meiotic divisions generate four haploid cells whose genetic complement consists of new combinations of parental genes. The reassortment of genes is a consequence of both crossing over and the random segregation of maternally and paternally derived chromosomes in each homologous pair. Meiosis therefore has important genetic consequences for sexually reproducing organisms.

SUMMARY

- In prokaryotes, cell division is by binary fission. The single circular molecule of DNA in a bacterial cell is attached to the plasma membrane. Following DNA replication, the attachment points of the two DNA copies are separated by growth of the plasma membrane and cell wall, which invaginate and partition the cell into two daughter cells.

- Cell division in eukaryotes involves two separate processes: division of the nucleus (mitosis or meiosis) and division of the cell (cytokinesis).

- In the cell cycle of eukaryotes, the G1, S and G2 phases constitute interphase, during which DNA and other molecules are synthesised. Cells about to divide enter the M (mitosis) phase, during which long strands of DNA condense into chromosomes. After nuclear division, the whole cell divides by cytokinesis (C phase).

- Chromosomes are essentially inert packages during mitosis. Their mitotically functional sites are kinetochores, which engage with spindle fibres to bring about alignment and subsequent separation of chromosomes into daughter cells.

- For mitosis, tubulin (the protein subunit of microtubules) is mobilised and assembled into spindle fibres. The spindle is formed from two sets of fibres that grow outwards from the poles. Some of these interact laterally with fibres from the other pole to create pole to pole fibres. These fibres define the axis of cell division. Others connect with kinetochores and are involved in moving chromosomes.

- During prophase, the nucleus is quiescent but chromosomes slowly condense. Centrioles move to opposite poles and form asters.

Disappearance of the nuclear envelope marks the start of prometaphase. Microtubules become inserted into the kinetochores and the chromosomes begin to move, first to the poles and then steadily to the middle of the spindle. By metaphase, the chromosomes are all in the centre of the spindle. Each centromere is attached by kinetochore fibres leading to opposite poles. The centromere splits and chromatids are drawn towards each pole during anaphase. During telophase, chromosomes clump tightly near the poles as a new nuclear envelope appears around them.

- Cytokinesis, cell division, is a separate event from mitosis. It may immediately follow mitosis or be delayed. In animal cells, actin filaments are mobilised into a contractile ring that pinches the cell in two after mitosis. In plants, the cytoskeletal microtubules form a phragmoplast and a preprophase band predicts the position of the new cell plate that divides the parent cell.

- Meiosis differs from mitosis in that it brings about a halving of the number of chromosomes in a cell. It normally precedes the formation of gametes and sexual reproduction. Meiosis involves two separate meiotic divisions and differs from mitosis in three important aspects:
  □ during meiosis I, homologous chromosomes pair precisely and exchange genetic material by crossing over;
  □ there is no DNA synthesis between meiosis I and II;
  □ meiosis II is a reductive division, halving the number of chromatids in the progeny nuclei.

QUESTIONS

1. How does a bacterial cell divide?

2. Describe the following structures and their role in cell division: chromatid, centromere, kinetochore, centriole, aster, microtubule.

3. What are homologous chromosomes?

4. What is meant by haploid and diploid? What are the possible advantages of diploidy?

5. State when each of the following occur during the mitotic cell cycle of eukaryotes: (a) DNA replication; (b) breakdown of the nuclear envelope; (c) chromosome aggregation in the middle of the spindle; (d) centromere division.

6. Why is chromatin extended during interphase? What is the significance of chromatin condensation for mitosis?

7. Describe how sister chromatids separate from one another and move to opposite poles during anaphase.

8. Compare and contrast cytokinesis in animal and plant cells.

9. What major events occur during meiosis I? What is its end result? What is the significance of meiosis II and how does it differ from a mitotic division?

10. Redraw Figure 8.15 so that the sets of homologous chromosomes during meiosis I have two or three chiasmata instead of one. Draw the possible combinations that can occur in the haploid cells resulting from meiosis II.

PART 2

GENETICS AND
MOLECULAR
BIOLOGY

INHERITANCE

There is a great range of variation between organisms but offspring tend to resemble their parents because they inherit certain features, **traits**, from them. The study of natural variation and heredity is genetics, the principles of which were first discovered by Gregor Mendel, a priest who grew different kinds of peas in his garden at the Augustinian monastery at Brünn, Moravia (now Brno in the Czech Republic; Fig. 9.1). Beginning in 1856, he studied the inheritance of a range of traits over several seasons.

INHERITANCE OF A SINGLE GENE

Mendel's experiments

The peas that Mendel grew are suitable for studies of inheritance because they can self-pollinate and are easy to cross artificially. For artificial crossing the young anthers of a flower are removed before their pollen is released. Pollen from a different plant can then be brushed on the stigma to fertilise the eggs in the ovules.

Mendel used strains of peas that had self-pollinated for many generations. This meant that the progeny looked identical to the parents and the strains were pure breeding for a range of different traits (Fig. 9.2).

Mendel succeeded in gaining an understanding of heredity where others had failed because in his peas he studied the inheritance of traits *one at a time*. For example, he studied seed colour as a single trait (Fig. 9.3). One pure-breeding strain had yellow seeds, another had green. When cross-pollinated, the progeny seed were always yellow, regardless of which parent had yellow or green seeds. In other words, Mendel observed that it did not matter which form of the trait was present in the pollen parent and which in the egg parent.

Mendel grew the yellow progeny seeds—the first filial generation, F_1—into plants, allowed them to self-pollinate and collected the progeny seeds—the second filial generation, F_2. This time, as well as

(a)

(b)

Fig. 9.1 (a) Gregor Mendel and **(b)** the Augustinian monastery in Brünn (now Brno in the Czech Republic), where he established the principles of heredity using garden peas. The plot where he grew his peas lies in the angle of the building, underneath the second floor window of his room

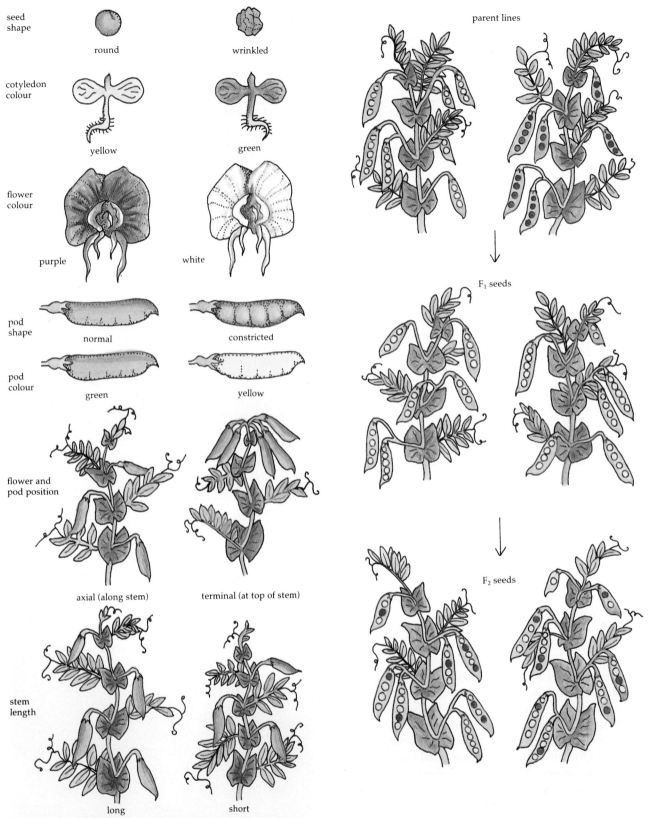

Fig. 9.2 Seven traits of garden peas studied by Mendel. Each trait occurs in two alternative forms including cotyledon colour (yellow or green), seed shape (round or wrinkled), flower colour (purple or white), pod shape (normal or constricted), pod colour (green or yellow), flower and pod position (along the stem or at the end) and stem length (long or short). In each case, the dominant form is listed on the left

Fig. 9.3 The results of Mendel's first type of experiment. He crossed two pure-breeding lines of peas, one with yellow seeds and one with green seeds. The progeny seeds resulting from such cross-fertilisations (F₁ seeds) were always yellow. These were grown into plants, which were allowed to self-fertilise to yield F₂ progeny seeds. On average, three-quarters of these seeds were yellow and one-quarter were green

getting yellow seeds, the green form reappeared. On average, three-quarters of the seeds were yellow and one-quarter were green.

Mendel explained his results by proposing that:

- yellow colour in seeds is determined by a factor Y and green colour by a factor y;
- the yellow trait is dominant and the green is recessive;
- pea plants carry two of these factors, which may be the same, YY or yy, or different, Yy;
- pea plants pass only one of these two factors on to their sperm or egg cells, that is, the factors separate (segregate) from each other;
- each gamete receives one or other of these factors with equal probability.

Monohybrid cross

We now recognise that these factors, Y and y, are alternative forms of a gene, called **alleles**. Individuals carrying two copies of the same allele are **homozygous** (YY or yy) and will produce gametes carrying only this allele. If self-pollinated, they will produce only homozygotes, that is, they will be **pure breeding**. Individuals carrying different alleles are **heterozygous** (Yy) and will, on average, produce gametes of which half carry one allele and half the other. If heterozygotes are self-pollinated ($Yy \times Yy$), they will produce both yellow and green offspring in the ratio 3:1 (Fig. 9.4).

The particular combination of alleles, for example, YY, Yy or yy, of an organism is its **genotype**, while its overall trait, for example, yellow or green, is its **phenotype**. Thus, in Mendel's experiments the yellow to green ratio of 3:1 is a phenotypic ratio. The genotypic ratio—homozygous YY (yellow):heterozygous Yy (yellow):homozygous yy (green)—is 1:2:1. Crosses such as this, which involve different alleles of a *single* gene, are **monohybrid crosses**.

Mendel demonstrated that these principles applied to the inheritance of six other traits in peas (Fig. 9.2), including seed shape, flower colour, pod shape, pod colour, flower and pod position, and stem length. He published the results of his experiments in 1866 but they were ignored until 1900 when others performed similar experiments and realised the importance of Mendel's findings some 34 years earlier.

Mendel recognised that individual traits are determined by discrete factors (genes) rather than being a blend of influences from each parent as earlier breeders had assumed. The factors remain discrete, like beads, rather than blending together like liquids. Mendel's first conclusion is often referred to as the **Principle of Segregation**. This can be stated as:

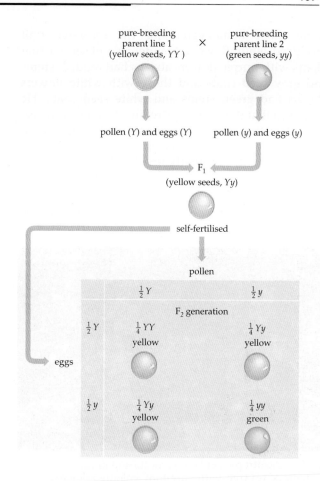

Fig. 9.4 Mendel's breeding program in which he followed the inheritance of seed colour in peas over two generations. This showed that the allelic forms of a gene controlling seed colour (Y and y) segregated from each other in equal numbers into the gametes formed by hybrid (F_1) plants. A matrix like this, which combines gametes into progeny, is a Punnett square, named after the British geneticist who first used it

alternative forms of an inherited trait are controlled by alternative forms of a gene (alleles), which segregate into gametes after meiosis with equal probability.

Mendel showed that heredity can be explained by discrete factors—alleles of a gene—which segregate into gametes in equal proportion. In a monohybrid cross involving alleles of a single gene, offspring will be produced in a phenotypic ratio of 3:1 and a genotypic ratio of 1:2:1.

Multiple effects of single genes

The gene for seed colour affects only one phenotypic trait, but sometimes a single gene may affect more than one characteristic. In his investigation of the inheritance of flower colour, Mendel noted that plants showing the dominant trait produced purple flowers, while those homozygous for the recessive trait produced white flowers. Mendel also observed that plants with purple flowers always had reddish stems and grey seed coats and those with white flowers always had green stems and white seed coats. He proposed that the factor controlling flower colour also controlled the colour of stem and seed coat. Geneticists today call this phenomenon, whereby one gene affects different, apparently unrelated characteristics, **pleiotropy** (Fig. 9.5).

> A pleiotropic gene affects different characteristics in the same organism.

Fig. 9.5 Pleiotropy can be seen occurring in this 'red damask' tea tree, with burgundy-coloured flowers, instead of white flowers, and dark-reddish leaves

CODOMINANCE AND BLOOD GROUPS

In our example of seed colour, only one allele, that associated with the dominant phenotype, is expressed in the heterozygote. However, for some genes, two different alleles can both be recognised as having an effect in heterozygotes.

In 1900, the same year that Mendel's work was rediscovered, Karl Landsteiner began experimenting with human blood. He separated blood into cells and serum and tested the effect of one person's serum on the red blood cells of another person. In some combinations, the red cells were clumped (agglutinated) by the serum, but cells were never clumped by their own serum. Landsteiner found that individuals could be placed into one of four groups, A, B, AB and O, depending on the pattern of interaction of their red cells with other sera.

We now know that the four blood groups are due to the presence of genetically determined antigens on the surface of red blood cells (Chapter 23) and that the ABO gene is located on chromosome 9. Group A individuals have A antigen on the plasma membrane of their red cells and anti-B antibodies circulating in

their serum (Table 9.1). Group B individuals have B antigen on their cells and anti-A antibodies.

The presence in the blood of antibodies to A and B antigens is *not* genetically inherited. Antibodies occur in response to contact with antigens in the environment. Some bacteria normally found in the gut have A-like and B-like antigens on their outer surface, which can provoke the immune system to produce anti-A and anti-B antibodies. However, if individuals are group AB, with both A and B antigens present on their red blood cells, the A-like and B-like antigens on the bacteria will be recognised as 'self' and no antibodies will be produced. On the other hand, if individuals are group O, with no AB antigens on their red blood cells, both antigens will be recognised as 'non-self', and anti-A and anti-B antibodies will be produced. Similarly, group A individuals will produce only anti-B antibodies, and group B individuals will produce anti-A antibodies.

Cells with particular antigens interact with matching antibodies in serum causing clumping. Thus, sera from group A individuals will clump cells from group B individuals but not cells from group A individuals, and vice versa. In AB individuals, red blood cells have both A and B antigens, with neither antibody present in their serum. Their cells are clumped by serum from either A or B individuals, but AB serum does not clump other red blood cell types. Finally, group O individuals have neither antigen on their red cells but both antibodies in their sera. Their red cells are not clumped by any other serum, while group O serum will clump all other red cell types— A, B and AB.

ABO blood groups are determined by three alleles, I^A, I^B and i. The I^A allele is responsible for producing the A antigen, I^B the B antigen, and the i allele produces neither. Thus, the genotype of individuals of type A phenotype are either I^AI^A or I^Ai, type B are I^BI^B or I^Bi, type AB are always I^AI^B and type O, the recessive phenotype, are always ii. Because the products of both alleles I^A and I^B are equally recognisable in the heterozygote, they are said to be codominant. **Codominance** is the full phenotypic expression of two alternative alleles in a heterozygote.

The A antigen is a complex carbohydrate with a chain of five sugars attached to the red cell membrane. The B antigen has the same length chain but has a different sugar on the outer end. In O individuals, the chain is only four sugars long. Thus, the ABO gene product is apparently an enzyme that adds the final sugar to the chain. The product of I^A adds one sugar (*N*-acetyl galactosamine), the I^B product adds another (galactose), while i does neither. This terminal sugar provides the specific shape that can combine with either type A or B antibodies.

It is important that the ABO blood group types are matched when giving blood transfusions. If they are not, then a large number of donor red cells bearing a foreign antigen may be introduced into serum containing antibody to that antigen. The incoming red cells will be clumped and the recipient may die. Alternatively, if the incoming red cells are, for example, type O (i.e. without A or B antigen), some problems may arise from the anti-A and anti-B antibodies present in the donor serum. These will clump recipient red cells if they have A or B antigen present. This effect is less severe than when donor cells are

clumped. The matching of ABO blood groups applies not only to blood transfusions. In organ transplantations, matching ABO and other antigen systems between donors and recipients is essential (Chapter 23).

Codominance is the full expression of two alternative alleles in a heterozygote.

Table 9.1 Characteristics of the human ABO blood group system, illustrating the antigens and antibodies present in each blood type

Blood group	Genotype	Antigens on red blood cells	Antibodies in serum	Blood
A	$I^A I^A$, $I^A i$	A	anti-B	
B	$I^B I^B$, $I^B i$	B	anti-A	
AB	$I^A I^B$	A, B	—	
O	ii	—	anti-A, anti-B	

Example of antigen–antibody interaction (red cell clumping):

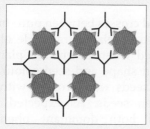

BOX 9.1   Rh BLOOD GROUPS AND PREGNANCY

In 1939 in the United States, a pregnant woman reacted violently to a blood transfusion from her husband. A week later she gave birth to a stillborn child. When tested, her blood was shown to contain antibodies that reacted both to her husband's blood and that of her child. It was suggested that the child had inherited from its father a factor that the mother lacked and against which she had developed antibodies. The antibodies were found to be anti-rhesus, named after the *Rhesus* monkey in which this blood group system was first identified. At least 85% of people are Rh positive and carry the rhesus factor; the remaining are Rh negative.

The discovery of the rhesus blood group system provided the explanation for a once common problem of pregnancy—*erythroblastosis fetalis* or haemolytic disease of the newborn. The Rh factor, which was absent in the mother in our example, is an antigen on the surface of red blood cells in Rh-positive people (the dominant phenotype) and is absent in Rh-negative people (the recessive phenotype), whose immune system recognises the Rh factor as 'foreign'.

Problems in pregnancy arise when an Rh-negative mother, *dd*, carries an Rh-positive baby, either *DD* or *Dd* (see figure). If the father is homozygous Rh positive (*DD*), this situation will occur in every pregnancy; if the father is heterozygous Rh positive (*Dd*), there is a 50% chance that a baby will be Rh positive. In ordinary circumstances, if an Rh-negative person has not been sensitised against the Rh factor, they will have no anti-Rh antibodies. However, during a problem pregnancy, and particularly during birth, some of the baby's Rh-positive red blood cells may enter the mother's circulation. The presence of the Rh factor on the baby's cells stimulates the immune system of the mother to produce anti-Rh antibodies. Usually this does not occur quickly enough to affect the current pregnancy, but the mother's immune system 'remembers' (Chapter 23). Upon a subsequent problem pregnancy, the mother, having been sensitised in this way, produces large amounts of antibodies very quickly after transfer of even a few of the baby's red cells. These antibodies pass without hindrance into the baby's circulation across the placenta and cause agglutination of the baby's blood. If this occurs late in the pregnancy, the baby may be born jaundiced; if seriously so, it may require a blood transfusion. If antibodies pass into the baby early in the pregnancy, the baby may die.

In many countries, erythroblastosis fetalis now rarely occurs. The blood group of all mothers is determined,

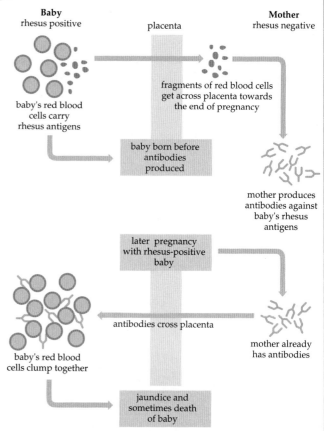

During pregnancy, an Rh-negative mother with an Rh-positive fetus may become sensitised to Rh antigen. This can cause problems in subsequent pregnancies with Rh-positive babies

and the blood group of her baby may be tested immediately after birth. If the results show that the danger combination exists, the mother is given an injection of anti-rhesus antibodies, passively immunising her (Chapter 23) against the rhesus factor. These antibodies quickly remove any Rh antigen that may have entered the mother's circulation and prevent her immune system from being sensitised. Passive immunity, used in this way for every problem pregnancy, has made erythroblastosis fetalis a problem of the past.

INHERITANCE OF COMBINATIONS OF GENES

Dihybrid cross

In addition to studying single traits, Mendel also studied the inheritance of *pairs* of different traits by carrying out **dihybrid crosses** between pure-breeding lines differing in two unrelated traits. He observed the progeny obtained when he crossed plants that differed in both seed shape and seed colour. Plants with round yellow seeds were crossed with plants with wrinkled green seeds. As expected, the F_1 generation showed both dominant phenotypes, yellow and round seeds. However, on raising these

F_1 plants and allowing them to self-pollinate, four types of F_2 progeny seeds were observed in the following numbers:

yellow round	315
yellow wrinkled	101
green round	108
green wrinkled	32
total	556

The results show that a dihybrid cross can produce new phenotypes (yellow wrinkled and green round) that are different from those of the parents (yellow round and green wrinkled).

We can summarise the dihybrid cross using the symbols Y for yellow, y for green, R for round and r for wrinkled seed:

Parents

$Y/Y;R/R$ × $y/y;r/r$

yellow round   green wrinkled

F_1 generation

$Y/y;R/r$ × $Y/y;R/r$

yellow round   yellow round

F_2 generation

$Y/Y;R/r$ & $Y/y;R/R$ & $Y/Y; R/R$ & $Y/y; R/r$

9 yellow round

$Y/Y;r/r$ & $Y/y;r/r$

3 yellow wrinkled

$y/y;R/R$ & $y/y;R/r$

3 green round

$y/y;r/r$

1 green wrinkled

Mendel realised that, if he looked at the two traits independently, about three-quarters of the F_2 seeds were yellow and one-quarter were green (416:140). Likewise, about three-quarters were round and one-quarter were wrinkled (423:133). These ratios were expected from his monohybrid crosses and his Principle of Segregation. Thus, he saw that seed colour and seed shape were inherited independently, the ratio 9:3:3:1 being effectively the product of two independent 3:1 segregations (Fig. 9.6). Mendel's results demonstrate that, during meiosis, the genes for seed colour and seed shape do not remain paired but segregate and recombine in different allelic combinations.

Back-crosses and test-crosses

Mendel tested his ideas by setting up another cross. Instead of allowing the F_1 plants ($Y/y;R/r$) to

self-fertilise, he crossed them back to the green wrinkled parent ($y/y;r/r$). A cross such as this between F_1 progeny and either of their pure-breeding parents is a **back-cross**. If it is to a parent with the recessive phenotype, then it is also a **test-cross**. In this case, the green wrinkled parent in the test-cross produces only one type of gamete—all yr. If the F_1 plant being crossed with the parent with the recessive phenotype produces the four types of gametes—YR, Yr, yR and yr—in equal proportions as predicted, then the progeny should occur in the proportions ¼ $Y/y;R/r$, ¼ $Y/y;r/r$, ¼ $y/y;R/r$ and ¼ $y/y;r/r$. Mendel observed the following numbers:

yellow round	55
yellow wrinkled	49
green round	51
green wrinkled	53
total	208

This is close to a 1:1:1:1 ratio as expected.

Mendel showed that this general pattern of inheritance applied to the inheritance of other pairs of traits tested in combination, even in cases where inheritance of three traits was examined. Mendel called this pattern of inheritance the **Principle of Independent Assortment**:

alternative forms of a gene controlling one trait assort into gametes independently of alternative forms of another gene controlling a different trait.

> When two different genes (each with two alleles) are inherited independently, the traits segregate in a 9:3:3:1 ratio in the F_2 generation or in a 1:1:1:1 ratio when back-crossed to a pure-breeding parent of the recessive phenotype (a test-cross).

GENES ARE ON CHROMOSOMES

Sex linkage

Mendel did not know how his factors (genes) were organised inside cells. However, soon after his work was rediscovered in 1900 it became clear that the transmission of genes from generation to generation closely paralleled the transmission of chromosomes (Chapter 8). Strong evidence that genes are located on chromosomes came when geneticists studied the inheritance of traits associated with sex chromosomes, that is, **sex-linked genes**.

BOX 9.2   KEY TERMS IN GENETICS

The following summarises the terminology that is used in genetics.

1. *Genes* are Mendel's discrete hereditary factors that determine traits.

2. *Alleles* are different forms of a gene.

3. *Locus* is the position of a gene on a chromosome; a locus may be occupied by any one of the alleles of a gene.

4. Organisms in which the members of a pair of alleles are different, for example, *Yy*, are *heterozygous*, and those in which both alleles are the same, for example, *YY* or *Yy*, are *homozygous*. A homozygous organism will be pure breeding for the trait type determined by the particular allele.

5. *Genotype* is the genetic make-up of an organism, that is, its specific allelic composition, for example, *Yy* genotype. *Phenotype* is the detectable properties or traits of an organism. It is the result of both the genotype and the environment. For example, a hydrangea grown in acidic soil has blue flowers, while a cutting from the same plant (same genotype) grown in alkaline soil has pink flowers (see figure). *Dominance* results in the expression of the same phenotype, such as yellow colour in seeds, by both the homozygous (*YY*) and the heterozygous (*Yy*) genotypes.

6. *Wild type* is the phenotype found in most individuals in a population.

7. A *monohybrid cross* involves crossing organisms that are heterozygous at one locus, for example, *Yy* × *Yy*.

8. A *dihybrid cross* involves crossing individuals that are heterozygous at two different loci. The cross involves the segregation of alleles of two genes.

9. *Autosomes* are chromosomes that have the same appearance in males and females. *Sex chromosomes* differ in appearance and number in males and females and are involved in sex determination. In organisms with an XX/XY sex-determining system, sex-linked genes occur on the X chromosome.

(a)

(b)

Flower colour in hydrangeas is influenced by the environment. Flowers are **(a)** pink if soil is alkaline, and **(b)** blue if soil is acidic

Genetic notation

In our discussion of Mendel's experiments on pea plants, we described the phenotypes of seeds in terms of the appearance of each character, for example, yellow round seeds or green wrinkled seeds. A geneticist would not do this. Instead, only phenotypes different from the wild type are specified (Table 1).

Table 1 Alternative phenotypic notations

Actual phenotype	Phenotypic notation
Yellow, round	Wild type
Yellow, wrinkled	Wrinkled
Green, round	Green
Green, wrinkled	Green, wrinkled

To summarise genotypes, capital letters are used for alleles associated with dominant phenotypes (e.g. *Y*) and small letters for alleles associated with recessive phenotypes (e.g. *y*). The full genotype notation *Y/Y;r/r* represents alleles of two separate genes, the semicolon indicating that the two genes assort independently. In a cross, such as *Y/Y;R/R* × *y/y;r/r*, the order of the two genes is consistent for the two parents. It is incorrect to present the cross in the order *Y/Y;R/R* × *r/r;y/y*.

Sometimes, alleles associated with the wild-type phenotype are shown as '+'. Thus, the notation +/+;+/+ is equivalent to *Y/Y;R/R* (Table 2).

Table 2 Types of genotypic notation

Type 1	Type 2	Type 3
YYRR	*Y/Y;R/R*	+/+;+/+
Yyrr	*Y/y;r/r*	+/y;r/r

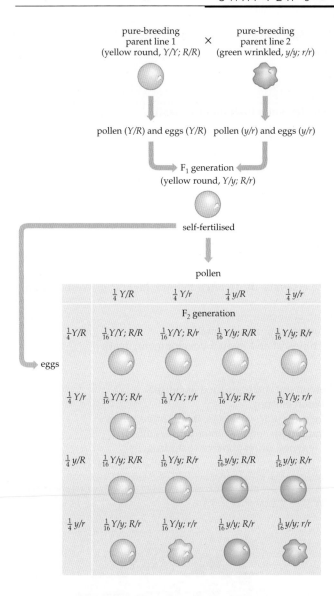

Fig. 9.6 Mendel's breeding program in which he followed the inheritance of both seed colour and seed shape in peas simultaneously. This showed that alleles of a gene controlling seed colour (*Y* and *y*) and alleles of a gene controlling seed shape (*R* and *r*) assort independently of each other into gametes of hybrid (F₁) plants, giving one-quarter of each combination

Most eukaryotes have two types of chromosomes, **autosomes**, which are the same in appearance and number in males and females, and **sex chromosomes**, which differ in appearance or number between males and females and are involved in sex determination. Genes that are responsible for human genetic variants that cause diseases may be autosomal or sex-linked (Table 9.2).

Autosomes are chromosomes that are the same in males and females. Sex chromosomes differ between males and females and are involved in sex determination.

In insects, such as *Drosophila*, and in mammals, the most common sex chromosome pattern is for males to have an X chromosome and a smaller Y chromosome and for females to have two X chromosomes. On average, half the sperm produced by a male carries an X chromosome while the other half carries a Y (Fig. 9.7). All eggs carry one or other of the two X chromosomes of the female. When these gametes come together at fertilisation, half the progeny will therefore be XX (females) and half XY (males).

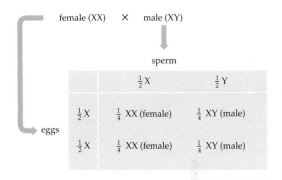

Fig. 9.7 Pattern of inheritance of sex chromosomes in humans. A male has an X and a Y chromosome, a female has two X chromosomes. Gametes have one of each. Sperm have either the X or the Y in equal numbers. Eggs have one or other of the two Xs in equal numbers. On average, half the progeny are male and half are female

In organisms with an XX (female) and XY (male) sex-determining system, the alleles of sex-linked genes are passed on to the next generation as if there are two in females and one in males. In other words, sex-linked genes are located on the X chromosome. In humans, one sex-linked trait is red-green colour blindness (Fig. 9.8). Most of us can distinguish red and green colours but about 8% of Caucasian males have difficulty in doing so. They have an abnormality in either the red-sensitive pigment or the green-sensitive pigment of their retina. Colour-blind males do not transmit this gene to their sons but they do to their daughters, who have a 50% chance of inheriting the gene.

This inheritance pattern can be accounted for by proposing that red-green colour blindness is a recessive phenotype, the gene for this trait is on the X chromosome; and there is no comparable gene on the Y chromosome (i.e. colour blindness is sex-linked). Males therefore can be either X^C (normal) or X^c (colour-blind) whereas females can be X^CX^C, X^CX^c (both normal) or rarely X^cX^c (colour-blind). Colour-blind males will always pass on the Y chromosome (without a colour vision gene) to their sons. However, their daughters will always receive the X^c gene on their father's X chromosome. If the daughter receives an X chromosome bearing the X^C allele from her mother, she will be heterozygous and thus a carrier.

Table 9.2 Examples of human traits and diseases determined by single autosomal or sex-linked genes (*a*)

Inheritance pattern	Frequency (1000 births)	Pathology
Autosomal dominant		
Huntington's chorea	0.1	Progressive dementia from mid life; death
Achondroplasia	0.04	Dwarfism
Myotonic dystrophy	0.05	Progressive muscular weakness
Neurofibromatosis	0.25	Tumours of peripheral and other nerves
Polycystic kidney disease	1.0	Progressive kidney failure
Osteogenesis imperfecta	0.05	Brittle bones; blue sclera
Autosomal recessive		
Phenylketonuria	0.1	Mental retardation unless diet controlled
Cystic fibrosis	0.4	Mucoid secretions, lung congestion
Beta-thalassaemia	0–10	Severe anaemia; frequency highest in Mediterranean and Oriental people
Sickle cell anaemia	0–10	Haemolytic anaemia, death often follows; highest in West African people
Albinism	0.1	No pigment; abnormal vision
Adrenal hyperplasia	0.1	Heterogeneous; abnormal genitalia
Mucopolysaccharidosis	0.05	Abnormal physical and mental development; death
Sex-linked (*b*)		
Red-green colour blindness	80	Red and green not distinguished as normal
Haemophilia A and B	0.2	defect in blood clotting
Muscular dystrophy	0.3	Progressive muscular weakness; death often follows
Fragile X-syndrome	0.9	Mental retardation; some women carriers affected
Testicular feminisation	0.02	Sterile XY females; unresponsive to androgens

(*a*) Frequencies per 1000 births are approximate and, except for haemoglobin diseases, relate to populations of northern European origin. Note that sex-linked diseases occur much more frequently in males. This is because in females the recessive phenotype is usually masked by the dominant phenotype, whereas in males, which have only one copy of the sex-linked gene, the recessive phenotype is expressed.
(*b*) Frequency is per 1000 *male* births.

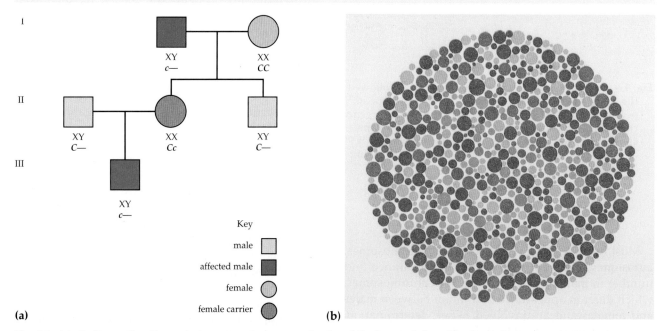

(a)

(b)

Fig. 9.8 (a) Pedigree (family tree) showing inheritance of colour blindness. Colour blindness in humans is caused by a defect in pigments of the retina that are sensitive to red or green light. This leads to an inability to distinguish red and green colours in the normal way. The genes controlling red–green colour vision are sex-linked, that is, on the X chromosome. An affected man always passes on the colour-blind *c* allele to his daughters but never to his sons. His daughter may pass the *c* allele on to her children with a 50% chance. In the pedigree shown, the son is colour-blind because he received his mother's X chromosome, bearing the *c* allele rather than that bearing the wild-type *C* allele.
(b) A test plate used for detecting colour blindness. Tests for colour blindness cannot be conducted with this material. For accurate testing, the original plates should be used

If a mother is a heterozygote carrier ($X^C X^c$), then her sons have a 50% chance of being colour-blind. If the father is also colour-blind ($X^c Y$), half the daughters will be affected. For a colour-blind homozygote mother ($X^c X^c$), all of her sons will be colour-blind and all of her daughters carriers. The inheritance of colour blindness from generation to generation follows exactly the inheritance of individual X chromosomes.

> Sex-linked genes are on the X chromosome in organisms with an XX/XY sex-determining system.

Y-linkage

If a character is passed from father to son and never observed in females, it is **Y-linked**, that is, the gene for that character must be on the Y chromosome. In mammals, the main Y-linked gene is the sex-determining gene or testis-determining factor. It has recently been cloned and is called SRY (**s**ex-**r**elated gene on **Y**). This gene switches development of the early fetus along the male pathway; if the Y chromosome (carrying this gene) is absent, the fetus becomes female. Once the particular sex-determining pathway has been activated, other genes that control male or female development take over. These are scattered on all the autosomes and a few are on the X chromosome. Other examples of Y-linked characters are rare.

> Y-linked genes are on the Y chromosome. Characters determined by Y-linked genes are passed from father to son and never occur in females.

Linkage on autosomes

As homologous chromosomes segregate at meiosis, so do the genes on these homologues. Therefore, genes located on a particular chromosome will be transmitted together, that is, they will be linked and will not follow Mendel's Principle of Independent Assortment. However, linkage between genes is never complete because of the exchange of segments between homologous chromosomes during crossing over (Chapter 8). Crossing over leads to recombination between genes in a pair of chromosomes.

A direct test for independent assortment is to test-cross a double heterozygote with the double recessive homozygote ($A/a;B/b \times a/a;b/b$). The four possible combinations of gametes from meiosis (AB, Ab, aB, and ab) are expected to form in equal proportions in the double heterozygote. Thus, the phenotypes of the test-cross progeny are expected in the proportions 1:1:1:1 if the two genes assort independently. For example, in the Australian sheep blowfly, *Lucilia cuprina*, a pure-breeding wild-type

strain has red eyes and straight bristles, and a pure-breeding mutant strain has white eyes and crooked bristles. When these two strains are crossed, the F_1 progeny have red eyes and straight bristles, indicating that these phenotypes are dominant. F_1 female flies are then test-crossed to males of the mutant strain (Fig. 9.9). (Note that a test-cross does not work in the opposite direction because there is no crossing over in male blowflies.) The results of this test-cross are shown in Table 9.3.

Table 9.3 Test-cross offspring

Phenotype	Number of individuals
Red eyes, straight bristles	96
White eyes, straight bristles	6
Red eyes, crooked bristles	7
White eyes, crooked bristles	91
Total	200

In this example, there are four phenotypes but not in the ratio 1:1:1:1 that we would expect from the Principle of Independent Assortment. Red eyes and straight bristles are inherited together as are white eyes and crooked bristles. However, if we compare eye and bristle traits separately, then the ratio of red eyes (96 + 7) to white eyes (91 + 6) is 103:97, which is approximately 1:1. Similarly the ratio of straight bristles (96 + 6) to crooked bristles (91 + 7) is 102:98, also close to the expected 1:1 ratio.

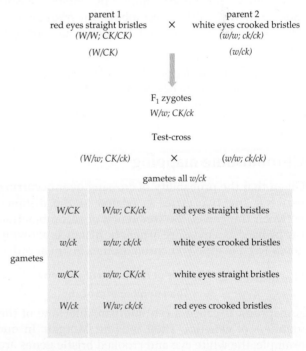

Fig. 9.9 A genetic explanation of the phenotypic ratios observed in test-cross offspring involving the F_1 generation. The original cross was between a pure-breeding wild-type (red eyes, straight bristles) strain of the Australian sheep blowfly and a pure-breeding strain of white eyes and crooked bristles

These data are consistent with the hypothesis that there is a single gene for eye colour and a single gene for bristle type, with the wild-type phenotype being dominant in both cases. The Principle of Segregation applies to the alleles of each gene, but the different genes do not assort into gametes independently of each other.

The data show that the phenotypes of the parental strains have been inherited together. Thus, red eyes and straight bristles are linked in inheritance, as are the inheritance of white eyes and crooked bristles. Only in a few cases (13 out of 200 or 6.5%) are the traits inherited in different combinations. These new phenotypes—red eyes crooked bristles, and white eyes straight bristles—are referred to as **recombinants**.

Because linked genes are on the same chromosome, the genetic notation for them is different from that for genes on separate chromosomes (genes that segregate independently). Linked genes are represented as:

$$\frac{A \quad B}{a \quad b}$$

where the lines represent two homologous chromosomes on which the alleles of the genes are located. The position of a particular gene on a chromosome is called a locus. The frequency of recombination will depend on the distance between the two genes. The closer the loci of two genes, the less often crossing over (recombination) occurs between them. Those sufficiently far apart may, in effect, assort independently, with a recombination frequency of 50% as would be expected for genes on separate chromosomes.

> Linkage is the tendency of two or more genes on the same chromosome to be inherited together.

Chromosome mapping

Given that the probability of crossing over occurring between any two genes is related to the distance between the genes, the frequency of recombination can be used as a measure of genetic distance, allowing genes to be mapped on chromosomes in linear order:

$$\text{Map distance} = \frac{\text{Number of recombinant progeny}}{\text{Total number of progeny}} \times 100$$

A map unit is called a **centimorgan**, after one of the founders of genetics, Thomas Hunt Morgan. In our example, the white eye and crooked bristle genes are 6.5 centimorgans apart.

Consider another gene, rusty brown colour (*ru*), which is linked to both white eye and crooked bristle. A test-cross involving white eye and rusty colour

shows a 15% frequency of recombination, indicating that white eye and rusty colour are 15 centimorgans apart. A further test-cross involving rusty colour and crooked bristles shows a 21.5% frequency of recombination, indicating that the genes for rusty colour and crooked bristles are 21.5 centimorgans apart. Thus, the genetic map is:

$$ck \longleftarrow 6.5 \longrightarrow w \quad\underset{\displaystyle \xleftarrow{\hspace{2cm}} 21.5 \xrightarrow{\hspace{2cm}}}{\longleftarrow 15 \longrightarrow} ru$$

or

$$ru \longleftarrow 15 \longrightarrow w \longleftarrow 6.5 \longrightarrow ck$$

The definite order of these genes can be determined using other linked genes as markers to construct a map of the relative positions of all the known genes on a chromosome. A map of some of the genes on chromosome 1 of the vinegar fly, *Drosophila melanogaster*, is shown in Figure 9.10.

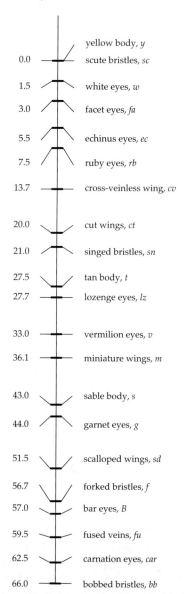

	yellow body, *y*
0.0	scute bristles, *sc*
1.5	white eyes, *w*
3.0	facet eyes, *fa*
5.5	echinus eyes, *ec*
7.5	ruby eyes, *rb*
13.7	cross-veinless wing, *cv*
20.0	cut wings, *ct*
21.0	singed bristles, *sn*
27.5	tan body, *t*
27.7	lozenge eyes, *lz*
33.0	vermilion eyes, *v*
36.1	miniature wings, *m*
43.0	sable body, *s*
44.0	garnet eyes, *g*
51.5	scalloped wings, *sd*
56.7	forked bristles, *f*
57.0	bar eyes, *B*
59.5	fused veins, *fu*
62.5	carnation eyes, *car*
66.0	bobbed bristles, *bb*

Fig. 9.10 (a)

(b)

Fig. 9.10 **(a)** Map of some of the genes on chromosome 1 (X chromosome) of *Drosophila melanogaster*. The positions of genes are relative to the gene closest to the end of the chromosome, and are in the map units, centimorgans. **(b)** *Drosophila melanogaster*, the vinegar fly, much studied as a model genetic organism for over 80 years

PHENOTYPIC INTERACTIONS: VARIATION IN MENDELIAN RATIOS

Incomplete dominance

Lack of dominance of one phenotype over another occurs in most organisms. For example, in snapdragons, *Antirrhinum*, flower colour is controlled by a single gene with two alleles. A cross between a homozygous red-flowered snapdragon and a homozygous white-flowered snapdragon will give a pink F_1 hybrid (Fig. 9.11). When pink F_1 plants are self-pollinated, they produce an F_2 generation containing a mixture of red-, pink- and white-flowered plants in the ratio 1:2:1.

Flower colour in snapdragons is referred to as **partially** or **incompletely dominant** because the phenotype of the heterozygote is intermediate between that of each homozygote. Incomplete dominance may occur if one allele produces one unit of a gene product while the other allele produces none. In this case, the homozygote for the active allele would produce two units of pigment, the heterozygote one and the other homozygote none.

> A character is incompletely dominant when the phenotype of the heterozygote is intermediate between that of each homozygote.

Novel phenotypes

We have already seen that the inheritance of combinations of gene pairs can be followed in individual breeding programs or pedigrees. For example, Mendel looked at the transmission of factors for both seed

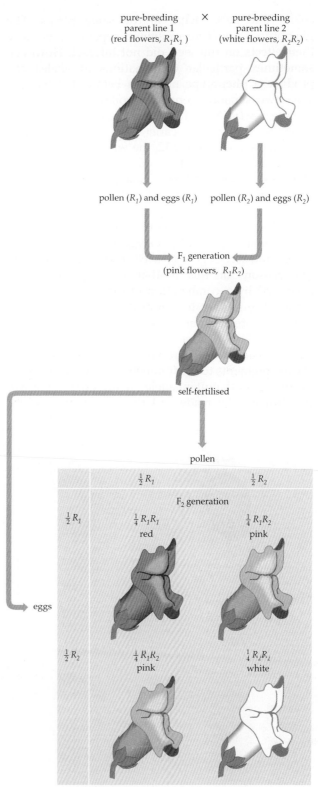

Fig. 9.11 Pure-breeding lines of snapdragons, *Antirrhinum*, show different flower colours—red, pink or white. When a red-flowered line is cross-pollinated with a white line (pollen parent), the F_1 generation is pink. When F_1 plants are self-pollinated, the phenotypes in the F_2 generation include all three colours. These data show that colours are controlled by two alleles, R_1 (red) and R_2 (white). Since heterozygous R_1R_2 plants have pink flowers, dominance is incomplete

colour and seed shape in the same plants. The phenotypes could be examined separately because their effects on the seed did not interact. However, sometimes particular combinations of alleles can produce a phenotype not seen before. Let us consider two separate genes controlling eye colour in *D. melanogaster* (Fig. 9.12). These are brown, with two alleles *Bw* and *bw*, and scarlet, with two alleles *St* and *st*. As shown in Figure 9.12, the wild-type eye colour in *D. melanogaster* is reddish-brown due to the presence of at least one *Bw* and one *St* allele. Flies with brown eyes are homozygous recessive for brown (*bw/bw;St/St* or *bw/bw;St/st*), while flies with scarlet eyes are homozygous recessive for scarlet (*Bw/Bw;st/st or Bw/bw;st/st*). However, flies that are homozygous recessive for both brown and scarlet genes (*bw/bw;st/st*) have white eyes. The reason for this is that wild-type colour is produced by a combination of red and brown pigments. The mutant brown allele (*bw*) is defective in producing red pigments so that the eye appears brown in brown homozygotes. The scarlet mutant allele (*st*) cannot make brown pigments, thus allowing the red pigments to predominate. When a fly is homozygous for both mutant alleles, neither red nor brown pigment can be made and the eye appears white.

Unexpected phenotypes may arise as a result of interaction between the phenotypes of different genes.

Epistasis

A different form of interaction between two different genes can occur if the phenotype controlled by one masks that of the other. This is the phenomenon of **epistasis**. For example, genes for eye colour in *Drosophila* cannot be expressed in a homozygous eyeless mutant! Epistasis has been studied in the plant *Arabidopsis thaliana*, a small weed species of the mustard family (Fig. 9.13). This plant is ideal for experiments because it is easy to grow in large numbers, it has only a six week generation time and many mutants are known. For example, one mutant with a recessive phenotype, *agamous* (without gametes), produces flowers that are abnormal. Petals form in place of the stamens and in place of the carpels, resulting in an attractive double flower (lacking male and female organs). Another mutant, also with a recessive phenotype, *leafy*, lacks flowers almost completely. F_1 plants heterozygous for both *leafy* and *agamous* have a normal non-mutant phenotype. When F_2 progeny are raised from self-fertilised F_1 plants, the expected phenotypic ratios are 9 wild type : 3 agamous : 3 leafy : 1 agamous leafy. However, it is not usually possible to determine whether or not plants of leafy phenotype are also agamous; since they have no flowers, the agamous phenotype cannot be observed. Thus, the *leafy* phenotype hides, or is epistatic to, the *agamous* phenotype. The modified ratio in the F_2 generation is therefore 9 wild type : 3 agamous : 4 leafy. Other types of gene interaction can occur and lead to different modified F_2 ratios.

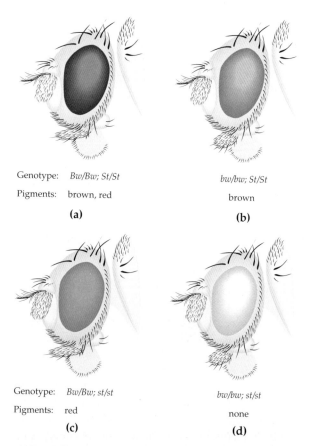

Genotype: *Bw/Bw; St/St*   *bw/bw; St/St*

Pigments: brown, red   brown

(a)   **(b)**

Genotype: *Bw/Bw; st/st*   *bw/bw; st/st*

Pigments: red   none

(c)   **(d)**

Fig. 9.12 Eye colour phenotypes of **(a)** wild type and two mutants **(b)** brown and **(c)** scarlet of *Drosophila*. **(d)** A different eye colour phenotype, white, occurs when the two mutants, both homozygous recessive, are bred together. This is an example of gene interaction

Fig. 9.13 (a)

(b)

(c)

Fig. 9.13 Flowering stems of the model plant, *Arabidopsis thaliana*, a small, easily grown species of the mustard family: **(a)** wild type with normal flowers and a seed pod; **(b)** *agamous* mutant (recessive phenotype) with many extra petals in place of stamens; and **(c)** mutant *leafy* (recessive phenotype), in which normal flowers do not arise. It is not possible to observe the agamous phenotype in leafy plants as the leafy phenotype masks it, that is, *leafy* is epistatic to *agamous*

Epistasis is an interaction between different genes such that the expression of an allele of one gene interferes with or masks the expression of an allele of another gene.

Phenotypes and environment

The effect of an allele can be masked or enhanced by variation in the environment. A well-known example of this is the coat colour mutation carried by Siamese cats (Fig. 9.14). The wild-type allele C results in fur being pigmented over all parts of an animal. In homozygotes for the mutant allele c^s, pigmentation is restricted to fur at the extremities of the face, ears, legs and tail. The mutant allele can produce pigment only at lower body temperatures, which occur at these extremities. At the higher temperatures on the main body surface, the product of the mutant c^s allele is almost inactive. Thus, expression of the c^s allele is affected by an environmental factor, in this case, body temperature.

Fig. 9.14 A Siamese cat homozygous for a mutant pigment allele, c^s. This allele is more active at lower temperatures, such as occur at the extremities of the animal's body

Variation in gene expression involves both expressivity and penetrance. **Expressivity** refers to the degree to which an allele is expressed in an individual, while **penetrance** refers to the proportion of individuals in a population carrying the allele that show a phenotypic effect. For example, in *D. melanogaster*, flies with the dominant phenotype *Curly* (*Cy*) have wings that curl upwards (Fig. 9.15). Curvature is caused by unequal contraction of the upper and lower epithelia of the wings during the drying period, after emergence of the adult from the pupal case. In a population of flies that are all heterozygous *Curly*, not all will show expression of the trait; some flies will have normal wings, like the wild type. If the temperature of the environment is 25°C, most flies will develop curly wings, but at 19°C, fewer flies will have curly wings. The extent of wing curling also varies depending on the temperature. If 80% of flies in a population of heterozygotes develop curly wings, then the gene is said to be showing 80% penetrance under these environmental conditions.

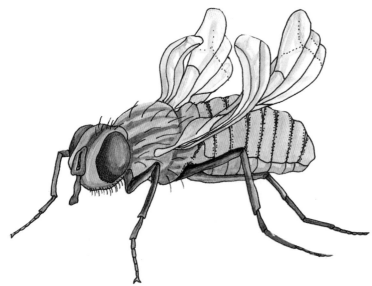

Fig. 9.15 Expression of the mutant *Curly* (*Cy*) in *Drosophila melanogaster*

Environmental effects on an individual organism may be so severe that it will resemble a mutant phenotype even though it is wild type. Such an organism is known as a **phenocopy**. For example, if we grow *Arabidopsis thaliana* on a medium containing the toxic metal cadmium, growth of the plant is stunted and the phenotype appears to be that of a stunted mutant named *dwarf* (*dw*). The plant grown on cadmium medium is a phenocopy of the dwarf mutant. However, it is genetically wild type and, if planted on cadmium-free medium, will grow normally. If crossed with a dwarf mutant, progeny will segregate as 3 wild type : 1 dwarf (mutant) in the F_2 generation. A phenocopy is not an environmentally induced mutation. It still has the wild-type '+' allele at the *dwarf* locus. The environment causes a phenotype that mimics the effect of a mutation.

The effect of a deleterious mutation may sometimes be overcome by changing the environment in which a mutant individual is raised. For example, the serious effects of the phenylketonuria mutation in humans (Table 9.2) can be prevented by modifying the diet of affected individuals. The wild-type product of this gene is an enzyme that converts the amino acid phenylalanine into tyrosine, an amino acid used in the synthesis of proteins and the pigment melanin. In recessive homozygotes, phenylalanine (and its phenylketone breakdown products) accumulate. This affects normal brain development leading to mental deficiency. As a pleiotropic effect of the mutant allele, affected individuals usually lack melanin pigment in their hair (Fig. 9.16).

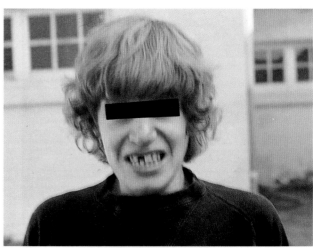

Fig. 9.16 A child with untreated phenylketonuria. The mutant results in a build-up of the amino acid phenylalanine and its breakdown products in the body tissues and the urine. Note that the child's hair has a low level of melanin pigment, a pleiotropic effect of this mutant

If the defect is detected early enough, an affected infant can be placed on a low phenylalanine, high tyrosine diet. Low phenylalanine intake reduces the amount of circulating toxic products, while the tyrosine makes up for the amino acid deficiency. A modified diet prevents brain damage occurring in early childhood and restores the individual to a phenotype not readily distinguishable from normal. This treatment is so successful that screening of all newborn infants is done routinely in Australia and other countries even though the frequency of affected individuals is only about 1 in 10 000 live births. A small sample of blood is tested for higher than normal levels of phenylalanine. Those babies are then tested further to determine if they are homozygous for the mutant form of this gene. If so, they are placed immediately on the modified diet.

> The effect of some genes on the phenotype of an organism can be masked or enhanced by variation in environmental effects. Expressivity is the degree to which a gene is expressed in an individual and penetrance is the proportion of individuals in a population carrying the gene that show the trait. A phenocopy is genotypically wild type but phenotypically mimics a mutant.

Quantitative characters

Up to now we have been examining the inheritance of genes by following their effects on characters that differ qualitatively between individuals. Qualitative characters can be used to group organisms into distinct phenotypic classes, for example, yellow or green seeds. Organisms also have quantitative characters such as height and weight in humans, milk yield in cattle, and yield and growth rate in cereals and legumes. **Quantitative characters** exhibit continuous variation and phenotype is measured along a continuous scale.

Quantitative characters are influenced by the combined action of a number of genes, each of which produces only a small effect. Grain colour in wheat (Fig. 9.17), for example, is controlled by two different genes A and B, each with two alleles A and a, and B and b. The A and B alleles produce red pigment; the a and b alleles do not. The effects of both the A gene and the B gene work additively in establishing seed colour, and a pure-breeding $AABB$ line has dark red seeds because it has four doses of red pigment. A pure-breeding $aabb$ has white seeds due to a complete absence of pigment. When a pure-breeding $AABB$ line is crossed with a pure-breeding $aabb$ line, the F_1 progeny are $AaBb$ and have medium red grains, with two pigment doses, one from A and one from B. The F_2 seeds from self-fertilised F_1 plants will display a range of five different colours ranging from dark red to white depending on how many doses of pigment are present in each grain (Fig. 9.17).

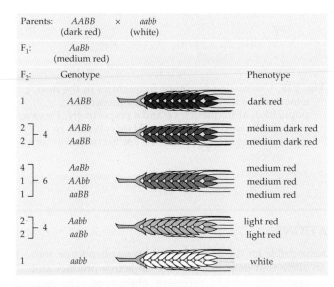

Parents:	$AABB$ (dark red)	×	$aabb$ (white)	
F_1:	$AaBb$ (medium red)			
F_2:	Genotype			Phenotype
1	$AABB$			dark red
2 ⎤ 4 2 ⎦	$AABb$ $AaBB$			medium dark red medium dark red
4 ⎤ 1 ⎥ 6 1 ⎦	$AaBb$ $AAbb$ $aaBB$			medium red medium red medium red
2 ⎤ 4 2 ⎦	$Aabb$ $aaBb$			light red light red
1	$aabb$			white

Fig. 9.17 Genetic control of a quantitative character, the amount of red pigment in wheat seeds. Two genes are involved, each with two alleles (A and a, and B and b, which act without dominance. AA produces twice as much red pigment as Aa, while aa produces none. The B allele works similarly and additively with the A allele. Thus, F_2 plants may have anything from none to four doses of pigment and range in phenotype from white to dark red

The amount of pigment in a wheat grain may also depend to some extent on its environment; for example, more pigment may be produced if a plant is in full sun than if it is in the shade. Thus, the amount of pigment in each of the five phenotypic classes of

F_2 grains may vary. If environmental variation is high, the amount of pigment in each of the five classes may overlap, resulting in a continuous variation in the character seed colour.

Many quantitative characters are controlled by more than two independent genes that act additively. Multiple individual genes each having a small additive effect on a quantitative character are **polygenes**.

When plant or animal breeders want to generate a new breed or variety, they select individuals carrying the properties they desire and breed from them. If the selected individuals are better because of the genes they carry, the overall performance of the selected line will improve. In some cases, a breeder can fix the desired genes by inbreeding to make the new line homozygous (pure breeding). In the F_2 wheat grains in Figure 9.17, for example, a variety with dark red grains could be established by selecting and inbreeding only from the dark red individuals. In a more realistic case, a tomato breeder would select plants that have the desired size, shape, growing period, response to fertilisers and resistance to fungus attack and inbreed from them. In this way, a new strain could be developed that has desirable polygenic characters.

> Quantitative characters show continuous variation and are controlled by both environment and polygenes. Polygenes are genes that each have a small additive effect on a character.

GENES IN POPULATIONS

Allele frequencies

For most genes there is one allelic form that occurs in the majority of a population. This is the wild-type allele (Box 9.3). However, to account for differences between members of a population, many genes must exist in more than one allelic form. The presence in a population of more than one allele of individual genes at levels higher than could be maintained by mutation alone is known as **genetic polymorphism**. Surveys have shown that many populations are polymorphic for 20–40% of genes.

We can estimate the frequency of different alleles of a gene in a population by examining the proportion of individuals that have particular phenotypes and therefore particular genotypes. For example, the human MN blood group is controlled by two codominant alleles, L^M and L^N. The heterozygote is blood type MN and the homozygotes are types M or N.

Consider a population of Aboriginal people from Elcho Island in Australia's Northern Territory (Table 9.4). To calculate the frequency of the L^M allele, remember that each member of the population with M type blood (28 individuals) is carrying two L^M alleles and each member with MN type blood (129) is carrying one. Therefore, the number of L^M alleles in the population is $2 \times 28 + 129 = 185$. The L^M **allele frequency** is the proportion of this allele relative to the total number of alleles in the population. For a population of 352 individuals with a total of 704 alleles (2×352), the L^M allele frequency is therefore $185 \div 704 = 0.26$. The L^N allele frequency is 0.74 (Table 9.4) since allele frequencies for a gene in a population must add to 1. Allele frequencies are usually abbreviated p and q, with $p + q = 1$.

Table 9.4 The frequency of MN blood group alleles in a population of Australian Aboriginal people from Elcho Island

| | Blood group | | | |
	M	MN	N	Total
Genotype	L^ML^M	L^ML^N	L^NL^N	
Number of individuals	28	129	195	352
Number of L^M alleles	56	129	0	185
Number of L^N alleles	0	129	390	519
Total number of alleles	56	258	390	704

Frequency of L^M allele = 185/704 = 0.26 = p.
Frequency of L^N allele = 519/704 = 0.74 = q.

Genetic polymorphism is the presence in a population of more than one allelic form of a gene. Wild type is the most common allele. Many populations are polymorphic for 20–40% of genes.

The Hardy–Weinberg principle

Frequencies of genotypes in a population can be calculated by dividing the number of individuals of each type by the total population size (Table 9.4). We can also calculate genotype frequencies expected in the next generation by using allele frequencies calculated for the present generation. Imagine that the next generation is a result of the random combination of gametes. Figure 9.18 shows the possible combinations of genotypes in a Punnett square. With a frequency p of gametes of one type and q of the other, the three genotypes arise in the next generation in the proportions p^2, $2pq$ and q^2. The proportions add to 1 and can be deduced from expansion of the binomial expression $(p + q)^2$. These genotype frequencies are called **Hardy–Weinberg frequencies** after G. H. Hardy and G. Weinberg, who independently discovered the principle in 1908.

For our MN blood group example, 0.26 (26%) of the gametes in the population will carry the L^M allele

		sperm	
		$p\,A$	$q\,a$
eggs	$p\,A$	$p^2\,AA$	$pq\,Aa$
	$q\,a$	$pq\,Aa$	$q^2\,aa$

Fig. 9.18 Calculating genotype frequencies in the next generation. Two alleles A and a are present in a population at frequencies p and q respectively. Progeny occur in the next generation in Hardy–Weinberg genotype frequencies: p^2 for AA, $2pq$ for Aa and q^2 for aa

BOX 9.3 WILD TYPES AND MUTANTS IN POPULATIONS

Generally, wild-type alleles are common in natural populations. Other alleles, such as those selected by early garden pea breeders to add variety to cultivated strains of peas, and those studied by Mendel, occur naturally but at low frequencies.

How do different forms of a gene arise? Once geneticists started following individual genes in breeding programs, they noticed that very occasionally genes changed from one form into another by the process of **mutation**. Thus, an allele can mutate, with the direction of change usually being reversible. Mutations may occur spontaneously and can be caused by certain agents (mutagens), such as X-rays and various chemicals (Chapter 32). Mutagens can be used to increase the frequency of mutations, for example, in plant breeding, although the inherited effect of a new mutation does not usually show up until a later generation, if at all. This is because most mutations change an allele that produces an active product and which is responsible for the dominant wild-type phenotype, into an allele that does not produce an active product and is associated with a recessive phenotype. The new mutant allele is only revealed in homozygous individuals. For example, to create mutations in flower colour, plant breeders expose seeds to a mutagen, germinate them and self-pollinate the plants; mutant forms will appear in the next generation.

By contrast, somatic effects of mutation may show up straight away in individuals in which the mutation has occurred. For example, cancer may arise after specific mutations occur in somatic cells. Survivors of the 1945 World War II atomic bombing of Hiroshima and Nagasaki, who received large doses of mutagenic radiation, showed greatly increased levels of leukaemia, cancer of white blood cells. However, to date, no significant changes have been detected in the genes that survivors passed on to their children.

(*p*) and 0.74 will carry L^N (*q*). Thus, in the next generation, 0.07 (p^2) will be blood type M, 0.39 (*2pq*) will be type MN, and 0.54 (q^2) will be type N. These Hardy–Weinberg genotype frequencies expected for the next generation are close to those of the current population.

Hardy–Weinberg genotype frequencies are expected to remain constant from generation to generation, given several assumptions:

- the interbreeding population is infinitely large;
- alleles mutate at the same rate;
- genotypes do not migrate into or out of the population at different rates;
- individuals mate at random;
- phenotypes have the same Darwinian fitness, that is, the chance of an individual producing fertile adult offspring is the same whatever the individual's phenotype.

If these assumptions are met, we would expect genotype frequencies to remain stable. If one or more of these assumptions is not met, allele frequencies in a population may change and the population may evolve.

Population size

The first assumption is that the population is extremely large. If a population is very small, or if it goes through periodic 'bottlenecks' of low numbers, some alleles in the population survive or are lost purely by chance. Random change in allele frequencies in small populations is **genetic drift** (Chapter 32). It leads to rapid fluctuations in allele frequencies, with the possibility that the frequency of a particular allele will be reduced to 0 or become fixed at 1, the population becoming homozygous. The remaining world population of cheetahs is homozygous for many genes, suggesting that small population size and genetic drift has led to homozygosity. Genetic drift also seems to account for high levels of homozygosity in small populations of the Australian bush rat, *Rattus fuscipes greyi*, on islands off Eyre Peninsula in South Australia (Fig. 9.19). Animals have been stranded on these islands by rising sea levels in the past and isolated for many generations.

Populations with low levels of genetic variation may survive but they are at risk if their environment changes and if genotypes fixed by chance are less well-adapted to the new conditions. In wildlife conservation, it is important to assess the level of genetic variation in natural populations to determine if variation is sufficient to allow them to respond to future environmental changes. Conserving genetic variation is particularly important for populations of Australian mammals that exist only in isolated fragments of their former wide distribution.

(a)

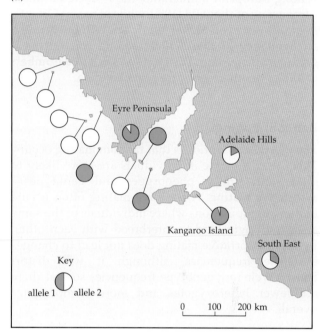

(b)

Fig. 9.19 **(a)** Bush rats, *Rattus fuscipes greyi*, from South Australia. **(b)** Thirteen populations of bush rats were screened for alleles of a gene controlling an enzyme present in heart and kidney cells. Allele frequencies are shown as pie diagrams. Three mainland populations and one large island population contained both alleles (i.e. they were polymorphic). Nine small island populations contained only one or other of the alleles, probably because of small population size and genetic drift

Random change of allele frequencies in small populations is genetic drift.

Mutation rate

The second assumption underlying Hardy–Weinberg genotype frequencies is that different alleles mutate at the same rate. Mutation ultimately provides genetic diversity, the raw material of evolution. However, mutation rates are usually very low and, even if different alleles mutated at different rates, this by itself is unlikely to lead to rapid change in populations.

Migration

The third assumption is that individuals with particular alleles do not migrate at different rates. Often this is not the case. Australia's human population has undergone great changes in genetic composition over time. In the Australian Aboriginal population, allele frequencies are quite different from the rest of the world. Immigrants from Europe, and more recently South-East Asia, have contributed to the diversity of the Australian population. Indigenous Aboriginal people, with the exception of a few individuals from the Cape York area, completely lack the I^B allele of the ABO blood group alleles (Box 9.4). Among European immigrants, the I^B allele occurs at a frequency of about 0.1 (10%), while in recent Oriental migrants it is even higher (0.2 or 20%). Thus, the overall frequency of the I^B allele among Australians is gradually increasing. The types and relative numbers of alleles of other genes are also changing.

Random mating

The fourth assumption is that **random mating** occurs. This is not the case if individuals are more likely to mate with similar phenotypes and genotypes— **assortative mating**. Assortative mating often occurs in human populations where individuals of the same ancestral origin tend to interbreed with each other. By itself, assortative mating does not lead to changes in allele frequencies, although it will distort Hardy–Weinberg genotype frequencies so that there are fewer heterozygotes and more homozygotes overall.

Darwinian fitness

Finally, constant Hardy–Weinberg frequencies will occur only if there are no differences in the **Darwinian fitness** of individual genotypes. Fitness is measured as relative reproductive success, and if one genotype leaves relatively more fertile offspring per head than others, the genotype frequencies in the next generation will not be in Hardy–Weinberg proportions.

For example, consider the case in which the recessive phenotype (homozygote) leaves no offspring. Relatively few of the recessive alleles will be transmitted to the next generation and their frequency will fall (all other factors being equal). In this way, deleterious alleles remain at low frequencies, with their loss usually being balanced by their creation through new mutation events. This is likely to be the case with severe human genetic diseases in which the affected individual does not reproduce, for example, before the cause of phenylketonuria was recognised and treated successfully by diet control.

Differences in fitness may also arise if environments change, leading to newly favoured phenotypes increasing in a population. The best known example of this is the evolutionary change in the peppered moth, *Biston betularia*, in industrial Britain (Fig. 9.20).

This moth rests during the day on the trunks of beech trees, relying on its camouflage to avoid being preyed on by birds. This worked well before the Industrial Revolution because the moth's wings had patterns closely matching the lichens covering the tree trunks. However, atmospheric pollution from factories established during industrialisation killed off the lichen on the trees and the peppered moths became clearly visible.

In pre-industrial times, a form of the moth with uniformly dark wings was seen occasionally. This resulted from the production of a dark pigment (melanin) by a dominant mutant gene (*D*). When lichen was present on the tree trunks, the dark phenotype was at a disadvantage because *DD* homozygote and *Dd* phenotypes were selectively eaten by predators leading to a relatively low contribution to the next generation (low Darwinian fitness). However, when the environment changed and the lichen disappeared it became advantageous to have dark wings matching the soot-covered bark. The dark, *DD* and *Dd*, phenotypes of the moths were now at a selective advantage and their Darwinian fitness was consequently higher than that of the *dd* form.

Selection acts on the phenotype and the frequency of the dark form was observed to increase from very low levels to 0.98 (98%) in the 45 years from 1850 to 1895. Thus, an ecological change led to an evolutionary change in the moth population. Reduced natural selection against the dark phenotype resulted in it increasing through the population.

Fig. 9.20 (a)

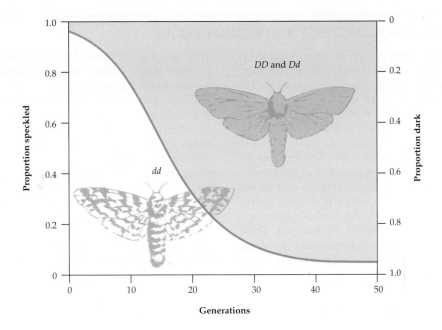

(b)

Fig. 9.20 Peppered moths, *Biston betularia*, from industrial Britain. **(a)** The speckled form (*dd*) and the dark melanic form (*DD* and *Dd*) are shown together on dark pollution-affected bark. **(b)** The predicted proportions of the two forms are shown over 40 generations of natural selection against the speckled form, if its relative reproductive success is only 80% of that of the dark form. After 40 generations, the dark form has greatly increased in number, representing more than 90% of the population

According to the Hardy–Weinberg principle, genotypes occur in populations in predictable frequencies, given several assumptions: large population size, random mating and no differences in allele mutation rates, phenotype migration rates or Darwinian fitness. Evolutionary change may result if these assumptions do not apply.

Applications of the Hardy–Weinberg principle

Calculations of Hardy–Weinberg genotype frequencies can be useful in predicting frequencies of alleles that are difficult to distinguish in the heterozygote because of dominance. For example, in the human genetic disease phenylketonuria, affected children are born in Australia at the rate of about 1 in 10 000 live births (a frequency of 0.0001). If the wild-type allele occurs in the population at the frequency p and the mutant allele at the frequency q, then we expect affected phenylketonuria zygotes to arise at the frequency q^2 under Hardy–Weinberg assumptions: that is, q^2 is 0.0001, and q is 0.01. Heterozygotes are likely to be present at the frequency $2pq$, or $2(0.01 \times 0.99) = 0.02$. In the Australian population, about 2% (1 in 50) of the population are likely to be carriers of a mutant phenylketonuria allele. There is a large reservoir of mutant phenylketonuria alleles in the Australian population, mostly carried silently in heterozygotes even though they

come together to form homozygous affected individuals in only about 1 in 10 000 children.

Balanced polymorphism

In some circumstances, natural selection can act to modify Hardy–Weinberg genotype frequencies without leading to an increase in the favoured phenotype. This occurs if there is **heterozygote advantage**. If the phenotype of the heterozygote is fitter in the Darwinian sense than the phenotypes of either homozygote (i.e. if it contributes relatively more to the next generation), then allele frequencies may change initially but will eventually reach an equilibrium point. The relative loss of one allele, for example, *A*, from the homozygote *AA* balances the relative loss of the other allele, *a*, from the homozygote *aa*. This leads to a stable **balanced polymorphism**.

Sickle cell anaemia is a well-known example of balanced polymorphism. The sickle cell allele results in a single amino acid change in the β chain of human haemoglobin, changing the tertiary structure of the protein. Individuals who are homozygous for this sickle cell allele suffer from severe life-threatening anaemia associated with distortion of their normally disc-shaped red cells into sickle-shaped forms. Heterozygous carriers are phenotypically normal under most conditions, although their red cells can be made to sickle under certain conditions in the

BOX 9.4   ALLELE FREQUENCIES OF ABO BLOOD GROUPS

ABO blood group alleles occur in polymorphic frequencies almost throughout the world. In northern Europeans, about 41% of individuals are group A, 11% are group B, 3% are group AB and 45% are group O. Allele frequencies can be calculated from these by assuming that Hardy–Weinberg genotype frequencies apply. The frequency of I^A is thus 0.26 (26%), I^B is 0.08 (8%) and i is 0.67 (67%). When world distributions are examined, the I^A allele is seen to occur at frequencies higher than this in parts of North America and among Aboriginal people in Australia, where it peaks at around 0.45 in Aboriginal people from the Central Desert and Murray–Darling Basin (see figure). By contrast, the I^B allele occurs at much higher frequencies in Asian populations (up to 0.30), and is mostly absent in native American and Aboriginal Australian peoples. Americans, at least those from South America, are almost monomorphic for the i allele.

These patterns of distribution reflect mutation, migration and genetic drift among human populations over many millennia. For example, the I^B allele may have arisen by mutation relatively recently in Asia and be moving into European and African populations through recent migrations. Those few I^B alleles in Australian Aborigines in the Cape York area may have come from recent Papua New Guinean immigrants. Genetic drift might have played a role in establishing the homogeneous type O in South American populations. These may have originated from relatively few immigrants and become fixed for the i allele by chance events.

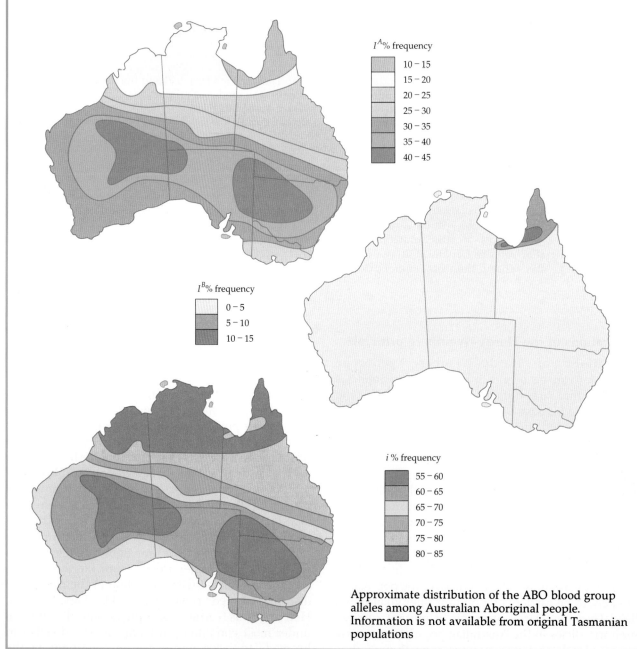

I^A% frequency
- 10 – 15
- 15 – 20
- 20 – 25
- 25 – 30
- 30 – 35
- 35 – 40
- 40 – 45

I^B% frequency
- 0 – 5
- 5 – 10
- 10 – 15

i % frequency
- 55 – 60
- 60 – 65
- 65 – 70
- 70 – 75
- 75 – 80
- 80 – 85

Approximate distribution of the ABO blood group alleles among Australian Aboriginal people. Information is not available from original Tasmanian populations

laboratory. The potentially lethal allele occurs at high frequencies (up to 10%) in certain African populations.

The distribution of the sickle cell allele closely matches the distribution of one form of malaria, an often-lethal parasitic disease (Fig. 9.21). Surveys of populations affected by malaria in West Africa showed that carriers of the sickle cell allele (heterozygotes) were less severely affected by malaria than were normal homozygotes. They had shorter bouts of fever, lower parasite counts in their blood and were hospitalised less often. It seems that red cells infected by the malarial parasite are rapidly removed from the blood in sickle cell heterozygotes. Thus, in malarial environments, sickle cell heterozygotes are at a selective advantage relative to both normal homozygotes (severely affected by malaria) and sickle cell homozygotes. Even though sickle cell homozygotes usually leave no offspring, the advantage gained by heterozygous carriers of the sickle cell allele is so high that the allele remains in the population at a balanced frequency of up to 10%.

The distributions of some other blood diseases are associated also with a high incidence of malaria. For example, thalassaemia is found in populations distributed from the Mediterranean through the Middle East, South Asia and South-East Asia and in parts of Africa where malaria is (or was) endemic (Fig. 9.21).

Thalassaemia also affects haemoglobin. In some alleles that cause β-thalassaemia, the production of the β chain is reduced; in others production is blocked. Homozygotes for some β-thalassaemia alleles are severely anaemic and may die unless they receive repeated blood transfusions. As a balancing selective agent, it seems likely that heterozygous carriers of β-thalassaemia are more resistant to malaria than are normal homozygotes, thus keeping the frequency of these otherwise deleterious alleles high.

In a malaria-free environment, the sickle cell allele and thalassaemia alleles do not confer a heterozygote advantage and will eventually be lost from the population. This is especially the case in emigrant populations such as Afro-Americans and Australians of Mediterranean and Asian origin. In fact, loss of the now deleterious alleles in Western populations is accelerating through prenatal diagnosis. In couples who are both carriers, testing can determine if the fetus is an affected homozygote who, if born, could expect a short life of low quality.

Polymorphisms can be maintained in balance if the heterozygote is fitter than either homozygote.

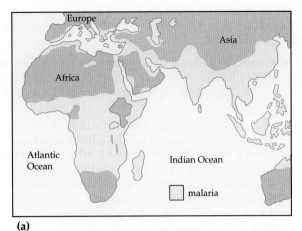

(a)

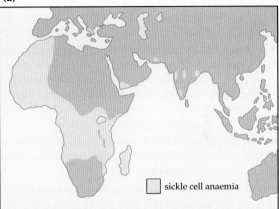

(b)

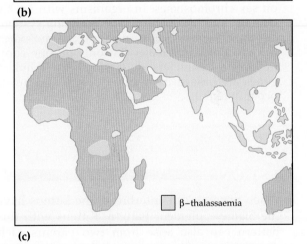

(c)

Fig. 9.21 (a) Former and present distribution of falciparum malaria, showing closely paralleled distribution of alleles for the human blood diseases (b) sickle cell anaemia and (c) β-thalassaemia. Sickle cell anaemia and β-thalassaemia homozygotes are severely anaemic and may not survive to adulthood. These deleterious alleles are maintained in malarial environments because heterozygous carriers are more resistant to malaria than normal homozygotes and leave relatively more offspring

SUMMARY

- Genotype is the genetic make-up of an organism and phenotype is an organism's observable traits, dependent on both genotype and environment.

- Phenotypic traits are often controlled by alternative forms of a gene (alleles). Individuals carry two alleles of each gene, which separate (segregate) into gametes. Half the gametes have one of the two alleles, half have the other. This is Mendel's *Principle of Segregation*.

- When gametes are formed, the segregation of alleles of one gene into gametes has no influence on the segregation of the alleles of another gene. This is Mendel's *Principle of Independent Assortment*.

- Phenotypes may be dominant, codominant or recessive. A dominant phenotype is expressed in a heterozygote. In codominance, there is full phenotypic expression of two different alleles in a heterozygote. A recessive phenotype is only evident in a homozygote.

- The expression of a gene often interacts with the expression of other genes (epistasis) as well as the environment.

- Most genes are on autosomes, chromosomes that look the same in the two sexes. Some genes are on sex chromosomes. In organisms with an XX (female) and XY (male) sex-determining system, sex-linked genes occur on the X chromosomes in two copies in females but only on one copy in males. Characters determined by Y-linked genes are passed from father to son and never occur in females.

- Absence of independent segregation of genes indicates linkage of genes on the same chromosome. Linked genes may be separated if crossing over occurs between them, resulting in recombinants. The frequency of recombination can be used to map genes on a particular chromosome.

- Traits that vary over a continuous scale may also be controlled by genes. Such quantitative traits result from the additive action of many genes (polygenes), each with only a small effect.

- Genes often occur in more than one form in a population (polymorphism). The proportions of each genotype can be predicted by calculating allele frequencies, based on the Hardy–Weinberg principle. Hardy–Weinberg phenotypic frequencies remain stable under certain conditions: the population must be large; individuals must mate at random; and genotypic frequencies must not be subject to the effects of genetic drift, differential mutation, migration or natural selection.

- Evolutionary change in a population can be viewed as resulting from changes in allele frequency. Such changes can result from genetic drift, differential migration of genotypes, and differences in Darwinian fitness between genotypes. However, if the heterozygote is fitter than either homozygote, genotype frequencies are maintained in a stable polymorphic balance.

QUESTIONS

1. When Siamese cats interbreed, the kittens have the Siamese pigment pattern. Kittens with this pattern can also arise from two parents each showing a non-Siamese pattern. Is the phenotype for Siamese pattern dominant or recessive? Explain your answer.

2. Polled cattle lack horns, a dominant phenotype arising as the result of the action of an autosomal gene *P*. Horned cattle are thus *pp* homozygotes. If a polled bull and a polled cow have a horned calf, what are the genotypes of the parents? If the same bull was bred with many horned cows, what proportions of polled and horned offspring are expected overall?

3. Consider your characteristics and those of your parents. How do Mendel's two principles explain the inheritance of maternal and paternal characteristics?

4. How are sex-linked genes inherited? Give an example.

5. If a gene occurs in three allelic forms, how many possible genotypes exist? In the absence of dominance, how many possible phenotypes will there be? If two of the alleles result in codominant phenotypes and the third in a recessive phenotype, how many phenotypes are possible?

6. A colour-blind man of ABO blood group type O and a woman with normal vision of blood type AB have two children. The older is a colour-blind boy of blood type A. The younger is a girl with normal colour vision also of type A. Draw up a pedigree (family tree) showing the genotypes of all family members.

7. Design an experiment to test if a new chemical increases the rate of mutation of the allele for

the recessive phenotype green seed colour in peas to the allele for the dominant phenotype yellow seed colour. How would you measure the rate of occurrence of new mutations in the opposite direction?

8. What is a polygene? Do polygenes control qualitative or quantitative traits? Explain.

9. About 1 in 2500 Australians are born with cystic fibrosis, a recessive autosomally inherited genetic disease. Estimate the frequency of the cystic fibrosis allele. What proportion of the Australian population are carriers? What proportion of couples are both carriers? What is the risk that such couples will have a child with cystic fibrosis?

10. Five general assumptions underly Hardy–Weinberg genotype frequencies. In real populations these assumptions may not apply. Which assumptions break down in populations showing (a) genetic drift, (b) assortative mating and (c) reduced reproductive success of one genotype?

GENES, CHROMOSOMES AND DNA

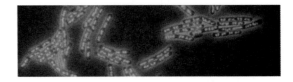

The idea that chromosomes are directly involved in inheritance was first formulated in the early twentieth century, after observations of the pairing of chromosomes during meiosis. In 1902, Walter Sutton formulated the *chromosomal theory of inheritance*, arguing that genes are on chromosomes and that chromosomes segregate during meiosis in a way similar to the factors of Mendel's model of inheritance. Sutton's theory, however, did not account for the observation that there are many more independently assorted traits than there are chromosomes. Later workers observed chiasmata and crossing over between pairs of homologous chromosomes (p. 149), which explained how genes on the same chromosome could assort independently.

The realisation of the link between chromosomes and inheritance led to the discovery of the molecular structure of genes and how genes function. DNA provides the genetic link between generations. For this role, not only must its base sequence be copied accurately during replication, but its integrity needs to be maintained by DNA repair mechanisms.

CHROMOSOMES AND DNA

In 1868, only three years after publication of Mendel's experiments, Friedrich Miescher isolated a substance, nuclein, from pus-soaked bandages of wounded soldiers. Pus contains large numbers of white blood cells, the source of nuclein. Subsequent painstaking work by others led to the idea that nuclein, which proved to be a mixture of DNA and proteins, was the material from which chromosomes were made. With this came the exciting realisation that if the precise composition of nuclein could be described, then it might be possible to explain inheritance at the molecular level.

For a long time DNA was regarded as too simple a molecule to convey the information of inheritance. DNA, with just four bases, was thought to act as some form of scaffold or support for proteins, which were believed to be the 'real' information carriers. Today, we know that the four bases of DNA encode the genetic message (Chapter 1).

Chromosomal DNA in the nucleus of a eukaryotic cell is the nuclear **genome** (Fig. 10.1). The genome of a prokaryote is its single circular chromosome. Chloroplasts and mitochondria (derived from prokaryotes, Chapter 3) have their own genomes, which are also circular chromosomes.

In addition to chromosomal DNA, prokaryotes can have **plasmids**, small circular double-stranded DNA molecules, often carrying genes conferring antibiotic resistance. In eukaryotes, plasmids may be present in the nucleus. Bacterial plasmids are useful in recombinant DNA technology (Chapter 12).

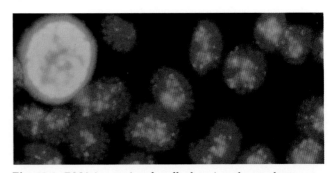

Fig. 10.1 DNA in a spinach cell, showing the nuclear genome (blue) and chloroplast genomes (red)

Genomes

The amount of DNA in a genome varies substantially between organisms, whether measured in terms of number of chromosomes or actual DNA content in number of bases (base pairs, bp; Table 10.1). The bacterium *Escherichia coli* has a genome of 4.7×10^6 bp (or 4700 kilobases, kb). In contrast, there are about 6×10^9 bp (6 000 000 kb) in a diploid human cell with 46 chromosomes (or half this amount in each sperm or egg, which are haploid).

While most organisms are diploid, some have more than two sets of chromosomes in each cell. A eucalypt somatic cell is diploid, whereas oilseed rape is tetraploid (four sets of chromosomes) and bread wheat is hexaploid (six sets of chromosomes).

> The genome is the total chromosomal DNA in a cell or organelle.

Table 10.1 Number of chromosomes of selected organisms (a)

Organism	Chromosome number	Base pairs (kb)
Bacterium, *Escherichia coli*	1	4 700
Yeast, *Saccharomyces cerevisiae*	18	28 000
Amoeba, *Amoeba*	50	
Garden pea, *Pisum sativum*	14	
Maize, *Zea mays*	20	9 000 000
Radiata pine, *Pinus radiata*	24	
Eucalypt, *Eucalyptus*	22	
Horsetail, *Equisetum*	216	
Adder's tongue fern, *Ophioglossum reticulatum*	1262	
Nematode, *Caenorhabditis elegans*	12	200 000
Vinegar fly, *Drosophila melanogaster*	8	330 000
Bandicoot, *Perameles*	14	
Human, *Homo sapiens*	46	6 000 000
Gorilla, *Gorilla gorilla*	48	
Salamander, *Amphiuma*	24	153 000 000
Horse, *Equus calibus*	64	

(a) For eukaryotes, the diploid number for the nuclear genome of somatic cells is given.

The arrangement of genes along the length of a chromosome can be described by either genetic or physical maps. As we saw in Chapter 9, genetic maps are usually obtained from mating experiments. They provide a description of the relative positions of identified genes in terms of recombination distances, but do not provide information on the position of unidentified genes. Although genetic loci may be mapped to a particular region of a chromosome, genetic maps do not necessarily identify the precise site of the locus. In contrast, a physical map describes the positions of genes on chromosomes in terms of molecular parameters (Fig. 10.2). Usually this physical description is provided in terms of the number of bases involved, measured in kilobases.

> Genetic maps describe the relative positions of identified genes in recombination distances, while physical maps define distances between loci in kilobases.

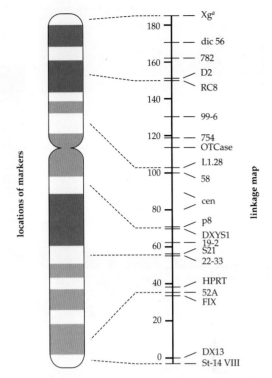

Fig. 10.2 A physical map of the human X chromosome. The marked locations indicate the positions of specific 'signposts', determined by restriction enzyme cleavage, which can be used to characterise the genome further. Distances between markers (left) are not directly related to the linkage map (right), in which distances are estimated by recombination

TYPES OF DNA SEQUENCES

Although a mammalian genome is about a thousand times larger than the *E. coli* chromosome, there is only 50 times more protein-coding information. The enormous variation in the amount of DNA in the genome of different organisms suggests that there is considerable DNA that does not code for polypeptides. In fact, as little as 2% of mammalian DNA codes for proteins and the functions of the remaining 98% of the DNA are only now being discovered.

Several types of DNA sequences are found in chromosomes (Fig. 10.3):

- genes, where a **gene** is all the DNA sequences necessary to produce a single RNA molecule and/or polypeptide;

- origin and terminus sequences involved in DNA replication;

- centromeres and telomeres of eukaryote chromosomes, and partitioning sequences of prokaryote chromosomes;

- repetitive DNA sequences, including both spacer DNA between genes and satellite DNA;

- transposable elements.

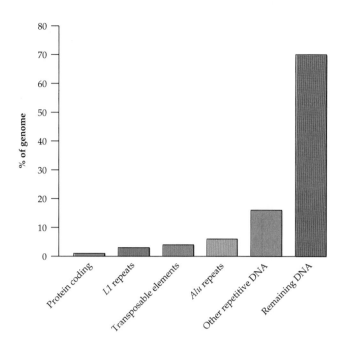

Fig. 10.3 Relative amounts of DNA sequences in the human genome. The proportion of protein-coding DNA is remarkably low, especially compared with the abundance of repetitive sequences. *L1* repeats have many long, dispersed repeated sequences

Genes

While crossing experiments suggest that genes are inherited as particulate factors (Chapter 9), a gene is no longer considered to be a single continuous sequence of protein-encoding DNA. A *gene* includes the coding region as well as regions of DNA immediately preceding and following it. The coding region of a gene includes individual coding sequences, **exons**, and intervening non-coding sequences, **introns**. Exons encode either polypeptide sequences or RNA, and are typically much shorter than introns (Fig. 10.4).

Most genes encoding proteins exist as single copies. However, some protein-encoding genes have been duplicated one or more times, and individual copies have been modified during the course of evolution. These related genes, for example, the genes that encode globin proteins, constitute a **gene family**. In contrast, specialised regions of chromosomes containing ribosomal RNA genes (rDNA), which provide coding information for the RNA components of ribosomes, are found in the nucleolus (Chapter 3) and exist in multiple copies. In most eukaryotes, there are more than 100 tandem (i.e. head to tail) repeats of rDNA. In certain toads, there are more than a million copies of rDNA, which make up about three-quarters of the genome. Many of these

help to code for the large number of ribosomes needed to produce yolk in egg cells. In contrast, in the *E. coli* chromosome, rDNA sequences are repeated seven times, constituting around 1% of the total genome.

Sequences associated with replication and division

The second type of DNA sequence is concerned with the process of DNA replication when a chromosome copy is made during cell division (Chapter 6). It includes a start sequence, the **origin**, and an end sequence, the **terminus**. In a circular prokaryotic chromosome, there is typically one origin, whereas along a eukaryotic chromosome, there are many origins of replication.

Telomere and centromere sequences are found only in eukaryotes. **Telomeres** are the DNA sequences found at the ends of chromosomes. They are important for the functional stability of linear chromosomes. In the absence of telomeres, chromosomes interact with one another, forming multiple structures, and are also sensitive to enzymic breakdown. A **centromere** is the site of attachment of a chromosome to the spindle, its position varying between chromosomes. Centromeres are involved in the separation of the two daughter chromosomes after DNA replication (Chapter 8).

The sequences of centromeres of the yeast *Saccharomyces cerevisiae* are among the best characterised. For each of the 16 yeast chromosomes, every centromere shows a unique sequence, but shares some similarity with other centromeres. For example, the yeast sequence for the centromere on chromosome 3 shows an 88 bp region high (93%) in adenine and thymine bases flanked by two shorter regions of conserved DNA.

Prokaryotic counterparts to the centromere are **partitioning sequences**, which are attached by specific protein linkages to the growing plasma membrane in the region where the cell is to divide.

Repetitive sequences

Apart from tandemly repeated genes, for example, those encoding ribosomal RNA, eukaryotic cells contain other small sequences of DNA that are repeated many times and have no known function. These are referred to as **repetitive DNA** (Fig. 10.3). As a result of these repetitive sequences (and other non-coding sequences) eukaryotic genes are spaced apart compared with prokaryotic genes. An example of repetitive DNA in the human genome is the *Alu* family, which is a class of approximately 300 bp of repetitive sequences present in about one million copies and representing about 6% of our total DNA.

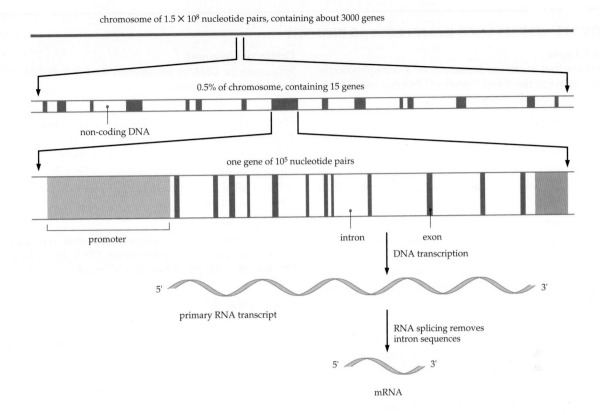

Fig. 10.4 A gene comprises a coding region as well as regions of DNA immediately preceding and following it. This scheme shows how genes are organised on a eukaryote chromosome. The promoter (regulatory region) is usually, but not always, located on the 5′ side of the gene. DNA-binding proteins bind to this region to initiate transcription. A long primary RNA transcript is produced, and later the intron sequences are removed (RNA splicing). The final mRNA transcript comprises the spliced sequences complementary to the exons

In the mouse and many other eukaryotes, the centromere is flanked by many repeats of short simple sequences of DNA, **satellite DNA**, which is mainly found in heterochromatin (condensed and dark staining parts of chromosomes). In insects, such as *Drosophila virilis*, over a third of the genome consists of satellite DNA. In this case, there are over 10 million copies of a few 7 bp sequences. The function of satellite DNA is unknown but it may have structural and organisational roles.

Transposable elements

A **transposable element** is a DNA sequence that is able to replicate and insert the copy into another location in the genome. These elements were discovered in maize by Barbara McClintock in the early 1940s and have become known as 'jumping genes'. Since they can replicate and transpose, these transposable elements are self-maintaining in the genome, and have become a valued tool in genetic engineering (Chapter 12).

> In eukaryotic chromosomes, DNA sequences include genes, sequences involved in DNA replication and division, repetitive DNA and transposable elements. In prokaryotes, chromosomes are simpler with fewer types of DNA sequences.

REPLICATION

In 1953, James Watson and Francis Crick first proposed a double-helical (duplex) structure for DNA. This model suggested a possible mechanism for the replication of DNA, based upon the complementary nature of the two strands of the double helix. The process involves **semiconservative replication**, that is, each of the two old strands pairs with a newly synthesised complementary strand (Fig. 10.5). DNA is synthesised by unwinding and separating the double helix into two single strands, and using each strand as a template for the synthesis of its complementary counterpart (Fig. 10.5): *template-directed synthesis*. Replication thus results in the production of two identical double helices from one original double helix. This leads to the formation of complete

BOX 10.1   TERMS USED IN MOLECULAR BIOLOGY

DNA ligase
Enzyme that catalyses formation of a phosphodiester bond between the 3′ and 5′ ends of DNA segments base-paired to template strand

DNA polymerase
Enzyme catalysing the addition of bases to a growing strand of DNA in a 5′ to 3′ direction

Exon
Segment of eukaryotic gene that is transcribed into mRNA

Exonuclease
Enzyme that hydrolyses terminal phosphodiester bonds of nucleic acid

Intron
Segment of eukaryotic gene that is transcribed but excised before translation

Lagging strand
DNA strand that grows in 3′ to 5′ direction, but is synthesised discontinuously in short fragments (5′ to 3′) that are later connected by covalent bonds

Leading strand
DNA strand that is synthesised continuously in 5′ to 3′ direction

Origin and terminus
Sequences of DNA at which synthesis is initiated and terminated respectively

Primer
Short sequence, usually of RNA, that pairs with a strand of DNA and provides free hydroxyl end for DNA polymerase to commence synthesis of nucleotide chain

Replication fork
Point of separation of strands of duplex DNA at which replication occurs

Restriction enzyme
A site-specific endodeoxyribonuclease that recognises short sequences of unmethylated DNA and cleaves the double strands either at the recognition site or nearby

Transposable element
DNA sequence able to insert itself into another position in the genome

BOX 10.2   ONE GENE, ONE PROTEIN

Many experiments have led to our current understanding of the relationship between genes and proteins.

Genes code for enzymes

A developmental study of eye colour in *Drosophila* in the 1930s led to the conclusion that genes are responsible for the production of enzymes. Two red-eyed mutants, *vermilion* and *cinnabar*, were used in the larval transplantation experiments outlined in the figure. Wild-type eye colour is brownish-red as a result of a combination of red and brown pigments. When an eye disc from a larva of either mutant was transplanted to the abdomen of an embryonic wild-type larva, it developed wild-type colour. When an eye disc from a *vermilion* larva was transplanted to the abdomen of a *cinnabar* larva, it also developed wild-type colour. However, when the reverse experiment was done, the transplanted *cinnabar* eye remained *cinnabar* when the *vermilion* larva became an adult. The conclusions from these studies were that the production of brown eye pigment involves at least two steps, which are controlled by different enzymes, and that the mutants lacked one or other of these enzymes.

The first step in the production of brown eye pigment involves an enzyme produced by the wild-type allele at the *vermilion* locus. It takes place in the general body tissue of a larva and produces a diffusible intermediate product. The second step involves the enzyme produced by the wild-type allele at the *cinnabar* locus. The intermediate product produced by the first step diffuses into eye cells where the second step takes place, producing brown pigment. Production of brown pigment therefore cannot occur when a *cinnabar* mutant eye is transplanted into a *vermilion* mutant larva because *vermilion* larval tissue is unable to make the intermediate product.

One gene: one enzyme

Neurospora is a fungus capable of synthesising virtually all of its nutritional needs, including amino acids and vitamins, from inorganic materials. Individual spores were irradiated to induce mutations and then grown until fruiting bodies formed. Each of the eight haploid ascospores in single fruiting bodies was grown separately on complete medium containing amino acids and vitamins. These cultures were then individually transferred to separate tubes of minimal medium containing only inorganic nutrients. Some continued

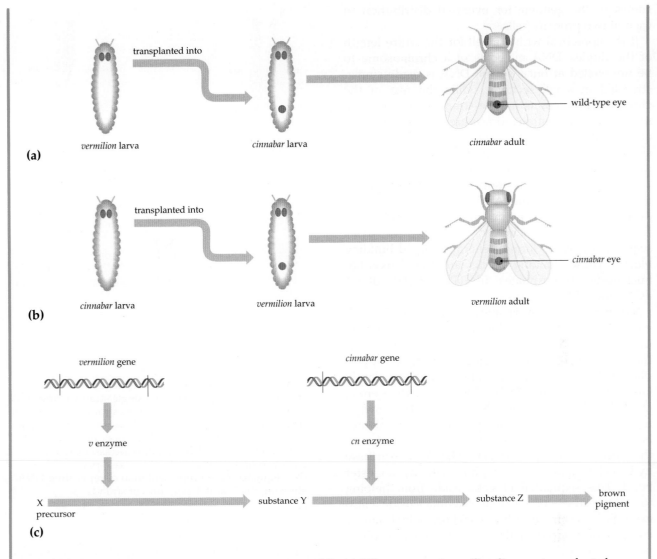

Transplantation experiments using eye discs of *Drosophila*. (a) When a mutant *vermilion* disc was transplanted into the embryonic abdomen of a *cinnabar* mutant larva, the disc behaved non-autonomously, developing into wild-type pigment. (b) The reciprocal experiment, a *cinnabar* disc transplanted into a *vermilion* mutant, resulted in autonomous development of *cinnabar* pigment. (c) This led the experimenters to deduce that in the biosynthesis of eye pigment, the vermilion gene product is produced before the cinnabar gene product

to grow while others did not. Those that did not grow were tested on different media to which a single amino acid, vitamin or other organic compound was added. Successful growth indicated which compound could not be synthesised and therefore, in which pathway a mutation had occurred. Using this procedure, many hundreds of nutritional mutations were identified, one for almost every enzymatically controlled reaction. Thus, *one gene codes for one enzyme*.

One gene: one polypeptide

Although enzymes are proteins, many proteins are not enzymes. A study of sickle cell anaemia and haemoglobin structure led to an expanded hypothesis:

one gene codes for one polypeptide. Pedigree studies of sickle cell anaemia showed that it is due to the mutation of a single gene with alleles for two codominant phenotypes (Chapter 9). Electrophoresis of haemoglobin from diseased and normal individuals identified two types of haemoglobin. Normal haemoglobin (Hb A) has a higher net negative charge than does sickle haemoglobin (Hb S). Heterozygotes have both Hb A and Hb S. Further biochemical study showed that Hb S has a *single* amino acid change (valine for glutamic acid) in the β-haemoglobin chain. (Adult haemoglobin is composed of two α and two β chains; Chapter 21.) Thus, a single gene provides the genetic information for the formation of one polypeptide, and a mutation resulting in the substitution of a single amino acid can cause a change in phenotype.

copies of the genome for eventual distribution to each of two progeny cells.

It is impractical within a cell for the entire length of the duplex DNA molecule of a chromosome to be unravelled at once, so the DNA is made single-stranded in small sections, while the rest of the chromosome remains double-stranded. These small sections of single-strandedness are where replication occurs, and involve the *duplex*, 'peeling apart' into single strands. In bacteria, replication occurs at the rate of approximately 500 nucleotides per second.

As a new strand is formed, adenine is placed opposite every thymine (A–T) and guanine opposite every cytosine (G–C), according to base-pairing rules (Fig. 10.5a). This is achieved by adding the bases sequentially in the form of energy-charged building blocks. These building blocks are the deoxyribonucleoside triphosphates (dNTPs): dATP, dCTP, dGTP and dTTP.

Synthesis of DNA is always from the 5' to 3' end, and therefore the deoxyribonucleoside triphosphates are laid down in this orientation in the newly synthesised strand, splitting off pyrophosphate (which is two phosphates covalently linked as an acid anhydride). Pyrophosphate is enzymatically cleaved to provide sufficient free energy to drive the reaction. The growing chain always has an exposed hydroxyl at its 3' end, ready for the entry of the next dNTP. The 3' deoxyribose hydroxyl is joined to the phosphate by an ester bond. This sequential process, aside from helping to attach the correct deoxyribonucleoside, also extends the sugar-phosphate backbone, which simply consists of alternating deoxyribose and phosphate (see Chapter 1).

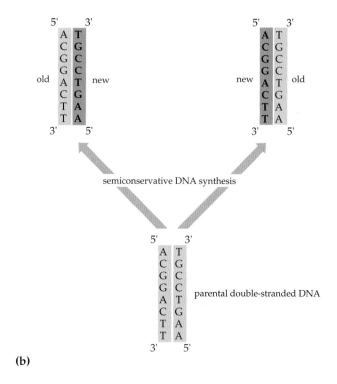

(b)

Fig. 10.5 **(a)** In semiconservative replication, double-stranded DNA separates, each single strand acting as a template for the synthesis of a new strand. In this three-dimensional diagram, the lower portion is still in a duplex and is base-paired. The upper portion is unwound and new DNA strands are associated with each parental strand by means of base-pairing.
(b) Sequence-based representation of replicating DNA showing the 5' and 3' polarities of strands

Replication begins when the double-stranded DNA molecule unwinds, separates and undergoes semiconservative duplication. Replication leads to the formation of complete copies of the genome for eventual distribution to each of two progeny cells.

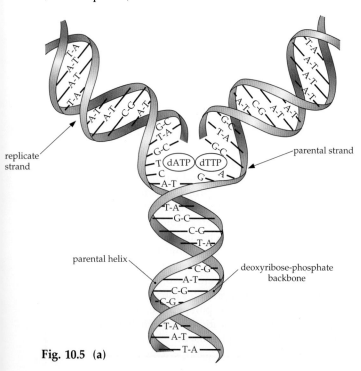

Fig. 10.5 (a)

REPLICATION IN PROKARYOTES

A bacterial chromosome, being simpler, provides a model system that has helped our understanding of more complex eukaryotic chromosomes. The circular chromosome of a prokaryote is about 1 mm in circumference when unwound. In the cell, it is compressed about a thousand-fold with the aid of special compression molecules, such as folding proteins and RNA, into the **nucleoid** (Figs 10.6, 10.7). Although highly condensed, this tiny genomic structure is in direct contact with the cytosol and is available for replication, transcription and repair.

BOX 10.3   BUILDING BLOCKS FOR DNA SYNTHESIS

Deoxyribonucleoside triphosphates (dNTPs) are the 'building blocks' for the biosynthesis of DNA, a polymer of deoxyribonucleoside monophosphate units containing four types of pyrimidine or purine bases (Chapter 1). Pyrimidines (dCTP and dTTP) and purines (dATP and dGTP) must therefore be synthesised in the cell in equal amounts. To balance the levels of purine and pyrimidine nucleotides, their synthesis is controlled by feedback mechanisms. Both biosynthetic pathways utilise simple precursors for assembly of dNTPs.

The balance between cellular concentrations of ribonucleoside triphosphates (ATP, GTP, CTP and UTP) and dNTPs (dATP, dGTP, dCTP and dTTP) is controlled by the enzyme ribonucleotide reductase, the specificity

and catalytic activity of which are subject to complex forms of regulation. In human lymphocytes that are dividing rapidly, ribonucleoside triphosphates are more abundant than dNTPs because the dNTPs are being used to maintain the rate of DNA synthesis for cell division.

A number of important anticancer drugs inhibit the synthesis of nucleotides in rapidly dividing cancer cells. Drugs such as methotrexate, have a selective toxicity for some cancers by blocking the synthesis of dNTPs. Other drugs, such as mercaptopurine, are incorporated into DNA in place of the natural dNTPs and disrupt the genetic code by pairing with 'wrong' nucleotides, thus causing fatal mutations in cancer cells.

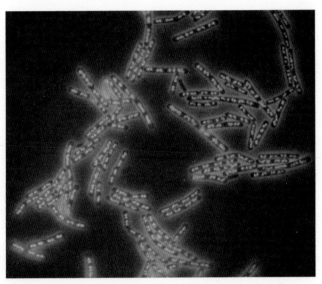

Fig 10.6 Fluorescence micrograph showing nucleoids (blue) containing DNA in the prokaryotic cells (*Bacillus subtilis*)

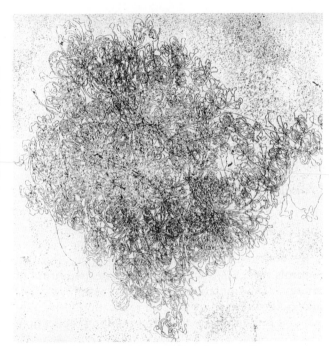

Fig. 10.7 Electron micrograph showing a partly unravelled nucleoid from the prokaryote *Bacillus licheniformis*

The nucleoid contains the circular DNA molecule of prokaryotes. It is compressed with the aid of folding proteins and RNA, and is located in the cytosol.

Initiation of replication

Bacterial chromosomes have a single origin. Replication begins at the origin, where unwinding takes place simultaneously in both directions. The precise DNA sequences of origins from many bacteria are known and contain recognition sites for specific *DNA-binding proteins*, which are part of the initiation machinery. Initiation of replication at the origin is the stage where replication is controlled.

Replication is initiated by a cut (nick) in at least one strand. The cut is made by a specific enzyme,

DNA gyrase. The DNA unwinds a **replication fork** is generated at each side of the origin and by initiation machinery. New strand assembly takes place at these replication forks as the double helix unwinds (Figs 10.8, 10.9). Replication forks move at an essentially constant rate.

Each replication fork contains a *leading strand* growing into the fork and a *lagging strand* growing out from the fork (Fig. 10.10). The two forks generated at the origin proceed bidirectionally (in opposite directions) around the circular bacterial chromosome, and eventually reach the terminus, where their movement is blocked (Figs 10.8, 10.10).

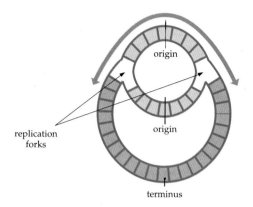

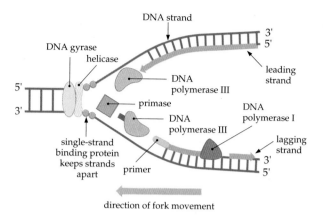

Fig. 10.8 DNA synthesis in circular chromosomes. Synthesis progresses in both directions from the single origin by means of replication forks around the circle until the growing forks meet at the terminus

Fig 10.10 The replication fork of *Escherichia coli*, including the various enzymes and accessory proteins that function in DNA synthesis. The enzymes gyrase, helicase and primase carry out DNA unwinding and priming

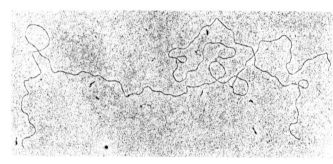

Fig. 10.9 Electron micrograph showing a replication fork in a DNA preparation from the prokaryote *Bacillus subtilis*

By controlling replication at initiation, the cell controls replication with a minimum expenditure of energy. The initiation machinery assembles and allows the creation of the two replication forks only when it is appropriate to start a round of replication.

> Replication begins at the origin. Unwinding of the double strands takes place simultaneously in both directions, forming two replication forks where new bases are added until the terminus is reached.

Enzymes needed for replication

DNA synthesis at the replication fork involves the co-ordinated participation of many enzymes. Separation of the duplex is achieved by ATP-driven enzymes (DNA helicases), which peel apart the base-paired strands (Fig. 10.10). Strands are kept apart by *single-strand binding proteins*, which bind to single-stranded DNA, preventing the strands from rewinding and ensuring that templates are available for building new complementary strands.

DNA normally occurs in a right-handed (positive) helix, with 10 base pairs per turn of the helix. Double-stranded DNA is usually supercoiled, with the double-stranded segments twisted around each other. The enzyme, DNA gyrase, undoes the supercoils and twists by making a transient cut and allowing twists to unwind before accurately renewing the phosphodiester bonds (Fig. 10.10).

At the replication fork, dNTPs are attached to the growing chain by **DNA polymerases**, enzymes with the ability to add bases to the 3'-OH group of DNA and to remove incorrect bases with its own *exonuclease* activity (see Box 10.1). In *E. coli*, DNA polymerase III is the major replication enzyme (Fig. 10.10), attaching bases to DNA in a 5' to 3' orientation (so synthesis must occur in this direction). For the leading strand, which points in a 5' to 3' manner in the same direction as the moving replication fork, this means that DNA polymerase continues to add nucleotides sequentially, keeping up with the replication fork. Since the two strands of DNA are antiparallel, the second strand must terminate in a free 5' end, and is not accessible to DNA polymerase. Within a replication fork, both strands continue to grow in the 5' to 3' orientation, but the second strand grows opposite to the direction of fork movement. This is the lagging strand, and its newly synthesised DNA comprises a series of short precursor fragments.

In a remarkably elegant process, DNA polymerase I, which has more than one enzyme activity, checks the added base and 'edits' any incorrect additions by means of its own 3' to 5' exonuclease activity. This editing removes any incorrect base together with its attached sugar and monophosphate, which diffuse away. DNA polymerase I replaces the base with the correct one. DNA polymerase I differs from DNA polymerase III in functioning to remove primers and fill gaps accurately (Table 10.2).

Table 10.2 Properties of DNA polymerases of *E. coli*

Property	DNA polymerase I	DNA polymerase III
Polymerisation 5′ to 3′	+	+
Exonuclease activity:		
3′ to 5′	+	+
5′ to 3′	+	−
Synthesis from:		
Intact DNA	−	−
Primed single strands	+	−
Primed single strands plus single-stranded binding proteins	+	+
In vitro chain elongation rate (nucleotides/min)	600	30 000
Molecules present per cell	400	10–20

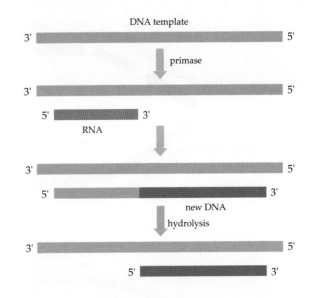

Fig. 10.11 Initiation of DNA synthesis in the lagging strand. The enzyme, primase, synthesises a short complementary stretch of RNA. This RNA is a primer for synthesis of new DNA. The RNA part of the new chain is hydrolysed, leaving a gap to be filled with DNA

Combined with the careful process of assembly, the editing process ensures that mistakes occur in less than about 1 in 10^8 cases. Thus, an entire bacterial chromosome of several megabases can be faithfully copied with only a small chance of insertion of an incorrect base.

> Bases are attached to a growing chain of DNA by DNA polymerases. These enzymes have different forms, and can both add bases and edit the DNA sequence.

DNA synthesis

DNA polymerase cannot start synthesis on its own, and only extends DNA if a free 3′-OH group is available at the end of the strand. Thus there is a need for a primer, a starting tag. A **primer** is a short sequence (usually of RNA), synthesised by primase. The primer allows DNA polymerase to commence synthesis from the 3′ end of the primer (Fig. 10.11). When it has finished its task, the primer is removed by an editing 5′ to 3′ exonuclease.

Since DNA polymerase can only assemble bases in a 5′ to 3′ direction relative to the newly synthesised strand, *continuous DNA synthesis* occurs on the leading strand in the same direction as the moving fork. Its synthesis proceeds more easily than, and slightly ahead of, the lagging strand.

The lagging strand faces in the 'wrong direction' for DNA polymerase action, and shows *lagging* or *discontinuous strand synthesis*. With the assistance of other proteins, primase lays down a primer on this discontinuous arm of the replication fork (Fig. 10.12). DNA polymerase III synthesises the lagging strand

away from the fork, to create an approximately 1–2 kb piece, known as an *Okazaki fragment*, named after its discoverer, Dr Reiji Okazaki. This synthesis eventually stops when the polymerase arrives at the 5′ end of the previously synthesised Okazaki fragment. DNA polymerase III finishes synthesising the Okazaki fragment and departs. The process then restarts at the site of the next primer.

Synthesis on the discontinuous strand therefore occurs in short bursts, and lags behind the more easily synthesised leading strand. The leading strand now consists of one unbroken chain, whereas the lagging strand is punctuated by regular breaks and attached short primers. DNA polymerase I enters at this point and, by the action of its own 5′ to 3′ exonuclease, erases each primer. DNA polymerase I is an unusual enzyme, because it then uses its polymerising ability to replace an erased primer with DNA. It is possible to do this because the 3′-OH end of the adjacent Okazaki fragment can be readily extended at this time. DNA polymerase I also uses its other 3′ to 5′ exonuclease, but only to check that it has attached the correct bases. This means that DNA (produced by the editing enzyme itself and therefore with no errors in base sequence) has replaced the primer. DNA polymerase I then moves to a recently synthesised Okazaki fragment and performs the same task.

The last event in completing this portion of the lagging strand is to join adjacent DNA fragments. This *ligation* (joining) is performed by DNA ligase and the result is that there is no break in the sugar-phosphate backbone of DNA (Fig. 10.12).

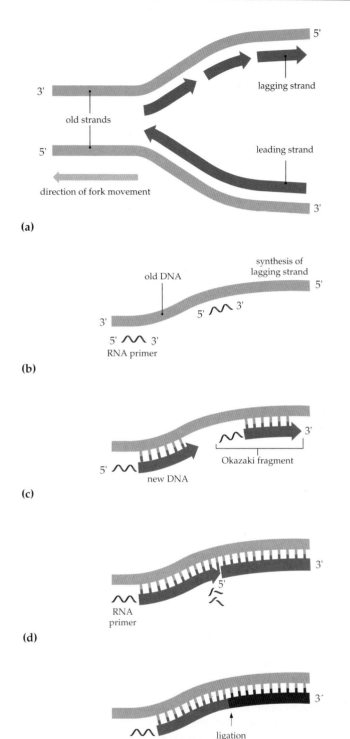

(a)

(b)

(c)

(d)

(e)

Fig. 10.12 **(a)** Organisation of the various elements in DNA synthesis in the leading and lagging strand at a replication fork. Catalysed by DNA polymerase III, synthesis of the leading strand depends on sequential addition of deoxyribonucleotides. The discontinuous lagging strand has several steps in its DNA synthesis: **(b)** RNA oligonucleotides copied from the DNA act as a primer; **(c)** DNA polymerase III elongates RNA primers with newly synthesised DNA; **(d)** DNA polymerase I removes 5′ RNA at end of adjacent fragment; and **(e)** adjacent fragments are linked by DNA ligase

Before reaching the terminus, the partially replicated chromosome looks somewhat like the Greek letter theta, θ, because it is a circle that now contains two replication forks (Fig. 10.8). It is known as a *theta structure*. One of the convincing demonstrations of a theta structure was obtained with the bacterium *Bacillus subtilis* (Fig. 10.13). The terminus in this bacterium stops the clockwise replication fork before the arrival of the other fork. This is achieved by the binding of a small protein to a precise DNA sequence in this region, which acts to prevent movement, ensuring that termination occurs followed by fusion of the two forks.

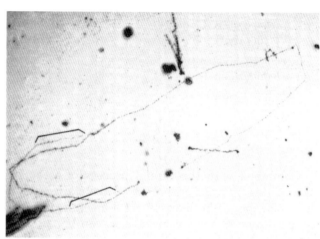

Fig. 10.13 Replicating circular chromosome of *Bacillus subtilis*. Brackets show newly synthesised regions adjacent to the replication forks

> DNA polymerase requires a primer (starting tag) made by an RNA polymerase (primase) before DNA synthesis commences. Two assembly mechanisms operate—continuous synthesis on the leading strand and discontinuous synthesis on the lagging strand.

REPLICATION IN EUKARYOTES

In eukaryotes, chromosomes have many origins (Fig. 10.14). Replication involves the formation at each origin of two replication forks, which are bidirectional with leading and lagging strands, in the same way as in prokaryotes.

There are a number of differences in replication between prokaryotes and eukaryotes. Eukaryotes have smaller Okazaki fragments, which are 100 to 200 bases, about a tenth of the size of those in most prokaryotes. In human cells, replication of the leading and lagging strands is performed by different DNA polymerases (alpha form for the lagging strand, and delta form for the leading strand).

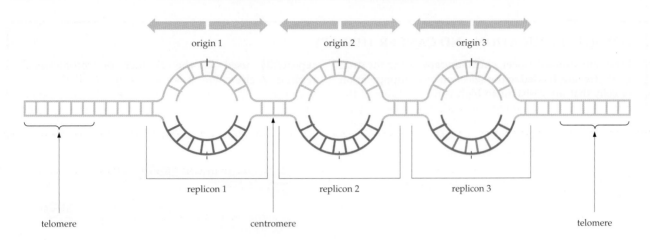

origin 1   origin 2   origin 3

replicon 1   replicon 2   replicon 3

telomere   centromere   telomere

Fig. 10.14 DNA synthesis in a chromosome of a eukaryote. Synthesis of DNA involves many origins of replication. Each origin is a separate replicon, and DNA synthesis is bidirectional from the origin

The size of the DNA molecule, and the fact that a replication fork in eukaryotes moves about 20 times more slowly than in prokaryotes, are major constraints in replication of eukaryotic DNA. In the hypothetical case of a eukaryotic DNA molecule that is 30 times the size of the *E. coli* chromosome, replication would take more than two weeks if it occurred bidirectionally from a single origin. These difficulties are overcome by having multiple origins of replication along each chromosome. Each chromosome thus has many **replicons**, unit regions of replication (Fig. 10.14). In contrast, a prokaryote chromosome has a single replicon (Fig. 10.8).

During the replicative part of the cell cycle (S phase, Chapter 8) each origin is used only once. As replication forks extend bidirectionally from an origin, they continue to move until they eventually join with replication forks from adjacent replicons, or until a replication fork approaches the end of the linear chromosome.

Replication of the genome of the vinegar fly, *Drosophila melanogaster* (165 Mb), takes about 3 minutes in early embryogenesis. There are thousands of origins, spaced about 10 kb apart (Fig. 10.15). In contrast, later in development, or in tissue culture where cells are grown *in vitro* in the laboratory, replication of the same genome takes hours rather than minutes. Replication of a mammalian genome *in vivo* takes about 8 hours.

Not all eukaryotic replicons are initiated at the same time. Collective banks or arrays of approximately 50 replicons appear to commence independently of other groups, so that a chromosome can be broadly divided into early, mid or late replicating regions. The time of replication appears to be at least partly due to whether the region contains active genes or not. Active regions containing frequently transcribed genes, such as those involved in the production of

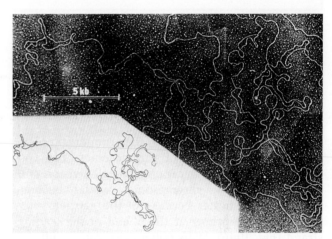

Fig. 10.15 Replication of *Drosophila melanogaster* chromosomal DNA. Many replication bubbles are present, due to multiple initiation sites

enzymes required for cell viability, tend to replicate early in S phase. In contrast, inactive regions tend to replicate late in S phase.

It is difficult to study replication processes when working with multicellular eukaryotes. For this reason, significant advances are being achieved by studying mutants of unicellular eukaryotes, such as the yeast *Saccharomyces cerevisiae*. Origin-binding proteins have been identified and this should lead to an understanding of the mechanism of initiation of replication. In mammals, clues are often obtained by observing events that go awry rather than when they proceed smoothly (Box 10.4).

Eukaryotic chromosomes have multiple origins of replication, from which replication is bidirectional. This speeds up total replication in large chromosomes.

BOX 10.4   REPLICATION AND CANCER THERAPY

In certain forms of cancer, chemotherapy, using the drug methotrexate, has led to the selection of a subpopulation of cells that are resistant to high concentrations of the drug. Methotrexate interferes with the action of an important enzyme involved in folate (vitamin B) metabolism. Coenzymes derived from folic acid are essential for formation of adenine and thymine in DNA synthesis. More rapidly growing cancer cells have a greater need for nucleotides, and die when starved accordingly.

Cancer cells have acquired resistance to methotrexate as a result of a mutation leading to a change in the structure of the enzyme, or by selective amplification of the gene encoding the enzyme, thereby generating many copies of the gene and increasing the intracellular concentration of the enzyme (see figure). In the latter case, up to several hundred copies of the gene can be generated. This can happen because there is an origin of replication near the locus of this gene, which is repeatedly used in the absence of chromosomal replication elsewhere. It is the selective (and aberrant) use of this origin that causes the eventual accumulation of the enzyme and resulting resistance of these cells to methotrexate, rendering this form of chemotherapy useless.

Selective amplification of the chromosomal region containing the dihydrofolate reductase gene (bottom). The normal chromosome (top) lacks the expanded long arm

Telomeres during replication

Completion of replication at each end of linear eukaryotic chromosomes presents a problem. Since DNA polymerases assemble DNA only in the 5′ to 3′ direction and need a primer, they cannot complete replication to the ends of chromosomes. You can imagine that, after many rounds of cell division and hence replication, chromosomes would become progressively shorter because of the gap left after removal of the primer at the 5′ end of the leading strand (Fig. 10.16). DNA polymerase cannot fill the gap because there is no 3′-hydroxyl group present as a starting point. The problem, however, does not arise because the integrity of chromosomes is maintained by the presence of the telomere, a terminal group of repeated DNA sequences. For example, in the unicellular ciliate *Tetrahymena*, there are many repeats of TTGGGG, and in humans there are many repeats of TTAGGG. These repeats are added to the ends of chromosomes by a special polymerase, telomerase. An associated RNA molecule provides the template to fill the gap and add more telomere sequences.

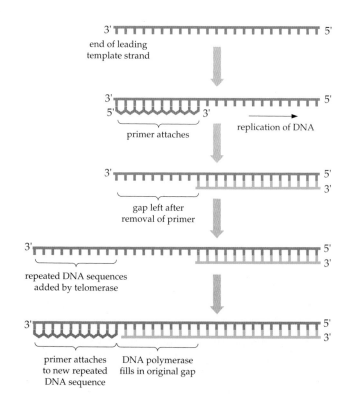

Histones during replication

Eukaryotic chromosomes are made up of roughly twice as much protein as DNA. As we saw in Chapter 3, DNA is condensed around histones. Histones are small basic proteins (11–21 kD) with a high concentration of side chains, such as lysine and arginine, which interact with the negatively charged sugar-phosphate backbone of DNA (Fig. 10.17).

Fig. 10.16 Completion of replication at ends (telomeres) of eukaryotic chromosomes. A gap is left after removal of primer at the 5′ end of the leading chain of DNA. Because of absence of the 3′-hydroxyl group, DNA polymerase cannot fill the gap; instead, telomerases, using an associated RNA molecule as a template, add repeated sequences to fill the gap

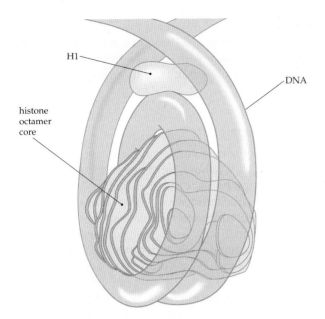

Fig. 10.17 Model of a nucleosome particle. DNA is wound around a histone octamer. This particle has been crystallised and examined at relatively low resolution using X-ray diffraction

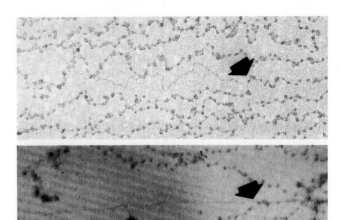

Fig. 10.18 Electron micrograph of a eukaryotic replication fork after cells were treated with cycloheximide to halt synthesis of new histones. Long strands are studded with beads (presumably old histones). One of the arms of the replication fork (arrow) has not been able to add any histones under these conditions, suggesting that old histones attach preferentially to one of the two arms

Histones are abundant in the nucleus, where most assemble in groups of eight to form a core upon which the DNA is bound. Double-stranded DNA of 146 bp in length is coiled slightly less than twice around each histone core, forming a **nucleosome core particle** (Fig. 10.17). Nucleosome core particles are linked by sequences of DNA of variable length, up to 80 bp. Together, linking DNA and nucleosome core particles are referred to as **nucleosomes**.

Origins of replication lack associated histones, making them more open to the action of enzymes such as nucleases. For this reason, origins are *nuclease hypersensitive sites*. The lack of histones presumably makes it easier for initiation proteins to recognise these regions and thereby facilitate replication. One type of histone, H1, located ouside of the nucleosome core particle, serves to attach the histone octamer firmly to the DNA. After replication forks have been formed, chromatin is substantially unravelled by dephosphorylating H1 histone, which disrupts interactions between nucleosomes.

The original histones remain associated with the leading strand of the replication fork, which remains substantially double-stranded during replication and so can associate with histones (Fig. 10.18). In contrast, the lagging strand contains many regions of single-strandedness. Histones bind more strongly to double-stranded than to single-stranded DNA. Newly assembled histones form 'beads on a string', and occur only on DNA on the lagging strand.

> During replication, original histones remain associated with the leading strand of the replication fork and new histones assemble on the lagging strand.

DNA REPAIR

DNA is susceptible to damage from environmental factors. DNA damage, which leads to mutations, can result from exposure to radiation (e.g. X-rays) or chemical mutagens. Degradation of DNA, such as spontaneous deamination of cytosine to form uracil, can also occur. Probably the most frequent event is depurination due to spontaneous hydrolytic loss of guanine or adenine from the sugar-phosphate backbone of DNA. Any damage to DNA can affect its ability to function and accurately pass on the full genetic message of inheritance. Correction of DNA sequence errors is necessary for the survival of organisms. Damage to a single strand of DNA can be repaired because genetic information is carried in both strands of the duplex DNA.

Recognition of errors in DNA

How does a cell recognise that damage has occurred to DNA? Although there are only four types of bases in DNA, some of these bases are chemically modified after incorporation, for example, they can be methylated. The pattern of modification is determined by the particular enzyme system involved. Thus, the source of DNA can be recognised by its specific modification pattern. For example, modification patterns enable a bacterial cell to determine whether DNA is self or foreign. Foreign DNA will not necessarily have the same pattern of modification, so that it will be susceptible to breakdown by **restriction enzymes**. In bacteria, these enzymes recognise the absence of methyl groups in the bases of the foreign DNA.

Systems that safeguard DNA

Mechanisms that repair damaged DNA have evolved and are primarily enzyme-based.

Photoreactivation

When DNA is exposed to ultraviolet (UV) light, adjacent pyrimidines, such as thymines, undergo a chemical rearrangement, bonding these adjacent bases to form *pyrimidine dimers* (Fig. 10.19a). This type of structure must be repaired because it distorts the double helix and interferes with transcription and with DNA replication. Visible light (blue) induces a

process of repair involving the photoreactivation enzyme, PRE (Fig. 10.19b). The enzyme associates with a dimer and with the absorption of a photon of light cleaves the bonds, thus reversing the effect of UV light.

Excision repair

There is a set of patrolling enzymes in a cell, especially in the nucleus, whose job is to respond rapidly to DNA damage by excising the damaged bases along with some nucleotides on either side (Fig. 10.19c). Sites of mutation are recognised by an enzyme complex, which creates a nick in the backbone of the DNA molecule. DNA polymerase I then fills in the gap by repair synthesis. The damaged portion of DNA is excised by the enzyme complex and the phosphodiester backbone is sealed by DNA ligase. Unlike photoreactivation, excision repair does not require light.

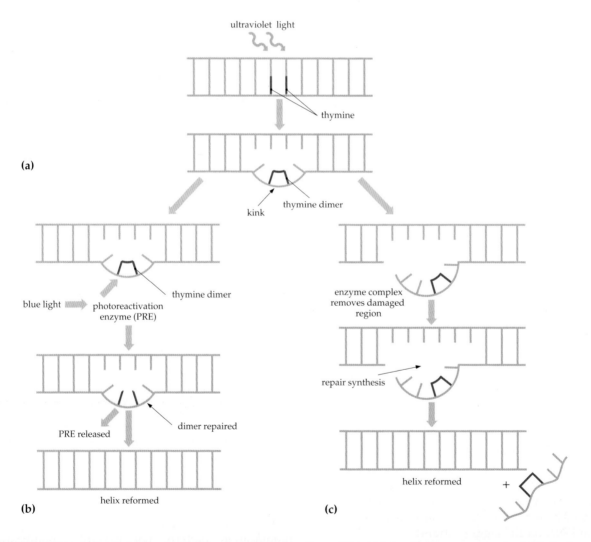

Fig. 10.19 DNA repair mechanisms. **(a)** Effects of a thymine dimer. Ultraviolet radiation can cause two adjacent thymines to link as dimers by formation of a covalent bond between them. The effect of a thymine dimer on the double helix is to make a kink in the duplex, which blocks DNA replication. **(b)** Photoreactivation repair mechanism of an ultraviolet-induced thymine dimer. **(c)** Excision repair mechanism of an ultraviolet-induced thymine dimer

A form of *xeroderma pigmentosum* (Table 10.3) is a rare skin disease in humans inherited as an autosomal recessive trait. The skin of an affected homozygote is very sensitive to sunlight and many sufferers die from skin cancer before the age of 30. The disease is caused by a defect in the endonuclease involved in excision repair.

Post-replication repair

Some pyrimidine dimers escape repair by either of the above mechanisms. During replication, DNA polymerase III passes over the dimer, leaving a gap. In the absence of repair, the DNA remains broken, preventing further rounds of replication. To overcome this, a recombination enzyme, rec A, interacts with both duplexes having an effect similar to crossing over (Fig. 10.20). As a result, each duplex contains one correct DNA strand. After completion of replication, the dimers are able to be repaired by the excision repair system.

Post-replication repair also can be triggered by the SOS response. There is a repressor molecule that normally represses the genes involved in producing enzymes needed for repair processes. However, binding of rec A to damaged DNA leads to cleavage of the repressor molecule, thus activating the repair system. Repair of DNA after an SOS response is error-prone because it includes a bypass system that permits incorrect replication across pyrimidine dimers. The proofreading properties of DNA polymerase are relaxed to permit this to occur.

> DNA can be damaged by environmental factors. Repair mechanisms involve a number of enzymes that remove the damaged part and replace it with a new sequence.

ORGANELLE GENOMES

For many years researchers knew that the nucleus contains genetic information, but it came as a surprise to discover that certain organelles contain DNA as well. Indications that this was the case first appeared when certain *petite* mutants of yeast were examined. Petite mutations result in small colonies of yeast as a result of impaired mitochondria, which leads to ineffective oxidative phosphorylation. These mutations did not map as nuclear mutations and, in fact, segregated independently of nuclear genes. This indicates that the petite mutation lies outside the nuclear genome. Subsequent studies revealed that mitochondria contain DNA and the mutations were identified in this organelle's genome. Mitochondria

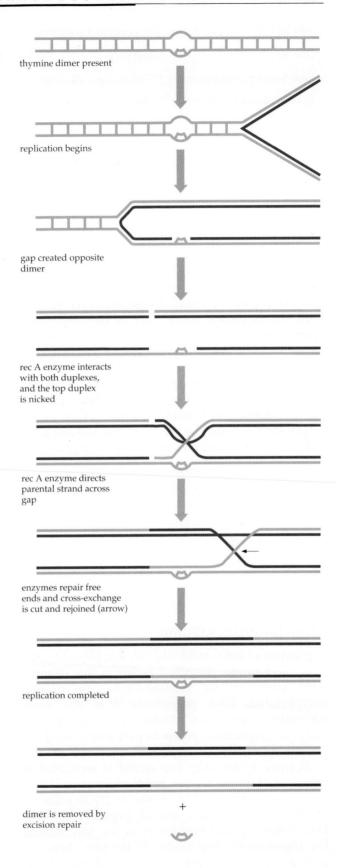

thymine dimer present

replication begins

gap created opposite dimer

rec A enzyme interacts with both duplexes, and the top duplex is nicked

rec A enzyme directs parental strand across gap

enzymes repair free ends and cross-exchange is cut and rejoined (arrow)

replication completed

dimer is removed by excision repair

Fig. 10.20 Mechanism of post-replication repair of an ultraviolet-induced thymine dimer

Table 10.3 DNA repair defects are associated with human diseases

Disease	Sensitivity	Cancer susceptibility	Symptoms and signs
Xeroderma pigmentosum	UV irradiation alkylation	Skin carcinoma	Skin photosensitivity
Ataxia telangiectasia	Gamma-irradiation	Lymphoma	Unsteady gait (ataxia); dilation of blood vessels in skin and eyes (telangiectasia); chromosomal aberrations
Fanconi's anaemia	Cross-linking agents	Leukaemia	General decrease in numbers of blood cells; congenital anomalies
Bloom's syndrome	UV irradiation	Leukaemia	Photosensitivity; defect in DNA ligase

and chloroplasts contain double-stranded circular DNA chromosomes, which vary in size depending upon the species. In humans, mitochondrial chromosomes are 16 569 bp, but in the yeast *S. cerevisiae* they are larger, at about 78 kb.

Sequencing has led to an appreciation of the compactness of organelle genomes, where much of the DNA has been shown to be coding. Genes code for products required by the organelle, and any extra coding information is provided by nuclear genes.

Mitochondrial genome

Replication of mitochondrial DNA is achieved by the basic mechanisms described earlier in this chapter, with some minor modifications. In human mitochondria, DNA polymerase δ is the major replication polymerase. Initiation occurs at a precise origin and replication proceeds part way around the circular genome. An important property of this replication is that only one strand is replicated at this time, leading to the production of a *D-loop*, consisting of displaced DNA (Fig. 10.21). Replication of the complementary strand begins when the D-loop has reached two-thirds of the way around the chromosome. This results in the formation of a replication bubble, with replication of one strand finishing much earlier than the other as a consequence of its earlier start. In fact, completion of the second strand consists of filling the gap resulting from the delayed synthesis of the second strand. This results in the production of two complete circular genomes.

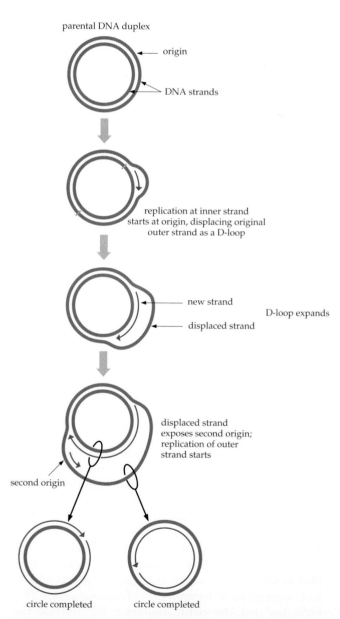

Fig. 10.21 D-loop expands an opening in a mammalian mitochondrial DNA chromosome for synthesis of a new strand of DNA. There are separate origins for the replication of each strand

Chloroplast genome

A chloroplast contains many copies of a circular chloroplast DNA molecule (cpDNA). Chloroplast DNA is about 120–160 kb in circumference and usually contains an inverted repeat (a segment that is in reverse sequence). Across the plant kingdom, cpDNA is highly conserved for the number and type of genes (about 160) and for their coding sequences (70–99% base conservation). Division of the chloroplast and replication of cpDNA are controlled by nuclear DNA independent of cpDNA expression.

Genes encoded by cpDNA fall into two main groups. The first group is concerned with chloroplast protein synthesis and includes chloroplast ribosomal RNA genes, a functional set of tRNA genes using normal codons and some of the chloroplast ribosomal proteins. Together with chloroplast ribosomal proteins and polymerases synthesised in the cytoplasm from nuclear-DNA-encoded mRNA, this chloroplast protein synthetic system makes products of the second group of genes, which encode chloroplast proteins concerned with photosynthesis. These photosynthetic proteins are complexes of proteins synthesised in both the chloroplast and cytoplasm, including Rubisco (ribulose bisphosphate carboxylase), photosystems I and II, the cytochrome complex and ATPase (Chapter 5). There are still a number of coding sequences for which functions are not known, and as yet no chloroplast genes have been identified that relate to non-photosynthetic functions.

The chloroplast genome of plants contains the residual genes of a cyanobacterial ancestral genome (Chapter 35) from which most of the functions have evolved to the nucleus in higher plants.

Chloroplasts and mitochondria contain DNA as small circular double-stranded molecules.

SUMMARY

- The genome is the total chromosomal DNA in a cell or organelle. DNA in cells is double-stranded, and each strand is arranged anti-parallel. Each chromosome comprises a single long DNA molecule.

- DNA sequences found in eukaryotic chromosomes include centromeres (one copy) which provide the site of attachment to the spindle and telomeres (two copies) at natural ends of chromosomes.

- Prokaryotes have only a single chromosome condensed with the aid of folding proteins and RNA to form a nucleoid. There is a single origin sequence for replication.

- Eukaryotic chromosomes have multiple origin sequences and multiple tandem (i.e. head-to-tail) repeats of the ribosomal RNA genes, as well as other repetitive sequences, including satellite DNA and other elements repeated up to a million times.

- DNA replication begins when the double-stranded DNA molecule unwinds and undergoes semiconservative duplication. Each parental strand of the pair remains unaltered, but a complementary strand is synthesised alongside each parental strand. Replication begins at the origin, where unwinding of the double strands takes place simultaneously in both directions, forming replication forks where new bases are added in the form of nucleotides until the terminus is reached. Nucleotides are attached to the growing chain by enzymes, DNA polymerases. These enzymes have different forms, one of which has many subunits for multitasking—both adding bases and editing the DNA sequence.

- DNA polymerases require a primer (starting tag) made by an RNA polymerase (primase) before DNA synthesis commences at the origin. Two assembly mechanisms operate—continuous synthesis on one strand, and discontinuous on the other.

- In eukaryotes, replication is bidirectional. Replication of the genome takes about 3 minutes in *Drosophila* embryo cells, but can take 8 hours in older cells.

- Telomere sequences are sited at the ends of chromosomes. The centromere has a defined sequence, which interacts with the kinetochore to attach a chromosome to spindle microtubules for chromosomal separation at cell division.

- Proteins associated with chromosomal DNA include histones and non-histone proteins. Histones are small basic proteins that form the spool upon which DNA is wrapped to form chromatin. Chromatin appears as a string of duplex DNA coiled around histone 'beads', the nucleosomes. During cell division, cyclical changes in the packaging occur, so that chromatin becomes more tightly packaged in mitotic chromosomes. Origins lack associated histones.

- DNA can be damaged by environmental factors. Repair mechanisms involve enzymes that remove the damaged part, while others replace it with a new sequence.

- Genomes of chloroplasts and mitochondria are circular DNA molecules.

QUESTIONS

1. Mammalian genomes are about 1000 times larger than those of bacteria, and yet there is only about 50 times more coding information. How can you account for this discrepancy?

2. How is the location of a centromere determined, and what is its role?

3. Why do you think human DNA can function with so many repetitive sequences?

4. (a) What are the names of the DNA sequences that start and end DNA replication? (b) Construct a flow diagram to show four major steps in DNA replication of a prokaryote genome.

5. (a) What is a replication fork? (b) How does replication differ in a prokaryote from a eukaryote? (c) What is the advantage of many sites of replication in a eukaryote chromosome?

6. Describe the various enzymes required for DNA synthesis in *Escherichia coli*.

7. (a) Why is the balance of dNTPs so critical for DNA synthesis and cell division? (b) How do cancer cells acquire resistance to methotrexate?

8. How do histones contribute to the construction of a eukaryote chromosome, and what happens to them during DNA replication?

9. (a) Why is the repair of damage to DNA so important? (b) Explain the mechanism underlying one DNA repair system.

10. In what ways are the genomes of mitochondria and chloroplasts similar to those of prokaryotes and different from the nuclear genome of eukaryotes?

G E N E   E X P R E S S I O N

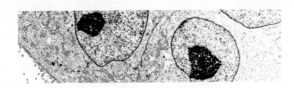

The central dogma of gene expression is that DNA carries genetic information, which is transcribed into RNA, and RNA is translated to produce a polypeptide (Fig. 11.1). During gene expression, information must be transferred accurately from DNA, through RNA, to the polypeptide chain because the three-dimensional folding of polypeptide chains and their specific cellular functions depend on the correct linear sequence of their constituent amino acids. Gene expression is the process by which information encoded in genes is decoded into amino acid sequences during synthesis of proteins.

Gene expression in cells is a two-stage process. Firstly, a DNA template directs the synthesis of an RNA molecule (Chapter 1). The gene sequence of one strand (the template) of duplex DNA is copied precisely to produce a single-stranded RNA molecule. This is the process of **transcription**. The information contained within the RNA is complementary to the DNA sequence; for example, the sequence 3'-ATGCAA-5' in a DNA template is transcribed as 5'-UACGUU-3' in messenger RNA (mRNA). Remember that complementary nucleic acid strands are antiparallel and that U substitutes for T in RNA (Table 11.1).

Secondly, the base sequence in mRNA specifies the amino acid sequence of a polypeptide during **translation**. This involves the participation of ribosomes and three kinds of RNA molecules—mRNA,

which codes for the polypeptide, transfer RNA (tRNA), which delivers the correct amino acids to the growing polypeptide, and ribosomal RNA (rRNA), a structural component of ribosomes, the workbench on which polypeptide assembly occurs (Chapter 3).

A gene is a DNA sequence that directs the synthesis of an RNA molecule (Chapter 10). While all polypeptides, and therefore the functional proteins that they form, are coded for by genes, not all genes code for protein. As we shall see below, some genes code for RNA molecules (tRNA and rRNA) that are used as tools during the process of protein synthesis.

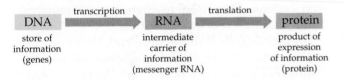

Fig. 11.1 Central dogma of gene expression. This simple scheme summarises the read-out of genetic information in living cells. Some RNA molecules (ribosomal RNA and transfer RNA) are key components of the machinery of protein synthesis and do not contain information specifying a polypeptide

Table 11.1 Base complementarity in DNA and RNA

Base in DNA template	Complementary base in RNA transcript
A	U
C	G
G	C
T	A

Gene expression involves the accurate transcription of information from DNA to RNA and the translation of an mRNA sequence into the amino acid sequence of a polypeptide.

THE GENETIC CODE

How does the sequence of nucleotides of DNA, and thus RNA, specify the order of amino acids in a polypeptide? What mechanisms are used to transcribe the four-letter nucleotide alphabet of DNA (A, T, G and C), firstly into complementary mRNA (U, A, C and G), and then to translate this sequence into the words (amino acids) of the protein language.

Only 20 amino acids are found in proteins; therefore, only 20 code words are theoretically required. Given the four-letter alphabet, a two-letter code would provide for a maximum of 16 amino acids, which is not enough. A three-letter code, which provides 64 possible combinations of triplets of the four nucleotides, is more than adequate.

Codons

In the **genetic code**, triplets of nucleotides are read as three-letter words, **codons**, which represent particular amino acids (Fig. 11.2). Experiments designed to assign meaning to triplets of nucleotides have shown that 61 different triplet codons specify particular amino acids. Almost every amino acid is specified by more than one codon and therefore the code is said to be *degenerate*. **Synonymous codons** are different codons that specify the same amino acid. Leucine, for example, is specified by six different codons; only tryptophan and methionine are specified by a single codon each (Fig. 11.2). Nevertheless, the code is unambiguous because each codon specifies only a single amino acid.

The degeneracy in the code is particularly evident in the third nucleotide of a codon, where changes often do not alter the amino acid that is specified.

Two codons with the same nucleotide at the first two positions are frequently synonyms, specifying the same amino acid. In Figure 11.2, about half the boxes belong to this category; for example, a CU combination encodes leucine regardless of the nucleotide in the third position. In the other half, the third nucleotide influences the code, but only as to whether it is a purine or a pyrimidine. The CA combination encodes both glutamine and histidine, depending on whether a purine or pyrimidine, respectively, occupies the third position.

In order to synthesise a protein, translation must start and stop at the appropriate codons. In almost every case, an AUG codon, specifying methionine (met or M), serves as the *start codon* (initiator codon). Occasionally, a GUG codon is the initiator and, when this occurs, GUG also specifies methionine, even though this codon specifies valine when found at positions other than the site of initiation. In any event, the first amino acid of a newly translated protein is always methionine. Translation ends at one of three *stop codons* (UAA, UAG and UGA), which do not specify an amino acid (Fig. 11.2).

Mutations that alter a codon to that specifying another amino acid are known as *missense mutations*. They lead to the production of an altered polypeptide. *Nonsense mutations* are alterations to codons such that they no longer specify an amino acid but have become stop codons. They lead to premature termination during translation of a polypeptide.

The genetic code is almost universal. The same basic codon dictionary is used to decipher all protein-coding RNAs in most organisms. However, slight variations in codon meanings do occur (Box 11.1).

> The common genetic code is made up of three-letter nucleotide code words (codons). Each group of three nucleotides may code for an amino acid or serve as a start or stop codon in the production of a polypeptide.

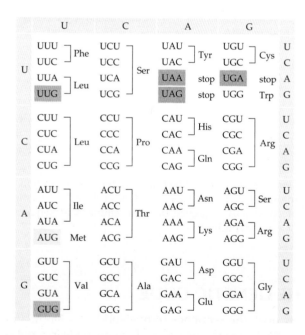

Fig. 11.2 RNA codes for amino acids. Termination or 'stop' codons are shown in pink, and the usual 'start' codon in green. The codons GUG or UUG (yellow) are occasionally used as starts. AUG, UUG and GUG usually code for methionine, leucine and valine respectively; but when these codons are used as start codons in bacteria, they direct the incorporation of N-formyl methionine, as the initiating amino acid for the polypeptide

Reading frames

In DNA and mRNA, codons are unpunctuated in the sense that successive triplets have no 'silent' letters or spaces between them. Consider the nine letter mRNA sequence, 5'-AUCGCCACU-3'. Triplets are read from the 5' end (i.e. from AUC) to the 3' end, so this mRNA codes for isoleucine-alanine-threonine by the triplets AUC GCC ACU. The nucleotides of any DNA sequence are read by a similar **reading frame** of triplets (Fig. 11.3). With regard to the growing polypeptide, triplet codons specify the sequence of amino acids reading from the N-terminus to the C-terminus (Chapter 1).

The nucleotides of a DNA sequence can form three different reading frames because codons are

BOX 11.1   IN SOME BIOLOGICAL SYSTEMS, A DIFFERENT CODE IS USED

The genetic code that is common to most organisms is the end product of millions of years of evolution. Across the range of organisms that use the common genetic code, there are variations in the frequency with which alternative codons are used. Codon usage may even differ within a single organism, depending on whether genes are expressed at high output or not. For example, of six alternative codons for leucine in bakers' yeast, *Saccharomyces cerevisiae*, UUG is used most frequently in highly expressed genes of the nucleus while the other codons, CUU, CUC and CUG, are hardly used at all. In contrast, genes that are moderately expressed use all six codons at about the same frequencies.

In several biological systems, the codon dictionary itself may vary from that of the common code. Thus, in mitochondria of most eukaryotes (except plants), UGA codes for tryptophan (not stop, Fig. 11.2) and AUA codes for methionine (not isoleucine). Further examples

are *Mycoplasma* (a prokaryote), in which UGA codes for tryptophan, and in eukaryotes such as the ciliated protists (e.g. *Paramecium*), in which UAA and UAG both code for glutamine (not stop). These deviations from the common genetic code tell us something of the evolution of eukaryotic cells and their organelles, providing evidence of the pathway along which evolution of a particular biological system has proceeded.

Two major issues confront biological scientists when considering these aspects of the genetic code. Firstly, what are the underlying chemical relationships, if any, that determined particular codons specifying given amino acids at the time cellular life began on earth and the common code became established? Secondly, for those modern systems that utilise a non-standard code, do such variant codes actually date from a time before the fixation of the common code? Or do they represent more recent changes?

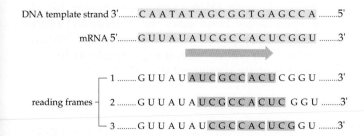

Fig. 11.3 The concept of reading frames. Bases are read sequentially in groups of three in the 5′ to 3′ direction (arrowed). In each strand of DNA there are three possible phases of the codon units by which the base sequence could be read to code for a protein (the reading frame). The complementary strand (not shown here) to the DNA template strand also could specify a set of three possible reading frames. A given region of double-stranded DNA usually only contains one reading frame that codes for a particular protein gene product

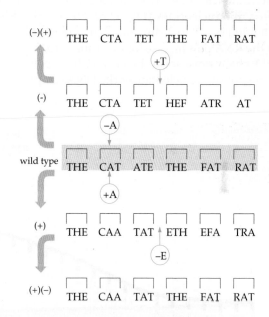

Fig. 11.4 Frameshift mutations. The addition (+) or deletion (–) of a nucleotide, here shown hypothetically, in the middle of a reading frame causes a frameshift mutation. All codons from that point on are altered. If there is both a deletion and an addition, the reading frame is not altered on either side of the region enclosed by the changed codons

unpunctuated (Fig. 11.3). Thus, since DNA is double-stranded, for each DNA segment there are six possible reading frames that could specify the amino acids making up a polypeptide. Each of the complementary strands of DNA can theoretically specify a set of three reading frames. In most cases, however, only one of the six potential reading frames in a gene sequence of duplex DNA actually encodes a protein; this is known as an *open reading frame*. In certain viruses, reading frames encoding proteins overlap, so that more than one protein can be produced from the same region of the DNA double helix.

Some chemical mutagens, for example, acridine dyes, cause **frameshift mutations** by the removal or addition of one or more nucleotide bases to a DNA sequence. The existence of this type of mutation

provided early evidence for the triplet nature of the code. When a single nucleotide was added or removed, the reading frame for all subsequent triplets was changed (Fig. 11.4). When two additions or deletions occurred, the reading frame was different again. However, if there was an addition *plus* a deletion, or the addition or deletion of *three* nucleotides, the original correct reading frame was restored in the remainder of the gene.

> A nucleotide sequence is read from the 5′ end to the 3′ end as a triplet 'reading frame'. It specifies the sequence of amino acids in the growing polypeptide from the N-terminus to the C-terminus.

TRANSCRIPTION

The synthesis of an RNA copy complementary to a DNA nucleotide sequence is the first stage of gene expression. The process involves transcription of the DNA 'template' using the enzyme **RNA polymerase**. For this to occur, the duplex strands of DNA temporarily separate and RNA polymerase moves along the template DNA (Fig. 11.5). RNA polymerase directs the synthesis of all three types of RNA—messenger, transfer and ribosomal. It functions in a similar manner to DNA polymerases in DNA replication (Chapter 10), adding nucleotides to the 3′ end of the growing chain, except that RNA polymerases use ribonucleotides rather than deoxyribonucleotides (Fig. 11.6). Hence, the RNA is extended in a 5′ to 3′ direction. The nucleotide chosen by the RNA polymerase for insertion into the growing RNA chain must form the correct base pair with the corresponding nucleotide in the DNA template (Table

11.1). RNA polymerases do not polymerise thymine nucleotides and uracil nucleotides are inserted into the RNA transcript in response to adenine residues in the DNA template. Genetic information within the DNA template is strictly maintained in the RNA transcript.

In contrast to DNA polymerases (Chapter 10), RNA polymerases do not require a short sequence of nucleic acid as primer to polymerise nucleotides. The RNA polymerase moves along the DNA double helix, without making transcripts, until it reaches and binds to a specific sequence, the **promoter**, which promotes initiation of transcription by the polymerase (Fig. 11.7). This occurs by the base-pairing of the first specified ribonucleotide (the promoter region itself is not transcribed); in our example, this is uridine triphosphate, with its complementary base adenine on the DNA template. The next specified ribonucleotide, guanosine triphosphate, is then brought in by the polymerase to base pair with its complement cytosine in DNA, and the first phosphodiester bond in the new RNA molecule is formed. As with DNA synthesis (Chapter 10), this phosphodiester bond links the 3′ hydroxyl residue of the already paired nucleotide to the 5′ hydroxyl residue of the incoming nucleotide and pyrophosphate is released (Fig. 11.6).

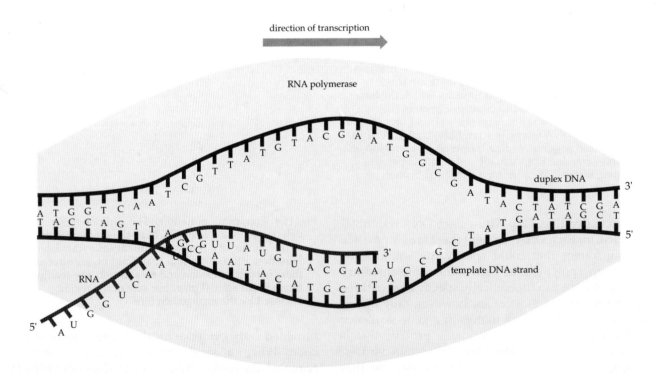

Fig. 11.5 Transcription of DNA by RNA polymerase. RNA is copied from only one strand (template) of a segment of DNA. Only a short region of the growing RNA chain remains attached, by hydrogen-bonded base-pairing, to the DNA template. A molecule of UTP is shown (U), ready to pair with next available base (A) in the template strand, being the next ribonucleotide to be added to the growing RNA chain. The continuously moving 'bubble' of unpaired DNA strands extends for about 17 nucleotides. RNA polymerase may cover a longer segment of DNA, certainly more than 50 base pairs

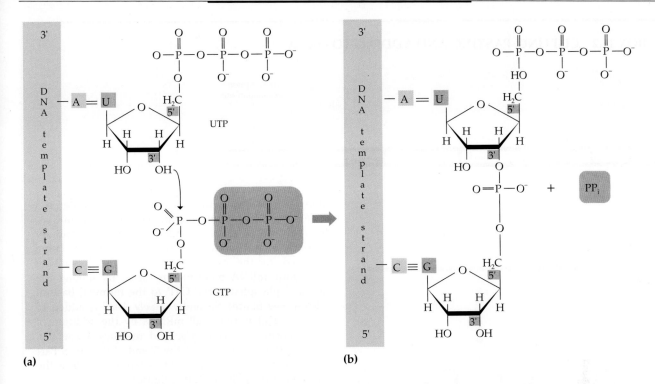

(a) (b)

Fig. 11.6 Formation of the first phosphodiester bond of transcription. In this example, the initiating nucleotide UTP is shown hydrogen bonded via its uracil base (U) to the adenine base (A) of its complementary nucleotide in the DNA template strand. The incoming guanine base (G) of ribonucleotide (GTP) forms hydrogen bonds with the next base cytosine (C) in the DNA template strand. Reaction occurs between the 3′-OH group of the ribose sugar ring in the first ribonucleotide (UTP) and the innermost phosphate of the incoming ribonucleotide triphosphate (GTP), leading to covalent linkage of the two ribonucleotides (3′–5′ phosphodiester) and removal of pyrophosphate (PP$_i$)

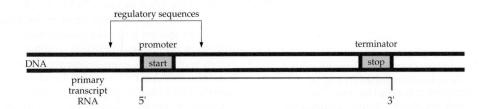

Fig. 11.7 The key features of a generalised transcription unit. The promoter and terminator represent start and stop signals in DNA, respectively, for the RNA polymerase that generates the primary transcript from a transcription unit. Regulatory sequences that control the rate of initiation of transcription are usually located upstream, but may be downstream, of the promoter sequence. In some genes, more than one regulatory sequence may be present, which allows fine control of the transcription of a given gene. The primary transcript may undergo post-transcriptional processing

Transcription continues with the sequential polymerisation of ribonucleotides complementary to template DNA. The product RNA does not remain base-paired to template DNA but dissociates as RNA polymerase moves along the template DNA (Fig. 11.5). This process continues until a specific termination signal on the DNA is reached. Like initiation, termination of transcription relies on a particular sequence of nucleotides in DNA. In this case, however, the sequence triggers not only the cessation of transcription but release from the DNA template of both RNA polymerase and the product RNA

transcript. The transcribed region of DNA, between promoter and termination sequences, defines a single **transcription unit** (Fig. 11.7) and its RNA product is a **primary transcript**.

In many cases, the primary transcripts of genes in eukaryotes, as well as in prokaryotes, undergo a series of maturation or processing reactions that produce a final product modified in several ways from the structure of the primary transcript (Box 11.2).

Transcription involves binding of RNA polymerase to the promoter sequence of DNA, initiation of synthesis, elongation of the RNA chain, termination and release.

BOX 11.2   CUTTING, PASTING AND ADDING TO mRNA

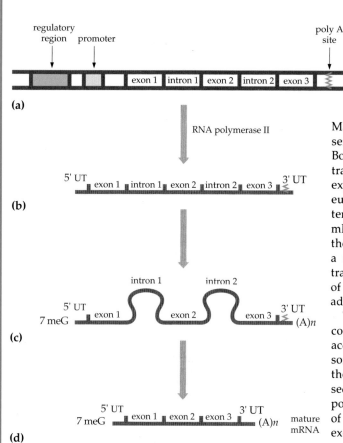

(a)

(b)

(c)

(d)

Post-transcriptional processing of eukaryotic mRNA transcripts. **(a)** A double-stranded DNA region containing a generalised transcription unit. For convenience, a single regulatory region is indicated, located upstream of the promoter. The coding region that specifies mRNA includes three exons interrupted by two introns. There are sites for addition of the 5' cap structure (7-meG) and the poly A tail [(A)n] at either end of the RNA transcript. The DNA regions that determine the transcription termination site are indicated. The transcription unit is transcribed by RNA polymerase II. **(b)** The resultant primary transcript, containing sequences corresponding to exons and introns, as well as the 5' and 3' untranslated regions (5'UT and 3'UT, respectively) that flank the outermost exons. The processing events that take place on the primary transcript, including the addition of 7-methyl guanosine (7-meG) at the 5' cap site and cleavage at the 3' end of the transcript, together with addition of a poly A tail of 100–250 nucleotides in length. **(c)** The final step in processing, namely, the splicing out of the intron sequences and joining of the ends of each adjacent pair of exons to yield a mature RNA. The removal of introns requires the activity of snRNPs (see text). **(d)** The final mature mRNA that is exported from the nucleus to the cytosol for translation on ribosomes

Many primary transcripts of mRNA genes undergo a series of modifications to produce functional mRNA. Both 5' and 3' ends of eukaryotic mRNA primary transcripts are processed in the nucleus, before being exported to the cytoplasm. Firstly, the 5' end of eukaryotic mRNA molecules is *capped* by linking the terminal phosphate of 5' GTP to the terminal base of mRNA (see figure). Secondly, a *poly A tail* is added to the 3' end of virtually all mRNAs by the addition of a sequence of polyadenylic acid residues. Generally, transcription proceeds past the 3' end of the gene. Part of the 3' end of the transcript is removed before the addition of the poly A tail (see figure).

The nucleotide sequence of an mRNA always corresponds precisely with its polypeptide end product, according to the rules of the genetic code. However, some eukaryotic genes are much longer than the mRNA they produce because the genes include additional sequences that interrupt the sequences that encode polypeptides. Interruptions are caused by the presence of non-coding introns between polypeptide coding exons in the DNA of a gene (Chapter 10).

After modification of the ends of the primary mRNA transcript, *splicing* may take place to remove non-coding introns interspersed between coding exons. During removal of introns, exon segments retained are spliced together, forming mature mRNA molecules, which pass into the cytoplasm (see figure). The accurate and efficient removal of introns requires the activity of small RNA molecules and helper proteins complexed together as snRNPs (**s**mall **n**uclear **ribon**ucleoprotein particles). Several different snRNPs are involved in the splicing process and form an organised structure on RNA during splicing. Splicing must be very accurate in order to generate a complete coding region containing an uninterrupted reading frame.

An alternative cutting–splicing mechanism has been found for rRNA primary transcripts in ciliates and slime moulds, and for primary transcripts of both rRNA and mRNA in yeast mitochondria. In these cases, the RNA transcripts are able to fold into specific conformations facilitating self-cutting and splicing without catalysis by protein enzymes. Such molecules, which use their own structure as substrate, are described as RNA enzymes or ribozymes. Some self-splicing RNAs derived from viroids have been modified in the laboratory to act as ribozymes that cut other RNA templates, thereby inactivating them. They are used in 'gene-shears' technology (Chapter 12).

Prokaryotes: transcription in the cytoplasm

In bacteria, a single RNA polymerase transcribes all genes. Since bacterial DNA is not enclosed by a membrane, newly synthesised RNA is immediately available in the cytosol for protein synthesis. As a result, even while being extruded from the RNA polymerase, the 5' end of a primary transcript of mRNA may interact with the protein synthesis apparatus (ribosomes; Fig. 11.8). In fact, during transcription, only the most recent 10–12 nucleotides laid down by RNA polymerase actually remain base-paired to the DNA template. The rest of the transcript simply peels away and is available for translation.

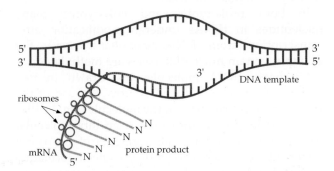

Fig. 11.8 Coincident transcription and translation in prokaryotes. Transcription is occurring as shown in Figure 11.5 and, at the same time, the newly synthesised mRNA has associated with ribosomes and translation of the protein product is underway

Many bacterial transcription units cover more than one polypeptide coding region and are known as **operons**. Resultant transcript mRNAs are **polycistronic**, that is, they contain more than one functional reading frame and code for more than one polypeptide. The protein-coding regions of polycistronic bacterial mRNAs are separated by non-coding nucleotide bases.

In prokaryotes, since bacterial DNA is not bound by a membrane, translation into a polypeptide can commence while mRNA is still being transcribed.

Eukaryotes: transcription in the nucleus

Transcription in eukaryotes differs in several respects from transcription in prokaryotes. Transcription of chromosomal DNA located within the membrane-bound nucleus must occur within this organelle. Also, the DNA of eukaryotic chromosomes is complexed with proteins (histones) to form chromatin (Chapter 10). Only during mitosis (Chapter 8) is chromatin condensed, giving the chromosomes their familiar and distinctive shape. During the remainder of the cell cycle, chromatin is much less condensed (euchromatin) and DNA is accessible to the RNA polymerases involved in transcription in eukaryotes.

The nucleolus (Fig. 11.9) is associated with the nucleolar organiser regions of chromosomes, which have many copies of rRNA genes organised as tandem repeats (i.e. head-to-tail). Here, transcription of rRNA occurs.

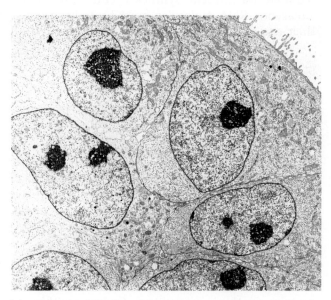

(a)

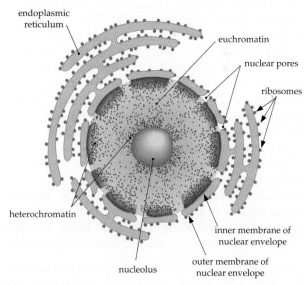

(b)

Fig. 11.9 Nuclear structure and transcription. **(a)** Electron micrograph of interphase nuclei of pig kidney cells. **(b)** In an animal cell nucleus, materials move between the nucleus and cytoplasm through nuclear pores. Ribosomal subunits are assembled in the nucleolus. Euchromatin is loosely packed chromatin that is transcriptionally active. In contrast, heterochromatin is tightly packed chromatin containing transcriptionally inactive DNA

In contrast to bacteria, which utilise a single RNA polymerase to transcribe all active genes, there are three RNA polymerases in the eukaryotic nucleus, each of which transcribes a specific set of genes. RNA polymerase I transcribes ribosomal RNA genes; RNA polymerase II transcribes the informational intermediates, messenger RNAs, which code for proteins; and RNA polymerase III transcribes transfer RNAs required for protein synthesis. After production, eukaryotic RNA molecules must be transported to their site of action, which is usually in the cytosol.

Transcription in eukaryotic cells occurs in the nucleus and involves three RNA polymerases.

The products of transcription

The products of transcription are the three major forms of RNA—mRNA, tRNA and rRNA. Each plays an important role in the synthesis of polypeptides.

Messenger RNA

Messenger RNAs carry information from DNA to the protein synthesis apparatus. Codons within an mRNA specify the exact order in which amino acids will be linked together. Typical eukaryotic mRNA is relatively small, containing 1000–2000 nucleotides. Such mRNAs contain information specifying a *single* polypeptide product. Only one reading frame is utilised in each mRNA molecule; they are **monocistronic**. Not every nucleotide of an mRNA molecule lies within the protein-coding region (Chapter 10). Most mRNAs have flanking regions, both upstream and downstream of the coding region, that are not translated.

In prokaryotic cells, mRNAs tend to be longer than their eukaryotic counterparts. This is because many bacterial mRNAs are polycistronic. The protein-coding regions of polycistronic bacterial mRNAs are separated by non-coding nucleotide bases. In some viruses (bacteriophages), reading frames may overlap.

Codon sequences of mRNA specify the order of assembly of amino acids during protein synthesis. In eukaryotes, each mRNA codes for a single polypeptide (monocistronic), while prokaryotic mRNAs may code for several proteins (polycistronic).

Transfer RNA

Transfer RNAs are small molecules, only 75–80 nucleotides in length. They are known as **adapter molecules** in protein synthesis because they bring amino acids and nucleotides together by covalently binding to an amino acid and hydrogen bonding with the nucleotides of mRNA. As a result, tRNAs align amino acids according to specific codons along the mRNA template. Each type of tRNA recognises only one amino acid and is conventionally named for this amino acid; for example, tRNA^{Leu} is specific for leucine. Enzymic mechanisms in the cell ensure that each tRNA is joined only to the appropriate amino acid.

Codon recognition is achieved by means of an **anticodon**, a specific trinucleotide sequence found at the end of a loop in each tRNA. The anticodon is *complementary* to the mRNA codon representing its amino acid. Complementary base-pairing enables tRNA to recognise the codon.

In both prokaryotes and eukaryotes, many nucleotides in tRNAs undergo modification after transcription from DNA (*post transcription*). For example, certain nucleotide bases are methylated.

In shape, tRNA molecules are drawn in two dimensions as clover-leaf structures because of the stem and loop structures in the molecule (Fig. 11.10a). Stem and loop structures arise from intramolecular base-pairing between small segments of the tRNA nucleotide sequence. This pairing results in formation of four double-stranded regions (stems) interspersed among single-stranded regions (loops). A particular loop (loop 2 in Fig. 11.10) is the anticodon loop, containing the anticodon sequence. Loops flanking the anticodon loop function in binding to the ribosome. The three-dimensional structure of tRNA shows that it is intricately folded (Fig. 11.10b).

Transfer RNA must be activated before it can fulfil its function as an adapter between an amino acid and the corresponding mRNA codon. This is achieved by **aminoacylation**, the covalent attachment of an amino acid to the adenosine residue at the 3′ end of tRNA (Fig. 11.11). For example, aminoacylated tRNA^{Leu} becomes leucyl-tRNA^{Leu} (Leu-tRNA^{Leu}). Such reactions are catalysed by aminoacyl-tRNA synthetases, with a different enzyme required to attach each amino acid to its specific tRNA. There are 20 such enzymes. When tRNA is charged with the amino acid corresponding to its anticodon, it is **aminoacyl-tRNA**.

There is a remarkable degree of specificity in this tRNA charging process. Each of the 20 aminoacyl-tRNA synthetase enzymes puts its amino acid on the tRNA molecule whose anticodon subsequently pairs with the codon specifying the relevant amino acid. Yet the enzymes do not recognise the anticodon itself: the enzyme specifically interacts with loop 1 (the DHU loop) of the tRNA.

The genetic code has 61 amino-acid-specifying codons. However, 61 different tRNAs are not required because a single tRNA may be capable of recognising

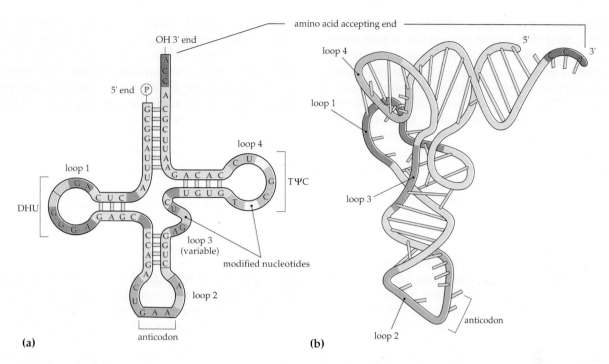

Fig. 11.10 Transfer RNA structure. **(a)** The base-paired regions in typical tRNA are shown in a two-dimensional clover-leaf structure. The loops are sometimes known by other names as follows: loop 1, DHU loop; loop 2, anticodon loop; loop 4, TΨC loop. The size of the variable loop 3 is different across a range of tRNAs. **(b)** An outline of the overall three-dimensional shape, as determined by X-ray diffraction

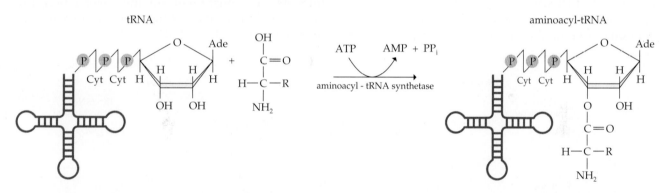

Fig. 11.11 Activation of tRNA by aminoacylation. The acylation of a tRNA molecule occurs in association with a specific aminoacyl-tRNA synthetase enzyme and requires energy from the hydrolysis of ATP. The end result is an aminoacyl-tRNA ready to participate in protein synthesis. The sequence CCA at the 3′ end of the tRNA (clover-leaf structure) is defined by the bases cytosine (Cyt) and adenine (Ade) joined to their respective ribose sugars. In the 3′-terminal nucleotide, the ribose ring is drawn in full to indicate that its 3′ hydroxyl becomes covalently linked to the carboxyl group of the amino acid. R indicates the side chain characteristic for each particular amino acid

more than one codon. Multiple codon recognition by a single tRNA can occur because base pairs, other than A–U and G–C, can form between the first nucleotide of an anticodon and the third nucleotide of a codon. Although these non-standard base pairs have weaker bonds than standard base pairs, the bonds are strong enough to allow specific codon recognition. This ability of tRNAs to wobble in their interaction at the third position of the codon decreases the number of tRNAs required to decode an mRNA. Eukaryotes have about 50 different tRNAs while prokaryotes contain up to 40.

Each tRNA must be charged with the amino acid corresponding to its anticodon (forming aminoacyl-tRNA) before delivering its specific amino acid to the growing polypeptide by base-pairing between its anticodon and the complementary mRNA codon.

Ribosomal RNA

Ribosomal RNA molecules combine with defined sets of proteins to form ribosomes (Chapter 2), which provide a non-specific workbench for the reactions of polypeptide synthesis.

The primary transcripts of rRNA genes are longer than the final rRNA product and are edited before being incorporated into the ribosome. Like tRNAs, each rRNA folds upon itself to form short double-stranded regions. The resultant stem–loop structures in each rRNA give the very large molecule a characteristic three-dimensional structure. This structure is important for the interaction with ribosomal proteins. Individual segments of rRNA may participate directly in the reactions of protein synthesis.

The size of rRNAs and ribosomes are commonly described by Svedberg (S) units, which describe the sedimentation properties of a particle when subjected to centrifugal force.

Each ribosome is composed of one large and one small subunit (Fig. 11.12). Prokaryotic ribosomes (50 S and 30 S subunits, together making a 70 S ribosome) are smaller than those of eukaryotes (60 S and 40 S subunits, together making an 80 S ribosome). (Note that, because the sedimentation property of a particle is a function of several parameters, including size and shape, the S value for an aggregate of particles is not simply the sum of the individual particles.) Each ribosome subunit contains a particular rRNA molecule or set of rRNAs. These rRNAs combine with a characteristic set of proteins to form the assembled ribosome subunit. Ribosomes of prokaryotes contain 23 S, 16 S and 5 S rRNA molecules while those of eukaryotes contain 28 S, 18 S, 5.8 S and 5 S rRNA molecules. When the large and small subunits of a ribosome combine to form a **monosome**, they undergo structural changes, forming a groove between them where the mRNA molecule fits (Fig. 11.13). Monosomes exist only when they are involved in the process of protein synthesis, otherwise they dissociate into their small and large subunits.

> Ribosomal RNA combines with a defined set of proteins to form ribosomes. In both prokaryotes and eukaryotes, each ribosome consists of a small and a large subunit.

TRANSLATION

In the process of **translation**, amino acids are linked in a precise sequence according to the sequence of nucleotides in an mRNA molecule. This process is carried out by ribosomes, with the assistance of tRNA. Prokaryotic ribosomes are smaller than those of eukaryotes (Fig. 11.12), but their fundamental activity in protein synthesis is the same.

Polypeptides are synthesised in three steps: initiation, which includes formation of the ribosome from its two constituent subunits; elongation, the ordered polymerisation of the amino acid chain; and termination (Figs 11.14, 11.15).

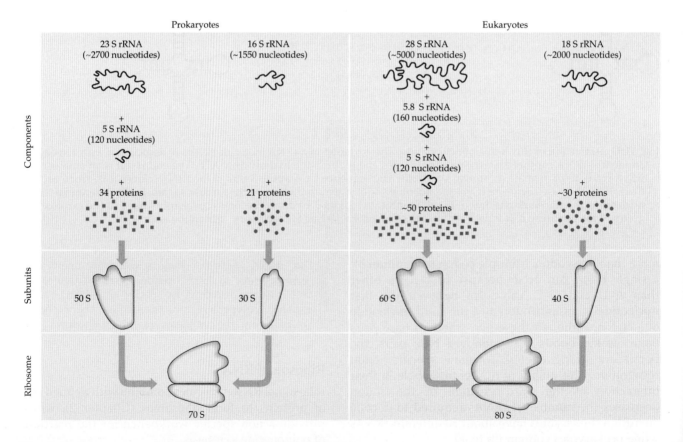

Fig. 11.12 Components of prokaryotic and eukaryotic ribosomes

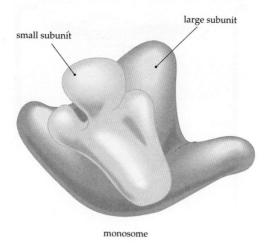

small subunit · large subunit · monosome

Fig. 11.13 Monosome of *Escherichia coli*. Three-dimensional model of the association of the small and large ribosomal subunits to form a functional monosome. The subunits undergo a conformational change upon association and form a groove into which an mRNA molecule fits

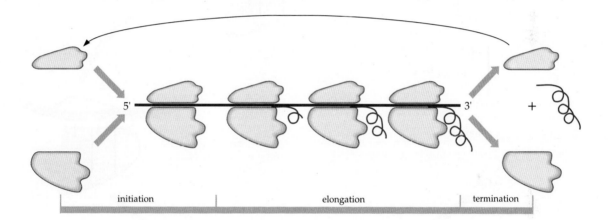

(a)

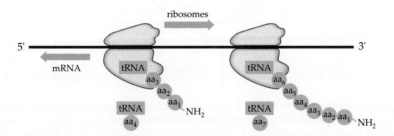

(b)

Fig. 11.14 (a) The three phases of protein synthesis. Initiation requires the formation of ribosomes from two subunits, which bind at the 5′ end of the mRNA. During the elongation phase the ribosome decodes the mRNA sequence to produce a growing polypeptide chain. At termination, the synthesis of the protein is completed and the ribosome dissociates into its two subunits, which may be recycled for further rounds of synthesis. The newly synthesised protein is simultaneously released. **(b)** Key events of protein synthesis. An mRNA molecule advances through a ribosome from its 5′ end to its 3′ end. The growing protein is extended from its N-terminal amino acid to its C-terminal amino acid. The building blocks of protein synthesis are amino acids activated by covalent binding to a specific tRNA molecule (catalysed by the enzyme aminoacyl-tRNA synthetase). Note that the growing polypeptide chain is always attached to a tRNA, which links it to the ribosome

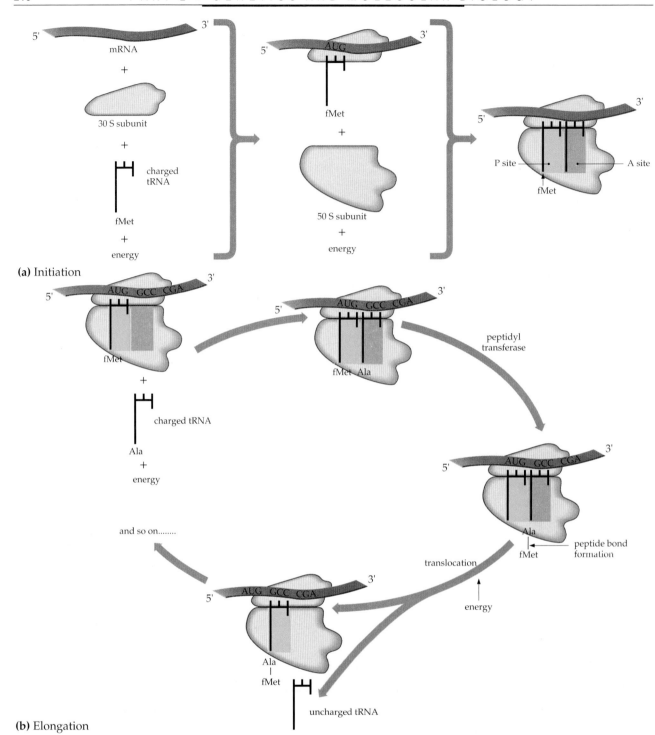

Fig. 11.15 Details of protein synthesis involving a bacterial ribosome (the process is essentially the same with eukaryotic ribosomes). **(a)** Initiation. Initiator tRNA, charged with methionine, binds to a site on the small ribosomal subunit. The small ribosomal subunit binds to an mRNA molecule near its 5′ end so that the mRNA AUG initiation codon pairs with the anticodon in the initiation tRNA. The large ribosomal subunit then joins the small subunit to form a functional initiation complex with two tRNA binding sites: A, which binds aminoacyl-tRNAs, and P, which binds peptidyl-tRNAs (tRNAs linked through their amino acid to the polypeptide being synthesised). Association of the large ribosomal subunit with the pre-existing complex causes fMet-tRNA$_i^{fMet}$ to move into the P site, ready for elongation to begin. **(b)** Elongation. The second aminoacyl-tRNA specified by the next mRNA codon (in this example, GCC coding for alanine) binds to the A site. When the A site is occupied, the two amino acids on the two tRNA molecules become joined by formation of a peptide bond. The reaction forming the new peptide bond is catalysed by the enzyme peptidyl transferase. Initiator tRNA loses its amino acid and the uncharged tRNA dissociates from the P site. The tRNA in the A site (now with two amino acids attached) translocates to the unoccupied P site and the mRNA moves through the ribosome to align the next codon with the now empty A site. The vacant A site now becomes occupied by another aminoacyl-tRNA as determined by the codon aligned to the A site and another cycle of elongation steps follows

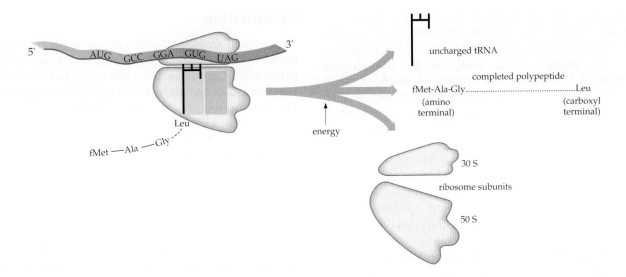

(c) Termination. Elongation continues until one of the three termination codons (UAA, UAG or UGA) is positioned in the A site. Translation ceases because there is no tRNA that recognises stop codons. The aminoacyl bond between the polypeptide and the peptidyl-tRNA is hydrolysed, releasing the polypeptide from the ribosome

> During translation, ribosomes assemble polypeptides according to information carried in mRNA codons.

Initiation

Protein synthesis begins with the assembly of an **initiation complex**, in which a ribosome binds to mRNA and the first specified aminoacyl-tRNA at the initiation codon of the mRNA (Fig. 11.15a). The first step in formation of this complex is the association of the small ribosomal subunit (30 S or 40 S), the initiator aminoacyl-tRNA and a molecule of GTP at the initiation codon of the mRNA. Assembly of these components requires a set of protein initiation factors and, in eukaryotes, energy released by hydrolysis of one molecule each of ATP and GTP. The first amino acid laid down during protein synthesis is always *methionine*, which is almost always specified by the AUG codon. Once the anticodon of the initiating methionine-charged tRNA has associated with the AUG initiation codon, the large ribosomal subunit (50 S or 60 S) binds to the complex and formation of the initiation complex is complete and ready for the elongation reactions to begin.

The formation of an intact ribosome establishes two tRNA binding sites. One site, the *A site*, binds aminoacyl-tRNAs, while the other site, the *P site*, accommodates peptidyl-tRNAs (tRNAs that are covalently linked through their original amino acid to all previously polymerised amino acids of the polypeptide being synthesised). Upon association of the large ribosomal subunit with the pre-existing complex, the initiating methionyl-tRNA moves into the newly created P site, ready for the elongation reactions to begin.

> Protein synthesis begins when the mRNA to be translated associates with the initiating aminoacyl-tRNA and a ribosome.

Elongation

Elongation of a polypeptide is a repetitive process involving sequential addition of amino acids onto the growing polypeptide chain (Fig. 11.15b). The elongation phase includes all reactions from synthesis of the initial peptide bond to the addition of the final amino acid. The process thus involves the formation of sequential peptide bonds as mRNA advances through the ribosome, matching codons of mRNA with anticodons of aminoacyl-tRNAs to place the correct amino acid at each position. The first amino acid laid down is at the amino (N-) terminus, and the polypeptide grows towards the carboxyl (C-) terminus. Addition occurs when the carboxyl group of the first amino acid is joined to the amino group of the second amino acid, and so on.

In order to form a peptide bond, there must be initially a peptidyl-tRNA in the P site on the ribosome, the A site being unoccupied. The appropriate incoming aminoacyl-tRNA then binds to the vacant A site on the ribosome, its anticodon matching the codon on mRNA in the A site. During the formation of the peptide bond, catalysed by the enzyme peptidyl transferase, the aminoacyl-tRNA in the A site is converted to a peptidyl-tRNA and is then shifted to the P site on the ribosome (translocation reaction; Fig. 11.15b). The tRNA that donated its amino acid during translation is displaced into the cytosol, where it can become reactivated by its specific enzyme (aminoacyl-tRNA synthetase). During the translocation reaction, as the new peptidyl-tRNA passes to the P site, it is still matched to the codon in mRNA. The mRNA has thus moved exactly one codon length along the ribosome, positioning the next codon at the vacant A site on the ribosome.

> Elongation of newly synthesised polypeptides involves the sequential addition of amino acids linked by peptide bonds. As the mRNA advances through the ribosome, translation of the mRNA codon sequence into a precisely defined chain of polymerised amino acids occurs.

Termination

The cycle of aminoacyl-tRNA delivery, peptide bond formation and translocation continues for each codon of the mRNA until one of the three stop codons (UAA, UAG or UGA) arrives at the A site on the ribosome. The positioning of a stop codon at the A site immediately stalls the process of translation because there is no tRNA that recognises such stop codons. This terminates polypeptide synthesis (Fig. 11.15c).

Specific *termination factors* then mediate the hydrolysis of the aminoacyl bond between the polypeptide and the peptidyl-tRNA. At the same time, this last tRNA to occupy the P site and the fully translated polypeptide are released from the ribosome. Once the P site of the ribosome is vacant, the ribosome dissociates from the mRNA and disassembles into its two subunits, using the energy released by the hydrolysis of a final GTP molecule. The individual free ribosome subunits can be recycled for another round of protein synthesis.

> Termination occurs when one of the three stop codons (UAA, UAG or UGA) is reached. The newly synthesised polypeptide is released from the ribosome.

BOX 11.3   PUTTING A SPANNER IN THE RIBOSOMAL WORKS

Many bacterial infections are treated by the use of antibiotics that inhibit protein synthesis and result in growth arrest or death of bacterial cells. These antibiotics selectively inhibit protein synthesis in bacterial cells rather than in surrounding eukaryotic cells. This specificity relies on structural and functional differences in the ribosomes of prokaryotic and eukaryotic cells. Certain steps of bacterial protein synthesis are inhibited by interaction between the antibiotic and specific protein components of the bacterial ribosome (Table). These antibiotics do not react with the equivalent protein components of eukaryotic ribosomes and thus they do not have an effect on protein synthesis in mammalian cells. Antibiotics have also proved useful tools for studying the mechanism of protein synthesis by allowing particular steps in the process to be defined.

Since the 1940s, penicillin, which inhibits the synthesis of bacterial cell wall components, and other antibiotics have become invaluable in treating infection. As a result of their widespread use, many bacterial strains have become resistant to these antibiotics, causing serious health problems in hospitals and clinics that harbour resistant bacterial strains.

In the 1970s, it was discovered how bacteria acquire widespread resistance to many antibiotics with various modes of action. Many bacterial cells carry multiple copies of *plasmids*, small circular DNA molecules (Chapter 10). Plasmids carry limited genetic information that may include genes encoding enzymes able to modify or destroy antibiotics. Thus, plasmids are able to confer the property of *drug resistance* to bacterial cells in which they reside, or into which they may be transferred.

Table Some selective inhibitors of bacterial protein synthesis

Antibiotic	Protein synthesis step blocked
Tetracycline	Binding of aminoacyl-tRNA molecules to the ribosome
Streptomycin	Pairing between aminoacyl and message codons
Chloramphenicol (Chloromycetin)	Peptidyl transferase reaction
Erythromycin	Translocation reaction

REGULATION OF GENE EXPRESSION

Gene expression is regulated so that a cell produces only those proteins that are needed. Thus, at any particular time, only a subset of the available genes will be expressed. For example, in photosynthetic organisms, genes leading to the formation of an active photosynthetic system are only expressed when the light intensity is sufficient for photosynthesis to occur.

Unicellular organisms regulate gene expression so that their metabolic and biosynthetic pathways change in response to changes in their nutritional and physical environment. For example, bacteria, such as *Escherichia coli*, growing in the presence of lactose, synthesise a set of enzymes that allow this sugar to be utilised. If lactose is replaced by another sugar, then a different set of metabolic genes are *induced* (expressed) and the genes required for lactose metabolism are *repressed*. Usually, enzymes involved in biosynthetic pathways for the production of molecules, such as amino acids, nucleotide bases and vitamins, are produced only as required. For example, if the amino acid tryptophan is available to bacteria, this source will be used and the genes required for synthesis of tryptophan will be repressed.

Unlike unicellular organisms, cells of multicellular organisms have an environment that remains relatively constant. Thus, an ability to change biosynthetic and metabolic reactions in response to environmental conditions is not required to the same extent as in unicellular organisms. In addition, many differentiated cell types of multicellular organisms, such as red blood cells, skin cells and lymphocytes, do not divide (Chapter 6). They die after a lifetime of specified length, during which they perform specific functions for the benefit of the organism as a whole. Nevertheless, no matter how many cell types make up a multicellular organism, every cell begins with the same DNA complement (Chapter 9). Through a series of regulatory events, only a small subset of genes are expressed at any one time, and new sets of genes are expressed sequentially as a cell differentiates to assume its particular functions.

Control points for gene expression

Regulation of gene expression may occur at several stages during the processes of transcription, translation and formation of a functional protein.

■ *Transcription*. When a protein is required, the gene encoding it may be transcribed at an increased rate. The resultant increase in mRNA levels would lead to a corresponding increase in protein production, providing control was not exerted during a later stage. Regulation at the transcription stage occurs in cells of both unicellular and multicellular organisms. Blocking transcription is most economical since both transcription and translation use energy. Far less energy is expended if an mRNA is simply not transcribed rather than if gene expression is blocked at a later stage.

■ *Stability of mRNA transcripts*. Some mRNAs are more stable than are others under certain conditions, so that they persist in the cell longer and produce a greater level of protein.

■ *Translation*. Even if a particular mRNA is produced continuously, controls can operate on the rate and/or frequency at which it is translated.

■ *Protein product*. Some proteins need specific signals before they become functional and this activation process may be controlled. Degradation of proteins may occur at different rates under different conditions, thus regulating the abundance of particular enzymes in a cell.

> Gene expression is regulated and a cell uses only a subset of the available genes. Regulation may occur at the point of transcription to produce mRNA, in the stability of mRNA, during translation, or even later in the lifetime of a protein.

Regulation of gene expression in prokaryotes

As mentioned earlier, prokaryotic DNA occurs in the same compartment as both RNA polymerase and ribosomes. Thus, mRNA transcripts begin to be translated while they are still being synthesised (Fig. 11.8). In prokaryotes, therefore, the most important regulatory point is at the transcription stage. Few prokaryotic mRNAs are subject to translational control.

At the 5′ end of the coding region, active prokaryotic genes possess a promoter upstream, which is directly accessible to RNA polymerase (Fig. 11.16a). Binding of this enzyme to initiate transcription is assisted by ancillary proteins. The promoter sequence is the key region concerned in major regulatory events, since binding of RNA polymerase to the promoter determines whether a gene will be transcribed or not.

In some instances, genes that encode enzymes or other proteins that are necessary for metabolism, cell growth and division, are expressed continuously regardless of the chemical environment. That is, they are expressed at a *constitutive* level. Other genes are *adaptive*; they are only expressed in a particular chemical environment. The expression of such genes may be modulated by products of *regulatory genes*, which may either induce (positive regulators) or repress (negative regulators) the transcription rate of structural genes (Fig. 11.17).

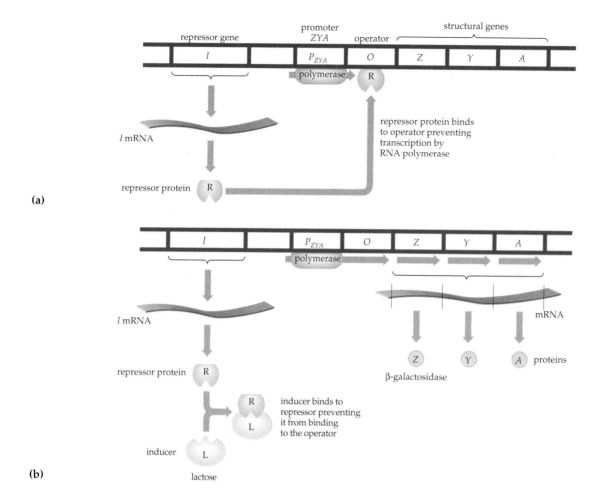

Fig. 11.16 Regulation of the lactose (*lac*) operon of *Escherichia coli*. **(a)** The repressor protein (R), specified by the repressor gene (*I*) and always made at a steady rate, binds to the operator region (*O*). The operator is a short sequence positioned between the promoter (*P*) and the region of the transcription unit that specifies mRNA coding for proteins (*Z, Y, A*). Binding of the repressor protein to the operator prevents RNA polymerase from binding to the promoter, thereby blocking transcription. **(b)** Lactose acts as an inducer by binding to the repressor protein. With inducer bound to it, the repressor protein can no longer bind to the operator. This allows transcription to be initiated at the promoter. The synthesis of proteins needed for lactose metabolism, encoded by the *Z, Y* and *A* genes, then takes place

The *lac* operon and negative control

One of the landmarks in our understanding of regulation of gene expression is the lactose (*lac*) operon of *E. coli* (Fig. 11.16). An operon is a single transcription unit that encompasses more than one polypeptide-coding region and therefore produces polycistronic mRNA. In general, operons contain structural genes.

The structural genes of the *lac* operon, *lacZ*, *lacY* and *lacA*, encode three enzymes necessary for lactose metabolism. The most important of these genes is *lacZ*, which codes for β-galactosidase. This enzyme is responsible for digesting lactose to its constituent sugars, glucose and galactose, both of which can be used by the cell as an energy source. The presence of lactose induces expression of *lacZ*, *lacY* and *lacA* by stimulating transcription of the polycistronic

mRNA. However, in the absence of lactose, synthesis of mRNA is repressed and, hence, the enzymes are not produced. Biologically, β-galactosidase is only useful if lactose is available in the environment.

How does lactose induce expression of these structural genes? Genetic tests identified two gene loci that are involved in regulation of the *lac* operon. Mutations at either locus cause constitutive expression of structural genes; that is, they are made at a steady rate at all times. One of these loci, *lacI* (for **i**nducibility), exerts its effect regardless of its position in bacterial DNA. The effect is attributed to production by *lacI* of a diffusible protein product (repress or protein). Thus, the *lacI* gene is *trans*-acting. The second regulatory locus, *lacO* (for **o**perator), could not be moved from its natural location immediately upstream of the structural genes without loss of function. The inference is that *lacO* does not produce a diffusible product; the locus is *cis*-acting. This locus lies between the RNA polymerase binding site of the operon promoter, *P*, and the operon structural genes (Fig. 11.16).

The detailed mechanism of regulation of the *lac* operon has been established on the basis of genetic and biochemical experiments. The *lacI* locus constitutively produces a protein repressor, which, in the absence of lactose, binds to the operator *O* locus and represses transcription of the operon by physically covering the site of transcription initiation. It prevents access of the RNA polymerase to the promoter. Thus, the repressor exerts *negative control* on transcription.

However, when lactose is present in the cell, it is metabolised to an effector molecule, the *inducer*. The inducer binds to the repressor protein forming an inactive complex. In this situation, the operator, *O*, is not covered by repressor and RNA polymerase can move through into the promotor, *P*, initiating transcription, and then transcribing the entire operon. Expression of *lacZ*, *lacY* and *lacA* leads to the efficient metabolism of lactose. If the level of lactose declines, free repressor will become available to bind to *O*, repressing the operon.

> The *lac* operon provides an example of negative control of transcription. In the absence of lactose, the *lac* repressor binds to the promoter, preventing transcription. When present, lactose binds to the repressor so that transcription occurs.

The arabinose operon and positive control

Positive control of transcription also occurs in bacteria. Regulation of the arabinose (*ara*) operon, responsible for supplying enzymes required for the breakdown of the sugar, arabinose, is regulated by the *araC* gene product. Unlike the *lac* repressor, the *araC* gene product stimulates transcription of the *ara* operon when bound to the inducer, arabinose. In the absence of arabinose, the *araC* gene product is inactive and transcription of the *ara* operon does not occur.

> The *ara* operon provides an example of positive control. When bound to arabinose, the *araC* gene product induces transcription.

Regulation of gene expression in eukaryotes

The precise control of expression of particular genes underlies cell differentiation in multicellular organisms (Chapter 16). Gene regulation also plays an important role in the responses of cells and tissues to short- and long-term changes in their environment. Chemical signals, such as cell–cell recognition factors and hormones, influence the expression of genes in the cells of multicellular organisms. Some of these signals affect gene expression by binding to the cell surface, while others must pass into the cytoplasm of the cell to elicit a response (Chapter 6).

One of the major control points for eukaryotic gene expression is when transcription is initiated. Transcription of most eukaryotic protein-coding genes is started by an initiation complex containing RNA polymerase II and several other protein factors, which must bind efficiently to the promoter to activate transcription. In addition, there are many protein transcription factors that play a role in modulating transcription of different eukaryotic genes by binding to specific DNA sequences in the promoter region. Many transcription factors vary in abundance between cell types and are likely to fulfil regulatory functions.

The action of some transcription factors has been shown to depend on specific signals. The effect of regulatory factors on transcription, in conjunction with their effectors, can be to either increase (positive regulators) or decrease (negative regulators) the transcription rate (Fig. 11.17).

Initiation of transcription by the initiation complex alone is unusual. Multiple promoter elements occur in the DNA region very close to the transcription start site of many genes. These are specialised DNA sequences that bind regulatory proteins that play a key role in transcription initiation. Many, if not all, eukaryotic genes require **activators**, which bind to promoter elements upstream of the gene, in order to be optimally transcribed. In yeast and other simple eukaryotes, one such promoter element is the *upstream activator sequence* (UAS). This sequence provides a single DNA binding site for a single activator protein and is sufficient to accelerate transcription in yeast.

A characteristic feature of eukaryotic genes is that they have *multiple* promoter elements (Fig. 11.17). Many of these respond to activators but others have a silencing function that depresses transcription of the gene. Such regulatory elements can operate so that the products of one gene can control activity of another gene and so initiate a cascade of events. In the absence of polycistronic mRNAs in eukaryotes, control of sets of genes through such cascades allows the effective regulation of a large number of genes by controlling a few 'master' genes. This is particularly important in the embryonic development of higher eukaryotes in which 'homeotic' genes control the formation of entire body parts or tissues (Chapter 16).

Another feature of genes of higher eukaryotes is the frequent occurrence of **enhancers**, activator-binding sites that may be far removed from the gene (but on the same chromosome), lying in either the

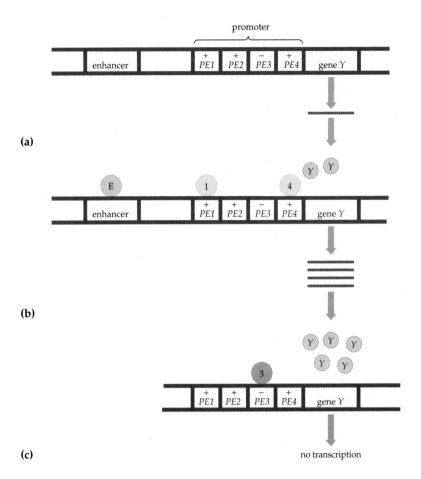

Fig. 11.17 Regulation of eukaryotic gene expression at the level of transcription usually occurs at more than one regulatory site. General activation of transcription across a broad region of DNA occurs at the enhancer, which may be 50 kb from the RNA start site, upstream of the gene (as shown here) or even downstream of the gene. The promoter may comprise several subsites, called promoter elements (*PE1*, *PE2*, etc.), at which different regulatory proteins bind. These proteins may act as activators (+) or repressors (–) of transcription. Three states are shown here. **(a)** In the absence of any proteins binding at regulatory sites upstream of the gene a basal level of transcription occurs. **(b)** Positive regulation occurs when proteins binding to specific elements activate transcription (e.g. promoter elements *PE1* and *PE4*). **(c)** Negative regulation occurs when protein binding to a particular element shuts down transcription; the promoter element *PE3* is called a silencer

upstream or downstream direction. Also, because of the complex organisation of chromatin, the activity of a gene in transcription is influenced by the extent of relaxation of the chromatin structure.

> Active eukaryotic genes characteristically possess promoter elements that may be activated by enhancer elements. Regulatory elements can operate so that the products of one gene can control activity of another gene, and so initiate a cascade of events.

Induction of *GAL* genes in yeast

The positive and negative regulation of inducible genes responsible for galactose metabolism in yeast provides a good example of transcriptional control in a eukaryote (Fig. 11.18). When yeast cells are grown in a medium containing galactose, expression of five particular genes, located on three chromosomes, is induced. All five *GAL* genes are expressed independently of one another. The expression of these genes is controlled by a positive regulator encoded by *GAL4* and a negative regulator encoded by *GAL80*, both of which are constitutively expressed. The *GAL4* product binds to the sequence upstream of each of the five inducible genes, inducing their expression. However, the *GAL80* product, which does not bind to DNA, represses expression by binding directly to the *GAL4* product. Thus, in the uninduced state, the five galactose-inducible genes are inactive due to the inability of *GAL4* to activate transcription, although bound to the UAS, because it is trapped in the *GAL4–GAL80* complex.

When galactose is introduced to the cell, an as-yet unidentified inducer, which is metabolically produced, causes the *GAL80* protein to dissociate from the *GAL4–GAL80* complex. This allows the bound

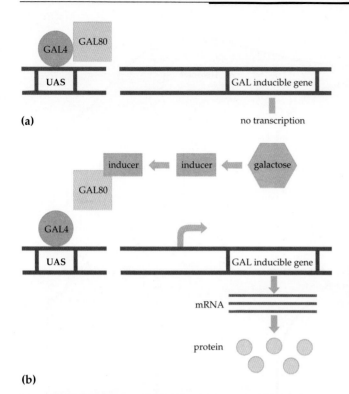

(a)

(b)

Fig. 11.18 Regulation of galactose (GAL)-inducible genes in yeast. A positive regulatory promoter element in yeast is often called a UAS (upstream activator sequence). Two regulatory proteins, GAL4 and GAL80, are formed constitutively. **(a)** Binding of the GAL4–GAL80 complex to the UAS for the galactose-inducible gene prevents transcriptional activation, which would occur if GAL4 were to bind to the UAS by itself. **(b)** An as-yet unidentified metabolic product, formed in yeast cells when galactose is present, binds to GAL80 and prevents it from binding to GAL4. This now allows GAL4 to function as an activator by binding to UAS by itself. Thus, the presence of galactose in cells allows the protein encoded by the galactose-inducible gene to be synthesised

GAL4 protein to activate the inducible genes. Thus, protein–protein interactions can also be important in the control of eukaryotic gene expression.

> Positive and negative regulation of galactose metabolising enzymes is achieved in yeast by regulatory elements.

PROTEIN TARGETING AND PROCESSING

Genes are not fully expressed until their protein product is transported from its site of synthesis to its site of action (*protein targeting*) and, if necessary, converted into a functionally active protein (*protein processing*). Eukaryotic cells, with many intracellular compartments, have highly specialised mechanisms that direct proteins to particular organelles or membrane systems of the cell (Fig. 11.19). Protein targeting is very accurate, such that the set of proteins taken up by any one of these organelles is highly specific. Proteins without a particular signal sequence or label remain in the cytosol.

Protein processing may include the formation of disulfide bridges between two cysteine residues within the same amino acid chain, cleavage of the amino acid backbone, or addition of one or more specific chemical groups. These modifications often occur during the targeting process. On reaching its correct destination, a polypeptide may need to associate with other proteins in order to form a functional unit, such as a multisubunit enzyme complex. This last stage is *protein assembly*. For example Rubisco (Chapter 5) is assembled in chloroplasts from two different polypeptide chains. Other enzyme complexes may contain more individual polypeptides, but many enzymes function as single polypeptide chains.

Targeting of polypeptides to different organelles has the following properties.
1. An 'address label' (or signal sequence) directing a polypeptide to a specific organelle is located within the polypeptide. Often the signal sequence is located before the N-terminus and is deleted once the cellular site has been reached and the label is no longer needed. Sometimes the signal sequence is located within the coding region and is not removed.
2. These labels are recognised and bound by specific receptors at the surface of the correct cellular site for the polypeptide.
3. Polypeptides that are to cross a membrane must have an unfolded conformation (Chapter 1).
4. Transport of polypeptides across membranes requires energy.
5. Import into organelles bounded by double membranes, such as mitochondria and chloroplasts, occurs at pores that are formed at junctions, or points of contact, between inner and outer membranes.

During protein targeting, processing and assembly, polypeptides are protected from damaging interactions with other molecules, or from prematurely folding on themselves too tightly, by protein *molecular chaperones* (Box 11.4).

> A signal sequence within a polypeptide directs it to a specific cellular site where it binds to a receptor (protein targeting). To form an active protein, a polypeptide may need to be modified or associated with other polypeptides (protein processing and assembly).

Polypeptides translated on free ribosomes

Polypeptides synthesised on 'free' ribosomes are released into the cytosol (Fig. 11.19). From there, they may pass to the nucleus and organelles such as

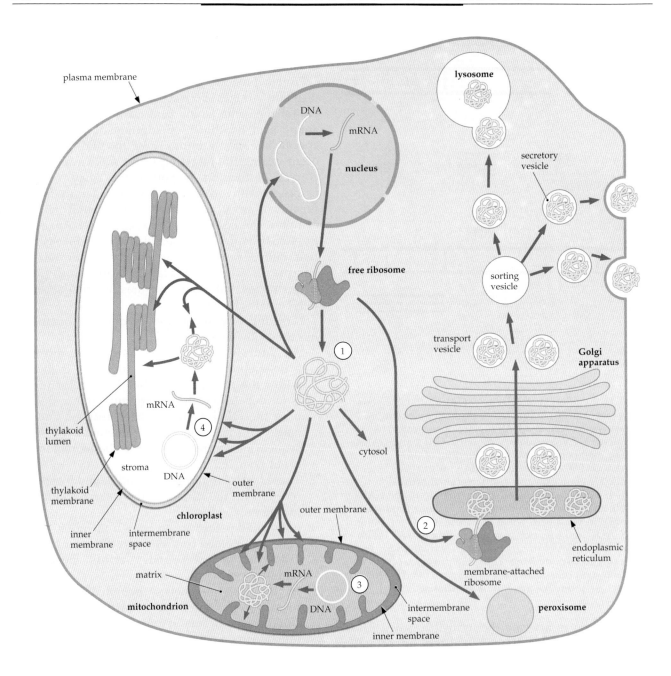

Fig. 11.19 Pathways of protein targeting in eukaryotic cells. Chromosomal genes in the nucleus encode mRNA molecules that move to the cytosol for translation by ribosomes. In one major targeting pathway (1), proteins synthesised on free ribosomes may remain in the cytosol, or be directed to one of the following membrane-enclosed organelles: nucleus, peroxisomes, mitochondria or chloroplasts. In the case of mitochondria, imported proteins are further targeted within the organelle to one of four possible destinations: outer membrane, intermembrane space, inner membrane or matrix. Similarly, proteins imported into chloroplasts are targeted to one of six possible destinations: outer membrane, intermembrane space, inner membrane, stroma, thylakoid membranes and thylakoid lumen.

   A second major pathway (2) involves ribosomes that become bound to the endoplasmic reticulum (ER). Proteins pass into the lumen of endoplasmic reticulum as they are synthesised. They may remain in the ER or move further to the Golgi apparatus, emerging within transport vesicles, then to sorting vesicles. Some proteins go to lysosomes, others to secretory vesicles for exocytosis. Some proteins may be embedded within a membrane that ends up enclosing a particular compartment or within the plasma membrane itself.

   Two further specialised protein targeting pathways are found in eukaryotic cells. In mitochondria (3), proteins are synthesised on mitochondrial ribosomes, encoded in mitochondrial DNA, and targeted to the inner mitochondrial membrane. In chloroplasts (4), proteins, encoded in chloroplast DNA and made on chloroplast ribosomes, may remain in the stroma, or be targeted to the thylakoid membranes or through into the thylakoid lumen

BOX 11.4 ROLE OF MOLECULAR CHAPERONES

As proteins are distributed to specific cell locations from their site of synthesis, they are exposed to the different environments of various intracellular compartments. A family of proteins, *molecular chaperones* (cpn), is involved in preventing interactions during the biosynthesis or transport of a polypeptide that could alter its functional structure or cause its inactivation.

Chaperones act, at appropriate stages, to maintain a polypeptide in an unfolded state or to mediate correct folding of a polypeptide. Chaperones do not themselves possess information specifying protein folding but rather act to prevent interactions that would otherwise result in misfolded protein structures (see figure). The mechanism by which this is achieved is thought to involve specific, non-covalent binding of the chaperones to protein surfaces.

The action of chaperones may be involved with movement of polypeptides across membranes and assembly of multisubunit protein structures (Chapter 1). For example, nuclear gene-encoded polypeptides targeted to mitochondria are synthesised in the cytoplasm and must be imported into mitochondria. One chaperone is involved in maintaining the polypeptide in a partially unfolded state suitable for movement across mitochondrial membranes. Once inside the matrix of a mitochondrion, the imported polypeptide is transferred to another chaperone that facilitates refolding of the polypeptide into a conformation suitable for assembly into an active protein. This is particularly important where polypeptides interact with other protein subunits leading to formation of an active enzyme. Chaperones themselves are not components of the assembled protein structures.

Chaperones also function to protect protein structure during environmental stresses, such as high temperature (heat shock). Such stresses commonly cause denaturation of proteins and formation of undesired and inactive aggregates. Chaperones limit the denaturation by protecting, and possibly even rescuing, proteins from the incorrect interactions that cause them to become denatured. In this way, cells can recover from limited stress.

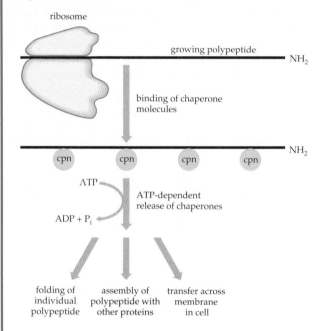

Role of chaperones in protein folding. As the N-terminal end of the growing polypeptide emerges from the ribosome, chaperone molecules (cpn) bind non-covalently to it. The binding of chaperones acts to keep the polypeptide in an unfolded state. Release of chaperones involves hydrolysis of ATP

mitochondria and chloroplasts. The set of proteins taken up by these structures is highly specific. For example, all proteins found within the nucleus are imported from the cytosol through large pores in the nuclear envelope. However, this is not a passive event. Receptors on the cytoplasmic face of the nuclear membrane, near the pores, recognise and bind only to polypeptides destined for the nucleus. After binding, hydrolysis of ATP drives the bound polypeptide through the pore. A key aspect of proteins targeted to the nucleus is that the label is entirely within the structure of the functional polypeptide and is not trimmed off after import. Such nuclear address signals may be anywhere within the polypeptide.

Another interesting example are mitochondria. These organelles make a small number of proteins, encoded by their own mitochondrial DNA, which are

involved in oxidative phosphorylation (Chapter 5). However, the vast majority of mitochondrial proteins are imported from the cytosol. Before being imported into the matrix, an incoming protein is kept unfolded by remaining bound to proteins that act as *molecular chaperones* in the cytosol (Box 11.4). Polypeptides are directed to mitochondria by a targeting sequence at the N-terminus. This sequence also signals import into mitochondria, at which time the sequence is usually trimmed off to form a shorter functional protein. Molecular chaperones maintain the protein in an unfolded state, or ensure proper folding and assembly, as required.

The majority of chloroplast proteins are also synthesised on free ribosomes. Less is known about the targeting of proteins to chloroplasts, but the mechanisms involved appear to be fundamentally similar to those that target proteins to mitochondria.

Import of proteins into the nucleus occurs with the aid of receptors associated with the pores in the nuclear membrane. Proteins imported into other organelles usually possess an N-terminal signal sequence that is trimmed off during passage into the organelle via membrane receptors. Molecular chaperones maintain proteins in an unfolded state, then correctly fold and assemble the imported proteins.

Polypeptides translated on membrane-bound ribosomes

Polypeptides synthesised on ribosomes bound to endoplasmic reticulum (ER) follow a different pathway from polypeptides synthesised on free ribosomes (Fig. 11.19). During translation, polypeptides are extruded across the ER membrane as they are assembled. Polypeptides entering ER have an N-terminal signal sequence of 16–30 residues with a high content of hydrophobic amino acids (Chapter 1). This signal sequence is translated by a ribosome during a short burst of translation while the ribosome is still free (Fig. 11.20). Specific receptors on the ER membrane recognise the signal sequence and insert it into the ER membrane, which binds the ribosome to ER. Translation of the remainder of the polypeptide proceeds, with the product being extruded across the membrane and into the ER lumen.

The N-terminal signal sequence embedded in the ER membrane may be trimmed from the newly synthesised polypeptide. This usually results in a free protein in the lumen of ER. These proteins usually proceed to the Golgi apparatus and then either to lysosomes, or to the plasma membrane and extracellular space via specialised membrane vesicles (Chapter 3). The N-terminal signal sequence is not trimmed from proteins destined to remain embedded in the membrane. Some of these membrane-bound proteins will become part of the plasma membrane of the cell.

Protein targeting is not exclusive to eukaryotic cells. Even though prokaryotic cells do not have an internal membrane system, proteins can be targeted 'out of the cell' across the plasma membrane by a similar but simpler process to that in eukaryotic cells. In prokaryotes, targeting and transport of the protein across the membrane occur after the protein has been completely synthesised on free ribosomes.

Proteins synthesised at membrane-bound ribosomes are extruded across the ER membrane as translation occurs. Most are transported to their various destinations within membrane-enclosed vesicles. Some will remain bound to the membrane.

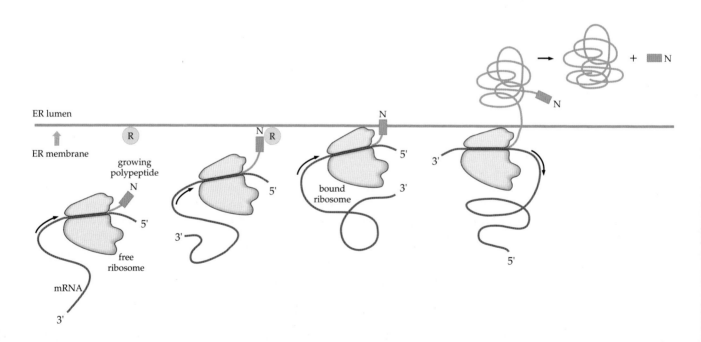

Fig. 11.20 Cotranslational transfer of proteins into the endoplasmic reticulum. The N-terminus of a protein that is to enter the ER system contains an N-terminal signal sequence that binds to a receptor (R) on the ER membrane surface. This interaction causes the large ribosomal subunit to bind to the membrane and thus a free ribosome becomes a bound ribosome. Interaction of the signal peptide (open rectangle) with the receptor causes the growing protein to be extruded through the ER membrane. During movement across the membrane, the N-terminal signal sequence is trimmed from the newly synthesised polypeptide. This usually results in a free protein within the lumen of the ER

SUMMARY

- Gene expression involves the flow of information from DNA to an intermediate molecule, mRNA (transcription), which then specifies the amino acid sequence of a polypeptide and directs the assembly of the protein at a ribosome (translation).

- The common genetic code is made up of three-letter code words (codons). Each group of three nucleotides of DNA may code for an amino acid.

- Transcription begins when RNA polymerase binds to the initiation signal sequence in DNA. The region of transcribed DNA between initiation and termination signals defines a single transcription unit. The RNA product directly synthesised within a transcription unit is a primary transcript. The primary mRNA transcript in eukaryotic cells is often much longer than the final mRNA that is translated in the cytoplasm and must be edited in several ways.

- In eukaryotic cells, mRNA generated by transcription moves from the nucleus to the cytoplasm where translation occurs. In prokaryotes, mRNA is synthesised in the cytosol and translation may commence before transcription is complete.

- Ribosomal RNA combines with a defined set of proteins to form ribosomes where polypeptide assembly takes place. Transfer RNA must be charged with the amino acid corresponding to its anticodon (aminoacyl-tRNA) before it can fulfil its function as an adapter between an amino acid and mRNA codon.

- Codons within mRNA specify the exact order in which amino acids will be linked together during protein synthesis. In eukaryotes, only one reading frame is utilised in each mRNA molecule (monocistronic). Bacterial mRNAs may contain two or more functional reading frames coding for different proteins (polycistronic).

- Elongation of the growing polypeptide involves the sequential addition of amino acids linked by peptide bonds as mRNA moves along the ribosome. When a stop codon is reached, termination and release occur.

- Gene expression is regulated and cells use only a subset of the available genes. Regulation may occur at the point of transcription to produce mRNA, in the stability of mRNA, during translation, or even later in the lifetime of a protein. The *lac* operon is an example of negative control of transcription of structural genes coding for enzymes metabolising lactose. The *ara* operon provides an example of positive control of transcription.

- Active eukaryotic genes characteristically possess a promoter element at which transcription is initiated. Such genes are activated by enhancer elements acting at a distance within a chromosome. Transcription may be controlled by the action of transcriptional regulators (positive or negative) that interact with DNA sequences at or near the promoter.

- Polypeptides are transported to their active site by protein targeting mechanisms. They may be structurally modified and, if necessary, associated with other polypeptides to form an active protein by protein processing and assembly mechanisms.

- Polypeptides formed on free ribosomes and destined for the nucleus contain information in the form of a signal sequence that is a permanent part of the polypeptide. By contrast, polypeptides destined for other organelles usually have an N-terminal signal sequence that is trimmed off during movement into the organelle.

- Proteins assembled at membrane-bound ribosomes are extruded across the ER membrane as translation occurs and often modified in the Golgi complex.

QUESTIONS

1. Describe the key features of the common genetic code.

2. How are DNA and RNA sequences involved in the two stages of gene expression?

3. What is the function of a promoter?

4. How do the mRNA molecules synthesised in eukaryotes differ from those synthesised in prokaryotes?

5. Why is precise excision of intron sequences from eukaryotic transcripts important?

6. Describe the structure and function of tRNA molecules.

7. What are the roles of the A and P ribosomal sites in protein synthesis?

8. What is the relationship of a codon to an anticodon?

9. (a) Describe in general terms how gene expression may be regulated in prokaryotes. (b) In the *lac* operon, when cells growing in non-lactose containing medium are switched to medium containing only lactose, what is the effect on gene expression?

10. In eukaryotes, what other sequence elements besides the promoter are important in regulating transcription?

11. What rules apply to the targeting of proteins to cellular organelles?

12. Why are molecular chaperones important in the movement of proteins around cells?

13. Compare the import of a protein into the mitochondrial matrix with cotranslational transfer of a protein into the endoplasmic reticulum.

RECOMBINANT DNA AND THE EUKARYOTIC GENOME

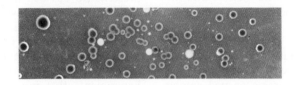

In the past decade, recombinant DNA technology has allowed molecular biologists to cut out small pieces of DNA (genes or fragments of genes) from a genome, isolate the fragments and multiply them in bacteria where the gene products can be expressed. As a result, genes can be sequenced readily, their positions on chromosomes can be located and their actions can be studied. Recombinant DNA technology is giving us new means of diagnosing, preventing and perhaps even curing human genetic diseases. Knowledge of the genomes of domestic animals and crop plants presents new opportunities for breeding superior varieties.

In this chapter, we will describe some of the new molecular tools that make it possible to explore the organisation and function of large and complex genomes, such as the human genome. We will consider how genes are organised, how they control cellular functions, and how genetic engineering makes it possible to manipulate gene function. Recombinant DNA technology has generated much public anxiety about the prospect of engineering 'genetic monsters' (see Box 12.1). However, when properly regulated, the technology has provided safe valuable research and diagnostic tools.

▌ Recombinant DNA techniques enable direct study of genes and manipulation of gene function.

RECOMBINANT DNA TECHNOLOGY

Enzymes have been discovered that cut DNA at specific sequences, producing segments of different sizes. These fragments can be separated by gel electrophoresis and spliced together in new combinations to create **recombinant DNA** (Box 12.2). Although cutting and splicing occur naturally as a result of genetic mechanisms, such as recombination, recombinant DNA technology produces new artificial associations

between DNA molecules or segments that are not normally associated, for example, between eukaryotic DNA and plasmids. Recombinant DNA technology makes use of a series of unique enzymes for the following sequence of steps.

1. Eukaryotic DNA molecules or segments (donor DNA) are isolated.

2. Donor eukaryotic DNA is joined to another DNA molecule, which acts as a vector, forming recombinant DNA.

3. The vector containing recombinant DNA is taken up by a bacterial cell (host).

4. The recombinant DNA is replicated repeatedly along with the host DNA and produced in abundance (cloned) in bacterial culture.

5. Cloned DNA can be isolated, purified and characterised.

6. Cloned DNA can be transcribed, mRNA translated and the gene product purified and characterised.

This technology permits recombinant gene products to be expressed in sufficient quantities for this to be an efficient means of producing proteins.

Restriction enzymes

The enzymes used to cut DNA are *restriction endonucleases*, a group of enzymes whose job in bacteria is to cut up DNA molecules carried into the cell by viruses (*restricting* the entry of foreign DNA). The names of restriction enzymes are based on the bacterium from which they were isolated. For example, the restriction enzyme, *Eco*RI, was isolated from *Escherichia coli*. Restriction enzymes cut double-stranded DNA at specific base sequences known as recognition sequences (Fig. 12.1). There are two classes: type I enzymes cut the DNA molecule non-specifically at a site close to the recognition sequence;

BOX 12.1   RECOMBINANT DNA: THE PUBLIC DEBATE

Methods for cutting and splicing DNA were first discovered and used to make recombinant molecules in the 1970s. Dr Paul Berg of Stanford University, one of the leaders in this new field, spliced foreign sequences into an animal virus. It was not long before Berg and other scientists became concerned that release into the environment of genetically modified organisms may have unforeseen consequences.

These scientists called for a moratorium on gene manipulation experiments until it could be shown that DNA technology posed no hazard to humans and other organisms. A great deal of discussion ensued among scientists about possible dangers of inadvertently creating disease organisms, adding a new organism to the natural environment or tinkering with the human genome. Many scientists noted, with some concern, that the *Escherichia coli* host to all these recombined vectors was originally isolated from the human gut. Would this mean that *E. coli* carrying recombinant plasmids with, say, a human cancer gene, could infect a careless research worker or pollute the water supply of a whole city? It did not take the media long to imagine frightening (and sometimes comic) scenarios of genetically engineered monsters taking over the world (see figure). The debate shifted from academic lecture theatres and scientific conferences to the front pages of newspapers.

The scientific debate resulted in a series of guidelines for the conduct of experiments involving recombinant DNA. At the time, these measures were hardly sufficient to allay public fears, but in retrospect, the guidelines were probably a little too cautious.

Relatively simple physical and biological containment rules apply to some kinds of recombinant DNA experiments, such as those involving DNA from organisms that could hybridise (cross) naturally. Cloning genes, such as of β-globin or insulin, into plasmids presents no obvious hazard, but must be performed under conditions that would prevent the escape of the recombinant organisms. This is done in laboratories designed for physical containment, and by selecting bacterial host cells that are weakened by mutation so that they cannot survive outside the laboratory. Experiments that involve splicing of genes that are known to make products dangerous to human health (e.g. bacterial toxins, human cancer genes) are subject to stringent regulation. Such experiments are permitted if they can be shown to be necessary for our understanding of human disease or the design of drugs and vaccines.

In Australia, a Genetic Manipulation Advisory Committee oversees conduct of all experiments using recombinant DNA technology, and control is exercised at a local level in each institution where genetic engineering experiments are being carried out. Trials have been permitted of genetically engineered bacteria that could prevent frost damage to plants, potatoes able to destroy attacking viruses, cotton plants that express their own recombinant insecticide (Chapter 15) and 'superpigs' with an extra set of growth hormone genes.

'Crack out the liquid nitrogen, dumplings . . . we're on our way.' A cartoon representing contemporary public fears of genetic manipulation

type II enzymes cut the DNA precisely within the recognition sequence. Restriction sequences of type II enzymes are *palindromic*, that is, their nucleotides read the same forward as backward. Some restriction enzymes cut at four-base sequences, others at five-base sequences; some have longer recognition sequences (Fig. 12.1). For example, if DNA from a virus is cut by *Eco*RI, the molecule will be cut at every position that contains the sequence GAATTC. A molecule with only one such sequence will be cut into two fragments. Each identical DNA molecule will be cut into the same two fragments. The lengths of fragments will depend on the location of the single recognition sequence in the DNA molecule.

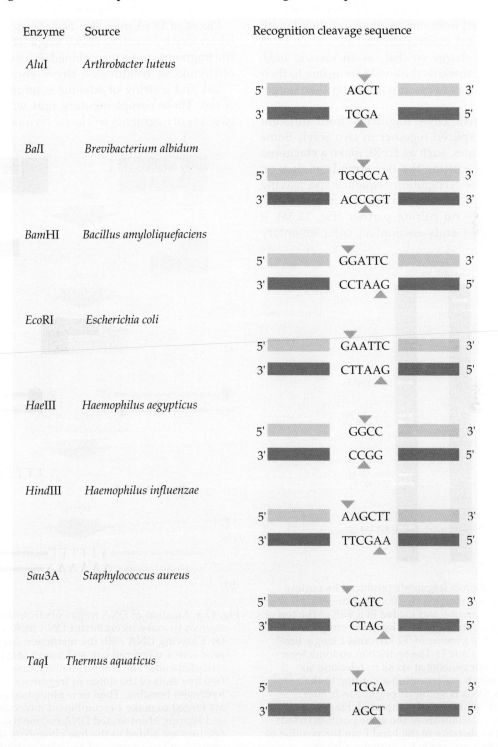

Fig. 12.1 Recognition sequences, cutting sites and source for some restriction enzymes. These restriction enzymes can recognise a six-base sequence (*Eco*RI, *Bam*HI) or a four-base sequence (*Hae*III) of DNA. Cutting sites on each strand are arrowed. Restriction enzymes can cut the two strands at the same site, producing a blunt-ended fragment (e.g. *Hae*III), or make staggered cuts to produce fragments with single-stranded sticky ends

If the same virus genome is cut with a different restriction enzyme, for example, *Bam*HI, it will be cut at every position that contains GGATTC. *Bam*HI fragments will overlap *Eco*RI fragments and thus cutting with both enzymes together will produce fragments whose lengths add up to those of the original *Eco*RI fragments, enabling their relative positions in DNA to be mapped. DNA fragments can then be separated from one another and their length determined by gel electrophoresis (Fig. 12.2). DNA has a negative charge so that, in an electric field, fragments move towards the anode according to their size, with small fragments moving faster than longer fragments.

Once separated, DNA fragments from different sources can be spliced together in two ways. Some restriction enzymes, such as *Eco*RI, make a staggered cut in the DNA, cutting the two strands at sites on either side of the recognition sequence and leaving short 'sticky' single-stranded regions at each end, which now have no pairing partner (Fig. 12.3a). If one of these free ends encounters complementary

bases on the free end of the other strand, the two may pair by formation of hydrogen bonds. Although not strong bonds, they may permit a new phosphodiester bond to form between the broken ends of the polynucleotide chain (Fig. 12.3a). This is *ligation* and is performed by enzymes in the cell whose normal job is to repair single-strand nicks in the backbone of DNA (Chapter 10).

Pieces of DNA may also be spliced together in the laboratory by creating sticky single-stranded tails on the fragments to be recombined. For example, a string of thymidine residues on the 5' ends forms a poly T tail, and a string of adenine residues forms a poly A tail. These complementary tails will pair and the two sets of fragments will ligate on mixing (Fig. 12.3b).

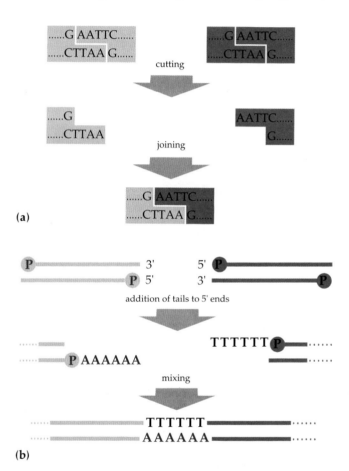

(a)

(b)

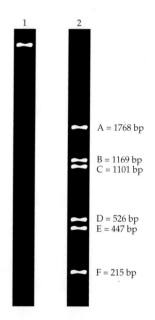

Fig. 12.2 Separation of fragments produced by cutting SV40 virus DNA with a restriction enzyme, *Hind*III. Viral DNA in solution was loaded in a well at the top of the gel (origin), and separated by gel electrophoresis. The uncut virus genome of 5.2 kb forms a single band near the origin (lane 1). The restriction endonuclease cuts the circular genome at six sites, releasing six fragments, which are separated according to their size in kilobases. DNA is visible as bright bands after staining with a fluorescent compound. Fragments of the same size accumulate at the same position (band) in the gel. The distance of the band from the position at which the DNA was loaded, compared with the distance travelled by marker fragments of known size, can be used to calculate the size of the fragment. Fragment size is usually expressed as base pairs (bp) of DNA

Fig. 12.3 Ligation of DNA fragments from different sources to make recombinant DNA molecules. **(a)** Cleaving DNA with the restriction enzyme *Eco*RI produces a staggered cut, and leaves short single-stranded ends. The complementary base sequences of two free ends of the different fragments may pair by hydrogen bonding. Then new phosphodiester bonds are forged to make a recombinant molecule. **(b)** Tailing and ligating blunt-ended DNA fragments. Adenine residues are added to the free phosphate (P) at the 5' ends of one fragment, and thymine residues to the other. When these fragments are mixed, the single-stranded A and T tails may form hydrogen bonds. Ligation then occurs, forming new phosphodiester bonds to make recombinant DNA molecules

Recombination of DNA from different sources is possible because recognition of complementary base sequences and repair of the backbone can be carried out by enzymes that do not distinguish the source of the DNA fragments. Two fragments cut by the same restriction enzyme can splice together equally well regardless of whether they were cut from a virus or a human genome.

> DNA molecules may be cut at specific sequences by restriction endonucleases. Fragments can be separated by gel electrophoresis and those from different sources spliced together to form recombinant DNA molecules.

DNA cloning

In order to make many copies (clones) of a specific length of DNA, the fragment is inserted into a self-replicating vector genome. Vectors include plasmids, bacteriophages, cosmids and yeast cells. As described in Chapter 10, plasmids, such as pBR322 (Fig. 12.4), are small double-stranded DNA molecules that occur independently of the chromosome in bacterial cells and can carry donor DNA of up to 10 kb. Bacteriophages, for example, λ (lambda) phage, can have the midpart of their chromosome removed and replaced with a fragment of donor DNA of up to 20 kb. Cosmids are hybrid vectors comprising both plasmid sequences containing antibiotic resistance genes and the terminal *cos* sequences of λ phage that are essential for assembly of chromosomes into phage heads. They allow the cloning of larger segments of donor DNA (up to 40 kb). Segments of yeast chromosomes (yeast artificial chromosomes, YACs) are vectors for large genomic fragments.

Replication of recombinant DNA occurs only in a living cell, so it is necessary to get the vector carrying the DNA to be cloned inside a host cell. Several microorganisms, especially bacteria and yeasts, have been employed as hosts for replication. A laboratory strain of *E. coli* (K12) is commonly used, particularly because it has been disabled and cannot survive outside the laboratory.

A circular plasmid is converted into a linear molecule by a single cut. A donor DNA molecule cut by the same enzyme may be joined to both ends of the plasmid and become inserted. Splicing is then allowed to take place, resulting in some molecules in which a vector DNA molecule has been spliced at both ends to a fragment of donor DNA, which now becomes part of the circle (Fig. 12.4). Alternatively, splicing can be accomplished by providing vector and donor DNA with complementary tails. Plasmid vectors contain plasmid signal sequences that initiate replication inside a bacterial cell (Chapter 10).

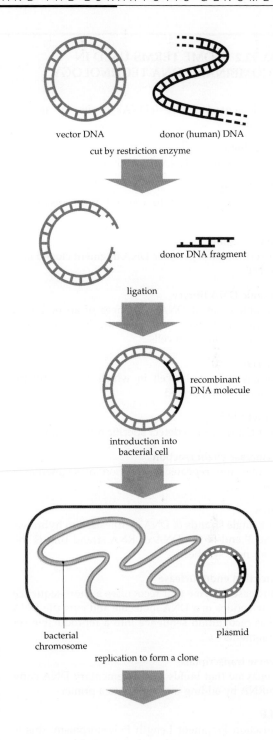

vector DNA   donor (human) DNA

cut by restriction enzyme

donor DNA fragment

ligation

recombinant DNA molecule

introduction into bacterial cell

bacterial chromosome   plasmid

replication to form a clone

Fig. 12.4 Cloning a human gene. Donor (human) DNA is cut by a restriction enzyme to leave sticky single-stranded ends. The vector DNA (plasmid or virus) is cut once with the same restriction enzyme to open out the circle. Ligation can then occur to insert a human DNA fragment into the vector. The recombinant DNA molecule is now introduced into a bacterial cell, which divides to form a clone, each cell of which carries a copy of the original human DNA fragment

BOX 12.2   SOME TERMS USED IN RECOMBINANT DNA TECHNOLOGY

cDNA
Complementary DNA (cDNA) copies of mRNA transcripts

cDNA clone
One recombinant cDNA cloned into host cell

cDNA library
cDNA clones spliced into a vector and transformed into host cells

Genomic clone
One recombinant genomic DNA fragment cloned into host cell

Genomic DNA library
A complete set of DNA sequences of an organism cut into fragments, spliced into a vector and transformed into host cells

Host cell
Typically a bacterial cell in which a recombinant vector can be replicated

Northern blot
A technique used to detect specific mRNA molecules

Polymerase chain reaction
Selective and repeated replication of segments of DNA

Primer
Short single strands of DNA synthesised to hybridise at the 3′ end of the DNA or RNA strand that is the site for new synthesis

Restriction endonuclease
A nuclease enzyme that recognises a short sequence of nucleotides in a DNA molecule and cuts the DNA at this recognition sequence into a reproducible set of fragments

Reverse transcriptase
An enzyme that builds a complementary DNA copy of mRNA by adding nucleotides to a primer

RFLP
Restriction Fragment Length Polymorphism: this is a variation in fragment length generated by presence or absence of restriction sites in different individuals

Southern blot
A technique used to detect specific DNA restriction fragments

Vector
A plasmid, bacteriophage, cosmid or yeast cell in which donor DNA can be spliced

Western blot
A technique used to detect specific protein gene products

Plasmid vectors must be introduced into *E. coli* by **transformation**, a process by which DNA is taken up into bacterial host cells. Bacterial cells are incubated with calcium chloride, which makes the outer membrane permeable, and then mixed with recombinant plasmids. As transformed bacteria grow and divide, recombinant plasmids are replicated by cellular enzymes along with normal plasmid genomes. Colonies containing many thousands of bacterial cells result from cycles of division and become visible on a culture plate. The cells of each colony are a clone of genetically identical cells, each cell descended from the original transformant and containing an identical copy of the recombinant plasmid. By isolating and growing this clone, any number of identical copies of the recombinant plasmid genome can be produced.

If the cloning vector is a viral chromosome, such as λ phage, donor DNA is spliced into viral DNA and these recombinant phages are allowed to infect bacterial cells. Once inside bacteria, the recombinant viral genomes take over the bacterial replication machinery, producing many identical copies of the recombinant phage DNA. Ultimately, the host cell bursts (lyses), releasing many progeny phages, all carrying the inserted DNA. These infect neighbouring cells, which in turn lyse, releasing phages that infect nearby cells. Multiplication of the phages and death of the bacteria result in a *plaque*, a clear area visible in the culture plate. Each plaque contains a clone of phages, descended from a single original virus and containing identical copies of the original recombinant phage genome. Thus, a single phage clone can be isolated and propagated indefinitely.

Many plasmid vectors are expression vectors, designed to permit expression of the donor gene or DNA fragment in bacterial colonies. They have been constructed with DNA that includes efficient promoter elements that can generate large amounts of mRNA complementary to the cloned sequences of donor DNA. Some vectors can express foreign proteins that are linked to prokaryotic proteins (fusion proteins), while other vectors can produce the foreign protein without such links.

> Fragments of DNA can be spliced into a self-replicating bacterial plasmid or phage vector. In this way, recombinant bacteria or phages can be isolated and grown to clone a single DNA fragment.

Identifying and isolating individual genes

Selecting recombinant clones

To identify and isolate an individual gene, it is necessary first to select out all the recombinant clones, that is, those that contain donor DNA, from those that do not. Splicing results in a mixture of products, with head-to-tail vectors and vectors that have re-ligated and closed upon themselves. Those that have not incorporated donor DNA need to be removed. This is achieved by suppressing the growth of bacteria containing non-recombinant plasmids. For example, one system involves bacteria containing the vector pBluescript II, a hybrid plasmid (phagemid) containing origin sequences from a phage DNA molecule (Fig. 12.5). Firstly, bacterial colonies containing plasmids are selected since they can grow on the culture plates that contain an antibiotic as only the plasmids have the antibiotic resistance gene. Secondly, blue- or white-coloured bacterial colonies are produced as culture plates and contain an artificial substrate, which gives a blue colour reaction in the presence of the enzyme β-galactosidase. The pBluescript vector contains part of the gene *LacZ*, which encodes an active β-galactosidase. Bacterial colonies containing the vector produce blue colonies. However, recombinant bacteria cannot produce the enzyme (or colour) and recombinant colonies are white. This occurs because the donor DNA is inserted with a restriction enzyme into the multiple cloning site (Fig. 12.5) and the insert interrupts and inactivates the *LacZ* gene. Thus, blue/white selection provides an efficient system for recovering recombinant colonies.

DNA libraries

Genomic libraries are thousands or millions of clones prepared from DNA of a particular animal or plant. The genome of the organism is isolated and cut into thousands or millions of fragments using restriction enzymes. Each fragment is small enough (only a few kilobases) to be incorporated into a vector and cloned (Fig. 12.6). Ideally, these libraries should contain clones of all the genes in an organism. Once a library has been prepared, almost any sequence of interest, whether single copy or repetitive DNA, can be cloned for study.

In eukaryotes, which have large genomes including much DNA with no coding function (Chapter 10), it is often useful to make a library representative only of the genes active in a tissue. This can be achieved using the mRNA transcripts that are active in a tissue as the starting material. Since RNA molecules cannot be spliced into DNA vectors, it is necessary to make **complementary DNA (cDNA)** copies of mRNA transcripts (Fig. 12.7). This is done using a special enzyme, reverse transcriptase, from RNA tumour viruses, which use the enzyme normally to make DNA copies of their RNA genomes during replication (Chapter 34). Reverse transcriptase uses the mRNA from the tissue as a template to make cDNA copies

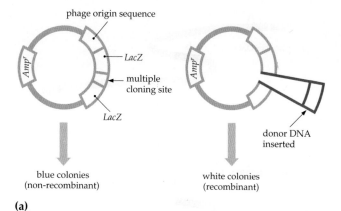

(a)

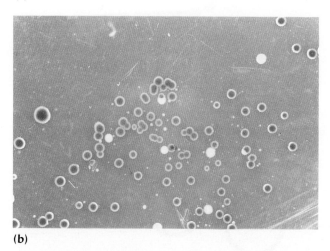

(b)

Fig. 12.5 Structure of the vector pBluescript II, which is a plasmid DNA molecule containing phage origin sequences (for DNA replication, making single-stranded DNA double-stranded). The antibiotic resistance gene is *Amp*^{r}, and blue/white selection is controlled by the *LacZ* gene. Blue-coloured bacterial colonies (when grown on culture plates containing artificial substrate) are produced when the gene is intact (as shown), and white colonies when the *LacZ* gene is interrupted (and inactivated) by insertion of donor DNA at the multiple cloning site

of all the mRNA transcripts (Box 12.3). Single-stranded poly A tails are added and these cDNA molecules are spliced into poly T tailed plasmid or virus vectors. The procedure also requires other enzymes, such as DNA polymerase I and S1 nuclease (Fig. 12.7). The recombinant vectors, each carrying one cDNA copy of an mRNA transcript, are inserted and cloned in bacteria. In this way, representative clones for all the 20 000 mRNAs expressed in a typical eukaryotic cell are assembled into a **cDNA library**.

Screening DNA libraries to select clones

The next step is to screen the genomic or cDNA library to select clones that contain the gene or DNA fragment of interest. Several methods have been developed. The *colony hybridisation* method involves the following steps (Fig. 12.8).

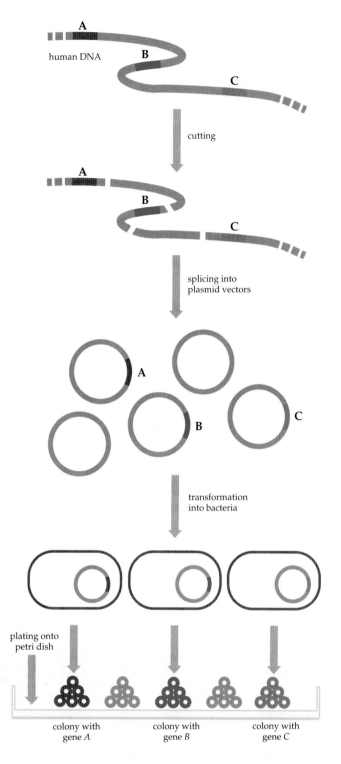

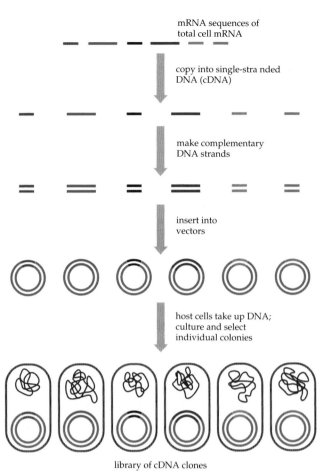

Fig. 12.7 Constructing a complementary (cDNA) library. In the first step, mRNA transcripts are copied into double-stranded DNA molecules. In the second step, the DNA copies are inserted into vectors, and taken up into host (bacterial) cells

1. Bacterial colonies containing recombinant plasmids from the library are grown on culture plates.
2. Samples of the colonies are transferred to a nitrocellulose filter (a 'lift').
3. A radioactive DNA or RNA probe is made to hybridise to a particular gene.
4. The probe is allowed to hybridise with colonies on a lift ('screening').
5. Hybridisation with colonies is detected by exposure to X-ray film.
6. Identified colonies are recovered from the original plate.

The basis of the method is that clones carrying sequences complementary to those of a specific probe are identified. One of the first probes used was mRNA. For example, two mRNAs code for two components of haemoglobin, α-globin and β-globin, and can be readily purified from precursor red blood cells. Purified mRNA from each is made radioactive for use

Fig. 12.6 Constructing a human genomic library. Human DNA is cut into fragments, containing genes *A, B, C* and so on. These fragments are all recombined with vector DNA (here a plasmid). The plasmids are transformed into bacteria, which are then plated on a culture plate. The transformed bacteria each grow into a colony. The great mixture of colonies, containing fragments of the whole human genome, is a library

as a probe and is allowed to hybridise (so that complementary base-pairing occurs) with the corresponding DNA from the clone library. The hybrid complexes are detected by their radioactivity (Fig. 12.8).

Genes that have an important general cellular function, and consequently whose DNA sequence is likely to have been conserved during evolution, can be cloned using the homologous gene from another organism as a probe. For example, many genes important in development have been cloned from the relatively small genome of *Drosophila melanogaster*. Radioactive single-stranded DNA, prepared from one of these cloned genes, can be used to probe, by cross-hybridisation, for similar sequences in human genomic or cDNA libraries so that the corresponding human gene can be isolated for study. For example, homeobox genes, detected in *Drosophila* by the transformation of body parts in mutants (Chapter 16), have homologues that are critical for early mouse (and presumably human) development.

Other screening methods have been developed to identify particular genes in genomic or cDNA libraries. *Differential screening* is one of the most useful. This method depends on the presence or absence of a gene, for example, in normal and mutant strains of an organism (in order to isolate the mutant gene), or in different tissues (in order to isolate genes expressed in only one tissue). In this case, the cDNA library is plated out and two 'lifts' of recombinant bacterial colonies (Fig. 12.9) are separately exposed to radioactively labelled cDNA from each of the tissues. For example, in isolating genes of oilseed rape plants, the library is screened against cDNAs from anthers and leaves (Fig. 12.9). By comparing the radioactive patterns of the colonies, recombinant colonies that are anther-specific can be selected and recovered from the plate. Oilseed rape plants are an important economic crop and understanding anther-specific genes will be useful in controlling seed production.

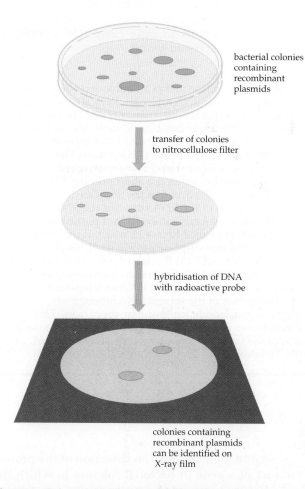

Fig. 12.8 Colony hybridisation method for screening DNA libraries. Bacterial colonies containing recombinant plasmids are grown in a culture plate. A 'lift' is prepared by transferring a portion of the colonies *in situ* onto nitrocellulose filter. The DNA on the filter is hybridised with a radioactive probe. The radioactive spots correspond to colonies containing the DNA of interest

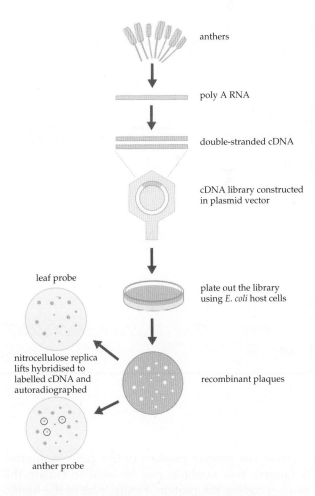

Fig. 12.9 Differential hybridisation for screening DNA libraries. In this example, a cDNA library to mRNA from anther is constructed, and clones specific to anther are selected by differentially screening the 'lifts' of recombinant plaques with an anther probe and a leaf probe. Those plaques detected only with the anther probe are selected and cloned for further analysis

BOX 12.3   GOING BACKWARDS: REVERSE TRANSCRIPTION

American scientists (and later Nobel laureates) Howard Temin and David Baltimore, worked separately on RNA tumour viruses that infect vertebrates (see Chapter 34). They discovered that these viruses, whose genetic information is stored as RNA, convert this information to DNA inside the cell as part of their reproductive cycle. The conversion is carried out by a special enzyme capable of using an RNA template to make a DNA copy (see figure). The enzyme is a DNA polymerase, *reverse transcriptase*, and was found to be associated with infectious RNA tumour virus particles. Since viruses harness the machinery of the cell in order to reproduce, they must use their RNA to 'go backwards' and make a double-stranded DNA copy of their genome inside the cell in order that their genes can be expressed. RNA tumour viruses are consequently now widely known as *retroviruses*. The DNA copy, the provirus, becomes integrated into a host cell's chromosome to become part of that host cell's genome essentially for the rest of the lifetime of that cell.

RNA tumour viruses not only provided the first glimpses to scientists of the process of reverse transcription in nature, but also were important in the development of our current knowledge of the molecular basis of cancer. Investigations on these viruses identified a class of **oncogenes**, whose products are able to change (transform) eukaryotic cells so that they grow in a way similar to tumour cells. Oncogenes are 'rogue' versions of natural growth-regulatory genes of cells. Expression of oncogenes leads vertebrate cells to abnormal growth behaviour, producing tumours or cancers. Many retroviruses cause cancer in animals, but very few are known to cause cancer in humans. The reason for this difference is not yet known.

Reverse transcriptase appears not only in retroviruses, but in mobile genetic elements within cells, in some bacteria, and in mitochondria of fungi and plants. One of the most powerful techniques utilised to analyse the molecular biology of complex cells is a recombinant DNA method in which mRNA is copied into DNA in the laboratory by means of purified retroviral reverse transcriptase, in order to make cloned DNA libraries (cloned sets of expressed genes).

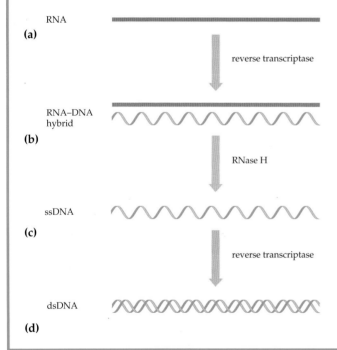

Reverse transcriptase makes a double-stranded DNA copy of a template RNA molecule. Each step is carried out by the enzyme reverse transcriptase. The diagram shows only the major nucleic acid molecules involved. **(a)** Synthesis of a single-stranded DNA copy of template RNA using reverse transcriptase. The complementary DNA strand remains attached to the RNA. This reaction requires a short priming RNA strand bound to the 3' end of the template RNA. **(b)** Degradation of the RNA strand of the RNA–DNA hybrid. This reaction is catalysed by a particular enzyme activity called RNase H, which is usually part of the reverse transcriptase protein itself. Single-stranded DNA (ssDNA) is released. **(c)** Synthesis of the second strand of DNA. Reverse transcriptase proceeds to utilise the single-stranded DNA as template for polymerisation of the second DNA strand from deoxyribonucleotide triphosphate precursors. This reaction also requires a priming strand bound to the 3' end of the template strand of single-stranded DNA. **(d)** The product is double-stranded DNA (dsDNA)

When the protein product of the gene of interest is known, two methods can be used to isolate the gene encoding the protein. Firstly, part of the amino acid sequence of the protein is determined by the technique of microsequencing. The nucleotide sequence encoding this peptide is deduced from the code table and then synthesised. The resultant oligonucleotide is radioactively labelled and used to screen the library.

A second method relies on detection of the protein product of a gene in bacterial colonies in which the gene has been inserted into specially constructed **expression vectors**. Specific probes, such as antibodies, are used to detect the colonies expressing the protein (which is the antigen to which the antibodies were raised). For example, a bacterial colony that contains a recombinant plasmid with a gene or sequence of special interest is detected by a positive colour reaction with an enzyme-labelled antibody probe. The chosen clone is transferred to a fresh culture plate. This clone can then be grown indefinitely and will continue to express the gene product.

Cloned genes or DNA fragments can be purified from bacterial colonies. Total DNA is extracted from the colonies and vector DNA separated from the much larger bacterial genome by centrifugation. The inserted fragment is cut out using the same restriction enzyme originally used to insert it and purified from vector DNA by gel electrophoresis. The result is an inexhaustible supply of a purified eukaryotic gene—something quite unobtainable by traditional methods.

> A DNA library is a collection of clones of DNA from an organism. A gene of interest may be identified in a library by binding to a specific DNA or RNA probe, or by the expression of a particular protein gene product. The clone containing the gene can be isolated and propagated indefinitely.

Isolating genes by PCR amplification

There is another useful technique for rapidly cloning a specific DNA sequence, without the need of a living cell—the **polymerase chain reaction (PCR)**. This technique selectively and repeatedly amplifies specific DNA sequences from a complex DNA mixture (Fig. 12.10). PCR makes use of the discovery that DNA replication can occur in a test tube in a reaction requiring DNA polymerase. PCR allows the synthesis of a selected part of the genome between two regions whose sequences are known. These known sequences are used to design two synthetic DNA oligonucleotides, one complementary to the 3' end of each strand of the DNA double helix (Fig. 12.10).

These oligonucleotides are added in excess to the denatured DNA at a temperature of 50–60°C. The oligonucleotides hybridise with the correct sequence in the DNA and serve as *primers* for DNA synthesis (Chapter 10). Synthesis begins when deoxynucleotides and a high-temperature-tolerant DNA polymerase (Taq polymerase from *Thermus aquaticus*, a bacterium from hot springs) is added (Fig. 12.10). At the end of the reaction, the mixture is heated to 95°C, separating the newly formed DNA duplexes. When the temperature falls again, each separated strand acts as a template for a new cycle of DNA synthesis. This is possible as excess primer remains. At each cycle of synthesis and separation, the number of copies of the sequence is doubled. Two advantages of PCR are that only small DNA samples are needed to start off the chain reaction and it is rapid, requiring only a few hours.

PCR forms the basis of many diagnostic procedures. It has particular use in forensic science (Box 12.4) for

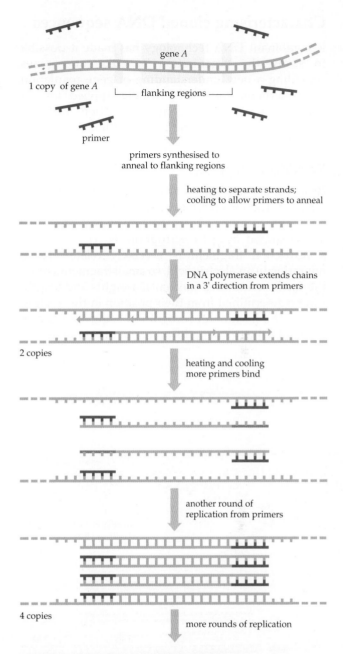

Fig. 12.10 Use of the polymerase chain reaction (PCR) to amplify a human gene. Short DNA primers complementary with sequences flanking the gene are synthesised. Human and primer DNA is heated to separate strands, then cooled. Primer strands bind to the regions on either side of the gene. DNA polymerase then extends the chain, replicating the gene from either direction. The cycle of heating and cooling is repeated, doubling the numbers of copies of gene in each cycle

identification based on traces of DNA in bloodstains, semen, or even a single hair left at the scene of a crime. PCR has even been used to amplify DNA sequences from traces of DNA in fossils of extinct organisms.

> The polymerase chain reaction (PCR) enables replication of specific DNA sequences from a complex mixture of DNA molecules in a test tube.

Characterising cloned DNA sequences

Recombinant DNA technology has made it possible to study the organisation of eukaryotic genes, providing a new understanding of gene regulation. Two methods have been widely applied for the analysis of eukaryotic genes: restriction mapping, sequencing, and DNA and RNA blotting.

Restriction mapping

Restriction mapping involves mapping the sites at which restriction enzymes cut cloned DNA. The DNA fragments resulting after restriction enzyme digestion are separated by gel electrophoresis. The length of a restriction fragment is inversely related to its migration in an electric field, so small fragments move the furthest and their molecular weights and lengths can be determined from their position in the gel (Fig. 12.11). The location of recognition sequences can be mapped using the lengths of fragments generated by

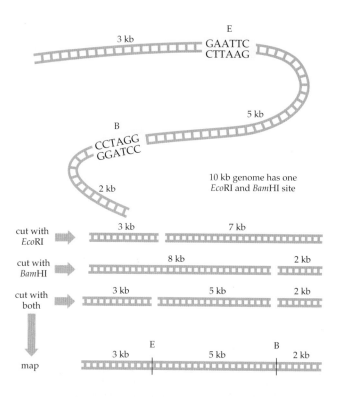

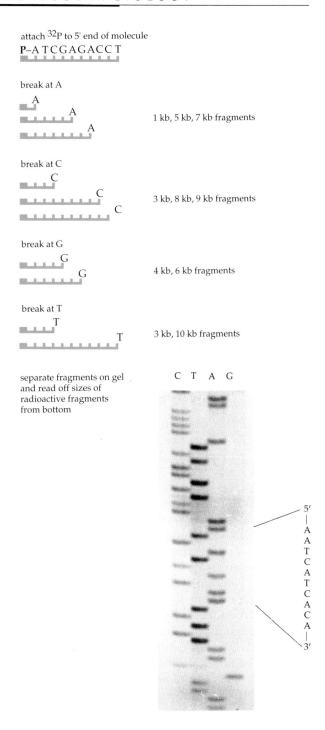

Fig. 12.11 Constructing a restriction map of the genome of a small virus. The viral DNA is 10 kb long and linear, and includes one *Eco*RI (E) and one *Bam*HI (B) site. When it is cut with *Eco*RI, fragments of 3 kb and 7 kb are seen on the gel. When the same genome is cut with *Bam*HI, fragments of 2 kb and 8 kb are released. When both are used together, the genome is cut into three fragments, of 2 kb, 3 kb and 5 kb. The 7 kb *Eco*RI fragment has disappeared, and been replaced by 2 kb and 5 kb fragments. This must mean that the *Bam*HI site is within the 7 kb *Eco*RI fragment, and must be 2 kb from the end of the molecule. Thus there is only one possible map

Fig. 12.12 Sequencing of a short DNA fragment. Firstly, the fragments are all tagged with radioactive phosphorus (^{32}P) at the 5′ ends. Then the chain is broken randomly at either A, C, G or T. The fragments generated in each of these four runs are separated on a gel, which is autoradiographed, showing the positions only of the radioactive fragments that include the 5′ phosphorus-32. The sequence can be read directly off the gel, starting at the bottom. Here the DNA molecule is broken at each of the four bases. This generates families of fragments, which are separated in the four lanes of the sequencing gel. The sequence can be read off from bottom to top

the restriction enzyme. For example, if a small linear viral genome is cut once by *Eco*RI, the sizes of the two fragments generated allow the position of the *Eco*RI site to be mapped with respect to the two ends of the molecule. If the same DNA molecule is cut once with *Bam*HI, the positions of the *Bam*HI site can be determined with reference to the ends of the molecule, and also to the *Eco*RI site. Restriction maps may be extended by piecing together the maps of overlapping cloned regions.

Nucleotide sequencing

The sequence of bases (A, C, G, T) within any short stretch of purified DNA can be determined using methods that rely on the generation of DNA fragments and their size separation by gel electrophoresis. Fragments are generated by cleaving single strands of DNA terminating at particular bases, for example, at A residues (Fig. 12.12). These fragments may be generated either by cutting DNA at a particular base or by synthesising new chains, incorporating altered nucleotides that terminate synthesis at a particular base. In both techniques, conditions are selected such that cutting or termination occurs only occasionally, generating a family of fragments that all end at the position of a particular base, for example, A residues (Fig. 12.12). By tagging their 5′ ends, labelled fragments will all start at a common origin and finish at one of the A positions (Fig. 12.12). In the same way, families of fragments that end at the T, C or G residues can be generated. The base sequence can be read simply from the position of the bands (Fig. 12.12). If neighbouring or overlapping clones are

sequenced, the sequences of large regions can be deduced.

DNA, RNA and protein blotting

DNA molecules produced by recombinant methods can be analysed by hybridisation to detect specific genes or restriction fragments by the technique of DNA blotting or 'Southern' blotting, named after its inventor, Edward Southern. DNA is purified and cleaved by a restriction enzyme and DNA fragments of different sizes separated by gel electrophoresis (Fig. 12.13). The fragments are then blotted onto a nitrocellulose membrane and denatured to make them single-stranded. A cloned DNA probe is radioactively labelled, made single-stranded and hybridised to the DNA on the filter. Unbound probe is washed away, and the bound probe detected using X-ray film (autoradiography). Of all the thousands of fragments into which genomic DNA is cut, the probe DNA will hybridise only to DNA fragments that have a sequence complementary to the probe. This is visualised as a band on the gel (Fig. 12.13).

A related technique can be used to confirm that a transcript in a cloned DNA sequence is represented in total RNA prepared from a tissue (Fig. 12.14). RNA is separated by gel electrophoresis and the bands transferred to nitrocellulose membrane. The RNA can then be hybridised with radioactively labelled cloned DNA and the bound probe detected by autoradiography. This method, known as Northern or RNA blotting (in contrast to Southern blotting), provides useful information on whether a particular gene is active in a tissue.

Cloned genes can be used as probes to detect complementary mRNA in RNA preparations from tissue extracts (Fig. 12.15). Several different tissue extracts can be screened for the expression of the gene. The tissue in which the gene is being transcribed can be identified.

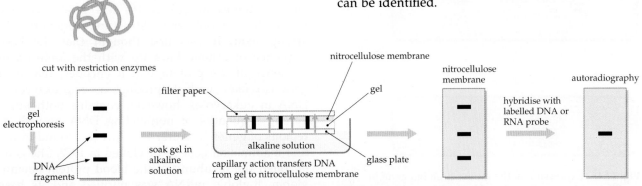

Fig. 12.13 DNA (Southern) blotting. DNA is cut by a restriction endonuclease. Fragments are separated by gel electrophoresis and blotted onto a nitrocellulose membrane. A radioactive DNA probe is allowed to hybridise with single-stranded DNA trapped on the membrane and binds only to the fragment containing the probe sequence. The position of the radioactive fragment in the gel is detected by a darkened band on an X-ray film placed against the membrane (autoradiography)

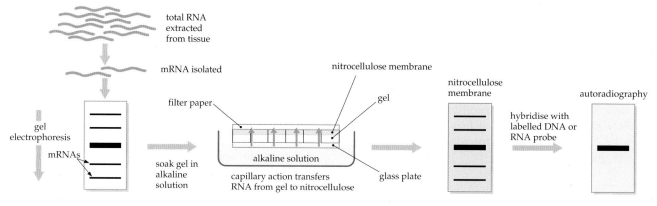

Fig. 12.14 RNA (Northern) blotting. RNA is extracted from the tissues and mRNA is isolated and separated according to size by gel electrophoresis and transferred to a nitrocellulose membrane. The membrane is incubated with a radioactive DNA or RNA probe, which detects the corresponding mRNA band. In this way, its molecular weight and quantity can be determined

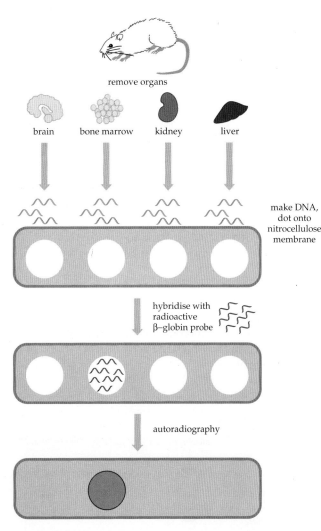

Fig. 12.15 Expression of the mouse β-globin gene in different tissues. Samples of brain, bone marrow, kidney and liver are removed from a dead mouse. RNA is prepared from different organs and dotted onto nitrocellulose membrane. The membrane is hybridised with radioactive β-globin DNA probe, and the probe bound to RNA is detected by autoradiography. The probe hybridises only to RNA from bone marrow, showing that the β-globin gene is being transcribed only in this tissue

Proteins, which are the products of genes, can be separated and visualised by gel electrophoresis and, as with nucleic acids, the protein bands can be transferred to membranes and detected by specific probes, such as antibodies (a procedure known as Western blotting).

Restriction maps are constructed using the lengths of fragments generated by restriction enzyme cutting to deduce the location of restriction sites. DNA sequences are deduced from the lengths of fragments generated by enzymes that cut at a particular base. DNA and RNA molecules can be analysed by hybridisation to detect specific base sequences.

Investigating gene structure in eukaryotes

The ability to isolate, map and sequence regions of large genomes in eukaryotes has produced new information about gene structure, function and arrangement. It was first thought that the base sequence of a gene lines up with the amino acid sequence of the protein that it encodes, that is, a gene is *colinear* with its product, as in prokaryotes. DNA in eukaryotes, however, contains both intervening sequences of non-coding DNA as well as regions of genes (Chapter 10). This was found in the first mammalian gene to be cloned in 1977. The gene encoded the β-subunit of the blood protein haemoglobin. β-globin mRNA was purified and its base sequence was shown to code for the known amino acid sequence of the β-globin protein. However, when a genomic clone of β-globin was obtained, it was possible to examine the gene itself, rather than just its RNA transcript. Restriction mapping showed that two parts of the gene were present, sited more than a kilobase apart in the genomic clone (Fig. 12.16). This

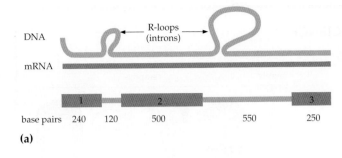

(a)

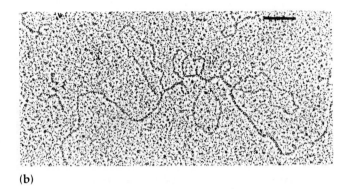

(b)

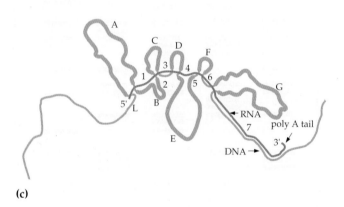

(c)

Fig. 12.16 **(a)** Structure of the β-globin gene. The sense strand of a cloned β-globin gene (blue) is hybridised to purified β-globin mRNA (red). Hybridisation occurs in three regions, separated by two loops of intervening sequence (R-loops). **(b)** Electron micrograph and **(c)** diagram of the ovalbumin gene, showing single-stranded DNA with R-loop structures obtained by hybridising the cloned ovalbumin gene to its mRNA. The gene is shown to contain seven large introns (A–G), which form R-loops, and eight exons (L,1–7), which bind to complementary regions of mRNA

was puzzling because the protein is only 146 amino acids long and should require a message of less than 500 base pairs! When a strand of genomic β-globin clone was hybridised to the globin mRNA, three separate regions of the gene showed complementary base-pairing with the mRNA and formed double-stranded regions (Fig. 12.16). Two other regions were left as single-stranded loops. Evidently these regions of the genome were not homologous to any sequences in the globin mRNA! These R-loops demonstrate that parts of the gene are not represented in mature mRNA (Fig. 12.16). These intervening sequences (introns) did

not take part in coding for the protein product and the β-globin gene is an interrupted structure: a string of exons and introns (Chapter 10).

Genes that code for enzymes in the same pathway, for example, those involved in production of the amino acid glycine, are distributed throughout the genome. Even genes that code for two subunits of a single protein may be located on different chromosomes. For example, the α-globin gene is located on human chromosome 16, whereas β-globin lies on chromosome 11. Even where families of related genes are clustered, such as β-globin-like genes (Fig. 12.17), they are all independently transcribed, unlike, for example, the genes of the *E. coli lac* operon (Chapter 11). Their location close together seems to be the result of their origin by tandem gene duplication and does not indicate co-ordinated control of gene activity.

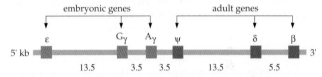

Fig. 12.17 Map showing the cluster of human genes related to β-globin and detected by a β-globin probe. The distances between genes (in kilobases) is marked. These genes are all the same size, have the same exon–intron structure and evolved from a single gene. β-globin is active from birth, and produces almost all of the product in adult blood. Delta(γ and α)-globin is also active in adults, but is expressed at a much lower level. The two (virtually identical) gamma(γ)-globin genes are active in the fetus, and the epsilon(ε)-globin gene in the embryo. The globin pseudogene (ψ) is inactive

Gene families

The availability of cloned probes has made it possible to detect genes with a similar sequence within genomic libraries. For example, a labelled β-globin probe detected not only the gene itself but several other very similar genes in the human genome (Fig. 12.17). Some of these genes produce proteins that have different respiratory functions in the developing human embryo and fetus. Unexpectedly two other genes were found among the globin genes that could not possibly function. While they had a similar structure and sequence to the functional β-globin-like genes, they had suffered mutations that rendered them untranslatable because they were interrupted by termination codons. These *pseudogenes* are therefore likely to be relics of once-functional globin gene copies. There are inactive copies of many genes in the eukaryote genome that now serve no coding function.

> The ability to purify, restriction map and sequence fragments of eukaryotic DNA has shown that eukaryotic genes have evolved into families with a related function.

BOX 12.4   DNA TECHNOLOGY IN FORENSIC SCIENCE

The ability to detect variation between individuals at the DNA level has been put to use in designing new techniques for solving crime. Whereas older techniques relied, for example, on the identification of blood group antigens or other proteins in body fluids using specific antibodies, the favoured material for identification is fast becoming DNA. There has been a good deal of media attention to cases in which a rapist has been positively identified by the DNA fingerprint left in semen on his victim, and the body of a murdered child was identified and her father identified, months after she was abducted.

Two factors increase the usefulness of DNA analysis. Firstly, sampling DNA is a good deal more straightforward than for proteins. DNA is a rather robust molecule, whose structure can survive drying out. Thus, even old bloodstains or dried or decomposed tissue may contain useable DNA. In addition, PCR techniques require so little material for amplification—potentially a single molecule—that even trace amounts may be sufficient.

Secondly, there is much more variation at the DNA level than at the protein level. This is because a great deal of eukaryotic DNA serves no coding function, and variation in its sequence is much more frequent, presumably because it is not under such stringent natural selection. In particular, variation in the number of tandemly repeated sequences may produce highly variable patterns of fragments, detected as 'DNA fingerprints'. A VNTR (variable number of tandem repeats) probe may detect up to 100 bands because these tandem repeats occur in different numbers at many different locations in the genome. The lengths of many or most of these fragments are also variable because the numbers of repeats at each position vary. With numerous alleles at 100 or more loci, the number of combinations is in the billions. Thus, banding patterns are virtually unique to an individual. Even related individuals are likely to have inherited different sets of bands, with the exception of identical twins (see figure).

DNA techniques also are potentially much more useful for establishing family relationships than the older method of blood-group matching. For example, in a case of disputed paternity of a child of blood type AB, a man with type O blood could be exonerated, but no *particular* individual of the correct blood group could be indicted. When individuals are tested for alleles at several different blood group loci, the chance of unrelated individuals having the same set of alleles is about 5%. However, the VNTR bands of a child will appear in the DNA either of the mother or the father; thus the father can be uniquely identified if the banding pattern of the child and mother are available.

Oddly enough, the legal profession has been extremely cautious in embracing these decisive and seemingly foolproof methods. One criticism raised is that, in unskilled hands, the technology can yield inconsistent banding patterns, in which some weaker bands may be obvious in some samples and undetectable in others.

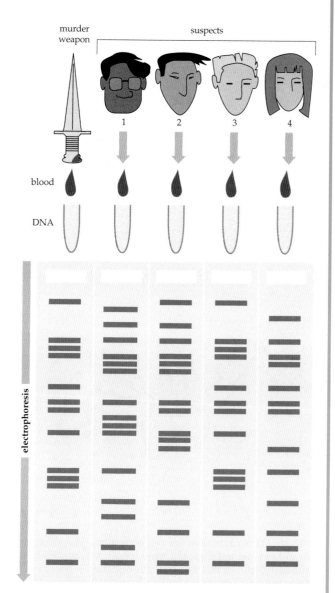

Find the murderer! DNA fingerprints were obtained from four suspects and compared with the DNA fingerprint from blood from a murder weapon. One of the suspects gives a perfect match

Another issue is that, although there are theoretically an almost infinite number of combinations of band positions, some alleles are much more frequent than others, so that certain combinations may be as frequent as 1 in 1000. This is still a much better prospect than the 5% chance that two unrelated individuals will have the same set of blood-group alleles. However, it is necessary to be cautious in interpreting results until the frequencies of these alleles have been determined for different human populations.

APPLICATIONS OF RECOMBINANT DNA TECHNOLOGY

Analysing genetic variation

Using DNA technology, we can identify base changes that cause changes in gene expression in mutant organisms. We can compare the differences in base sequence between wild-type alleles that produce variants of normal proteins (protein polymorphisms). Some base changes produce amino acid substitutions causing no change in protein function, while others produce no amino acid change. For example, comparisons of the sequence of a gene may be made between normal and affected family members in order to identify a mutation that results in a genetic disease. This has been done for a number of haemoglobin abnormalities and we now know which base changes produce protein that does not function normally.

It is routine to use PCRs to detect even single base differences between alleles. This can be done by deliberately constructing short primers that hybridise perfectly to one allele, but not to another. Binding of the primer and amplification of the specific gene will take place only in a reaction where the primers match perfectly.

Another method to study genetic variation is to compare the restriction fragment lengths within a cloned region (Fig. 12.18). If a base change has occurred and disrupted a restriction sequence or created a new one, there will be one fewer or one more cut. The sizes of fragments will therefore be different between individuals, resulting in **restriction fragment length polymorphisms (RFLPs)**, detectable because the number and position of the radioactively labelled bands will be altered, giving different restriction patterns (Fig. 12.18).

Other methods of comparing sequences in DNA from different individuals rely on hybridisation between complementary strands. Where there has been a base change, DNA will not pair; this base mismatch can be detected by chemicals or enzymes, and the mutation site pinpointed. Variation in the genomes of different individuals can be detected with probes that detect clusters of repeated sequences rather than single unique sequences. These short sequences are often clustered in small groups of different sizes at several regions of the genome. A number of fragments of various sizes, containing variable numbers of tandem repeats (VNTR), will be detected by Southern blotting, using the repeated sequence as a probe. The number of tandem repeats is very variable, so that these probes detect a variable set of bands by Southern blotting. Some of these are so variable that individuals each show unique DNA fingerprints, which are so specific they are used in forensic investigations (Box 12.4).

> Differences in base sequence can be identified by cloning and sequencing normal and mutant alleles. Mutations that alter numbers of restriction sites are detected as restriction fragment length polymorphisms (RFLP) by Southern blotting. Repeated sequence probes can detect extremely variable numbers of tandem repeats.

Mapping genes

Classically, segregation of parental differences in a phenotypic character is observed among offspring and the amount of *recombination* between genes used as a measure of their physical distance apart, as explained in Chapter 10. This method has served well for eukaryotes, such as fungi, *D. melanogaster* and several plant species. However, *linkage analysis* is difficult in vertebrates, especially mammals, and almost impossible for humans. We can now use differences detected at the DNA level. Linkage maps can be constructed using RFLPs as markers for human chromosomes. PCR-typing of individual sperm

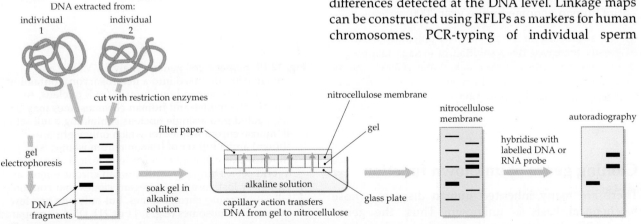

Fig. 12.18 Detection of restriction fragment length polymorphisms (RFLPs). DNA from two different individuals is extracted and digested with restriction enzymes and analysed by Southern blotting. The blot is probed by an RFLP, that is, an oligonucleotide of known sequence or a cDNA clone, which will hybridise with a DNA band that may vary between the two individuals

provides the potential to detect recombination in large numbers of gametes rather than limited number of progeny.

Mapping may also be accomplished by hybridising cells rather than mating organisms. Somatic cells from different species can be fused into cell hybrids. These hybrid cells eliminate chromosomes (and the genes they carry) from one of the parental species, in a kind of segregation event, making it possible to correlate retention of molecular markers, and hence particular genes, with retention of particular chromosomes. This technique provides a rapid way to locate (map) genes to chromosomes. For example, human cells can be fused with mouse cells (Fig. 12.19). Mouse–human cell hybrids eliminate certain human chromosomes and their genes. Human and mouse β-globin genes can be distinguished by their specific Southern blot pattern when probed with a radioactive β-globin probe. After analyses of many hybrids, some are detected that retain a human chromosome 11 and all retain the human β-globin gene. Hybrids that lack the human chromosome 11, all lack the human gene. The β-globin gene must, therefore, be on chromosome 11.

More direct methods of physical gene mapping, which do not require allelic differences between parents, nor a mating event, are now available. Instead, a radioactive gene probe is applied directly to a preparation of human chromosomes. Figure 12.20 shows how this method can be used to map the human β-globin gene. This *in situ hybridisation* method gives a physical location for any cloned gene. If enough gene markers are used, genetic and molecular mapping methods can produce detailed maps.

The use of these new mapping methods has led to a rapid increase in our understanding of complex eukaryotic genomes, such as the human genome. Restriction mapping and DNA sequencing methods are being used with the aim of characterising the entire human genome by the year 2000.

> The availability of a host of DNA polymorphisms has greatly increased the resolution of linkage mapping in humans and other higher eukaryotes. Genes can be assigned to chromosomes in mammals and also mapped directly by *in situ* hybridisation.

Cloning genes of unknown function

There are many inherited human diseases whose biochemical basis is unknown. Thus, the genes responsible for the disease cannot be cloned by identifying abnormal protein products. Instead, the gene is first mapped by looking for linkage between the disease and a set of DNA markers whose chromosome location is already known. When linkage

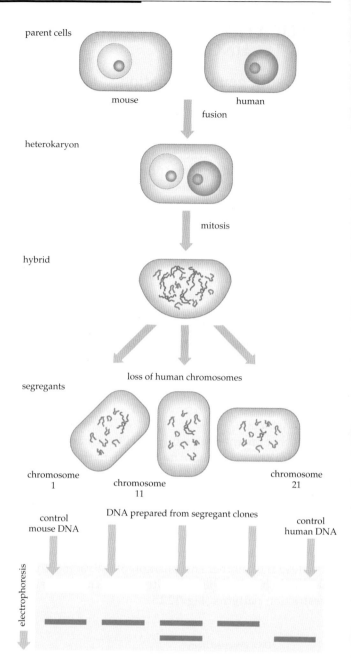

Fig. 12.19 Somatic cell genetic analysis. Mouse and human cells are fused into a heterokaryon, containing one mouse (blue) and one human (red) nucleus. At mitosis, the mouse and human chromosomes may be segregated into a single nucleus containing a full set of 40 mouse chromosomes (of which only eight are shown) and a full set of human chromosomes (six shown). This cell may proliferate to form a clone of hybrid cells. During growth, human chromosomes are randomly lost, forming segregant clones that contain a full set of mouse chromosomes, but only one or a few human chromosomes (e.g. 1, 11 or 21). DNA is prepared from each clone, and cut with a restriction enzyme. The presence or absence of the human β-globin gene is determined from the restriction fragment length on a Southern blot. There is found to be a correlation between the human β-globin and human chromosome 11. Thus, β-globin is located on chromosome 11

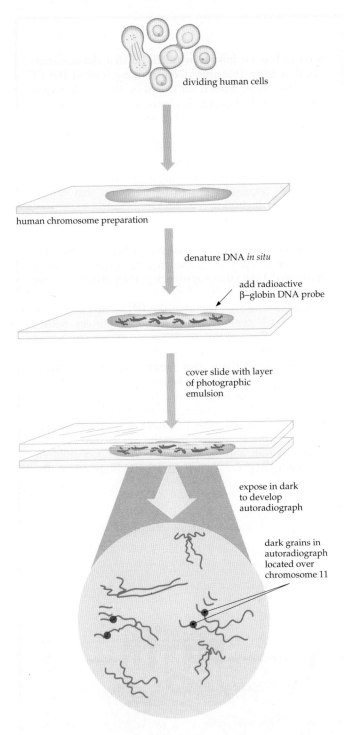

Fig. 12.20 Mapping the human β-globin gene by *in situ* hybridisation. Dividing human cells are spread onto a microscope slide. The DNA is denatured within the fixed chromosomes by alkali treatment. Then radioactive single-stranded cloned β-globin DNA is added. This probe will hybridise to chromosomes only at the site of the β-globin genes. This site may be revealed by covering the slide with photographic emulsion. The radioactive probe will produce dark grains in the film immediately over the site of hybridisation. These grains are found over the short arm of chromosome 11, the site of the β-globin gene

is detected, the disease can then be assigned to a chromosome region. The position of the unknown gene is further defined by measuring recombination between DNA markers closer and closer on either side. Eventually, the DNA between these markers can be cloned and screened for the presence of functional genes. This can be done by looking for signals that are common to all genes, for example, promoter and control sites, splice sites and sequences that are translatable into amino acid sequences (open reading frame). Finally, the base sequence of the gene may be compared in normal and affected individuals. This allows us to pinpoint the lesion in DNA that produced the abnormal product leading to the disease state. This approach has been successful recently in isolating the gene that is responsible for cystic fibrosis (Box 12.5).

Sometimes genetic abnormalities can be caused by major rearrangements of chromosomes. Missing pieces (deletions) and exchanges of large chromosome regions (translocations) can be observed in cytological preparations under the microscope, and may provide a guide to the location of an unknown gene. For example, a study of the DNA of a number of patients with deletions or rearrangements of the Y chromosome led to the successful identification of the gene that controls male sexual development in humans (Box 12.6).

> DNA technology, combined with new gene mapping methods, can be used to pinpoint a gene responsible for a disease, allowing the region to be cloned and the gene identified.

Genetic engineering

The techniques of **genetic engineering** make use of recombinant DNA technology for applications in the pharmaceutical, agricultural and veterinary industries. This technology has permitted genes to be modified and transgenic organisms to be produced.

There are several ways in which genetic engineering is proving useful. Firstly, genes encoding particularly valuable human proteins can be transformed into bacterial expression systems, which produce genetically engineered **recombinant proteins** with an amino acid sequence identical to that in natural proteins in human tissues. Many proteins, for example, hormones and enzymes, are needed for therapeutic purposes in medicine. Although it is desirable to use natural proteins, supply is usually limited to a small amount that can be purified from donated blood or cadavers (where problems have arisen with contamination by rare viruses). For example, recombinant human insulin is now available for diabetics instead of insulin extracted from pigs.

BOX 12.5   THE CYSTIC FIBROSIS GENE

The disease cystic fibrosis (CF) becomes apparent at a young age and affected children suffer persistent chest infections. The autosomal recessive pattern of its inheritance is well known; affected children have normal parents who both are carriers of the mutant *cf* allele. CF is the most common genetic disease in Caucasian populations, with the frequency of carriers being as high as 1 in 20. The pattern of abnormal mucus secretion, respiratory infections and early death is also well known. However, the primary biochemical cause of the disease was unknown until the advent of molecular cloning.

The first step was to locate the *CF* gene by following the inheritance of the disease in families and correlating it with the presence or absence of variants of other genes and DNA sequences. This turned out to be a most difficult task, with many false leads. In 1989, with a great deal of co-operation between scientific groups in London and Toronto, the *CF* gene was shown to be linked to a gene (*MET*) known to lie on chromosome 7. The location of *MET* on chromosome 7 was not known precisely but it narrowed the search to DNA markers located on this chromosome.

The DNA of the CF-containing region was cloned by **chromosome walking**. In this technique, overlapping DNA clones are selected to 'walk' along the chromosome to an adjacent gene of interest. The DNA marker closest to *CF* was used to screen a genomic library for clones that overlapped it (Fig. a). This clone, in turn, was used to screen for more overlapping clones. These clones were used to test for linkage with CF, so that the scientists could confirm that they were walking toward the *CF* gene, not away from it. In this way, the whole region between the flanking markers was cloned, and the region most likely to contain the *CF* gene identified.

How do we recognise the *CF* gene when its size, sequence and function are entirely unknown? The large amount of non-coding DNA in the human genome makes this step far from straightforward. Fortunately, there are sequences shared by all genes that can be recognised, including promoter and control sequences, splicing signals and a coding region (open reading frame). Also, important proteins are likely to be shared by other mammals, so it might be expected that the sequence of such an unknown gene would be highly conserved. Using these approaches, the putative *CF* gene was identified in 1989.

A promising sign was that the DNA sequence of the putative *CF* gene would code for a protein with a structure that could serve as a membrane-bound regulator of ionic exchange, that is, an ion channel (Chapter 4). Perhaps this would explain the claggy mucus and abnormal salt balance of sweat in affected children. Even more promising was the finding that the putative *CF* gene in children with CF had several base changes compared with the same gene in unaffected children. This suggested that the ion channel protein must have an abnormal function in affected children, explaining why the electrolyte balance is abnormal. The

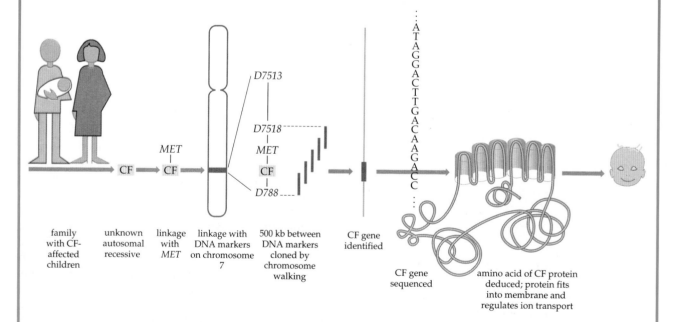

| family with CF-affected children | unknown autosomal recessive | linkage with *MET* | linkage with DNA markers on chromosome 7 | 500 kb between DNA markers cloned by chromosome walking | CF gene identified | CF gene sequenced | amino acid of CF protein deduced; protein fits into membrane and regulates ion transport |

(a) Identifying the *CF* gene. From knowing little about the gene except its autosomal recessive mode of inheritance, it is possible to identify it using genetic and molecular mapping techniques. The *CF* gene can be mapped in relation to other markers, its chromosomal location determined and its physical position progressively narrowed down. The region containing *CF* can be cloned and the gene recognised and characterised. Its DNA sequence can be used to deduce the amino acid sequence of the protein it makes, and the form and the function of the protein product deduced

discovery and isolation of the *CF* gene opened the way for a study of the structure and function of the *normal* protein product of the *CF* gene (CF transport protein). Chains of non-polar amino acids allow the protein to lie buried in the cell membrane, and lengths of polar amino acids give it its capacity to make a channel, which selectively allows passage of sodium and potassium ions through the membrane.

But does this discovery offer any hope to the children who suffer and die from cystic fibrosis, and to their distressed parents? Immediate application to diagnosis of the disease was possible, even before the gene itself was cloned. DNA diagnosis involves showing that RFLPs within (or linked to) the abnormal allele were present in the DNA of an individual. This made it possible to tell whether both parents carried the abnormal allele (Fig. b). It also made *prenatal diagnosis* possible, using DNA from embryonic cells of the placenta (chorionic villus sampling) or shed into the amniotic fluid (amniocentesis). The parents can then elect to abort an affected fetus.

The possibility of using the cloned *CF* gene to cure affected children has been discussed for decades but, until recently, seemed to belong in the realm of science fiction. However, exciting experiments have recently been performed on rats and mice to show that an adenovirus (the common cold virus), containing a normal *CF* allele, can deliver the *CF* gene to the lungs where it infects and transforms lung cells. Transformed cells can produce enough CF transport protein to correct the ion channel defect and alleviate the symptoms of the disease. These experiments raise hopes that humans with CF can be treated in the same way.

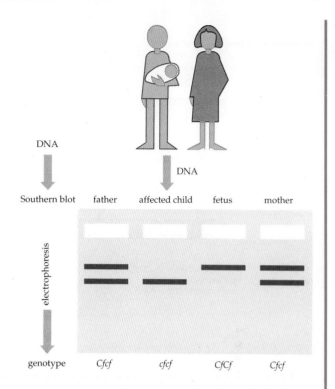

(b) Two normal parents have an affected infant. The mother is again pregnant and the parents are anxious to know whether their second child will be normal. Southern blot analysis shows that in the affected child both abnormal *cf* alleles are contained in a 7 kb fragment. The parents each have this fragment, as well as a 9 kb fragment that must contain the normal *CF* allele. DNA from the developing fetus can now be compared. There is good news for the couple, for their baby has inherited the normal allele from both parents and will therefore be normal

BOX 12.6 FINDING AND ISOLATING THE HUMAN SEX-DETERMINING GENE

In humans, as in other mammals, sex is determined by the presence or absence of the Y chromosome (Chapter 9). It has long been supposed that this chromosome carries a critical gene that operates a switch in a five-week old human embryo, which turns undifferentiated gonads into testes. Once testes are formed, they produce hormones that influence other aspects of male development. The *testis determining gene, TDF*, has been inferred for a long time, but its sequence and how it functions as a genetic switch were unknown until recently.

Cloning and characterisation of *TDF* involved examination of a number of patients with abnormal sexual development. The position of *TDF* was narrowed down by gene mapping in XX males (who had only part of a Y chromosome) obviously carrying *TDF*, and XY females who therefore had a part of the Y lacking *TDF*. By aligning the regions of the Y present in XX males and absent in XY females (deletion mapping, Fig. a), two groups of research workers, one in London and one at Boston, attempted to clone *TDF*.

In 1988, the US group announced the cloning of a candidate *TDF* gene. Their deletion mapping seemed to pinpoint *TDF* to a small region on the short arm of the Y chromosome. A search of this region at Boston identified a new gene, called *ZFY* (for zinc-bound polypeptide fingers, Y chromosome) because it coded for a protein that would bind DNA with zinc-bound polypeptide fingers. The sequence of *ZFY* was highly conserved on the Y chromosome in other primates, cats, mice and other placental mammals, as would be expected of *TDF*. It seemed as if the search for *TDF* was over. However, gene mapping at La Trobe University in Australia showed that *ZFY* was present in the DNA of kangaroos, but was not on the Y chromosome. This immediately cast doubts on the proposed role of *ZFY* in sex determination because the real *TDF* gene would be expected to lie on the Y chromosome in all mammals.

This unexpected finding led to further mapping and cloning. The London group, involving two young Australian scientists, studied other XX male patients and found a few who had only the very tip of the Y

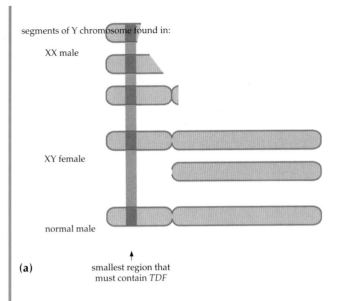

segments of Y chromosome found in:

XX male

XY female

normal male

(a)

smallest region that
must contain *TDF*

chromosome. This tiny region did not include *ZFY* but must have included *TDF* since the patients were male. Thus, deletion analysis pinpointed a small region of the Y chromosome that must contain *TDF*. This region was studied in great detail and a sequence, *SRY* (for sex region, Y gene), was found to be present in men and absent in women. This *SRY* gene is male-specific and located on the Y chromosome in all other mammals tested, including kangaroos.

SRY has all the hallmarks of a sex-determining gene. Its sequence is altered in at least some individuals who have a Y chromosome and yet are female, presumably because they have a mutant *SRY* gene. The final proof was delivered by recent work, involving another young Australian scientist working in a different group in London. The group injected the cloned mouse *SRY* gene into a female mouse embryo with two X chromosomes and no Y. The transgenic embryo developed into a mouse, named 'Randy', who looked and acted like a normal male mouse (Fig. b).

(a) Deletion mapping of the human Y chromosome. XX males who have only a small part of the Y chromosome all retain the region outlined in red. XY females who have even large parts of the Y all lack this region. The *TDF* gene can be pinpointed to this region of the Y chromosome. Analysis of this 30 kb region led to the recognition of the *SRY* gene. **(b)** An XX mouse embryo was injected with mouse *SRY* gene, which has caused male development. The transgenic 'Randy' (right) has normal testes and male genitalia indistinguishable from his normal male litter mate (left), but is sterile because he lacks other genes on the Y chromosome required for sperm production

(b)

Genetically engineered blood clotting factor is used to treat boys with haemophilia; the recombinant product may be preferable to natural clotting factor purified from donated human blood because it is free of the risks of blood-borne diseases, such as AIDS.

Secondly, the genetic characteristics of laboratory and domestic animals can be altered so that they are more useful. A number of methods have been developed for inserting foreign genes into animal cells. Cloned DNA can be taken up by cells in a manner analogous to DNA transformation in bacteria. Fertilised mouse eggs can be micro-injected with purified cloned DNA (Fig. 12.21), resulting in the incorporation of foreign genes into the mouse genome. These **transgenes** are inherited by all the cells that are descended from the egg, and may be expressed if they have been incorporated along with their normal control signals. Injecting cloned growth factor gene into sheep or pigs can make them grow much faster (Fig. 12.21) and yield meat with a lot

(a)   **(b)**

Fig. 12.21 Transgenic mice. **(a)** Cloned DNA is injected into a fertilised mouse egg, through a fine glass micropipette. One or many copies of the gene may become inserted into one of the mouse chromosomes, and may be expressed. **(b)** The insertion of the rat growth hormone into the fertilised mouse egg produced 'Supermouse' (left), shown with his normal litter mate

less fat. In Melbourne, a research company is actively investigating inserting blue pigment genes into a rose, which would have novelty value for the world's largest cut flower crop (Chapter 15).

Gene or *DNA therapy* in humans is in the early stages of development. The idea is that, rather than replacing the protein missing from patients with a particular genetic disease, genetic engineering could be used to replace the faulty gene itself. While the technology for inserting foreign genes into cells and embryos is well advanced, methods for incorporating genes into specialised cells that lack a particular gene product need to be refined. For example, in the future, the β-globin gene may be inserted into the red blood cell precursors of sickle cell anaemia patients, and the normal allele of the cystic fibrosis gene into the lungs of cystic fibrosis patients. A helium-propelled microprojectile delivery system that makes use of DNA-coated microparticles is available. These can be 'shot' into tissues, such as muscle, that are readily accessible. This technology gives a high rate of successful transformation and provides the basis for a new generation of DNA-based vaccines.

> Recombinant DNA technology can be used to change the genetic make-up of organisms. Bacteria can be transformed to produce recombinant human proteins. Animals and plants may be genetically modified by the insertion of cloned genes.

SUMMARY

- The eukaryote genome, especially the human genome, has been studied with recombinant DNA techniques, which are used to isolate genes or small regions of chromosomes.

- DNA molecules can be cut at specific sequences by restriction endonucleases. The fragments are separated by gel electrophoresis and sizes (in kilobases) determined. Fragments from different sources can be spliced together to form recombinant DNA molecules.

- Fragments of a eukaryotic genome can be cloned by splicing them into self-replicating bacterial plasmids or viral vectors to form a genomic DNA library that contains fragments of all the DNA of the genome. Growth and isolation of a single bacterial or viral clone allows a single DNA fragment to be propagated and purified.

- Complementary DNA (cDNA) copies (clones) of mRNA transcripts can be spliced into plasmids or viral vectors, yielding a cDNA library, a mixture of the genes active in a tissue. A gene of interest may be recognised among this library by the binding of its DNA in bacterial colonies or plaques to that of a specific DNA or RNA probe, or by expression of a particular protein product. The clone containing the gene of interest may then be isolated and propagated indefinitely.

- The polymerase chain reaction (PCR) offers a rapid means of isolating and replicating purified eukaryotic genes and DNA fragments. PCR makes use of the replicative machinery of the cell to make copies of small, defined regions of the eukaryote genome by amplifying from short synthetic primer DNA sequences.

- Any cloned fragment can be subjected to restriction mapping and DNA sequencing. Restriction maps are constructed using the lengths of fragments generated by restriction enzyme cutting to deduce the position of restriction sites. DNA sequences are deduced from the lengths of fragments that all terminate at a particular base.

- The ability to purify, restriction map and sequence fragments of eukaryotic DNA has made it possible to examine the structure of genes in higher organisms.

- New DNA technology permits us to look directly at the genome to identify changes in base sequence. This can be done by cloning and sequencing normal and mutant alleles. Changes that alter the numbers of restriction sites can be detected as restriction fragment length polymorphisms (RFLPs) by Southern blotting. Repeated sequence probes may detect extremely variable numbers of tandem repeats.

- Cloned genes can be used as probes to detect complementary mRNA in tissue extracts, or in intact tissue. These techniques can be used to determine when and where, in the developing body, a particular gene is being transcribed.

- The availability of a host of DNA polymorphisms has greatly increased the resolution of linkage mapping in humans and other higher eukaryotes. Somatic cell genetic techniques have also allowed thousands of genes to be assigned to chromosomes in mammals. Genes can also be mapped directly by *in situ* hybridisation.

- DNA technology, combined with new gene mapping methods, can be used to pinpoint a gene responsible for a disease, allowing the region to be cloned and the gene identified.

- Recombinant DNA technology can be used to change the genetic make-up of organisms. Bacteria may be modified to produce a valuable human protein. Flowering plants and animals may also be genetically modified by the insertion of a cloned gene. Transgenesis may be used to improve domestic plant or animal species, and in the future may be used to treat genetic disease.

QUESTIONS

1. (a) What information is contained in a genomic DNA sequence of a eukaryote that is not found in a cDNA sequence? (b) How would you decide which vector to use in constructing a genomic library of a eukaryote?

2. In making cDNA clones of DNA, what steps are catalysed by (a) reverse transcriptase, (b) DNA polymerase I and (c) S1 nuclease?

3. The action of restriction endonucleases produces three different kinds of ends on the DNA molecules that are cleaved. Illustrate these by means of diagrams.

4. Several human genes produce protein products that have potential therapeutic uses. These genes contain a number of introns but the cloned DNA needs to be inserted into an expression vector

recombinant protein expressed in *E. coli*. How is this possible when prokaryotes will not excise introns from mRNA?

5. **(a)** What is the special function of Taq polymerase in the polymerase chain reaction? **(b)** How is DNA sequence amplification achieved?

6. Which recombinant DNA technique has most application in forensic science and why?

7. How would you isolate a gene responsible for a human disease in which no abnormal protein has been discovered?

8. Explain how molecular markers and chromosome walking can be used to isolate eukaryotic genes.

9. **(a)** What techniques are available for genetic transformation in humans or human cells? **(b)** Limitations apply to the use of gene transformation in gene therapy of inherited diseases. Why?

10. List some of the issues that are important to human society in terms of the ethics of the use of recombinant DNA technology.

PART 3

REPRODUCTION

AND

DEVELOPMENT

REPRODUCTION

One of the most striking characteristics of living organisms is that they have the capacity to produce others of their own kind. For an individual, the process of reproduction ensures genetic continuity; that is, at least some of an organism's genes continue to exist in its offspring. More generally, reproduction of individuals ensures the continuation of the species.

Organisms reproduce asexually, sexually or by both means. In **asexual reproduction**, new individuals are **clones**, that is, they have one parent only and are genetically identical to that parent, for example, when a unicellular alga divides by mitosis or a bacterium divides by binary fission (Chapter 8). In **sexual reproduction**, a new individual forms from the fusion of haploid gametes from two different organisms. At some stage before gamete formation, meiosis occurs, during which homologous chromosomes exchange genetic material producing new combinations of parental genes (Chapter 8). By allowing DNA from one organism to be combined with that from another, sexual reproduction results in unique individuals and generates genetic variation within populations.

ASEXUAL REPRODUCTION

Some plants, fungi and animals typically reproduce asexually, with sexual reproduction occurring rarely or not at all. The fact that these organisms have survived for millions of years indicates that asexual reproduction (possibly punctuated by rare bouts of sexual reproduction) is a very successful strategy for survival.

Asexual reproduction involves the production of new individuals through mitotic cell divisions. An advantage of asexual reproduction is that it allows offspring to form without the need to get two individuals of different sex into reproductive

condition at the same place and the same time. A disadvantage is that, because offspring are genetically identical to the parent and to each other, genetic diversity is limited. This limits the ability of organisms to adapt to new or changing environments.

In the life cycles of many parasites, there are stages involving prolific asexual reproduction. These parasites include rust fungi, which cause diseases of plants, such as wheat; invertebrates, such as tapeworms; and protists, such as the parasite that causes malaria. The asexual stage of the life cycle dramatically increases the number of infective individuals, which enhances the chance of finding a new host. Asexual reproduction can occur by several different means, including vegetative growth and parthenogenesis.

> In asexual reproduction, new individuals arise from a single parent by mitotic cell division and are genetically identical to their parent.

Vegetative growth

As any gardener knows, many plants can be propagated asexually from small parts of a parent plant. For example, strawberry plants and spinifex grass (Fig. 13.1) have runners or **stolons**, long stems that grow along the surface of the soil. The stolon produces leaves and roots at regular intervals (at each node) and can be subdivided into a number of new plants. The weeping willow, *Salix babylonica*, reproduces asexually and disperses widely by means of broken branchlets that float along rivers and take root on riverbanks (Fig. 13.2). Known also as Napoleon's willow, the tree was introduced into eastern Australia from a single clone from the island of St Helena (where the Emperor Napoleon was exiled). In Australia, this willow reproduces only asexually because the introduced clone was female and there are no male trees.

Fig. 13.1 Spinifex grass, *Spinifex hirsutus*, has long stems (stolons) that grow along the surface of the soil, producing new plantlets (leaves and roots) at each node

Fig. 13.2 Weeping willow trees, *Salix babylonica*, reproduce asexually by means of broken branchlets that float along rivers, such as the Murray River, dispersing widely and taking root on riverbanks. These are also known as Napoleon's willow and reproduction has been asexual since only a single female clone was introduced

(a)

(b)

Fig. 13.3 (a) The potato develops new tubers from swollen regions of the stem. **(b)** The common reed, *Phragmites australis*, spreads rapidly by suckering through aquatic habitats

Underground storage organs of plants, such as corms (gladioli), bulbs (daffodils) and tubers (potatoes; Fig. 13.3), may also function in vegetative propagation. **Rhizomes** are underground stems that give rise to new shoots. They are characteristic of bracken and other ferns, many grasses and irises. 'Suckers' are new shoots that arise from roots. Trees and shrubs that sucker, such as reeds (*Phragmites australis*; Fig. 13.3), wattles (*Acacia*) and blackberries (*Rubus*), can spread quickly into a vacant patch of habitat after disturbance.

Agriculturists and horticulturists clone plants with particularly desirable characteristics, such as the red-flowering form of *Eucalyptus ficifolia*, for commercial use by means of modern methods of tissue culture. Tissue culture is a propagation technique in which one or a few cells are treated so that they give rise to a whole new plant. Instead of planting bulky potato tubers, in future, farmers will be able to sow 'artificial seeds' made of clusters of potato cells grown in tissue culture and coated with a protective covering.

Regeneration

Regeneration involves the production and differentiation of new tissues and is normally responsible for replacement of damaged and missing parts of the body. It is highly developed in many invertebrate animals, such as hydras, flatworms, annelids and echinoderms. In some of these animals, new individuals can arise from separated body parts. Starfish, for example, can regenerate a new individual from a single arm provided that part of the central disc is present (Fig. 13.4). In some species of aquatic annelids, the capacity to regenerate missing body parts has evolved into a means of reproduction by fragmentation. Periodically, these worms simply break into several parts, each of which then regenerates the missing parts to form a complete worm.

Reproduction by regeneration is little more than an extension of the normal processes of growth and tissue repair. It involves cellular replication by mitosis followed by differentiation of the tissues but no rearrangement of genetic material.

Fig. 13.4 A single arm of the sea star, *Linkia mulifora*, regenerating many new arms to form an entire new individual

Budding

Budding involves the development of a new individual from outgrowths of the body wall of the parent. This is typical of *Hydra* and other polyps, such as corals, and some plants (Fig. 13.5). Hydras produce

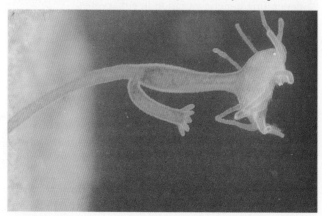

Fig. 13.5 (a)

(b)

Fig. 13.5 Asexual reproduction by budding. **(a)** In *Hydra*, a new individual develops from outgrowths of the body wall of the parent by producing a bud that branches from the side of the body. This is typical of *Hydra* and other polyps such as corals. **(b)** *Kalanchoe* generates tiny plantlets by budding along the margins of its leaves. These plantlets drop to the ground and take root

buds that branch from the side of the body. They may break off to begin life independently, eventually growing to normal adult size. Alternatively, a bud may remain attached and, together with other buds, contribute to the formation of a colony.

Asexual reproduction can occur by means of vegetative growth, regeneration or budding.

Parthenogenesis

In bees, wasps and ants, fertilised eggs develop into diploid females, while unfertilised eggs become males (Fig. 13.6). Unfertilised eggs become activated and develop by **parthenogenesis**, which is initiated by a mitosis of the haploid egg nucleus that is not followed by cytokinesis. The two nuclei produced then fuse to form a diploid nucleus and the egg develops as if it had been fertilised.

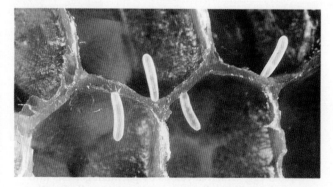

Fig. 13.6 In bees, fertilised eggs develop into female workers and unfertilised eggs into males. Bees breed in tree holes and here, eggs in cells can be seen

In some species of aphids, during spring and summer, females produce eggs that develop parthenogenetically into new offspring. A female might have up to 10 parthenogenetic cycles in a season, producing three to 100 offspring per cycle. The offspring mature rapidly and within a few days may also be producing eggs. Thus, under favourable conditions, population numbers can rise dramatically. Most offspring produced in this way are female, but some are winged males, which can fly to new host plants. These males can mate with females to produce fertilised eggs, which are more able to survive harsh winter conditions than parthenogenetic eggs. Surviving fertilised eggs hatch in the next spring to provide a new, genetically varied population.

Parthenogenesis occurs in a few species of vertebrates, such as the common Australian gecko (see Chapter 32). The whiptail skink from eastern North America also has secondarily adopted a unisexual way of life. Only females are present in the population and eggs develop by parthenogenesis. Intriguingly, this species retains part of a previous sexual pattern of reproduction: ovulation only occurs after a stereotyped sequence of courtship behaviour in which another female takes a male behavioural role.

Apomixis

Apomixis is asexual reproduction involving a form of parthenogenesis in flowering plants (Fig. 13.7).

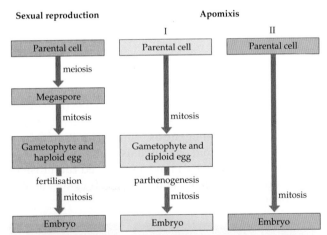

Fig. 13.7 Comparison of the developmental mechanism of sexual reproduction and apomixis in flowering plants. In sexual reproduction, the parental spore-producing cell is centrally located in the ovule and undergoes meiosis to produce a megaspore, which develops into the female gametophyte containing the female gamete, the egg. After fertilisation, the egg develops by mitotic divisions into the diploid embryo. In the two types of apomixis shown here, a diploid cell of the ovule adjacent to the sexual pathway develops by mitosis as follows—type I: into a diploid female gametophyte containing a diploid egg, which develops by parthenogenesis into an embryo (dandelions and grasses); type II: directly into a diploid embryo (*Citrus*), which competes with the sexually derived embryo to produce a seed

Developing diploid eggs do not undergo meiosis but begin to divide by mitosis to form an embryo. Apomictic embryos are genetically identical to the parent plant and develop directly into a seed rather than a plantlet. In many dandelions and grasses, sexual reproduction is completely replaced by apomixis. In kangaroo grass, *Themeda triandra*, and in *Citrus*, both sexual reproduction and apomixis can occur side by side in the same flower. In the case of *Citrus*, cells near a sexually produced embryo are triggered to divide and produce diploid embryos. These somatic embryos compete with the single embryo produced by sexual reproduction and, being more numerous, usually form the seed.

Because of the absence of meiosis, apomictic progeny do not show any new gene combinations and hence potential adaptive variation. They provide a means for rapid multiplication of individual plants that are successful in their present environment.

> Parthenogenesis is a form of asexual reproduction in which egg cells develop into embryos without fertilisation.

SEXUAL REPRODUCTION

Sexual reproduction is characteristic of nearly all eukaryotes. It often involves a considerable cost to the parents in energy and building materials for the production of gametes and to ensure mature male and female gametes occur together. However, its most important advantage is the redistribution of parental genes into offspring. Advantageous genetic combinations may result from the random assortment of parental characters during meiosis (Chapter 9) and enhance the chances of survival of individuals. Sexual reproduction also generates variation within populations, which has been exploited in selective breeding of domesticated plants and animals.

> Sexual reproduction involves redistribution of parental genes into offspring and is the principal mechanism for generating genetic variation in eukaryotes.

Life cycles

A **life cycle** is the sequence of stages in the growth and development of organisms from zygote to reproduction, that is, from one generation to the next. Life cycles of sexually reproducing organisms show a pattern of alternation between diploid and haploid stages, with fertilisation alternating with meiosis (Fig. 13.8). However, the relative amount of time spent in diploid and haploid stages varies greatly between species.

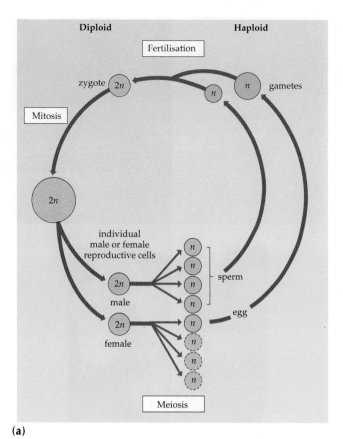

(a)

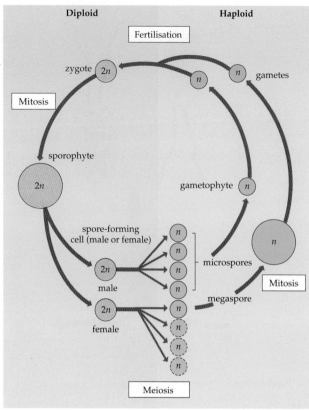

(b)

Fig. 13.8 Life cycles of sexually reproducing organisms. **(a)** In animals, after meiosis, reproductive cells produce either four haploid sperm or one egg (the others degenerate, shown by broken circles). Two gametes fuse to form a diploid zygote, which then undergoes mitotic divisions to produce the embryo, the beginning of the diploid life stage. **(b)** In plants, there are from two to many mitotic divisions after meiosis before gamete production occurs. Meiosis produces female (often only one, with the other three products degenerating) and male haploid spores (four), each of which undergoes cell division by mitosis to produce the female or male gametophytes, which produce haploid gametes by mitotic divisions

In animals, fertilisation follows on directly from meiosis and therefore the haploid generation is simply the gamete stage (Fig. 13.8a). Plants, in contrast, have from two to many mitotic divisions after meiosis before gamete production occurs (Fig. 13.8b). In the life cycles of plants, meiosis produces female and male haploid spores, each of which undergoes cell division by mitosis to produce a multicellular haploid stage, the female or male **gametophyte**. These gametophytes produce haploid gametes by mitotic divisions. Two gametes fuse to form a diploid zygote, which then undergoes mitotic divisions to produce the diploid life stage, the **sporophyte**. In flowering plants, the sporophyte is the diploid vegetative structure that produces flowers. At some stage, diploid cells of the sporophyte undergo meiosis and produce haploid spores, so completing the life cycle (see Chapter 37 for different kinds of plant life cycles).

Life cycles of sexually reproducing organisms show a pattern of alternation between diploid and haploid stages.

Types of gametes

Sexual reproduction involves the fusion of gametes, which may be similar in structure, or distinctly different (Fig. 13.9). **Isogamy**, in which gametes are similar in appearance, is the simplest form of sexual reproduction in eukaryotes; for example, certain algae and fungi have this reproductive pattern. Although the gametes both look the same, at the molecular level, they are of two mating types, '+' and '−', and can only fuse with the opposite type.

In the unicellular green alga, *Chlamydomonas*, the two mating types—mt^+ and mt^-—are determined by recognition proteins located at the tips of the flagella, where initial fusion between cells of opposite mating types occurs (Fig. 13.9). A given cell is either mt^+ or mt^-. After meiosis, two of the four product cells are mt^+ and two are mt^-. Gamete formation occurs only when the alga grows under reduced nutritional conditions, suggesting that sexual reproduction is confined to less favourable environments. In good conditions, reproduction is by asexual means, binary fission.

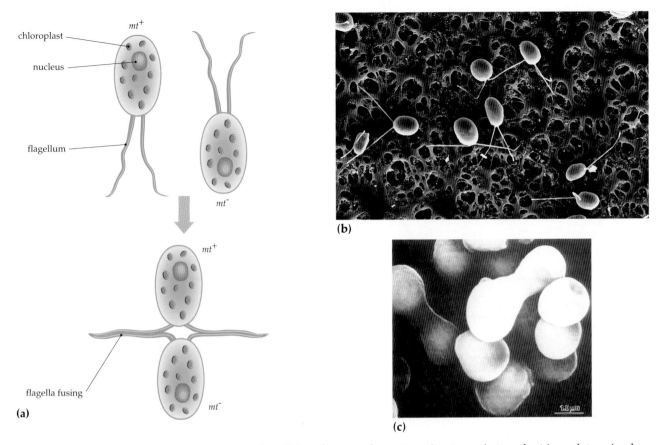

Fig. 13.9 Isogamy. **(a)** In the unicellular green alga, *Chlamydomonas*, the two mating types (*mt⁺* and *mt⁻*) are determined by recognition proteins located at the tips of the pair of posterior flagella, where initial fusion between cells of opposite mating type occurs **(b)**. **(c)** In the yeast, *Hansenula*, sexual reproduction occurs when haploid cells of opposite mating types make contact and form aggregates of cells

BOX 13.1   CONJUGATION IN PROKARYOTES

Sexual reproduction occurs only in eukaryotes. Conjugation is a form of gene transfer in prokaryotes that is different to sexual reproduction but similar in outcome (see figure). Conjugation requires cell–cell contact, with DNA being transferred directly from one bacterium to another. In bacteria, DNA transfer is initiated by a plasmid, a fragment of the bacterial chromosome (Chapter 10). Forty years ago, Joshua Lederberg and Edward Tatum discovered that plasmids can pass from one bacterial cell to another. Only cells containing a particular type of plasmid (F, for fertility factor), which may be integrated into the bacterial chromosome, can effect gene transfer. This plasmid has genes for both DNA replication and the formation of a pilus, the hollow tube that forms a conjugation bridge between two pairing cells.

The cell containing the F plasmid is the donor (male) and the other, the recipient (female). Transfer proceeds when the F region of the chromosome associates with the pilus and initiates DNA synthesis through the bridge between the paired cells. In this way, a copy of the entire bacterial chromosome of the donor may be transferred across to the female cell. There, the single strand inserts and recombines with the DNA of the female cell. Later, DNA replication followed by binary fission produces a new recombinant cell with genes derived from both parents. Gene transfer is not always complete because the pilus may break apart before all the DNA strand has passed through.

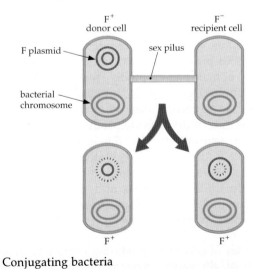

Conjugating bacteria

In the yeast, *Hansenula*, sexual reproduction occurs when haploid cells of opposite mating types make contact and form aggregates of cells (Fig. 13.9). Recognition involves sex-specific molecules on the surface of the cell walls, rather than contact between flagella as in *Chlamydomonas*.

Anisogamy involves the fusion of gametes that are different in appearance. In anisogamous systems, the smaller of the pair of gametes is considered male. Male gametes, **sperm**, are present in large numbers and

their function is associated with motility and dispersal. The large, non-motile female gametes, **eggs**, are produced in smaller numbers and contain food reserves. Most eukaryotes, including algae, fungi, plants and animals, are anisogamous (Fig. 13.10).

Isogamy involves the fusion of gametes that are morphologically similar but of opposite ('+'/'−') mating types. In anisogamy, male and female gametes differ in appearance.

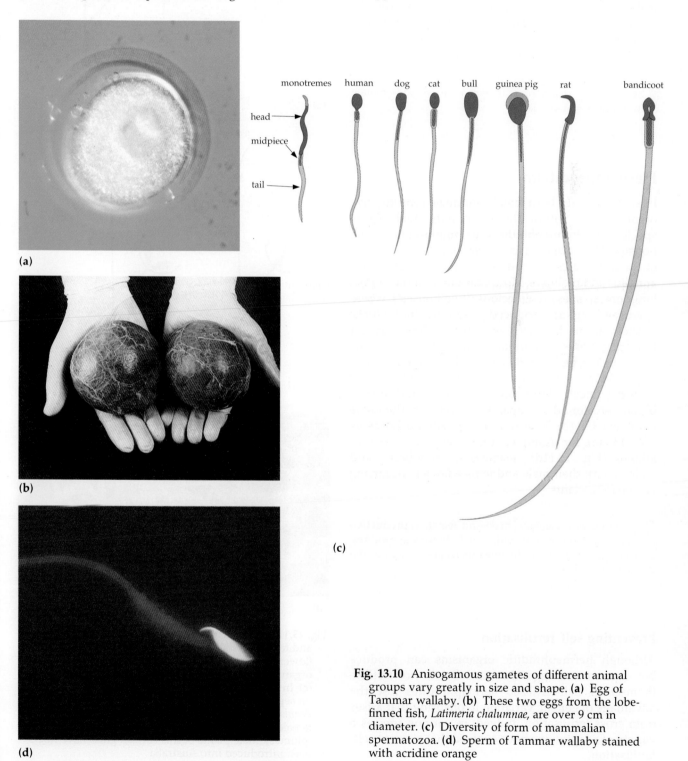

Fig. 13.10 Anisogamous gametes of different animal groups vary greatly in size and shape. **(a)** Egg of Tammar wallaby. **(b)** These two eggs from the lobe-finned fish, *Latimeria chalumnae*, are over 9 cm in diameter. **(c)** Diversity of form of mammalian spermatozoa. **(d)** Sperm of Tammar wallaby stained with acridine orange

Sex

Sex refers to the distinction between individuals of different mating types, usually referred to as male and female. The chemical signals used between isogamous mating types '+' and '−' are considered to be the earliest example of sex. In anisogamous organisms, individuals that produce only one form of gamete are often morphologically, physiologically and behaviourally distinct. Organisms producing only sperm are **male** and those producing only eggs are **female**. The distinctions in appearance and behaviour between male and female individuals of a species are known as secondary sexual characteristics.

> Sex is the distinction between male and female individuals of a species.

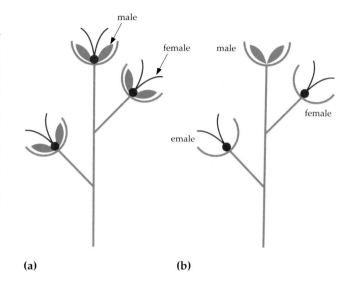

(a)   (b)

Hermaphroditism

In some animals, both female and male reproductive organs occur within the same individual. These species are **hermaphrodites** or **monoecious** ('one house'). This situation is common in a wide range of invertebrates including sponges, flatworms, annelids and molluscs. More commonly in the animal kingdom, species are **dioecious** ('two houses'), where sperm and eggs are produced by separate individuals.

These terms are used somewhat differently for flowering plants. In hermaphroditic plants, the male and female organs are housed within a single flower (Fig. 13.11a), whereas in monoecious plants, for example, maize, birch and oak, male and female organs are found in separate flowers on the same plant (Fig. 13.11b). Some flowering plants are dioecious (Fig. 13.11c), for example, alpine pepper, asparagus, kikuyu (Fig. 13.11d), seagrasses, strawberry and willow, with their male and female flowers occurring on separate plants.

> In dioecious organisms, male and female reproductive organs occur in separate individuals. In hermaphrodites, male and female reproductive organs are present in the same individual.

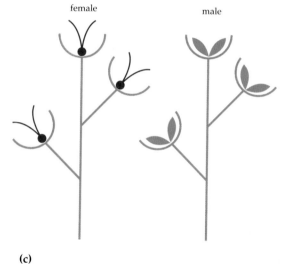

(c)

(d)

Preventing self-fertilisation

Although hermaphroditic organisms can produce both sperm and eggs, they usually do not fertilise themselves. Self-fertilised hermaphrodites lose the advantage of genetic variation in offspring that arises from cross-fertilisation. Indeed, in various species a variety of adaptations have evolved that prevent self-fertilisation.

Fig. 13.11 (a) In hermaphroditic plants, the male (blue) and female (red) organs are housed within a single flower. (b) In monoecious plants, male and female organs are found in separate flowers on the same plant. (c) In dioecious plants, male and female flowers occur on separate plants. (d) Female plant of kikuyu grass, *Pennisetum clandestinum*, a common lawn grass. The female flowers are evident by the white stigmas (plumes). Male plants of this African grass have not been introduced into Australia

In hermaphroditic animal species, where eggs and sperm are produced simultaneously, self-fertilisation may be prevented by anatomical separation of the gametes. In the earthworm, male and female reproductive openings are well separated. The worms have a complex mating procedure with another individual in which they fertilise each other's eggs (Chapter 39).

In many marine hermaphrodites that release both eggs and sperm simultaneously, the eggs become fertile some time after release, by which time the sperm are generally no longer capable of fertilisation. This mechanism ensures cross-fertilisation.

Most hermaphroditic flowering plants have a self-recognition system that prevents self-fertilisation. Self-incompatibility is determined by specific recognition genes sited in the interacting male and female partner tissues, and is discussed more fully below (p. 269).

Many adaptations have evolved to limit self-fertilisation in hermaphroditic organisms.

Changing sex

One reproductive strategy involves an individual alternating between male and female sex so that it does not produce both eggs and sperm at the same time. In some species, there is a single change of sex. An organism may start off as male, converting to female at some later stage, **protandry**, or as female, with a later conversion to male, **protogyny**.

In salmon, fish start life as males and over several years gradually get bigger until they exceed a threshold size, at which time their gonads transform into ovaries (Fig. 13.12). There is a clear advantage in this pattern of reproduction. Sperm are small, so even small males can produce sufficient sperm to fertilise vast numbers of eggs. However, because eggs are large and yolky, the larger a female is, the greater the number of eggs she can produce.

Fig. 13.12 Salmon are protandrous. After reaching a threshold size, they change from male to female and begin producing thousands of yolky eggs

Protandry also occurs in hermaphroditic flowers, such as eucalypts (Fig. 13.13). Upon flower opening, the filaments of the stamens extend and the male organs, anthers, open to release their pollen. Several days later, the flower becomes female as the female organ, the pistil, matures and is able to be pollinated.

Fig. 13.13 Protandry occurs in the red-flowering eucalypt, *Eucalyptus ficifolia*. Upon flower opening, the red filaments of the stamens extend and anthers open to release the yellow pollen. After three to four days, the flower becomes female as the pistil matures, and glistens with a liquid exudate, as shown here.

A coral reef fish, the blue wrasse, provides an example of protogyny. The large dominant male controls a harem of smaller, drab-coloured females. The male alternates his colour between green and blue. About an hour before spawning, the blue colour dominates and he commences mating (Fig. 13.14). This involves elaborate courtship behaviour followed by spawning with each female in turn. His colour then changes back to green. If the male is lost from a group, the largest female will undergo sex reversal and change colour to become the male with control of the harem. Protogyny in the blue wrasse maximises reproductive output, since all but one of the individuals is female and producing eggs, and the single male is able to fertilise eggs produced by all females in its harem.

Fig. 13.14 Blue wrasse are protogynous. This large dominant male is blue in colour, signalling his readiness to fertilise the eggs spawned by all the females in his harem

Some flowering plants, for example, wattle trees, are protogynous (Fig. 13.15). Initially, flower buds open with the pistil protruding, ready to receive pollen. Pollen release from male anthers is delayed until the next day, thus changing the sex of the hermaphroditic flower.

> In protandry and protogyny, self-fertilisation is limited because individuals do not produce eggs and sperm at the same time.

Fig. 13.15 Protogyny occurs in the wattle tree, *Acacia retinodes* (wirilda), whose flowers are grouped in heads of 20 flowers. Flower buds open with the pistil protruding, ready to receive pollen (female phase). The male phase occurs the next day, when the flower heads are covered in anthers and resemble golden spheres. In this flowering shoot, female and male phases are present at the same time

SEXUAL REPRODUCTION IN ANIMALS

Most animals reproduce sexually, although some species can reproduce either sexually or asexually, depending on environmental conditions. In a suitable environment, rotifers multiply asexually until crowding occurs or conditions deteriorate (Fig. 13.16).

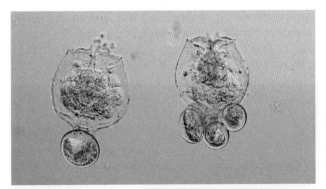

Fig. 13.16 Two female rotifers, *Brachionus angularis* Gosse, from a Murray River billabong. In this genus, most females reproduce parthenogenetically, as seen in the female on the left, which is carrying a single, mitotically produced (female, 2*n*) egg. Increasing population density stimulates the production of sexually producing females, which produce several smaller eggs by meiosis. One of these females carrying five smaller (male, *n*) eggs is shown on the right

These circumstances stimulate a switch to sexual reproduction, generating a variety of starter clones, which then multiply asexually and compete for available resources. In corals (Box 13.2), asexual reproduction ensures growth locally, while sexual reproduction, which includes a free-swimming larval stage, provides a dispersal mechanism as well as a means of increasing the genetic diversity of offspring.

The *primary* sex organs in animals are **gonads**, which produce gametes. Sperm are produced in the **testes** and consist primarily of a nucleus and a small amount of cytoplasm. They are usually highly motile, propelled by means of flagella or cilia. Eggs are produced in the **ovaries** and are usually large, non-motile and endowed with a protein- and lipid-rich material, **yolk**, which forms a reservoir of nutrient to support development of the embryo.

Animals usually have a number of *secondary* sex organs, including glands producing secretions that protect or nourish the gametes, and ducts that carry gametes from the gonads. These ducts may be highly specialised and serve storage, protective or nutritive functions. Primary and secondary sex organs together constitute an animal's reproductive system. A secondary but important role of gonads is the production of hormones that regulate development of gametes and secondary sex organs, produce sexual differences in appearance and behaviour, and act as pheromones or sex-attractants (Chapter 25).

> In animals, eggs and sperm are produced in ovaries and testes respectively. These gonads also produce hormones that regulate secondary sexual characters and behaviour.

Methods of fertilisation

Complex physiological and behavioural mechanisms are often used to bring sperm and egg into close proximity to allow fertilisation to occur.

Many marine animals achieve fertilisation by simply shedding their gametes into the sea. Whether sperm and egg meet is a matter of chance, but two things occur to increase the likelihood of successful fertilisation. Firstly, the number of gametes released is often enormous. A single spawning oyster can release millions of eggs but the majority of the tiny larval offspring is unlikely to survive and develop into an adult. Secondly, within a population, gamete release tends to be synchronised. Environmental cues, such as water temperature, tides and day length, control the reproductive cycle, so that most individuals in a region reach reproductive condition at the same time. When one animal starts to spawn, pheromones released along with gametes will stimulate nearby individuals to spawn, which in turn release more pheromone, resulting in co-ordinated spawning over a wide area. During mass spawnings

BOX 13.2 SYNCHRONISED SPAWNING

Reproduction of many marine organisms on coral reefs is triggered by environmental cues. Annual, lunar, tidal and daily rhythms modulated by temperature appear to regulate their biological clocks. Corals produce one of the biggest and most spectacular examples of this in their synchronised, annual mass spawning on the Great Barrier Reef, Queensland (see figures). In colonies of the large plate coral, *Acropora tenuis*, spawning occurs during spring and early summer, one night or so after a full moon. Spawning reaches its peak on the fourth, fifth and sixth nights during a period of neap tides and rising sea temperatures. Within one day, fertilised eggs develop, forming into swimming larvae, planulae. After a few days at the surface, the larvae descend to find a suitable site to form a new colony. Although millions of larvae are produced, almost all are eaten by predators. Of the few remaining, only a tiny fraction find a suitable substrate, settle and develop into adulthood.

The precise timing after a full moon ensures nights when the sea is characteristically calm. With neap tides, there is little tidal movement, so that less water will move away from the reef. Thus, there is less chance that the developing larvae will be carried away from suitable habitats.

Since multiple species of corals all spawn together, there must be precise mechanisms for species recognition between sperm and egg. Self-matings between male and female gametes from the same colony are uncommon and matings between male and female gametes from different species do not occur.

Flowering of seagrasses is also related to monthly and annual lunar cycles. The Australian sea nymph, *Amphibolis antarctica*, which is distributed on the southern and western seaboards of Australia, is dioecious and flowering occurs only once a year, in early summer. *Thalassia hemprichii* grows on the seaward side of wave-cut platforms of tropical reefs around the Indian Ocean and Northern Australia. It flowers regularly once each month in summer. Flowers emerge during the period of neap tides and pollen is shed coincident with the following period of spring tides.

In certain tropical reef fishes, spawning can occur as frequently as once a day. Reproductive behaviour of the male moon wrasse, *Thalassoma lunare* is centred around female spawning time, which occurs at high tide each day during summer. As the high tide starts to fall, surface currents sweep all the fertilised eggs away from the reef and its predators.

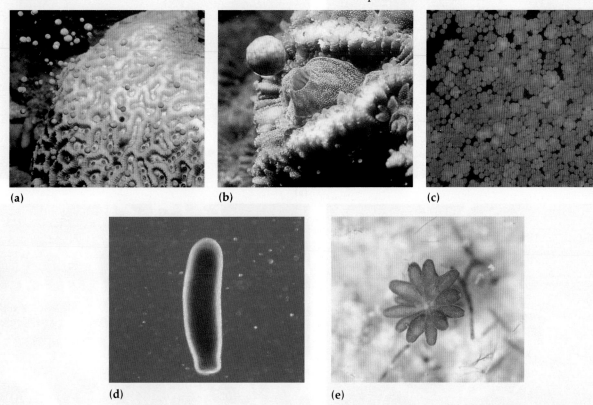

(a)   (b)   (c)

(d)   (e)

Mass spawning of corals on the Great Barrier Reef. (a) Release of bundles, containing sperm and egg from a brain coral, *Platygyra* sp. (b) Detail of release of bundles, containing sperm and egg bundle from mouth of a coral polyp, *Goniastrea palauensis*. The bundle is 3–4 mm across, containing hundreds of eggs as it is launched into the sea. (c) On reaching the sea surface, the egg and sperm bundles of the large plate coral *Acropora tenuis* float and begin to break up, ready for fertilisation. (d) Two to three days later, the larva of *Acropora* is cylindrical, 2–3 mm in length. (e) After about 3 months, the larva has attached to the substrate and laid down its skeleton, and grown into a tiny coral polyp that is capable of budding new polyps

of coral on Australia's Great Barrier Reef, the number of gametes shed is so great that, for a time, the sea turns milky (Box 13.2). An additional benefit of mass spawning is that predators rapidly become sated by the sudden glut of planktonic food and consequently a higher proportion of gametes and larvae escape predation.

Other species rely less on chance to get sperm and egg together. In most frogs, for example, males and females attract each other with special mating calls. Once together, the male climbs onto the back of the female, clasping the female behind the head. Males often develop special 'nuptial pads' on the side of their forelimbs to facilitate this grip (Fig. 13.17). The grip provides a behavioural cue for the female to lay her eggs and, as the eggs emerge, the male sheds sperm over them. This behaviour, **amplexus**, facilitates fertilisation by providing fresh sperm in close proximity to the eggs as they are laid.

These examples involve fertilisation of eggs in the external environment. External fertilisation requires a moist environment since unprotected sperm and eggs are prone to drying with consequent loss of viability. An alternative to this, internal fertilisation, is normally found in terrestrial animals but has also evolved in many aquatic species. Many variations are seen but usually sperm are placed directly into the female's reproductive ducts, often by a specialised organ, the **penis**. The penis is connected by a duct to the testes. There is usually a complex behavioural sequence, **copulation**, associated with penetration of the female reproductive duct to effect sperm transfer.

Copulation using a penis is not the only means of internal fertilisation. A wide variety of animals use packets of sperm, **spermatophores**. In some mites, the transfer of spermatophores is a chance affair. As the spermatophores mature, they are simply deposited by the male, and females pick them up whenever they

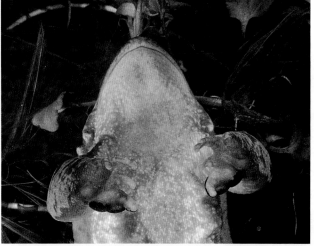

(a)

(a)

(b)

(b)

Fig. 13.17 (a) The forelimbs of the male common frog, *Rana temporaria*, showing dark nuptial pads used to clasp the female firmly during amplexus. (b) During amplexus, the male spotted grass frog, *Limnodynastes tasmaniensis* (on top), sheds sperm onto the freshly laid eggs

Fig. 13.18 (a) A spermatophore of the bush cricket. (b) Sedentary barnacles, *Balanus perforatus*, have an extra long penis capable of reaching considerable distances for fertilising neighbouring eggs

happen to come across them. In many insects and birds, transfer follows a highly ritualised behavioural sequence (see Chapter 28).

More bizarre methods of fertilisation are seen in some leeches, which are dioecious. Copulating leeches become entangled and one injects a spermatophore directly through the body wall of the other, using the penis rather like a hypodermic needle. Sperm then migrate through the leech's body from the injection site to fertilise eggs in the ovary.

For species with a sedentary lifestyle, the ability to exchange gametes directly with others is limited. This limitation has been overcome in barnacles, for example, which have developed an extremely long penis and so can fertilise others some distance away (Fig. 13.18b).

> Fertilisation involves a variety of adaptations that ensure that mature gametes come into close proximity. External fertilisation usually involves the co-ordinated release of gametes into the environment; internal fertilisation follows transfer of sperm into the female reproductive tract.

Reproductive strategies

After fertilisation, the zygote undergoes a sequence of divisions and developmental changes that ultimately result in the formation of a new adult animal (Chapter 14). The nature of development is related to the reproductive strategy of a particular species. Some animals show **indirect development**, with one or more intermediate larval forms before the adult form is attained. Others show **direct development**, in which the offspring hatch or are born in miniature adult form.

Indirect development

In many animals showing indirect development, fertilisation follows the random release of gametes into the environment. Since vast numbers of gametes are released, the amount of yolk that the female can invest in each individual egg is limited. Each egg has sufficient yolk to sustain only a brief period of embryonic development. The embryo then hatches into a larval form, which feeds and grows, accumulating food resources, before metamorphosing into the adult form.

Indirect development is not restricted to aquatic animals with external fertilisation. For example, butterfly eggs hatch into caterpillars, which feed voraciously until sufficient reserves have been stored to support change into the adult form (Fig. 13.19a). The caterpillar secretes a silky cocoon about itself, within which it undergoes metamorphosis into a butterfly. In fact, all insect eggs develop into various larval stages, **instars**, which feed and grow before metamorphosis.

Similarly, in frogs, eggs hatch into aquatic, herbivorous tadpoles (Fig. 13.19b). After a period of feeding and growth, a tadpole undergoes metamorphosis into a frog.

(a)

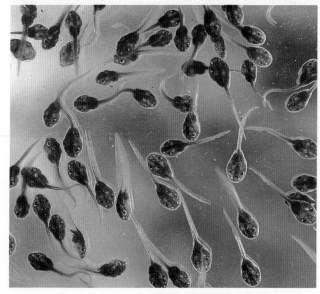

(b)

Fig. 13.19 Indirect development in animals involves larval stages such as (**a**) an emperor gum caterpillar, *Antheraea eucalypti*, and (**b**) tadpoles

Direct development

Animals showing direct development are typically those where females produce smaller numbers of large yolky eggs containing sufficient food reserves for more elaborate development. Turtles and crocodiles lay clutches of eggs in sandbanks beside the sea or a river (Fig. 13.20). When an egg hatches, the offspring is essentially a small adult, able to crawl up from the buried nest to the surface, to find its way to the water and to feed independently. In birds, hatchlings also have adult form but most are more dependent than turtles and crocodiles. Most young birds depend on their parents for food and shelter, sometimes for an extended period.

Fig. 13.20 Direct development in a yolky egg gives rise to tiny crocodiles, *Crocodylus porosus*

Mammals also show direct development but their eggs contain relatively little yolk. Extended development is sustained by the transfer of nutrients directly from the mother to the embryo, either via the placenta or in milk during suckling.

> Some animals release large numbers of eggs with relatively little yolk, which typically undergo indirect development. Other animals have fewer eggs with large nutrient reserves, or direct transfer of nutrients, which allow direct development.

Development in mammals

In eutherian mammals, the developing embryo is retained within the mother's body in a specialised region of the oviduct, the uterus, for the major part of its development, **gestation** (Box 13.3). Here the embryo receives the protection of the mother's body and an excellent supply of nutrients from both uterine secretions and maternal blood. A specialised organ, the **placenta**, facilitates this nutrient exchange.

At birth, young eutherians may be essentially independent. For example, guinea pigs are highly advanced at birth, with fully developed teeth and able to survive on solid foods (Fig. 13.21a). They can be weaned immediately after birth. Large herbivores, such as antelope or horses, have young that can stand and run soon after birth, though they are still dependent on milk for at least a while. At the other extreme, many shrews give birth to offspring at a very early stage of development, and even mice and rat pups are virtually helpless at birth (Fig. 13.21b). Most eutherians lie between these two extremes, with newborn offspring (neonates) dependent on parents for food and shelter for a short period relative to their overall life expectancy.

The other two groups of mammals, monotremes (platypus and echidna) and marsupials (Box 13.4),

(a)

(b)

Fig. 13.21 Different eutherian mammals are born at very different stages of development, for example, (a) independent newborn guinea pigs and (b) immature newborn mice

show quite different patterns of development. Monotremes are oviparous (egg-laying) and marsupials are viviparous (bearing live young) but, in both cases, embryos emerge at a very early stage of development. This short gestation followed by a prolonged lactation contrasts with most eutherians, which have longer internal development and shorter lactation.

Monotremes produce large, relatively yolky eggs. These are fertilised internally and undergo part of their embryonic development in the female reproductive tract before they are laid (Fig. 13.22). After about 10 days incubation in the nest, during which time the embryo is nourished by yolk, the eggs hatch. The small, nearly helpless offspring emerge and remain dependent on milk from the mother for a substantial part of development.

In marsupials, very early development involves nutrient supply via a placenta but, like monotremes, the offspring are poorly developed at birth. Newborn honey possums weigh less than 10 mg, and the offspring of the red kangaroo weighs only half a gram. These offspring undergo a major part of their development within the mother's pouch, totally dependent on milk for their nutrients.

BOX 13.3   HUMAN REPRODUCTION

As in other mammals, human reproductive systems are adapted for internal fertilisation and internal embryonic development. Sperm are produced by meiosis in seminiferous tubules in the testis (see Fig. a). Several hundred million sperm are produced daily and these pass into the epididymis, a convoluted tube lying beside the testis, where they complete their maturation and are stored. In humans, for normal sperm formation, the testes must be kept several degrees below normal body temperature, so they are held outside the abdominal cavity in the scrotum. A specialised set of muscles hold the testes closer to or further away from the body to regulate temperature. At ejaculation, muscular contractions cause sperm to pass through the vas deferens; then various secondary sex glands, including the prostate and Cowper's glands, add their secretions, producing a buffered nutrient-rich seminal fluid. The seminal fluid then passes into the urethra and through the penis.

Seminal fluid passes into the female tract during copulation. The female reproductive system comprises paired ovaries, where, normally, a single egg is released each month. The ovulated egg is collected by the specialised ends of the oviduct, the fallopian tube (see Fig. b). It is usually in the fallopian tube that an egg meets and is fertilised by sperm. The zygote then passes into the uterus, a region of the oviduct specialised for nurturing embryo development. The young embryo embeds in the wall of the uterus and forms the placenta, an organ specialised for nutrient exchange. The lower portion of the reproductive tract, the vagina, serves as a conduit for the sperm and as a birth canal for the developed young.

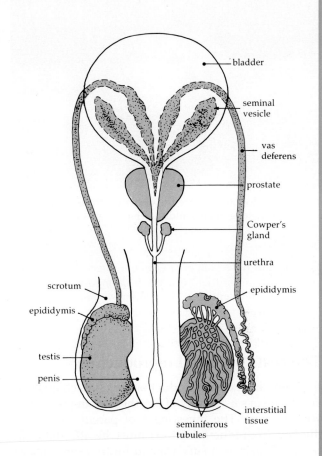

(a)

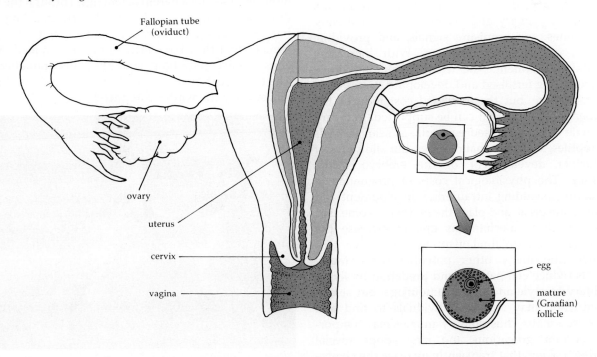

Diagrams showing **(a)** the human male reproductive system and **(b)** the female reproductive system

Fig. 13.22 Monotremes are the only oviparous mammals. This is the egg of an echidna, which are usually laid in pairs

Fig. 13.23 This male green and gold bellfrog, *Litoria aurea*, expends energy in calling to attract a mate. While calling, he risks attracting the attention of potential predators

> Mammals have yolk-poor eggs but direct development by means of nutrient transfer from mother to embryo during development. Monotremes are oviparous, marsupials and eutherians are viviparous.

The costs of sexual reproduction

Engaging in sexual reproduction involves changes in an organism's usual way of life; these changes are not without cost. The investment of parents in reproduction (reproductive effort) must be affordable in terms of their own survival and future reproductive success (the cost) and must increase the chance of survival of their offspring (the benefit).

Reproductive costs vary greatly between species and can include devoting energy to the production of gametes, primary and secondary sexual organs, pheromones, synchronising signals, and providing parental care to the offspring. With increasing parental investment, the chances of individual gametes being fertilised and developing successfully might increase, but the increased cost reduces the number of offspring that can be supported. In regard to female costs, increased egg yolk, as seen in birds and reptiles, saps maternal reserves, but allows faster or further development of the embryo before hatching. The physiological costs of pregnancy in eutherians, including intrauterine development and suckling, are great, and place the mother at some risk. However, these mechanisms greatly increase the chance that an individual offspring will survive.

There are many other potential costs. Many animals reduce competition and predation by living a solitary lifestyle and using camouflage; but sexual reproduction may require individuals to find and recognise a mate. Thus, solitary animals may temporarily become gregarious and may adopt special breeding colours that transiently increase the chance of finding a mate at the expense of increased risk of predation. Frogs begin calling, which attracts both

potential mates and predators (Fig. 13.23); the colourful breeding plumage of some birds attracts the attention of potential mates but conflicts with their usual protective camouflage.

Aggressive tendencies in predatory animals may need to be inhibited. It is pointless for a male and female to come together for reproduction if predatory instincts pre-empt the reproductive process and one individual kills the other.

The costs of reproduction are often very high. In the red deer, with the approach of the breeding season, or rut, adult males grow antlers that are sometimes massive. They are used in dominance-related fighting and are shed after the rut. Males expend considerable energy and risk severe injury in fighting to secure a harem of females. Through the rut, dominant males are so involved with guarding the harem, warding off intruders and mating with the females that they have little or no time to feed. As a consequence, their body condition falls dramatically. The toll is such that the chance that a dominant buck will survive the winter is reduced.

Fig. 13.24 Male grey kangaroos fight to gain mating access to females. This behaviour involves considerable cost to the dominant male

BOX 13.4 REPRODUCTION IN THE RED KANGAROO

Marsupials from the kangaroo and wallaby family (Macropodidae) have an extraordinary ability to control embryonic development. Like all marsupials, gestation is rather short, and females mate and become pregnant again just after giving birth. Since milk production (lactation) lasts much longer than gestation, it is necessary to delay development of the new embryo until the pouch becomes vacant. Macropodids have developed the ability to completely halt development at a very early stage (see Fig. a). This delay in development, **embryonic diapause**, is controlled by the suckling of the young in the pouch.

Suckling (see Fig. b) stimulates release of the hormone prolactin from the pituitary gland. Prolactin promotes milk secretion by the mammary gland and it also acts on the ovary to suppress production of the hormone progesterone. At weaning, the suckling stimulus wanes and prolactin levels decline, allowing increased progesterone production by the ovary. This increased progesterone stimulates the uterus to become more secretory, and the change in the uterine environment leads to reactivation of the diapausing embryo. Development of the embryo then leads to birth of a new young about a month later, and mating will follow soon after, producing a new diapausing embryo.

In red kangaroos, reactivation of the diapausing embryo occurs before the young is fully weaned, so a female may have an older joey out of the pouch but still being suckled, a newborn young in the pouch, and an embryo in diapause in the uterus. At this stage, remarkably, the mother will be producing two different types of milk simultaneously. The newborn young will get a low lipid, high carbohydrate milk, while the young 'at foot' feeding from a different teat will get large volumes of milk high in lipid and low in carbohydrate. This 'production line' approach allows very rapid population growth when conditions are good, even though the mothers have only a single young at a time.

(b) Newborn Tammar wallaby, suckling in pouch

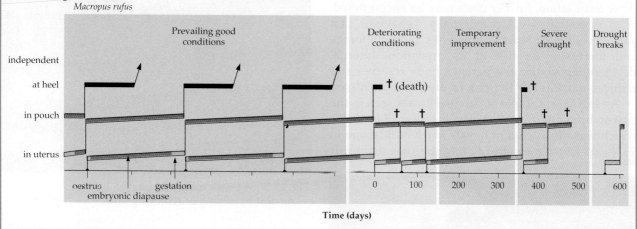

(a) Diagram illustrating patterns of reproduction in the red kangaroo under different environmental conditions

Male kangaroos also fight strenuously to secure their mating rights, using teeth, forefeet and hind-feet, and sometimes leaning back on their tail to deliver a two-footed attack (Fig. 13.24). Fights tend to be ritualised, with relatively few, highly damaging blows, and the skin is heavily thickened in the regions most prone to attack.

An extreme price is paid for sexual reproduction by males of the marsupial mouse *Antechinus* (Fig. 13.25). Both males and females are sexually mature in

their first year. Breeding occurs in mid-winter with a mating period of two to three weeks. By the time the females produce their litters, all the males are either dead or dying. The death of males occurs as a result of the following sequence of events. Levels of the male hormone testosterone increase dramatically in males as the photoperiod shortens with the onset of winter. High levels of testosterone cause increased territorial aggressiveness and fighting between males, and enlargement of the adrenal glands. The males are highly stressed during the mating period. Extremely high levels of adrenal hormones suppress the immune system, causing actual breakdown of immune tissue. As a result, their wounds do not heal, and they are unable to fight parasites or bacteria and quickly die from infection. If, however, males are kept isolated in captivity, this sequence of events does not occur and they can live for several years.

Fig. 13.25 The marsupial mouse, *Antechinus*. The mottled appearance of this male indicates loss of condition as a result of stress towards the end of his first and only mating season

Parental care

After fertilisation, further costs may be encountered. With eggs that develop externally, behaviour that protects offspring reduces a parent's ability to forage and increases the risk of its predation. Internal development drains the parent's body reserves and reduces mobility, so the risk of predation increases.

An extreme example of parental care is seen in the gastric brooding frog, *Rheobatrachus silus* (Fig. 13.26), which was discovered in forests north of Brisbane in 1974 and which may already have become extinct. It

Fig. 13.26 Young froglets emerging from the mouth of a female gastric brooding frog after developing directly in their mother's stomach

shows remarkable reproductive behaviour: eggs are shed and fertilised during amplexus, as in other amphibians but, after this, rather than leaving the eggs to develop alone and unprotected, the female swallows them. In the stomach, digestive secretions cease and the eggs settle into the stomach wall where they are protected and apparently absorb nutrients from the parent. This gastric brooding appears to last six to seven weeks, during which time the female does not eat. When the young frogs are ready, they are regurgitated through the mouth. Thus these animals have external fertilisation but internal development.

In birds, fertilisation is internal but the fertilised egg undergoes the majority of its development externally, within the egg. Parental care is often needed continuously. Most species brood their eggs, with parents taking turns so that each can go and feed during the incubation period. Other behaviours may also be used. The megapodid birds, such as Australia's brush turkey, build a mound from twigs, soil and leaf litter, into which they place their eggs (Fig. 13.27). Heat from decomposition of the leaf litter keeps the eggs warm but the male parent frequently tends the nest, adding or removing material to control the temperature of incubation.

Fig. 13.27 A brush turkey maintains the temperature of its nest by adding or removing litter

Sexual reproduction imposes significant costs in terms of increased energy use, decreased food intake, increased risks to survival, and in developing and maintaining specialised anatomical and physiological adaptations.

SEXUAL REPRODUCTION IN FLOWERING PLANTS

Reproductive mechanisms in flowering plants ensure some degree of selectivity in fertilisation and embryo development. These mechanisms optimise **pollination**, the transfer of pollen from the anthers to the pistil. This transfer is carried out by animals, and air and water currents, and enables the female gametophyte to screen out undesirable pollen (see Chapter 15). Since plants are essentially immobile, this is the only way in which selectivity of fertilisation partners can operate. The selectivity operates in most species to ensure that cross-fertilisation between different plants occurs and, in a few species, to ensure that only self-fertilisation occurs.

> Reproductive mechanisms in flowering plants enable them to optimise pollination and for the pistil to screen out undesirable pollen.

Cross-fertilisation

Cross-fertilisation involves the transfer of pollen from one plant to the female organs of a different plant of the same species. It maintains maximum genetic variability of offspring and is an advantage in unpredictable or changing environments. Cross-fertilisation is ensured by physical separation of the sexes in dioecious plants, by temporal separation as in protandry and protogyny (p. 259), and by a genetic mechanism, self-incompatibility.

Self-incompatibility

In most flowering plants, self-fertilisation is not possible because of **self-incompatibility**, which prevents self-pollen from fertilising eggs of the same plant. The pollen is, however, perfectly capable of fertilising eggs in the pistil of most other plants of the same species. Self-incompatibility is regulated by the S gene, which has up to 50 alleles. If the same S allele is present in both pollen and pistil, fertilisation will not be successful (Fig. 13.28). One example is the cultivated sweet cherry, *Prunus avium*. Each cultivar has its own particular pair of S alleles and is asexually propagated by grafted cuttings. By reciprocally crossing three different cultivars in all combinations, we can illustrate the mechanism of self-incompatibility (Fig. 13.29). Cultivar Bedford is S_1S_2, Napoleon is S_3S_4 and Van is S_1S_3. When pollen of variety Bedford (haploid S_1 or S_2) lands on the pistil of Van (diploid S_1S_3), only half the pollen—that containing the S_2 pollen—will grow successfully. S_1 pollen is recognised as self and its growth is therefore arrested. All pollinations should succeed when Bedford and Napoleon are crossed as their pollen and pistils have no S alleles in common. In the most common type of self-incompatibility, growth of self-pollen tubes is arrested soon after their entry into the pistil (Fig. 13.28).

Many other flowering plants, such as apple, pear and macadamia plants, are self-incompatible. Information obtained from reciprocal crosses of the

kind described for cherry is important for orchard management to ensure maximum yields from effective pollination. Orchards are designed to ensure that cross-pollinating varieties are interplanted according to designs that optimise pollination, for

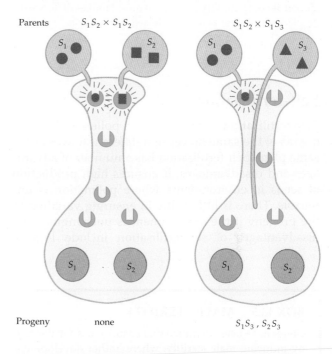

Fig. 13.28 Mechanism of self-incompatibility. Fate of pollen (S_1 or S_2) on a self-pollinated pistil (S_1S_2, left) and the same kind of pistil cross-pollinated with S_1 or S_3 pollen (right)

		Female		
		Bedford S_1S_2	Van S_1S_3	Napoleon S_3S_4
Male	Bedford S_1 or S_2	—	S_1S_2 S_2S_3	S_1S_3 S_2S_3 S_1S_4 S_2S_4
	Van S_1 or S_3	S_1S_3 S_2S_3	—	S_1S_3 S_1S_4
	Napoleon S_3 or S_4	S_1S_3 S_2S_3 S_1S_4 S_2S_4	S_1S_4 S_3S_4	—

Fig. 13.29 Offspring from self-incompatible matings. The checkerboard shows reciprocal crosses between three different cultivars of sweet cherry. The female parent (top), the pistil, is diploid and carries two S alleles. The male parent (left), haploid pollen grains, has half the grains carrying one S allele and half the other S allele in each anther. Pollen tubes carrying an S allele in common with the stigma are arrested in the style and cannot produce progeny

example, by appropriate placement of hives of honey bees at flowering time. Self-incompatibility is not restricted to fruit trees, occurring in most families of flowering plants including most wattles, *Acacia*, and most grasses.

> Self-incompatibility is found in most families of flowering plants. Its effect is to exclude self-fertilisation.

Self-fertilisation

Self-fertilisation occurs between pollen and female organs of the same flower or a different flower of the same plant. Self-fertilisation has a number of advantages and disadvantages. It ensures high production of seeds in environments where pollination is unreliable. There is little value in ensuring variation in the progeny if the environment is unchanging. The disadvantages of self-fertilisation include loss of

genetic variability and the appearance of lethal recessive gene combinations (inbreeding depression) manifested by reduced yield. For example, there is a 30% decrease in seed yield after a single generation of self-fertilisation in maize; and in some cases, embryo abortion within a few days of self-fertilisation, for example, in *Cacao*, the cocoa tree.

In some plants, self-fertilisation is ensured by **cleistogamy**, where flowers remain closed and anthers open within the closed petals (Fig. 13.30). Examples include the violet, *Viola*, and *Impatiens*. Some species of violet have normal cross-fertilising flowers (they open normally) in spring but cleistogamous flowers without petals in summer when conditions for cross-pollination are less favourable. Cleistogamy is frequent in plants that flower underground, for example, in several Australian annuals, the grass *Amphicarpum*, the weed *Emex* and the daisy *Gymnarhena*. In wallaby grass, *Danthonia spicata*, both closed cleistogamous flowers and open cross-fertilising flowers are present.

BOX 13.5   MALE STERILITY

Genes have been identified that ensure cross-fertilisation by inducing **male sterility**, where anther development and pollen production are blocked. Male sterility genes can be recessive (*ms*) or dominant (*Ms*) nuclear genes located on chromosomes, or cytoplasmic genes located in the mitochondrial genome. These genes occur at a low frequency in most plant populations. However they have proved very valuable experimental tools in plant breeding for the production of F_1 hybrid seeds. Since the female parent used in the hybrid cross is male sterile, only the female parts of flowers are functional. This ensures that all seeds produced by the female parent *must* be cross-fertilised and thus F_1 hybrid in origin.

Genetic engineering has also been used to create male sterile lines of crops for hybrid seed production. A Belgian company has spliced a 'killer' gene coding for the enzyme ribonuclease to a promoter of a gene expressed only in the anther and transformed it into oilseed rape plants (see Chapters 12 and 15). The killer enzyme is expressed in the precise organ (anther) and at a precise time (anther development), aborting the developing pollen. At the University of Melbourne, a research group has spliced an antisense version (coding region reversed) of another gene expressed only in the anther, and when transformed, the new gene causes male sterility at a precise time in pollen development (see figures).

(a)

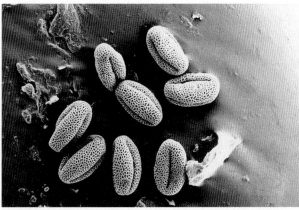

(b)

Control of male sterility. **(a)** Shrivelled pollen from transgenic *Arabidopsis* plant containing an antisense version of an anther-specific gene. This procedure inactivates expression of the gene and sterility shows that it is essential for normal pollen development. **(b)** Normal wild-type pollen for comparison.

(a)   (b)

Fig. 13.30 **(a)** Flowers of the violet open in spring. **(b)** In autumn, cleistogamous flowers of the violet develop in which the petals remain closed and self-fertilisation occurs within

Self-fertilisation limits heterozygosity, meaning that the number of homozygous gene combinations is increased in the progeny. After about eight generations of self-fertilisation, a pure line is created that is almost homozygous and breeds true. This is the classic method used to produce new varieties in plant breeding. Cross-fertilisation between such varieties leads to an increase in heterozygosity and thus variation. When two varieties of tobacco were crossed, leaf yield increased by 2% and plant height by 6% in the first generation. This illustrates the phenomenon of **hybrid vigour**, which occurs when two pure lines (homozygous) are crossed. The resulting hybrid seed is much more vigorous than either parent, often giving an increase in yield of as much as 30%. This is the basis for F$_1$ hybrid seed production in agriculture.

> Self-fertilisation enables genetic continuity in plants that are successful in their environment. Seed production is enhanced through cleistogamy in environments where cross-fertilisation is unreliable.

Reproductive success

Reproductive success is the ability of a plant to parent as many viable embryos as possible. Female reproductive success depends on viable ovules. The female gametophyte is usually viable for only a short period of time. Under some weather conditions,

ovules can abort before fertilisation occurs. For example, in cherry and peach, there are two ovules in each ovary (Fig. 13.31). The larger one is viable when the flower first opens, then it aborts and the second ovule becomes receptive. This mechanism extends the window of opportunity for fertilisation, especially during unfavourable weather. It also ensures that there are never two seeds in a cherry stone!

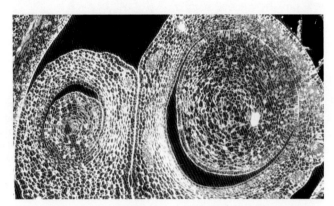

Fig. 13.31 Ovule viability. Ovary of a peach, showing a large viable receptive ovule (right), and smaller ovule (left), which will become viable later when the ovule on the right has degenerated. This device ensures a longer period of ovule viability for fertilisation

Some plants such as the cherry and avocado have only one or very few ovules in each ovary. Ovaries develop into large fruits which provide a substantial source of nutrition for the new offspring. Other plants, such as orchids, have several thousand ovules in each ovary. However, the seeds have almost no nutrient store and require an association with a particular fungus as a source of nutrients for growth.

Male reproductive success is dependent on pollen quantity and viability. The number of pollen grains produced in each anther is generally constant for a species but varies widely between species. For example, an exotic pasture grass, rye-grass (*Lolium perenne*), has 5400 grains per anther, while Australian grasses have less pollen: kangaroo grass (*Themeda triandra*) has 2000 grains and wallaby grass (*Danthonia caespitosa*) has only 400 grains per anther.

Pollen viability is defined as the ability of pollen to carry out successful fertilisation. This is difficult to estimate experimentally. Pollen quality, that is, whether the grains are living or not, is readily estimated by germinating a sample of pollen, which can take 24 hours (Fig. 13.32a), or by a rapid microscopic test, which takes only a few minutes (Fig. 13.32b).

Pollen grains from grasses and pumpkin retain high quality for only a few hours. Others from cabbage and the native climber, *Pandorea jasminoides*, slowly lose their viability over four or five days, after which only 8–10% of pollen is living. Pollen of *Banksia*

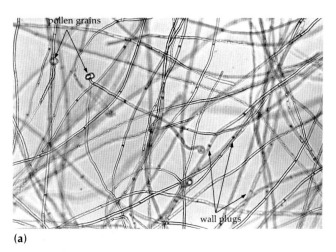

(a)

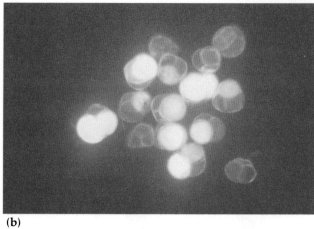

(b)

Fig. 13.32 Pollen viability. **(a)** Pollen germination test *in vitro*. Pollen grains of wild tobacco were cultured in a liquid medium and the extent of pollen tube growth was photographed after four hours. **(b)** Pollen quality can be tested using fluorescence microscopy, a procedure that takes only a few minutes to perform. This micrograph shows pollen of a hybrid *Rhododendron*, in which many of the clusters of four grains are sterile. In this procedure, pollen takes up a non-fluorescent compound, which is broken down by enzymes within the grain to form a fluorescent polar compound. This is retained within intact plasma membranes (green fluorescence), but diffuses out of dead grains, which have porous plasma membranes (no fluorescence, black)

spinulosa has 50% living pollen after eight days, although pollen from *Banksia menziesii* lives for only one day. Pollen of *Iris* can be stored for years at room temperature! Eucalypt pollen is remarkably heat tolerant and can remain viable even at temperatures of 45°C, when most other types of pollen are killed by temperatures over 30°C. Most pollen grains have to be stored frozen in liquid nitrogen at −176°C to maintain viability for prolonged periods for use in plant-breeding programs.

Each plant allocates its resources to its essential functions, which are vegetative growth, maintenance and reproduction. Maintenance costs include such activities as regulation of water movement, metabolism and avoidance of predation and pathogens. The cost of reproduction includes dispersal and selection of pollen, fertilisation and parental care. Reproductive effort represents the proportion of the total budget that is devoted to reproduction and hence into maturing embryos. Most organisms devote sufficient reproductive effort to maximise the chance of their offspring reaching maturity.

Reproductive success is the ability of a plant to parent as many viable embryos as possible.

SUMMARY

- In asexual reproduction, new individuals arise from a single parent by mitotic cell division and are genetically identical to that parent. Asexual reproduction can occur by means of vegetative growth, regeneration or budding.

- Parthenogenesis is a form of asexual reproduction in which egg cells develop into embryos without fertilisation.

- Sexual reproduction involves redistribution of parental genes into offspring and is the principal mechanism for generating variation in eukaryotes.

- Life cycles of sexually reproducing organisms alternate between diploid and haploid generations. In animals, the haploid stage is usually simply the gametes formed by meiosis, whereas the haploid stage in plants is a multicellular gametophyte and gametes are formed by mitosis.

- Isogamy involves the fusion of gametes that have a similar appearance but are of opposite (+/−) mating types. In anisogamy, male and female gametes differ in appearance, typified by large yolky eggs and smaller motile sperm.

- Sex is the distinction between male and female individuals of a dioecious species; that is, where male and female reproductive organs occur in separate individuals. In hermaphrodites, male and female reproductive organs are present in the same individual.

- Many mechanisms have evolved to limit self-fertilisation in hermaphroditic organisms. In protandry and protogyny, self-fertilisation is limited because individuals do not produce eggs and sperm at the same time.

- In animals, eggs and sperm are produced in the ovaries and testes respectively. These gonads also produce hormones that regulate secondary sexual characters and behaviour.

- Fertilisation involves a variety of adaptations, which ensures that mature gametes come into close proximity. External fertilisation usually involves the co-ordinated release of gametes into the environment; internal fertilisation follows transfer of sperm into the female reproductive tract.

- Some species release large numbers of eggs with relatively little yolk and these typically undergo indirect development. Other species have fewer eggs with large nutrient reserves, or direct transfer of nutrients, and this allows direct development.

- Mammals have yolk-poor eggs but direct development by means of nutrient transfer from mother to embryo during development. Monotremes are oviparous, marsupials and eutherians are viviparous.

- Sexual reproduction imposes significant costs in terms of increased energy use, decreased food intake, increased risks to survival, and in developing and maintaining specialised anatomical and physiological adaptations.

- Reproductive mechanisms in flowering plants enable them to optimise pollination and for the pistil to screen out undesirable pollen.

- Cross-fertilisation maintains heterozygous gene combinations, thus increasing genetic variation. Self-incompatibility, found in most families of flowering plants, is a genetic mechanism to ensure cross-fertilisation. In plant breeding and biotechnology, male sterility is a genetic mechanism used for hybrid seed production.

- Self-fertilisation enables genetic continuity in plants that are successful in their present environment but reduces yield in normally cross-fertilising crop plants. Seed production is enhanced through cleistogamy in environments where cross-fertilisation is unreliable.

- Reproductive success is the ability of an organism to parent as many viable embryos as possible. Reproductive effort represents the proportion of an organism's total energy and nutrient budget that is devoted to reproduction and hence into maturing embryos. Most organisms devote sufficient reproductive effort to maximise the chance of their offspring reaching maturity.

QUESTIONS

1. List the advantages and disadvantages of (a) asexual reproduction and (b) sexual reproduction.

2. What are the major differences between the life cycles of animals and flowering plants?

3. (a) Why can parthenogenesis in the whiptail skink (p. 254) be considered a degenerate form of reproduction? (b) What processes of sexual reproduction are avoided to produce seeds by apomixis?

4. The mechanisms involved in bringing gametes together for fertilisation are different in plants and animals. Use examples to explain how these mechanisms are related to an organism's lifestyle.

5. Most hermaphroditic plants and animals prevent self-fertilisation. Describe the various ways by which they do this.

6. Explain why changing sex is an advantage for sexual reproduction in both salmon and the blue wrasse.

7. (a) What is self-incompatibility? (b) By means of a checkerboard, deduce the S alleles of the offspring when a self-incompatible plant, A (S_1S_2), is crossed reciprocally with another plant, B (S_2S_3).

8. How is mass spawning of corals triggered? What advantages and disadvantages does it have for the organisms?

9. Explain why animals that produce large numbers of eggs often undergo indirect development, while those that produce fewer eggs often undergo direct development.

10. Outline some of the potential costs of sexual reproduction for (a) animals and (b) plants.

ANIMAL DEVELOPMENT

Multicellular animals begin life as a single cell. Within a short period that may be days, weeks or, at most, months, this cell develops into a complex animal, such as an echidna, a mayfly or a wedge-tailed eagle. This transformation is particularly striking because it is compressed into a relatively brief period. In most animals, body form is basically established during *embryogenesis*, the phase between fertilisation and hatching or birth.

Three main processes are involved in development:

1. **cell proliferation**—the production of millions of new cells by repeated mitotic divisions of the zygote (Chapter 8);

2. **differentiation**—specialisation of these cells into various types;

3. **morphogenesis**—the ordered assembly of these cells to form complex organs.

Cell differentiation is brought about by the selective switching on and off of specific genes in the nuclei of different cells (Chapter 16). Morphogenesis also involves regulation of gene activity but the important aspect of this process is the three-dimensional rearrangement of cells. For example, there is no difference between your arms and legs in terms of the *types* of cells present (each contains skin, muscle, connective tissue, bone and blood cells). The difference lies in their three-dimensional organisation.

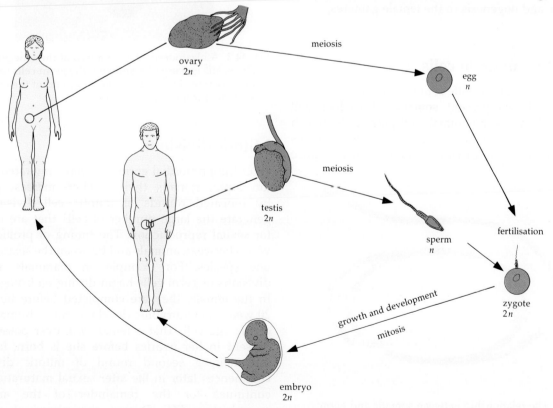

Fig. 14.1 The human life cycle. Fusion of the haploid sperm and egg in the act of fertilisation produces the diploid zygote. This cell develops into the multicellular embryo and eventually to the adult diploid human. Germ cells within the gonads of the adult (testis in man and ovary in woman) develop into the haploid gametes, completing the life cycle

Development is a progressive process, with complexity arising by degrees, building upon earlier events. For this reason, it is only possible to understand one event in the context of that which has gone before. Development proceeds under the instructions of genetic information. However, the connection between genes and the end product of their activity—the parts of the body—involves interactions between molecules, organelles, individual cells, groups of cells, tissues and organs. To understand the control of development we need to examine these interactions.

> Animal development is a progressive process, involving cell proliferation, cell differentiation and morphogenesis.

THE FORMATION OF GAMETES: GAMETOGENESIS

Embryonic development in animals begins with the fusion of sperm and egg to form a zygote (Fig. 14.1). However, since the early stages of development depend upon the properties of the sperm and particularly the egg, we should begin with an examination of **gametogenesis**, the sequence of events involved in the production of the mature gametes. **Spermatogenesis** refers to the formation of the male gametes, and **oogenesis** to the female gametes.

Migration of germ cells

Early in the embryonic life of most animals, two broad types of cells are present: **somatic cells**, which will form the body of the animal; and **germ cells**, which alone have the ability to develop into gametes (Fig. 14.2). In some cases it is possible to recognise

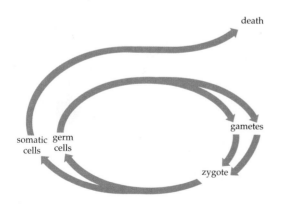

Fig. 14.2 The relationship between somatic and germ cells. Somatic cells die along with the organism, whereas germ cells become gametes, which fuse with another gamete to form the beginning of a new generation

germ cells very early in embryonic development as they separate from somatic cells (Fig. 14.3). At this time they are called **primordial germ cells**. Usually only a relatively small number of these primordial germ cells is formed initially and, strangely, they first appear far removed from their final location, which is the gonads (ovary or testis). They make their way to that distant site reliably, moving through a complex terrain, in some cases by a process of active migration. How germ cells find their way is poorly understood.

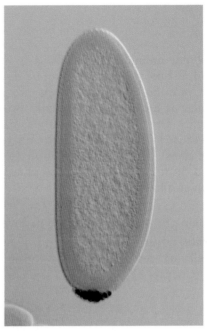

Fig. 14.3 A blastoderm stage embryo of the vinegar fly, *Drosophila melanogaster*, showing the germ cells (brown) in a discrete cluster at the posterior end of the embryo, separated from the somatic cells

Mitotic divisions

After the primordial germ cells have migrated to the developing gonads, their numbers increase greatly by repeated rounds of mitotic cell divisions to generate the large number of cells that are needed for sexual reproduction. The timing of proliferation varies between animals and between the sexes of any one species. For example, in mammals, mitotic divisions of germ cells begin during embryogenesis. In the female, they are completed before birth (by 20 weeks of embryonic development in humans): all of the egg cells that a female will ever possess are present in her ovaries before she is born. In male mammals, a second round of mitotic divisions commences later in life after sexual maturation and continues for the remainder of the animal's reproductive life. During the phase of mitotic divisions, female germ cells are **oogonia** (singular, **oogonium**) and male germ cells are **spermatogonia** (singular, **spermatogonium**) (Fig. 14.4a,b).

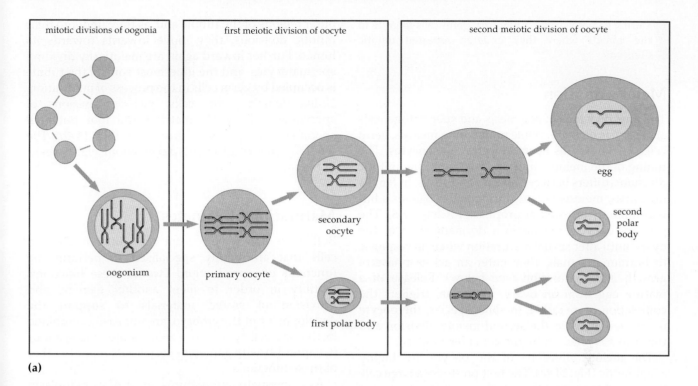

(a)

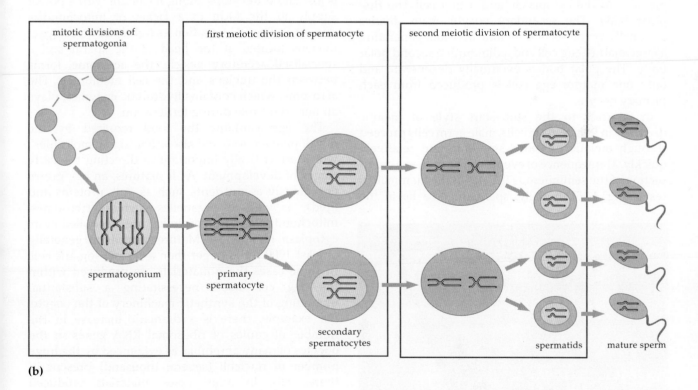

(b)

Fig. 14.4 (a) The sequence of events during oogenesis. A series of mitotic divisions generates a pool of diploid oogonia. These cells then undergo meiosis. The first meiotic division is unequal, generating one secondary oocyte and a small polar body, both of which are haploid. Each of these cells then undergoes the second meiotic division. The secondary oocyte divides unequally to produce another polar body and a haploid egg cell or ovum, and the first polar body divides equally to produce two additional polar bodies. (b) The sequence of events during spermatogenesis. Repeated mitotic divisions of the spermatogonia are followed by the two divisions of meiosis. The primary spermatocyte divides equally to produce four haploid spermatids. Differentiation of the spermatid into a mature sperm cell follows

During gametogenesis, primordial germ cells migrate to the gonads where they undergo repeated mitotic divisions.

Meiotic division

The final divisions of oogenesis and spermatogenesis are *reduction divisions*. While undergoing meiosis, germ cells are known as **spermatocytes** and **oocytes**. The timing of the meiotic division, like that of the mitotic divisions, differs between species and sexes. In female mammals, meiosis begins during embryogenesis but is commonly arrested at prophase I (Chapter 8). The primary oocytes remain in a dormant state in the ovary until after sexual maturation when, in response to hormonal signals, they enter an active phase of growth and maturation (see below). Release of a mature egg from an ovary, **ovulation**, triggers the completion of meiosis I. In some species, the oocyte is arrested again in the second meiotic division and this arrest is released by the act of fertilisation.

The meiotic divisions of the oocyte are highly asymmetric (Fig. 14.4a). The first produces a large cell, the **secondary oocyte**, which contains the vast majority of the cytoplasm, and a tiny cell, the **first polar body**. The secondary oocyte again divides unequally at the second meiotic division, generating a large mature egg cell and a diminutive second polar body. The polar bodies eventually degenerate and only one mature egg cell is produced from each primary oocyte.

In contrast to the stop–start style of meiotic divisions in female germ cells, male germ cells proceed through meiotic divisions smoothly and relatively quickly. The sequence of events can be seen in a cross-section of the seminiferous tubules of the mammalian testis (Fig. 14.5). Initially, spermatogonia lie in the outer regions of the tubule and, as they undergo their mitotic divisions, they move inwards towards its lumen. Further inward again are meiotically dividing spermatocytes, and the innermost zone of the tubule is occupied by germ cells in the process of maturation. Unlike female germ cells, meiotic divisions of spermatocytes are symmetrical and four potential sperm cells are produced from each (Fig. 14.4b). The whole sequence of events takes about eight weeks in humans.

Maturation of gametes

Both male and female germ cells differentiate into cells that are highly specialised to perform the functions of reproduction. Two of these functions, motility in order to meet another gamete, and provision of stored materials to support the development of the embryo, are not easily combined in the one cell type. Most animals are anisogamous (Chapter 13), with gametes specialised for one or other of these functions.

As a spermatocyte matures, most of its cytoplasm is lost and it develops a long flagellum with a power supply in the form of a jacket of mitochondria (Fig. 14.6). The DNA condenses tightly into a compact nucleus located at the head of the sperm and a specialised secretory vesicle, the **acrosome**, forms between the nucleus and the cell membrane. The acrosome, which contains hydrolytic enzymes, plays an important role during fertilisation.

The egg contains the food required by the developing embryo and also other 'signal molecules', which are critically important in directing the early stages of development. As it matures, an egg grows enormously as nutrients such as yolk proteins and lipids, as well as organelles such as ribosomes, mitochondria and microtubules, are stockpiled in its cytoplasm for later use. Mature egg cells are generally at least 1000 times larger than a typical somatic cell. In some cases these materials are produced within the egg cell itself, necessitating a substantial upgrading of the synthetic machinery of the oocyte. For example, there is a dramatic increase in the number of copies of ribosomal RNA genes in the oocytes of many amphibia, as evidenced by the huge number of nucleoli (several thousand) present in these cells. In other cases, materials produced elsewhere in the female's body are transported into the eggs. For example, virtually all of the yolk in the bird egg is first produced in the liver of the female and is then delivered to the oocytes in her ovaries via the circulatory system. In mammals, the oocyte is surrounded by layers of cells, called **follicle cells**, and one of the functions of these cells is to regulate the movement of substances such as yolk from the bloodstream into the oocyte (Fig. 14.7).

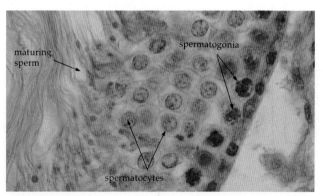

Fig. 14.5 Cross-section through a portion of the seminiferous tubule of the mouse, showing different stages in spermatogenesis. The dark-staining nuclei in the outermost region of the tubule (right hand side) are spermatogonia undergoing mitosis. The lighter nuclei further to the left are spermatocytes in various stages of meiosis. The small, condensed nuclei of spermatids and the long tails of mature sperm can be seen adjacent to the lumen of the tubule (left hand side)

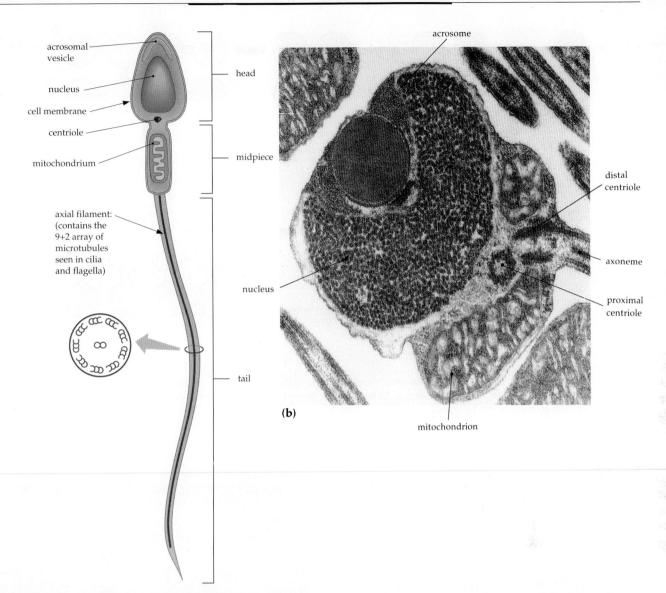

Fig 14.6 (a) The structure of a human sperm cell. Three regions can be recognised: the head contains the nucleus, the centriole and the acrosomal vesicle; the midpiece consists of a jacket of mitochondria; while the tail consists of a bundle of microtubules organised in the typical arrangement of a flagellum. (b) Electron micrograph through the head of a sperm from the pin-cushion starfish

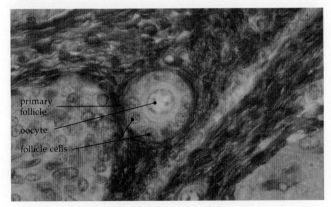

Fig. 14.7 A section through the ovary of a mouse. A primary follicle lies in the centre: it consists of a single oocyte, surrounded by a layer of smaller follicle cells. The follicle cells absorb materials from the circulatory system and transfer them to the oocyte

In many invertebrates, including some insects, annelids and molluscs, the oocyte is connected to a number of **nurse cells** by cytoplasmic bridges: the nurse cells and the oocyte are in fact sister cells, produced by repeated mitotic divisions of an oogonium in which cytokinesis (Chapter 8) is incomplete (Fig. 14.8). After these mitotic divisions are over, one of the cells develops into an oocyte and commences meiosis, while the others become nurse cells. RNA produced in the nurse cells is transported into the oocyte through the cytoplasmic bridges.

Mature egg cells also contain molecules that will later influence directly the developmental fate of embryonic cells. The synthesis of these signal molecules, and their precise location within the egg, takes place during oogenesis and therefore is directed

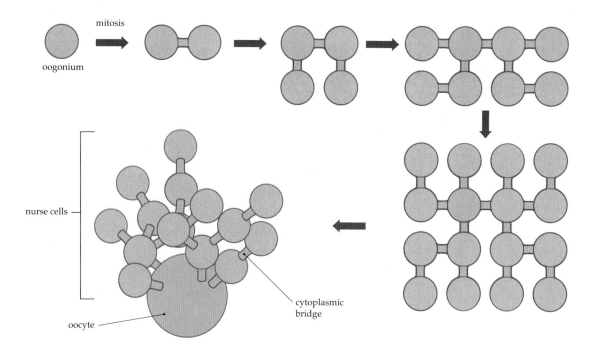

Fig. 14.8 In some insects, the oocyte and its associated 15 nurse cells are products of four rounds of mitotic divisions of a single oogonium. The cells remain connected to each other by cytoplasmic bridges, allowing materials to move from the nurse cells to the oocyte by direct cytoplasmic transfer

by maternal genetic information. This *maternal inheritance* is discussed in more detail in Chapter 9.

As an egg cell matures, it becomes surrounded by extracellular membranes (envelopes), which are secreted by the oocyte and/or follicle cells. These membranes provide protection after the egg is released from the sheltered environment of the female reproductive tract. In most species, a thin yet tough layer, the **vitelline membrane** (or **zona pellucida** in mammals) lies closest to the plasma membrane of the egg. A variety of other envelopes may surround the vitelline membrane: the harsher the environment into which an egg is released, the more highly developed these are (Fig. 14.9).

> After meiosis, germ cells mature into eggs and sperm. Oogenesis includes a phase of dramatic growth associated with storage of nutrients and signal molecules for use during embryogenesis.

FERTILISATION

Fusion of gametes, **fertilisation**, marks the beginning of the new organism. It involves two processes, **egg activation** and **nuclear fusion**. Firstly, the egg, which towards the end of oogenesis entered a metabolically dormant state, is activated by the fusion of egg and

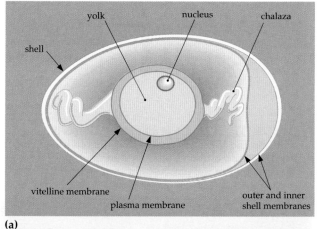

(a)

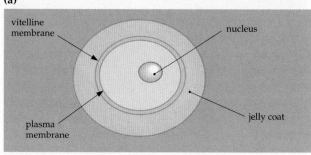

(b)

Fig. 14.9 Envelopes associated with the egg cell in (**a**) the chicken and (**b**) the frog. Chicken embryos, which develop in a terrestrial environment, have a more extensive and protective set of membranes than do frog embryos, which develop in water

sperm plasma membranes to begin synthetic activity again. Secondly, fertilisation leads to the fusion of the sperm and egg nuclei (called **pronuclei**), thereby creating a diploid zygote.

In the sea urchin, contact between sperm and the jelly coat that surrounds the egg triggers the acrosomal vesicle in the head of the sperm to release its contents (Fig. 14.10). These contents include hydrolytic enzymes, which soften the jelly coat and the vitelline membrane. A long process, formed at the head of the sperm cell as a result of the breakdown of the acrosome, bores through the jelly coat to the vitelline membrane. Here species-specific recognition occurs involving binding between molecules on the sperm plasma membrane and complementary molecules on the vitelline membrane. The acrosome process then breaks through the vitelline membrane and fusion of the egg and sperm cell membranes takes place. The initial contact between egg and sperm cell membranes triggers a cascade of changes in the egg. These begin with a change in permeability of the egg plasma membrane to Na⁺ and a rise in the intracellular Ca^{2+} concentration, and culminate in the activation of DNA and protein synthesis. The cell has commenced along the path of embryonic development.

Fusion of egg and sperm plasma membranes allows the sperm pronucleus to pass into the egg cytoplasm and fuse with the egg pronucleus. Firstly, however, the sperm pronucleus must locate the egg pronucleus, a not inconsiderable task considering the size of the egg cell. In humans, a period of 20 hours elapses between entry of the sperm pronucleus into the egg and nuclear fusion. With pronuclear fusion, a diploid zygote is formed. These events have been used in legal definitions of the beginning of a new life. For example, in Australia, the Victorian state government had until recently defined pronuclear fusion, as opposed to the initial membrane contact between egg and sperm, as marking the beginning of a new human individual: after this time, experimental manipulation of the human embryo was prohibited by law.

Curiously, in many animals (with the notable exception of mammals), the contribution from the sperm, the set of *paternal* genes, plays little if any role in early embryonic development. This can be demonstrated in sea urchins and frogs by artificially activating an egg, such as by pricking its plasma membrane or changing the pH of the surrounding water. The activated egg undergoes development to the gastrula stage (see later). However, development proceeds no further.

> Fertilisation involves fusion of the plasma membranes of sperm and egg, which activates the egg cell, and fusion of sperm and egg pronuclei, which restores the diploid number of chromosomes.

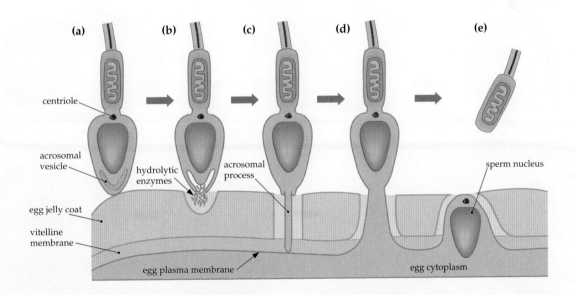

Fig. 14.10 Fertilisation in the sea urchin. (**a**) Contact between the sperm and the egg jelly coat leads to (**b**) the breakdown of the acrosomal vesicle. The hydrolytic enzymes thus released soften the jelly coat and (**c**) allow the acrosomal process to penetrate through the jelly coat and vitelline membrane. (**d**) Fusion of the plasma membranes of the sperm and egg activates the egg and (**e**) allows the sperm nucleus to move into the egg cytoplasm. The tail and midpiece of the sperm are left behind

MAJOR EVENTS OF EMBRYONIC DEVELOPMENT

Fertilisation sets in train the process of embryonic development, which appears very different in different animal species. Nonetheless, certain events or phases are found in all animal groups, namely cleavage, gastrulation, organogenesis and growth. The separation of embryogenesis into these phases is arbitrary because one stage leads imperceptibly into the next and they overlap in time in different parts of the embryo. However, the division is useful for outlining the developmental sequence.

Cleavage

Most egg cells are too large for efficient movement of molecules within the cell, and between the cell and the environment. During **cleavage**, which is triggered by activation, the zygote undergoes a series of rapid mitotic divisions. Immediately after each cytokinesis, cells enter the S phase and then move directly into the M phase; G1 and G2 phases are eliminated (Chapter 8). As a result, the size of the embryonic cells, **blastomeres**, progressively decreases during cleavage, while the overall size of the embryo remains unchanged. The reduction in cell size is important as it reduces the cytoplasmic volume, restoring the surface area:volume ratio of blastomeres to a more normal somatic cell value.

In most cases, early cleavage divisions follow a rigid pattern and a regular time sequence. Each cleavage furrow is oriented in a particular plane with respect to other divisions. As cleavage proceeds this regularity in pattern and time is usually lost, although in some animals, divisions are rigidly stereotyped right through to hatching. For example, in the nematode *Caenorhabditis elegans*, developmental biologists have been able to trace the origin and fate of every cell from zygote to adult worm to produce a complete cell lineage.

While all animal embryos undergo cleavage, the way in which it occurs varies greatly. One of the most important factors influencing the pattern of cleavage is the amount of yolk in the egg. Yolk displaces the spindle to an off-centre position and it also presents a physical barrier to the passage of the cleavage furrow, slowing down or stopping cytokinesis. In those animals with a small amount of yolk, such as the sea cucumber, yolk is evenly distributed within the egg and the mitotic spindle is centrally located. Early cleavage divisions therefore divide the egg into equal-sized blastomeres and subsequent divisions proceed at about the same rate in all parts of the embryo. As a result the blastomeres in the **blastula**, as the embryo is known at the end of cleavage, are of approximately equal size (Fig. 14.11). The sea cucumber blastula shows a feature seen in many other animal groups—a fluid-filled central cavity, the **blastocoel**.

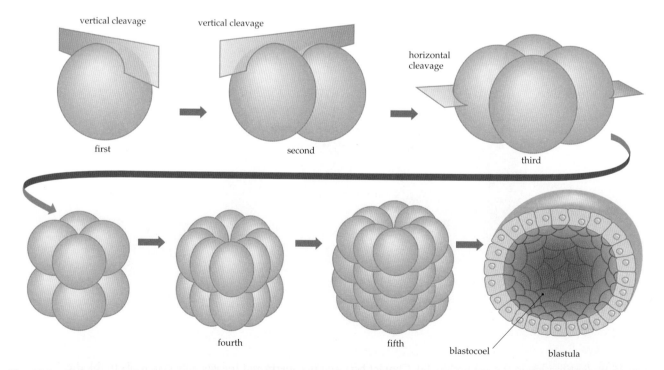

Fig. 14.11 The pattern of cleavage divisions in the sea cucumber embryo. The first and second divisions are vertical and at right angles to each other, the third is horizontal, the fourth vertical and the fifth horizontal. Precisely the same sequence of divisions takes place in every embryo. The end product of cleavage, the blastula, is a ball of equal-sized cells surrounding a central cavity, the blastocoel

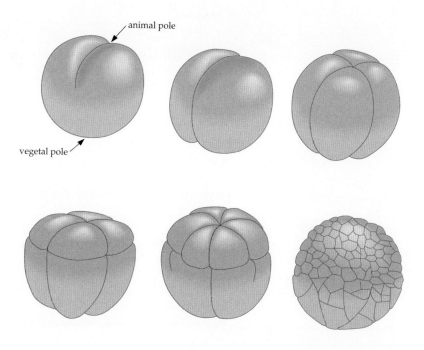

Fig. 14.12 Cleavage in the frog. The first two vertical cleavage divisions produce equal-sized blastomeres, but the third horizontal cleavage is displaced towards the animal pole because of the high yolk content in the vegetal hemisphere. As a result, the top tier of four blastomeres is smaller than the bottom tier. Later cleavage divisions in the yolk-filled vegetal region proceed more slowly than those in the animal hemisphere, resulting in a blastula in which the vegetal cells are larger than animal cells

Amphibians have eggs with an intermediate amount of yolk that is distributed in a graded fashion, ranging from a high concentration at one end of the egg, the **vegetal pole**, to a low concentration at the opposite end, the **animal pole** (Fig. 14.12). In amphibians, cytokinesis involves the entire egg but divisions proceed more slowly in the vegetal region than in the animal region due to the higher yolk concentration. As a result, more divisions occur in the animal region and thus blastomeres are smaller in the animal region of the blastula than in the vegetal region.

Birds and insects have yolk-rich eggs. In birds, most of the egg cytoplasm and the spindle are displaced to a thin, yolk-free disc on top of a large mass of yolk (Fig. 14.13). Cytokinesis is confined to this superficial layer of cytoplasm and at the end of cleavage the embryo takes the form of a cap of cells, a **blastodisc** or **blastoderm**, lying on top of an uncleaved yolk mass. In insects, cleavage produces a superficial layer of cells (blastoderm) enclosing a central yolk mass.

Development in eutherian mammals, for example, the mouse, shows a different pattern of cleavage. The eggs of most mammals are virtually yolk-free and

almost all of the nutrients required by an embryo are provided by the placenta. Like the sea cucumber, cleavage divisions in mammals are equal but the orientation of early cleavage planes is quite different from that seen in other animals. The first cleavage is vertical but at the second division, one blastomere divides vertically, while the other divides horizontally (Fig. 14.14). Furthermore, mammalian embryos cleave much more slowly than do other animals and even the early divisions are often asynchronous. In mice, divisions are about 12 to 24 hours apart, whereas a sea urchin zygote develops into a tiny free-swimming larva in about 12 hours. The blastula of mammals is also somewhat different from that of other animals with yolk-poor eggs in that it possesses a cluster of cells, the **inner cell mass**, at one end of the blastula, which is called a **blastocyst** in mammals (Fig. 14.14). Only the cells of the inner cell mass give rise to embryonic tissues. The surrounding shell of trophoblast cells contributes to the formation of the chorion, which fuses with the uterine wall to form the placenta. Cleavage in marsupials proceeds in a different manner again (Box 14.1), a consequence of the increased amounts of yolk in the eggs of these mammals.

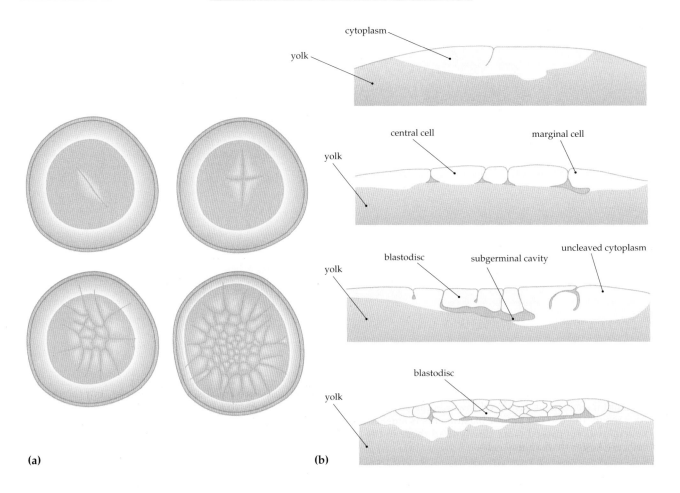

Fig. 14.13 Cleavage in the chicken showing **(a)** views from above the blastodisc and **(b)** views in longitudinal section. Cleavage furrows divide up the yolk-free patch of cytoplasm but do not penetrate the mass of yolk underneath. This results in a disc of cells, the blastodisc, separated from the uncleaved yolk mass beneath by the subgerminal cavity

As a result of cleavage, the contents of the egg cytoplasm are segregated into individual blastomeres. Because many components of the egg cytoplasm are not uniformly distributed throughout the egg, different blastomeres receive different types of molecules. This is visible in some species in which there are differently pigmented regions of the egg cytoplasm. As we shall see later, certain cytoplasmic materials influence the developmental fate of the blastomeres that receive them.

In many animals, the genes of the new individual do not participate in the events of cleavage. Cellular processes taking place during this stage, which are largely associated with mitosis, depend entirely upon maternal gene products stored in the egg during oogenesis. This idea is supported by two experimental observations. Firstly, fertilised sea urchin and frog eggs that have had their nucleus removed continue to cleave normally, showing that the presence of the nucleus is not necessary; and secondly, sea urchin embryos treated with the transcriptional inhibitor actinomycin-D cleave normally, showing that new synthesis of RNA is not needed (Chapter 11).

> Cleavage produces a multicellular embryo and restores a more usual nucleus:cytoplasm ratio in blastomeres. The pattern and timing of cleavage is affected by the amount of yolk in the egg.

Gastrulation

The blastula is structurally simple, consisting of a sphere or, in some cases, an irregular mass, of similar cells, often with a single internal cavity, the blastocoel. Contrast this form to the body plan of most mature animals, which have several tissue layers, many organs and two internal cavities. Furthermore, there is a change in symmetry: a blastula is generally radially symmetrical, whereas most animals are bilaterally symmetrical. During the next stage of embryogenesis, **gastrulation**, the embryo undergoes a whole-scale rearrangement of its cells to form a **gastrula**, an embryonic stage that displays many of the basic features of the definitive body plan of the adult animal.

BOX 14.1   CLEAVAGE IN MARSUPIAL EMBRYOS

Unlike eutherian mammals, marsupial eggs have three external envelopes, the zona pellucida, mucoid layer and shell (see figures). These envelopes enclose the marsupial embryo in its own microenvironment within which the embryo is constructed. The shell is porous to uterine secretions that nourish the embryo until it implants in the uterine wall and forms a placenta at a relatively advanced stage of organogenesis.

Marsupial eggs are rich in yolk, which becomes localised to a particular region of the zygote cytoplasm during the activation process of fertilisation. This yolk-rich cytoplasm is separated from the remainder of the cytoplasm during the first two cleavage divisions (see figure). Marsupials show a variety of cleavage patterns related to the patterns of yolk localisation in the zygote and of yolk segregation during cleavage. In Australian marsupial mice and native cats, yolk is segregated into a single yolk mass at the first cleavage division.

The first three divisions are equal and vertical but the fourth is unequal and horizontal, producing two tiers of eight blastomeres. The tier of smaller blastomeres lies in the hemisphere containing the yolk mass. In marsupials, blastomeres do not clump together to form an inner cell mass as in many eutherian mammals. Instead, blastomeres line the zona pellucida to form the single-layered epithelium of the unilaminar blastocyst. A number of eutherian mammals such as primitive insectivores, pigs and goats also have a blastocyst with no inner cell mass. Even though these mammals have no inner cell mass in a morphological sense, they do have a specific set of blastomeres, which go on to give rise to the tissues of the embryo proper. These cells, like the inner cell mass of mammals such as the mouse, are exposed to a different microenvironment to the rest of the embryonic cells, which may determine their particular developmental fate.

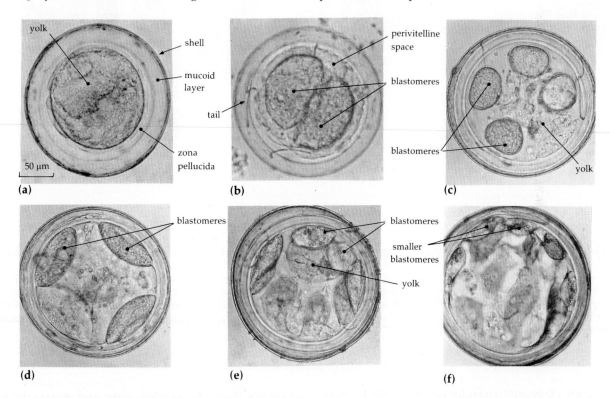

Stages in development of embryos of the brown *Antechinus*. (a) The zygote is surrounded by the zona pellucida, mucoid layer and shell. The yolk mass has become localised at one pole of the zygote. (b) Two-cell stage viewed from the yolk-poor pole. The two blastomeres have not yet moved apart and are concealing the yolk mass, which was pinched off from the blastomeres during the first division. The perivitelline space is opaque because it is filled with small yolk-like particles, which have been extruded by the blastomeres. The tail of the spermatozoan is trapped in the mucoid layer. (c) Oblique view of an arrested four-cell stage showing the four approximately equal-sized blastomeres lying apart from each other and slightly below the yolk mass. (d) Late four-cell stage when the four blastomeres are flattening on the zona pellucida before the third division.
(e) An oblique view of the eight-cell stage at the end of the vertical third division. Six of the blastomeres have stretched longitudinally from the yolk mass towards the opposite pole. The remaining two blastomeres have completed the division and rounded up but have not yet flattened longitudinally. (f) Sixteen-cell stage viewed from the side. The yolk mass is out of focus at the top of the figure. The fourth division is an unequal one and divides the blastomeres into an upper tier of eight smaller blastomeres, which lie over the yolk mass, and a lower tier of eight larger blastomeres

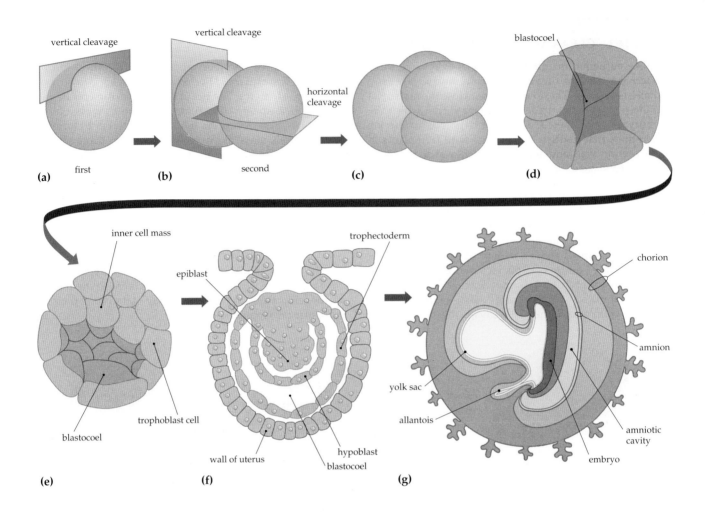

Fig. 14.14 Cleavage and tissue formation in the mouse. The first division (**a**) is vertical but the second (**b, c**) is vertical in one of the blastomeres and horizontal in the other. Furthermore, these divisions are not necessarily synchronous. Cleavage results in a single layer of trophoblast cells surrounding a blastocoel, with a cluster of cells, the inner cell mass, at one end. (**d, e**) The embryo subsequently sinks into the wall of the uterus and, as it does, the inner cell mass separates into various layers. (**f**) Some of these give rise to the embryo proper, while others contribute, along with the trophectoderm cells, to the development of the allantois, amnion and chorion, a system of membranes which protects the embryo (**g**)

As with cleavage, gastrulation proceeds in very different ways in different animal groups. We shall compare gastrulation in an echinoderm, the sea urchin, and a vertebrate, the frog, *Xenopus laevis*.

Sea urchin gastrulation

The sea urchin blastula is spherical in shape and consists of a single layer of closely packed epithelial cells around a large central blastocoel. The first sign of gastrulation is a flattening of the vegetal pole region to form the **vegetal plate**, caused by a lengthening of epithelial cells in this region (Fig. 14.15). Then, a small group of cells, **primary mesenchyme cells**,

detaches from the vegetal plate and moves to the interior of the embryo. These cells migrate along the inner surface of the blastula towards the animal pole and ultimately give rise to the skeleton of the embryo. Next there is a change in shape of the epithelial cells of the vegetal plate from a columnar to pyramidal form. At the same time, the vegetal plate starts to buckle inwards (**invaginates**). A new cavity, the **archenteron**, forms as a result of this invagination.

As invagination continues, a new group of mesenchyme cells, **secondary mesenchyme cells**, moves from the vegetal plate region into the blastocoel. These cells send out long processes (filopodia), which contact the inner surface of the

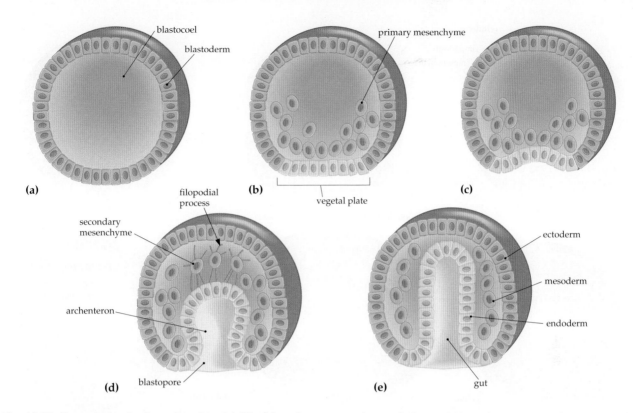

Fig. 14.15 Gastrulation in the sea urchin. (**a**) The blastula consists of a single-layered blastoderm surrounding a central blastocoel. (**b**) The vegetal plate flattens and, at the same time, primary mesenchyme cells detach from the vegetal plate and move into the blastocoel. (**c**) The vegetal plate begins to invaginate, creating a new cavity, the archenteron, which becomes the gut. (**d**) Secondary mesenchyme cells detach from the blastoderm and migrate into the blastocoel. Some send out long processes, filopodia, which adhere to the invaginating vegetal plate and to the inner wall of the blastoderm. (**e**) The invaginating vegetal plate contacts the blastoderm on the opposite side. The primary germ layers, ectoderm, endoderm and mesoderm are now evident

blastula in the animal hemisphere region. The vegetal plate continues to invaginate toward the animal pole, partly as a result of contraction of the filopodial processes and partly as a result of rearrangement of cells in the wall of the archenteron. Secondary mesenchyme eventually fills the blastocoel space.

The opening on the surface leading to the archenteron is the **blastopore** and forms the posterior end of the gut, the anus. The tip of the invaginating archenteron eventually touches the cells near the animal pole and the mouth of the sea urchin larva forms at this site. Gastrulation in the sea urchin therefore also marks a change from radial to bilateral symmetry with the appearance of the anteroposterior axis.

By the end of gastrulation three discrete tissue layers, the **germ layers**, can be identified:

1. **ectoderm**—the outer layer of cells, consisting of that part of the blastoderm that did not migrate inwards or invaginate;

2. **mesoderm**—the intermediate layer of cells formed from mesenchyme cells or by an outpocketing of the archenteron (see below);

3. **endoderm**—the innermost layer, the wall of the archenteron.

In the body of the adult animal, ectoderm gives rise to the outer body covering and the nervous system, endoderm to the lining of the gut and associated organs, and mesoderm to the tissues and organs that lie in between.

Another characteristic feature of the body plan of the sea urchin is the presence of a second body cavity, the **coelom** (a cavity that is entirely enclosed by mesoderm). In the sea urchin, the coelom develops from an outpocketing of the archenteron, which then expands to form inner and outer mesoderm layers (Fig. 14.16). In essence, the body plan of a coelomate animal is a tube within a tube: an outer tube of ectoderm and mesoderm, forming the body wall, and an inner tube of mesoderm and endoderm, enclosing the gut cavity. The space between, the coelom, is completely lined by mesoderm.

Xenopus gastrulation

Gastrulation in the frog *Xenopus laevis* appears very different to that in the sea urchin. The small size of the frog blastocoel and the presence of many large,

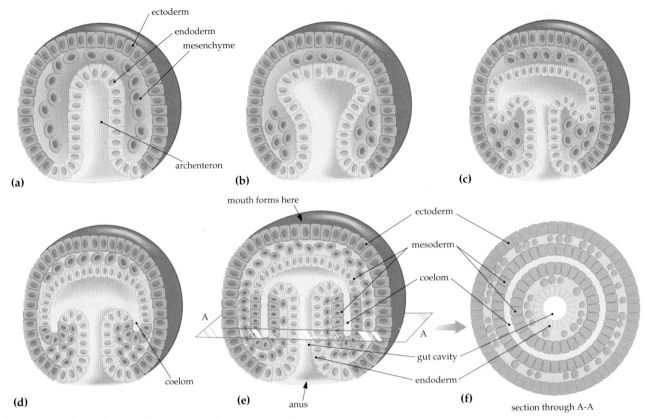

Fig. 14.16 (a)–(e) Coelom formation in the sea urchin. The coelom forms from an outpocket of the archenteron, which grows posteriorly and eventually separates from the archenteron. The inner and outer walls of this outpocket contribute to the mesoderm, which therefore lines the coelom on both sides. (f) A cross-section through the embryo at plane A–A, shows the definitive body plan of the sea urchin, with three germ layers and two body cavities

yolk-filled cells in the vegetal hemisphere make gastrulation by simple invagination impossible. Instead, presumptive mesoderm (mesoderm-to-be) rolls into the interior of the embryo by a process known as **involution**, and presumptive endoderm is internalised by overgrowth of ectodermal cells, a process called **epiboly**.

In the frog embryo, the anteroposterior axis and the dorsoventral axis are evident at the beginning of gastrulation. The very first cells to undergo involution lie in the animal hemisphere just beneath the surface of the embryo along the future dorsal midline. These cells will give rise to the **notochord**, a longitudinal rod-like structure characteristic of all chordate embryos (Chapter 40). They also play a central role in the formation of the central nervous system and other dorsal structures (see below). They move deeper into the embryo by first migrating in a vegetal direction, then rolling under their neighbours and moving towards the animal pole (Fig. 14.17). A lip, the **dorsal lip of the blastopore**, forms at the site where the presumptive notochord cells roll inside the embryo and this lip marks the boundary of the blastopore. Later, more lateral and then ventral cells roll inwards and spread out, eventually forming a continuous sheet of mesoderm

under the outer layer of the embryo, and completing the formation of the blastopore. As in the sea urchin, the blastopore marks the site of the anus and the mouth breaks through at the opposite end of the archenteron; this pattern is characteristic of deuterostomes (echinoderms and chordates, Chapter 40).

While involution of presumptive notochord and mesoderm is taking place, yolk-rich cells in the vegetal region of the blastula are also being internalised. They come to lie in the interior of the embryo as a result of the lateral expansion of the superficial sheet of animal hemisphere cells. By the end of gastrulation, these animal hemisphere cells enclose the entire embryo, forming the ectoderm, and the internalised yolk-rich vegetal cells form the endoderm. The archenteron, lined by endodermal cells, forms as the presumptive endoderm moves inwards. A large mass of yolk-filled endodermal cells lies ventral to the archenteron and mesoderm lies in between endoderm and ectoderm. Soon a split appears within the mesoderm and expands to form the coelom.

Gastrulation results in the formation of:
- the germ layers (ectoderm, mesoderm and endoderm);
- new body cavities (archenteron and coelom);
- in many cases, the basic axes of body symmetry.

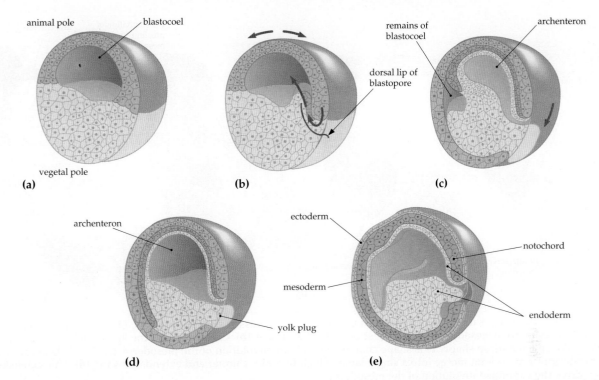

Fig. 14.17 Gastrulation in the frog *Xenopus*. (**a**) The blastula. (**b**) Involution of presumptive notochord cells around the dorsal lip of the blastopore. (**c**) Involution continues, the archenteron forms and the blastocoel is crowded out. Epiboly of the presumptive ectoderm takes place. (**d**) The endoderm has been internalised. (**e**) The completed gastrula showing the three germ layers

Mechanisms of morphogenesis

The tissue and cell movements that generate the germ layers and internal cavities in the gastrulae of frogs and sea urchins are different: invagination and movement of single cells are important in the sea urchin whereas involution and epiboly play the dominant role in the frog. However, the end product of gastrulation in each case is the same: a three-layered, bilaterally symmetrical gastrula with an archenteron and a coelom. Furthermore, the cellular processes that underly the movements of gastrulation are similar in the two species, the most important being change in cell shape and change in cell adhesion. For example, the detachment of primary mesenchyme cells from the blastoderm of the sea urchin blastula is accompanied by a change in shape of these cells from a columnar to a tear-drop shape and by a disappearance of the cellular connections that link these cells with their neighbours (Fig. 14.18).

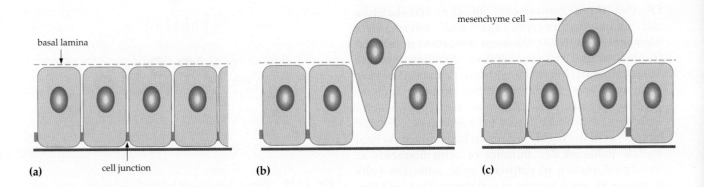

Fig. 14.18 Cellular changes underlying early gastrulation in the sea urchin. (**a**) Epithelial cells of the vegetal region of a late blastula. The cells are held together by intercellular junctions and an underlying basal lamina. (**b**) Cellular junctions and the basal lamina break down, freeing an epithelial cell, which changes from columnar to tear-drop shape. (**c**) The cell becomes a primary mesenchyme cell as it breaks away from the vegetal plate

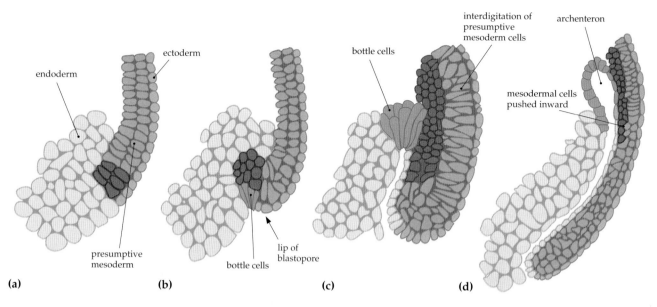

Fig. 14.19 Cellular changes underlying involution in *Xenopus*. (a) A cross-section through the blastopore region of the blastula. Presumptive mesodermal cells lie just beneath the surface. (b) Certain cells at the lip of the blastopore, 'bottle cells', take on a tear-drop shape. This shape change may initiate involution of the mesoderm. (c) The cells in the layers of presumptive mesoderm merge into a single layer, which becomes thinner and extends forward. (d) This extension may drive the continued involution of the mesoderm

Similarly, involution in the frog embryo is initiated by elongation and rearrangement of cells at the dorsal lip of the blastopore (Fig. 14.19), involving changes in cell shape and in cell adhesion. Similar changes underly gastrulation movements in other animals and, as we shall see below, the complex morphogenetic changes occurring during organogenesis.

Gastrulation is influenced by the structure of the blastula and varies widely between species; however, similar cellular processes are involved.

The cytoskeleton and cell shape

Of the three components of the cytoskeleton, microfilaments, microtubules and intermediate filaments (Chapter 3), the most important in relation to cell shape changes during development are microfilaments. Microfilaments are composed of long actin rods associated with other proteins such as myosin. Myosin molecules can interact with actin filaments to produce forces for movement or shape change, as in the highly organised filaments of skeletal muscle (Chapter 27). Bundles of actin filaments at the apical surface of epithelial cells, adhesion belts (Chapter 6), are a feature of cell sheets that undergo folding. The invagination of the vegetal plate during gastrulation in the sea urchin may be generated by a wave of constriction of the band of microfilaments found at the apex of these epithelial cells (Fig. 14.20).

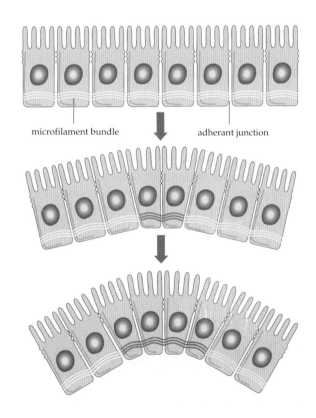

Fig. 14.20 When bundles of microfilaments in adhesion belts in certain epithelial cells contract, they cause a decrease in the diameter of the cells at that point. If these cells retain adhesive contacts with their neighbours (adherant junctions), the epithelial sheet as a whole will bend

In many animal cells, a network of actin microfilaments found just beneath the plasma membrane of the cell has been implicated in cell shape change and particularly in the movement of whole cells. In migrating cells, bundles of microfilaments project into extensions of the plasma membrane, **filopodia** and **lamellipodia** (Fig. 14.21). These structures appear to act as feelers for exploring the environment and for pulling the cell forward. We have already seen how filopodia are involved in movements of mesenchyme cells in the sea urchin gastrula.

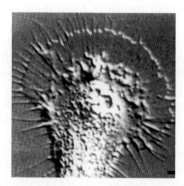

Fig. 14.21 The growing tip, or growth cone, of a nerve cell from the mollusc *Aplysia* in tissue culture; a thin veil, or lamellipodium, extends around the perimeter of the growth cone from which long, thin filopodia project

Cell–cell recognition and adhesion

In an embryo, migrating cells move along particular pathways and come to rest at precise locations. To do this they must be able to sense their environment and respond to it, either by adhering to surrounding cells or by sending out lamellipodia or filopodia in a directed manner. The importance of cell adhesion in morphogenesis is not confined to cell migration. For example, cells are held together within an epithelial sheet partly by adhesive interactions with their neighbours. When a cell leaves a sheet, as do mesenchyme cells during sea urchin gastrulation, a change in this adhesive property must have occurred.

Adhesive relationships may be quite specific. If a germ layer is treated with enzymes to dissociate its cells, and isolated cells from different germ layers are mixed together, they reaggregate according to cell type (Fig. 14.22). This *in vitro* phenomenon has been used as an assay to identify cell–cell adhesion molecules.

In recent years biologists have begun to identify the molecules involved in cell–cell recognition and adhesion. Cell-adhesion molecules found to date are glycoproteins projecting from the outer surface of cell membranes (Fig. 14.23). Some projecting molecules are identical, while others are complementary. Two broad classes, Ca^{2+}-dependent and Ca^{2+}-independent, have been identified. The tissue distribution of cell–cell adhesion molecules within embryos and changes in their distribution are consistent with their involvement in morphogenesis (Box 14.2).

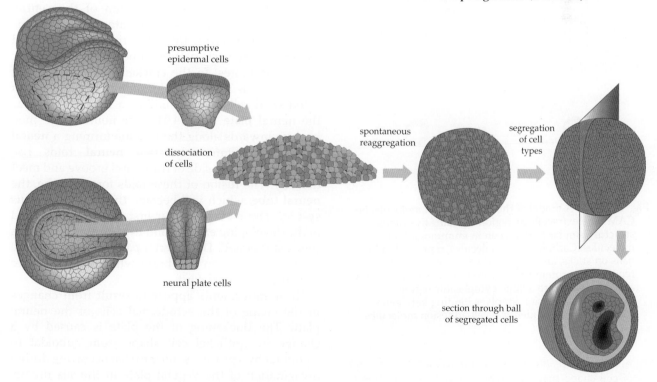

presumptive
epidermal cells

dissociation
of cells

spontaneous
reaggregation

segregation
of cell
types

neural plate cells

section through ball
of segregated cells

Fig. 14.22 When amphibian cells from different germ layers are combined and grown *in vitro*, they spontaneously separate out into aggregations of similar cell type

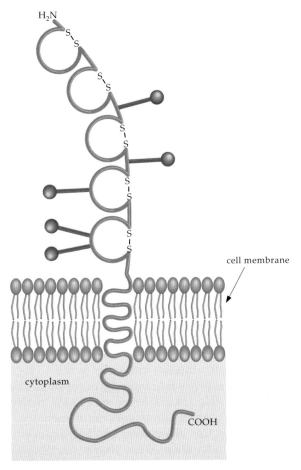

(a)

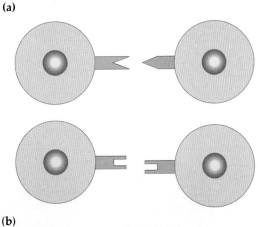

(b)

Fig. 14.23 **(a)** A model of the cell-adhesion molecule, N-CAM. The extracellular region contains five loops, a structure similar to that seen in immunoglobulin molecules. Carbohydrate molecules, represented by balls on sticks, are associated with this extracellular region. The protein passes through the cell membrane and continues into a large cytoplasmic region.
(b) Cell–cell adhesion involves binding between complementary or identical cell-adhesion molecules

Actin filaments often play an important role in changes of cell shape that accompany morphogenesis. Specific adhesive interactions, mediated by cell-adhesion molecules, underly many morphogenetic processes.

Organogenesis

While the basic body plan of an animal is laid down during gastrulation, the morphogenetic processes that mould tissues into complex three-dimensional organs such as the heart, lungs and limbs, take place during the next phase of embryogenesis, **organogenesis**. Like gastrulation, this is a rapid and highly dynamic phase of development. It takes place in humans between 3 and 6 weeks after conception, during which time the fetus is particularly susceptible to damage. Many agents known to cause birth defects, such as the drug thalidomide and the virus causing rubella (German measles), have their effects during this period.

Though the final shapes of organs are extraordinarily diverse, a limited number of morphogenetic mechanisms are involved in their formation. These mechanisms include thickening and folding of epithelial sheets; disaggregation of tissues into individual cells and migration of those cells to new sites in the embryo; localised cell proliferation; and localised cell death. We shall examine the involvement of these morphogenetic mechanisms in the formation of two vertebrate organs, the nervous system and limbs.

Neurulation

The nervous system is the first major organ system to develop in a vertebrate embryo. **Neurulation**, the process by which the primordium of the central nervous system forms, so dominates the embryonic landscape that, during this stage, the embryo is called a **neurula**. The basic pattern of neurulation is similar in all vertebrates. In the chicken, the first stage of neurulation is seen as a thickening of the sheet of ectodermal cells in the midline, a region known as the **neural plate** (Fig. 14.24). The neural plate then buckles inwards along the midline forming a **neural groove**. Simultaneously, two **neural folds** rise upwards on either side of the neural groove and meet dorsally. The fusion of these folds forms a tube, the **neural tube**, which will become the central nervous system. These events occur first at the anterior end of the developing embryo and then continue towards the posterior end. The anterior part of the neural tube bulges out to form the brain while the remainder becomes the spinal cord.

These movements appear to result from changes in the shape of the ectodermal cells of the neural plate. The thickening of the plate is caused by a change in epithelial cell shape from cuboidal to columnar by a process similar to that occurring during invagination of the vegetal plate in the sea urchin. The formation of the neural tube is an example of morphogenesis by thickening and folding of an

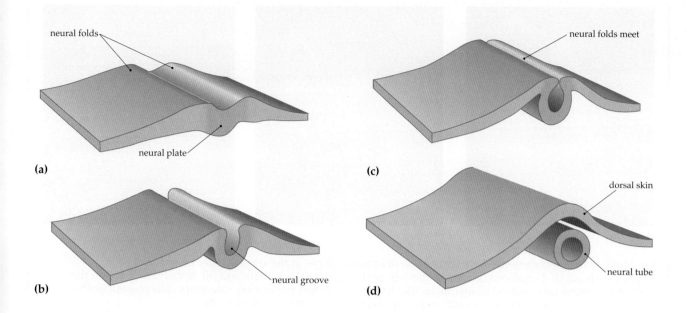

Fig. 14.24 Neurulation in the chicken. (**a**) Neural folds form on either side of the dorsal midline. (**b**) The folds rise upwards, creating a neural groove. (**c**) The folds continue to rise up and meet in the dorsal midline. (**d**) The neural tube separates from the overlying ectoderm, which forms the dorsal skin

epithelial sheet. Development of other organs, such as the pancreas, lungs, trachea and eye, involve similar mechanisms.

As the neural tube separates away from the overlying ectoderm, cells, originally from the crests of the neural folds, transform into mesenchyme and migrate away from the junction of the tube and the ectoderm. These are known as **neural crest cells** and they play an important role in the further development of vertebrates. They migrate *through* the tissues of the embryo as individual cells and lodge at certain sites where they give rise to a variety of organs. For example, some do not migrate far and form groups of sensory nerve cells (ganglia). Others migrate further and form autonomic nerve cells, part

of the adrenal glands, some sensory organs (such as hair cells, Chapter 7), and the sheaths around nerve axons. In addition, they are a major contributor to the soft and hard connective tissues of the skull, face, jaws, pharynx, gills and major arteries. During the evolution of vertebrates, these structures enabled an increased uptake of oxygen and higher metabolic rates, faster swimming, better spatial orientation and detection of prey, and an increased ability to respond quickly and accurately to sensory information. The appearance of organs derived from neural crest cells largely allowed vertebrates to increase in size and adopt an active, predatory lifestyle in contrast to the slow-moving, filter-feeding invertebrates from which they evolved.

BOX 14.2 THE NEURAL CREST

The development of the neural crest is a spectacular example of morphogenesis by cell rearrangement and cell migration, and it highlights the role of cell adhesion in these events.

Cell-adhesion molecules and migration

Tests performed on cultured presumptive nerve tissue from chicken embryos show that cell–cell adhesion declines in the neural crest cell population at the time of onset of cell migration. Transmission electron microscopy shows that the cells lose specialised intercellular adherens junctions (Chapter 6) at this stage.

Likewise, specific molecules belonging to known cell-adhesion molecule families (p. 291) decrease on neural crest cells. One important group is the cadherin family of calcium-dependent adhesion molecules, which are associated with adherens junctions. Using antibodies against various cadherins as markers, epidermal ectoderm cells can be shown to possess E-cadherin, whereas neural ectoderm cells possess N-cadherin. These enable cells with like cadherins to adhere. Neural crest cells initially show N-cadherin while they are part of the neural epithelium but this is lost when they become mesenchyme (see Fig. a).

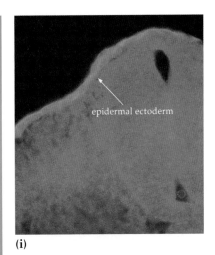
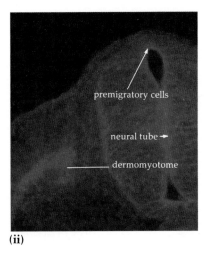
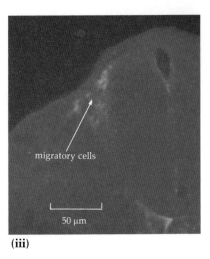

(i)   (ii)   (iii)

(a) Multiple immunolabelling of a single section of a chick embryo at the stage of early neural crest cell migration shows (i) E-cadherin (green) on the cells of the epidermal ectoderm and (ii) N-cadherin (red) on the cells of the neural tube and dermomyotome. (iii) In the neural crest population, migratory cells possess neither E- nor N-cadherin, while those cells still at the premigratory stage possess N-cadherin

If the onset of migration of neural crest cells is due to a loss of cell–cell adhesion, artificially decreasing cell–cell adhesion before the onset of migration should provoke premature migration. Cadherins can be selectively inactivated by exploiting their extraordinary sensitivity to protease enzymes in the absence of calcium. Applying this technique to cultured neural tissue from chicken embryos bears out the prediction that inactivation of cell-adhesion molecules will stimulate the onset of cell migration. Premigratory neural crest cells commence migration immediately when briefly treated with protease.

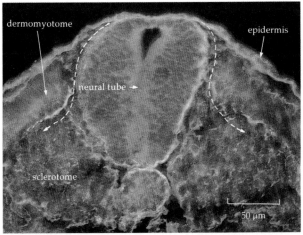

(b) Triple immunolabelling of a section of a chick embryo, with the major pathway of neural crest cell migration indicated by dotted lines. The preferred pathway between the dermomyotome and sclerotome shows a green colour, indicating a predominance of the adhesive ECM molecule fibronectin. The potential pathway between the dermomyotome and epidermis is not followed at this stage, and its blue colour indicates a predominance of an anti-adhesive ECM molecule. The red colour labels N-cadherin

Migration routes and signposts

Transplanting labelled neural crest mesenchyme cells to different regions in embryos shows that, although neural crest cells have highly developed migratory abilities, the pattern of their migration is controlled by the surrounding microenvironment. The early stage of neural crest cell migration occurs into cell-free spaces. Microenvironmental structures that could provide stable cell adhesion and a substrate for this migration include the extracellular matrix (ECM). Transmission electron microscopy has shown that neural crest cell extensions contact and adhere to the ECM, and antibody labelling has revealed that glycoproteins of the ECM include many adhesive molecules, such as fibronectin, laminin and collagens, which form fibrous meshworks. In addition, adhesion-inhibiting molecules, such as chondroitin sulfate proteoglycans, have also been found in the ECM. These are distributed broadly in and around the neural crest cell migration pathways, suggesting some role for the ECM in defining those pathways (see Fig. b).

ECM molecules can be purified and used as substrates for growth of neural crest cells in tissue culture. These experiments show, for example, that chicken neural crest cells adhere to fibronectin, laminin and collagens, and that migrating cells can accurately follow ECM molecule pathways (see Fig. c). Neural crest cells adhere to the ECM via cell-surface receptors, which link extracellular ECM molecules and the intracellular cytoskeleton.

Other cells are also potential substrates for the migration of neural crest cells. It is suspected that neural crest cells entering blocks of mesodermal cells known as somites may use the somite cells themselves as a migratory substrate since there is little ECM present.

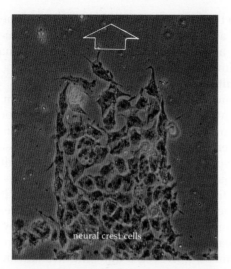

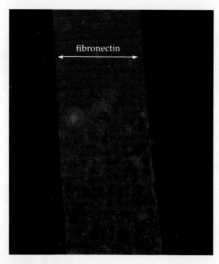

(c) (i) In tissue culture, neural crest cells, shown in phase contrast microscopy, migrating in the direction of the arrow, are confined to a narrow band

(ii) Immunofluorescent microscopy of the same field shows that cells are following a track of the adhesive ECM molecule, fibronectin

In addition, electron microscopy has revealed cell–cell adhesions between neural crest and somite cells, but the cell–adhesion molecules involved in this are not yet known.

Preventing cell adhesion inhibits migration

Cell adhesion can be disrupted using antibodies against the adhesive region of a ligand (e.g. an ECM molecule) that prevent the cell receptor from locking onto it. Alternatively, antibodies can be produced against the receptor molecule with similar effect. The receptors can also be blocked using small soluble peptides mimicking the ligand. Microinjection of such reagents known to inhibit cell–fibronectin adhesion in cell culture has been used successfully to curb neural crest cell migration in chicken embryos (see Fig. d). This evidence confirms the importance of adhesive interactions in neural crest cell migration.

The end of migration

Neural crest cells eventually cease migrating. The best studied examples involve the formation of neural ganglia, where the cells form clumps within expanses of non-crest cells with which they were previously interspersed. Antibody labelling has shown the renewed appearance of cell-adhesion molecules such as N-cadherin on cells in these ganglionic clusters and a reduction of fibronectin receptors. These observations are consistent with increased adhesion between neural crest cells and reduced adhesion between these cells and their surroundings, which could explain the reversion to cell aggregation and cessation of movement.

The overall conclusions drawn from these studies on neural crest systems are, firstly, that complex morphogenetic changes can be in part controlled by alterations in a relatively small number of cell-adhesion molecules, ECM molecules and adhesion receptors. Secondly, these molecules act in concert and may have overlapping functions. Thirdly, since there is

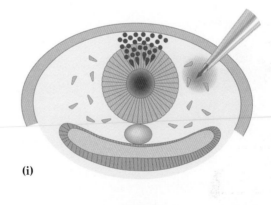

(i)

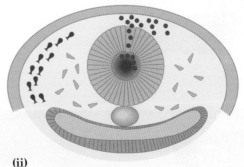

(ii)

(d) Scheme showing that **(i)** an adhesion-perturbing antibody against the fibronectin receptor microinjected into the migration pathway of neural crest cells (black) on one side of the head of a chick embryo **(ii)** curtails later migration into this region. The uninjected side serves as a control. This provides strong evidence for the role of cell adhesions in cell migration

considerable similarity between neural crest cell behaviour and behaviour of other cells at different stages of development, regulation of cell adhesion is likely to be involved in morphogenesis in these other situations.

Limb formation

The formation of the vertebrate limb involves several different morphogenetic mechanisms. A limb first appears as a small bud on the flank of the embryo. It consists of a core of mesodermal cells covered by a layer of ectodermal cells. The ectoderm at the tip, the apical ectodermal ridge, is thicker than elsewhere and stimulates underlying mesodermal cells to divide. The limb bud elongates as a result of this localised proliferation of mesoderm (Fig. 14.25).

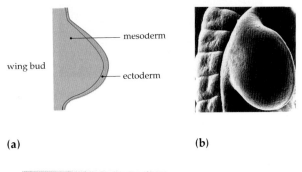

wing bud — mesoderm

ectoderm

(a)   (b)

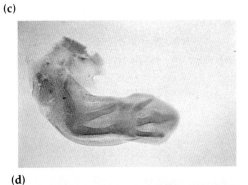

(c)

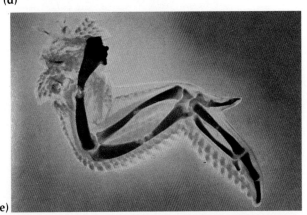

(d)

(e)

Fig. 14.25 The morphogenesis of the chicken limb bud.
(a, b) A small bud of mesoderm, surrounded by a layer of, ectoderm, develops on the flank of the embryo.
(c)–(e) As the limb grows, mesodermal cells condense together to form the rudiments of bones

As the limb elongates, the divisions of mesodermal cells at the base of the bud slow and some of these cells aggregate to form cartilage plates that are the rudiments of the limb bones (Fig. 14.25). Simultaneously, other mesodermal cells condense around this cartilage, eventually to form muscles. Recently, it has been found that the mesoderm that produces the muscle does not arise within the limb. Rather, it migrates into the limb bud as single cells that split off from blocks of mesoderm, somites, which surround the developing neural tube.

Initially, a limb bud is paddle-shaped, with no sign of the digits characteristic of a mature limb. In amphibians, the rudiments of digits arise by a process of local proliferation of mesodermal cells. In birds and mammals, however, a different mechanism operates. Here, digits are sculptured from the paddle by a process of selective cell death: mesodermal cells in the zones between the digits start to degenerate and then die, thus separating the digits (Fig. 14.26). This cell death appears to be programmed because mesodermal cells from the interdigit zones placed into tissue culture initially appear healthy but then die on schedule, at the same time as interdigit zone cells left in the limb bud.

Fig. 14.26 A chick limb bud at the 'paddle-bud' stage. Cells within the dark blue regions die later in development, giving rise to the shapes of digits

The basic set of morphogenetic mechanisms evident during nervous system and limb bud development are used in different combinations and in different situations to produce the vast array of forms seen in the organs of animal embryos. These tissue changes are, in turn, generated by changes in a limited set of cellular properties, including some changes, such as in cell shape and cell adhesion, that were involved earlier in development, during gastrulation.

Morphogenetic changes during organogenesis are generated by thickening and folding of epithelial sheets; disaggregation of tissues into individual cells and single cell migration; localised cell proliferation; and localised cell death.

Cell differentiation

As organ rudiments take shape, another important developmental process, **cell differentiation**, is taking place within the cells composing them. Differentiation involves changes in the pattern of gene expression of the cell (Chapter 16). However, before a cell switches on the genes that reveal its final differentiated character, changes have already taken place that restrict it to a particular fate. This process is **determination**. There are often no obvious changes in the appearance or biochemical activity as a cell becomes determined. It is possible to assess whether determination has taken place by transplanting cells to a different site in a developing embryo. If the transplanted cells develop in a similar way to their new neighbouring cells, the transplanted cells were undetermined at the time of transplantation. If, however, the transplanted cells continue to develop as they would have done at their original site, despite their new location, they *were* determined at the time of transplantation. The series of experiments shown in Figure 14.27 shows that presumptive neural ectoderm in amphibians becomes determined during the gastrulation stage.

The path to differentiation usually involves a series of steps in which the developmental potential of a cell is *progressively* determined. For example, while dorsal ectoderm becomes determined to form neural tissue during gastrulation, the precise type of neural tissue to be formed by individual cells is determined much later.

Growth and maturation

The basic structure of an organism is laid down very early in development, for example, in the human embryo, when it is only about 12 mm long. The remainder of development largely involves gradual growth of component parts. Further cell proliferation, differentiation and expansion occurs, along with an increase in the extracellular matrix. Growth, like all developmental processes, is tightly regulated, but the mechanisms by which this occurs are not understood.

Some animals hatch or are born in a relatively immature state and acquire many of their organs during post-embryonic development. Examples include the larvae of marine annelids, molluscs and echinoderms. In other groups, such as some insects and amphibia, the larvae are relatively well-developed organisms that are radically different from the mature adult in both body form and lifestyle. In this case, the transformation from larva to adult, metamorphosis, involves dramatic changes: existing

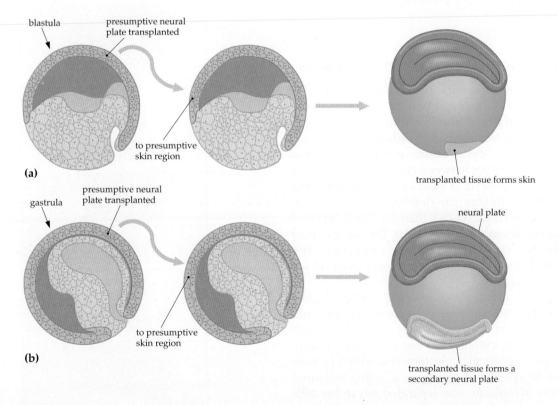

Fig. 14.27 An experiment to demonstrate determination in the frog embryo. **(a)** At the early gastrula stage, the presumptive neural plate tissue is undetermined and will give rise to skin if transplanted to the presumptive skin region of another gastrula. **(b)** If the same experiment is carried out at the late gastrula stage, the presumptive neural plate, which is now determined, gives rise to a secondary neural plate

cells degenerate or are modified, new cells are produced and differentiate, and substantial structural changes occur.

> The later stages of embryogenesis involve growth and maturation as a result of cell proliferation and differentiation, and an increase in extracellular matrix materials.

REGULATION OF DEVELOPMENT

Reproduction almost always produces 'normal' offspring. For example, the frequency of abnormalities in newborn humans is very low. The reason is that developmental events occur sequentially and are very tightly controlled. Just how these processes are regulated has occupied the attention of biologists for well over a century and is today one of the most active research fields in biology.

Cells alter their activities during development in highly reproducible ways because they respond to specific signals emanating either from within the cell or from the extracellular environment. We will consider some of what is known about the nature of these signals, their sources, the timing of events and the nature of the resulting cellular responses, by examining two developmental systems.

Cytoplasmic specification in sea squirts

One of the earliest events during embryogenesis is the segregation of the cytoplasm of the egg cell into cells of the blastula, the blastomeres. Since egg cytoplasm is not homogeneous, different blastomeres inherit different cytoplasmic components, thus setting up differences between cells in different parts of the early embryo. Evidence that **cytoplasmic determinants** influence the developmental pathway taken by the cells that receive them has been obtained in a number of embryonic systems. One of the most intensively studied systems concerns the determination of the muscle cell lineage in larval sea squirts (ascidians).

Early this century, pioneering embryologists observed that clear differences in pigmentation could be seen in different regions of eggs of ascidians, such as *Styela*. These cytoplasmic regions were reliably segregated to particular blastomeres, which in turn gave rise to specific tissue types: a yellow-coloured cytoplasm, **myoplasm**, was separated out to the cells of the mesodermal lineage that produce muscle; a grey crescent portion went to the notochord and neural tube; the clear animal cytoplasm·went to the ectoderm, which became the larval epidermis; and

the grey vegetal region went to the endoderm, which gave rise to the gut (Fig. 14.28). It was suggested that factors associated with the differently coloured cytoplasmic regions might determine cell fate; that is, myoplasm might contain a 'muscle determination' factor. Indeed, if blastomeres containing yellow myoplasm are removed from an embryo early in cleavage and allowed to develop in isolation, they produce cells that differentiate into precisely the same cell type, muscle, that is produced by these blastomeres in the intact embryo (Fig. 14.29). Testing the progeny of isolated myoplasm-containing blastomeres for muscle specific proteins, such as acetylcholinesterase, also shows that these

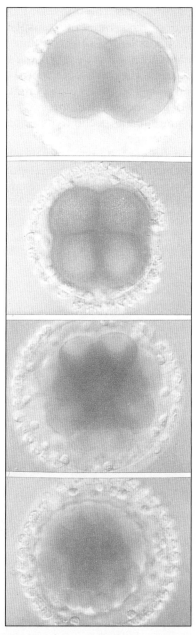

Fig. 14.28 The separation of different cytoplasmic regions in the egg of the ascidian *Styela* to different blastomeres during cleavage

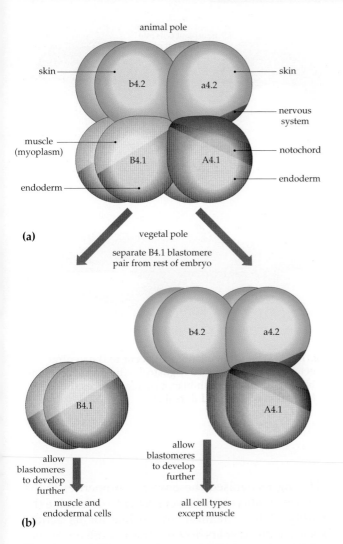

(a)

(b)

Fig. 14.29 (a) The normal fates of different regions of the eight-cell *Styela* embryo. (b) When separated from the rest of the *Styela* embryo, the B4.1 blastomeres give rise to muscle, mesenchyme and endodermal cells; this is the same range of cell types that is generated by these blastomeres in the whole embryo during normal development

blastomeres are determined for a muscle fate very early in development. Most other blastomeres show similar patterns of development, producing the same cell types in isolation that they produce in the whole embryo. The early sea squirt embryo seems to be a mosaic of parts that undergo independent determination and differentiation; thus, this mode of development is known as **mosaic development**.

These findings are consistent with the idea that determination of the muscle lineage is due to the presence of myoplasm. Direct evidence in support of this hypothesis requires a different experiment, in which some myoplasm is relocated to blastomeres that normally have a different fate (such as skin), in another region of the embryo. If the hypothesis is correct, these cells should express proteins characteristic of muscle cells.

This decisive experiment had to wait some 80 years after the original discovery of myoplasm and was carried out in 1982 by J. R. Whittaker, an American biologist. He pressed a glass needle against the vegetal blastomeres of the eight-cell stage *Styela* embryo, displacing some of the myoplasm, which normally is restricted to the vegetal B4.1 blastomere pair, into the animal b4.2 blastomeres. In addition to their normal ectodermal progeny, the animal blastomeres went on to generate muscle cells and expressed the muscle-specific biochemical marker, acetylcholinesterase (Fig. 14.30). Hence the presence of myoplasm imposed a new developmental fate on the animal cells that inherited it.

Many questions remain about the involvement of myoplasm in the specification of muscle cell lineage. The biochemical identity of myoplasm, or indeed any other cytoplasmic determinant, is unknown. We have little idea of how the myoplasm comes to lie in its characteristic position within the egg. Also, as we do not know what biochemical changes take place in

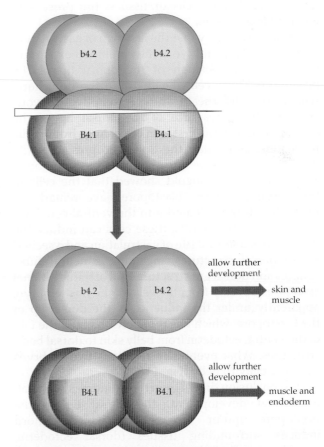

Fig. 14.30 When a *Styela* embryo is pushed with a glass needle during the third horizontal cleavage division, some of the myoplasm in the blastomere B4.1 pair moves into the animal hemisphere blastomeres b4.2. Some of the progeny of these b4.2 cells subsequently adopt a muscle fate

the blastomeres when they become determined for a muscle fate, we can only guess at the role of the myoplasm in this process. Experiments using transcriptional inhibitors suggest that determination of muscle type requires transcription but little is known beyond this fact.

> Cytoplasmic determinants, parcelled out to blastomeres during cleavage, may influence the developmental fate of cells that receive them. Cell determination in ascidian embryos appears to involve such factors.

Primary induction and mesoderm induction in amphibia

Amphibians have long been favoured subjects for embryological research because of the ready availability, accessibility and large size of their eggs and embryos. Spemann and Mangold's study of the determination of the central nervous system in the amphibian embryo, carried out in 1924, provides a classic example of cell fate being regulated by interactions *between* cells or tissues, the process of **embryonic induction**.

We saw previously that the first sign of gastrulation in the amphibian embryo is the involution of a group of cells at the position of the dorsal lip of the blastopore. These cells develop into the **notochord**, which lies underneath the presumptive nervous system. The appearance of the dorsal lip of the blastopore is also the first sign of a change from radial to bilateral symmetry: the dorsal lip prescribes the position of the future dorsal midline of the embryo.

Spemann and Mangold showed that the cells of the dorsal lip of the blastopore have remarkable properties. If transplanted into the ventral ectoderm of another early gastrula, these cells can induce the formation of a second site of gastrulation and a second set of dorsal structures, including a neural tube, somites and dorsal gut structures. All of the additional structures develop from the cells of the host embryo, apparently under the influence of the dorsal lip of the blastopore, which has therefore changed the fate of the ventral ectoderm from belly skin to dorsal body structures. What eventually results is two embryos attached belly to belly (Fig. 14.31).

The conclusion drawn from these findings was that, in normal development, the dorsal lip of the blastopore and/or the presumptive notochord induces surrounding tissues (dorsal ectoderm, mesoderm and endoderm) to form an organised dorsal body side and hence it became known as the **organiser**. The action of the organiser represents an early example of embryonic induction, and hence it was called **primary induction**. As we shall see below, this is a misnomer because it is actually preceded by other inductive events.

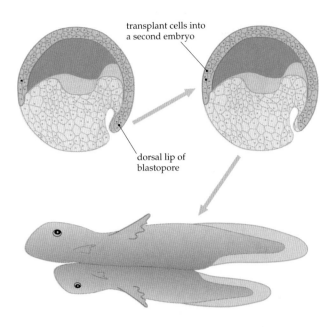

Fig. 14.31 When transplanted adjacent to the ventral ectoderm of another early gastrula, the dorsal lip of the blastopore of an amphibian gastrula induces the formation of a second dorsal axis

Having recognised the powerful properties of the organiser, embryologists were eager to determine the identity of the chemical responsible for its action. Despite some 60 years of effort, there is still no answer to this question. However, considerable progress has been made in understanding the events that lead to its appearance. The organiser has been found to be the product of an earlier inductive interaction between animal and vegetal cells in the equatorial region of the blastula. After removal from the embryo and culture in isolation, animal hemisphere cells produce only epidermis, and vegetal hemisphere cells produce only gut tissue. However, if animal and vegetal hemisphere cells are cultured together, mesodermal tissues, including notochord and muscle, which are normally produced solely from equatorial cells, also appear (Fig. 14.32).

A strong candidate for a mesoderm-inducing factor has recently been identified. This molecule, activin-A, is produced within the blastula of the South African clawed toad, *Xenopus laevis*, and, when injected into a ventral vegetal blastomere, it causes the formation of a second organiser and dorsal axis.

> Determination of dorsal tissues, such as the central nervous system, in the amphibian embryo involves an interaction with a neighbouring tissue, the organiser. The organiser is itself the product of an earlier inductive interaction between animal and vegetal hemisphere cells.

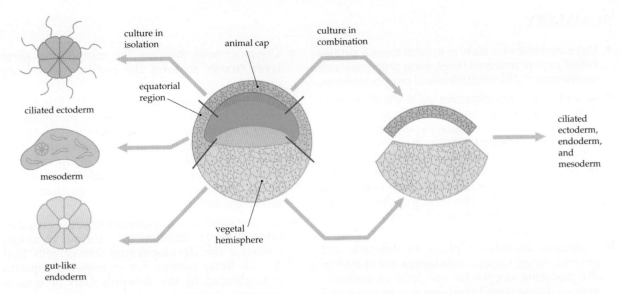

Fig. 14.32 Mesoderm induction in amphibia. When cultured in isolation, tissue from the animal cap, the equatorial region and the vegetal hemisphere of the blastula give rise to ciliated ectoderm, mesoderm and gut-like endoderm, respectively. However, if the equatorial region is removed and presumptive endoderm is cultured in combination with presumptive ectoderm, mesodermal tissue also develops

Cytoplasmic and intercellular signalling

Two cases of developmental control that provide examples of regulatory signals arising from either within or outside the cell have been considered. Factors within the egg cytoplasm can play an important role in the early determination of the fate of blastomeres that inherit those factors. As the embryo develops and new anatomical relationships between cells arise, extracellular factors (intercellular induction) take on an increasing role in controlling development.

In most animals, both modes of signalling are used at some stage of embryogenesis, although their relative importance varies across different animal phyla. Cytoplasmic specification is dominant in ascidians, nematodes, annelids and molluscs, which show mosaic development. Intercellular signalling is more important in mammals. Even at the four- and eight-cell stage, mammalian blastomeres are **totipotent**, they retain the ability to form an entire embryo. Mammalian embryos show **regulative development**; they can develop normally, despite the removal of some cells at an early stage in development.

> Factors within the egg cytoplasm play an important role in early development while extracellular factors become increasingly important as development proceeds.

SUMMARY

- Development of a mature animal from a single-celled zygote involves three main processes: cell proliferation, differentiation and morphogenesis.

- Gametes, the reproductive cells which fuse to form a zygote, are products of a sequence of events that includes migration of germ cells to the gonads; repeated mitotic divisions of germ cells to generate a stockpile of potential gametes; meiotic divisions to halve the chromosome number; and maturation to produce specialised egg and sperm cells. The timing of these events varies between species and between the sexes of any one species.

- Oogenesis includes a phase of dramatic cell growth, when various substances are stored in the maturing oocyte for use later in embryogenesis. These stored products may be produced within the oocyte itself or by surrounding maternal cells.

- Fertilisation involves penetration of the egg cell plasma membrane by the sperm, which activates egg cell metabolism and initiates embryogenesis, and fusion of the sperm and egg pronuclei, which restores the diploid number of chromosomes.

- Cleavage is a rapid series of mitotic cell divisions which restores the nucleus:cytoplasmic ratio and forms the blastula. The pattern of cleavage is affected by the amount of yolk in the egg.

- Gastrulation involves rearrangement of blastomeres to form the gastrula, which has three germ layers (ectoderm, mesoderm and endoderm), new body cavities (archenteron and, if present, coelom) and, in many cases, the basic axes of body symmetry.

- Organogenesis involves the assembly of germ layer tissues to form the rudiments of body organs.

- Gastrulation and organogenesis involve a limited set of morphogenetic mechanisms, including: thickening and folding of epithelial sheets; disaggregation of tissues into individual cells that migrate to new sites in the embryo; localised cell proliferation; and localised cell death. Movement of cells involves changes in cell shape and in cell adhesion.

- In some animals, cytoplasmic factors, segregated into different blastomeres during cleavage, influence the developmental fate of cells that receive those factors. For example, myoplasm is implicated in the determination of muscle tissue in sea squirts.

- Determination of dorsal tissues in the amphibian embryo, such as the central nervous system, involves an interaction with a neighbouring tissue, the organiser. The organiser and mesoderm are the products of an earlier inductive interaction between animal and vegetal hemisphere cells.

- In general, factors within the egg cytoplasm play an important role in early development and extracellular factors become increasingly important as development proceeds. Cytoplasmic specification is dominant in ascidians, nematodes, annelids and molluscs, which show mosaic development. Intercellular signalling is paramount in mammals, which have totipotent blastomeres and regulative development.

QUESTIONS

1. **(a)** Explain why mutations that take place in muscle cells during the life of an organism are not passed down to the next generation. **(b)** If a mutation were to occur in a single oogonium that had just completed its last mitotic division, how many egg cells would carry the mutation? **(c)** Answer part **(b)** for a spermatogonium.

2. What are the roles of **(a)** follicle cells and **(b)** nurse cells in oogenesis?

3. **(a)** List four consequences of cleavage. **(b)** In what way do cleavage divisions differ from conventional mitotic divisions?

4. Explain why the blastomeres in the animal hemisphere of the frog blastula are smaller than those in the vegetal hemisphere.

5. 'Gastrulation usually establishes the basic body plan of an animal.' Illustrate this statement with reference to sea urchin and frog embryos.

6. List the four principal mechanisms of morphogenesis in animals and give an example of each.

7. **(a)** How are microfilaments involved in the folding of epithelial sheets? **(b)** How do changes in cell shape and cell adhesion underly the morphogenetic movements of neural crest cells?

8. **(a)** Explain what is meant by mosaic development. **(b)** Why does this phenomenon provide evidence for a role for cytoplasmic factors in cell determination?

9. Outline a direct experimental test of the hypothesis that a cytoplasmic factor such as the myoplasm is directly involved in determination of cell fate.

10. Explain the term 'induction' and give two examples of this phenomenon in amphibian development.

11. Using mammalian development as an example, explain **(a)** totipotency and **(b)** regulative development.

PLANT DEVELOPMENT

During reproduction, plants, like animals, pass through single-celled stages in their life cycles. In animals, gametes are produced directly by meiosis but, in plants, meiosis produces haploid spores, which then develop into male or female gametophytes (Chapter 13). In flowering plants, the male gametophytes are pollen grains, which are physically transported to the female by air, water or animal vectors. Pollen lands on the stigma and the pollen tube transfers the male gametes to the female gametophyte, the embryo sac, contained within an ovule. Within a short period, fertilisation occurs and the ovule develops into a seed containing the embryo. The seed is the dispersal unit. In this chapter, we will examine each of these developmental stages in flowering plants.

Growth and development of a flowering plant involves repeated mitotic divisions of the single-celled zygote to produce new cells, differentiation of these cells into the diverse types found in the sporophyte, and morphogenesis (ordered assembly) of these cells into the three-dimensional arrangements of mature organs of the sporophyte. These developmental processes are brought about by the switching on and off of specific genes in the nuclei of different cell types (Chapter 16), as part of specific programs of development, and result in the formation of different organs such as the shoot or root. Plants are usually able to produce new organs throughout their life. Plant cells retain the ability to produce a complete new plant. This property of totipotency is the basis of plant tissue culture and genetic engineering technologies.

is no evidence for the existence of such a germ line in flowering plants. On the contrary, cells of the sporophyte form the reproductive organs and, within each organ, reproductive cells are determined by their central position. These cells undergo meiosis to form haploid spores, which then divide mitotically to produce the male and female gametophytes. In order to facilitate comparison with the processes in animals (Chapter 14), gametogenesis is applied here in its widest sense to include the development of both the gametophyte and gametes.

The reproductive organs of flowering plants occur within the **flower**. Flowers are short shoots with transformed 'floral leaves', which grow into four whorls (Chapter 16). The two outer whorls are the sepals, which are leaf-like, green and leathery but can be reduced to scales or may even be absent in some plants (Figs 15.1, 15.2), and the petals, which are usually brightly coloured and scented. Some petals have honey guides, lines that direct insects and other animal pollinators to nectary glands, which are sited on the floral axis (receptacle). Floral nectaries produce nectar (Figs 15.1, 15.2), a liquid secretion containing sugars, amino acids and other rewards for floral visitors.

The two inner whorls comprise the reproductive organs, the whorl of stamens and of pistils. Stamens are the male reproductive organs and consist of anthers borne on slender filaments. The pistil is the female reproductive organ, which lies centrally on the receptacle (Fig. 15.1). The whorls of sepals and petals surround and protect the reproductive organs. The evolution of the flower and the diversity of modern flowering plants is described later in Chapter 37.

THE FORMATION OF GAMETES: GAMETOGENESIS

You will recall that gametogenesis in animals involved a special line of germ cells, which are distinct very early in development and whose exclusive function is to undergo meiosis and form the gametes. There

Development of the male gametophyte

Within the anther, the male program of reproduction takes place. Meiosis and subsequent mitotic divisions lead to the development of the male gametophyte and differentiation of male gametes.

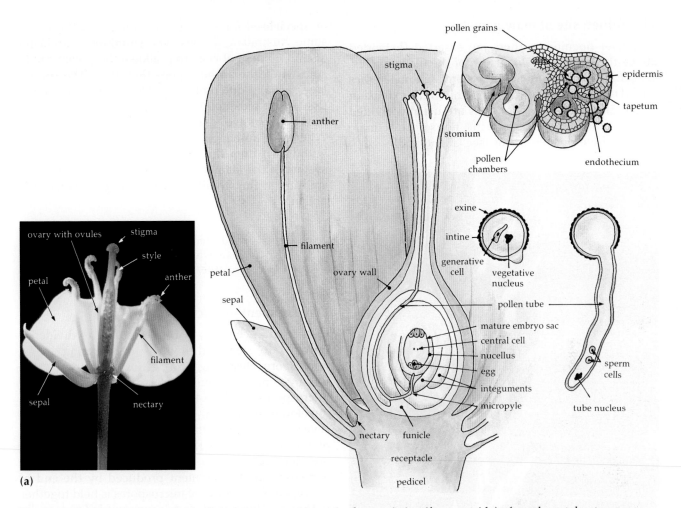

Fig. 15.1 The principal organs of a flower are arranged in whorls, consisting (from outside) of sepals, petals, stamens (comprising filament and anther, which contain numerous pollen grains) and pistil (comprising stigma, style and ovary, which contain rows of many ovules). (**a**) Longitudinal section of a flower of oilseed rape *Brassica napus*. (**b**) Diagram of longitudinal section of flower. (**c**) Cross-section of an anther, showing the pollen grains and tissues of the wall, including tapetum, endothecium and epidermis, as well as the stomium which splits when the anther opens. (**d**) Mature pollen grain with pair of sperm cells within the vegetative cell. (**e**) Germinating pollen grain with pollen tube containing a pair of sperm cells and tube nucleus

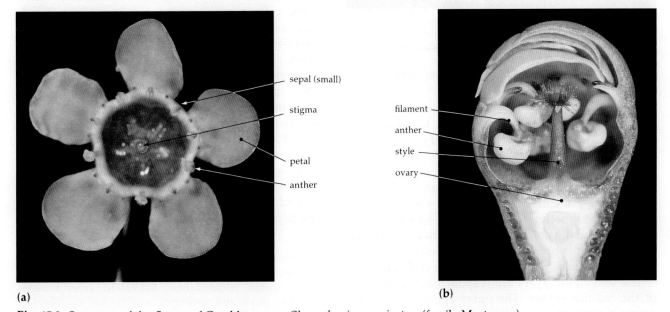

Fig. 15.2 Structure of the flower of Geraldton wax, *Chamaelaucium uncinatum* (family Myrtaceae)

The anther: site of male gametes

At flower opening, anthers are exserted (extended above the petals) by elongation of the filament (Fig. 15.3). Each anther contains two elongated pollen sacs joined together by a connective, which is an extension of the filament. The pollen sacs are usually divided into two chambers, which house the male gametophytes, pollen grains (see Chapter 13).

Fig. 15.3 Flowers of the rye-grass *Lolium perenne* exsert their anthers when flowers are open. The green glumes part and the anthers elongate on their slender filaments, so that the anthers hang from the flower. Pollen is released from slits at the base of anthers into air currents

The anther wall typically contains four layers of tissues when viewed in cross-section (Fig. 15.1c). The innermost layer is the **tapetum**, which forms the nurse tissue for the developing male gametophytes. Tapetal cells, usually binucleate, are metabolically very active and secrete enzymes, nutrients and wall materials. Usually, the tapetum is composed of secretory cells that surround the pollen sac. Outside the tapetal layer is a layer that is prominent in the developing anther and then usually degenerates during development.

The outer two layers are responsible for opening of the mature anther. The outer epidermis usually serves a protective function but can be pigmented

or specialised for scent production, as in flowers of some Australian wattles. The **stomium** is a cluster of thin epidermal cells that allows the pollen sac to split when the anther opens (Fig. 15.1). The second layer usually differentiates into the **endothecium**, a tissue that aids splitting of the pollen sac. Its cells possess massive, bar-like secondary thickenings. When the anther is ready to split open, water is withdrawn from the endothecium into the vascular system of the filament. This results in the pollen sac splitting at the stomium for subsequent dispersal of pollen.

> Anthers are specialised for differentiation, nutrition, protection and dispersal of pollen. The tapetum has a nutritive function, while endothecium and stomium are responsible for anther opening.

Development of microspores

Cells located centrally within each anther divide by mitosis to form diploid spore-forming cells, **microsporocytes** (Fig. 15.4). Most of these will undergo meiosis, giving rise to tetrads of haploid **microspores** (Fig. 15.4), unicellular structures that develop into pollen grains. Microspores undergo several changes. At the beginning of meiosis, most of the organelles in the microsporocyte cytoplasm are destroyed, with new organelles required for gametophyte development produced by the end of meiosis II. Each tetrad of microspores is held together within a callose wall and is surrounded externally by the tapetum. Callose has different structural properties from cellulose, although both are polymers of glucose. Callose prevents the passage of molecules larger than sugars into microspores, thus sealing off microspores from parental influences.

The end of meiosis is signalled by the dissolution of the callose wall, carried out by callase (1,3 β-glucanase), an enzyme secreted from the tapetal cells. This releases the microspores into the pollen sac. After release, the microspores become spherical and commence their major growth period. Microspores show exponential growth, increasing in volume about three times before growth levels off at maturation. This major growth period is marked by the appearance of a large vacuole within the microspore. The nucleus and cytoplasm become peripheral, giving the microspores the characteristic appearance of a 'signet ring' (Fig. 15.4a). The microspore is unicellular until the end of the growth period (vacuolation), when typically the first mitosis occurs, forming the pollen grain.

> Microspores are formed after meiosis. Each tetrad is enclosed in a callose wall. At the end of meiosis, microspores, released from the tetrad, enter a major growth phase and develop a large central vacuole. The microspore becomes a pollen grain at first mitosis.

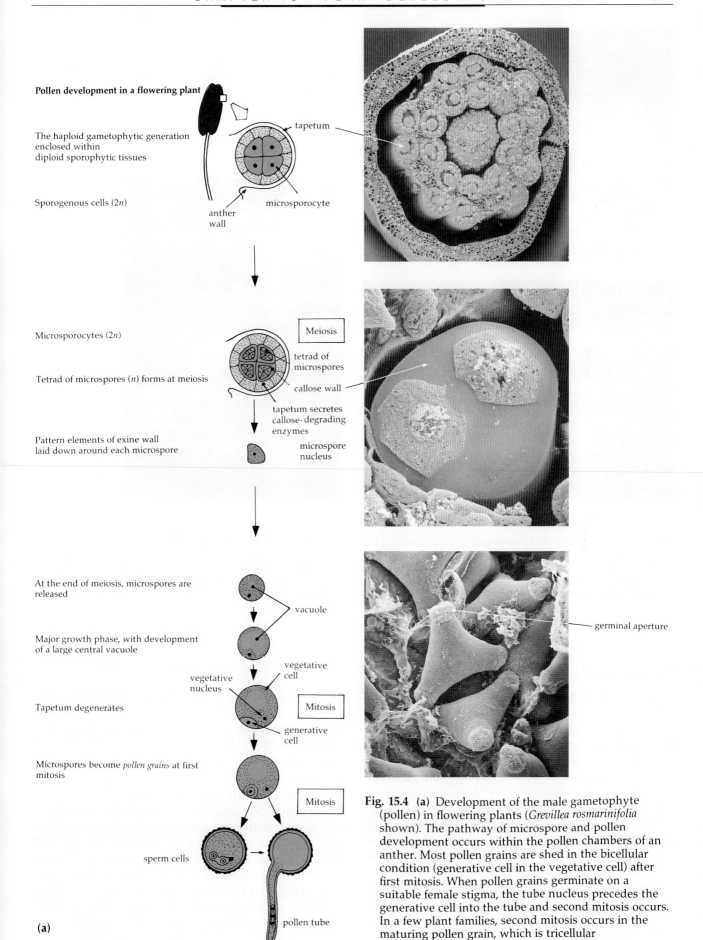

Pollen development in a flowering plant

The haploid gametophytic generation enclosed within diploid sporophytic tissues

Sporogenous cells (2*n*)

anther wall

tapetum

microsporocyte

Microsporocytes (2*n*)

Meiosis

tetrad of microspores

callose wall

Tetrad of microspores (*n*) forms at meiosis

tapetum secretes callose-degrading enzymes

Pattern elements of exine wall laid down around each microspore

microspore nucleus

At the end of meiosis, microspores are released

vacuole

Major growth phase, with development of a large central vacuole

vegetative cell

vegetative nucleus

Mitosis

Tapetum degenerates

generative cell

Microspores become *pollen grains* at first mitosis

Mitosis

sperm cells

pollen tube

germinal aperture

(a)

Fig. 15.4 (a) Development of the male gametophyte (pollen) in flowering plants (*Grevillea rosmarinifolia* shown). The pathway of microspore and pollen development occurs within the pollen chambers of an anther. Most pollen grains are shed in the bicellular condition (generative cell in the vegetative cell) after first mitosis. When pollen grains germinate on a suitable female stigma, the tube nucleus precedes the generative cell into the tube and second mitosis occurs. In a few plant families, second mitosis occurs in the maturing pollen grain, which is tricellular

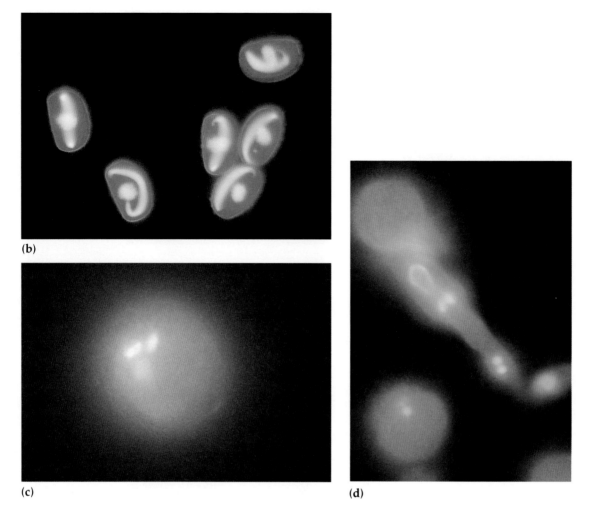

Fig. 15.4 **(b)** Bicellular *Tradescantia* pollen treated with a DNA fluorochrome so that the nuclei fluoresce blue, showing elongate generative nuclei and spherical vegetative nuclei. **(c)** Tricellular pollen of oilseed rape treated with the same fluorochrome, showing bright blue sperm nuclei and fainter vegetative nucleus. **(d)** The pollen tube of oilseed rape transports the pair of sperm cells and tube nucleus within the living cytoplasm at the tube tip. The pair of sperm nuclei and tube nucleus at the tip fluoresce blue, while yellow fluorescence shows the callose wall sealing off the tip from the rest of the grain

Formation of pollen

Pollen is formed when microspores undergo an asymmetric mitosis, cutting off a small **generative cell** from a larger **vegetative cell** (Fig. 15.4a). At first, the generative cell is attached to the perimeter wall but becomes cut off and centrally sited within the vegetative cell, as a *cell within a cell*. Pollen maturation then proceeds with loss of the vacuole and accumulation of storage reserves of lipid, carbohydrate and protein. Mature pollen grains of most species are shed from the anther in this *bicellular* condition (Fig. 15.4b). When pollen lands on a stigma surface, a pollen tube forms. The nuclei and cytoplasm of the pollen grain enter the tube. The generative cell divides within the pollen tube tip to produce two sperm cells (Figs 15.4a, c). The nucleus of the former vegetative cell has become the **tube nucleus**,

which regulates pollen tube function. In other cases, pollen is shed in the *tricellular* condition, where mitotic division of the generative cell occurs in the maturing pollen grain (Fig. 15.4b).

In a pollen grain and tube, sperm cells are usually linked to each other and to the vegetative (tube) nucleus to form the **male germ unit** (Fig. 15.5). The two sperm cells come to lie adjacent to each other, end to end, within a common sac, the inner pollen tube plasma membrane. One or both sperm cells possesses extensions of the plasma membrane that surrounds or penetrates the tube nucleus (Fig. 15.5). This arrangement may ensure transfer of male DNA along the cytoskeleton of the pollen tube tip. The unit may also lead to the correct physical placement of sperm cells for double fertilisation (p. 317).

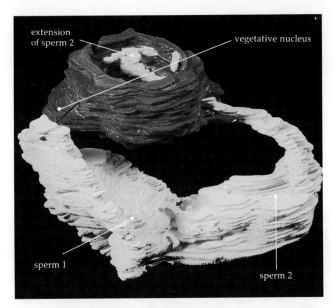

Fig. 15.5 The male germ unit provides a means for sperm transport in the pollen tube. Three-dimensional reconstruction of a male germ unit in a pollen grain of oilseed rape, *Brassica campestris*. The pair of sperm cells are linked together and one has a long extension that penetrates the vegetative nucleus in mature pollen

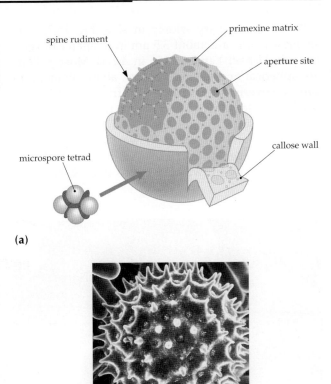

(a)

(b)

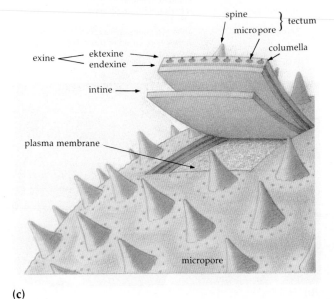

(c)

Fig. 15.6 The primexine establishes pollen wall pattern formation. (**a**) Tetrad of microspores of the morning glory, *Ipomoea*, dissected away to reveal the surface pattern of the developing microspore, showing the pattern of the primexine matrix, spine rudiments and sites of germinal apertures. (**b**) Scanning electron micrograph of mature pollen of *Ipomoea* showing the prominent spines encircling the many germinal apertures (pores). (**c**) Typical two-layered exine showing roof layer, tectum, supported by rod-like columellae over a foot layer, endexine, and the smooth inner intine layer

The pollen grain wall

The *primexine* is a wall laid down external to the microspore plasma membrane but within the callose wall (Fig. 15.6). It is a matrix within which the intricate pattern of the future pollen wall, the **exine**, is initiated. This pattern is due to the outer *exine* wall and the number and location of *germinal apertures*, which form slits or circular openings (pores) in the exine (Fig. 15.6). The exine is made of **sporopollenin**, a polymer that is tough and plastic-like and extraordinarily resistant to biodegradation. There is no known enzyme in flowering plants that can degrade sporopollenin. Chemically, sporopollenin resembles lignin (Chapter 6) and is found only in exines and spore walls of certain algae and fungi. The exines preserved in fossil deposits provide valuable clues to the history of plant life in past geological eras (Chapters 31 and 41).

The inner layer of the pollen grain wall is the **intine**, made largely of pectic polysaccharides (Fig. 15.6c). This layer meets the surface of the grain only at the germinal apertures. When pollen germinates, the intine bulges out and the pollen tube grows out by tip extension from one of the apertures. The pollen tube wall comprises an outer pectic layer and an inner callose layer.

Pollen grains vary widely in size (Fig. 15.7). The smallest grains are about 3.5 μm in diameter (forget-me-not, *Myosotis*) and 6.5 μm in *Acacia*. Most pollens are spherical and average in size, about 35 μm, and can be covered in a variety of intricate and group-specific patterns. However, some range in size up to about 300 μm in pumpkin and the Turk's cap, *Malvaviscus*, the national flower of Hawaii. An Australian seagrass, *Amphibolis*, holds the record for the longest pollen, which is more than 3000 μm in length and filamentous with hooks at the ends (Fig. 15.7). Some pollen grains are compound, for example, occurring as tetrads in which the four microspores of a meiotic tetrad do not separate from each other at the end of the meiosis (Fig. 15.7).

Most pollen is bicellular, containing a generative cell enclosed in a vegetative cell. The generative cell divides to form two sperm cells after germination within the growing pollen tube. In tricellular pollen, the generative cell divides within the pollen grain.

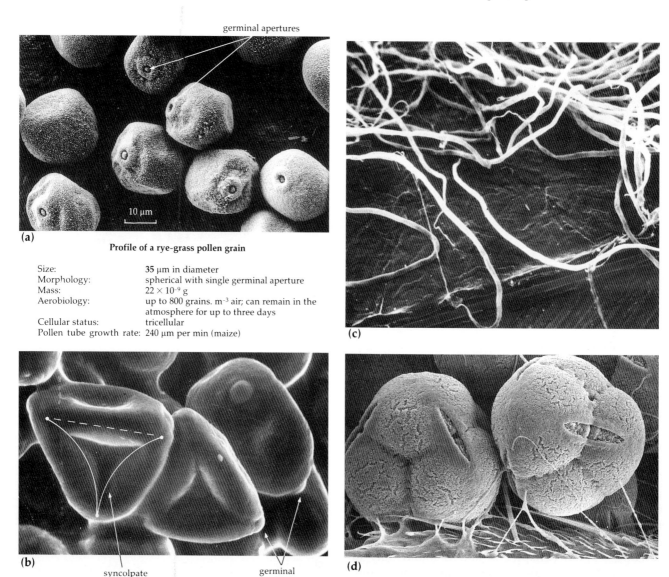

Profile of a rye-grass pollen grain

Size:	**35** μm in diameter
Morphology:	spherical with single germinal aperture
Mass:	22×10^{-9} g
Aerobiology:	up to 800 grains. m^{-3} air; can remain in the atmosphere for up to three days
Cellular status:	tricellular
Pollen tube growth rate:	240 μm per min (maize)

Fig. 15.7 Morphological features of different types of pollen. **(a)** Scanning electron micrograph of rye-grass pollen grains showing spherical pollen with a single aperture (pore), some covered by a cap of exine. **(b)** Triporate pollen of *Eucalyptus* (family Myrtaceae) adheres together by pollen coat materials in the open anther. Three apertures occur in slits (colpi) at each corner of the triangular grains. Eucalyptus grains have a characteristic furrow pattern linking the colpi (syncolpate) (magnification 1600×). **(c)** Filamentous pollen of the Australian sea nymph, *Amphibolis* (family Cymodoceaceae), which is 3200 μm in length and lacks an exine, so that it can bend in seawater currents. **(d)** Structure of compound pollen grains, in which all the microspores in a meiotic tetrad remain joined together at maturity. Pollen grains of *Rhododendron* (family Ericaceae) adhering by means of sporopollenin threads to hairs on the legs of a honey bee. These grains are compound, held in tetrads containing the meiotic products of a single microsporocyte and, in this case, are arranged rhomboidally with one grain overlying three others. The slit-like germinal apertures form across adjacent grains

BOX 15.1 GRASS POLLEN AEROBIOLOGY, HAYFEVER AND ALLERGIC ASTHMA

Grass flowers open soon after sunrise and again in the afternoon. Their golden pollen is released into air currents for dispersal. Spore traps show that grass pollen is the major single component in the air in spring and early summer in cool temperate climates such as Melbourne (Fig. a). Studies of Melbourne schoolchildren who are sensitive to grass pollen show that asthma attacks occur coincidentally with the highest peaks of grass pollen.

Rye-grass *Lolium perenne* pollen is the most abundant pollen of all grasses. It produces two different proteins that trigger hayfever and allergic asthma in sensitive individuals (allergens, see Chapter 23). The function of pollen allergens has only just begun to be understood through the isolation and cloning of genes encoding these proteins (Chapter 12) at The University of Melbourne. The genes can be sequenced and the triplets of nucleotides used to predict the amino acid sequence of the allergen.

The major allergen has proved to be a 263 amino acid protein, with a coding region whose sequence begins at the N-terminus: **I-A-K-V-P** . . . (in the single letter amino acid code, Chapter 1). The protein has an 'export' signal sequence, as seen in secreted enzymes such as amylase, suggesting that this protein is targeted to the surface of pollen grains. Pollen grains impact on the eyes and nose, where they trigger the symptoms of hayfever.

A second protein that has proved to be a major allergen has also been isolated. This allergen differs in its N-terminal amino acid sequence, which begins **A-D-A-G-Y** . . . and shows no similarities with other known proteins in its amino acid sequence. This protein has a signal sequence suggesting that it is targeted to chloroplasts. Antibody labelling has confirmed that this allergen is located in starch granules produced in amyloplasts (a type of chloroplast, Chapter 3).

The implications of finding a potent allergen in starch granules are important. When placed in rainwater, grass pollen is known to burst by osmotic shock (Fig. b). In this way, rain causes the release of allergen-containing starch granules from grass pollen into the atmosphere, with each pollen grain releasing hundreds of granules. During the grass pollen season, air samples in Melbourne have a staggering 54 000 granules in each cubic metre of air on days after rainfall, compared with only 1000 on sunny days. These starch granules are small enough (0.5–2.5 μm in diameter) to be respirable. They can enter the bronchi, where they may trigger allergic asthma. Wheezing is caused by constriction of the bronchi (Fig. c).

In fact, epidemics of asthma sweep south-eastern Australia in November of most years within 24 hours after major thunderstorms. In 1989, there was a 10-fold increase in the number of asthmatics admitted to Melbourne hospitals after such a thunderstorm and laboratory studies of a sample of patients showed that all were sensitive to grass pollen. When lung function was tested with aerosols containing isolated starch granules from rye-grass pollen, these same patients showed bronchial constriction in every case.

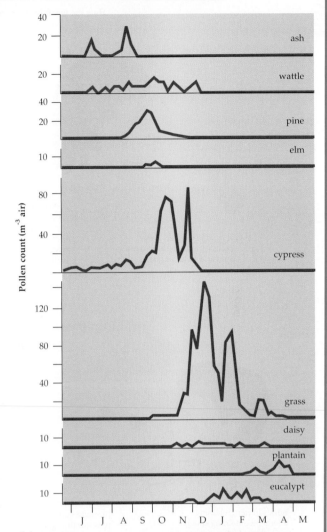

(a)

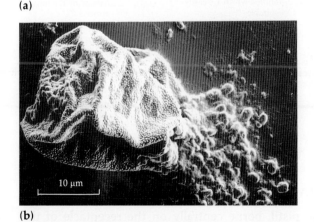

(b)

Association of atmospheric pollen content with hayfever and allergic asthma. **(a)** Pollen calendar for a cool temperate city (Melbourne) showing the seasonal progression of different types of pollen detected in spore traps. Many of the types are exotic to Australia, including ornamental trees and agricultural grasses. **(b)** Grass pollen grains burst by osmotic shock in rainwater, releasing many hundreds of starch granules

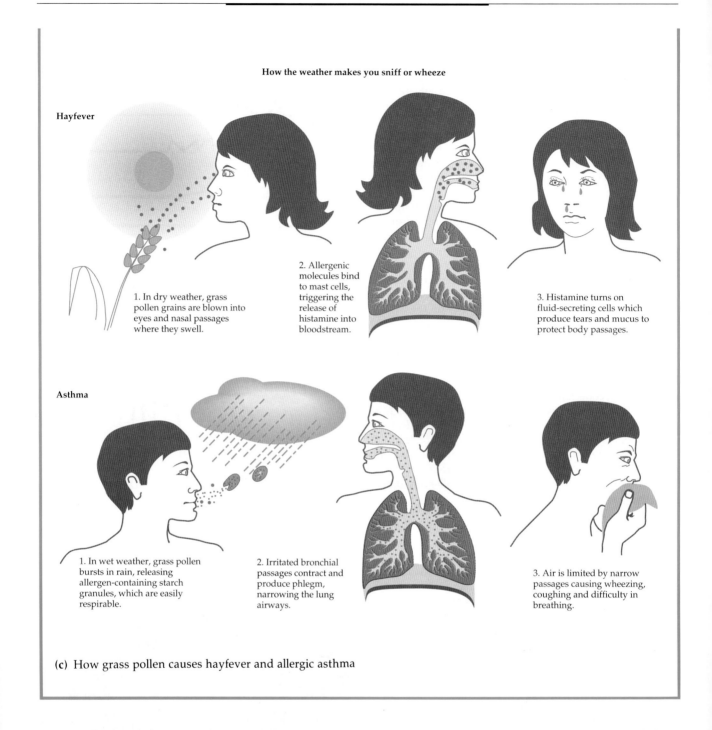

How the weather makes you sniff or wheeze

Hayfever

1. In dry weather, grass pollen grains are blown into eyes and nasal passages where they swell.

2. Allergenic molecules bind to mast cells, triggering the release of histamine into bloodstream.

3. Histamine turns on fluid-secreting cells which produce tears and mucus to protect body passages.

Asthma

1. In wet weather, grass pollen bursts in rain, releasing allergen-containing starch granules, which are easily respirable.

2. Irritated bronchial passages contract and produce phlegm, narrowing the lung airways.

3. Air is limited by narrow passages causing wheezing, coughing and difficulty in breathing.

(c) How grass pollen causes hayfever and allergic asthma

Development of the female gametophyte

The **pistil**, borne centrally on the receptacle of the flower, receives and screens pollen and houses one to many female gametophytes. A pistil comprises stigma, style and ovary (Fig. 15.1). A pistil can possess one to several *carpels*—leaf-like units containing ovules. Carpels can be single and separate (in which case a pistil is a carpel) or joined together (Chapter 37).

The **stigma** is at the tip of the pistil and its epidermal cells may be smooth or form elongate hairs, papillae, which serve to trap pollen and begin the screening process. A receptive stigma is one that is ready for pollination and the papillae, if present, are fully expanded. The stigma surface may be either dry or covered by a copious exudate (Fig. 15.8). Dry-type stigmas are usually covered with a membrane-like sticky layer, the pellicle. The pellicle has several functions in the interaction between pollen grains and stigma (see p. 315).

stigma papillae

style

(a)

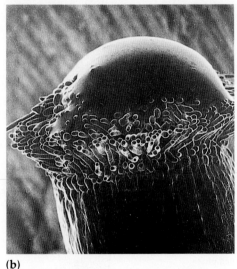

(b)

Fig. 15.8   Scanning electron micrographs of stigmas. **(a)** Dry-type stigmas of *Gladiolus* showing the row of papillae along the edge of the style. **(b)** Wet-type stigmas of *Eucalyptus*, showing pollen grains germinating in the pool of exudate that covers the papillae

Beneath the stigma is the **style**, which forms a transmitting pathway or, in some cases, a central stylar canal, for pollen to grow to the ovary. The ovary at the base contains the ovules, each comprising a central nucellus (nurse tissue) within which the female gametophyte will develop (Fig. 15.9). During development, a pair of integuments usually forms, enveloping and protecting the nucellus: the inner and outer integuments. The micropyle, a funnel-shaped entrance to the female gametophyte, develops at the tip of the ovule (Fig. 15.9). The micropyle is located at the side of the egg-shaped ovule, as in some eucalypt **hemitropous ovules** (Fig. 15.9). Commonly, the ovule then bends back on itself and the micropyle comes to lie adjacent to the funicle when mature, as in **anatropous** ovules such as passionfruit.

Within each ovule, a single diploid cell that lies centrally in the nucellus enlarges to form the diploid spore-forming cell, the **megasporocyte** (Fig. 15.10). The megasporocyte divides by meiosis, producing four haploid megaspores. These are located in a row that stretches towards the micropyle. Commonly, the three megaspores nearest the micropyle abort and the remaining megaspore is the first cell of the gametophyte generation (Fig. 15.10). In animals also, meiosis produces only one oocyte (Chapter 14). In a few plants, such as the lily, all four megaspore nuclei participate in formation of the female gametophyte.

The female reproductive organ, the pistil, bears a stigma to receive and screen pollen grains, a style with transmitting tissue or central canal for growth of pollen tubes and a basal ovary containing ovules.

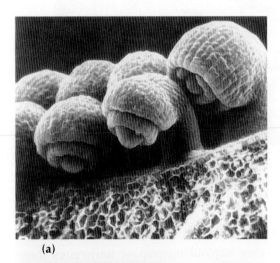

(a)

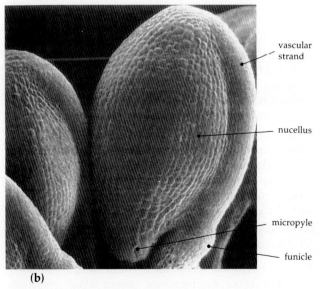

vascular strand

nucellus

micropyle

funicle

(b)

Fig. 15.9 **(a)** Developing and **(b)** mature ovules of *Passiflora* (passionfruit), in which the micropyle is sited at the base of the ovule (anatropous)

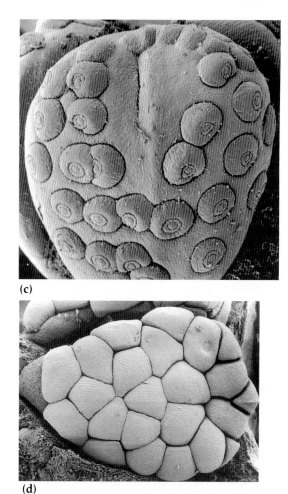

(c)

(d)

Fig. 15.9 Scanning electron micrographs of
(c) developing and (d) mature ovules of *Eucalyptus*, in
which the micropyle is sited on the side of the ovule

Embryo sac formation

The surviving haploid megaspore undergoes three
rounds of mitotic divisions to form a female
gametophyte, the **embryo sac**, within each ovule
(Fig. 15.10). Cell walls then develop, dividing the
eight-nucleate embryo sac into a seven-celled female
gametophyte. The **egg cell**, the female gamete, occurs
centrally at the micropylar end (Fig. 15.10). Two
synergids, which function to receive the pollen tube,
lie one on either side of the egg and three nutritive
antipodal cells lie at the opposite end of the sac. A
central cell is formed when one nucleus from the
micropylar end and another from the opposite end
associate as **polar nuclei** and later fuse to form a
diploid nucleus. The central cell is the largest in the
embryo sac, occupying its mid-portion. This is the
condition when the embryo sac is mature, awaiting
fertilisation.

> The embryo sac is formed by three mitotic divisions,
> producing an eight-nucleate (seven-celled) structure.
> This female gametophyte develops an egg, two
> synergids, central cell (with two polar nuclei) and three
> antipodal cells.

Development of embryo sac

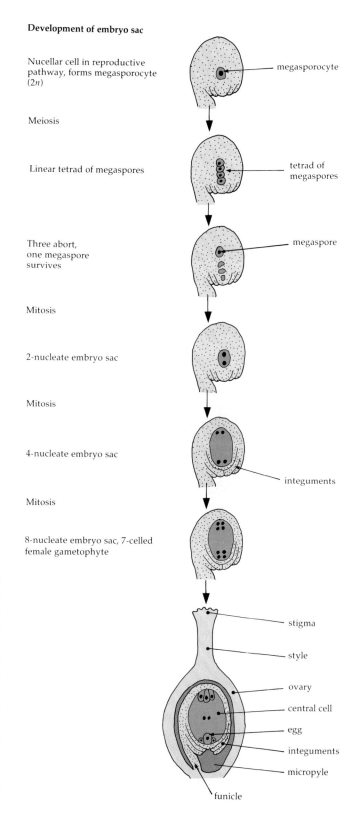

Nucellar cell in reproductive
pathway, forms megasporocyte
(2*n*) — megasporocyte

Meiosis

Linear tetrad of megaspores — tetrad of megaspores

Three abort,
one megaspore
survives — megaspore

Mitosis

2-nucleate embryo sac

Mitosis

4-nucleate embryo sac — integuments

Mitosis

8-nucleate embryo sac, 7-celled
female gametophyte

stigma
style
ovary
central cell
egg
integuments
micropyle
funicle

Fig. 15.10 Pathway of development of the female
gametophyte in flowering plants and the meiotic and
mitotic divisions involved

FERTILISATION

Pollen grains alight on a stigma and set in train a series of interactions that, if successful, may lead to fertilisation (Fig. 15.11). In the case of dry-type stigmas, such as grasses, pollen grains attach directly to the stigma pellicle. If chemical recognition occurs, fluid passes out from the stigma into the pollen grain. As a consequence, the pollen grain becomes hydrated and swells in volume. Finally, the grain germinates by producing a pollen tube through a germinal aperture. In the case of wet-type stigmas, such as eucalypts (Fig. 15.8b), the pollen grains become immersed in the stigma exudate that surrounds the papillae, where they swell and germinate.

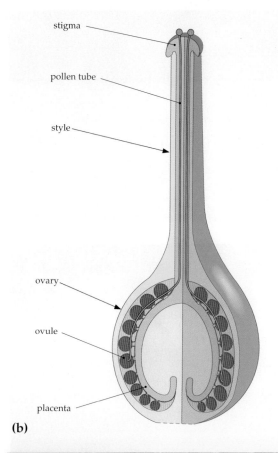

(b)

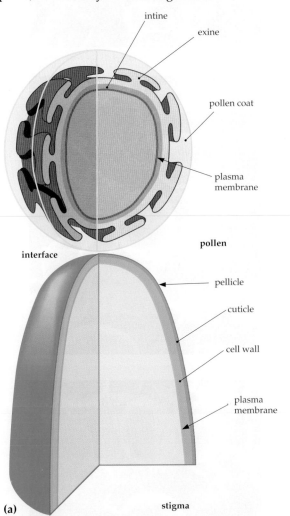

(a)

(c)

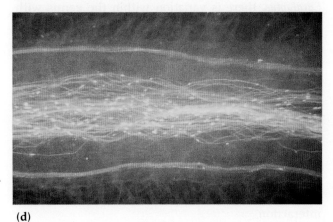

(d)

Fig. 15.11 Events of pollen–pistil interactions. **(a)** When the pollen grain arrives at the stigma surface (dry types), surface informational molecules of the pollen coat interact with the stigma surface. The pellicle initiates signals which permit *attachment* and release of fluid for pollen *hydration* in successful pollinations. **(b)** Pathway of pollen tube in pistil of wild tomato (dry type). **(c)** Fluorescence micrographs showing pollen tubes in pistil of wild tomato after compatible pollination. Pollen grains have germinated and tubes pass through the stigma (red zone). **(d)** Passage of tubes through part of the style

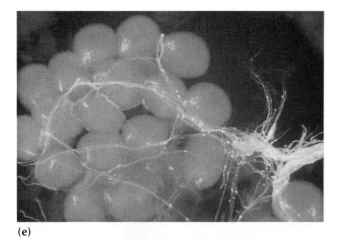

(e)

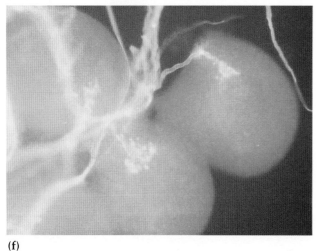

(f)

Fig. 15.11 (e) Arrival of tubes in the ovary and passage of rope of tubes passing ovules (red spheres). (f) Entry of a pollen tube into each ovule via a micropyle

(a)

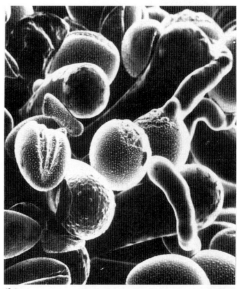

(b)

Screening of foreign pollen

The fate of foreign pollen varies enormously depending on the plant group. Species in some genera such as *Grevillea* hybridise readily, whereas others such as *Banksia* hybridise rarely. In large genera such as *Eucalyptus*, there is variation within the genus with some species crossing readily and others failing completely.

The pollen grain interacts with the pistil at fertilisation (Fig. 15.11). At each step of the interaction, the pistil can screen out and arrest development of incompatible pollen and pollen tubes. Foreign pollen alighting on a stigma often fails even to germinate because of a mismatch of stigma fluids and remains dehydrated on the stigma surface (Fig. 15.12). In some cases, pollen may germinate, producing a pollen tube that is unable to penetrate the stigma surface and is arrested at this early point in the interaction.

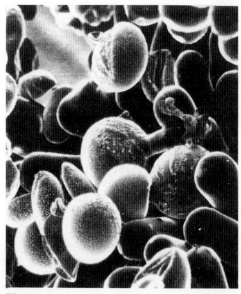

Fig. 15.12 (c)

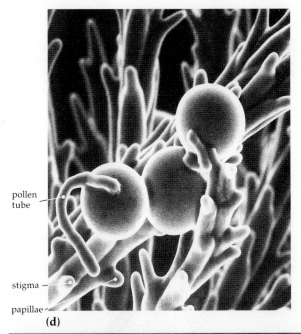

(d)

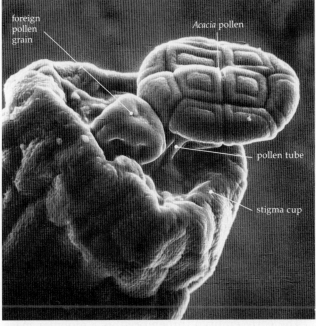

(e)

Fig. 15.12 Screening of foreign pollen on stigmas of *Gladiolus*. **(a)** Foreign pollination: lily pollen on *Gladiolus* stigmas (different families), showing that the lily pollen remains dehydrated. **(b)** Intergeneric pollination: *Crocosmia* pollen on *Gladiolus* stigmas (both members of family Iridaceae) showing *Crocosmia* pollen hydrates and germinates but pollen tubes are unable to penetrate stigma surface. **(c)** Compatible pollination: *Gladiolus* pollen on *Gladiolus* stigmas, showing pollen germinating and tubes penetrating stigma surface. **(d)** Grass pollen germinating and producing pollen tube on feather-like stigma papillae. **(e)** Compound disk-like pollen of *Acacia* germinating in the stigma cup, also containing a foreign pollen grain (dehydrated). The *Acacia* pollen grains have germinated and pollen tubes can be seen emerging from under the disk, and penetrating the stigma

Successful hybrid crosses have pollen tubes that penetrate the stigma and enter the ovules, whereas many unsuccessful combinations result in abnormal pollen tubes with swollen tips that do not reach the ovary. The major barriers in the pistil are the upper style and the ovary. However, some crosses that show normal pollen tube growth and ovule penetration still fail to result in hybrid seed, indicating that barriers to hybridisation exist in the ovule.

Compatible interactions

Pollination typically involves a large number of pollen grains. These compete in pollen tube growth through the style to be first to reach the ovary and fertilise the ovules. Pollen grains in compatible matings between the individuals of a species will hydrate and germinate. The pollen tube penetrates the stigma surface and grows into the style, passing through the gelatinous walls of transmitting tissue or mucilage of the stylar canal.

In bicellular pollen systems, the time interval from pollination until fertilisation may be 16 to 48 hours or even longer, reflecting the slow metabolism in this pollen. Pollen tubes in tricellular systems, which have a high metabolic rate, can grow more than 20 times faster than those in bicellular systems. The time from pollination until fertilisation is only 20 minutes in grasses and 45 minutes in sunflowers.

The male reproductive cells and tube nucleus are transferred within the growing pollen tube tip (Fig. 15.1e). As the tube grows through the style, the cytoplasm in the pollen tube tip is sealed from the remainder of the pollen tube by a succession of callose wall plugs. During growth through the style, the tube nucleus usually precedes the sperm cells, lying just behind the growing tube tip.

Events of double fertilisation

During passage of the pollen tubes down the style, one of the synergids in the embryo sac degenerates (Fig. 15.13a, b). The first pollen tube to arrive penetrates the micropyle, entering the ovule. On reaching the embryo sac, the pollen tube passes into the degenerated synergid. The tube tip ruptures, releasing the pair of sperm cells and tube nucleus (Fig. 15.13c). In this position, there are no cell walls and male or female gametes are bounded only by their plasma membranes. One sperm fuses with the egg cell to form the diploid embryo, the first cell of the new sporophyte generation (Fig. 15.13d). The other sperm fuses with the central cell to form the **triple fusion nucleus**. This nucleus divides mitotically to produce nutritive **endosperm**. Thus, the process of double fertilisation in flowering plants involves true fertilisation (fusion of sperm and egg) plus an accessory fusion (between sperm and central cell).

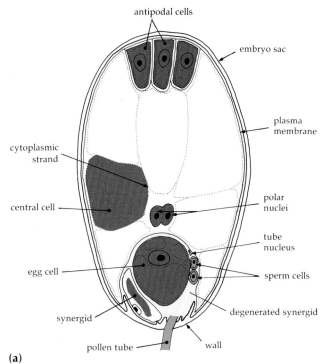

(a)

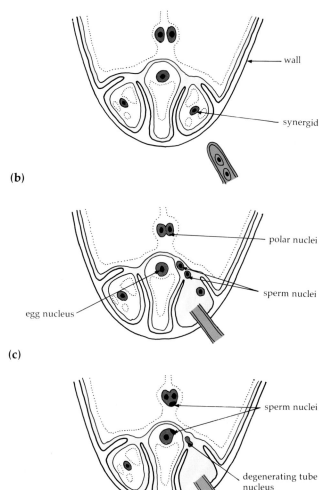

(b)

(c)

(d)

Fig. 15.13 Cellular processes in a typical double fertilisation event in flowering plants. **(a)** Embryo sac at fertilisation seen in longitudinal section. View of the micropylar end of the embryo sac, showing the sequential events of double fertilisation. **(b)** Pollen tube containing sperm cells and tube nucleus approaches the embryo sac (details of ovule omitted for clarity). **(c)** Pollen tube pushes through the wall of the embryo sac, enters the now degenerated synergid, and releases sperm cells into space between the plasma membranes of the egg and central cells. Cellular fusion follows. **(d)** Nuclear fusion follows when smaller sperm and egg nuclei fuse, and larger sperm and polar nuclei fuse. The tube nucleus degenerates

Double fertilisation appears a complex process, and yet it has increased the rate and timing of reproduction in flowering plants compared with other vascular plants. For example, double fertilisation in grasses can take place in only 20 minutes compared with single fertilisation in conifers, which requires many months (Chapter 43).

Fertilisation in flowering plants begins when pollen alights on the stigma, and pollen tubes emerge and grow through the stigma and style to the ovary and penetrate the ovules. There, double fertilisation occurs. One of the pair of sperms in the pollen tube fuses with the egg to produce the diploid zygote, the other with the central cell to produce the triploid endosperm.

MAJOR EVENTS OF EMBRYONIC DEVELOPMENT

A zygote undergoes rapid cell divisions to form an **embryo**, a young plant formed within a seed by the processes of embryogenesis. The first mitosis is typically in a transverse plane. The lower cell subsequently forms a filament of cells, the **suspensor**, terminating in a basal cell (Fig. 15.14a). The suspensor provides nutrition to the developing embryo. In some plants, repeated mitoses of the upper apical cell gives rise to the embryo (Fig. 15.14b), while in others both cells contribute to embryo formation.

The early embryo (proembryo) is a globular mass of about 40 cells (Fig. 15.15). As the **cotyledons** (seed leaves) develop, the embryo becomes heart-shaped and then torpedo-shaped, establishing the axes of the embryo. The growing meristem of the shoot forms the **epicotyl** (the part of the axis above the cotyledons that forms leaves and lateral branches). The **hypocotyl** lies immediately below the cotyledons. The root meristem of an embryo grows in the opposite direction to the epicotyl, forming the **radicle**, the primary root of the seedling (Fig. 15.15c). This structure is at the suspensor end of the original embryo. Root and shoot meristems continue to function throughout the life of the plant, showing a potentially unlimited pattern of growth.

In cereals and grasses, a single fertilised ovule develops into a grain. The integuments form the outer

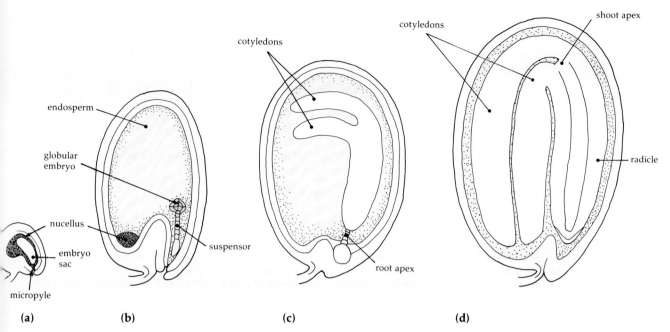

Fig. 15.14 Scheme showing early stages of embryogenesis in the dicotyledon *Arabidopsis*. (**a**) Ovule at the time of fertilisation, showing embryo sac within nucellus. (**b**) Globular embryo and suspensor in ovule, surrounded by endosperm with nucellus now pushed to opposite end of embryo sac. (**c**) Torpedo-shaped embryo and suspensor in ovule. (**d**) Mature seed 0.5 mm in length containing a pair of cotyledons and radicle

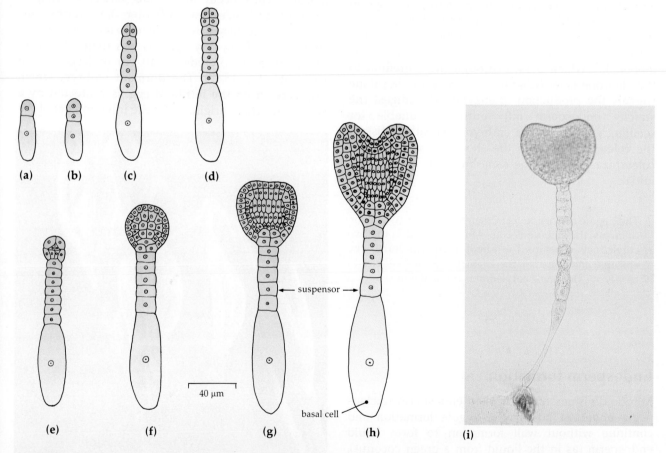

Fig. 15.15 Cellular structure of embryo and suspensor at different stages of development viewed in longitudinal sections. (**a**) Two-celled proembryo, after first (horizontal) mitosis. (**b**) Three-celled proembryo. (**c**) Eight-celled proembryo after first longitudinal mitosis of apical cell and enlargement of basal cell at tip of the suspensor. (**d**) Eleven-celled proembryo. (**e**) Sixteen-celled proembryo showing initiation of the epidermal layer. (**f, g**) Globular stages of embryogenesis. (**h**) Heart-shaped stage of embryogenesis. (**i**) Heart-shaped embryo and suspensor of a related plant, *Brassica*

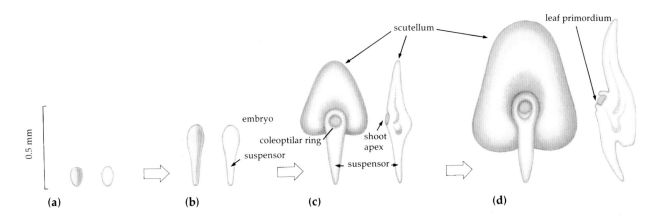

Fig. 15.16 Early stages of embryogenesis in the kernel of maize (monocotyledon) showing the origin of the single cotyledon, the scutellum. The diagrams show the embryo in face view (left) and in longitudinal section (right): **(a)** proembryo; **(b)** globular embryo and suspensor; **(c)** coleoptilar embryo showing origin of scutellum and coleoptilar ring around shoot apex; **(d)** later embryo showing development of leaf primordium (meristem) within the coleoptilar ring

protective layers (bran), surrounding the embryo and endosperm. The single cotyledon, the **scutellum** (Fig. 15.16), is an interface organ between embryo and endosperm. In wheat or barley grains, endosperm forms an outer layer, the **aleurone layer**, which functions in seed germination. In barley grains, the germinating embryo produces a signal hormone, gibberellin, which passes through the scutellum to the aleurone layer (Chapter 24). There the hormone triggers the production of enzymes that digest the starchy endosperm into nutrients suitable for seedling growth. In the embryo, a sheath of tissue, the **coleoptile**, protects the shoot initial, while the **coleorhiza** performs the same function for the root initial.

> The zygote divides to form the embryo and suspensor. In dicots, the embryo becomes heart-shaped as the two cotyledons develop. The shoot meristem forms the epicotyl and the root meristem forms the radicle. In monocots, the single cotyledon is the scutellum.

Endosperm formation

Mitotic divisions of the triploid endosperm nucleus begin about 24 hours after zygote formation and continue without wall formation to form liquid endosperm (as in the liquid from a green coconut). At about the time the proembryo is at the globular stage, the endosperm usually becomes cellular. It loses its liquid appearance as it forms cell walls around each nucleus and expands to form the major nutritive tissue of the seed, rich in storage reserves.

Fruit and seed development

After fertilisation, an ovary grows and develops into a fruit. Fruits occur in a wide variety of shapes, textures and sizes (see Chapter 37). During this process, the fertilised ovules develop into **seeds**. Seeds vary in size from tiny spheres less than 0.1 mm in diameter, as in orchids, to 10 cm or larger, as in coconuts (Fig. 15.17). The dormant embryo, which originates from the fertilised egg, is protected by a seed coat, which originates as the integuments of the

Fig. 15.17 (a)

(b)

Fig. 15.17 (a) Several thousand seeds form in the fruit of the orchid, *Phaius tancarvillae*. **(b)** Germinating coconut showing the embryonic shoot emerging through hard white cellular endosperm, brown shell (seed coat) and fibre

ovule. The **seed coat** or **testa** is often hard and is composed of sclerenchyma (Chapter 6).

In some plants, the triploid endosperm is consumed before seed maturity and development depends on food stored in the cotyledons (Fig. 15.14), which become swollen and filled with storage reserves. The small seeds of some orchids are truly non-endospermic because there is no sperm fusion with the central cell. They rely on a symbiotic relationship with a fungus to obtain nutrition for early development (Chapter 18).

> Seeds result from fertilised ovules and have a seed coat derived from the integuments of the ovule and an embryo from the fertilised egg. Endosperm is the nutritive tissue in many mature seeds; other seeds store food reserves in the cotyledons.

Seed maturation and dormancy

As seeds mature, development of the embryo is arrested. The onset of dormancy is marked by a lowering of metabolic rate of the embryo. Dormant seeds are dry, about 10% water content (normally more than 85%). The seed coat is impervious to water and/or oxygen and mechanically resistant to embryo enlargement. In some seeds, germination is prevented for some time by chemical inhibitors. Dormancy

ensures that the seeds do not germinate while conditions are unsuitable for early plant development.

Seeds of many tropical plants do not show any dormancy, as conditions are usually suitable for germination throughout most of the year because of the absence of marked seasonal variation.

> Embryo development is arrested in most mature seeds (dormancy).

Triggering seed germination

The life span of seeds varies widely, from a few days (e.g. poplar seed) to a thousand years or more. In Japan, viable seeds of a water plant, *Nelumbo*, have been shown by radiocarbon dating of associated fragments of wood to be about 3000 years old.

Re-initiation of embryo development is triggered by environmental cues (see Chapter 24). Germination cannot take place until water and oxygen reach the embryo and this is initiated by fracture of the seed coat.

The radicle is the first organ to emerge through the seed coat; the epicotyl follows. Pea and cereal seeds show **hypogeal germination**, where the hypocotyl remains short, the cotyledons do not emerge from the seed and the epicotyl reaches the surface by its own growth (Fig. 15.18a). In contrast, bean and *Acacia* seeds show **epigeal germination**, in which the hypocotyl elongates into a U-shape, pushing cotyledons and epicotyl above ground (Fig. 15.18a). As a consequence, when cotyledons are brought above ground in this way, they emerge from the encasing seed coat and function as leaves for a short time.

> Seed germination cannot take place until water and oxygen reach the embryo, usually by fracture of the seed coat. The radicle and epicotyl emerge directly (hypogeal germination) or the hypocotyl pushes the cotyledons above ground (epigeal germination).

ORGANOGENESIS

Growth and development of new organs occurs from the tips of shoots and roots. There is a specialised region of cells, the **apical meristem**, at the extreme tip of every shoot and root of a flowering plant. The function of these clusters of cells is to undergo rapid mitotic cycling. At each division, two daughter cells are produced, one of which remains as part of the meristem and the other differentiates and contributes to the tissues that form part of the mature body of the plant. The first apical meristems are the root and shoot meristems of the embryo. Additional apical meristems form at the tips of new shoots and roots as the plant grows.

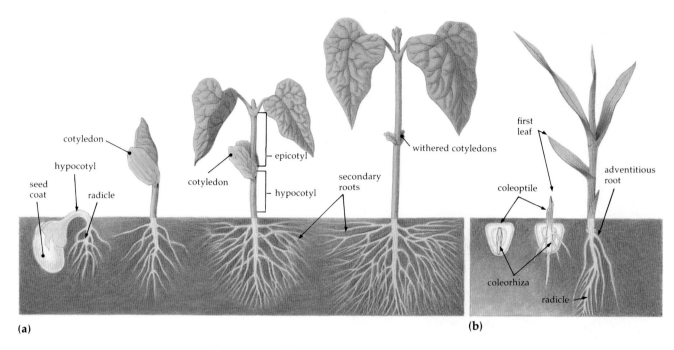

(a)   **(b)**

Fig. 15.18 Types of seed germination. **(a)** Epigeal germination in which the elongating hypocotyl (U-shaped) pulls the cotyledons up into the air. **(b)** Hypogeal germination in which the cotyledon remains underground, the epicotyl grows upwards into the air and the radicle grows downwards from the seed

The shoot apex

The shoot apical meristem produces new leaves. The apex of the shoot is a spherical dome of meristematic cells, which divide to produce leaf primordia on the flanks of the dome (Fig. 15.19a, b). The apex consists of two distinct regions of cells, the tunica and corpus. The **tunica** (*covering*) consists of one to three layers of cells that cover the apical dome. The **corpus** (*body*) is an area of cells below the tunica, occupying a central position in the meristem (Fig. 15.19c). Tunica and corpus divide in different ways to produce different tissues. Cells of the tunica divide in a regular manner to produce an outgrowth that becomes a leaf

primordium. Cells of the corpus divide in an irregular manner to form the mature tissues of the stem. Divisions of both tunica and corpus lead to the production of a leafy shoot.

Axillary buds, which give rise to side branches, arise separately from a lateral meristem. Lateral meristems develop in the axil (angle between stem and leaf) of a leaf primordium (Fig. 15.19c).

In contrast to flowering plants, growth in ferns is derived from divisions of a single, large, **apical cell** situated at the tip of a stem or root. All tissues in a fern are derivatives of an apical cell (Chapter 37).

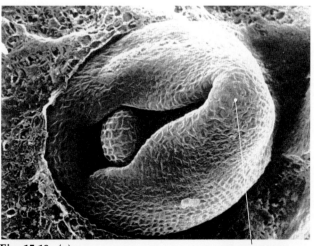

Fig. 15.19 (a)   leaf primordium

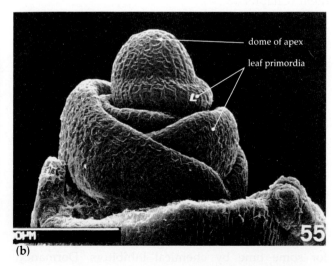

(b)

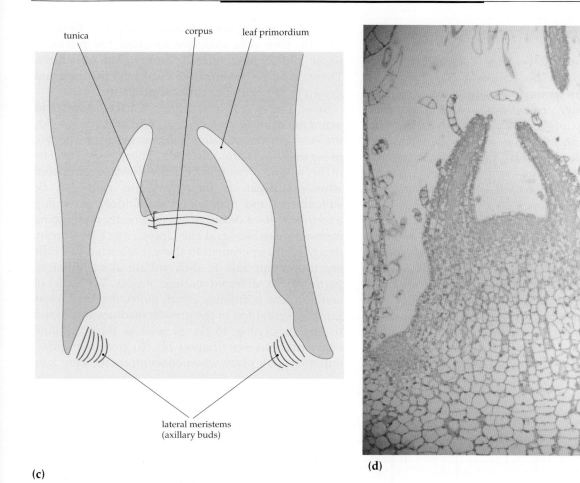

(c)

(d)

Fig. 15.19 Characteristic features of the shoot apical meristem of wheat viewed by scanning electron microscopy. **(a)** View of apex from above, with dome covered by expanding leaf primordium. **(b)** Side view showing apical dome with leaf primordia present on flanks of the apex. **(c)** Diagram showing tissues of shoot apex. **(d)** Lateral meristems in shoot apex of *Coleus* as seen in longitudinal section (magnification 10×)

The root tip

Root apical meristems of flowering plants do not have a tunica–corpus structure. They contain a zone of rapidly dividing cells, which gives rise to all the mature tissues of the root. Roots do not produce lateral structures equivalent to the leaves on a stem. As we will see later (Chapter 17), root hairs arise by elongation of an epidermal cell, while lateral roots arise deep within the tissues of more mature parts of the root, with no involvement of the root apex. Root tips produce a structure, the **root cap** (Fig. 15.20), that has no equivalent in the shoot. The root cap is a protective covering for the meristem as it grows through the soil. Root cap cells are produced from the meristem by cells in front of the apex.

The tissues of stems and roots produced by apical meristems are primary tissue. Later, in woody plants, secondary tissue, forming bark and wood, develops from specialised meristems once the primary tissues are mature (see Chapter 17).

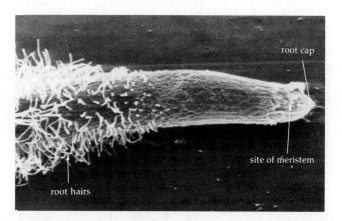

Fig. 15.20 Scanning electron micrograph showing the structure of the root tip of wheat

Apical meristems of shoots and roots are clusters of rapidly dividing cells that give rise to all the cells of the plant.

Totipotency of plant cells

The fate of plant cells becomes determined during development. When development has proceeded beyond a certain point, cells are committed to a particular pathway, for example, a leaf primordium will only develop into a leaf (Fig. 15.19). Cells dividing and differentiating at the apical meristem are channelled into the 'leaf' pathway. A peg-like primordium elongates and the leaf lamina differentiates by increased cell division and elongation in this region.

However, in contrast to animal cells, plant cells are **totipotent**. A single cell possesses the full potential to regenerate an entire organism. This is shown by the experiments of F.C. Steward from Cornell University in 1958, who first demonstrated the totipotency of carrot cells in sterile culture. He took transverse sections of a carrot root and, under sterile culture conditions, cut out tiny pieces of secondary phloem tissue. These were then cultured in a shaking flask in a liquid medium containing nutrients and

minerals, plus green coconut milk to stimulate growth. Free cells sloughed off from the pieces of tissue and divided, forming embryo-like structures. These could be transplanted onto solid medium and grown in the light into new carrot plants.

Small pieces can be cut from tobacco stems and maintained by *in vitro* culture. Within a few weeks, massive growth of callus cells occurs from the site of wounding at the cut ends. These cells are part of the 'wound program' of cell division that is initiated when an organ is injured. Callus cells can be subcultured and maintained indefinitely, providing a source of plant cells that are constantly undergoing mitosis. Folke Skoog, at the University of Wisconsin, designed an experiment to show that such callus cells are totipotent, able to differentiate along different pathways in different culture media. The ratio of certain plant hormones (small molecules that act as growth regulators) in the growth medium was found to be crucial (Fig. 15.21), as well as their absolute concentrations (see Chapter 24). No growth or differentiation took place when concentrations were low.

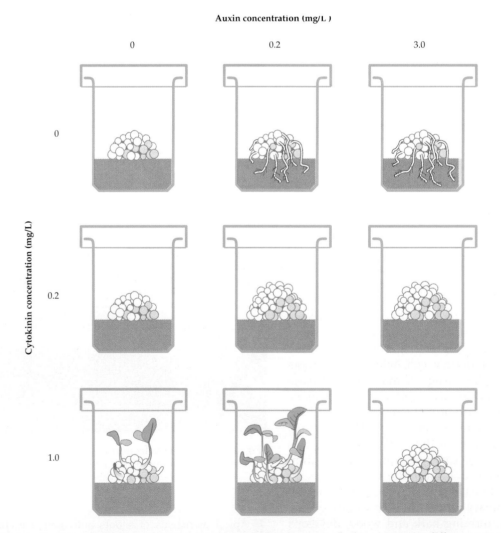

Fig. 15.21 Developmental programs of tobacco callus when cultured in flasks containing different concentrations of the plant hormones: auxin (indole acetic acid, IAA) and cytokinin (kinetin)

However, when the hormone auxin was high, root formation occurred on the clumps of cells (*root program*). When the hormone cytokinin was high and auxin low, shoot formation was induced (*shoot program*). With moderate amounts of both hormones, growth of the callus clumps occurred without differentiation (*callus program*).

Thin cell layers of epidermis peeled from tobacco stem can be cultured and the cells induced directly to enter different programs of differentiation, for example, into vegetative buds, roots or callus. If the epidermal cells came from flowering shoots, then flowers can also be induced.

These experiments suggest that single plant cells, if in a suitable environment of nutrients and light, can undergo cell division and growth to generate an entire organism.

> Plant cells are totipotent, possessing the full potential to regenerate an entire organism.

Tissue and single cell culture

Cell culture techniques have made use of microbiological culture methods, especially adapted for plant cells. Their use has resulted in dramatic advances in regenerating plants from individual cells, organs or tissues. Such techniques are an essential component of genetic engineering programs for gene transfer.

Tissue culture can involve micropropagation directly from fully organised structures, for example, apical and lateral meristems and young axillary shoots. 'Meristem culture' was established to make use of heat treatment to eliminate viruses from meristem regions. Increasing the temperature in meristem culture allows plant cells to divide faster than viruses can multiply, thus eliminating viruses from the tissue. This process has been widely used to produce strawberries and potatoes free of virus diseases.

During the past 20 years, callus cultures and suspension cultures have been widely used for mass propagation of plants. However, tissue culture of callus cells has resulted in increases in chromosome number and changes in gene expression and mutations. Plants regenerated from callus cultures often showed structural and physiological abnormalities. In some cases, where crops were regenerated using these methods, there was a disastrous effect on the commercial viability of the crop. Mass propagated strawberries remained vegetative after field planting in the United States; after five years in the field in Malaysia and New Guinea, oil palms grown from tissue culture of roots produced sexually modified flowers with drastically reduced kernel yields. Methods involving induction of callus tissue are now avoided and techniques giving rise directly to somatic embryos are now used.

Hybrid embryo rescue

Interspecific hybridisation provides the potential to generate new genetic combinations but, even with successful cross-fertilisation, the hybrids may form non-viable seeds. *In vitro* culture of developing embryos from fertilised ovules of hybrids is a useful method to rescue hybrid embryos. For example, hybrids between clovers (*Trifolium ambiguum* × *T. repens*) were given embryo rescue treatment in New Zealand with the aim of providing a source of new pasture legumes. When each of the parent clover species is self-fertilised, the embryos progress normally through globular, heart-shaped, torpedo-shaped stages until mature embryo formation. However, in interspecific hybrid embryos, growth in the first two days after pollination is more rapid than in the parents, but then slows, with development within the seed being arrested at the heart-shaped stage, when the embryos have about 1000 cells. This developmental arrest is due to the action of deleterious genes, which appear to act by blocking the normal nutrient supply to the growing embryo.

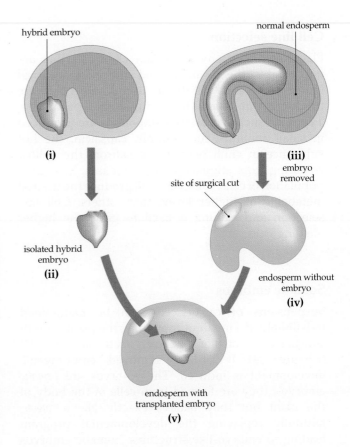

Fig. 15.22 Hybrid embryo rescue. (**a**) Transplantation of a hybrid embryo into a de-embryonated normal endosperm for culture until maturity

Fig. 15.22 (b) New clover hybrid raised by embryo rescue in New Zealand: parents are *Trifolium ambiguum* (left) and *T. repens* (right), with the new hybrid (centre)

Several rescue techniques have been successfully used. Nurse cultures are initiated with hybrid embryos being cultured on a normal endosperm (Fig. 15.22a). Embryos can also be cultured directly on artificial media, in which the required nutrients are provided. These embryo rescue techniques successfully produced new hybrid clovers (Fig. 15.22b).

Cell line selection

Plants contain many secondary chemicals that are important in the cosmetic, food and pharmaceutical industries. Nearly all these useful molecules are produced in secondary metabolic pathways. Typically, such secondary compounds are produced in small amounts, for example, capsaicin, the hot substance in chilli peppers, or saffron, the yellow colouring from stigmas of wild crocuses. In a large population of cells, some cells will produce the desired metabolite at higher levels than others. Cell line selection and cloning is used to grow the higher yielding lines.

Somatic embryos

Suspensions of carrot cells can be maintained indefinitely in a medium containing the plant growth substance 2,4-dichlorophenoxy acetic acid (2,4-D) (Chapter 24). If 2,4-D is removed, embryogenic development is induced. The embryos are *somatic* embryos; they are formed from cells of the body of the plant but they behave exactly like a *zygote*, faithfully replaying the developmental program leading to embryo-like structures. Somatic embryos can also be induced from epidermal cells of seedling hypocotyls when cultured in the presence of particular hormones, especially cytokinins (Fig. 15.23).

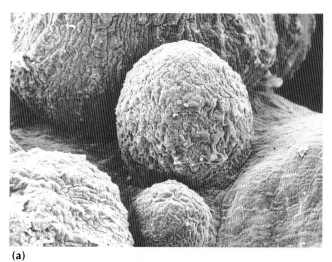

(a)

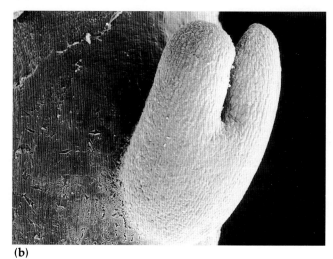

(b)

Fig. 15.23 Embryo culture: scanning electron micrographs of developing somatic embryos developing from hypocotyl of clover *Trifolium*. **(a)** Globular embryos. **(b)** Heart-shaped embryos

Applications in genetic engineering

Tissue culture techniques are a part of genetic engineering technologies. Firstly, the desired genes (e.g. disease-resistance genes) need to be isolated (see Chapter 12). Secondly, these genes need to be physically transferred into the host plant cell. The process by which the genetic make-up of a single cell is altered is *transformation* (see Chapter 12). Micropropagation techniques can then be used to regenerate a new plantlet with the altered phenotype from this cell.

Transformation is typically achieved using a vector to transfer the gene (Fig. 15.24). These are usually *plasmids*, small DNA molecules in the cytosol of bacteria that carry one or more genes and can replicate themselves in bacteria. The most common vector is the Ti plasmid of *Agrobacterium tumefaciens*, crown gall disease, which produces tumour-like growths in plants (but for transformation, the tumour-inducing

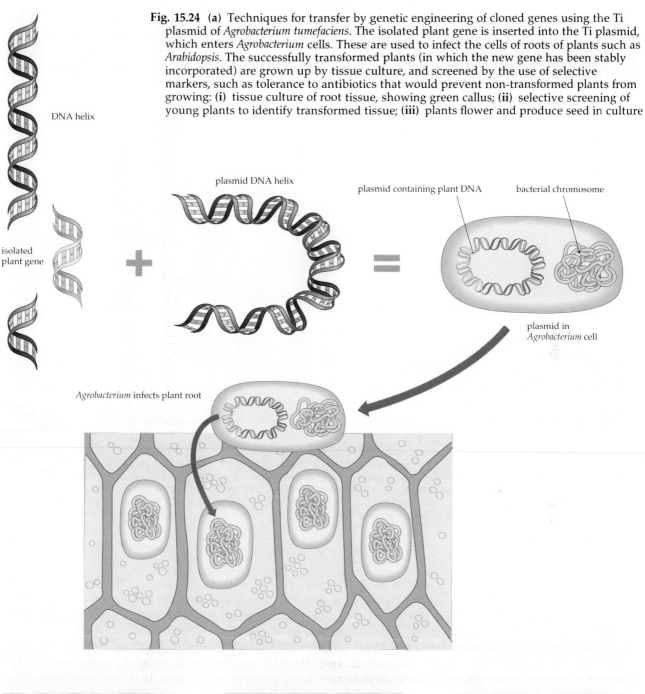

Fig. 15.24 (a) Techniques for transfer by genetic engineering of cloned genes using the Ti plasmid of *Agrobacterium tumefaciens*. The isolated plant gene is inserted into the Ti plasmid, which enters *Agrobacterium* cells. These are used to infect the cells of roots of plants such as *Arabidopsis*. The successfully transformed plants (in which the new gene has been stably incorporated) are grown up by tissue culture, and screened by the use of selective markers, such as tolerance to antibiotics that would prevent non-transformed plants from growing: **(i)** tissue culture of root tissue, showing green callus; **(ii)** selective screening of young plants to identify transformed tissue; **(iii)** plants flower and produce seed in culture

DNA helix

isolated plant gene

plasmid DNA helix

plasmid containing plant DNA

bacterial chromosome

plasmid in *Agrobacterium* cell

Agrobacterium infects plant root

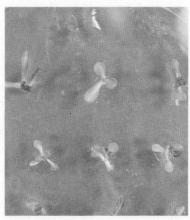

(a) (i)   **(ii)**   **(iii)**

(b)   **(c)**

Fig. 15.24 **(b)** Growth of transformed *Arabidopsis* plant by tissue culture. **(c)** Testing of *Arabidopsis* plant for selectable marker. The plant gene being transferred is expressed only in anthers, as seen here in this flower using a marker enzyme that gives a blue colour after incubation in a suitable substrate. The presence of the blue colour in the anthers confirms that the gene has been successfully transferred

gene is disabled.) The critical feature is that the bacterium transfers part of the Ti plasmid to the plant nucleus (known as T-DNA), thus effecting transformation. The Ti plasmid is efficient at transferring one or two genes (~15 kilobases) into plant cells.

The use of *Agrobacterium* for transformation is confined to certain dicotyledons, including tobacco and *Arabidopsis*. A vector is not always needed for transformation. Direct gene transfer can be achieved by using a biolistic DNA delivery system, which uses helium gas under pressure to fire DNA-coated tungsten microprojectiles into plant tissues.

Genetic engineering has some major advantages over traditional methods of cross-breeding. It is rapid and as it involves the transfer of only a few exotic genes, the balance of the host plant genome is not usually disturbed. This means that the special characteristics of the cultivar are conserved. There also appears to be few restrictions on the source of the gene, as even viral genes have been expressed in plants. Examples include genes encoding viral coat proteins in the *coat protein cross-protection* approach,

a technique that provides protection against viral infection (see Chapter 12), and use of bacterial genes encoding an insecticidal protein, known as BT toxin, from *Bacillus thuringiensis*.

One disadvantage is that the number of genes available remains small, especially compared with the large number of diseases that threaten crops. Also the tissue culture techniques needed for transformation and regeneration have still to be developed for many crop plants, especially cereals. The insertion of exotic DNA into plant chromosomes is apparently random, so position effects can occur when the expression of a gene is modified by adjacent genes. Genetic engineering is particularly well developed at present in crops such as potatoes and tomatoes, members of the family Solanaceae.

Tissue and single cell culture techniques have resulted in the generation of plants from individual cells, organs or tissues. Such techniques are essential for gene transfer into host plant cells using a vector.

BOX 15.2   ETHICS OF GENETIC ENGINEERING: FIELD TRIALS OF TRANSGENIC COTTON IN AUSTRALIA

The cotton industry in Australia currently generates $900 million annually in exports but the cost of pesticides used to ensure the crop amounts to $250 million, creating economic, environmental and ecological problems. The pesticides are needed to control the cotton budworm, the larvae of which attack the leaves and flowers. The CSIRO Division of Plant Industry, in collaboration with Monsanto Chemical Company, have engineered cotton plants with bacterial crystal protein genes that express BT toxin. The protein acts in the gut of the insect and is potent and specific towards the budworm larvae.

Release of transgenic plants for field trials in northern New South Wales began in 1992 and has been carefully monitored by the Genetic Manipulation Advisory Committee, the regulatory body in Australia. For field trials, the transgenic cotton plants are grown within a buffer zone of normal cotton varieties. This is necessary to prevent the dispersal of exotic genes in pollen from the transgenic cotton plants by pollinating bees and accidental cross-hybridisation with native cotton species that grow in the area. Experiments have shown that cotton is largely self-fertilised and cross-fertilisation is 20% between adjacent rows, but falls to less than 1% only 7 metres from the transgenic plants. Fortunately, experiments have shown that cultivated cotton does not cross-hybridise with native species of cotton (*Gossypium*).

The use of genetically engineered cotton is expected to lead to a reduction in the use of chemical pesticides, reducing the financial burden on growers and the chemical burden on the environment. Non-target insects and vertebrates, including humans, are relatively safe from the introduced gene, which is expressed at the level of 0.2% of total plant protein. For these positive reasons, there is considerable optimism for the future of transgenic cotton, especially when the ethical issues involved both in genetic engineering and in the use of chemical pesticides are considered.

SUMMARY

- In flowering plants, anthers are specialised for differentiation, nutrition, protection and dispersal of pollen. The tapetum has a nutritive function, while endothecium and stomium are responsible for anther opening.

- Microspores are formed within an anther at meiosis, each tetrad being embedded within a callose wall. At the end of meiosis, microspores are released from the tetrad and enter a major growth phase, developing a large central vacuole. The microspore becomes a pollen grain after the first (asymmetric) mitosis.

- The pollen grain is the male gametophyte. Most pollen is bicellular and contains a generative cell enclosed within a vegetative cell. The generative cell divides to form two sperm cells after pollen germination within the growing pollen tube in the style. In tricellular pollen, the generative cell divides within the pollen grain.

- The female organ, the pistil, bears a stigma to receive pollen, a style with transmitting tissue or central canal for growth of pollen tubes and a basal ovary containing ovules.

- Ovules of flowering plants consist of a female gametophyte (embryo sac) containing the female gamete, the egg, enclosed by a nucellus and one or two integuments.

- The embryo sac is formed from a diploid megasporocyte, which undergoes meiosis to produce four haploid megaspores. Typically, three abort and the remaining megaspore differentiates by three mitotic divisions, producing an eight-nucleate embryo sac. A seven-celled female gametophyte is formed, containing an egg, two synergids, central cell with two polar nuclei and three antipodal cells.

- Fertilisation in flowering plants begins when pollen alights on the stigma, and pollen tubes emerge and grow through the stigma and style to the ovary and penetrate the ovules. There, double fertilisation occurs. The pair of sperms are usually linked together and to the tube nucleus as the male germ unit. One of the pair of sperms in the pollen tube fuses with the egg to produce the diploid zygote, the other with the central cell to produce the triploid endosperm.

- The zygote divides to form the embryo and a suspensor that acts as a nutritive link. In dicotyledons, the embryo develops into a globular, then heart-shaped and finally torpedo-shaped structure as the two cotyledons differentiate. In monocotyledons, the single cotyledon is the scutellum.

- Seeds result from fertilised ovules and have a seed coat derived from the integuments of the ovule and an embryo from the fertilised egg. Endosperm is the nutritive tissue in the mature seed; other seeds store food reserves in the cotyledons.

- Embryo development is arrested in the mature seed when it enters a period of dormancy, which is marked by a dehydrated state and low metabolic rate.

- Seed germination cannot take place until water and oxygen reach the embryo, usually by fracture of the seed coat. The radicle and epicotyl emerge (hypogeal germination), while the hypocotyl pushes the cotyledons above ground (epigeal germination).

- At the tips of every shoot and root are specialised apical meristems, clusters of rapidly dividing cells, that give rise, directly or indirectly, to all the cells of the plant.

- Plant cells are totipotent, possessing the full potential to regenerate an entire organism. Tissue and single cell culture techniques have resulted in an ability to generate plants from individual cells, organs or tissues.

- Regeneration is essential for gene transfer in biotechnology. Isolated genes are physically transferred into the host plant cells using a plasmid vector, usually by infection with *Agrobacterium* or by biolistic delivery.

QUESTIONS

1. (a) What is the male gametophyte of a flowering plant? (b) Where is a microspore? (c) What is the function of the callose wall during microsporogenesis?

2. (a) What is the female gametophyte of a flowering plant? (b) Where is it located? (c) By means of diagrams, show how it forms by meiotic and mitotic divisions.

3. What are the special features of the location of the egg and central cells within the embryo sac? Draw a diagram to illustrate your conclusions.

4. Describe the sequence of events from when a pollen grain lands on the surface of a compatible stigma until double fertilisation.

5. Plants grow from meristems. (a) What organs are derived from the tunica? (b) What is the main difference between apical meristems of ferns and flowering plants? (c) The shoot apical meristem of grasses is cone-shaped. What are the peg-like structures that accumulate along its length?

6. Do flowering plants go through embryonic stages equivalent to those of animals? Compare and contrast embryo development in flowering plants with that in animals (Chapter 15).

7. Why is the cereal grain really a fruit?

8. By means of labelled diagrams, distinguish between epigeal and hypogeal forms of germination.

9. In a dicotyledon embryo, what is (a) an epicotyl, (b) a hypocotyl and (c) a radicle? (d) What is the scutellum of a monocotyledon embryo?

10. In 1992, the CSIRO began field trials of genetically engineered cotton plants containing a bacterial gene expressing a pesticidal toxin that is expected to displace chemical sprays. (a) List five advantages that this new technology will have for cotton growers in Australia. (b) What are the ethical issues that need to be considered before plants containing exotic genes are released commercially?

GENES AND DEVELOPMENT

In most eukaryotes, each new organism begins life as a single cell, a zygote. As we saw in Chapter 14, in multicellular organisms, this cell divides repeatedly by mitosis to produce a highly organised adult composed of many different types of cells and tissues. During very early development, all embryonic cells are similar. As development proceeds, a stage is reached when cells begin to become different from one another and specialised for particular functions. The process of cell specialisation is differentiation, which is under the control of genes. The genome of a zygote contains all the genes required to produce every type of cell found in the adult. However, in differentiated cells, only some genes are active. The gene for haemoglobin is active as red blood cells develop; in nerve cells, genes control the production of neurotransmitter molecules; in lymphocytes, genes that produce a particular antibody may be active; and in flower petals, genes that produce coloured pigments are expressed.

In this chapter we will examine the temporal, spatial and regulatory roles of genes in the control of the processes of differentiation and development.

> Differentiation is the process that leads to the expression of a particular pattern of genes in a specialised cell.

GENES INVOLVED IN DEVELOPMENT

We can identify two general ways in which genes are involved in development.

1. *Regulation of gene expression.* In this case, gene expression is regulated so that the amount of gene product varies at different stages of development or in different embryonic tissues. For example, some genes are active in most cells of an organism. These *housekeeping genes* produce the macromolecules that are required for basic cellular functions, such as provision of energy, repair and maintenance of organelles, and cell division. Housekeeping genes do not regulate the activity of other genes.

2. *Regulation of other genes.* Some genes regulate the activity of other genes, thereby governing the sequence of events during development and the pattern of the final product. Many regulatory genes are themselves regulated by the products of other genes.

Recent research into the control of development has led to the discovery of a group of **homeotic genes**, whose products appear to regulate the activity of other genes during development. For example, in *Drosophila melanogaster*, a single mutation of one gene will cause a leg to grow in place of an antenna (p. 346). Since the production of a leg requires the action of many genes, the single mutation must have occurred in a homeotic gene that controls the set of leg-forming genes.

Researchers have now shown that all homeotic genes contain a short sequence of DNA in which the sequence of bases is almost identical—the *homeobox region*. The homeobox region codes for a protein that binds strongly to DNA, allowing the protein to regulate other genes. Interestingly, this same sequence of bases has been identified in a wide variety of organisms.

> Some genes are regulated so that the amount of gene product varies at different stages of development. This regulation is carried out by other genes.

CONTROL OF GENES

We know from studies, such as grafting experiments with the alga *Acetabularia* (Box 16.1), that the nucleus controls the phenotype of a cell. But what controls the nucleus? During development in animals, early gene control appears to be due largely to *intracellular* regulatory factors located in particular regions of the cytoplasm of the egg (Chapter 14). These substances affect cellular processes or interact with genes themselves, so that the cell proceeds along a particular path of development. Later, *extracellular* signals from other differentiating cells within the organism, or from the environment, become important. In each case, the phenotypic response to these signals is specified by the genetic information carried in the nucleus of the responding cell.

A similar pattern of control is seen in plants. Intracellular regulatory factors are produced in the developing egg cytoplasm as a result of maternal gene expression. Extracellular control in plants may be due to hormones, for example, the flowering hormone produced in leaves is able to transfer across graft junctions to stimulate the development of flowers in an apical meristem that would otherwise have produced a shoot.

Intracellular control

As a zygote commences development, mRNA transcripts in the egg cytoplasm, which are derived from maternal genes during oogenesis, are translated. The genes that produce these mRNA transcripts are *maternal-effect genes*; they encode proteins that control the early development of the embryo. For example, in many animals, these proteins control cleavage, which can occur even in a zygote that has had its nucleus removed (Chapter 14). During gastrulation, maternal-effect proteins can function as intracellular signals, interacting with the genome of the cells in which they are located to switch certain genes on and others off. In *D. melanogaster*, substances stored in the egg cytoplasm play a role in laying down the body pattern. They specify the basic body symmetry of the larva, that is, the posterior–anterior and dorsal–ventral axes (pp. 344–5).

> Control of early development is due in part to intracellular (cytoplasmic) factors produced by maternal-effect genes.

BOX 16.1   THE ROLE OF THE NUCLEUS

The role of the nucleus in controlling cell phenotype was clearly demonstrated in a series of grafting experiments carried out in the 1930s using *Acetabularia*, a large single-celled marine alga, which can reach a length of 1–2.5 cm. Each cell consists of three principal regions: cap, stalk and rhizoids (Fig. a). The rhizoids are branched root-like cell extensions that anchor the small plant, and the nucleus stays within the rhizoids for most of the life cycle. If the cap is cut off, a new cap regenerates over the next few days. If the new cap is removed, another is formed, and so on.

Two closely related species with different-shaped caps are *A. mediterranea*, which has an umbrella-like cap, and *A. crenulata*, which has a cap that is deeply dissected. If the cap of either species is removed, a new cap of the same shape grows back (Fig. b). In a reciprocal experiment, caps were removed from individuals of each species, and then the stalks were cut and transplanted onto rhizoids of the other species (Fig. c). The first caps that grew were intermediate in form between the two species. When these were removed, the next and any further regenerated caps had the form of the rhizoid

What is the explanation for these results? Cap regeneration requires the synthesis of many specific proteins and lipids within the cytoplasm of the stalk. The first regenerated caps were intermediate in shape because instructions (mRNA) from the original parent remained in the stalk cytoplasm and these were translated along with new mRNAs coming from the nucleus in the rhizoid. When the intermediate caps were removed, the next cap formed was, in each case, that of the rhizoid species because regeneration was directed only by mRNA from the nucleus in the rhizoid. Thus, the nucleus located in the rhizoid responds to the absence of a cap and provides mRNA instructions, not only as to how the cap is to be made, but what type it will be.

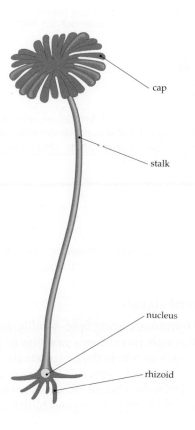

cap

stalk

nucleus

rhizoid

(**a**) The major regions of the single-celled alga, *Acetabularia mediterranea*. The nucleus is always associated with the rhizoids

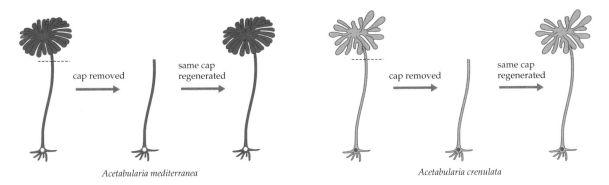

(b) Grafting experiments in *Acetabularia*. Cap removal and reformation. If the cap of *A. mediterranea* or *A. crenulata* is cut off, then another cap grows. It is always the same type as the original species

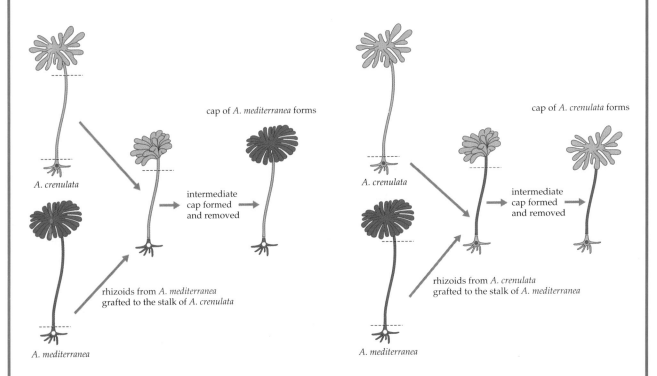

(c) A grafting experiment in *Acetabularia*. The caps are removed from both species of *Acetabularia*, and then the stalk of *A. mediterranea* is grafted to the rhizoid of *A. crenulata* (or vice versa); the first cap that develops in each case is intermediate and when this is removed the cap that is formed is that appropriate to the rhizoid containing the nucleus

Extracellular control

Later in development, extracellular signals from other differentiating cells or from the environment play an increasingly important role in regulating gene activity. Embryonic induction in animals (Chapter 14) and the action of many hormones (Chapter 25) are examples of extracellular signals originating within the organism that regulate gene activity. Other molecules and ions in the vicinity of the cell membrane, physical factors, such as temperature, movement and light, may selectively influence gene activity during development, particularly during the later stages.

Hormonal signals

Steroid hormones, being lipid-soluble, are able to pass easily through membranes and many pass directly into the nucleus where they regulate the transcription of specific genes (Chapter 7). Their action is blocked by the DNA-dependent RNA synthesis inhibitor, actinomycin D. Other hormones bind to a cytoplasmic receptor and the hormone–receptor complex passes into the nucleus where it appears to interact with non-histone proteins (Chapter 10) to determine the pattern of genes transcribed.

In insect larvae, the steroid hormone ecdysone controls the genetic switches that regulate metamorphosis into the sexually mature adult. Ecdysones stimulate moulting, growth and differentiation of tissues into adult structures. For example, during pupation in the silkworm, the 'silk' gene is switched on by ecdysone to a very high level of expression, resulting in the production of the silk needed to form the pupal case (p. 338). On the other hand, juvenile hormone, as its name implies, stimulates a pattern of gene expression that maintains the larval state (Chapter 25).

Environmental factors

Sometimes the signal that alters gene expression in a cell originates outside the organism itself, for example, temperature, food supply, movement and light may selectively influence gene activity.

The cellular slime mould, *Dictyostelium discoideum*, provides an example of environmental control of development. Slime moulds spend much of their life as single-celled amoebae living in the soil and ingesting bacteria, but under adverse conditions they aggregate to form a multicellular complex in which the cells adhere, communicate with each other and change their shape (Figs 16.1, 7.1). Co-operation between a large number of amoebae results in the survival of some of them.

When there are plenty of bacteria to eat, single amoebal cells grow and divide. When the local food supply is exhausted, stressed cells release a signal chemical that causes other amoebae to aggregate around them. The aggregate of cells forms a mound, containing up to 100 000 amoebae, which first extends upwards, then falls on its side forming a small 'slug-like' organism, which migrates through the soil for a few days before moving to the soil surface and developing into a fruiting body. About 30% of the cells make up a basal disc and stalk, and the remaining 70% form thick-walled spores, able to germinate into single amoebae after dispersal.

The chemical involved in this signalling process in *D. discoideum* is cyclic AMP (cAMP), a common signal molecule in cells (Chapter 7). When cAMP binds to a receptor on the membrane of amoebae during aggregation, the cells move towards the source of the signal, and then release cAMP themselves, thus amplifying the message. cAMP finally activates a range of genes, bringing about slug and fruiting body formation. The cAMP signal is broken down by the enzyme, phosphodiesterase. This type of signalling occurs during many aspects of cell development, including the action of some hormones (Chapter 7).

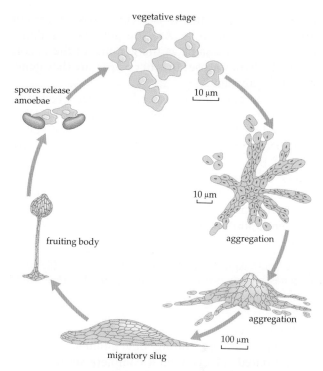

Fig. 16.1 Life cycle of the slime mould, *Dictyostelium discoideum*. When the food supply becomes limiting, the individual amoebae aggregate in response to a signal and form a 'slug'. The slug consists of two types of cells: cells that make the stalk of the fruiting body, which occupy the anterior of the slug and extend around the outside; and cells that become spores, which occupy the posterior part of the slug. The slug migrates and ultimately forms a stalked fruiting body, which releases spores that can germinate under appropriate conditions to form new amoebae

The genetic control of this process in *D. discoideum* has been demonstrated through the isolation of mutants that affect development in different ways. For example, mutations that affect the size, texture or pigmentation of the fruiting body, or mutations that inhibit aggregation or stop development at the slug stage, enable identification of gene products that are essential to the orderly progression of this developmental system.

> Development is regulated by extracellular signals, including hormones, such as ecdysone in insects, and environmental factors, such as the availability of food and levels of cAMP in slime moulds.

REGULATING GENE EXPRESSION DURING DEVELOPMENT
Genomic equivalence

As mentioned earlier, differentiated cells produce some proteins in common and other proteins specific to each cell type. The question arises as to whether differentiation is due to selective loss of genetic information from cells. Are the genomes of differentiated cells the same as the genome of the zygote from which they arose? In other words, are they genomically equivalent?

Genomic equivalence of cells has been tested in a number of ways. Biochemical analysis of DNA extracted from different cells of an organism shows that DNA type and content are the same in most cells of that organism (excluding germ cells) at all stages of development. These results indicate that there is no wholescale loss of genes during differentiation and is consistent with the notion that differentiation involves the differential expression of genes.

> Genomic equivalence means that cells have the same genes, although they may be differentially expressed in different tissues.

Other experiments have shown that genes are not irreversibly altered during development and that differentiated cells contain a complete set of potentially functional genes. Differentiated plant cells are totipotent, that is, they are capable of giving rise to a complete organism under appropriate conditions. For example, in the carrot, *Daucus carota*, at least some differentiated cells in root tissue can be induced to develop into a mature plant. When phloem cells are isolated from roots and grown in tissue culture, they divide and form groups of cells with 'embryo-like' properties (Fig. 16.2). In the presence of certain hormones, some of these groups of cells will give rise to differentiated plantlets with stems, leaves and roots (Chapter 15).

Similar experiments have been attempted with animal cells, but it has not been possible to induce a fully differentiated somatic cell to behave like a zygote. However, it has been possible to show that, at the level of the nucleus, the process of differentiation is reversible. In two species of frog, the grass frog, *Rana pipiens*, and an African frog, *Xenopus laevis*, nuclei were transferred from a differentiated cell into an egg of the same species with its nucleus removed, either surgically or by irradiation (Fig. 16.3). Some of the eggs with transplanted nuclei continued to develop through a number of stages and, occasionally, normal tadpoles were produced.

We can conclude from these experiments that at least some nuclei of differentiated cells are totipotent and that the genome has not changed irreversibly as a consequence of differentiation. Genes that were inactive in one situation were able to be expressed in a different cellular environment. These experiments also show that gene expression is not irreversibly programmed from the outset along a particular developmental pathway.

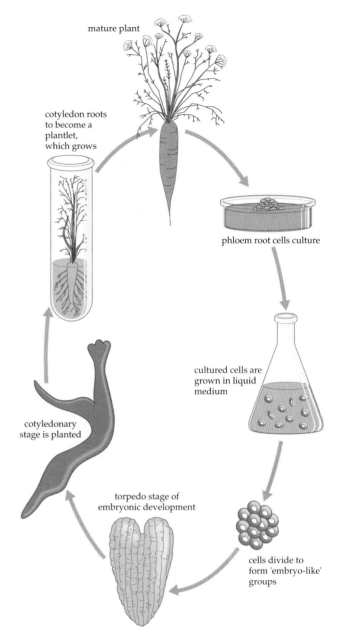

mature plant

cotyledon roots to become a plantlet, which grows

phloem root cells culture

cultured cells are grown in liquid medium

cotyledonary stage is planted

torpedo stage of embryonic development

cells divide to form 'embryo-like' groups

Fig. 16.2 Development of a carrot plant from differentiated root cells. Phloem cells from the root are isolated and cultured, and grown in liquid medium. As the cells divide they become differentiated and form 'embryo-like' groups (somatic embryos). Transfer of the somatic embryos to appropriate solid media stimulates them to form plantlets, which can then be grown into mature plants

> Some differentiated plant cells are totipotent, that is, they are capable of giving rise to a complete organism under appropriate conditions. Nuclei from some differentiated animal cells can support full embryonic development.

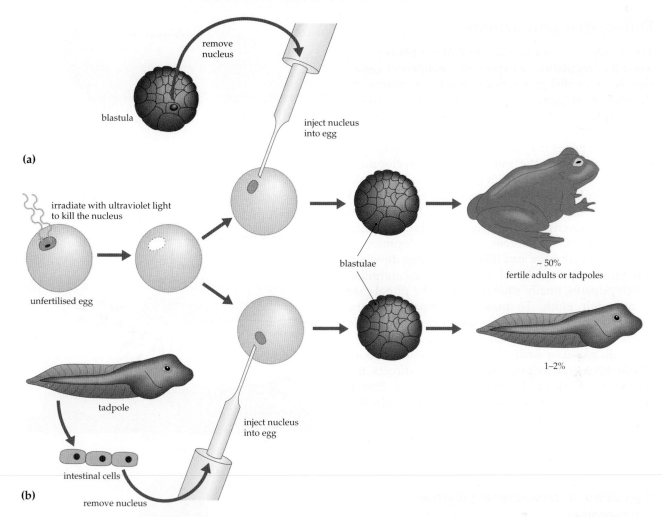

Fig. 16.3 The nucleus of a differentiated amphibian cell can program development. Fertilised eggs are enucleated by killing the nucleus with ultraviolet light. They are then injected with differentiated nuclei. The success of these in programming development depends on the developmental stage of the transplanted nuclei. **(a)** If the nucleus is taken from the early developmental stage, the blastula, then there is a high probability that the transplanted nucleus could direct development to produce fertile adults. **(b)** If, on the other hand, the nucleus is taken from a differentiated tissue, such as tadpole intestinal cells, most transplants are unsuccessful, but in a few cases the transplanted nucleus could direct development to the tadpole stage, indicating that the necessary genes are present and active

Differentiation and determination

Differentiation leads to the production of many morphologically distinct cell types from the original zygote. This process is highly co-ordinated and gene expression must be precisely regulated.

Differentiation involves a series of pathways that lead from the fertilised egg to mature cells in each specialised tissue. The particular pathway that will be followed is decided by a series of changes that takes place in each cell during development. At a particular time, a cell becomes committed to a certain 'fate', a process called **determination**. The time of determination can be revealed by transplantation experiments (Chapter 14). Once differentiated, cells do not normally change into a different cell type.

Cells usually become determined (committed) to differentiate along a particular pathway long before any morphological change is apparent. For example, stem cells in the bone marrow of mammals give rise to many types of blood cells, including red blood cells, which transport oxygen; lymphocytes, which produce antibodies; and megakaryocytes, which produce platelets (Chapter 6). In the early stages of proliferation, although progenitor cells are difficult to distinguish, they are already committed to different fates.

Determination is the commitment of a cell to a particular developmental pathway. Cells may become 'determined' long before any morphological change is apparent.

Differential gene activity

The differences between differentiated cells occur as a result of regulation of expression of different genes, that is, *differential gene activity*, during development. Regulation of gene expression may occur during transcription or translation, or post-translationally (Chapter 11).

Different genes may be expressed in different cells of an organism at the same stage of development or within the same cell at different times. For example, at a particular stage of development, many mRNA transcripts can be made from the same DNA segment and each transcript can be used to produce multiple copies of a specific protein. In the development of the silkworm, *Bombyx mori* (Fig. 16.4), an egg develops through a series of larval stages into a caterpillar, which pupates, finally emerging from the pupal case as an adult moth. To make the pupal case, the caterpillar must produce large quantities of 'silk', the major component of which is a protein, *fibroin*. In each silk gland, a single fibroin gene makes about 10 000 RNA transcripts, each of which directs the synthesis of about 100 000 molecules of fibroin produced over four days. The silk gene is controlled by the steroid hormone, ecdysone.

> Gene expression may be regulated at the transcription stage, for example, the silk gene, or at later stages.

Regulation of transcription: polytene chromosomes

Ecdysones are a group of hormones that can induce particular genes to be transcribed rapidly during development. The effect of this hormone on chromosomes can be visualised using the 'giant'—*polytene*—chromosomes that are contained in some cells of certain insects, for example, the salivary glands of larvae. Polytene chromosomes are interphase chromosomes (Fig. 16.5) and so are much more elongated than metaphase chromosomes. They are formed as a result of repeated duplications of DNA without any cell division. Maternal and paternal homologues are generally perfectly paired throughout their length so that the chromosome resembles a multistranded cable. In *Drosophila*, there are nine duplications, thus:

$$2^9 = 512 \times 2 \text{ strands as it is diploid}$$
$$= 1024 \text{ strands}$$

The midge, *Chironomus*, has an extra round of replication and therefore has 2048 strands.

Polytene chromosomes have a series of light and dark bands and it has been possible to assign specific gene loci to specific dark bands. Each gene locus apparently occupies a single band. There are approximately 5000 bands in the polytene chromosomes of *Drosophila*, so they may represent the full complement of an individual's genes. In all

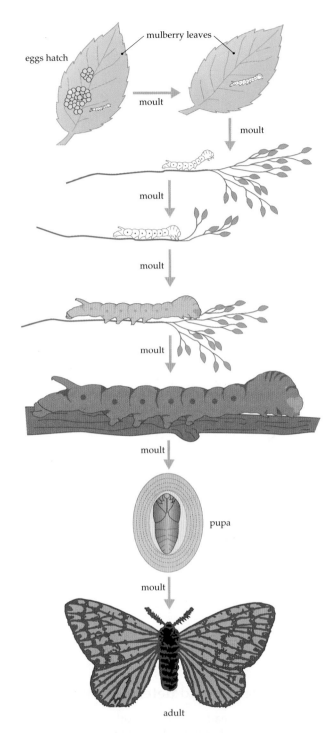

Fig. 16.4 The normal pattern of development in the silkworm, *Bombyx mori*. The series of moults that leads to the development of the adult is triggered by the insect hormone, ecdysone

cases, the relative positions of genes and their map distances apart, as determined by genetic analysis, correlate well with the positions of bands identified as the sites of particular gene loci.

The width of a polytene chromosome varies along its length. At some places, the chromosome can be quite swollen, up to three or four times the usual diameter (Fig. 16.5). These regions are *puffs* and are regions of RNA synthesis. This can be shown by supplying a cell with radioactive uridine (the nucleotide unique to RNA) and using auto-radiographic detection methods to demonstrate that the label is incorporated in the region of the puff. Therefore, if dark bands represent genes, then puffed bands are genes that are being actively transcribed. Interestingly, in an organism such as *Drosophila*, the location of puffs on polytene chromosomes varies from one cell to another, indicating that different genes are active in various cell types. Even in the one type of cell, the puffing pattern varies over time, indicating that different genes are active at different times.

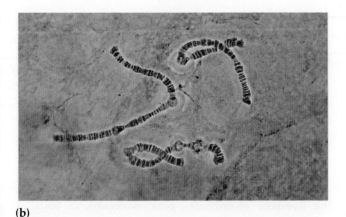

(a)

(b)

Fig. 16.5 **(a)** Part of a polytene chromosome of the midge *Chironomus*, unpuffed (top) and showing puffs (centre and bottom), where active transcription is occurring **(b)** Polytene chromosomes of the Australian midge, *Chironomus duplex*. These chromosomes have been stained to show the banding pattern and the puffs present

A good example of differential gene activity is seen during insect moulting, that is, when the cuticle is shed. Each time an insect moults, a predictable pattern of puffing occurs. Moults are triggered by increasing levels of ecdysone. If ecdysone is injected into a larva of the genus *Chironomus* just after it has completed a moult, when its endogenous levels of ecdysone are low, a new moult can be induced. The normal pattern of moulting puffs then begins. The first puff appears within about 30 minutes and then all the other 'premoult' puffs appear in sequence, until the entire puffing pattern is completed. It appears that the injection of ecdysone into a larva turns on a series of genes involved in the moulting process. This is confirmed by the increase in RNA seen after ecdysone injection, which is not found in the cells of untreated larvae (Fig. 16.6).

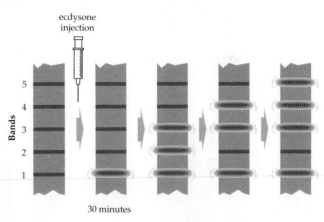

Fig. 16.6 Induction by ecdysone. The insect hormone, ecdysone, turns on and off the transcription of a specific sequence of bands (genes) involved in insect moulting in polytene chromosomes if injected into a larva just after a moult, when the levels of the hormone are low

The evidence for differential transcription has been strengthened by demonstrating a correlation between the presence of a particular puff and the presence of a specific gene product. There are a large number of species of *Chironomus* but two of them, in particular, differ distinctly. *Chironomous pallidivittatus* has granules of a specific protein in some cells of its salivary gland, while the related species, *Ch. tentans*, does not. Furthermore, *Ch. pallidivittatus* shows an extra puff towards the end of chromosome 4 that is missing in chromosomes of *Ch. tentans*. These two species can be crossed and it has been found that both the extra puff and the granular protein are inherited as single gene traits. Chromosome mapping shows that the locus of the gene for granular protein is at the position of the extra puff. When salivary gland proteins are separated by gel electrophoresis, that is, on the basis of charge and size in a gel matrix, there are a number of common proteins and one that is specific to *Ch. pallidivittatus*. In hybrids between the two species, which have a puff only in the chromosome derived from *Ch. pallidivittatus*, an intermediate level of the specific protein is found (Fig. 16.7).

In the midge *Chironomus*, there is a correlation between a specific polytene chromosome 'puff' and transcription of a gene producing a particular salivary protein.

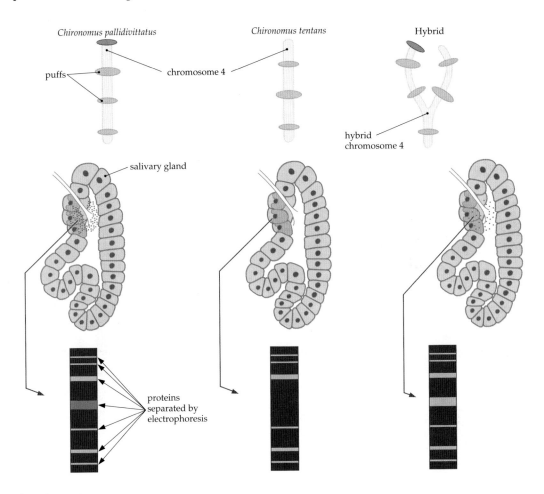

Fig. 16.7 Correlation between the presence of a particular chromosome puff and the production of a particular protein in some salivary gland cells in chironomid midges. *Chironomus pallidivittatus* produces a puff and a particular protein that is not present in *Ch. tentans*. A hybrid between the two species produces the particular puff in the chromosome inherited from *Ch. pallidivittatus* and a small amount of protein. Note the protein and the puff are present only in the few cells that contain the puff

BOX 16.2   SPORE FORMATION IN BACTERIA

The bacterium, *Bacillus subtilis*, is an organism that can be grown readily and is easy to manipulate. In response to adverse conditions, such as nutrient limitation, this bacterium initiates a developmental program that results in sporulation, that is, the formation and release of a spore inside the bacterial cell (see figure). Sporulation involves passage through a series of well-defined morphological stages, and results in lysis of the cell and release of the spore. The spores are highly resistant to normally lethal conditions, such as heat or desiccation.

During sporulation, a cell develops two compartments, the larger *mother cell* and the smaller *forespore*. The two compartments inherit identical chromosomes from the last round of mitotic DNA replication within the cell. The forespore develops into the mature spore, while the mother cell contributes to the development of the spore, and is disrupted when the mature spore is released. The

forespore therefore represents a 'germ-line' cell because it will give rise to subsequent progeny (Chapter 14). The mother cell, on the other hand, is similar to 'somatic cells', which do not contribute genes to the next generation.

Gene expression in these two compartments is differentially regulated, each compartment expressing a different set of developmental genes. However, the two developmental programs are not completely independent; communication between the mother cell and the forespore co-ordinates gene expression in both compartments.

Spore development requires the expression of 80 or more developmental genes:

■ *spo* genes, spore-forming genes identified and defined by mutations that block development of the spore at a particular stage but have little or no effect on vegetative growth;

- *ger* genes, required for normal germination of the mature released spore;

- *cot* genes, required for the polypeptide components of the spore coat;

- *ssp* genes, which code for a family of small acid-soluble proteins located in the core of the spore.

Studies of the regulation of these genes show that spore development is correlated with an ordered program of gene expression in which sets of genes are activated in a sequential compartment-specific manner.

A major feature of the developmental process is the formation of the septum that divides the cell into the two compartments. At an early stage, invagination of

the bacterial membrane encloses a chromosome and some cytoplasm. Then, a thick and tough surface layer of protein is laid down on the outside of the membrane. At the same time, transcription of genes normally used in vegetative growth is shut down. As differentiation continues, we see that the spore contains a few ribosomes, a copy of the bacterial chromosome and the proteins necessary for germination. The germination process converts spores to a condition in which they can make vegetative proteins and thus reverse the shutdown process and make viable vegetative bacteria capable of growth and division. Differential gene expression in the two compartments of the cell is controlled by different forms of RNA polymerase (and particularly the σ-factor responsible for the initiation of transcription; Chapter 11). Different forms of this factor are present in the mother cell and the forespore.

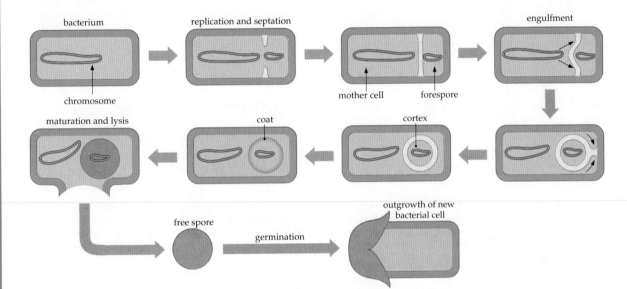

Morphological stages of sporulation in the bacterium, *Bacillus subtilis*. The spore forms as a result of nutrient depletion and the original bacterial cell forms two compartments, the mother cell and the forespore, which develops into the spore. At the completion of spore formation, the mother cell disrupts and releases the spore

BOX 16.3   DEVELOPMENT IN *DROSOPHILA*

The life cycle of *Drosophila* consists of a series of distinct stages (Fig. a). Development is indirect; the larva undergoes a complete change of form, metamorphosis, during which many tissues degenerate while others differentiate, to form the adult, or *imago* (Chapter 14).

Although larvae look nothing like adult flies, they contain groups of cells, imaginal discs (from 'imago'), the precursors of adult structures. Imaginal discs are pouches of epithelial tissue that invaginate and differentiate at metamorphosis. Each imaginal disc occupies a defined position in the larva and will form a specific structure in the adult, for example, a wing or a leg (Fig. b). When a larva develops into a pupa and metamorphosis begins, larval cells break down and cells in the imaginal discs proliferate to form adult structures.

Although imaginal disc cells appear relatively undifferentiated in the larva, their fate is already determined so that, if transplanted before metamorphosis, they will form the determined structure. For example, if you transplant a wing imaginal disc into another larva, an extra wing will develop. Occasionally, however, an antennal imaginal disc will give rise to a leg, or the second thoracic segment will give rise to an extra wing. This change in the fate of an imaginal disc is known as *transdetermination* and it can be genetically manipulated. Mutant strains of *Drosophila*, which show disturbances of their body plan, occasionally arise. The genes involved may specify the formation of imaginal discs, their position in the larva or their role in the adult.

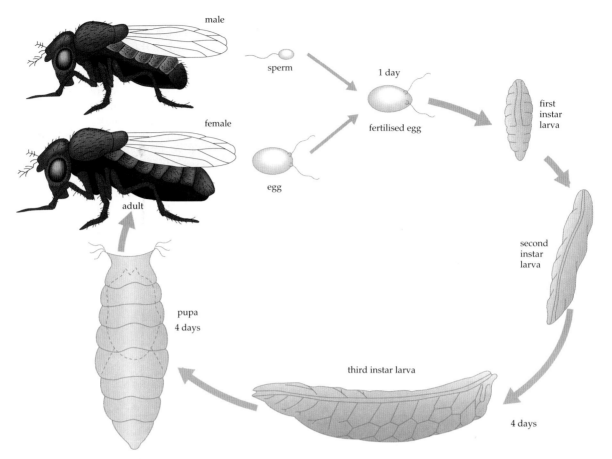

(a) Developmental stages of *Drosophila* from egg to adult fly. The times of development are for fly stocks reared at 25°C. After the egg is fertilised, the zygote develops into a sexually immature form, a larva. Each larva undergoes two moults, increasing in size between each. The larva then undergoes a final moult to become a pupa and the cuticle hardens so that the pupa is completely encased. The insect undergoes metamorphosis within the pupa to emerge as an adult

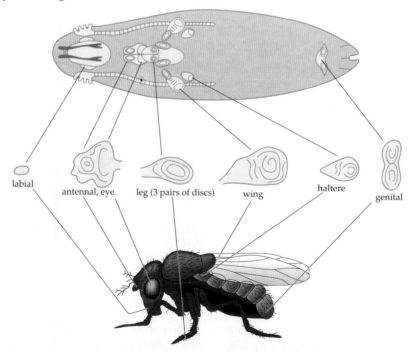

(b) Locations of imaginal discs in a *Drosophila* larva and the adult structures derived from them, showing the general morphology of the discs late in larval development

GENE REGULATION AND DEVELOPMENT OF BODY FORM

In recent years, we have begun to identify some of the genes involved in development of body form and to gain an understanding of how they act.

Genes and *Drosophila* development

Many developmental genes have been identified in *Drosophila*. We will consider those that have effects on the body plan of the insect. Both larval and adult *Drosophila* have a segmented body plan, comprising a head, three thoracic segments and eight abdominal segments (Fig. 16.8). Developmental genes modify both the fate of segments and the relationships between them.

Early development in *Drosophila* takes place within the egg—the zygotic nucleus divides and a series of 13 mitotic divisions takes place (Fig. 16.9). The first

nine of these divisions take place without cytokinesis (cell division) and occur very quickly, each division taking only 5–10 minutes. Therefore, a cluster of nuclei is formed within the egg, remaining at the centre of the egg. Most of the nuclei then begin to migrate to the periphery of the egg where cell membranes are formed around each of the nuclei. During this phase the germ line is formed from about 10 cells, the pole cells, which are segregated at the posterior end of the embryo.

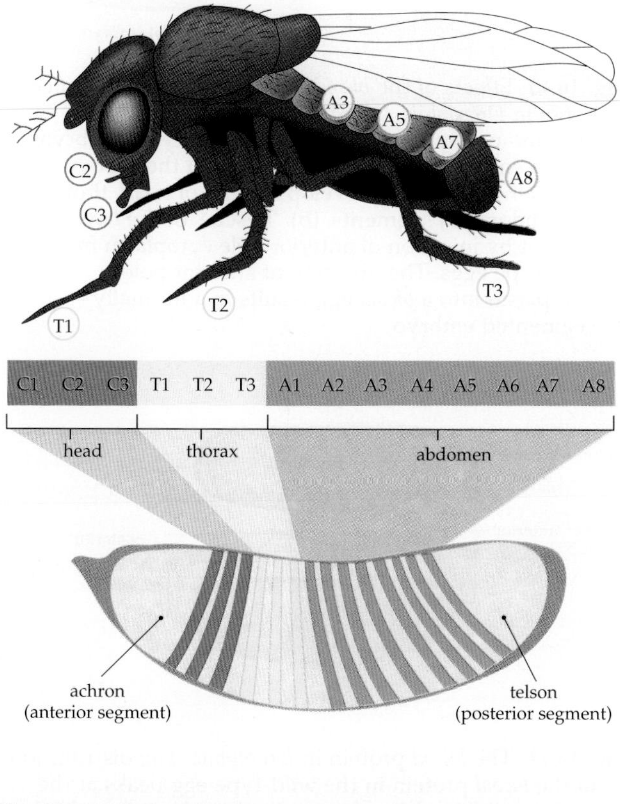

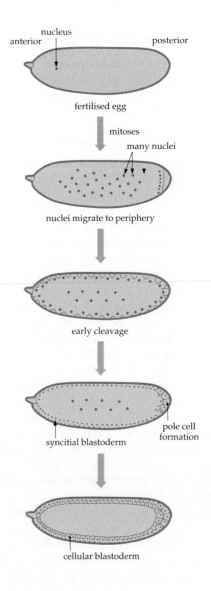

Fig. 16.9 Early developmental stages in the *Drosophila* embryo

Fig. 16.8 Relationship between larval and adult segmentation in *Drosophila*. The segmented body plan is laid out in the embryo. Segments C1–C3 produce the head, while segments T1–T3 produce the thorax. Each thoracic segment carries a pair of legs. The wings develop on the second thoracic segment (T2) and the halteres (flight balancers) on the third thoracic segment (T3). Segments A1–A8 give rise to the abdomen

The earliest stages of development are controlled by maternal-effect genes (p. 333). Mutations in maternal-effect genes are readily identified in genetic experiments. If we carry out reciprocal crosses (*m* is a recessive maternal-effect mutant and + is the wild-type or normal allele), then:

female $\dfrac{m}{m}$ × male $\dfrac{+}{+}$ ⟶ $\dfrac{+}{m}$ (abnormal development)

and

female $\dfrac{+}{+}$ × male $\dfrac{m}{m}$ ⟶ $\dfrac{+}{m}$ (normal development).

The progeny of both crosses are genetically identical but the homozygous $\dfrac{m}{m}$ females produce eggs that are not capable of normal development, while homozygous $\dfrac{m}{m}$ males produce normal sperm. The wild-type gene products from the female are necessary for normal development of the zygote.

> Products of maternal-effect genes are located in egg cytoplasm and are essential for normal development of the zygote.

Specifying the anterior–posterior axis

In *D. melanogaster*, analysis of maternal-effect mutants shows that maternal-effect genes produce substances, stored in the egg cytoplasm, which establish the polarity of the *Drosophila* egg before fertilisation takes place. They determine which parts of the egg are dorsal or ventral and which are anterior or posterior and thus determine the basic axes of the embryo.

If a small amount of the cytoplasm is removed from the anterior region of certain insect eggs, the embryo develops with its head and thorax missing. This observation led to the idea that some factor, an **anterior determinant**, which specifies both the head and thoracic segments, is released from the anterior pole of the egg. A search for mutants whose phenotype resembled embryos produced by removal of the anterior pole cytoplasm led to the discovery of the *bicoid* mutant (Fig. 16.10a). The *bicoid* mutation can be reversed (or 'rescued') by injecting anterior pole cytoplasm from a normal wild-type embryo into a *bicoid* egg, suggesting that the wild-type product of the *bicoid* gene may be the anterior determinant (Fig. 16.10b).

The *bicoid* gene was isolated (cloned into a vector) and used to detect its mRNA and protein product, and their positions in the embryo. After fertilisation, *bicoid* mRNA is confined to the anterior tip of the egg. The *bicoid* protein formed by translation of this mRNA diffuses away from the anterior pole and a concentration gradient is established. The peak of this gradient is at the anterior pole and the low point about two-thirds of the way to the posterior pole (Fig. 16.11). During this phase, the embryo is a syncytium. This means that the *bicoid* protein has free access to nuclei in the regions in which it occurs and can interact with genes in them.

Injection of pure *bicoid* mRNA into a mutant *bicoid* egg causes a full set of head and thoracic structures

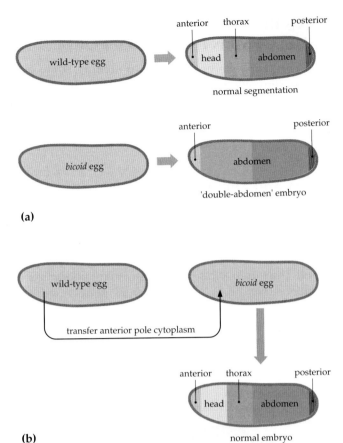

(a)

(b)

Fig. 16.10 Effects of the *bicoid* mutation (a maternal-effect gene) in *Drosophila*. **(a)** Structures produced in the anterior–posterior axis in normal (wild-type) embryos and in embryos of the *bicoid* mutant. In the mutant, two abdominal segments develop instead of head, thorax and abdominal segments. **(b)** 'Rescue' of the *bicoid* mutant by injection of anterior pole cytoplasm from wild-type eggs. The injection of anterior pole cytoplasm into a *bicoid* egg results in a normally segmented embryo

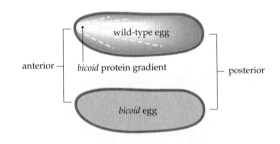

Fig. 16.11 The *bicoid* protein in *Drosophila*. The distribution of the *bicoid* protein in the wild-type egg peaks at the anterior pole and forms a gradient to about two-thirds of the distance to the posterior pole. This gradient is absent in the eggs of the *bicoid* mutant

to form in that embryo and the site of head formation corresponds to the site of injection (Fig. 16.12). Over 20 genes and their products are involved in the

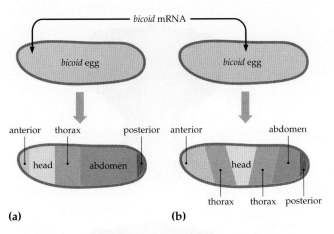

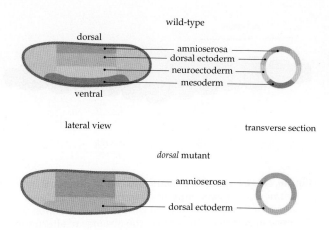

Fig. 16.12 The effects of injection of wild-type *bicoid* mRNA into mutant *bicoid* eggs. Injection of *bicoid* mRNA causes anterior structures to form, with the site of injection defining the location of the head. **(a)** When the mRNA is injected into the anterior end of the egg, a normal set of segments forms; **(b)** whereas when it is injected into the middle of the egg, head structures form at the middle with a duplicated set of thoracic structures on either side

Fig. 16.13 The *dorsal* mutant of *D. melanogaster*. The *dorsal* mutant lacks structures found in the ventral regions of wild-type embryos

appearance of the anterior body segments, so the *bicoid* protein does not itself cause these anterior segments to form but regulates other genes lower in the hierarchy of interactions. The *bicoid* gene is maternally inherited and the *bicoid* mRNA first appears in nurse cells, which are connected to the oocyte by cytoplasmic bridges (Chapter 15). It is subsequently transported across these bridges and accumulates in a spherical region at the anterior pole of the oocyte. This localisation of *bicoid* mRNA is also genetically controlled.

> The *bicoid* gene is an example of a gene that regulates other genes and activation of *bicoid* transcription in nurse cells is an example of transcriptional control.

Specifying the dorsal–ventral axis

Gene action during oogenesis also leads to the establishment of the dorsal–ventral axis. The genes involved were identified by looking for mutant embryos with defects of the dorsal–ventral axis formation, and the gene *dorsal* appears to be involved in this process. In embryos carrying mutant (inactive) dorsal genes, structures that normally appear on the ventral side of the embryo, such as the ventral nervous system and the mesoderm, are replaced by dorsal structures, such as dorsal epidermis (Fig. 16.13).

> Maternal-effect genes establish the polarity of the *Drosophila* egg before fertilisation takes place. They determine the anterior–posterior and dorsal–ventral axes of the future embryo.

Segmentation genes

The next set of genes to act on the developing embryo, the **zygotic genes**, are the developmental genes of the zygote genome. These begin to be expressed just before cellularisation of the embryo. Also known as *segmentation genes*, zygotic genes extend development beyond the pattern established by maternal genes and act to determine the position and pattern of the segments in the embryo. Segmentation genes act more or less sequentially to produce increasing levels of organisation and fall into three classes (Fig. 16.14). The first segmentation genes to act are the *gap* genes, which act in regions along the embryo to interpret the posterior–anterior information and establish a spatial organisation that will lead to segmentation. Mutations in *gap* genes result in an embryo having one or more missing segments, so that gaps appear in the normal structural pattern of the embryo. The other two classes of segmentation genes act on all segments. For example, the *pair-rule* genes control differentiation of the embryo into discrete segments; mutations in pair-rule genes cause deletion of every second segment. The mutation *even-skipped* affects even-numbered segments, while *odd-skipped* affects only odd-numbered segments. Finally, there are *segment-polarity* genes that react to signals to determine the pattern of development within each segment of the embryo. Mutations in segment-polarity genes often lead to the normal number of segments, but part of each segment may be deleted and replaced by a symmetrical mirror-image duplication of the portion that is retained.

> Zygotic (segmentation) genes extend the developmental program beyond that established by the maternal-effect genes.

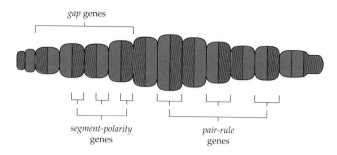

Fig. 16.14 Groups of segments affected in different segmentation gene mutants. Mutants in *gap* genes cause the absence of several contiguous segments, those in *pair-rule* genes cause the absence of alternating segments or parasegments, and those in *segment-polarity* genes fail to undergo the normal anterior–posterior differentiation within each parasegment

Homeotic genes

Finally, a set of genes designates the various segments along the length of the larva and adult. These homeotic genes, when mutant, result in a transformation of one body part into another. Therefore, some very striking changes can occur in adults that are mutant for these genes, for example, *antennapedia* mutants have legs on the head where the antennae would normally be (Fig. 16.15). Another example is the mutant *bithorax*, in which the halteres (flight balancers) are converted into a second pair of wings on an extra thoracic segment (Fig. 16.16).

The normal *antennapedia* gene designates the structure of the second thoracic segment and the legs on it, so the *antennapedia* mutation leads to a gain of function in which the dorsal part of the head becomes a functional thoracic segment and the antennae become legs. Thus, homeotic mutations appear to be changes in control genes involved in determination.

Many of the homeotic genes in *Drosophila* are found as gene clusters. One of these clusters is the *bithorax* complex and comprises at least eight genes that affect the identity of the larval segments. The role of the *bithorax* complex is revealed by a study of larval development. A mutation in one of the genes of the complex *bithoraxoid* leads to a fly with legs on the first abdominal segment, and the first abdominal segment of the larva would be a repeat of the third thoracic segment (Fig. 16.15).

The homeotic genes of *Drosophila* were first identified as alterations in the phenotype produced by mutant alleles. Upon DNA sequence analysis of a number of homeotic genes, it became apparent that a short sequence (approximately 180 base pairs) is characteristic of many homeotic genes. This sequence has been termed the **homeobox**. The homeobox

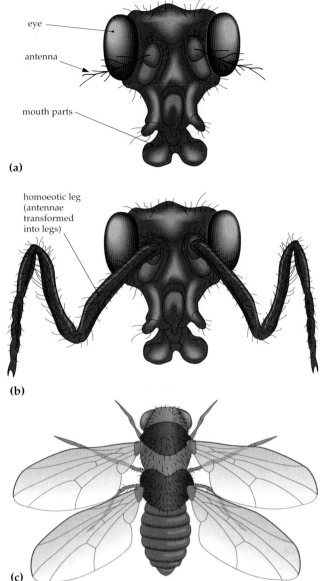

Fig. 16.15 Examples of homeotic mutants that transform one body part to another. **(a)** The head of a wild-type *Drosophila*. **(b)** *Drosophila* in which the antennae are transformed into legs (*antennapedia* mutant). **(c)** An adult *Drosophila* with four wings produced by mutations in the *bithorax* complex. The mutants convert the third thoracic segment into the second thoracic segment, and the halteres normally present on the third thoracic segment become converted into a second pair of wings

sequence has been detected in a wide range of organisms, including yeast and humans, and has been implicated in the control of development.

> Homeotic genes act to control the developmental pattern within individual compartments of the embryonic organism and, as such, are an example of genes that control development.

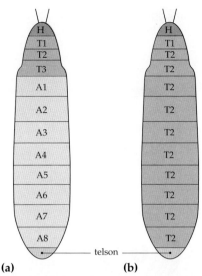

Fig. 16.16 (a) Segmentation pattern in a wild-type *Drosophila* larva. There is a head (H) comprising the archon and C1–C3, three thoracic segments (T1–T3), eight abdominal segments (A1–A8), and a telson. (b) Segmentation pattern in a mutant larva that lacks the complete set of *bithorax* genes. The head, segment T1 and the telson are normal but all the other segments develop as T2 segments

Fig. 16.17 Flowers of wild-type plant *Arabidopsis thaliana*, showing whorls of sepals, petals, stamens and carpels

Flower development in *Arabidopsis*

While much of our understanding of how homeotic genes act to regulate development has come from genetic studies of segmentation in *Drosophila*, homeotic genes also play a regulatory role in plants. During flower formation, these genes control the identity of the whorls of floral organs as they develop at the floral meristem. The model plant, *Arabidopsis*, has been used to study genes that regulate flower development because of its small genomic size, rapid generation time from embryo to seed set and ease of use in generating mutants.

Mutants are produced by treating seeds with a chemical mutagen, such as ethylmethane sulfonate, at concentrations that kill up to 15% of seeds. Seedlings of this M_1 (mutation 1) generation are grown to flowering, self-pollinated and their seeds grown (M_2 generation). Floral mutations are visible in the M_2 generation in a small proportion of plants. Most of the mutants obtained are recessive and only produce a visible phenotype in the M_2 generation when in the double-recessive condition.

The war of the whorls

In *Arabidopsis* flowers, there are four whorls of organs. Outermost are the four sepals (*whorl 1*, Fig. 16.17), followed by the four petals (*whorl 2*) internal to and alternating with the growing sepals, then a whorl of six stamens (*whorl 3*) and finally a central pistil

formed from two carpels (*whorl 4*). These are precisely sited in relation to each other because they are formed sequentially during development: sepals initiated first and carpels last.

Study of floral mutants has identified a class of *whorl identity genes* (Fig. 16.18). These are homeotic genes, that is, regulatory genes whose actions determine which whorl develops. This can be seen in the floral mutants. For example, the *agamous* mutant has flowers that do not develop stamens and carpels. Instead, they have sepals (whorl 1), petals (whorl 2), petals (whorl 3) and sepals (whorl 4), forming 'double' but sterile flowers (Fig. 9.13). Another mutant, *apetala 2*, lacks sepal and petal whorls, which are replaced by carpels and stamens, that is, flowers have carpels (whorl 1), stamens (whorl 2), stamens (whorl 3) and carpels (whorl 4). These examples show that mutations always affect not just one whorl but a *pair of adjacent whorls*.

A genetic model has been proposed to explain these observations in *Arabidopsis*. The four whorls of a wild-type flower are specified by three sets of homeotic genes, which are active in three overlapping regions of a floral meristem (Fig. 16.19). These regions, A, B and C, each express a specific set of organ identity genes. The set of genes active in each region specifies two adjacent whorls. If one region is missing as a result of mutation, the organs specified will differ from wild-type. A specific combination of genes specifies each whorl because the regions are overlapping. Sepals can only be specified by region A organ identity genes acting alone; petals arise if there is

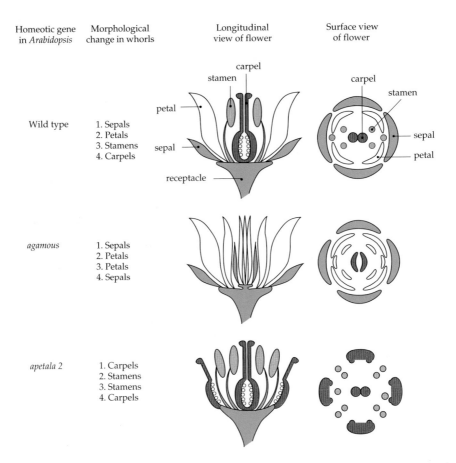

Fig. 16.18 Flower development patterns in wild-type and two whorl identity mutants of *Arabidopsis*. In each mutant, the two outer or inner whorls have been switched as indicated. The *agamous* mutant flowers are shown in Figure 9.13

overlapping expression of region A and B genes; stamens from overlapping expression of region B and C genes; and carpels only from region C gene expression.

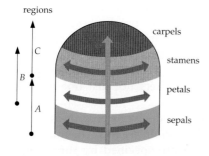

Fig. 16.19 Whorl identity genes regulate the sequential development of floral organs in flowers of *Arabidopsis*. These genes are active in three overlapping regions of the floral meristem (arrows on left). As the floral meristem develops, sepals are initiated by a set of genes in region A, petals are initiated by set of genes in A or B, stamens by set of genes in B or C, and carpels by set of genes in C only. Regions A and C are mutually antagonistic, and a mutation in, for example, region A means that all the floral organs are specified by region C, and vice versa

In order to explain the phenotypes of many mutants that have been observed, we need to assume, in addition, that the activities of regions A and C are mutually antagonistic. This means that when a mutant is in region A (i.e. gene set A is missing), genes from region C are expressed in all four whorls. *Apetala 2* is therefore a mutation of a homeotic gene in region A, and this mutation changes the pattern of development of the outer two whorls of sepals and petals. When the mutant is in the C region, genes from region A are expressed in all four whorls. *Agamous* is a mutation of a homeotic gene in region C, and this mutation changes the development of the inner two whorls of stamens and carpels.

Interactions between these homeotic genes controlling different regions of the flower have been termed the 'war of the whorls'.

Genes that regulate flower development have been identified by analysis of homeotic mutations affecting flower structure. Mutations affect the identity of floral organs in adjacent whorls, switching them to alternative types of organs.

SUMMARY

- Differentiation is the process that leads to the expression of a particular pattern of genes in a specialised cell. Some genes are regulated during development, so that the amount of gene product varies. Other genes regulate processes and timing, and the activity of other genes.

- Early development is controlled by intracellular (cytoplasmic) factors produced by maternal-effect genes. Development is regulated by extracellular signals including hormones, such as ecdysone in insects, and environmental factors, such as availability of food and levels of cAMP in slime moulds.

- Genomic equivalence means that cells have the same genes, although they may be differentially expressed in different tissues.

- Differentiated plant cells can be totipotent, that is, capable of giving rise to a complete organism under appropriate conditions. In animals, a differentiated somatic cell cannot be induced to behave like a zygote. Under certain circumstances, nuclei from differentiated animal cells can direct embryonic development.

- Determination is the commitment of a cell to a particular developmental fate. Cells may become 'determined' before any morphological change is apparent.

- Differential gene expression can result from differential transcription. In the midge *Chironomus*, there is a correlation between a specific polytene chromosome 'puff' and transcription of a gene producing a particular salivary protein.

- The insect *Drosophila melanogaster* is used to study the effects of genes on development. Developmental mutants can easily be isolated and the effects of these mutations on the normal segmentation pattern of the insect determined.

- Products of maternal-effect genes are located in egg cytoplasm and are essential for normal development of the zygote. Maternal-effect genes establish the polarity of the *Drosophila* egg before fertilisation takes place. They determine the anterior–posterior and dorsal–ventral axes. Zygotic (segmentation) genes extend the developmental program beyond that established by the maternal-effect genes.

- Homeotic genes act to control the developmental pattern within individual compartments of the embryonic organism and, as such, are an example of genes that control development.

- The model plant, *Arabidopsis*, is used to study genes that regulate flower development. Mutants can be readily isolated and genes identified because of the small genome size. Homeotic mutants show a switch in position of the floral organs to positions normally occupied by other organs.

QUESTIONS

1. Distinguish between the terms differentiation and determination during animal development.

2. (a) Why are plant cells considered to be totipotent? (b) While animal cells cannot be considered to be totipotent, what is the significance of experiments in which the nucleus from differentiated cells can be transplanted and shown to direct embryonic development?

3. We assume that all somatic cells in an organism contain equivalent genetic information. (a) How then do we get different tissues formed? (b) What experimental systems might you use to investigate this process?

4. How is the fate of embryonic cells or tissues decided?

5. In *Drosophila*, the larval stage can be considered as a raft for the imaginal discs. Explain this statement.

6. (a) What is a 'puff' in polytene chromosomes in the salivary glands of certain insects? (b) What role does differential transcription play in development?

7. Give two examples of specific gene activity during development.

8. Provide a molecular explanation for the events of sporulation in bacteria.

9. (a) What is a homeobox region? (b) Describe the effect of homeotic mutations, for example, in *Drosophila*.

10. (a) Draw diagrams to show the whorls of organs in flowers of two different homeotic mutants in *Arabidopsis*. (b) Explain the simple model of the role of homeotic genes in regulating floral development.

PART 4

REGULATION OF
THE INTERNAL
ENVIRONMENT

STRUCTURE OF PLANTS

Many structural features of plants are related to the requirements of living on land. These include preventing desiccation, acquiring nutrients and water from the soil, and providing physical support in an aerial environment. Plants have an aerial shoot system for photosynthesis and an underground root system for absorption and anchorage. Vascular plants, such as ferns, conifers and flowering plants, have a transport system linking the shoot and root.

While there are basic similarities in structure common to all vascular plants, there are also differences between taxonomic groups and between individual plants adapted to different environments. For example, eucalypts are easily recognisable by their characteristic branching and foliage patterns (Fig. 17.1a). The crown of the tree is open with branchlets in clusters, and leaves that hang vertically. In contrast, a palm tree has a single trunk topped with large umbrella-like leaves (Fig. 17.1b).

In this chapter, we will consider the structure of vegetative organs of flowering plants, that is, stems, roots and leaves. The tissues and cells that make up these organs have been described in Chapter 6. In Chapter 18, we describe the physiological functions of plants.

(a)

(b)

Fig. 17.1 Characteristic form of trees. **(a)** A eucalypt, here *Eucalyptus papuana*, has an open crown with branchlets in clusters. **(b)** An Australian rainforest palm, in contrast, has a non-branched trunk topped with large compound leaves

STEM STRUCTURE

Aerial shoots consist of **stems** with attached leaves. The stem provides support, contains vascular tissues for water movement and sugar transport and, in some plants, is an important storage organ. The shoot grows from an apical meristem, which produces new cells and leaf primordia (Chapter 16). Leaves are attached at **nodes**, while the portions of stem between successive nodes are **internodes** (Fig. 17.2).

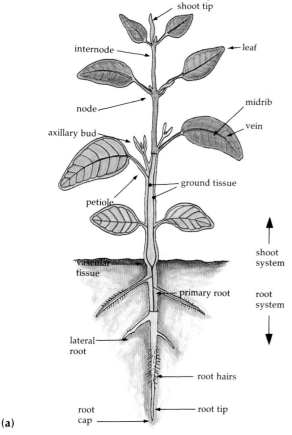

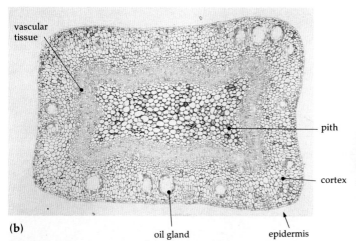

(b)

Fig. 17.2 The structure of a flowering plant.
(a) Arrangement of the organs of a eucalypt, with primary stem and root (portion cut away to show vascular bundles). (b) Cross-section of a stem of *Eucalyptus preissiana*, showing major tissue types, including the ring of vascular tissue

Primary growth

Vascular bundles are the conspicuous structural features in young stems of eucalypts, wattles and beans, which are flowering plants called dicotyledons (Chapter 37). Together with parenchyma, the vascular bundles form a central cylinder, the **stele** (Fig. 17.3a). Bundles diverge regularly from the stele and pass through the outer tissues into the leaves. Vascular

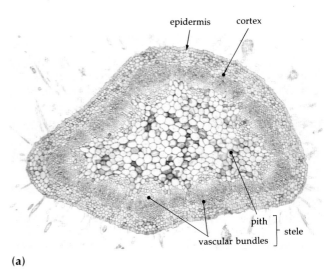

(a)

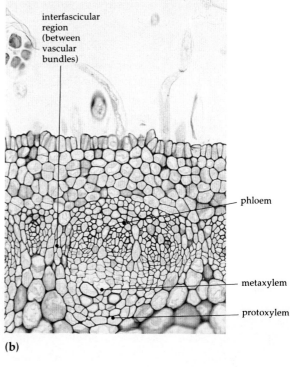

(b)

Fig. 17.3 Primary stem structure in the dicotyledon *Acacia*. Cross-section showing (a) ring of vascular bundles dividing the stem into cortex and pith; and (b) vascular bundle showing protoxylem and metaxylem elements forming an endarch xylem. Phloem is to the outside of the xylem

bundles of the stem of a dicotyledon are usually arranged in a single ring, although double rings occur in some plants (e.g. pumpkin, *Cucurbita*). The vascular ring divides the stem into outer **cortex** and inner **pith**. The cortex contains parenchyma, usually with chloroplasts. The pith also contains parenchyma, often with prominent intercellular spaces, and usually disappears during stem growth.

Vascular bundles consist of xylem and phloem tissues (Chapter 6). The first-formed and earliest maturing xylem, **protoxylem**, develops as thin strands towards the centre of the stem. To the outside of protoxylem, larger, thicker walled xylem elements, **metaxylem** develop. This pattern of primary xylem development, in which new xylem is added to the outside of the protoxylem, is known as **endarch xylem**. Primary phloem forms outside the xylem on the same radius (Fig. 17.3b).

In contrast to dicotyledons, vascular bundles in the stems of monocotyledons, such as grasses, lilies and palms, are scattered throughout the ground tissue, so that there is no division into cortex and pith (Fig. 17.4). Each vascular bundle comprises central phloem tissue surrounded by xylem.

> Stems of dicotyledons have a prominent ring of vascular bundles containing endarch xylem. Stems of mono-cotyledons have vascular bundles scattered in ground tissue.

Secondary growth

The mature stem of most herbaceous (non-woody) plants consists only of primary tissues as we have just described. However, in woody dicotyledons (shrubs and trees), there is another, quite different type of growth and cell production, secondary growth, which involves **cambium**, a secondary meristem. Unlike an apical meristem, which produces new cells behind it and continually adds length to a shoot (axial growth), cambium produces sheets of new cells *laterally*, thus adding girth (radial growth). There are two types of cambium, **vascular cambium**, which produces wood, and **cork cambium**, which produces bark. Together these meristems produce the majority of tissue in the stems of dicotyledonous shrubs and trees. In 'woody' tissues of monocotyledons, such as coconut palm, there is no vascular cambium and no secondary growth; strengthening of the stem results from many individual vascular bundles.

Vascular cambium

In most trees and shrubs, vascular cambium is a continuous multilayered cylinder of living meristem between the cortex and pith, and is responsible for the production of the secondary vascular tissues (Fig. 17.5). Like primary vascular tissue, secondary vascular tissues consist of a number of cell types, most prominently tracheids and vessels of **secondary xylem**, and sieve cells of **secondary phloem**. Cell divisions of vascular cambium are generally periclinal, that is, the plane of division is parallel to the stem surface. When a cambial cell divides, daughter cells form to the inside and the outside of the cambium.

The vascular cambium contains two types of meristematic cells: fusiform and ray initials (Fig. 17.5c). **Fusiform initials** are greatly elongated cells aligned longitudinally in the stem, producing xylem vessels, tracheids and sclerenchyma fibres to the inside, and phloem sieve cells to the outside. **Ray initials** produce parenchyma cells that are aligned radially in the stem and aggregated into clusters, forming wood rays. Ray parenchyma cells are the only living cells in the mature secondary xylem, and maintain links with the vascular cambium. Parenchyma cells of the xylem can remain functional for several years after they are produced.

Secondary growth is most rapid in spring, and the xylem vessels produced are large and thin-walled.

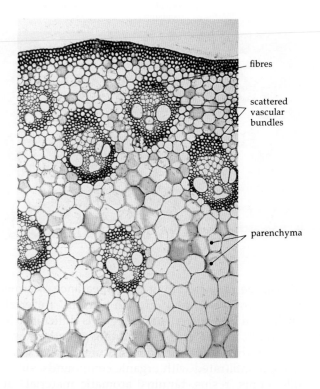

fibres

scattered vascular bundles

parenchyma

Fig. 17.4 Stem structure of a monocotyledon. Cross-section of a stem of maize, *Zea mays*, showing scattered vascular bundles, with outer bundles surrounded by fibres. In contrast to dicotyledons, there is no division into cortex and pith

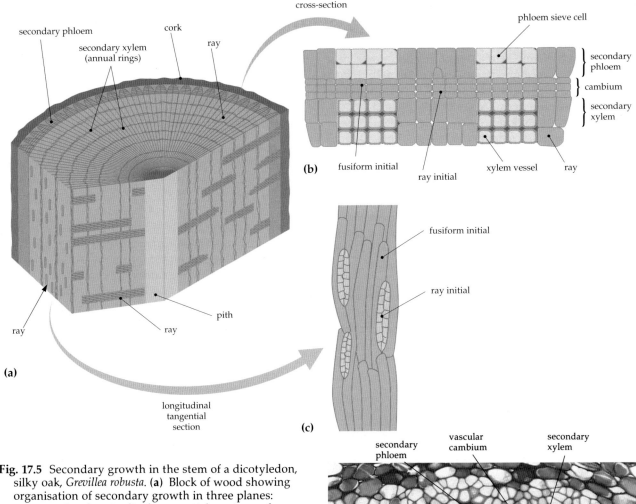

Fig. 17.5 Secondary growth in the stem of a dicotyledon, silky oak, *Grevillea robusta*. (**a**) Block of wood showing organisation of secondary growth in three planes: cross-section, longitudinal–tangential section and longitudinal–radial section. (**b**) Cambium and secondary xylem and phloem formation shown in cross-section. Fusiform initials of the cambium give rise to xylem vessels and phloem sieve cells. Ray initials give rise to ray cells, parenchyma cells that elongate radially. (**c**) Organisation of the fusiform and ray initials seen in longitudinal–tangential section. (**d**) Cross-section of secondary xylem and phloem in the stem of a eucalypt

As the growing season nears its end, environmental factors, such as reduced water supply, shorter day lengths and lower temperatures, slow growth so that smaller and thicker walled xylem is produced. Eventually growth ceases until the following spring. Consequently, there is an abrupt contrast between the last-formed xylem of one season and the new season's growth, which is visible in wood as annual rings (Fig. 17.5a). The age of many trees can be estimated by counting these annual rings, although allowances have to be made for drought years when growth may be interrupted and resumed later in the same season. Also, climatic changes during a tree's life can be predicted from the appearance of growth rings. However, where growth seasons are erratic, such as in the Australian mallee, it is difficult to age eucalypts, for example, by counting growth rings.

When the secondary xylem matures, xylem vessels become infiltrated with organic compounds, such as oils, gums, resins, tannins, aromatic materials and pigments. At this time, the parenchyma cells of the rays die, and the central region of xylem loses its water-conducting capacity and becomes **heartwood**. The outer sheath of xylem, **sapwood**, continues to function. The strength of heartwood compared with sapwood makes it useful in the building industry.

Cork cambium

Periderm is a protective tissue that replaces the epidermis in older stems and forms corky tissue (Fig. 17.6a). Periderm develops from the cork cambium (phellogen), which consists of a series of small, temporary meristems. Individual meristems first form as disc-shaped sheets of cells just underneath the epidermis in the outer stem cortex. Like other meristematic cells, they are continually dividing. Toward the outside of a stem, they produce **cork**, or phellem (Fig. 17.6b), and to the inside, small amounts of thin-walled parenchyma. Cork consists of closely packed dead cells whose cell walls are heavily impregnated with suberin, a very resistant and impermeable material. Suberin protects the delicate inner tissues of stems and roots, forming the outer bark (Fig. 17.6b). Bark is a general term for all tissue outside the vascular cambium, and includes secondary phloem, cork cambium and cork. The bark of 'cork oak', *Quercus suber*, is impermeable to gases and liquids, and hence one of its uses is as stoppers for wine bottles.

Each individual cork meristem is short-lived. New discs of meristematic cells continually form deeper in the cortex. As these new cambia produce cork, older tissues to the outside are shed from the plant, for example, as ribbons of bark in the ribbon gum, *Eucalyptus viminalis*. Eventually all of the original primary tissues of the cortex are discarded, and all subsequent cork cambia arise in the no-longer functional outer layers of the secondary phloem. The reason that large quantities of secondary phloem never accumulate in the stem is because older phloem is continually shed by cork cambial action.

Like leaves (p. 368), young stems have special sites for gas exchange, stomata, in the epidermis. Because stomata of the stem are shed with the cortex during the earliest activity of cork cambia, new sites of gas exchange between the atmosphere and inner living tissues of a woody stem are needed. These are **lenticels**, tiny regions of the periderm consisting of loosely arranged cells with an extensive network of intercellular spaces (Fig. 17.6c). Lenticels often appear as raised areas or dots on stems and the skins of fruits; in old stems, they occur in furrows in the bark.

> Cambia are secondary meristems, sheets of cells forming secondary tissues. The vascular cambium produces wood, and cork cambium produces periderm (corky tissue of bark).

(a)

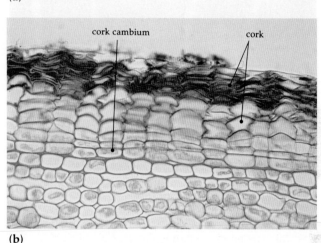
(b)

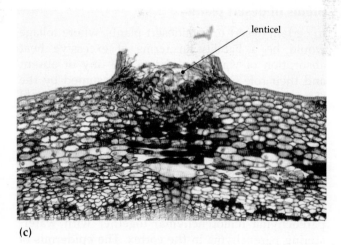
(c)

Fig. 17.6 The periderm or outer protective layer of plants. **(a)** Bark is the protective covering on the trunk of this eucalypt, *Eucalyptus peltata*. **(b)** Cross-section of bark showing the periderm and cork cambium. **(c)** A lenticel in the bark of a tree forms an aeration pore

Special functions of stems

Underground stems

Not all stems are aerial; some grow underground as horizontally growing *rhizomes*, from which aerial branches or leaves arise. Rhizomes often function in vegetative propagation, for example, in grasses (Chapter 14). Underground stems also function as food storage organs. A potato tuber, for example, is actually a swollen underground stem, complete with scale leaves and buds, produced in a single growing season. The edible parts of ginger (a monocotyledon;

Fig. 17.7) and Jerusalem artichokes (a dicotyledon related to sunflower) are long-lived perennial rhizomes, with many short thick nodes bearing papery scale-like leaves. The *corms* of gladioli are an example of swollen underground stems, bearing buds and membranous leaves on the upper surface, and adventitious roots below (see p. 360).

Fig. 17.7 Ginger tuber, showing the short thick nodes bearing papery scale-like leaves

Fig. 17.8 Stems of she-oak, *Allocasuarina*, are cladodes, and function as photosynthetic organs. Leaves are reduced to scales at each node

Stems of desert plants

In certain desert or semidesert plants, where foliage would be a liability in terms of excessive heat absorption or water loss, leaves are tiny or absent and their role in photosynthesis is assumed by the stem. In she-oaks, *Allocasuarina*, leaves are reduced to small, scale-like structures in a whorl at the nodes of stems, **cladodes** (Fig. 17.8), which are the photosynthetic organs. The cladodes have stomata located in grooves that extend the length of the internodes. Some desert plants, such as many cacti, are entirely leafless. Swollen, succulent stems of plants in the family Euphorbiaceae have special photosynthetic parenchyma (chlorenchyma) together with water-storing parenchyma in the cortex. The epidermis of these plants is often multilayered and covered by a thick cuticle.

Stems of marsh and aquatic plants

Plants inhabiting saline environments, which, like deserts, induce water stress in plants, have developed a similar growth form. Leaves of the saltmarsh plant, *Sarcocornia*, are reduced to scales (Fig. 17.9) and the fleshy stem cortex consists of large palisade cells, which store water as well as carry out photosynthesis. Marsh soils are also waterlogged. Oxygen levels are

Fig. 17.9 Stems of the saltmarsh plant, *Sarcocornia* (samphire), are thick and fleshy for water storage

low and plant stems often contain extensive aerenchyma, consisting of parenchyma cells with large gas-filled intercellular spaces (Fig. 17.10).

In submerged aquatic plants, the epidermal wall is covered by a thin cuticle, which allows gas exchange directly with the surrounding water. Epidermal cells are rich in chloroplasts and carry out photosynthesis in addition to leaves. In seagrasses, the vascular system is greatly reduced and the xylem is not lignified.

Stems show specific adaptations to the environment in which the plant lives. Some underground stems are storage organs, stems of desert plants can be photosynthetic, while stems of aquatic plants are adapted for living submerged in water.

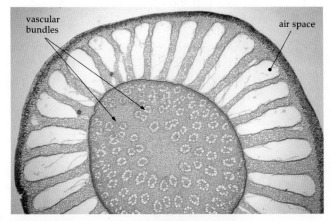

Fig. 17.10 Stems of the marsh plant rush, *Juncus*, have large air spaces in the cortex formed by cell breakdown. The central area contains scattered vascular bundles typical of monocotyledonous plants

ROOT STRUCTURE

Roots are the underground organs of vascular plants and grow from apical meristems at their tips. They grow downwards by tip growth in response to gravity. Unlike stems, roots lack nodes and internodes. They do, however, have lateral branches, which originate from internal tissues. Roots and their surfaces have several important functions (Fig. 17.11), including nutrient and water uptake from the soil, anchorage and support, synthesis of plant hormones, and storage of nutritional reserves (Chapter 18). Roots of some plants are also modified for special functions, including aerial roots for oxygen uptake in salt marshes and swamps, clasping roots in climbing plants, prop roots for support and contractile roots to pull the plant firmly into its substrate.

The first root of a plant develops from the radicle of the embryo. As this root enlarges it produces lateral roots, which in turn divide to form a branched root system. This type of root system, centred around a main taproot, is the common type in dicotyledons

Fig. 17.11 The structure and functions of a root are affected by both developmental and environmental stimuli. Here we see the root of a radish, growing downwards by tip growth. Unlike stems, it lacks nodes and appendages, and develops root hairs and later lateral root branches. The internal (developmental) and external (environmental) factors that control root structure and development are shown

(Fig. 17.12a, b). In some plants, roots can also arise from deep within the tissues of stems. They are **adventitious roots**, which are common at the nodes of grasses and other monocotyledons such as palms. In these plants, the taproot degenerates early and the functional root system is formed entirely from adventitious roots (Fig. 17.12c).

A mature plant develops a large number of roots with enormous total length and surface area. A single four-month old rye plant (*Secale cereale*) may possess 14 million roots with a total length of 630 km (about the distance from Melbourne to Adelaide). A single cubic centimetre of soil from under a Kentucky bluegrass plant (*Poa pratensis*) may contain a total root surface area of 400 cm².

> The main root produced from the radicle is the taproot; lateral roots develop to form a highly-branched root system. Adventitious roots arise from stems.

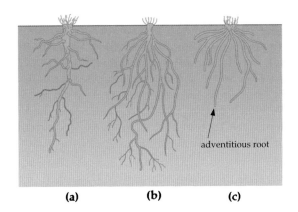

Fig. 17.12 Types of root systems: **(a)** and **(b)** show primary roots with lateral roots and **(c)** shows adventitious roots, which are characteristic of grasses such as **(d)** annual meadow grass, *Poa annua*, in which many lateral roots form at the base of the stem

Primary root structure

A root is covered by an epidermis on the outside, with a well-developed cortex and an inner vascular cylinder (Fig. 17.13). A unique feature of roots is the root cap at the tip, which protects the apical meristem (Chapter 16).

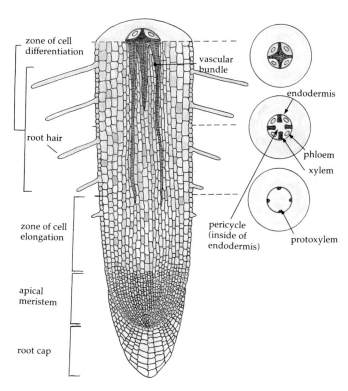

Fig. 17.13 Organisation of roots. Tissues in a root of a dicotyledon, showing the root cap, apical meristem, zone of cell elongation, and zone of cell differentiation. Cross-sections of the root are shown for three regions

Root cap

The **root cap** (Fig. 17.14) comprises large parenchyma cells, which secrete a polysaccharide, mucigel, containing a mucilaginous matrix and sloughed-off living cells. Mucigel acts as a lubricant, enabling the root to grow between soil particles without damage. It also binds soil particles together and may function as a culture medium for bacteria and other microorganisms associated with the root.

Root cap cells frequently contain masses of starch-containing amyloplasts. These are typically sited at the lower side of each cell, where they have accumulated under the force of gravity (Fig. 17.14). It has been suggested that these starch granules act as statoliths, that is, cellular gravity sensors (Chapter 24).

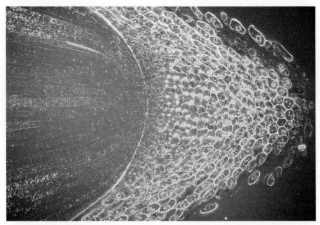

Fig. 17.14 Root cap of maize viewed in longitudinal section. Starch granules lie to one side of each cell (arrow), suggesting they may act as the cellular sensors of gravity. Can you guess in which direction this root was growing?

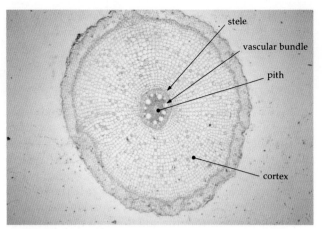

Fig. 17.15 Root of a monocotyledon, *Canna*, viewed in cross-section, showing the central vascular bundles (polyarch) and pith

Epidermis and root hairs

The extensive root surface area, required for uptake of water and nutrients from soil, is enhanced by a zone of numerous fine root hairs, finger-like extensions of the epidermal cells just behind the growing tip (Figs 17.11, 17.13). Root hairs usually possess a thin cuticle, which offers little resistance to the uptake of water and minerals. The soil immediately surrounding this absorptive portion of the root is the **rhizosphere**. In this region, interactions take place between the plant and its soil environment, and with free-living and symbiotic micro-organisms. Behind this root hair zone, lateral roots may develop.

Vascular cylinder

Most roots, including dicotyledons, have vascular tissue at the centre; there is no pith as in stems. Roots have **exarch xylem**, in which xylem forms from the outside towards the inside of the root. A stele is formed with a star-shaped central core of metaxylem with protoxylem points (protostele, Fig. 17.13). Bundles of primary phloem form separately and alternate with the protoxylem on different radii; in stems they are on the same radii.

In monocotyledon roots such as wheat or rice, metaxylem does not completely fill the centre of the root. The xylem is **polyarch**, divided into many ridge-like projections (archs) of protoxylem and metaxylem vessels surrounding a large pith (Fig. 17.15). Phloem alternates with the protoxylem points.

Pericycle

The vascular cylinder is surrounded by the **pericycle** (Fig. 17.16), a layer of one or more cells thick from which lateral roots arise. Lateral roots grow from the

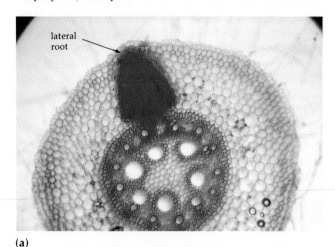

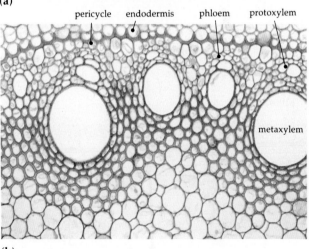

Fig. 17.16 (a) Thick section of a root showing the development of a lateral root (stained red). Cell divisions within the pericycle lead to the development of a new root meristem, which grows and pushes through the cortex and epidermis. (b) Endodermis in the root of a monocotyledon, maize, *Zea mays*. Semi-thin section of the root, showing the thick suberised radial and inner tangential wall (stained red), overlying the pericycle, the outer layer of vascular tissue. Phloem alternates with the xylem, which shows both protoxylem and metaxylem elements

pericycle through the cortex, physically forcing and digesting their way to the root surface and the soil (Fig. 17.16a).

Endodermis

Immediately outside the pericycle is the **endodermis**, a specialised layer one cell thick that plays a critical role in water uptake (Fig. 17.16b). The four radial walls of each endodermal cell are impregnated with a strip of suberin, which makes the cell wall impermeable and water-resistant (Fig. 17.17). This suberin layer in the endodermis is known as the **Casparian strip**. Most water movement from the soil through the epidermis and cortex of the root takes the path of least resistance through cell walls and intercellular spaces, the **apoplastic pathway**. Because this route does not involve crossing plasma membranes, the plant does not exert any physiological control over water movement. Once water reaches the endodermis, however, it is prevented from continuing along the cell wall pathway by the Casparian strip. Instead, it must pass across the plasma membrane and into the cytoplasm of the endodermal cells, the **symplastic pathway**.

Because of the presence of the Casparian strip, ions must be actively transported through living cells of the stele before they can enter the non-living cells of the xylem (Fig. 17.17). Inward ion movement causes osmotic changes and the water potential (Chapter 4) in the stele decreases. Water enters, increasing turgor

pressure in the stele. The endodermis is important because it enables a plant to control the intake of water and ions. The Casparian strip prevents backflow of water from the vascular tissue into the root cortex.

In most of the root, the endodermis also prevents the loss of sucrose from the phloem along the apoplastic pathway and sugar unloading occurs through the symplastic pathway. However, near the root tip, where the endodermis has yet to develop, there is no barrier and sugars are unloaded along the apoplastic pathway.

As a root ages, the endodermis becomes progressively modified to maintain turgor pressure in the stele. Suberin is laid down, covering all the primary wall except for sites of plasmodesmata. Additional thickening with cellulose is accompanied by lignification, which further toughens and seals the endodermis except for the plasmodesmata. The outer tangential wall of the endodermis is not usually suberised. In addition, some endodermal cells opposite the xylem are passage cells, remaining thin-walled and allowing some movement of materials. Despite suberisation, endodermal cells remain living cells.

Since the endodermis is internal to the cortex, its position leaves the cortex exposed to ions diffusing from the soil along the apoplastic pathway. In some roots (e.g. *Citrus*), an **exodermis** is formed at the junction of the epidermis and cortex. This layer appears to have many of the properties of an endodermis, but the Casparian strip is less well developed.

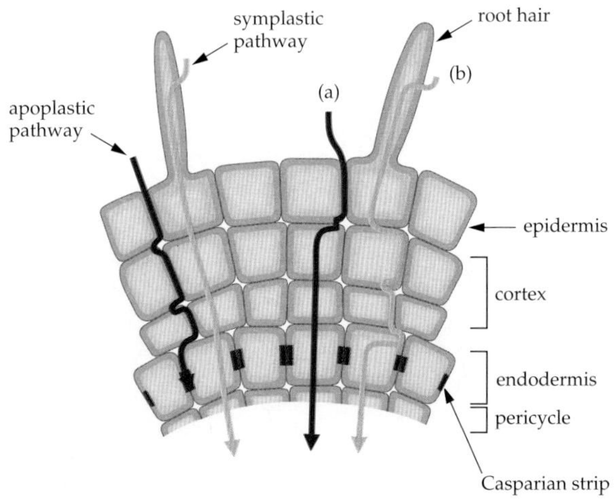

> In most roots, vascular tissue forms a solid central core, surrounded by pericycle and endodermis. Endodermis is a specialised cell layer that controls water uptake.

Fig. 17.17 The endodermis and paths of water and mineral movement in the root epidermis and outer cortex. There are several possible routes: through cell walls (apoplastic) with a switch at the Casparian strip; through cytoplasm (symplastic) after entry through root hair; **(a)** shows the effect of a switch from an apoplastic pathway to symplastic pathway; **(b)** shows the effect of a switch from a symplastic pathway to an apoplastic pathway

Secondary growth

Secondary growth of roots, like stems, is characteristic of dicotyledonous flowering plants and other woody plants, such as pine trees. It results from the formation of secondary xylem and phloem from a vascular cambium in the stele. Periderm forms at the cortex as bark. Because of the presence of bark, roots with secondary growth function primarily in transport and not water absorption; water uptake from the soil only occurs near the tips of primary roots where there are root hairs. As a result of secondary growth, adjacent roots that come in contact with one another sometimes become united. Such root grafts commonly occur between adjacent trees and are one way that infectious diseases are transmitted from tree to tree in forests.

BOX 17.1   HOW PALMS GROW INTO TREES

There are few plants as distinctive or striking as palm trees, which are a common feature of tropical Australian rainforests. Their majestic unbranched trunks and dense crown of large, frond-like leaves makes them a favoured landscape tree in public gardens, beachfront reserves and stately avenues. Palms belong to the family Arecaceae. Together with several other distant relatives, such as *Pandanus* (screw palm), yucca and bananas, they are referred to as tree-forming (arborescent) monocotyledons. The fact that they can grow into trees is quite remarkable because, like other monocotyledons, they do not produce a vascular cambium and so cannot increase their stem diameter by secondary growth. All of the cells and tissues in a palm tree trunk, including the xylem and phloem, are produced by the apical meristem of the shoot and must remain functional for the life of the tree. This is in stark contrast to woody dicotyledons, in which the primary tissues soon degenerate and the structural and metabolic functions of the stem are assumed by secondary tissues.

When a palm seedling emerges from a seed, the apical meristem is relatively small and enlarges gradually as the stem grows. Mature palm trees have massive apical meristems, up to 30 cm across. This results in a trunk of increasing diameter, which is narrowest at its base. Such a trunk is inherently unstable and is unable to support a large tree. The narrow base also forms a bottleneck, severely restricting the amount of water that can be conducted into the stem. The seedlings of most species of palm, such as coconuts and dates, grow downward into the soil for some distance before then growing toward the surface. The widening of the apex is achieved during this period of underground growth, ensuring that the narrow base of the trunk is firmly embedded in the soil (Fig. a). A few palms and arborescent monocotyledons, such as *Pandanus*, overcome the problems of instability and water transport by a different but equally unique solution. Adventitious roots arise from the stem and grow down to the ground (Fig. b). These prop roots are thick rigid organs that stabilise the base of the tree, but their name belies their other major function. They effectively bypass the narrow base of the stem by delivering water from the soil directly to the thicker portions of the trunk.

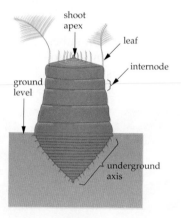

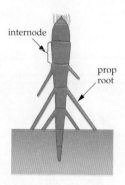

(a) Most palms have an underground stem that progressively widens with short internodes; growth in height is achieved by development of longer internodes formed after the diameter of the adult stem is attained

(b) Screw palm, *Pandanus*, with prop roots. Growth in height begins early, and the widening stem is supported by prop roots

Special functions of roots

Roots of aquatic plants

Plants grow in many different environments, and their roots show specific modifications, especially those associated with aquatic habitats. For example, the mangrove *Rhizophora* has upright aerial roots, **pneumatophores** (Fig. 17.18a), which are exposed at low tide and function in gas exchange for aerobic respiration. They are negatively gravitropic and grow upwards toward the light. Pneumatophores have a cortex containing aerenchyma through which, at low tide, oxygen diffuses from the air down to roots buried in the anaerobic mud. Similarly in rice, cortical root cells break down to form large air spaces (Fig. 17.18b),

which are required for gas exchange when the soil is flooded.

Aquatic plants, especially those that are totally submerged, suffer little water loss by transpiration, so they have a substantially lower requirement than terrestrial plants for water uptake. As a consequence, their root systems are generally simple in structure. Seagrass roots, in particular, have very weakly developed vascular tissues, lacking lignified xylem. Like mangroves, they also have cortical aerenchyma with intercellular spaces for gas exchange because the concentration of oxygen in water is relatively low.

Other types of aerial root are found in epiphytic orchids, which grow on other plants for support (Chapter 43). The epidermis, known as the **velamen**, covers all but the absorptive tip of the orchid root and is thick and multilayered, preventing water loss (Fig. 17.18c).

Roots and salinity

Root structure plays a role in determining salt tolerance in river red gums, *Eucalyptus camaldulensis*. River red gum roots have an exodermis (Fig. 17.19), a suberised layer beneath the epidermis that develops as roots mature. In trees susceptible to salinity, there is always a zone of non-suberised hypodermal cells near the root apex. In salt-tolerant trees, suberised exodermis develops earlier and near the root tip, minimising the zone of non-suberised cells. The

(a)

(b)

(c)

Fig. 17.18 Special functions of roots. **(a)** Aerial prop roots and pneumatophores, the latter functioning in gas exchange, of the mangrove *Rhizophora stylosa*. **(b)** Appearance and structure of the root of rice viewed by scanning electron microscopy. The cross-section of the root shows epidermis, cortex, endodermis (E), pericycle (PE), xylem (X) and phloem (P). The cortical cells have broken down to form larger air spaces (L) needed for gas exchange in waterlogged paddy fields. **(c)** Aerial roots of an epiphytic orchid covered by a thick outer layer, the velamen

presence of the exodermis is correlated with the root's ability to exclude chloride ions, and this barrier thus acts to protect the plant from saline conditions.

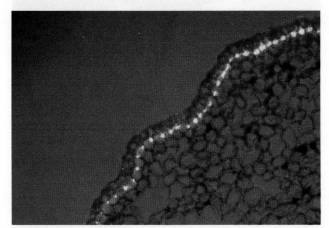

Fig. 17.19 Root tip of river red gum, *Eucalyptus camaldulensis*, in longitudinal section, showing the presence of exodermis, a suberised endodermal-like layer surrounding the cortex. This layer develops close to the root tip, enabling some plants to be salt tolerant since salt cannot pass across the exodermis

Storage roots

Adaptations associated with food storage often involve a swollen taproot and an unusual type of secondary growth. Carrot, *Daucus carota*, has a vascular cambium that produces a moderate amount of secondary xylem interspersed with a large amount of storage parenchyma. Roots of the sweet potato, *Ipomoea batatas*, develop additional vascular cambia that mainly produce parenchyma to form the swollen tuber, an organ of vegetative reproduction that forms roots below and stems above. The fleshy root of beetroot is also a product of multiple, concentric vascular cambia. Each new cambium forms a ring of parenchyma (dark bands) alternating with vascular tissue (light bands) to form the fleshy texture and familiar pattern of beetroot slices.

> Roots of some plants are modified for growth in aquatic or saline conditions. Storage roots involve unusual secondary growth.

Root adaptations and nutrient supply

A wide range of plants, especially legumes (family Fabaceae) and wattles (family Mimosaceae), have **root nodules** (Fig. 17.20a). Root nodules contain nitrogen-fixing bacteria, which convert inert gaseous nitrogen from the atmosphere to organic nitrogenous compounds that are then available to the host plant (Chapter 44).

Soil bacteria, for example, *Rhizobium* in legume roots or *Frankia* in non-legume roots such as the she-oak,

Allocasuarina, enter roots via root hairs. They then form an infection thread that passes through several cell layers and into the root cortex. The infection stimulates rapid cell division of the root resulting in the formation of a root swelling or nodule, which joins with the conducting system of the host plant. Infected cells contain numerous *bacteroids*, which are modified bacterial cells formed within the root nodule cells. The nodules have a distinct pink colour (Fig. 17.20b) due to the presence of the transport protein haemoglobin (Chapter 20), which is usually produced in the host cell to maintain a high flux of oxygen.

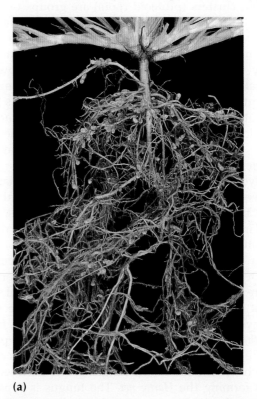

(a)

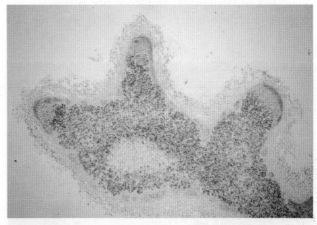

(b)

Fig. 17.20 Structure of nitrogen-fixing root nodules. (a) The roots of white clover show the presence of nodules, in which symbiotic, nitrogen-fixing bacteria live. (b) In cross-section, the infected cells (central in the nodule) are a distinct pink colour because haemoglobin is usually produced in the host cell

Coralloid roots are found in the roots of she-oaks, alder (*Alnus*) and cycads (e.g. *Macrozamia*). These result from root hair infection by the bacterium *Frankia* and cyanobacteria, producing a coral-like growth of roots, which grow upwards with a characteristic dichotomous branching structure. The cells of apical meristems are uninfected, while growing cortical cells are invaded by thin filaments of the micro-organisms. The cortical cells develop vesicles in which nitrogen fixation takes place. The infected outer cortex contains periderm and tannin-filled cells, while the inner cortex and vascular system are uninfected.

Root clusters (proteoid roots) are groups of hairy rootlets that form dense mats at the soil surface in *Banksia* and *Dryandra* (family Proteaceae) and in some legumes (*Viminaria* and *Davesia*, family Fabaceae). They result from massive proliferation of rows of lateral rootlets, up to 1000 per centimetre, along the parent root (Fig. 17.21). Masses of proteoid roots have a white glistening appearance and feature extensive development of vascular tissue xylem inside the root. Their function is to enhance nutrient uptake in nutrient-poor soils.

Mycorrhizae are examples of symbiotic associations between roots and fungi (Chapter 36). Such symbiosis aids the root in the absorption of nutrients, particularly phosphorus, from the soil, and mycorrhizae are often essential for successful plant growth. They are a feature of most terrestrial plants, particularly epiphytic orchids and heath plants (Chapter 41). There are two types of mycorrhizae. The fungal symbiont of *ectomycorrhizae* produces a thread-like mass or mycelium (Chapter 36) on the root surface, *fungal sheath*, which is partly in the rhizosphere and partly in the root (Fig. 17.22a). When the fungal mycelium penetrates the cortex, the fungus grows between the cells forming the *Hartig net*. The fungus is typically restricted to the cortex, and does not penetrate the endodermis. Infected roots are short and usually

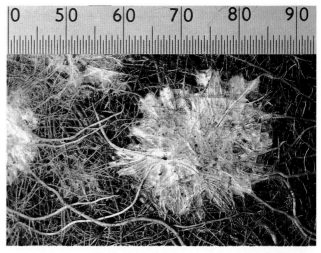
(a)

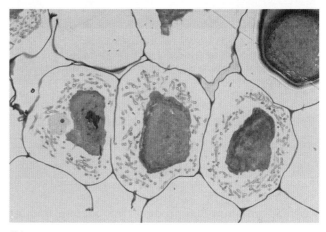
(b)

Fig. 17.22 Mycorrhizae—a fungal–root association.
(a) Ectomycorrhizae produce a fungal sheath on the surface of pine roots, partly in the rhizosphere and in the root, which grows between the cells of the cortex but does not penetrate the endodermis.
(b) Endomycorrhizae occur in orchid roots, in which the coils of hyphae are intracellular, forming pelotons (blue-staining regions) in the outer tissues

dichotomously branched. In contrast, in *endomycorrhizae*, particularly common in orchid roots, the plasma membrane of a host plant cell surrounds the fungal hyphae, which form *intracellular* coils of hyphae (Fig. 17.22b).

Roots show several modifications associated with increasing the supply of nutrients: root nodules and coralloid roots fix nitrogen, while root clusters (proteoid roots) and mycorrhizae improve phosphorus uptake.

LEAF STRUCTURE

Leaves are the prominent photosynthetic organs of vascular plants. They have an enormous diversity of shapes and sizes depending on the species and the environment in which a plant is growing. Most leaves

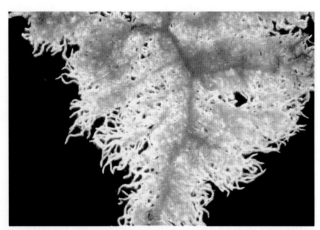

Fig. 17.21 Root clusters (proteoid roots) of *Banksia* result from a proliferation of lateral rootlets, enhancing nutrient uptake

consist of a narrow stalk (petiole) and a thin, flat, expanded blade (lamina) and are attached to stems at nodes (Fig. 17.2). The structure and orientation of leaves maximises the interception of sunlight while minimising water loss, and allows carbon dioxide to be extracted from the atmosphere. Leaves also have secondary roles such as the storage of toxic compounds, which deter herbivores (Chapter 19). Some leaves are modified to perform particular functions, such as protection in the form of bud-scales and spines, carbohydrate storage in fleshy bulbs and support as tendrils.

Leaf arrangement and life span

Leaves are arranged on the stem in an ordered and regular pattern. In most plants this pattern is a spiral, but leaves in pairs, whorls or some other configuration are also common. The pattern of leaf arrangement, **phyllotaxy**, is largely determined when leaf primordia are formed at the shoot apex, and ensures minimum shading of one leaf by another and maximum exposure to sunlight.

Individual leaves have a limited life span, generally about one growing season. Evergreen plants, which bear leaves all year round, such as most Australian trees, continuously shed their oldest leaves and replace them with new ones, thus maintaining a full complement of leaves on the tree at all times. In northern Australia there are some deciduous trees that shed their leaves at the onset of the dry season (Fig. 17.23). This is usually at the end of the growing season, so that a complete set of foliage is renewed the following spring. In the forests of Europe and North America, which are dominated by deciduous oaks, birches, beeches and maples, trees shed their leaves in autumn before the onset of the cold winter.

Leaves of herbaceous plants typically wither on the plant and decay, but leaves of woody plants are actively shed. Shedding occurs as a result of changes in a special **abscission zone**. This is located at the base of the leaf petiole and comprises several layers of small cube-like cells, which secrete enzymes that digest cell walls and cause leaf fall (abscission). This process is regulated by hormones such as ethylene, and by several environmental factors including drought, low temperature, low light intensity and short day lengths (Chapter 24).

> Leaves are photosynthetic organs and have a limited life span: evergreen leaves are shed continuously while deciduous trees shed their leaves before the onset of a cold or dry period.

Leaf shape and orientation

Leaf size ranges from several metres in palms to tiny scales of a few millimetres. Leaves with a single lamina are **simple leaves**, while those divided into many leaflets, each with their own stalk, are **compound leaves** (Fig. 17.24). Clover leaves are an example of

Fig. 17.23 Young trees of *Brachychiton megaphylla* shed their large leaves at onset of the dry season in Kakadu National Park, Northern Territory

Fig. 17.24 Organisation of compound leaves of a legume, showing the main rachis with a swollen pulvinus, which controls the angle of the leaf to the stem at the base, and the many leaflets attached to each secondary rachis forming the bipinnate leaf

compound leaves that have three leaflets arising from the same point (*trifoliate*). Compound leaves of silver wattle, *Acacia dealbata*, have leaflets borne in rows and are *bipinnate* (Chapter 41). In bipinnate leaves of legumes, the base is differentiated into a **pulvinus**, a motility organ that can change the position of the leaflet. In *Albizzia*, the pulvinus operates so that each leaflet is held open horizontally during the day, and closed in a vertical position at night as a kind of sleep movement. In most Australian wattles, compound leaves are replaced by phyllodes as the plant grows. A **phyllode** develops by expansion of the petiole and rachis (midrib of a compound leaf) to form a photosynthetic lamina (Chapter 41).

In most plants, leaves are displayed horizontally and are *dorsiventral*, that is, with a definite upper (dorsal) and lower (ventral) surface. Some plants, however, have leaves that are held vertically. In these leaves both surfaces are the same, and the leaf is *isobilateral* or *bifacial*. Leaves of adult eucalypts are typically pendant and hang vertically from a branch, while those of the geebung, *Persoonia*, stand erect. Both are adaptations for equal illumination on both sides of the blade while at the same time maintaining an orientation that reduces heat absorption.

Juvenile and adult leaf structure

Many plants produce a number of different leaf types during growth and maturation from a seedling to an adult. Juvenile and adult leaves can be markedly different in size, shape, colour, arrangement and even anatomy. Figure 17.25 shows these differences in *Eucalyptus calcicola* from Western Australia. The juvenile leaves are shiny, bright green and lack a petiole (Fig. 17.25a). They are arranged in opposite pairs on the stem, and are held horizontally for increased light interception. In contrast, the adult plant has the typical isobilateral, petiolate leaf of

eucalypts (Fig. 17.25b). The juvenile leaves of the blue gum, *Eucalyptus globulus*, are waxy and blue-grey and strikingly different from those of the adult. Some eucalypts, such as spinning gum, *Eucalyptus perreniana*, reach full size, flower and set fruit, but their foliage never undergoes a transition to adult leaves. This results in a mature tree bearing only juvenile leaves. Such foliage is attractive and used in the horticultural export trade.

> Simple leaves have a single lamina while compound leaves have many leaflets. Juvenile leaves can differ from adult foliage in colour and form.

Leaf structure and organisation

A leaf blade typically comprises upper and lower epidermis, enclosing a mesophyll containing chloroplasts. Vascular bundles enter the petiole from the stem and branch out into the lamina to form a network of **veins**, the conducting tissue of the lamina. The leaves of dicotyledons usually have a distinct central vein (midrib, Fig. 17.2), prominent lateral veins and a network of smaller veins that form a web-like reticulate pattern. Leaves of most monocotyledons have veins running parallel to each other along the axis of the blade.

Epidermal cells

The leaf epidermis is typically a single layer of cells that are rectangular in profile (Fig. 17.26). The epidermal cells are covered with a thick cuticle and particles of wax, and contain specialised pores, **stomata**, which allow uptake of carbon dioxide from the atmosphere (Chapter 5). Each stomatal apparatus consists of two kidney-shaped **guard cells** that are

(a)

(b)

Fig. 17.25 Juvenile and adult foliage of *Eucalyptus calcicola*. **(a)** Shiny green juvenile leaves are heart-shaped and held horizontally on the stem, without a petiole. **(b)** Adult leaves are waxy with a petiole, and are pendant

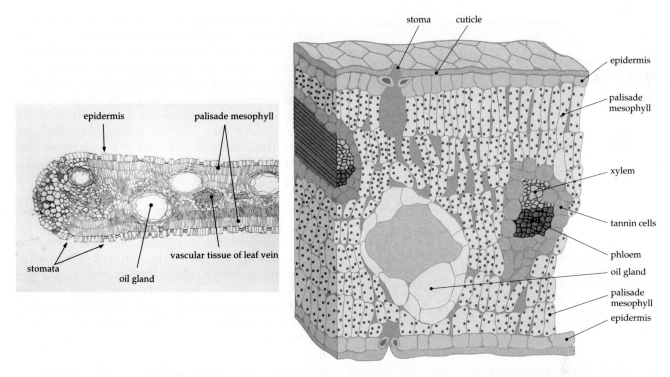

Fig. 17.26 Structure of an isobilateral leaf of *Eucalyptus preissiana*, viewed in cross-section. Palisade mesophyll is developed on both sides of the leaf. Epidermal cells are thickly cutinised on both sides of the leaf. The cuticle partially extends over the stomata, which are sunken beneath the epidermis

joined at the ends of their abutting walls. These walls are impregnated with suberin, and when the guard cells fill with water and become turgid, the differential thickening of their walls causes a gap to open between them, the stomatal pore, exposing internal cells of the leaf to the atmosphere. The stomatal opening is very small, typically 7–40 μm long, and 3–12 μm wide when fully open. Guard cells contain chloroplasts, but other epidermal cells are devoid of photosynthetic pigments.

The number of stomata on leaf surfaces varies. On the lower surface, numbers of 15–1000 per mm^2 are typical, the lower number referring to plants with large stomata. For example, cabbage, a leafy vegetable, has about 600 per mm^2. In Australian plants, the number of stomata varies from about 28 per mm^2 in geebung, *Persoonia*, to 100–340 per mm^2 in eucalypts.

As well as stomata, there may be one or more types of specialised cells present in the epidermis, such as leaf hairs (Fig. 17.27), secretory gland cells and crystal-containing cells. Grass leaves possess several unique cell types. Dumbbell shaped cells with silicate crystals in their walls form in rows on the epidermis, giving many grasses a knife-edge quality and abrasive texture, which deters grazing animals. Bubble-shaped bulliform cells are long and thin-walled, and contain a large vacuole (Fig. 17.28). Change in turgor of these cells appears to cause the lateral rolling and unrolling of a grass leaf in response to environmental conditions.

Mesophyll

Mesophyll is the ground tissue of leaves. It consists of chloroplast-containing parenchyma cells, which are the sites of photosynthesis. In dorsiventral leaves, the shape of mesophyll cells is different in the upper and lower halves of the leaf. Toward the upper surface, mesophyll cells are elongated, with their long axis perpendicular to the leaf surface; this is the *palisade*

Fig. 17.27 Fluorescent micrograph of the surface of a leaf of kikuyu grass, a monocotyledon, showing parallel veins, stomata (closed) and small hairs. Patterns of epidermal cells can also be seen

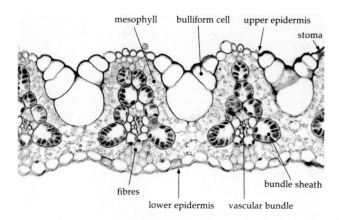

Fig. 17.28 Structure of the leaf of love grass, *Eragrostis brownii*, a monocotyledon, viewed in cross-section. The vascular bundle (xylem above phloem) is surrounded by a prominent bundle sheath, comprising large cells containing chloroplasts, indicating a C_4 pathway of photosynthesis. There are large bulliform cells on the upper surface responsible for unrolling the leaf

mesophyll. These mesophyll cells have the highest numbers of chloroplasts and intercept most of the sunlight. Toward the lower surface, mesophyll cells are more or less regular in shape (isodiametric) and loosely packed, forming the *spongy mesophyll*. The extensive intercellular space in spongy mesophyll allows free diffusion of carbon dioxide from the stomata, which are located predominantly in the lower epidermis.

Isobilateral adult leaves of eucalypts have palisade mesophyll developed at both surfaces, with a small amount of spongy mesophyll restricted to the centre of the leaf. Eucalypt leaves are also equally thickly cutinised on both sides (Fig. 17.26). The cuticle arches over the stomata so that they are sunken beneath the epidermis, which reduces water loss. The vascular bundles are surrounded by tannin cells, which may serve as an ultraviolet filter.

Leaves of many plants, especially the myrtle family Myrtaceae, which includes eucalypts and clove, and the mint family Labiatae, have oil glands. In *Eucalyptus*, oil glands are lined with cells that secrete oils, which accumulate in a central cavity (Fig. 17.26). Each gland can fill a space as large as the entire width of the upper or lower palisade. Oils are a mixture of toxic compounds that make the leaves less palatable to herbivorous insects, but they are also the feature that make herbs and spices valuable culinary plants. Some aromatic flavourings are extracted from glandular hairs, for example, menthol and peppermint oil from the mint *Mentha piperita*.

Vascular system of leaves

Veins (vascular bundles) of the leaves of flowering plants form an interconnected branching system. Leaf vascular bundles contain xylem located toward the upper surface and phloem toward the lower surface (Figs 17.28, 17.29). The minor veins in dicotyledons and the veins of monocotyledon leaves are surrounded by a **bundle sheath**, a layer of tightly packed parenchyma cells (Fig. 17.28). These cells are usually elongated in shape (in the direction of the vein axis) and may contain chloroplasts. In most grasses, the bundle sheath cells link with fibres that connect with the epidermis to provide increased rigidity (Fig. 17.28). In some tropical grasses, which have the C_4 pathway of photosynthesis (Chapter 5), the bundle sheath cells are enlarged and contain many chloroplasts (Fig. 17.28). In some plants, especially mountain and alpine peppers, *Tasmannia*, (family Winteraceae), the sheath comprises sclerenchyma, which makes the leaf more rigid.

In legume leaves, paraveinal mesophyll cells are linked to form a network connecting the mesophyll to the phloem (Fig. 17.29). They are *transfer cells* with conspicuous ingrowths of their cell walls that greatly increase the surface area of the cell. They also have plasmodesmata, which allow direct connections between neighbouring cells. Paraveinal mesophyll cells form a pathway for transfer of sugars from photosynthetically active tissue to the vascular system in one direction, and distribution of xylem-transported nutrients, for example, amides, nitrates and ureides, to leaf tissues in the other direction. Paraveinal mesophyll cells are the site for storage of nitrogen reserves as proteins. Clover leaves, for example, are rich in protein and are important in the diet of grazing animals.

> The leaf lamina consists of palisade and spongy mesophyll, both of which contain chloroplasts. The leaf epidermis has specialised cells, including stomata for gas exchange.

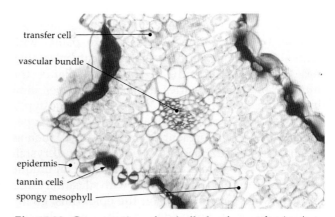

Fig. 17.29 Cross-section of a phyllode of a wattle, *Acacia cardiophylla*. Spongy mesophyll is the photosynthetic tissue, and many of the cells are transfer cells with wall ingrowths (red-stained patches). The transfer cells form part of the paraveinal mesophyll cell system that connects the phloem with the photosynthetic tissue. The layer of cells beneath the epidermis are tannin cells (stained dark purple)

Modifications of leaf structure

Environmental effects

Leaf size and shape are usually similar within a species, which indicates that the major underlying developmental processes are under genetic control. However, there are a number of aspects of leaf morphology and anatomy that are directly influenced by the environment. For example, differences in leaf size and structure are evident in parts of the same plant exposed to different illumination levels (high and low light, or sun and shade). Shaded leaves are generally larger in size than are leaves exposed to full sunlight. Leaves growing in full sunlight have substantially more palisade mesophyll than similar leaves growing in shade. The structure of chloroplasts also differs in sun and shade leaves (Box 5.2).

The thickness of cuticle deposited on the epidermis is strongly influenced by the environment in which the plant grows. Plants grown outdoors can have a leaf cuticle 10 times thicker than that of similar plants grown in a glasshouse. The cuticle is needed to prevent water loss, protect the photosynthetic surface from the sun's rays and prevent the entry of pathogens.

Leaf adaptations

Plants growing in particular environments show special adaptations of their leaves. For example, plants in arid and semiarid environments often have small, scale-like leaves (Fig. 17.8). Unusual pore-like structures, **hydathodes**, occur on leaves of certain rainforest plants, along their margins or at the tips. These permit water to be extruded by guttation, in which water is forced out of the hydathodes by root pressure (hydrostatic pressure in the xylem). Guttation typically occurs early in the morning when root pressure is high, such as in the calla lily. It helps maintain water movement and therefore ion transport through a plant that is growing under conditions of high relative humidity when transpiration is low. A considerable number of tropical plants also have leaves with 'drip tips' (Fig. 17.30a), an adaptation that allows runoff of rainwater in wet tropical rainforests.

Succulents such as pigface, *Disphyma*, have swollen leaves that contain water-storage tissue—large parenchyma cells filled with a central vacuole containing hydrophilic colloids. Succulence is an adaptation for drought tolerance in dry habitats.

Some climbing plants have leaves modified as tendrils, which twine around the substratum for support. For example, the climbing rattan palms of Australian and South-East Asian tropical rainforests (Fig. 17.30b) have whip-like extensions of the midrib covered with spines that enable the plants to climb to the upper canopy of the forest (and inflict pain on the unsuspecting visitor who walks by!). In the grape vine, *Vitis*, the entire leaf is a tendril, while in the garden pea, *Pisum*, only the terminal leaflet is modified (Fig. 17.30c).

Many Australian flowering plants, particularly those that live in dry habitats, have stiff needle-like leaves. For example, the needle-like leaves of *Hakea* and Geraldton wax, *Chaemaelaucium uncinatum*, have an epidermis that is heavily cutinised with sunken stomata (Fig. 17.30d). Palisade mesophyll is arranged in concentric layers around the leaf and contains characteristic sclereids that are dumbbell-shaped, while the central spongy mesophyll contains isolated vascular bundles and areas of sclerenchyma. These features combine to confer rigidity and drought resistance on the leaf. Some plants in dry environments also have leaves covered by wax plates, giving them a dull glaucous appearance.

Surface hairs or trichomes can have several functions and are important in certain environments. Their physical presence reduces water loss by creating a barrier to evaporation, as in the case of 'woolly' leaves that are covered in dense hairs. They may also provide an unusual source of nutrition, as in carnivorous plants, where the hairs secrete a mucilage to trap insects and surface glands secrete enzymes that digest the captured prey. They may secrete chemicals to deter herbivores. For example, the substances secreted by glandular hairs in potatoes, tomatoes and sunflowers give protection from attack by aphids, caterpillars and other herbivores. Tomato leaves have six types of trichomes; one of these (type IV) can secrete more than 22 different small molecules (sesquiterpene lactones). These molecules have the remarkable property that they can be detected by the nervous system of insects through special hairs on the insect's surface, causing the insect to leave. These chemicals thus act as natural insect repellents.

> Plants growing in particular environments show special leaf adaptations. In dry environments, leaves may be needle-like, have stomata sunken in pits and be covered with a thick cuticle and special wax plates. Hairs or trichomes on leaves may secrete small molecules as insect repellents.

Fig. 17.30 Adaptations of leaves. **(a)** Drip tips on the leaves of the giant stinging nettle tree ensure that excess rainwater runs off the leaves. **(b)** Rattan palms have hooked tendrils, which enable them to climb over rainforest vegetation. **(c)** Tendrils on compound leaves of the garden pea, *Pisum*. **(d)** Needle-like leaf of Geraldton wax, *Chamaelaucium uncinatum*, viewed in cross-section, showing epidermis covered by a thick cuticle with stomata, and spongy photosynthetic mesophyll

SUMMARY

- Plants have an aerial shoot system, consisting of stem and leaves, for the function of photosynthesis, and a root system for the functions of absorption and support.

- Stems of dicotyledons have a prominent ring of vascular bundles containing endarch xylem, while stems of monocotyledons have vascular bundles scattered in ground tissue.

- Cambia are secondary meristems, the sheets of cells forming secondary tissues of woody stems. The vascular cambium produces wood and cork cambium produces the corky tissue of bark.

- Stems of many plants have special functions related to the environment in which they live. Some stems are storage organs and others may contain enlarged parenchyma cells specialised for photosynthetic and water-storage functions.

- The main root produced from the radicle is the taproot; lateral roots develop to form a highly branched root system. In most roots, including dicotyledons, vascular tissue forms a solid central core within the cortex, bounded by the endodermis and pericycle. In roots of monocotyledons such as grasses, the vascular tissue forms a ring surrounding the central pith.

- Roots of some plants are modified to grow in aquatic or saline conditions. Storage roots involve unusual secondary growth. Some plants have roots specialised for improving nutrient supply. Root nodules, coralloid roots and mycorrhizal roots involve symbiotic associations with micro-organisms.

- Leaves have a limited life span: evergreen leaves are continuously being shed while deciduous trees shed their leaves in autumn before the onset of cold conditions, although some eucalypts shed their leaves before the dry season.

- Simple leaves have a single lamina while compound leaves have a number of leaflets. The leaf lamina consists of palisade and spongy mesophyll, both of which contain chloroplasts. The leaf epidermis has specialised cells, including stomata for gas exchange. Plants produce different types of leaves as they develop from seedling to adult.

- Plants growing in particular environments show special leaf adaptations. In dry environments, leaves may be needle-like, have stomata sunken in pits and be covered with a thick cuticle and special wax plates. Hairs or trichomes on leaves may secrete small molecules as insect repellents.

QUESTIONS

1. By means of diagrams, show two different types of xylem organisation in stems.

2. Describe the function of cork cambium. What is periderm and what is bark?

3. What is the function of a root cap? Why is it covered in mucilaginous slime?

4. What is the function of the pericycle in a root?

5. How does the endodermis limit flow of water through a root?

6. (a) What do root hairs and leaf hairs have in common? (b) What are the functions of each type of hair?

7. How can you distinguish the upper side from the lower side in a dorsiventral leaf?

8. (a) Describe the structure of an isobilateral leaf. (b) In what environment would you expect to find such a leaf? (c) Give an example.

9. (a) What is the main difference in leaf venation between a dicotyledon and a monocotyledon? (b) In what ways does the *Eucalyptus* adult leaf differ from that of a grass? (c) What are possible functional aspects of these differences?

10. (a) What are the possible advantages of different developmental leaf stages in eucalypts? (b) Relate Figure 17.25 to the environment you might expect a seedling and a mature tree to encounter.

11. (a) What is a stem growth ring? (b) How can the study of the growth rings of a fossil tree from Antarctica help scientists determine past climate?

PLANT NUTRITION,
TRANSPORT AND WATER USE

In addition to light, carbon dioxide and water required for photosynthesis (Chapter 5), plants use other raw materials to synthesise the organic compounds they need for growth. These nutrients are inorganic elements, 13 of which are essential, especially nitrogen (N), phosphorus (P), potassium (K) and magnesium (Mg). Compared with animals (Chapter 19), the nutrient requirements of plants are relatively simple.

A unicellular green alga, such as *Chlamydomonas* (Chapter 35), has little need for a transport system: nutrients, such as nitrate, phosphate, calcium and carbon dioxide, are dissolved in the water that surrounds the cell. The distances over which these nutrients move are sufficiently small that simple diffusion is adequate. In contrast, multicellular plants obtain nutrients from their environment by root uptake generally from soil. These nutrients are required throughout the plant and, in plants that are more than a few centimetres tall, they are distributed by a transport system (Chapter 17). In a vascular land plant, such as a tall eucalypt tree, mineral nutrients absorbed from the soil solution are transported in the xylem from the roots up the stem to more than 30 m above ground. Similarly, the products of photosynthesis are transported in the phloem from the leaves to all parts of the plant.

Land plants operate under a further constraint, one that is at least as significant as the need to transport inorganic nutrients and the products of photosynthesis. Sometimes referred to as the 'dilemma' of land plants, this constraint arises from the conflicting demands of having to obtain carbon dioxide from surrounding air while at the same time ensuring that cells remain hydrated. When stomata of leaves are open to allow carbon dioxide to enter for photosynthesis, water loss by evaporation is inevitable. In other words, growth, which requires a supply of atmospheric carbon, cannot occur without some water loss. To manufacture 1 g of dry weight,

a plant uses 200–1000 g of water! This dilemma limits the types of plants that can grow in particular environments, especially arid habitats where supply of fresh water is limited. The ways by which plants get around this dilemma are diverse and are discussed in this chapter.

> In addition to light, carbon dioxide and water, plants require inorganic nutrients, which they obtain from their environment and transport to all parts of the plant. Because photosynthesis requires uptake of carbon dioxide, growth in plants is inevitably accompanied by evaporative water loss.

NUTRITION OF PLANTS

Although plants are composed largely of water and organic compounds containing the elements carbon, hydrogen and oxygen, plants also synthesise compounds, such as amino acids and vitamins, thus requiring nitrogen, calcium, zinc and other elements (Table 18.1). All plants appear to have very similar nutrient requirements, although the amounts required vary between individual plants and between species. In general, **macronutrients** are required in large amounts and **micronutrients**, **trace elements**, in small amounts.

Essential minerals

The 13 mineral elements identified as essential for plant growth were discovered from the early 1800s to the 1940s. Experiments designed to identify an essential element involve growing plants either in sand or water culture (hydroponics), and not in soil. Controlled amounts of highly purified nutrients from chemical sources are then added. Complete nutrient

Table 18.1 Essential elements required by plants *(a)*

Element	Form absorbed by plants	Relative concentration in dry tissues
Macronutrients		
Nitrogen (N)	NO_3^-, NH_4^+	1000
Potassium (K)	K^+	250
Calcium (Ca)	Ca^{2+}	125
Magnesium (Mg)	Mg^{2+}	80
Phosphorus (P)	$H_2PO_4^-$, HPO_4^{2-}	60
Sulfur (S)	SO_4^{2-}	30
Micronutrients		
Chlorine (Cl)	Cl^-	3.0
Boron (B)	H_3BO_3	2.0
Iron (Fe)	Fe^{2+}, Fe^{3+}	2.0
Manganese (Mn)	Mn^{2+}	1.0
Zinc (Zn)	Zn^{2+}	0.30
Copper (Cu)	Cu^+, Cu^{2+}	0.10
Molybdenum (Mo)	MoO_4^{2-}	0.001

(a) Some plants also require sodium (Na) and silicon (Si).

Fig. 18.2 Seedlings of *Eucalyptus viminalis* grown on alkaline calcareous soil are severely chlorotic (yellow) due to the low availability of iron at a high soil pH

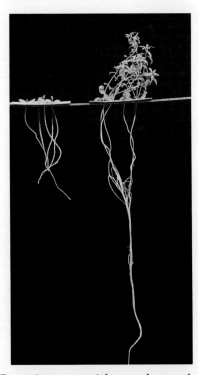

Fig. 18.1 Boron is an essential trace element for plants, but not animals. These flax plants have been growing in a nutrient solution for three weeks without boron (left) or with the addition of 0.1 ppm boric acid, H_3BO_3 (right). Glass-distilled water cannot be used in these experiments because it is contaminated with boron from borosilicate glassware used in the construction of stills

solutions include the elements shown in Table 18.1. Various ions can be omitted and the resulting plant deficiency symptoms observed (Fig. 18.1). Plants growing on soil deficient in iron, for example, typically become yellow (chlorotic) and may die (Fig. 18.2).

Mineral nutrients in the form of ions serve a *general* function in plants in that they affect the ionic balance of cells and the regulation of water balance. Minerals also have *specific* functions; for example, Mg^{2+} is a component of chlorophyll, K^+ affects the conformation of a number of proteins, and Ca^{2+} is important for maintenance of the physical properties of membranes and as a component of primary cell walls. A key role of trace elements is their catalytic function as part of enzymes that are essential in metabolism. The trace elements Mn, Mo, Cu, Zn and Fe are involved in redox reactions (Chapter 2); iron, which can exist in the forms Fe^{2+} and Fe^{3+}, is a constituent of cytochromes, which are involved in the electron transport reactions of photosynthesis and respiration (Chapter 5).

> Thirteen mineral elements are essential for the growth of plants. N, P, K, Mg, Ca and S are required in large amounts and Fe, Cl, Mn, Cu, Zn, B and Mo are required in trace amounts.

Where do essential nutrients come from?

Plants obtain essential minerals from soil (Fig. 18.3a), which is formed from the weathering of underlying rocks (the lithosphere). However, the supply of nutrients to plants does not come solely from this source. Consider the supply of phosphorus necessary for the growth of a forest tree (Fig. 18.3b). Within

BOX 18.1   AUSTRALIA: A TRACE ELEMENT DESERT?

Scientific research identifying the trace element requirements of plants and animals has been of great benefit to Australia. Before the 1930s, there were large tracts of bushland that were unsuitable for agriculture. Even when the major deficiency of phosphorus in soil was alleviated by the application of superphosphate, crops simply would not grow and grazing animals suffered nutritional deficiencies. Australian soils are geologically old and many nutrients important to plants have been leached far below the soil surface.

Pioneering research in South Australia and Western Australia showed that the soils of many areas were deficient in the trace elements Cu, Zn, Mo and Mn. It was demonstrated in the early 1940s that application of as little as 5 g of Mo per hectare greatly improved the growth of clovers in many Victorian pastures. These legumes supplied the necessary nitrogen to pasture grasses, which then supported a greatly increased number of grazing animals.

In other areas, animals did not thrive without additions of trace elements to the soil because of deficiencies in Cu and cobalt (Co) (see figure). Once the soils of these areas had been treated with very small

amounts of these trace elements, together with superphosphate fertiliser, it was possible to establish farms.

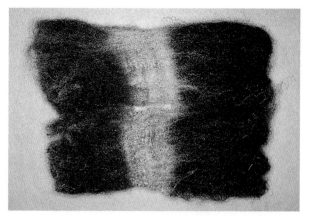

A sheep with black fleece fed on pasture growing on a copper-deficient soil. Copper deficiency in the sheep is indicated by the band of pale colour in the fleece

---

a forest ecosystem, pools of phosphorus exist within living trees and understorey shrubs, in dead plant litter on the ground and in the soil. The total phosphorus in this ecosystem moves between these pools as bacteria decompose litter and plants absorb the released nutrients. This is an example of a nutrient cycle (Chapter 44). The supply of phosphorus

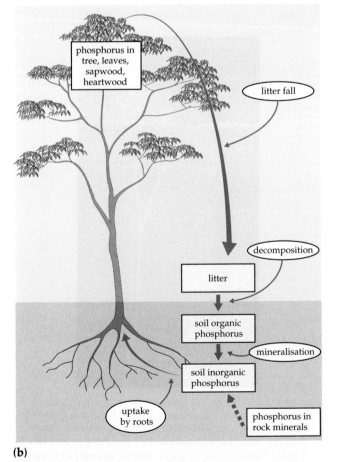

(a)   (b)

Fig. 18.3 (a) Terra rossa soil (red earth) derived from limestone at Buchan, eastern Victoria, is relatively fertile. **(b)** Plants not only obtain nutrients from the lithosphere, but depend on nutrient cycling. The cyclic movement of phosphorus in a forest involves annual leaf fall, decomposition of litter by soil organisms and return of nutrients to the soil

available for tree growth is more dependent upon the amounts of phosphorus in the various pools and the rates of movement between them, than on the very slow release of phosphorus to the soil from the underlying rocks.

Another exception is nitrogen, which is absent from the lithosphere. Besides being the most important of the essential elements (Table 18.1), nitrogen is the major gas of the atmosphere. Nitrogen (N_2) is 'fixed' to form ammonium (NH_4^+) and eventually nitrate (NO_3^-) by certain bacteria (Chapter 44), which is then absorbed by plant roots. These bacteria may live free in the soil or in association with plants, such as the *Rhizobium* bacteria in the root nodules of legumes, or actinomycete nitrogen-fixing organisms in the roots of casuarinas (Box 18.2).

> Plants obtain essential nutrients primarily from soil. Nutrients cycle from plants, to decomposers, to soil and back into plants. Nitrogen in the atmosphere becomes available to plants as nitrate and ammonium by nitrogen-fixing bacteria.

BOX 18.2   DOING A LOT WITH LITTLE: HOW AUSTRALIAN PLANTS COPE WITH LOW NUTRIENTS

Most plants adapted to infertile environments, such as Australian native plants (Chapter 41), have high rates of uptake of available soil nutrients. Some achieve this through specialised root structure, for example, by having many fine roots and/or root hairs, or by having a high ratio of roots to shoots. Some may even be able to exploit immobilised forms of key nutrients such as phosphorus. Red bloodwood, *Eucalyptus gummifera*, for example, grows on coastal sands that are low in nutrients, and may be able to utilise insoluble iron and aluminium phosphates. There is evidence that exudates from the plant's roots may convert insoluble forms of phosphate to soluble forms, which the roots can absorb.

Other more common ways of enhancing nutrient absorption usually involve some form of interaction between roots and soil micro-organisms. The most widespread microbial symbioses are those between roots and mycorrhizal fungi, which are extremely common in woody plants. Many different fungi are involved and both external (ectomycorrhizal) and internal (endomycorrhizal) associations occur (Chapter 17).

So important are mycorrhizal associations that it is common practice to inoculate many exotic agricultural and forestry crops with fungal spores at the time of planting to ensure optimum rates of growth. Development of mycorrhizae stimulates a considerable increase in root surface area, thus greatly increasing the absorption surface for the uptake of minerals. It is also probable that mycorrhizae play an important role in nutrient cycling, because the fungal hyphae readily penetrate litter and compete with other decomposers for phosphorus and other nutrients.

Another form of specialised microbial association, proteoid roots (Chapter 17), is more restricted in its occurrence. Proteoid roots are highly absorbent and their growth is stimulated by particular micro-organisms in the root zone. They are particularly well developed in the family Proteaceae (e.g. many species of *Banksia*, *Grevillea* and *Hakea*). This family is commonly found in Australia and South Africa, particularly in heaths growing on sandy or rocky soils of very low nutrient status (see figure); the phosphorus contents of some of these heath soils are the lowest in the world.

Heath plants are adapted to soils of low fertility, such as this sand sheet, near the Gippsland Lakes, Victoria

For most plants inhabiting low-nutrient environments, the ways of coping with restricted soil reserves are two-fold. As well as maximising rates of nutrient uptake, these plants also are very conservative in their internal usage of nutrients. This may be achieved in part by matching rates of growth to rates of uptake. Very low growth rates and an evergreen perennial habit are common characteristics of plants growing in nutrient-deficient environments. The relatively small, hard (sclerophyllous) and long-lived leaves that are typical of many Australian plants are generally capable of producing much more plant material (biomass) for a given input of phosphorus or other key nutrients than are the larger, softer and shorter-lived leaves of deciduous trees more common in Northern Hemisphere forests. Economies of use are also achieved by internal recycling (e.g. withdrawal of nutrients from leaves before they are shed, a strategy common to many eucalypts) and internal storage (e.g. in stems, trunks, lignotubers and roots) to allow retranslocation of nutrients when needed by actively growing plant parts.

BOX 18.3   ORGANIC VERSUS ARTIFICIAL FERTILISERS

The present high yields of agricultural crops depend to a great extent on 'artificial' fertilisers. These are chemicals manufactured by the fertiliser industry and used to augment nutrients, particularly nitrogen, phosphorus and potassium, which may be below optimal levels in the soil for plant growth. Artificial fertilisers include urea, ammonium sulfate, superphosphate and potassium chloride. In contrast, growers who produce 'organically grown' food (see figure) do not use artificial fertilisers but use organic manures and compost, a farming method used for centuries.

An ammonium ion (NH_4^+), however, is chemically the same whether it is derived from animal manure or from ammonium sulfate. Similarly, a phosphate ion derived from a rotting bone does not differ chemically from one derived from superphosphate. Plant roots cannot distinguish between these different sources of ions in the soil solution. It is doubtful, therefore, that the source of the inorganic essential nutrients affects the quality of plant produce.

If there is a difference in quality, for example, between hydroponically grown and organically grown tomatoes or lettuces, it is not related to the origin of the nutrient ions in the soil solution, but may relate to other factors, such as water supply or care of the crop. Although there is much evidence that the structure and water-holding capacity of many soils is improved by the use of organic manures and compost, the case against the use of artificial fertilisers as inferior suppliers of essential nutrients is unsubstantiated.

Organically grown tomatoes

PATHWAYS AND MECHANISMS OF TRANSPORT

Non-gaseous substances that are transported into and within plants are water and solutes such as ions, amino acids and sucrose. There are a number of pathways:

- between solutions in the external environment and plant roots;
- between cells either along the apoplastic pathway or the symplastic pathway via plasmodesmata (Fig. 17.17);
- between compartments within a cell, for example, between vacuole and cytoplasm, which are separated by a membrane, the tonoplast (Fig. 18.4);
- long-distance transport in xylem and phloem.

Transport mechanisms include both passive and active processes (Chapter 4). Passive processes in plants include diffusion, mass flow and osmosis, and active processes include uptake or movement of ions and sugars with the expenditure of energy. *Mass-flow transport* occurs when solutes are carried in a flow of solution driven by gradients of hydrostatic pressure, as in a hose pipe. Such gradients of pressure occur in xylem and phloem, and are responsible for the movement of water and sugars.

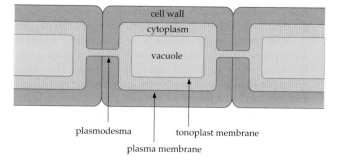

Fig. 18.4 Diagram of the compartments in vacuolated plant cells. The apoplast is the volume outside the plasma membrane, comprising cell wall and intercellular spaces, while the symplast is the intracellular volume, comprising the nucleus, cytoplasm and large vacuole. Transport can take place through the apoplast and/or symplast (via the plasmodesmata)

> In plants, cell-to-cell movement of water and solutes occurs via apoplastic and/or symplastic pathways. Long-distance transport occurs in xylem and phloem. Transport mechanisms include passive processes, such as diffusion, mass flow and osmosis, and active processes, which require energy.

WATER TRANSPORT

The movement of water molecules from soil into roots, or from leaf mesophyll cells through stomata to the atmosphere, takes place *down* gradients of water potential (ψ, psi), that is, down a gradient of decreasing free energy (Chapter 4). The free energy of water molecules in the soil is normally greater than in the plant, which, in turn, is greater than the free energy of water molecules in the air (Fig. 18.5). Hence, water moves from soil into roots and up a plant.

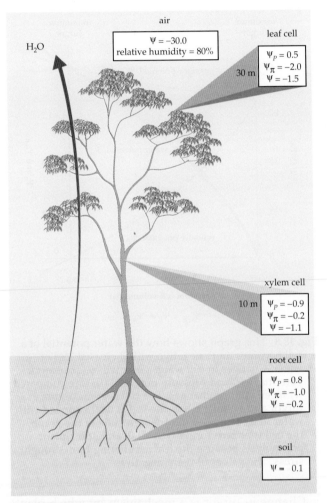

Fig. 18.5 Gradients of free energy of water, indicated as water potential (ψ), in the pathway of water flow from the soil, through the plant to the atmosphere. Values of ψ become progressively more negative from soil to atmosphere. In the vascular system, the xylem sap is under tension so that pressure potential (ψ_P) becomes lower at the top of the tree. Also, the xylem sap is more dilute than that in cell vacuoles and hence osmotic potential (ψ_π) is higher (less negative) in the xylem compartment compared with that in root and leaf cells. The units of ψ, ψ_π, and ψ_P are in MPa

Values of ψ in soil, plant or air are expressed relative to the energy status of pure water at the same temperature and pressure. For pure water, ψ is set

at 0, just as the temperature of a water–ice mixture on the Celsius scale is set at 0°C. The units in which ψ is measured are those of pressure (MPa, megapascals), which is also equivalent to energy per unit volume. (One megapascal equals approximately 10 bars on other scales of pressure.)

Factors that affect water potential

In soil and within a plant, ψ is less than 0, that is, it has a negative value. This is because the free energy of water molecules is decreased below that of pure water by the presence of solutes as well as solids that adsorb water molecules. For example, a 1 M sucrose solution has a water potential, $\psi_{solution}$, of −3.5 MPa (Fig. 18.6). The free energy of the water molecules in this case is reduced below 0 by the presence of the sucrose molecules.

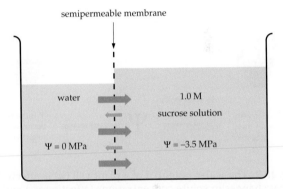

Fig. 18.6 The net flow of water through a semipermeable membrane and down a gradient of water potential (ψ), from pure water to a 1 M sucrose solution

Water potential is affected by hydrostatic pressure. It is decreased when a fluid is under tension (negative pressure), such as occurs in xylem, whereas it is increased by the application of positive pressure, such as occurs in a turgid plant cell. A plant cell wall has elastic properties and it can swell and contract as the water content and volume of the cell change. In this way, the elastic wall of a turgid cell exerts a positive hydrostatic or turgor pressure on the cell contents.

Thus, the water potential of a plant cell (ψ_{cell}) is the sum of the positive effect of cell turgor pressure (pressure potential, ψ_P) and the negative osmotic effect (osmotic potential, ψ_π) of solutes:

$$\psi_{cell} = \psi_P + \psi_\pi$$

If the ψ_π of a solution in a plant cell is −1.2 MPa and ψ_P is 0.5 MPa, then ψ_{cell} is −0.7 MPa.

This equation needs some modification because an assumption is made that solutes cannot penetrate a semipermeable membrane, which, by definition, is permeable only to water molecules. Few, if any, membranes are impenetrable to solutes. So ψ_π needs to be qualified by a factor, σ (sigma), which is a measure of a membrane's leakiness to solutes. For a completely semipermeable membrane, σ has a value of 1, and 0 for a membrane that presents no barrier to movement of solutes. The equation now becomes:

$$\Psi_{cell} = \psi_P + \sigma\psi_\pi$$

In summary, gradients of water potential, and therefore direction of water flow between cell compartments separated by membranes, are influenced by both osmotic potential and pressure potential (Fig. 18.7). In open-ended xylem vessels lacking plasma membranes, and in phloem sieve tubes with open sieve plates, σ = 0, and thus $\Psi_{cell} = \psi_P$. Fluid movement in xylem and phloem is by mass flow, driven only by gradients of hydrostatic pressure.

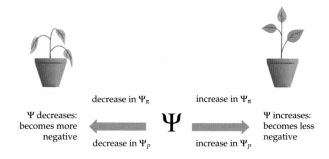

decrease in Ψ_π increase in Ψ_π

Ψ decreases: Ψ increases:
becomes more   **Ψ**   becomes less
negative   negative

decrease in Ψ_P increase in Ψ_P

Fig. 18.7 The water potential of a plant cell (ψ) is affected by changes in solute concentrations (affecting ψ_π) and hydrostatic pressure (affecting ψ_P)

In plants, water potential (ψ) is negative and is the sum of osmotic potential (ψ_π) and pressure potential (ψ_P).

Water uptake by plant cells

How do these properties of water movement relate to the ability of a plant to take up water from soil and to maintain a high degree of hydration? We have shown that the water potential of a cell determines the capacity of that cell to take up or lose water from its surroundings. When plant cells exchange water with their environment, they shrink or swell to a limit imposed by the cell wall. These changes are accompanied by changes in pressure potential (ψ_P) and solute concentration, and therefore osmotic potential (ψ_π) . If cell water content and volume increase, the solutes are diluted, ψ_π decreases and ψ_{cell} becomes less negative (Fig. 18.8). When a cell is fully turgid, $\psi_P = \psi_\pi$ and $\psi_{cell} = 0$. The fully turgid cell has no capacity to absorb water, even from distilled water.

Consider, however, what happens to a leaf cell on a hot day. Water vapour lost through stomata exceeds water taken up by roots. Cells lose water and their volume decreases. As water content and cell volume decrease, ψ_P decreases, ψ_π increases and ψ_{cell} becomes more negative. At the point of incipient plasmolysis, there is no wall pressure ($\psi_P = 0$), and $\psi_{cell} = \psi_\pi$. Thus, the capacity of the leaf cell to absorb water from surrounding cells, such as the xylem, is greatly increased and water moves into the cell.

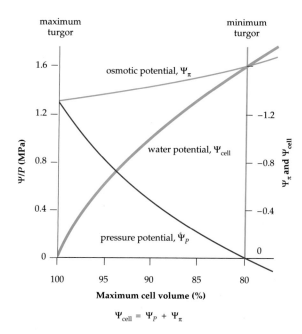

$$\Psi_{cell} = \Psi_P + \Psi_\pi$$

Fig. 18.8 This graph shows how the water potential of a cell becomes more negative with increased loss of water. Changes in cell water potential (ψ_{cell}), osmotic potential (ψ_π) and pressure potential (ψ_P) are compared with cell volume changes. The cell is at maximum turgor and volume when the leaf is well-supplied with water (left). However, on the right side of the graph, the cell is at zero turgor ($\psi_P = 0$) and at incipient plasmolysis because the leaf is suffering from drought stress. The values of ψ are negative

Change in turgor is a mechanism by which plants adjust their water-retaining and water-absorbing capacities in drought conditions. When water availability increases, as at night when the rate of evaporation from leaves is low, cells have the capacity to take up water and regain turgidity (Fig. 18.9a). Changes in turgidity have many important consequences: for example, when turgor has been lost, growth by cell expansion cannot occur and, in leaves, stomatal closure takes place, restricting photosynthesis. With a loss of turgor, thus pressure potential, tissues also lose mechanical strength or rigidity and leaves will wilt and may suffer permanent damage (Fig. 18.9b). Most plants die if their water potential drops below about −4 MPa.

(a)

(b)

Fig. 18.9 Wilting of leaves of a cucumber plant. **(a)** Leaves are wilted at noon on a hot day when their cells have lost turgor. **(b)** Leaves have regained turgor at 8 pm the same day

When a cell loses water and decreases in volume, pressure potential and osmotic potential both decrease, with the result that water potential also decreases.

Osmotic adjustment in plant cells

Loss of turgor and wilting can occur very rapidly, in a matter of minutes, resulting in equally rapid decreases in cell water potential. When plants are subjected to *persistent* soil water deficits over days or weeks (i.e. drought stress), they respond in a different way. In this case, the amount of solutes in cell vacuoles increases, which has the effect of decreasing osmotic potential and reducing cell water potential without adversely affecting cell turgor and growth. This response to drought is **osmotic adjustment**. Osmotic adjustment allows growth and photosynthesis to continue in drier conditions. Solutes that contribute to increased osmotic pressure of the vacuole include ions, sugars and organic solutes, such as the amino acid proline. Osmotic adjustment also occurs in plants growing in wet but

saline soils. This is achieved by accumulation in cell vacuoles of Na^+ and Cl^- from the soil solution. Salt-tolerant plants (halophytes, Fig. 18.10) may accumulate sufficient inorganic ions in their cells to generate an osmotic potential as low as -2.5 to -3.0 MPa.

Fig. 18.10 *Disphyma australe* is a halophyte (salt-tolerant plant). Plants growing in wet saline soils are able to osmotically adjust their cells to retain and absorb water

The capacity of a cell to retain and to absorb water is increased by osmotic adjustment. Plants respond to longer term soil water deficits or salinity by accumulating solutes in the vacuole, which allows turgor and growth to be maintained in dry or saline conditions.

Transpiration

Transpiration is the loss of water by evaporation from leaves, and is a process that requires energy from incoming solar radiation to vaporise water (Fig. 18.11).

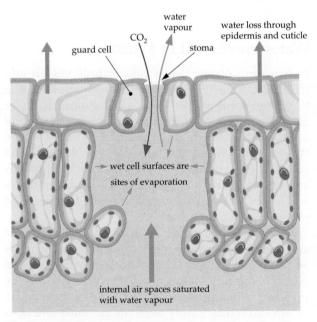

Fig. 18.11 Pathway of the movement of carbon dioxide and water vapour into and out of a leaf

Water vapour passes through both stomatal pores and, to a lesser extent, the leaf cuticle. In a well-watered growing plant, transpired water is replaced by water drawn up from the roots through the xylem. On a sunny day a single leaf from a herb, such as a sunflower, may lose an amount equivalent to its entire water content in just three hours. In mountain ash trees, *Eucalyptus regnans*, growing in a moist environment, the amount of water transpired on a summer day can be as much as 300 L. On a hot day, a hectare of wheat can lose up to 100 tonnes of water each day, which is equivalent to 10 mm of rainfall. Transpiration is a physical process and its rate is highly dependent upon climatic conditions.

The overall gradient of water potential from soil to air is maintained during the day by the negative water potential of the surrounding air (Fig. 18.5). Even at a relative humidity of 90% and temperature of 20°C, air has a water potential of approximately −14 MPa, which is much more negative than the water potential of plant cells. Moving along this gradient, water enters the root system in the tip region of young roots where root hairs are in close contact with the soil solution. Water and solutes move radially from the root epidermis, across the cortex, to the endodermis and pericycle, and then into the xylem elements. Water moves through the apoplast by mass flow but, at the endodermis, the apoplastic pathway is obstructed by the Casparian strip (Chapter 17). At this point, the whole flux of water and solutes must travel through the symplast of cells via plasmodesmata to enter the stele of the root (Fig. 17.17).

> Transpiration in plants is the loss of water vapour by evaporation from leaf surfaces.

Water movement in xylem

In flowering plants, xylem cells include vessels and tracheids. A vessel consists of elements joined end to end to form a tube that may be quite short or may extend up to 15 m long, as in one species of oak tree. The diameter of large vessels is about 50 μm. Water and solutes, but not gas bubbles or solid particles, pass through pits in the walls of both vessels and tracheids. Both vessels and tracheids contain a continuous column of xylem sap, extending from the root to the veins of a leaf.

Xylem sap is a dilute solution of inorganic ions and some organic nitrogen compounds (amino acids and amides, Fig. 18.12). The column of sap in the xylem of a transpiring plant is under negative hydrostatic pressure. When the stem of a transpiring plant is cut, xylem sap retreats away from the cut ends and does not leak out. There are, however, exceptions. Aborigines living in the Australian desert discovered

that, when cut, mallee roots will provide a source of water (Fig. 18.13). Mallee roots release sap because it is not under tension; root water potential is very high (close to 0) and root xylem vessels are of large diameter, offering little resistance to flow. Water flows out when cut roots are upended.

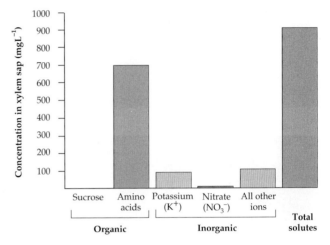

Fig. 18.12 Chemical composition of xylem sap from white lupin, *Lupinus albus*

Fig. 18.13 Xylem sap is tapped from roots of mallee gums by Aborigines at the Yalata reserve, South Australia

Root pressure

Under certain conditions, such as very moist soil and high atmospheric humidity, when the transpiration rate is zero or very low, root xylem sap may be under positive pressure. Pressure produced by roots, **root pressure**, is responsible for the exudation of sap from the cut stems of some vines, walnut trees and North American sugar maple in early spring. It does not account for the ascent of xylem sap in tall plants. Root pressure is not well understood, but possibly an osmotic flow of water is generated in roots due to an internal accumulation of solutes.

▌Positive pressure occurs in the roots of some plants under certain conditions.

How does water reach the top of tall trees?

So far, we have established that the driving force for mass flow of sap in xylem is a gradient of water potential, and that this gradient, which is increasingly negative towards the air, 'pulls' the xylem fluid upwards. This idea assumes that there is a continuous water column in narrow xylem elements, and that this column exists under tension without rupturing.

Atmospheric pressure is only sufficient to hold a column of water up to a height of 10 m. However, water columns much higher than 10 m, as in a tree 100 m high, can be maintained in very narrow tubes such as xylem vessels. This is because molecular forces will hold water molecules together and to the sides of a tube (capillary attraction), resulting in a column of water under tension. Removal of water by evaporation in leaves at the top of the column results in a flow of water from the roots through a plant. This explanation for how water rises to the top of a tall tree is called the **cohesion theory**.

The cohesion theory hypothesis received support from the experiments of John Milburn, a plant physiologist, who used acoustic devices to 'listen' to stems of water-stressed plants. When the tension in a column of water becomes very high, as in a plant suffering from drought stress, the column sometimes breaks, **cavitates**. The spaces formed in the column, momentarily a vacuum, fill with water vapour, forming an air bubble, or embolism (Fig. 18.14).

Milburn noted that high tension in the xylem produced audible 'clicks', which became more frequent as the degree of water stress was increased. The clicks were the sound made by the breaking of the column of xylem sap under tension, resulting in an embolism. The spread of an embolism is restricted by bordered pits, which act as valves in the vessel walls. Plants recover from these embolisms at night, when the tension in the xylem decreases, possibly assisted by root pressure.

▌A continuous column of water ascending in xylem of trees is dependent on strong adhesion of water molecules to the walls and intermolecular cohesive forces that maintain the flow.

Water movement from leaves

Leaf mesophyll cells obtain water from leaf veins and are surrounded by air spaces, which may occupy up to 50% of the volume of a leaf (Fig. 18.11). Because these air spaces are in contact with wet mesophyll cells, they are saturated with water vapour (relative humidity 100%). The movement of water vapour from leaves meets two points of resistance. The first is the *cuticle*, which offers the greatest resistance, and *stomata*, through which 90% of the flux of water vapour passes. The resistance due to stomata depends upon the mean aperture of all the stomatal pores on the leaf surface; when the stomata are closed, resistance is high and vice versa. The second point of resistance is the *leaf boundary layer*, that is, the layer of still air just outside the leaf surface (Fig. 18.15). Water vapour tends to accumulate around a leaf increasing humidity relative to the atmosphere. The thickness of the boundary layer depends on the shape of the leaf and the presence of leaf hairs, which trap water vapour. Wind increases transpiration rate by decreasing the thickness of this boundary layer.

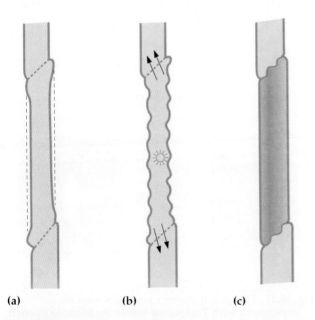

(a)   (b)   (c)

Fig. 18.14 Cavitation in a xylem vessel. (a) Under tension, xylem sap strains the wall of the vessel, pulling it inwards. (b) A cavitation bubble forms and the xylem walls vibrate, resulting in a 'click'. Water moves out of the vessel element, which (c) becomes filled with air

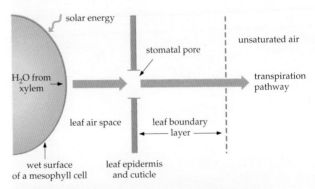

Fig. 18.15 Transpiration. Water evaporating from a mesophyll cell meets two points of resistance to movement: the cuticle and stomatal pores, and the leaf boundary layer, where there is a gradient of water vapour concentration to the outside air. These two resistances affect transpiration rate

Stomata: turgor-operated valves

Although a leaf of *Eucalyptus globulus* may have more than 30 000 stomata per square centimetre, the proportion of the leaf surface area occupied by stomatal pores is only about 1% (Fig. 18.16). Despite this, the amount of transpiration from a leaf with open stomata is nearly the same as evaporation from an open wet surface; gas movement through small pores is remarkably efficient. By controlling the size of stomatal pores, a plant can regulate the exchange of carbon dioxide, oxygen and water vapour. Regulation of stomatal pores allows a trade-off between carbon dioxide uptake required for photosynthesis and potentially damaging water loss.

The size of a stomatal pore is dependent upon the turgor of the two guard cells, which may be likened to two sausage-shaped balloons with less extensible walls on the inner surfaces. When the guard cells become turgid, they expand outwardly, opening the pore, which can have a diameter of up to 10–20 μm. When they lose turgor, water moves out of the guard cells to the surrounding subsidiary and epidermal cells. The guard cells deflate, closing the pore. Guard cells are therefore *turgor-regulated valves*.

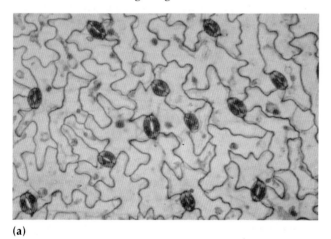

(a)

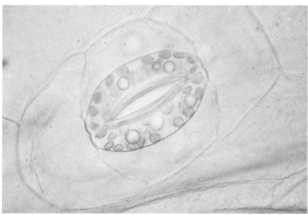

(b)

Fig 18.16 Stomata in the leaf epidermis of a garden pea. **(a)** Stomata are stained with neutral red, a dye that penetrates cells with natural openings. **(b)** A living, half-closed stoma showing chloroplasts in the guard cells

Control of stomatal opening and closing

What are the causes of changes in guard cell turgor that are associated with stomatal opening and closing? Increase in guard cell turgor (pressure potential) is caused by uptake of water by osmosis from surrounding cells. This occurs when the water potential of a guard cell is more negative than that of adjacent cells. Stomatal opening, therefore, involves a decrease in guard cell water potential, which is brought about by importing solutes into the guard cell (Fig. 18.17).

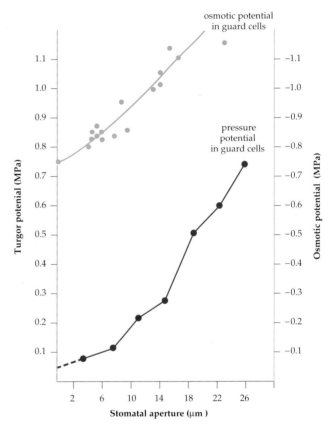

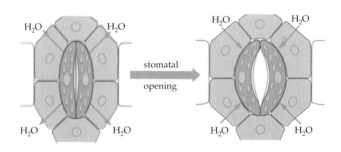

Fig. 18.17 This graph shows the results from an experiment with *Tradescantia* leaves. As stomata opened, measurements were made of the size of the stomatal aperture, and the osmotic and pressure potentials in guard cells. Stomata opened as osmotic potential became more negative, causing guard cells to take up water (increasing turgor or pressure potential)

Stimuli that induce stomatal opening activate membrane-bound H^+–ATPase antiports (Chapter 4). These antiports actively transport H^+ out of the guard cells (Fig. 18.18), generating an H^+ gradient that is coupled to the import of K^+ and Cl^-. Guard cells of open stomata contain about 20 times more K^+ ions than do those of closed ones (Fig. 18.18). There are also increases in concentration of the anion, malate, which together with Cl^- balances the cation, K^+.

Malate is generated within the guard cell by the activity of the enzyme, PEP carboxylase (Chapter 5). An outward flux of these ions results in guard cell osmotic potential becoming less negative so water moves out, turgor is lost and stomata close.

Stomatal opening and closing is the result of movement of solutes, particularly K^+ and Cl^-, into and out of guard cells. This results in changes in osmotic potential, water potential and thus cell turgor.

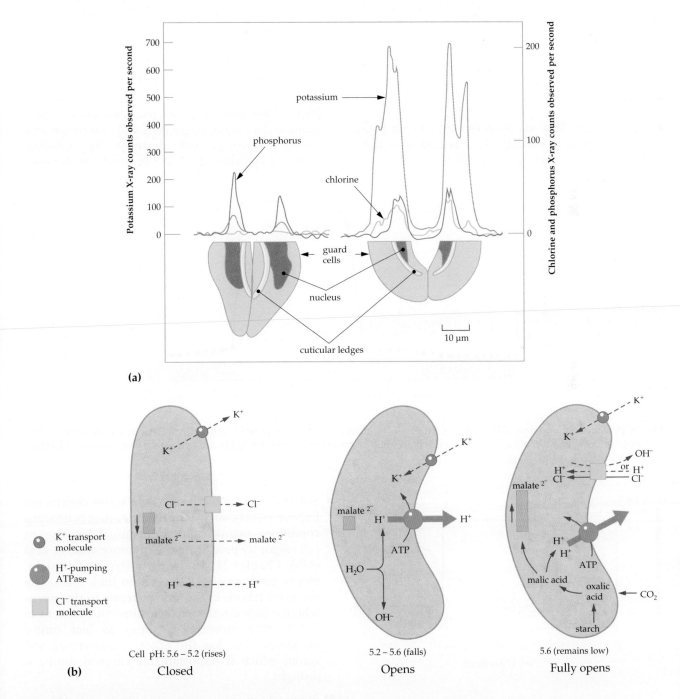

(a)

(b)   Closed   Opens   Fully opens

Cell pH: 5.6 – 5.2 (rises)   5.2 – 5.6 (falls)   5.6 (remains low)

Fig. 18.18 Changes in amounts of various ions in closed and open stomata. **(a)** Profiles across stomata of the elements potassium, chlorine and phosphorus were obtained by scanning electron microscopy using an electron probe, which detects each element by emission of X-rays. **(b)** Ion movements into and out of guard cells during stomatal closure and opening depend on proton-pumping of ATPase, which provides the proton gradients that are coupled to other secondary active transport mechanisms for K^+ and Cl^-

Diurnal stomatal rhythms

When transpiration and stomatal behaviour are measured in plants growing in well-watered conditions, a *diurnal stomatal rhythm* is observed: stomata are open by day and closed by night. CAM plants (Chapter 5) do the reverse, opening their stomata by night, closing them early in the day and, in moist conditions, reopening them in the afternoon. This stomatal behaviour is related to the ability of CAM plants, such as pineapple and succulent plants, to assimilate carbon dioxide at night, as well as by day (Chapter 5).

Environmental stimuli that affect stomatal opening and closing are: light, intercellular carbon dioxide concentration, air humidity and plant and soil water deficits.

- *Light intensity.* Stomata open rapidly when a plant that has been in the dark is provided with light (Fig. 18.19). Blue light is most effective.

- *Carbon dioxide concentration.* Stomata are sensitive to CO_2 concentration in the surrounding air. This CO_2 signal appears to be sensed by the plasma membranes of the guard cells. Decreases in CO_2 concentration cause stomata to open and increases cause stomata to close. In C_3 plants that are photosynthesising, the CO_2 concentration inside a leaf may be 100 ppm less than that in air, which is sufficient to cause stomata to open during the day. At night, when there is no photosynthesis, leaf respiration causes an increase in internal CO_2 concentration sufficient to decrease stomatal aperture. In CAM plants, assimilation of CO_2 at night causes CO_2 concentration to decrease and stomata to open at night (Fig. 18.20).

- *Air humidity.* If air flowing over a leaf changes from high to low humidity, the transpiration rate of a plant will increase, guard cells will lose turgor and stomatal aperture will decrease. By reducing stomatal aperture, the leaf can restrict water loss.

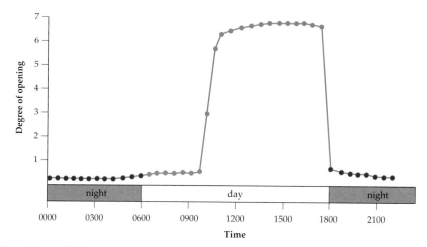

Fig. 18.19 Stomatal opening and closing in response to changes from dark to light and light to dark in a cabbage leaf. The response is rapid, stomata closing completely in about 30 minutes when the lights are turned off and opening in about 60 minutes when the lights are switched on

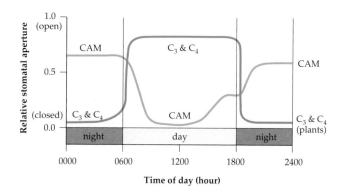

Fig. 18.20 Stomatal opening and closing in C_3 and C_4 plants compared with CAM plants. The graph shows the results of experiments under standard growth conditions with a day length of 12 hours

- *Soil and plant water deficits.* When water deficits are imposed *slowly*, such as during the onset of drought conditions, plants respond to gradual decreases in cell turgor by producing the hormone abscisic acid (ABA, Chapter 24). In the early stages of drought, plants generate an ABA signal in their roots. The hormone travels in the xylem stream to the leaves where it induces stomatal closure. ABA content of leaves may increase 40 times as leaf turgor decreases. This has the effect of conserving soil water, which is important for survival during a drought.

> Stomatal opening and closing is affected by environmental stimuli, including light intensity, carbon dioxide concentration, air humidity, and soil and water deficits. During drought, plant roots generate a hormonal signal (ABA), which induces leaf stomatal closure.

TRANSLOCATION OF ASSIMILATES

The raw materials for the growth of all parts of a plant come from the transport, **translocation**, of assimilates, the end products of photosynthesis. The main *sources* of assimilates are leaves. Other non-photosynthetic tissues act as *sinks*, that is, importers of assimilates. A storage organ, however, which imports and exports carbohydrates, can act as both a source and a sink. In a wheat plant, it has been shown that carbohydrates in the grain, the sink, are mainly derived from the photosynthetic activity of the two upper leaves, the source. Thus, sinks tend to be supplied by the nearest source. The growth of an apple, for example, depends greatly on assimilates from the nearest leaves.

Pathway of assimilate transport

How are assimilates transported from a source to a sink? Evidence of assimilate transport in phloem comes from observing ringbarked trees. In the widespread clearing of forests in Australia in the nineteenth and early twentieth centuries, settlers who wished to establish farms in place of forests often ringbarked trees (Fig. 18.21). Ringbarking involves cutting through the bark to the wood around the base of a tree trunk. This causes the tree to die gradually, often over one to two years. Experiments show that leaves do not wilt after ringbarking, indicating that transport in xylem is unimpaired. Above the ring, cell division and growth occur and sucrose levels are higher than below the ring. This indicates that assimilates move downwards in a tissue of the bark. Thus, ringbarked trees die because all the tissues below the cut portion, including the roots, are starved of assimilates. These observations implicate phloem as the assimilate-transporting tissue of bark.

Evidence also comes from experiments using radioactive carbon dioxide. When leaves photo-synthesise in the presence of carbon-14 labelled carbon dioxide, which is radioactive and emits soft β-radiation, the pathways by which assimilates move in the plant can be traced. Figure 18.22 shows the distribution of carbon-14 labelled assimilates in a single source leaf. The assimilates moved out of the source leaf through the stem. The sinks to which assimilates moved are the roots, the youngest expanding leaves and the shoot tip. No movement occurred into fully expanded leaves. A cross-section of a leaf petiole cut two hours after feeding the leaf with radioactive carbon dioxide, showed that the carbon-14 label was confined to the phloem, within sieve cells and companion cells.

> Assimilates are transported in phloem from their source in leaves to sinks in other parts of the plant.

Fig. 18.21 Ringbarking of trees kills the phloem and hence the tree. This forest in north-east Victoria was cleared by ringbarking to create grazing land

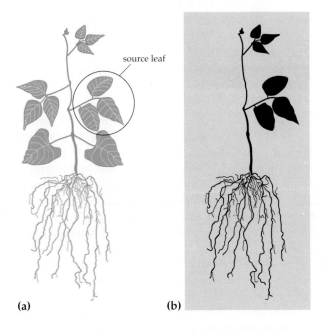

source leaf

(a)   (b)

Fig. 18.22 Transport of carbon-14 labelled assimilates from a single source leaf of French bean, *Phaseolus vulgaris*. (**a**) The source leaf has been fed radioactive carbon dioxide for 4 hours. In this experiment, autoradiography was used for locating carbon-14 in the plant: radioactive carbon dioxide produces an image on unexposed photographic film. (**b**) An autoradiograph of the plant shows the distribution of the carbon-14 label from the source leaf to sinks. Three expanded leaves show no carbon-14 label, and thus are not sinks

What substances are transported in phloem?

Phloem sap can be collected from cut stems or from aphids, which are sucking insects that use their mouthparts (stylets) to penetrate phloem sieve tubes for sugar (Fig. 18.23). Because there is a positive hydrostatic pressure in sieve tubes (up to 3–4 MPa), the sugar solution flows into the insect's stylet. If the stylets of feeding aphids are cut, samples of phloem sap exude out and can be collected for analysis.

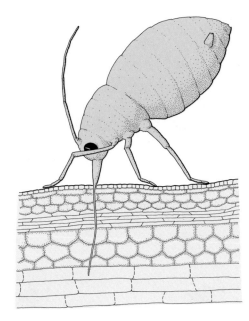

Fig. 18.23 Phloem-feeding aphids are used to tap the contents of sieve tubes

Phloem sap is a concentrated solution of solutes (Fig. 18.24), predominantly sucrose at a concentration of about 0.5 M. However, the trisaccharide raffinose, rather than sucrose, occurs in white ash, and mannitol occurs in apple and apricot. Phloem sap is a sweet liquid, which when fermented is the basis of many alcoholic beverages. For example, the phloem sap of the century plant, *Agave americana*, is collected and fermented in Mexico to make pulque and tequila.

Sucrose in the sieve tubes is not always directly derived from current assimilation in leaves, but may also come from the breakdown of stored carbohydrates. For example, about 20% of the carbohydrates in the grain of wheat are derived from the remobilisation of stores in the upper parts of the flowering stem. In the tubers of Jerusalem artichoke, *Helianthus tuberosus*, carbohydrate comes from redistribution of storage reserves in the stem, and not directly from the current products of photosynthesis.

▌ In most plants, phloem sap is under positive pressure and contains a high concentration of sucrose.

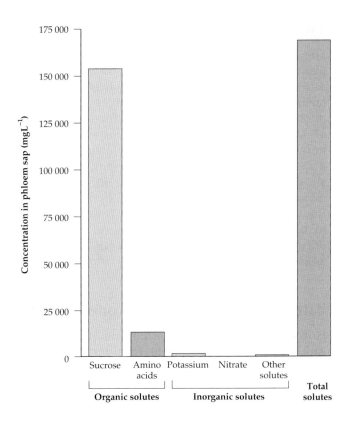

Rate of phloem transport

The rate of phloem transport can be measured by tracing the speed of a pulse of radioactively labelled assimilates as it passes down a petiole and stem, away from a leaf. Transport in phloem (unlike xylem) involves living cells and the rate of transport is sensitive to temperature. The rate, for example, slows when a leaf petiole is cooled to 0°C. When the petiole is killed by heat treatment, movement of assimilates down the petiole stops (Fig. 18.25).

Fig. 18.25 Translocation can be blocked by heat-killing a small section of phloem in the vascular tissue of a petiole. This does not prevent transport in the xylem

The concentration of sucrose and the hydrostatic pressure in sieve tubes is greater near a source than at a sink and a gradient exists between them. CSIRO scientists found that in wheat, for example, the rate of translocation varies from 39 cm per hour down the leaf sheath to 87 cm per hour in the stem. These rates of transport greatly exceed (by a factor of about 10^4) those predicted by a simple diffusion model. Diffusion, therefore, cannot be the mechanism by which sucrose is transported in the phloem.

> The rate of transport in phloem ranges from approximately 40 cm to 100 cm per hour.

Mass-flow mechanism of transport

Sucrose transport from a source, such as mesophyll cells of a leaf, to a sink, such as growing cells at a root tip, involves three processes:

- *phloem loading*, that is, active transport from the mesophyll to the sieve tubes in the veins of a leaf;
- *long-distance transport* of sucrose in the stem and root phloem;
- *phloem unloading*, that is, passive transport from the sieve tubes to the cells at the root tip.

The pathway of the loading of sucrose, from a photosynthesising mesophyll cell to sieve tubes, is shown in Figure 18.26. Sugar moves from the photosynthesising cell into a chain of paraveinal mesophyll cells, transfer cells (Chapter 17), to the phloem of a vascular bundle. Transfer cells have many projections and folds on their inner walls, which increase the surface area for transport.

The concentration of sucrose in a photosynthesising cell is approximately 50 mM. In contrast, concentrations in the sieve element at the site of loading may be 10 to 20 times higher. Phloem loading, therefore, involves transport against a concentration gradient, and is an active process.

The loading of sucrose is accompanied by a pH increase. This indicates that proton transport is associated with active sucrose uptake. The site of this sucrose pump is thought to be the plasma membrane of the sieve element. The driving force for the active movement is a membrane-bound ATPase. The discharge of the proton gradient is coupled to a protein (sucrose translocator) that transports both sucrose and H^+ into the sieve element. Such a process may also be coupled to the transfer of K^+ inwards, accounting for the high concentration of both sucrose and K^+ in the phloem sap. Phloem unloading takes place passively *down* a concentration gradient of sucrose, and transfer cells are often present at unloading sites.

Movement of phloem sap in sieve tubes between sites of loading and unloading is by *mass flow* (Fig. 18.27), that is, driven by a gradient of hydrostatic pressure. As a consequence of the increased

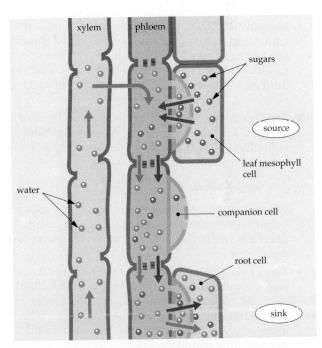

Fig. 18.27 Mass-flow transport of phloem sap between source and sink. As sucrose is actively loaded into phloem, water also enters by osmosis, creating a hydrostatic pressure that moves the sap

concentration of sugar at the loading site, water potential decreases and water enters the sieve tube by osmosis. This increases hydrostatic pressure in the source end of the transport column so that it is greater than at the sink and movement occurs in the direction of source to sink. This model accounts for transport of solutes in phloem of different vascular bundles to occur in different directions.

> Sucrose is transported in sieve tubes by a pressure-driven, mass-flow mechanism.

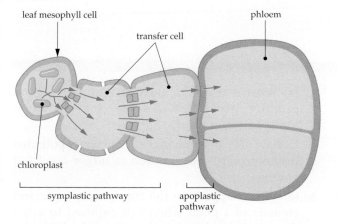

Fig. 18.26 Pathway of active sucrose transport leading to phloem loading

SUMMARY

- In addition to light, carbon dioxide and water, plants require inorganic nutrients (minerals) for growth, which they obtain from their environment and transport to all parts of the plant. Because photosynthesis requires uptake of carbon dioxide through open stomata, growth in plants is inevitably accompanied by evaporative water loss.

- Thirteen mineral elements are essential for the growth of plants: N, P, K, Mg, Ca and S are required in large amounts and Fe, Cl, Mn, Cu, Zn, B and Mo are required in trace amounts. These minerals are obtained primarily from soil, and cycle from plants to decomposers and back to the soil. The source of nitrogen, however, is the atmosphere and becomes available to plants as nitrate and ammonium due to the action of nitrogen-fixing bacteria.

- In plants, cell-to-cell movement of water and dissolved nutrients occurs via apoplastic and/or symplastic pathways. Long-distance transport of water and solutes occurs in xylem and sugars are transported in phloem. Transport mechanisms include passive processes, such as diffusion, mass flow and osmosis; or active processes, which require energy.

- The movement of water in plants is down gradients of water potential. Water potential (ψ) is negative and is the sum of osmotic potential (ψ_π) and pressure potential (ψ_p):

$$\psi_{cell} = \psi_p + \sigma\psi_\pi$$

When a plant cell loses water and decreases in volume, pressure potential and osmotic potential decrease, with the result that water potential decreases.

- The capacity of a cell to retain and to absorb water is increased by osmotic adjustment. Plants respond to longer term soil water deficits by accumulating solutes in the vacuole, which allows turgor and growth to be maintained in dry or saline conditions.

- Plants lose water by evaporation from leaf surfaces, the process of transpiration, which requires solar energy to vaporise water. In a transpiring plant, xylem sap is under negative hydrostatic pressure or tension. In some plants when transpiration is low, a positive pressure can occur in roots (root pressure).

- Water moves up xylem to the top of a plant by mass flow, which is driven by the gradient of water potential from soil to air. The rise of a continuous column of water to the top of a tall tree (above the point supported by atmospheric pressure) is dependent upon strong adhesion of water molecules to the walls of narrow xylem elements and intermolecular cohesive forces that maintain the water column.

- By controlling the size of stomatal pores, a plant can regulate the uptake of carbon dioxide and the loss of water. Stomatal aperture is dependent on the turgor of guard cells, which is affected by changes in water potential caused by changes in solute concentrations, particularly K^+ and Cl^-.

- Stomata have a diurnal rhythm, which is affected by environmental stimuli, including light intensity, carbon dioxide concentration, air humidity, and soil and water deficits. During drought, plant roots generate a hormonal (ABA) signal, which induces leaf stomatal closure.

- Assimilates are transported from their source in the leaf via phloem to sinks in other parts of a plant. Mature (source) leaves do not import assimilates from another source.

- At a source, sugar is actively loaded into phloem against a concentration gradient. Water also enters the phloem by osmosis, creating a positive hydrostatic pressure. This pressure drives the mass movement of phloem sap in sieve tubes to sites of unloading. At a sink, sugar is unloaded by passive diffusion.

QUESTIONS

1. What essential inorganic nutrients does a plant require for growth? Which of these nutrients are required in trace amounts? What are some of the general functions of inorganic nutrients?

2. Describe, with examples, two ways in which some Australian plants are able to survive on low-nutrient soils.

3. (a) Trace the two possible pathways of a water molecule moving from the soil into the stele of a root. (b) By what processes do roots take up ions?

4. (a) What is transpiration? (b) Why is it a good practice to remove some leaves when transplanting a plant?

5. Explain in terms of changes in cell water potential (ψ) how guard cells function as turgor-regulated valves.

6. (a) What happens to the transpiration rate and stomata of a plant when it is exposed to a hot dry wind? (b) What is the role of the hormone ABA during prolonged periods of water stress?

7. Explain why salt makes soil water less available to a plant and has similar effects to a drought.

8. (a) Explain how soil water can be transported to the top of a tall tree. (b) What special structural features of xylem enable this to occur?

9. What is the composition of phloem sap and how is it different from that of xylem sap?

10. Explain the mechanisms by which assimilates are translocated in phloem from a source to a sink.

11. Explain how (a) radioactive tracers and (b) aphids have been used to study the transport and distribution of carbohydrates in plants.

HETEROTROPHIC NUTRITION IN ANIMALS

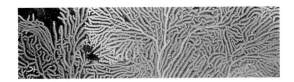

Animals are heterotrophs—unable to manufacture organic compounds from inorganic molecules, they obtain them by eating other organisms. During the evolution of animals, many different kinds of digestive systems able to break down large molecules into molecules small enough for absorption have arisen. Some animals also depend on symbiotic relationships with other organisms for certain nutrients. For example, some corals and the giant clams found throughout the Great Barrier Reef contain algae, zooxanthellae, in their tissues (Fig. 19.1a). Zooxanthellae are photosynthetic and synthesise carbohydrates, some of which are released and utilised by the host's cells. Other animals, such as the koala (Fig. 19.1b), have bacteria in their gut, which produce enzymes that digest cellulose (p. 400). Still other animals, such as intestinal tapeworms and marine pognophoran worms (Fig. 19.1c), have no gut at all, and rely totally on other organisms for their nutrition.

(b)

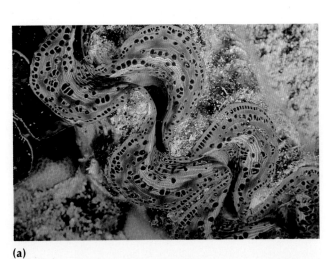

(a)

(c)

Fig. 19.1 Animals have many different ways of obtaining nutrition. **(a)** This giant reef clam filters microscopic organisms from the water, but it also has autotrophs (algae) incorporated into the blue mantle tissue, which provide additional nutrients. **(b)** The koala eats *Eucalyptus* leaves and relies on gut micro-organisms to break down cellulose. **(c)** These pognophoran worms at a deep-sea thermal vent have no gut and obtain all necessary nutrients from chemotrophic bacteria inside their body

The activities associated with obtaining and processing food do not occur in isolation from other behaviour. Feeding in some animals is limited by environmental influences, such as day length, tidal level or temperature. For example, animals may confine food gathering and processing activities to times when the risk of being caught by a predator is lowest, or when a food source is most abundant (Chapter 28). The study of nutrition and digestion in a broad context is known as nutritional ecology.

WHAT NUTRIENTS DO ANIMALS NEED?

Besides organic compounds, such as carbohydrates and lipids, which are needed as a source of chemical energy, animals require particular chemical compounds—amino acids, fatty acids, vitamins, minerals and water. These nutritional requirements may vary diurnally, seasonally and over an animal's lifetime. Both nutrient content and quality of food are important. Some nutrients cannot be stored in an animal's body, so these must be eaten regularly and there is no value in eating them in excess amounts. Also, because different metabolic pathways are interconnected, an abundant nutrient may be poorly utilised if another nutrient is in short supply.

For most animals, the major sources of energy are carbohydrates (sugars, starches and fibre) and lipids (fats and oils). These substrates can be oxidised to yield energy for the synthesis of ATP (Chapter 5). Compounds that normally serve other needs in an animal also may be oxidised for energy. For example, amino acids required for protein synthesis also yield energy upon oxidation.

Essential amino acids are those that cannot be manufactured by an animal and so must come from a dietary source. Many animals obtain some essential amino acids from micro-organisms that live symbiotically in the host gut or tissues. Different amino acids are essential for different animals. The actual amino acids required depend on an animal's metabolism and body form. For example, sheep have a greater need than do many other animals for the sulfur-containing amino acid methionine because it is an important component of wool.

While particular requirements vary, almost all animals, except some insects, require certain fatty acids in their diet. Fatty acids form part of the phospholipids of cell membranes and contribute to the structure of other biologically active compounds, such as hormones.

Vitamins are usually described as organic compounds that are required by animals in small amounts for normal growth and maintenance. Vitamin C (ascorbic acid) is an example of a vitamin that can be synthesised by most animals, but humans, other primates and fruit bats, for example, lack one of the necessary enzymes. Vitamins may be water soluble (B group, C) or fat soluble (A, D, E, K). Water-soluble vitamins are transported as free compounds in the blood. They cannot be stored and need to be replenished continually from a dietary source by those animals that cannot synthesise them. In contrast, fat-soluble vitamins are transported as complexes with lipids or proteins and can be stored in varying amounts. Consequently, deficiencies of fat-soluble vitamins are unusual, at least in adults. In excess, some vitamins can be toxic; for example, several starving polar explorers were poisoned by excess vitamin A after eating the livers of seals and sled dogs!

Mineral elements are essential nutrients for all animals. Macroelements are those that are needed in relatively large amounts: sodium, chlorine, potassium, calcium and phosphorus. Sodium is the major cation and chlorine the major anion in body fluids; calcium is a major component of bone; phosphorus is a central element in ATP and in genetic material. Microelements are required in much smaller amounts and are often components of particular enzymes. Microelements include cobalt, iron, copper, molybdenum, zinc and iodine.

Although not strictly a nutrient, water is required by all animals and is obtained from a variety of sources. Animals produce metabolic water in their bodies during the complete oxidation of carbohydrates and fats (Chapter 5). Many animals drink, especially vertebrates, and water is naturally present in most food. Even apparently dry seeds contain 10–20% water, depending on the prevailing humidity.

> Animals need organic compounds for chemical energy, amino acids, fats, vitamins, minerals and water.

Mineral nutrition of animals: a comparison with plants

Although the mineral requirements of animals are similar to those of plants (Chapter 18), there are important differences. Animals are unable to use inorganic nitrogen as a nitrogen source, and they need more sodium and chlorine than do plants. Plant tissues are rich in potassium and, normally, not in sodium. Animals do not have a boron requirement, while vertebrates, such as farm animals, need sources of iodine, selenium and cobalt in their diet. Sometimes the vegetation on which farm animals are grazing is deficient in supplying adequate amounts of copper to the grazing animal, even though the plants may contain sufficient of this element for their own growth.

BOX 19.1   READING THE LABELS

There is an increasing trend among consumers to expect food manufacturers to provide details of the nutritional content of their product on the packaging. Tables that list the composition of most foods consumed by Australians are available. However, interpretation of this information is not always straightforward.

Sometimes the analyses do not refer to a food as it is normally consumed, but are given on a dry weight basis (that is, after deducting the amount of water naturally present). Some information given is not useful. Consider, for example, the energy yield (total energy content) of foods, which is measured by burning a small sample. Total energy content is the value reported on food packets as energy yield. A kilogram of sawdust, for example, has a similar total energy content to a kilogram of prime hay. However, while 60% of the hay might be able to be digested by a kangaroo, sawdust is practically indigestible. Thus, energy content is not a useful measure; it is content of *available* energy rather than total energy content that is relevant.

Protein content is also sometimes difficult to interpret. Rarely is the true protein content of a food measured. More often the total *nitrogen* content of a food is measured and then a conversion factor applied. Resultant values are correctly called 'crude protein' levels. This is usually sufficient for the types of food found in human diets,

but many animal foods contain non-protein nitrogen-containing compounds, so estimates of crude protein may only approximate true protein. Of equal importance is the composition of the protein. Because amino acids cannot be stored, foods that are deficient in one or more amino acids may be of lesser nutritional value than smaller amounts of a complete protein food (one containing all essential amino acids).

Many of our processed food products show carbohydrate content. If these merely state the percentage of carbohydrate, they are not very useful. Dietitians advise that we reduce our sugar intake and increase intake of fibre, which is largely the insoluble constituents of plant cell walls. Yet both sugar and fibre are carbohydrates. Analyses that divide carbohydrates into different groups on the basis of their nutritional characteristics would be more useful.

Finally, all animals need minerals and trace elements in varying amounts. Again, it is availability rather than total mineral content of a food that is important. The calcium contained in many cereals is chemically bound in such a way that it is not easily absorbed.

The key to understanding whether a particular food is an adequate source of nutrients is to consider not what a food contains but, rather, how available the nutrients are to an animal.

NUTRIENT COMPOSITION OF FOODS

Plant tissues

In general, the nutrient composition of plant tissues is more variable than animal tissues (Fig. 19.2). For example, the nutritional quality of north Australian tropical grasses declines markedly between the beginning and end of the northern dry season (Fig. 19.3). Most plant tissues are poor sources of protein, which has led to suggestions that protein availability is the major limiting factor for herbivores. However, optimal nutrition depends not on a single nutrient group, but on the right mix of nutrients. For example, proteins will be poorly utilised by herbivores if their diet does not contain adequate amounts of carbohydrates.

Although plant tissues eaten by animals are predominantly carbohydrate, the type of carbohydrate varies. Monosaccharides, disaccharides and starch are digested relatively easily but some of the more complex polysaccharides, such as cellulose and pectins of plant cell walls, are not. These components of plant fibre are important in digestion because of their ability to form bulk. Fibre takes up water and swells, and many plant cells retain their

shape and size, even after their contents have been digested. In humans, one effect of dietary fibre is to stimulate movement of food by gut muscle. Diets of highly processed foods are low in fibre and may cause constipation, which can lead to a number of serious bowel diseases. On the other hand, for some herbivorous animals that eat high-fibre diets, the bulky fibre filling the gut can limit the absorption of nutrients (p. 405).

The mineral content of plant tissue may vary greatly between different geographic regions. In general, animals do not show specific appetites for mineral elements, apart from sodium. A hunger for salt (sodium chloride) has a major effect on the behaviour of many animals (including humans). For example, on the higher slopes of the Snowy Mountains region of Australia, much of the sodium has been leached out of the soil, and animals, such as wombats, eastern grey kangaroos and feral rabbits, are chronically exposed to sodium shortage. These animals show a range of mechanisms for retaining sodium in their bodies, and will avidly chew on any material that contains sodium (Fig. 19.4).

The nature and availability of plant matter in aquatic and terrestrial environments are different. In aquatic environments, most autotrophs are algae, which produce cellulose and storage carbohydrates

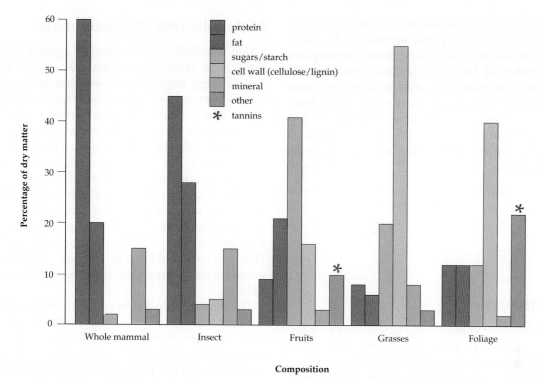

Fig. 19.2 The composition of some foods eaten by animals

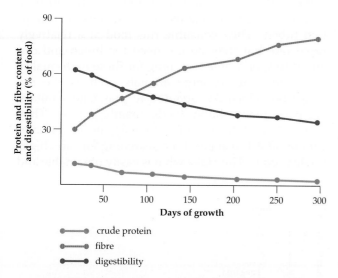

Fig. 19.3 The composition of many plant foods changes seasonally. Spear grass, *Heteropogon contortus*, in tropical Australia declines in protein content and increases in fibre content as it ages. The combination of these means that the nutritive value (measured as percentage digestibility) also declines throughout the year

Fig. 19.4 Wild mammals have a specific appetite for sodium in the alpine regions of Australia, where sodium is sparse. The blocks impregnated with sodium chloride were eaten while those soaked in potassium, calcium or magnesium chloride were ignored

but, unlike most land plants, have no secondary thickening. In aquatic environments, plants are physically supported by water and even vascular plants lack secondary cell walls. This means that aquatic plants and algae are softer than land plants and easier for an animal to digest. Seagrasses, for example, are much easier to crush than terrestrial grasses. Dugongs, major consumers of seagrasses, have horny pads for crushing, whereas horses have hard enamel teeth with well-developed sharp ridges for chewing grasses.

Terrestrial vascular plants have complex cell walls and often lignified woody tissues (Chapter 17), which are difficult to digest. Rupturing plant cell walls makes the entire cell contents available. This requires hard jaws or teeth that can grind against each other. Terrestrial herbivores, such as molluscs, arthropods and vertebrates, all have mechanisms incorporating hard tissues that enable the animal to bite or chew pieces of plants. The digestive actions of fungi tend to make plant detritus more accessible. An added bonus for detritivores (animals that feed on detritus, dead organic matter), is that any fungi and bacteria feeding on the decaying wood will also be consumed.

> Because of a lack of secondary thickening, tissues from aquatic plants are generally easier to digest than are those from terrestrial plants. Plant matter is high in carbohydrates, low in protein and has variable mineral content.

Animal tissues

Most animal tissues contain large amounts of protein, which is relatively easy to digest, but little carbohydrate. Given that carbohydrates, glucose in particular, are important sources of energy for most animals, an absence of dietary carbohydrates could pose serious problems. The brain, for example, uses glucose as its main energy source and a constant supply is critical. In the absence of dietary sources of glucose, animals produce glucose from other substrates, such as fats and proteins. The mineral composition of animal matter is essentially constant.

> Animal tissues are rich in protein, poor in carbohydrates and have relatively constant mineral composition. They are easy to digest.

HOW MUCH FOOD IS REQUIRED?

The nutrient requirements of a particular animal depend on its metabolic rate, age and reproductive state. Metabolic rate varies with level of activity, body mass and prevailing environmental conditions. Activity is probably the most important factor affecting metabolic rate: increased activity increases metabolic rate in both ectothermic and endothermic organisms. In addition, many animals show pronounced circadian changes in activity and metabolic rate.

For endothermic animals (Chapter 29), basal metabolic rate is the rate of metabolism of a quiet inactive animal in a thermoneutral environment. Metabolic rate increases when ambient temperature rises or falls below an endothermic animal's thermo-neutral range. For ectothermic animals, standard metabolic rate is that of a quiet inactive animal and is temperature-dependent; increasing ambient temperature increases metabolic rate.

Metabolic rate and body mass

When metabolic rate is plotted against body mass for a wide range of mammals, it can be seen that daily energy requirements increase with increasing size. However, plotting metabolic rate *per kilogram body mass* shows an opposite trend: the energy requirements of an animal increase exponentially as body mass falls (Fig. 19.5). Thus, an elephant requires more energy per day than a mouse; but a mouse requires more energy *per unit of body mass*. To put it another way, five tonnes of mice require a lot more energy in a day than a five tonne elephant! Mass specific food intake (food intake per unit body mass) also increases exponentially as body mass declines. This pattern holds true for all vertebrate and invertebrate groups that have been examined (Fig. 19.5).

The relationship between body mass and energy requirements may be related to the types of diets that animals select. For example, consider the Australian macropod marsupials—kangaroos, wallabies, and rat kangaroos. More than 60 species, ranging in size from 0.6 kg to 90 kg, have a variety of diets (Fig. 19.6). The diets of small animals, such as the musky rat kangaroo, are composed of somewhat scarce but nutritious items, such as insects, or contain easily digestible material, such as starch-rich tubers. They consume this food at a relatively rapid rate but they do not need too much and can afford to spend time searching for these high-quality items. In contrast, large animals such as the red kangaroo, which has adapted to grassland habitats, eat poor-quality bulky foods, mainly grasses. They need a large total amount of energy every day and cannot afford to spend time searching for rare, high-quality foods. They take what is easily obtainable and

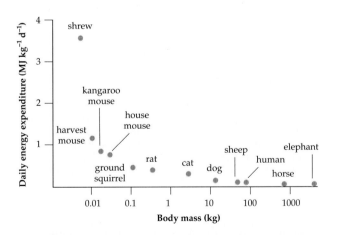

Fig. 19.5 Although large animals use more energy each day than small animals, the energy expenditure of animals per unit mass of tissue (plotted here) increases exponentially as animals get smaller. This is the mouse–elephant curve. Similar relationships exist for other animals, such as insects and marsupials

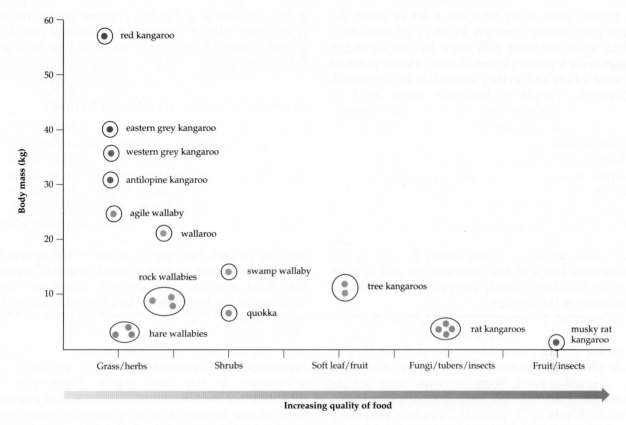

Fig. 19.6 The relationship between the nutritional quality of preferred foods and the body mass of some kangaroos, wallabies and rat kangaroos. Generally, smaller bodied species eat better quality foods, although this relationship is not perfect

eat a lot of it. Not *all* kangaroos fit this pattern. Some have diets that, in theory, would appear to be so poor that they should suit only larger animals. The challenge for nutritional ecologists is to understand how, for example, small animals such as hare wallabies survive on a poor diet.

> Small animals need more energy per kilogram of body mass than do large animals. Small animals usually eat better quality foods.

THE DIGESTIVE PROCESS

Selecting or catching food is only the first step in obtaining necessary nutrients. Large organic molecules, such as fats, proteins and carbohydrates, must be broken down into molecules small enough to be transported across the gut epithelium. To do this within a reasonable period of time requires the action of digestive enzymes, often preceded by physical breakdown of food, which increases the surface area available for enzymatic attack.

Physical digestion

Physical breakdown of food into small particles by grinding or chewing occurs in different regions of

the gut in different animals. It may take place upon ingestion, after storage or during chemical digestion. For example, the action of teeth and jaws in vertebrates, and the rasping action of the radula in snails (Chapter 39), break up food items before they enter the gut. In some birds, a highly muscular gizzard, containing swallowed stones or shells, is used to grind food before it enters the stomach. Ruminants (p. 404) repeatedly regurgitate and rechew plants to break down cell walls.

> For many animals, physical breakdown of food into small particles is necessary for efficient chemical digestion.

Enzymatic digestion

Although there are many different feeding mechanisms, the process of digestion is basically similar in all animal groups. Digestion involves breakdown of complex molecules by hydrolytic enzymes usually secreted into the gut lumen. Not surprisingly, there is a good correspondence between the types of food an animal eats and the types of digestive enzymes it produces. For example, starlings do not eat fruits rich in sucrose and they lack the enzyme sucrase. People who are vegetarian tend to have higher amylase activity (which splits starches to sugars) than those who eat a lot of meat. An interesting human example involves galactosidase, which splits the main milk sugar lactose. Although this enzyme is present in most babies, it is not secreted in some adults and so they are unable to digest milk adequately. People of European origin tend to continue to consume and digest lactose, whereas many of Asian and African origin are intolerant to lactose in milk.

Digestive enzymes are catalysts and their rate of reaction is affected by temperature and pH of the medium in which they operate (Chapter 2). For example, human salivary amylase works best in an environment with a pH of 6.5, which means that it acts well in the mouth. However, another amylase (pancreatic amylase), which enters the gut in the duodenum, has a higher optimum pH and acts on starches that have already been exposed to the acid conditions of the stomach.

In contrast to many other enzymes, digestive enzymes tend to have a lower level of specificity (Chapter 2). They tend to be specific for a type of molecule, such as protein or carbohydrate, rather than for a specific bond. Some enzymes may act on particular bonds within a molecule, such as terminal peptide bonds of a protein. Complete enzymatic breakdown of food usually involves the sequential secretion of different digestive enzymes along the length of the gut. In vertebrates, nervous and hormonal control mechanisms are responsible for ensuring that this occurs.

> The composition of digestive enzymes largely reflects the composition of the normal diet. Digestive enzymes tend to have a lower level of specificity than do other enzymes.

Control of digestive secretion in humans

Saliva secreted into the buccal cavity lubricates food for its passage through the gut, contains enzymes, such as amylase, that are involved in initial chemical digestion, and may contain toxins for use in defence or attack. Secretion of saliva is generally associated with ingestion of food. The Russian physiologist Ivan Pavlov conditioned dogs to salivate by ringing a bell (Chapter 28), showing that salivation was under nervous control. Enzyme secretion in the stomach and duodenum are under hormonal control (Chapter 25). Food reaching the stomach stimulates the secretion of the hormone **gastrin** from mucosal cells in the stomach (Fig. 19.7). Gastrin circulates in the blood and stimulates the release of hydrochloric acid and pepsinogen from the stomach mucosa. **Pepsinogen** is a **zymogen**, an inactive precursor of a protease, in this case, **pepsin**. Pepsinogen is activated to pepsin by either acid or existing pepsin. Cells secrete zymogens rather than active proteases to prevent damage to the cells themselves.

As the acidic stomach contents move into the small intestine, they stimulate intestinal cells to release another hormone, **secretin**. Secretin stimulates the pancreas to release hydrogen carbonate ions, which

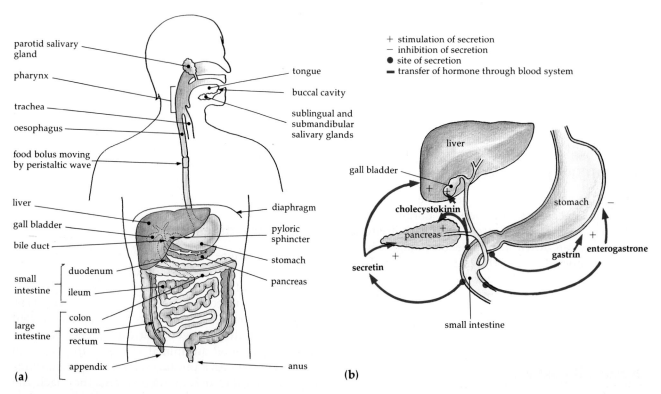

Fig. 19.7 The digestive system in humans: **(a)** major features of the gastrointestinal tract; **(b)** hormonal control of human digestive secretions

neutralise the acidity of the partially digested food. It also stimulates the release of bile from the gall bladder. Bile emulsifies fat, increasing the surface area of fat droplets. This is necessary because fat-digesting enzymes (lipases), like all digestive enzymes, are water-soluble and work in an aqueous environment. Without bile, lipases would have little access to fat. Another hormone, **cholecystokinin**, released from the duodenum, stimulates the pancreas to release other zymogens, including **trypsinogen**, which is activated to the protease **trypsin**. A third hormone, **enterogastrone**, inhibits the continual release of stomach acid and enzymes by a feedback system so that these are present only when required.

> Enzymatic digestion is the co-ordinated, sequential breakdown of large molecules by enzymes secreted along the length of the gut. In most animals, it involves both hormonal and nervous control.

The time that digestive enzymes have to act depends on the speed at which food moves through the digestive tract. Some materials, such as cellulose, are resistant to hydrolysis and need a long time for digestion, whereas others, such as sucrose and starch, are readily hydrolysed. Increasing the length of the gut and expanding regions for storage allow longer periods for enzymatic breakdown. However, if food remains in the gut for long periods and yields fewer nutrients than new food would provide, there is no nutritional advantage. The right balance between digesting food for long periods or making room for new food depends largely on the metabolic rate of an animal. Small animals, with high mass-specific metabolic rates, tend to feed on foods that are digested relatively easily and that pass through the gut relatively rapidly, whereas in larger animals with lower mass-specific metabolic rates, the opposite tends to occur (Fig. 19.8).

> The speed at which food moves through the digestive tract governs the time that digestive enzymes have to act upon it.

Once foods have been digested into small molecules, such as amino acids and monosaccharides, they are absorbed across the epithelium of the gut wall. This may occur by simple diffusion, but more often it occurs against a concentration gradient by active transport (Chapter 4). Separate protein channels transport particular types of compounds. There is a close correlation between the natural diet of an animal and the numbers and types of transport channels in the gut. For example, fruit- and nectar-eating birds have a greater capacity to take up glucose than they do amino acids, whereas carnivorous birds have a greater capacity for amino acid uptake.

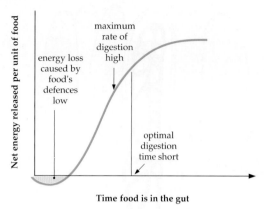

(a) High-quality food

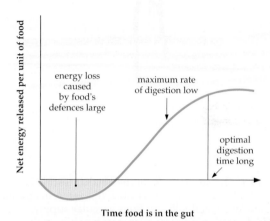

(b) Low-quality food

Fig. 19.8 An idealised model showing predicted patterns of digestion for **(a)** low- and **(b)** high-quality foods. The amount of energy released from a food will initially be negative because the animal needs to expend energy in chewing. Initially, digestion is rapid because of release of material from the cell contents but, thereafter, it becomes progressively harder to digest the structural material. The maximum rate of digestion is lower on the lower quality food but the time that the food is kept in the gut is greater. If food is filling the gut, then the animal cannot continue to eat more. Animals must trade-off the time taken to further digest food already in the gut against the possibility of finding and eating new food

The folding and finger-like projections (villi) of the gut mucosa, and microvilli on epithelial cells, which can be seen in the small intestine (Fig. 19.9), increase the surface area of gut epithelium. This increases the number of transport channels and enables more rapid absorption of nutrients. The surface area of the small intestine is related to an animal's energy requirements. For example, in the wild, mammals require about 20 times as much energy per day as do lizards of similar size. Comparison of the intestines from similar-sized mammals and lizards shows that, although the overall dimensions of the small intestines are similar, the amount of folding of the

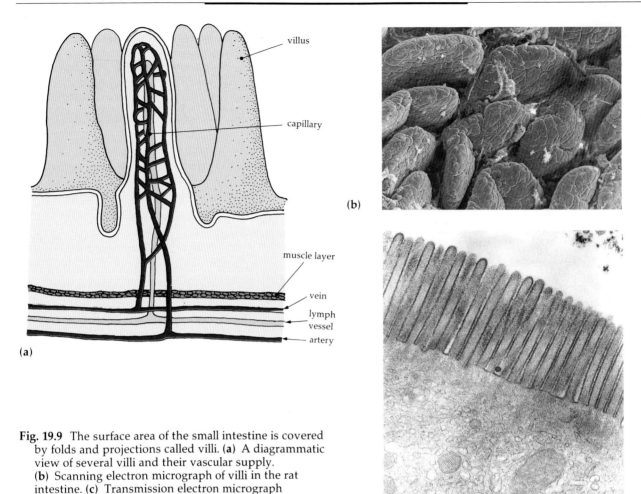

Fig. 19.9 The surface area of the small intestine is covered by folds and projections called villi. **(a)** A diagrammatic view of several villi and their vascular supply. **(b)** Scanning electron micrograph of villi in the rat intestine. **(c)** Transmission electron micrograph showing microvilli on the epithelial surface of the villi

lining of a mammal's small intestine is much greater, increasing its surface area by up to 50 times. This allows mammals to take up nutrients much faster than lizards and so fuel their greater metabolic rate.

The size and absorptive capacity of the gut can change in response to change in diet or when energy requirements change, such as during lactation or exposure to cold. For example, during winter times and during lactation, most mammals seem to grow more intestine; some parts of the gut can become 40% larger. This allows animals to eat more food, while extracting the same proportion of nutrients from it.

> Nutrients are actively absorbed from the gut lumen by special transport proteins. The type and number of these transport channels and the absorptive surface area of the intestine reflect an animal's dietary requirements.

Dealing with plant cell walls

Many herbivorous animals ingest the structural parts of plants. Cell walls are difficult to digest and impede access to the soluble sugars and proteins of cell contents. Many animals reject the cell wall fraction or pass it rapidly through the gut, utilising only the easily digested parts of the plant. For example, near Townsville at the beginning of the rainy season, flying foxes eat leaves of leguminous trees. They chew the leaves, swallow the juices (soluble proteins, sugars and starches) and then spit out the fibrous portion. Giant pandas swallow entire bamboo leaves but they have a short uncomplicated gut through which fibre passes rapidly. Consequently, pandas eat large amounts of plant material and extract only the proteins and carbohydrates from cell contents.

> Plants provide a diet of readily digestible cell contents surrounded by a tough cell wall.

Digesting cellulose

The cell walls of plants are a rich source of energy for animals that can utilise cellulose. Cellulose is the most abundant carbohydrate on earth, but very few animals can produce the necessary enzymes to hydrolyse cellulose molecules. To release the energy of the chemical bonds of cellulose, animals need a group of enzymes known broadly as **cellulases**. It used to be thought that no animals produced their own cellulases, but we now know that several Australian insects, including some termites and the wood cockroach, and some land snails can produce cellulase. Nonetheless, it is likely that all animals (certainly all vertebrates) that digest cellulose rely primarily on cellulases produced by symbiotic micro-organisms (Box 19.3). Since larger animals do not need to acquire energy as rapidly as smaller animals, they can afford the time necessary for symbiotic micro-organisms to slowly digest the cellulose.

Digestion of cellulose by microbial fermentation occurs in nearly all vertebrate herbivores and a wide range of invertebrate herbivores. The microbes are usually housed in an expanded region of the gut, which protects them from being continually lost with the passage of food, and provides an anaerobic environment at relatively constant temperature and pH. The microbes secrete cellulases, which hydrolyse cellulose into its constituent sugars. These sugars are then rapidly used by the microbes, which release short-chain fatty acids (acetic acid, propionic acid and butyric acid) as wastes. It is these latter molecules that are then absorbed by the host animal and used as a source of chemical energy.

> Cell walls are a rich source of energy if a herbivore has the enzymes to digest them. Cellulose is usually digested by microbial symbionts living in the gut. Gut microbes hydrolyse cellulose to glucose which they use. They release short-chain fatty acids, which can be absorbed by the host and used for energy.

There are two broad patterns of microbial fermentation. In foregut fermentation, the microbial population is situated anterior to the true stomach (the region where acid and pepsin are secreted). Examples include all but one of the kangaroos, sheep, cattle and their allies (deer, antelope, goats), some primates and tree sloths from South America. The other pattern, hindgut fermentation, is more common and the microbes are located after the true stomach. Animals that show this pattern include termites, koalas, ringtail possums, wombats, horses and related forms, many rodents, and many herbivorous fishes. Generally, foregut fermentation is restricted to relatively large animals, while hindgut fermentation spans a wide range of body size.

BOX 19.2   HOW DOES THE STOMACH PROTECT ITSELF AGAINST DIGESTION?

Biologists have long wondered why enzymes in the stomach can digest meat and other foods and yet, generally, do not digest the stomach lining itself. Painful gastric ulcers are evidence that breakdown of protection can occur and medication for peptic ulcers is one of the major sources of income for the world pharmaceutical industry. Many of these medicines are intended to counter the natural acidity of the stomach, but new approaches are continually sought. The discovery of Australian frogs that incubate their young in the stomach (the gastric brooding frog, Fig. 14.26) prompted speculation that the mechanisms used by these frogs to turn off stomach secretions could be useful in the treatment of ulcers. However, prolonged suppression of acid production in the stomach has undesirable side-effects in humans, such as weight gain and the formation of cancer-promoting agents in the gut.

During investigation of the mechanism that naturally protects the cells lining the stomach, Australian scientists have found a surface-active phospholipid (SAPL) in the stomach, chemically similar to some corrosion inhibitors that protect metal against acid. Further experiments have shown that these SAPL molecules make the stomach wall hydrophobic but that this property is lost when exposed to aggressive ulcer-inducing agents, such as bile salts and alcohol. Some bacteria, for example, *Helicobacter pylori*, may also be involved in inducing gastric ulcers. All these agents tend to remove the SAPL by dissolving it. It has now been shown that these SAPL molecules protect the stomach itself against the digestive processes.

Antiulcer drugs currently in use are very effective in curing ulcers in the short term but the rate of relapse is 80–100% over two years if medication is discontinued. Lamellar structures similar to the SAPL observed in the stomach have been found in bananas, eggs and unpasteurised milk. The use of these simple foods might prevent ulcers at a fraction of the cost of the medication currently prescribed.

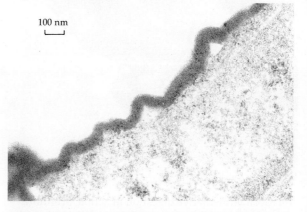

100 nm

The gastric mucosal barrier. A transmission electron micrograph from the epithelial surface of an oxynic duct in the stomach of the rat. In this region, there is no mucous lining, and the pH is about 1. This is the most corrosive environment in the stomach. The layered structures protecting the stomach epithelium are clearly visible. Some invasive micro-organisms use similar surfactant coatings to protect themselves from digestion. The bar represents 100 nm

BOX 19.3   MICROBES IN THE GUT

Most animals harbour a range of different micro-organisms in the gut, including protists, bacteria and even fungi. For example, in the eastern grey kangaroo, there are about 5×10^3 protists per millilitre of digestive fluid and about 3×10^9 bacteria. Protists, although in lower numbers, form up to 40% of the microbial biomass of the gut. The smaller bacteria consist of a diverse group ranging from those that digest cell wall carbohydrates (cellulose and pectin) to those that specialise on soluble sugars. Others require substrates (e.g. formic acid) that are formed entirely by other bacteria.

Many of these micro-organisms in the gut attach themselves to particles of plant matter and then secrete extracellular enzymes to digest away parts of the plant. The microbes generally attach to broken tissues and so initial mechanical disruption of food by the host's teeth, or by other means, is very important.

Some microbes in the gut can also detoxify plant toxins. For example, cattle in northern Australia could not eat the otherwise-nutritious legume *Leucaena leucocephala* because it contains a toxic amino acid, mimosine. Australian scientists observed that, in Hawaii, goats were able to eat this plant without problems and suggested that the Hawaiian animals had a specific bacterium that could break down the toxin. This proved correct and, after some digestive fluid from Hawaiian goats was brought back to Australia, all animals treated were able to eat *Leucaena* safely. The bacterium was spread naturally and *Leucaena* is now an important part of our northern grazing industries.

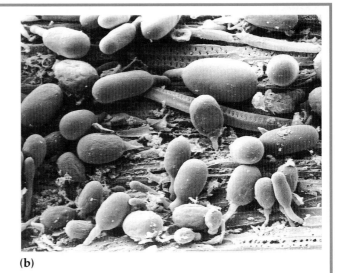

(b)

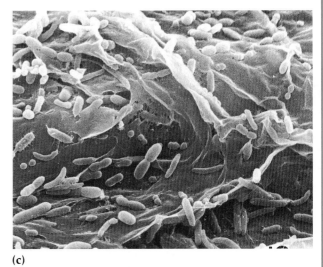

(c)

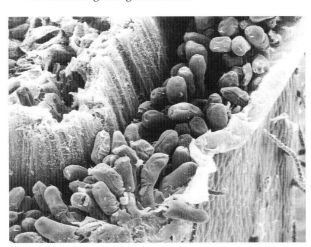

(a)

Microbial fermentative digestion in action. Scanning electron micrographs showing **(a)** protists on the cut end of a lucerne stem (densely packed between the epidermis and the vascular bundle) after 15 minutes in the rumen of a sheep; **(b)** sporangia of anaerobic fungi found on fragments of lucerne in the rumen of sheep; **(c)** bacterial populations on pieces of *Eucalyptus* leaf from the caecum of a brushtail possum; **(d)** bacteria attached to plant tissues by means of extracellular material (from caecum of brushtail possum)

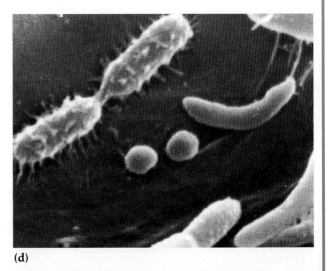

(d)

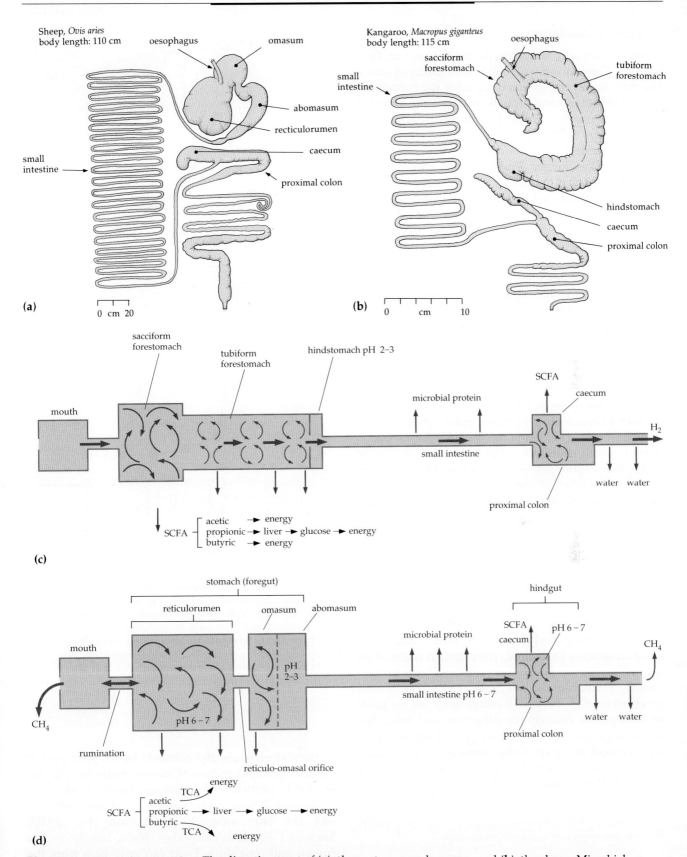

Fig. 19.10 Foregut fermentation. The digestive tract of (**a**) the eastern grey kangaroo and (**b**) the sheep. Microbial fermentation occurs in the forestomach of both species. Note the difference between the structure of the stomachs. The kangaroo stomach is dominated by a tubiform region whereas the sheep stomach is dominated by the sacciform reticulorumen. In both species, there is a second, smaller area of microbial fermentation in the caecum. Schematic representation of the digestive processes occurring in the digestive tract of (**c**) the eastern grey kangaroo and (**d**) the sheep. Small arrows indicate mixing of ingesta; bold arrows indicate one-way flow

Foregut fermentation

In the eastern grey kangaroo, *Macropus giganteus*, food is chewed and mixed with copious amounts of saliva. The first part of the stomach is a large sac containing billions of bacteria and protists, and partially digested food particles (Fig. 19.10). The microbes attack the food and the soluble part (sugars and starches) is quickly digested and absorbed by them. Digesting the fibre takes longer. The microbes attach to the fibre (Box 19.3) and break the chemical bonds between the glucose molecules that make up cellulose. They then further metabolise the glucose for their own use. So far, all the nutrients in the grass have been captured by the microbes and there is little return for the kangaroo.

However, the microbes in the kangaroo stomach excrete fatty acids that the kangaroo absorbs and uses as an energy source (Fig. 19.11). Bicarbonate ions in saliva help to buffer these fatty acids so that the pH of the microbial broth stays relatively constant. The end result is that while cellulose is not utilised directly by the host animal, the microbes convert it into usable molecules.

This seems to be an efficient process with benefits for both kangaroo and microbes, but it is not without some cost to the kangaroo. As we said, the kangaroo cannot directly digest the cellulose but it *could* directly digest the soluble sugars of the grass. However, the microbes have first access to these and the kangaroo must rely on using the microbes' organic wastes. Energy-transforming processes are never 100% efficient (Chapter 2); the kangaroo gets about 75% of the original energy value of the soluble sugars.

> Foregut fermentation allows for extensive degradation of cellulose, but soluble sugars and starches are first used by microbes.

Similarly, proteins are first available to microbes before becoming available to the host. Proteins in grass are quickly fermented. The nitrogen is absorbed by the microbes as ammonium and used for microbial growth. However, the microbes themselves are an excellent source of essential amino acids for the host. As food passes through the digestive tract, microbes are washed out of the saccular part of the stomach into the final, acid part of the stomach and are a high-quality source of complete protein that is digested in the small intestine.

> In foregut fermenters, dietary protein is used by microbes and then the host digests the microbes, which are a rich source of essential amino acids.

Ruminant foregut fermenters

Sheep and cattle, which form the basis of our extensive livestock industries, and related wild herbivores, are foregut fermenters that **ruminate**; that

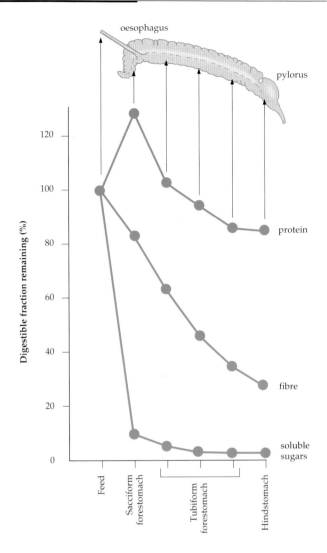

Fig. 19.11 A schematic diagram of the stomach of the eastern grey kangaroo showing the disappearance of certain components of the diet as food passes through the stomach. Note how the soluble sugars are fermented and disappear immediately. The fibrous part of the diet is fermented and digested more slowly throughout the stomach. Protein content actually increases because the micro-organisms that attach to the food are themselves rich in protein

is, they regurgitate the contents of the first part of the stomach and rechew it, further reducing the size of food particles. Ruminants have a stomach composed of four parts: rumen, reticulum, omasum and the true stomach, abomasum (Fig. 19.10). Microbial fermentation takes place largely in the rumen and reticulum. Food can only pass from the reticulum to the omasum after it has been reduced to a certain size by rumination. In sheep, for example, only particles less than 1.0–1.5 mm are likely to pass into the omasum through the reticulo-omasal orifice. This means that a ruminant animal keeps food in its fermentation chambers for prolonged periods, giving the microbes plenty of time to digest the cellulose. However, this may become a disadvantage if the quality of the pasture becomes very poor. As grasses mature, the amount of soluble sugars and proteins declines and the proportion of indigestible fibre increases markedly (Fig. 19.3). This means that an animal must spend more and more time rechewing ingested food before it is small enough to pass into the omasum. Eventually, processing indigestible material can take so long that an animal can starve, gaining no further nutrients from ruminating but, because its stomach is still full, unable to ingest new food with available soluble nutrients.

All animals that eat fibrous diets are limited to some extent by the gut-filling effect of dietary fibre, and the rumen system can be a digestive bottleneck under certain conditions. This problem is not faced by kangaroos, which do not have to rechew food to pass it out of the stomach. As pastures decline in quality, kangaroos simply eat more and digest less (Fig. 19.12).

Hindgut fermentation

The nature of microbial fermentation in the koala is similar to that of foregut fermenters such as the eastern grey kangaroo, except that fermentation occurs in the hindgut (Fig. 19.13). Chewed eucalypt leaves are mixed with saliva but the saliva has a very different composition from that found in foregut fermenters. Soluble sugars and proteins are hydrolysed by the koala's own digestive enzymes and glucose and amino acids are absorbed in the small intestine. The microbial populations in koalas and kangaroos act on different substrates. In koalas, microbes utilise mainly cell walls using similar digestive reactions to those in the kangaroo. Cell walls are broken down, with glucose being absorbed by the microbes; the fatty acids they produce are absorbed and utilised by the koala. The major difference between the two systems is that microbes washed out of the koala hindgut are not able to be digested as a high-quality protein source but are lost in the faeces.

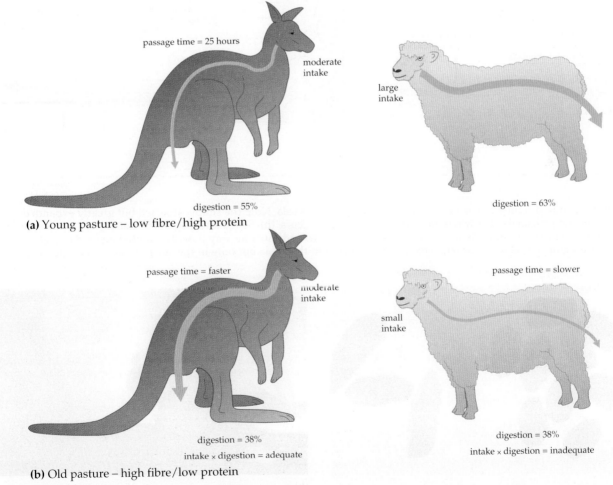

(a) Young pasture – low fibre/high protein

passage time = 25 hours
moderate intake
digestion = 55%

large intake
digestion = 63%

(b) Old pasture – high fibre/low protein

passage time = faster
moderate intake
digestion = 38%
intake × digestion = adequate

passage time = slower
small intake
digestion = 38%
intake × digestion = inadequate

Fig. 19.12 When pasture quality is good, eastern grey kangaroos eat and digest less than sheep. When pasture quality is poor, kangaroos eat at the same level but have a quicker passage of food through the gut. In contrast, sheep cannot increase their food intake because they must continually rechew the tough food before they can pass it from the gut. In effect, sheep have a rate-limiting step (the reticulo-omasal orifice) whereas kangaroos do not. Kangaroos can usually maintain themselves longer on poor quality pastures

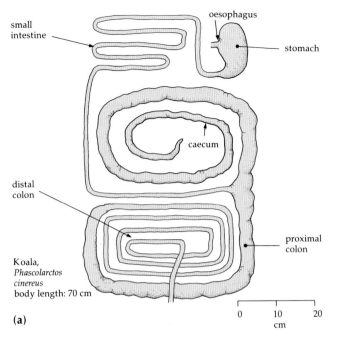

Koala,
*Phascolarctos
cinereus*
body length: 70 cm

(a)

Common ringtail possums make use of the rich source of microbial protein in their faeces. During the day, when they are normally resting in nests or tree hollows, they stop eliminating normal faeces and start to pass out unchanged caecal contents (Fig. 19.14). This material is very rich in microbes and the animal eats the pellets as they are expelled. The microbes are then digested and the nutrients absorbed in the stomach and small intestine. This behaviour is known as caecotrophy or coprophagy, and is possible because of a complicated system of particle sorting that occurs in the possum hindgut. This behaviour is restricted to relatively small animals such as rabbits and many small rodents. Without doubt, the ability to use this extra source of protein is one reason rabbits were able to colonise vast areas of forests and grasslands in Australia.

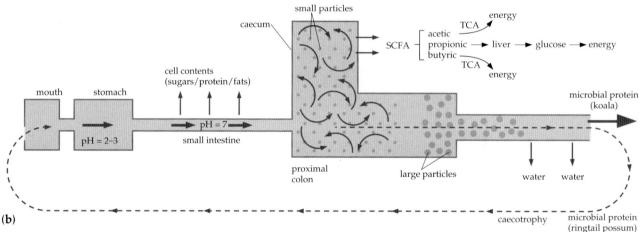

(b)

Fig. 19.13 Hindgut fermentation. (a) The digestive tract of the koala. Note the simple stomach but greatly expanded caecum and proximal colon where microbial fermentation occurs. (b) Schematic diagram of the digestive processes in the gut. Small arrows indicate mixing of ingesta; bold arrows indicate one-way flow. The dotted line indicates the reingestion of special faecal pellets which does not occur in the koala but only in the smaller common ringtail possum

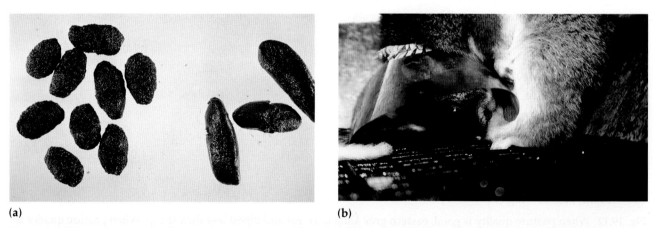

(a)   (b)

Fig. 19.14 The common ringtail possum ingests special faecal pellets (called caecotrophs) during the daylight hours. (a) The special pellets on the left contain higher levels of protein and B-group vitamins than do the normal faecal pellets on the right. (b) The pellets are eaten directly from the anus. The possum here is wearing a plastic collar to try to prevent the behaviour so that samples of special pellets could be collected for analysis

Hindgut fermentation is similar to foregut fermentation, except that most animals cannot utilise high-quality microbial protein. Some small mammals do so by eating special faecal pellets.

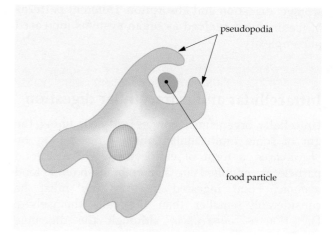

(a)

BOX 19.4   THE POISONED PLATTER

Along with useful nutrients, herbivores also often ingest potentially harmful or toxic compounds found in plants. These chemicals are known as secondary plant compounds because they were thought to have no role in the primary metabolism of plants. We now believe that many of them have a deterrent effect on herbivores and that they are important in disease resistance.

Tannins and phenolics are one of the most important groups. Tannins can precipitate proteins. When 'tanning' an animal hide, plant tannins react with the skin's proteins to form leather. When an animal eats a tannin-rich plant, released plant proteins may be precipitated and so their availability to the animal is reduced. Alternatively, tannins can bind with digestive enzymes and so inhibit the digestion of other food constituents. For example, common brushtail possums are less able to digest *Eucalyptus* leaves containing active tannins than leaves in which the tannins have been deactivated. By developing varieties of grain that contain higher than normal levels of tannins, plant breeders have turned the deterrent properties of tannins to their advantage. Such 'bird-resistant grains' are less attractive to crop-raiding birds, such as corellas and galahs.

Other plant compounds exert more direct toxic effects. Some plants in Western Australia contain a compound called sodium monofluoroacetate which, as compound 1080, is used around the world as a poison for vertebrate pests. The native mammal fauna of Western Australia has been exposed to sodium monofluoroacetate for millions of years and have developed a remarkable resistance to its effects. Toxicity testing showed that the lethal dose for a brushtail possum in Western Australia was about 100 times that of the same species in eastern Australia, where plants do not contain 1080. This provides land-management authorities in Western Australia with a wider safety margin when controlling feral animals such as foxes, cats and pigs. Recent efforts to re-establish small marsupials in areas from which they have disappeared depend on controlling these introduced animals.

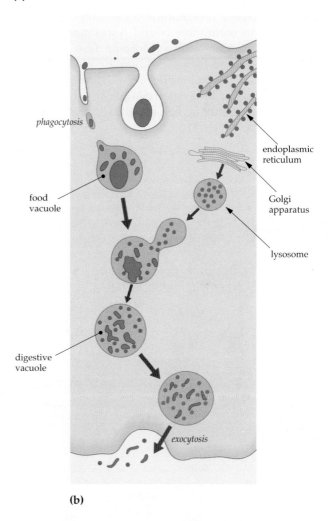

(b)

Fig. 19.15 (a) Amoeba engulfing a food particle with pseudopodia. (b) Intracellular digestion of a food particle in a protozoan. Food material that the cell takes in by phagocytosis is enclosed in a food vacuole, which fuses with a lysosome containing digestive enzymes. Digestion takes place within the composite structure thus formed (digestive vacuole), and the products of digestion are absorbed across the vacuolar membrane. The vacuole eventually fuses with the cell membrane and then ruptures, expelling digestive wastes to the outside

EVOLUTION OF DIGESTIVE SYSTEMS

The evolution of multicellularity allowed division of labour between specialised cells and led to the evolution of organs involved in food procurement, storage, digestion and absorption. Different patterns of feeding also evolved as organ systems increased in complexity.

Intracellular and extracellular digestion

Unicellular organisms, sponges and cells lining the gut of some multicellular animals ingest food by *phagocytosis*, a form of endocytosis where a food particle is engulfed in a membrane-bound food vacuole. Food ingested in this way must be considerably smaller than the cells themselves. Digestion is *intracellular*, although the digestive process is separated from the cytoplasm by a membrane (Figs 19.15, 19.16). Enzymes produced by the endoplasmic reticulum and the Golgi apparatus are stored in membrane-bound lysosomes (Chapter 3). A food vacuole and a lysosome fuse to form a digestive vacuole in which complex molecules are broken down into molecules small enough to cross the membrane into the cytoplasm (Fig. 19.15). Undigested residue is then expelled from the cell by exocytosis (Chapter 4).

Most animals use *extracellular* digestion, which involves the secretion of digestive enzymes onto potential food either externally, as in many spiders and starfishes (p. 410), or into the lumen of a gut, as in vertebrates. Chemical breakdown occurs and the resultant simple molecules are absorbed across the epithelial surface and into the animal. For this reason, food in the lumen of the gut is, in a digestive context, still considered to be outside the animal.

Simple digestive cavities

Corals, anemones and jellyfishes (cnidarians), and free-living flatworms (turbellarians), have a simple, sac-like gut into which enzymes are secreted and in which extracellular digestion occurs. In cnidarians, enzymes are diluted by considerable amounts of water also taken into the cavity. Tiny food particles may also be engulfed directly by endocytosis and digested intracellularly. Waste products are released back into the cavity and then out through the mouth. Digestive systems of this kind are relatively inefficient because both food and waste products pass through the same opening and, in cnidarians, digestion is slow due to dilution of enzymes (Fig. 19.17).

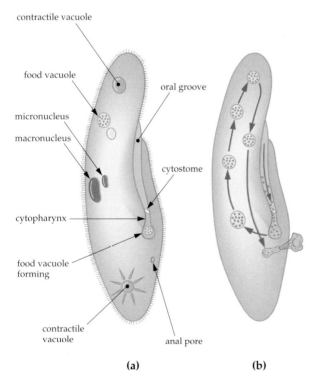

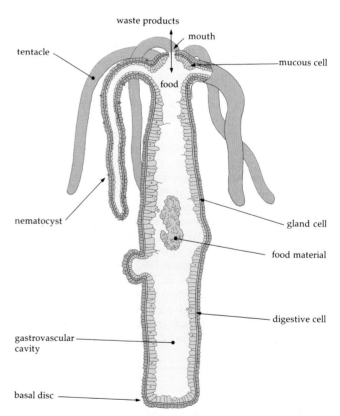

Fig. 19.16 **(a)** Major structures of a paramecium. **(b)** A food vacuole formed at the lower end of the cytopharynx, separates and moves toward the anterior end of the cell while enzymes are secreted into it. Digestion takes place as in amoebae and the products of digestion are absorbed into the general cytoplasm. The vacuole then moves toward the posterior end, attaches to the anal pore and expels digestive wastes. The vacuole undergoes several changes in size and appearance as it moves

Fig. 19.17 Gastrovascular cavity of *Hydra*. The cavity contains food material. The gland cells secrete digestive enzymes and the digestive cells absorb the products of digestion. The nematocysts are specialised cells that help capture prey

Some flatworms have many complex blind sacs opening from a muscular pharynx (Fig. 19.18). This increases the absorptive surface area and shortens the diffusion pathway from the gut to other cells of the animal. The muscular pharynx is capable of ingesting small animals or tearing off pieces from a larger, usually dead, organism.

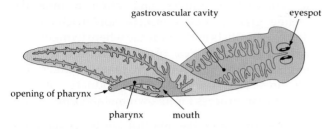

Fig. 19.18 A planarian, showing the highly branched gastrovascular cavity and muscular pharynx extruding through the mouth

Gut parasites, such as tapeworms, which live in the digestive tract of another animal, are continually surrounded by food molecules that are the end products of extracellular digestion by the host animal. Tapeworms are highly specialised flatworms that have lost their digestive tract and simply absorb required nutrients directly through their body wall.

Two openings: one-way movement of food

The next major advance in the evolution of digestive systems is seen in simple worms such as nemerteans and nematodes (Chapter 38). These worms have a *one-way* digestive tract, with a mouth at one end and an anus at the other. A one-way gut allows the independent elimination of wastes and regional specialisation involving sequential secretion of different enzymes. The gut is usually only one cell thick, with no muscular lining. Food is forced along the tract by external body pressure or pressure of following food. One group, the rotifers, have jaws that are able to grind food, although it is not clear how effective they are.

A one-way gut with a mouth and anus allows regional specialisation of the gut for more efficient digestion and waste elimination.

Muscular gut wall: coelomates

The evolution of a coelom (Chapter 15) in more complex groups, such as annelids (Fig. 19.19),

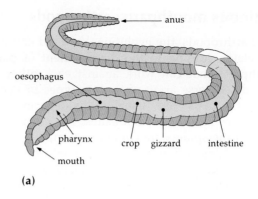

(a)

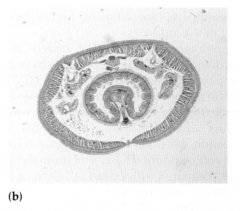

(b)

Fig. 19.19 (a) Digestive system of an earthworm showing the one-way system where food is ingested by the mouth and waste products are eliminated by the anus. (b) Cross-section of an earthworm in the intestinal region. The typhlosole, which projects into the cavity of the intestine, greatly increases the surface area available for absorption of food

arthropods, molluscs, echinoderms and chordates, led to a marked increase in length of the gut and development of a muscular gut wall, with its own blood supply. Waves of sequential contraction and relaxation, peristalsis, move food through the gut independently of whole body movements. Different rates of food movement are possible depending on the size of food particles and how difficult they are to digest. There is a capacity to store ingested food for later digestion, allowing feeding to be discontinuous, leading to an increased capacity to utilise localised food sources. Water absorption from the gut lumen is particularly important for terrestrial animals.

This basic pattern of gut structure and function is found in all complex animals. With increased specialisation of the gut, animals are able to eat and digest a wider range of foods, including larger or hard to digest items.

Increased gut specialisation allowed animals to eat and digest a wider range of foods.

Chitinous mouthparts: arthropods

In the arthropods, the evolution of a hard chitinous exoskeleton leading to the development of paired mouthparts that can work against each other has allowed a diverse array of feeding types to evolve. The arthropods are usually divided into the chelicerates and the mandibulates. The chelicerates, of which the spiders are common examples, tend to pierce and inject digestive enzymes into their prey and then suck out the predigested food; their gut is short. The aquatic mandibulate crustaceans are generally filter-feeders, predators or scavengers and their mouthparts are adapted for ingestion and crushing of the food.

The predominantly terrestrial insects show the greatest diversity of feeding types; they have crushing jaws or piercing mouthparts (Chapter 38). The prey of carnivorous insects includes soft-bodied worms and larvae from the soil, other arthropods, which have a resistant exoskeleton, and vertebrates with an endoskeleton. Most herbivorous insects feed on sap or plant cell contents. The size of an insect, such as an aphid, compared with a plant cell allows it to pierce the phloem tubes or other cells with specialised mouthparts (Fig. 18.23). The mouthparts of an aphid for sucking plant sap are in many ways similar to those used by a female mosquito for sucking vertebrate blood (Fig. 19.20). Plant cell contents are readily digestible and so an aphid's gut is short and uncomplicated as it is in carnivorous insects.

Insects that feed on plant tissue and ingest cell wall material have cutting or grinding mandibles, often with specially hardened cutting edges. At moulting, the cuticle, including the mandibles, is lost and then replaced. Consequently, a new set of mandibles is produced at each moult. This is particularly important to those insects that eat abrasive foods, such as grass, because mechanical processing of plant tissue wears the leading edges of the mandibles. Many of these insects do not digest cellulose but extract the cell contents and then eliminate the residue. These animals tend to have relatively short guts combined with complex crushing and grinding mandibles and often a muscular grinding gizzard. Other insects hold bacteria and colonies of protists in their caecae where fermentation of the cell wall takes place; their gut is more complicated and gut retention time is longer. The host insect utilises the by-products of this fermentation process for its own metabolism.

> Insects with an exoskeleton that can be modified into complex mouthparts have developed an enormous array of dietary specialisations.

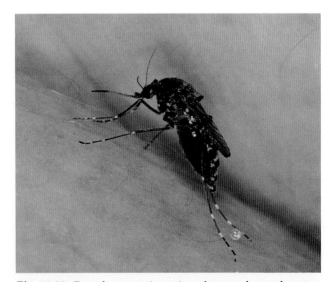

Fig. 19.20 Female mosquito using elongated mouthparts to suck blood

Jaws and teeth: vertebrates

In vertebrates, jaws supporting hard enamel-covered teeth provide an efficient mechanical processing system that allows improved access to large, hard or tough foods. The earliest vertebrates to evolve were the fishes (Chapter 40). Today, cartilaginous and bony fishes (teleosts) include specialist filter-feeders, detrital bottom feeders and carnivores. Herbivores, which consume algae and seagrasses, are only found among the bony fishes. The contrasting jaw and gut structures associated with the two dietary extremes, carnivory and herbivory, are well illustrated when the carnivorous flathead is compared with the herbivorous luderick. The flathead has sharp grasping needle-like teeth, a large but simple stomach and a very short almost straight intestine. The luderick has flat crushing teeth and a long coiled gut.

The skull structure of higher bony fishes is complex with a series of hinged bones supporting the teeth, which can be moved relative to other parts of the skull (Figs 19.21, 19.22). This kind of bone movement is known as kinesis and its development has produced a diverse range of specialist feeders that can manipulate often strange food items, particularly in coral reef fishes (Chapter 40).

All adult amphibians appear to be carnivorous. Frog larvae, tadpoles, with long coiled guts feed primarily on algae. At metamorphosis into the carnivorous adult, the gut shortens markedly, illustrating a general pattern seen in many other animals—herbivores have longer intestines than carnivores (Fig. 19.23).

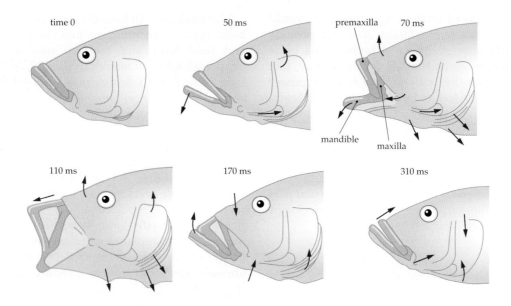

Fig. 19.21 Feeding in *Serranachromis*, a cichlid fish. The mouth at rest is closed and conforms to the streamlined shape of the body. The mouth can be rapidly opened to a very large gape and the jaws protruded to grasp prey. The mouth is then rapidly closed. Only the premaxilla and mandible support teeth. The maxilla acts as a lever to raise the premaxilla. Herbivorous fish or fish that prey on slow moving animals do not have such wide gapes but have powerful jaw-closing muscles

Fig. 19.22 Movable or extendible mouthparts are an important feeding adaptation in many fishes. This slingjaw wrasse, *Epibulus insidiator*, can rapidly extend the whole jaw and catch unsuspecting prey

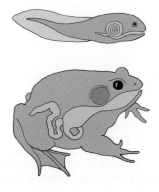

Fig. 19.23 Gastrointestinal tract of an adult frog and tadpole. The much-coiled intestine of the herbivorous tadpole is far longer relative to the size of the animal than is the intestine of the adult frog

Ectothermic reptiles have lower energy requirements than endothermic birds and mammals. This would seem to allow herbivorous reptiles the time required to ferment plant cell walls. However, this advantage may be offset to some extent by their lower body temperature, at least at night, and by poor mechanical preparation of food since reptiles have relatively uncomplicated teeth and poor chewing mechanisms.

Aquatic reptiles are mainly carnivorous, although a few, such as marine iguanas and green turtles, consume algae and seagrass. Some carnivores swallow the prey whole without dismemberment. Snakes disarticulate the lower jaw to accommodate the broadest part of the prey. However, feeding in this way means that a large amount of non-nutritive material may also be ingested.

Birds are constrained in what they eat by their mode of locomotion. Flight requires a lot of energy and there is a high cost in having heavy organ systems. Many birds consume high energy foods such as insects, nectar or fruit. A relatively large gut filled with a heavy mass of fermenting vegetation would prohibit flight, and fermentation does not provide energy rapidly enough. The few herbivorous birds tend to be larger ground-loving forms or may, like the emu and the other ratites, have lost the power of flight altogether. Modern birds have lost their teeth, presumably a weight-saving adaptation, but the cost is a loss of mechanical processing capacity. For predators such as owls, which swallow prey whole, feathers and bones ingested are regurgitated after the easily digestible material has been processed. In some grain-eating species, such as chickens, a muscular part of the stomach known as the gizzard has developed a mechanical processing role (Fig. 19.24).

Birds generally eat high-energy easily digestible foods to fuel their high metabolic rates and to keep weight low.

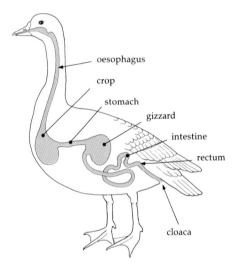

Fig. 19.24 Digestive system of a bird. The crop stores food and the chamber for mechanical break-up, the gizzard, is located posterior to the acid-secreting stomach, which means that the food is pretreated before it enters the gizzard. Note the similarity to the earthworm

Mammals

Mammals maintain high constant body temperatures, and have complicated very hard teeth held in jaws that can move independently in precisely controlled ways. Their teeth have become an integral part of the digestive system, and both teeth and gut have become specialised for particular diets. The parallels between tooth structure and digestive function in mammals of diverse phylogenetic origins, but similar diets, are striking. However, if one component of an animal's diet needs specialised teeth or gut, the animal tends to show these adaptations, even though the dietary component may be only a minor part of its diet. It is partly for this reason that one can be misled when attempting to deduce diet from functional anatomy.

Mammalian teeth have become specialised into four main types with different functions. Incisors grasp and hold food, and tend to remain similar and unspecialised in all mammals (Fig. 19.25). Canines are long conical teeth, which may be used for stabbing and gripping prey, and which are usually well developed in carnivores and lost in herbivores. Post-canine teeth, the cheek teeth, are involved with mechanical processing and the lower teeth work against the upper teeth in either cutting or grinding actions. The first set of cheek teeth, premolars, together with incisors and canines, are deciduous teeth and are replaced in the adult animal by permanent, usually larger teeth. The posterior teeth, molars, are not replaced, only one set being produced in the life of most mammals. For animals with an abrasive diet, tooth wear may reduce effective feeding (Fig. 19.26).

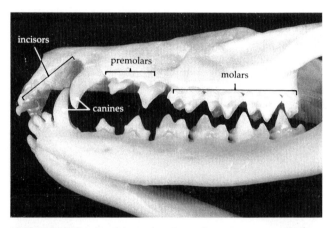

Fig. 19.25 The dentition of an insectivorous mammal, the marsupial native cat. The incisors, canines, premolars and molars are indicated. The molars have two functions: to puncture and then crush insects with a rigid exoskeleton

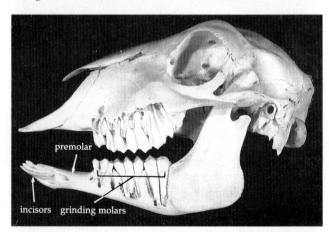

Fig. 19.26 The dentition of a mammalian herbivore, the sheep. The grinding molars occupy most of the jaw. The teeth have high crowns, which are held in the body of the jaw and have continually growing roots. This means that for a good part of the animal's life the enamel crown is continually being replaced as it is worn away by abrasive foods

Small mammals require and are generally restricted to high-energy foods that are quick to digest. Their main source of high-energy food is energy-rich plant products, such as nectar, pollen, fruit and flowers, and arthropods, which have an exoskeleton of relatively indigestible chitin. The smallest mammalian herbivores utilise plant tissue with little cell wall material and with energy-rich starch-filled cell contents. They have efficient teeth with relatively large surface areas, powered by large jaw muscles. The gut is relatively simple and short, usually only several times the body length. In general, larger mammalian herbivores incorporate more structural material in their diet than smaller mammals.

Small insectivores (carnivores for which insects form a major part of the diet) use teeth to penetrate the tough cuticle of their prey and expose the nutrient-rich haemolymph. The molar teeth of virtually all small insectivorous mammals are well suited to this, initially puncturing and crushing the exoskeleton followed by fine shearing of the inner tissues. Large carnivores cannot survive on insects, unless they are concentrated in large numbers as are ants and termites, because it takes too long to catch them in sufficient quantities. Consequently, large terrestrial mammalian carnivores are limited to preying upon other vertebrates that have an endoskeleton. Among carnivores such as dogs, cats and marsupial lions, scissor-like carnassial teeth have evolved independently several times (Fig. 19.27). These teeth are adapted for shearing chunks of meat off the bone. Some carnivores, such as the canids and especially the hyenas, have specialised molars for crushing bone. Cats, which tend to be specialist flesh-eaters, have lost nearly all teeth except the canines and carnassials (Fig. 19.27). The toothed whales and most seals are predators, mostly on large single prey items such as fish or other mammals. The gut of carnivores remains relatively unspecialised with a simple stomach, short intestine and often a reduced or missing caecum.

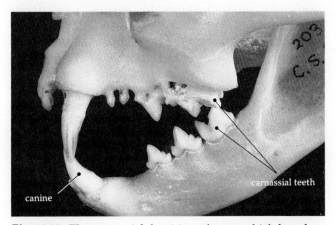

Fig. 19.27 The carnassial dentition of a cat, which largely consumes other vertebrates with an endoskeleton. The carnassial tooth is specialised for shearing flesh only

> Small mammals tend to consume easily digested foods, such as nectar, fruit and insects. Small insectivorous mammals have simple guts and complex teeth for puncturing and crushing insect exoskeletons. Carnivorous animals have short digestive systems, while herbivores have long, often complex digestive systems.

Most herbivores utilise only the cell contents, thereby avoiding less digestible cell walls that require slow enzymatic breakdown or fermentation. Fermentation of plant cell walls is enhanced by high temperature, large body size (accommodating a large gut and contents) and efficient initial mechanical processing of food. Large mammals are thus well adapted to process structural plant tissue. Herbivorous mammals typically have broad crushing teeth adapted to an abrasive fibrous plant diet.

Herbivores such as the koala (p. 405), which rely on the cell contents being exposed and digested in the small intestine before the food reaches the bacteria in the hindgut, also need very efficient teeth. Herbivores that use bacteria in the stomach as a means of fermenting plant cell walls rely on their teeth to damage cell walls for bacterial attack rather than to expose the cell contents for digestion. If the cell wall is degraded, then the contents will spill out and become available in any case.

The advantage of size can be seen in the common ringtail possum and koala, which both consume eucalypt leaves. The possum is small and selects a diet lower in fibre than does the larger koala. Both animals eat the same leaves but the possum, with its smaller mouth, can bite out parts of the leaf between major veins. It thus avoids lignified fibre, and the mesophyll and epidermal cells that it ingests are relatively easy to crush between its teeth. The koala with its larger mouth size cannot avoid the lignin and must deal with it in its diet. It has teeth with more pronounced cutting edges to process the resistant fibre bundles.

Omnivorous animals can utilise a variety of different food types. We might presume that omnivores are not as efficient at every task as specialists, but are compensated by having a greater variety and abundance of resources available to them. The problem with this argument is that it does not take into account processing time within the animal and the limitations of gut fill. If an animal cannot process food fast enough, and hence efficiently enough, to satisfy its requirements, an abundance of food may not be an adequate recompense. It is interesting that omnivores tend not to be found among the smallest mammals where rate of energy acquisition is most important. Those that do occur are likely to eat high-energy easily digestible plant and animal products, such as nectar and insects (these animals tend to be considered as insectivores).

Filter feeding: eating tiny particles

This method of feeding is represented in most animal phyla. It limits an animal to tiny particles suspended in water and it is no surprise that filter feeding is predominantly found in organisms that feed in aquatic systems (Fig. 19.28). Vertebrates that filter

(a)

(b)

Fig. 19.28 Sifting the seas—many animals feed by filtering microscopic plant and animal life from large volumes of water. **(a)** Baleen whales feed using large sieve-like plates. **(b)** Corals intercept the water current in a variety of ways—this gorgonian coral intercepts the current

small particles of food from large volumes of water include mammals such as the baleen whales (e.g. the humpback whale of the Australian east coast) and birds such as flamingos.

Filter feeding presents a number of difficulties, particularly for intertidal animals. For example, intertidal animals can feed only when they are submerged. Perhaps not surprisingly, individuals from the high intertidal zone consume more prey and absorb more nutrients in a given time period than do those from the low intertidal zone. Generally, filter-feeding animals ingest whatever material happens to pass by; the only discrimination possible is on the basis of the size of a particle. The wide range of prey ingested (e.g. bacteria and plankton, and also inorganic particles) means that the animals need a wide range of digestive enzymes to be able to use what is captured.

Filter feeding is one of the most common feeding mechanisms among sessile or mobile aquatic invertebrates. Most invertebrates lack the structures to reduce the size of large prey. In both freshwater and oceans, detritus particles, and living and dead phytoplankton (such as diatoms), are abundant. The mechanism for trapping small particles varies from sheets of mucus (e.g. tunicates or 'sea squirts') to modified tube feet (sea cucumbers). One of the most widespread is the use of cilia, which may set up local currents to trap microscopic food particles in a mucous stream to pass this towards the mouth. This type of pattern is commonly found in bivalves such as oysters and mussels.

Generally, shallow oceanic waters contain more particulate matter than the open ocean but filter-feeders must cope with a wide variation in food supply. Studies of larvae of the crown-of-thorns starfish have shown just how important food supplies can be for filter-feeding animals. The natural levels of particles in Great Barrier Reef waters are generally too low or marginal for the development of the crown-of-thorns' larvae. However, at certain times there are sufficient phytoplankton to ensure a moderate survival of larvae and this may contribute to subsequent outbreaks of the starfish on the Great Barrier Reef.

SUMMARY

- Heterotrophic animals are unable to manufacture their own organic compounds and must obtain them from other organisms. All animals need energy, nitrogen-containing compounds, fats, vitamins, minerals and water.

- Plant tissues are rich sources of carbohydrates but generally poor sources of protein. Animal tissues are the reverse.

- Large animals eat more food than do small animals but small animals actually need more food per unit body mass. This means that small animals often feed on better quality diets than do large animals.

- Digestion is a stepwise process that usually starts with the physical breakdown of food, providing a large surface area for the action of digestive enzymes that break down complex molecules. The type of digestive enzymes present usually reflects the composition of the normal diet.

- Enzymatic digestion normally involves the sequential secretion of different digestive enzymes along the length of the gut. These processes are controlled by both nervous reflexes and hormonal actions.

- Not all foods can be digested by an animal's own enzymes. Plant cell walls are a potentially rich source of energy but most animals need enzymes from symbiotic microbial organisms to hydrolyse the cellulose and hemicellulose. The host animal benefits from the waste products (mostly short-chain fatty acids) excreted from the micro-organisms.

- In some animals (e.g. kangaroos and sheep), the microbes are located before the true acid-secreting stomach. This is foregut fermentation. In hindgut fermenters (e.g. possums and horses), the microbes are located posterior to the region of acid secretion.

- Foregut fermenters benefit by being able to digest some of the actual microbes, which provides them with a rich source of essential amino acids. However, the microbes digest any sugars and proteins in the diet and although some of these nutrients will eventually be made available to the host animal, the transformations are inefficient.

- Hindgut fermenters benefit by being able to directly use the sugars and proteins in the diet but are unable to digest microbial protein, except by eating faeces, coprophagy. This is a common behaviour in small herbivorous mammals.

- Digestive systems became more complex as animals evolved into large multicellular forms.

- The simplest digestive systems involve engulfing of food particles into a membrane-bound vacuole and intracellular digestion.

- Extracellular digestion, in contrast, involves the secretion of digestive enzymes onto food. In the simplest forms, this occurs in simple sac-like guts but the development of guts with two openings allows for foods and wastes to be separated and for sequential digestion along the gut.

- Animals with coeloms developed longer guts with muscular gut walls and their own blood supply. These larger guts meant the food could be ingested and stored for later digestion. Increasing gut specialisation meant that animals could feed on a wider range of foods.

- Since many foods are hard, physical processing is necessary before digestion can occur. The chitinous mouthparts of insects and the jaws of vertebrates have allowed an enormous range of dietary specialisations.

- Similar design patterns are found in a wide range of animals. Carnivores tend to have short and relatively simple digestive systems whereas herbivores have long, large and complex guts.

- Filter feeding is practised by a wide range of animals and clearly demonstrates the close match between the nature of the food and the design of feeding and digestive systems.

QUESTIONS

1. Explain why 'total carbohydrate' is unhelpful when it appears in a statement of food composition.

2. Explain the terms 'essential amino acids' and 'essential fatty acids'.

3. Why do small animals tend to eat better quality foods than do large animals?

4. What are zymogens and why are they important?

5. Describe some of the advantages and disadvantages of microbial fermentation in the foregut and the hindgut of herbivorous mammals.

6. Relate differences in the structure of the gut of sheep and kangaroos to the likely responses of each to poor-quality pastures.

7. Outline the dietary differences faced by aquatic herbivores and terrestrial herbivores.

8. What are the advantages of a one-way digestive tract?

9. Why are there few birds that rely on fermentation of plant cell walls as their major energy source?

10. Grasses are rich in fibre and abrasive particles. How have the teeth of grazing mammals adapted to these independent factors?

11. What problems are faced by blood-sucking and sap-sucking insects? How do they overcome the response of their prey?

GAS EXCHANGE
IN ANIMALS

Gas exchange supports cellular respiration by supplying a fuel, oxygen (O_2), and removing a by-product, carbon dioxide (CO_2; Chapter 5). Gas exchange involves the movement of these gases between the environment and mitochondria, the site of cellular respiration. In unicellular organisms, O_2 reaches mitochondria simply by diffusing through the plasma membrane and cytosol. In multicellular organisms, diffusion is usually inadequate and a variety of mechanisms that ensure an adequate supply of O_2 and removal of CO_2 have evolved. The nature of these mechanisms reflects the size of an animal, its level of activity and whether or not the external respiratory medium is water or air. Animals inhabit a diverse array of habitats. Usually, terrestrial animals obtain their O_2 from the atmosphere, while aquatic animals extract O_2 from water. Some animals, such as frogs, breathe water as larvae and air as adults.

| Gas exchange involves the supply of oxygen for cellular respiration and the removal of carbon dioxide.

AIR AND WATER AS RESPIRATORY MEDIA

As respiratory media, air and water have very different properties. Compared with air, water has a greater density, is more viscous and presents a greater resistance to diffusion of both O_2 and CO_2. In a respiratory context, the most important property is how much O_2 or CO_2 can be contained in a given volume of air or water. The concentration of O_2 in a litre of air is 20.9% by volume. For water, the concentration (C) depends on the solubility of O_2 (amount physically and chemically bound) and the **partial pressure** of O_2 (the proportion of total pressure of a gas mixture provided by O_2) as shown in the equation:

$$C_{O_2} = \beta_{O_2} . P_{O_2}$$

where β is a constant related to solubility (**capacitance** of water for O_2), and P_{O_2} is the partial pressure of O_2 in water. A similar relationship holds for CO_2.

Figure 20.1 shows the concentrations of O_2 and CO_2 plotted as a function of their partial pressures in distilled water and air. The slope of the line of concentration versus partial pressure corresponds to the capacitance of the fluid for the gas. Two important properties are illustrated in this figure: the capacitance of air for O_2 and CO_2 is identical, and the capacitance of water for O_2 is far lower than for CO_2. Thus, water, in equilibrium with air, contains some 20–40 times less O_2 than does the air. The precise amount of O_2 contained in water also depends on factors such as temperature, salinity and distance from the air–water interface. An increase in either temperature or salinity decreases the capacitance of water for gases and hence decreases gas concentrations. On the other hand, because of the higher capacitance of water for CO_2, CO_2 diffuses through water some 20–40 times faster than does O_2. These differences in the capacity of water for O_2 and CO_2 have significant consequences for gas exchange.

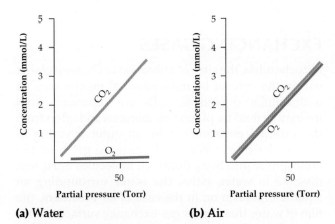

Fig. 20.1 Concentrations of O_2 and CO_2 versus their partial pressures in (**a**) water and (**b**) air. The slope of the line is related to the capacitance for a gas. The capacitance of CO_2 in water (solubility) is far greater than that for O_2, while the capacitance for both gases is the same in air

The capacitance of air for O_2 and CO_2 are the same, whereas the capacitance of water for O_2 is 20–40 times less than for CO_2. Increased temperature and salinity reduce the capacitance of water for dissolved gases.

There are several other factors that influence O_2 availability in aquatic habitats. In the sea, despite the low content of O_2, there is usually enough O_2 at the surface to meet the aerobic demands of animal life. The same is true for deeper regions of the ocean, where lower temperatures raise the capacitance for O_2 and replenishment is assisted by ocean currents. The O_2 content of freshwater can also be maintained provided that there is sufficient water circulation to achieve effective aeration. In situations where water circulation is reduced or lost, as in swamps, ponds or tidal pools, O_2 deprivation may occur. However, this may be offset by the production of O_2 by photosynthetic organisms. The balance between removal of O_2 by cellular respiration and addition of O_2 by photosynthesis can vary daily and seasonally.

Except for caves and burrows, where air circulation is reduced, the composition of air is relatively stable. The amount of O_2 available in air does, however, depend on altitude. With increase in altitude, total atmospheric pressure, and thus the partial pressure of O_2, decreases. Although the proportion of O_2 remains the same, O_2 concentration decreases with increased altitude.

When compared with aquatic dwellers, air breathers live in environments that tend to be more stable and contain more O_2. However, unlike water breathers, air breathers are faced with the risk of water loss and desiccation (Chapter 22).

The composition of air is relatively stable and O_2 content is high compared with water; however, air breathers often face problems of desiccation.

EXCHANGING GASES

Mitochondria, the sites of utilisation of O_2, are isolated within the cytosol by mitochondrial membranes. In multicellular organisms, cells are separated from interstitial fluid by plasma membranes and often from the external environment by an outer covering of skin, scales or exoskeleton. In order to get from the atmosphere into body fluids, O_2 molecules must first dissolve in water, either the water surrounding an aquatic organism or, in the case of air breathers, the film of water that covers gas-exchange surfaces. Once in solution, O_2 diffuses across successive membrane barriers provided there is a favourable partial pressure gradient. Net diffusion at any of the barriers will cease if the partial pressure difference becomes zero. Cellular respiration reduces intramitochondrial P_{O_2}

towards zero, so there is clearly an overall gradient of P_{O_2} from the atmosphere to the inside of mitochondria.

The rate of diffusion of a gas across a membrane not only depends upon the partial pressure gradient but also on properties of the membrane. Fick's law shows that the amount of gas transferred per unit time across a membrane (*conductance, dM/dt*) is directly proportional to its *permeability* (chemical and physical properties, *P*), *surface area* (*A*), and the *difference in partial pressure* of the gas on either side of the membrane (*Δp*), and inversely proportional to its *thickness* (diffusion distance, *d*).

$$\frac{dM}{dt} \; \alpha \; \frac{PA\,\Delta p}{d}$$

In diseases in which fluid accumulates in the alveoli, such as emphysema and pneumonia, diffusion distance between air and blood becomes so great that the rate of O_2 exchange is inadequate. This causes breathing to become laboured and the patient is unable to tackle even moderate exercise without great distress. The nature of the diffusion barrier is also important in gas exchange. For example, vertebrate embryos are often encased in protective structures (shells) that pose potential problems for gas exchange (Box 20.1).

Diffusion across a membrane is directly proportional to the permeability, surface area and partial pressure gradient, and inversely proportional to the thickness of the membrane.

SURFACE AREA AND VOLUME

It has been estimated that, for a spherical animal with a metabolic rate appropriate for a single-celled organism surrounded by air-equilibrated water, the maximum diameter that can be supported by diffusion alone is about 2 mm. There are several reasons for this limitation, the most important of which is the animal's surface area-to-volume ratio. This is a simple consequence of geometric relationships—as an object increases in size, its surface area-to-volume ratio decreases.

This relationship results because volume (*V*) is related to the cube of the length (*l*), whereas surface area (*A*) is related to the square of the length. As a result, surface area is proportional to volume to the two-thirds power. Thus, if one species is twice the length of another in each direction, then it is likely to have eight times the volume but only four times the surface area. This can be seen graphically by taking a cube and doubling its length in each dimension (see p. 420). The resulting cube will contain eight cubes of the original size but only have four times the external surface area.

BOX 20.1 GAS EXCHANGE ACROSS THE SURFACE OF EGGS

Fish and amphibian eggs are often surrounded by gelatinous capsules, while those of most reptiles, all birds and monotremes are surrounded by shells (Fig. a). The rate at which gas diffuses through an egg covering, its **conductance**, is described by Fick's law.

The capsules of fish and amphibian eggs are composed mainly of water across which dissolved gases diffuse slowly. Under these circumstances, it is impossible to achieve high gas conductances; thus, these embryos are small (always less than 1 g) with low metabolic rates and gas-exchange requirements. Bird eggs, on the other hand, are surrounded by a shell pierced by thousands of tiny air-filled pores through which gas diffusion occurs easily. Such shells permit high rates of gas exchange, so embryos can be larger, up to several kilograms in flightless birds (and dinosaurs). Embryos of this size depend upon internal transport systems (Chapter 21) to achieve adequate rates of O_2 delivery and CO_2 removal to cells. In bird eggs ranging in size from the humming bird to the emu, conductance and metabolic rate increase in parallel. The result is that levels of O_2 and CO_2 inside the shell are about the same in all birds.

Conductance is also adapted to match conditions prevailing in unusual incubation environments. The Australian mound-building birds (mallee fowl and brush turkey) lay their eggs underground where humidity and CO_2 levels are high, and O_2 availability is low (Fig. b). In compensation for this rather stuffy subterranean atmosphere, the shells of these birds are unusually thin, with very high conductances. The incubation mound protects the fragile shells from breaking and gas tensions inside the shell are similar to those in chicks of other species.

Given the rigid construction of a bird's egg, shell conductance of most bird eggs remains fairly constant throughout incubation. However, conductance of frog egg capsules changes during development. Recent studies of the eggs of the Australian terrestrial toadlet *Pseudophryne bibronii*, which lives and lays its eggs under leaf litter, have shown that the embryo gradually absorbs water from the environment into the space beneath the capsule. This makes the capsule swell, decreasing its thickness and therefore greatly increasing its conductance (Fig. c). The fact that conductance increases with increasing metabolic rate means that the oxygen level near the embryo is kept high and constant throughout development.

(b) An unhatched egg and a hatchling chick of the Australian brush turkey found inside an incubation mound. The mound is constructed from decaying leaves and twigs from the forest floor. The eggs lie about 60 cm below the surface and chicks require about two days to dig their way out

(a) Pathways of oxygen and carbon dioxide diffusion through the jelly capsule in an amphibian egg (left) and an avian eggshell (right). Respiratory gases diffuse slowly through the aqueous layers of the capsule and support relatively low metabolic rates in amphibian embryos. Inside the capsule, gas exchange occurs across the embryo's skin and gills. In contrast, high rates of metabolism in bird embryos are supported by rapid diffusion through air-filled pores in the shell and shell membranes. Oxygen is taken up, and carbon dioxide is given off, by blood flowing through the chorioallantois, the respiratory membrane that grows from the embryo. The inner and outer shell membranes separate at the blunt end to form an air cell where the embryo begins to breathe shortly before hatching

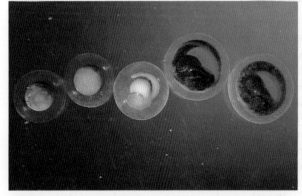

(c) Eggs of the Australian toadlet, *Pseudophryne bibronii*, at 3, 6, 15, 37 and 54 days of incubation, showing the development of the embryos and the increase in fluid volume within the jelly capsule

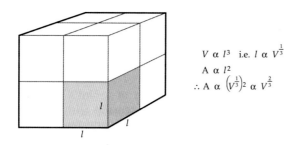

$$V \, \alpha \, l^3 \quad \text{i.e. } l \, \alpha \, V^{\frac{1}{3}}$$
$$A \, \alpha \, l^2$$
$$\therefore A \, \alpha \, \left(V^{\frac{1}{3}}\right)^2 \, \alpha \, V^{\frac{2}{3}}$$

In other words, each 10-fold increase in edge length reduces a cube's surface area-to-volume ratio by a factor of 10. This means that the bigger an animal is the less surface area it has available for exchange of respiratory gases (or heat for that matter). As an animal grows, the mass of metabolically active cells (the prime determinant of O_2 *demand*) rapidly outstrips the surface area available to *supply* O_2. In animals thicker than about 2 mm, the surface area is insufficient and diffusion is too slow for adequate O_2 to be supplied by diffusion alone. In general, if an active animal is to grow much larger than a few millimetres in size, it must rely on some other method of enhancing the diffusion process.

We know from Fick's law that diffusion across a membrane can be improved by altering one or other of the variables in the equation. Thus, an increase in the partial pressure difference, a decrease in the diffusion distance or an increase in the surface area will all improve the delivery of O_2. During the evolution of animals, a variety of mechanisms incorporating changes in one or more of these factors has arisen.

> As an object increases in size, its surface area-to-volume ratio decreases. Animals larger than a few millimetres in size have evolved ways of enhancing exchange with the environment.

VENTILATION AND CONVECTION

A small organism such as an amoeba is limited in obtaining sufficient O_2 by the presence of a layer of stagnant water around it, the **boundary layer**. As O_2 is taken up from this layer, P_{O_2} adjacent to the amoeba decreases, thus decreasing the partial pressure gradient for diffusion across the membrane. Replenishment occurs as O_2 diffuses into the boundary layer from surrounding water. However, far more effective replenishment occurs if the boundary layer is stirred or flushed away by currents in the pond water.

Rather than relying upon natural currents to renew the external media at the gas-exchange surface, many organisms generate their own currents by means of cilia or flagella, or make use of various paddle-like appendages to promote water flow. In animals, this

is commonly associated with restriction of gas-exchange surfaces to a reduced region of the body surface. This frees other regions to become specialised for locomotory, protective or sensory functions.

The bulk flow of a fluid (air or water) is known as **convection**. When convection of the external medium at the site of gas exchange is generated by the animal, it is referred to as **ventilation**. Ventilation occurs in all larger animals and is responsible for renewing the external medium, thereby maintaining a favourable partial pressure gradient and maximising the rate of diffusion at the site of gas exchange.

The value of convection in maintaining a favourable partial pressure gradient applies equally well to the internal side of the gradient—body fluids. Many animals possess some means of circulating body fluids (Chapter 21), although an efficient internal transport system is first seen in the segmented worms, annelids (Chapter 39). In terms of gas exchange, internal convection, **perfusion**, is responsible for transporting O_2 from the site of gas exchange to cells, thereby maintaining the P_{O_2} gradient across the respiratory surface. For CO_2 a similar situation occurs in reverse: internal convection moves CO_2 from cells to the site of gas exchange, thereby maintaining favourable gradients for the loss of CO_2 at both sites.

Thus, in terms of gas exchange, the evolution of larger animals was dependent upon the development of two systems (Fig. 20.2):

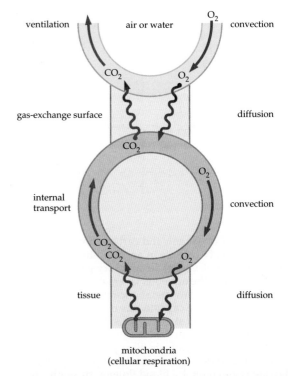

Fig. 20.2 The pathways for O_2 and CO_2 in a metazoan. O_2 is transferred from the environment to the mitochondria whereas CO_2, a waste product, is transferred in the opposite direction. Whether transfer is by convection or diffusion is shown on the right of the diagram

- a gas exchange system, often involving ventilation of the respiratory medium;
- an internal transport system.

Both systems depend upon convection *and* diffusion to transfer gases. Diffusion is responsible for conveying both O_2 and CO_2 between the two systems and from the transport system to mitochondria. Circulatory systems are considered further in Chapter 21.

> Ventilation renews the external medium at the gas-exchange surface, thereby maintaining favourable partial pressure gradients for O_2 and CO_2. Internal convection maintains partial pressure gradients for O_2 and CO_2 across the gas-exchange surface, and transports O_2 and CO_2 to and from cells.

GAS-EXCHANGE ORGANS

Water breathers

Ventilation of the gas-exchange surface is particularly important in water because of its low O_2 content and diffusibility of O_2. Furthermore, the thickness of the boundary layer adjacent to the exchange surface decreases with increasing flow of water past the gas-exchange surface (Fig. 20.3). Maintenance of a good convective flow of water past the gas-exchange surface both supplies fresh oxygenated water and reduces the boundary layer.

The simplest multicellular animals are the sponges (phylum Porifera). While they have no special organs for gas exchange, their body is permeated with pores, canals and chambers through which a water current is established to bring in food and O_2, and to remove waste products, including CO_2 (Fig. 20.4). Specialised cells lining the inside walls, choanocytes, establish the water current by the beating of their flagella. The beating of the flagella has no particular synchrony so the flow of water is regulated by regulating the size of the osculum and closing the incurrent pores.

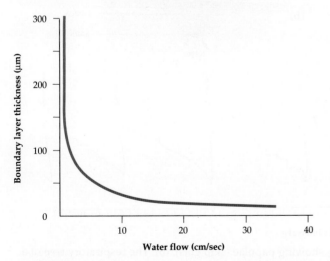

Fig. 20.3 The relationship between the boundary layer formed and the speed of water past a surface

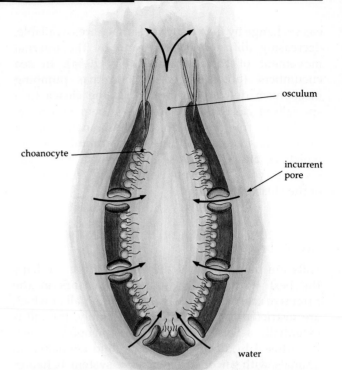

Fig. 20.4 The flow of water through a sponge, generated by the beating of the choanocytes, is regulated by opening and closing the osculum and incurrent pores

Cutaneous exchange

Aquatic oligochaetes depend on the exchange of gases across the general body surface—**cutaneous exchange**. Replacement of the surrounding water is either actively maintained or relies upon natural convection. A good example of active ventilation is found in *Tubifex*, a freshwater oligochaete. This worm lives in burrows but achieves ventilation by the rhythmic waving of its well-vascularised posterior end. The frequency of waving increases as the amount of O_2 in the surrounding water decreases. Annelids also possess a circulatory system for convection of blood carrying O_2 and CO_2 between tissues and skin.

As larger animals evolved, protective outer surfaces developed, which had a reduced ability to exchange gases. This, together with a decreasing surface area-to-volume ratio, led to the evolution of ventilatory organs specialised for gas exchange. Nevertheless, many fishes and amphibians rely on cutaneous exchange, although often only as a supplement to the main form of gas exchange.

> Cutaneous exchange is limited in larger organisms by protective outer surfaces and a low surface area-to-volume ratio.

Simple gas-exchange structures

In echinoderms, a variety of structures are involved in gas exchange. Sea stars (asteroids) have many tiny outgrowths of the body wall (papulae), which aid gas exchange by increasing the surface area available, decreasing diffusion distance and by the internal movement of coelomic fluid (Fig. 20.5a). In sea cucumbers (holothurians), a primitive pumping mechanism delivers sea water via the cloaca to a specialised gas-exchange surface, known as a rectal respiratory tree (Fig. 20.5b, c). The branches of the respiratory tree end in thin-walled ampullae, across which O_2 and CO_2 diffuse into coelomic fluid. The respiratory tree is actively ventilated by contractions of the cloaca.

Invertebrate gills

Gills are outgrowths of the body surface. Perhaps the best example of a simple gill occurs in the primitive crustacean family Anaspididae, all of which are restricted to Tasmania. In this family, the gill is essentially a flattened appendage (pleopod), which is perfused by haemolymph, the fluid circulated in animals with an open circulatory system (Chapter 21), circulating through channels around the edge and across the centre (Fig. 20.6a). The gill is exposed and

ventilation is created by the rhythmic beating of swimming appendages.

In other aquatic crustaceans and molluscs, the gills hang freely in the water. In these animals, the total respiratory surface area has been increased by increasing either the number or complexity of the gills, or both. Most species, however, possess gills enclosed within a **gill chamber**. While the gill chamber protects the delicate gills from physical damage, active ventilation of the gill is a necessity if a good supply of oxygen is to be achieved.

▪ Ventilation is necessary for gills housed in a gill chamber.

In the majority of molluscs, except for cephalopods and land-dwelling gastropods (pulmonates), ventilation of the gill chamber (mantle cavity) is created by cilia covering the surface of each gill and the mantle itself. For crustaceans with gills housed within chambers (branchial chambers), ventilation of the gills is maintained by rhythmic beating of a specialised pair of appendages, the scaphognathites (Fig. 20.6). In both cases, the flow of water is on the whole *unidirectional* in that it has separate routes of entry and exit from the gill chamber.

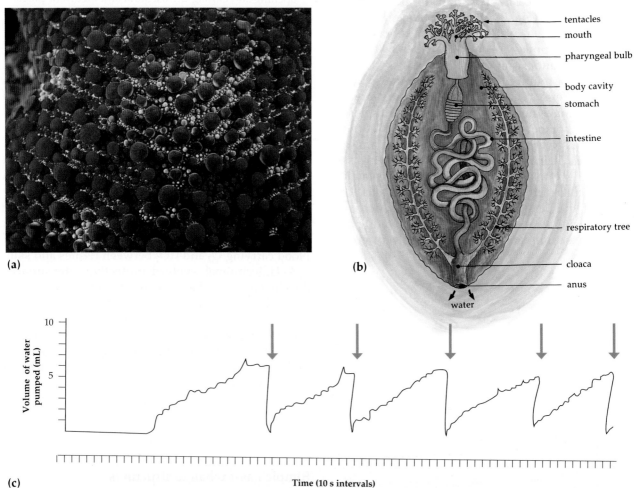

(a)

(b)

- tentacles
- mouth
- pharyngeal bulb
- body cavity
- stomach
- intestine
- respiratory tree
- cloaca
- anus

water

(c)

Volume of water pumped (mL)

Time (10 s intervals)

Fig. 20.5 **(a)** Close-up view of the centre surface of a sea star, showing papulae (skin gills). **(b)** The respiratory tree of a holothurian. A number of small contractions of the cloaca, represented by each small upward movement on **(c)** the chart, results in water being drawn into the respiratory tree. Water is then expelled in a single expiratory contraction

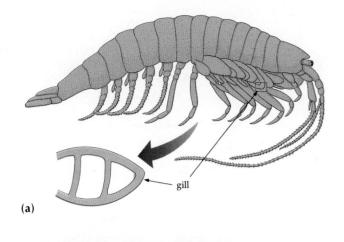

(a)

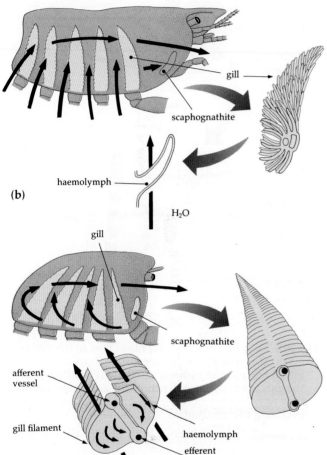

(b)

gill

afferent vessel

gill filament

haemolymph

efferent vessel

(c)

The relationship between water flow (ventilation) and haemolymph flow (perfusion) for three types of crustaceans is shown in Figure 20.6. An enormous advantage for gas exchange exists for the crab in which ventilation draws water inwards over the gill filament in the opposite direction to that of haemolymph perfusing within the filament. Such an arrangement, **countercurrent flow**, greatly increases the efficiency of gas exchange when compared with water flowing in the same direction as haemolymph—**cocurrent flow**.

The benefit of countercurrent flow arises because haemolymph entering the gill filament has a low P_{O_2} and first encounters water leaving the gill, which has already given up some of its O_2 (Fig. 20.7a).

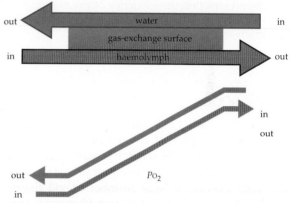

(a) Countercurrent flow

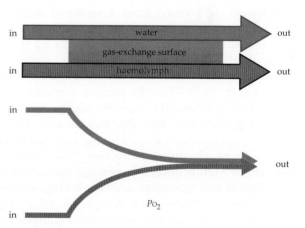

(b) Cocurrent flow

Fig. 20.6 Representative gills of three types of crustacea: **(a)** Anaspididae; **(b)** crayfish; and **(c)** crab. Ventilation in Anaspididae is achieved by the beating of the swimming appendages, whereas in the two decapod crustaceans, in which the gills are housed within brachial chambers, ventilation (blue arrows) is unidirectional and achieved by the rhythmic beating of the scaphognathites. The relationship between ventilation and perfusion is shown for the two decapod crustaceans. In the crab, the flow of water is in the opposite direction to the flow of haemolymph, permitting countercurrent gas exchange

Fig. 20.7 Functional models for **(a)** countercurrent flow and **(b)** cocurrent flow. The advantage of countercurrent flow, in which water and haemolymph flow in opposite directions through the gas exchanger, is that the partial pressure of oxygen (P_{O_2}) of the haemolymph leaving the gas exchanger is in near equilibrium with the P_{O_2} of the water entering. With cocurrent flow, water and haemolymph reach an equilibrium and leave the gas exchanger at an intermediate P_{O_2}

Nevertheless, the partial pressure gradient results in O_2 diffusing from water into haemolymph. As the haemolymph continues through the filament, it meets fresher water, with progressively higher O_2 levels; therefore, the partial pressure gradient between water and haemolymph is maintained, and O_2 diffusion continues into haemolymph. The result is that haemolymph leaves the gill with almost the same partial pressure of O_2 as the surrounding freshwater. By contrast, in cocurrent flow, with water flowing in the same direction as the haemolymph, equilibration of O_2 would eventually occur at a partial pressure somewhere between those of the incoming haemolymph and water (Fig. 20.7b).

> The directions in which water and blood (or haemolymph) flow relative to one another has a profound effect on the efficiency of gas exchange, with the greatest efficiency being achieved by a countercurrent arrangement.

Vertebrate gills

Gill structure in fishes is similar in principle to that in the crab except that the surface area is greatly increased and the gills are better vascularised. Most fishes have four gill arches on either side of the mouth, each arch bearing a large number of gill filaments (Fig. 20.8). The surface area of each filament is further enlarged by the addition of vertically stacked lamellae on both the dorsal and ventral surfaces. Each lamella is perfused with blood from a series of branchial arches derived from the ventral aorta.

Most fish actively ventilate their gills by means of a double-action pump involving sequential contractions and expansions of buccal and opercular cavities (Fig. 20.9). The gills present a resistance to water flow between the two chambers, effectively separating the two, and the mouth and pair of opercula operate as valves. During inspiration the mouth opens, the opercula remain closed and the floor of the mouth lowers, increasing the volume of both buccal and opercular cavities. Water is drawn through the mouth into the buccal cavity and through the gill curtain into the opercular chamber. On expiration, the mouth closes, the opercula open, and the floor of the mouth is raised. This movement forces water from the buccal cavity into the opercular cavity and out through the open opercula. By co-ordinating the movement of mouth and opercula, the fish maintains a hydrostatic pressure difference across the gills, which drives water across them. This is an open-flow or unidirectional ventilatory system. Blood and water flows at the level of the secondary lamellae (the gas-exchange structures) are countercurrent, with the expected advantage for exchange efficiency (Fig. 20.9).

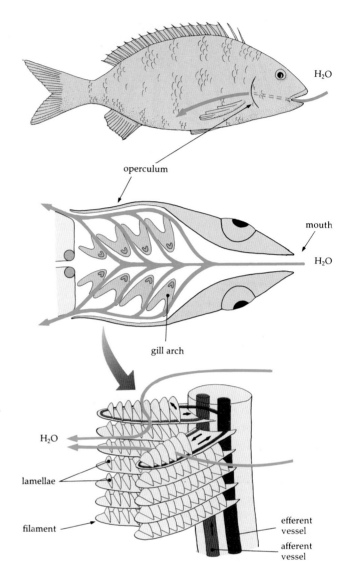

Fig. 20.8 Direction of water and blood flow in the gills of a fish. The section of a fish head from above shows the gill arches and the pathway of water. Water flow past the gills is unidirectional, permitting a countercurrent flow to exist between the water and the blood perfusing the gill lamellae

> Fishes have a unidirectional ventilatory system and countercurrent flow in the secondary lamellae.

Another advantage of a unidirectional water flow pattern is that some fishes can effectively ventilate their gills simply by opening their mouth and opercula while swimming. This **ram ventilation** is observed in many fast-swimming fishes, such as sharks and tuna. In fact, tuna are unable to pump ventilate their gills and as a result must swim continuously. Ventilation in ram ventilators is powered solely by the powerful locomotory

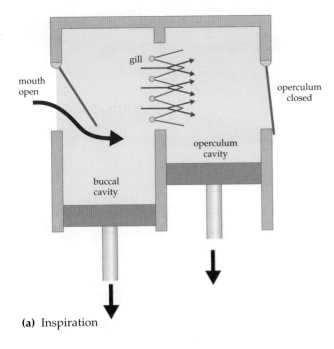

(a) Inspiration

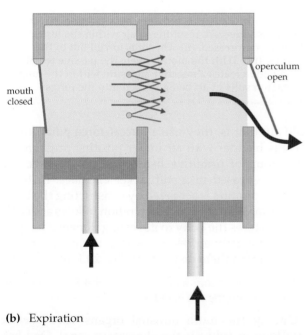

(b) Expiration

Fig. 20.9 A model showing how teleost fishes ventilate their gills showing **(a)** inspiration and **(b)** expiration. The flow of water is indicated by blue arrows and the movement of the floor of the mouth is shown by red arrows

musculature of the body and does not require additional energy expenditure by buccal and opercular muscle systems. Ram ventilation has the advantage in that it removes the braking effects due to drag created by the cyclic opening of the mouth and opercula. Fishes that ram ventilate can accelerate to faster swimming speeds without increasing the energetic cost of swimming.

The transition from water to air breathing

Gills are thin-walled vascularised outgrowths of the body surface and are not particularly well suited to function in air. Their physical weight, together with the powerful effect of surface tension, would cause them to collapse from lack of support and stick together in air, reducing the surface area available for gas exchange. To prevent gill collapse, many intertidal and terrestrial crabs strengthen their gills with a chitinous covering over part of the gill surface and increase the spacing between the gill filaments. These changes reduce the effective surface area available for gas exchange, but this is easily compensated for by the greater availability of O_2 in air.

In terms of gas exchange, terrestrial crabs are adapted to cope with the aerial environment. The advantage of O_2-rich air is offset to some extent by difficulties associated with water loss from the gills. Because of their large surface area and the short diffusion distances, aerial gas-exchange surfaces are particularly vulnerable to water loss. In terrestrial crabs, gills are housed within a chamber to minimise water loss and to provide physical protection for the gills. In general, internally housed gas-exchange surfaces were a prerequisite for terrestrial existence.

Some terrestrial crabs, such as robber crabs, *Birgus* (Fig. 20.10), have very reduced gills that play only a minor role in gas exchange. In *Birgus*, gill reduction is associated with an increase in the size and vascularisation of the gill chamber and some blood vessels in the walls of the gill chamber are extended into tufted outgrowths, increasing the surface available for gas exchange. Even though O_2 and CO_2 diffuse more readily in air, the chamber must be ventilated. This is usually achieved by movement of the scaphognathites.

In air breathers, gas-exchange surfaces are usually housed internally for physical protection and to minimise water loss.

Fig. 20.10 A robber crab, *Birgus latro*

In vertebrates, the evolutionary transition from aquatic to aerial gas exchange probably began in fishes and occurred independently in a number of lineages. Air breathing offered an advantage for fishes in habitats with a shortage of O_2 in water. Survival probably depended upon a fish's ability to gulp a bubble of air and draw upon this as a source of O_2.

In some teleost fishes, **gas bladders**, which evolved originally as outgrowths of the alimentary tract and were generally used for buoyancy, were also used as gas-exchange organs. Most air-breathing fishes, for example the primitive teleost *Amia calva* (Fig. 20.11a),

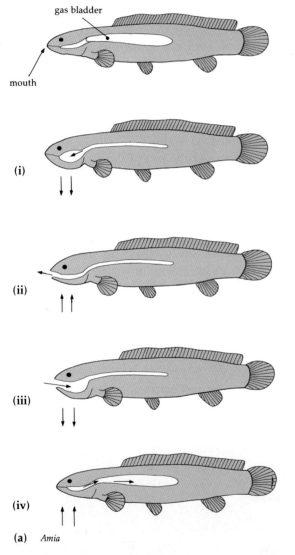

(a)   *Amia*

Fig. 20.11 Pattern of air flow in two types of air-breathing fish. **(a)** *Amia* uses a buccal-force pump. **(i)** With the mouth closed air is drawn from the gas bladder into the buccal cavity by lowering of the mouth floor. **(ii)** With the mouth open, buccal cavity elevation forces air out. **(iii)** The floor of the mouth is again lowered, drawing air into the buccal cavity. **(iv)** The mouth closes and buccal cavity elevation compresses the air in the buccal cavity which is then forced under positive pressure into the gas bladder

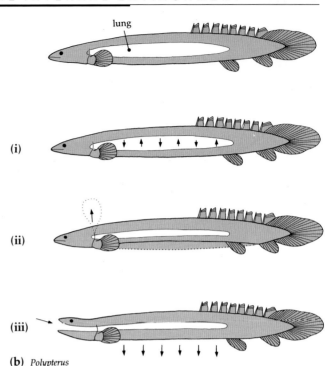

(b)   *Polypterus*

Fig. 20.11 (b) *Polypterus* breathes by aspiration. **(i)** The body is squeezed, resulting in air within the lungs being compressed. **(ii)** The air is forced out of the lungs. **(iii)** With the mouth open, the passive recoil of the body creates a negative pressure within the lung which results in air being sucked into the lungs

gulp air; that is, they use a **buccal-force pump** to fill the gas bladder with air under positive pressure. In one group of primitive fishes, the polypterids, the body is encased in a stiff jacket of scales. In these fishes, expiration is achieved by constricting the body to force air from the lung; inspiration follows as elastic recoil restores the body to its resting dimensions. This creates a negative pressure within the lungs and air is drawn into them (aspirated) (Fig. 20.11b).

> Most air-breathing fishes use a buccal-force pump to force air into the gas bladder.

One of the more unusual organs used for air breathing in fishes is the alimentary canal. One fish that uses intestinal gas exchange is the Japanese weatherloach, *Misgurnus anguillicaudatus*. Although it possesses gills, this small (2–3 g) fish periodically shoves its head through the surface of the water, swallows air and passes it through the alimentary canal before expelling the gas through the vent (Fig. 20.12a). Gas exchange occurs in the posterior 60% of the gut, which has an epithelial layer so thin that the diffusion distance is considerably less than that of gills. Its use of the intestine for gas exchange depends on the partial pressure of O_2 in the surrounding water. As water P_{O_2} decreases, frequency of air gulping increases, increasing gas exchange across the intestine and supplementing gill gas exchange (Fig. 20.12b).

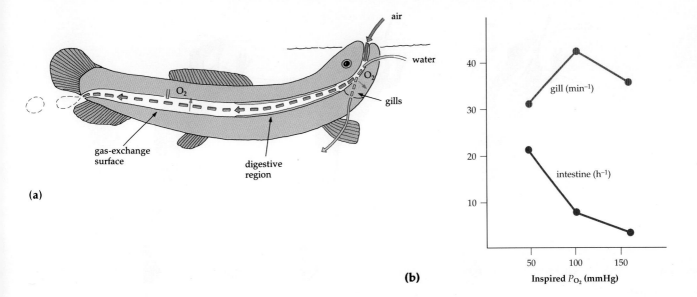

Fig. 20.12 (a) Intestinal air breathing in the Japanese weatherloach supplements gas exchange across the gill.
(b) Frequency of gill or intestinal ventilation as a function of ambient P_{O_2}

Air breathing: lungs

The onset of aerial breathing in fishes paved the way for early tetrapods and the beginning of the vertebrate invasion of the land (Chapter 40). The air-breathing organs of vertebrates, lungs, develop as outgrowths of the gut.

Amphibians

Most amphibians can and do breathe through their skin; characteristically the skin is the major site for CO_2 loss and the lungs are for O_2 absorption. The simple sac-like form of the amphibian lung appears to have departed little from that of its ancestors. The gas-exchange surface is usually well vascularised with a network of capillaries, the degree of vascularisation varying with the overall importance of the lung in gas exchange.

Amphibians, like most air-breathing fishes, ventilate their lungs by a buccal-force pump. Several breathing patterns have been described in frogs and toads (anurans). In general, they depend upon co-ordinated action of the nostrils and glottis in relation to changes in volume of the buccal cavity, which functions like a bellows to inflate the lungs. The totally aquatic South African clawed toad, *Xenopus laevis*, ventilates by first allowing its lungs and buccal cavity to empty passively through open glottis and nostrils. It then closes the glottis, lowers the floor of the mouth and draws fresh air into the buccal cavity. Finally, it closes the nostrils, opens the glottis and forces air under positive pressure into the lungs (Fig. 20.13a).

A somewhat different pattern is observed in the frogs *Rana* and *Bufo*. These frogs ventilate the buccal cavity continually by pumping air in and out through the nostrils with the glottis closed. Intermittently, ventilation of the lung commences with a large buccal inspiration drawing fresh air into the buccal cavity. The glottis then opens and air moves passively from the lungs and mixes with buccal gas, with a little escaping through the nostrils as they close. The gas within the buccal cavity is then forced back into the lungs as the buccal cavity is compressed. The nostrils then open and the glottis closes, and ventilation of the buccal cavity recommences (Fig. 20.13b).

Pumping air into and out of internally housed lungs results in ventilation that is periodic and tidal. With **tidal ventilation**, it is inevitable that some air that has been in contact with the gas-exchange surface will be rebreathed, in contrast to open-flow ventilation where only fresh medium is passed across the gas-exchange surface. In terms of O_2 exchange, rebreathing results in mixing fresh air with air that has already lost some O_2; thus lung P_{O_2} is reduced. However, this presents no problem given the far greater quantity of O_2 in air than in water, and can be compensated for by reducing the diffusion distance and increasing the surface area available for gas exchange. More importantly, as a result of tidal ventilation and rebreathing, CO_2 accumulates in the lungs, which is significant in relation to the control of ventilation (p. 439).

> Amphibians use the buccal cavity to force air into the lungs under positive pressure. Tidal ventilation inevitably involves rebreathing air, which lowers P_{O_2} and raises P_{CO_2} in the lungs.

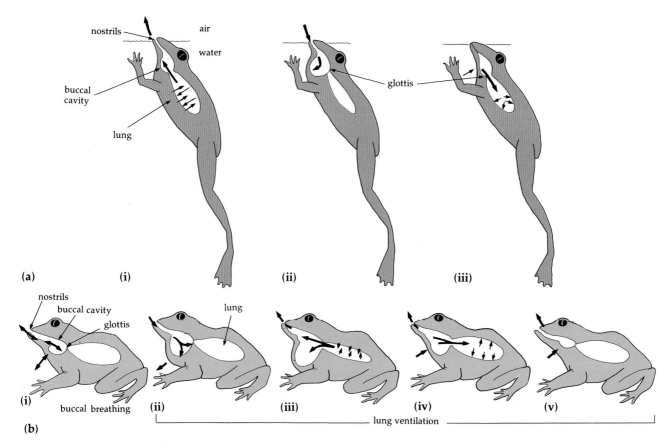

Fig. 20.13 Amphibian breathing patterns. **(a)** *Xenopus laevis*. **(i)** Lungs are emptied out through the open glottis and nostrils. **(ii)** With glottis closed, the floor of the buccal cavity is lowered and air is drawn into the buccal cavity through the nostrils. **(iii)** Nostrils close and the glottis opens; then the floor of the buccal cavity is raised forcing air under positive pressure into lungs. **(b)** *Bufo*. **(i)** Buccal breathing. With the nostrils open, repeated lowering and raising of the floor of the buccal cavity moves air in and out of the buccal cavity continually. **(ii)**–**(v)** Lung ventilation. **(ii)** With glottis closed, a large buccal inspiration draws air into the buccal cavity through open nostrils. **(iii)** The glottis then opens and air moves from the lungs into the mouth with a little escaping as nostrils close. **(iv)** With the nostrils closed, the buccal cavity floor is elevated and air is forced under positive pressure into the lungs. **(v)** Nostrils then open, the glottis closes and buccal ventilation recommences

Reptiles

Unlike amphibians, where ventilation is achieved using a force pump, reptiles use an **aspirating pump**, that is, a pump that draws air into the lungs under negative pressure. Aspiration is achieved in much the same way as in polypterid fishes, by actively moving the body wall out, resulting in a negative pressure that draws air into the lung (inspiration) and passively or actively expelling air from the lung (expiration). Unlike polypterid fishes, where body stiffness required for aspiration is supplied by thick scales, body stiffness in reptiles is supplied by ribs. As in amphibians, ventilation is tidal.

Reptilian lung structure is relatively simple, although it differs between groups in regard to the number of chambers and distribution of the gas-exchange surface. For example, in a simple sac-like lung as found in the Australian central netted dragon, *Ctenopherous*, most of the gas-exchange surface is located in the anterior section of the lung (Fig. 20.14). In some species of reptiles, for example, goannas, the gas-exchange surfaces are raised and further subdivided in order to increase surface area. This results in an increase in the overall thickness of the lung wall.

The physical consequences of such differences in lung structure among reptiles affect the mechanics of lung ventilation. Airflow resistance is inversely related to the radius raised to the fourth power, so small changes in the radius of subdivisions have a large influence on resistance. In addition, an increase in the overall thickness of the lung wall makes the lung less **compliant**, so a greater pressure will need to be developed to achieve a given change in volume. Increased airway resistance and/or decreased compliance contribute to the metabolic cost of ventilation in that more energy is required to ventilate the lungs.

Aspiration involves the expansion of a relatively stiff chamber, creating a negative pressure that draws air into the lungs. The mechanics of lung ventilation are governed by airway resistance and compliance.

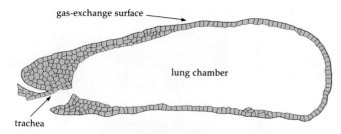

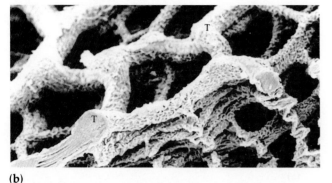

(b)

Fig. 20.14 The lung of the Australian central netted dragon, *Ctenophorous nuchalis*, **(a)** showing the distribution of the gas-exchange surface. **(b)** Scanning electron micrograph showing the gas-exchange surface stretching from the very thin external wall of the lung to an internal network of supporting muscular trabeculae (T)

Mammals

In mammalian lungs, a single tube, the trachea, branches repeatedly until it terminates in an array of cup-shaped chambers, the **alveoli** (Fig. 20.15). Gas exchange occurs in alveoli and also in **alveolar ducts**, which, being thin-walled and better ventilated than the alveoli, are at least as important as alveoli in gas exchange. Alveoli have extremely small diameters, for example 40 µm in mice and 250 µm in humans. The walls of alveoli comprise only capillaries and a small amount of tissue. The mammalian lung is little more than a dense capillary network encased by a thin epithelial layer and exposed on all sides. The diffusion distance between air and blood is very small; in humans it is 0.62 µm but it is as small as 0.25 µm in the smallest mammal known, the Etruscan shrew.

Alveoli are lined with a thin film of liquid and the resultant **surface tension** produces a force that tends to cause some to collapse. Consider two bubbles of different radii that are joined together (Fig. 20.15c). Pressure inside each bubble is directly proportional to surface tension and inversely proportional to radius. Thus, the smaller bubble has a higher internal pressure, which forces its air into the larger bubble, causing the smaller bubble to collapse. The same principle applies to alveoli, but alveolar collapse is

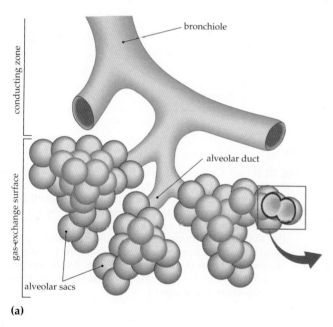

(a)

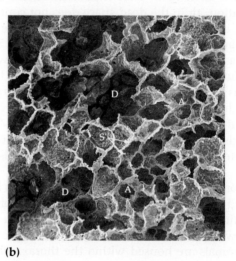

(b)

Fig. 20.15 (a) The terminal portion of the mammalian lung showing the cup-shaped alveoli. **(b)** A scanning electron micrograph of the lung of a red kangaroo, showing individual alveoli (A) opening onto an alveolar duct (D). Alveoli are separated by a thin septa (S). **(c)** If two bubbles of different radii (r) are joined together, surface tension (T) in the smaller bubble will create a higher internal pressure (P), resulting in air moving from the smaller bubble to the larger one and leading to the inevitable collapse of the smaller bubble

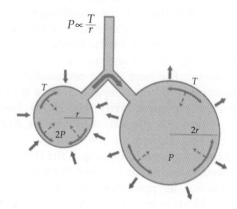

(c)

prevented by the presence in the liquid film of a phospholipid known as **surfactant**, which greatly reduces the surface tension of the lining fluid layer. Surfactant is synthesised and stored in special cells (type-II alveolar cells; Fig. 20.16a) and is secreted into the fluid layer that covers the alveolar epithelium. Of more importance is the fact that surfactant reduces surface tension to a greater degree as the area of the surface layer (the radius of the alveolus) decreases (Fig. 20.16b). Not only does this prevent alveoli from collapsing but permits continuous differences in their dimensions, a requirement for ventilation. Premature infants whose lungs fail to secrete surfactant are unable to inflate their lungs due to the high surface tension: the resultant 'respiratory distress syndrome' can be treated by application of synthetic surfactant.

> Surfactant lowers the surface tension at the air–water interface, preventing the collapse of the alveoli during expiration.

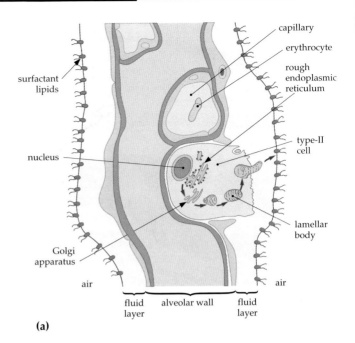

(a)

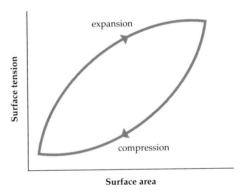

(b)

Fig. 20.16 **(a)** Cross-section of the alveolar wall showing capillaries and a type-II cell. The synthesis, storage and secretion pathway for surfactant in a type-II cell is shown. On secretion, the surfactant lipids form a layer on the surface of the fluid lining the alveolar wall.
(b) Relationship between surface tension and surface area. Surfactant prevents alveolar collapse by altering the surface tension of the surface film. As the surface area of the alveolus decreases, the surface tension in the film layer decreases

In humans, 23 successive branchings occur between trachea and alveolar ducts. The volume of these conducting airways represents a **dead space**, in which air remains after expiration. At the end of inspiration, air in the dead space does not contribute to gas exchange. This volume of dead space is by no means insignificant. In a dog, it is about 150 mL, whereas in a whale it may be up to 300 L. On average it accounts for about one-third of each breath. If dead space is artificially increased, as when a diver breathes through a snorkel, the build-up of stale air can be so great that a diver may lose consciousness. Modern snorkels have been designed to have the smallest possible internal volume in order to minimise this risk.

In compensation for the low compliance of a highly divided lung, mammals have a more efficient breathing mechanism than simple movement of the body wall, as is found in many reptiles. The lungs of mammals are housed within the thorax, which is separated from the abdomen by the **diaphragm** (a muscular and tendonous sheet). Expansion of the rib cage and contraction of the diaphragm create a negative pressure within the thorax, which draws air in through the airways to fill the lungs.

Except in large mammals, such as elephants and possibly horses, the lung is not physically attached to the thoracic wall but is separated by a thin fluid-filled space, the **pleural cavity** (Fig. 20.17). Negative pressure within the pleural cavity keeps the lung surface closely applied to the thoracic wall and prevents the lung from collapsing. If the thoracic wall is punctured, allowing air to enter the pleural space, a state known as pneumothorax, the lungs will collapse. Inspiration is an active muscular process in mammals; expiration generally results from passive elastic recoil of the thoracic wall. Some mammals synchronise ventilation with locomotion (see Box 20.2).

> Mammals have stiff lungs with low compliance that are ventilated with the aid of a diaphragm. Inspiration is achieved with a negative pressure.

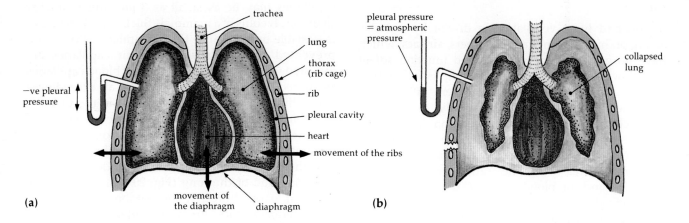

Fig. 20.17 The thoracic cavity of a mammal. Ventilation is achieved by movement of the ribs and diaphragm. **(a)** The lungs are held in close contact with the chest wall and prevented from collapse by the negative pressure within the pleural cavity. This negative pressure is the result of the ribs, tendency to recoil out and the lungs, tendency to recoil in (arrows). **(b)** Puncture of the thoracic wall results in this negative pressure being lost and the inevitable collapse of the lungs

BOX 20.2   BREATHING AND RUNNING

In some animals, the mechanical basis of lung ventilation during exercise differs from that found during resting. Studies of frequency of limb movement in running mammals in relation to frequency of breathing have shown that, at least in some gaits, six types of mammals (e.g. dogs and horses) take one breath per stride. Trained athletes were found to take one breath every two strides (but they could abruptly change breathing frequency). This phenomenon is known as 'locomotory-respiratory coupling'. Such coupling has been also described in fishes, and flying and running birds.

Locomotory-respiratory coupling is an example of *entrainment*, the phase-locking of two oscillating systems. Little is known of the mechanisms controlling entrainment in animals. There is some evidence for the role of information from mechanoreceptors in joints, but in cases where limb frequencies are synchronised with breathing in a 1:1 ratio, mechanical explanations have been proposed. The first, seen in kangaroos, is a *visceral piston mechanism*, in which bouncing of the abdominal viscera with each hop displaces the diaphragm, pushing air into and out of the lungs. Another mechanism that has been proposed is flexion and extension of the back

in a galloping animal. Flexion of the lumbar region tends to push the abdominal viscera and diaphragm forwards, compressing the lungs and causing exhalation; extending the back achieves the opposite.

One of the clearest examples of locomotory-respiratory coupling occurs in the hopping wallabies (see figure). This was first observed when researchers from Flinders Medical Centre and the University of British Columbia observed that, as Tammar wallabies, *Macropus eugenii*, hopped on a chilly morning, a jet of vapour would leave the nostrils with each hop. Their proposition was confirmed when, on a motor-driven treadmill, hopping wallabies showed a phase-locked 1:1 ratio between respiratory and locomotory frequencies at all hopping speeds. The data were explained by a piston mechanism, with inspiration beginning just as the animal leaves the ground. The animal's lungs are fullest just before it reaches the highest point in the hop. Kangaroos and wallabies have a central tendon in the diaphragm that may aid the piston displacement. In addition the viscera are rather loosely slung, which would help them move within the abdominal cavity.

Locomotory-respiratory coupling. **(a)** At the beginning of a hop, lungs are compressed by the weight of the viscera. **(b)** Leaping forward, viscera attached to the diaphragm are thrust backward by inertia, helping to draw air into the lungs. **(c)** At landing, viscera move forward, pushing air out of the lungs

Birds

In the bird lung, ventilation is driven by muscles acting on highly compliant chambers, **air sacs**, which have no gas-exchange surfaces and function solely as bellows (Fig. 20.18a). The high compliance enables rib movements alone to produce very efficient air movement in bird lungs. Additionally, because the gas-exchange regions are separated from the air sacs, the gas-exchange region is immobile. Because the volume of the gas-exchange region does not vary, its surface area-to-volume ratio can be greatly increased and air–blood diffusion distances can be extremely short.

The gas-exchange region in birds is composed of a parallel series of tubes, the **parabronchi**. The lumen of each parabronchus gives rise to narrow conducting tubes, which in turn give rise to numerous **air capillaries** (Fig. 20.18b). The diameter of these air capillaries may be as small as 3 μm (much smaller than the alveolus of a mammal). Such a small diameter is possible only because the air capillary is rigid and prevented from collapsing. The air capillaries are invested with an intricate network of blood capillaries separated from the air capillaries by an exceedingly small diffusion distance (e.g. 0.12 μm in the pigeon).

In birds, airflow through the parabronchi of the gas-exchange region is unidirectional. During inspiration, most air passes along the bronchus and into the posterior air sacs; the remainder enters the posterior side of the lung (Fig. 20.19a). Once in the lung, air passes through the parabronchi and into the anterior air sacs. Air is prevented from entering the junction between the bronchus and the anterior side of the lung during inspiration by pressure differences that create zones of no flow. At the onset of expiration, the pressure differences change (Fig. 20.19b). Air can now pass between the anterior side of the lung and the bronchus, but is prevented from passing between the junction of the bronchus and the posterior side of the lung. As a result, air from the posterior air sacs is directed into the posterior side of the lung through the parabronchi where it joins air being squeezed from the anterior air sacs.

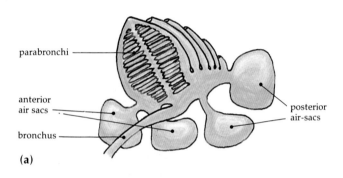

parabronchi

anterior air sacs

bronchus

posterior air-sacs

(a)

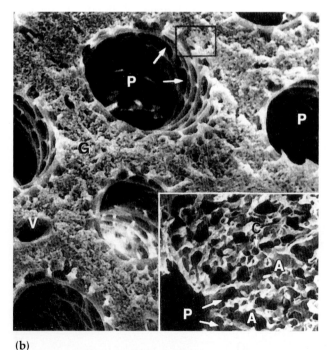

(b)

Fig. 20.18 The bird lung and its fine structure. (a) The gas-exchange region (parabronchi) and the air sacs. (b) The lumen of a parabronchus showing the air capillaries. The space between the air capillaries is filled with a network of blood capillaries

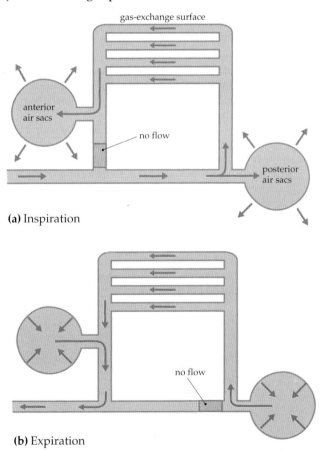

gas-exchange surface

anterior air sacs

no flow

posterior air sacs

(a) Inspiration

no flow

(b) Expiration

Fig. 20.19 The flow of air through the bird lung. Zones where airflow is prevented because of pressure differences are shown as darker regions. The patterns of **(a)** inspiration and **(b)** expiration are explained in the text

This air, which has passed through the gas-exchange region of the lung, is directed through the anterior junction of the bronchus and on out of the system. Thus, airflow through the gas-exchange region is unidirectional and occurs during both inspiration and expiration.

The unidirectional airflow through parabronchi and the fact that blood flow in the blood capillaries is towards the lumen of the parabronchus suggest the possibility of countercurrent flow between air and blood. However, closer examination shows that this is not the case (Fig. 20.20). Movement of air in air capillaries only occurs by diffusion. Starting at the entrance to a parabronchus, the partial pressure of O_2 is essentially the same as outside air. Due to uptake of O_2 as air progresses through the parabronchus, P_{O_2} falls in subsequent air capillaries and the amount of O_2 extracted by the blood progressively declines. Blood leaving the parabronchus is a mixture drawn from all capillaries and usually has a P_{O_2} greater than that of the air leaving the parabronchus. This pattern of blood and airflow has been termed **cross-current exchange**.

> Birds have highly compliant chambers (air sacs) used to ventilate stiff gas-exchange surfaces. Airflow through the bird lung is unidirectional and occurs during both inspiration and expiration. Cross-current exchange is more efficient than cocurrent exchange but less efficient than countercurrent exchange.

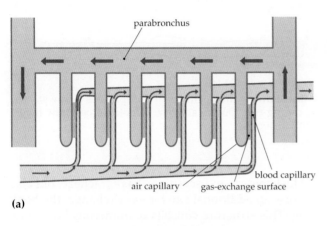

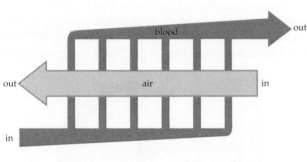

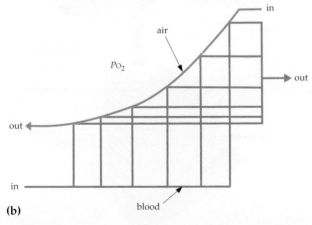

(a)

(b)

Fig. 20.20 **(a)** The arrangement of the air capillaries and the blood capillaries in the bird lung. **(b)** A functional model for cross-current flow. In cross-current flow, capillary blood flow is perpendicular to the main flow of air. In most cases, this permits the blood leaving the gas exchanger to have a higher partial pressure of O_2 (P_{O_2}) than the air leaving the organ

Air breathing: tracheae

In insects, which are successful terrestrial arthropods, air passes internally to cells through tubular ingrowths of the body surface, **tracheae**. Tracheae usually arise at openings, **spiracles**, of which there are normally 10 pairs on the thorax and abdomen (Fig. 20.21a). Tracheae branch repeatedly and terminate in blind-ending tubules, **tracheoles**, which contain liquid (Fig. 20.21b). In some insects with exceptionally long legs, possible diffusion problems over such distances are circumvented by having additional spiracles on the legs. Spiracles can be opened or closed, which assists in minimising water loss. Tracheae taper to a diameter of 0.5–0.1 μm and deliver O_2 to within a few micrometres of the cell, often indenting the cellular membrane and, in some cases, passing near to mitochondria. Tracheal volume varies greatly between species. In locusts, the tracheal system may occupy almost half of the insect's body volume. Tracheal systems of varying complexity are also found in most spiders and scorpions (arachnids) and in terrestrial isopods (crustaceans).

Diffusion is usually adequate for the renewal of gas along tracheae, especially for small insects or the inactive stages of larger species. However, the blind tips of tracheoles are filled with liquid through which exchange occurs between tracheoles and cells. With such small diameters, one might expect a tendency for the tracheoles to collapse due to surface tension, as occurs with alveoli. This does not occur because tracheole walls are strengthened with chitin.

Osmotic forces (Chapter 4) are used to reduce diffusion distance within tracheole liquid during exercise. During increased muscle activity, there is a release of metabolites into interstitial fluid, raising tissue osmotic pressure. Water is drawn osmotically

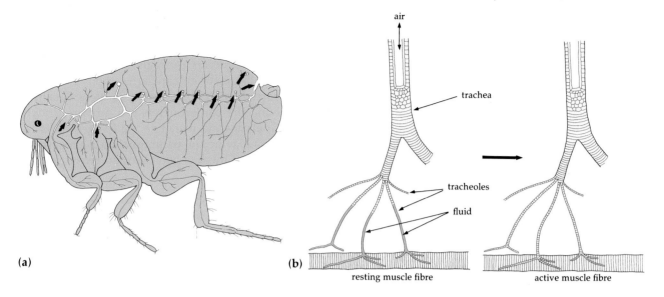

Fig. 20.21 Insect tracheal system. (a) The tracheal system of the flea showing the openings of the spiracles (arrows). (b) The terminal portions of the tracheoles are filled with fluid during rest. During activity, osmotic forces draw fluid from the tracheoles and permit air to advance further down the tracheoles

from tracheoles to the tissue. Given the greater diffusion of gases in air, such a mechanism has obvious advantages. The removal of fluid from the tracheole during increased activity permits gaseous air to travel further down the tracheole, reducing the diffusion barrier (Fig. 20.21). At the end of activity, the metabolites are dispersed and water moves back into the tracheoles.

> In insects, air passes directly to tissues through a network of tracheae and is generally renewed by simple diffusion.

Gases may also move through the tracheal system by bulk flow. Numerous large insects actively ventilate by movements of the abdomen. Expiration is brought about by muscular activity, whereas in many species inspiration relies upon elastic recoil of the cuticular exoskeleton. Many flying insects make use of flight movements to ventilate tracheae supplying the flight muscles through open spiracles.

Ventilation in insects is usually tidal, although some beetles (order Coleoptera) are equipped with two pairs of large-diameter tracheae that run the length of the thorax. From these, numerous secondary tracheae branch and invade the flight muscles. During flight, air is ducted through the large tracheae as a result of ram ventilation and the flow is therefore unidirectional.

Not all insects have spiracles. In some aquatic insects, the tracheal system has no external connection. Whether their tracheal systems are sealed or open, aquatic insects show a number of modifications to the tracheal system that permit gas exchange in water (Box 20.3).

Most spiders and scorpions have tracheae, but also possess an additional site for gas exchange, the **book lung**. This structure consists of numerous horizontal air spaces, connected to the outside air through a spiracle, which alternate with haemolymph spaces, giving the appearance of pages in a book and hence the term book lung (Fig. 20.22). Such an arrangement

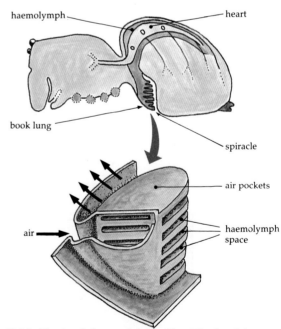

Fig. 20.22 The book lung of the spider. The book lung consists of numerous air spaces, which alternate with haemolymph spaces (see inset). The air spaces are connected to the outside by a spiracle. Oxygen diffuses from the air spaces to the haemolymph, returning to the heart. Once in the heart this oxygenated haemolymph is distributed to all parts of the body

BOX 20.3   GAS EXCHANGE IN AQUATIC INSECTS

Aquatic insects have, in terms of evolution, returned to water from the land and their tracheal systems are modified to permit air-filled tracheae to operate in water. There are essentially two patterns of gas exchange: the air-filled tracheal system may be in open contact with some form of gas source or it may be completely internal.

Modification of a spiracle into a tube that pierces the surface of the water permits some aquatic insects, such as mosquito larvae (Fig. a), to form a direct link with air above the water surface. When the larva is suspended at the surface, air moves by diffusion through the tube and into the tracheal system. Interestingly, the larvae of one genus of mosquito, *Mansonia*, replenish their air supply by piercing aquatic plants with the tip of their spiracle tube and drawing air from aerenchyma tissue (Fig. b).

When they visit the surface, diving beetles and bugs trap bubbles of air and their tracheae open into these bubbles. As O_2 is absorbed by the insect from the bubble, an O_2 diffusion gradient is established and more O_2 diffuses in from the surrounding water. The high solubility of CO_2 in water means that CO_2 is easily lost to the surrounding water and the P_{CO_2} inside the bubble is negligible. Therefore we might expect that an insect with a bubble would never need to surface. However, as a result of a decrease in the partial pressure of O_2 within the bubble, the partial pressure of nitrogen (N_2) increases, favouring loss of N_2 from the bubble to the water. Also, hydrostatic pressure increases with depth. The deeper a beetle dives, the higher the gas pressure inside the gas bubble. This increases the loss of N_2 and O_2 from the bubble. As a result, the bubble inevitably collapses and the insect must surface to replenish its air supply.

If the bubble were prevented from collapsing, the animal would not need to surface as O_2 would continue to diffuse from the water into the bubble (N_2 being maintained in equilibrium with the water). Some diving bugs and beetles prevent bubble collapse by covering part of their surface with very fine submicroscopic hairs that trap air between and beneath them. Physical factors,

(i)

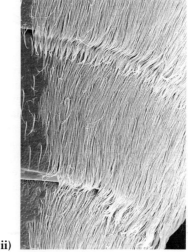

(ii)

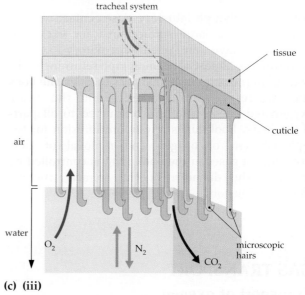

(c) (iii)

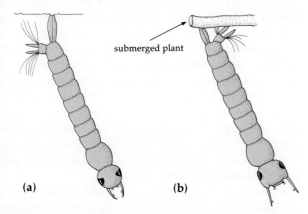

(a)   (b)

(a) A mosquito larva obtaining air at the water surface.
(b) Another mosquito, *Mansonia*, obtains O_2 by piercing air-filled plant tissue

(c) (i) General view and (ii) fine view of the ventral surface of an aquatic beetle, *Simsonia tasmanica*, showing the fine hairs of the plastron. (iii) Diagram illustrating function of plastron

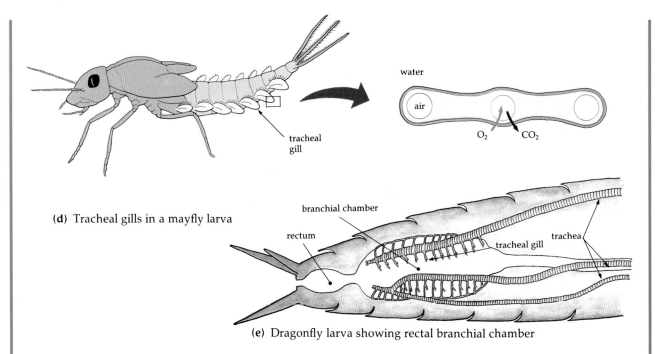

(d) Tracheal gills in a mayfly larva

(e) Dragonfly larva showing rectal branchial chamber

such as surface tension, prevent hydrostatic compression of the bubble. The air is in contact with a spiracular system that has fine hairs to prevent water entering. Such a structure is known as a **plastron** and is well-developed in the beetle *Simsonia tasmanica* (Fig. c). In these bugs, the majority of the ventral and dorsal surface is covered in many submicroscopic hairs, up to 2.5 million/mm². Because the bubble is prevented from decreasing in size, bugs with a plastron can descend to depths of several metres.

In some aquatic insect larvae, the gas-filled tracheal system is sealed, having no openings to external air supplies. Abdominal appendages, tracheal gills, form the external gas-exchange site, which relies on diffusion between gas contained in the tracheal system and the surrounding water (Fig. d). Ventilation over the exposed gills is created by water passing over the body or by undulation of the body. In dragonflies, the gills are housed in an enlarged region of the rectum, a 'branchial chamber' (Fig. e), and are ventilated by muscular pumping actions.

brings haemolymph into very close contact with air and provides a large surface area for exchange. Arachnids have an open circulatory system. Nevertheless, almost all haemolymph returning to the heart from the abdomen must pass through the book lung where it is oxygenated. From the heart, oxygenated haemolymph is conducted to all parts of the body. Book lungs rely upon diffusion for the replenishment of oxygen and removal of CO_2. Exchange of gases in the book lung is controlled by changing the diameter of the spiracle, thereby affecting diffusion.

GAS TRANSPORT
Transport of oxygen

Because of its low solubility, the content of O_2 in water and physiological fluids is low. Most animals therefore improve the transport of O_2 within the body by the addition of specialised O_2-carrying molecules,

respiratory pigments, such as haemoglobin and haemocyanin. These pigments (coloured molecules) substantially increase the maximum amount of O_2 that can be carried by a fluid—O_2-**carrying capacity**—and hence increase the transport of O_2.

Respiratory pigments are found in the blood of animals with closed circulatory systems (e.g. nematodes, annelids, cephalopods and vertebrates), the haemolymph of animals with open circulatory systems (e.g. arthropods and molluscs) and, sometimes, in coelomic fluid (e.g. some annelids and echinoderms). The pigment may be either directly dissolved in the fluid or contained within specialised cells (corpuscles, erythrocytes or coelomocytes) suspended in the fluid. Pigment-containing cells are found in some annelids, molluscs, echinoderms and all vertebrates except the ice fish.

In mammals, haemoglobin is located entirely within specialised cells, **erythrocytes**, which do not contain a nucleus or mitochondria. Packaging haemoglobin into specialised cells avoids problems that would be caused by the presence of this osmotically active material in plasma, and provides the pigment with a stable and regulated environment with the appropriate ions and enzymes.

▎ Respiratory pigments increase the O_2-carrying capacity of a fluid.

Oxygen-carrying capacity

When a pigment is present, the total O_2 content of blood is the sum of that transported in solution and that bound to the pigment (Fig. 20.23). Besides having the ability to bind with O_2, a respiratory pigment must be able to release O_2 under certain conditions. The ability of a pigment to load and unload O_2 (reversible binding) is partly dependent on local P_{O_2}.

The relationship between P_{O_2} and total O_2 content is known as the **oxygen equilibrium curve**, sometimes called the oxygen dissociation curve. Examination of such a curve for haemoglobin reveals that the amount of O_2 bound to a pigment is not a simple relationship (Fig. 20.23). The S-shaped (sigmoid) curve demonstrates that, in a region of low P_{O_2}, haemoglobin contains very little O_2 and that this rapidly increases with increasing P_{O_2} until at high P_{O_2} the pigment is fully saturated and the curve reaches a plateau.

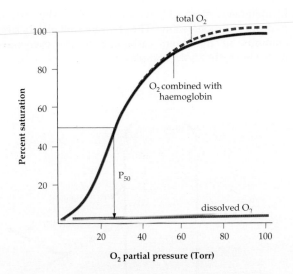

Fig. 20.23 The amount of O_2 in the blood is dependent upon the partial pressure of oxygen. The amount of O_2 binding to a respiratory pigment, in this case haemoglobin, usually exceeds that in solution to an extent where it is the major source of O_2

▎ A respiratory pigment must be able to combine reversibly with O_2. The amount of O_2 bound to a respiratory pigment increases with increases in the partial pressure of O_2.

Not all O_2 equilibrium curves are sigmoid because the shape for a given pigment depends on the number of sites to which O_2 can bind (Fig. 20.24). Where there

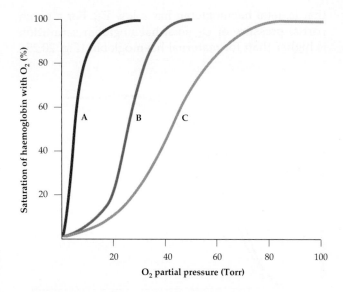

Fig. 20.24 The shape of the O_2 equilibrium curve for respiratory pigments reflects the number of binding sites for oxygen. A single binding site results in a hyperbolic curve (curve A), while more than one binding site results in a sigmoid curve (curves B and C); the more binding sites, the steeper the curve. The steeper the curve the lower the P_{O_2} for saturation

is a single binding site, as in lamprey haemoglobin or in myoglobin (a respiratory pigment found in muscle), the curve is hyperbolic (curve A). A sigmoid curve (curves B and C) results from the fact that binding of O_2 at one site increases the affinity for O_2 at the remaining sites (co-operative binding). The advantage in a pigment having a steep dissociation curve is that such a pigment will change from loading to unloading with minimal change in P_{O_2}. This tends to maximise diffusion gradients for O_2 exchange at gas-exchange organs and tissues.

▎ Co-operative binding, which results in a sigmoidal oxygen equilibrium curve, permits the loading and unloading of O_2 over relatively small changes in P_{O_2}.

Oxygen affinity

Different pigments have different affinities for O_2, expressed in terms of the P_{O_2} at which the pigment is half saturated, P_{50}. A pigment has low affinity for O_2 if it requires a high P_{O_2} to become saturated and vice versa. The functional significance of pigments with different O_2 affinities can be illustrated by considering mammals during development. In eutherian mammals, the developing fetus must take up O_2 from the blood of its mother. Since O_2 is also used by maternal tissues, the partial pressure of O_2 in maternal blood falls and therefore fetal O_2 uptake must occur at lower partial pressures of O_2 than those for the mother taking up O_2 from air. This is possible because the O_2 equilibrium curve for fetal haemoglobin lies to the left of the maternal curve, that is, fetal haemoglobin has a low P_{50}. For a given partial pressure of O_2, fetal haemoglobin saturation is higher than for maternal haemoglobin (Fig. 20.25).

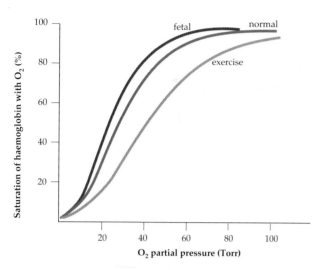

Fig. 20.25 The oxygen equilibrium curve for haemoglobin. The curve for fetal haemoglobin is shifted to the left of the maternal (normal) curve showing that fetal haemoglobin has a greater affinity for oxygen than does maternal haemoglobin. The rightward shift of the normal curve caused by exercise results in a lowering of the oxygen affinity of haemoglobin, favouring unloading of oxygen

The affinity of haemoglobin for O_2 is partly dependent upon the prevailing P_{O_2}. Oxygen affinity can also be altered by the presence of various molecules such as organic phosphates, by pH and by temperature, which effectively shift the O_2 P_{50} horizontally. In most cases, O_2 affinity decreases as pH decreases. This is the **Bohr effect** and it assists in both loading O_2 in lungs or gills (where pH is slightly higher) and unloading it in the tissues (where accumulated CO_2 tends to make the region slightly more acidic). Oxygen affinity of mammalian haemoglobin is reduced by CO_2 binding reversibly with the pigment (at a different site on the molecule to the O_2 binding site): the **Root effect**. An increase in temperature also reduces O_2 affinity by directly affecting the reaction between haemoglobin and O_2, and to a small extent by affecting pH.

The functional significance of the effects of pH, CO_2 and temperature on O_2 affinity becomes clear if we consider an actively metabolising tissue. Such a tissue produces heat and CO_2, and may also affect pH by the production of lactic acid (Chapter 5). As blood passes through an active tissue, these products enter the blood and lower the O_2 affinity of haemoglobin, which results in more O_2 being unloaded.

The oxygen affinity of a respiratory pigment decrea. with a decrease in pH and with an increase in CO_2 temperature.

Transport of carbon dioxide

Carbon dioxide, a by-product of cellular respiration, must be removed. The main pathways for CO_2 transport in mammals are given in Figure 20.26. After its formation in the cell, CO_2 diffuses rapidly into blood. In air-breathing vertebrates, some CO_2 dissolves in plasma but the majority diffuses from plasma into erythrocytes. Regardless of whether CO_2 is in plasma or erythrocytes, most of it is hydrated to form carbonic acid (H_2CO_3), which in turn is dissociated into hydrogen carbonate (HCO_3^-) and hydrogen ions (H^+). The hydration step proceeds very slowly and, despite the fact that dissociation of H_2CO_3 is a rapid reaction, few HCO_3^- are formed in plasma. In erythrocytes, the enzyme **carbonic anhydrase** greatly speeds up the hydration reaction. As a result, a large HCO_3^- concentration gradient exists between an erythrocyte and plasma, and HCO_3^- diffuses from erythrocytes to plasma. To maintain electrical balance within an erythrocyte, Cl^- moves in the opposite direction. At the lungs, opposite gradients exist and HCO_3^- diffuses out of erythrocytes and is exchanged for Cl^-, which moves in. The exchange of Cl^- between plasma and erythrocyte, known as the **chloride shift**, permits substantial exchange of HCO_3^- between the two.

In mammals, most CO_2 is transported in the form of hydrogen carbonate ions. The chloride shift allows large amounts of hydrogen carbonate to be exchanged between erythrocytes and plasma.

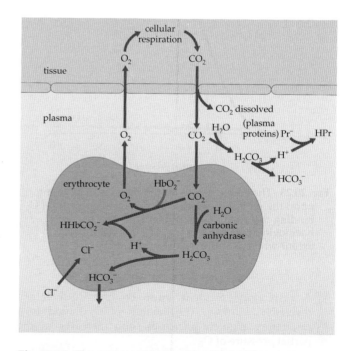

Fig. 20.26 The main pathways describing CO_2 transport in mammals. The diagram shows movement of CO_2 into blood from tissues. The reverse occurs as CO_2-laden blood passes through the lungs

In mammals, some of the CO_2 entering an erythrocyte binds directly with haemoglobin forming carbaminohaemoglobin. The more deoxygenated the haemoglobin, the better is its ability to take up CO_2, a shift known as the **Haldane effect**. Up to 10% of CO_2 can be transported this way. A side product of this reaction is the production of H^+. To prevent changes in blood pH, these H^+, together with those produced with HCO_3^-, must be removed. Haemoglobin can take up or give off H^+ at particular sites; that is, haemoglobin can act as a **buffer**. In addition, the more deoxygenated the haemoglobin the better it binds with H^+. Buffering of H^+ in plasma is achieved by plasma proteins.

Thus, CO_2 diffusing into capillaries of tissues is transported either as HCO_3^-, carbaminohaemoglobin or dissolved in solution. Furthermore, almost all the H^+ produced from the HCO_3^- and carbamino-haemoglobin reactions is buffered by haemoglobin. In the lungs, the above reactions are reversed as CO_2 moves from the blood to the alveoli (Fig. 20.26). As a result of the buffering capacity of haemoglobin, the pH difference between arterial and venous blood is generally very small.

In invertebrates, the mechanisms involved in CO_2 transport appear similar to those in vertebrates. The enzyme carbonic anhydrase is found in the gills of polychaetes, gastropods, cephalopods and some crustaceans where it assists in CO_2 elimination. However, in aquatic organisms, CO_2 transport is more complex because the ions involved, HCO_3^- and H^+, can be exchanged directly across gills using active membrane transport.

Temperature also affects CO_2 transport. An increase in temperature reduces the amount of H^+ that is buffered by haemoglobin. As a result there is an accumulation of H^+, which slows down the CO_2 hydration process, producing less HCO_3^-. The effects of temperature on H^+ buffering are particularly important in poikilothermic animals whose body temperatures vary with environmental fluctuations.

CONTROL OF VENTILATION

Ventilation with respect to oxygen concentration is shown in Figure 20.27 for a number of water and air breathers. It can be seen that ventilation in water breathers is some 10–30 times higher than that of air breathers. Also shown is the effect on ventilation for a mountaineer at sea level and at two elevated altitudes.

Given the high ventilation rates of water breathers and the high solubility of CO_2 in water, CO_2 is easily removed as a waste product from water breathers. As a consequence, the difference in partial pressure

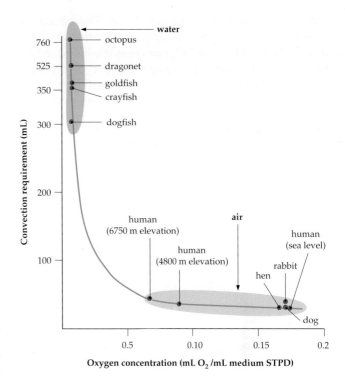

Fig. 20.27 Convection requirements against concentration of oxygen for a number of air and water breathers

of CO_2 between inspired and expired water is very small and, because CO_2 diffuses easily, internal partial pressures for CO_2 are not very different from external partial pressures. The main problem for water breathers is to obtain sufficient oxygen and it is not surprising that they regulate ventilation in response to changes in O_2.

Air breathers face a different problem. Tidal volume can be reduced because of the higher availability of O_2 in air compared with water. As a consequence, such animals tend to have higher internal levels of CO_2. The main difficulty for air breathers is to control the level of CO_2 to protect against pH imbalances. Ventilation in air breathers, although sensitive to changes in O_2, is found to be particularly sensitive to CO_2 or pH (which is directly affected by CO_2 levels).

Water breathers need to ventilate around 10–30 times more than do air breathers because of the low concentration of oxygen in water. Ventilation in water breathers is sensitive to changes in oxygen. Ventilation in air breathers is particularly sensitive to changes in carbon dioxide or pH.

In order to use O_2 or CO_2/pH levels to regulate ventilation, an animal must have some means of detecting these. Chemoreceptive tissue responsible for the predominant effects of CO_2 and pH on ventilation in mammals is located in the medulla region of the brain. These central receptors respond exclusively to changes in the pH of the cerebral interstitial fluid that surrounds them. The effects of CO_2 on these receptors is apparently mediated by its effects on pH. Since CO_2 can cross the blood–brain barrier but H^+ cannot, changes in pH of cerebral interstitial fluid reflect changes in the partial pressure of CO_2. The response of central receptors overrides input from other chemoreceptors, such as those in the carotid bodies.

In mammals, peripheral chemoreceptors sensitive to O_2 and CO_2/pH levels are found in the **carotid bodies**, tiny organs located at the bifurcation of the common carotid arteries into internal and external branches. The carotid body has an extensive capillary network and contains a specialised tissue composed

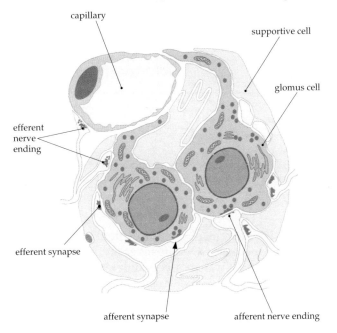

capillary

supportive cell

glomus cell

efferent nerve ending

efferent synapse

afferent synapse   afferent nerve ending

of glomus cells, and efferent and afferent nerve endings (Fig. 20.28). Glomus cells contain a variety of neurotransmitter substances, which, when released in response to changes in partial pressures of O_2 or CO_2, or change in pH, regulate the rate of firing of the chemoreceptor nerve endings. An increase in P_{CO_2} or a decrease in pH or P_{O_2} result in an increased rate of firing of the chemoreceptors. This is transmitted to the central nervous system and brings about an increase in ventilation. Carotid bodies are primarily concerned with transmitting information about O_2 levels, but also respond to changes in CO_2 and pH.

> Central chemoreceptors respond to changes in pH (and thus P_{CO_2}). Peripheral chemoreceptors respond primarily to change in P_{O_2}.

In fishes, oxygen-sensitive chemoreceptors are located in the gills. An external receptor responds to changes in ambient oxygen levels whereas the internal receptors monitor blood oxygen content and flow. Both receptors can initiate compensatory changes in breathing and/or heart rate.

There are other receptors important in the control of ventilation, such as mechanoreceptors in fish gills that monitor the water flow and are responsible for altering gill geometry and adjusting breathing pattern; lung stretch receptors in air-breathing vertebrates, which provide feedback about changes in lung volume, pressure or wall tension; and mechanical receptors that are involved in defence reflexes that protect the respiratory tract.

Output from all receptors is fed via neural pathways to a central pattern generator that adjusts the breathing pattern to maintain homeostasis or to integrate breathing movements with other activities, such as feeding, talking or locomotion.

SUMMARY

- Gas exchange in animals involves the supply of O_2 for cellular respiration and the removal of CO_2, a product of cellular respiration.

- Water contains 20–40 times less O_2 than air, which has important consequences for gas exchange.

- Gas exchange involves gaseous diffusion across membranes. It is dependent upon the concentration gradient, and the permeability, thickness and surface area of the membrane. Convection of external medium (ventilation) and internal medium (perfusion) maintains a favourable concentration gradient.

- Gills are outgrowths of the body surface that, when housed in gill chambers for protection, require ventilation.

- Air breathers have gas-exchange surfaces housed internally to minimise water loss. In most cases, this results in tidal breathing. Amphibians ventilate lungs using a positive pressure buccal pump. Reptiles, birds and mammals generate a negative pressure within the lungs, by expanding the thorax, and in mammals lowering the diaphragm, to draw air into the lungs.

- The alveoli of mammalian lungs have a tendency to collapse because of the surface tension at the air–liquid interface. This tendency can be overcome by surfactant.

- Airflow through the gas-exchange region of a bird lung is unidirectional and is cross-current to flow in the blood capillaries.

- Insects conduct air directly to the tissues by means of tubular tracheae, the tips of which are fluid-filled. Air in tracheae is generally replaced by diffusion. During exercise, fluid is osmotically withdrawn from tracheole tips, decreasing the diffusion barrier.

- Blood transport of O_2 generally involves a respiratory pigment, which reversibly binds with O_2 according to the partial pressure of O_2. The affinity of haemoglobin for O_2 decreases with a decrease in pH and an increase in CO_2, temperature or organic phosphates.

- In air-breathing vertebrates, most CO_2 is hydrated to form carbonic acid, which dissociates and is transported as hydrogen carbonate ions. The enzyme carbonic anhydrase, located in erythrocytes, speeds up the hydration step. Haemoglobin acts as a buffer to counteract the increase in H^+ that occurs with the dissociation of carbonic acid into bicarbonate ions.

- In general, ventilation in water breathers is more sensitive to P_{O_2} whereas ventilation in air breathers is particularly sensitive to P_{CO_2}. On the whole, changes in P_{O_2} are detected by peripheral chemoreceptors, whereas central chemoreceptors respond to changes in pH (as a consequence of changes in P_{CO_2}).

QUESTIONS

1. What are the two systems that are responsible for the supply and removal of O_2 and CO_2 in most animals? What two processes are fundamental to the operation of these two systems?

2. When compared with air, water is more viscous and dense, and is lower in O_2 content. As a result, what design constraints have been imposed on gas exchange in water breathers?

3. Describe and explain with examples the difference between aspiration and buccal-force pump ventilation.

4. Contrast the mechanisms and efficiency of ventilation in a bird and a mammal.

5. Why is surfactant important in lungs?

6. Surface tension forces at an air–water interface tend to collapse structures with a small radius. Discuss ways in which different groups of animals have overcome this problem while at the same time maintaining viable gas-exchange surfaces.

7. In air breathers, gas-exchange surfaces provide an avenue for water loss. Reduction of water loss was an important factor in the evolution of animals able to inhabit terrestrial environments. How is this reflected in the gas-exchange organs of terrestrial animals?

8. Explain the difference between cocurrent, countercurrent and cross-current exchange.

9. Why are respiratory pigments important? What factors affect the transport of O_2?

10. (a) Explain the reason for differences in the shape of oxygen equilibrium curves. (b) What advantages are offered by a respiratory pigment with a steeper curve and a lower P_{50}?

11. (a) Outline the transport of CO_2 in mammals. (b) How does the transport of CO_2 affect pH regulation in a mammal?

12. Water and air breathers differ in their convection requirements. What relevance is this to the control of ventilation?

CIRCULATION

Cells must exchange materials with their surroundings to stay alive. As we have seen in previous chapters, oxygen (O_2) must be supplied to mitochondria within cells, and carbon dioxide (CO_2) must be eliminated; nutrients and waste products of cellular metabolism must similarly move into and out of cells. These substances are transported between the exchange surfaces and the cells of the body. Within cells, movement occurs by diffusion and by a form of intracellular circulation, cytoplasmic streaming. Diffusion also occurs from cell to cell, but it is a slow process that can transport substances effectively only over relatively short distances, usually less than a millimetre or so. Thus, the only animals that rely exclusively on diffusion are small or thin, for example, flatworms. Larger animals require circulatory systems that speed the internal transport of materials.

Circulation involves convection, the bulk movement of fluid and any substances it contains. It brings dissolved substances close enough to cells or exchange surfaces so that diffusion can be effective, and it enhances diffusion by reducing the thickness of the so-called boundary layer, which is a stagnant layer of fluid near the exchange surface through which diffusion occurs.

Circulatory systems have three general functions.

- They provide mass transport of substances and blood cells throughout the body.

- They transport heat between different parts of the body, for example, to or from the external surface.

- They allow transmission of force, which is used, for example, for locomotion by many worms and molluscs.

A cardiovascular system that provides sufficient convective transport of O_2 almost invariably is more than adequate in providing for the other functions. Therefore, levels of respiratory gases in the blood are important factors in regulation of cardiovascular function. In animals where the circulatory system is not used for respiratory gas transport, the volume of circulating fluid can be less. For example, insects, which use tracheae to supply O_2 to their cells (Chapter 20), have a blood volume that is proportionally only one-fifth that of crustaceans, which use blood for transport of O_2 and CO_2.

> Circulatory systems provide internal convective transport of substances, cells and heat, and transmission of force.

TYPES OF CIRCULATORY SYSTEMS

We normally think of a circulatory system circulating blood, but there are many animal systems that circulate other body fluids. Extracellular fluid compartments include coelomic spaces, vessels (blood and lymphatic) and interstitial spaces between cells. The arrangements of these fluid compartments, and the patterns of fluid movement through them, are extremely diverse, especially among invertebrates.

Circulatory systems are often described as being *open* or *closed*, referring to whether or not the circulated fluid is always contained within a system of vessels. Of course, to carry out their exchange functions, circulatory systems are never actually 'closed'. Open circulatory systems, found in many invertebrates such as crustaceans (Fig. 21.1a) and some molluscs, are those in which a heart or hearts pump fluid through vessels that finally open into interstitial spaces. The fluid then percolates among the cells and makes its way rather slowly back to the heart, often by way of gills that oxygenate the fluid on its return journey. Thus, in open systems, the circulated fluid is indistinguishable from interstitial fluid.

Closed circulatory systems occur in a few invertebrate groups, for example, annelid worms (Fig. 21.1b) and cephalopod molluscs, and in vertebrates. In these animals, blood pumped from the heart (or hearts) passes around the body within a series of branching vessels leading to thin-walled capillaries and then into veins, which return blood to the heart.

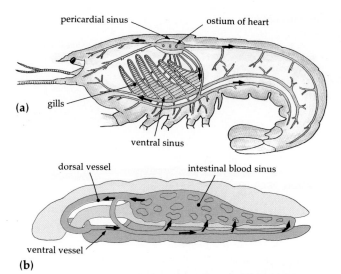

pericardial sinus   ostium of heart

(a)   gills

ventral sinus

dorsal vessel   intestinal blood sinus

ventral vessel

(b)

Fig. 21.1 **(a)** Open circulation in a lobster. Blood enters the heart from the pericardial sinus through ostia and is pumped through the body in large vessels. It leaves the vessels, percolates through the tissues and enters the ventral sinus, leading back to the heart by way of the gills. **(b)** Closed circulation in a primitive annelid worm. Blood tends to move anteriorly in the dorsal vessel and posteriorly in the ventral vessel. It surrounds the gut and exchanges material in a large intestinal sinus before returning to the dorsal vessel

In closed systems, therefore, the circulated fluid, blood, is anatomically separated from interstitial fluid throughout its circuit.

Differences in the organisation of circulatory systems generally reflect differences in the level of activity of animals. It is interesting to consider the costs and benefits of having a closed circulatory system. Open circulatory systems do not circulate fluid very rapidly, so only low rates of metabolism can be supported. More active animals with higher metabolic rates are better served by closed systems in which blood flows rapidly throughout the body within a system of smooth vessels. However, in a closed circulatory system more energy is required because the heart must pump more strongly to overcome the increased resistance to flow as blood moves through small vessels. The real advantage in having a closed circulatory system lies in the fact that, when blood is anatomically separated from interstitial fluid, it can develop different properties. For example, blood may contain respiratory pigments, such as haemoglobin, which dramatically increase its O_2-carrying capacity. This allows a considerable energy saving as much less blood needs to be circulated to carry the same amount of O_2.

> In open systems, the circulated fluid is indistinguishable from interstitial fluid. In closed systems, the circulated fluid, blood, is enclosed in a system of vessels and is distinct from interstitial fluid.

VERTEBRATE CIRCULATORY SYSTEMS

From a comparative point of view, the general arrangement of major arteries in different vertebrate groups is believed to reflect changes that occurred during their evolution. Similarly, the changes that occur during development of the circulatory system in embryonic vertebrates suggest a history of evolutionary change (Box 21.1). Within vertebrates, there is a trend towards reduction in the number of major arteries. In early vertebrates, the circulatory system is thought to have included a series of gill vessels (arches) supplied with blood directly from the heart (Fig. 21.2). This primitive arrangement is only slightly different in modern fishes, with the loss of some of the anterior gill arches. With the evolution of lungs in terrestrial vertebrates, gills were no longer necessary. Most gill arches were progressively lost; those that remained became involved with other functions. Some became the major systemic arteries, such as the aorta, and others supply blood to the lungs (Fig. 21.3).

> During the evolution of vertebrates, there has been a trend towards reduction in the number of major arteries.

There were also dramatic changes in the pattern of blood flow through the heart. In fishes, the heart is a series of four chambers (**sinus venosus**, **atrium**, **ventricle**, and **conus arteriosus**). The chambers are separated by valves that prevent reverse flow; the chambers fill and empty in sequence, moving the blood forward (Fig. 21.4). The sequence of contraction begins in the sinus venosus, but most of the propulsive force originates in the highly muscular ventricle. Circulation in fishes is a single circuit, with the heart, gills and body tissues supplied with blood in that order (Fig. 21.2). Thus, gill capillaries are perfused with blood under relatively high pressure, which tends to favour fluid loss across the lamellar epithelium, much as happens in glomerular capillaries of the kidney (Chapter 22). Also, passage through gill capillaries lowers pressure in the rest of the circuit, so that blood flow is fairly sluggish throughout the rest of the body.

Although living lungfishes are not ancestral to amphibians, their hearts have characteristics that offer insight into what might have been the condition in early amphibious fishes. In the heart of the Australian lungfish, *Neoceratodus*, for example, the atrium is almost completely divided by a septum and the ventricle and conus arteriosus are also partly divided by septa (Fig. 21.5). A large fibrous plug attached to the septa acts as a valve between the atrium and ventricle. Although incomplete, the dividing septa tend to separate blood from different origins. Surprisingly little mixing of oxygenated and

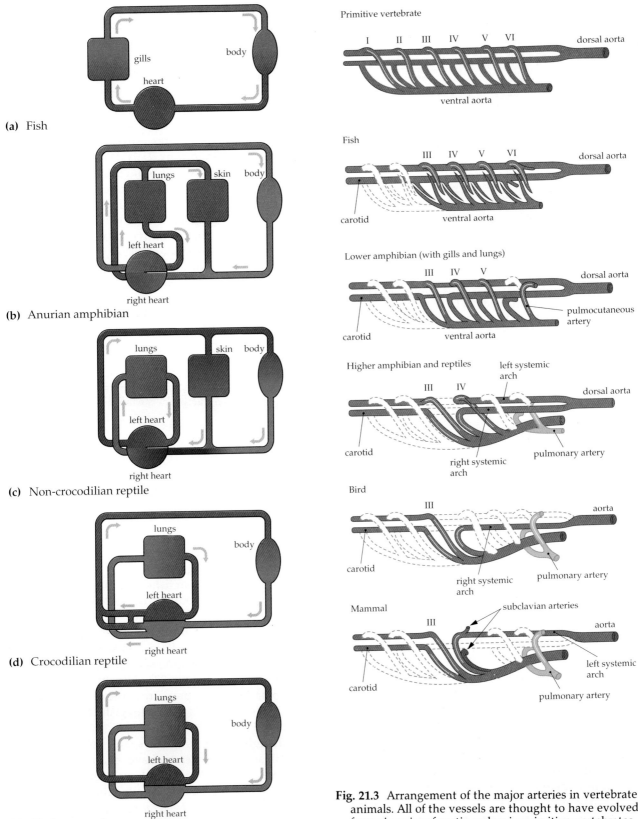

Fig. 21.2 Patterns of circulation in vertebrate animals.
(a) Fish have a single circuit with complete separation of deoxygenated and oxygenated blood.
(b–d) Amphibians and reptiles have a partly divided double circuit. (e) Birds and mammals have a completely divided double circuit

Fig. 21.3 Arrangement of the major arteries in vertebrate animals. All of the vessels are thought to have evolved from six pairs of aortic arches in primitive vertebrates. Originally these arches perfused the gills, but beginning with the first air-breathing amphibians, arch VI became the pulmonary artery. Notice that the aorta in mammals is derived from the left aortic arch IV, while in birds it is derived from the right aortic arch IV, indicative of independent origins of birds and mammals from reptilian ancestors that had both arches

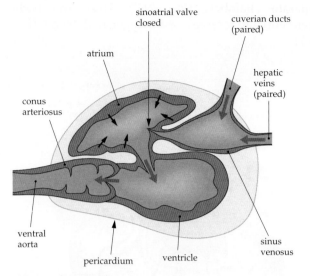

(a) Atrial contraction

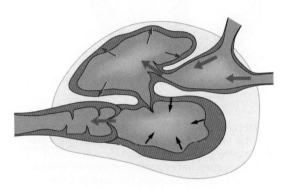

(b) Ventricular contraction

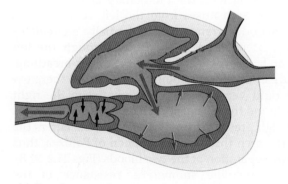

(c) Conal contraction

Fig. 21.4 The sequence of contraction in a fish's heart propels blood from the atrium, to the ventricle, to the conus arteriosus, to the ventral aorta: **(a)** atrial contraction; **(b)** ventricular contraction; **(c)** conal contraction. The pericardial compartment around the heart chambers is rigid, so contraction of one of the chambers tends to expand the others

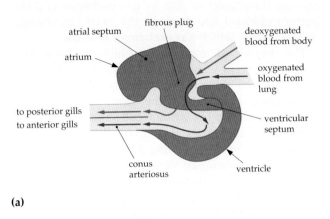

(a)

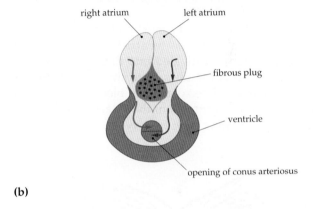

(b)

Fig. 21.5 (a) Longitudinal and **(b)** transverse sections through the heart of the Australian lungfish, *Neoceratodus forsteri*. The blood from the body and lung divides on either side of the atrial septum and flows past the fibrous plug that acts as a valve between the atrium and ventricle. A partial ventricular septum hangs down from the ventricle roof. Oxygenated and deoxygenated bloods are largely separated as they enter the divided conus arteriosus

deoxygenated blood occurs in the ventricle because the ventricular chamber is like a sponge, and blood flowing into either side tends not to mix.

Functional separation of blood through the heart of a lungfish makes gas exchange more efficient. Poorly oxygenated blood returns from the body to the right side of the atrium. From there it passes through the ventricle and out through channels leading to the posterior gills, where it is oxygenated. Meanwhile, oxygenated blood from the lung enters the left side of the atrium, passes through the ventricle and out through another channel to the anterior gill arches. The anterior arches are simple tubes without gill capillaries, so oxygen-rich blood from the lung can be distributed to the body without the danger of loss of oxygen to the surrounding water. Lungfishes can use their lungs or gills for gas exchange and the pattern of blood flow through the heart can change, depending on which gas exchanger is in use.

The hearts of modern amphibians have two separate atria, the right atrium collecting deoxygenated blood from the body (and some oxygenated blood from the skin) via the sinus venosus and the left atrium collecting oxygenated blood from the lungs (Fig. 21.6). Both atria drain through one valved opening into a single ventricle. Although these animals have no trace of a ventricular septum, mixing in the ventricle is minimised by folds in its internal spongy wall and by a spiral fold in the conus arteriosus. Thus, deoxygenated blood is moved preferentially toward the pulmocutaneous arteries bound for the lung and skin where it can obtain oxygen, and most oxygenated blood is moved to the body (Fig. 21.2).

As in amphibian hearts, most reptilian hearts have two atria and a single ventricle. However, the ventricle is partially divided into three chambers by two ridges. Recent studies have shown that mixing of oxygenated and deoxygenated blood is minimised because, during contraction, the ridges close against the walls, isolating the two types of blood into

separate chambers (Fig. 21.7). Thus, the body preferentially receives well-oxygenated blood (Fig. 21.2).

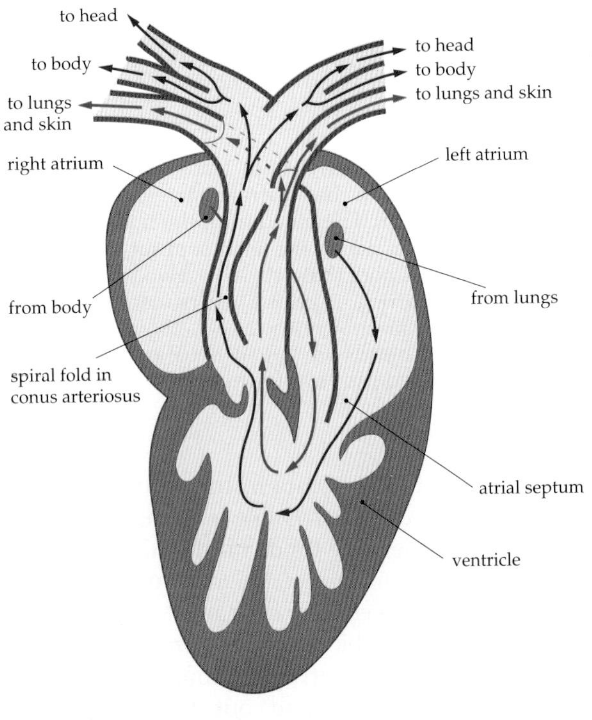

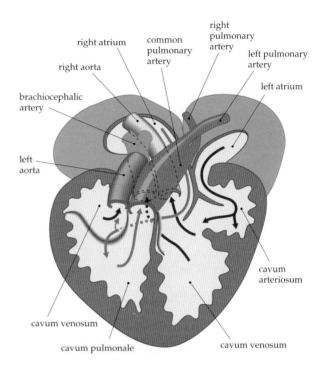

Fig. 21.7 Pattern of blood flow through a turtle heart. Although the ventricle consists of three incompletely divided chambers, oxygenated and deoxygenated blood remain somewhat separate because of laminar flow and a valving effect during the cardiac cycle

Crocodilians are unique among reptiles in having both atria and ventricles completely divided. However, the left aortic arch originates from the right ventricle along with the pulmonary artery, and the right aortic arch arises from the left ventricle (Fig. 21.8). This odd arrangement potentially recirculates deoxygenated blood to the body through the left aortic arch. However, when a crocodile is breathing, pressure is lower in pulmonary vessels and therefore in the right ventricle. The valve between the right ventricle and the left aortic arch remains closed. Blood under high pressure from the left ventricle enters the left aortic arch through the foramen of Panizza, thus ensuring separation of the two bloods (Figs 21.2, 21.8). When a crocodile submerges, resistance of the pulmonary vessels increases, as does right ventricular pressure, which opens the valve into the left aortic arch. Thus, blood is able to bypass the lung and pass back to the body (Fig. 21.8).

The conclusions of these and other recent studies are that mixing of oxygenated and deoxygenated blood through the heart in reptiles and probably amphibians is regulated and that it is advantageous under certain circumstances. Fishes have relatively

Fig. 21.6 The three-chambered heart of a frog. Oxygenated and deoxygenated bloods are separated in two atria, but there is potentially some mixing of blood in the ventricle. However, mixing is minimised by separation of blood in the spongy walls of the ventricle and also by streaming of blood in the conus arteriosus. Deoxygenated blood is sent preferentially to the lungs and skin

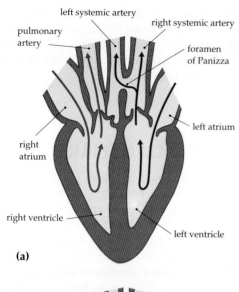

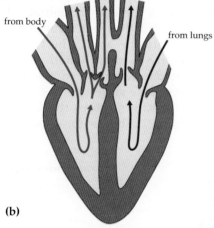

Fig. 21.8 The crocodilian heart. (a) Normally there is no mixing of oxygenated and deoxygenated blood because the valve at the beginning of the left systemic artery remains closed, and oxygenated blood flows through the foramen of Panizza. (b) During long dives, however, contraction of the pulmonary artery causes right ventricular pressure to rise and open the valve. Thus, some of the blood from the body bypasses the lung and returns to the body through the left systemic arch

low metabolic rates, but they breathe water that is very low in O_2 content, so they need to breathe fairly continuously. Amphibians and reptiles occupy an interesting position. They can breathe air, which is rich in O_2, but they are ectothermic and have relatively low metabolic rates. This means that they do not *need* to breathe continuously to provide sufficient O_2. When they are not breathing, such as during long periods under water, there is no advantage in using energy to pump all of the blood through the pulmonary circulation. But when they do breathe, there is an advantage in passing deoxygenated blood through the lungs at an increased rate to pick up O_2. Hence, the ability to shunt blood either towards or away from the

pulmonary circulation is an advantage for amphibians and reptiles.

The same argument does not hold for birds and mammals. Although they also breathe air, birds and mammals are endothermic and usually ventilate their lungs continuously to provide sufficient O_2 for their needs. There is no mixing of oxygenated and deoxygenated blood in the four-chambered hearts of birds and mammals (Fig. 21.9) and all blood that

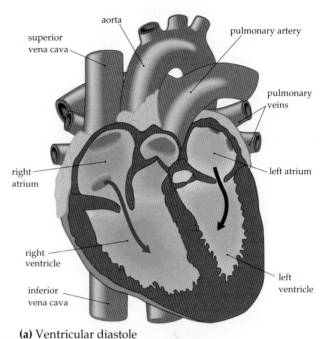

(a) Ventricular diastole

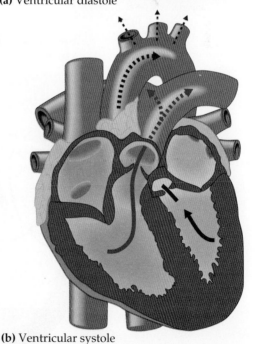

(b) Ventricular systole

Fig. 21.9 Circulation through the mammalian heart during (a) ventricular relaxation (diastole) and (b) contraction (systole). There is complete separation of deoxygenated blood in the right heart from oxygenated blood in the left heart. The heart is essentially the same in birds and mammals

passes through the systemic circulation must then pass through the pulmonary circulation (Fig. 21.2). Average flow rates in both sides of the circulation are equal.

In both birds and mammals the sinus venosus has been reduced to a small node of tissue in the right atrium, the **sinoatrial node**, which is the pacemaker region of the heart, and a single aorta leads directly from the left ventricle (Box 21.1). Interestingly, the aortae of birds and mammals are derived from different aortic arches of their reptilian ancestors, birds retaining the fourth right arch and mammals retaining the fourth left arch (Fig. 21.3).

> In fishes, blood flows through a single circuit, from heart to gills to body. In amphibians and most reptiles, atria are divided, flow is regulated to some degree through an undivided ventricle and blood may be shunted towards or away from the lungs. In birds and mammals, there is complete separation of oxygenated and deoxygenated blood, and all blood passes through the pulmonary circuit.

Heart muscle itself requires a supply of nutrients and oxygen. It receives these via the **coronary arteries**, which, in fishes, arise from the arteries leaving the gills. Coronary arteries are not found in amphibians and are only weakly developed in reptiles but their

heart muscle has a spongy construction, which permits blood to permeate the tissue. In contrast, the heart walls of birds and mammals are so compact that a coronary blood supply, which arises from the base of the aorta, is essential.

THE MAMMALIAN HEART

The mammalian heart consists of four separate chambers that alternately contract, **systole** (pronounced sis-toh-lee) and relax, **diastole** (pronounced dye-as-toh-lee; Fig. 21.10). Blood from the body enters the right atrium, moves through the **tricuspid valve** into the right ventricle, and then through the **pulmonary semilunar valve** to the lung. At the same time, blood returning from the lung enters the left atrium, goes through the **mitral valve** into the left ventricle, and then out through the **aortic semilunar valve** into the aorta. The tricuspid and mitral valves are strengthened by fibrous attachments, **chordae tendinae**, to the ventricular wall, but the semilunar valves are not. The closing of the two sets of valves causes the familiar heart sounds, 'lubb-dupp'. 'Lubb' is due to simultaneous

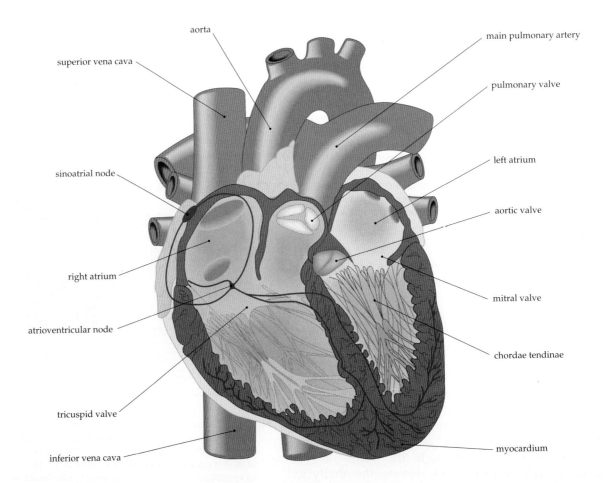

Fig. 21.10 The human heart

closure of the tricuspid and mitral valves when the ventricles begin to contract. The semilunar valves open silently and blood is forced into the two major arteries. At the end of ventricular systole, pressure in the arteries causes the semilunar valves to abruptly close, and the 'dupp' sound is heard. Clear crisp heart sounds indicate competence of the valves, but a 'fifft' sound indicates that a valve is leaking.

In humans, atrial contraction does not contribute greatly to filling of the ventricles. One reason is that there are no valves to prevent blood from passing back into the veins as there are in fishes. Another is that the ventricles largely fill themselves by creating a suction during diastole, when the relaxing ventricular walls passively expand. Indeed, in the human heart, if disease causes the atria to stop contracting entirely, the ventricles continue to fill almost normally and there is little change in circulation. It is perhaps more surprising that an *adequate* circulation can be maintained even if the right ventricle also ceases to function. The wall of the right ventricle is thinner than that of the left and pulmonary pressure is much less than systemic pressure.

The heart wall is composed of three layers of cardiac muscle, collectively called the **myocardium**. Cardiac muscle fibres are similar to skeletal muscle fibres in being striated and covered with an electrically active membrane, the sarcolemma. As in striated muscle (Chapter 27), excitation of the sarcolemma causes the muscle fibre to contract. The bundles of muscle fibres run spirally over the ventricles and when they contract, blood is squeezed from the bottom of the heart (the apex) up toward the major arteries.

> In mammalian hearts, atrial contractions push some blood into ventricles, which otherwise fill passively. Atrioventricular and semilunar valves prevent back flow.

Electrical activity

Contraction of the atria and ventricles is highly co-ordinated. Both atria contract together and then, after a short delay, the ventricles contract. The heart cycle is initiated in the right atrium at the **sinoatrial node**, or *'pacemaker'* (Fig. 21.10). This structure is derived from the sinus venosus, which initiated the sequence of contraction in early vertebrate hearts. The sinoatrial node is a small group of non-contractile cardiac muscle cells that show spontaneous rhythmic electrical activity. The heartbeat in vertebrates is **myogenic**, that is, it originates in the heart muscle itself. In contrast, many invertebrate hearts are **neurogenic**, that is, they will contract only in response to stimulation by a nerve.

The walls of pacemaker cells initiate action potentials, which are conducted through the atria (Fig. 21.11). Like a ripple in a pond, a wave of depolarisation sweeps over the atria at a rate of about

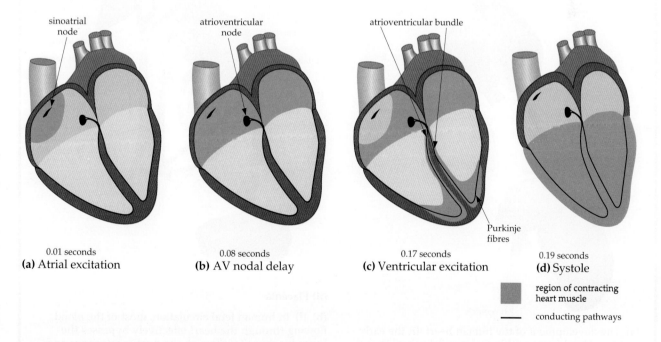

| 0.01 seconds | 0.08 seconds | 0.17 seconds | 0.19 seconds |
| **(a)** Atrial excitation | **(b)** AV nodal delay | **(c)** Ventricular excitation | **(d)** Systole |

region of contracting heart muscle

—— conducting pathways

Fig. 21.11 The pattern of electrical excitation in the human heart. **(a)** The sequence begins at the sinoatrial node (pacemaker) and **(b)** sweeps over the atria in about 0.08 seconds, causing atrial contraction. The wave of excitation encounters the atrioventricular node at about 0.03 seconds but it is delayed for 0.13 seconds before **(c)** being rapidly conducted by the atrioventricular bundle and Purkinje fibres to the apex of the ventricles. Ventricular contraction begins at 0.17 seconds, well after the atrial contraction is completed and **(d)** systole occurs at 0.19 seconds

BOX 21.1   EMBRYONIC CIRCULATION IN MAMMALS AND BIRDS

There are obvious similarities between the phylogenetic (evolutionary) and ontogenetic (embryonic) development of the circulatory system in vertebrates (Chapter 38). In mammals, for example, the embryonic heart begins as a simple straight tube. During growth and differentiation, the tube bends upon itself, forming an S-shape, and gradually becomes divided into

pulmonary (right side) and systemic (left side) circuits by the appearance of intra-atrial and intraventricular septa (Fig. a). However, the separation of embryonic

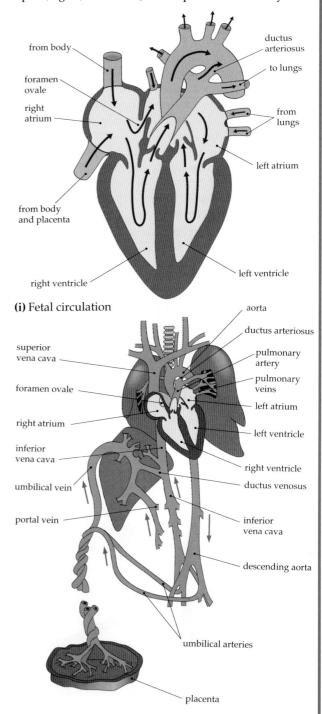

(i) Fetal circulation

(ii) Placenta

(a) The development of the human heart. In the early embryo, blood forms within two parallel tubes. These fuse into the four-chambered linear heart, which begins pumping blood. Later the heart flexes upon itself, and divides into two halves again, resulting in two, two-chambered hearts in parallel

(b) (i) In human fetal circulation, most of the blood flowing through the heart effectively bypasses the lungs by shunting through the ductus arteriosus and foramen ovale. (ii) Circulation to the placenta occurs through two umbilical arteries branching from the femoral arteries, and it returns via the umbilical vein, which feeds into the liver and ductus venosus

circuits is not complete. At this stage, the lungs are not functional and all exchanges occur across the placenta (Fig. b). The fluid-filled lungs offer a high resistance to blood flow through pulmonary vessels and therefore most blood ejected from the right ventricle is forced through the **ductus arteriosus**, which connects the pulmonary artery directly to the aorta. Meanwhile, blood returning to the heart from the body and placenta into

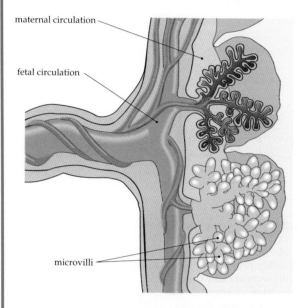

maternal circulation

fetal circulation

microvilli

(b) (iii) At the placenta, fetal blood enters small microvilli where it comes into equilibrium with maternal blood

the right atrium moves across to the left atrium through a one-way valve, the **foramen ovale**. These two pathways enable blood to bypass the lungs and supply the most oxygenated blood to the body.

At birth, however, dramatic changes occur. When a newborn mammal inflates its lungs with its first gasps, pulmonary vascular resistance drops and more blood flows to the lungs. As this blood returns from the lungs, pressure in the left atrium increases, closing the foramen

ovale. The ductus arteriosus gradually constricts and closes (as do the umbilical vessels to and from the placenta). Failure of the ductus arteriosus or the foramen ovale to close results in mixing of blood between the pulmonary and systemic circuits, and circulation of inadequately oxygenated blood explains the 'blue' appearance of human babies born with these problems.

Mammals have a single extraembryonic circulation to the placenta but, in embryonic birds, there are two circulatory loops—one goes to the yolk sac and the other to the chorioallantois (Fig. c). Yolk sac circulation is mainly involved with uptake of nutrients from the yolk while the chorioallantoic circulation largely involves gas exchange and control of fluid volumes in the egg compartments. Near hatching, circulatory changes occur in a way similar to those in mammals, only much more slowly. The yolk sac is gradually absorbed into the body cavity of the chick over several days. During the last day of incubation in chickens, the bird begins breathing gas that has collected in the air cell inside the shell. As it gradually aerates its lungs, blood flow slowly shifts away from the chorioallantois toward the pulmonary vessels. At hatching, therefore, the lungs are fully functional and chorioallantoic circulation has all but stopped.

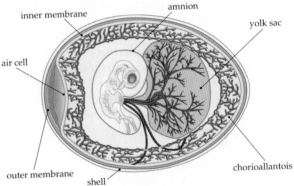

inner membrane

amnion

yolk sac

air cell

outer membrane

shell

chorioallantois

(c) Circulation in the embryonic chick. The chorioallantois is a well-vascularised embryonic membrane that functions in gas exchange and solute regulation while the yolk sac blood vessels supply nutrients to the embryo

30 cm/sec, causing contraction of both atria and pushing some blood into the ventricles. Depolarisation is prevented from entering the ventricles by a non-conducting layer, except at the **atrioventricular node**, a group of specialised muscle cells located at the junction between the right atrium and ventricle. At the atrioventricular node, the speed of conduction of the action potential is slowed to about 5 cm/sec due to the particular structure of the muscle cells (they have very small diameters and few electrical connections). The slow passage through the atrioventricular node allows time for the atrial contraction to finish before ventricular contraction begins.

After leaving the atrioventricular node, the action potential is conducted very quickly (150–250 cm/sec) through the **atrioventricular bundle** (bundle of His), which runs down the interventricular septum towards the apex of the heart. In this case, the conducting muscle fibres have large diameters and many electrical connections resulting in rapid conduction velocities. On its way, the atrioventricular bundle first divides into two main branches, which lead to both ventricles, and then into many **Purkinje fibres**, which invade ventricular muscle. This anatomical arrangement ensures a synchronous ventricular contraction, which begins at the apex of the heart.

The electrical activity of heart muscle is similar to that of skeletal muscle (see Chapter 27). When stimulated, the membrane produces an action potential during which its polarity reverses briefly (depolarisation) and then returns to normal (repolarisation). In skeletal muscle, an action potential is complete in a fraction of a millisecond, but the cardiac action potential lasts for about 250 milliseconds, time enough for the heart to complete its contraction and eject blood (Fig. 21.12).

> The mammalian heartbeat is highly co-ordinated. Contraction is initiated in the pacemaker region, conducted through atria and, via the atrioventricular node, into the ventricles. Rapid conduction through the atrioventricular bundle triggers a co-ordinated contraction of the ventricles.

The electrocardiogram

The electrical activity that sweeps over the heart during contraction is strong because it involves a large number of cells operating in unison. Some electrical current generated by the heart is passively conducted throughout the salty fluids of the body. This 'leakage' can be detected by electrical leads placed on the body surface and is the basis of the **electrocardiogram** (ECG or EKG), which is of enormous diagnostic value for cardiac physicians (Fig. 21.12). The ECG is usually divided into three parts: the P wave, QRS complex, and T wave. The P wave represents atrial depolarisation, the QRS complex results from ventricular depolarisation, and the T wave comes from ventricular repolarisation. Atrial repolarisation

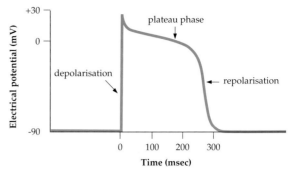

(a) Action potential

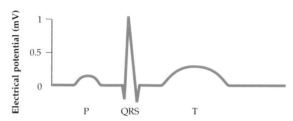

(b) Electrocardiogram

Fig. 21.12 (a) An action potential recorded in a single ventricular muscle fibre. The cell membrane is initially polarised at about −90 millivolts. When stimulated, it quickly reverses polarity, remains depolarised for 250 milliseconds and then quickly repolarises. **(b)** These electrical changes can be detected by electrodes placed in contact with the skin, yielding the typical electrocardiogram

is obscured by the QRS complex. Abnormalities in the timing of this sequence can reveal damage to the conducting pathway, which results in poor co-ordination of cardiac events and reduced pumping ability.

BOX 21.2   CIRCULATION AND THERMOREGULATION

Circulating blood can transport heat between different parts of the body and is instrumental in regulating body temperature (Chapter 29). In amphibians and reptiles that bask in the sun, heat from warmed skin is transferred to the body core, much as a solar water heater provides hot water to a house. Basking birds and mammals can also obtain solar heat this way, but they usually use their circulatory systems to control heat loss.

When a mammal is overheated, blood flow is increased to the skin and other evaporative surfaces where excess heat can be lost; when metabolic heat needs to be retained, circulation to the skin and appendages is greatly reduced. Many native people living in cold environments show cardiovascular adaptations to cold weather. Australian Aborigines undergo pronounced vasoconstriction of skin vessels and sleep comfortably on very cold nights. The hands of Alaskan Inuit (Eskimos) are often exposed to freezing temperatures without the distress and pain felt by Europeans. Periodic bouts of vasodilation, during which a pulse of warm blood flows through their extremities, protects them from frostbite.

Circulation to the limbs in some birds and mammals is adapted to conserve heat by countercurrent exchange, where heat is transferred between two bloodstreams flowing in opposite directions. As arteries enter a limb, they run parallel with veins returning blood to the body. Warm arterial blood loses most of its heat to cooler venous blood, which carries heat back into the body rather than out into the limb. The best examples of this mechanism come from aquatic birds and mammals with fins or flippers. Blood flows to the fins of marine mammals in central arteries that are completely surrounded by veins (Fig. a). When in cold water, arterial heat transfers to the veins and is retained. However, when an animal is in warm water, metabolic heat may need to be lost. At such times, the central veins close and superficial veins open, eliminating the countercurrent exchange and allowing venous blood to lose heat to the water.

In some animals, instead of having single arteries surrounded by veins, the arteries divide into a series of small parallel vessels, which interdigitate with branching veins before entering the limb. This arrangement increases the surface area available for countercurrent exchanges so that transfer may be virtually complete. In cross-section, the vessel walls of this structure resemble a net (Fig. b), so it is usually called a **rete** or *rete mirabile*, which means 'wonderful net'.

Rete structures also occur in the cranial circulation of some mammals and birds where they maintain brain temperature if the animal is overheated. Certain 'warm-blooded' fast-swimming fishes, such as tuna and some sharks, also use retes to isolate their warm active muscles from the cold blood returning from their gills.

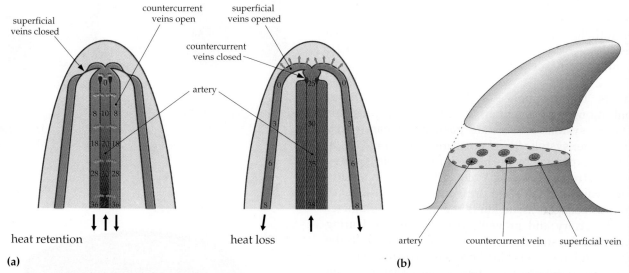

(a) Countercurrent heat exchange in the fin of a cetacean. When heat needs to be retained, warm blood from the body core loses its heat to the venous blood returning to the body in countercurrent veins that surround each artery. When heat needs to be lost, the blood returns in the superficial veins, where the heat is transferred to the sea water. The fin circulation acts as a 'thermal window', which can be opened or closed to heat. (b) Transverse section through a fin, showing the location of vessels described in (a)

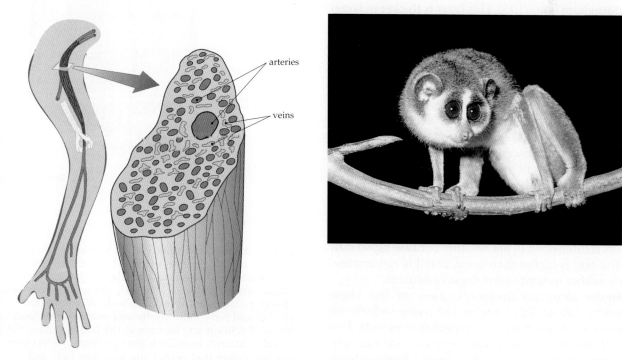

(c) Cross-section through the rete in the limb of a small mammal (the loris). The arteries and veins interdigitate and facilitate countercurrent heat exchange in the vessels at the base of the limb

CONDUCTING VESSELS

Blood circulates through the body in arteries, arterioles, capillaries, venules and veins. In addition, there is a system of lymphatic vessels, which returns tissue fluid to the circulation (p. 458).

Arteries carry blood away from the heart, either to the body in the systemic arteries or to the lungs in the pulmonary arteries. Artery walls are relatively thick, with four layers of tissue, including an inner endothelial layer surrounded by a thin connective tissue layer, a thick layer of smooth muscle and elastic connective tissue, and a thinner fibrous outer layer (Fig. 21.13). Elastic membranes may separate these layers. The abundance of elastin helps smooth out pulsations and maintain a high blood pressure. During systole, arteries are passively expanded by the higher systolic blood pressure. Thus, some of the energy expended by the heart during systole is stored in the stretched arterial walls rather than in greatly increasing arterial blood pressure. The outer layer of fibrous collagen tissue limits the maximum diameter of the artery and prevents over-expansion. During diastole, the stored energy is transferred back to the blood and maintains blood pressure. The result is that pressure in the arterial system remains relatively high and fluctuations are strongly damped (Fig. 21.14). If the walls become less elastic with age or disease, blood pressure fluctuations can become extreme.

Veins collect blood from capillaries and return it to the heart. They have a thin layer of smooth muscle between the outer connective tissue and inner endothelium (Fig. 21.13). This layer is thicker in larger veins, particularly in veins of the legs, which may become as muscular as arteries. Interestingly, when veins are transplanted to replace diseased coronary arteries, they grow to resemble arteries in a few weeks.

Venous blood pressure is low and is not sufficient on its own to raise blood back to the heart against gravity. Fortunately, return of venous blood is assisted by external compression of veins due to contraction of adjacent muscles, and intravenous valves that prevent reverse flow (Fig. 21.13). Veins are distensible and, if venous pumping is prevented, for example, in soldiers standing completely motionless at attention, blood may pool in the legs and consequently reduce blood flow to the head. This effect is intensified in hot weather because blood flow to the skin is higher than usual, and it is not unusual for a soldier to faint under these conditions.

Under most circumstances, most of the blood capacity (about 75%) lies in the veins, which are consequently referred to as **capacitance vessels**. This blood can be mobilised when required, for example, to compensate for blood loss due to haemorrhage. Contraction of veins increases venous return to the heart and therefore cardiac output, which in turn

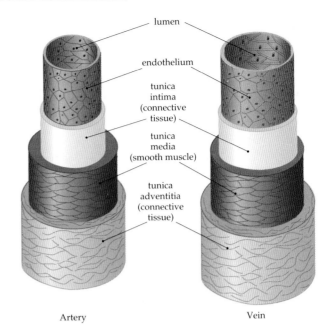

Artery   Vein

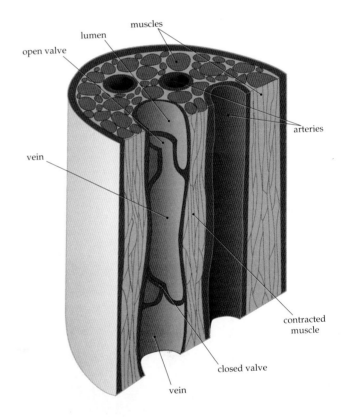

Fig. 21.13 Arteries and veins consist of four layers of cells. Outside of a single layer of endothelial cells is the tunica intima (collagenous and elastic fibres), the tunica media (elastic and muscle fibres), and tunica adventitia (tough collagenous fibres). The walls of arteries are thicker and have more elastic and muscle fibres than those of veins at any location in the body because the blood in arteries is under higher pressure. Veins often contain valves that permit one-way flow only and create a pumping action when external muscles compress the segment between adjacent valves

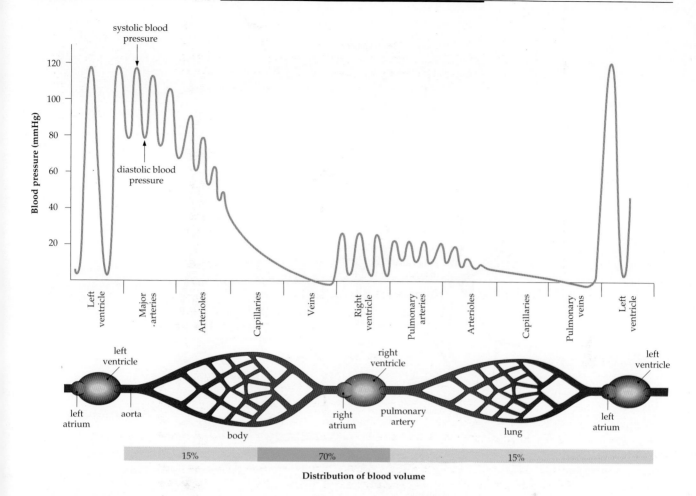

Fig. 21.14 Blood pressure and volume in the systemic and pulmonary circulation. Large pressure changes during systole and diastole in the ventricle are damped by the elastic arteries. The major drop in blood pressure occurs in the arterioles and capillaries. There is not much change in pressure along the veins because they are numerous and offer little resistance to blood flow. Although the capillaries normally contain only 5% of the total blood volume, they represent over 90% of the vascular surface area

raises arterial blood pressure. There is insufficient blood to perfuse the entire vascular bed at the same time. Blood flow is distributed to different organ systems according to need (p. 459). For example, after a meal blood flow to the gut is increased, generally at the expense of flow to skeletal muscle. The reverse is true during exercise. The brain and heart, however, must receive adequate blood flow at all times.

> Arteries carry blood away from the heart. Their elastic muscular walls store some pressure energy and reduce fluctuations in arterial blood pressure. Veins return blood to the heart. At any time, most blood is held in the veins (capacitance vessels).

The microcirculation

Vessels between the smallest arteries and venules constitute the **microcirculation** (Fig. 21.15). These include arterioles, which are completely surrounded

by smooth muscle, metarterioles, which are surrounded by discontinuous smooth muscle, capillaries, which have no muscle except for precapillary sphincters, and non-muscular venules, which are a single layer of endothelial cells surrounded by a layer of collagen.

Arterioles are very small arteries (about 30 μm diameter) that lead to capillaries. Arterioles impose the major resistance to blood flow through the circulation and, although smooth muscle in the main arteries partly controls distribution of blood to various organs, most vascular control occurs at the level of arterioles. For this reason, arterioles are often referred to as **resistance vessels**. Under nervous, hormonal and local control, arterioles can constrict completely, preventing blood flow to a particular capillary bed, or dilate to several times their normal size, greatly increasing flow.

Capillaries are the sites of exchange between blood and tissues. They are numerous (approximately 10^{10}

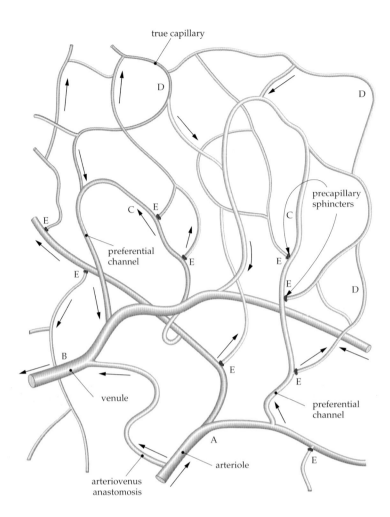

Fig. 21.15 The microcirculation. Blood enters the tissue in fine, muscle bound, arterioles (A) and is collected in venules (B). At rest, most of the blood travels through preferential channels (C), but with increased cellular activity, precapillary sphincters (D) open and allow blood to flow through the true capillaries (E). In some tissues, blood can bypass the exchange area through an arteriovenous anastomosis

of them in humans) and quite small (about 1 mm long and 8–10 μm in diameter). They are so narrow that red blood cells must pass through in single file. The total surface area of capillaries in humans is about 6000 m² and end to end, they would stretch about 100 000 km, or more than twice around the earth.

> The microcirculation consists of the small vessels lying between arteries and veins. Muscular arterioles and capillary sphincters control blood flow to vascular beds. Capillaries are the sites of exchange between blood and interstitial fluid.

Capillary walls are single endothelial cells, like the lining of the vessels leading to them (Fig. 21.16). The primary mode of capillary exchange is simple *diffusion*. The rate of diffusion depends inversely on diffusion distance and the cells of capillary walls are particularly

thin (1–3 μm). Small lipid-soluble molecules, such as O_2, CO_2, glucose and urea, diffuse directly through the lipid walls of the endothelial cells. The second mode of exchange across the capillary is *pinocytosis* (Chapter 4). In many capillaries, electron microscopy reveals an abundance of tiny vesicles of 60–70 nm (1 nm = 10^{-9} m) in the endothelial cytoplasm (Fig. 21.16), which are thought to be responsible for the slow exchange of certain substances including large particles and lipid-insoluble materials.

The third mode of exchange is *filtration*, where pressurised fluid is forced through various openings in the capillary wall, such as gaps between adjacent cells, and specialised regions known as fenestrae (Fig. 21.16). Carried with the fluid are substances small enough to pass through the openings. The relative area involved is quite small, less than a thousandth of the surface area of the capillary walls. However,

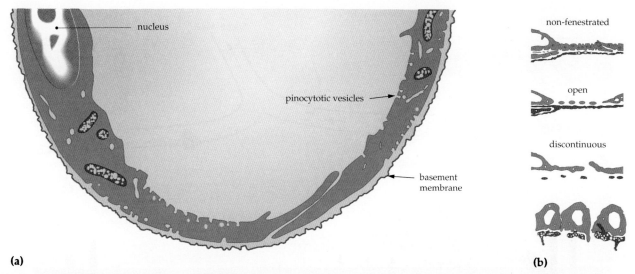

Fig. 21.16 Capillary in cross-section. (**a**) Diffusion occurs through the thin endothelial walls of the capillary. Larger molecules can move across by pinocytosis in vesicles, and water can be filtered through gaps in the lining. (**b**) The occurrence of fenestrae (windows) and discontinuities in the wall is related to the requirement to exchange large molecules

this varies from tissue to tissue. Particularly permeable capillaries are found in tissues where filtration of fluid and rapid transport of large molecules is important, for example, in the kidney glomerulus, intestinal villi, spleen, bone marrow, liver and endocrine glands. On the other hand, non-fenestrated capillaries are found in muscle, skin, fat and nervous tissue, where little filtration occurs and small molecules are exchanged mainly by diffusion or active transport. The maximum size of particles that can move through fenestrations is about 4 nm. Thus, red blood cells and large plasma proteins, principally albumin, are not filtered, but smaller molecules, for example, amino acids and vitamins, are filtered. The filtered fluid enters interstitial fluid spaces where exchange with cells takes place. Tissue fluid balance is maintained because most filtered fluid is reabsorbed into capillaries by osmosis and that which is not is removed by the lymphatic system.

Understanding of the balance between filtration and reabsorption is attributed to Ernest Starling, who proposed that *net* fluid movement between capillary and interstitial fluid is determined by the balance between two forces—*hydrostatic* (fluid) pressures and *colloid osmotic* (oncotic) pressures across the capillary wall (Fig. 21.17). Hydrostatic pressure and colloid osmotic pressure of interstitial fluid (HP_{IF} and COP_{IF}) are relatively constant along the length of the capillary. However, as blood flows along a capillary, its hydrostatic pressure (HP_B), due to the pumping action of the heart, decreases from about 30 mmHg at the arteriolar end to about 15 mmHg at the venular end. Colloid osmotic pressure due to large proteins ('colloids') retained in the blood (COP_B), is about 25 mmHg and changes little along the length of a capillary.

At the arteriolar end of a capillary, the balance of forces across the capillary wall results in a net outward filtration pressure and fluid leaves the blood. At the venular end, largely as a result of the decrease in HP_B along the capillary, net filtration pressure is into the capillary and fluid is reabsorbed. Filtration and reabsorption do not necessarily occur equally in the same capillary. Some capillaries may largely filter while others largely absorb, the balance depending on the hydrostatic pressure gradient in each one. Overall, however, in any tissue, most filtered fluid is reabsorbed.

Two important features of circulation put the Starling principle into perspective. Firstly, it has been estimated that, of the 8000–9000 L of blood pumped from the human heart each day, only 20 L (0.25%) are filtered from capillaries into tissues. With regard to exchange of respiratory gases, nutrients and wastes, therefore, filtration is of negligible importance. Secondly, because of the imbalance between net outward and net inward filtration pressures, not all filtered fluid is reabsorbed back into capillaries. About 2–4 L (10–20% of filtered fluid) remains and is collected from tissues by the lymphatic system and returned to the blood via the great veins near the heart. Failure of the lymphatics to return excess filtrate to the blood results in an increase in interstitial fluid and swelling of tissue, **oedema**. For example, in tropical regions, filarial nematode worms can block lymphatic vessels in humans and cause enormous swelling of the affected parts, a condition known as elephantiasis (Chapter 38).

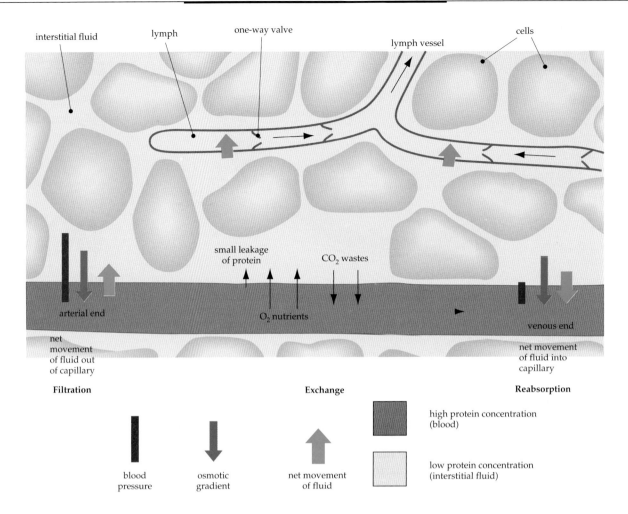

Fig. 21.17 Filtration and reabsorption in the capillary. Balance of pressures across capillary walls and resultant movement of fluid between blood plasma and interstitial fluid. Net filtration occurs due to the high blood pressure at the arterial end of the capillary. As blood pressure decreases along the capillary, the colloid osmotic pressure eventually exceeds blood pressure, and osmotic reabsorption occurs. Dissolved substances diffuse along their concentration gradients, and are also carried with filtered and reabsorbed water. Excess interstitial fluid is returned to the blood vascular system via the lymphatic channels

Oedema is particularly dangerous in the lung because increased interstitial fluid may leak into the alveoli and greatly increase the diffusion barrier between air and blood (Chapter 20). In normal circumstances, however, excess filtration in the lungs is unlikely because pulmonary blood pressure is much lower than systemic blood pressure. Thus, in the lung, HP_B is normally less than COP_B and any water that gets into the lung is immediately absorbed back into the blood. However, in the case of heart failure involving the left ventricle, a backup of blood in pulmonary vessels causes a rise in pulmonary HP_B and fluid may filter into the lungs.

Net movement of fluid across the capillary wall occurs in response to the *net* filtration pressure, which is the result of balance between colloid osmotic and hydrostatic pressures of interstitial fluid and blood. Most of the fluid that moves out at the arterial ends of capillaries is reabsorbed.

The lymphatics

Mingled among cells in interstitial spaces are primary lymphatic capillaries. Like blood capillaries, they are simple endothelial channels but, in contrast, they are blind-ended. Interstitial fluid enters lymphatic capillaries through gaps between endothelial cells, some of which appear to have valve-like flaps that prevent reverse flow back into the tissues. Lymph fluid is moved as a result of contraction of smooth muscle in walls of larger lymph vessels and by indirect external pressure on lymph vessels due to contraction of adjacent skeletal muscles. In some fishes and amphibians, lymph flow is also assisted directly by muscular 'lymph hearts'. As in veins, one-way valves cause lymph to flow in one direction only. In humans, lymph passes through many lymph nodes (Chapter 23), finally entering the venous system from either the right lymphatic duct or the thoracic duct. Some tissues lack lymphatic vessels (for example, the central nervous system, bone marrow and lung), but they are abundant in other tissues (for example, gut, skin and upper airways).

The lymphatic system has a particularly important role in tissue fluid balance. There is a small leakage of protein from capillaries into the interstitial fluid. If allowed to accumulate in interstitial fluid, COP_{IF} would soon rise, reducing the net reabsorption pressure and causing oedema. In terms of tissue fluid balance, the main role of the lymphatic system is to return these escaped plasma proteins, along with excess interstitial fluid, back to the blood.

The lymphatic system also transports other materials. For example, it is the major route of lipid transport from the small intestine to the blood and it carries certain products of the liver (Chapter 19). Another important function of the lymphatic system is its role in defence, which is considered in Chapter 23.

> Fluid that is not reabsorbed back into capillaries, and proteins that leak from capillaries, are returned to the circulation by way of the lymphatic system.

REGULATION OF BLOOD FLOW AND PRESSURE

The vertebrate circulatory system is essentially a pump that propels blood around a circuit of vascular tubing. Whether it is a single circuit, as in fishes, or a completely double circuit, as in the systemic and pulmonary circuits of birds and mammals, the heart works against friction in moving blood through the vessels. By analogy with the flow of current in an electrical circuit, we can describe blood flow by Ohm's law, which states that current is equal to the ratio of voltage over resistance. In our case, rate of blood flow (V_b) is equal to the pressure gradient along the vessel ($P_{b\,(artery)} - P_{b\,(vein)}$) over the resistance of the vessels (R; Fig. 21.18).

In mammals and birds, **cardiac output** (V_b) is the amount of blood pumped by the left ventricle into the aorta every minute (mL/min). In contrast, in vertebrates with undivided ventricles, cardiac output refers to the entire output of the heart. Blood pressure (P_b) is traditionally measured in units of millimetres of mercury (mmHg). V_b and ($P_{b\,(a)} - P_{b\,(v)}$) are dependent on the peripheral resistance (R, mmHg·min/mL), which is regulated by controlling the diameter of arterioles. In the homeostatic condition, any changes in overall arteriole diameter are matched by reciprocal changes in cardiac output so that blood pressure is regulated at a more or less constant level.

Looking more closely at the 'plumbing' of the circulatory system (Fig. 21.19), we can see that it consists of many organs usually arranged in parallel rather than in series. The liver is one exception in

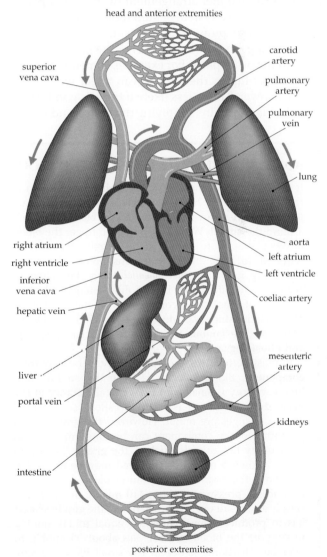

Fig. 21.19 Diagram of the double circulation of mammals. The vascular beds are arranged in parallel, except for the intestine and liver, which are connected in series by the portal vein

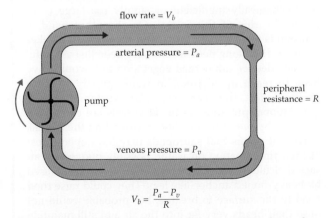

$$V_b = \frac{P_a - P_v}{R}$$

Fig. 21.18 Blood flow in the circulatory system depends on the difference in blood pressure between the main arteries and veins, and also on the resistance to blood in the smaller vessels (peripheral vessels)

that part of its circulation comes from the gut. Parallel blood supply ensures that organs are supplied with oxygenated blood delivered at high pressure by the systemic arteries. It also permits changes in blood flow to individual organs to be somewhat independent of flow through the whole system. If organs were arranged in series, all would be obliged to receive the same blood flow, and those first in line would receive the freshest blood.

> Blood flow through vessels is governed by blood pressure and resistance to flow in the vessels. Blood supply to most organs is arranged in parallel, rather than in series, providing fresh blood to most organs and total flow independent of flow to individual organs.

Cardiovascular regulation in vertebrates

Despite the parallel arrangement of blood supply to most organs, any system of fluid reticulation supplied by a common pump tends to suffer from the fact that a change in flow to one part of the system affects flow to the other parts. For example, someone turning on a water tap in the kitchen can affect water flow to the shower. This is less likely if there is good water pressure in the mains supplying the house, and would be even less likely if mains water pressure were regulated to stay the same in the face of changing demands. Circulatory systems in vertebrates have both these features: they generate relatively high pressures and have negative feedback mechanisms that regulate arterial blood pressure to compensate for redistribution of flow. As a result, in ordinary circumstances, an increase in blood flow to one organ need not decrease flow to the rest of the body.

A negative feedback system consists of three parts:

- *Sensors* (receptors), which are able to detect change or disturbance and pass this information on.
- An *integrating centre*, which evaluates the information and directs it.
- *Effectors*, which produce an appropriate compensatory response that reduces the change.

The body has no receptors capable of measuring blood flow. However, there are several types of sensors that provide information on variables directly related to changes in blood flow. For example, stretch receptors in vessel walls and in the atria detect blood pressure, and chemoreceptors in arteries and the brain detect changes in O_2, CO_2 and pH. Information from sensory receptors throughout the body is integrated in the brainstem and appropriate

---

BOX 21.3   BERNOULLI, GRAVITY AND BLOOD FLOW IN DINOSAUR NECKS

It takes energy to move fluid through tubing. Much of the work that the heart performs is used to generate *pressure energy* to overcome friction as blood is moved around the circuit. Work is also needed to get the blood moving, that is, to impart *kinetic energy*, and work is required to raise blood to organs above the heart, increasing its *gravitational potential energy*. Fluid moves according to differences in *total* fluid energy, which is the sum of these three components of energy (Fig. a). This relationship was described in 1738 by Daniel Bernoulli, the famous Swiss mathematician, physicist and physiologist.

In most vertebrates, pressure energy is the important component. Blood velocity is generally so low that kinetic energy is small, and the vertical distance of the head above the heart is relatively short, so potential energy changes are minor. However, the gravitational potential energy component can become important in particularly tall animals, such as the giraffe (Fig. b). In a fully grown 5 m giraffe, the head may be about 150 cm above the heart. The column of blood in the neck produces a pressure of about 110 mmHg at the heart, simply as a result of gravity. Therefore, the giraffe's heart has to produce a mean pressure equal to 110 mmHg to support the blood column, *plus* about 90 mmHg to overcome frictional losses in the small blood vessels. Direct measurements confirm that the mean blood pressure in the giraffe's aorta is about 200 mmHg. In comparison to many mammals in which mean aortic pressure is about 100 mmHg, the giraffe certainly has a high blood pressure. To produce such a pressure, the walls of the giraffe's left ventricle are extremely thick.

Giraffes, at about 4–5 m, are the tallest living animal, but some of the long-necked dinosaurs (the sauropods, often called brontosaurs) were much taller. Modern reconstructions show giants such as *Brachiosaurus* or *Seismosaurus* towering 12–14 m above the ground and feeding from tall trees. Similar calculations to those for giraffes provide an estimate of required blood pressure at the heart of 500–750 mmHg. Some physiologists believe that such pressures are impossible because the heart muscle would have to be so thick that it would be mechanically inefficient or simply too large to fit inside the ribs of the animal. On the other hand, if sauropods were not terrestrial and they floated in lakes or rivers, their long neck would have been just as useful to reach deeper submerged vegetation as it would have been to reach up to trees. In water, gravity is not a problem for the cardiovascular system because gravitational pressures in blood vessels are essentially matched by pressures in water surrounding the animal.

Thus, it seems that sauropods may have spent most of their time floating on water, possibly buoyed by air sacs that filled spaces in their vertebrae, and stabilised by heavy-boned anchor-like legs. They could raise their head to the surface to breathe, but probably could not raise their heads very far into the air and still maintain blood flow to the brain. These animals doubtless came on land, perhaps to lay eggs, but they almost certainly did not forage in tall trees as often depicted.

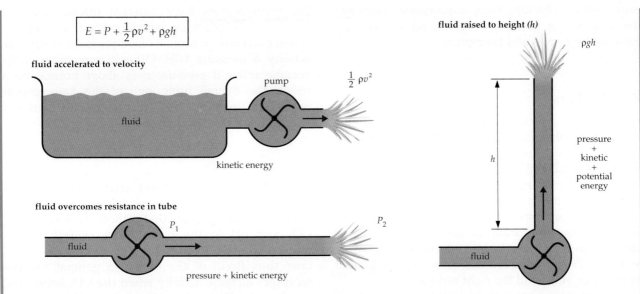

$$E = P + \frac{1}{2}\rho v^2 + \rho gh$$

fluid accelerated to velocity

pump

$\frac{1}{2}\rho v^2$

fluid

kinetic energy

fluid overcomes resistance in tube

P_1

fluid

P_2

pressure + kinetic energy

fluid raised to height (h)

ρgh

h

pressure
+
kinetic
+
potential
energy

fluid

(a) The Bernoulli equation. The energy that a pump must produce to move fluid is the sum of three components: the pressure energy (*P*), the kinetic energy of fluid motion ($1/2\rho v^2$), and the gravitational potential energy (ρgh). In the equation, ρ is the density of the fluid, *v* is the velocity, *P* is the hydrostatic pressure, *g* is the gravitational acceleration constant, and *h* is the height of the fluid column. In blood circulatory systems, the kinetic energy component is small because the velocity is low. Sufficient fluid pressure is important, however, to overcome the resistance to flow in the tubing. The heart must work against the pressure developed in the vertical blood column above it

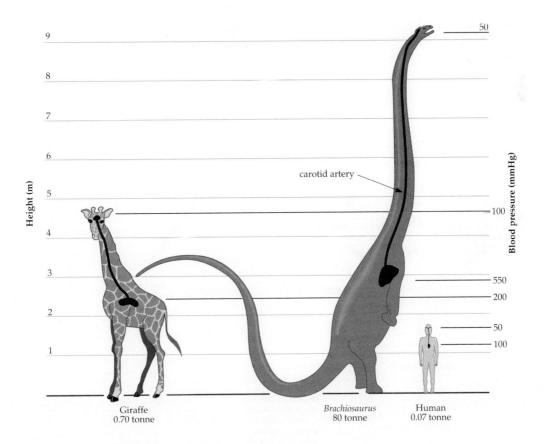

carotid artery

| Giraffe 0.70 tonne | Brachiosaurus 80 tonne | Human 0.07 tonne |

(b) Blood pressures measured at the levels of the heart and head in a human and an adult giraffe, compared with that calculated in a sauropod dinosaur

responses by effectors, such as pacemaker cells and the muscles of the heart and blood vessels, are initiated by nerves and hormones.

Baroreceptors

Regulation of arterial blood pressure is exceedingly complex, involving many different sensory receptors, neural pathways, hormones, and also local effects (Fig. 21.20). Here it is only possible to present a simplified account. Stretch receptors in the cardiovascular system are known as **baroreceptors**. In mammals, arterial baroreceptors are mainly located in the carotid sinus, at the bifurcation of internal and external carotids, and along the carotid arteries and the aortic arch (Fig. 21.21). There are also baroreceptors located in the large veins and the right atrium.

Baroreceptors appear to be simple nerve endings that divide repeatedly and invade the muscular wall. When blood pressure rises, the wall stretches and the nerve endings are deformed, resulting in increased rate of firing of action potentials (Chapter 26). Baroreceptors have different threshold levels. Some baroreceptors are *tonically active*; they generate action potentials at normal blood pressures, but cease activity if pressure falls. Other baroreceptors only become active if pressure rises above normal. As a result, the overall firing rate of the baroreceptor population increases and decreases with increases and decreases in blood pressure.

Chemoreceptors

Associated with the carotid arteries and aorta in mammals are the carotid and aortic bodies, which contain chemoreceptors that respond to levels of O_2, CO_2 and pH in the blood. Also, central chemoreceptors in the medulla respond to changes in pH (and thus P_{CO_2}) of the blood. In general, chemoreceptors increase activity when the CO_2 level rises, or when the O_2 level or pH fall. In addition to their effects on ventilation (Chapter 20), stimulation of chemoreceptors increases heart rate and blood flow to the lungs, improving gas exchange. They also influence systemic blood flow and pressure.

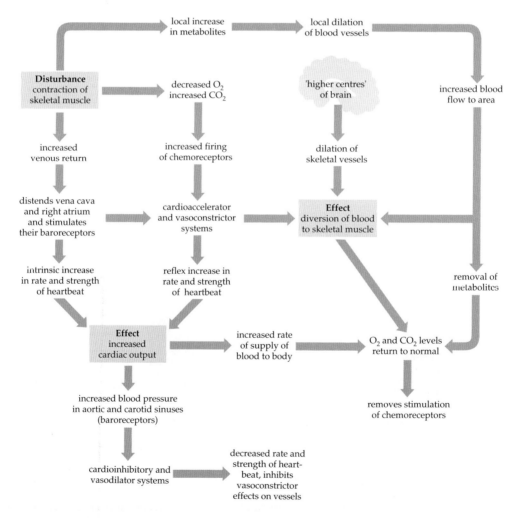

Fig. 21.20 Example of some of the pathways involved in the homeostatic control of blood flow and pressure during exercise (the disturbance) in a mammal

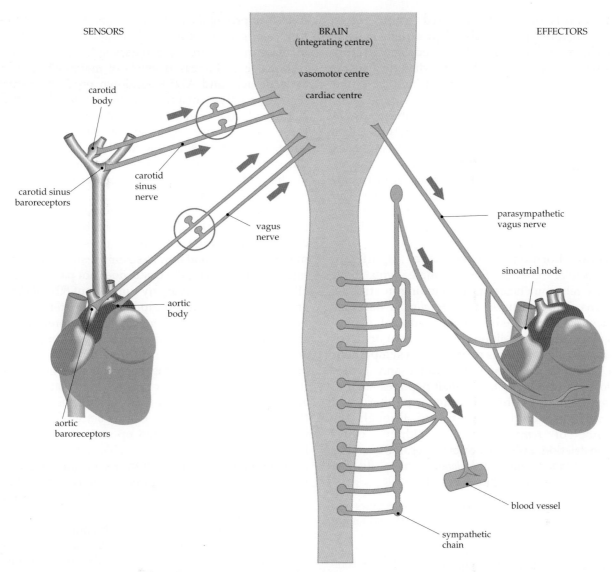

Fig. 21.21 Major neuronal feedback systems for the regulation of blood pressure. Receptors detect stretch in the arterial walls and levels of blood gases and pH. Sensory nerves relay this information to the brain, where it is processed in the vasomotor and cardiac centres. Effector nerves can change the power and rate of the heartbeat and also alter peripheral resistance by affecting smooth muscles in the arterioles

Integration

Information from baroreceptors is sent to the **vasomotor centre** and the **cardiac centre** in the brainstem (medulla and pons). The vasomotor centre controls the state of contraction of smooth muscle of arterioles, and the cardiac centre controls the rate and force of beating of the heart. Nervous impulses from chemoreceptors pass to the brainstem where they primarily influence the respiratory centres controlling breathing, but they also have an impact on the cardiovascular centres.

Responses

In vertebrates, both cardiac output and regional blood flow are regulated by the autonomic nervous system, including parasympathetic and sympathetic nerves (Chapter 26), by circulating hormones and by local (intrinsic) factors. In general, if arterial blood pressure drops, output from the vasomotor and cardiac centres causes an increase in activity of sympathetic nerves and a decrease in activity of parasympathetic nerves. This causes most peripheral arteries and arterioles to contract further, increasing overall peripheral resistance. Increased sympathetic input and decreased parasympathetic (vagal) input to the heart cause the pacemaker to produce action potentials more quickly, increase the speed of conduction through the atrioventricular node and increase the force of ventricular contraction. Increased sympathetic nerve activity to venous blood vessels causes them to contract and to force more venous blood towards the heart. Increased venous return stretches baroreceptors in the right atrium, causing a reflex increase in heart rate. Increased venous return also stretches cardiac muscle fibres directly, causing them to contract more forcefully, raising the volume of blood pumped. All of these responses raise blood pressure. If arterial blood pressure rises above normal, opposite responses occur to again return blood pressure towards normal. These nervous responses, particularly those involving parasympathetic nerves, are extremely fast, and compensatory adjustments can occur within the time of a single heartbeat. This is important for blood pressure regulation during postural changes; if they were slower, we would faint every time we stood up.

Several hormones also have significant cardiovascular effects. For example, you have probably felt your heart start pounding in stressful situations. This is caused by *adrenaline,* a hormone produced and released from the adrenal glands, particularly under conditions of excitement or stress. Like *noradrenaline,* which is released from sympathetic nerves, adrenaline speeds the heart and causes arterioles to constrict. Other small peptides such as *vasopressin,* produced by the posterior pituitary, and *angiotensin,* produced in the blood through the influence of *renin* from the kidney, cause vasoconstriction. Still other substances, for example, *kinins, histamines,* and *prostaglandins,* have distinct effects on the heart and blood vessels.

> Negative feedback control involves baroreceptors and chemoreceptors, integration in both the vasomotor and cardiac centres of the brain, and sympathetic and parasympathetic outputs that control the timing and strength of the heartbeat and the state of contraction of muscles in vessel walls. Regulation is also affected by blood-borne hormones.

Regulation of blood flow to tissues

The rate of blood flow to particular capillary beds is controlled by intrinsic mechanisms (local factors within the tissue itself), and extrinsic mechanisms (influences from outside the tissue), including nerves and hormones. In different tissues, different mechanisms may predominate. For example, coronary blood flow is almost entirely under intrinsic control, whereas control of flow to skeletal muscle or skin has a strong neural component.

Intrinsic control: autoregulation

Blood flow through capillaries is largely under the control of precapillary sphincters (Fig. 21.15), which are responsive to local conditions in tissues. There is an inherent myogenic activity (tone) in the smooth muscle of precapillary sphincters and the level of this tone tends to increase if a vessel is stretched by increase in pressure. Precapillary sphincters are also responsive to increased levels of metabolites, such as lactic acid and ATP breakdown products. When metabolic activity of a tissue increases, levels of CO_2, lactic acid, ADP and AMP increase, and any one of these may cause opening of the precapillary sphincters, increasing blood flow to the area. Vasoconstrictor and vasodilator substances produced in the endothelium also have powerful effects on vascular smooth muscle.

Local control of blood flow, **autoregulation**, is also responsible for the distribution of blood within a capillary bed. When a tissue is at rest, it receives a minimal rate of blood flow to sustain resting metabolic rate. Blood perfusing the tissue does not pass through all capillaries because, at any time, about 80% of them are closed by their precapillary sphincters. Blood normally flows in larger capillaries, **preferential channels** (Fig. 21.15), which remain open continually and allow blood to pass through at a velocity appropriate for exchanging gases and substrates with the tissues. Preferential channels are relatively far apart, but exchange with them is sufficient to meet the needs of resting tissue. During activity, however, precapillary sphincters open up and allow blood to flow through true capillaries. This not only allows a greater blood flow through the tissue, but it ensures that blood passes more closely to active cells, thus decreasing the diffusion distance and increasing the rate of exchange.

The distribution of blood flow through the lungs appears to be dependent on local factors, but in the opposite direction to that in systemic tissues. Although the lungs of mammals receive all blood flowing from the right ventricle, blood tends to flow preferentially through areas of the lungs containing the freshest air. Here, under the influence of high levels of O_2, pulmonary arterioles dilate rather than constrict. In lower vertebrates with incompletely divided ventricles and intraventricular shunting (Chapter 40), the fraction of blood sent to the lungs can be changed according to the level of O_2 in the lung and the requirements of the animal.

Extrinsic control

Extrinsic regulation of vessel diameter is brought about by nerves and hormones. Arterial vessels are innervated by sympathetic nerves (Chapter 26), which are tonically active, releasing noradrenaline (and possibly other neurotransmitters), usually producing a vasoconstrictor tone. However, in the same vascular bed, the response can change direction depending on the level of activity in the nerves. In skeletal muscle, for example, vasodilation occurs at low levels of sympathetic input when noradrenaline stimulates a set of receptors (β-receptors) on arterial smooth muscle. However, at higher levels of sympathetic nerve activity, noradrenaline begins to stimulate a different set of receptors (α-receptors), which cause vasoconstriction. In some species, there are also cholinergic vasodilator nerves innervating vessels supplying skeletal muscle.

Blood flow to specific organs can also be modified by neural output from higher (conscious) centres of the brain. Familiar examples include blushing of the skin with emotion, skin blanching with shock, and erection of genital organs during sexual arousal.

Hormonal regulation of vascular smooth muscle involves circulating hormones such as adrenaline, vasopressin, angiotensin, renin, kinins, histamines and prostaglandins.

> Blood flow within a tissue is controlled by arterioles and precapillary sphincters, which are responsive to both intrinsic mechanisms (stretch, metabolites and other locally released vasoactive substances) and extrinsic mechanisms (effects of nerves and hormones).

BLOOD

Blood has numerous functions, but its most significant roles are the transport of respiratory gases (Chapter 20), nutrients, waste products and hormones. It is also central to the defence mechanisms of the immune system (Chapter 23), and regulation of fluid balance in tissues (Chapter 22) and body temperature (Chapter 29). The formation of mammalian blood is described in Chapter 6. Here we will examine some of its general properties.

In vertebrates, whole blood consists of two fractions: **plasma**, a clear, slightly yellowish fluid (containing a significant amount of protein, chiefly albumins and globulins) and suspended cells or fragments of cells (Fig. 21.22), produced in the bone marrow. When whole blood is spun in a centrifuge, the cells pack at the bottom of the tube and their volume as a proportion of the total volume is the **haematocrit**. In healthy humans, haematocrit is around 40–45%. Red blood cells (erythrocytes), which transport O_2, are the most numerous. White blood cells (leucocytes) are less common (about 1 per 600 erythrocytes), but their roles in protecting the body against disease and removing cellular debris are essential and are considered in detail in Chapter 23. Finally there are **thrombocytes**, which, in lower vertebrates, are small nucleated cells but, in mammals, are cell fragments called **platelets**. Platelets are disc-shaped membrane-bound cell fragments, about half the diameter of an erythrocyte. They do not contain a nucleus or DNA. Huge cells, **megakaryocytes**, in the bone marrow, pinch off thousands of platelets

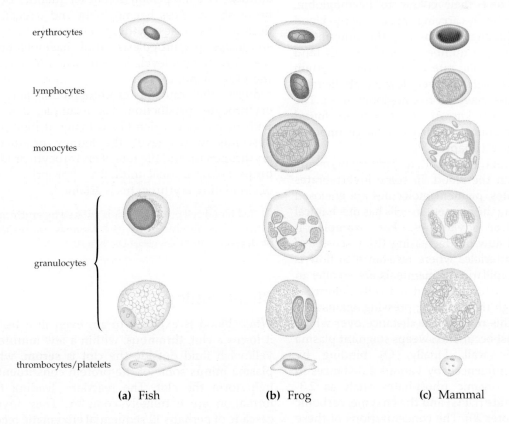

erythrocytes			
lymphocytes			
monocytes			
granulocytes			
thrombocytes/platelets			
	(a) Fish	**(b)** Frog	**(c)** Mammal

Fig. 21.22 Blood cells from **(a)** a fish, **(b)** a frog and **(c)** a mammal

during their life. During their 10-day life after release into the circulation, platelets may take part in wound healing and clot formation (see below).

Vertebrate blood contains cells suspended in a fluid plasma. In mammals, haematocrit is about 45%.

Erythrocytes

In humans, erythrocytes are round discs, about 7–8 µm in diameter, with thick edges, giving them a biconcave appearance. Although only 2.1 µm in maximum thickness, if all the red blood cells in your body (about 30 million million) were stacked like plates, the stack would measure 63 000 km, or about one and a half times the circumference of the earth. Their combined surface area is about 3840 m^2, or about half the area of a soccer field. Mammalian erythrocytes are the smallest among vertebrates, lack nuclei in the mature form, and have low metabolic activity. On the other hand, erythrocytes of other vertebrates are larger (up to 78 $\times$ 46 µm in a salamander), nucleated and have higher metabolic activity. The significance of these differences is unclear. However, smaller cells have a larger relative surface area for gaseous exchange, and a low metabolism reduces the amount of oxygen consumed by the cells themselves on their way to the tissues.

Erythrocytes owe their colour to haemoglobin, which increases the O_2-carrying capacity of the blood (Chapter 20). Among vertebrates, the amount of haemoglobin and the number of erythrocytes tend to reflect the metabolic rate of the animal. Certain Antarctic fishes, which have very low metabolic rates in the frigid water, have no haemoglobin at all. The highest haematocrits known, about 60%, are found in some diving mammals, which store large amounts of O_2 in their blood (Chapter 29).

In many invertebrate animals, respiratory pigments circulate freely in the blood. In some invertebrates and all vertebrates, pigment molecules are enclosed in cells. Packaging the pigment in cells has mechanical and physiological advantages. For example, it facilitates blood flow by decreasing the viscosity of blood in small arterioles where resistance to flow is highest. Some capillaries of mammals are smaller in diameter than erythrocytes and red cells deform as they pass through the capillary, pressing against the capillary wall. This reduces the distance over which gas diffusion must occur and sweeps stagnant plasma away from the wall. Finally, O_2 binding by haemoglobin is influenced by various substances in the blood (e.g. organic phosphates such as 2,3-diphosphoglycerate [DPG] and the enzyme, carbonic anhydrase; Chapter 20). The concentrations of these substances inside the cell can be regulated at higher levels than would be possible if they were all freely dissolved in the blood.

In mammals, erythrocytes are formed in bone marrow (Chapter 6). They are incapable of dividing or repairing themselves and they simply wear out. In humans, they circulate for about 120 days. Before they actually break down, old erythrocytes are phagocytised in the spleen, liver and bone marrow. Proteins are broken down into amino acids and recycled, and the iron from haemoglobin is retained for synthesis into new haemoglobin. Part of the haemoglobin molecule is converted by the liver into bilirubin, which leaves in the liver bile.

In humans, about two and a half million erythrocytes are formed and destroyed every second. However, the number of erythrocytes in the blood remains remarkably constant, reflecting a highly efficient regulatory system. The hormone *erythropoietin* is continually produced by the kidney (an organ that usually receives a high and constant blood flow in order to carry out its excretory role) and released into the blood. When erythropoietin reaches bone marrow, it stimulates red cell production (erythropoiesis, Chapter 6). The amount of erythropoietin released depends on the oxygenation of the blood. If blood has a low O_2 content, erythropoietin production increases. Low blood O_2 can result from **anaemia** (low levels of haemoglobin), respiratory or cardiovascular disease, or living at high altitude. When a person donates a quantity of blood, he or she has less haemoglobin and oxygen levels decline. This causes erythropoietin production, which stimulates erythropoiesis until haemoglobin and therefore oxygen levels have returned to normal. On the other hand, excessive oxygenation of the blood reduces the rate of erythropoietin, and hence erythrocyte, production. For example, if someone with haemoglobin-rich blood living at high altitude descends to sea level, the kidney produces less erythropoietin and the rate of erythrocyte production drops below normal until the appropriate number of circulating erythrocytes is attained.

Red blood cell production is regulated by erythropoietin, which is produced by the kidneys in response to decreases in O_2 levels of the blood.

Blood clotting

When blood is exposed to air, even in a test tube, it forms a clot, **thrombus**, within a few minutes. The yellowish fluid outside the clot is **serum**, which is plasma minus some of the protein constituents that help form the clot. The reactions leading to clot formation are extremely complex. They involve a cascade of perhaps 12 sequential enzymatic reactions, with 'failsafe thresholds' at each stage to prevent accidental triggering of clot formation. Each reaction converts an inactive form of an enzyme into an active form, which is capable of catalysing the next reaction of the cascade and producing hundreds of active products. Thus, the number of products is multiplied at each step. The final reaction involves the enzyme **thrombin**, which converts soluble plasma protein, **fibrinogen**, into fine strands of insoluble **fibrin**. These fibres form a meshwork that traps erythrocytes and platelets to form a clot (Fig. 21.23). The enzyme sequence of the clotting cascade has been studied in humans with variations of the genetic disease **haemophilia**, which prevents the blood from clotting normally.

In damaged tissue, factors other than clotting are also involved in preventing blood loss. Firstly, traumatised vessels usually contract, dramatically reducing blood flow through the vessel and therefore blood loss. When a vessel is broken, the endothelial layer is breached, exposing collagen fibres in the deeper layers. Platelets immediately attach to the collagen and each other. Thrombin, which forms as outlined above, has three effects on platelets. It causes them to:

- become sticky and attach to fibrinogen;

- become fragile and release ADP, which enhances linking with fibrinogen;

- form long pseudopod-like processes, which attach to adjacent platelets.

Thrombin converts fibrinogen to fibrin, which further strengthens the clot. The gap in the vessel quickly fills with a mat of fibrin and aggregated platelets. Thrombin apparently stimulates platelets to contract, squeezing serum from the clot and drawing the broken edges of the vessel together. Platelets also release a substance that makes the surrounding blood vessels contract, further reducing the blood flow to the area.

Tiny breaks in vessel walls occur all the time as a result of normal activity. These are rapidly sealed by platelets, which plug holes usually without preventing blood flow through the vessel. Platelets are so important in preventing blood loss that, when their production is inhibited, as in some leukemias, numerous tiny bruises appear over the body.

> Blood clotting involves a series of enzyme-controlled reactions leading to production of the insoluble protein, fibrin. Platelets attach to the exposed edges of a wound, to each other, and to fibrin, and contribute greatly to closing damaged vessels.

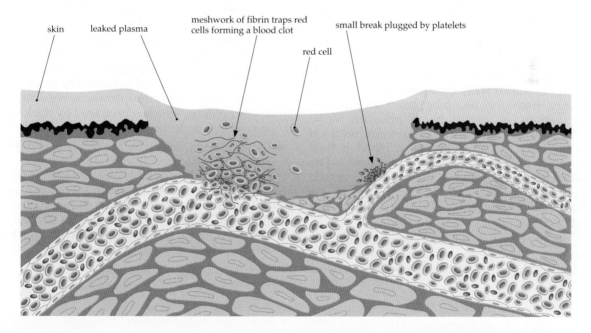

Fig. 21.23 Clot formation in a broken blood vessel. Platelets adhere to the collagen exposed in the damaged tissue and secrete their contents. Other platelets adhere to them and help close the wound. Thrombin is also produced and it converts fibrinogen to strands of fibrin, which entangle red blood cells, forming a clot

SUMMARY

- Circulatory systems are internal convective systems that transport materials between the environment and cells, and from one part of the body to another.

- In open systems, the circulated fluid is indistinguishable from interstitial fluid. In closed systems, the circulated fluid, blood, is enclosed in a system of vessels and is distinct from interstitial fluid. Closed systems favour more rapid flow rates and fluid with higher O_2-carrying capacity.

- Fishes have circulatory systems with a single circuit—heart, gills and body—with a low pressure systemic blood supply. In lungfishes, amphibians and reptiles, mixing of oxygenated and deoxygenated blood in the ventricle is regulated by septa, spongy heart muscle and/or valves. Complete separation of oxygenated and deoxygenated blood and high pressure systemic flow is found in birds and mammals.

- In vertebrates, cardiac contraction is myogenic, that is, initiated in the striated heart muscle itself. Excitation is conducted through the heart by a pathway of specialised cardiac muscle fibres, including the sinoatrial node (pacemaker) in the right atrium, the atrioventricular node, which delays conduction into the ventricles, and the atrioventricular bundle and Purkinje fibres, which conduct the signal rapidly throughout the ventricles.

- Arteries carry blood away from the heart. Their elastic, muscular walls store some of the pressure energy from the pumping heart and reduce fluctuations in arterial blood pressure. Veins collect the blood from the tissues and return it to the heart. At any time, most blood is held in the veins (capacitance vessels). The microcirculation consists of the small vessels lying between arteries and veins. Capillaries are the sites of exchange between blood and extracellular fluid.

- Net movement of fluid across the capillary wall occurs in response to the *net* filtration pressure, which is the result of balance between colloid osmotic and hydrostatic pressures of interstitial fluid and blood. Fluid, carrying dissolved materials, moves out of the arterial ends of capillaries and most of this fluid is reabsorbed back into the capillaries, carrying with it dissolved waste products from the cells.

- Fluid that is not reabsorbed into capillaries, and proteins that leak from capillaries, are returned to the circulation by way of the lymphatic system.

- Regulation of both cardiac output and regional blood flow is the result of both intrinsic and extrinsic mechanisms. Negative feedback control involves baroreceptors and chemoreceptors, integration in the vasomotor and cardiac centres of the brain, and sympathetic and parasympathetic outputs that control the timing and strength of the heartbeat and the state of contraction of muscles in vessel walls. Regulation is also affected by blood-borne hormones.

- Blood flow within a tissue is controlled by circular muscle fibres in arterioles and precapillary sphincters. These smooth muscles are responsive to both intrinsic (stretch, metabolites and other locally released vasoactive substances) and extrinsic mechanisms (effects of nerves and hormones).

- Vertebrate blood contains cells suspended in a fluid plasma. In mammals, haemoglobin-containing erythrocytes (red blood cells) occupy about 45% of the blood volume. The production of erythrocytes, erythropoiesis, is regulated by erythropoietin, which is produced by the kidney in response to decreases in O_2 levels of the blood.

- Blood clotting involves a cascade of enzyme-controlled reactions leading to the production of strands of the insoluble protein fibrin. Platelets attach to the exposed edges of a wound, to each other, and to fibrin. Under the influence of thrombin, platelets undergo a series of changes, assist in closing the wound and reducing blood flow in adjacent vessels.

QUESTIONS

1. Why are circulatory systems an inevitable consequence of the evolution of large animals?

2. Distinguish between open and closed circulatory systems.

3. Describe the sequence of electrical and mechanical events that occur during the cardiac cycle of a mammal.

4. Explain how heart sounds and an electrocardiogram can be useful in diagnosis of heart disease.

5. (a) Compare the functions of arteries, veins and capillaries. (b) What are the mechanisms of exchange of materials at the capillary level?

6. If an animal is wounded and loses blood, some of the lost blood volume is regained from the interstitial fluid. Describe the mechanism that underlies this shift in fluid balance.

7. How are blood pressure and flow regulated in mammals **(a)** in the circulatory system as a whole, and **(b)** at the level of individual tissues?

8. What is the haematocrit, and how is it regulated?

9. Describe the processes that prevent blood loss from a wound.

10. What circulatory changes occur **(a)** around the birth of a mammal or **(b)** during the hatching of a bird?

11. Describe the circulatory mechanism that prevents loss of heat from the feet of a penguin.

EXCRETION AND WATER BALANCE

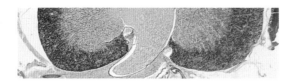

There is considerable exchange of water and solutes, including waste products, between an animal's body fluids and its environment. This occurs as a result of food and fluid intake, and respiratory, urinary and faecal losses. In most animals, it is essential for the normal functioning of cellular processes that the ionic and osmotic composition of body fluids is maintained relatively constant, despite the constant fluxes of water and solutes into and out of the animal.

Homeostasis is the maintenance of relatively constant internal conditions with respect to one or more variables, such as salt or water levels, body temperature or levels of oxygen in the blood. Homeostasis of the water and solute composition of body fluids is brought about by controlling the intake of substances from the environment and regulating the excretion of solutes and water. **Excretion** refers to the removal of substances that once formed part of the body. Excretion is not to be confused with **elimination**, which is the removal of unabsorbed food in faeces (Fig. 22.1). Unabsorbed food has never been part of the metabolic pool of the body.

Excretory systems generally produce a specialised fluid, **urine**, which is expelled outside of the animal.

In some animals, such as insects, reptiles and birds, urine is mixed with faeces in the hindgut before excretion. There are also other avenues for excretion. Carbon dioxide is excreted across respiratory surfaces (skin, gills or lungs; Chapter 20). Salts and water are excreted in sweat and other skin secretions. The gut also plays a role in excretion; for example, bile pigments produced in the liver as metabolic breakdown products of haemoglobin are secreted into the gut for excretion.

> Body fluid solutes and water are regulated by excretion, the removal of substances that once formed part of the body. Elimination voids unabsorbed food.

The excretory systems of animals contribute to the regulation of the internal environment in three main ways:

- maintenance of normal solute composition;
- control of normal body water content;
- excretion of metabolic waste products and unwanted substances that may have been inadvertently absorbed, such as dyes and toxins.

EXCHANGE OF WATER AND SOLUTES

Various ions and other solutes are invariably present in the body fluids of animals and are an essential part of the internal environment in which all biological processes occur. Life evolved in sea water and the solute composition of the body fluids of animals reflects this marine origin. The solutes found in body fluids include inorganic ions, such as sodium, potassium, chloride, calcium, magnesium, sulfate and phosphate, a variety of organic molecules, such as glucose, amino acids and proteins, and nitrogenous wastes, including ammonia, urea and uric acid. The concentrations of particular solutes in solution are usually expressed as molar concentrations, whereas

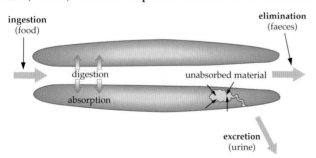

Fig. 22.1 Excretion is the discharge of water and solutes from the body fluids; these materials have been absorbed into the body pool of water and metabolites. Elimination is the discharge from the digestive tract of unabsorbed materials that have never been part of the body pool of metabolites. Note that there are other avenues for excretion (e.g. CO_2 by the lungs, sweat by the skin) and that some materials are secreted into the gut for excretion (e.g. bile pigments)

the total concentration of all dissolved solutes is expressed as the osmolar concentration (Box 22.1).

There is a continual exchange of solutes and water across the plasma membranes of animal cells (Chapter 4). Solutes may be passively exchanged by diffusion down a concentration gradient, and many are exchanged more rapidly or against a concentration gradient by active transport. Water molecules readily move through plasma membranes by osmosis (Chapter 4), the passive diffusion along an osmotic concentration gradient (from low to high osmotic concentration). There is no active transport pump for water molecules.

Occasionally, water moves across epithelia against a considerable osmotic gradient. This is not accomplished by active transport of water; there is no active 'water pump'. Movement is a passive osmotic consequence of active solute transport. For example, the blowfly *Calliphora* has rectal pads, which can withdraw water from rectal contents against a considerable osmotic gradient (Fig. 22.2). The epithelium of rectal pads actively reabsorbs ions from the rectal lumen and water is passively withdrawn by osmosis. Ions are then actively removed from the reabsorbed fluid, leaving it relatively dilute.

In aquatic animals, solutes are lost or gained by passive diffusion across the body surface, depending on the concentrations of the body fluids and external medium, and water is gained or lost by osmosis (Fig. 22.3). There may also be active transport of salts into or out of the body fluids, for example, across gills. In addition, water and solutes are gained by drinking and eating, and lost in urine and faeces. Water is also produced by cellular metabolism, but this is generally an insignificant source of water.

For a terrestrial animal, loss of water due to evaporation is an ongoing problem. Animals adapted to terrestrial environments therefore usually have outer body surfaces that are relatively impermeable to water. This limits the exchange of ions or water across the general body surface. However, water is lost across respiratory surfaces, and both water and solutes are gained across the gut wall and lost as urine and in faeces. There may also be loss of solutes and water by skin secretions, such as sweat in mammals.

Exchange of water and solutes occurs across many surfaces, including excretory organs, gut, lungs, gills and skin.

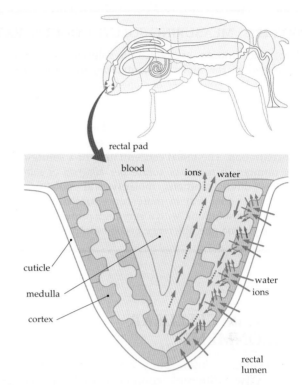

Fig. 22.2 The rectal pads of some insects are able to absorb a hypo-osmotic fluid from the rectal contents by a complex arrangement of active ion pumps. Ions are actively absorbed from the rectal lumen and water passively follows by osmosis; ions are subsequently removed from the absorbate by active transport,

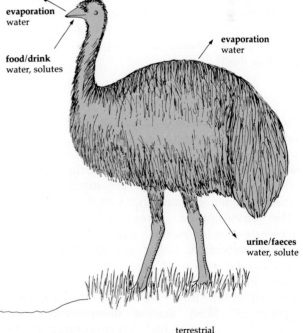

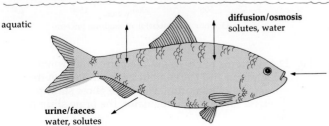

Fig. 22.3 The primary avenues for water and solute exchange by aquatic and terrestrial animals

---

BOX 22.1   MEASURING SOLUTE CONCENTRATIONS

The concentration of solutes in solution is usually expressed as the **molar concentration**, the number of moles of that solute per *litre of solution* (molarity, M). A mole of solute is 6.02×10^{23} molecules, or the number of grams of solute equal to its gram molecular weight. For low concentrations, either millimolar (mM) or micromolar (μM) units are more convenient than molar concentration. The **molal concentration** of a solution is the number of moles of solute per *kilogram of solvent*. In very dilute solutions, molarity and molality are practically equal, but in more concentrated solutions especially in solutions, with a very high protein content, such as plasma, the difference becomes important.

The total concentration of all osmotically active particles is the **osmotic concentration**. It is measured as osmols per litre (osmolarity, Osm) or osmolals per kilogram of water (osmolality), where an osmole is a mole of solute particles, regardless of the actual nature of the solutes. For solutes that dissociate when dissolved in water, the osmotic concentration is not the same as the molar concentration. For example, in water, sodium chloride (NaCl) dissociates into two osmotically active particles, the ions Na^+ and Cl^-. So when 1 mole of sodium chloride is dissolved to 1 L of solution, the molar concentration is 1 M NaCl but the osmotic concentration is 2 Osm.

THE EXTRACELLULAR ENVIRONMENT

In multicellular animals, body water and solutes are located in two different 'compartments': the *intracellular* compartment, consisting of fluid inside cells, and the *extracellular* compartment. The composition of the extracellular environment is typically different from the intracellular fluid (Chapter 4). However, in animal cells, the osmotic concentration of extracellular fluid is the same as intracellular fluid. In animals with a closed circulatory system (Chapter 21), extracellular fluid is further compartmentalised into the fluid inside the circulatory system (plasma) and the fluid outside the circulatory system and between the cells (interstitial fluid).

> Water and solutes in animals are distributed in intracellular and extracellular compartments.

The ion concentrations of extracellular fluids in some aquatic animals are similar to the ion concentrations in the external environment because these animals are unable to regulate internal ion concentrations. Such animals are **ionoconformers** because their ion concentrations conform to those of the environment. They include sponges, coelenterates, echinoderms and the hagfish, a primitive vertebrate. Other aquatic animals are **ionoregulators**, with mechanisms that enable them to maintain their extracellular ion concentrations at values different from those of the external environment. Most marine invertebrates are ionoregulators, although their ion concentrations do not differ greatly from those of sea water. Marine fishes and freshwater animals are also ionoregulators, maintaining extracellular ion concentrations that are considerably different from the surrounding water.

Animals also differ in relation to the osmotic concentrations of their extracellular body fluids. Some

aquatic animals simply allow the osmotic concentration of their extracellular fluids to follow that of their surroundings; these animals are **osmoconformers**. Most marine invertebrates, such as the ghost shrimp, *Callianassa*, are osmoconformers (Fig. 22.4). In contrast, many aquatic animals regulate the osmotic concentration of their extracellular fluids despite considerable variation in the osmotic concentration of the surrounding water; these animals are **osmoregulators**. The brine shrimp, *Artemia*, for example, osmoregulates its body fluids at about 300

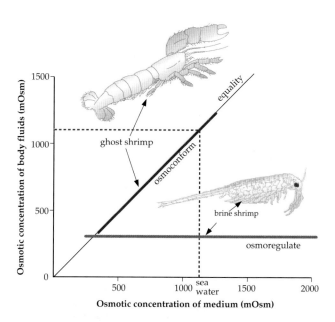

Fig. 22.4 Patterns of change in the body fluid osmotic concentration and the osmotic concentration of the medium, for an osmoconformer, the ghost shrimp, *Callianassa*, and an osmoregulator, the brine shrimp, *Artemia*

mOsm over a very wide range of external osmotic concentrations. The brine shrimp has salt pumps on its appendages, which pump NaCl out of the body fluids to regulate the ionic composition (see p. 479).

> Some aquatic animals are ionoconformers but most are ionoregulators. Some aquatic animals are osmoconformers, but many are osmoregulators.

PATTERNS OF OSMOREGULATION AND OSMOCONFORMATION

Marine animals

Life evolved in the oceans, and the ionic and osmotic characteristics of the first animals were compatible with sea water as the external medium. The osmotic concentration of modern marine protists and many invertebrates reflects this. Most marine protists and invertebrates are osmoconformers. The extracellular environment of marine protists and simple animals, such as sponges, is sea water itself, and for many invertebrate animals, the extracellular fluid is very similar to sea water. For example, the extracellular fluid of the horseshoe crab, *Limulus*, has about the same Na+, K+ and Cl- concentrations, and the same osmotic concentration, as sea water (Table 22.1). Only one vertebrate, the hagfish *Myxine*, is both an ionoconformer and an osmoconformer. The intracellular fluid osmoconforms to the extracellular fluid (or external environment) but invariably is ionoregulated at very different Na+, K+ and Cl- concentrations to the extracellular fluid (or external environment).

Most marine vertebrates regulate both the ionic concentrations and osmotic concentration of their body fluids. For example, the ion and total osmotic concentrations of marine bony fishes (teleosts) are about one-third those of sea water (Fig. 22.5). These fishes lose water by osmosis from their body fluids to sea water and continually gain ions by diffusion across the skin and gills. Water loss is replenished by drinking sea water, but this also increases salt gain. Ion pumps located on the gills excrete the excess salt, thereby maintaining a constant ionic and osmotic concentration of the body fluids.

It might seem likely that marine animals that are osmoconformers would also be ionoconformers and that those that are osmoregulators would also be ionoregulators. In many cases this is true, but a

Table 22.1 Comparison of the ionic and total osmotic concentrations of sea water, a marine protist, *Uronema*, and a marine crustacean, the horseshoe crab, *Limulus*

	Sea water	Marine protist (intracellular)	Horseshoe crab Intracellular	Horseshoe crab Extracellular
Na+ (mM)	480	84	29	445
K+ (mM)	10	134	129	12
Cl- (mM)	517	16	43	514
Osmotic concentration (mOsm)	950	950	953	953

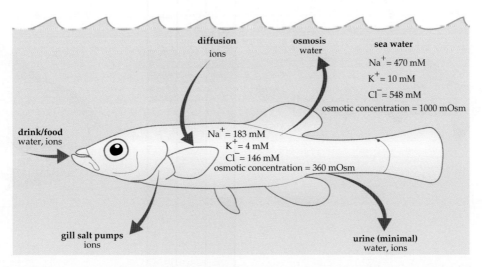

Fig. 22.5 A marine teleost is an ionoregulator and osmoregulator at a body fluid osmotic concentration of about 360 mOsm. Consequently, there is a constant influx of ions by diffusion and a loss of water by osmosis. Water is replaced by drinking and from food, and excess salts are removed by active gill pumps. There is some loss of water and ions by the production of small amounts of urine

number of marine osmoconformers are quite efficient ionoregulators. Many marine invertebrates regulate the concentration of some ions, for example, sulfate ions for buoyancy control, and are therefore capable of limited ionoregulation. Of the marine vertebrates, elasmobranchs, the coelacanth (a lobe-finned fish) and the crab-eating frog accumulate high levels of urea in their body fluids, thereby osmoconforming with sea water and avoiding the need to drink. Urea is used as a convenient osmotic solute to balance the osmotic concentration of their body fluids with that of sea water. However, these animals are ionoregulators of Na^+, K^+ and Cl^-, and thereby maintain the ionic composition of body fluids within the range encountered in other vertebrates, except hagfish (Box 22.2).

Most marine invertebrates, but only one vertebrate, the hagfish, are both ionoconformers and osmoconformers. Marine teleosts are ionoregulators and osmoregulators. Some marine animals are osmoconformers but also ionoregulators.

BOX 22.2   UREA IN MARINE IONOREGULATORS/OSMOCONFORMERS

A number of marine vertebrates are osmoconformers to the osmotic concentration of sea water, but regulate their extracellular ion levels considerably lower than those found in sea water. For example, chondrichthyean fishes (elasmobranchs—sharks, skates and rays—and the rat fishes) have almost the same osmotic concentration as sea water (1100 mOsm). The body fluid osmotic concentration is actually a little higher than the osmotic concentration of the sea water, and so some water diffuses into these fishes by osmosis.

The extracellular Na^+ and Cl^- concentrations of these chondrichthyean fishes are much lower than those of sea water. There is an 'osmotic gap' of about 500 mOsm between the total ion concentration and the osmotic concentration. These fishes synthesise and actively accumulate high concentrations of urea (which is normally a nitrogenous waste product of terrestrial animals) and lower concentrations of trimethylamine oxide as balancing osmolytes to fill the osmotic gap. Consequently, these fishes are said to be ureo-osmoconformers because urea is a major osmolyte that allows osmoconformation with sea water. The high urea concentrations of chondrichthyean fishes would be toxic to many other vertebrates, but their metabolic physiology is adapted to function normally with high urea levels.

The ureo-osmoconforming strategy of chondrichthyeans is also used by two other marine vertebrates. The coelacanth, *Latimeria chalumnae*, a relict fish species related to the evolutionary line leading to tetrapod vertebrates (Chapter 40), is also a ureo-osmoconformer. The crab-eating frog, *Rana cancrivora*, is an unusual amphibian because it can survive in sea water. Adult crab-eating frogs are ureo-osmoconformers. Their tadpoles also survive in sea water, but are not ureo-osmoconformers; rather, they are ionoregulators and osmoregulators, like marine teleost fishes. The crab-eating frog is not, however, completely marine; its eggs require freshwater to develop, and its tadpoles need freshwater to metamorphose successfully.

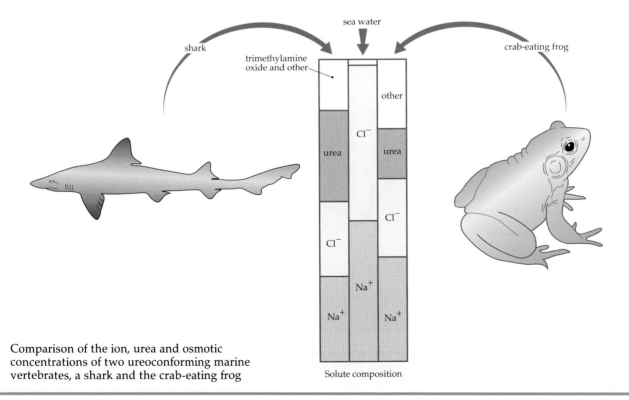

Comparison of the ion, urea and osmotic concentrations of two ureoconforming marine vertebrates, a shark and the crab-eating frog

Solute composition

Hypersaline animals

While many animals are able to live in sea water, only a select few can survive in **hypersaline** water. A number of invertebrates can live in hypersaline water, such as natural salt lakes formed by salt leaching from the soil and water evaporation, and the evaporation ponds of salt works. The brine shrimp, *Artemia*, can survive in water over five times as concentrated as sea water. Some teleost fishes can live in salt lakes, estuaries and pools, where sea water is concentrated up to two to three times by evaporation.

Animals living in hypersaline water *must* be ionoregulators and osmoregulators. An excessive build-up of ions in their extracellular body fluids would disrupt the normal cellular functions and, like marine teleost fishes, they constantly lose water by osmosis. Water loss is replenished by drinking the salty water. Salts that are ingested or diffuse through their body wall are actively excreted by salt pumps located on gills in fishes, and on the appendages of brine shrimps (see p. 479).

> Animals living in hypersaline water must be ionoregulators and osmoregulators.

Freshwater animals

Animals that live in freshwater gain water and lose salts to their environment. They *must* be ionoregulators and osmoregulators, because their cells require some ions and other solutes for normal function. However, many freshwater animals have remarkably dilute body fluids compared with marine animals. For example, the osmotic concentration of intracellular fluid of many freshwater protists is about 50 mOsm, and the intracellular and extracellular fluids of some freshwater animals are also very dilute (sponges, 55 mOsm; molluscs, 40–100 mOsm; worms, 100–200 mOsm).

Freshwater teleosts have about the same extracellular composition as marine teleosts, with a total osmotic concentration of about 300 mOsm (Fig. 22.6). These freshwater fishes gain water by osmosis and drinking, and surplus water is excreted as urine. Salts lost by diffusion across the skin and gills, and in urine, are replenished by ions absorbed by the gut and by active uptake of ions across the gills. The few freshwater elasmobranchs that have evolved are ionoregulators and osmoregulators, like freshwater teleosts. Unlike marine chondrichthyeans, they do not accumulate urea in their body fluids, excreting ammonia instead.

Terrestrial animals

Terrestrial animals invariably ionoregulate and the body fluid osmotic concentration is generally regulated at about 300–600 mOsm. The ion balance of terrestrial animals is relatively straightforward because there is no significant exchange across the body surface by diffusion or active transport. Ions are gained in food and drink and lost in urine and faeces.

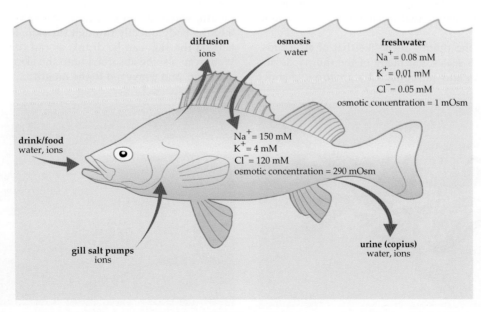

Fig. 22.6 Freshwater teleost fishes have higher body fluid ion and osmotic concentrations than does their freshwater medium. Water is gained by osmosis, by drinking and in food, and ions are lost by diffusion and in urine. Water is excreted by producing a copious amount of dilute urine. Ions are actively absorbed by gill salt pumps, and are obtained in food

Terrestrial animals lose water by evaporation from their body and lung surfaces, and in their urine and faeces (Fig. 22.3). This water loss is replenished by drinking, by water contained in food and by water produced through cellular metabolism (Chapter 5). For most terrestrial animals, drinking is the most important avenue of water gain. One lizard, the Australian thorny devil, *Moloch horridus*, has evolved a remarkable mechanism to facilitate drinking; its skin acts as a 'blotter' to absorb rain and dew and transfer the water to its mouth (Box 22.3). Water loss must be minimised because of the limited availability of water in the terrestrial environment. Consequently, terrestrial animals have physiological mechanisms and behaviours that minimise evaporative, urinary and faecal water loss.

Many desert-adapted animals are able to minimise their evaporative, urinary and faecal water losses to such an extent that they can survive by relying mainly on metabolic water production as their water source. For example, the Australian hopping mouse, *Notomys alexis*, a desert seed-eating rodent, does not have to drink, although it will drink if free water is available. Under dry conditions, the hopping mouse can obtain sufficient water from the seeds it eats and from metabolic water production to cover water losses (Table 22.2).

Terrestrial animals are ionoregulators and osmoregulators. Drinking is an important means of water gain, and water loss is minimised.

Table 22.2 Routes of water gain and loss for the Australian hopping mouse, *Notomys alexis*, which can survive without drinking

Avenue for water gain and loss	Amount of water (mL per day)
Water gain	
Drinking	0.0
Food	0.7
Metabolic water	0.9
Total gain	1.6
Water loss	
Evaporation	1.21
Urine	0.33
Faeces	0.06
Total loss	1.6
Net gain/loss	0

Some terrestrial arthropods, including insects, arachnids (ticks) and crustaceans (isopods), supplement their normal avenues for water gain (food

---

BOX 22.3   WATER UPTAKE BY THE THORNY DEVIL

Desert animals must take maximum advantage of the desert's infrequent rainfall. Many desert reptiles lick rainwater or fog condensate from their skin, but the Australian thorny devil lizard, *Moloch horridus*, has a remarkable skin texture that facilitates drinking of dew and rainwater.

The skin of the thorny devil, like that of all lizards, is covered with scales, but many of the thorny devil's scales are small and surrounded by a narrow channel

about 10 μm wide. These interscalar channels attract water by capillary action, and the skin absorbs water like a piece of blotting paper. The interconnected network of channels directs this water towards the mouth, where it can be drunk. Water is not absorbed into the body directly through the skin. Rainwater that falls on the skin can be drunk, as can condensed dew. Water can also be absorbed onto the skin from wet soil after rain, and conveyed to the mouth.

(a)   **(b)**

(a) The thorny devil has **(b)** numerous small scales, surrounded by narrow channels, which absorb water by capillarity

and drink) by absorbing water directly from unsaturated air. For example, some isopods can absorb water vapour from air if the relative humidity is greater than 93%. Mealworm larvae, *Tenebrio molitor*, can absorb water from air with a relative humidity of 88% or higher. Desert silverfish, *Thermobia*, can absorb water from air as dry as 45% relative humidity. A variety of mechanisms are used by terrestrial arthropods to absorb water vapour. Some secrete concentrated salt solutions, which are hygroscopic and absorb water vapour from air. Some couple water uptake with ion uptake across their rectum, and are able to absorb water vapour from air in their hindgut (p. 483).

Some terrestrial arthropods are able to absorb water vapour from unsaturated air.

EXCRETING WASTE PRODUCTS

Cellular metabolism produces a variety of waste products that must be removed. Water and carbon dioxide are the only waste products from the complete aerobic metabolism of carbohydrates and lipids. It is this metabolic water, which enters the general body water pool, that can be an important avenue of water gain for animals such as *Notomys*. Carbon dioxide is readily excreted across the body surface, gills or lungs.

The metabolism of protein forms **ammonia** (NH_3) by deamination of amino acids to ketoacids (Fig. 22.7). Ammonia is very toxic to animals, but it is also very soluble in water. Consequently, aquatic animals are able to excrete ammonia readily across their body or gill surface. The ammonia is diluted in the external water and thus does not accumulate to toxic levels.

If terrestrial animals excreted their nitrogenous waste as ammonia across their body surface, it would accumulate to harmful levels. For this reason, ammonia is usually converted to a less toxic form. Many terrestrial vertebrates, such as amphibians and mammals, and invertebrates, such as earthworms and pulmonate snails, convert ammonia to **urea** (CON_2H_4), a less toxic yet very soluble nitrogen waste product. A biochemical cycle of reactions, called the urea cycle, condenses ammonia with carbon dioxide to form urea in a process that requires energy. Some terrestrial animals, such as insects, reptiles and birds, convert the ammonia to an even less toxic and highly insoluble nitrogen waste product, **uric acid** ($C_5H_4O_3N_4$), by a series of metabolic reactions called uricogenesis. This conversion involves a number of specific enzymes and intermediate reactions, and an even greater expenditure of energy. A few terrestrial animals, such as spiders and scorpions, excrete **guanine** ($C_5H_5ON_5$), a nitrogenous waste product related to uric acid.

Ammonia, urea, uric acid and guanine are the primary nitrogenous waste products of animals.

The metabolism of nucleic acid bases is another source of nitrogenous wastes. In the case of pyrimidines (thymine, cytosine and uracil), ammonia is produced. Purines (adenine and guanine) are degraded by a complex series of reactions to guanine (spiders and scorpions), uric acid (terrestrial snails, reptiles and birds), or urea (many mammals). Many aquatic animals further degrade urea to ammonia for ready excretion into their aquatic environment.

There are relative advantages and disadvantages associated with each of these ways of disposing of nitrogenous waste. Ammonia is very soluble and is the usual nitrogenous waste in aquatic animals

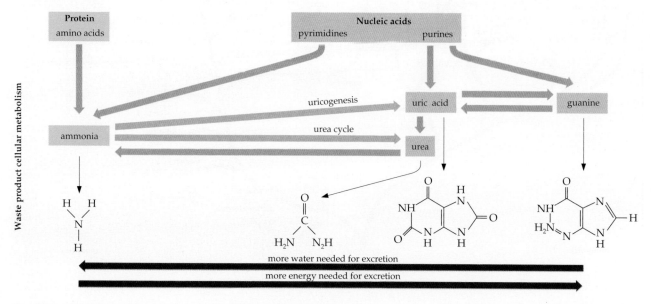

Fig. 22.7 Metabolic pathways for the formation of various nitrogenous waste products from protein and nucleic acid metabolism ('purple' arrows) and the synthesis of complex nitrogenous wastes from ammonia ('yellow' arrows)

because its toxicity is not a problem and no energy is expended in its synthesis. Urea is less toxic than ammonia and the excretion of nitrogen as urea requires less water than ammonia because each molecule of urea contains two nitrogen atoms (whereas an ammonia molecule contains only one nitrogen atom). However, the synthesis of urea from ammonia requires the expenditure of energy, so there is a metabolic cost with urea excretion. Uric acid excretion conserves even more water than urea excretion. Uric acid contains four nitrogen atoms per molecule; it is highly insoluble and non-toxic. However, its synthesis requires even more energy expenditure than urea synthesis.

A few terrestrial animals excrete the purine, guanine, which is almost insoluble and can be excreted as a semisolid paste with little water loss. Guanine is produced at considerable metabolic cost, but confers an advantage that can be useful in avoiding water stress. Purine waste products are present in urine, which is emptied into the hindgut,

where excess water can be reabsorbed. A pasty urine is then eliminated with the faeces.

The pattern of nitrogenous waste excretion seen in a particular animal group generally reflects the availability of water (Fig. 22.8). Not surprisingly, most aquatic animals excrete ammonia—**ammonotely**. A few aquatic animals excrete urea—**ureotely**. Elasmobranchs and the lobe-finned coelacanth use urea as a major body fluid solute for osmoregulation (Chapter 40). Only a few terrestrial animals excrete ammonia, often as a vapour across the body surface rather than in solution as urine. For example, some terrestrial isopods, insects and snails excrete gaseous ammonia across their body surfaces or via faeces.

In contrast, animals that live in dry environments often excrete urea or purines—**purinotely**. This is a consistent trend, apparent in both invertebrates and vertebrates (Fig. 22.8). Most terrestrial amphibians and mammals excrete urea. Lungfishes periodically enter a period of dormancy out of water, aestivation, during which they produce urea rather than ammonia

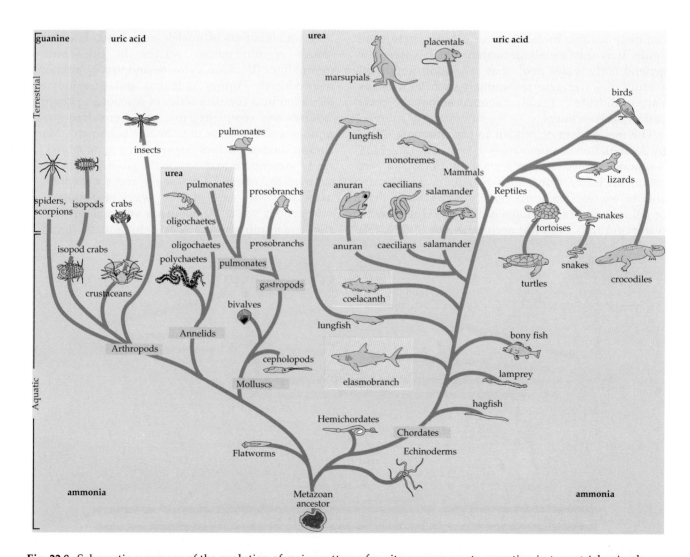

Fig. 22.8 Schematic summary of the evolution of major patterns for nitrogenous waste excretion in terrestrial animals

as nitrogenous waste. Terrestrial worms and some snails excrete urea. Most terrestrial insects and snails, reptiles and birds, and some crabs excrete uric acid—**uricotely**. Spiders and scorpions excrete guanine.

> Aquatic animals are generally ammonotelic, whereas few terrestrial animals are. There is a consistent trend towards ureotely or uricotely in terrestrial animals.

EXCRETORY SYSTEMS

There is considerable diversity in the structure of excretory organs, but they can be classified into two general types.

- Solute pumps, located at the surface of the animal, which transport specific solutes across the body surface.
- Internal tubular excretory organs, which form a liquid urine that is expelled to the exterior.

Surface epithelial pumps are highly specialised for a particular solute. They are important in regulating the exchange of a few specific ions, such as Na^+ and Cl^-, but are generally not used to eliminate a wide variety of different ions, organic solutes, metabolic waste products or water. Consequently, most animals have an internal tubular excretory organ that performs the functions of excretion: maintenance of normal solute composition of body fluids, regulation of the body water content, and excretion of nitrogenous wastes.

Some freshwater protists, such as the amoeba and paramecium, and most cells of simple freshwater sponges, excrete water and solutes by the action of an intracellular organelle, the **contractile vacuole**. These cells face a continual osmotic uptake of water, which must be excreted from the cell if it is to maintain a constant volume. A series of small tubules, the spongiome, collects intracellular fluid and delivers it to a spherical vesicle, which contracts and expels the fluid from the cell through a pore. The rate of contractile vacuole activity is inversely proportional to the osmotic concentration of the external medium; that is, water is excreted faster in dilute media.

> Contractile vacuoles excrete excess intracellular fluid to regulate cell volume.

Surface excretory pumps

Certain regions of the outer epithelial layer of many animals are specialised for active transport of solutes,

either into or out of the animal. For example, brine shrimp living in hypersaline solutions gain salts by diffusion and drinking. They have epithelial salt glands, which actively excrete Cl^-, and Na^+ moves passively along with the Cl^- to maintain electrical neutrality (Fig. 22.9).

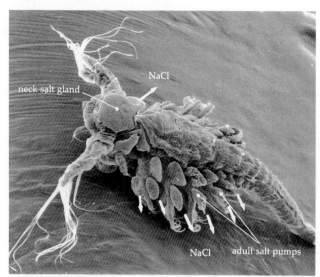

Fig. 22.9 The epithelial salt pumps of immature brine shrimp, *Artemia*, are located in a neck salt gland, whereas the salt pumps of adult brine shrimp are located on the appendages

Freshwater and marine teleosts also maintain salt balance using salt pumps located on their gills. These pumps actively excrete Cl^- (and Na^+ passively follows) in marine species, and actively absorb Cl^- (and Na^+ passively enters) in freshwater species. Amphibian skin actively transports Na^+ into the body fluids (and Cl^- passively follows).

If the movement of ions results in a change in the osmotic gradient across the epithelium, water will also tend to move across by osmosis. Thus, through active transport and regulation of permeability of an epithelium to water, Na^+, Cl^- and water can be passively transported across a surface.

> Surface epithelial salt pumps excrete or absorb specific ions for ionoregulation.

Tubular excretory organs

Surface epithelial pumps regulate the exchange of a few specific ions and are not used to eliminate a wide variety of substances. This is usually carried out by internal tubular excretory organs, which have a number of excretory roles (Fig. 22.10). There are three basic types of tubular excretory organs in animals—nephridia, coelomoducts and Malpighian tubules (Fig. 22.11). All excretory tubules form a fluid filtrate at their inner end by **filtration** of coelomic fluid or blood,

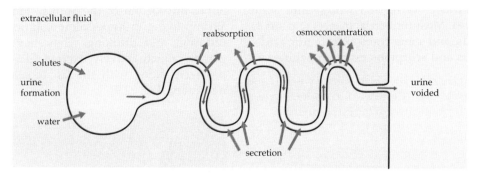

Fig. 22.10 Tubular excretory organs form urine, subsequently modify it by reabsorption and secretion, and then void it to the exterior

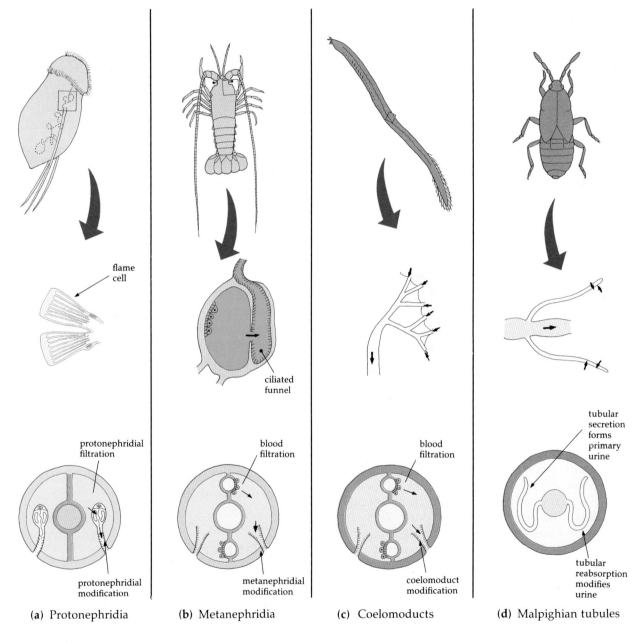

(a) Protonephridia   (b) Metanephridia   (c) Coelomoducts   (d) Malpighian tubules

Fig. 22.11 Different forms of tubular excretory organs. **(a)** Protonephridia have a terminal flame cell, which transports fluid from the coelomic cavity into the tubule by ciliary action. **(b)** Metanephridia have a ciliated funnel that transports coelomic fluid into the tubule. **(c)** Coelomoducts are similar to metanephridia in structure and function. **(d)** Malpighian tubules of arthropods form a filtrate by the active secretion of K^+ into the tubule; the tubular fluid is emptied into the hindgut

or by active transport of potassium ions (K^+), leading to passive solute and water flow. As this fluid passes along the excretory tubule, its composition may be substantially modified by reabsorption of solutes and water, or by solute secretion. Active **reabsorption** of useful solutes and water is important because the initial fluid has essentially the same solute composition as blood plasma, and contains many useful ions, organic solutes and vitamins. Active **secretion** of specific waste products into the urine occurs. In some animals, to conserve water, the excretory tubules also **osmoconcentrate** urine by the active addition of solutes and/or osmotic withdrawal of water.

> Excretion of a wide variety of substances is carried out by tubular excretory organs, which form, modify and expel urine.

Nephridia

Nephridia develop as ingrowths from the body surface towards the body cavity (Fig. 22.11). They may be blind-ended (protonephridia) or may terminate at a ciliated opening, the nephridiostome (metanephridia). Protonephridia are present in many simple invertebrates, for example, rotifers. The end cells of the blind-ended protonephridia have numerous cilia, which draw fluid into the tubule via perforations in its wall. The continual beating motions of the cilia resemble the flickering of a flame, hence the common name 'flame cell' for these terminal cells. The terminal cells also form the perforations that allow coelomic fluid to be drawn into the tubule lumen by a negative hydrostatic pressure produced by the ciliary action.

Metanephridia are found in more complex invertebrates. The metanephridium of crustaceans, for example, has a terminal ciliated opening, the nephridiostome, that is continuous with the coelomic space. Coelomic fluid is drawn into the metanephridial lumen by ciliary action. The coelomic fluid is formed elsewhere in the coelom, by filtration of fluid out of blood vessels.

Coelomoducts

Coelomoducts are present in many invertebrates, and also primitive vertebrates, such as the hagfish. Nephrons of the vertebrate kidney are derived from coelomoducts. In contrast to nephridia, coelomoducts develop outwards, from the coelomic lining towards the external body surface, but they resemble metanephridia in general structure. Coelomic fluid is drawn into a ciliated funnel-like opening, the coelomostome, then passes along the tubule and is eliminated from the coelomopore. Coelomic fluid is formed elsewhere in the coelom, by filtration of fluid from blood vessels. In advanced vertebrates, the roles of the fluid-forming blood vessels and the coelomoducts are combined into a single structure, consisting of a vascular glomerulus and the nephron tubule, which is the functional unit of the kidney.

Malpighian tubules

Many arthropods, such as insects and some spiders, have **Malpighian tubules**. These are linear or branched blind-ended tubules that open into the digestive tract at the junction of the midgut and hindgut. Fluid and solutes are drawn into the blind ends by the active transport of K^+ from the extracellular fluid into the tubule; water and other solutes passively follow K^+ into the tubule. The 'urine' formed by the Malpighian tubules is emptied into the hindgut, where it is subsequently modified by the rectal epithelium, then excreted with the faeces.

EXCRETION IN INVERTEBRATES

Protists and primitive animals, such as freshwater sponges, rely on ion pumps for ionoregulation and contractile vacuoles for volume regulation by excretion of excess intracellular fluid. Multicellular invertebrates, such as flatworms, nemerteans, pseudocoelomate worms and some annelids, have protonephridia. These collect body fluid to form a primary filtrate, which is modified to form urine. For example, a freshwater rotifer, *Asplanchna*, excretes urine (40 mOsm) that is more dilute than its body fluids (80 mOsm) but more concentrated than lake water (18 mOsm). The larvae of many higher invertebrates also have protonephridia.

Many of the more complex invertebrates have metanephridia rather than protonephridia, particularly as adults. For example, annelid worms have metanephridia and/or coelomoducts, which excrete a very dilute urine. Coelomic fluid is filtered from one body segment of earthworms into the metanephridium of the more posterior segment (Fig. 22.12). The osmotic concentration of the urine declines as it passes through the tubule because of solute reabsorption (mainly active Na^+ and passive Cl^- reabsorption).

The metanephridial excretory organs of crustaceans are **antennal glands** (or green glands). These paired structures of the head consist of a coelomic end sac, which forms the urine by filtration of fluid from blood vessels; a labyrinth chamber and excretory tubule which reabsorb solutes and form a dilute urine; a bladder, which stores the urine and further reabsorbs solutes to dilute the urine; and an excretory pore through which the dilute urine is excreted (Fig. 22.13).

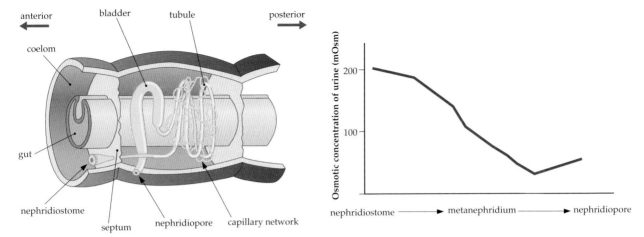

Fig. 22.12 The excretory system of the earthworm, *Lumbricus*, consists of paired segmentally arranged metanephridia, which collect coelomic fluid from the more anterior body segment, modify the urine by active solute reabsorption, then void it to the exterior via the nephridiopore. The osmotic concentration of the urine decreases as it passes along the tubule because of active solute reabsorption

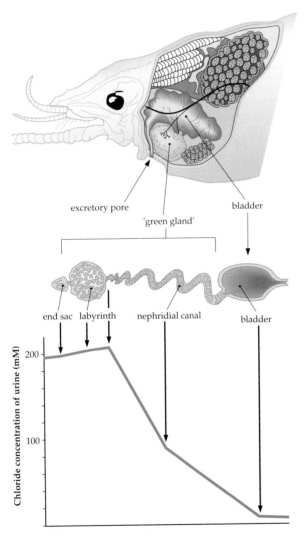

Fig. 22.13 The crustacean antennal gland is a pair of metanephridia, each with a coelomosac that forms urine by filtration. The urine is modified as it passes along the tubule by active solute reabsorption, and the osmotic concentration declines

Many arthropods have both metanephridial excretory tubules, **coxal glands**, and Malpighian tubules (Fig. 22.14). For example, primitive spiders have two pairs of coxal glands, consisting of a thin-walled spherical sac that collects haemolymph, an excretory tubule that modifies the urine, and an excretory pore on the coxal part of the leg. More advanced spiders rely primarily on two branched Malpighian tubules for excretion.

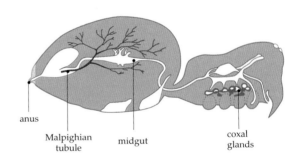

Fig. 22.14 Primitive spiders have two pairs of metanephridial coxal glands and branching Malpighian tubules

Insects have a variable number of Malpighian tubules for excretion. The urine is formed in the blind ends of the tubules and is modified as it passes along the tubules by the reabsorption of ions and the precipitation of uric acid. The urine empties into the hindgut, where it is further modified in the rectum by reabsorption of ions, other solutes, such as glucose and amino acids, and water to form a very dry urate waste.

The rectal reabsorption of water is coupled in some insects with active pumping of K^+ into the blind ends of the Malpighian tubules (Fig. 22.15). This creates a concentration gradient that assists in the osmotic removal of water from the rectal contents, thus

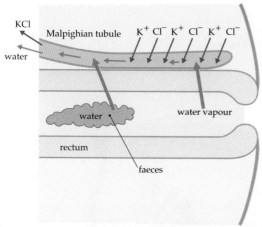

Fig. 22.15 The hindgut of the mealworm larva, *Tenebrio molitor*, fuses with the blind ends of the Malpighian tubules to form a complex structure, the cryptonephridial system. This system reabsorbs water from the faeces in the hindgut, and can absorb water vapour from air inside the hindgut at relative humidities greater than 88%

forming very dry faeces. This mechanism also allows the absorption of water vapour from air in the rectum, down to a relative humidity of about 88%.

> Protists and freshwater sponges have ion pumps for ionoregulation, and contractile vacuoles for water regulation. Primitive multicellular animals use protonephridia for excretion. More complex invertebrates have metanephridia, coelomoducts or Malpighian tubules for excretion.

EXCRETION IN VERTEBRATES

The primary excretory organ of vertebrates is the **kidney**, but various other organs can be involved with solute and water regulation, including the skin, gills, gut and salt glands.

Aquatic vertebrates excrete ions and water using gill salt pumps and their kidney. Freshwater fishes and amphibians are in constant danger of becoming overhydrated by osmotic influx of water from the dilute medium, and hypoionic by diffusional loss of ions. Their kidneys excrete the excess water as a copious flow of dilute urine, and salt pumps on the gills (fish) or skin (fishes and amphibians) absorb salts from the dilute medium. The kidneys of freshwater fishes are relatively large, reflecting their need to have a high urine flow. In contrast, marine teleost fishes gain ions by diffusion; these ions are excreted by gill or skin salt pumps. They lose water by osmosis and as urine, although the urinary water loss is low because the kidneys are small and they often lack glomeruli (urine is formed by secretion, not filtration). Marine elasmobranchs and coelacanths excrete ions via their kidneys and a special rectal gland located in the posterior gut.

Terrestrial vertebrates rely primarily on their kidneys for ionoregulation and osmoregulation, although some reptiles and birds have salt glands for ion excretion.

> The principal excretory organ of vertebrates is the kidney.

Kidneys

Vertebrate kidneys are paired organs that range in shape from compact ovoid structures to elongated more diffuse organs, especially in fishes (Fig. 22.16). Each kidney consists of thousands, or even millions, of individual nephrons. The nephrons of primitive vertebrates, such as hagfish, are coelomoducts, having a ciliated funnel that collects coelomic fluid into the nephron tubule. The nephrons of more advanced vertebrates are structurally modified so that the urine formed by the nephron is not collected from the coelomic space, but is formed by filtration of blood from a tuft of capillaries, the **glomerulus**. Some of the nephrons of amphibians have both ciliated coelomostomes and glomeruli (Fig. 22.16).

In glomerular nephrons, the proximal end of the nephron is dilated and invaginated to form a hollow double-walled cup; the outer layer of cells forms the renal or Bowman's capsule (Fig. 22.17). The renal capsule envelopes the tuft of glomerular capillaries and urine is formed by glomerular nephrons through filtration of glomerular blood. Each glomerular capillary is covered by podocyte cells, which are highly modified inner cells of the capsule (Fig. 22.17). The capillary walls and podocytes form the filtration membrane.

The glomerular filtrate is subsequently modified by reabsorption and secretion. In nephrons of mammals and some birds, the tubular fluid is also osmotically concentrated. The nephron tubule, which extends from the renal capsule, consists of an initial proximal convoluted tubule and a subsequent distal convoluted tubule. The terminal part of the tubule, the collecting duct, conveys fluid from the distal convoluted tubule to the ureter. The amphibian nephron well illustrates the structure of the nephron tubule (Fig. 22.17). Nephrons of fishes lie in a straight line and have an intermediate segment located between the proximal and distal convoluted tubules. In mammalian and some bird nephrons, the intermediate segment turns a hairpin loop, the **loop of Henle**, resulting in descending and ascending segments lying parallel and close together. This arrangement greatly improves the kidney's ability to reabsorb water from urine and is extremely important for terrestrial mammals living in arid environments.

> The vertebrate nephron has a renal capsule, which encloses the glomerulus, proximal and distal convoluted tubules, and a collecting duct.

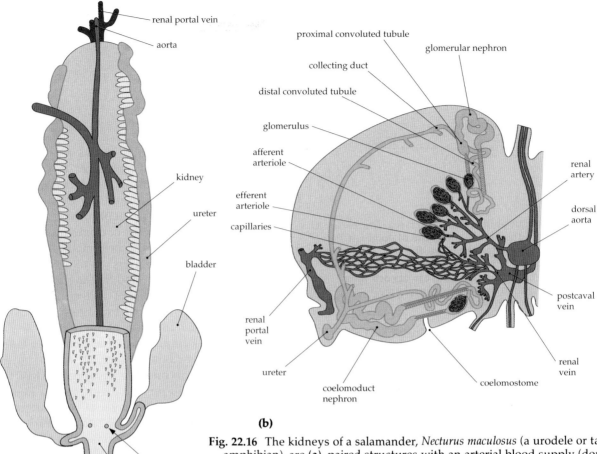

Fig. 22.16 (a) The kidneys of a salamander, *Necturus maculosus* (a urodele or tailed amphibian), are (a) paired structures with an arterial blood supply (dorsal aorta and renal arteries), a venous portal blood supply (renal portal veins) and venous drainage (renal veins and postcaval vein). (b) The kidneys contain two types of glomerular nephrons, those with a coelomostome and glomerular nephrons, typical of tetrapods that lack a coelomostome

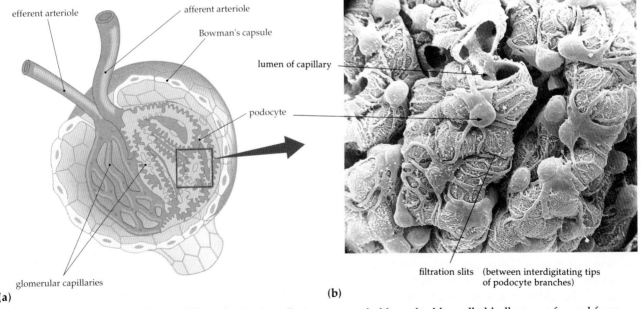

Fig. 22.17 (a) The glomerulus is a spherical tuft of capillaries surrounded by a double-walled hollow cup formed from the terminus of the nephron tubule. Bowman's capsule is formed from the outer layer of capsule cells. (b) The inner layer of capsule is formed from highly modified podocytes, which support the glomerular capillaries and form filtration slits, as shown in this SEM of the glomerulus of a rat kidney

Filtration

Glomerular filtrate is formed by the hydrostatic filtration of fluid from the glomerular capillaries into the renal capsule. Glomerular capillaries are perforated by numerous tiny holes (fenestrae) that allow ready loss of fluid and solutes from the capillaries. The podocytes of the renal capsule have numerous branches (pedicels); the interdigitated pedicels of neighbouring podocytes support the glomerular capillaries and form filtration slits (Fig. 22.17). The capillary fenestrae, the basement membrane of the capillary, and the filtration slits form the filtration membrane, across which blood plasma is forced by the blood hydrostatic pressure to form the glomerular filtrate.

Blood pressure in glomerular capillaries is relatively high, similar to arterial blood pressure, because the efferent arteriole is narrower than the afferent arteriole. This high pressure forces fluid through the glomerular filtration membrane. The larger blood constituents, for example, blood cells (8–1000 μm diameter) and proteins (molecular weight greater than 60 000 daltons), are retained in the capillaries; otherwise, glomerular filtrate has essentially the same composition as blood.

A complex balance of forces determines the overall filtration pressure that forms glomerular filtrate. The main force causing filtration is the high blood pressure within the glomerular capillaries. There is also a hydrostatic pressure in the Bowman's capsule, as a result of filtered fluid entering the capsule, which is required to force fluid along the nephron tubule. The intracapsular hydrostatic force is considerably lower than that of the glomerular capillaries. In addition, the difference in protein concentration between the glomerular capillary blood and the glomerular filtrate also affects filtration because proteins are essentially impermeable to the filtration membrane. The high protein concentration of glomerular blood effectively creates a 'suction pressure' (blood colloid osmotic pressure), which tends to draw water into the capillaries by osmosis. The glomerular filtrate also has a colloid osmotic pressure tending to draw water from blood into the Bowman's capsule but, because of the low protein concentration of the filtrate, glomerular filtrate colloid osmotic pressure is negligible. The overall filtration pressure due to the balance of these forces is:

$$\left(\begin{array}{c}\text{glomerular}\\\text{hydrostatic}\\\text{pressure}\end{array}+\begin{array}{c}\text{filtrate}\\\text{colloid}\\\text{osmotic}\\\text{pressure}\end{array}\right)-\left(\begin{array}{c}\text{capsule}\\\text{hydrostatic}\\\text{pressure}\end{array}+\begin{array}{c}\text{blood}\\\text{colloid}\\\text{osmotic}\\\text{pressure}\end{array}\right)$$

In humans, for example, the balance of forces is an overall filtration pressure of about 10 mmHg (Fig. 22.18).

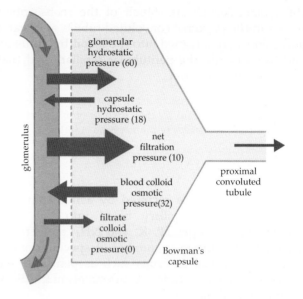

Fig. 22.18 The net filtration pressure that filters fluid from the glomerular capillaries into Bowman's capsule is the sum of the hydrostatic and colloid osmotic pressures acting across the filtration membrane. The approximate pressures (in mmHg) for the human nephron are shown

Formation of the glomerular filtrate is due to the balance of hydrostatic and colloid osmotic forces across Bowman's capsule.

Reabsorption

The rate of glomerular filtration is much higher than the rate of urine excretion. For example, in a resting human, about 1200 mL of blood (about one-quarter of the cardiac output) enter the kidneys per minute; renal plasma flow is about 60% of blood flow, that is, 720 mL min⁻¹. About one-fifth of the renal plasma flow is filtered to form glomerular filtrate; the glomerular filtration rate is about 125 mL min⁻¹ (or 180 L per day)! Obviously most of this glomerular filtrate is reabsorbed. Urine flow rate is actually less than 1% of the glomerular filtration rate, about 1 mL min⁻¹; over 99% of the glomerular filtrate, including most of its solutes, is reabsorbed. Each day, the human kidney reabsorbs about 178 L of water, 1200 g of salt, and 250 g of glucose. The kidney has a high metabolic rate relative to its weight, reflecting its active absorption and secretion of solutes. It has a low metabolic rate relative to its blood flow, because the high blood flow is required for glomerular filtration, not just the supply of nutrients and removal of waste products.

Reabsorption allows the recovery of important ions and nutrients, such as glucose and vitamins. The chemical composition of body fluids is precisely regulated by the control of solute reabsorption from the glomerular filtrate. Much of the reabsorption occurs in the proximal convoluted tubule, and water and solutes are returned into the blood by diffusion and osmosis into the peritubular capillary bed that surrounds the nephron.

| Most of the glomerular filtrate, particularly water and important body fluid solutes, such as amino acids and glucose, are absorbed by the nephron tubules.

Secretion

Some substances are secreted by active transport from peritubular capillary blood into the renal tubules. For example, H^+, K^+ and NH_4^+ are actively secreted. In addition, a number of specific organic molecules, as well as drugs such as penicillin, are actively secreted. Secretion occurs mainly in the distal convoluted tubule.

Marine fishes typically have small kidneys and their nephrons are often modified to minimise urine flow. For example, the glomeruli are often small to minimise the urine production. The nephrons of some marine fishes are **aglomerular**; they lack glomeruli altogether. Urine is formed in these aglomerular kidneys only by the active secretion of solutes, followed by the passive osmotic inflow of water and other solutes into the tubules (as in insect Malpighian tubules).

Secretion of H^+ is an important mechanism for the regulation of blood and urine pH. Carbon dioxide in the epithelial cells of the distal convoluted tubule combines with water in the presence of carbonic anhydrase to form hydrogen carbonate and H^+, which is actively transported into the renal tubule in exchange for Na^+. The excretion of H^+ in this manner is made more effective by the presence of buffers, such as HPO_4^{2-} and $H_2PO_4^-$ in the renal tubular fluid. In addition, the H^+ combines with NH_3 to form NH_4^+, thus 'trapping' the H^+ in the urine. By buffering and trapping H^+ in the urine, excess metabolic H^+ can be excreted by the urine at only moderately low pH.

| Ions such as H^+, K^+ and NH_4^+ are actively secreted by the nephron.

Osmoconcentration

The kidneys of mammals and some birds are able to form a hyperosmotic urine that has a higher osmotic concentration than body fluids. Many mammals, in particular, are able to excrete very concentrated urine, as high as 5000–9000 mOsm. This is important since terrestrial animals rely on food and fluid intake, and metabolism, for water gain. The most important of these, drinking, is often not possible. For example, because of infrequent rainfall, many desert animals are almost never able to drink water. It is therefore essential for these animals to minimise

their urinary water losses, while at the same time maintaining excretion of their solute wastes.

Mammalian kidneys have two types of nephrons. *Cortical nephrons* are located near the outer cortex of the kidney. *Juxtamedullary nephrons* are located nearer the inner part of the kidney, the medulla, and have a long loop of Henle that penetrates to the inner medulla (Fig. 22.19). Cortical nephrons are important for filtration, reabsorption and secretion, but not for osmotic concentration of the urine. In addition to these processes, juxtamedullary nephrons also provide the mechanism for the osmotic concentration of urine. Since active removal of water is not possible, urine osmoconcentration must utilise active transport of solutes and passive osmotic removal of water (Chapter 4). The nephrons of the vertebrate kidney can osmotically reabsorb water from urine to make it hyperosmotic to body fluids. The epithelium of insect rectal pads can also osmotically reabsorb water from a hyperosmotic solution, although by a very different mechanism (Fig. 22.2).

| Mammals and some birds can osmotically concentrate their urine using juxtamedullary nephrons, which have long loops of Henle.

The long loops of Henle of juxtamedullary nephrons descend into the renal medulla, making a sharp hairpin bend at the tip of the medulla, and ascending back into the cortex. The loop of Henle establishes an osmotic concentration gradient in the renal medulla by the countercurrent exchange of solutes and water between the descending and ascending limbs. There are marked structural and functional differences between the thin descending limb of the loop of Henle and the thick ascending limb.

The ascending portion of the loop of Henle has a relatively thick epithelium and actively transports Cl^- out of the nephron. Na^+ passively follows the Cl^- out of the nephron for electrical neutrality, but water is not able to follow because the tubule wall is impermeable to water. The effect of these Cl^- pumps is therefore to increase the solute concentration outside the ascending limb of the loop of Henle and to dilute the fluid remaining inside the ascending limb, as it passes along the tubule. In contrast, the thin descending limb lacks Cl^- pumps and is permeable to water. Because it runs in close contact with the ascending limb, but in the opposite (counter) direction, there is countercurrent exchange between the descending and ascending limbs. As fluid passes through the descending limb, it encounters regions of greater osmotic concentration because of the solutes pumped out of the ascending limb. Consequently, water is drawn by osmosis out of the descending limb and some solutes may enter. The osmotic concentration of fluid in the tubule increases

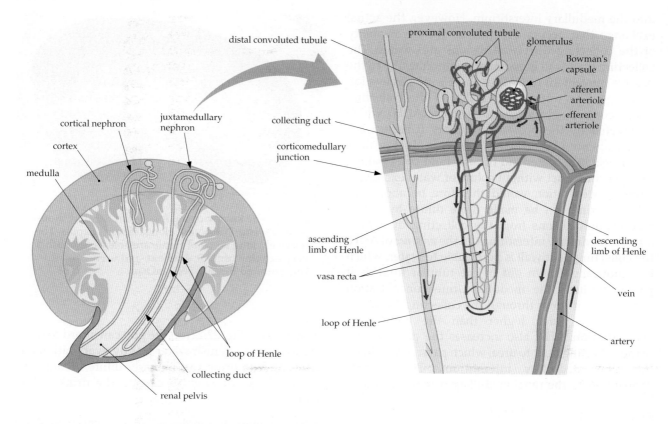

Fig. 22.19 The kidney of a rodent illustrates the arrangement of cortical and juxtamedullary nephrons in the renal cortex and medulla, and the structure of the mammalian nephron

as it passes down the descending limb and becomes progressively more dilute as it passes along the ascending limb. A countercurrent arrangement of the medullary capillaries (the vasa recta) prevents breakdown of this renal medullary osmotic gradient by the blood supply.

> The role of the loop of Henle is to establish an osmotic concentration gradient in the renal medulla.

The net effect of Cl$^-$ pumps and countercurrent exchange in the loops of Henle is to establish an osmotic gradient in extracellular and intracellular fluids of the renal medulla. This gradient increases from normal body fluid concentration at the cortex–medulla boundary to a high osmotic concentration at the renal medulla tip, where the loops of Henle make their hairpin bends (Fig. 22.20). However, the fluid leaving the ascending tubule has a lower osmotic concentration than the normal body fluids.

Osmoconcentration of urine occurs as fluid flows through the collecting ducts, which pass through the osmotic concentration gradient of the renal medulla to their openings into the renal pelvis. In the medulla, the osmotic gradient between the medullary extracellular fluid and the collecting duct fluid provides the potential for ions to diffuse into the collecting duct and for water to move by osmosis

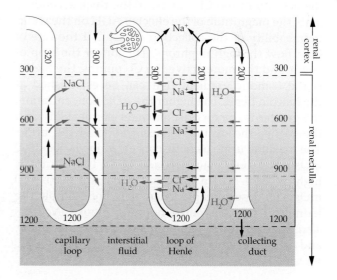

Fig. 22.20 The ascending limb of the loop of Henle actively transports Cl$^-$ out of the tubule. The descending limb is permeable to water and solutes. Countercurrent exchange of water and solutes between the ascending and descending limbs establishes an osmotic concentration gradient in the renal medulla. This osmotic gradient is used to osmoconcentrate urine as it passes through the collecting duct. Blood in the blood vessels to this part of the nephron also is osmotically concentrated, by passive equilibration of the blood with the medullary osmotic gradient. Active transport is shown by red arrows, passive transport by blue arrows

into the medullary interstitium. However, the actual exchange that occurs depends on the permeability of the collecting duct wall to ions and water. The collecting duct wall is essentially impermeable to the passive movement of ions, so there is no passive ion exchange. The water permeability of the collecting duct is controlled by *antidiuretic hormone* (ADH), which is secreted by the posterior pituitary (Chapter 25). In the *absence* of ADH, the collecting duct wall is impermeable to water and tubular fluid passes along the collecting duct without solute or water exchange. Urine concentration would therefore be about 200 mOsm. Urine can be made more dilute by active reabsorption of ions from the collecting duct fluid into the medullary interstitium. In the *presence* of ADH, the collecting duct wall is permeable to water, which is osmotically drawn out of the tubular fluid as it passes through the medulla. Urine concentration increases as it flows through the collecting duct, and can become much higher than the body fluid concentration. ADH also increases the permeability of the collecting duct to urea, which passively diffuses into the medullary interstitium. This further contributes to the renal medullary osmotic gradient.

Osmotic concentration of urine occurs in the collecting ducts.

The degree to which urine can be osmotically concentrated depends on the length of the loop of Henle, the activity of the Cl^- pumps of the thick ascending limb, the magnitude of the effect of ADH on the water permeability of the collecting ducts, and the rate of fluid flow through nephrons. The longer the loop of Henle and the more active the Cl^- pumps, the higher is the osmotic gradient that can be established at the hairpin bend. The higher the ADH concentration, the greater is the increase in water permeability and the higher the osmotic concentration of the urine. The maximal urine osmotic concentration is 500–1000 mOsm for some mammals and birds, but can be as high as 2000 mOsm for some birds and 8000–9000 mOsm for desert-adapted mammals. For example, the kidney of the Australian hopping mouse, *Notomys alexis*, has relatively long loops of Henle (Fig. 22.21), and it can excrete urine as concentrated as 9000 mOsm.

Salt glands

Some crocodiles, turtles, snakes, lizards and birds have **salt glands**. These salt glands excrete a salt solution of either NaCl or KCl, which is more concentrated than their body fluids. For example, the herring gull has a salt gland in the orbit of the eye that can secrete a solution of primarily NaCl at about 1600 mOsm. This is more concentrated than sea water

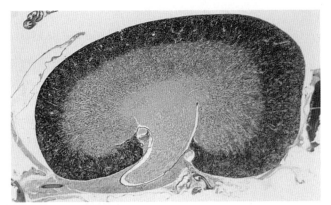

Fig. 22.21 The kidney of the Australian hopping mouse, *Notomys alexis*, has a very elongate renal papilla. The relatively long loops of Henle are able to establish a very extreme osmotic gradient, and urine can be concentrated to 8000–9000 mOsm

(1100 mOsm; Fig. 22.22) so the gull can afford to drink sea water to replenish water lost by evaporation and urine, and its urine can be fairly dilute because the kidney does not have to excrete the drinking salt load.

Some reptiles and birds can excrete concentrated salt solutions using salt glands.

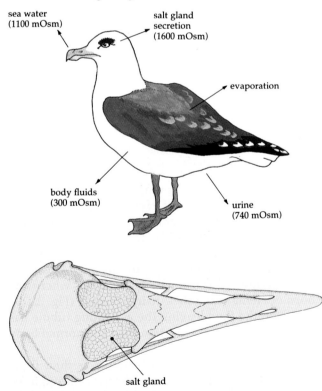

Fig. 22.22 The orbital salt gland of the herring gull, *Larus*, excretes a concentrated NaCl solution (up to 1600 mOsm), which is more concentrated than sea water (1000 mOsm). This enables the gull to drink sea water to replenish the water lost in urine and faeces, and by evaporation

SUMMARY

- The excretory systems of animals are responsible for regulating the water and solute composition of body fluids, and removing waste products.

- Metabolism of protein and nucleic acids produces nitrogenous solutes that are usually excreted. Ammonia is the nitrogen waste for most aquatic animals. Urea, uric acid and guanine are the nitrogenous wastes commonly produced by terrestrial animals.

- Water exchange between an aquatic animal and its environment occurs by osmosis through the body and gut walls, by evaporation, and by urine and faecal loss. Ion exchange occurs by diffusion and active transport across the body and gut walls, and by urine and faecal loss. For terrestrial animals, drinking, food consumption, urine and faecal losses are the primary avenues for water and ion exchange; considerable amounts of water are often lost by evaporation.

- Most marine invertebrates and one vertebrate, the hagfish, have similar extracellular ion and osmotic concentrations to sea water. All other marine vertebrates are ionoregulators; many also are osmoregulators. Freshwater animals invariably are ionoregulators and osmo-regulators. Terrestrial animals gain water and ions by drinking and in their food, and lose water and ions in urine and faeces, and water also by evaporation.

- Many freshwater protists and cells of freshwater sponges have contractile vacuoles for excretion of intracellular fluid. There are two general types of excretory organs in animals: surface epithelia with ion pumps, and tubular excretory organs, such as nephridia, coelomoducts, nephrons and Malpighian tubules. Tubular excretory organs form urine, and subsequently modify its composition by reabsorption and secretion. Nephrons of mammals and some birds can also osmotically concentrate urine.

- Excretory organs of vertebrates include kidneys, skin, gills and salt glands. The functional unit of the kidney of higher vertebrates consists of a glomerulus capillary tuft and a nephron tubule. The balance of hydrostatic and colloid osmotic pressures between the glomerulus and Bowman's capsule forms a glomerular filtrate, which is modified as it passes through the nephron by reabsorption of ions, other solutes and water, and by secretion of specific wastes.

- The juxtamedullary nephrons of mammalian and some bird kidneys have long loops of Henle. The loop of Henle establishes an osmotic gradient in the medulla of the kidney. This osmotic gradient allows the osmotic concentration of urine as it flows through the collecting ducts. The water permeability of the collecting ducts and therefore the concentration of urine is controlled by antidiuretic hormone.

- Some reptiles and birds have salt glands, which excrete highly concentrated salt solutions.

QUESTIONS

1. Compare the advantages of ammonia, urea and uric acid as nitrogenous waste products in (a) aquatic animals and (b) terrestrial animals.

2. What are the principal differences in the solute concentrations of intracellular and extracellular fluids of animals?

3. Define ionoconformation, ionoregulation, osmoconformation and osmoregulation. Generally speaking, which animals are ionoconformers and osmoconformers, which are ionoregulators and osmoconformers, and which are ionoregulators and osmoregulators?

4. How do brine shrimp, which live in hypersaline solutions, solve their ionic and osmotic problems?

5. How is urine formed by protonephridia, metanephridia and Malpighian tubules? How does the composition of the initial urine differ from that of the body fluids?

6. What processes are used to modify the composition of urine after it is formed by excretory tubules? How do these processes affect ion, organic solute and water loss?

7. Compare and contrast the structure and function of the excretory organs of earthworms, crustaceans and insects.

8. What is the structure of a mammalian nephron? What is the primary physiological function of each part of the nephron?

9. Explain how the loop of Henle establishes an osmotic concentration gradient in the renal medulla. How is this osmotic gradient used to osmotically concentrate urine?

10. Desert mammals such as the Australian hopping mouse, *Notomys alexis*, never have to drink. What are the physiological adaptations of these animals for maintaining solute and water homeostasis?

11. Compare and contrast how water is apparently transported against an osmotic gradient in rectal pads of *Calliphora*, and reabsorbed against an osmotic gradient by mammalian nephrons.

THE IMMUNE SYSTEM

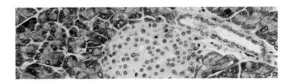

All organisms must be able to defend themselves to some degree against damage stemming from their environments. Damage to an organism can occur at three levels:

- *physical damage*, such as occurs through predation or accidental injury;

- *damage resulting from extreme environmental conditions*, such as excessive or limiting levels of water, oxygen or salts, or extremes of pH or temperature;

- *damage from macromolecules, micro-organisms and parasites*. It is at this level that the immune system is of key importance in defence.

The role of the immune system is to provide protection against viruses, bacteria, fungi and other small parasites, and any other macromolecular organic matter that threatens the integrity of the organism. As life has evolved, so too have immune systems, and the increasing complexity of organisms is reflected in the possession of more complex and refined immune systems. In this chapter we shall deal mainly with the mammalian immune system, primarily because it is among the most complex and advanced but also because it is the best understood. Immunity in non-mammalian animals and in plants will be discussed at the end of the chapter.

There are four important concepts around which much of this chapter revolves.

- Substances that an immune system reacts against, for example, molecules on the surface of a bacterium, are **antigens**.

- The cells in the body that carry out many of the functions of the immune system are **lymphocytes**. There are two main types: B lymphocytes and T lymphocytes.

- B lymphocytes make and release **antibodies**, which are important defence molecules that react with and destroy antigens.

- T lymphocytes regulate immune responses and can directly attack and kill foreign organisms or infected cells.

NON-SPECIFIC DEFENCE MECHANISMS

Organisms also have defences that are not part of an immune system but which provide non-specific protection against foreign organisms (Fig. 23.1). At the simplest level these include 'external' defences, such as enzymes in secretions that can damage or

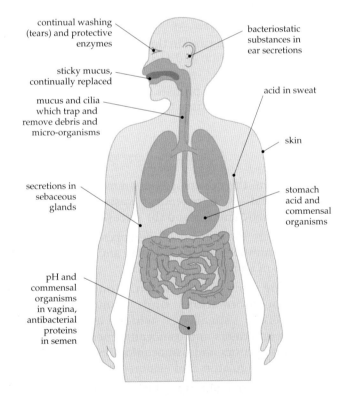

Fig. 23.1 Simple non-specific defence systems. Some of the external physical and chemical defence mechanisms that prevent foreign organisms from gaining access to the body. Remember that the inside of a mammalian body—warm, wet and rich in nutrients—is an ideal environment for the growth of many micro-organisms, so the best first line of defence is just not to let them in

Table 23.1 Key terms in immunology

Term	Definition
Antigen	Any molecule (usually organic) that can be recognised by (i.e. bind to) one of the specific molecules (antibodies or T-cell receptors) of the immune system. These include self-antigens and non-self (foreign) antigens. Antigens are not necessarily immunogenic.
Immunogen	An antigen that stimulates immune responses—either antibody production or the generation of effector T cells.
Epitope	The actual portion of an antigenic molecule that is recognised by a receptor. A large protein antigen may have dozens or hundreds of different epitopes, each of which could induce a unique response.
Antibody	The protein molecule produced by B cells in response to antigen and which reacts very specifically with that antigen. Each different antibody is the result of a random genetic rearrangement event.
Immunoglobulin (Ig)	Another term for antibody.
Lymphocyte (lymphoid cell)	Mononuclear cells that are the predominant cells in immune organs such as lymph nodes, spleen, thymus, Peyer's patches and tonsils.
B cell (or B lymphocyte)	Lymphocyte that makes antibodies; produced in the bone marrow, spleen or gut lymphoid tissue.
T cell (or T lymphocyte)	Lymphocyte that matures in the thymus, recognises antigen by means of the T-cell receptor, and functions independently to kill micro-organisms. It is also crucial in controlling B-cell responses.
T-cell receptor (TCR)	The glycoprotein molecule on the surface of T cells that recognises an antigen. Like antibodies, each of these molecules is the result of a genetic rearrangement event.
Auto-	From the same or a genetically identical individual (e.g. in autoimmune, autoantigen, autograft).

destroy micro-organisms (e.g. lysozymes in saliva and tears), low pH in the stomach, sticky mucus on many surfaces, and the presence in the gut and genital tract of commensal organisms that are able to prevent the growth of pathogenic organisms. The last example explains some of the gastrointestinal side-effects of antibiotics that kill harmless or beneficial gut flora, while allowing more dangerous micro-organisms to grow.

Internal non-specific defences include molecules that increase in number during the course of an infection and protect against a range of organisms. These substances include molecules of the **complement system** (Box 23.3), which cause bacterial rupture, **lysis**, and enhance the removal of bacteria by phagocytic cells, and **interferons**, which, among other effects, induce resistance to viral infection in some cell types. To some extent these non-specific mechanisms can be seen as a first line of defence. They are always present at low levels and are rapidly activated early in an infection or after tissue damage. They reduce or slow down the development of infections during the several days that it takes for the immune system to develop a specific response.

> Non-specific defences play an important role early in infections. They include antimicrobial substances in secretions, and molecules that slow the growth of micro-organisms and aid in their removal.

IMMUNE RESPONSES

Immune responses can be distinguished from non-specific defences by their specificity. Immunisation with a polio vaccine results in immunity to polio, but not to other diseases. A key part of specific immunity is that it involves *immunological memory*. When a person is immunised against polio, a *primary response* occurs. The immune system 'remembers' this and, if the person comes into contact with the polio virus again, a *secondary response* occurs, which is usually bigger and faster than the primary response and sufficient to prevent infection the second time. However, if the same individual is immunised later against a different organism (e.g. rubella virus, which causes measles), this results in a new primary response that is unaffected by the previous response to polio. Immune responses are not always *absolutely* specific to a particular organism; protection may be conferred upon closely related organisms, as in the famous example where infection with the cowpox virus also provides immunity against the smallpox virus (Chapter 34).

Why is the immune system so complex?

Consider the following situation. During the lifetime of an individual, the immune system has the difficult task of protecting against a possibly infinite and certainly unpredictable variety of foreign molecules or organisms. No organism can possibly predict which foreign organisms it may encounter or which new ones may evolve during its lifetime. The task is even more difficult because a complex organism is itself composed of a very large number of cells and molecules that must *not* be subject to immune attack. Inappropriate immune responses against these 'self' molecules (autoimmune responses) can have very damaging consequences. An immune system therefore needs a large array of very *specific* defences, which operate at the level of molecular recognition.

How can an immune system meet these requirements? One way would be to have millions of genes coding for millions of defensive molecules that do not react with self. But, even in mammalian genomes, with an estimated 100 000 genes for an entire organism, the millions of additional genes required would take an impossible amount of genetic space. Evolution has resulted in a novel solution to the problem. A relatively small number of genetic elements are randomly shuffled, or rearranged, in such a way that a vast number of different defence

molecules is generated. This solution is very efficient in that it uses little genetic space; but it has the disadvantage that the range of defence molecules generated is unpredictable and will include many that may react with self components. As we shall see, many of the complexities of the immune system result from the subsequent need to screen the unpredictable range of defence molecules in order to remove any that are potentially self-destructive.

> The immune system must be able to respond to an infinitely large and unpredictable array of foreign molecules, although only a relatively small number will be encountered in the lifetime of any individual, but not to 'self' molecules.

CELLS OF THE IMMUNE SYSTEM

Cells that take part in immune responses can be separated into three categories (Fig. 23.2):

- *specifically acting cells*—T and B lymphocytes;
- *non-specifically acting cells*—phagocytic cells, macrophages and granulocytes;
- several groups of *partly specific cells*—that is, cells that can kill tumour cells or virus-infected cells but not normal cells.

Most cells found in blood, bone marrow and lymphoid tissues are derived from multipotent haemopoietic stem cells that reside mainly in the bone marrow (see Box 6.2). Further development of T cells occurs after migration of stem cells to the thymus; B-cell development occurs in various tissues in different species, but in the bone marrow in humans.

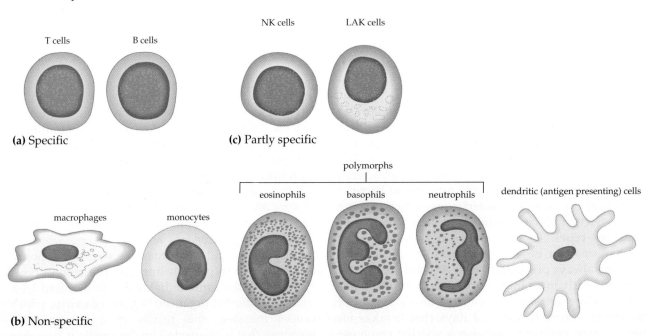

Fig. 23.2 Cells of the immune system: **(a)** specific; **(b)** non-specific; and **(c)** partly specific

Lymphocytes and specificity

Lymphocytes are small round mononuclear cells (Fig. 23.3), generated in large numbers each day. They make up the majority of white blood cells and cells in the organs of the immune system. T and B lymphocytes, although superficially similar in appearance, can be distinguished on the basis of their development and functions. Individual lymphocytes are different from their neighbours because each carries a distinct antigen receptor that is the result of gene rearrangement (Box 23.1). Thus, only one or a very small number of cells has a particular arrangement and hence particular surface receptor.

In principle, each receptor recognises a different antigen, although among a large number of randomly generated receptors, it is likely that several different receptors could interact with the same antigen. A normal laboratory mouse has approximately 500 million lymphocytes bearing at least several hundred million different specificities referred to as its 'immune repertoire'.

During an immune response, each cell that is able to react with an antigen proliferates, generating a clone of daughter cells, each with the same receptor, each capable of responding to the antigen. This clonal expansion boosts the response to the stimulating antigen by increasing the number of cells reacting to it. The key features of this **clonal selection** are outlined in Figures 23.4 and 23.5.

> Each lymphocyte carries a different surface receptor. During an immune response, those few cells that interact with an antigen proliferate, increasing the number of cells reacting to that antigen. This process is known as 'clonal selection'.

Cellular and humoral immunity

There are two kinds of specific response to foreign antigens—cellular and humoral (Fig. 23.6). Both can be triggered, but one or other tends to predominate and be most effective in any particular infection. The cellular immune response is most effective against fungal and other parasites, and intracellular viral infections, whereas the humoral response is most effective against the extracellular phases of bacterial and viral infections.

Cellular immunity involves active destruction of foreign organisms or of the body's own virally infected cells by T cells. T cells regulate most reactions in the immune system, including antibody formation, and self–non-self discrimination.

Humoral immunity is mediated by soluble antibody molecules secreted by B cells in the serum or other body fluids. Once formed, antibodies act independently of B cells, and pure cell-free serum retains its humoral activity. Unlike cellular immunity, antibodies are reasonably stable outside the body.

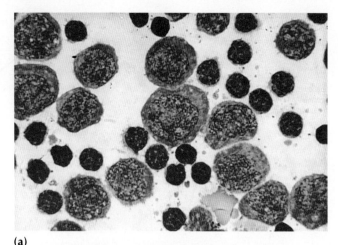

(a)

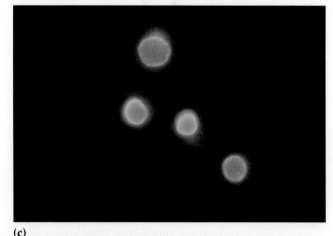

(b)

(c)

Fig. 23.3 The lymphocyte. (**a**) A smear of a population of lymphocytes taken from the efferent lymph leaving a stimulated lymph node. The small round cells with little cytoplasm are small lymphocytes, the large ones are activated, dividing lymphocytes responding to antigen. There are both T and B cells among the small and large cells but they cannot be distinguished by visual criteria alone. (**b**) An electron micrograph of a small lymphocyte, with little cytoplasm and few outstanding features or organelles. Again this could be a T or B cell. (**c**) A group of lymphocytes labelled on the surface with an antibody to a molecule called Thy 1, which is only expressed on T cells. The antibody is marked with a green fluorescent tag, so only T cells show up as green. B cells in the picture are not visible as they are not labelled

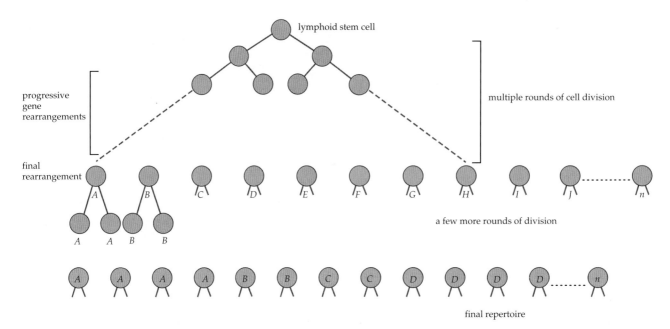

Fig. 23.4 Generating the preimmune repertoire. The timing of gene rearrangement and cell proliferation gives rise to a large number of small clones of cells, each clone expressing a different antigen receptor (Ig for a B cell, TCR for a T cell). The final repertoire is the range of clones available before any response to antigen has occurred. *A, B, C . . . n* are receptors of different specificities

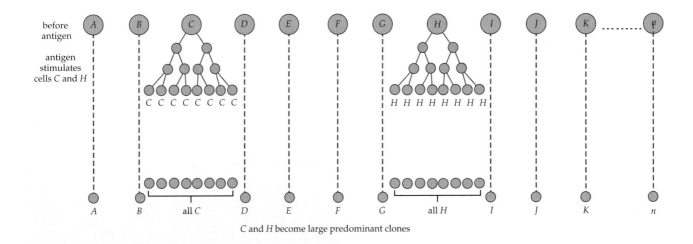

Fig. 23.5 Clonal expansion and the immune repertoire. Some of the cells from within the preimmune repertoire will eventually encounter the antigen that their surface receptor recognises, or can interact with. These cells will be stimulated to proliferate and carry out effector functions. The proliferation will result in an increase in the number of these cells and they will make up a larger proportion of all the lymphocytes. If the same antigen is encountered again, there will already be large numbers of cells present, and only activation will be required. All the time required for proliferation will be saved, allowing much faster responses on secondary stimulation

Antibodies are present in breastmilk and are important in protecting a baby from infection. Antibodies can also be effective across species and are used for providing 'passive' immunity against diseases such as hepatitis.

There are two distinct kinds of immune response that are usually effective against different kinds of micro-organisms: a cellular response, mediated by T cells, and a humoral response (in serum), mediated by antibodies produced by B cells.

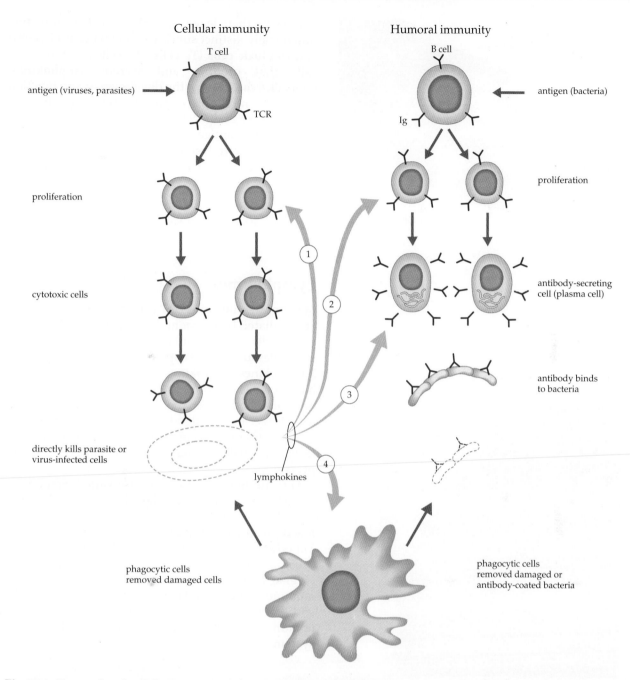

Fig. 23.6 Humoral and cellular immunity—the two main arms of immune responses. The molecules released by activated T cells, called lymphokines, are important in helping both T and B cells to proliferate (1 and 2), and B cells to make antibody (3). Lymphokines also stir up the macrophages (4), increasing the speed of removal of debris

T cells and the thymus

T cells mature in the thymus (Fig. 23.7). If the thymus is absent, as happens in certain genetic abnormalities, or removed early in fetal development, no T cells are made and profound **immunodeficiency** results. Stem cells enter the thymus and proliferate rapidly. During this time the genes coding for the T-cell receptor (TCR) for antigen are being rearranged (Box 23.1) and, eventually, TCR proteins are expressed on the cell surface. At this point, all cells must be screened

so that only useful cells, that is, those cells that do not recognise self-antigens, continue development. Only about 1% of cells survive screening. Those that do survive go on to mature, finally leaving the thymus and populating the lymphoid organs; the remaining 99% die within the thymus.

> T cells are produced in the thymus in very large numbers. Only a few useful, non-self-reactive T cells survive the screening processes and are exported to the lymphoid organs.

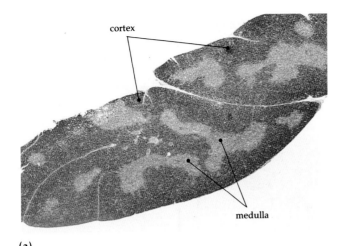

(a)

thymic lymphoid cells (thymocytes)

epithelial cells

macrophages and dendritic cells

(b)

Fig. 23.7 The structure of the thymus. **(a)** A section of a mouse thymus. The light-stained central areas are medulla, which contains mainly mature cells waiting to migrate out to the lymphoid organs, such as spleen and lymph nodes. The darker outer zone of densely packed cells is the cortex where most of the proliferation and differentiation occurs. **(b)** Detailed structure of the thymus, with a meshwork of fixed epithelial cells supporting scattered bone-marrow derived macrophages and dendritic cells, and all the spaces packed full of thymic lymphocytes, often called thymocytes. The areas left open in the diagram for clarity would normally be full of thymocytes

T cells that leave the thymus fall into two functionally distinct subclasses: **helper cells** (T_H cells) and **cytotoxic cells** (T_C cells). T_H cells are regulatory cells that produce and secrete **lymphokines**, molecules that control the development and function of other T and B cells, as well as accessory cells such as macrophages. T_C cells are effector cells that, when stimulated by antigen and lymphokines produced by T_H cells, directly lyse or kill target cells recognised by the T_C cells on the basis of their particular antigen.

> There are two types of T cells: helper cells (T_H cells), which mainly regulate other immune cells, and cytotoxic cells (T_C cells), which kill foreign organisms or infected body cells.

Lymphokines

Many of the functions of T cells, particularly T_H cells, are mediated by hormone-like molecules released by these cells after stimulation by an antigen. These molecules, usually glycoproteins, are variously called lymphokines, cytokines or interleukins. Some lymphokines act only in the local environment, so only other cells near the T cell are affected. Others may travel in the bloodstream, stimulating cells at distant sites or exerting generalised effects that cause fever or 'malaise', nausea or changes in blood pressure.

The effects of lymphokines are varied. Some help B cells make antibodies, others help T cells develop. Still others stimulate the production or activation of macrophages and granulocytes, red cells or platelets. Yet others increase the permeability of blood vessels, causing swelling that may accompany inflammation. Overall, lymphokines are key players in almost every immune response, relaying signals between T cells and other cells, and communicating between the immune system and the endocrine and nervous systems.

B cells and their development

B-cell development parallels that of T cells but it occurs at different sites. In humans, B cells develop mainly in the bone marrow. There is extensive proliferation as the cells progressively differentiate and, during this time, the genes that encode antibodies (also known as **immunoglobulins** or Ig's) are rearranged (Fig. 23.8).

Mature B cells manufacture antibodies that are expressed on their surface and function as receptors. New B cells die within a few days if they do not interact with their specific antigen, which is the fate of the majority of B cells produced. When a membrane-bound antibody receptor binds to its specific antigen in the presence of stimulated helper

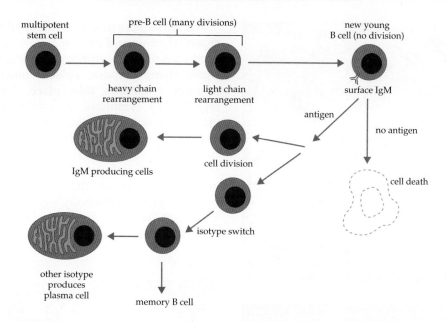

Fig. 23.8 B cell development, showing some of the steps in development from a multipotent stem cell

T cells, the receptor stimulates further differentiation of the B cell. Some B cells will become **memory cells**; others differentiate further into **plasma cells**, non-dividing antibody factories that produce and release large quantities of antibody molecules.

> In humans, B cells are made in the bone marrow. They have antibodies on their surface that, on binding with their specific antigen in the presence of helper T cells, trigger further differentiation into memory cells or plasma cells.

Phagocytic cells

About a hundred years ago, Elie Metchnikoff made the first observation of white blood cells (leucocytes) engulfing micro-organisms and he called the process **phagocytosis** or cellular eating. There are two main types of phagocytic cells—mononuclear phagocytes (macrophages and monocytes), and polymorpho-nuclear granulocytes, usually known as polymorphs or granulocytes.

Macrophages and monocytes

These cells play an important role in body maintenance. They engulf and digest dead and damaged cells, old red cells, debris, antibody-coated micro-organisms and damaged fatty particles or molecules (Fig. 23.2). They participate in wound healing, and in both acute and chronic inflammation. Almost all of these activities increase when the cell becomes activated through contact with lymphokines released by T cells undergoing an immune response.

Polymorphonuclear granulocytes

There are three types of polymorphonuclear granulocytes, so-called because of their multilobed nucleus and their many cytoplasmic granules (Fig. 23.2). The most common of these encountered under normal conditions are *neutrophils*, which are predominantly phagocytic cells. *Eosinophils* are essentially killing cells, which are especially important in the removal of larger parasitic animals, such as flukes and worms. However, the destructive activities of eosinophils can also cause tissue damage at sites of allergic reactions, so they are potentially dangerous. Circulating *basophils* and stationary *mast cells* are mainly involved in acute inflammation and allergy.

Dendritic cells

There are many cells in the immune system, including macrophages and B cells, that can present antigens to T and B cells. Most effective of all are **dendritic cells**, a poorly understood but ubiquitous cell type, which are able to break down foreign molecules and present them to lymphocytes. Dendritic cells, so-called because they have many long thin processes (dendrites) that spread out between surrounding cells (Fig. 23.9), are located in all lymphatic tissues, including the thymus, where they play an important role in the screening process. They are also found in blood, in skin (Langerhans cells) where they are key players in many inflammatory reactions, and at mucosal surfaces in contact with the outside environment, such as the gastrointestinal and respiratory tracts.

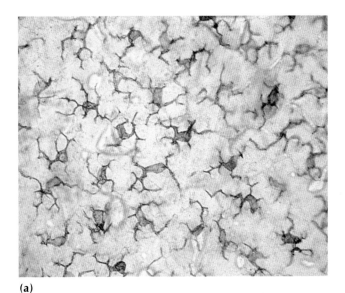

(a)

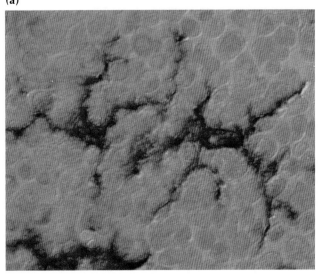

(b)

Fig. 23.9 Dendritic cells play a sentinel role in peripheral tissues, in particular at mucosal surfaces in contact with the external environment, where their prime function is surveillance for incoming antigens, which are subsequently transported to central lymphoid organs for presentation to the immune system. **(a)** Epidermal Langerhans cells stained for ATPase. **(b)** Airway intraepithelial dendritic cells stained for MHC

There are several cell types that assist lymphocyte functions. Phagocytic cells (macrophages and granulocytes) remove damaged or foreign cells and can kill parasites. Macrophages, B cells and dendritic cells can present foreign molecules to lymphocytes.

SECONDARY LYMPHOID TISSUES AND THE IMMUNE SYSTEM

To understand how the immune system functions in maintaining the health of an individual, it is necessary to appreciate the structure and behaviour of the immune system in the whole body (Fig. 23.10).

The immune system is made up of *primary lymphoid organs* (thymus and bone marrow), which are largely producers of new lymphocytes, and *secondary lymphoid organs* (lymph nodes, spleen, tonsils, etc.) where immune responses take place. These are linked together by lymphatic vessels and blood vessels. Since a wound, infection or dying cell might be anywhere in the body, the immune system needs to survey the entire organism continually. Surveillance is achieved by keeping the cells of the immune system on the move around the body. Lymphocytes are the body's great travellers, and the highways for their travelling are the network of lymphatic vessels.

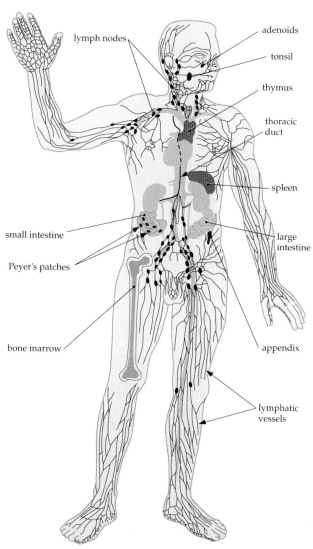

Fig. 23.10 The immune system of the human. This comprises primary lymphoid organs, which produce new lymphocytes (thymus and bone marrow), secondary lymphoid organs where organised immune responses occur (lymph nodes, spleen, adenoids, tonsils, appendix, Peyer's patches), the network of blood and lymphatic vessels that link all the tissues and, of course, the lymphocytes themselves, which migrate around the body and reside in the lymphoid tissues

BOX 23.1   GENERATION OF DIVERSITY AND GENE REARRANGEMENT

The complex rearrangement events that generate diversity among antibody and TRC molecules are probably unique to those molecules. Each Ig and TCR molecule is made up of two protein chains, each bearing a variable portion that contributes to the antigen-binding site. The potential diversity of a population of whole molecules is thus the product of the potential diversity of each chain. Part of this diversity comes from the variety of variable (V) genes that can be used in each chain, but much more comes from the way the genetic elements are joined. Between the V and constant (C) genes are small elements known as diversity (D) and joining (J) elements. Immense variation can occur as different combinations of V, D and J segments are combined.

The approximate contribution of each of these sources to total variation for Ig and TCR molecules is shown in the table. As you can see, there may be 10^{11} possible antibody molecules and 10^{15} possible TCR molecules. Since mammals contain 10^8–10^{12} lymphocytes, it is obvious that there are many more possibilities than will ever occur in any one individual. The figure shows a scheme of these events for a typical rearrangement process.

Genetic sources of diversity in Ig and TCR molecules
(a)

| | Ig | | TCR | |
	Light chain	Heavy chain	α chain	β chain
V genes	250	250–1000	100	25
D segments	0	10	0	2
J segments	4	4	50	12
N region addition (b)	none	V–D, D–J	V–J	V–D, D–J
Possible combinations (c)	$\sim 10^{11}$		$\sim 10^{15}$	

(a) Number of gene segments in DNA of genes for the two chains of immunoglobulin and the TCR.
(b) Random insertion of nucleotides at the point of joining between segments.
(c) Estimated number of different protein chains that could be generated by different combinations of V, D and J segments, along with N-region effects.

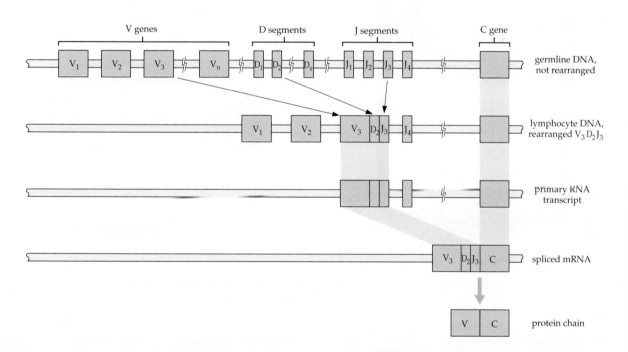

Rearrangement events in a hypothetical Ig or TCR chain. The gene segments used in this case were V_3, D_2 and J_3. During the rearrangement of the DNA, the section between V_3 and D_2 (V_4–n and D_1) was excised, as was the section between D_2 and J_3 (D_3–n and $J_{1,2}$). The primary RNA transcript includes everything between the rearranged V gene and the C gene, but this RNA is later spliced to bring VDJ beside the C gene. Note that some chains only have V, J and C segments (Ig light chain and TCR α chain), while others have V, D, J and C (Ig heavy chain and TCR β chain). Some chains also have several C genes to select from, but these do not affect diversity of the antigen-binding site

The lymphatic network

The network of lymphatic vessels has two important roles:

- to clear proteins and fluids from tissues;
- to transport cells.

The finest lymphatic vessels join up to form larger and larger vessels, eventually flowing into a small number of major lymphatic vessels, which return fluid back to the blood (Fig. 23.11). Lymphatics also pick up lymphocytes and macrophages that are wandering through tissues, and collect micro-organisms or debris from inflamed or infected sites.

Lymph nodes, located at numerous points along the lymphatic vessels, act as filters and are also highly organised lymphoid tissues capable of producing a vigorous immune response. Although an infection results in local inflammation and influx of immune cells, all the cells and debris and antigens from this site will eventually be carried by the lymphatics to the local lymph node where the strongest immune response, well organised and distant from the inflammation, occurs. Hence, an infected hand results in enlarged and tender glands (lymph nodes) in the armpit, the site of the main lymphatic filters for the lymphatics in the hand and arm. Other lymphoid tissues, such as tonsils, Peyer's patches and adenoids, function in a similar way (Fig. 23.12).

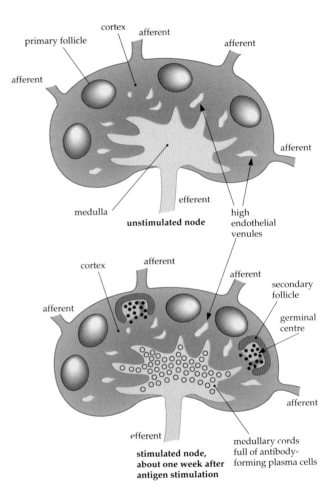

Fig. 23.12 Lymph node structure. B-cell areas (follicles and medulla) are indicated in red, while the T-cell areas (cortex) are blue

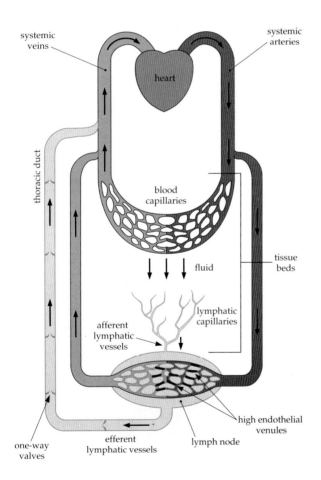

Fig. 23.11 Lymphocyte recirculation. Blood circulation is shown in red, lymphatic circulation in blue. The arrows show the direction of flow of fluid and cells. Blood circulation is driven by the heart. Afferent lymphatic vessels collect fluid and cells through osmotic and fluid pressure, while lymph moves through the efferent vessels by squeezing past one-way valves as the muscles move and contract. Hence, prolonged inactivity in a standing or sitting position can lead to swollen hands or feet as lymph accumulates and there is insufficient muscle movement to massage the lymph up the long lymphatic vessels of the arms and legs

Even in the absence of infection there is a major traffic of lymphocytes from blood, through specialised blood vessels directly into the lymph nodes (Figs 23.13, 23.14). After leaving the bloodstream, these cells migrate through the nodal tissue back into the lymphatics and to the blood again. The majority of lymphocytes are constantly recirculating—going round and round this circuit, about one trip per day, and passing through many lymph nodes each week.

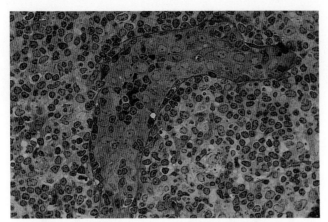

Fig. 23.13 A histological section of a high endothelial venule in the cortex of a lymph node. Note the lymphocytes migrating between the high endothelial cells

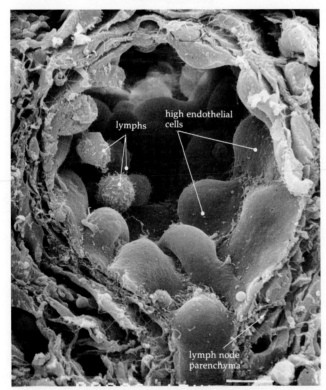

Fig. 23.14 Scanning electron micrograph of the inside of a blood vessel in a lymph node. Small round lymphocytes (lymphs) adhere to the dome-shaped high endothelial cells that make up the walls of this high endothelial venule, a specialised blood vessel, which allows the emigration of lymphocytes from the blood into lymphoid tissues. The lymphocytes adhere to the endothelial cells, then migrate between them, into the body (the parenchyma) of the node, and then on out into the efferent lymph to continue their recirculation back to the blood

Antibodies formed in an activated lymph node will be carried to the blood through lymphatic vessels and dispersed around the body. Similarly, immune-activated cells and memory cells will eventually leave the node and impart immunity to all parts of the

body. Some of these will leave the blood again at the site of inflammation, carrying effector cells right to the cause of the trouble.

Blood is filtered in a similar way by the *spleen* (Fig. 23.15). Should an infection get into the bloodstream, the spleen will filter out the foreign organisms and co-ordinate an organised immune response involving T and B cells and antigens.

> Lymphocytes continually migrate around the body, circulating from blood to lymphoid tissue and back to blood through lymphatic vessels. This allows lymphocytes to monitor a wide range of sites around the body. Secondary lymphoid organs act as filters and sites where immune responses can be organised.

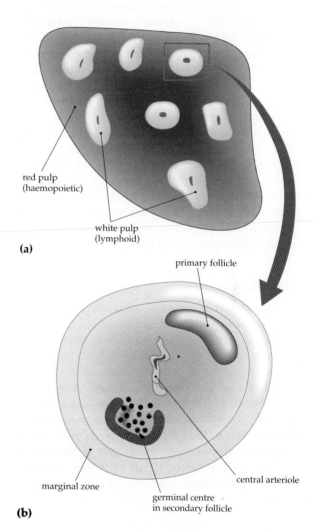

Fig. 23.15 (a) The spleen is made up of red pulp, coloured red because of the high erythrocyte content, and islands of white pulp, coloured white because there are few erythrocytes, mainly lymphocytes. (b) White pulp enlarged. T-cell area is shown in blue, the B-cell areas in red. The white pulp always has a blood vessel running through the centre and a marginal zone, rich in antigen-presenting cells, around the outside. As in lymph nodes, primary follicles mature into secondary follicles and germinal centres upon antigen stimulation. In the spleen, plasma cells are often found scattered through the red pulp

MOLECULES OF THE IMMUNE SYSTEM

The important molecules of the immune system are *antigens*, molecules against which the immune system will react, *antibodies*, produced by the immune system to help get rid of foreign antigens, *T-cell receptors*, molecules on the surface of T cells that interact with antigen, and *major histocompatibility complex (MHC) molecules*, which help to present antigens to the immune system.

Antigens

An antigen is a molecule that can react with and bind to the variable region of an antibody (immunoglobulin) or T-cell receptor molecule. If this binding stimulates a lymphocyte to make an immune response, then the antigen is also an **immunogen**. B cells can respond to a wide variety of antigens, including proteins and carbohydrates. T cells respond predominantly to protein antigens.

Proteins are the best understood and most diverse antigens, and the smallest molecules that can function as antigens are peptides of about 10–15 amino acids. Molecules any smaller than this are invisible to antibodies or TCRs. A typical large protein molecule of hundreds or thousands of amino acids will contain many 'minimal' peptides, or **epitopes**, that are potentially antigenic (Fig. 23.16).

Protein antigens

Large immunogenic protein antigens are broken down by antigen-presenting cells, such as macrophages, into small peptides. These peptides are the units displayed or presented by cell-surface MHC molecules to T cells. T cells are stimulated only by MHC-bound peptides and they recognise peptides derived from protein antigens that may *not* have been located on the surface of the intact molecule. Antibodies, on the other hand, are much more likely to recognise epitopes from the *surface* of the whole antigen molecule because the B cells that produce them can be stimulated directly by whole antigen molecules.

Carbohydrate antigens

The second important group of antigens are carbohydrate antigens. As for proteins, only large complex polysaccharides function as antigens. In general, carbohydrates are more likely to stimulate B cells to make antibodies than they are to stimulate T cells. Some important bacterial and viral antigens are polysaccharides. Another well-known group of carbohydrate antigens is the human *ABO blood group antigens*. Most humans (except those of blood type

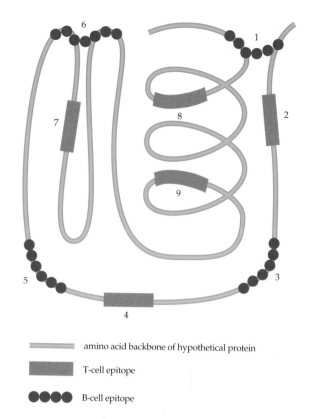

amino acid backbone of hypothetical protein

T-cell epitope

B-cell epitope

Fig. 23.16 Hypothetical protein molecule showing T- and B-cell epitopes (the actual small piece of an antigen which the antibody or TCR binds to), which can be quite distinct. B-cell epitopes, those bound by antibody molecules, are usually on the surface of a molecule and may be dependent on the structure of the molecule (e.g. epitopes 1 and 6), or on a string of amino acids, often one which stands out on the surface (epitopes 3 and 5). T-cell epitopes are entirely dependent on the short sequence of amino acids, since they are always presented after the amino acid chain is broken into small bits. They can therefore be internal to the whole molecule (epitopes 7, 8 and 9), and cannot be dependent on the structure of the whole molecule

AB) have antibodies against one of these antigens in their serum (see Box 23.2).

> Both proteins and carbohydrates can be antigens. T-cell receptors and B-cell antibodies 'see' antigens quite differently. Antibodies react with an antigenic molecule as a whole, whereas TCR's only recognise small peptides that are presented by MHC molecules on the surface of antigen-presenting cells.

Antibodies

The specificity of the immune response resides in the B and T lymphocytes. Each has its own distinct but related molecule that mediates this specificity. For B cells this molecule is an antibody and for T cells it is the T-cell receptor.

Antibodies are either membrane-bound or exist free in solution. For any particular B cell, both forms of

BOX 23.2 BLOOD GROUP ANTIGENS

Blood group antigens are predominantly polysaccharides and are found on the surface of red blood cells. The most important and best known blood group antigens are the human ABO and Rhesus antigens. Individuals who have group A blood have A antigen on their red cells. Those with group B blood have B antigen on their red cells. Those with group AB have both and those with O blood have neither antigen on their surface.

In principal, transfusion of blood between individuals of different blood groups would eventually result in an immune response against the antigens that differed from those on the recipient red blood cells. However, everyone with blood group A makes antibodies directed against blood group B, and vice versa. Group O individuals make antibodies against both antigens, while group AB individuals make neither (see table).

The apparently spontaneous existence of antibodies against group A or group B antigens probably results

Human ABO blood groups

Blood group	Red cell antigen	Antibody present in serum	Transfusion status
O	O	Anti-A and Anti-B	Universal donor
A	A	Anti-B	Selective
B	B	Anti-A	Selective
AB	AB	None	Universal recipient

from the common occurrence of similar antigens in bacteria. So, for example, antibodies are produced when a person is exposed, usually early in life, to bacteria with these antigens in their structure, but they would not develop against any antigens similar to the person's own antigens.

---

antibody have exactly the same antigen-binding region. Antibodies are *glycoproteins*, all of which have a similar underlying structure (Chapter 1). Each molecule is made of two identical *heavy* chains and two identical *light* chains (Fig. 23.17). Each antibody molecule therefore has *two* identical antigen-binding regions at the ends of its two arms.

Both heavy and light chains can be divided into distinct regions: a *variable* region, the portion coded for by the rearranged part of the DNA (Box 23.1), which contains the antigen-binding site, and a *constant* region which, as the name suggests, is relatively constant from antibody to antibody. The variable region is the part that binds to the antigen (e.g. on the surface of a bacterium), but it is the constant region that carries out most of the other functions, such as killing the bacterium or promoting clearance by macrophages. Variation in the heavy chain constant regions is used to divide immunoglobulins into various classes (or isotypes), known as IgM, IgD, IgG, IgA and IgE, which are found in different sites and have different functions.

> Antibody molecules function as antigen receptors on B cells and as effector molecules in all body fluids. They have a variable region, which recognises and binds to antigen, and a constant region, which is responsible for most other functions.

The T-cell receptor

Unlike antibodies, which may be membrane-bound or free in body fluids, TCR molecules are always membrane-bound. Like antibodies, TCRs are composed of variable and constant regions, with the

antigen-binding site a product of the joining of the two chains (Fig. 23.18). Antibodies and TCRs have many features in common, especially the genetic rearrangement events that generate their diverse specificities (Box 23.1). In contrast to antibodies, which have two antigen-binding sites, each TCR has only one antigen-binding site. Differences between TCRs and antibodies are shown in Table 23.2.

Table 23.2 Differences between antibodies and TCR molecules

Feature	Antibody (immunoglobulin)	T-cell receptor
Number of chains	2 heavy, 2 light	1 α, 1 β (or 1 γ,1 δ)
Antigen-binding sites	2	1
Number of subtypes	10 (IgM, IgG, etc.)	2 (αβ, γδ)
Localisation	Cell surface and soluble	Cell surface
Polymerisation	Up to 5 polymers	No
'Sees' soluble antigen	Yes	No
Variable part	Yes	Yes
Constant part	1–4 per chain	1 per chain

The TCR functions primarily as a tool for interaction between T cells and other cells at their cell surfaces. The TCR is one of a large group of molecules involved

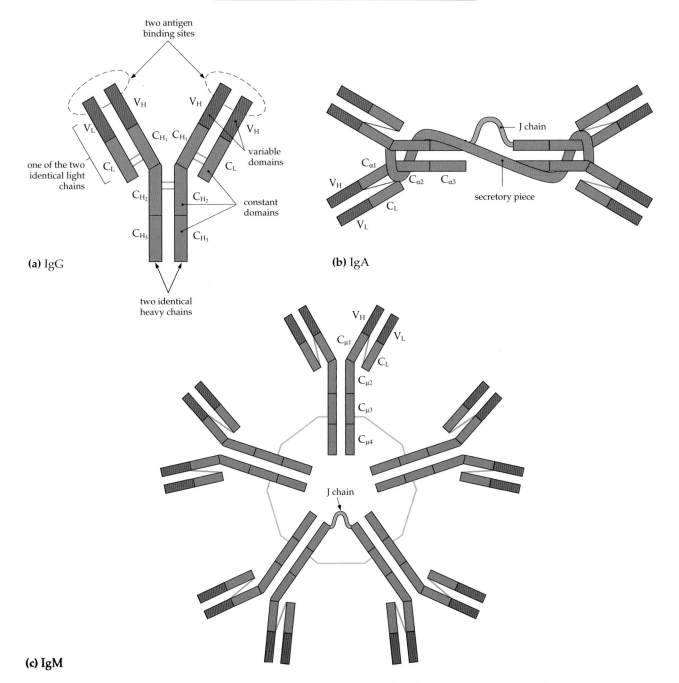

Fig. 23.17 The structures of antibody molecules. (**a**) Schematic diagram of an IgG molecule, one of the common antibodies in blood. The variable domains are in red, the constant domains in blue. The C domains are quite similar to each other, but small structural differences allow each domain to have specialised functions. (**b**) Schematic diagram of the dimeric IgA molecule. The secondary piece functions in the transport of the dimer across mucous membranes. The J chain holds the two monomers together. (**c**) Schematic diagram of a pentameric IgM molecule. Again the monomers are held together by a J chain. Because IgM has 10 antigen-binding sites, it cross-links antigens more easily, enabling better clearance

in this interaction or in the responses that follow, but the specificity of the interaction is determined by the variable region of the TCR itself.

The TCR does not interact with free antigen. In fact, it only recognises antigen in the form of peptides bound to the surface of molecules of the MHC (see below). Once the TCR recognises a foreign peptide/MHC complex, a complicated rather poorly understood series of events happens, resulting in a

response within the T cell. The response involves activation, proliferation and differentiation of the T cells, for example, to become killer cells or to produce lymphokines.

TCRs are always bound to the T-cell surface. Each TCR has only one antigen-binding site, unlike antibodies, which have two.

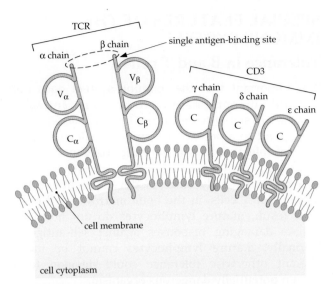

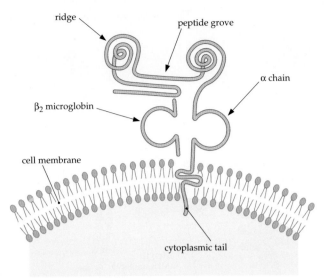

Fig. 23.18 Schematic diagram of the T-cell receptor and the associated chain of the CD3 molecule, which is essential for the appearance of TCR chains in the membrane, and for the transduction of signals when TCRs bind to antigens. Notice that the domains of the CD3 chains outside the membrane are similar to the C domains of the TCR molecule or of immunoglobulin

The major histocompatibility complex

MHC molecules are the only molecules that can present (show) antigen to T cells. Since T cells regulate most reactions in the immune system, including antibody formation and self–non-self discrimination, MHC molecules play a key role in every aspect of immunity. B cells, as we have seen, can see antigens directly without needing MHC molecules to present them, but since B cells usually need help from T cells, the production of antibodies is also dependent on MHC molecules. Rejection of tissue grafts is another immune reaction in which the role of T cells and MHC molecules is central.

Antigens detected by T cells are small peptides having only eight to 20 amino acids derived from whole proteins. These peptides are only recognised as antigens by T cells when they are displayed on the end of the MHC molecules (Fig. 23.19). The peptide, together with the two neighbouring portions of the MHC molecule, is the structure recognised by the TCR (Fig. 23.20). The peptide is 'picked up' as the MHC molecule is synthesised inside the cells. Peptides may be derived from other molecules being synthesised by the cell (e.g. viral antigens), or from molecules external to the cell that have been phagocytosed and broken down.

The genes coding for MHC molecules are part of a *multigene complex*, which is a region in the genome containing many kinds of related genes. In the case of the MHC complex, some of the genes are associated with tissue graft rejection, and most of these occur in many *allelic* forms, with up to 100 alleles present in a given population. As the name implies, this is

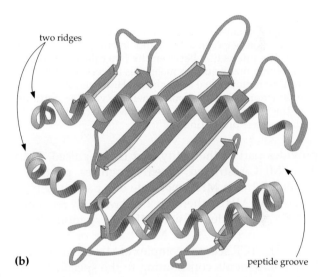

Fig. 23.19 **(a)** Structure of the MHC class I molecule, attached in the cell membrane. It has two chains, a large α chain and a smaller β₂ microglobulin chain. **(b)** View of the peptide groove and ridges from the surface, giving a more accurate picture of how the amino acid chain (i.e. the ribbon) coils and criss-crosses to make the platform and ridge structure

the most important set of genes in the determination of graft rejection. If two individuals have different alleles for *any* of these many MHC genes, then tissue grafts between them will be rejected as foreign material.

MHC restriction

A key concept of modern immunology is the concept of **restriction**. Most interactions of T-cell receptors with antigen are *MHC restricted*, which simply means that the antigen is recognised in association with an MHC molecule and not alone. This is not surprising since we know that the antigen is presented by MHC molecules, but the key point is that the MHC *itself*

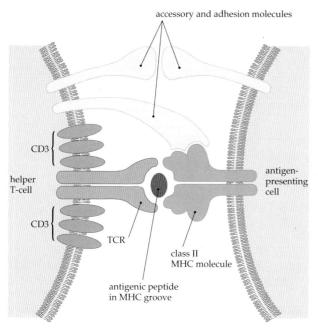

Fig. 23.20 Schematic representation of the TCR-mediated interaction between a T cell and an antigen-presenting cell. The interaction between the two cells is complex, involving many surface molecules (accessory and adhesion molecules) on each cell. However, the interaction is initiated only when the variable portion of the TCR recognises an antigen/MHC structure on the antigen-presenting cell

provides part of the entity that is recognised. Thus, if an MHC-restricted T cell recognises peptide X presented by an MHC molecule of a particular type, for example MHC^a, then this T cell will only recognise (MHC^a + peptide X), not (MHC^a + peptide Y) or (MHC^b + peptide X), or MHC^a alone or peptide X alone. The TCR sees the complex of MHC and peptide, both components contributing to the specificity and both being required to trigger the particular T cell (see Table 23.3).

Table 23.3 MHC restriction and the specificity of the TCR

Antigen recognised by TCR	Antigen being presented	T-cell response
MHC^a + peptide X	MHC^a + peptide X	Yes
MHC^a + peptide X	MHC^b + peptide X	No
MHC^a + peptide X	MHC^a + peptide Y	No
MHC^a + peptide X	MHC^a alone	No
MHC^a + peptide X	peptide X alone	No
MHC^a + peptide Y	MHC^a + peptide Y	Yes
MHC^a + peptide Y	MHC^a + peptide X	No

MHC molecules present to T cells a sample of both internal synthesised antigens and external scavenged antigens. The MHC molecule is part of the structure that the T cell recognises and interacts with.

SPECIAL FEATURES OF THE IMMUNE SYSTEM
Tolerance in B and T cells

It is important that the immune system does not respond inappropriately to certain antigens, particularly self-antigens. If the immune system fails to respond to a potentially immunogenic antigen, a state of **tolerance** is said to exist. Tolerance occurs when lymphocytes come in contact with self-antigens at a particular early stage of development—in the thymus for T cells, in the bone marrow for B cells. As a result, mature lymphocytes do not normally initiate damaging responses against self-antigens. Normally, mature lymphocytes cannot be made tolerant otherwise tolerance could develop to a foreign potentially dangerous organism.

Both T and B cells can be made tolerant if they contact self-antigens at critical early stages of their development. Normally, mature lymphocytes cannot be made tolerant.

Autoimmunity

The processes that give rise to self-tolerance result in the removal, inactivation or suppression of most cells with the potential to react against self-antigens. This process is not perfect and, in some individuals, immune responses develop against the individual's *own* antigens, giving rise to **autoimmune diseases**. These diseases are usually chronic and may strike people at almost any stage of life. They are second only to cancer in their cost to the health-care system.

Autoimmune diseases affect many tissues, including lungs, kidney, stomach, skin, joints, adrenal glands, thyroid or pancreas (Fig. 23.21). Damage to the pancreas results in a reduction in the capacity of the tissue to make insulin, resulting in diabetes. Autoimmune diseases are usually treated with anti-inflammatory drugs (especially in the case of arthritis) or, in severe cases, with immunosuppressive drugs to dampen the autoimmune response. This, of course, carries the risk of infection because other immune responses are also depressed.

THE NATURE OF IMMUNE RESPONSES
Humoral responses

Imagine bacteria entering the bloodstream through a cut or scratch and beginning to proliferate. Macrophages at the tissue site or in the spleen break down some of the bacteria and recycle some bacterial peptides back to the cell surface on MHC molecules.

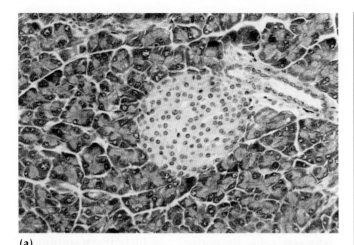

(a)

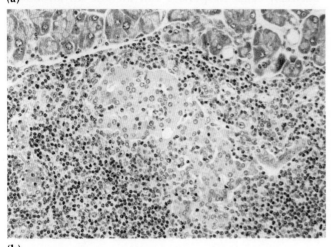

(b)

Fig. 23.21 Autoimmune diabetes. **(a)** A section of a
normal pancreatic islet, which is the part of the
pancreas that makes insulin, embedded in the exocrine
pancreas. **(b)** Severe invasion of the area of the islet
with lymphocytes (small round darkly stained cells),
and a breakdown in the structure of the islet.
Eventually this islet and others like it will be too badly
damaged to make enough insulin to regulate blood
sugar levels, and diabetes will result. Notice that there
is essentially no invasion of the exocrine tissue. The
invading T cells are quite specific for the islets

T_H cells in the spleen or local lymph node that
recognise these peptides respond to stimulation by
dividing and releasing lymphokines. At the same
time, B cells come into contact with bacteria and those
that have a surface antibody that recognises a
bacterial surface antigen are stimulated. If the B cells
also receive a signal from lymphokines made by T_H
cells, they will proliferate and begin to release
antibodies. These antibodies will bind to the bacteria
and, in the presence of complement (Box 23.3), kill
them and/or make it easier for macrophages and
polymorphs to dispose of them. This will provide
further stimulus for T_H cells, further lymphokines and
more B cells to respond. As increasing numbers of
cells circulate through the spleen, more cells reactive

BOX 23.3   THE COMPLEMENT SYSTEM

Complement is the collective term for a series of about
20 serum proteins that exist in the fluids of the body
in an inactive state. If the first component in this
series is turned into the active form, a complex
cascade of reactions occurs, in which each component,
once activated, activates the next.

The first component can be directly activated by
substances on the outer surface of certain bacteria
or parasites, by some serum proteins bound to a
variety of micro-organisms, and by antibodies that
are bound to their antigens (antigen–antibody
complexes). Complement can thus be activated
directly by pathogens or more specifically by
antibody. Several of the components along the
cascade, when activated, have important effects on
the immune system. The final components can form
a molecular complex, which punches holes in the
membranes of cells (microbes, parasites, or the host's
own cells) to which it is bound.

The activation of the complement cascade has three
main consequences:

- activation of a range of cell types involved in
  immune responses and inflammation, including
  macrophages, granulocytes and mast cells;

- facilitation of the engulfment by phagocytic cells
  of microbes or cells with complement bound to
  them;

- lysis of cells with complement bound to their
  surface (by hole-punching).

In addition to these effects, complement plays
a role in the induction of local inflammation, increasing
vascular permeability at sites of infection, inducing
contraction of smooth muscle (blood vessel walls), and
attracting and holding granulocytes at sites where
complement is bound.

to the antigens on this particular bacterium can
become stimulated and more B cells are available to
produce antibody. Soon some B cells will become
plasma cells, pumping out huge amounts of antibody
against the bacteria. Eventually, the increasing levels
of antibody will swamp remaining bacteria, resulting
in their elimination by macrophages and polymorphs.

Once the bacteria are gone, there is no more antigen
to stimulate the T or B cells, so they eventually stop
proliferating and making lymphokine or antibody,
and the system goes back to a resting state, the threat
of bacterial invasion averted. However, the
proliferation results in many more T cells able to
recognise the particular bacterium, and many more
B cells that can make antibody to it, scattered
throughout the lymphoid tissues as memory cells.
This is the immunised state. Should the same
bacterium come along again, a second immune
response (secondary or memory response) will be
much quicker and more effective.

When both B cells and helper T cells are stimulated by a foreign antigen, T_H cells will help the B cells make antibody, which will help kill and remove the pathogen. Once the antigen has been cleared, there is no further stimulus to the T or B cells and the response subsides.

Cellular responses

If the initial infection is a virus that proliferates within body cells, some viral antigens would appear as peptides in MHC molecules on the surface of infected cells. Antigens of this kind are more likely to stimulate T_C cells, which attack and kill any cell with viral antigen on the surface. Some free viral antigens might also be taken up by macrophages, and peptides from them presented by MHC molecules to T_H cells, stimulating them to respond and further helping the T_C cells to react. The stimulated T cells will also release a range of lymphokines that will attract and stimulate other cells, and may cause local inflammation or fever. Some B cells might also respond to the viral antigens and, in concert with the T_H cells, make antibodies against some viral components. Since the whole virus is inside the cell, there is little opportunity for B cells to be directly stimulated by it, or for antibodies to reach it, since antibodies cannot normally get inside cells. However, antibodies may prevent the spread of virus from cell to cell.

Again, proliferation of T_C cells means that more cells will be available to protect the individual from a subsequent infection and, if antibodies have been made, they might also protect against initial infection while the virus is outside the cells.

Cellular responses are also largely responsible for rejection of foreign tissues (Box 23.4). Here the foreign MHC antigens of the grafted tissue, be it heart, skin or kidney, stimulate both T_H and T_C cells, giving rise to cytotoxic cells, which can destroy the graft, or to lymphokines, which induce macrophages and granulocytes to cause inflammation, also destructive to the graft.

When antigens, such as viral antigens, are synthesised by cells and presented on MHC molecules, protective responses involve direct killing of infected cells by T_C cells, activation of phagocytic cells and inflammation due to lymphokines produced by stimulated T cells.

Regulation of immune responses

As we have seen, immune responses can take several courses, depending on the nature of the stimulus, but in all cases there is a great deal of co-operation between cells. Responses can include an antibody response, a T_C response and/or a T_H cell response. Helper T cells can make different lymphokines in different circumstances. The balance of these various responses will vary widely, although as you might expect, a response completely restricted to one component would be unusual. A great deal of research is underway to understand the subtle processes that alter the balance of these responses, since every disease is different in terms of the kinds of immune response that best provide protection or recovery. For example, it is no use making a vaccine that stimulates a strong antibody response if T_C cells are needed.

The regulation of immune responses is not always perfect, and in some conditions responses may overshoot or undershoot, causing damage to self-tissue, or not clearing the pathogen effectively. Secondary damage to surrounding tissues caused as a result of an immune response is **immunopathological** (Box 23.5), as distinct from damage caused by the pathogen that caused the response. In some cases, the immune response does more damage than the disease during its attempts to remove a relatively benign foreign organism.

BOX 23.4 TISSUE TRANSPLANTATION

Skin grafts, widely used for cosmetic surgery, usually involve transfer of skin from one part of the body to another. The tissue is 'self' and so there is no immune response against it; it is not rejected. Had the skin come from another non-identical individual (anyone but an identical twin), it would have been rejected within a week or two due to an immune response mounted against the MHC molecules it carried. However, with improvements in surgical techniques, tissue matching (MHC and blood-group typing) and immuno-suppressive drugs, which dampen the rejection process, a number of tissues can now be transplanted between individuals with a good chance of success.

In Australia, there are 500–600 kidney transplants per year (mostly cadaver donations), with a one-year success rate of 80–90%. There are about 100 heart transplants and 50 liver transplants, both with about the same 80–90% one-year success rate. Another tissue frequently transplanted is the cornea, in an operation that is now fairly routine and with a success rate (when the recipient eye is not inflamed or severely damaged) well over 95%.

A major difficulty in transplantation is the risk of infection when immunosuppressive drugs are used. These suppress the response to pathogens as well as to the transplant, so immunosuppressed patients are often kept isolated to reduce the risk of infection. Much current research is aimed at finding ways to suppress immune responses to particular antigens (i.e. on the transplant) without causing generalised immuno-suppression.

Multiplicity of immune responses

Pathogenic organisms may contain many antigenic molecules with tens or even hundreds of potential T- and B-cell epitopes. Usually only some of these will actually stimulate a response during a normal immune reaction. Some may stimulate B cells and others may stimulate T cells. In the overall immune response to any micro-organism, there may be dozens of different antibodies being produced in varying levels, as well as T-cell responses to many different epitopes. Which antigens are responded to, and the nature of the responses to them, can be of vital importance in determining the outcome of an infection. A typical immune response will have many different components, some of which may be protective, while others may be not useful or even harmful, stimulating allergy or other immuno-pathology.

> There are many possible arms to the immune responses, and many potential antigens in most foreign organisms. Different antigens from the same organism might stimulate different parts of the immune response.

Stress and immune responses

These days there is considerable talk of 'the stresses of living in a fast-moving society' and the effect this can have on health. How does stress or, more broadly, mental state affect the immune system? There is good evidence that acute stress, such as sudden accident or injury, can have a positive effect on an immune response that follows it. However, when either mental or physical stress is prolonged there can be marked reductions in the ability to develop immune responses, partly through prolonged exposure to stress hormones (Chapter 25).

There are more subtle and complex interactions between the nervous system and the immune system. For example, if an immunosuppressive drug is administered to an experimental animal in association with a particular taste stimulus (e.g. saccharine), then the taste and drug become associated in the animal's mind. Later, if only the saccharine is given, some of the immunosuppressive effects of the drug are reproduced in the complete absence of the drug (Fig. 23.22). Furthermore, in humans, some immune reactions, in particular, allergic responses, can be induced by *placebos*. In other words, if the patient thinks the allergic substance is present, even if it is not, some symptoms, at least, will appear. The mind is causing the immune system to respond to an antigen that is not there. The mechanisms for this are as yet unclear, but lymphocytes have receptors for a number of molecules produced by the nervous and endocrine systems, and stimulated T cells make

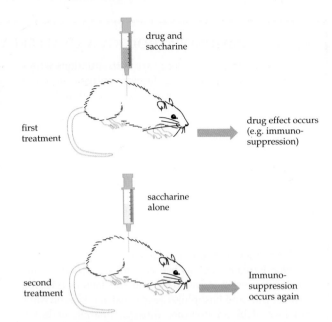

If the first treatment was either drug alone or saccharine alone, then the saccharine given at the second treatment does not lead to a drug effect

Fig. 23.22 The brain can reproduce drug effects. In this experiment, the taste of saccharine becomes associated with a particular drug effect. Later, the taste alone can reproduce the effects of the drug, presumably by acting through the central nervous system. Other groups in this experiment show that if the first treatment was just the drug or just the saccharine, then nothing happens at the second treatment with saccharine

a number of molecules, in addition to lymphokines, that can have effects on the central nervous system. This is an exciting and relatively new area of research that holds promise of interesting future treatments for a variety of diseases.

> Acute stress may stimulate immune responses but chronic stress can be inhibitory. Inhibition can be caused by stress hormones and through interactions between the immune system and the central nervous system.

Immunity to bacteria, viruses and fungi

Viruses are diverse and cause a wide range of diseases in many organs (Chapter 34). Each type of virus grows only in certain types of cell. Influenza virus, for example, only grows in respiratory epithelium, while polio virus grows in the nervous system. Infections may be acute, chronic, latent (present but not easily detected), recurrent or subclinical (detectable virus but no obvious disease). The role of the immune system in many of these conditions is unclear, although it plays a vital role in acute infections, such as influenza. Early in an infection, non-specific molecules provide some level of resistance. If general *viremia* occurs, antibodies can be important, but once

BOX 23.5 IMMUNOPATHOLOGY AND ALLERGY

Immunopathology and allergy are both situations where an immune response to a foreign antigen becomes excessive and results in damage to surrounding tissues, or causes life-threatening physiological responses.

Allergic responses occur when initial contact with an antigen (for example, pollens or house dust) results in production of IgE, which binds to the surface of mast cells. Later contact with the same antigen results in the antigen binding to the IgE on the mast cells, stimulating mast cells to release substances that cause acute inflammation. Since mast cells are common in the respiratory tract, it is here that allergic reactions often occur, resulting in constriction of the airways and, in extreme cases, suffocation. An allergen that gets into the bloodstream, such as bee venom, also has dramatic effects in allergic individuals, and the factors released by mast cells and basophils cause constriction of airways and loss of blood pressure through dilation of blood vessels. Both these potentially life-threatening symptoms are reversed by adrenaline, which is the usual treatment for acute allergic attack.

Other immunopathologies result when immune responses become disregulated or antigens persist in tissues, resulting in prolonged immune responses, inflammation and damage. For example, contact with certain chemicals results in the binding of the chemical to molecules in the skin. Immune responses to the bound chemical result in local inflammation as T cells release lymphokines, which stimulate macrophages and other inflammatory cells. The antigen is not easily cleared because it is bound to self-tissues, so the immune response is prolonged and difficult to contain and considerable tissue damage may ensue.

Infection by parasites may also cause immunopathology. Macrophages or other cells can become persistently infected with intracellular parasites and constantly activated by surrounding T cells. Hyperactive macrophages multiply and enlarge to cause granulomas, which are densely packed large swellings in the skin or internal organs. This response may ultimately be protective if the parasite is cleared, but persistence of the parasite can result in chronic granulomas that may become life-threatening.

Antibodies too can cause damage, not only by binding to foreign antigens expressed on tissues, but also because antibody attached to its antigen (an immune complex) can form precipitates that cause damage to blood vessels and kidneys. This most often occurs with persistent antigens, such as self-antigens or antigens from uncleared infections. Precipitates form when certain concentrations of antigen and antibody exist in the serum. Precipitated complexes can activate complement causing local inflammation at sites of accumulation.

BOX 23.6 IMMUNODEFICIENCY AND AIDS

The immune system is critical for the protection of an individual against infection. In the absence of immune protection, fatal infection almost invariably follows. In a clinical setting, the immune system may be inhibited by immunosuppressive drugs after transplantation or by radiation or drugs used in the treatment of cancer. In these cases, great care and use of antibiotics will usually see the patient through until the immune system recovers. There are, however, a number of diseases in which babies are born with defective immune systems. Di George syndrome is a T-cell deficiency resulting from the absence of a thymus. Severe combined immunodeficiency (SCID) results when genetic abnormalities result in the failure of both T and B cells to develop. In the past, children with these diseases died soon after birth due to severe viral, bacterial or fungal infections. Even the best modern medicine cannot save these children indefinitely without the help of an immune system. In recent years, some of these diseases have been successfully treated by transplanting normal MHC-matched bone marrow into the children, replacing the abnormal marrow and allowing the immune system to develop from the stem cells in the normal marrow transplant.

There are also a number of diseases that result in damage to the immune system. The best known example is AIDS (acquired immunodeficiency syndrome), a disease caused by the human immunodeficiency virus (HIV). This virus is normally difficult to transmit since its target cells are mainly lymphocytes, which are not accessible from the outside. Direct contact of body fluids through blood transfusion or as a result of certain sexual practices, such as anal intercourse, which almost inevitably results in minor bleeding, is necessary. Once inside the body, the virus binds to a molecule called CD4, which is predominantly expressed on T_H cells. After binding to the CD4 on the surface of the T_H cell, the virus, like all retroviruses (Chapter 34), enters the cell and integrates into the DNA, essentially becoming a part of the cell. It then uses the cell's own replication mechanisms to multiply, eventually damaging or killing the T_H cell. Over a long period of time, the virus slowly spreads through the immune system, severely reducing the number of T_H cells, making the immune system less and less effective. With time, secondary infections become more and more serious, eventually causing the death of the patient. In the absence of these secondary infections, HIV would not itself cause lethal disease.

The great difficulty in treating patients with AIDS comes from the fact that the virus is so completely integrated into the host cell genome and becomes part of the cell itself. There is little that is unique to the virus that could be the target of a drug. Furthermore, the very cells that would help initiate an immune response are themselves being destroyed.

the virus is inside its target cell, T_C cells are probably essential. Thus, the required proportions of antibody or T cells varies with the particular disease and the stage it is at. Antibodies may be useful in preventing spread or reinfection by attacking the virus while it is outside the cells, but T cells are probably required to clear an existing infection (Fig. 23.23). Some viral diseases, such as polio, measles and mumps, are easily controlled by vaccination. Others, such as herpes or AIDS (Box 23.6) are not yet easily controlled.

Within a host, *bacteria* are mainly extracellular and so are easier to deal with than are viruses. The main sources of protection are intact skin and mucosal surfaces. When these are breached, rapid proliferation of bacteria may initially outrun the immune system. Rapid but relatively non-specific defences, such as phagocytosis and killing by macrophages, polymorphs and complement (see Box 23.3), are crucial at this stage. Later, antibodies are involved through direct lysis mediated by complement, and because bacteria coated with antibody are more efficiently removed by macrophages, which finally remove the bacteria.

Fungi mostly cause superficial infections and do not usually get inside tissues or blood of healthy people. Common fungal diseases include athlete's foot, ringworm and candida, a common disease of female genital mucosa. Immunity to these diseases is not well understood and, being largely external, the organisms are out of reach of T cells and antibodies. Some resistance, probably T-cell mediated, does develop and is manifested through local release of lymphokines, which activate macrophages and polymorphs to clear infectious material.

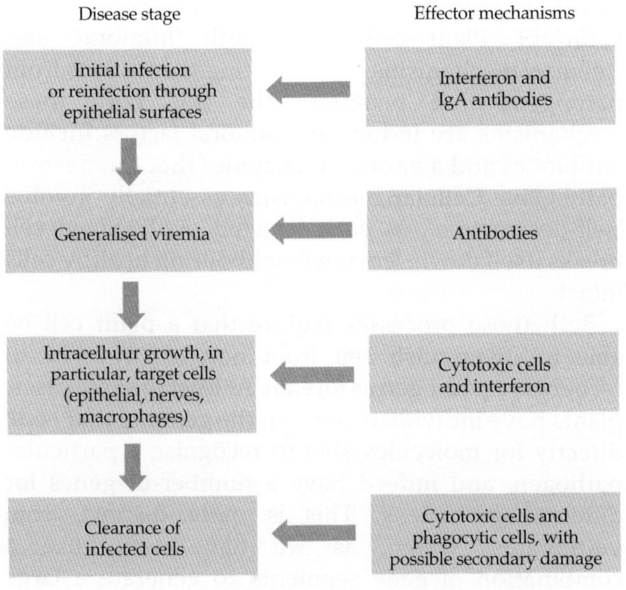

Immunity to protist and animal parasites

Protist and animal parasites are more difficult for the immune system to deal with because they are larger and more complex than are viruses or bacteria; some intestinal worms can be metres long. Also, their life cycles are often complex, with many rapidly changing stages, and a variety of mechanisms that subvert or avoid the immune system have evolved. This difficulty is apparent from the high levels of chronic infection in some human populations, particularly in the tropical world. Diseases such as malaria and sleeping sickness (protist parasites, Chapter 35) and schistosomiasis and hookworm (roundworm parasites, Chapter 38)) each affect over 100 million people worldwide. Not surprisingly, resistance to these parasites involves almost every aspect of the specific and non-specific elements of the immune system.

T cells are important in controlling parasites, particularly because the lymphokines they release stimulate macrophages, polymorphs and mast cells into states where they can kill parasites. Antibodies can also play a vital role, simply 'smothering' a parasite, or attracting macrophages or complement to cause lysis. In many cases, a combination of all these mechanisms is necessary to rid the body of a parasitic organism.

Immunity to tumours

There is a widely held belief that cancers are tissues that have escaped the control of the immune system; that is, the surveillance function of lymphocytes has broken down. This, however, is not the case: tumours are made up of our own tissue, admittedly growing in an unregulated manner, and are generally recognised as 'self' by the immune system and ignored. The major fatal human cancers of the lung, gut, breast and urogenital tract proliferate unchecked because their cells are *not immunogenic*, not because the immune system has failed.

Nevertheless, there may be some instances where a tumour could express antigens that are 'not self' and hence be potential targets for immune attack. These include viral antigens, since some tumours are thought to be virus-induced, or antigens normally expressed only in the fetus, which are re-expressed in tumours of adults. These could possibly be the targets of immune therapy since they are unique to the tumour in the adult, and if an immune response to them could be induced, it would not damage other tissues in the individual.

Disease stage	Effector mechanisms
Initial infection or reinfection through epithelial surfaces	Interferon and IgA antibodies
Generalised viremia	Antibodies
Intracellular growth, in particular, target cells (epithelial, nerves, macrophages)	Cytotoxic cells and interferon
Clearance of infected cells	Cytotoxic cells and phagocytic cells, with possible secondary damage

Fig. 23.23 Immune mechanisms that protect against viral infections as a viral disease progresses from initial infection to recovery

EVOLUTION OF IMMUNE RESPONSES

Animals

Clearly, every organism must have mechanisms for keeping itself distinct from its environment or from other organisms. These mechanisms become more complex as organisms become more complex. Invertebrates do not have organised lymphoid tissues or lymphocytes or any truly specific immune responses. However, all multicellular invertebrates have some kind of roving phagocytic cell that is able to engulf and destroy damaged or foreign material (Fig. 23.24). Some invertebrates can reject **allografts** (tissues from other individuals of the same species), and some have molecules able to bind to foreign material and enhance phagocytosis. In molluscs, annelids and arthropods, these molecules may be

induced by damage or infection. In no invertebrate studied to date is there evidence of any kind of memory or secondary response.

Each of the above reactions requires that the defensive molecules or cells are able to distinguish self from non-self, even though specificity is lacking in the sense of being able to distinguish between different non-self molecules. It is not clear how this recognition occurs, although it is likely that substances released from damaged self-cells play some role. Removal of foreign invaders in this case may involve removal by engulfment of any damaged self tissue.

Essentially all birds and mammals (including marsupials and monotremes) have all the characteristics of the immune system that we have described in this chapter (Fig. 23.25). With the exception of the more primitive fishes, most vertebrates have something resembling the mammalian system, with antibodies or similar molecules, as well as spleen, thymus and T and B cells. Nonetheless, there is a gradual increase in the complexity and refinement of immune systems as we move through the vertebrates.

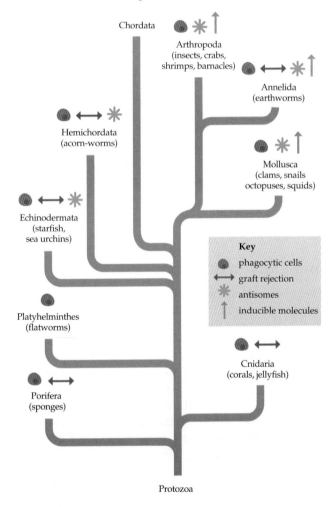

Fig. 23.24 Schematic diagram showing the increasing complexity of 'immune' responses as organisms become more complex. Antisomes are molecules that are able, in a general way, to attach to foreign (non-self) material, making destruction or clearance easier. They function like antibodies but are not highly specific, and may or may not be inducible (i.e. produced when needed in the event of an infection)

Immunity in plants

Like all other living organisms, plants must have mechanisms that protect them against invasion and, indeed, a number of complex processes have evolved in plants to provide such protection. Plants differ from most animals in that they have no circulatory system or migrating cells, so immune mechanisms must be local; each cell must fend for itself.

The first line of protection is, of course, the cell wall but, as in animals, physical barriers are never sufficient. Plant cells have both 'humoral' and 'cellular' mechanisms for protecting themselves from pathogens that penetrate the cell. Both these mechanisms are inducible. Humoral factors include antibiotics and a variety of enzymes that can destroy pathogens. Cellular mechanisms essentially involve 'self-destruction', as each infected or damaged cell breaks itself down, leaving neighbouring healthy cells intact.

Both these processes require that a plant cell be able to distinguish self from non-self, that is, to recognise a pathogen as foreign. At least in some cases, plants have individual genes in the genome that code directly for molecules able to recognise a particular pathogen, and indeed have a number of genes for different pathogens. This is quite distinct from vertebrates, which, as we have seen, use a combination of gene segments to generate a large random repertoire of defence molecules. In that case, there is no pre-existing gene for any particular

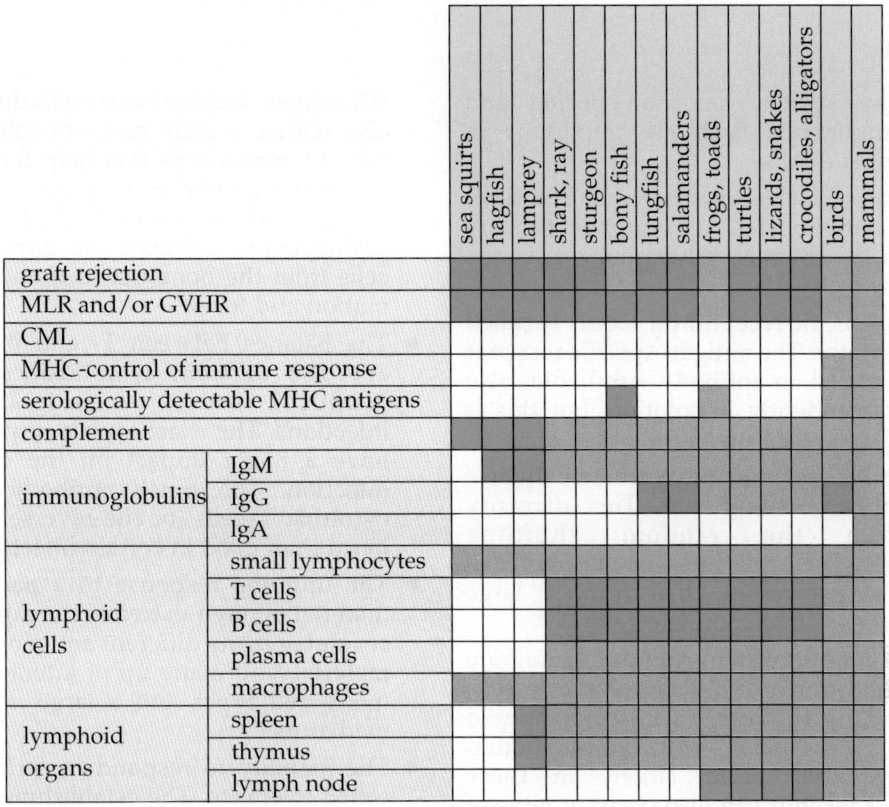

		sea squirts	hagfish	lamprey	shark, ray	sturgeon	bony fish	lungfish	salamanders	frogs, toads	turtles	lizards, snakes	crocodiles, alligators	birds	mammals
graft rejection															
MLR and/or GVHR															
CML															
MHC-control of immune response															
serologically detectable MHC antigens															
complement															
immunoglobulins	IgM														
	IgG														
	IgA														
lymphoid cells	small lymphocytes														
	T cells														
	B cells														
	plasma cells														
	macrophages														
lymphoid organs	spleen														
	thymus														
	lymph node														

Fig. 23.25 Changes in immune responses with increasing complexity in the Chordata. Green boxes indicate the presence of immune functions similar to those of mammals. Pink boxes indicate the presence of atypical or only partly developed traits. White boxes show cases where the particular trait is absent, or there is no information available

pathogen, although there may be a rearranged gene in one, or a number of clones of B or T cells. Obviously, plants cannot have genes for all pathogens, since they would have the same problem as animals in using up too much genetic space, and in being unable to deal with new pathogens. We must assume that they have evolved matching genes for important common pathogens, but must have some additional, as yet unknown, mechanisms for distinguishing self from non-self more generally.

SUMMARY

- The immune system has non-specific and specific components. Both are important in defence against disease.

- Cells that carry out specific immune functions are called lymphocytes. T lymphocytes are produced in the thymus, B lymphocytes in the bone marrow.

- The antigen-specific receptor on T cells is called the T-cell receptor. The antigen-specific receptor on B cells is called an antibody. Antibodies also function independently in solution, but this is not true of T-cell receptors.

- Antigen-specific receptors are extremely diverse in their antigen-binding site. This diversity comes from the random shuffling (rearrangement) of a relatively small number of genetic elements, rather than from a large number of pre-existing complete genes.

- Cells of the immune system continually migrate around the body in blood and lymphatic vessels, passing regularly through organised lymphoid tissues, such as spleen and lymph nodes. Immune responses occur largely in these organised tissues, but can also occur at sites of inflammation.

- Antigens for T cells are mostly proteins, and these must be broken down and presented to the T cell in the form of short peptides attached to an MHC molecule. Antigens for B cells can be proteins or carbohydrates, and the whole or part of the antigen can interact with an antibody.

- The primary role of B cells is to make antibodies in large quantity, to kill directly or assist in the killing of foreign infectious agents by macrophages and granulocytes.

- T cells have more diverse functions. They can kill antigen-bearing target cells directly, but they also release a wide range of soluble mediators called lymphokines that help B cells proliferate and make antibodies; help other T cells respond to antigen-stimulated macrophages and granulocytes; enhance the formation of blood cells from the bone marrow; and cause inflammation and fever.

- The balance between T- and B-cell responses and the different lymphokines that can be produced varies considerably between infections. The exact form a response takes can have a major impact on the outcome of an infection. Too much antibody and too few cytotoxic T cells (or the reverse) may do more harm than good in certain infections.

- The immune response to a particular foreign material is often extremely complex, usually to several or many different antigens on the foreign material, and made up of interactions between many cell types and a large array of soluble mediators.

- The inability to respond to certain antigens is called tolerance. The establishment of tolerance to self-antigens is an important aspect of the development of lymphocytes. A breakdown in self-tolerance can result in destructive immune responses against self tissues, resulting in autoimmune disease.

- The evolution of the immune system parallels the evolution of more complex life in general. All organisms need some form of defence against infection; however, specific immune responses, organised tissues and specialised cells and molecules exist only in vertebrates. The immune system seems to be at its most complex in birds and mammals.

QUESTIONS

1. What are the important characteristics of the array of micro-organisms that the immune system must deal with? How does the immune system cope with this?

2. What are the differences between specific and non-specific components of immune responses?

3. Explain how the process of gene rearrangement occurs, and how the timing of it during the development of lymphocytes gives rise to a wide range of different TCR and antibody molecules?

4. What is immune tolerance? Under what circumstances is it desirable, and when should it be avoided?

5. How and why do lymphocytes migrate around the body?

6. What are the essential differences between T-cell and B-cell antigens?

7. What are the main functions of MHC molecules?

8. Explain some ways in which immunological memory might work.

9. How do T cells regulate other cells during immune responses? Which other cells do they regulate?

10. Discuss three ways in which the immune system can go wrong and do damage to self tissues.

11. Explain why patients die of AIDS.

12. Why is transplantation between individuals so difficult and how are these difficulties overcome?

13. Name three characteristics of lymphocytes that make them different from most (but not necessarily all) other cell types in the body.

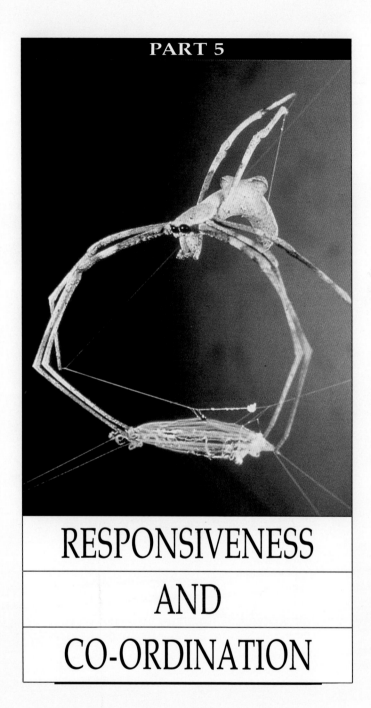

PART 5

RESPONSIVENESS

AND

CO-ORDINATION

PLANT HORMONES
AND GROWTH RESPONSES

Plants compensate for their relative immobility by modifying their growth in response to stimuli from the environment in which they live. For example, shoots grow and bend towards a light source, roots grow down into soil and flowers are produced at times of the year suitable for seed production. Thus, environmental factors, such as light, day length, temperature and gravity, can exert dramatic effects on plant growth and development. Nevertheless, some developmental changes are largely programmed genetically and hence occur regardless of environmental factors.

Responses of plants to both internal and external influences may involve changes at the molecular level, through altered gene expression; the cellular level, where cells may either undergo division or growth and differentiation; and the whole plant level, where the growth pattern of a root or shoot is altered or new organs are produced. Many plant responses that are triggered by environmental stimuli are brought about by the action of hormones. **Plant hormones** are molecules that affect plant growth and development at very low levels, sometimes at concentrations below one part per billion. This may involve action at some distance from the site of synthesis, but frequently plant hormones are produced in the same tissue in which they exert a response. However, whether growth responses are regulated primarily by changes in the *levels* of plant hormones or by changes in the *sensitivity* of tissues to these substances is still a matter of debate among plant physiologists. Both mechanisms probably occur.

Our understanding of the action of plant hormones has been aided in the last 10 years by the use of modern methods of analysis, such as gas chromatography, mass spectrometry and immunological assays, and by the use of single-gene mutants that markedly alter growth and development compared with wild-type plants (Fig. 24.1). If exogenous (external) application of a particular hormone to such a mutant induces development of the wild-type

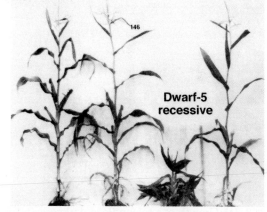

Fig. 24.1 Effect of the plant hormone, gibberellic acid (GA), on normal and dwarf corn. Left to right: normal plant (control); normal plus GA; dwarf plant (control); dwarf plus GA

phenotype then, by implication, that hormone is probably involved in the normal process of growth and development.

In this chapter, we will discuss five main groups of plant hormones—auxins, gibberellins, cytokinins, abscisic acid and ethylene—the physiological processes that they influence, and the growth responses that are triggered by environmental stimuli.

> Plant growth and development are controlled by many environmental and genetic factors. Many responses are brought about by the action of hormones, molecules active at very low concentrations.

AUXIN
Phototropism

Auxin was the first plant hormone to be discovered. In 1881, Charles Darwin and his son Francis performed experiments on coleoptiles, the sheaths enclosing young leaves in germinating grass seedlings

(Fig. 24.2). They exposed coleoptiles to light from a unidirectional source and observed that they bend towards the light. This phenomenon is termed **phototropism**. A trophic response is one in which growth is directed by an environmental factor and usually involves bending of the organ involved. By covering various parts of the coleoptile with light-impermeable metal foil, the Darwins discovered that light is detected by the coleoptile tip, but that bending occurs below the tip in the zone of cell elongation. They proposed that a messenger is transmitted in a downward direction from the tip of the coleoptile.

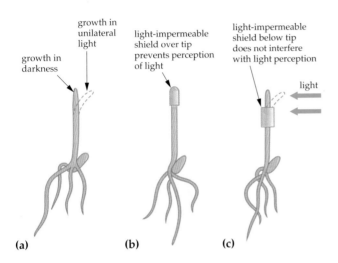

Fig. 24.2 **(a)** Growth of a coleoptile in darkness and unilateral light. **(b)** The tip of a coleoptile is the site of perception of unilateral light. If the tip is covered with aluminium foil there is no response. **(c)** Covering the zone below the tip does not prevent bending towards the light source

In the 1920s, Friedrich Went, a plant physiologist from The Netherlands, showed that a chemical messenger diffuses from coleoptile tips. When small agar blocks containing messenger from coleoptile tips were transferred to the cut ends of similar coleoptiles, they caused the coleoptiles to bend (Fig. 24.3). Went later

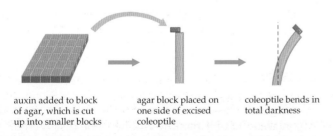

auxin added to block of agar, which is cut up into smaller blocks

agar block placed on one side of excised coleoptile

coleoptile bends in total darkness

Fig. 24.3 Went's experiment showing that application of auxin to one side of a coleoptile with the tip removed will show a growth response in darkness

proposed that the messenger substance is a growth-promoting hormone, which he named *auxin*, that becomes asymmetrically distributed in the bending region. He suggested that auxin is at a higher concentration in the shaded side, promoting cell elongation, which results in a coleoptile bending towards the light.

Since these experiments were carried out, the existence of auxin and its tendency to move downwards from the tip of a coleoptile have been confirmed, and Went's original experiment with agar blocks has been repeated with similar results. However, doubts now exist about the exact role of auxin in coleoptile bending. Determinations of auxin distribution using modern analytical techniques have shown that approximately equal levels of auxin occur on the illuminated and shaded sides of coleoptiles and stems. Furthermore, it has been reported that the shaded side contains lower levels of growth-inhibiting substance(s). This, of course, would exert an effect similar to an accumulation of growth-promoting auxin on the shaded side. The original method of determining the growth-promoting potential of illuminated and shaded tissue, by means of bioassays of coleoptiles, would not distinguish between a *low* level of auxin and an *accumulation* of growth-inhibiting substances. Thus, bending of coleoptiles may be regulated by auxin combined with growth inhibitors.

Auxins are growth-promoting hormones; they induce bending of coleoptiles towards light, an effect that may also involve growth inhibitors.

Indole-3-acetic acid

The main auxin occurring naturally in plants is **indole-3-acetic acid** (IAA), which is a small molecule consisting of an indole group with an acetic acid side chain (Fig. 24.4). When applied to intact plants, IAA does not elicit large responses, but is very effective in promoting growth of portions of stem or coleoptile segments. Concentrations of IAA as low as 0.2 parts per million (approximately 10^{-7} M) will produce a response (Fig. 24.5). IAA exerts this effect by promoting elongation of cells rather than by increasing the number of cells present in the segment.

Within a plant, IAA may be synthesised primarily from the amino acid tryptophan. The level of active IAA in tissue can be influenced by the formation or hydrolysis of conjugates, which are composite molecules comprising IAA bonded to another molecule, such as a sugar. The level of IAA in a particular organ can also increase as a result of transport from another part of the plant. IAA shows *polar transport*, that is, the hormone is transported from the top to the bottom of stem segments much more readily than in the reverse direction.

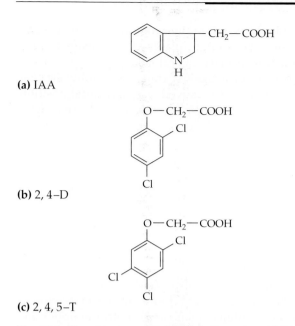

(a) IAA

(b) 2, 4–D

(c) 2, 4, 5–T

Fig. 24.4 The structure of **(a)** indole-3-acetic acid (IAA) and the synthetic auxins **(b)** 2,4-dichlorophenoxyacetic acid (2,4-D) and **(c)** 2,4,5-trichlorophenoxyacetic acid (2,4,5-T). These synthetic auxins are used as herbicides because at high levels they disrupt normal growth

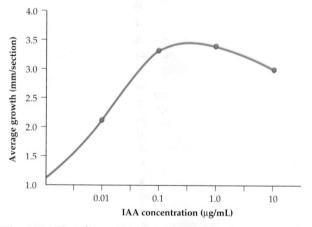

Fig. 24.5 The effect of IAA on elongation of segments of pea epicotyl

Several other naturally occurring auxins have been discovered, including 4-chloro-indoleacetic acid, which has been identified in extracts from certain legumes. There are also many synthetic auxins, not found in nature, that are used in agriculture as herbicides (Fig. 24.4) or as root-promoting substances in plant propagation (Fig. 24.6). For example, both 2,4-dichlorophenoxyacetic acid (2,4-D) and 2,4,5-trichlorophenoxyacetic acid (2,4,5-T) are well-known herbicides that were used during the Vietnam war. 2,4,5-T was a component of Agent Orange, used as a defoliant in jungle warfare. Unfortunately, dioxins, which are among the most toxic substances known, were present as trace contaminants in herbicides containing 2,4,5-T and the use of these substances is now banned.

Fig. 24.6 The effect of an auxin on root initiation. Leaves of bean cuttings were either untreated (i.e. a control, left), or treated with 5 mg/L (middle) or 50 mg/L (right) of the synthetic auxin naphthalene acetic acid (NAA). Increasing concentrations of NAA induce the formation of roots on the stems

Auxin and cell elongation

One mechanism proposed for the effect of auxin on cell elongation is the *acid growth hypothesis*. The hypothesis is that auxin stimulates release of H^+ from within a cell (Table 24.1), which results in a loosening of bonds in the cell wall, making it more flexible. As a result of turgor pressure, cells elongate. This hypothesis explains two experimental observations: excised stem segments floating on a solution of low pH show enhanced elongation; and the pH of a solution in which auxin-treated stem segments are floating becomes lower.

Auxin may also affect plant development by regulating gene expression. For example, in strawberry fruits, high levels of a specific mRNA are associated with a block in fruit development. Auxin within the plant represses the level of this specific mRNA, allowing fruit to develop normally.

> The acid growth hypothesis for the action of auxin on cell elongation suggests that release of H^+ from cells into the walls causes cell-wall loosening. This leads to cell expansion under turgor pressure. Auxin may also regulate gene expression.

Auxin and apical dominance

Many plants consist of a single dominant shoot because the apical bud inhibits the growth of axillary buds further down the stem. On removal of the apical bud of such a shoot, axillary buds will grow out to form branches. However, if IAA is applied to the cut surface of a stem after removal of the apical bud, lateral bud growth remains suppressed. IAA mimics the effect of the apical bud and, since shoot tips are important sites of IAA production, IAA is therefore implicated in the maintenance of apical dominance. This is supported by the increased apical dominance

Table 24.1 Biochemical and molecular responses to auxin

Growth response	Response time (min)	Tissue
Increases in:		
Respiration rate	5	Maize coleoptile
mRNA sequences	5	Soybean
	10–20	Garden pea
H⁺ extrusion	7–8	Maize coleoptile
Membrane potential	12	Oat coleoptile
β-glucan synthase activity	10-15	Garden pea
K⁺ uptake	30	Oat coleoptile
Protein synthesis	30	Tobacco protoplast
Decreases in:		
ATP:ADP ratio	5	Oat coleoptile
Cytosolic pH	5	Maize coleoptile

shown in certain plants genetically engineered to overproduce IAA. For example, when a gene for IAA production was transferred from *Agrobacterium tumefaciens* (a bacterium widely used in genetic transformation) into petunias, IAA levels were increased and the degree of branching decreased. But how can IAA, a growth promoter, act as a growth inhibitor? It may do so by diffusing downwards from the apical bud and influencing cells in the nodes adjacent to the lateral buds to produce a second factor, which suppresses the growth of the lateral buds.

▌ IAA can substitute for the apical bud in the maintenance of apical dominance.

Gravitropism

In **gravitropism**, roots bend towards the earth's gravitational field while stems and coleoptiles grow away from it. Gravitropic bending of a root or shoot placed in a horizontal position results from differential growth on upper and lower sides. In roots, the root cap detects the stimulus of gravity, whereas actual bending occurs in the elongation zone behind the root cap. In shoots, however, there does not seem to be such a clear demarcation: the capacity to detect the direction of gravity appears much more uniformly spread along a coleoptile or stem.

Detection of gravity appears to involve the sedimentation of plastids (statoliths), usually amyloplasts, which contain starch granules (Chapter 17). Evidence for the role of amyloplasts comes, for example, from observing plants grown in spacecraft where they are exposed to only a weak gravitational field. These plants have roots that are orientated randomly as

are amyloplasts within their root cap cells. Also, some starch-deficient mutant plants show impaired gravitropism. In a normal root or rhizoid growing vertically, amyloplasts settle against transverse cell walls (Fig. 24.7). When a root is moved to a horizontal position, amyloplasts settle on the side of cells that were originally vertical. The movement of amyloplasts may result in a gradient of growth hormones within cells, causing growth and bending of the root in the direction of the force of gravity.

A role for auxin in mediating gravitropism was proposed in the 1930s by the Russian plant physiologist Cholodny. Support has come from recent work on soybean hypocotyls, which show an asymmetrical distribution of auxin-regulated mRNAs correlated with the gravitropic response. However, there is no apparent auxin gradient in the direction of the force of gravity and the auxin level in coleoptiles placed horizontally is similar on both the upper and

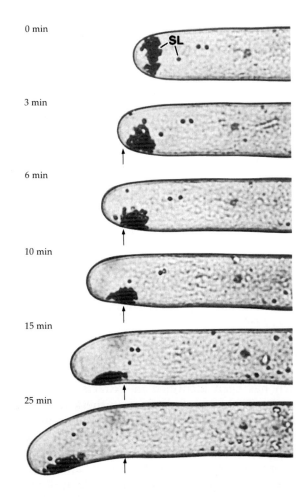

Fig. 24.7 Time-lapse photographs of a rhizoid of *Chara*, a green alga related to land plants (Chapter 37). The rhizoid, originally vertical, was displaced to a horizontal position. The photographs show sedimentation of statoliths (SL) followed by bending downwards of the rhizoid after 25 minutes. The arrow indicates the same point on the cell wall in each photograph

lower sides. Therefore, rather than alter the *level* of auxin, gravity may somehow influence the *sensitivity* of tissue to endogenous auxin and thus affect the production of mRNA.

There is evidence that Ca^{2+} is also an important messenger in gravitropic responses. It has been suggested that, in horizontally orientated roots, a downward movement of Ca^{2+} towards the lower side is responsible for the subsequent bending of the root.

> Gravitropism results in bending of stems and roots in response to gravity. In roots, the response seems to involve auxin and Ca^{2+} gradients.

GIBBERELLINS

Gibberellins are another group of plant hormones that influence growth, including stem elongation, seed germination, mobilisation of starch reserves in seeds and leaf expansion. They occur in low concentrations in vegetative tissues (about 10 parts per billion) but in higher concentrations in developing seeds (more than 1 part per million).

More than 80 different gibberellins have been isolated from a range of plants and fungi. All are small molecules (diterpenes) of similar basic molecular structure, with 19 or 20 carbon atoms (Fig. 24.8). They were first recognised in the 1930s as a product of a pathogenic fungus, *Gibberella fujikuroi*, which infects rice plants, making them tall and spindly. The molecular structure of the active compound, gibberellic acid (GA), was identified later in 1956. The major endogenous active gibberellin that causes stem elongation in many plants is GA_1 (Fig. 24.8).

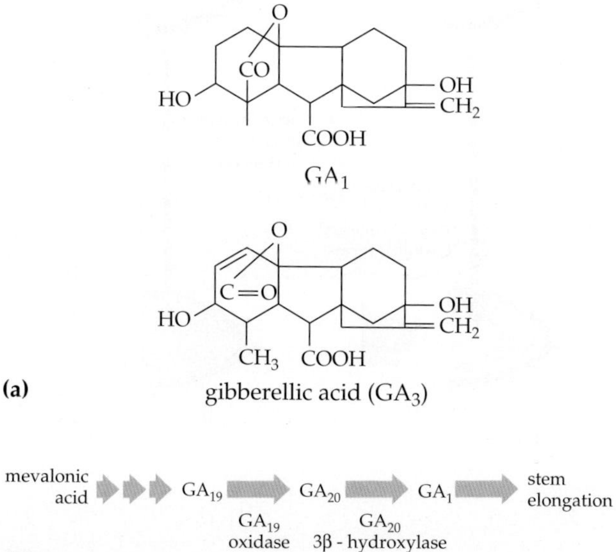

GA_1

gibberellic acid (GA_3)

(a)

mevalonic acid ➤➤➤ GA_{19} ➤ GA_{20} ➤ GA_1 ➤ stem elongation

GA_{19} oxidase GA_{20} 3β - hydroxylase

(b)

Fig. 24.8 **(a)** Structure of two active gibberellins, GA_1 and GA_3. **(b)** The major biosynthetic pathway in the vegetative tissue of many higher plants leading to the active gibberellin, GA_1, which is responsible for shoot elongation

Gibberellins and stem elongation

When Mendel's dwarf (*le*) pea plants (Chapter 9) are treated with 1 μg of GA, normal stem elongation and growth are restored (Fig. 24.9). Dwarfness is due to the *le* mutation, which blocks the last step in the biosynthetic pathway leading to GA. Stems and leaves of tall (wild-type) peas contain about 15 ng of GA_1 per gram of fresh weight, while shoots from dwarf *le* peas contain only about 1 ng per gram. However, dwarf *le* peas do contain high concentrations of GA_{20} (about 70 ng per gram), the precursor of GA_1. Since tall plants have only a third of this, the *le* mutation seems to result in an inability to convert GA_{20} to GA_1 (Fig. 24.8). This was confirmed experimentally when GA_{20} was labelled with a radioactive isotope and fed to dwarf and tall peas. Only tall peas produced significant quantities of labelled GA_1, indicating that the *le* mutation blocks the step converting GA_{20} to GA_1. The response of dwarf peas to gibberellin involves both cell division and elongation. The response is dependent on the level of GA_1, the biologically active form of the hormone, and not on the total level of gibberellins.

Fig. 24.9 The response of dwarf peas to applied GA_1. The control plant is on the right. The plant second from the right has been treated with 0.01 μg GA_2, with increasing GA_1 doses in two-fold steps to 24 μg of GA_1 (plant on left)

Gibberellins are also involved in bolting, the rapid shoot elongation of many rosette plants before flowering. This process is normally induced by either the onset of longer day lengths or the completion of a period of exposure to low temperatures (see p. 533). For example, in spinach, transfer from short-day lengths to long-day lengths results in the accumulation of GA_1 and its immediate precursor GA_2 due to increased activity of the enzyme, GA_{19} oxidase (Fig. 24.8).

Evidence for the role of gibberellins in breaking seed dormancy and stimulating germination has come from studying *Arabidopsis* mutants (Fig. 9.13). Seeds of mutants deficient in gibberellins do not germinate, even when provided with light, water and prechilling. The only treatment that results in germination is application of exogenous active GA. Release from dormancy in wild-type *Arabidopsis* appears to involve an increase in the sensitivity of seeds to endogenous GA. It has also been suggested that light, often required for germination, may increase the biosynthesis of endogenous gibberellins.

Gibberellins and mobilisation of seed storage products

In cereal crops, the endosperm of grain contains carbohydrate and protein reserves upon which the developing embryo depends for energy and nutrition until it commences photosynthesis. The reserves in the endosperm must therefore be mobilised and transported to the embryo (Fig. 24.10). The breakdown of endosperm starch and proteins into smaller more easily transported molecules, such as sugars and amino acids, results from the action of a range of

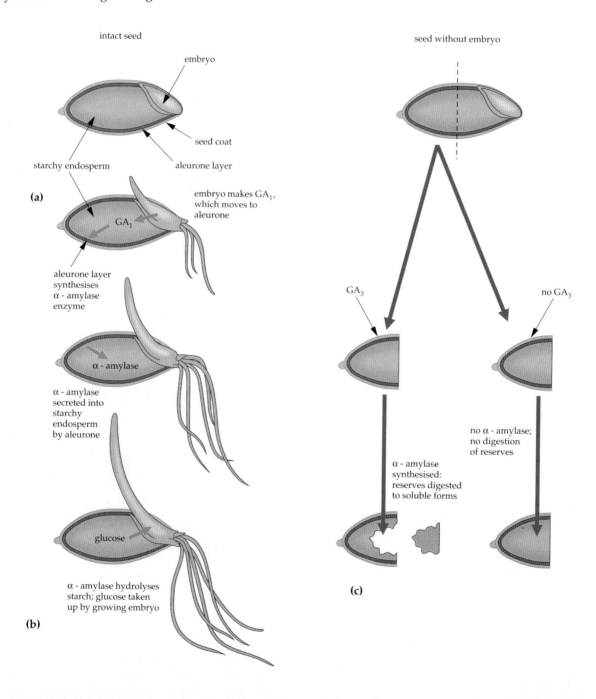

Fig. 24.10 Longitudinal sections through barley grains indicating the **(a)** major tissue types, **(b)** growth responses and **(c)** the action of gibberellin and the embryo on α-amylase production

---

BOX 24.1   SYNTHETIC GROWTH RETARDANTS

Synthetic growth retardants, such as paclobutrazol, reduce vegetative plant growth and have important applications in agriculture and horticulture. Paclobutrazol, which is a substituted pyrimidine, acts by blocking the plant's synthesis of gibberellin. Treatment of plants with such growth retardants can result in reduced lodging (falling over) in small grain crops; reduced need for pruning, and increased flowering and fruiting in orchard trees; reduced growth in ornamental trees (e.g. for use near power lines); and more compact shoot growth in herbaceous potted plants. Less frequent mowing of turf grasses also may be a potential application for these compounds.

Fenarimol is another substituted pyrimidine, related to paclobutrazol, but with more activity against fungi than plants. In fungi, it blocks the biosynthesis of ergosterol, an essential requirement for growth, and is thus used as a fungicide to control plant diseases. At high doses, however, it can affect the growth of plants as well. In apple trees, overuse of fenarimol causes reduced growth of leaf petioles, stunted shoot growth, flower and fruit abortion, all symptoms of reduced gibberellin biosynthesis.

---

hydrolytic and proteolytic enzymes, including α-amylase, which breaks down starch molecules into maltose (a disaccharide consisting of two glucose molecules).

In barley, enzymes are produced specifically by cells in the outermost layer of endosperm, the aleurone layer. The secretion of these enzymes is under the control of the embryo, since if the embryo is removed from the seed before soaking the seed for germination, no endosperm hydrolysis takes place (Fig. 24.10). GA_1 is the messenger that moves from the embryo to the aleurone layer, inducing the transcription of various genes, increasing the level of mRNAs coding for α-amylase production.

Thus a plant hormone (GA_1) produced in one tissue (the embryo) may regulate gene expression in another tissue (in this case, the aleurone layer). Another plant hormone, abscisic acid (p. 525), also regulates the transcription of these genes, but has the opposite effect to GA_1, resulting in reduced mRNA levels. An understanding of this system has allowed gibberellins to be used commercially in the brewing industry, which depends on the breakdown of endosperm reserves for the malting process.

> Gibberellins are a group of growth-promoting hormones that stimulate both cell division and elongation, thus influencing stem growth and bolting. In germinating cereal seeds, GA_1 produced in the embryo moves to the aleurone layer, where it activates genes that control the synthesis of the enzyme that converts starch to sugar.

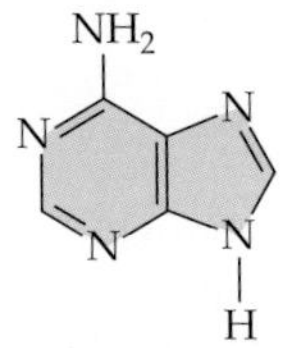

(a) Adenine (6-amine purine)

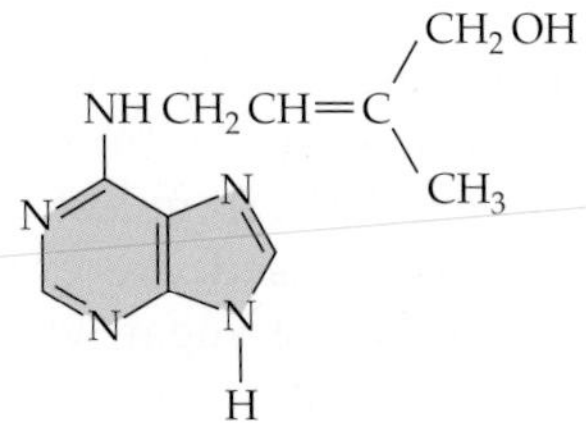

(b) Zeatin

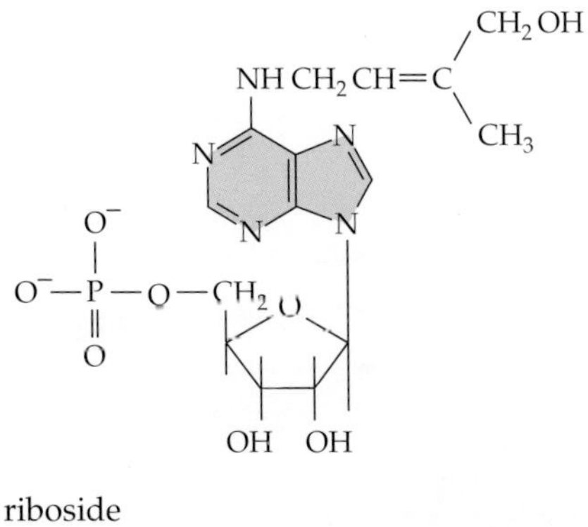

(c) Zeatin riboside

Fig. 24.11 Structure of (a) the purine adenine and the cytokinins (b) zeatin and (c) zeatin riboside

CYTOKININS

Cytokinins are plant hormones identified by their ability to stimulate cell division in plant tissue cultures. Most cytokinins are derivatives of the purine adenine (Fig. 24.11). Roots and developing fruits are major sites of cytokinin synthesis in flowering plants. **Zeatin**, isolated from corn seed, is the most active known naturally occurring cytokinin.

Cytokinins may regulate senescence (yellowing) of leaves on an intact plant. In experiments where leaves are removed from certain plants and floated on water containing cytokinin, leaves take longer to senesce than do leaves in control treatments not provided with hormone. In other words, the longevity of cut leaves is enhanced by cytokinin.

Cytokinins may also be involved in the regulation of the growth of axillary buds since bud growth is promoted by direct application of cytokinin (Fig. 24.12). Exogenous cytokinin and auxin are therefore antagonistic in their effects on axillary bud growth.

Fig. 24.12 The effect of a cytokinin on axillary bud growth. The bud of the plant on the right was treated with 300 parts per million of the synthetic cytokinin, kinetin, three days previously. The plant on the left is a control

Cytokinins are used commercially to induce organogenesis in tissue cultures. Small pieces of plant material, for example, from leaf or stem tissue where cell division has ceased, are excised and placed on a medium containing sugar, vitamins, various salts, and various concentrations of auxin and cytokinin.

Under these conditions, the piece of plant material may give rise to a mass of callus (meristematic) cells (Chapter 15). From this callus, shoots and roots may be regenerated. Within certain concentration ranges, a high cytokinin-to-auxin ratio results in the production of new shoots, whereas a high ratio of auxin to cytokinin is conducive to root formation (Fig. 24.13). Cytokinins are also widely used in micropropagation, where axillary or apical buds of shoots are used as the starting material for the production of many new cloned shoots, each of which may give rise to a new plant. Native Australian plants that have been propagated in this way include river red gum, *Eucalyptus camaldulensis*, kangaroo paws, *Anigozanthos*, and Christmas bells, *Blandfordia grandiflora*.

In *Agrobacterium tumefaciens*, a gene has been identified that controls an important step in cytokinin production. This gene has been transferred to tobacco plants, and the transgenic plants have increased levels of certain cytokinins and increased lateral bud outgrowth. The reason scientists transfer genes from an organism such as *A. tumefaciens* is largely because there are no known single-gene mutants in flowering plants that specifically cause large changes in the levels of cytokinins. However, this is not the case in mosses. One moss, *Physcomitella patens*, has a cytokinin-producing mutant with much higher levels of certain cytokinins than the wild type.

Cytokinins are plant hormones that stimulate cell division in tissue cultures and may be involved in the control of growth of lateral branches and leaf senescence.

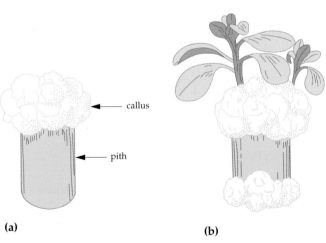

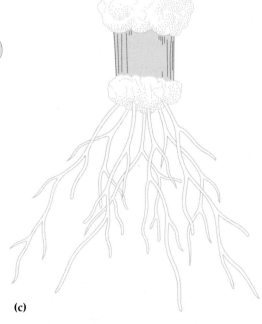

Fig. 24.13 The effects of various levels of a synthetic cytokinin (kinetin) and of auxin (IAA) on the growth of shoots or roots from tobacco callus cultured on nutrient agar: **(a)** kinetin 0.2 mg/L, auxin 0.18 mg/L; **(b)** kinetin 1 mg/L, auxin 0.03 mg/L; and **(c)** no kinetin, auxin 0.18 mg/L

ABSCISIC ACID

In addition to growth stimulators, such as auxins, gibberellins and cytokinins, plants also have growth inhibitors. These assist them to tolerate or avoid periods of adverse conditions, such as drought, salinity and low temperatures. Responses to stressful environments (Chapter 29) may involve modification at a morphological level (e.g. leaf drop and the formation of dormant buds in deciduous trees), changes at a physiological level (e.g. closure of stomata) or changes at a biochemical level (e.g. increase in frost resistance). The search for regulatory compounds involved with plant responses to adverse conditions occurred during the 1950s and 1960s and resulted in the isolation of the growth inhibitory compound abscisic acid.

Abscisic acid contains 15 carbon atoms (Fig. 24.14) and is synthesised from mevalonic acid via violaxanthin, a carotenoid intermediate. This pathway was indicated by work with maize mutants with blockages at various steps in the pathway for carotenoid synthesis. These mutants are almost white and possess markedly reduced levels of ABA.

ABA and drought resistance

Three mutations of tomato are known that cause plants to be ABA deficient so that even the mildest water stress results in a tendency for wilting to occur. There is a close correspondence between the severity of the wilting phenotype and ABA concentration (Table 24.2). Addition of ABA can cause the mutants to revert to wild type, confirming that this hormone controls the tendency to wilt.

Due to ABA deficiency, the stomatal aperture remains relatively large in mutants and thus these plants have much higher rates of transpiration. In wild-type plants, leaf ABA levels rise dramatically after water stress, resulting in rapid closure of stomata. In stomatal guard cells this increase in ABA level may be 20-fold. An even quicker response (in terms of stomatal closure) may be mediated by the redistribution of ABA already present within leaves. These results suggest that modification of drought tolerance in breeding programs may be possible by selecting for modified ABA levels or sensitivity. This could be of benefit to arid countries such as Australia.

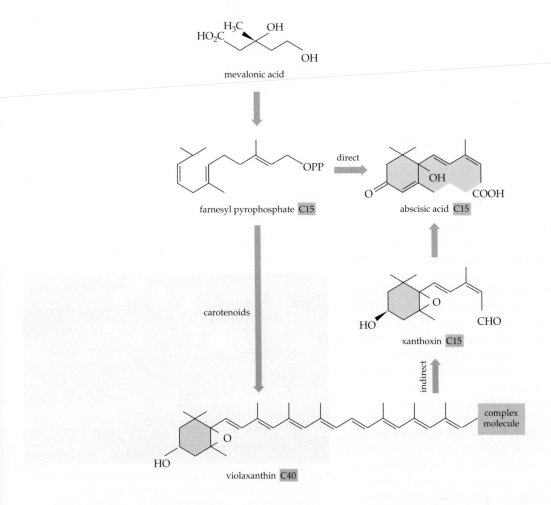

Fig. 24.14 The two possible biosynthetic pathways leading to abscisic acid: the direct pathway from farnesyl pyrophosphate and the indirect pathway, which occurs via the breakdown of carotenoid pigments

Table 24.2 Abscisic acid levels regulate wilting in tomatoes, as shown by specific mutants (a)

Genotype	Phenotype	ABA (ng.g⁻¹ fresh weight)
Flc/Flc; Not/Not; Sit/Sit	Normal	110
Flc/Flc; not/not; Sit/Sit	Mildly wilty	52
flc/flc; Not/Not; Sit/Sit	Wilty	23
Flc/Flc; Not/Not; sit/sit	Wilty	9
flc/flc; Not/Not; sit/sit	Wilty	10
flc/flc; not/not; Sit/Sit	Extremely wilty	4
Flc/Flc; not/not; sit/sit	Extremely wilty	3

(a) Mutants: *flacca* (*flc*), *notabilis* (*not*) and *sitiens* (*sit*). Plants were grown in a controlled environment at 95% relative humidity for six weeks.

ABA and frost tolerance

Frost tolerance is correlated with elevated ABA levels. For example, ABA levels increase transiently during exposure to cold temperatures in potatoes, while an ABA-deficient mutant of *Arabidopsis* will not become cold-tolerant. Cycloheximide, an inhibitor of protein synthesis, can prevent increase in frost resistance, suggesting that the synthesis of new proteins is involved in the response.

There is now clear evidence that ABA mediates responses to environmental stress by regulating gene expression. Certain genes are up-regulated by ABA while others are down-regulated. Understanding the genetic and biochemical regulation of plant responses to environmental stress is an important part of research programs aimed at maximising crop production under periodically unfavourable conditions.

Seed dormancy

Once a seed matures and development of the embryo ceases, dehydration follows (Chapter 15). The seed may then either be dormant or quiescent. A dormant seed will not germinate immediately when rehydrated. A quiescent seed will germinate if rehydrated because of its higher metabolic state. For many plants seed dormancy is advantageous since at the time of seed release conditions are often inappropriate for growth and survival of young seedlings. This is particularly so in climates where seed dispersal in late summer or autumn is followed by winters of sufficient severity to kill young recently germinated seedlings.

Dormancy is often broken by a period of exposure to low temperature, often less than 5°C, over a period of weeks or months in the presence of moisture (Fig. 24.15). A period of cold, which breaks dormancy and causes germination, is known as **stratification** or prechilling. By the time the prechilling requirement has been met, winter will have passed and then germination may proceed in the more favourable conditions

of spring. The optimum temperature for the breaking of dormancy (1–5°C) is quite different from the optimum temperature for subsequent germination and seedling growth (usually 15–25°C).

Just as it may be disadvantageous for seeds to germinate during winter, problems for emergent seedlings may also occur if seeds germinate while they are buried too deeply in soil, covered by litter or beneath an extremely dense canopy. In these cases, a seedling may exhaust its food reserves before its leaves are able to begin harvesting energy from sunlight. Many species have a mechanism to prevent such premature germination: their seeds are **photodormant**, that is, they will not germinate unless they are exposed to appropriate levels of red light. The effect of light on germination appears to be under the control of phytochrome. In addition to the effects of low temperature and light, dormancy in some seeds is reduced by the effect of water, which may leach out inhibitors. In desert plants, seeds often do not germinate after a light shower, but only after steady rain that washes out sufficient inhibitor. This ensures that seedlings only emerge when enough rain has fallen to provide adequate moisture for growth. Seed dormancy is also reduced with the passage of time, possibly as inhibitors break down.

Comparative studies of mutant and wild-type plants have been important in discovering the role of ABA in the control of seed dormancy. For example, dormancy of freshly harvested seeds of wild-type *Arabidopsis* can be broken by exposure to 2°C for several days followed by exposure to light at 24°C, in the presence of water. Such treatment results in the germination of 40–70% of seeds. Without the cold treatment, *fresh* seeds will not germinate. In contrast, seeds of the mutant *aba* form, which is deficient in abscisic acid, do not show dormancy. Mutants insensitive to abscisic acid, which appear incapable of responding to their own (endogenous) abscisic acid, also have reduced seed dormancy. However, even

Fig. 24.15 The seeds of high-altitude plants, such as this snow daisy *Celmisia asteriifolia*, growing in the Australian alps, require the cold of winter to break dormancy, ensuring that seed germination occurs under more favourable conditions in spring or summer

in wild-type plants of *Arabidopsis*, dormancy diminishes with time in dry storage, so that after a period of several weeks up to 75% of seeds are capable of germinating in the presence of light and water without a cold treatment. It thus appears that *development* of dormancy requires a certain level of abscisic acid but that the hormone is not responsible for *maintenance* of dormancy over time.

> Abscisic acid is a growth-inhibitory hormone that affects plant responses to stress, for example, drought and frost, through modified gene expression. The development of seed dormancy requires the presence of abscisic acid and dormancy is often broken by a period of low temperature (stratification) or by red light.

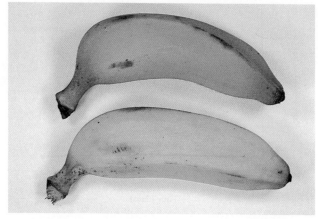

Fig. 24.16 Fruit ripening in banana. Green unripe and yellow ripe fruits of twin bananas. Ethylene is used commercially to ripen bananas which have been transported in the unripe state

ETHYLENE

Alone among plant hormones, **ethylene** (C_2H_4) is a gas of low molecular mass. Its effects on plant growth have been known since last century, when it was observed that gas leaks from underground pipes caused leaf abscission in shade trees growing along streets. However, it was not until the 1960s that ethylene was shown to be produced naturally by higher plants and also to influence plant development at hormonal levels (as low as 0.01 μL/L). The gaseous nature of ethylene means that it readily diffuses through intercellular spaces from the site of production and that it may diffuse out of a plant completely. This allows a rapid equilibrium to be established between production and loss, with consequently rapid changes in the plant's response.

Applied ethylene influences a wide range of developmental processes from seed germination to shoot elongation, flowering, fruit ripening, abscission and senescence. However, plants seem to be able to grow and reproduce without ethylene since mutants insensitive to ethylene and plants grown in the presence of specific inhibitors of ethylene synthesis grow normally. Ethylene appears to be involved primarily in plant responses to certain environmental stresses and wounding.

Fruit ripening

As fruits ripen under natural conditions they undergo an integrated set of changes, including a decline in organic acids, an increase in sugar content, softening and often a colour change (Fig. 24.16). These all involve active metabolic processes and, in many fruits, coincide with a period of increased respiration, the respiratory **climacteric**. During the climacteric there is also a dramatic increase in ethylene production. In some fruits (e.g. melon), this natural increase in ethylene precedes the rise in respiration but in others it coincides with or follows it, suggesting that ethylene levels may not be the endogenous controlling factor in fruit ripening. Rather, changes in tissue sensitivity to ethylene or a fall in the level of a ripening inhibitor may be involved.

Applied ethylene can initiate the climacteric in certain fruits and is used commercially to ripen tomatoes, avocados, melons, kiwi fruit and bananas. This allows fruit to be harvested green and ripened 'on demand'. It also explains the age-old saying that 'one bad apple can spoil the barrel', since the bad apple produces ethylene, which triggers ripening in the other fruit. The commercially available compound, ethephon, which breaks down to release ethylene once inside a plant, is sprayed on tomatoes to ensure uniform ripening. In contrast, inhibitors of ethylene action, such as CO_2, are used to retard ripening of stored fruit.

Shoot growth and flowering

Applied ethylene also influences shoot growth. In dark-grown seedlings, which are etiolated, application of ethylene can result in reduced elongation of the stem, bending of the stem (i.e. the loss of the normal gravitropic response) and swelling of the epicotyl or hypocotyl. The combination of these responses is referred to as the triple response and is frequently observed when many seedlings are grown together in a confined, unventilated space or a laboratory with gas outlets.

One of the best examined and most startling results of ethylene application is the rapid elongation of the shoots of some aquatic plants such as deep-water rice. These plants elongate rapidly when submerged (Fig. 24.17). Elongation is caused by an increase in the level of ethylene, which builds up because diffusion of the gas out of the plant is retarded in a submerged shoot. The endogenous role of ethylene

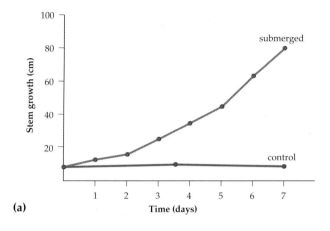

(a)

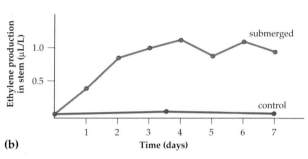

(b)

(c)

Fig. 24.17 Effect of submergence in water on **(a)** stem growth and **(b)** ethylene production within the stem of deep-water rice plants. Control plants were not submerged. **(c)** Deep-water rice (right) can elongate at rates up to 30 cm per day during flooding and can grow in water up to 4 m deep; plants here were growing in 1.5 m of water. Irrigated rice (left) usually grows at 0.1–0.2 m water depths

can be demonstrated by the application of ethylene antagonists, such as silver ions (Ag⁺), which inhibit the elongation response.

Ethylene can induce flowering in plants of the pineapple family and this is an important commercial application. Ethylene also promotes flower senescence in plants such as petunias, carnations and peas. Alternatively, flower senescence can be inhibited by treatment with the ethylene antagonist, Ag⁺ (Fig. 24.18). This is of practical benefit to the horticulture trade, where it is used to prolong the shelf-life of certain cut flowers.

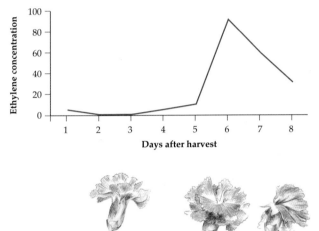

(a)

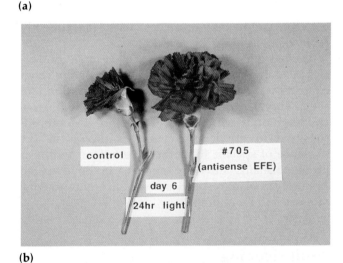

(b)

Fig. 24.18 Senescene in cut flowers of carnation.
(a) Petals senesce when levels of ethylene produced in the petals rise about six days after flowers are cut.
(b) Improved post-harvest life of the flowers is achieved through genetic engineering. A Melbourne plant biotechnology company, Calgene Pacific Pty Ltd, has 'silenced' expression of a gene associated with ethylene production in carnation petals, so that the flowers remain fresh for many days longer. Here, flowers of a normal carnation (left) are compared with those of a genetically engineered carnation (right) on day 6

Interactions of auxin with ethylene

High auxin levels can induce ethylene production. Consequently, responses to high doses of auxin can produce ethylene responses such as stem swelling and reduced stem elongation. Such effects are perhaps the clearest indication of how plant development and responses depend on interactions between hormones and should be taken into account when examining both ethylene and auxin responses.

> Ethylene is a gaseous plant hormone that influences a wide range of developmental processes from seed germination to shoot elongation, flowering, fruit ripening, abscission and senescence.

(a)

PHOTOPERIODISM AND CONTROL OF FLOWERING

The cycle of night and day is an ever-present environmental stimulus that regulates the growth and development of certain plants. Response to the length of light and dark periods in a 24-hour cycle, **photoperiodism**, allows plants to reproduce synchronously and in the appropriate season. For example, in temperate climates, fruits and dormant seeds are produced before the onset of unfavourable winter conditions.

Unequivocal evidence of the importance of photoperiod (day length) in flowering was obtained in 1920, when it was discovered that certain varieties of soybeans always flowered on the same day at a particular latitude regardless of when they were planted. Similarly, it was shown that, in the glasshouse, a new tobacco variety, Maryland Mammoth, did not flower during summer but only during winter. Several environmental factors were investigated, including temperature, nutrition and soil moisture, to see what controls flowering. It was found that both soybean and tobacco would flower provided that the photoperiod was reduced by a few hours relative to summer conditions.

(b)

Short-day and long-day plants

There are two major types of responses to photoperiod: plants may be short-day plants or long-day plants. Short-day plants flower when the photoperiod becomes less than a particular day length, called the *critical day length*, which is usually between 12 and 14 hours (Fig. 24.19). Thus, short-day plants are typically autumn-flowering plants, such as the chrysanthemum, and are frequently of tropical or subtropical origin. By comparison, long-day plants flower when the photoperiod exceeds a critical day length and typically include many spring and early summer

Fig. 24.19 (a) Effect of photoperiod on flowering in the short-day plant, *Pharbitis nil*, Japanese morning glory. The plant at left (with flower) was grown under conditions of short days while the vegetative plant on the right was grown under conditions of long days. (b) Effect of photoperiod on flowering in the long-day plant, *Brassica campestris*, rapeseed. The plant grown under conditions of short days (right) has remained vegetative while exposure to long days (left) has caused flowering, accompanied by the pronounced stem extension known as bolting

flowering plants of temperate origin (Box 24.2). Plants that do not show a photoperiod response for flower initiation are *day-neutral* plants.

It is actually the length of the *dark period*, rather than the length of the light period, that determines when flowering occurs. Soybean, for example, is a short-day plant that will not flower unless it receives greater than 10 hours of darkness (Fig. 24.20). (A period of high-intensity light is also required by short-day plants in order to meet their photosynthetic requirements.) In long-day plants, the reverse occurs. The dark period must be shorter than some critical length. In both long- and short-day plants the interruption of the dark period by a short period of light (as little as 1 minute in some particularly sensitive short-day plants, but usually an hour or more in long-day plants) negates the effect of the dark period (Fig. 24.21). Consequently, long-day plants are really short-night plants and short-day plants are long-night plants.

Environmentally induced flowering responses are frequently under simple genetic control; for example, in garden peas, a single mutation (*Sn* to *sn*) converts a long-day type into a day-neutral type. In some species that have a wide geographic range, different varieties (ecotypes) have evolved that are suited to local environmental conditions. Kangaroo grass, *Themeda triandra* (Fig. 24.22a), includes short-day, long-day and day-neutral ecotypes. Short-day ecotypes occur in tropical regions in New Guinea and the Northern Territory from latitudes 6–15°S (Fig. 24.22b). Long-day ecotypes extend from southern New Guinea down the east coast of Australia to temperate regions in Tasmania (43°S). In the Australian Alps,

BOX 24.2   EXAMPLES OF PLANTS WITH SPECIAL REQUIREMENTS FOR FLOWERING

Short-day plants
Chenopodium rubrum
Chrysanthemum morifolium
Xanthium strumarium
Coffee arabica (coffee)
Glycine max (soybeans)
Cannabis sativa (hemp)
Oryza sativa (rice) (*a*)
Gossypium hirsutum (cotton)
Zea mays (corn) (*a*)
Fragaria (strawberry)

Day-neutral plants
Nicotiana tabacum (tobacco)
Pisum sativum (peas)
Oryza sativa (rice) (*a*)
Zea mays (corn) (*a*)
Lycopersicon esculentum (tomatoes)
Vicia faba (broad beans)
Solanum tuberosum (potatoes)

Long-day plants
Avena sativa (oats)
Arabidopsis thaliana
Lolium temulentum (rye-grass)
Hordeum vulgare (barley)
Pisum sativum (peas)
Triticum aestivum (wheat)
Spinacia oleracea (spinach)
Brassica rapa (turnip)
Vicia faba (broad beans)

Vernalisation
Chrysanthemum morifolium
Pisum sativum (peas)
Triticum aestivum (winter wheat)
Arabidopsis thaliana
Hordeum vulgare (winter barley)
Spinacia oleracea (spinach)

(*a*) Particular varieties.

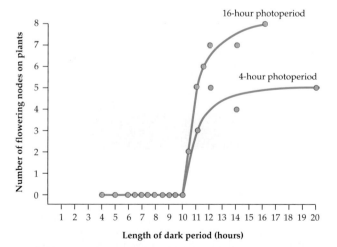

Fig. 24.20 The effect of different periods of darkness on the flowering of the short-day plant, soybean, *Glycine max*, exposed to either 4-hour or 16-hour light periods. Plants need more than 10 hours of darkness to flower and so are really long-night plants

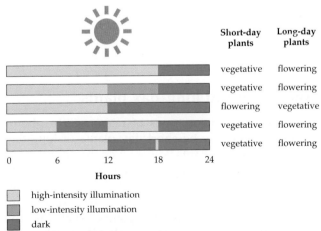

Fig. 24.21 The flowering behaviour of short-day and long-day plants subjected to various photoperiods. A critical photoperiod of 13 hours is assumed for both response types

(a)

where harsh winters occur, kangaroo grass requires vernalisation (p. 533) and long days before flowering. Ecotypes from dry inland sites, where rainfall is sporadic, are day-neutral, which may be advantageous if they are to flower and fruit as soon as adequate water is available, regardless of season.

Some plants are very sensitive to photoperiod, allowing very precise timing of the onset of flowering. For example, a variety of rye-grass, *Lolium temulentum*, and the Chicago strain of cocklebur, *Xanthium strumarium*, flower in response to a single cycle of the appropriate photoperiod. Some plants can even detect changes in photoperiod as small as 15–20 minutes.

> Plants in which flowering is affected by photoperiod are either long-day or short-day plants. The length of the dark period, rather than the length of the light period, is the controlling factor. Plants not affected by day length are day-neutral.

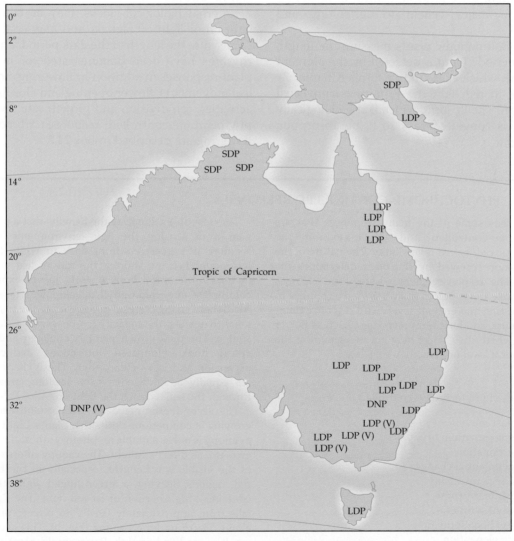

(b)

Fig. 24.22 (a) The Australian native grass *Themeda triandra* includes several ecotypes, which differ in their environmental requirements for flowering. **(b)** Distribution of the photoperiod ecotypes: day-neutral (DN); long-day plants (LDP); short-day plants (SDP); and plants responding to vernalisation (V)

Detection of photoperiod

The photoperiod response is detected by leaves. If leaves, but not the shoot apex, are exposed to the appropriate photoperiod, flowering will occur. When the shoot apex, but not the leaves, is exposed to the inductive photoperiod, no flowering occurs. This simple result implies that a signal must pass from the leaves, where induction of flowering occurs, to the apex, where flower initiation occurs. Grafting studies also indicate that a transmissible signal is involved. In some cases the signal appears to be a flower-promoting signal, often referred to as the hypothetical hormone **florigen** but, in other instances, flower-inhibiting substances appear to be involved. Plant physiologists have been able to transfer the promoting signal between long-day and short-day plants, but its chemical identity has not been determined, despite nearly 50 years of intensive research. Discovering the hormone(s) responsible for flowering remains one of the major challenges of plant science research, since artificial control of flowering would be of major benefit to agriculture and horticulture.

How do photoperiodic plants measure the length of the dark period? Light is detected by the pigment phytochrome, which exists in two forms (Chapter 7). By exposure to red light the P_r form is converted to the P_{fr} form. The P_{fr} form, which is the biologically active form, is converted back to P_r by absorption of far-red light:

$$P_r \underset{\text{far-red light}}{\overset{\text{red light}}{\rightleftarrows}} P_{fr} \text{ (active)}$$

During sunlight much of the phytochrome in a plant exists as P_{fr}. In darkness, as in the night, P_{fr} levels decline. Thus, the relative levels of the two forms of the pigment give a plant a way of detecting the light–dark transition. However, the simple decline of P_{fr} levels in darkness is not the mechanism by which plants measure the *length* of the photoperiod.

One suggested timing mechanism is an *hour-glass mechanism*, where the amount of some substance (other than P_{fr}) provides a measure of the length of the dark period between two light periods. A second idea is that *endogenous circadian rhythms* act as the timing mechanism. It is suggested that rhythms of approximately 24 hours occur within plants, even in continuous light or continuous dark. These rhythms, or internal biological clocks, may be controlled by light-on or light-off signals, and may enable a plant to measure the length of the dark period. Endogenous rhythms have been demonstrated for many other plant responses in addition to flowering, and include leaf movements, flower opening and closing, enzyme activities, cell division and growth rate. Examples of phytochrome-controlled responses in addition to flowering are described in Box 24.3.

BOX 24.3 PHYTOCHROME-CONTROLLED RESPONSES

The table lists some of the many responses that are phytochrome-controlled and which exhibit the characteristic red/far-red reversibility. The nature of the last irradiation received by the plant determines the response. The responses include short-term changes with response times of less than 5 seconds, changes in gene expression and enzyme levels with response times of a few hours, and long-term responses, such as flower initiation, where the change may not be evident at the morphological level until many weeks later.

Table

Examples of phytochrome-controlled processes
Chloroplast movement in algae
Protonema growth in bryophytes
Spore germination
Electrical potential changes
Membrane permeability
Anthocyanin synthesis
Chlorophyll synthesis
Hypocotyl hook formation
Internode elongation
Flower initiation
Seed germination

Two responses illustrate the benefits that a plant may gain from sensing the light environment. When a seedling is transferred from dark to white light, its internodes become shorter, the rate of leaf expansion increases, the apical hook uncurls, leaves expand and chloroplast and chlorophyll development takes place. All of these changes are advantageous to the plant, since it has to develop the capacity to utilise light as an energy source. The dark-grown plant is etiolated and has an apical hook, elongated internodes, reduced leaf expansion and lacks chlorophyll. These characteristics are equally advantageous to a seedling growing up through the soil to reach the light (Fig. a).

The second example deals with established plants growing in competition with other plants. Light beneath a canopy is not as suitable for photosynthesis (and hence growth), as direct sunlight. The canopy filters out most of the visible wavelengths, especially in the blue and red regions, leaving a green-tinged light that also contains a high portion of far-red light. The P_{fr}:P_r ratio falls from approximately 0.6–0.7 in direct sunlight to about 0.4–0.5 under a canopy (Fig. b). This change can result in modified growth. For example, plants adapted to growth in potentially shaded environments respond by producing longer internodes, smaller and thinner leaves and altered chlorophyll levels. These changes are of adaptive value under conditions of shade. In addition, changes in P_{fr} levels due to shading may allow plants to detect the proximity of other plants. Thus, phytochrome pigment not only allows plants to sense qualitative changes in the light environment (such as from light to darkness) but also to perceive quantitative changes in light quality.

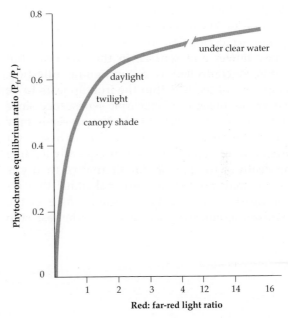

(a) The effect of darkness on pea plants 15 days old. The plant on the left was grown in complete darkness while the plant on the right was exposed to natural light

(b) The relationship between the ratio of red to far-red light and the proportion of phytochrome in the P_{fr} form. Values are given for conditions of shade, twilight, natural daylight and underwater

Phytochrome pigment in leaves enables plants to detect light and darkness. This pigment interacts with an internal clock mechanism to measure the length of the dark period.

VERNALISATION AND FLOWERING

A second major environmental cue that induces flowering in many species is a period of low temperature. Induction of flowering by low temperature is termed **vernalisation**. Like long-day plants, plants with a vernalisation response frequently flower in spring. Some of these plants grow as compact rosettes and their stems only elongate, that is, bolt (p. 521), when the plant is about to flower (Fig. 24.19b). Some plants require both a particular photoperiod and

vernalisation to stimulate flowering (Box 24.2). In others, a particular photoperiod or vernalisation is not essential but hastens flowering.

Temperatures that are effective in vernalisation range from −1°C to 9°C, and are usually required for at least four weeks. The site of detection of low temperature is usually the shoot apex, but may be in the leaves in some plants. Where leaves are sensitive to a period of low temperature, a graft-transmissible signal can be demonstrated as for photoperiodic effects. Vernalisation can be reversed (devernalisation) by relatively short periods of high temperatures (two to four days at 25–40°C) provided that these conditions immediately follow the cool period.

The flowering of certain plants is promoted by exposure to a period of low temperature: vernalisation.

MONOCARPIC SENESCENCE

In annual plants, once flowering and fruiting has taken place, the death of the entire plant normally follows. This is an example of monocarpic senescence. This phenomenon can present a dramatic sight when whole fields of wheat or maize die *en masse* within a few days. Biennial plants grow vegetatively for one year, overwinter and then flower once before dying.

The most unusual plants that undergo monocarpic senescence are long-lived species, such as bamboo (Fig. 24.23), which may thrive in the vegetative state for in excess of 30 years before flowering and fruiting once, then senescing. All the bamboo plants in one area may flower and senesce in the one year, but how this is controlled is an intriguing question. Development of seeds within the fruit appears to be crucial for the onset of monocarpic senescence since it is retarded by removal of fruit. Young seeds may simply outcompete vegetative parental tissue for essential but limited nutrients, consequently leading to the death of the parent. An alternative theory is that seeds may produce a hormonal signal that results in death of the vegetative plant. Whatever the mechanism, monocarpic senescence is the last step in growth and development and is a genetically controlled and programmed event.

Fig. 24.23 Bamboo plant in flower

SUMMARY

- The development of a plant from seed to maturity involves seed germination, vegetative growth, flowering, fruit development, senescence and seed dormancy. All these phases are strongly influenced by environmental and endogenous factors that may act by changing the level of, or tissue sensitivity to, a hormone.

- Experiments in which a hormone is applied to an intact plant, or to an excised part of a plant, have provided a basis for determining the endogenous roles of plant hormones. However, single-gene mutations that block the biosynthesis of a hormone have proved crucial in determining how plant growth and development is controlled by plant hormones.

- The best known plant hormones are auxins, gibberellins, cytokinins, abscisic acid and ethylene, all of which are effective at very low concentrations.

- Auxin (IAA) promotes growth of excised stem or coleoptile segments. The primary action of auxin is on cell elongation and may involve the release of H^+ from cells into walls, softening the walls and allowing cells to expand (acid growth hypothesis). Auxin also regulates gene expression.

- Over 80 gibberellins have been identified. The active members of this group are growth-promoting hormones that stimulate both cell division and cell elongation. They influence stem growth, cause stem bolting and promote germination. In germinating cereal seeds, GA_1 produced in the embryo moves to the aleurone layer where it activates genes that control the synthesis of the enzyme that converts starch to sugar.

- Cytokinins are plant hormones that stimulate cell division in tissue cultures and appear to be involved in the control of growth of lateral branches and leaf senescence.

- Abscisic acid is a growth-inhibitory hormone that affects plant responses to stress and modifies gene expression. It is involved in the control of stomata and transpiration, frost resistance and seed dormancy. Seed dormancy is often broken by a period of low temperature (stratification) or by red light.

- Ethylene is a gaseous hormone that influences a wide range of developmental processes, from seed germination to shoot elongation, flowering, fruit ripening, abscission and senescence. In fruit ripening, ethylene production is associated with a period of increased respiration, the respiratory climacteric.

- In many plants, flowering is controlled by photoperiod, but the length of the dark period, rather than the length of the light period, is the controlling factor. The pigment phytochrome enables leaves to sense light and darkness. The measurement of the length of the photoperiod appears to depend on the interaction of phytochrome with an internal clock mechanism. Flowering may also be promoted by exposure to a period of low temperature—vernalisation.

QUESTIONS

1. Define the terms 'hormone' and 'tropism'. List the major types of plant hormones.

2. What is the best known naturally occurring auxin in plants? Explain the acid growth theory for the action of auxin on cell elongation.

3. Explain how roots probably detect gravity.

4. The mobilisation of food reserves in cereal crops such as barley is an excellent example of hormonal control in plants. Explain whether this is a valid statement.

5. Of what value to a plant is the capacity to synthesise a growth-inhibitory substance? Give an example of the role of abscisic acid as such a growth inhibitor.

6. What advantages are conferred on a plant by the gaseous nature of the hormone ethylene? Give an example of the role of ethylene in the response of certain plants to flooding.

7. Synthetic hormones are used in agriculture. What, for example, are 2,4-D and 2,4,5-T and what are their uses?

8. A plant is described as a long-day plant with an absolute requirement for vernalisation. Explain in simple terms what this means.

9. In order to show that a certain developmental trait is controlled by a given hormonal substance, it is insufficient to simply demonstrate that application of that substance affects the characteristic in question. What additional evidence is necessary?

10. How have the techniques of genetic engineering become valuable to studies of plant hormones? What are the benefits of using single-gene mutations to study the control of plant growth and development?

11. In an herbaceous species of plant, a mutant has arisen that branches more than does the wild-type. Cytokinins are thought to promote such branching. Possibly the shoot of the mutant is more sensitive than the wild-type to root-produced cytokinins. Alternatively, the mutant may produce more cytokinins in its roots, which move into the shoot, than does the wild-type. Explain how a grafting experiment may differentiate between these two possibilities.

HORMONAL CONTROL

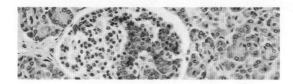

Hormonal control systems, with widely differing degrees of complexity, are found in representatives of a range of organisms, including protists, fungi, plants and all animals. With the increasing complexity of hormonal control systems that has occurred during evolution, the chemical nature of hormones appears to have been strongly conserved. Hormones with the same, or closely related, chemical structures have been found in all animals so far studied. However, the biological roles of these hormones vary widely between different animal groups.

As we saw in Chapter 24, **hormones** are chemical messengers secreted by cells of an organism in response to specific stimuli. Their presence in the watery 'internal environment' modifies the activity of cells in various organs as a result of interaction with specific receptors, so providing an appropriate co-ordinated response to the stimulus.

ANIMAL HORMONES

The word 'hormone' (meaning 'to stir up') was introduced by the physiologist E. H. Starling in 1905, to categorise a substance released by cells in the duodenum during digestion. This substance, which he named secretin, is released into circulating blood of mammals in response to the presence of acidic stomach contents entering the duodenum after a meal (Fig. 25.1). The presence of secretin in blood stimulates another digestive organ, the pancreas, to secrete an alkaline juice into the duodenum. This neutralises the acid, providing a suitable environment for further digestion of food entering from the stomach. This study introduced the concept of non-neural control of biological functions.

The existence of animal hormones was first recognised in mammals, and the organs that produced them were called 'glands of internal secretion' or **endocrine glands** to distinguish them from glands of external secretion, **exocrine glands**, such as salivary glands, sweat glands and digestive glands. Initially,

the term 'hormone' was restricted to the regulatory products of discrete endocrine glands that are released into circulating blood and that exert their action at a distance from the site of their secretion. However, it is now clear that hormones are secreted by a wide variety of tissues, not necessarily organised into endocrine glands, and that they may reach their site of action by simple diffusion.

Historically, hormones have been considered to differ from another class of chemical messengers found in animals, **neurotransmitters**, which are released at the synapse between the terminals of a nerve cell axon and its effector cell (Chapter 26). While hormones and neurotransmitters both interact with specific receptors to produce their effects, neurotransmitters usually act only locally, at the site of release. However, given our increasing knowledge of secretion of hormones by nerve cells, **neurosecretion**, the distinction between hormones and neurotransmitters is no longer clear.

> Animal hormones are secreted in response to specific stimuli. They diffuse or are transported through the extracellular environment to their site of action where they modify the activity of particular cells to provide an appropriate co-ordinated response to the stimulus.

Sites of action

One way to group hormones is in terms of the distance over which they travel to exert their effect (Fig. 25.2).

Effects on distant cells: endocrine hormones

Endocrine hormones are usually secreted into circulating blood, so that a hormone can reach its target organ rapidly and at an adequate concentration. In some invertebrates, endocrine hormones reach their target organs via circulating haemolymph or by diffusion through extracellular fluid.

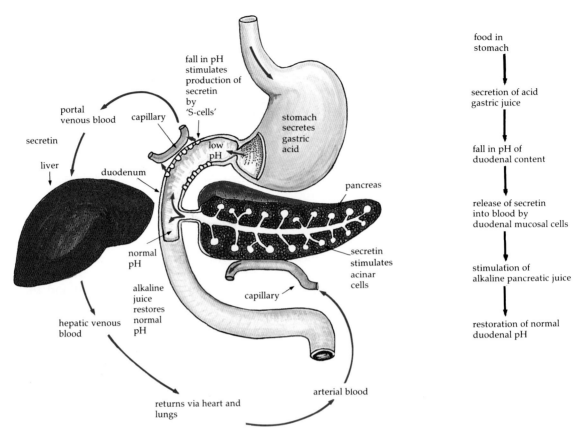

Fig. 25.1 Diagram illustrating a simple form of hormonal control of a physiological function, using, as an example, the first hormone to be recognised—secretin

Effects on nearby cells: paracrine hormones

Paracrine hormones usually act over very short distances and travel to their site of action by diffusion through extracellular fluid. However, they may also diffuse into blood and exert an effect on more distant organs.

Effects on the same cell: autocrine hormones

Some cells can be stimulated to secrete **autocrine hormones**, which interact with receptors on their own surface to produce a response, usually cell division.

Effects on another organism: pheromones

To the above three groups can be added an additional form of chemical communication. **Pheromones** are released into the *external environment* and provide chemical communication *between* individuals, through senses such as smell and taste. They are usually highly volatile compounds that are released by exocrine glands and are detected at extremely small concentrations by chemical receptors on the surface of the recipient, for example, the nasal epithelium of mammals, or the antennae of insects. There are also less volatile and more persistent pheromones that

may be excreted in urine, faeces or saliva, as well as from specialised surface glands. These pheromones are deposited on objects as territorial or mating signals, and to induce reproductive activity. These effects serve to synchronise reproductive activity and maximise reproductive success, particularly in dispersed and solitary species.

> Endocrine, paracrine and autocrine hormones and pheromones are grouped according to the distance over which they travel to exert their effect.

General functions

Hormones do not appear to initiate any unique cellular activities. Rather, they *modify* the rates of existing activities, usually through the induction or repression of enzymes within cells. In some cases they act at the nucleus of a cell to influence the activity or expression of genes; in other cases they influence the permeability of cells to solutes, or the activity of cytoplasmic enzymes.

Through their control of a wide variety of physiological processes, the major actions of hormones are to influence reproduction, growth and development, metabolism, osmoregulation and mineral exchange.

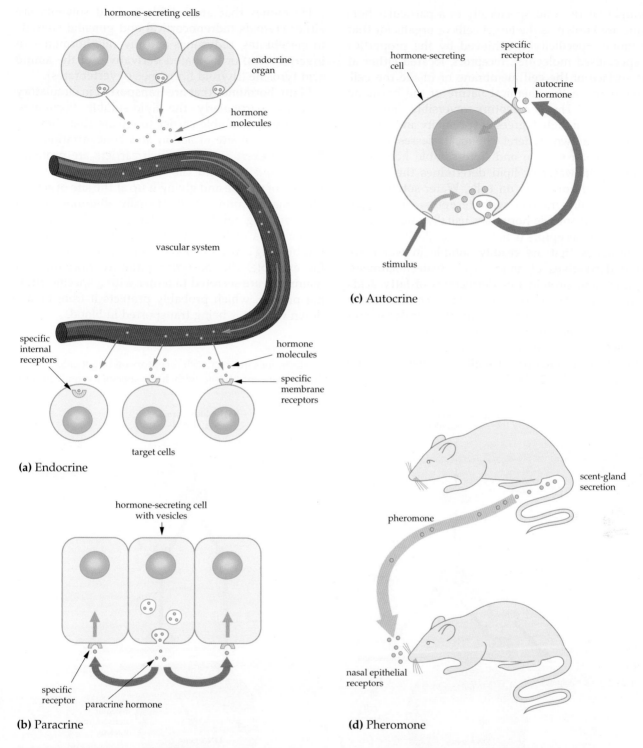

Fig. 25.2 Classification of hormones in terms of the distances over which they act: **(a)** endocrine; **(b)** paracrine; **(c)** autocrine; and **(d)** pheromone

A most important consequence of these actions is the ability hormones confer on an organism to respond appropriately to changes in its environment, particularly in the long term.

> Hormones have regulatory biological functions that enable organisms to respond appropriately to changes in their environment.

MECHANISMS OF HORMONE ACTION

As with plant hormones, animal hormones exert their effects at very low concentrations, of the order of 10^{-9}–10^{-12} M. Although the term hormone implies excitation, the effects of hormones on cells may, in fact, be either stimulatory or inhibitory. Cells or organs that respond specifically to a particular hormone are known as the **target cells** or **organs** for that hormone. Specificity is achieved by the properties of specialised molecules, **receptors**, located either at the surface of the cell membrane or inside the cell, depending on the chemical nature of the hormone (Chapter 7). These receptors reversibly bind the hormone with high degrees of specificity and affinity.

There are two general chemical classes of hormones—water-soluble and lipid-soluble hormones. Solubility in water or lipid determines the way in which a hormone acts on cells. Water-soluble hormones usually interact with receptors at the cell surface; lipid-soluble hormones usually interact with intracellular receptors (Chapter 6; Fig. 25.3).

Hormones that are readily soluble in water are either derivatives of amino acids (catecholamines, peptides and proteins) or derivatives of fatty acids (eicosinoids). There is recent evidence that some peptide hormones, such as mammalian prolactin and growth hormone, may, after binding, move with their transmembrane receptor to the nucleus where they influence cell function, usually by affecting gene transcription (Fig. 25.3a).

Hormones that are soluble in lipid solvents are either steroids (adrenocortical and gonadal steroids in vertebrates, ecdysones and juvenile hormones in invertebrates) or iodinated derivatives of the amino acid tyrosine (thyroid hormones of vertebrates).

Many hormones that are transported in circulatory systems, particularly the lipid-soluble hormones, need a water-soluble carrier to ensure that they are carried to the site of action at a concentration sufficient to exert an effect. Such carriers are proteins. They reversibly bind the hormone, taking it up at the site of release and giving it up at the site of action. They may be non-specific (usually albumins, with a high binding capacity) or highly specific to a particular class of hormone (usually globulins, with a low binding capacity). Some water-soluble hormones, for example, the posterior pituitary hormones of mammals, are secreted together with a specific binding protein, which probably protects it from breakdown while it is being transported in blood.

> Hormones interact with receptors on target cells in very low concentrations, with high specificity and affinity.

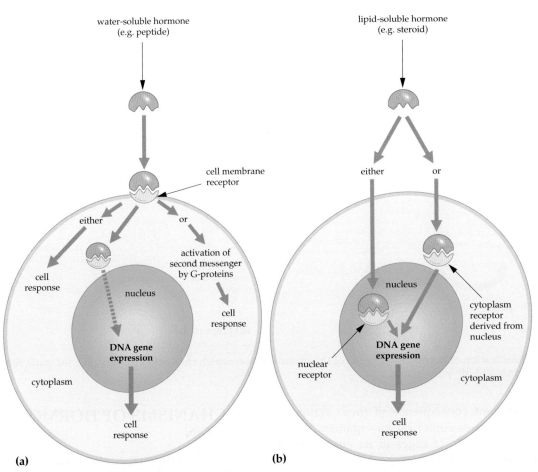

Fig. 25.3 Different mechanisms of action of the two major chemical classes of hormones: **(a)** water-soluble and **(b)** lipid-soluble hormone (see also Chapter 6). Water-soluble hormones interact with receptors on the external cell surface, whereas lipid-soluble hormones pass through the cell membrane and interact with nuclear-derived intracellular receptors

HORMONE CONTROL SYSTEMS

There are two distinct ways in which the secretion of hormones in animals can be controlled by environmental conditions. One is to relay information from the external environment to the hormone-secreting cells via the nervous system, either by direct innervation or by **neurosecretory cells**, specialised nerve cells that secrete hormones. The other is for the hormone-secreting cells themselves to respond *directly* to changes in the local environment. Both mechanisms can be integrated to provide a complex and precise hormonal control of biological function.

From an evolutionary point of view, neural control of hormonal secretion seems to have been the first to appear. Neurosecretory cells are the only hormone-secreting cells to be found in the simplest invertebrates, where their hormones influence growth and maturation to adulthood. As animals evolved further, this basic form of hormonal control was retained, with the addition of non-neural endocrine organs and with increasing complexity.

Neurosecretion

Neurosecretory cells derived from the embryonic source of nerve cells are an important source of hormones. They have been identified by selective staining reactions in most species of animals so far investigated. Neurosecretory cells may be scattered diffusely through all parts of the nervous system or they may be aggregated into discrete, highly vascularised **neurohaemal organs** (Fig. 25.4).

Neurosecretory cells look like nerve cells and are innervated by the axons of other nerve cells via contacts (synapses) with their dendrites (Chapter 26). However, the cell bodies of neurosecretory cells are often larger than in most neurons and their cytoplasm contains large secretory granules. The axon of a neurosecretory cell is usually densely packed with secretory granules of the order of 1000–2000 nm in diameter. These granules are membrane-bound vesicles containing hormone (or its precursor), which is usually a protein. Characteristically, neurosecretory axons do not form synapses with other nerve cells.

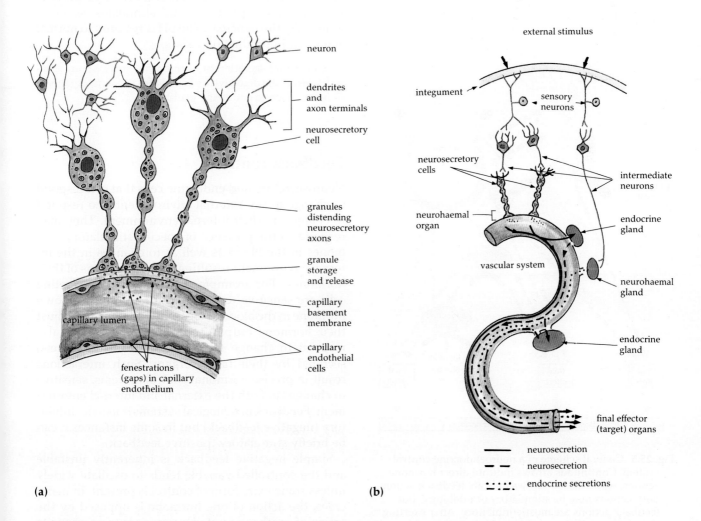

(a)　　　　　　　　　　　　　　　　　(b)

Fig. 25.4　(a) Structure of a neurohaemal organ. **(b)** Neurosecretions and endocrine secretions may act either directly on a final effector organ, or they may stimulate or inhibit secretion from another endocrine organ

Instead, they release their hormone by exocytosis (Chapter 4) into the local environment, usually in the vicinity of a blood capillary (Fig. 25.4). Neurosecretions then influence effector organs either *directly* or *indirectly* through the stimulation of other hormone-secreting cells (Fig. 25.5).

> Neurosecretion involves the release of hormones by exocytosis from nervous tissue into the local environment or a blood vessel.

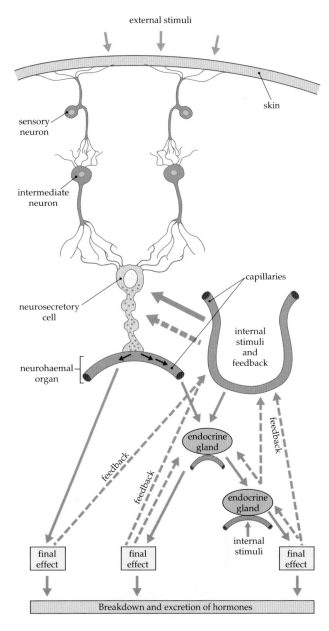

Fig. 25.5 General scheme of a neuroendocrine control system. Continuous arrows indicate direct hormone actions, while dotted arrows indicate feedback actions. Both actions may be stimulatory or inhibitory, but feedback actions are mostly inhibitory. After exerting their actions, the hormones are broken down and excreted from the body

Non-neural hormone secretion

Non-neural hormone-secreting cells are derived mostly from tissues in the digestive and excretory systems. These cells are usually organised into discrete endocrine glands, which respond either indirectly, through the nervous system, or directly to changes in the external environment. Changes in the external environment detected by the nervous system may be relayed to an endocrine gland either by direct innervation of the endocrine gland or by a multistage 'cascade' of stimulating hormones (Fig. 25.5).

Some non-neural endocrine glands themselves monitor the local concentration of the variable they control, responding directly to a change in its concentration by secreting hormones that activate mechanisms to reduce the change. When the concentration returns to normal, secretion of the hormone declines. Such endocrine glands are usually not influenced by other hormones. This is an example of **negative feedback control**; secretin is the classic example of such control. The regulated variable is the pH of the duodenal contents. If these become acid, secretin is produced and stimulates secretion of alkali by the pancreas. The pH is restored to normal and secretin production ceases.

> Non-neural secretion of hormones usually occurs from tissues of the digestive and excretory systems, organised into a discrete endocrine gland.

Feedback control

Neurosecretory and endocrine cells that can respond to change in the external environment also respond to changes in their internal environment. They may respond to the presence of specific stimulatory hormones in the blood as well as to changes in the internal environment resulting from the actions of their hormones. For example, many hormone-secreting cells are sensitive to the concentration of their own hormone in the blood or local environment and adjust their hormone output accordingly. They may also respond to change in concentration of hormone(s) secreted by their target organs. Such interactions result in precise hormonal control systems, sensitive to changes in both the external and internal environment. Feedback in biological systems is usually inhibitory (negative feedback) but in some instances it can be briefly stimulatory (positive feedback).

Simple negative feedback is inherently unstable and the controlled variable tends to oscillate widely unless some extra form of control is present. In many cases, the action of one hormone is opposed by the action of another, secreted in response to an opposite change in the same variable and having the opposite effect, so reducing the size of oscillations. Control of blood glucose or blood Ca^{2+} concentration in vertebrates (p. 557) is achieved in this way.

> Hormone release is usually under negative feedback control.

HORMONE-SECRETING ORGANS OF INVERTEBRATES

Isolated neurosecretory cells are found in the nervous system of hydrozoans (phylum Cnidaria) where their secretions influence *growth* and *regeneration* (Fig. 25.6). With the first appearance of circulatory systems, in annelids, scattered associations between neurosecretory cells and blood vessels (neurohaemal associations) are found.

In polychaete and oligochaete worms, neurosecretions are *inhibitory*, preventing the morphological changes that lead to sexual maturity, known as **epitoky** (Fig. 25.7). Some environmental change reduces the activity of the neurosecretory cells and so permits the genetically programmed development to sexual maturity. This is a simple example of the modulatory effect of a hormone on an existing biological mechanism. In leeches, by contrast, neurosecretory control of reproduction is *stimulatory*. Destruction of the brain of a leech causes a marked reduction in size and number of the gamete clusters in the gonads, which are restored by injection of an extract of leech brain. In both cases, it seems that the secretion of hormone is continuous and control by external events is to reduce secretion.

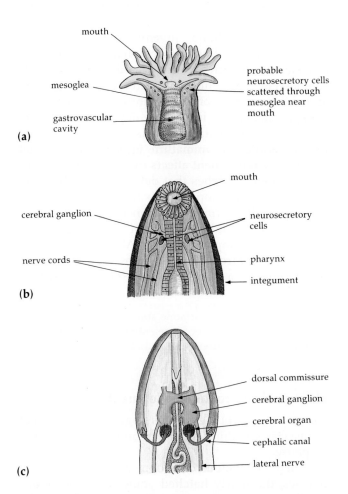

Fig. 25.6 Location of neurosecretory cells in (a) an anemone (phylum Cnidaria), (b) a flatworm (phylum Platyhelminthes) and (c) a ribbon worm (phylum Nemertinea). This illustrates the evolution of neurosecretory organisation from scattered cells to specific cells in neural ganglia, and then to discrete neurohaemal organs associated with neural ganglia

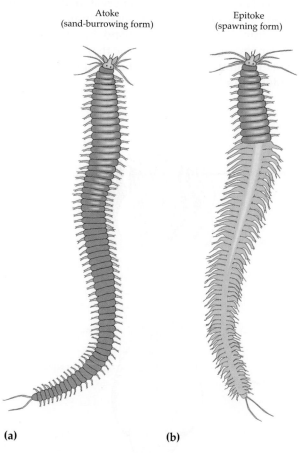

Fig. 25.7 Hormonal control of development and reproduction in a polychaete worm (phylum Neridae). A juvenile hormone produced by neurosecretory cells in the cerebral ganglion normally prevents development from (a) non-reproductive (atoke) to (b) the reproductive (epitoke) form. Production of the neurosecretion is under the influence of day length, light intensity and the lunar cycle, presumably mediated through the eyes

Aggregations of neurosecretory cells into discrete neurohaemal organs occur in arthropods (crustaceans, spiders and insects), together with discrete non-neural endocrine organs. (Insect endocrine systems are considered in more detail below.) In centipedes and millipedes, neurosecretory cells have been found in all areas of the central nervous system and the principal site of termination of their axons is the **cerebral gland**, which may take the form of a group of small lobes or a compact body. The cerebral gland is connected to the brain by a nerve containing the neurosecretory axons. Secretory granules tend to accumulate in the axon terminals, which lie close to haemolymph sinuses.

In molluscs, non-neural endocrine control seems to be dominant. Although cells resembling neurosecretory cells are present in some parts of the brain, their functions are unknown. Cephalopods (squids and octopuses) have a non-neural endocrine gland, the *optic gland*, which is controlled by the nervous system. The optic gland is innervated directly by neurons in the nearby *optic lobe* of the brain (Fig. 25.8). When stimulated, the optic gland becomes highly vascular and produces a hormone that stimulates the gonads to produce gametes; that is, the hormone is **gonadotrophic**.

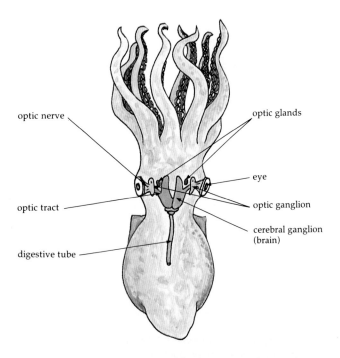

Fig. 25.8 Diagrammatic view of the eyes and brain of an octopus, to show the location of the non-neural optic glands. These are close to the optic tract, which joins the optic ganglion to the cerebral ganglion. Visual stimuli from the eye travel via the optic nerve to the optic ganglion. Nerve cells in the optic ganglion form synapses with cells in the optic gland, which *inhibit* secretion of the optic gland hormone. The optic gland hormone causes enlargement of the gonads and increase in gamete production (i.e. it is gonadotrophic). This is a mechanism by which visual stimuli can affect reproductive activity

Insects

In insects, both neural and non-neural endocrine glands are present. The principal neurohaemal organs are the **corpora cardiaca**. These are small paired structures located posterior to the brain and close to the dorsal aorta and oesophagus. The corpora cardiaca are highly vascularised with haemolymph sinuses and contain many terminals of neurosecretory axons swollen with neurosecretory granules. The chemical nature of these neurosecretions is uncertain but they are probably proteins. Depending on the species, they may be involved in colour change, moulting and metamorphosis, initiation of regeneration, gametogenesis, parturition, control of metabolism and control of salt and water balance.

The non-neural endocrine glands in insects are the **corpora allata** and the **ecdysial glands**, present in the head or, in the case of the ecdysial glands, the upper thorax of some species (in which they are named the **prothoracic glands**). The corpora allata secrete a lipid-soluble hormone, **juvenile hormone**, which stimulates laying down of nymphal structures. The ecdysial glands secrete a steroid, **ecdysone**, also lipid-soluble (Fig. 25.9), which stimulates moulting, growth and differentiation of tissues into adult structures.

Reproduction in insects is dependent on hormonally induced metamorphosis to produce sexually mature adults. In immature insects, the internal hormonal environment affects the direction and timing of genetic switches that determine the direction and order of development to the sexually mature adult. In adults, hormones influence the development of the gonads and secretion of pheromones that influence courtship and mating. All these events are under the control of the neurosecretory system, responding in a species-specific way to environmental cues (Fig. 25.10).

In insects, juvenile hormone stimulates laying down of nymphal structures; ecdysone stimulates moulting and differentiation of tissues into adult structures.

Hormonal control of development

Hormones control the various stages of development in insects—moulting, pupation and emergence, and diapause.

In **hemimetabolous** development, as seen in cockroaches, the newly hatched young resemble adults in all except size, presence of wings and maturity of the gonads. These young undergo a succession of moults, producing a succession of **nymphal instars**, by which they acquire adult characters, mainly during the final moult. **Brain hormone**, stored in the neurohaemal organs, **corpora cardiaca**, can be released into the blood as a result of external stimuli, including change in temperature, humidity and distension of

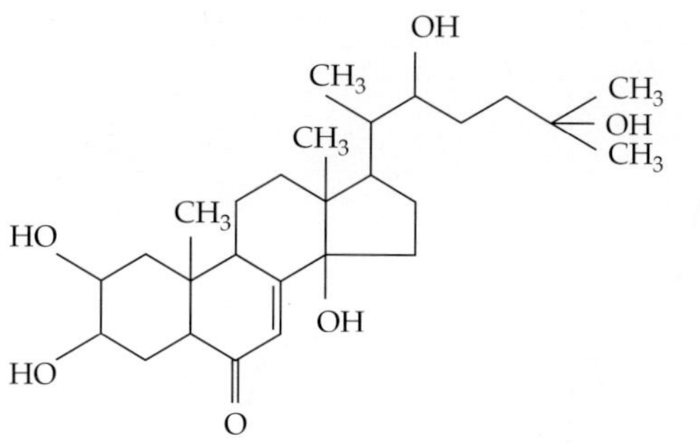

(a) Ecdysone

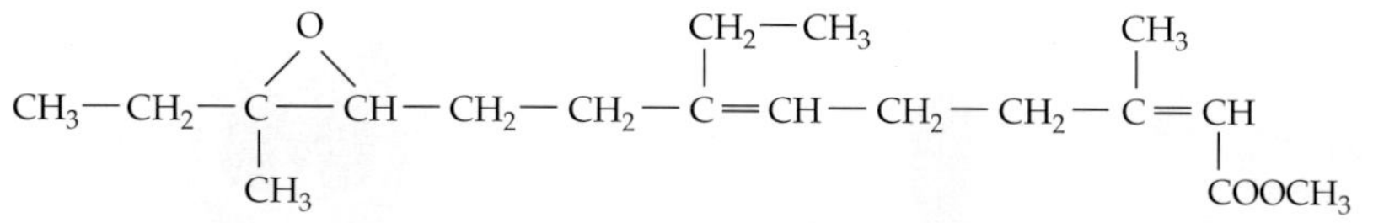

(b) Juvenile hormone

Fig. 25.9 Chemical structures of two non-neurosecretory hormones in arthropods: **(a)** ecdysone (a steroid) and **(b)** juvenile hormone (methyl 10-epoxy-7-ethyl-3,11-dimethyl-2,6-tridecadienate), a derivative of fatty acids

the gut by food. It acts on the **prothoracic glands** to produce the moulting hormone **ecdysone** (Fig. 25.10). These processes are inhibited by **juvenile hormone**, secreted by the paired **corpora allata**, which are also under neuroendocrine control.

If development is **holometabolous**, as in moths and butterflies, young have the form of worm-like larvae bearing little resemblance to adults. After a succession of moults, induced in the same way as in hemimetabolous insects, the last instar transforms into a chrysalis or pupa, which is quiescent and incapable of feeding. During this quiescent stage, the larva undergoes complete structural reorganisation to assume the adult form, known as an **imago**, which emerges when the pupal case bursts. These events are genetically programmed but, again, the sequence and timing of the expression of different sets of genes is under hormonal control. The hormones involved are the same as in hemimetabolous insects.

The integration of neural and non-neural hormone secretion to produce growth and maturation in a hawk moth is illustrated in Figure 25.10a. Juvenile hormone from the corpora allata promotes the retention of larval characters, while a neurosecretion from the corpora cardiaca stimulates secretion of ecdysone, the moulting hormone, by the prothoracic gland. This promotes differentiation to the sexually active adult form. The relative rates of secretion of these hormones are influenced by environmental factors, such as day length, temperature and humidity.

The switch-over from the pupal state to emergence of the adult, **eclosion**, is produced by a neurosecretion from cells in the medial part of the brain of a pupating insect. It collects in the corpora cardiaca and is released into the blood at the time appropriate for emergence. Secretion of this hormone appears to

depend on the functioning of a 'biological clock' in the brain, which detects the duration of light and dark through photoreceptors. The appropriate photoperiod for emergence is genetically determined and eclosion can occur only within the time 'window' or 'gate' set by the biological clock. Once the pupating insect has developed to a stage appropriate for emergence, the gate is opened. The insect can now emerge when eclosion hormone is secreted, that is, the switch is turned on. Hence, adults emerge at a particular time of day or night, according to the genetic program of the species.

Hardening and darkening of the adult cuticle after emergence is caused by another neurosecretion, **bursicon** or **tanning hormone**, which is produced by brain neurosecretory cells. It is released into the fused thoracoabdominal ganglia about the time of transition from larval to adult form. A similar substance has been found in crustaceans, such as crayfish, and presumably plays a part in hardening of the exoskeleton.

In insects, expression of genetically determined adult sexual characters is under the control of opposing stimulatory and inhibitory hormones, themselves controlled by neurosecretory cells of the nervous system. In pupating insects, emergence is controlled by a hormonal switch, linked to a neural clock that detects the duration of photoperiod.

At some stage in their life cycle, many insects undergo a period of dormancy, **diapause**, in which growth and differentiation practically cease. The duration of diapause varies considerably, depending on seasonal and environmental factors, particularly photoperiod. Its function appears to be to make use of environmental cues to ensure that young are

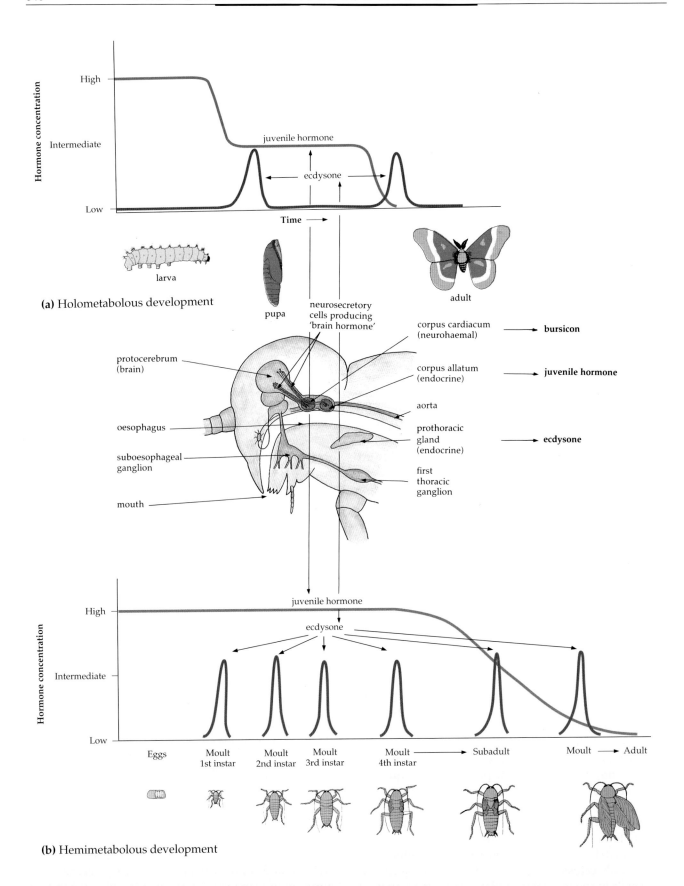

(a) Holometabolous development

(b) Hemimetabolous development

Fig. 25.10 Hormonal control of development of **(a)** holometabolous insects and **(b)** hemimetabolous insects. Juvenile hormone is a long-chain fatty acid that prevents adult development, possibly acting on cell-surface receptors. Edysone is a steroid that acts on nuclear chromosomes to initiate gene transcription. Bursicon, which hardens the exoskeleton, is a protein released by neurosecretory cells in the brain or ventral ganglia immediately after emerging from the pupa

hatched at a time when their chances for survival are maximal or, in the adult, to permit survival when environmental conditions become hazardous. Diapause is under the control of a hormone secreted by two neurosecretory cells in the suboesophageal ganglion.

Hormonal control of reproduction

Once the adult reproductive state has been reached, reproduction itself is also under hormonal control. However, hormonal control of reproductive functions in adult insects is not well understood. In the majority of insects, reproduction occurs cyclically over a life span that may extend from only a week to several months. In some insects, the adult, on emergence, immediately mates and the female deposits her eggs. The mouthparts of such adults are non-functional, so that the insects cannot feed and all the energy required for reproduction is derived from stores laid down in the larval or nymphal stage. Having reproduced within the first few hours of emergence, these adults then die.

Gonadal functions of insects undergoing cyclic reproductive activity are influenced by secretion from the corpora allata, which is essential for deposition of yolk in the eggs of females and for formation of spermatophores in many species. The hormone subserving this **gonadotrophic** function of the corpus allatum is juvenile hormone, which, as described above, has a different function in the immature insect.

In the adult female, juvenile hormone acts in conjunction with another unidentified neurosecretion, which mobilises energy reserves, principally fat tissues, to make available lipids, proteins and carbohydrates for incorporation into the developing oocytes. This process, **vitellogenesis**, is common to all egg-laying animals and the neurosecretion is **vitellogenic hormone**. In mosquitoes, vitellogenesis is stimulated by the ingestion of protein, for example, blood.

Finally, gonadotrophic stimulation of the ovary in the female induces the secretion of an **ovarian hormone**, which feeds back to the neurosecretory cell and inhibits further gonadotrophin secretion. In this way a cyclic reproductive pattern can be produced.

Because changes in the external environmental cues will determine the reproductive state of a mature insect at any particular time, these presumably reflect the optimum time for fertilisation and egg-laying. In some species, the act of copulation provides the final stimulus for gonadotrophin-induced maturation of eggs before they are fertilised some time after mating.

> When insects become adult, juvenile hormone becomes gonadotrophic, stimulating development of eggs. Energy reserves for incorporation into developing eggs are mobilised by vitellogenic hormone.

Pheromones

Many insects depend on the secretion of pheromones to attract members of the opposite sex or to induce mating behaviour (Chapter 28). This is readily demonstrated in the cockroach, in which release of a volatile chemical by the sexually mature female attracts males and induces courtship behaviour. Removal of the gonads does not affect mating behaviour, but removal of the corpora allata from females prevents secretion of the pheromone and thus courtship and mating. Removal of the corpora allata from males has no effect on courtship and mating, which, therefore, must be triggered only by the pheromone. Similar dependence on pheromone secretion for the attraction of mates and induction of courtship and mating is found in some moths and butterflies and is almost certainly

---

BOX 25.1 CONTROL OF METABOLISM IN ARTHROPODS

Control of metabolism is obviously essential for normal growth, development and successful reproduction. This is achieved by interaction between hormonal control systems.

Insects secrete a **hyperglycaemic hormone**, which increases the sugar concentration of extracellular fluid. It is a peptide, secreted by the corpora cardiaca, which acts on the enzyme systems of the 'fat body', where glycogen is stored. Here, it causes breakdown of glycogen to **trehalose**, the main carbohydrate of insects, which is liberated into the haemolymph. There is also evidence that the corpora allata exert some sort of inhibitory control over the breakdown of fats in the fat body, and that both a secretion of the corpora allata and a brain neurohormone are involved in the control of protein metabolism.

In crustaceans, the level of sugar in the blood appears to be regulated by a hyperglycaemic hormone present in the eye stalk. After removal of the eye stalk, the blood sugar concentration decreases while the glycogen content of the hypodermis (from which the exoskeleton is derived) is increased. Chitin, one of the major chemical constituents of the exoskeleton, is derived from glycogen, so that control of blood sugar may be important during moulting. In intact crustaceans, injection of aqueous extracts of the whole eye stalk, or of the 'x'-organ/sinus gland complex, causes an increase in blood sugar level. Increase in blood sugar level is also produced by exposing the animal to an adverse environment, which may also induce moulting.

widespread among insects. The volatile pheromones can exert their influence over surprisingly long distances.

Other pheromones may have inhibitory effects on reproduction and some have multiple functions. The classic example is the chemical 9-keto-decanoic acid, which is secreted by the mandibular glands of the queen in a hive of honeybees. This is ingested by the workers and inhibits both the development of the ovaries and the behaviour that would otherwise lead to the construction of more royal cells for the rearing of new queens. This substance also acts as a sex attractant during the nuptial flight. The secretion of pheromones is probably always under hormonal control through gonadotrophin secretion.

> Reproduction in insects depends on external environmental cues that determine the nature of the internal hormonal environment through neurosecretion. These cues may be day length, temperature, humidity or chemical agents detected by surface receptors.

Crustaceans

In crustaceans (crayfish, prawns and yabbies), the neurohaemal organs are the sinus gland in the eye stalk, the pericardial organs located within or attached to the tissue surrounding the heart, and the post-commissural organ, about which little is known, behind the oesophagus (Fig. 25.11).

The **sinus gland** is easily recognisable as a blue-white opalescent spot in the eye stalk below the eye. It is a highly vascular reservoir for neurosecretions contained in the terminals of axons from neurosecretory cells located mainly in the **'x'-organ** in the largest of the eye-stalk ganglia. In some species, additional x-organ neurosecretory cells are found on the surface of the eye stalk, in the sensory pore x-organ. Neurosecretory axons terminate on a thin membrane lining a blood sinus, into which they can

release their contents. The sinus gland hormones (also known as eye-stalk hormones) influence, either directly or indirectly, colour change, movement of retinal pigments in the eye (to adapt to changing levels of light), moulting, reproduction, water balance and blood glucose concentration. The **pericardial organs** receive neurosecretory axons from cells in the thoracic and other segmental ganglia and also contain neurosecretory nerve cell bodies. These organs influence the rate of the heartbeat and, probably, the flow of blood through the gills.

The non-neural endocrine glands of the crustaceans are the 'y'-organ, androgenic glands and, in females, the ovaries. The **y-organ** is located in the head, usually at the base of the first antenna. However, some species have an *exocrine* **antennal gland** at this location and, in these, the y-organ is located near the mouth. It secretes a steroidal moult-stimulating hormone, crust-ecdysone, which closely resembles the insect hormone (Fig. 25.9). The **androgenic gland** is associated with the genital duct and secretes a protein that affects all primary and secondary sex characteristics. These endocrine glands are under the influence of neurosecretions produced by neurosecretory cells, and are stored and released at the neurohaemal organs, so that the events associated with reproduction, growth and differentiation are geared to changes in the external environment.

HORMONE-SECRETING ORGANS OF VERTEBRATES

In vertebrates, there is a greater diversity of non-neural hormone-secreting organs, derived mostly from the embryonic digestive tract. Most of these are controlled by neurosecretions, but some act independently, responding directly to changes in blood and tissue fluid composition.

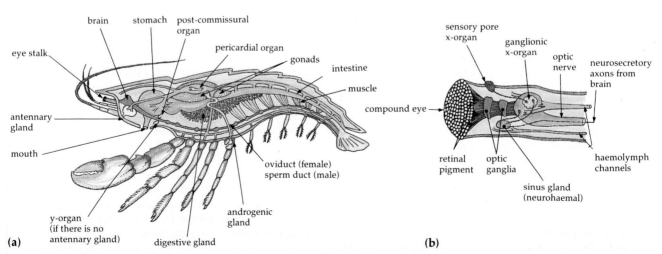

Fig. 25.11 (a) Hormone-secreting organs of a decapod crustacean. If there is no antennary gland, the sinus gland takes over its role. (b) Enlargement of the eye stalk

Neurohaemal organs

The major neurosecretory region is found at the base of the brain in a mass of nervous tissue, the **hypothalamus**, surrounding the third brain ventricle (Fig. 25.12). The hypothalamus is an important neural control centre, receiving information from most parts of the nervous system and providing central control of many functions by the autonomic nervous system.

The pituitary gland, also known as the **hypophysis**, comprises an outgrowth of neural tissue (the pituitary stalk) from the base of the hypothalamus to which a non-neural endocrine gland, derived from an outgrowth of the digestive tract, is closely attached (Fig. 25.13). The two structures together form a distinct appendage to the base of the brain, which is embedded in a cavity in the base of the skull. The

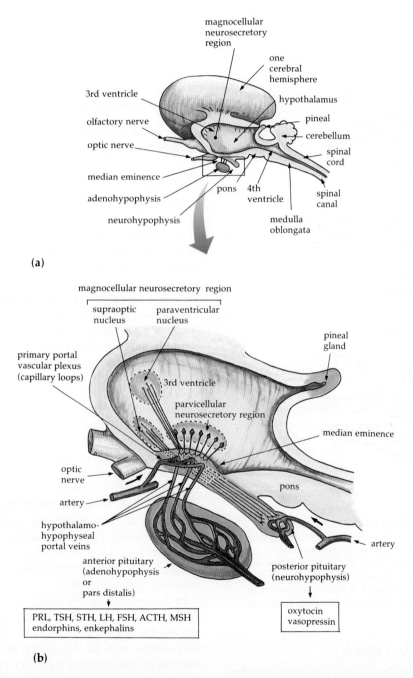

Fig. 25.12 (a) Diagram showing the location of the major neurohaemal organs in the lateral walls and floor of the third brain ventricle of a mammal (the hypothalamus). The neurosecretory regions are highly vascular. **(b)** In the region of the median eminence, an array of capillary loops, derived from nearby arteries, coalesce into the hypothalamo-hypophysial portal veins, which supply the hormone-secreting cells of the anterior pituitary gland. Neurosecretions released at the median eminence reach the anterior pituitary gland at high concentration and stimulate or inhibit the secretion of anterior pituitary hormones

neural element contains two major neurohaemal organs, the **posterior pituitary gland** and the **median eminence**. Collectively, these two organs are termed the **neurohypophysis**. The non-neural anterior part is the **anterior pituitary gland** or **adenohypophysis** (Fig. 25.12).

Two other important neurohaemal structures in vertebrates are the **pineal gland** and the **chromaffin organs**.

> In vertebrates, the pituitary gland has distinct neurohaemal and non-neural endocrine regions. Neurosecretory tissue is found in the hypothalamus, neurohypophysis (posterior pituitary gland and median eminence), pineal gland and chromaffin organs.

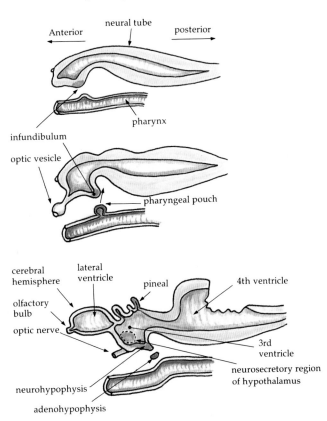

Fig. 25.13 Diagram to illustrate the embryonic development of the neural tube in vertebrates, in this case a reptile, and the location of the central neurosecretory structures. An outgrowth of the nearby digestive tract buds off to form the adenohypophysis (anterior pituitary gland) and the neurohypophysis forms from a downgrowth of the developing neural tube

The posterior pituitary gland

In most mammals and birds, the posterior pituitary gland is a distinct appendage connected to the brain by a **neural stalk** or **infundibular process**. In other species it may be only a minor bulge. The neurosecretory cells that send axons to the posterior pituitary gland are located bilaterally in two nuclei in the walls of the third ventricle (The supraoptic and paraventricular nuclei). The axons, which contain dense neurosecretory granules, pass into the neurohypophysis and terminate in the vicinity of highly permeable blood capillaries.

In monotremes and eutherian mammals, the posterior pituitary hormones are the 5-amino-acid peptides (pentapeptides) **vasopressin** and **oxytocin**, which differ from each other in amino acid composition but retain the same cyclic pentapeptide configuration (Fig 25.14). Vasopressin influences blood pressure and water balance, and oxytocin affects reproductive functions. All the vertebrates so far studied, including marsupials, have similar pentapeptides with minor differences in amino acid composition, such as **vasotocin**, the principal posterior pituitary hormone of birds and reptiles.

> The posterior pituitary secretes peptides that affect water balance and reproduction.

The median eminence

In all vertebrates except teleost fishes, axons of a set of neurosecretory cells in the inferior and lateral walls of the third ventricle terminate in the vicinity of a dense group of capillary loops in a neurohaemal organ, the median eminence. This is a small highly vascularised prominence on the floor of the third ventricle, just anterior to the posterior pituitary gland. Capillaries emerging from the median eminence coalesce to form a leash of veins, the **hypothalamo-hypophyseal portal system**. These veins transport the neurosecretions released at the median eminence directly, at high concentration, to an adjacent endocrine gland, the anterior pituitary gland. There, the veins divide into a second network of capillaries, which perfuse the secretory cells of the anterior pituitary (Fig. 25.12).

The neurosecretions from the median eminence act specifically to stimulate or inhibit the secretion of particular anterior pituitary hormones, providing a chemical link between the neural and non-neural elements of the hormonal control system. This arrangement of a neurosecretory system with a direct vascular link to a non-neural endocrine gland is analogous to the corpus cardiacum–corpus allatum relationship in insects.

In teleost fishes, the axons of the neurosecretory cells pass directly to the anterior pituitary where they release their secretions in the vicinity of specifically responsive cells. In this group of animals, the neural control of the anterior pituitary is thus of a *paracrine* rather than *endocrine* nature.

Pineal gland

The pineal gland is an outgrowth in the midline of the roof of the third brain ventricle, between the

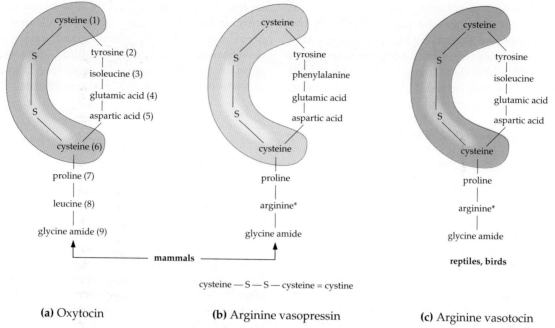

cysteine — S — S — cysteine = cystine

(a) Oxytocin (b) Arginine vasopressin (c) Arginine vasotocin

Fig. 25.14 Chemical structures of the major neurohypophyseal hormones of mammals: **(a)** oxytocin; **(b)** vasopressin; and **(c)** vasotocin. Note the minor changes in amino acid composition that confer differences in biological activity

cerebral hemispheres and the cerebellum (Fig. 25.15). It varies greatly in size and complexity between vertebrates. In many vertebrates, associated tissue becomes differentiated into a separate organ called the parapineal organ (in fishes), frontal organ (in amphibians) or parietal organ (in reptiles). In fishes, amphibians and some reptiles (but not snakes) the parapineal, frontal or parietal organ and even the pineal gland itself contain *photoreceptor cells.*

Particularly in lizards, the whole complex takes the form of a well-differentiated eye with structures analogous to a cornea and lens, which serves a definite photosensory function (Fig. 25.15). Hence, it is sometimes referred to as the 'third eye'.

In mammals and birds the photosensory cells have become modified into **pinealocytes**, and these have a neurosecretory function. The major secretory product is **melatonin**, which is derived from the

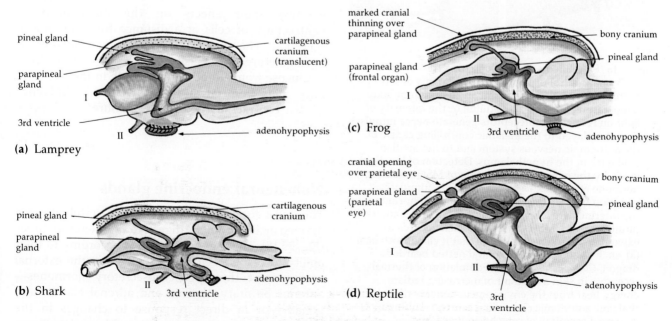

Fig. 25.15 Arrangement of the pineal/parapineal complex in four different species of vertebrates (I: olfactory nerve; II: optic nerve): **(a)** lamprey; **(b)** shark; **(c)** frog; and **(d)** reptile. Note how, in some reptiles, the parapineal becomes a functioning eye. The neurosecretion of the pineal gland is melatonin

neurotransmitter, serotonin. One function of melatonin is to aggregate the pigment cells of the skin (melanophores), so causing blanching of the skin in many vertebrates (Fig. 25.16). Melatonin secretion is inhibited by light and stimulated in the dark, which demonstrates the link between the photoreceptor in the 'third eye' and control of pigmentation through neurosecretion.

Another important effect of melatonin, particularly in birds and mammals, is inhibition of reproductive activity in response to changes in environmental photoperiod. This is one mechanism by which seasonal reproduction is controlled (Chapter 13).

Melatonin secretion by pinealocytes appears to be under direct nervous control and the pinealocytes

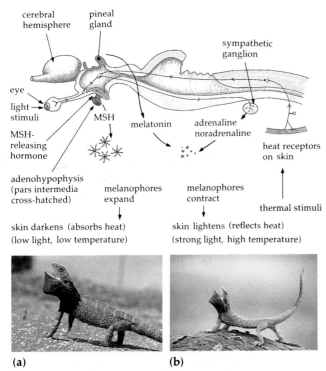

cerebral hemisphere
pineal gland
sympathetic ganglion
eye
light stimuli
MSH
melatonin
adrenaline noradrenaline
heat receptors on skin
MSH-releasing hormone
adenohypophysis (pars intermedia cross-hatched)
melanophores expand
melanophores contract
thermal stimuli
skin darkens (absorbs heat) (low light, low temperature)
skin lightens (reflects heat) (strong light, high temperature)

(a)   (b)

Fig. 25.16 Simplified scheme illustrating the neural and hormonal control of skin pigmentation in a desert reptile, which helps it to maintain an optimum body temperature. Thermal skin stimuli or the intensity of light on retinal photoreceptors generate nerve impulses that are relayed to the thalamic controlling centre of the autonomic nervous system and to the median eminence in the hypothalamus. Detection of low light levels by the eyes causes secretion of MSH-releasing hormone at the median eminence. This stimulates secretion of MSH (melanocyte stimulating hormone) by the intermediate lobe (pars intermedia) of the anterior pituitary gland. MSH causes skin melanocytes to expand and darken the skin, so that it can absorb heat, **(a)** such as shown in this central netted bearded dragon, *Ctenophorus nuchalis*. Stimulation of thermal receptors in skin as a result of increasing radiant energy heat from the environment activate the thalamic autonomic controlling centre, resulting in increased secretion of catecholamines from sympathetic nerve endings. These cause contraction of melanocytes and blanch the skin, so that it can reflect heat, **(b)** again shown in *C. nuchalis*

secrete their product directly into blood capillaries, without the intervening extension of an axon. The nerve supply to the pineal gland varies between vertebrates. In mammals and birds, the innervation appears to be exclusively from the sympathetic division of the autonomic nervous system.

> The pineal gland secretes melatonin, which influences pigmentation of the skin in a wide variety of vertebrates and reproductive activity in birds and mammals. It is under the influence of light through photoreceptors and the nervous system.

Chromaffin organs

Chromaffin organs are groups of cells containing small cytoplasmic granules that take up chromate stains, hence the name. These cells aggregate around capillary vascular sinuses in the vicinity of the kidneys (Fig. 25.17) and are innervated by pre-ganglionic fibres of the autonomic nervous system (Chapter 26). They are considered to be post-ganglionic sympathetic neurons that have been modified to become neurosecretory cells.

In mammals, chromaffin tissue forms the **adrenal medulla**, which is surrounded by a cortex of non-neural steroidogenic cells to form the **adrenal gland**. Adrenal chromaffin tissue secretes the catecholamines adrenalin, noradrenalin and dopamine, all derived from the amino acid tyrosine. These are, in fact, classified as neurotransmitters because they are also released by the terminals of sympathetic axons and are involved in neurotransmission in the central nervous system. However, when released into the bloodstream from organs such as the adrenal medulla, they behave as hormones, exerting their effects on the metabolism or contractility of cells at a distance from the site of secretion. The major effects of catecholamines as hormones are to mobilise energy reserves in the form of glucose from stored carbohydrate in the liver and fatty acids from fat stores, as well as to influence cardiac contractility and blood pressure.

Non-neural endocrine glands

The non-neural endocrine glands of vertebrates are derived mainly from the digestive and excretory tubes (Fig. 25.18). They may be grouped according to their ability to respond either to changes in the external environment through neurosecretory hormones—anterior pituitary, thyroid and adrenal glands, and gonads—or in direct response to changes in the internal environment—parathyroid and ultimobranchial glands, islets of Langerhans and gastrointestinal mucosa.

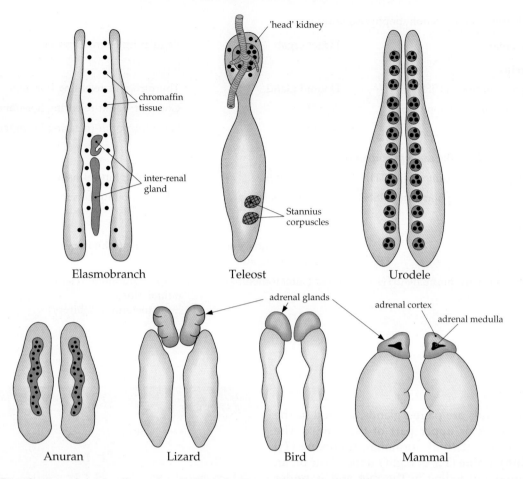

Fig. 25.17 Distribution of chromaffin and steroidogenic tissues in different vertebrates. In mammals, the two types of tissue have combined to form a discrete endocrine gland, the adrenal gland, with a cortex and medulla of quite different origins and subserving distinct functions

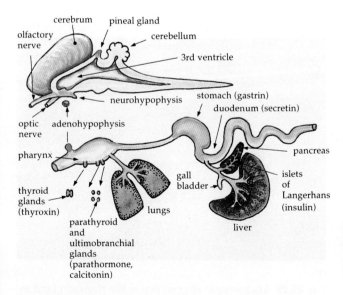

Fig. 25.18 Origins of some non-neural endocrine organs from the developing digestive tract of vertebrates

Anterior pituitary gland

The anterior pituitary gland secretes seven peptide or glycoprotein hormones that stimulate the activity of other endocrine glands and influence metabolism, growth and reproduction, either directly or by way of secretions of other hormones (Table 25.1). The endocrine target organs are the thyroid glands, gonads and adrenal cortices.

The secretion of most, possibly all, anterior pituitary hormones is subject to **hormonal feedback control**, by which the presence in the blood of the anterior pituitary hormone itself or the *target organ* hormone inhibits the secretion of either the releasing hormone at the hypothalamic level, or the adenohypophyseal hormone at the site of its secretion.

> The anterior pituitary secretes protein-based hormones that either control the secretion of other endocrine glands or act directly on cells of the body. Secretion is under the influence of stimulatory or inhibitory neuro-secretions from the hypothalamus, which perfuse the anterior pituitary via the portal vessels of the median eminence.

Table 25.1 Nature of the adenohypophyseal hormones

Class and name	Target organ	Regulating neurosecretion
Glycoproteins		
Thyrotrophic hormone (TSH)	Thyroid gland	Thyrotrophin-releasing hormone (TRH)
Follicle-stimulating hormone (FSH)	Gonads	Gonadotrophin-releasing hormone (GnRH)
Luteinising hormone (LH)	Gonads	Gonadotrophin-releasing hormone (GnRH)
Peptides derived from the same precursor molecule		
Adrenocorticotrophic hormone (ACTH)	Adrenal cortex	Corticotrophin-releasing hormone (CRH)
Melanocyte-stimulating hormone (MSH)	Pigment cells	MSH-inhibiting hormone (MIH)
Large peptides		
Prolactin (PRL)	Mammary glands, gonads and secretory epithelia	Dopamine (inhibitory)
Somatrophic (growth) hormone (STH)	Liver (somatomedins)	Growth hormone-releasing hormone (GRH) (stimulatory) Somatostatin (inhibitory)

Thyroid glands

The thyroid glands are paired glands, located in the front of the neck or upper part of the thorax. The production of the thyroid hormones, **thyroxine (T_4)** and **triiodothyronine (T_3)**, is highly dependent on an adequate intake of iodine in the diet and is under the neurosecretory control of TSH, which itself is subject to negative feedback control by the thyroid hormones (Fig. 25.19).

The two thyroid hormones are lipid-soluble and penetrate cell membranes, T_4 more easily than T_3. They bind to a protein in the cytoplasm of most cells in the body, which acts as an intracellular store and releases the hormone to diffuse into the nucleus, where it binds to the 'true receptor' and converts it into a **genetic transcription factor**. This initiates the production of enzymes that are essential for normal growth and maturation of immature individuals and stimulates metabolic rate of cells. Furthermore, both hormones directly stimulate mitochondria, and T_3 has a direct effect on cell membranes, increasing their permeability to amino acids and glucose.

The role of adenohypophyseal thyrotrophic hormone and thyroid hormone secretion in growth and metamorphosis of an anuran tadpole to an adult frog is shown in Figure 25.20, which also illustrates how the secretion of such hormones is dependent upon the development of the vascular link between the hypothalamus and the adenohypophysis.

The membrane actions of T_3 resemble catecholamine actions and produce short-term effects, measured in minutes or hours. Signs of excessive thyroid hormone activity, a condition known as **thyrotoxicosis** in humans, include elevated metabolic

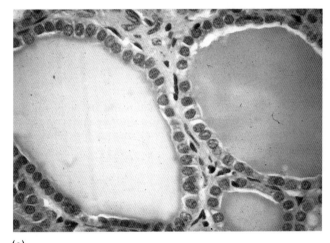

(a)

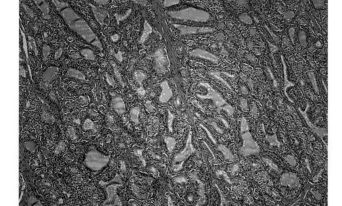

(b)

Fig. 25.19 Microscopic appearance of the thyroid gland in: (a) the unstimulated 'resting' state and (b) the TSH-stimulated active or 'secretory' state

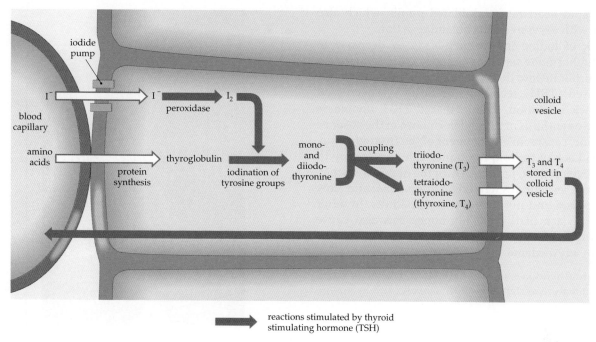

reactions stimulated by thyroid
stimulating hormone (TSH)

Fig. 25.19 (c) The mechanism of thyroid hormone production by vesicle cells. Circulating iodine is 'trapped' by an iodide pump in the cell membrane. Lack of iodine in the diet or interference with iodine pumping by chemicals present in some vegetables can impair thyroid hormone production. Amino acids provide the substrate for producing thyroglobulin. Thyroid hormones are stored in the 'colloid' contained within the thyroid vesicles

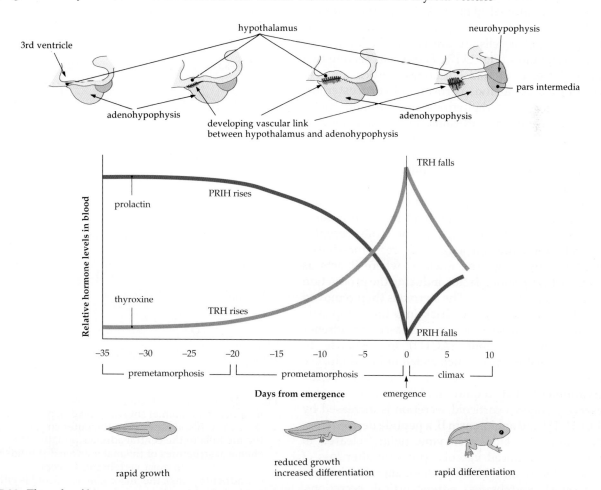

Fig. 25.20 The role of hormones in growth and development of the tadpole of a frog. Prolactin secretion by the anterior pituitary declines due to secretion of prolactin-inhibitory hormone (PRIH) from the hypothalamus. At the same time, thyroxine-releasing hormone (TRH) secretion rises and stimulates thyroxine secretion by the thyroid gland

rate, rapid heart rate and dilated pupils, which also occur with excessive catecholamine action.

Lack of iodine in the diet leads to decrease in thyroid hormone secretion because there is no iodine to incorporate into thyroglobulin. This, in turn, causes increased TRH secretion because of decreased feedback inhibition. The increased TRH, unable to increase synthesis of thyroid hormones in the absence of iodine, causes an enlargement of the thyroid gland, **goitre**. Lack of thyroid hormones can impair growth and development, particularly of the nervous system, in young vertebrates (**cretinism** in humans) and slow metabolism and nervous activity in adults. In adult humans, lack of thyroid hormones can also cause abnormal accumulation of water in subcutaneous tissues, **myxoedema**.

> Thyroid hormones stimulate growth and metamorphosis in immature vertebrates and metabolic rate in mature vertebrates.

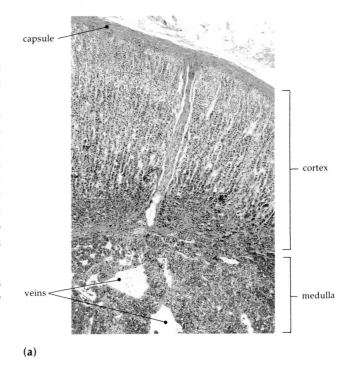

(a)

Adrenal cortices

In mammals and birds, the adrenal cortex is the outer layer of the **adrenal gland** (located at the anterior, cephalic, end of the kidneys). Less well-defined groups of similar cells are present in the kidneys of other vertebrates (Fig. 25.17). These cells secrete two major classes of steroid hormones (corticosteroids): **glucocorticoids**, such as cortisol and corticosterone, which influence carbohydrate and protein metabolism; and **mineralocorticoids**, such as aldosterone, which influence Na^+ and K^+ balance (Fig. 25.21). The chemical nature of the principal corticosteroids is very similar among vertebrates, with only minor variations between groups.

Like thyroid hormones, corticosteroids diffuse readily into cells where they bind to a cytoplasmic protein, which, in this case, is the 'true' receptor derived from the nucleus. The receptor–steroid complex moves into the nucleus, where it acts as a genetic transcription factor, inducing the production of cytoplasmic enzymes. The steroid is then removed and the receptor migrates back into the cytoplasm.

Glucocorticoid secretion is increased by adrenocorticotrophic hormone (ACTH) in response to metabolic disturbances, such as fall in blood glucose concentration, and to stressful disturbances of the environment that require mobilisation of energy reserves. Mineralocorticoid secretion is increased by both ACTH and **angiotensin II**, a peptide derived from the action of a kidney enzyme, renin (released in response to reduced vascular pressure after loss of body sodium), on a blood protein, **angiotensinogen**.

In most vertebrates, adrenocortical secretions, particularly mineralocorticoids, are essential for survival. Death usually follows removal of the adrenal

Fig. 25.21 (a) Section through the adrenal cortex and medulla of a mammal showing the different zones of the cortex. Blood flows from the outer cortex through the medulla to the central adrenal vein. (b) The chemical structures of the major adrenal steroids found in vertebrates. In most vertebrates (except chondrichthyians), the major glucocorticoid is either cortisol (fishes and mammals other than rodents) or corticosterone (reptiles, birds and rodents). The major mineralocorticoid in all species is aldosterone

glands and is usually a consequence of loss of body Na^+, with dehydration and circulatory failure. Many vertebrates, such as rats, red kangaroos and possibly North American opossums, can survive complete removal of both adrenal glands provided they have access to sufficient salt to replace that lost through the kidneys as a result of lack of mineralocorticoids. Echidnas appear to be unaffected by removal of both adrenal glands provided the external environment is stable. However, when exposed to a cold environment, an adrenalectomised echidna is unable to increase thermogenesis and maintain body temperature. The ability to thermoregulate is restored by injection of cortisol. This is a good example of the *permissive* nature of some hormone actions; that is, a lack of the hormone does not appear to influence body functions in the normal state but its presence is essential for regulatory responses to a disturbance of the environment.

The glucocorticoid hormones, as their name implies, influence sugar metabolism. They are important in the induction of enzymes that synthesise glucose from non-carbohydrate sources, such as amino acids, a process known as **gluconeogenesis**. They also cause breakdown of cell proteins to their constituent amino acids. The combination of these two actions results in the production of glucose, as a source of energy, at the expense of tissue protein. Two important tissues susceptible to this *catabolic* action are muscle and lymphoid tissues, both of which reduce in mass if high glucocorticoid levels are sustained in the blood.

Among the lymphoid cells affected by glucocorticoids are thymus-derived lymphocytes (T cells, Chapter 23), so that prolonged increase in glucocorticoid secretion can impair the ability of an animal to withstand infections. Other actions of the glucocorticoids include inhibition of release of the tissue factors that cause the inflammatory response to injury. Because of this, glucocorticoids can delay wound healing. All the adverse effects of glucocorticoids can occur as a consequence of prolonged exposure to a stressful environment.

Glucocorticoids also enhance the actions of catecholamines on cardiac contractility and blood pressure, a property useful in the management of shock in severely injured humans.

> The adrenal cortex secretes two major classes of steroid hormones (corticosteroids)—glucocorticoids, which influence carbohydrate and protein metabolism, and mineralocorticoids, which influence Na^+ and K^+ balance.

Gonads

In addition to their basic function of germ-cell production, ovaries and testes, located near the genital tracts, contain hormone-secreting cells. In females,

theca interna cells of ovarian follicles secrete the steroidal **oestrogens** (oestradiol, oestrone and oestriol). In males, Leydig cells of testes secrete the steroidal **androgens** (testosterone and dihydrotestosterone). These steroidal hormones diffuse into the cells of their target organs and bind to a specific nuclear receptor, which then becomes a genetic transcription factor. In some tissues, such as brain cells in mammals, testosterone is converted to an oestrogen (oestradiol) and then combines with an oestrogen receptor in the nucleus. Dihydrotestosterone, however, acts more directly on nuclear gene transcription.

Oestrogens and androgens influence growth, development of male and female sexual characteristics and behaviour, development of gametes, and ovulation (Fig. 25.22). Their effects on metabolic exchange are anabolic, that is, they favour increase of tissue mass. They are under the control of the two adenohypophyseal gonadotrophic hormones. Follicle-stimulating hormone (FSH) stimulates thecal cells of ovarian follicles to proliferate and secrete oestrogens. Luteinising hormone (LH) acts in concert with FSH to cause ovulation. LH then stimulates thecal cells of the ovulated follicle to proliferate and form a **corpus luteum**, which secretes a further steroid hormone, **progesterone**. Progesterone prepares the reproductive tract either to secrete an appropriate coating for the eggs or to receive an embryo and maintain its smooth muscle quiescent through pregnancy (Fig. 25.22).

In males, FSH stimulates spermatogenesis and LH stimulates the secretion of the steroidal androgens. Mammalian gonads also produce a peptide hormone **inhibin**, which can inhibit secretion of LH. Control of secretion of this hormone is not yet fully understood. Such a hormone could prove useful for male contraception.

> The ovaries and testes secrete oestrogens and androgens in females and males respectively. These steroidal hormones influence growth, development of male and female sexual characteristics and behaviour, development of gametes, and ovulation.

Parathyroid and ultimobranchial glands

When present, these glands are located in the neck or upper thorax. They are outgrowths of the **branchial clefts** of the embryonic pharynx (Fig. 25.18) and form discrete clusters of cells usually close to or incorporated in the thyroid glands (Fig. 25.23). In mammals, ultimobranchial cells are incorporated into the thyroid glands and are called **thyroidal C-cells**. Both glands are concerned with the regulation of extracellular Ca^{2+} concentration and their functions are best understood in birds and mammals.

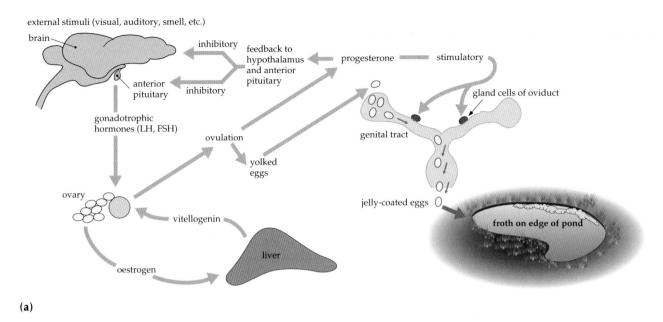

(a)

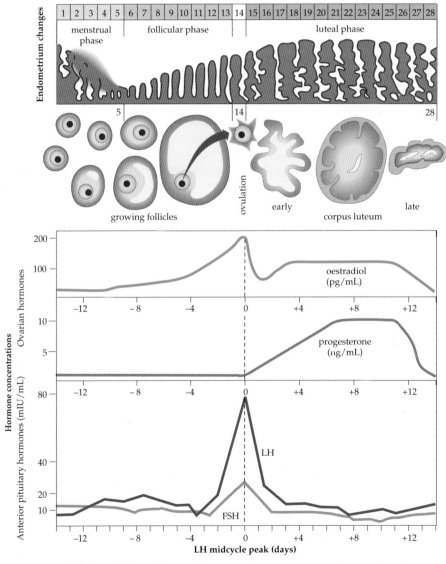

(b)

Fig. 25.22 (a) Outline of hormonal control of ovulation and vitellogenesis in an amphibian. (b) The relationship between hormone concentrations, follicular development and changes in the uterine endometrium during the menstrual cycle in a human female

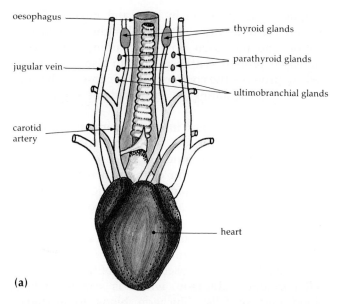

oesophagus

jugular vein

carotid artery

thyroid glands

parathyroid glands

ultimobranchial glands

heart

(a)

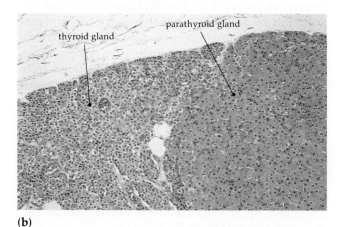

thyroid gland

parathyroid gland

(b)

Fig. 25.23 **(a)** The location of the thyroid, parathyroid and ultimobranchial glands in the neck of a bird. Similar relationships are found in other species. **(b)** Microscopic appearance of a parathyroid gland in a human, showing the close association with the thyroid gland

The parathyroid glands secrete **parathormone** (PTH) and the ultimobranchial glands secrete **calcitonin**. These are peptides that bind to cell-membrane receptors and activate intracellular messengers to stimulate existing enzyme activity. Their three main sites of action are bone, the intestine and the kidneys. PTH increases Ca^{2+} concentration in blood plasma by mobilising bone Ca^{2+}, increasing Ca^{2+} uptake by the gut and promoting Ca^{2+} reabsorption by renal tubules. Calcitonin has the opposite effects. It depresses plasma Ca^{2+} concentration by promoting its uptake by bone and increasing its excretion by the kidney. In addition to its direct actions, PTH also stimulates synthesis of *Vitamin D*, which also enhances Ca^{2+} absorption from the gut and Ca^{2+} mobilisation from bone.

These glands respond directly to changes in plasma Ca^{2+} concentration. Increase in Ca^{2+} in blood plasma increases calcitonin secretion and decreases PTH

secretion, so that Ca^{2+} is removed from the blood. Decrease in plasma Ca^{2+} concentration has the opposite effect, so that Ca^{2+} is added to the blood.

The opposing actions of calcitonin and parathormone produce a very precise control of extracellular Ca^{2+} concentration, which is necessary because of the important role Ca^{2+} plays in neuromuscular transmission (Chapter 27). In humans, a small fall in extracellular Ca^{2+} concentration causes a marked increase in neuromuscular irritability, a condition known as **tetany**, which can lead to convulsions and death. The occasional development of tetany after removal of a goiterous thyroid gland in humans led to the discovery of the parathyroid glands and their role in Ca^{2+} balance. In those vertebrates that have one, it is the bony skeleton that is a major reservoir of exchangeable Ca^{2+} and a major site of action of Ca^{2+}-regulating hormones. For example, excessive PTH secretion leads to demineralisation of bones, which become weak and prone to fracture.

Ultimobranchial glands are well defined in both cartilaginous and bony fishes. The Mexican cavefish, *Astyanax mexicanus*, is normally exposed to dim light for a short time during the day. If it is kept in complete darkness, the ultimobranchial bodies enlarge and the bony skeleton becomes decalcified. It was this observation, made in 1954, that led to the later realisation that the ultimobranchial glands were concerned with regulation of Ca^{2+} balance. It also shows that, in this species of fish at least, secretion by the ultimobranchial gland can be influenced by a change in the external environment—light.

Parathyroid glands evolved in tetrapods, where together with the ultimobranchial glands, they appear to influence Ca^{2+} exchange in much the same way as described for mammals. In birds and some reptiles, the parathyroid and ultimobranchial glands have an additional function related to the production of large eggs with calcified shells. The large amount of Ca^{2+} required, especially in birds, appears to be derived from bone under the influence of PTH. The calcification of the egg is facilitated by oestrogens from the female gonads. Oestrogens also increase Ca^{2+} mobilisation from bone and, in addition, stimulate the liver to produce a phospholipoprotein from the liver, **vitellogenin**, which strongly binds Ca^{2+} and is deposited in the egg. Oestrogens are also important in regulation of Ca^{2+} exchange in human females, depletion of bone Ca^{2+}—**osteoporosis**—being a fairly common consequence of reduction in gonadal oestrogen production after menopause if Ca^{2+} intake is low.

Parathyroid and ultimobranchial glands secrete parathyroid hormone and calcitonin respectively. These hormones regulate extracellular calcium concentration through opposing actions on uptake, mobilisation and excretion of calcium.

Islets of Langerhans

These cells are associated with the outgrowth of the digestive tube that forms the exocrine **pancreas** (Fig. 25.18). However, the islet cells have only an endocrine function, releasing their secretions into the circulating blood (Fig. 25.24). There are three major cell types in these islets—alpha, beta and gamma—which secrete respectively the peptides **insulin**, **glucagon** and **somatostatin**. Insulin is a most important regulator of carbohydrate metabolism in vertebrates. It is a protein that occurs, with minor chemical modifications, in most animals. It has even been found in unicellular organisms, including bacteria, although its functions, if any, in such organisms are unknown.

Insulin is secreted in response to a rise in blood glucose concentration, for example, after ingestion of carbohydrate. It acts by binding to a transmembrane receptor on its target cells, inducing a cascade of events that lead to an increase in the permeability of the cell membrane to glucose and amino acids, and the induction of metabolic enzymes. This results in increased glucose utilisation and increased storage of glucose as glycogen in liver and muscle. These actions cause a fall in extracellular glucose concentration and increased uptake of amino acids with increased protein synthesis. Insulin also acts on fat cells to promote the conversion of glucose to glycerol and acetate, resulting in increased fat production (lipogenesis).

The human disease **insulin-dependent diabetes mellitis** is characterised by high blood glucose concentration and excretion of glucose in the urine, and is due to destruction of the pancreatic islet β-cells, possibly by an autoimmune reaction after a viral infection. The disease can be controlled by administration of insulin.

The effects of insulin are **hypoglycaemic** (blood glucose-lowering) and **anabolic** (favouring synthesis of tissue). They are opposed by one of the other islet cell hormones, **glucagon**, which is secreted when blood glucose falls. Glucagon binds to a transmembrane receptor leading to a decrease in glucose oxidation by cells and an increase in breakdown of glycogen to glucose in liver and muscle. It also stimulates production of gluconeogenic enzymes, which synthesise glucose from amino acids derived from structural proteins. These effects are **hyperglycaemic** (blood glucose-raising) and **catabolic** (favouring tissue breakdown).

While there are many other hormones that participate in the control of blood glucose and metabolism, such as glucocorticoids from the adrenal cortex and somatotrophin from the adenohypophysis, the blood glucose level is normally accurately controlled by the interaction between insulin and glucagon effects (Fig. 25.25).

Islet cells respond directly to changes in extracellular glucose concentration. Insulin secretion increases with rising extracellular glucose concentration, causing increased uptake of glucose by cells, which counters the rise. Glucagon secretion increases when extracellular glucose falls, causing mobilisation of glucose stores in the liver, so countering the fall. In this way, insulin and glucagon control extracellular glucose levels in much the same way as calcitonin and PTH control extracellular Ca^{2+} concentration.

> Insulin and glucagon are secreted by the alpha and beta cells respectively of the islets of Langerhans in the pancreas. Insulin increases tissue uptake and storage of glucose, and its secretion is stimulated by rising blood glucose concentration. Glucagon has the opposite effects. Between them they control blood glucose concentration.

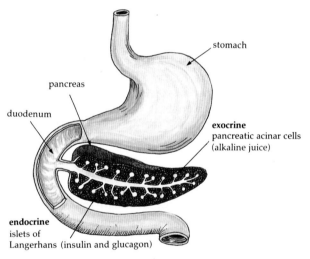

stomach

pancreas

duodenum

exocrine
pancreatic acinar cells
(alkaline juice)

endocrine
islets of
Langerhans (insulin and glucagon)

(a)

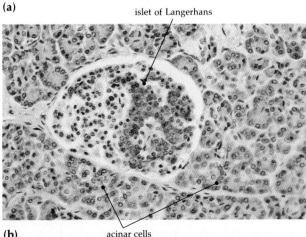

islet of Langerhans

(b)   acinar cells

Fig. 25.24 **(a)** Location of the islets of Langerhans in the head of the exocrine pancreas. **(b)** An islet containing insulin-containing β-cells, stained brown, surrounded by acini

Gastrointestinal mucosa

The cells that secrete gastrointestinal hormones are scattered in loose aggregations throughout the gut mucosa. They are bipolar, the side on the mucosal

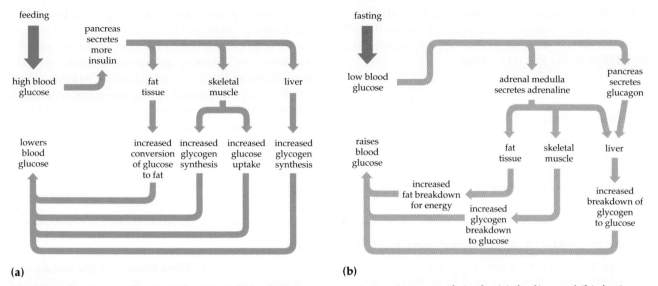

Fig. 25.25 Summary of hormonal control of blood glucose concentration in mammals in the **(a)** feeding and **(b)** fasting states

surface detecting a change in the chemical environment provided by the contents of the digestive tract, while the internal side secretes the hormone into local blood vessels in response to an appropriate change in gut content (Fig. 25.26).

These hormones control the secretion of digestive juices. There are three major hormones in this group, **secretin**, **gastrin** and **pancreozymin/cholecystokinin**. They are all released into the portal vein to be carried in the blood through the liver to their target organs,

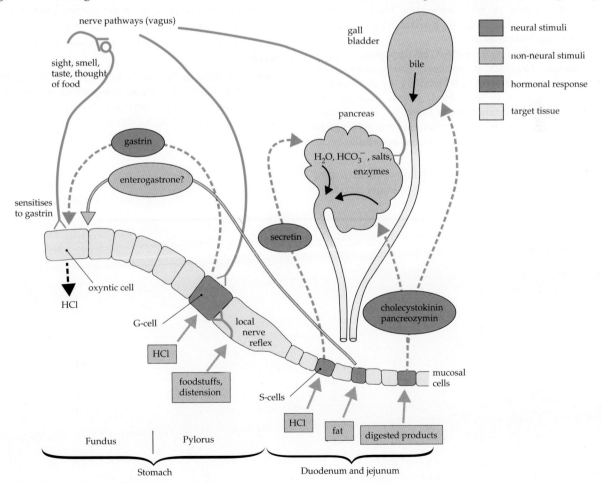

Fig. 25.26 The major hormone-secreting regions of the digestive tract and the role of gastrointestinal hormones in controlling the composition of the digestive juices

where they optimise the composition of the digestive juices to ensure adequate digestion of ingested food, and control the motility of gut smooth muscle and blood flow to the gut (Fig. 25.26).

Polypeptide growth factors

This is a group of polypeptide hormones, secreted by a wide variety of cells, that act as **mitogens**, that is, they stimulate mitosis of cells in the environment in which they are released. Some examples of these

are epidermal growth factor, nerve growth factor and transforming growth factor. They bind with high affinity to the extracellular domain of specific protein receptors on plasma membranes and activate intracellular messengers to stimulate mitosis. They therefore have important functions in the regulation of growth and differentiation. It is now thought that viruses can induce cancer by producing analogues of these locally acting growth factors. A considerable research effort is being made to understand the functions of these hormones with a view to controlling cancers.

BOX 25.2   AUTOCRINE AND PARACRINE HORMONES

These hormones constitute a diverse group of biologically active agents released by a wide variety of tissues in response to local stimuli, which act on cells in the vicinity of their release.

Eicosanoids

Eicosanoids are the most widely studied members of this group of hormones and most of the knowledge applies to eutherian mammals. They are all derived from the unsaturated fatty acid **arachidonic acid** and all have 20 carbon atoms in their molecule. There are four subgroups—prostaglandins, prostacyclins, thromboxanes and leukotrienes.

All mammalian cells, except erythrocytes, contain eicosanoids. They act at extremely low concentration in the same environment in which they are synthesised. They also appear in the blood, where they can be measured, but their functions as blood-borne hormones are uncertain. In mammals, eicosanoids are involved in a wide variety of functions, all mediated by cellular cyclic AMP. These include:

- induction of local increase in blood flow and increased capillary permeability in response to injury of cells— the *inflammatory response*;

- stimulation of nerve fibres subserving the sensation of pain;

- induction of fever by affecting the thermoregulating region of the brain;

- participation in blood pressure control by acting on vascular smooth muscle;

- participation in the induction of blood clotting at a site of injury;

- control of several reproductive functions, including induction of labour, lysis of the corpus luteum and some aspects of reproductive behaviour;

- participation in regulation of the sleep–wakefulness cycle.

Although little is known about their evolutionary history, it is likely that these compounds are widely distributed in vertebrates and are important in reproduction and in mediating local responses to injury.

SUMMARY

- Different hormones interact to achieve control of growth, maturation, reproduction, metabolism and the inorganic composition of the body fluids.
- Neurosecretion is the earliest form of hormonal control to have evolved in animals, and is probably the major form of hormonal control in invertebrates.
- Neurosecretory hormones are concerned with growth, maturation, reproduction and metabolism.
- Non-neural hormone-secreting organs controlled by the nervous system evolved later and add to the effectiveness and range of actions of the neurosecretory control of body functions.
- Reproduction in insects depends on external environmental cues that determine the nature of the internal hormonal environment via neurosecretion. These cues may be day length, temperature, humidity, or chemical agents detected by surface receptors. The effects of neurosecretions on the functions of the corpora allata appear to be important. In some species, the level of food intake can initiate gonadotrophin secretion, or determine whether or not the environmental cues are effective.
- Control of energy exchange in arthropods is effected mainly by neurosecretions from either the eye stalk or corpora cardiaca, which mobilise energy reserves in response to environmental change.
- Hormonal control systems in vertebrates have become increasingly complex and affect most physiological functions. However, the basic ground plan of neurosecretory response to environmental change, first evident in lower invertebrates, is retained.
- The increasing complexity of vertebrate physiological control systems is the result of the evolution of additional hormonal mechanisms responding to and controlling variables in the internal environment.

- In vertebrates, neurosecretory tissue is found in the neurohypophysis and non-neural endocrine tissue in the adenohypophysis. The neurohypophysis secretes peptides affecting water balance and reproduction. The adenohypophysis secretes protein hormones that either control the secretion of other endocrine glands or act directly on all cells of the body.
- The pineal gland, which is under the influence of light through photoreceptors, secretes melatonin, which influences pigmentation and reproductive activity in many birds and mammals.
- Thyroid hormones stimulate growth and metamorphosis in immature vertebrates and metabolic rate in mature vertebrates.
- Parathyroid and ultimobranchial glands secrete parathormone and calcitonin respectively, which, through opposing actions on uptake, mobilisation and excretion of calcium, regulate extracellular Ca^{2+} concentration.
- The adrenal cortex secretes corticosteroids—glucocorticoids, which influence carbohydrate and protein metabolism—and mineralo-corticoids, which influence Na^+ and K^+ balance.
- The ovaries and testes secrete oestrogens and androgens in females and males respectively. These steroidal hormones influence growth, development of male and female sexual characteristics and behaviour, development of gametes, and ovulation.
- Insulin and glucagon are secreted by the alpha and beta cells respectively of the islets of Langerhans in the pancreas. Insulin increases tissue uptake and storage of glucose and glucagon has the opposite effects. Between them, they control blood glucose concentration.

QUESTIONS

1. Describe two basic differences in the modes of action of water-soluble hormones and lipid-soluble hormones.

2. With reference to hormone actions, what is meant by the terms autocrine, paracrine and endocrine?

3. What is meant by neuroendocrine control? Describe (a) a neurosecretory cell and (b) a neurohaemal organ.

4. In vertebrates, explain the developmental origins of the neurohypophysis and adenohypophysis.

5. What are the general properties, target organs and physiological functions of (a) neurohypophyseal hormones and (b) adenohypophyseal hormones?

6. (a) What is the significance of dietary iodine to thyroid function? (b) What physiological processes do thyroid hormones influence?

7. (a) What is the functional difference between a mineralocorticoid and a glucocorticoid? (b) What is the effect of ACTH on the adrenal gland?

8. (a) Where are the ultimobranchial and parathyroid glands? (b) Draw a flow diagram to show how the hormones they secrete interact to control blood Ca^{2+} levels.

9. What are the endocrine cell types in the islets of Langerhans and what are their functions?

10. What are the hormonal effects of removing the eye stalks from (a) an immature and (b) a mature sexually quiescent crustacean? Give the reasons for these effects.

11. What are the functions of (a) eclosion hormone and (b) bursicon in pupating insects?

12. What is meant by the term vitellogenesis? Which hormones stimulate vitellogenesis in (a) insects and (b) vertebrates?

13. Draw a diagram to illustrate the hormonal control of ovulation in a mammal.

14. (a) What are the major functions of progesterone in mammals? (b) What are the functions of prolactin in mammals and birds?

NERVOUS SYSTEMS

Co-ordination of function is achieved in all animals by the passage of information within cells and between cells, tissues and organs. This is accomplished in two main ways—by means of hormones, as explained in Chapter 25, and by the nervous system. The nervous system has specific roles in integrating information about the state of an organism, and in providing rapid and precise signalling between parts of the body. In both simple and complex organisms, the nervous system is responsible for stereotyped activities, such as avoidance behaviours and movements of internal organs. In mammals, it provides remarkable computational power, which is expressed in such phenomena as visual memory and fine movement control. The human nervous system has evolved further to allow speech and abstract thought.

Communication in nervous systems depends on the functions of **neurons**, cells specialised for conducting information. Neurons, together with supporting **glial cells**, form the nervous system (Table 26.1).

Table 26.1 Features of two cell types found in nervous systems

Neurons	Glial cells
■ Electrically excitable	■ Provide mechanical support for neurons
■ Receive inputs from the environment (sensory neurons) or from other neurons (interneurons and motor neurons)	■ Provide electrical insulation for neurons
	■ Maintain the extra-cellular environment of neurons
■ Provide output to other neurons or to effectors (mostly muscle and gland)	■ Guide neuronal development and repair

NEURONS: FUNCTIONAL UNITS OF NERVOUS SYSTEMS

The function of a neuron is to receive information, process that information, and transmit the processed information to other cells. Most neurons have three major structural regions (Fig. 26.1a): **dendrites**, processes that are generally short and receive inputs; the **soma**, the cell body containing the nucleus; and an **axon**, a long process that carries the output. Neurons are often quite long cells, for example, neurons whose axons reach your feet have their cell bodies located in the spinal cord at about the level of your kidneys, and axons in the arms of giant squids run for more than 10 m. Some neurons in simple animals, such as jellyfish (phylum Cnidaria), are unspecialised in structure and pass information in any direction (Fig. 26.1b). In more complex animals, neurons of many shapes have evolved with regions specialised for receiving, processing or passing on information (Fig. 26.1b). Transduction of information by sensory cells has been considered in Chapter 7.

Neurons receive, process and transmit information. Glial cells provide both physical and nutritional support, and insulation.

Types of neurons

In simple animals, a single neuron can carry out several functions. It can detect incoming stimuli, transfer information between other pairs of neurons and activate muscles. This arrangement is not very specific for processing and routing signals, and is not conducive to integration of responses. The neurons of most animals are more specialised; they can be grouped into three functional classes: sensory neurons, interneurons and motor neurons. *Sensory neurons* are associated with specialised detectors, known as sensory receptors (Chapter 7). Receptors that detect signals from the external environment, such as sensory neurons for touch or smell, are known as *exteroreceptors*. Those that monitor internal states, such as blood pressure or core temperature, are *enteroreceptors*. Sensory neurons pass their information to other specialised neurons.

Interneurons receive information from one group of neurons and pass it on to another. They have major

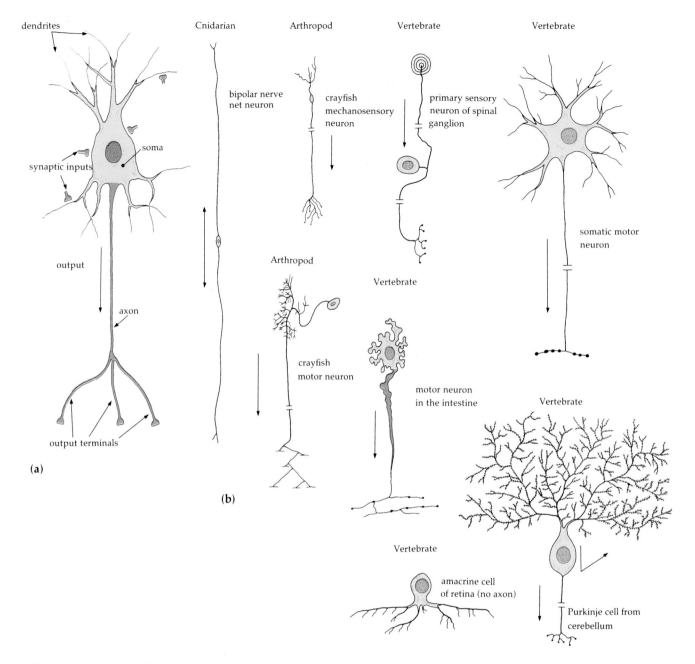

Fig. 26.1 **(a)** A generalised representative neuron. Most neurons have a series of short processes, called dendrites, that receive information via synaptic inputs from other neurons. Excitation of the cell body (soma) causes an electrical event, the action potential, to travel along the axon towards the output terminals of the neuron, which release chemical transmitters that influence other neurons or effectors such as muscles or glands. **(b)** A comparison of the morphology of a number of different neurons from a variety of animals and locations within nervous systems. Arrows show direction of travel of information

roles in the integration of information, that is, sorting, comparing, combining and discarding information from various sources, and in providing appropriate outputs.

Motor neurons provide the final output of the nervous system to the muscles and glands that it controls. The cells or tissues innervated by motor neurons are called *effectors*.

Pathways carrying information *towards* the central nervous system are called *afferent pathways*, whereas those carrying it *away* are called *efferent pathways* (the same terms are used for other structures, such as blood vessels: afferent means passing towards a point or area, efferent means passing away from it). Thus, sensory neurons are often referred to as afferent neurons and motor neurons as efferent neurons.

Sensory neurons detect signals in the internal and external environment. Interneurons receive, process and transmit information from one group of neurons to another. Motor neurons influence the activities of muscles and glands.

Neurons conduct electrical signals

The manner in which information passes along a neuron depends on the passage of electrical currents across its plasma membrane. A current will travel when there is a difference in voltage between two points, and a conducting path along which a current can flow. In a neuron, electrical communication depends on the voltage difference across the plasma membrane, the inside of a neuron being negative with respect to the outside when the neuron is at rest (Chapter 4). Local changes in membrane voltage may be conducted away passively, that is, their amplitudes decrease with distance and time (Box 26.1).

> Information in the form of electrical signals is conducted along the membranes of neurons.

BOX 26.1 PASSIVE ELECTRICAL PROPERTIES OF NEURONS

The plasma membrane of nerve cells, like all plasma membranes, is a protein and lipid bilayer that incorporates integral membrane proteins (Chapter 3). Membranes have electrical capacitance and resistance, which are depicted in the electrical equivalent circuit in Figure a. Electrical charge and the insulating lipid together create the capacitance component. The charge is present as ions in aqueous solutions on either side of the membrane and partly as polar groups on the membrane surface. Pores, composed of integral proteins, allow ions to traverse the membrane but offer some opposition to their passage; this is the membrane resistance represented in Figure a. There is also resistance to flow of current through the cytoplasm in nerve cell processes.

If the voltage across the membrane is changed at a point, the current generated will leak both across the membrane at that point and lengthways along the process before leaking across the membrane at other points. The greater the resistance of the membrane, per unit length (R_M, ohm/cm) relative to the cytoplasmic resistance, per unit length (R_L, ohm/cm), the greater the tendency of the current to follow the length of the process. The measure of the conductance along the process in relation to leakage across the membrane is the length constant, more generally called the **space constant**, which is represented by λ, the distance for an applied voltage to fall to $\frac{1}{e}$ (about 37%) of its original value. The space constant, λ, is equal to $\sqrt{R_M/R_L}$. The capacitance per unit length of the membrane (C_M, coulomb/volt. cm) delays the effect of an applied current on the voltage across a nerve fibre membrane because the current must first displace capacitative charge before it generates the full voltage change. The rate of discharge of the capacitance is measured as the **time constant**, τ, which is the time to change the difference between the initial and final state by a factor of $1-\frac{1}{e}$ (about 63%); τ equals $R_M \cdot C_M$. The electrical properties of the membrane determine how quickly the peak of an electrical event is conducted passively in time and space. This speed is λ/τ.

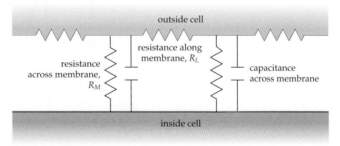

(a) Electrical model of a membrane (R_M represents the resistance to flow of current across the membrane and R_L the resistance to flow of current through the cytoplasm)

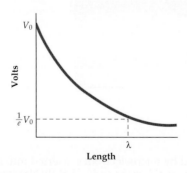

(b) Length–voltage relationship for an axon: charge leaks through the membrane (V_0 is voltage at zero length; λ is length constant)

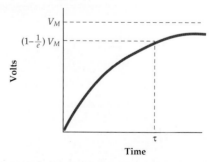

(c) Time–voltage relationship: an imposed current takes time to change the voltage because it must displace capacitative charge (V is maximum voltage attained; τ is time constant)

The voltage difference between the inside and outside of a neuron depends on transmembrane proteins in the neuronal membrane that function as ion pumps, using energy from ATP (Chapter 4). These pumps transport Na^+ to the outside and K^+ to the inside of the cell, creating and maintaining concentration gradients for these ions. In the resting state, the neuronal membrane is more permeable (leaky) to K^+ than to Na^+; K^+ moves down its concentration gradient from inside to outside, leaving the inside of the neuron electrically negative compared with the outside (Box 26.2). The polarisation of a membrane can be measured using a microelectrode inserted through the neuronal membrane (Fig. 26.2). Such measurements show that the electrical polarisation of most neuronal membranes is about 80 mV; that is, the cell has a resting membrane potential of −80 mV.

The surface membrane of a neuron is electrically polarised, with the inside negative compared with the outside.

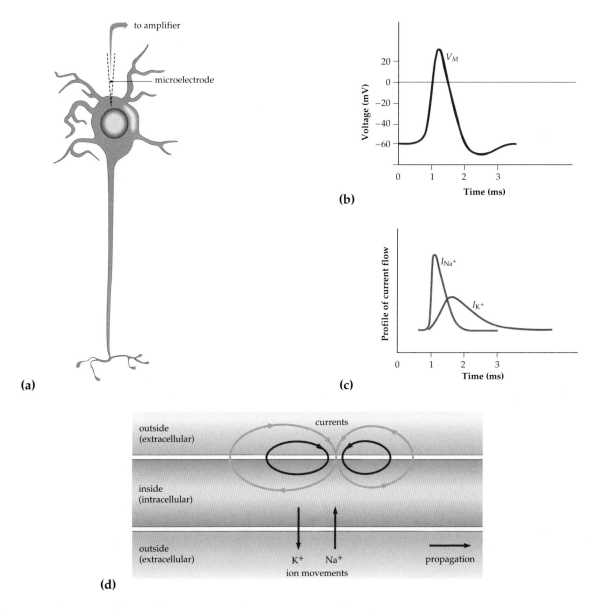

(b)

(c)

(d)

Fig. 26.2 The neuronal action potential. **(a)** Action potentials are recorded by microelectrodes inserted into neurons. **(b)** The action potential consists of a brief depolarising potential, with a rapid onset and a slightly slower decline, followed by a slight undershoot. **(c)** The voltage changes are the consequence of brief changes in permeability to Na^+ and K^+, producing an inward current (I_{Na^+}) and an outward K^+ current (I_{K^+}). **(d)** The movement of Na^+ and K^+ during the action potential causes local current circuits. The electrical disturbance propagates along the axon. After the action potential has passed, the usual ionic gradients are restored by the Na^+–K^+ pump

Active responses

When the membrane potential inside the neuron becomes less negative—depolarised—a local feedback can amplify the potential change, giving rise to an event called an *active response*. Active responses eventually die away with distance from their points of initiation.

An active response involves charged particles (ions) moving through channels in the membrane. Two cations, sodium (Na^+) and potassium (K^+), are important for active responses.

Sufficient depolarisation to generate an active response can be caused physiologically by excitation from another neuron or a sensory stimulus, or by

---

BOX 26.2 THE NEURON MEMBRANE POTENTIAL

A voltage difference exists across the plasma membrane because it is differentially permeable to ions and separates solutions of different ionic compositions (Chapter 4). Each type of ion tends to move down its concentration gradient, but the passages of different ions are hampered to different extents. It is this tendency of ions to move across the membrane that creates the voltage difference.

Firstly, consider K^+. The plasma membrane is permeable to K^+ and the concentration of K^+ inside the cell is greater than that outside ($[K]_i > [K]_o$), which means that there will be a net diffusional movement of K^+ outwards. Outward K^+ movement makes the cell interior more negative; the membrane is thus equivalent to a battery with its negative pole inside and its positive pole outside the cell.

In solution, ions behave like an ideal gas; that is, the work that the gas or ions in solution can do by moving down a concentration gradient depends on the concentration difference. From the gas laws, the diffusion energy available for movement from concentration [K] to concentration zero is

$$RT \log_e [K] \text{ joules/mole}$$

where R is the gas constant (in J mol^{-1} degree^{-1}). For a gas, the equation is usually expressed in terms of pressure, which is proportional to the concentration of gas molecules; the concentration of solute molecules is proportional to diffusional pressure.

The difference in diffusion energy across the membrane (driving K^+ out) is the difference in the diffusional energy of the inside and outside solutions:

$$RT \log_e [K]_i - RT \log_e [K]_o$$

This difference in diffusional energy must equal the electrical energy that the battery, created as a result of the concentration difference, can deliver. Battery energy (the work that a battery would need to do to cause movement of a single ion) is equal to the strength of the battery (V; voltage) multiplied by the charge on the ion (qZ; elementary charge × valency), that is,

$$VqZ \text{ volts coulomb}$$

We need to put this in molar terms. To do this we use Faraday's constant, F (the number of coulombs in a mole), giving

$$VFZ \text{ volts coulomb mole}^{-1}$$

Because the two energies—the battery energy and the energy for diffusion—must be equal,

$$VFZ = RT \log_e[K]_i - RT \log_e[K]_o$$
$$= RT (\log_e[K]_i - \log_e[K]_o)$$
$$= RT \log_e \frac{[K]_i}{[K]_o}$$

That is, the voltage across the membrane is

$$V = \frac{RT}{ZF} \log_e \frac{[K]_i}{[K]_o}$$

This is known as the Nernst equation and V is the equilibrium potential for K^+, E_K; that is, the potential that would be generated by the distribution of this ion if the membrane were not permeable to any other ions. Because the ability of K^+ to diffuse across the membrane determines E_K, it is also called the diffusion potential for K^+. In thinking about this equation, remember that the logarithm of a number greater than 1 is positive, the logarithm of 1 is zero and the logarithm of a number less than 1 is negative. A positive value of V means the outside cell is positive and the inside cell is negative.

Now imagine a membrane permeable to Na^+, K^+ and Cl^-. Each will provide a driving potential due to differences in its concentrations across the membrane and charge. The effect of the driving potential (equilibrium potential) of any ion on the resting membrane potential will depend on its ability to cross the membrane—only ions that can cross contribute to the diffusion potential.

In fact, the membrane of the cell resists the movements of all ions to some extent. The effect of this resistance is to reduce the effective diffusional energy supplied by a given concentration of ions. This is equivalent, in effect, to lowering the concentration. To take into account membrane resistance, we should replace [K] by $[K]_i$/resistance. In practice, we use the inverse of resistance to flow of ions, a term known as conductance (g_K). Thus, the Nernst equation can be rewritten as

$$E_K = \frac{RT}{F} \log_e \frac{g_K[K]_i}{g_K[K]_o}$$

Of course, the g_K terms cancel out, but considering that there are other ions being 'pushed' by their concentration differences through different conductances, adding in the terms for these ions we get

$$E_M = \frac{RT}{F} \log_e \frac{g_K[K]_i + g_{Na}[Na]_i + g_{Cl}[Cl]_o}{g_K[K]_o + g_{Na}[Na]_o + g_{Cl}[Cl]_i}$$

This is known as the Goldman equation. Note the inversion of the Cl^- concentrations because the Cl^-

charge, and hence the electrical equivalence of its diffusional drive, is opposite to Na^+ and K^+. Simply looking at the equation shows that the greater the value of g_K relative to other conductances, the closer the Goldman equation comes to the Nernst equation for K^+, in other words, the closer the membrane potential (E_M) approaches the K^+ equilibrium potential (E_K).

Test yourself on the questions below. Cover the right column, answer the questions on the left and then compare your responses with those provided.

Questions	Answers	Questions	Answers
What happens to E_M if $[K]_o$ increases?	E_M becomes more positive (i.e. the membrane depolarises).	In this last case, what happens to overall conductance?	It increases.
What happens if g_{Na} increases?	E_M becomes more positive, it moves towards E_{Na}.	What would be the relative effects on E_M of the diffusion potential for a fourth ion at rest after all the other conductances (but not its conductance) is increased?	The effect of the fourth ion would be diminished.
What happens to E_M if all the conductances increase, but the ratios between them stay the same?	E_M does not change.		

a current passed through a microelectrode (Fig. 26.2). The potential at which depolarisation is sufficient to trigger an active response is the *threshold potential* for that membrane. At this potential, conformational changes occur in certain transmembrane proteins, increasing their permeability to external Na^+ (and occasionally Ca^{2+}). These proteins are voltage-dependent channels because the voltage across the membrane determines whether they are open or closed (Chapter 7). When they open, Na^+ diffuses down its concentration gradient, making the inside of the neuron even less negative. This change in membrane potential causes additional voltage-dependent Na^+ channels to open. The active response terminates when the membrane depolarises sufficiently for a different set of voltage-dependent channels to open, in this case, K^+ channels, allowing K^+ to leave the cell. This moves the potential back towards its resting level.

The distance that an active response spreads along a neuron process and the time it lasts depend on both passive and dynamic characteristics of the membrane. In some neurons, voltage change spreads rapidly, with little decrease in amplitude over large areas of membrane; in others, it affects only the membrane immediately around the stimulus site. Different parts of the membrane of the same neuron can have different characteristics. Local active responses are important in transmission of information over short distances.

> The active response of a neuronal membrane is dependent on the dynamic properties of the voltage-dependent channels, and time and space constants of the membrane.

Action potentials

Active responses are not suitable for conducting information along processes of neurons that are from several millimetres to metres long. Long-distance communication involves the conduction of an electrical event that does not die out with distance. This event is the **action potential** or spike. Action potentials travel the full lengths of neuron processes without loss of amplitude because they regenerate themselves at successive points. The regenerative nature of action potentials allows information to be carried over long distances without distortion or loss.

In neuronal membranes that generate action potentials, the number of voltage-dependent Na^+ channels opening at the threshold potential increases rapidly and in a non-linear fashion. This creates a positive feedback loop in which more channels opening induce yet more to open so that the potential inside the membrane becomes rapidly positive. The positive feedback loop is broken when high levels of depolarisation cause the Na^+ channels to close and the voltage-dependent K^+ channels to open. All this happens within a few milliseconds, giving rise to a rapid inward Na^+ current followed by a short-lasting outward K^+ current (Fig. 26.2).

For a brief period after each action potential, the membrane becomes less excitable. This period of reduced excitability, the refractory period, limits the frequency at which action potentials can occur in a neuron to about 100 per second. The depolarisation caused by an action potential spreads into adjacent areas of membrane in the same way as local active responses. When this occurs in an axon, the depolarisation is effectively channelled along the cable, opening voltage-dependent Na⁺ channels in the adjacent region of the neuronal membrane where the action potential is regenerated. Sequential activation regenerates the action potential at contiguous points along the axon, that is, it is propagated or *conducted*.

> Some membranes are capable of producing regenerative active responses called action potentials.

Conduction of action potentials

The speed of conduction of action potentials along axons determines, to a large extent, how quickly animals can respond to their environment. There is an advantage in rapid conduction in those parts of nervous systems involved in fast responses that affect survival, for example, escape from predators or prey capture. Axon conduction speeds vary from about 0.5 to 120 ms⁻¹ and depend on morphological characteristics of the axon that determine its effectiveness as an electrical cable, particularly its diameter and the insulating properties of the plasma membrane and closely adjacent glial cells. The more efficient the insulation and the greater the diameter, the faster the conduction.

Based on these factors, two ways of increasing conduction velocity in neurons have evolved. Vertebrate axons that conduct at speeds greater than 5–10 ms⁻¹ are invariably wrapped in an insulation of glial cell membranes forming a coat called myelin (Fig. 26.3). The myelin sheaths are interrupted by small regions of bare axon, 1–1.5 mm long, called nodes of Ranvier. Action potentials skip from node to node, in a process known as **saltatory** (meaning leaping) **conduction**. In many invertebrates, rapid conduction is achieved by means of large axon diameters. In the squid, axons more than 1 mm in diameter supply mantle muscles, which contract during the jet propulsion used for escape and prey capture (Fig. 26.3).

> Speed of conduction of an action potential along an axon is increased by increase in diameter and/or insulation of the axon.

Transmission of information to other cells

Neurons transmit information directly to other neurons, or to muscles, at small areas of close contact, **synapses**. Transmission can increase activity in the postsynaptic cell, *excite*, or reduce it, *inhibit*. At *chemical synapses*, which are most common, there are narrow separations about 20–50 nm wide between presynaptic axon terminals and the surfaces of the post-synaptic cells they influence. Signals are transmitted across these gaps via the release of chemical signalling substances, **neurotransmitters** (Fig. 26.4).

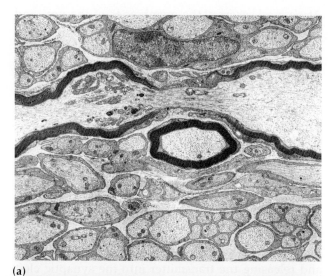

(a)

(b)

Fig. 26.3 (a) Fast conducting axons in vertebrates are wrapped in many layers of glial cell plasma membrane, the myelin sheath, as shown in this electron micrograph. A gap in the myelin (a node of Ranvier) can be seen at the arrow. (b) Much of the original work on the function of neurons was carried out on giant axons from the squid. These axons are responsible for producing the very rapid jet propulsion shown by these animals under certain circumstances

presynaptic axon

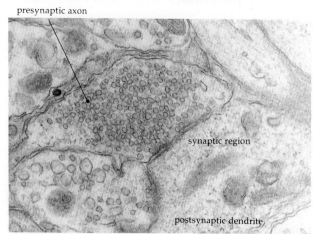

synaptic region

postsynaptic dendrite

Fig. 26.4 A chemical synapse as it is seen in the electron microscope. The major features are the presynaptic accumulation of vesicles and the pre- and post-synaptic membrane thickenings

At *electrical synapses*, the membranes of the pre- and post-synaptic cells are apposed and communicating junctions are present, allowing electrical signals to pass across directly (Chapter 6). Some electrical synapses pass signals in both directions, others only one way.

Within the presynaptic nerve ending at a chemical synapse are small membrane-bound structures, *synaptic vesicles*, containing chemical transmitter molecules. When an action potential arrives at the pre-synaptic nerve ending, it causes the opening of voltage-dependent Ca^{2+} channels. Ca^{2+} then rushes down its concentration gradient into the nerve ending. This triggers a series of biochemical events that culminate in the synaptic vesicle membrane coalescing with the presynaptic plasma membrane and releasing the transmitter into the synaptic cleft by exocytosis (Chapter 4). Embedded in the post-synaptic membrane are specialised receptors to which the transmitter chemical binds. This binding triggers a post-synaptic response. For example, activation of receptors could cause muscle to contract, glands to secrete, or neurons to be excited or inhibited. It is important to note that the same transmitter chemical can cause excitation of one cell and inhibition of another, depending on the receptor and the post-receptor mechanism on which it acts (Chapter 7).

One recently discovered chemical transmitter does not act in the way just described. This is nitric oxide, which is not stored but is rapidly synthesised when Ca^{2+} enters the presynaptic nerve ending. The nitric oxide then rapidly diffuses through the presynaptic membrane, across the synaptic cleft and into the post-synaptic cell, where it stimulates responsive enzymes.

> Signals, which can excite or inhibit the post-synaptic cell, are transmitted by a neuron either by release of a neurotransmitter or by direct electrical communication.

Synaptic potentials

By changing the permeability of the post-synaptic membrane to one or more ions, a neurotransmitter can change the post-synaptic membrane potential. When the membrane potential becomes more negative, the neuron is *hyperpolarised*; that is, the voltage difference between the inside and the outside (membrane polarisation) is increased. In this circumstance, activation of the neuron is *inhibited*. Conversely, when the membrane potential of a neuron is made more positive, the neuron is *depolarised*; that is, the membrane potential is brought closer to the threshold level for action potential initiation. The neuron is thus *excited*. Muscle cells are similarly inhibited or excited by membrane potential changes.

During chemical neurotransmission, a series of events links the arrival of the neurotransmitter to

the generation of a post-synaptic potential (Fig. 26.5). Firstly, the transmitter selectively binds to a receptor in the post-synaptic membrane. This binding is brief, usually lasting less than a millisecond, and causes a conformational change (alteration in shape or electrical charge distribution, usually both) in the receptor protein. This alteration of the receptor protein can elicit an electrical change in a number of ways (Chapter 7). The receptor protein may be an ion channel, with the conformational change leading to altered ionic permeability, or the receptor may be linked to an ion channel whose conformation is changed by the altered receptor. Alternatively, receptor changes may trigger intracellular messenger systems, such as the adenylate cyclase system or the inositol triphosphate system, which, via protein phosphorylation, change the permeability of ion channels. The transmitter nitric oxide does not act on a membrane receptor but passes through the membrane to act on an intracellular enzyme, usually guanylate cyclase.

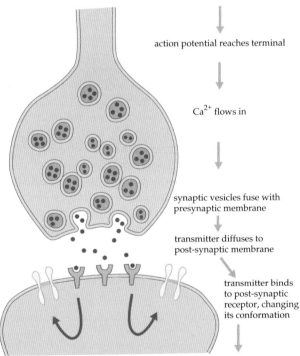

action potential reaches terminal

Ca^{2+} flows in

synaptic vesicles fuse with presynaptic membrane

transmitter diffuses to post-synaptic membrane

transmitter binds to post-synaptic receptor, changing its conformation

binding to the receptor triggers conformational changes in ion channels, altering membrane permeability

Fig. 26.5 Transmitter release and action at a chemical synapse

The electrical event caused by transmission at an excitatory synapse between two neurons is a depolarisation of the post-synaptic neuron and is known as an excitatory post-synaptic potential (*epsp*; Fig. 26.6). The response at an inhibitory synapse is almost always a hyperpolarisation, an inhibitory post-synaptic potential (*ipsp*; Fig. 26.6).

The dominant ionic permeability change underlying an *epsp* is usually an increase in Na^+ conductance (movement of Na^+ across the membrane), but a decrease in K$^+$ conductance can also cause an excitatory potential change (Box 26.2). An *ipsp* can result from a transient increase in K$^+$ permeability or an increase in Cl$^-$ conductance. Although the equilibrium potential for Cl$^-$ (E_{Cl}) is close to the resting membrane potential, a significant increase in Cl$^-$ permeability increases the influence of E_{Cl} on the membrane potential (Box 26.2). As a result, the concomitant opening of channels for other ions has less effect than usual and inhibition can occur with little change in membrane potential. One of the most important inhibitory transmitters of the vertebrate central nervous system, gamma-aminobutyric acid (GABA), acts via the opening of Cl$^-$ channels.

Many different molecules function as neurotransmitters (Box 26.3). Each transmitter can produce either an excitatory or inhibitory response according to the nature of the post-receptor mechanism (Chapter 7).

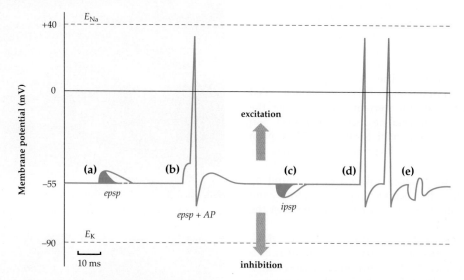

Fig. 26.6 Depiction of the types of electrical events that occur in neurons. Presynaptic inputs cause excitatory post-synaptic potentials (*epsp*'s) and inhibitory post-synaptic potentials (*ipsp*'s) in the post-synaptic neuron. (**a**) An *epsp* that does not reach threshold to evoke an action potential. The time course of the current underlying the *epsp* is indicated by the shading. (**b**) An *epsp* that reaches threshold to generate an action potential (*AP*). (**c**) An *ipsp* with its current indicated by shading. (**d**) Action potentials. (**e**) An *ipsp* prevents an action potential that invades the region of the *ipsp* from reaching threshold; only a small depolarisation is now caused by the action potential. The equilibrium potentials for Na$^+$, E_{Na}, K$^+$, E_K, resting membrane potential and zero potential are indicated. The values of these potentials are representative; they differ slightly between neurons

---

BOX 26.3   SOME IMPORTANT NEUROTRANSMITTERS

Acetylcholine
Excitatory transmitter to skeletal muscle in vertebrates, inhibitory transmitter to vertebrate heart. Excitatory transmitter at many neuro-neuronal synapses.

Adrenaline
Excitatory transmitter to the heart, blood vessels and various viscera in vertebrates. In mammals, the same role is served by the closely related compound, noradrenaline.

Glutamate
An amino acid that is a commonly used excitatory transmitter in the brain. In arthropods, glutamate is an excitatory transmitter to muscle.

γ-Aminobutyric acid (GABA)
An amino acid that is an inhibitory transmitter in the vertebrate brain and at some arthropod neuromuscular junctions.

Serotonin
A transmitter to heart muscle and in the ganglia of some invertebrates. A transmitter in the vertebrate brain.

Peptides
Peptides, which are chains of three to over 30 amino acids, are involved in transmission in many invertebrate and vertebrate species. Over 50 neurotransmitter peptides have been identified. Closely related or identical peptides are found in neurons of taxonomically widely separated species (e.g. hydra and humans).

Nitric oxide
A free radicle that is a transmitter to some visceral muscles of vertebrates, and may be involved in memory formation.

BOX 26.4   ANIMAL AND PLANT NEUROTOXINS

Nervous systems play an important role in movement, co-ordination and survival of most animals. It is therefore not surprising that chemicals that cause nervous systems to malfunction have evolved as offensive or defensive weapons in both plants and animals. For example, two Australian animals, the red-back spider, *Latrodectus mactans hasselti* (Fig. a), and the tropical taipan, *Oxyuranus scutellatus* (Fig. b), produce neurotoxins that cause an excessive outpouring of transmitters from neurons. As a consequence, bites from these animals can cause paralysis in both prey and predators. Many plants produce alkaloids and other chemicals that poison or deter animal predators. The Australian corkwood, *Duboisia* (Fig. c), contains in its bark a substance called hyoscine, which prevents certain receptors for the neurotransmitter acetylcholine being activated. An almost identical toxin is found in the unrelated European plant, deadly nightshade, *Atropa belladonna*. It has the name belladonna (beautiful lady) because, when placed in the eye, the toxin causes dilation of the pupil by interfering with the acetylcholine receptors and preventing activation of the muscles that constrict the pupil in the eye.

Another interesting neurotoxin is strychnine, which comes from the seeds of an Indian tree, *Strychnos nux-vomica*. It was shown by the Australian neuroscientist and Nobel prize winner, J. C. Eccles, that strychnine acts by blocking inhibitory synaptic potentials in the central nervous system, causing powerful convulsions. Strychnine blocks receptors for the neurotransmitter glycine.

The golden orb-weaving spider, *Argiope*, one of the spiders responsible for the beautiful, intricate orb webs we find in our gardens, produces a toxin that rapidly paralyses insects by blocking the excitatory receptors on their muscles. The toxin only works, however, on channels already opened by the excitatory transmitter, glutamic acid. This means that the more an insect

struggles to escape, the more severe the paralysis becomes. The insect, still alive but paralysed, is wrapped in silk and hung up for storage. The poison in its system ensures that it will reparalyse itself whenever it starts to struggle. Thus, the spider has a secure source of fresh food.

(b) The tropical taipan, *Oxyuranus scutellatus*

(a) The red-back spider, *Latrodectus mactans*

(c) The corkwood tree, *Duboisia hopwoodii*

In general, as with hormones (Chapter 25), evolution has been highly conservative and all groups of animals utilise a selection from a common group of neurotransmitters.

> Neurotransmitters bind to receptors in the post-synaptic membrane, causing post-synaptic potentials that may be excitatory or inhibitory.

Integration of information by neurons

There are two major levels of integration in the nervous system: integration that occurs at the level of a single neuron and integration that involves a number of neurons. A deceptively simple act, such as catching a ball in flight, involves integration of information involving thousands of neurons. Even at the level of a single neuron, integration of information is complex. The role that any one synaptic input plays in generating an output by the post-synaptic cell depends on the *activity* of that synapse, the *location* of the synapse on the post-synaptic neuron, and the *timing* of input activity in relation to the activity at other synapses on the same neuron.

The location of synapses on the post-synaptic neuron may be axodendritic, axosomatic or axo-axonal. The space constant of the post-synaptic membrane, together with the active response characteristics of the surrounding membrane, determine how far the effect of synaptic potential changes will travel. In general, synapses on distal dendrites have relatively smaller individual influences on the membrane potential of the cell body than do more proximal synapses. The summation of effects of numerous excitatory and inhibitory axodendritic and axosomatic synapses contribute to the net membrane potential of the cell body and hence to the firing pattern of the neuron.

The first part of the axon, the axon hillock or proximal process (Fig. 26.1a), is a key region because it is the site at which the active response characteristics of the membrane permit the generation of an action potential for transmission into the axon. Inhibitory inputs at the axon hillock can block the propagation of an action potential and therefore cancel out the effects of even massive excitatory inputs onto the dendrites or cell body. In neurons where the cell body is situated on the end of a process, away from the dendrites and axons, as occurs in vertebrate sensory neurons and in many invertebrate motor neurons (Fig. 26.1b), the membrane of the cell body plays little or no part in the signal processing function of the neuron.

In some cases, an axon terminal forms an inhibitory synapse on the terminal of a second axon, near its synapse. Activation of the inhibitory synapse can reduce transmitter output from the terminal of the second axon and so decrease or even block transmission across that synapse. This process is known as presynaptic inhibition.

The combination of synaptic inputs, together with the characteristics of the neuron membrane, determine the final levels of depolarisation in different parts of a neuron. This, in turn, determines its output in terms of rate of action potential firing and amount of neurotransmitter release.

> Integration occurs both at the level of a single neuron and in assemblies of neurons. The influence of each synapse on signal output depends on its level of activity, its location on the post-synaptic cell and the electrical characteristics of the membrane.

EVOLUTION OF NERVOUS SYSTEMS

The basic properties of neurons from all animals appear to be the same. The evolution of more flexible and complex behaviours, however, has been accompanied by an increasing complexity of the nervous system.

Almost all cells have some sort of potential difference between the inside and the outside. The development and use of this potential difference for signalling and for control of effectors appears to have occurred early in evolution. Intracellular recordings from single-celled protists, such as *Paramecium*, reveal a resting potential similar to that of cells in multicellular animals. Moreover, potential changes are part of cell-signalling in protists. Depolarisation of the plasma membrane in *Paramecium*, for example, causes a reversal in the direction of the ciliary beat used for locomotion.

In primitive metazoans, such as cnidarians, a number of levels of electrical communication are found. For example, some muscle cells are in direct electrical communication with each other via communicating junctions (Chapter 6). Although the process is slow and rather imprecise, because excitation tends to move outwards in all directions from the source like ripples in a pond, it does play a role in the co-ordination and behaviour of the whole organism. Cnidarians and platyhelminthes also have nerve nets. These are networks of neurons connected by synapses to form diffuse networks (Fig. 26.7). Signals can pass through the network in many directions but tend to conduct radially and decrease gradually with distance. These simple methods of conduction have the obvious disadvantages of being slow and diffuse. It is therefore not surprising to find that, although a number of groups of animals have retained nerve nets for local responses, they have also evolved faster, more directional systems to complement them. Large jellyfish, for example, have defined neuronal pathways consisting of larger neurons dedicated to

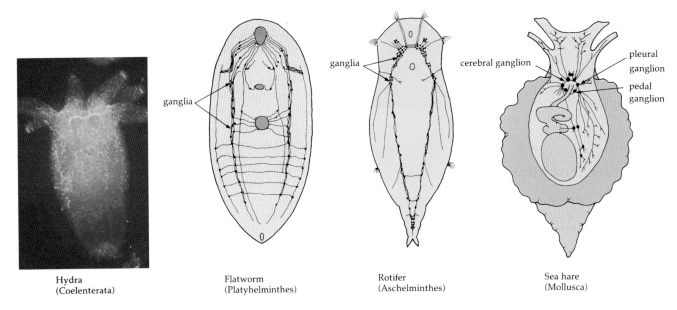

Hydra
(Coelenterata)

Flatworm
(Platyhelminthes)

Rotifer
(Aschelminthes)

Sea hare
(Mollusca)

Fig. 26.7 Invertebrate nervous systems. These examples show the increasing complexity of nervous systems in more complicated animals and the formation of anterior aggregations of nerve cells (encephalisation). In coelenterates, illustrated here by hydra, isolated neurons form a network without any significant aggregation of nerve cell bodies. In addition to a nerve net, platyhelminthes also have nerve cells gathered into longitudinal trunks with ladder-like cross connections, and concentrations at the front of the animal. In the rotifer, groupings of nerve cells in primitive ganglia can be seen. In the drawings of higher invertebrates, molluscs and arthropods, individual nerve cells are no longer represented. In these animals, tens of thousands of nerve cells are gathered in ganglia, in the head, viscera and foot of the sea hare, and in the head of the cuttlefish and grasshopper

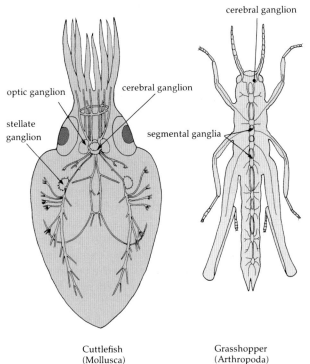

Cuttlefish
(Mollusca)

Grasshopper
(Arthropoda)

co-ordinating the swimming bell; these neurons are not part of the nerve net.

Other trends in neural properties associated with more complex behaviour are apparent. As we have seen, integration of electrical information occurs ultimately at the membrane level. The time and space constant characteristics of neuronal membranes place physical limits on the distances over which this integration can occur. It is therefore effective for cells that exchange and integrate information to be close together (Fig. 26.7). Flatworms (phylum Platyhelminthes) have simple aggregations of nerve cells along the through-pathways down each side of the animal. Transverse pathways link these with the equivalent groupings on the other side, producing a ladder-like nervous system. Similar, even more highly organised groupings of neurons, known as ganglia, occur in all other metazoan phyla. In many invertebrates, local information can be processed within ganglia as simple voltage changes without the need for action potentials (this information is said to be amplitude- or analogue-coded), but is converted into an action potential code for transfer over greater distances, to other ganglia or to effectors, such as muscles (this information is said to be frequency- or digitally-coded).

A number of organisms, such as flatworms and annelids, move through the environment with a preferred end forward. This allows specialisation of locomotory structures for more efficient movement. In these animals, the advantages of localising food sensing and gathering structures at the anterior end are immediately apparent. The anterior concentrations of neurons that accompany this trend, **cerebral ganglia**, are the forerunners of brains. The simplest ones receive information from chemical and movement sensors, and often primitive light detectors, and may direct co-ordination of feeding structures and body movement. The progressive aggregation of nerve cells in groups at the anterior end of an animal is **encephalisation** (Fig. 26.7).

Over the course of evolution of arthropods, the ganglia on either side of the animal have moved closer together and eventually fused. In metamerically segmented animals (Chapter 39), a segmental ganglion is responsible for co-ordinating the movements within each segment and for passing essential information to ganglia of adjacent segments and, where necessary, to the cerebral ganglia.

The nervous system of echinoderms is interesting because these animals are bilaterally symmetrical as larvae but metamorphose into radially symmetrical adults. Because of the radial symmetry, the nervous system shows little tendency to be concentrated in one place. In starfish, the nervous system has equally sized ganglia at the base of each arm joined by a nerve-fibre ring around the mouth.

The animal kingdom can be divided on the basis of the early development of its members into two large groups or 'superphyla'—the protostomes and the deuterostomes (Chapter 38). The layout of the nervous system in the two groups is fundamentally different. Protostomes, exemplified by arthropods, have a dorsal brain connected by a pair of connectives passing either side of the oesophagus to a suboesophageal ganglion and a ventral nerve cord with segmental ganglia. Deuterostomes, including chordates, have a dorsal brain and a dorsal segmentally ganglionated nerve cord. The parallel evolution of encephalisation and coalescence of local ganglia in the two groups suggests that the evolutionary advantages were considerable.

> Although the basic properties of neurons are the same in all species, increasingly complex behaviour is accompanied by increases in nervous system complexity, including the formation of ganglia linked by nerve cords, and by encephalisation.

Complex nervous systems: vertebrates

From simple to complex animals, the total number of neurons increases dramatically; there are about 300 in a rotifer, about 500 000 in an arthropod, 10^8 in some cephalopod molluscs and 10^{10} in mammals. This vast increase in numbers is associated with increased behavioural complexity and adaptability. Most neurons in vertebrates have their cell bodies located in the brain and spinal cord, which together constitute the *central nervous system*. As vertebrates evolved, the brain became larger in relation to the spinal cord.

In the most primitive chordates, such as the lancelet amphioxus (Chapter 40), the pattern of nervous system organisation typical of vertebrates can be seen. In such species, the central nervous system develops by the infolding of the dorsal ectoderm and differentiation of the cells that surround the canal formed by this infolding (Chapter 14). Further differentiation occurs, particularly at the head end, resulting in the typical hollow dorsal central nervous system of vertebrates. The major regions of the vertebrate brain are present in primitive eel-like fish called cyclostomes, where the brain is divided into forebrain, midbrain and hindbrain, and is continuous with the spinal cord.

In addition to the central nervous system, vertebrates have numerous peripheral ganglia. Many of these ganglia are involved in control of the viscera, such as the heart and the digestive system. Signals are relayed between the central nervous system and the organs of the body via the *peripheral nervous system*.

> Vertebrates have a central nervous system, which contains the majority of their neurons, and a peripheral nervous system comprising numerous ganglia and connecting nerves.

The mammalian brain

The dominant features of mammalian brains are the external layers—the cerebral and cerebellar cortices (sing. cortex; Fig. 26.8). The *cerebral cortex* is part of the forebrain. In humans, it is very highly developed

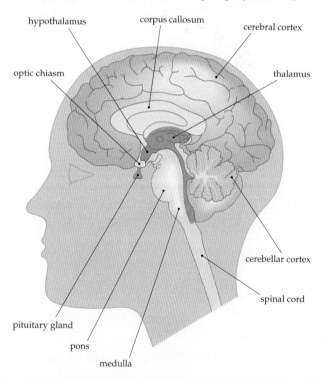

Fig. 26.8 Major structural divisions of the mammalian brain, in this case, the human. The brain is drawn with its left side cut away, so the deeper structures are not hidden by the cerebral cortex. Its appearance with the cortex intact is shown in Figure 26.9

and its growth outstrips the rest of the brain so much that it folds to maintain close apposition to the underlying brain stem. In mammals such as the rabbit (Fig. 26.9), rat or guinea pig, the cerebral cortex is smooth and may not completely obscure the upper brain stem from a lateral view. These species also tend to have prominent olfactory bulbs, in relation to the rest of the brain.

The cerebral cortex is concerned with the control of movement and with what are known as higher nervous functions. In humans, these include learned behaviours, memory and recall, pattern recognition, emotional responses, abstract thought and language. Many attempts have been made to localise these higher functions in the cerebral cortices; this has been only partly successful.

The major regions of the human cortex, and the functions these regions subserve are shown in Figure 26.10. In some cases, there is lateralisation of cortical function, meaning that one side of the brain predominates in controlling that function. Motor commands from the cortex are lateralised and crossed; that is, the cortex on the right side commands muscles on the left and vice versa. A most remarkable lateralisation is that of human speech. If part of the left temporal lobe is injured, then speech can be entirely lost, whereas no deficit is caused by a comparable injury to the right side. Patients with speech loss due to left lobe injury can understand written or spoken speech perfectly well and can make correct word associations, as demonstrated by their ability to give written responses to questions.

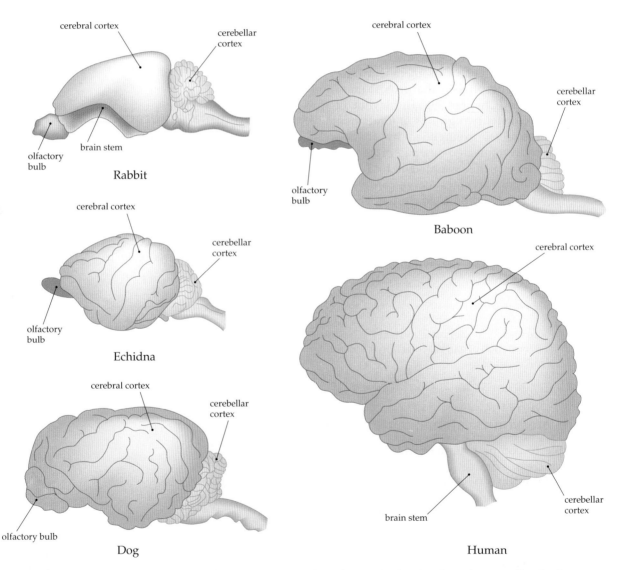

Fig. 26.9 Differences in structure of mammalian brains seen in lateral view. In the simplest of mammalian brains, represented here by the rabbit, the cerebral cortex is smooth; it only partly obscures the upper brain stem and does not cover the cerebellar cortex. The olfactory bulb is prominent. As brains become more complex, convolutions of the cerebral cortex progressively develop, the upper brain stem and cerebellar cortex become enveloped by the cerebral cortex and the olfactory lobes become relatively less prominent. In humans, the angle of the lower brain stem changes to accommodate upright posture

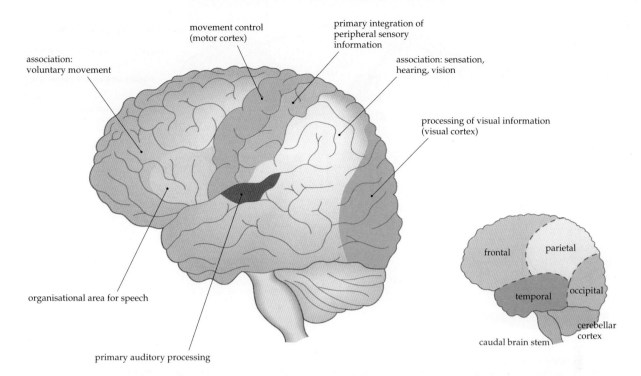

Fig. 26.10 Lateral view of the human brain, indicating the positions of some of the major functional regions of the cerebral cortex and the four cortical lobes

The *cerebellar cortex* (or cerebellum) is a part of the hindbrain. It plays a crucial role in the co-ordination of balance and movement. The cerebellum is one of the brain regions most obviously susceptible to alcohol, whose earliest effects are thus on balance and eye–hand co-ordination.

Underlying the cerebral and cerebellar cortices is the *brain stem*, which is derived from structures that arose earlier in vertebrate evolution. It is composed of the remainder of the forebrain, the small midbrain and the hindbrain. The major regions of the brain stem are the *thalamus*, which is concerned with relaying and processing information related to loco-motor activity between the cortex and lower centres; the *hypothalamus*, which is concerned with integrating visceral activities, such as appetite and body tempera-ture control; the *pons*, which includes auditory processing circuits; and the *medulla*, which includes centres for control of specific viscera, such as the cardiovascular and digestive systems. Twelve pairs of *cranial nerves* emerge from the brain stem; through these information is received from or passed to more peripheral sites.

The brain stem is continuous with the spinal cord, which lies within the vertebral column and has a very characteristic structure when viewed in cross-section (Fig. 26.11). Except for the head and some structures in the neck, the majority of motor com-mands pass out from the central nervous system via the spinal cord. Moreover, a lot of sensory informa-tion, including all sensations from the limbs and from the surface of the trunk, enters via the cord. Every-thing that occurs in the spinal cord is able to be controlled or modified by higher brain centres. This includes spinal reflexes, such as the tendon jerk reflex, which occurs when the tendon below the knee is struck.

> The main regions of the mammalian brain are the cerebral and cerebellar cortices, the thalamus and hypothalamus, and the pons and medulla.

Higher functions of brains

All species that show behaviour involving complex computations, such as those required for learning and language, have large arrays of highly ordered neurons characterised by repeating patterns. The best-known examples outside the vertebrates are those found in insects (Fig. 26.12) and octopuses. The brains of these animals have far fewer cells available for such computations than does a vertebrate but can never-theless generate some surprisingly sophisticated behaviour. Octopuses, for example, can learn to dis-tinguish abstract symbols to gain a reward or avoid a punishment. Bees communicate the direction, dis-tance and quality of food sources to other members of their hive using an elaborate dance language that encodes the information (Chapter 28). They also remember a map of the locality around their hive and fly directly back to it if released anywhere within the area covered by the map.

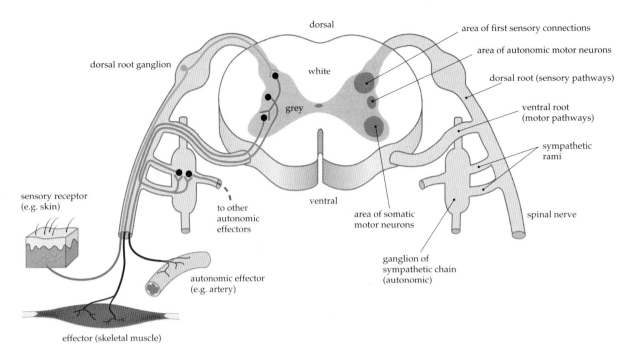

Fig. 26.11 Representation of a cross-section through the mammalian spinal cord. The spinal cord has a central grey region, consisting of nerve cell bodies, their dendrites, initial parts of axons, and synaptic inputs, and a white region, consisting of axons that pass up and down the cord. At each vertebral level, bundles of axons, the dorsal and ventral roots, connect with the spinal cord. These join to form spinal (segmental) nerves, small branches from which join the sympathetic chains at thoracic and upper lumbar levels

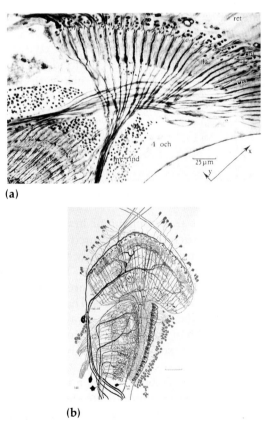

(a)

(b)

Fig. 26.12 Highly ordered patterns of neuronal organisation are seen in animals capable of complex behaviours, such as learning and language.
(a) Photograph and **(b)** composite drawing showing the complex organisation of nerve cells in the visual system of the insect brain

The brains of vertebrates are many orders of magnitude larger in terms of neuronal numbers, and the complexity of their activity reflects this. Their higher level functions can be divided into two groups, those of thought (cognition) and those of emotion (affect). In line with this, psychiatric conditions are divided into affective and cognitive disorders. The cerebral cortices have major roles in all higher nervous functions. Because these functions can be localised to specific cortical areas, or groups of areas, and can be altered by drugs that modify synaptic transmission. It is a tenet of modern neurobiology that what we refer to as 'mind' is the outcome of the functioning and interaction of neural circuits of the cortex. Examples of cognitive functions include the comprehension and formulation of speech, learning and memory, anticipation of future consequences of existing conditions, and abstract thought. The formulation of speech is localised in an area of the left frontal lobe, known as Broca's area, whereas the site of speech comprehension is in the temporal lobe. Certain emotions are also centered in the temporal lobe, and patients with temporal lobe epilepsy may have disorders of affect. The older parts of the brain, in evolutionary terms, for example, the thalamus and the pons, are also important in storing and implementing emotional responses. In the immensely complex cortex, with its multiple inter-connections between areas, regions associated with higher functions often have parallel and serial inter-connections. This has made it particularly difficult to define precisely which areas are associated with particular neural processes.

| Higher nervous system functions include thought (cognition) and emotion (affect).

NEURAL CIRCUITS

In order to understand how nervous systems produce co-ordinated behaviour, it is necessary to unravel individual neural circuits. This is usually done either by tracing a circuit backwards from the motor side or by following the path forward from the sensory side. Extensive studies of this kind have been carried out in invertebrates where the smaller number of neurons makes access and interpretation easier. The nervous systems of molluscs, such as the sea hare, *Aplysia* and the snail, *Helix,* and arthropods, such as locusts, *Locusta,* crayfish, *Procambarus,* and the vinegar fly, *Drosophila,* are the most often investigated (Fig. 26.7). *Drosophila* is particularly important because our detailed understanding of its genetics and development provides a bridge to understanding these aspects of nervous system biology. In many invertebrate species, precisely equivalent neurons can be identified reliably from animal to animal and the physiological, chemical and structural attributes of these neurons investigated (Box 26.5). Identifiable

neurons provide the opportunity to examine general principles of circuit operation, which aids in establishing guidelines for the vastly more difficult task of analysing vertebrate circuits.

Vertebrate neural circuits often consist of thousands or millions of neurons. Even the simplest of reflex circuits usually involves far more neurons than could be included in a diagrammatic representation. In order to explain such circuits, stylised diagrams are drawn in which large numbers of neurons are represented by single morphologically simplified neurons.

The neural circuit for the light reflex controlling the mammalian iris is shown in Figure 26.13. This is typical of the way that such circuits are depicted. Approximately 1 billion photoreceptors are represented by a single photoreceptor and the integration of photoreceptor information, which determines the pattern of firing of the retinal ganglion cells, is ignored. On the output side, several hundred pre- and postganglionic autonomic neurons are involved to provide a graded and eucentric change in pupillary diameter.

| In many invertebrates, specific neurons can be identified reliably and the physiological, chemical and structural attributes of these neurons investigated. Vertebrate neural circuits often consist of thousands or millions of neurons.

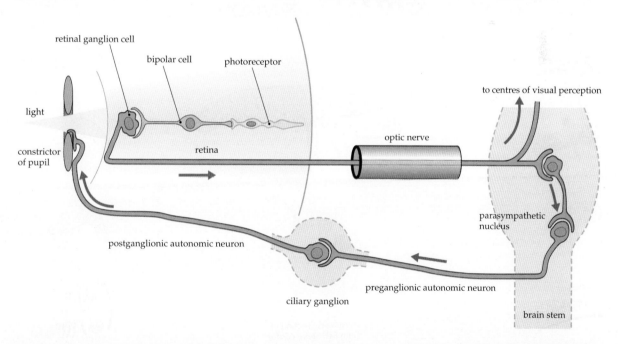

Fig. 26.13 Representation of a neuronal circuit: the light (pupilloconstrictor) circuit. Neuronal circuits are usually very much simplified when represented in diagrams. In reality, wherever there are neuroneuronal or neuroeffector junctions (six in this circuit), there is integration of information. Wherever a single neuron is drawn, it represents many parallel pathways. The light reflex is an example of a simple negative feedback circuit; if the amount of light falling on the retina is increased, the pupil constricts, and if the light level drops, the pupil will dilate. Neurotoxins that block transmission from the final pupilloconstrictor neurons cause the pupil to enlarge. Neurons are usually represented schematically, as in this diagram, as a round cell body, without depicting dendrites, a line for the axon and a Y-shaped structure at the synapse

BOX 26.5   RAPID ESCAPE RESPONSES

Invertebrates and vertebrates have evolved rapidly responding nerve networks for avoiding predators. If you have ever tried to catch a crayfish (yabby) in a net you will know that they usually try to escape by flicking their tails and propelling themselves away rapidly. Neurobiologists have discovered that this escape is controlled by two sets of giant neurons with axons running the entire length of the nerve cord (Fig. a). They are termed lateral or medial giant axons according to their position in the nerve cord. The medial giant axons are activated by abrupt visual or touch stimuli to the animal's head. They conduct action potentials rapidly (15–20 ms^{-1}) backwards along the body and directly stimulate neurons activating a selected group of muscles in the legs and tail. The muscles in the legs cause them to flex in alongside the trunk, reducing the drag on the body as it moves through the water, and the tail muscles produce a strong abdominal flexion that propels the animal backwards (Fig. a). The lateral giant fibres function similarly but with distinct and interesting differences. They are activated by mechanical stimuli from behind, particularly abrupt waterborne vibrations, such as occur when something lunges at the crayfish

from behind. They conduct action potentials forwards to a different set of muscles that cause the animal to somersault away from the stimulus (Fig. a). In this case, the leg muscles are not activated; their position does not affect the efficiency of the somersault.

Vertebrates also have large neurons with functions dedicated to escape. Some fishes have a pair of giant cells, called Mauthner neurons, after the biologist who first described them (Fig. b). These cells are activated by a mechanical stimulus to the side of the fish; which one of the pair is stimulated is determined by the site of the stimulus. Water movements with an intensity equivalent to those associated with the predatory strike of another fish are sufficient to trigger the cells. When a Mauthner cell fires, it initiates a series of complicated movements involving the trunk, fin, eye, jaw and opercula muscles. Because the axon of each Mauthner neuron crosses over to the other side of the body, the initial body contraction is on the side away from the stimulus so that the fish's body rotates between 30° and 100° about its centre of mass in 15–40 msec. This rapid turn away from the stimulus positions the fish for a powerful acceleration out of the area (Fig. b).

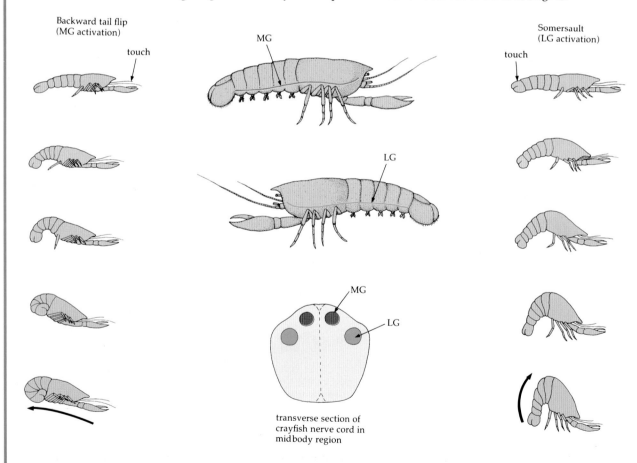

(a) There are two sets of giant neurons in the crayfish nerve cord: a medial pair and a lateral pair called the medial giant (MG) axons and lateral giant (LG) axons respectively. The MG axons are activated by a stimulus to the anterior end of the animal and cause abdominal contractions that cause the animal to move backwards. The LG axons are activated by a sharp stimulus to the tail and activate muscles that cause a somersault. The silhouettes of the animals are shown at 10 ms intervals

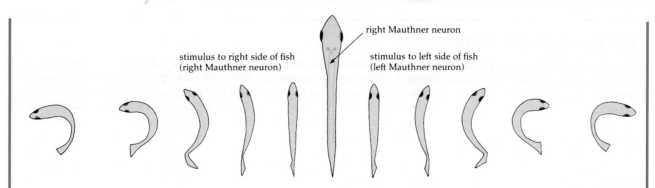

stimulus to right side of fish
(right Mauthner neuron)

right Mauthner neuron

stimulus to left side of fish
(left Mauthner neuron)

(b) Giant-sized Mauthner neurons in many varieties of fishes are responsible for an abrupt body turn away from a lateral stimulus. Body profiles here are shown at approximately 5 ms intervals

FUNCTIONAL DIVISIONS OF NERVOUS SYSTEMS

Motor control systems

Motor or somatic control systems are concerned with the control of posture and movement and direct all movements under voluntary control. Movement can be initiated by the central nervous system or by reflexes via peripheral sensory inputs. The system is more or less hierarchically arranged. At the lowest end of the hierarchy are simple reflexes that pass through segmental ganglia of invertebrates or restricted regions of the spinal cord of vertebrates. These include such reflexes as the closing of the claws in crabs in response to stroking of hairs on the chelae, and tendon reflexes in vertebrates that cause the leg to 'jerk' in response to a tap to the tendon just below the knee. These are monosynaptic reflexes in which sensory neurons connect directly with motor neurons. Monosynaptic reflexes are not common and most connections between the sensory and motor side, including most spinal reflexes, are disynaptic or polysynaptic (Fig. 26.11).

The next level of organisation is responsible for integration of movements involving sets of muscles acting around one joint. Simple hinge joints are controlled by opposing sets of muscles (Chapter 27), flexor muscles that bend the joints and extensor muscles that straighten them. Most arthropod joints are of this type so that the muscles are capable of producing only one sort of movement and the circuitry is relatively simple. Vertebrate joints are usually more complicated because the muscles can produce different actions depending on which other muscles are active at the same time and the integration required is necessarily more complicated. An example is the shoulder joint, which can extend in several directions and rotate.

Limb movements involve co-ordination of several or many joints. Integration for this is more complicated again. It is still carried out in local ganglia or spinal segments but more of them interact and there is direction from the cerebral ganglion or brain. When you lift an object, by bending the elbow, for example, integrated muscle action is needed. In this case, both flexor and extensor muscles of the wrist are contracted so that the wrist is held firm, the elbow flexors are contracted and the elbow extensor muscles are inhibited from contracting. More and more local centres are activated and co-ordinated as other limbs become involved. In both insects and cats, for example, reflexes producing strong extension in one leg can produce flexion in an adjacent one. If you step on a nail, there will be a reflex withdrawal of the injured foot and, at the same time, a reflex extension of your other leg, bracing it to take all your weight.

In both vertebrates and invertebrates, parts of rhythmically co-ordinated patterns, such as those involved in breathing, walking, flying and swimming, can be produced by neural circuits in the absence of input from the cerebral cortex or cerebral ganglion, although the decision about whether to turn them on or off may come from these higher centres. Flight is one of the most complicated motor behaviours. The basic flight rhythm is produced by rhythmic generators but their output is constantly modified by inputs from other centres, such as those co-ordinating wind- and groundspeed detection, gravity, and balance in the yaw, pitch and roll planes (Chapter 27).

Co-ordination involving consciously driven patterns of movement, such as those used to hit the keys on the piano with correct order, force and interval in response to the music sheet in front of your eyes, require the highest levels and most widely distributed neuronal interactions. This is because inputs from higher brain centres and sensors in different areas all over the body have immediate and constantly changing involvement. For production of the optimal output, adjustments are being made continuously at a number of levels.

Motor control systems are hierarchically organised. They are concerned with posture and movement, and direct all movements under voluntary control.

BOX 26.6   BATS AND MOTHS

In 1793 an Italian biologist, Lazzaro Spallanzani, showed that owls cannot avoid objects when flying in complete darkness whereas some bats can. We now know that these bats emit ultrasonic sound pulses (frequencies of 40–90 kHz) while flying, and detect the extremely faint echoes that return from objects in the immediate vicinity. Bats that can do this have ears specialised for detecting the direction of the returning echoes (Fig. a) and auditory pathways in the nervous system specialised for selecting the frequencies necessary for echo analysis. The bat determines the distance to an object by measuring the time taken for the echo to return; echoes from distant objects take longer to arrive than do those from close ones. The echo detection system is sufficiently accurate that the bats can use it to hunt flying insects that make up their diet, and much of the investigative work has been done on bats trained to catch mealworms tossed into the air.

(a) The mouse-eared bat, *Myotis adversus*, has specialised ears for detecting faint echoes from ultrasonic pulses emitted when the bat is hunting

Observations on some of the insects hunted by insectivorous bats shows that they have specialisations for avoiding bat predation. Some moths in the family Noctuidae (Fig. b) have ears on either side of the thorax that are tuned to the bat emission frequency. When noctuid moths detect the sound of a hunting bat, they exhibit characteristic avoidance behaviour. The details have been analysed by playing bat calls over loudspeakers set up in the open, and observing the behaviour of free-flying moths to the onset of the artificial bat calls (Fig. c). The experimenter can control the amplitude of the signal and also decide when to activate it. When the signal is faint, equivalent to a distant bat, the moth turns so that the source of the sound is directly behind

it, and flies away. If the signal is loud, representing a bat close by, the moth flutters erratically to the ground, or folds its wings and drops vertically (Fig. c).

A series of experiments of this type have shown that the point at which a moth's behaviour switches from steady flight out of the area to rapid evasive action is roughly equivalent to the point at which a bat can detect the presence of the moth. This suggests that at least some aspects of moth hearing evolved specifically in response to the selective evolutionary pressure imposed by hunting bats.

(b) The noctuid moth, *Heliothis armigera*, which forms part of the bat's diet

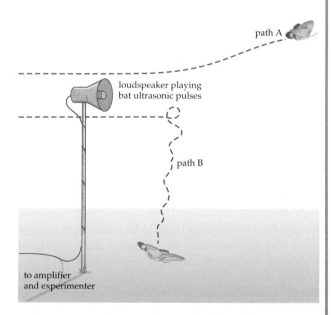

(c) Experiment showing that moths detect bat ultrasound and respond to it by altering their behaviour. Bat calls were played over a speaker set up at night in a clearing. A spotlight at the edge of the clearing allowed the experimenter to see moths approaching the speaker. A low-intensity sound, equivalent to a distant bat, caused moths to fly out of the area (path A). A high-intensity sound, equivalent to a bat close by, caused moths to take rapid evasive action (path B)

Sensory systems

The environments of the earth contain energy of many different kinds. Most of it is of little relevance to animals (e.g. radiowaves). Animals have sensory receptors that enable them to respond selectively to those aspects of their environment that impact on their ability to survive and reproduce. The types of sensory receptors present, and the sensitivity, discrimination and range of these receptors, differs substantially between animals. For example, a wombat has lower visual sensitivity than a kookaburra; dogs use chemical scents extensively for communication and have far greater olfactory discrimination than humans; some moths have chemoreceptors that can detect one molecule of pheromone; some bats emit and can hear ultrasonic sounds beyond the range of most animals, except the moths they hunt with ultrasound (Box 26.6). Many animals respond to parts of the energy spectrum undetectable to humans. For example, a number of animals, including snakes, can detect and use infra-red radiation, bees see ultraviolet light, some fishes detect weak electric currents, and sensitivity to magnetic fields occurs in a wide range of animals including birds and insects (Chapter 7).

The key elements of sensory reception are receptors specialised for detecting particular forms (modalities) of environmental energy: parts of the electromagnetic spectrum (light, heat), mechanical displacement (movement, sound) and chemicals (odours, tastes). Sensory receptors distributed widely over the surface or within the body are the basis for the *general senses* (e.g. touch, pain and joint position). Aggregations of receptors into specialised organs provide for the *special senses* (e.g. sight and hearing). Those receptors that detect internal states, such as blood (or fluid) pressure and fluid chemistry (e.g. oxygen and carbon dioxide tension) are known as *visceral receptors* or enteroreceptors.

The special senses are characterised by receptor organs in which the neural receptor cells are associated with each other and with accessory cells of various types that assist in the collection and primary processing of a specific stimulus energy (Fig. 26.14). In the case of light, photoreceptor cells that contain light-sensitive pigments (Chapter 7) may be associated with lenses and pigment cells controlling the light pathway. In the case of chemoreception, cells sensitive to different chemicals may be grouped and the groups supported inside structures that maximise the probability of contact. For sound, the vibration-sensitive neurons may be attached to larger vibrating structures that select, filter and sometimes amplify the frequencies that are important to the animal.

> Sensory receptors monitor conditions in an animal's internal and external environment and provide information that enhances the animal's ability to survive and reproduce.

Vision

There is evidence that light-sensitive pigments appeared very early in animal evolution because the photosensitive pigments, called opsins, are common to all light-detecting animals. Light is detected when it interacts with visual pigments that respond by changing their conformation; the conformational change is used to trigger a cascade of events (Chapter 7), that eventually results in an electrical signal and the particular reaction of that species to light.

The most primitive type of eye, an eyespot or pigment-cup eye, is found in almost all the major animal phyla. It consists of a small patch of light-sensitive receptors in an open cup of screening pigment, which limits the direction from which incident light can reach the receptors to excite them. Some estimates suggest that simple eyes of this type have evolved independently at least 40 times. Without a device to restrict the acceptance angle of individual receptors, pigment-cup eyes do not resolve images well enough for pattern recognition. The development of an optical system that focuses the light arose much later in evolution and in fewer groups of animals. About 10 different ways of forming an image have evolved and these fall into one of two categories: *simple eyes*, single-chambered eyes constructed on the same principle as a camera, and *compound eyes*, in which a number of lens systems form images that are then combined physically or by neural integration to provide pattern information (Fig. 26.15). The most elementary simple eye is found in the cephalopod mollusc *Nautilus*, which has a pinhole instead of a lens and no cornea. All other simple eyes, including our own, use a lens, a cornea or both to form an image on the photoreceptors of the retina. The commonest and probably most primitive form of compound eye is the *apposition eye*. Each lens unit of the eye, called an ommatidium, forms an image onto a photopigmented area. Sections of the image are taken through layers of neurons that reconstruct one image for the whole visual field. Thus, the final image is constructed by the nervous system from 'pixels' contributed by a number of ommatidia.

Interesting visual specialisations are encountered. For example, a number of animals, including cats and some deep-sea fishes, have surfaces at the back of the eye that reflect light back into the photoreceptor layer after it has passed through it once. This increases the light sensitivity of the eye but at the expense of resolution. Eagles and jumping spiders, both active hunting species, have an additional lens immediately in front of the retina so that the optics function like a telephoto lens. Bees and some birds can detect ultraviolet light, invisible to most animals, and ultraviolet lines are an important component of the colour patterns on the petals of bee-pollinated flowers (Chapter 37). A number of animals, including birds and arthropods, can detect the polarisation of the

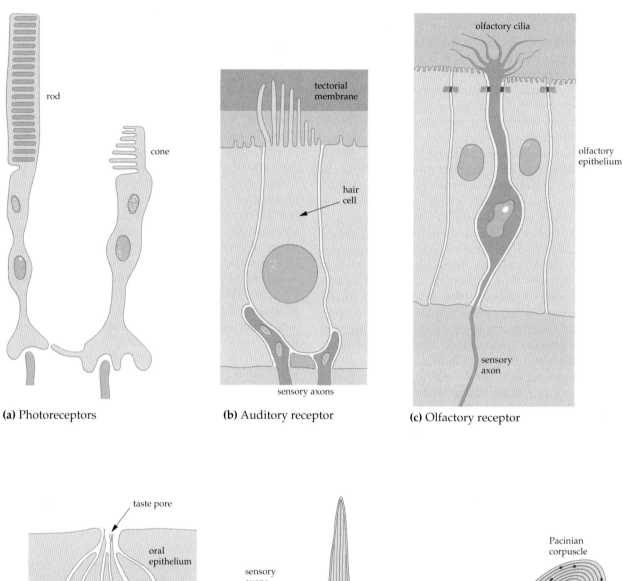

(a) Photoreceptors

(b) Auditory receptor

(c) Olfactory receptor

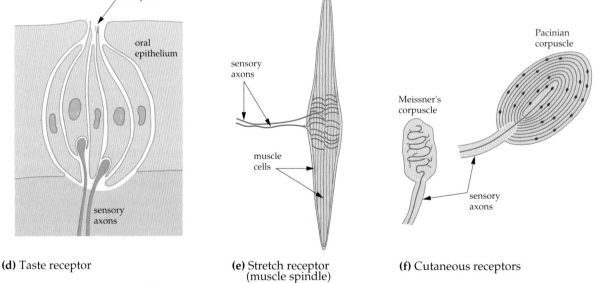

(d) Taste receptor

(e) Stretch receptor (muscle spindle)

(f) Cutaneous receptors

Fig. 26.14 Examples of sensory receptors. **(a)** Photoreceptors are elongated cells with disc-like structures at one end. Within these discs are light-sensitive pigments. The interaction of light with the pigment causes an electrical change, causing the release of a chemical transmitter to activate neurons in the retina (see also Chapter 7). **(b)** The receptors for sound are hair cells where cilia are embedded in the tectorial membrane, which sound waves move relative to the hair cell. **(c)** Olfactory neurons have chemosensitive cilia that penetrate the mucus of the nasal passages. **(d)** Taste buds lie in the oral epithelium, principally of the tongue. They contain chemosensitive cells that form excitatory synapses with taste sensory axons. The ends of axons of the two cutaneous axons that are illustrated here are encapsulated. **(e, f)** Mechanical distortion of the stretch receptor or the cutaneous corpuscles causes their axons to be excited

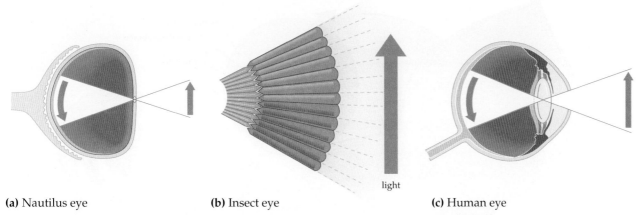

(a) Nautilus eye **(b)** Insect eye light **(c)** Human eye

Fig. 26.15 Comparison of mechanisms of visual detection. **(a)** The mollusc, *Nautilus*, has an eye cup, open to the sea through a small hole. The eye acts like a pinhole camera. It thus has the ability to form an image, but suffers from poor light-gathering ability. **(b)** The compound eye of the insect has thousands of individual detectors, like small eyes. Each of these provides an input to the nervous system. These eyes sacrifice resolution: the resolution can be no better than the area of each unit. **(c)** The vertebrate eye, in this case a human eye, has a focusing lens and a fine-grained retina, giving greater light sensitivity than *Nautilus* and better resolution than the insects. All these groups use the same chemical, *cis*-retinal, to detect photons of light

light coming from the sky, which can be used as a navigational clue.

Hearing

Hearing is a specialised form of mechanoreception. Sound travels through air, water and solids as pressure waves (small compressions and rarefactions). Animals detect these with mechanoreceptors specialised for detecting and magnifying minute vibrations. Naked hairs, such as those on the cerci of insects such as cockroaches, can detect air currents. Very sensitive ones can detect airborne sound vibrations but only at low frequencies. Hearing organs that detect a range of frequencies have mechanical devices that amplify the movements generated by those frequencies useful to the animal to the point where mechanoreceptors can detect them. The commonest mechanical amplifier for airborne sounds is a membrane or tympanum. The tympanal organs of insects are found on the legs, body or wings, while those of vertebrates are found in the ears.

In most terrestrial vertebrates, vibrations of the membrane are further amplified by bony levers and transmitted to a fluid-filled canal (Fig. 26.16). Within the canal are hair cells that have evolved from those that detect water movements around fish and which are sensitive to waterborne vibrations (Chapter 7).

Chemoreception

Chemoreception is the detection of specific chemicals in the environment. It occurs throughout the animal kingdom and is thought to provide the basis for the evolution of inter- and intraspecific communication between simple organisms. Aquatic animals commonly have chemoreceptors over the body surface, with concentrations in some places for specific purposes, such as food or mate detection. Some chemoreceptors are extremely sensitive; catfish can detect some amino acids at concentrations less than 1 part in a million. Terrestrial animals have specialised organs for olfaction (smell), detecting airborne chemicals, and taste, detecting those derived from food. In addition to receptors for common amino acids and sugars, which would be useful to most animals searching for food, many animals have evolved receptors for specific chemicals emitted by prey or predators or pheromone chemicals released by others of their species signalling states such as sexual readiness. Receptors in the antenna of the male silk moth, for example, can detect single molecules of the appropriate sexual pheromone released by the female.

Mechanoreception

A range of receptors inform animals about their mechanical interactions with their environment and provide information to the brain about such things as joint position and tension in the walls of viscera, such as the lungs or stomach. Cutaneous mechanoreceptors include stress detectors in animals with chitinous exoskeletons, detectors of pressure in animals with elastic integuments and detectors that monitor the deflection of hairs. Some mechanoreceptors have specialised transducer structures, such as sensory corpuscles or scolopales at the end of the sensory axon (Fig. 26.14), whereas in others, the transduction of movement to an electrical signal occurs in the endings of the axonal membrane.

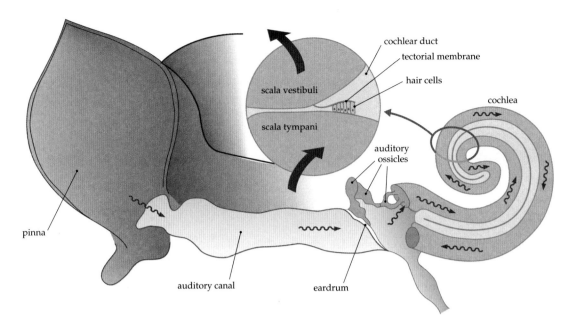

Fig. 26.16 How sound is detected by the mammalian ear. The energy available in soundwaves needs to be concentrated into a smaller area so that it is sufficient to excite the hair cells. This is done by an external ear (pinna) that funnels soundwaves into the auditory canal at the end of which they cause the eardrum to vibrate. Via a series of small bones (the auditory ossicles: the hammer, the anvil and the stirrup), which act as levers, the vibration is conducted to a fluid-filled canal, the cochlea, in which sit rows of hair cells (inset). The intrinsic frequency of vibration of the membrane on which the hair cells sit varies along the cochlea. This allows different tones to be distinguished. The inset is a diagram of the cochlea in cross-section. Soundwaves run up the scala tympani and down the scala vestibuli, as indicated by the arrows. In so doing, they vibrate the membranes within the cochlea and distort the hair cells. This distortion causes a chemical to be released from the hair cell, which excites the adjacent sensory nerve ending

Pain

All animals with nervous systems avoid encounters with noxious external stimuli. However, it is not clear how complicated a nervous system has to be before an animal experiences something akin to pain felt by humans. Vertebrates that vocalise in response to stimuli that humans interpret as painful certainly behave as if their experience is similar. At present we have no way of answering questions of the general type 'Do worms feel pain?'.

In humans, painful sensations generally come from the body surfaces, which are supplied by three types of pain receptors. Mechanical pain receptors respond to strong stimuli, most effectively from sharp objects. Heat pain receptors respond when the skin is heated to 45°C or more. Polymodal pain receptors respond to noxious mechanical, heat or chemical stimuli. For all pain sensations, the nerve endings are believed to be activated by chemicals released from the damaged or irritated tissue. The signal is carried to the spinal cord or brain stem by the axons of sensory neurons. A series of relay neurons then carry the information to the cerebral cortex, which must be functional for pain to be perceived. Psychological and behavioural influences on pain are pronounced. Individuals show marked differences in pain threshold and the same individual in different circumstances

can experience pain differently. A striking example is the suppression of pain during vigorous physical activity or in a crisis situation.

Drugs related to opium suppress pain, apparently by mimicking the actions of peptides, known collectively as opioid peptides, which are transmitters in the central nervous system. The best known of these transmitters are enkephalin and endorphin.

Visceral control

Organs whose functions are not consciously controlled are the visceral organs, including the heart, stomach, spleen and pancreas, blood vessels, specialised glands, such as sweat glands, and smooth muscles of the eye. Complex animals, both vertebrate and invertebrate, have a specialised neural network for this purpose, which is termed the visceral or autonomic nervous system (Fig. 26.17). It integrates closely with the endocrine system (Chapter 25) to co-ordinate the involuntary or regulatory physiological functions of the body and to stabilise the internal environment. The major functions influenced by the autonomic nervous systems are listed in Box 26.7.

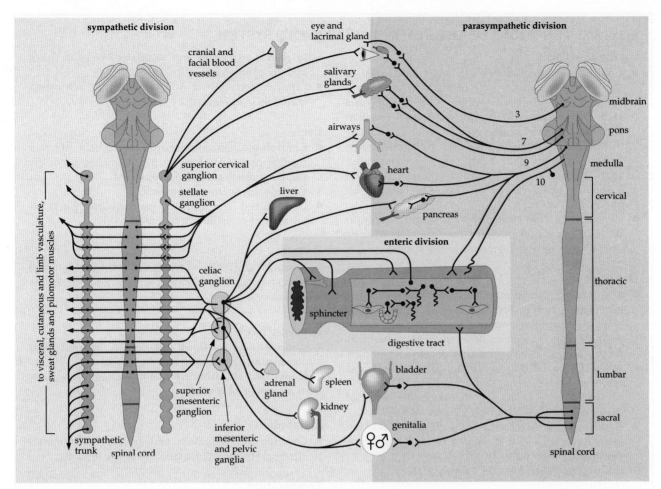

Fig. 26.17 Diagrammatic representation of the autonomic nervous system. This system has three divisions through which organs of the thoracic, abdominal and pelvic cavities, glands of the head and skin, peripheral blood vessels and ocular smooth muscles are controlled. For clarity of presentation, the brain stem and spinal cord are represented twice, on the left to show the sympathetic outflows and on the right to show parasympathetic pathways. The enteric division of the autonomic nervous system is embedded in the wall of the digestive tract. Enteric reflexes are modulated, and sometimes overridden, by inputs from the sympathetic or parasympathetic divisions. The cranial nerves that include autonomic motor pathways are numbered (3, 7, 9 and 10). Visceral sensory neurons connecting with the digestive tract and within the enteric division of the system are marked by asterisks. There are many other visceral sensory neurons that are not included in this diagram

> The autonomic nervous system innervates the visceral organs of the body; their functions are not consciously controlled.

The vertebrate autonomic nervous system

The vertebrate autonomic nervous system consists of a central portion within the brain stem and spinal cord, and a peripheral part comprising ganglia and connecting nerves. Autonomic axons that leave the central nervous system, with the exception of those innervating the adrenal gland, make synaptic connections with nerve cell bodies that are grouped in peripheral ganglia. Autonomic neurons whose axons project to peripheral ganglia from cell bodies in the central nervous system are known as *preganglionic neurons* and the neurons with which they

connect are *postganglionic neurons*. Autonomic ganglia are small centres for integration of neuronal information. Their output consists of patterns of action potentials in axons that supply smooth muscle, cardiac muscle, glands or other autonomic neurons.

The peripheral nerves and ganglia are classified into three subsystems, known as the *sympathetic, parasympathetic* and *enteric* divisions of the autonomic nervous system (Fig. 26.17). The sympathetic division is that part of the autonomic nervous system whose motor pathways emerge from thoracic and lumbar parts of the spinal cord. The parasympathetic division consists of those pathways arising from the brain stem and from the sacral spinal cord. The enteric nervous system is the division of the autonomic nervous system that is embedded in the walls of digestive organs, notably the stomach, small intestine, colon,

BOX 26.7   FUNCTIONS OF THE AUTONOMIC NERVOUS SYSTEM

Cardiovascular

- Controls the rate and strength of the heart beat of the heart. The heart muscle of many invertebrates, such as crabs, requires specific excitatory nervous stimuli to initiate each contraction. A small ganglion of around nine to 15 cells controls the rate and order of contraction of the muscle fibres of the crab heart. In vertebrates, the heart beat is generated spontaneously by the cardiac muscle but the frequency and strength of beat is influenced by autonomic nerves.
- Controls the relative distribution of blood flow to different organs by changing the diameters of their arteries of supply at key points in the system.

Digestive

- Controls mixing and propulsion of food through the gut. Some invertebrates, such as crabs, have complex stomachs that grind and filter food. The complex co-ordination of the dozens of small muscles required for this is done by an aggregation of some 30 neurons called the stomatogastric ganglion. The strength and frequency of the mixing and propulsive movements of the muscle layers in the digestive tract of vertebrates is controlled by autonomic nerves.
- Controls digestive secretions into various parts of the digestive tract and transport of electrolytes, hence controlling water balance.
- Controls secretion from the salivary glands.

Respiratory

- Controls variation in diameter of major airways of lungs in vertebrates, and the degree of contraction or relaxation in gills and respiratory trees in fishes and invertebrates.
- Controls secretion of mucus and other lubricating materials over respiratory surfaces.

Eye

- Controls mechanisms for adjusting the light intensity reading photoreceptors, such as the diameter of the pupil in vertebrate and octopus eyes.
- Controls lubricatory secretions.
- Is involved in focusing mechanisms, such as movement or distortion of lenses.

Excretory

- Controls emptying contractions of waste storage reservoirs, such as bladders, and filtration pressures in kidneys.

Reproductive

- Controls contractions of organs storing and conducting eggs, sperm and embryos.
- Is involved in intromission and ejaculation in males of a number of species.

Metabolic

- Controls the formation and release of hormones affecting the overall metabolism of the organism.

Temperature

- Controls systems used to cool or warm animals, such as cutaneous blood flow and sweating in vertebrates and spiracular opening in insects.

---

pancreas and gall bladder. The enteric nervous system is unique in that it contains complete reflex circuits and thus it can function when all neural connections with the central nervous system are severed. The enteric nervous system is innervated by the sympathetic and parasympathetic divisions of the autonomic nervous system.

Many autonomic pathways are tonically active, which means that some groups of autonomic neurons fire action potentials continually and therefore control the level of activity of the organ they innervate. This is rather like having the accelerator of a car partly on; by the use of a single controlling device, a greater or lesser activity can be obtained. Sympathetic nerves control the diameter of arteries in this way. If you are sitting at rest at a comfortable temperature, the arteries to your skin will be slightly constricted due to sympathetic vasoconstrictor neurons being active. If the room becomes warmer, these vasoconstrictor neurons will decrease activity, the arteries will dilate and the increased flow of warm blood will lose heat to the surrounding environment. This heat loss can be enhanced by evaporative cooling due to increased sweat production, also caused by changed autonomic activity. Conversely, if you become cold, sympathetic neurons to skin arteries become more active, constricting the vessels and reducing blood flow to the skin. Vascular changes of other organs are similarly controlled.

Other examples of tonic activity are vagal (parasympathetic) neurons to the heart, which in humans maintain heart rate below its intrinsic or 'free-running' level (Chapter 21), and the parasympathetic neurons that keep the pupil partly constricted at normal daylight levels (Fig. 26.13).

Together with the endocrine system, the autonomic nervous system regulates the activities of internal organs to produce a stable internal environment.

SUMMARY

- The plasma membranes of almost all cells show the basic physical properties and ionic selectivity that are the basis for evolution of electrical signalling within a specialised type of cells, neurons.

- Nervous systems are made up of networks of neurons and are capable of providing precise and rapid co-ordination of cellular function, movement and behaviour.

- Sensory neurons convert other forms of energy to electrical signals, allowing the nervous system to monitor changes in the external and internal environments. Interneurons pass information from neuron to neuron and aid in the integration of responses. Motor neurons influence the excitability of tissues, such as muscles or glands.

- Conduction along the processes of neurons is electrical; electrical signals can be conducted passively or actively over short distances but must be actively conducted over longer distances. Regenerating active responses that are conducted over long distances are called action potentials.

- Transmission from neurons to other cells is usually chemical but, in some cases, is electrical.

- The same range of chemical transmitters is found across the animal kingdom although different phyla and classes have characteristic groupings of these.

- The more complex and adaptable the behaviour of an organism, the more extensive and complex is its nervous system.

- The motor outputs of the vertebrate nervous system occur by way of two subsystems, the somatic system that is under voluntary control, and the autonomic system that automatically adjusts the activity of internal organs.

QUESTIONS

1. Draw a representative neuron and label its parts. What are the meanings of the words afferent and efferent?

2. (a) What are the structural features of a chemical synapse? (b) Describe the events involved in the release of neurotransmitter.

3. What causes the inside of a neuron electrically negative with respect to the outside?

4. Indicate the major structural components of the mammalian central nervous system.

5. What are the functions of the cerebral and cerebellar cortices? What is meant by higher nervous functions?

6. Compare the visual mechanisms of different types of eyes.

7. What special sensory mechanisms are available to the platypus and echidna, bees, pit vipers and bats?

8. Describe the series of changes in ionic permeability underlying the action potential.

9. What types of conductance changes underly epsp's and ipsp's and why do these lead to excitation and inhibition?

10. Explain the differences between somatic and autonomic components of the vertebrate nervous system.

11. What do you think would happen to the activity of vasoconstrictor neurons to skeletal muscle during exercise, or to the intestine during digestion?

12. What trends in nervous system organisation accompany increasing behavioural complexity?

ANIMAL MOVEMENT

Locomotion refers to movement of an animal from one place to another. The majority of animals are capable of movement throughout life, with locomotion being of fundamental importance for survival and reproduction. In the case of *sessile* marine animals, such as sponges, tunicates and barnacles, motility may be limited to the early planktonic stage of development, where it facilitates the dispersal of offspring.

Locomotion requires force to be exerted on the surrounding environment and the resultant movement is always in accordance with Newton's three *Laws of Motion*. The modes of locomotion displayed by animals are extremely diverse, but the production of forces upon which locomotion depends can be broadly divided into two categories—muscular and non-muscular.

Muscular movement involves elongated cells that are specialised for contraction, and is the principal method of force generation in higher animals. Once contracted, a muscle cell cannot extend itself. It must be extended by an external force before being able to contract again. Non-muscular locomotion is limited to small animals or the motile larval stages of larger organisms. In the majority of these, locomotion is achieved using flagella or cilia, which are motile organelles (Chapter 3).

The mechanisms of force production in both types of locomotion are examined later in this chapter. Firstly, we will look at some of the different forms of locomotion used by animals in water, on land and in the air.

> All forms of locomotion require force to be exerted on the surrounding environment. Locomotion can be broadly divided into two categories—muscular and non-muscular.

AQUATIC LOCOMOTION

Animals use many methods to exert a force on the surrounding water, accelerating the water in one direction and the body in the opposite direction. Some mechanisms are effective only if an animal is small, others operate best for large animals. We shall examine aquatic locomotion, not in a phylogenetic framework, but instead by grouping animals according to their method of swimming. This approach is used because similar ways of moving in water have evolved in animals from widely separated phyla. But firstly we need to consider the property of buoyancy in aquatic animals.

> Aquatic locomotion involves an animal exerting a force on the surrounding water. This accelerates water one way and the body in the opposite direction.

Buoyancy

Buoyancy is the tendency for objects to float. An animal that rises when at rest is positively buoyant, whereas one that sinks is negatively buoyant. If there is neither a tendency to rise nor sink, then an animal is neutrally buoyant. Neutral buoyancy occurs where the downward-directed force, due to gravity acting on the mass of the body, is opposed by an equal and opposite force produced by upthrust from the surrounding water. Many aquatic organisms are, or are very nearly, neutrally buoyant. Neutral buoyancy allows animals to hover in mid-water without having to swim constantly. In addition, a neutrally buoyant animal requires less **power** (work per unit time) to swim at a particular speed than if it were negatively buoyant. Neutral buoyancy allows energy savings in both these circumstances.

Positive buoyancy reaches an extreme in *Physalia*, the Portuguese man-of-war or 'bluebottle' of our warm seas. Each 'bluebottle' is actually a colony of many individuals, one of which is highly modified and forms the conspicuous gas-filled sac that acts as a float.

Negative buoyancy occurs when the overall density of an animal is greater than the density of the surrounding water. All benthic organisms, those that live on the bottom of oceans and lakes, are negatively buoyant. If the animal ceases to swim, it sinks.

Nautilus, the only surviving genus of the subclass Nautiloidea (Fig. 27.1) is found in coastal waters of the South-west Pacific. This animal has a spiral shell, divided into numerous chambers. All of these chambers are gas-filled, with the exception of the largest chamber at the mouth of the shell, which is occupied by its body. The gas-filled chambers are much less dense than the surrounding water, compensating for the greater density of the shell and body tissues. As a result, the animal is neutrally buoyant. Many teleost fishes use a gas-filled sac, the swim bladder, to control buoyancy. The amount of gas in the swim bladder can be adjusted to achieve the necessary buoyancy. Even a moderately small swim bladder makes a relatively large difference to the overall density of a fish, as it increases body volume without appreciably increasing body weight.

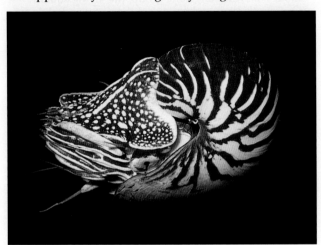

(a)

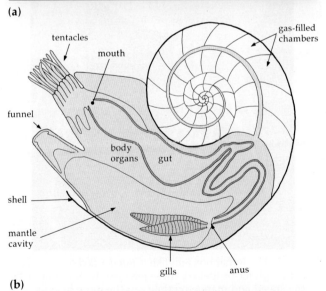

(b)

Fig. 27.1 *Nautilus*, the only surviving shelled cephalopod. This animal achieves neutral buoyancy by secreting gas into the spaces within its shell. Swimming is achieved by expelling water from the funnel, with the direction of swimming dependent on which way the funnel is pointing during expulsion. When feeding, it swims forwards, tentacles extended to capture prey items, but it swims backwards at other times

Certain sharks approach the condition of neutral buoyancy by having very large livers that contain lipids that are less dense than sea water. These lipids compensate for other more dense body tissues, which results in the overall body density being close to that of sea water. Many sharks are slightly more dense than sea water and therefore have a tendency to sink. Even so, they are found throughout the water column, from the surface waters to the sea bed. They counteract the tendency to sink by using fins, particularly their pectoral fins, as **hydrofoils**. These act like the wings of aircraft and produce an upward-directed force, known as **lift**. The caudal (tail) fin also produces some lift and provides the motive force. Thus, a swimming shark has the potential to cruise in mid-water without sinking, whereas a stationary shark will sink.

> Highly mobile aquatic animals are usually neutrally buoyant. This permits individuals to occupy any level in the water column with minimum energy expenditure. Neutral buoyancy is achieved by incorporation of low-density substances in the body.

Non-muscular locomotion

Cilia and flagella

Cilia and flagella are long thin extensions of cytoplasm able to undergo vigorous bending movements (Chapter 3). Typical flagellate locomotion occurs when a wave of bending travels from the tip of a long flagellum to its base, or from the base to the tip, forcing water in the opposite direction. Animals that have shorter (and more numerous) cilia tend to be larger than flagellates and can move faster. In cilia, the locomotory forces are produced by a powerstroke, where a rigid cilium is moved by bending at its base alone. The cilium then becomes limp and is drawn forward in preparation for the next powerstroke (Fig. 27.2). The co-ordinated beating of numerous cilia can be seen as a visible wave propagated across a field of cilia.

Non-muscular locomotion involving cilia and flagella is only practical for animals of small size. Many of the single-celled protists use these structures to produce propulsive forces, and the majority of invertebrate phyla (with the apparent exception of the Arthropoda and some of the more obscure phyla that comprise the Lesser Protostomes) contain species that have a ciliated larval stage.

Muscular locomotion

Jet propulsion

The cephalopod molluscs, octopuses, squids, cuttlefish and *Nautilus* have the ability to move by jet propulsion. This involves a forceful contraction of

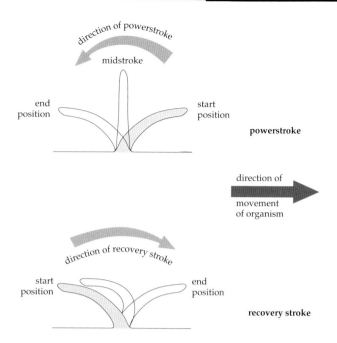

Fig. 27.3 Cuttlefish use undulations of their fins for steering and propulsion at low speeds, but resort to jet propulsion for fast locomotion. Squids use similar modes of swimming. One group, the 'flying squids', *Onycoteuthidae*, can shoot out of the water and glide for considerable distances

Fig. 27.2 Beating of a cilium. The powerstroke provides the motile force and is followed by a recovery stroke. Ciliated animals use numerous beating cilia to move through fluids. Larger animals often use cilia to move fluids within their bodies or across their surface. Our respiratory tract has numerous cilia, which move fluids and debris away from the lungs

circular muscles in the wall of the mantle cavity, expelling water rapidly through the funnel and accelerating the animal in the opposite direction (Fig. 27.3). These animals move forwards or backwards depending on which way they point the funnel, but they generally travel backwards when moving fast.

Cephalopods are not alone in using jet propulsion. Many jellyfish, such as the Indo-Pacific sea wasp or box jellyfish, *Chironex fleckeri*, found in the warm coastal regions of Australia, produce thrust by contracting circular subumbrellar muscles (Fig. 27.4), which results in the ejection of a pulse of water from the bell. They are not high-performance swimmers, primarily because of the small amount of muscle involved in the contraction process.

Most bivalve molluscs are sedentary. However, scallops and file shells can hop and swim by clapping their shells together. They may do this to escape from predators, particularly starfish, or may simply clap their way to a new site on the sea bed. Muscles attached to the internal surfaces of the shells provide the power by pulling the shells shut, ejecting water rapidly. An elastic hinge ligament, which is compressed when the valves are pulled shut, acts as the antagonist (Fig. 27.5). Thus, when the muscles relax, the energy contained in the compressed hinge causes the valves to spring apart.

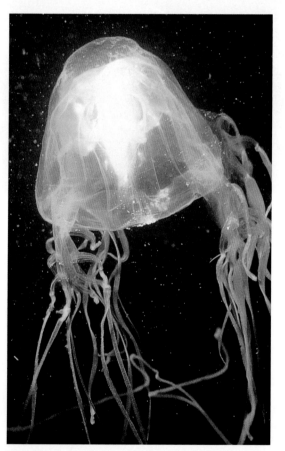

Fig. 27.4 The lethal box jellyfish, *Chironex fleckeri*. Although jet-propelled, these jellyfish do not approach the speed and manoeuvrability of the more heavily muscled cephalopods

Rowing

In rowing, limbs push against the surrounding water. In the backwardly directed powerstroke, the extended limbs present a large surface area to the water,

(a)

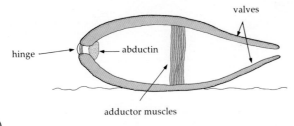

(b)

Fig. 27.5 Scallops have a block of a rubber-like protein called abductin in the hinge region of the shell valves. Adductor muscles close the valves, squeezing the abductin. Muscle relaxation allows the compressed abductin to 'spring' the valves open

rather like the blade of an oar. The forward recovery stroke presents a smaller profile to the water so that, over the whole limb cycle, the net effect is to move water backwards, thus propelling the animal for-wards. Water beetles achieve this by having hairs that fan out during the powerstroke, increasing the size of the effective 'oar'. The hairs lie flat during the recovery stroke, reducing the profile presented to the water.

Plesiosaurs, aquatic reptiles of the dinosaur era, may have used a form of rowing that involved their four large paddle-shaped limbs. Alternatively, the paddle may have had a cross-section much like that of a bird's wing and plesiosaurs may have 'flown' underwater. Another hypothesis is that they used a combination of the two techniques, resulting in paddle movements similar to those of sea lions (Fig. 27.6). The simple fact that the question is under debate illustrates the plausibility of rowing as a locomotor style that can be used by animals differing in size by orders of magnitude.

Swimming using body undulations

The majority of fishes use trunk musculature and the caudal fin to propel themselves through the water. Locomotion powered by fins other than the caudal

fin is found in many groups of fish. The weedy seadragon, *Phyllopteryx taeniolatus*, manoeuvres by propagating a wave along its dorsal fin; trigger fish, *Balistes*, use both the dorsal and anal fins; while skates and rays use highly modified pectoral fins (Fig. 27.7), the latter reaching its extreme manifestation in the huge manta rays.

Aquatic mammals, such as whales, dolphins, porpoises and manatees (dugongs), use large tail flukes, and others, such as seals and sea lions, use modified hind limbs in a similar way to tail flukes. A major difference between fishes and aquatic mammals is that fishes employ a side-to-side sweep of the tail while mammals use vertical oscillations. Eels and snakes have slender elongated bodies and swim by lateral wave-like movements, which travel backwards along the body, the amplitude of the wave increasing as it travels towards the tail. This is known as anguilliform locomotion (Fig. 27.8).

> Methods of aquatic locomotion include ciliar and flagellar motion, jet propulsion, rowing, swimming by body or fin undulation, and underwater flying.

AERIAL LOCOMOTION

The majority of animals that fly use muscle-powered wings and are capable of sustained flight. Other flying animals either have different sources of power, or use unpowered flight, such as gliding.

> Aerial locomotion can be divided into muscle-powered flight and unpowered or gliding flight.

Unpowered flight

Gliding flight is more economical in energetic terms than flapping flight. Muscular involvement is limited to the maintenance of a particular posture. The wings, or their equivalent, are outstretched and held motionless relative to the body, apart from constant small adjustments in response to changing air conditions or flight requirements. Some animals are only capable of gliding for a few seconds, others for hours; some have maximum gliding speeds in the order of metres per second, others in the order of centimetres per second. In all cases, in order to stay aloft, an animal has to produce enough upward-directed force, *lift*, to counteract the effect of gravity (Box 27.1). This lift depends predominantly upon the speed of flight, the size and geometry of the wing, or aerofoil, and how the wing is tilted relative to the direction of travel—the *angle of attack*.

> An aerofoil is used to generate lift in all types of flight. Lift opposes the effect of gravity. The size, shape and orientation of the aerofoil determines flight performance.

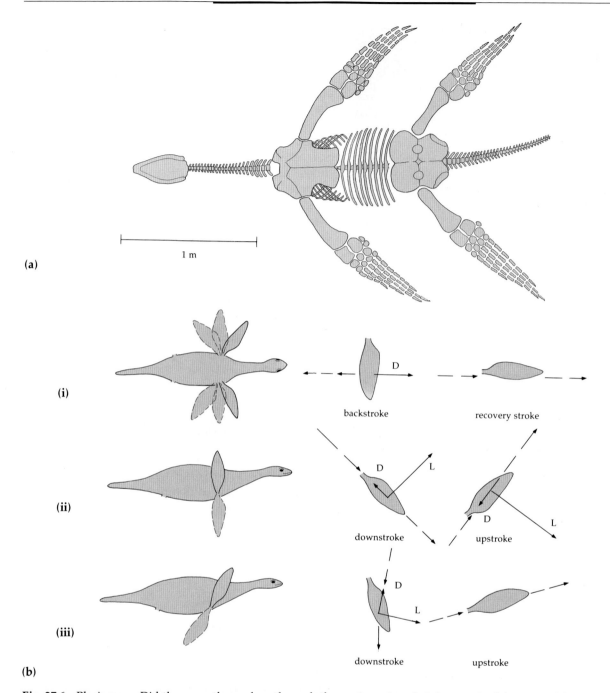

Fig. 27.6 Plesiosaurs. Did they row themselves through the water, using their large paired fins as paddles, or were they able to 'fly' underwater, using their limbs as hydrofoils? **(a)** Skeleton of a long-necked plesiosaur shown from below. **(b)** Three possible swimming techniques: **(i)** rowing; **(ii)** and **(iii)** underwater flight. Diagrams on the left show how flippers would have been moved relative to the body and those on the right show the drag (D) and lift (L) forces acting during the downstroke and upstroke (only the foreflippers are shown)

Many of the larger sea birds and birds of prey (raptors) with large wingspans can stay aloft by gliding for long periods of time. The wandering albatross, *Diomedea exulans*, has a wingspan of about 3.2 m and the wedge-tailed eagle, *Aquila audax*, has a span of 2.5–3.0 m. Predictions made using flight equations suggest that the narrow wings of the albatross, compared with those of the eagle, would result in a higher wing loading and thus a greater gliding speed (Box 27.1). Gliding performance for a variety of animals is shown in Figure 27.9.

Most, if not all, gliding animals can alter the size, shape and angle of their wing or flight membrane to modify gliding performance. Gliding speeds are generally reduced by postural adjustment immediately before landing to reduce the chance of

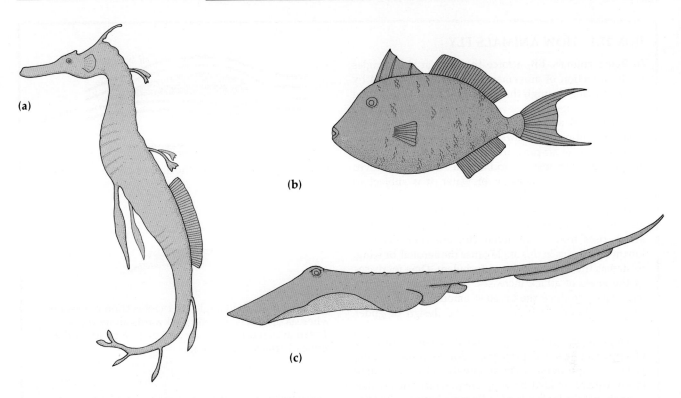

Fig. 27.7 Swimming using fins other than the caudal fin. **(a)** The weedy seadragon uses its dorsal fin to manoeuvre. **(b)** Undulations of the dorsal and anal fins power the low-speed swimming of trigger fish. **(c)** Waves of motion along the highly enlarged pectoral fins of rays provides the thrust for swimming. The caudal fin and tail may act as a rudder, or may be modified as a defensive weapon as in the stingrays

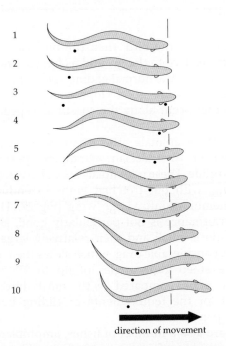

direction of movement

Fig. 27.8 This is anguilliform locomotion, which is often used by long thin animals when swimming. The head does not deviate to either side by very much, but by the time the wave of contraction reaches the tail it has a very large amplitude. The wave is produced by co-ordinated contraction of muscle blocks on each side of the body. The dots indicate wave crests and can be seen to move backwards along the body. This sequence is taken from 10 photographs taken at 50-ms intervals

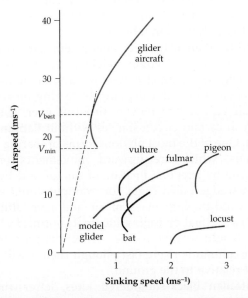

Fig. 27.9 Gliding performance of some animals and glider aircraft. Sinking speed is the rate at which height is lost during flight and is related to the airspeed. Stalling will occur below the minimum gliding speed (V_{min}). A line drawn as a tangent to the curve from the origin of the graph gives the 'best glide ratio', that is, the speed (V_{best}) at which maximum range can be achieved for any given starting height

BOX 27.1   HOW ANIMALS FLY

In flying animals, **lift**, a force that acts at right angles to the direction of movement, is usually generated by wings moving through the air. A cross-section of a wing shows the upper and lower surfaces to be different (asymmetrical) and the flow of air over the upper surface is faster than over the lower one (Fig. a). This produces an area of lower air pressure above the wing and higher air pressure beneath. In addition, as an object moves through a fluid, such as air or water, it is subject to **drag** forces due to friction.

The angle of attack (Fig. b) affects the amount of lift and drag a wing creates. Lift increases as the angle of attack rises from 0° to about 20°, but then declines. Further increases in this angle cause the aerofoil, or wing, to stall as the airflow becomes turbulent. Drag increases as the angle of attack increases. Narrow wings with a large span produce the greatest amount of lift for the smallest amount of drag. Hence the design of glider aircraft.

Simple analysis of gliding flight has revealed a number of important general principles. For example, gliding speed is proportional to the square root of wing loading (body weight divided by wing area), which means that animals with a high wing loading are better suited for faster flight than are those with low ones. Another important parameter is the **aspect ratio** (tip-to-tip length divided by average width). Animals with large aspect ratios are able to glide at shallow angles relative to the

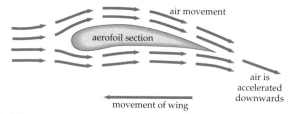

(a) Schematic diagram of the airflow past a wing that is producing lift

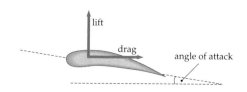

(b) Lift acts at right angles to the direction of motion, whereas drag acts directly backwards along the direction of motion. The lift and drag depend on the angle of attack of the wing

ground. This is another reason why glider aircraft have extremely long thin wings. They can fly at glide angles as small as 1.3° to the horizontal, compared with about 3° for the wandering albatross, which has the smallest glide angle of all birds.

---

injury, whereas increased gliding speeds can assist prey capture. The peregrine falcon, *Falco peregrinus*, folds its wings back and then enters a fast dive, or *stoop*, to catch birds in mid-air.

Soaring is an extremely energy efficient form of gliding. In normal gliding, height is slowly lost and eventually the animal comes to ground or has to regain height by active flapping. In soaring, an animal uses air currents to remain airborne, even gaining height in certain circumstances without flapping its wings. One of the most obvious forms of soaring uses **thermals**, which are upward movements of air produced by irregular heating of the ground by the sun. A wedge-tailed eagle, for example, could enter an upward-moving mass of air and gain altitude. Within a thermal, an eagle glides in its normal fashion, losing height relative to the surrounding air, but if the air column is rising fast enough then it will gain height relative to the ground.

Australian black-shouldered kites, letter-winged kites and nankeen kestrels are adept at slope soaring (in addition to their ability to hover using powered flight). They are often seen facing into the wind, remaining stationary relative to sloping ground, and moving their wings only fractionally. These birds are soaring in wind that is angled upwards relative to the horizontal. If this angle is equal to the normal

gliding angle of the bird, then the bird can remain stationary relative to the ground as long as the wind speed is equal to the normal gliding speed (Fig. 27.10). Many sea birds are good slope soarers, using the updrafts that occur along the coastline, especially at cliffs.

A variety of animals other than birds, including a number of Australasian marsupial species, are known to glide. These marsupials move between trees by gliding, with outstretched limbs extending the gliding membrane into a large 'wing' (Fig. 27.11). Their flight performance is not particularly good, as their aspect ratio is small and their relatively large body produces considerable drag, which slows them down. They use steep glide angles of up to 27° which, although poor compared with most birds, are sufficient for the requirements of gliding between trees in open forest.

There are also a number of fishes, amphibians and reptiles capable of unpowered flight. Flying fish belong to the family Exocoetidae, and one species in particular, *Exocoetus volitans*, has greatly enlarged pectoral fins that it uses as wings. These fish swim until a high enough speed is attained, at which point they break through the water surface, extend their pectoral fins and glide. There are a number of 'flying frogs', *Rhacophorus*, in the Malaysian region that glide,

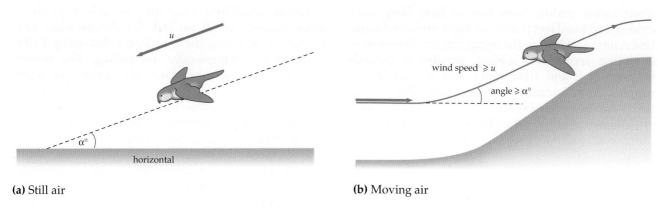

(a) Still air   **(b)** Moving air

Fig. 27.10 Hovering kites and the kestrel are often observed hanging motionless in the air while looking for prey. **(a)** If a bird can glide downwards through still air at a speed u and glide angle $\alpha°$, then **(b)** it will be able to remain stationary relative to the ground if it faces into the wind, moving at speed u, as long as the wind is directed upwards by at least angle $\alpha°$

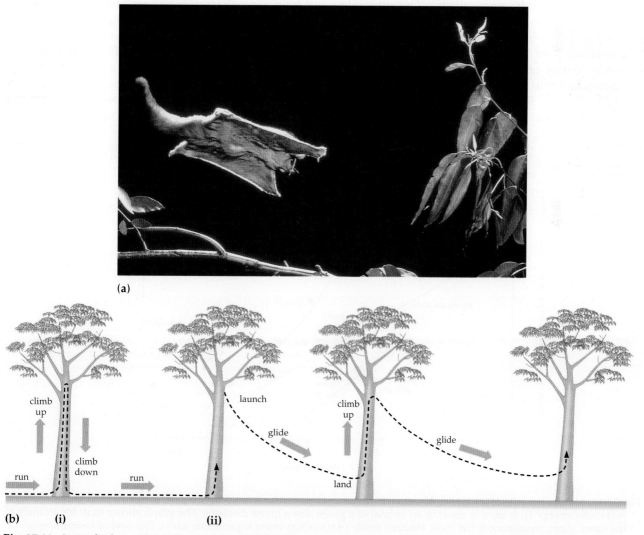

Fig. 27.11 Australia has at least five species of glider. These range in size from the large greater glider, *Petauroides volans*, weighing about 1200 g, to the diminutive feathertail glider, *Acrobates pygmaeus*, of about 15 g. **(a)** This is a sugar glider, *Petaurus breviceps*, about to land. **(b)** To move between trees it is energetically more expensive to **(i)** climb down, run to the next, and climb up, than it is to **(ii)** glide between trees. Gliders use the second type of locomotion for tree-to-tree movement when foraging. Powered flight in the vertebrates may have evolved via gliding in tree-living animals. This arboreal or 'trees-down' theory was proposed by Darwin in 1859. An alternative theory is that active flight evolved in ground-living animals via a run-jump-glide-fly progression. This is the cursorial or 'ground-up' theory

albeit rather poorly, from tree to tree. They have enlarged webbed feet that act as flight surfaces. South-East Asia is also host to the flying dragon, *Draco volans*, a gliding reptile. Its gliding membrane extends between fore and hind limbs and has additional support from laterally extended ribs. The extinct pterosaurs are probably the most well-known group of gliding reptiles, the largest of which was *Quetzalcoatlus northropi*, which had a wingspan of about 12 m.

Powered flight

In flapping flight, strong muscular contraction is needed to power wing movements. Active flight is much more complex than gliding flight. In most birds, it is the downstroke that produces most of the useful lift and thrust, which support body weight and propel the animal forwards respectively. The upstroke normally generates little or no lift (and may even produce negative lift), but it does return the wings to a position from which the next downstroke can be made. Wings tend to be fully extended during the downstroke to generate maximum lift and are brought closer to the body during the upstroke to reduce drag forces.

Various small birds (and bats) are able to produce large amounts of lift on *both* the downstrokes and upstrokes, enabling them to hover. Hovering flight is, however, energetically demanding. The power requirements are proportionately higher for large animals and, consequently, hovering is effectively limited to small animals. Dragonflies and many other insects are adept hoverers, as are the hummingbirds of the Americas. In general, birds up to about the size of pigeons and doves are able to hover for short periods of time, but larger birds cannot generate sufficient power. Although active flight is extremely energy-demanding, requiring a high output of power, it is an efficient way of moving a unit mass over a unit distance. For a given body size, active flight is a far cheaper way to travel than walking or running, although swimming is cheaper still (Fig. 27.12).

Birds (Table 27.1) and bats provide good examples of how wing characteristics are linked with ecology. Normal flight behaviour, foraging behaviour and habitat preference can be shown to be strongly influenced by the size and shape of bat wings. Species with low wing loading are very agile and are able to change direction rapidly, whereas those with high wing loading are not. More agile species often occupy 'cluttered' environments and are capable of hawking

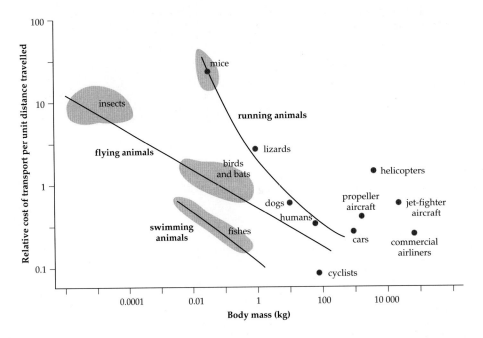

Fig. 27.12 Metabolic costs of transport. Measurements of oxygen consumption can be used to reveal how much metabolic energy (J) is used in moving an animal of a given size a given distance. The graph shows that, for animals of the same mass, swimming is the most efficient mode of transport (note logarithmic axes). The cost of swimming is lower than flying, which is in turn lower than legged locomotion. (These curves generally refer to animals that are specialised for the particular form of locomotion represented here.) Swimming is efficient because the body is supported by water. In terrestrial locomotion, energy is consumed in order to counteract the force of gravity and, in addition, energy is required to accelerate and decelerate legs during a stride cycle. Notice that all the curves slope in the same direction. Large animals use less energy per kilogram body mass than do small animals when moving the same distance. Bicycle travel is very efficient, primarily because no work is done against gravity. Energy is primarily used in overcoming air resistance

Table 27.1 Relationships between flight mode, wing characteristics and flight performance of various birds

| Flight and foraging modes | Example | Observed wing characters | | | Flight performance predicted for the observed wing characters based on aerodynamic theory |
		Aspect ratio	Wing loading	Wingspan length	
Cruising and hawking in open spaces	Swifts, swallows, hawks, falcons	High	High	High	Relatively slow and enduring flight. Relatively high manoeuvrability
Cruising flight in search of prey in open spaces	Albatrosses, harriers, gulls, pelicans	High	Low	High	Slow enduring flight, soaring ability
Soaring or cruising flight in search of prey, often perching	Sea-eagle, wedge-tailed eagle	Low	Low	High	Slow flight, soaring ability
Perching and hawking within vegetation	Flycatchers	Low	Low	Low	Slow, highly manoeuvrable, but rather expensive flight
Foraging in open water	Ducks, geese, swans	High	High	Low	Rapid enduring flight adaptations to migration and commuting, low manoeuvrability. Usually need taxiing run for take-off

for insects in the complex three-dimensional space. Less agile bats tend to occupy more open habitats where good turning ability is less crucial.

> Flight is energetically expensive, but is the cheapest mode of transport per unit distance for terrestrial animals.

TERRESTRIAL LOCOMOTION

Movements performed by applying force against the ground include burrowing, crawling, walking, running and jumping.

Locomotion without legs

Amoeboid movement

Single-celled organisms, such as *Amoeba*, and motile cells within higher animals, such as some vertebrate white blood cells, use this slow form of locomotion, which involves cytoplasmic streaming. Progression is made by extending finger-like projections (pseudopodia) forwards and attaching them to a surface. The fluid endoplasm then streams forwards ahead of this adhesion point through a tube of more solid cytoplasm, the outer ectoplasm. At the leading edge it is transformed to the stiffer gel-like ectoplasm. At the rear, the cell 'shrinks' as the ectoplasm is converted

back to endoplasm, which then streams forwards (Fig. 27.13).

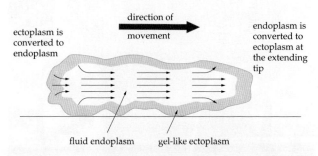

Fig. 27.13 Microfilaments in the cytoplasm are thought to power amoeboid locomotion. This diagram of a vertical section through a crawling *Amoeba* demonstrates how these types of cell move by relocating the material that forms their 'tail end' to the 'front' of the cell. Cells may be free-living, as the amoebas, or may be motile cells within an organism, such as phagocytic white blood cells in our own bodies

Peristalsis

Worms move on or through soil by a form of crawling that uses peristaltic contractions. Their bodies are divided up into segments, each of which has two sets of opposing muscles (circular and longitudinal) in the body wall. As each segment is fluid-filled its volume is constant. Contraction of longitudinal muscles shortens a segment and increases its diameter, and stiff hairs (setae) are extended to anchor with the substrate. When the circular muscle contracts, diameter reduces, the setae are retracted and the segment elongates. Progression is made by successive waves of contraction of each muscle layer, extending the body segment and anchoring it in turn (Fig. 27.14). The direction of these co-ordinated contraction waves can be reversed, enabling worms to crawl backwards.

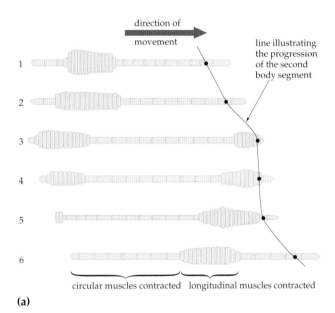

(a)

(b)

Fig. 27.14   **(a)** Worms generally burrow through the soil, moving by alternately contracting circular and longitudinal body wall muscles. The six positions shown represent body positions after short intervals of time. **(b)** The giant earthworm, *Megascolides* (class Oligochaeta), of Australia reaches lengths of about 3 m

Pedal waves

Molluscs, such as snails, slugs and chitons, seemingly glide across the ground without obvious movement

of their broad flat foot. Some species of snail, particularly small snails and those that live on soft substrates, move by ciliary propulsion, but most use waves of muscular contraction of the foot, to propel themselves along. These pedal waves can easily be seen by observing snails crawling up the inside of an aquarium. The mucus that coats the ventral surface of the foot has peculiar physical properties. Beneath non-moving parts of the foot, the mucus appears to behave like a solid and effectively sticks the foot to the ground. This enables other portions of the foot to pull against these 'anchored' regions. The mucus beneath the actively extending regions of the foot has markedly different properties. It behaves like a viscous fluid, enabling these moving portions of the foot to slide over the ground with ease.

> Crawling includes amoeboid movement, peristaltic and pedal wave forms of locomotion.

Locomotion with legs

Stability

Legs raise an animal above the ground, greatly reducing friction during locomotion. When animals are stationary, legs provide support and static stability. Stability increases as the height of the centre of mass of the body above the ground decreases, as feet become relatively larger and as the area enclosed by those feet that are in contact with the ground increases. Humans are bipedal, walking upright on their hind limbs. When we stand still, it is essential to maintain our centre of gravity directly above the area bounded by the outlines of our feet, or we topple over. We accomplish this without conscious effort by feedback mechanisms involving sight, balance, muscles and special receptors in the joints and tissues of the body (Chapter 26). Animals with four or more feet in contact with the ground are more stable as it is harder to 'accidentally' cause the centre of gravity to move outside of the support base. In a similar way, squat animals with outstretched limbs are more stable than are tall animals with limbs held beneath the body. This latter point is of particular importance to animals of low mass, such as insects and spiders, where there is a considerable risk of being blown over by air currents. Similar factors affect aquatic legged animals, such as crabs, which are prone to being overturned by water currents. In both instances, the limbs tend to project away from the body enough to ensure stability without compromising the animal's ability to move.

In legged animals, movement is accomplished by lifting individual legs, moving them forwards and placing them back on the ground. Animals with four or more legs move their legs in a particular order to maximise stability and to minimise the chances of one leg interfering with another. Quadrupeds move their feet in the order shown in Box 27.2 when walking slowly. Insects tend to use an alternating tripod **gait**, where three legs are moved while the remaining three legs form a stable tripod (Fig. 27.15). These examples illustrate stable gaits, where the centre of mass remains within the triangle of support at all times. Stability becomes less of a problem as the number of legs increases, but for multilegged animals the potential for 'tripping over' their own legs becomes greater. Precise co-ordination of leg movement prevents interlimb interference, as demonstrated by the wave of leg movement seen in centipedes and millipedes.

> Support and stability are important for legged animals. To ensure stability when walking, limbs are moved in a precise sequence. This varies according to the animal and the number of legs possessed.

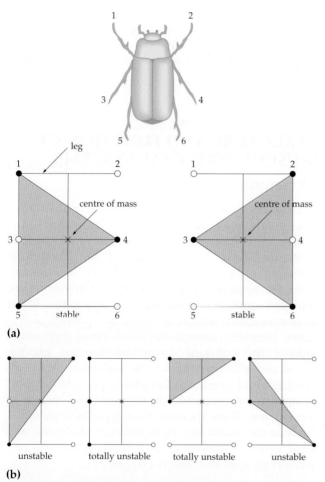

Fig. 27.15 (**a**) Insects use a stable tripod gait where three of their six feet are in contact with the ground at any one time. Feet on the ground are represented by solid circles. The centre of mass stays well within the triangle formed by the supporting legs indicated by the dashed lines. (**b**) The animal would topple over if it used these combinations of foot–ground contact as the centre of mass falls outside or on the outer border of the triangle of support

Walking and running

Animals often change gait as they increase speed. Horses and many other quadrupedal mammals walk, trot and then gallop as they increase speed. Kangaroos use a curious pentapedal form of locomotion when moving slowly, with the tail acting as a 'fifth limb', but use a bipedal hopping gait at higher speeds. The reasons for altering gait with speed appear to be linked to, firstly, the energetic cost of locomotion and, secondly, simple physical constraints that would otherwise limit speed.

We will examine the first reason further, using the example of horses that are trained to change their gait on command. Experimental analysis of the metabolic energy cost of travel shows that the cost of moving unit distance varies with speed within each gait (Fig. 27.16). There is a particular speed, within

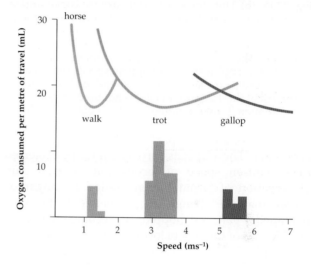

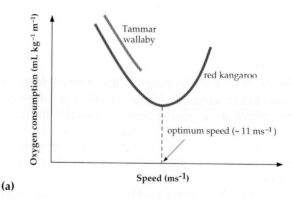

Fig. 27.16 Energetic cost of horse and kangaroo locomotion. (**a**) The oxygen cost to move unit distance against speed for a horse. The histograms show the speeds that were used when the horse moved freely. Results from experiments with large kangaroos suggest that their oxygen consumption varies with the speed of locomotion. The curve is similar to that of the trotting horse, suggesting that there is an optimum speed at which these kangaroos will travel. Small kangaroos, of less than about 5 kg, appear to have an oxygen consumption independent of speed, but why this occurs is still a mystery

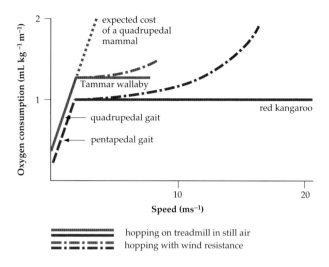

hopping on treadmill in still air
hopping with wind resistance

Fig. 27.16 (b) How the steady-state oxygen consumption for medium-sized red kangaroos varies with speed and gait. The effect of wind resistance is more important in big kangaroos in comparison to small wallabies. This is because the aerodynamic drag acting on them is relatively large at the high hopping speeds they can attain. Unlike the situation seen in the red kangaroo, the tail of the Tammar wallaby is not used in slow locomotion. It uses a form of quadrupedal locomotion

the walk, trot and gallop, at which the energetic cost is minimal. These speeds correspond to the preferred freely chosen speed of locomotion in each gait. Observation of locomotor energetics in kangaroos shows that the change in gait between the slow pentapedal gait and hopping is accompanied by a levelling off of the cost of locomotion (Fig. 27.16). This is possibly due to an increased importance of energy storage and return in tendons. Thus, it appears that changing gaits can minimise the metabolic energy required to move a unit distance.

Jumping

Animals may use jumping as the normal form of locomotion (a kangaroo hop is, in effect, a jump) or, more often, as a mechanism of escape. In most instances, a forceful and rapid extension of limbs, powered by contraction of skeletal muscle, propels the individual at high speed away from its original position.

Fleas are able to leap about 0.1 m, which may seem poor when compared with humans who can raise the centre of mass (located at about the level of the navel) about 1 m, but it represents a leap of about 100 times body height compared with our half body height jump. In general terms, the actual height of a jump depends on the speed and angle of take-off, and the interrelationship between the animal's mass and its air resistance. Small animals tend to have large surface areas relative to their volume (or mass) and have greater drag than larger and heavier animals.

Therefore, for any given speed of take-off, the small animal will slow down faster than the large one and not jump as high.

Click-beetles, *Elateridae*, also jump, but not using their legs. They throw themselves into the air by rapidly flexing their body at a joint between two thoracic segments. This movement involves a catapult mechanism whereby muscles contract over a relatively long period of time, straining part of the thorax. A catch mechanism releases all of the stored energy 'instantaneously' to produce a relatively large leap. Catapult-type mechanisms, where energy is stored in stretched or compressed tissues (of the limbs), also power the jumps of grasshoppers, locusts and fleas. In the Orthoptera (grasshoppers and crickets), the energy is stored in the cuticle of the limb and the tendon (apodeme) of the leg extensor muscle, whereas fleas utilise the properties of an elastic material known as resilin.

> Jumping animals use fast powerful movements, often in the form of a quick extension of the supporting limbs, in order to throw themselves into the air. In many cases, energy is stored in stretched or compressed structures, and then released 'instantaneously' to produce large forces.

STRUCTURE AND FUNCTION OF MUSCULOSKELETAL SYSTEMS
Skeletons

In neutrally buoyant aquatic animals, the downward force of gravity is counteracted by an upthrust due to displaced water. As a result, in many cases skeletal structures have become adapted for locomotion rather than for support against gravity. If these animals are removed from their aquatic environment, and consequently lose this support, they may collapse under their own weight, or will be incapable of movement. Terrestrial animals need to support themselves against the constant downward-directed gravitational acceleration without the assistance of a supporting fluid medium. Skeletons provide support against gravity, in addition to their involvement in locomotion. There are three main types of skeletons—hydrostatic skeletons, rigid internal skeletons and rigid external skeletons. Each type, as we will see, has advantages and disadvantages.

> The three main skeletal forms are hydrostatic skeletons, and rigid external and internal skeletons.

Hydrostatic skeletons

Fluid-filled body compartments form the basis of the hydrostatic skeleton in various invertebrates (cnidarians, flatworms, nemerteans, nematodes and

BOX 27.2 ANIMAL GAITS

A **gait** can be defined as a distinctive form of locomotion. Gaits can thus be defined for aquatic, aerial and terrestrial locomotion, although gaits are most commonly considered in association with the latter.

As an example, consider the changes in gait that occur as a horse gradually increases speed. The first and slowest gait is the walk, this is followed by the trot, the canter and finally the fastest gait, the gallop. Galloping is not just a faster version of walking, but a distinctly different form of locomotion. It is often easiest to describe a gait diagrammatically or by using numerical values of various parameters, such as the *relative phase* of a foot (defined as the time at which it is set down expressed as a fraction of the total stride time). The relative phases used by walking, trotting, cantering and galloping horses are shown in the Figure. When walking and galloping, each foot hits the ground at a different time; when trotting, the diagonally opposite feet hit the ground simultaneously; and in the canter only one diagonal pair of feet hit simultaneously. This is why the sounds of trotting, cantering and galloping horses have quite different rhythms.

Camels and long-legged dogs often use a modified trot, the pace, where the limbs on one side of the body are moved in phase with each other. This reduces the likelihood of the fore and hind limbs interfering with one another. Small animals (possums, gliders) tend to use a bound or half-bound instead of a gallop (Figure).

A gait may change because of a simple inability to move faster in the one currently used. Thus, it is analogous to changing gear in a car. Experimental evidence suggests that gaits change before the top speed is reached in any one gait. This may be linked to the observation that horses choose speeds of locomotion within a gait that are the most energetically economical. It has also been noticed that the forces on limb bones rise with speed within a gait, but fall at gait transitions. It may be that changing gaits is a method of ensuring that the skeleton is not exposed to potentially damaging forces during normal locomotion.

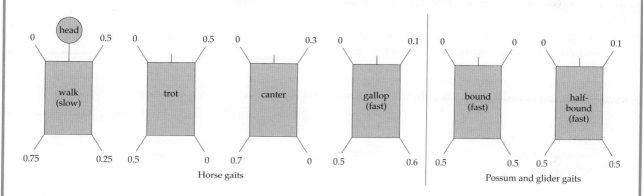

Diagram showing the relative phases of each foot for different gaits in horses and possums and gliders. The front left limb starts the cycle (or stride) and always has a relative phase of zero

---

annelids). Since water is non-compressible, a shortening of a cylindrical compartment results in an increase in its diameter. As we saw in annelids (p. 601), circular and longitudinal muscles of each segment act on a hydroskeleton to produce peristaltic locomotion. Locomotion may involve the propagation of undulatory waves of contraction along the body producing eel-like swimming, if in a fluid medium, or a crawling progression, if on a solid substrate (see below). Animals that have hydrostatic skeletons are generally small. This is due, in part, to the lack of protection afforded by the skeleton and because it provides only limited mechanical support, especially in the terrestrial environment.

> Hydrostatic skeletons are fluid-filled compartments, often under pressure. Movement is mediated by muscle contraction. This skeletal form offers little protection, provides limited support against the effects of gravity and is basically limited to 'small' animals.

Exoskeletons

Exoskeletons are found only in invertebrates. They are formed when hard skeletal material is deposited on the external body surface. The chemical composition of the skeleton varies considerably in different invertebrates. For instance, the shells of gastropod molluscs are composed of calcium carbonate and an organic material in the ratio of about 2:1, whereas in arthropods, including insects, crustaceans and spiders, the principle component in the exoskeleton, or cuticle, is the polysaccharide, chitin.

As animals with exoskeletons grow they need to either continuously enlarge the skeleton to accommodate the growing soft tissues (e.g. limpets and snails) or periodically replace it with a new larger one (moulting in crabs and spiders). Exoskeletons may be 'one-piece' (a snail shell), 'two-piece' (bivalve molluscs, such as clams), or 'multipiece' as in the case of the exoskeletons of the arthropods. A one-piece skeleton does not play a direct role in locomotion, but provides good protection for soft body parts. Two-piece skeletons are used also primarily for protection as they are usually capable of completely enclosing the soft parts, but they may also be involved in locomotion, as seen in the jet-propelled scallops. In arthropods, the entire body is covered by a chitinous exoskeleton and movement is possible because of joints between each rigid section of the body. The arthropod skeleton is secreted by the epidermis and consists of an outer epicuticle, which acts as a waterproofing membrane, an intermediate exocuticle, a tough layer of tanned chitin and protein, and an inner endocuticle of untanned chitin and protein (Fig. 27.17). The stiff exocuticle layer is absent at the joints, allowing flexibility.

Exoskeletons provide excellent protection and they can act as almost impermeable barriers to water. This is important in reducing the risk of desiccation in land animals and is likely to have been an important factor in the highly successful colonisation of land by insects. Disadvantages of exoskeletons include the weight of the skeleton, the limitation it imposes on body size and, importantly, the vulnerability of an animal immediately after moulting when the new exoskeleton is soft. Very heavily armoured invertebrates are marine, where their effective weight is less due to support from the surrounding water.

> Exoskeletons provide good protection but are often heavy and/or limit animal size. Some forms are replaced periodically (via moulting), whereas others are enlarged as the individual grows.

Endoskeletons

Vertebrate animals have an endoskeleton. It is constructed primarily of bone, except in elasmobranch fishes (the sharks and rays) where it is cartilaginous. The functions of the skeleton are numerous: it is involved in movement, mechanical support of the body, protection of parts of the body, mineral storage and regulation (Chapter 25), and in the production of red and white blood cells (haemopoiesis, Chapter 6).

The major constituents of bone are collagen and a complex form of calcium phosphate (hydroxyapitite). Collagen, a structural protein, is strong in tension and gives bones a degree of flexibility, whereas the inorganic component, hydroxyapitite, is strong in compression and gives bone much of its strength.

New bone is usually produced by deposition on a pre-existing surface or in a non-calcified substrate (cartilage) during normal growth and development of the vertebrate skeleton. The material that forms the bone shaft is compact and dense in comparison with the 'honeycomb' configuration found in the expanded ends of bones (Fig. 27.18), with these two

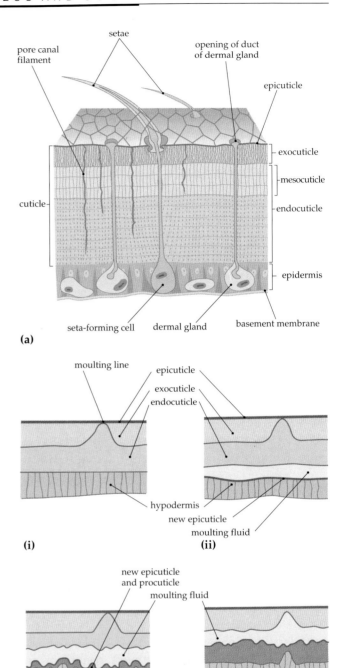

Fig. 27.17 (a) Diagrammatic section through the integument of an arthropod. (b) Representation of the moulting sequence in an arthropod. (i) Normal appearance of the exoskeleton in intermoult. (ii) The epidermal layer (hypodermis) separates from the exoskeleton and secretes a new epicuticle and moulting fluid. (iii) The untanned endocuticle is eroded by the moulting fluid and a new procuticle is formed. (iv) The animal is encased in the old and new skeleton immediately before moulting. The old skeleton breaks along predetermined lines and the animal emerges with its new pliable exoskeleton. Uptake of water or air enlarges the animal before hardening of the cuticle occurs

morphologies being related to the functional requirements of bony structures. Woven-fibred bone differs from most forms of new bone in that it does not require a surface or non-calcified model for deposition. It is found in regions of skeletal repair, and is the type of bone involved in the callus formation after bone fracture.

Bone is a living tissue and, as such, is capable of adaptive change. Large changes occur during growth, for example, the long bones of the forearm of a human adult are much larger than are those of an infant, both in terms of bone lengths and diameters. Even when full adult stature is achieved, bones are still capable of change. New bone layers can be deposited or old bone removed at the surfaces of bones, and the internal bony structure can be altered by a process known as secondary remodelling. Thus, the skeleton can adapt to new situations where the pattern and/or magnitude of forces acting on bones is altered. For example, professional players of racquet sports often have thicker and heavier bones in their racquet arm than in the other arm. This bone hypertrophy is in direct response to the large repetitive forces the racquet arm is exposed to. When a player retires from the game, within a moderate period of time the skeletal asymmetry declines as a result of remodelling and net bone resorption in the bones of the racquet arm. When an injured arm is immobilised in a plaster cast, net resorption and remodelling occurs in response to the abnormally low levels of bone loading. Similarly, large reductions in the strength and mineral content of astronauts' bones after prolonged periods of weightlessness have been recorded. These are partially due to changes in the mechanical environment, although in these cases the causes of bone loss are complex and the hormonal status of individuals also plays an important role.

> Internal skeletons provide good support, allow growth to be continuous and are remodelled according to use throughout life.

Joints

If motion is to occur between two or more rigid structures, whether they form part of an exoskeleton (Fig. 27.19) or an endoskeleton (Fig. 27.20), there has to be a joint present (Box 27.3). Muscles and skeletal components interact to produce movement, with muscles producing force and the skeleton acting as levers, articulating with one another at joints (Box 27.3).

In humans, joints are classified according to their structure and function. The three *functional* groups are immovable, such as the joints between the bones of the skull (sutures); slightly movable, such as the discs between the vertebrae of the spinal column; and freely movable joints, most of the joints in the arms and legs. *Structurally*, joints can be grouped into fibrous joints, where tough collagenous fibres link two bones; cartilaginous joints, where the articulating bones are bound together by cartilage; and lastly, synovial joints.

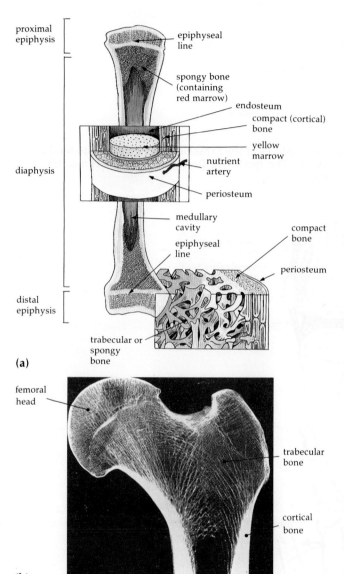

Fig. 27.18 (a) Diagram of the structure of a typical mammalian bone (human tibia). (b) Section through the head of a human femur. Mammalian long bones have an approximately tubular shaft (diaphysis) composed of compact (cortical) bone and containing bone marrow. At each end of the diaphysis there are expansions of the bone, the epiphyses. Within each epiphysis there is a complex 'honeycomb' of supporting struts, the trabeculae. The orientation of this trabecular bone is thought to be intimately related to the forces that the bone is subjected to in normal life. Increases in bone length are achieved by growth at the epiphysial plates, which remain unossified until full skeletal maturity is reached. Changes to the thickness of the walls of bones occur by the deposition or removal of bone at their external (periosteal), or internal (endosteal) surfaces

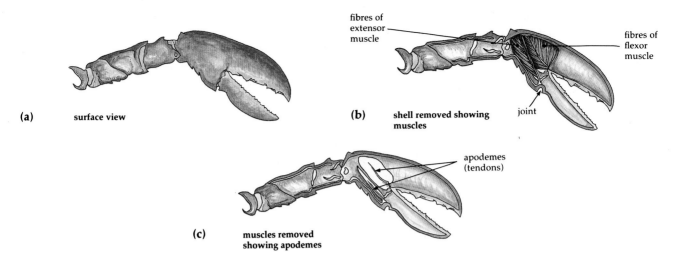

(a)   surface view

(b)   shell removed showing
            muscles

(c)   muscles removed
            showing apodemes

Fig. 27.19 In a crab claw, the muscle fibres originate over the inner surfaces of the exoskeleton and insert onto tendon-like apodemes that increase the surface area available for insertion. The apodemes direct vectors of the contractile force onto the site of action where the apodeme attaches to the skeleton

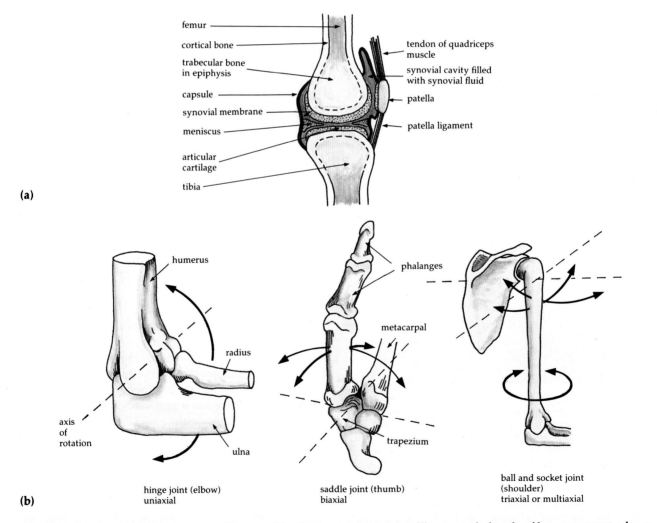

Fig. 27.20 **(a)** Synovial joint structure, illustrated by the human knee joint. The expanded ends of bones support a layer of smooth articular cartilage and the whole joint is bounded by a fibrous joint capsule, the inside of which is lined with a synovial membrane that secretes synovial fluid. This serves to lubricate the joint surfaces, resulting in an almost frictionless articulation. Ligaments (not shown) hold the joint surfaces together and passively 'control' and limit joint movement. The menisci help to transfer forces across the knee joint. **(b)** Joint movement is often described using easily understood terms. Joints in the forelimb illustrate this

BOX 27.3 BONES AS LEVERS

A lever is a rigid structure that rotates about a **fulcrum** when a force is applied. In a joint, muscles produce the force, bones act as levers and the joint itself is the fulcrum (Fig. a).

The precise arrangement and proportions of these components are important in determining how the system operates. A hypothetical change in the site of insertion of a muscle alters the lever system of the human forearm (Fig. b). With the arrangement found in *Homo sapiens*, the insertion of the tendon of biceps brachii onto

the radius is close to the elbow joint. This results in the muscle having to generate large amounts of tension to support even quite modest loads held in the hand. This may be viewed as being a disadvantage. However, the advantage of this arrangement is seen when the speed and range of movement is considered. Small length changes of the biceps brachii muscle result in large displacements of the forearm and these movements can occur quickly. A muscle inserting more distally on the radius would significantly change the properties of

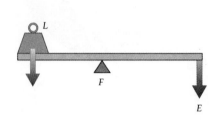

(i) First-class lever

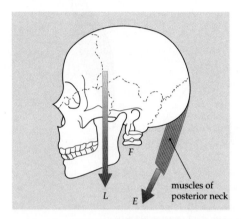

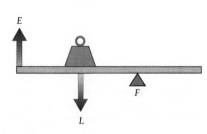

(ii) Second-class lever

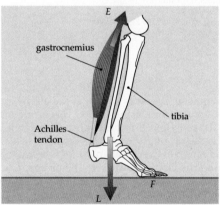

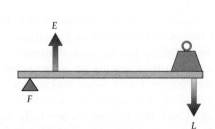

(iii) Third-class lever

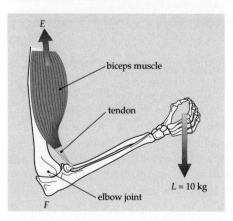

(a) Bones as levers. There are three classes of levers which are defined by the relative positioning of a pivot (fulcrum, *F*), a muscular force (effort, *E*) and a resistance or passive load (*L*). **(i)** A first-class lever has the fulcrum between the effort and the load. **(ii)** A second-class lever has the load between the effort and the fulcrum. **(iii)** A third-class lever has the effort between the fulcrum and load

the lever system. The muscular tension required to support loads carried in the hand will be considerably less than in the above case. Thus, larger loads could be carried, or the same loads carried with less effort. This arrangement of bones, joints and muscle would have the disadvantages, in comparison with the true situation, of having a smaller range of motion and a reduced speed of forearm flexion.

The lengths of limb bones are generally related to function. The echidna, for example, has short limb bones, which, together with the associated musculature, can perform highly efficient and powerful digging movements. By analogy, this system of levers is like that shown in Figure b (ii). Kangaroos, on the other hand, use a lever system more like that shown in Figure b (i) and are adapted for moving at speed by virtue of their elongated hind limbs. Long legs allow for longer strides to be taken, which can result in greater speeds being attained.

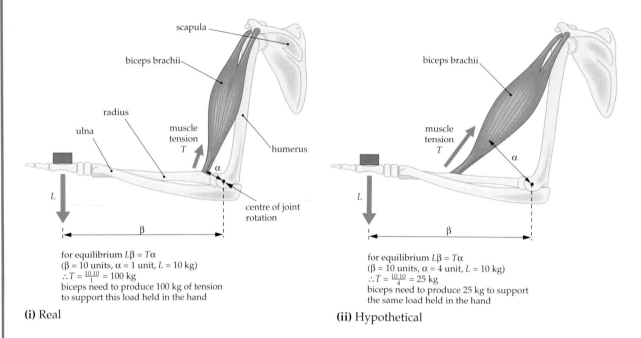

for equilibrium $L\beta = T\alpha$
($\beta = 10$ units, $\alpha = 1$ unit, $L = 10$ kg)
$\therefore T = \frac{10.10}{1} = 100$ kg
biceps need to produce 100 kg of tension to support this load held in the hand

(i) Real

for equilibrium $L\beta = T\alpha$
($\beta = 10$ units, $\alpha = 4$ unit, $L = 10$ kg)
$\therefore T = \frac{10.10}{4} = 25$ kg
biceps need to produce 25 kg to support the same load held in the hand

(ii) Hypothetical

(b) The biceps brachii produces flexion at the elbow joint. **(i)** The muscle inserts, via a tendon, onto the radius of the forearm at a point close to the elbow. This arrangement is well adapted for rapid movement as a small shortening of the muscle will flex the forearm through a large arc. **(ii)** If the muscle were attached to a more distal point, the degree of flexion would be much less than in (i), for the same shortening of the biceps brachii. The latter arrangement is better adapted for forceful movements of the forearm, demonstrated by balancing moments around the elbow. Tension in the biceps, T, acting at distance α from the elbow, acts in a clockwise direction and has to equal the load, L, acting at a distance β from the elbow in an anticlockwise direction if equilibrium is to be maintained. It is obvious that in the situation illustrated in (i), the muscle is in a poor position to support large loads held in the hand

In synovial joints the articulating bone surfaces, which are covered with cartilage, are located in an articular cavity containing **synovial fluid** (Fig. 27.20). This fluid acts as an extremely effective lubricant, resulting in almost frictionless movement. The geometry of the articulating surfaces effectively determines the types of movements available at any particular joint. In addition, the shape of the surfaces, the position of **ligaments** around the joint and the muscles and tendons that cross the joint, are all involved in determining how much movement is possible at that articulation. Ligaments are, with a few exceptions, strong bands of a collagenous material attached to bone on either side of joints. They keep the joint surfaces pressed against each other and are positioned so as to allow specific movements of adjacent bones. The strength and positioning of ligaments act to stabilise joints, and damage to ligamentous structures often results in a large loss of stability. If you badly twist your ankle, for example, the stretched or torn ligaments will take a long time to repair. Until the ligaments are fully healed, further ankle injury is likely because of compromised joint stability.

Skeletons provide the means to transfer forces generated by muscles to the environment. In the case of rigid skeletons, the bones or body segments are hinged at the joints and act as levers.

Size and the skeleton

Many valuable insights into 'how and why' animals have evolved into the forms we see today can be achieved by the application of mechanical engineering principles to biology—the discipline of **biomechanics**. The skeleton provides us with numerous examples of the interrelationships between skeletal structure and function. One such example is the interaction between animal size and skeletal structure.

The bones of small mammals tend to appear rather delicate, whereas large mammals have relatively robust skeletons. Imagine a small macropod marsupial, such as the potoroo, magnified so it becomes the same size as a red kangaroo. The bones of the hind limb will be about the same length in both animals, but the diameter of these bones will be different. The femora (thigh bones), for example, would have a smaller diameter in an 'enlarged potoroo' than in the red kangaroo. The reason is simple. For animals of the same shape, a limb bone's cross-sectional area will increase as the square of the linear dimensions and its mass (derived from the volume) will increase by the cube of the linear dimensions (Fig. 27.21). The stress (force per unit area) applied to the limb bones can be seen to have increased in the 'enlarged potoroo', and would be undesirable. This is because of a reduction in the difference between a load that the bones could withstand and one that would break them. As a consequence, large animals have disproportionately thicker limb bones to ensure that bone stresses remain at safe levels.

Another noticeable effect of body size on the skeleton is that many large animals tend to support their bodies on relatively straight legs, whereas small animals use flexed-limb (crouching) postures. Humans, elephants and large dinosaurs exemplify the former limb posture, and mice, rats and rabbits the latter. A straight-legged posture is an adaptation that reduces both the muscular energy costs of standing (try standing for some time with your legs very bent!) and locomotion. It also results in the forces that act on limb bones well aligned with the long axis of the bones during locomotion and standing still. This is advantageous as bone is strongest when loaded in this way (compression). If the forces were less well aligned, the limb bones would have to be bigger if they were to be as strong. The disadvantageous aspects of straight-legged postures include reduced manoeuvrability and ability to accelerate rapidly, as the limbs are almost fully extended when at rest. Animals that rest with highly flexed limbs are able to produce long and powerful extensions of the limbs to produce rapid body acceleration. This is why human sprinters start a race from a crouched position on racing blocks—to improve their acceleration.

Muscle

Muscle tissue has a unique property in that it can contract. It is the force produced by shortening that powers locomotion. In vertebrates there are three types of muscle tissue—skeletal, cardiac and smooth muscle—each of which is adapted to perform

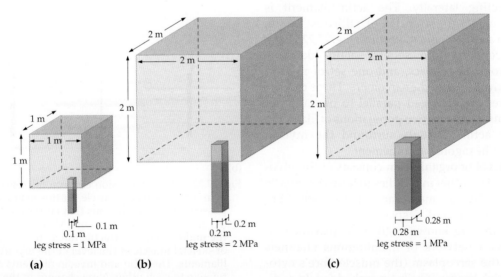

Fig. 27.21 (a) A hypothetical 'one-legged animal'. Its body tissues have a uniform density of 1000 kgm^{-3}. It has a volume of 1 m^3 and thus has a body mass of 1000 kg. The leg has a cross-sectional area of 0.01 m^2. The stress (force per unit area) acting in the leg is (1000 g/0.01) $\simeq$ 1.10^6 newtons per square metre (or 1 MPa), where g is gravitational acceleration. **(b)** If the 'animal' doubles each of its linear dimensions, it would increase its mass to 8000 kg and its leg would now be 0.04 m^2, resulting in a stress of 2 MPa. This doubling of stress in the leg increases the likelihood that it may break. Consequently, in real animals, leg bones of large animals are proportionally thicker than are those of small animals in order to keep bone stresses low. **(c)** The leg size required to keep the stress at 1 MPa

particular tasks within the body. They are all capable of converting chemical energy into mechanical energy in order to produce force, but it is skeletal muscle that is directly involved in locomotion and maintenance of posture. Some invertebrates use smooth muscle to power locomotion, although the majority use forms of striated (skeletal) muscle.

> All vertebrates and many invertebrates use striated (skeletal) muscle to power their locomotor movements.

Skeletal muscle structure

Skeletal muscle cells, or muscle fibres, are roughly cylindrical, contain many peripherally located nuclei and have alternating light and dark bands crossing them, called striations.

The intracellular structure of a skeletal muscle cell can be examined using transmission electron microscopy. Photomicrographs taken at high magnification show a regular arrangement of filaments contained within intracellular structures called **myofibrils** (Fig. 27.22). The myofibrils are tubular and are divided into compartments by discs (Z-discs), which lie perpendicular to the long axis of the myofibril. The distance between two adjacent discs is known as a **sarcomere**. Within each myofibril there are two types of myofilament, thick and thin. Analysis shows that each thick myofilament is composed of the protein **myosin** and the thin myofilaments are predominantly the protein **actin**. A myosin filament consists of a bundle of hundreds of myosin molecules, each of which is composed of two identical proteins. Each protein is shaped like a golf club, with the 'shafts' lying parallel to the myosin filament with the double heads projecting laterally. The actin filament is composed of two strands of fibrous actin, tropomyosin and troponin molecules (Fig. 27.22). Each sarcomere contains two sets of actin filaments extending from the Z-discs, and one set of myosin filaments in the centre of the sarcomere. These thread-like structures are arranged parallel to one another and, in regions where the two populations of filaments overlap, the actin filaments surround the myosin filaments in a hexagonal arrangement.

The next level of organisation consists of the myofibrils grouped together in bundles to form the muscle fibres (cells). The cell membrane of a muscle fibre is the sarcolemma, and a specialised endoplasmic reticulum in muscle, known as the sarcoplasmic reticulum, forms a network of membranous channels throughout the sarcoplasm (the muscle fibre's cytoplasm) in which the myofibrils are embedded. In addition, there is a specialised membrane system known as the transverse (T) tubule system, formed by the invagination of the sarcolemma, which runs to the core of the fibre (Fig. 27.23). Although the T-tubules and the sarcoplasmic reticulum are in contact where

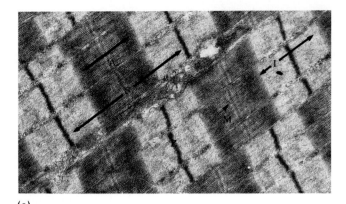

(a)

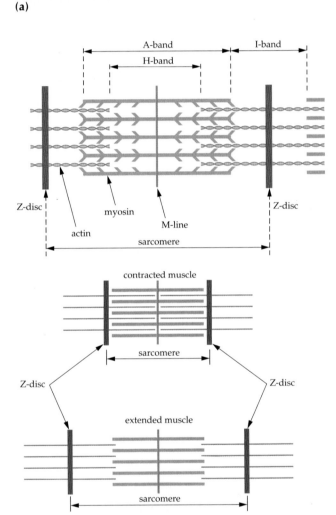

(b)

Fig. 27.22 (a) Transmission electron micrograph of mammalian skeletal muscle. At this magnification it is possible to see the regions that correspond to the actin (A) and myosin (M) filaments. This muscle is in about the middle of its contractile range. (b) Representation of skeletal muscle at the level of the myofibrilar filaments. The actin and myosin filaments can slide relative to each other. Note changes in the width of bands between contracted and extended muscle

the two sets of tubules cross, there is no direct connection between them; therefore there is no exchange of fluid between the two types of tubules.

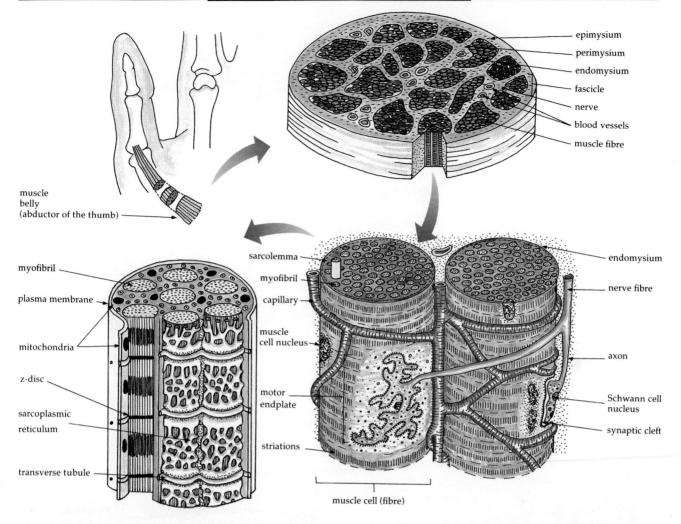

Fig. 27.23 Schematic diagram to show the hierarchical structure of muscle from the whole belly to individual microfibrils, and the structural relationship between the myofibrils and the surrounding sarcoplasmic reticulum and transverse (T) tubule system

Connective tissue binds the components of the muscle together, supports the capillary and nerve networks of the muscle and acts to transfer the force from the muscle fibres to the skeleton. The connective tissue sheath that surrounds individual fibres is the endomysium. Groups of muscle fibres (also termed fascicles) are surrounded by the perimysium, and the muscle belly as a whole is bounded by the epimysium (Fig. 27.23).

Muscles involved in movement of the skeleton insert into the skeleton either directly, or indirectly via tendons (apodemes in arthropods). It is common to find a tendon at either end of a muscle belly, attaching it to bones. The attachment of a muscle that moves less during contraction is called the origin, whereas the one that moves most is called the insertion. In muscles associated with moving the limbs, the origin is usually the proximal attachment point and the insertion is the distal attachment point.

Cardiac muscle is the tissue of which the heart is predominantly made. The muscle cells join in an irregular manner to form a branching network. This form of striated muscle is very resistant to fatigue and, unlike skeletal muscle, has a mechanism by which electrical potentials can pass from cell to cell. It is innervated by the autonomic nervous system.

Smooth muscle (also called visceral muscle) is found in numerous sites throughout the vertebrate body: in the walls of blood vessels, bronchioles, the bladder, ureters, uterine tubes, the ductus deferens and the digestive tract. Individual smooth muscle cells have a single nucleus and are elongated, and usually pointed at each end. This type of muscle produces relatively slow but forceful contractions and can produce force over a wide range of lengths. Innervation is by the autonomic nervous system.

How skeletal muscle works

The mechanism of contraction is explained by the **sliding filament model**. In this model it is proposed that the actin filaments, which slide relative to the myosin filaments, move towards the middle of the sarcomere and pull the Z-discs, to which they are

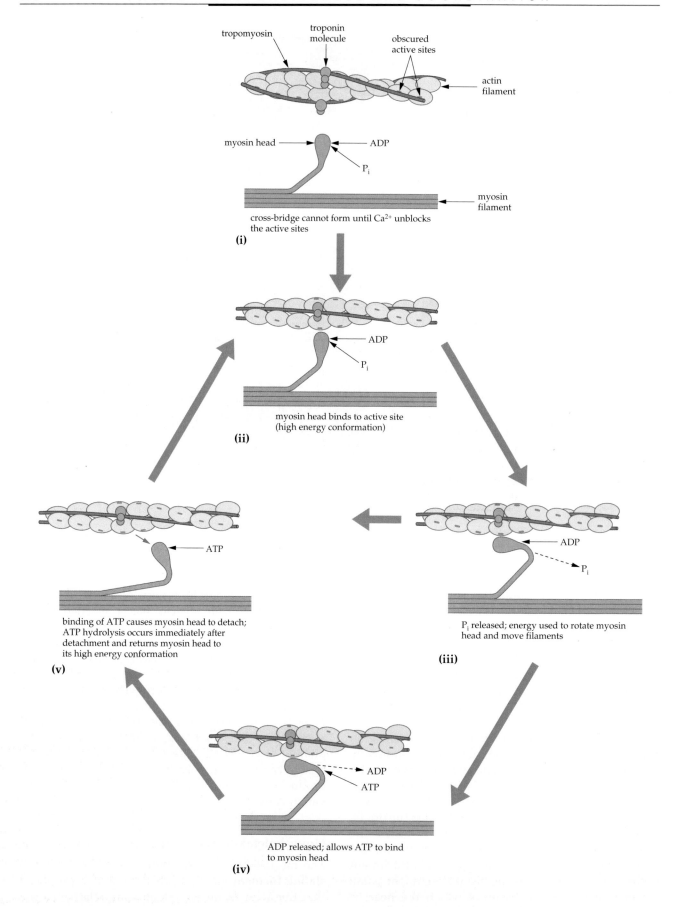

Fig. 27.24 Schematic representation of the mechanism of muscle contraction

attached, towards one another. This results in a shortening of the sarcomere and hence the muscle contracts. Note that individual filaments do not change length or contract but simply slide relative to one another. The whole process requires an input of energy, which is derived from the breakdown of adenosine triphosphate (ATP).

The basis for the sliding filaments is the interaction between the heads of myosin molecules and specific sites on adjacent actin filaments, for which myosin heads have a high affinity (Fig. 27.24). When a muscle is in its resting state, these active sites on the actin filament are obscured by tropomyosin, a regulatory protein associated with the actin filament. In this condition, the myosin heads are unable to form attachments or cross-bridges with the actin. Muscle contraction can only occur when these binding sites are exposed, and this requires the presence of calcium ions (Ca^{2+}), which are normally stored within modified endoplasmic reticulum, the sarcoplasmic reticulum. Upon neural stimulation great enough to produce muscle contraction, an action potential spreads across the muscle cell membrane and along the transverse (T) tubule system formed by invagination of the plasma membrane. The T-tubules conduct the action potential deep into the fibre, causing a wave of depolarisation to cross the membrane of the sarcoplasmic reticulum. This increases membrane permeability to Ca^{2+}, thus releasing these ions into the sarcoplasm. The Ca^{2+} binds to troponin and causes a conformational change in the tropomyosin–troponin complex, which uncovers the binding sites, allowing contraction to occur.

Myosin heads now bind to actin filaments forming cross-bridges. A subsequent change in shape of the myosin heads results in the actin filament being pulled towards the centre of the sarcomere. (It is at this stage of the cycle that mechanical work occurs, that is, the muscle shortens and/or generates force.) The binding of ATP to a myosin head causes it to dissociate from the actin filament and to return it to its original 'primed' condition before reattachment to another binding site further along the actin filament (Fig. 27.24). This cycle, which lasts for tenths or hundredths of a second, can then be repeated, resulting in further muscle shortening. The energy needed to fuel one complete cycle is ultimately derived from the hydrolysis of one ATP molecule. Muscle relaxation also requires an input of ATP, but in this case it is used in the active transport of Ca^{2+} back into the sarcoplasmic reticulum. A reduction in the Ca^{2+} concentration within the sarcoplasm allows the tropomyosin–troponin complex to re-establish its position, blocking the actin binding sites once again, and preventing the establishment of new cross-bridges.

After death, intracellular levels of ATP decline. As a result, Ca^{2+} leaks into the muscle cells and causes the uncovering of active sites. Cross-bridges form and remain until the tissues start to decay, there being insufficient ATP available to dissociate the myosin heads from the active sites. The development of muscle rigidity after death is termed rigor mortis.

The force a muscle fibre can produce is related to the length of its sarcomeres. If the sarcomeres are in a highly shortened or a highly extended state, then the muscle fibre cannot generate much force (Fig. 27.25). In addition, the rate at which muscles can contract is limited by the rate at which cross-bridges can attach, pull, release and reattach. The faster the rate of shortening of a muscle, the smaller the force it can produce. At the fastest rate of shortening a muscle can produce no force. On the other hand, when a muscle is successfully stimulated but is held

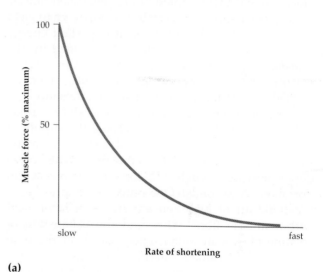

(a)

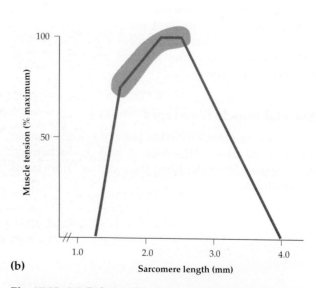

(b)

Fig. 27.25 **(a)** Relationship between the speed of muscle contraction and the force it can produce. Muscles that contract quickly cannot generate much force. **(b)** The length–tension curve for skeletal muscle. Most muscles are thought to operate in the shaded region of the curve

at a constant length (isometric contraction), it produces a large force. In life, muscles usually operate between these two extremes with muscles shortening while producing reasonable forces.

> Skeletal muscle derives its energy ultimately from the splitting of ATP. The relative motion between actin and myosin myofilaments within muscle cells produces shortening of individual cells, resulting in muscle contraction.

Muscle energetics

The direct source of energy for muscular contraction is ATP and is produced by aerobic or anaerobic respiration (Chapter 5). The amount of ATP stored in a resting muscle cell is low, but other energy stores are present within the cell. Glucose is converted into glycogen and this is stored in the cell. When muscle activity occurs, glycogen can be broken down to yield glucose. Creatine phosphate is an important energy storage compound in muscle and is formed by the transfer of a high-energy phosphate from ATP to creatine. It can be utilised for the rapid conversion of ADP to ATP in conditions of sudden demand.

Using the example of human runners, short periods of intense activity (sprinting) exhaust the supply of ATP present in resting muscle almost immediately. Anaerobic respiration and creatine phosphate breakdown produce enough ATP for further contraction (for about 20 seconds). One outcome of sprinting is the build-up of high concentrations of lactic acid within muscle fibres. This, and the exhaustion of creatine phosphate stores, effectively limit the time over which sprinting can be maintained. Aerobic respiration is used to produce ATP in long-distance running, where the muscular effort is sustained at submaximal levels for long periods of time. Under these conditions, fatty acids form a more important energy source than does glucose, although glucose is also metabolised.

Skeletal muscle fibre types

Muscle fibres, even within the belly of a single muscle, are not all alike. Histological and histochemical observations have identified three distinct fibre types in mammals.

1. *Slow oxidative (type I) fibres*, also called slow-twitch fibres. These contain numerous mitochondria and appear red due to their high vascularity and large myoglobin content. They have a great capacity for generating ATP by oxidative (aerobic) activity and are difficult to fatigue as the high blood flow to the fibres delivers oxygen and nutrients at a rate sufficient to keep pace with the breakdown of ATP. They have a relatively slow contraction velocity due to the slow rate of splitting of ATP and produce relatively little tension.

2. *Fast oxidative (type IIA) fibres*, also called fast-twitch fatigue-resistant fibres. These are red fibres with a fast contraction speed resulting from an ability to split ATP at a rapid rate. They have a moderate capacity for both oxidative and glycolytic (anaerobic) activity.

3. *Fast glycolytic (type IIB) fibres*, also called fast-twitch fatigable fibres. These are large fibres with a small myoglobin content and few capillaries, giving them a white appearance. They are adapted to generate ATP by the anaerobic process of glycolysis, using local glycogen stores. These stores are rapidly exhausted and there is a rapid build-up of lactic acid, causing these fibres to fatigue rapidly.

The relative proportions of these fibre types in a muscle belly can often be related to the normal tasks of that muscle. Type I fibres are ideal for maintaining posture and are found in relatively high proportions in the soleus (calf) muscle, which helps us maintain our balance when standing, and in neck muscles that hold the head up. In contrast, type IIB fibres are found in high proportions in arm and shoulder musculature, which are often used for rapid and powerful movements. Type IIA muscles are found in large proportions in the legs of sprinters.

Fibre types are probably determined genetically and an analysis of elite athletes shows that the percentage of fibre types differs from the normal population according to their chosen sport. Long-distance runners and cross-country skiers have a large proportion of type I fibres for endurance, whereas sprinters have more type II fibres for speed and power. Similar observations can be made between different species of animal according to the lifestyles and behaviours commonly used.

SUMMARY

- Many different methods of moving have evolved in animals in order to feed, avoid various dangers and find a mate for reproduction.

- Aquatic locomotion is achieved by an animal exerting a force on the surrounding water. This accelerates water one way and the body in the opposite direction. Many highly mobile aquatic animals are neutrally buoyant, which permits them to occupy any part of the water column from the surface to the bottom. They achieve neutral buoyancy by incorporating low-density substances in the body, either in discrete structures (e.g. the fish swim bladder) or within the general body tissues.

- Methods of aquatic locomotion include ciliar and flagellar motion, jet propulsion, rowing, swimming by body undulation or fin undulation, and underwater flying.

- Aerial locomotion can be divided into muscle-powered flight and different forms of unpowered, or gliding, flight. All types of flight use an aerofoil, such as wings, skinfolds or webbed feet, to generate lift, an upward force that opposes the effect of gravity. The size, shape and orientation of the aerofoil determines flight performance and can be used to predict how an animal will fly.

- Flight is energetically expensive per unit time, but works out to be the cheapest mode of transport per unit distance.

- Methods of terrestrial locomotion by a very diverse group of animals include walking, jumping, burrowing and crawling. Crawling includes amoeboid movement, peristaltic and pedal wave forms of locomotion.

- Support and stability are important for legged animals whose bodies are raised up from the ground. To ensure continuous stability when walking, it is necessary for limbs to be moved in a particular sequence depending on the animal and the number of legs it possesses.

- Jumping animals use fast powerful movements, often in the form of a quick extension of supporting limbs, to throw the body into the air. In many cases, energy is stored in stretched or compressed structures and then released 'instantaneously' to produce large forces.

- Hydrostatic skeletons consist of fluid-filled compartments, often under pressure. Movement is mediated by muscle contraction. This skeletal form is basically limited to 'small' animals. It offers little protection and provides limited support against the effects of gravity.

- Exoskeletons provide good protection, but are often heavy and/or limit animal size. Some forms are replaced periodically (by moulting), whereas others are enlarged as the individual grows.

- Internal skeletons allow growth to be continuous, provide good support and are important in blood cell production and the maintenance of blood Ca^{2+} levels.

- Skeletons provide the means to transfer forces generated by muscles to the environment. In the case of rigid skeletons, the bones or body segments act as levers, being hinged at the joints.

- All vertebrates and many invertebrates use striated (skeletal) muscle to power their locomotor movements.

- Skeletal muscle derives its energy ultimately from the splitting of ATP. The relative motion between actin and myosin myofilaments within muscle cells produces shortening of individual cells, resulting in muscle contraction.

QUESTIONS

1. Define buoyancy. Explain how it might be achieved and the possible benefits gained by neutrally buoyant animals.

2. Describe how jet propulsion is used in the swimming of squids and jellyfish.

3. What is 'lift' and how is it generated by the wings of a gliding bird?

4. Why is the ability to hover limited to small animals?

5. Define 'gait'. Using the horse and kangaroos as examples, explain how their gaits change with speed.

6. How does an *Amoeba* move?

7. What are the main differences between internal and external skeletons? How and why may these differences affect the final size and rate of growth of animals?

8. Why do we have joints and why aren't all joints of the very mobile 'ball-and-socket' type?

9. Explain how the sliding filament model of muscle contraction operates.

10. What are the advantages and disadvantages of muscle fibres that use aerobic and anaerobic respiration? Give examples of the functional use of the different fibre types.

ANIMAL BEHAVIOUR

During spring, male satin bowerbirds, *Ptilonor-hynchus violaceus*, build bowers consisting of two parallel rows of vertical twigs. The area immediately around the bower is decorated with flowers, feathers, snail shells and sometimes items such as bottle tops and toothbrushes (Fig. 28.1). Each male displays to attract females to his bower, and mating usually takes place at or near the bower.

Understanding why a male bowerbird takes so much trouble to attract the attention of a female is one of many questions that is addressed in **ethology**, the study of animal behaviour. Ethology is not simply documenting the activities of animals; it aims to explain *why* animals behave in the ways they do. Niko Tinbergen, one of the founders of modern ethology, believed that ethology should try to explain the causation, development, adaptive value and evolution of any behaviour under investigation. In response to the question 'Why do male bowerbirds construct and decorate bowers?', for example, we might suggest answers in terms of:

- *Causation.* The hours of daylight increase during springtime, and this might trigger changes in hormone levels that stimulate the male to build bowers.

- *Development.* Males have learned the behaviour from their parents or neighbours.

- *Adaptive value.* Males build bowers in order to attract females for breeding; males without bowers are unable to attract and mate with females.

- *Evolutionary history.* Complex bower-building behaviour may have evolved from more simple nest-building behaviour in the ancestors of bowerbirds.

These explanations fall into two main categories. *Proximate explanations* are those that consider the causation of a particular behaviour and *ultimate explanations* are those that are concerned with the evolution and function of a particular behaviour. A proximate explanation for bower-building might emphasise the learning abilities of bowerbirds or their

Fig. 28.1 A male satin bowerbird, *Ptilonorhynchus violaceus*, beside a bower, which he constructs to attract females

responses to objects of different colours, while ultimate explanations consider the effects of bowers and decorations on the reproductive success of males. Proximate and ultimate explanations are not alternative explanations but rather answers to different kinds of questions. This chapter mostly addresses the evolution of behaviour and therefore emphasises ultimate explanations. Nevertheless, it is often necessary to consider causal or proximate explanations in order to understand the mechanisms underlying ultimate explanations. The 'internal' mechanisms by which behaviour is controlled, such as neural pathways and the influence of hormones, are the subject of Chapters 25–27.

GENETICS AND THE EVOLUTION OF BEHAVIOUR

An animal's behaviour is an integral part of its phenotype. It has a genetic basis and consequently is under the influence of *natural selection* (Chapter 32) and sexual selection (p. 629). For example, a marsupial mouse, *Antechinus*, that forages in exposed areas during the day may be more likely to be captured by a predator than if it were to forage under the cover of bushes at night. Similarly, an *Antechinus* that forages for food that yields less energy than the animal expends obtaining it will not survive. It is therefore possible to attribute survival value to an animal's behaviour in the same way that it is attributed to an animal's anatomical or physiological characteristics. Frequently, an animal's behaviour reflects a balance between several alternative behaviours; an *Antechinus* that spends excessive time hiding from predators rather than foraging may be as much at risk of death, through starvation, as is one that ignores the risk of predation. Understanding how animals balance these conflicting demands is a common theme in contemporary studies of animal behaviour.

Natural selection can act only on genetic differences between individuals within a population. Consequently, an animal's behaviour, like other aspects of its phenotype, can evolve only if there is, or was in the past, variation in the behaviour of individuals in the population; if these differences in behaviour are heritable or genetic in origin; and if some behaviours provide greater survival or reproductive success than others. The genetic basis of animal behaviour can be illustrated by examining genetic markers, by artificial selection experiments and by comparing populations with genetic differences.

Genetic markers

Male and female vinegar flies, *Drosophila*, usually copulate for about 20 minutes. However, duration of copulation can be altered by subjecting flies to agents that produce genetic mutations. In one mutant, the male does not disengage from the female after the usual time, while in another mutant the male disengages from the female after only 10 minutes and thus fails to fertilise any eggs. These differences in mating behaviour are the result of changes in the sensory receptors and muscles of the flies, which can be attributed directly to genetic mutations.

Selection experiments

It is possible to select artificially for particular behaviours in the same way that plant or animal breeders select for physical attributes of flowers, crops or domesticated animals. For example, some males of the field cricket, *Gryllus integer*, call to attract females, while others remain silent and attempt to intercept females that are attracted to the calling males. It is possible to produce populations of males that call very frequently by allowing only frequently

calling males to breed. Similarly, a population of almost silent males will result if frequently calling males are prevented from breeding.

Populations with genetic differences

Individuals from populations that are geographically separated may differ in their morphology and behaviour, often as a result of different selection pressures imposed by different ecological conditions. For example, the garter snake, *Thamnophis elegans*, is found in separate populations on the coast and inland in south-western United States. Inland snakes are mostly aquatic and feed on frogs and small fish, while coastal snakes are terrestrial and feed mainly on slugs. Food-choice experiments with newborn snakes reared in the laboratory showed that individuals derived from coastal populations had a stronger preference for slugs than did snakes from inland populations. Furthermore, offspring obtained by matings between individuals from coastal and inland populations had intermediate preferences for slugs. These experiments indicate that the food preferences of snakes from the two populations has a genetic basis and is associated with local ecological conditions.

> All forms of animal behaviour have a genetic basis, which means that, like anatomical and physiological characteristics, behaviour is under the influence of natural selection.

LEARNING AND THE DEVELOPMENT OF BEHAVIOUR

The undignified tricks that some animals are obliged to perform in circuses and zoos are examples of learning. But learning is an integral, and frequently important, feature of the behaviour of animals under natural conditions. Any change in an individual's behaviour that is due to its experience can be referred to as **learning**. One advantage of learning is that it provides a greater potential for changing behaviour according to changes in an individual's circumstances. For example, a worker honeybee, *Apis melliphera*, can learn the location of her hive and the location and quality of several flower crops during her brief three-week life span. This means that she can fly directly to the most productive crops rather than randomly 'discover' them every time she leaves the hive.

The behaviour of an organism capable of learning is unlikely to be constant throughout its life but will change over time. Thus, it is important to distinguish learning from the *development of behaviour*. For example, barnacles (Cirripedia) have a wide range of behaviours that are specific to their age. A barnacle changes from a planktonic, swimming, grazing nauplius to a swimming and then crawling cyprid, which seeks out a settlement site, attaches itself to the substrate and metamorphoses into a sessile filter-feeding adult, which behaves in a different way again. Newly hatched chicks of the domestic hen peck more accurately at grains on the ground as they grow older. Although this may be partly due to experience, the improvement will also occur irrespective of how frequently the chicks are allowed to peck at food. In other words, older chicks are more accurate as a result of their maturity, regardless of how much practice they have had. The reduction in play behaviour with increasing development, common in many animals, including humans, is another example of a change in behaviour with age.

Imprinting is a feature of the early development of some kinds of behaviour. When a young gosling hatches it will recognise the first moving object as something to follow for the next few weeks. Usually, this will be its mother, and consequently the gosling will *imprint* on its mother. However, a chick will imprint on a human if the human is the first moving object the chick sees (Fig. 28.2). Imprinting usually occurs only during a *sensitive period*, the exact duration of which varies between species.

Fig. 28.2 These young geese have imprinted on the famous ethologist Konrad Lorenz, and behave as if he were their mother

While certain behaviours simply develop with age, others clearly include an element of learning. For example, the songs of many birds are learned from their parents or other nearby birds. A young bird's first attempt at song usually occurs some months after hatching. This song, called a subsong, resembles the adult song in length, pitch and tonal quality, but is variable, imprecise and lacks specific features that are typical of the adult song. For example, a young white-crowned sparrow, *Zonotrichia leucophrys*, raised in isolation, will be able to sing the subsong but will fail to develop the normal adult song. In contrast, young birds that are raised with their parents or other birds will learn to sing the adult song (Fig. 28.3).

There are several ways in which animals learn behaviour and these can be roughly placed into two categories: *associative* and *non-associative learning*. The difference relates to whether the animal learns the behaviour in association with some **stimulus** or property of the environment. The classic example of associative learning is Pavlov's conditioning experiments with dogs. Pavlov had noticed that dogs salivate when presented with meat or a meat-smelling substance. In one experiment, Pavlov rang a bell every time he presented a dog with meat. The dog eventually learned to associate the sound of the bell with the meat, and Pavlov was able to make the dog salivate simply by ringing the bell. Learning by trial and error is another form of associative learning. For example, white-winged choughs, *Corcorax melanorhamphos*, obtain their food by digging into leaf litter and soil. This method of foraging is both time- and labour-intensive, and difficult for the young bird to master; juvenile choughs can take up to eight months before they learn the technique and can forage independently. *Habituation* is a form of non-associative learning. For example, fish may fail to show an escape response to a shadow passing overhead if this stimulus is repeated every few minutes.

Clearly, some behaviour is learned while other behaviour is not. However, it is by no means the case that complex behaviour is always learned and simple behaviour is not. For example, the pallid cuckoo, *Cuculus pallidus*, is a nest parasite; it lays its eggs in the nests of other species, especially the superb fairywren, *Malurus cyaneus*. The cuckoo chick usually hatches before the wren chicks, and shortly afterwards ejects the wren eggs from the nest, an activity that requires considerable dexterity. The behaviour of the cuckoo chick could not be learned from either its cuckoo parents or its foster parents. Similarly, young ogre-faced net-casting spiders, *Deinopis*, not only construct a web between their legs, but also recognise prey and snare them with the net. This behaviour cannot have been learned from parents or other adult spiders (Fig. 28.4) because they never come in contact with each other.

For many years there was controversy among ethologists over whether most animal behaviour is *innate* (determined by the genotype) or *learned* (determined by environmental experiences). However, these views represent two extremes of a continuum; most animal behaviour is the result of an interaction between genetic instructions and learning from the environment.

> Learning refers to changes in an individual's behaviour that are due to experience, and it provides a great potential for changing behaviour under different circumstances. In contrast, the development of behaviour involves changes that occur as a result of age, not experience.

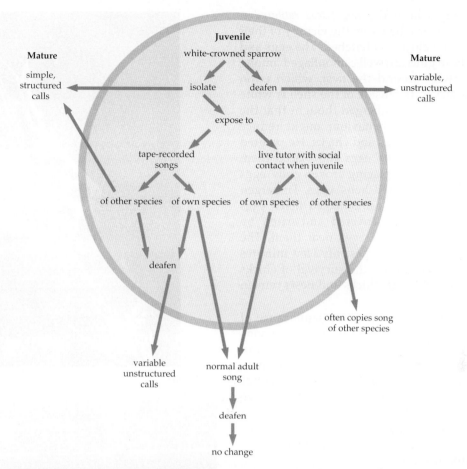

Fig. 28.3 This diagram illustrates the results of numerous experiments that have examined the factors responsible for the development of song in the white-crowned sparrow, *Zonotrichia leucophrys*

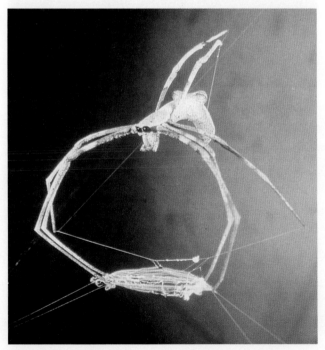

Fig. 28.4 The ogre-faced net-casting spider, *Deinopis subrufa*, builds a silk web between its legs and then casts this web onto a passing insect. The insect is unable to escape from this sticky 'net'

UNDERSTANDING COMPLEX BEHAVIOUR

There are three methods of obtaining data that are used to understand the evolutionary basis of animal behaviour: intraspecific comparisons, manipulative experiments, and interspecific comparisons.

The most common way in which animal behaviour is studied is by comparison of the behaviour of individuals within a species. Although this approach has provided numerous insights, interpreting the observed patterns is often difficult. For example, observations of wallabies may reveal that they have a different foraging behaviour in grassland and woodland habitats. If the two habitats vary in both the kinds of available food and the risk of predation, it would not be possible to decide which of the two factors is responsible for the differences in the feeding behaviour of the wallabies. Indeed, it may be another, unknown, factor that is responsible.

Experimental manipulations provide a less ambiguous indication of the selective pressures that may have led to a particular behaviour, but they are more difficult to undertake. A simple but effective experiment was performed by Niko Tinbergen, who wanted to know why black-headed gulls, *Larus ridibundus*, removed the empty eggshell from the vicinity of their nests shortly after a chick had hatched. The eggs and hatchlings of these gulls are well camouflaged among the grass and twigs around the nest, which is on the ground. However, the inside of the shell is white and highly conspicuous. Tinbergen thought that the parents' eggshell-removal behaviour might reduce the risk of predation on eggs and chicks. He painted hens' eggs to resemble gulls' eggs and placed them throughout the colony. By placing broken gulls' eggs beside some of these painted hens' eggs, Tinbergen was able to show that the camouflaged hens' eggs were more likely to be discovered and eaten by predators if they had broken eggs near them. The parents' removal of eggshells takes only a few minutes each year, but it is crucial for the survival of chicks because it ensures that both chicks and nests remain inconspicuous.

Most ideas about the evolutionary significance of particular behaviours are derived from comparisons *between* species. The generality of an explanation of why a certain behaviour is found in one species may be derived from an understanding of why it is not found in another species. For example, a comparative study could examine whether the ability to sing complex songs is associated with territorial behaviour (p. 626) by comparing the song of territorial species with non-territorial species.

> Comparison of the behaviour of individuals within a species, and between species, and experimental manipulations are three ways in which the behaviour of animals is studied.

OBTAINING FOOD

Animals obtain their food in extraordinarily diverse ways (Fig. 28.5). Some species, such as the cheetah, *Acinonyx jubatus*, and jumping spiders (Salticidae) are *free-ranging* or *active foragers*, searching for and then pursuing their prey. Other species, such as the death adder, *Acanthophis antarcticus*, or praying mantids (Mantodea) are *sit-and-wait predators*, remaining relatively immobile and attacking only those prey items that venture within striking distance. These predatory strategies represent two extremes and the foraging behaviour of many species may include elements of both. Active foragers also include animals that move considerable distances in search of pollen, seeds and other plant matter.

Despite the diversity of ways in which food is obtained, the foraging behaviour of animals has an important common feature. In general, animals do not forage in a haphazard way, but rather individuals make *foraging choices* that maximise their survival.

(a)

(b)

(c)

Fig. 28.5 Examples of the ways in which animals obtain food: **(a)** a praying mantid feeding on a grasshopper; **(b)** a cheetah chasing a springbok; and **(c)** a sea slug, *Cyerce nigricans*, grazing on turtle weed

These choices (or decisions) might include the proportions of time spent resting and foraging, whether to forage in one place rather than another, the size and kind of food to eat or the time spent foraging in one location. It is convenient to think of a foraging animal as a 'decision maker', although this does not necessarily imply that this is a conscious decision. Rather, natural selection will favour those 'decisions' that represent the best balance of costs and benefits associated with each choice: some food items may be time-consuming to obtain but rich in nutrients, and some locations may have a rich supply of food but may entail greater risk of predation. Viewed in this way, it is possible to evaluate these costs and benefits mathematically and predict which decisions might be expected under different circumstances. This approach is the basis of **foraging theory** (Box 28.1).

Honeybees, *Apis mellifera*, foraging on nectar must, at some stage, return to their nest and empty their crops. Interestingly, bees frequently return to the nest with less than the maximum load of nectar that they can carry in their crops. In fact, a bee carrying a large load expends more energy flying than does one carrying a small load. Thus, while the gross quantity of nectar increases as the bee continues to forage, the net return on a foraging trip will decrease as a result of the additional energy expended in carrying the load. Consequently, the behaviour of bees is expected to reflect a balance between the benefits of obtaining additional amounts of food against the energetic costs of carrying it. Experiments that involved the manipulation of the distances bees had to travel to artificial flowers showed that this was the case: bees returned to the hive with smaller loads when they were forced to fly greater distances to the artificial flowers.

This kind of analysis has been remarkably successful in predicting the foraging behaviour of animals, both in the laboratory and in natural field populations. However, while providing insight, the approach has some limitations. In particular, earlier studies made the common assumption that animals forage in a way that maximises their rate of food acquisition. This means that other influences, such as the risk of predation, were ignored, which may not be a realistic assumption for many foraging animals. For example, the risk of predation may vary between food sites, such that a site with high-quality food may also have a higher risk of predation than, for example, a site with low-quality food. Thus, the foraging animal may have to balance the relative costs and benefits of different sites according to the quality of food and risk of predation at each site. North American grey squirrels, *Sciurus carolinensis*, are sometimes faced with this problem and field experiments show that they apparently balance these costs and benefits. If food items (biscuits in these

---

BOX 28.1   FORAGING THEORY: A SIMPLE PREY–CHOICE MODEL

Foraging animals are often confronted by many kinds of prey, and an important decision is whether to specialise on only one kind of prey (at the exclusion of others) or to take every type of food as it becomes available. The decision will depend upon the costs and benefits of each type of prey. For example, some prey may be easy to find and process but provide little nutrition. Foraging on this prey may exclude the opportunity of obtaining other prey types that provide greater returns. The costs and benefits of prey types, and their influence on the foraging decisions of an animal, can be evaluated mathematically.

Imagine a foraging animal that can feed on two types of prey that contain E_1 and E_2 kilojoules of energy, and where handling times (e.g. the amount of time taken to husk a seed) are h_1 and h_2 seconds. The profitabilities (or benefits) of these prey therefore provide E_1/h_1 and E_2/h_2 energy returns for the time taken to process each. The animal searches for a total of T_s seconds and encounters these two prey types at rates of m_1 and m_2 prey per second. If the foraging animal eats both types of prey, it will obtain the following energy in T_s seconds:

$$E = T_s(m_1E_1 + m_2E_2)$$

The total time to obtain this food will be the searching time plus the handling time:

$$T = T_s + T_s(m_1h_1 + m_2h_2)$$

The forager's rate of food intake is simply the total energy divided by the total time:

$$E/T = (m_1E_1 + m_2E_2)/(1 + m_1h_1 + m_2h_2)$$

Note that the foraging time, T_s, has cancelled out.

If the forager is to specialise on one prey type, then the energy gain (E/T) from foraging only on this prey must be greater than if it were to forage on both types. In other words, if it is to forage only on prey type 1, then E/T from eating prey type 1 must be greater than E/T from eating both prey items:

$$(m_1E_1)/(1 + m_1h_1) > (m_1E_1 + m_2E_2)/(1 + m_1h_1 + m_2h_2)$$

This can be simply rearranged to give:

$$1/m_1 < (h_2 - h_1)E_1/E_2$$

Thus, a forager should specialise on prey type 1 if the time taken to encounter it is less than the difference in handling times multiplied by the ratio of kilojoules obtained for each. Notice that if prey type 2 takes less time to handle than prey type 1, an animal should never specialise on prey type 1 irrespective of the encounter rate or energy content of this prey.

experiments) were placed near trees in which the squirrels could seek refuge from predators, they usually took each biscuit back to the trees and ate them there. But when the biscuits were placed further from the trees, the squirrels brought only large biscuits back to the trees, and ate the smaller biscuits in the open field. The energetic costs of travelling to and from the trees in order to feed on food that has low energetic returns outweighs the costs associated with increased risk of predation. These decisions may be further complicated by an animal's energetic requirements. A small animal with low food reserves may have to forage in an area of higher predation risk if alternative, less risky sites, provide insufficient food.

> Animals forage for food in ways that maximise the benefits of particular kinds of food against the costs of obtaining these food items.

AVOIDING BEING EATEN

Natural selection has favoured numerous anatomical, physiological and behavioural characteristics that are *defence mechanisms* against predators and parasites. Essentially, these adaptations increase the costs to a predator of detecting, identifying, attacking, pursuing, handling and consuming prey. An obvious antipredator behaviour is the ability to flee quickly from predators, thus increasing a predator's pursuit time and energy expenditure. Greater escape speed may mean that the cost of capturing the prey exceeds the energetic return and the predator should pursue another victim. However, escape from predators is not the only way that animals avoid predation. Many insects produce a noxious chemical when attacked, which deters the predator, and lizards may lose part of their tail, which may distract the predator while the lizard escapes.

Some species are well camouflaged or **cryptic** and blend into the background substrate; this deception works only if the animal remains motionless or moves in an appropriate way. Leaf-insects (Phylliinae), for example, look like leaves, and the deception is even more convincing when they move slightly, as if blown by the wind. Being cryptic increases the time taken for the predator to locate the prey, and consequently it may search for other victims.

Many animals resemble other unrelated species. These resemblances may be visual, chemical, behavioural or acoustic and are usually referred to as **mimicry**. There are two main types of mimicry: Batesian and Mullerian. Batesian mimics are palatable (or edible) species that resemble an unpalatable species, while Mullerian mimics are slightly unpalatable species that resemble more strongly unpalatable species. For example, very few predators feed on ants because they are distasteful, and many species

of arthropods mimic ants, thereby reducing the probability of predation (Fig. 28.6). Although mimicry is frequently a mechanism that reduces the probability of predation for the mimic, in some species it has evolved as a mechanism by which predators capture their prey. This is sometimes called *aggressive mimicry*. Bolas spiders (Araneidae) swing a droplet of adhesive (the bolas) attached to the end of a silk thread at their prey (Fig. 28.7). Bolas spiders prey exclusively on male moths, which are attracted to the spider because the bolas contains a chemical substance that mimics the mate-attracting pheromone of a female moth. Interestingly, the spider does not swing the bolas continuously, but only when a moth flies nearby. Apparently, the swinging behaviour of the

Fig. 28.6 Many different kinds of animals resemble ants in appearance and behaviour, such as these juvenile long-horn grasshoppers. These *ant-mimics* may be less likely to be victims of predators, because few animals are predators of ants

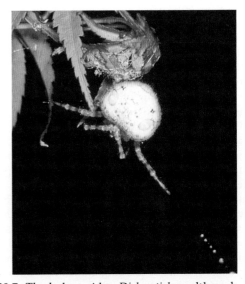

Fig. 28.7 The bolas spider, *Dichrostichus*, although belonging to the orb-web-building family Araneidae, does not build a web but instead captures male moths on the end of the sticky bolas. The moths are attracted to the bolas because it contains a chemical that mimics the sex-attracting pheromone produced by female moths

spider is elicited in response to the vibrations generated by the actions of the moth's wings.

The relatively immobile and nutritious larvae of butterflies are particularly vulnerable to enemies, and a number of defence mechanisms have evolved in this taxa: larvae may be cryptic, hiding during the day and only feeding at night; or they may be highly distasteful by incorporating toxins from the plants they feed on. One intriguing solution has evolved in the lycaenid, or blue butterflies. The larvae of many species of this large family are tended by ants that swarm over each larva, touching it with their antennae. Numerous experiments with the Australian imperial blue butterfly, *Jalmenus evagoras* (Fig. 28.8), which is tended by the small meat ant, *Iridomyrmex*, have demonstrated that large numbers of larvae perished from predation and parasitism if the ants were artificially removed. The ants tend the larva, rather than attack and eat it, because the larva produces carbohydrates and amino acids harvested by the ants. The relationship is *mutualistic* (Chapter 43); the lycaenid gains protection from the ants, which gain a profitable source of food. The larvae may also produce chemicals that mimic part of the ants' communication system, perhaps eliciting tending, rather than aggressive, behaviour in the ants. Indeed, workers of the Australian green tree ant, *Oecophila smaragdina*, carry young larvae of the lycaenid, *Liphyra brassolis*, into their nest. Once in the nest, the lycaenid starts feeding on the ant larvae and pupae, but the parasitic relationship is somehow maintained by the young butterfly larvae fooling the ants.

Fig. 28.8 The larvae of the imperial blue butterfly, *Jalmenus evagoras*, an Australian lycaenid whose larvae and pupae are tended by various species of *Iridomyrmex* meat ants. The ants provide protection against predators and parasitoids, and in return obtain food in the form of carbohydrates and amino acids

Living in groups

Many animals reduce the risk of predation by living in groups. There are several ways in which group living can reduce the risk of predation. The most obvious is that if a predator makes a single attack on a group of prey, and all individuals within the group are equally likely to be caught, then the probability that any individual will be captured is simply the inverse of the group size. This probability is 0.5 for a group of two prey individuals, 0.1 for a group of 10 individuals and 0.01 for a group of 100 individuals. The dilution effect of living in groups is clearly reduced if groups are more conspicuous than solitary individuals, or if predators make multiple attacks on individuals within the group. Furthermore, group members may not gain equal protection if the predator selects particular kinds of individuals, such as juveniles or unhealthy individuals.

An individual at the centre of a group can be protected by others if the predator takes prey from the edge, rather than the middle of a group. This may explain why many group-living animals, such as birds and fishes, commonly form tight flocks when a predator approaches. Such behaviour may also produce a *confusion effect* because it is more difficult for the predator to focus on any one individual in a tight group of rapidly moving individuals. In some species, members of a group may actively defend themselves from predators by *mobbing*. Goannas, *Veranus*, are often mobbed or harassed by sulfur-crested cockatoos, *Cacatua galerita*, if they get too close to the birds' nests, and the goanna usually leaves without destroying the eggs.

Group living may also provide an advantage in detecting predators. The success of many predators depends upon surprise attack, but the chance of a successful attack is reduced if the predator is detected early. An individual could avoid predation by remaining constantly alert, but this is impossible if other activities, such as pecking for food, are incompatible with looking out for predators. Prey individuals must therefore partition their time between looking out for predators and feeding. Early detection of the predator, and hence escape, depends upon how much of the time the potential prey is alert, or vigilant, for predators. By living in a group, each individual can take advantage of the vigilance behaviour of the others. Studies of many species of birds and mammals have shown that a predator is more quickly detected by individuals in groups than by those foraging alone.

If there are so many benefits to living in groups, why do many animals live alone? The answer is that there are costs associated with living in groups. One of the most important is that group living increases the opportunity for competition over resources, such as food.

Swift locomotion, camouflage, mimicry and interactions with other species are examples of defence mechanisms against enemies, such as predators or parasites. Another common way of reducing the risk of predation is by living in groups.

COMPETITION AND TERRITORIAL BEHAVIOUR

If bread is thrown to ducks by two people on opposite sides of a pond, the ducks are likely to distribute themselves between the two sources of food according to the rates at which the bread is thrown into the pond. If one person throws in twice as much food as the other, then that person is likely to attract twice as many ducks as the other. This can be called an **ideal free distribution** of ducks; *free* because the ducks are free to choose between resource sites, and *ideal* because each duck goes to the place that provides the highest returns. For the ducks, the site with the higher rate of food delivery clearly provides more food, but if all the ducks foraged there, each of them would obtain less food because of competition. Clearly, it would pay some ducks to move to the other site where there is a lower delivery rate but less competition. Eventually, in an ideal free distribution, the ducks distribute themselves in the pond in proportion to the food delivery rate.

In this kind of competition, access to the resource depends only upon the number of other individuals competing for it. Later arrivals to the resource are not excluded, and no one individual attempts to monopolise the entire resource.

In some species, it is very common for one individual to exclude actively others from the resource. This may be achieved by establishing a **territory**, or area that contains the resource, and defending it against other individuals (Fig. 28.9).

Territories often have clearly defined borders; birds, for example, may sing along the borders of their territory and chase off any intruder that crosses that border. The kinds of resources defended within a territory vary; for example, male frogs and butterflies may defend areas in which females lay eggs, birds defend nesting sites, and ants defend areas that contain rich food resources.

Fig. 28.9 Magpies, *Gymnorhina tibicen*, are vigorous defenders of their territories

Territorial behaviour

The reason why some species are territorial and others are not may be found in terms of the costs and benefits of defending a territory. Territory-holders do not have to share their resource, but instead must endure the costs of excluding others, including the time and energy required to chase off intruders. Both costs and benefits are likely to increase with larger territories, but not necessarily at the same rate. Thus, there may be a territory size that maximises the net benefit (Fig. 28.10). Larger territories may increase benefits, but at some point the additional benefits will be outweighed by higher costs.

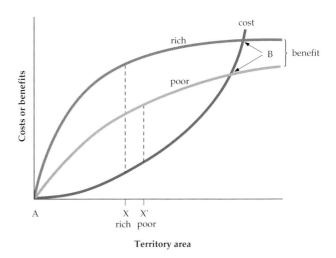

Fig. 28.10 The costs and benefits of defending a territory increase with territory size but at different rates. The benefits of defending territories of different sizes will depend upon the quality of the resources within the area; smaller territories in quality habitats may provide as much benefit as larger territories in poor quality habitats. There is likely to be a territory size X and X' at which the net benefit (benefits minus costs) is maximised. The resource is economically defendable between A and B

Territories are not always defended by one individual. In many species, a male and female may jointly defend a territory, and in some species, such as the Australian grey-crowned babbler, *Pomatostomus temporalis*, family groups defend territories. Joint territoriality is frequently associated with breeding, and not necessarily related to the economic defence of resources for each individual.

> Animals may defend areas containing food, nesting sites or other resources, thereby forming a territory. These territories are only defended if the benefits of exclusive use of the resource exceed the costs, in time and energy, of excluding others.

ANIMAL CONTESTS

Competition between individuals of the same species for particular resources, such as food, nesting sites and sexual partners, frequently involves direct conflict or *contest* between individuals. Some contests, such as fights between male red kangaroos, *Macropus rufus*, are obvious (Fig. 28.11), and in some species may result in serious injury. For example, between 5% and 10% of musk ox bulls, *Ovibos moschatus*, die each year as a result of injuries sustained through fights. However, not all contests between individuals are as dramatic or obvious. Sea anemones, while appearing to be sedentary, often engage in fights using specialised tentacles, called acrorhagii, as weapons against each other.

Animal contests that result in physical injury are not common, and most disputes are resolved before there is serious injury to either contestant. Examination of the costs and benefits of different kinds of contests reveals why this is the case. Winning a contest will provide benefits in terms of the resources obtained, but may also incur costs in terms of time

Fig. 28.11 A contest between two male red kangaroos, *Megaleia bufa*. The males grab each other with their front legs and attempt to kick each other with their hind legs while using their tails for balance. The claws on the hind legs are capable of causing a serious injury to the opponent

or injury. For example, if the costs of injury to a contestant that defends its resource are greater than the value of that resource, the contestant should relinquish the resource before escalation of the fight. The relative costs and benefits of the different behaviours can be examined using ideas developed in a mathematical technique called **game theory**, which is also used commonly in economics (see Box 28.2). Game theory models show why escalated contests involving physical injury are unlikely to be maintained within a population.

In reality, the outcome of contests between individuals depends upon a variety of factors, including the fighting ability of the contestants, the value of the resource to each contestant, and which party currently owns the resource. Clearly, it would be to a contestant's advantage to be able to assess the fighting ability of another individual before engaging in combat, and there are numerous examples of contestants advertising their fighting ability, by singing, calling, roaring and so on, before the fight escalates. These displays may escalate into physical contests if the value of the resource exceeds the costs of fighting, perhaps in terms of the likelihood of injury. However, if the risk of injury to the attacker exceeds the value of the resource, then the attacker may leave.

> Competition between individuals for scarce resources can sometimes result in direct conflict or physical contest. These contests can result in physical injury, but most disputes are resolved before this occurs.

COURTSHIP AND MATING BEHAVIOUR

A crucial requirement for many sexually reproducing animals is finding and mating with a partner of the same species. Males and females usually come together as a result of one sex advertising its location to the other using chemical, auditory and visual cues: female moths produce a sex pheromone that attracts males; male frogs attract females by calling; and many species of male fireflies attract females by producing flashes of light. An important feature of these sex-attracting mechanisms is that they are *species-specific*. Thus, species of frogs have calls that are distinct from the calls of other species that are found nearby, and the chemical composition of moth sex pheromones differs between species. These differences ensure that the advertising cue attracts individuals of the same species, thereby reducing the possibility of either sex wasting time or energy attempting to fertilise eggs with the sperm of another species. This is particularly important for the female, who may lose her entire reproductive output, whereas the male can at least attempt to mate with another female.

BOX 28.2   HAWKS AND DOVES: FIGHTING BEHAVIOUR AND EVOLUTIONARILY STABLE STRATEGIES

Assume that two contestants can engage in a contest in one of two ways. A 'hawk' strategy means an individual will fight until its opponent either dies or is too seriously injured to continue, and a 'dove' strategy means an individual only displays and never engages in serious fights. When a 'hawk' encounters a 'dove', the 'hawk' always wins because the 'dove' always retreats without fighting. When a 'hawk' encounters a 'hawk', they fight and we assume that each wins on half the occasions. Similarly, a 'dove' encountering another 'dove' always displays and wins on half the occasions. Although imaginary, these two 'strategies' represent the two possible extremes of fighting behaviour.

We can investigate whether each of these strategies can be maintained in the population by assigning scores to the outcomes (winning or losing) and the costs (of injury and time wasted in a display) of fights between individuals. As an illustration, we can assign the following imaginary numbers that reflect some measure of fitness as a result of the conflict. Let the value of the resource equal 40. Therefore, the winner scores +40 and the loser 0. The cost of injury equals −60; and the cost of displaying equals −10. We will further assume that hawks and doves reproduce their own kind, and the number of offspring produced is in proportion to their overall scores. A higher score at the end of the contest is the equivalent of producing more offspring. The exact values do not matter; the game could be analysed using algebra. The pay-offs to the attacker for each strategy played against the other strategies can be shown as a matrix:

	Opponent	
	Hawk	Dove
Attacker		
Hawk	$0.5(40) + 0.5(-60)$ $= 10$	+40
Dove	0	$0.5(40 - 10) + 0.5(-10)$ $= +10$

It is now possible to examine what would happen in the population if individuals played different strategies. If all individuals played 'dove', then the average pay-off for each individual would be +10. However, a mutant 'hawk' individual introduced into this population would be at a great advantage, and the 'hawk' strategy would rapidly spread. But once the population comprises mostly hawks, doves would be at an advantage because the average pay-off for the 'hawk' strategy is −10, which is less than 0, the average pay-off when a dove encounters a hawk.

Clearly, it is best to be a hawk in a population comprising mostly doves and a dove in a population consisting mostly of hawks. In each situation, selection will favour the rarer strategy. Nevertheless, a stable mixture of hawks and doves can occur, when the average pay-off for hawks equals that for doves. The stable mixture can be calculated quite simply. Let h be the proportion of hawks. The proportion of doves will then be $(1 - h)$. The average pay-off is the pay-off for each type of fight multiplied by the probability of meeting that type of opponent:

average pay-off to hawks, $H = -10h + 40(1 - h)$
average pay-off to doves, $D = 0h + 10(1 - h)$

The stable mixture of hawks and doves occurs when $H = D$, that is,

$$-10h + 40(1 - h) = 0h + 10(1 - h)$$

which is true when $h = 0.75$. Thus, the population is stable when three-quarters of the population are hawks, and one-quarter doves. This mixture could arise in two ways. The population could comprise individuals that played *pure* strategies of either hawk or dove, with three-quarters of the population being hawks and one-quarter doves. Alternatively, the population could comprise individuals that played *mixed* strategies of hawks and doves, with every individual playing hawk three-quarters of the time and dove one-quarter of the time. Either scenario would result in an **evolutionarily stable strategy**, which is a strategy that cannot be beaten by any other strategy in the game. The important point about this very simple model is that the behaviour of one contestant can depend critically upon the behaviour of other individuals within the population.

Courtship

In many animals, **courtship behaviour** takes place after the two sexes have located each other. Courtship behaviour varies dramatically between species, ranging from little more than the male moving directly towards the female, to an elaborate display of vocalisations and body movements. Courtship behaviour may provide additional confirmation that a male and female are of the same species if there are particular aspects of the behaviour that are species-specific, such as the leg-waving behaviour of many jumping spiders. The female's response to a male's courtship behaviour can also indicate to the male whether the female is ready to mate.

The courtship displays of many animals are far more complex and elaborate than might be expected if their sole function were to indicate species identification and readiness to mate. For example, the courtship display of the raggiana bird of paradise,

Paradisea raggiana, involves an extraordinary combination of colour, movement and vocalisations (Fig. 28.12). The male Prince Rupert's blue bird of paradise, *Paradisea rudolphi*, sings from a perch and then slowly rotates backwards. Once he is hanging upside down, he shakes himself with repeated movements, extending his plumage and making a soft monotonous song.

The extraordinary courtship displays of many animals have most likely evolved as a result of **sexual selection** (Box 28.3). Sexual selection favours those characteristics in males that improve their chances of mating success through the process of either *male–male competition* or *female choice*. The former process does not involve courtship, but the latter refers to female discrimination between males on the basis of the male's secondary sexual characteristics, which may include courtship displays. The more spectacular courtship displays occur in birds: a male Australian musk duck, *Biziura lobata*, courts a female by leaning back and spraying her with water using his legs; male superb fairywrens, *Malurus cyaneus*, present females with a flower petal; and male superb lyrebirds, *Menura novaehollandiae*, spread their elegant tails above their heads while courting the female.

Fig. 28.12 A male raggiana bird of paradise, *Paradisea raggiana*. The female's plumage is quite dull in comparison. The colourful plumage and extravagant courtship behaviour of the male has evolved through sexual selection by female choice

BOX 28.3 SEXUAL SELECTION AND SECONDARY SEXUAL CHARACTERISTICS

The theory of sexual selection was first proposed by Charles Darwin, who was interested in explaining the extravagant and often brightly coloured adornments that are characteristic of the males of many species, such as the peacock, *Pavo cristatus* (Fig. a). The problem for Darwin was that these adornments seemed unlikely to have evolved through natural selection (Chapter 32) as a result of improved survivorship for the male. In order to understand why these adornments may have evolved (and why it is mostly males that possess them) it is necessary to appreciate the fundamental difference between the sexes: males produce sperm and females produce eggs. A single egg is more expensive to produce than a single sperm, and a male has the potential to fertilise far more eggs than a female can produce. This means that while female fecundity depends on her ability to produce eggs, male reproductive success depends on the number of mates he can secure. Females are a limiting resource for male reproductive success because there are more males capable of mating than there are females available.

The difference between the sexes means that males compete with each other for access to female partners. The rewards for male success in this competition, in terms of reproductive success, are high and hence

(a) The stunning appearance of the male peacock, *Pavo cristatus*, is thought to have evolved as a result of positive feedback between female choice and elaborate plumage in males

selection for male ability to secure matings is strong. Selection of characteristics that increase mating success is referred to as *sexual selection*. Sexual selection can occur in two ways: either by *intrasexual selection*, which favours the ability of one sex (usually, but not always, the male) to compete directly with other members of the same sex; or by *intersexual selection*, which favours characteristics in one sex that attract the other sex.

Intrasexual selection favours any trait that increases a male's competitive ability, and it is often referred to as male–male competition. The most obvious traits are weapons for fighting, such as horns on stag beetles and many ungulates (Fig. b). However, there are other, more subtle mechanisms that have evolved through male–male competition. Invertebrates have a particularly diverse range of mechanisms to prevent females, once mated, from mating with other males. For example, males of the Australian big greasy butterfly, *Cressida cressida*, deposit a substance over the genital opening of the female while mating with her. The substance hardens to form a permanent structure, known as a sphragis, that prevents other males mating with her, and thereby ensures that only his sperm fertilises her eggs.

Intersexual selection occurs when one sex shows a preference for certain individuals of the other sex. The sex that provides the greatest material investment in offspring is the one that shows the preference. In most species, the female provides the greater investment, and hence intersexual selection is often referred to as *female choice*. Females may select males on the basis of their

ability to provide resources, such as food and oviposition sites, and on the basis of their genetic qualities (in addition to their ability to provide resources). This form of female choice is thought to be responsible for the evolution of the elaborate adornments and courtship displays of males of many species (Fig. a).

The evolution of secondary sexual adornments as a result of female choice occurs through a positive feedback mechanism. Initially, females may have had slight preferences for males possessing a certain characteristic. The characteristic might be colourful feathers in birds, and a male with more colourful feathers will have a mating advantage. Thus, this male will be preferred over other males, and a female that prefers to mate with such a male will produce sons that have more colourful feathers (and hence a mating advantage) and daughters that have her preference for such males. A positive feedback between female preference and male characteristics continues indefinitely, unless the mating advantages of the characteristic are exceeded by a decrease in male survival. This opposing effect of natural selection on male survival may explain why the extravagant tail feathers of some birds are not even more extravagant. While there is good evidence that females have preferences for certain kinds of males, it is still not clear how the feedback loop started.

(b) A red-deer stag, *Cervus elaphus*, and a male stag beetle; both have armaments that are used in contests with other males over access to females

In some species, courting males provide females with a **nuptial gift**. These gifts may be a prey item or a 'package' containing material synthesised by the male. Males of several species of birds provide the female with a prey item (Fig. 28.13), and this may reflect the ability of the male to provide for the offspring later. Male lygaeid bugs improve their reproductive success by providing females with a nuptial gift that contains nutrients that the females use to produce more or larger eggs (Fig. 28.14). Courting males of most species of empiid midges can only mate with the female if they present her with an insect gift that is wrapped in silk. However, in one species, the male cheats by presenting the female with an empty silk cocoon. Although there is no nutritional value to this gift, the male must still present it if he is to mate with the female.

Fig. 28.13 Male azure kingfishers, *Alcedo azurea*, offer prey items to the female during courtship

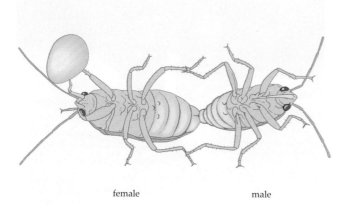

female   male

Fig. 28.14 The male lygaeid bug collects and predigests an edible seed as a nuptial gift for his partner. After giving his gift to the female, she permits him to copulate while she consumes the gift

Courtship may also provide a mechanism for reducing the risk of cannibalism in **sexually cannibalistic species**, where the female attempts to consume the male. Males of some orb-weaving spiders attract the female onto a specially constructed mating thread, where the risk of cannibalism is lower than mating on the orb-web because the female is suspended from only one thread. However, courtship can only function as a means of reducing cannibalism in those species in which the cannibalism occurs before mating. It is therefore ineffective in *post-mating* sexually cannibalistic species, such as nudibranchs, praying mantids and several spiders (most notably the red-back spider, *Latrodectus*). It is often stated that sexual cannibalism in praying mantids is necessary for the male to copulate effectively. In fact, males are perfectly capable of mating successfully without losing their lives. Sexual cannibalism in these species most likely occurs when the female has been deprived of food and devouring the male may increase her fecundity.

Competition between males does not always cease once copulation has taken place; there are many male **mate-guarding behaviours** that prevent other males from fertilising the female's eggs. Mate-guarding may occur before or after mating has taken place. Pre-mate-guarding occurs if the sperm from the first male to mate with the female fertilises most of the eggs, whereas males guard females after mating if the sperm from the last male to mate fertilises most of the eggs. For example, in many damselflies, a male will guard a female by grasping her thorax with the tip of his abdomen in the so-called tandem or wheel position. This prevents other males from subsequently mating with the female and thus protects his sperm. Males of the linyphiid spider, *Linyphia*, destroy the female's web before mating takes place. The silk of the female's

web contains sex pheromones that attracts other males, so, by destroying her web, the male reduces the possibility that other males find and mate with her.

Males often exhibit elaborate and complicated courtship behaviours after a female has been located. These behaviours can enable each sex to determine whether they are of the right species; provide the female with a basis from which to choose between males; and, in some species, may reduce the risk of sexual cannibalism.

MATING SYSTEMS AND PARENTAL CARE

In many animals, one or both parents may provide some form of **parental care** for their offspring, such as feeding them, or protecting them from predators. Whether the female alone, the male alone, or both parents care for the offspring depends on the species. For example, in seahorses, it is the male that provides parental care by carrying the fertilised eggs in a special brood pouch (Fig. 28.15).

Fig. 28.15 Male seahorses, *Hippocampus*, brood care for their developing offspring by maintaining and feeding them in a special pouch. Here the offspring are being born

Which parent provides parental care seems to have an important influence on the *mating system* of the species. Mating systems are defined according to the number of partners each sex may have during its lifetime or during the mating season. There are four main categories of mating system: **monogamy**, where the male and female form a relatively long-lasting pair-bond; **polygyny**, in which a male mates with several females, but a female mates with only one male; **polyandry**, the reverse of polygyny; and **promiscuity**, where both males and females may mate with several individuals.

The relationship between the nature of parental care and the mating system, shown in Table 28.1, reveals a conflict of interest between the sexes. Essentially, the sex that does not provide care deserts its partner and mates with other individuals. In mammals, females are predisposed to providing care: the offspring have a prolonged gestation and, after birth, the mother must provide the infant with milk. There is little parental care that the male can provide, other than protecting the female and her young and, in general, mammals are polygynous.

Fig. 28.16 Male and female wandering albatrosses may court for several years before mating and attempting to produce their first clutch of eggs. They will then remain mates for the rest of their lives

Table 28.1 General association between vertebrate mating systems and parental care

Parental care	Mating system	Taxonomic example
Male and female	Monogamous	Many birds
Only females	Polygynous	Most mammals
Only males	Polyandrous/ promiscuous	Some fishes

In most species of birds, the number of offspring that can be reared successfully depends upon the amount of food that is provided by the parents. Since two parents can provide more food than one, both parents could increase their reproductive success by staying together. If one parent deserted, its reproductive output may not only be reduced but it would also have to spend additional time finding a new partner. Thus, in birds, monogamy and care provided by both parents is the general rule. These monogamous relationships can last a lifetime in some species, particularly many seabirds (Fig. 28.16).

Guarding and fanning eggs and brood are the primary forms of parental care shown by fishes and these can be done by either parent alone. The question of which sex provides care seems to depend partly upon whether the species has internal or external fertilisation. If fertilisation is internal, the female cares for the offspring because the male can desert immediately after mating, and the mating system is polygynous. If fertilisation is external, the female can desert before fertilisation has taken place and the eggs may be cared for by the male, resulting

in a polygamous mating system. Why the male cares for the eggs is not clear, but it may be because the female lays eggs in the male's territory and thus defence of the eggs, particularly from predation and cannibalism, is incidental to defending the territory in order to attract other females.

There are many exceptions to the patterns described in Table 28.1, and their explanations are still controversial. Nevertheless, it is generally agreed that if females can be easily defended from the sexual attentions of other males, then selection may favour polygyny. The ability of males to defend more than one other female will depend upon the distribution of sexually receptive females in both space and time.

> The number of mating partners that is typical for each sex defines a species' mating system. The different kinds of mating systems for different species seem to depend, in part, on which sex provides most of the care of the offspring.

SOCIAL ORGANISATION AND CO-OPERATIVE BEHAVIOUR

In most species, individuals live a generally solitary life, only briefly pairing up with a member of the opposite sex for mating and reproduction. In others, individuals may form loose aggregations, as a result of their attraction to a particular resource, such as a water hole or rich supply of food, rather than to each other. Vinegar flies, *Drosophila*, that congregate around rotting fruit are an example of this kind of aggregation. The size and membership of these aggregations usually varies over time, and although individuals within these groups may frequently interact with each other, these interactions are likely to be competitive in nature.

Aggregations, flocks, or schools of individuals can become established because individuals benefit more by being in groups than being alone. One advantage of living in a group is that individuals gain protection against predators (p. 625). Another advantage is that individuals in groups may gain protection from the cold (Fig. 28.17). Group living may also provide benefits in terms of obtaining food. Individuals can take advantage of the food-finding abilities of others or, by foraging together, may increase the chance of capturing prey. The latter is a common feature of the foraging behaviour of social carnivores, such as lions, hyenas and hunting dogs; some individuals may drive prey towards others that are concealed by vegetation, or may share in the task of chasing the prey until it is exhausted.

Fig. 28.17 Emperor penguins huddle in groups to reduce the chill factors of Antarctic winds

More complex *social behaviour* is found among individuals that form essentially permanent groups. Individuals in such groups usually **co-operate** in finding, hunting and processing food, defending themselves against predators and competitors, and in rearing young. Social behaviour occurs in a diversity of animals, including spiders, insects, birds and mammals. Individuals of the Australian social thomisid spider, *Diaea*, co-operate in building nests by weaving together *Eucalyptus* leaves with their silk. Nests may contain over 50 individuals, which shelter in the nest by day and co-operate in capturing insect prey that walk on or near the nest at night. Individuals probably live together in social groups because their co-operative activities result in individuals obtaining more prey, or because they are less likely to be victims of predators than if they lived alone.

The composition of *social groups* varies between species, resulting in considerable diversity in the ways in which individuals interact with each other. Interactions between individuals within a *Diaea* nest, for example, do not seem to vary, perhaps because the interests of each individual are similar. In contrast, the interactions within groups of primates, such as chimpanzees or baboons, are considerably more variable, leading to a more complex social organisation.

The relationships between individuals within a group primarily reflect the conflicts and congruences of interests that arise within these groups. Social groups usually comprise one or a few breeding individuals, infants, juveniles and in many cases a number of other adults that do not raise their own offspring (see below). There are exceptions: some groups do not contain non-breeding adults and, in some species, males may, during part of their lives, form bachelor groups. The number of breeding individuals varies, ranging from one male and one female, through one male and several females, to a few males and numerous females. The principal reason for the conflicts of interests within a group is the difference between individuals in the opportunity for breeding.

Dominance relationships between individuals in a group arise when one individual gains access to the contested resource at the expense of another. Domestic chickens, for example, have a linear dominance hierarchy in which one individual is the most dominant in the group and is capable of displacing all others. Below her is a second-ranking bird who can dominate all but the top-ranking bird, and so on until the most subordinate bird, which is displaced by all other birds. Chickens are unusual in having this kind of dominance hierarchy. Primates, for example, may have more complex hierarchies, such that individual A may dominate individual B and B dominate C, but A may not necessarily dominate C. An important feature of dominance hierarchies is that, while there may be considerable antagonistic or aggressive behaviour between individuals during the establishment of the hierarchy, this behaviour rarely continues.

Although individuals within a social group co-operate in various ways, some adults in the group do not themselves reproduce, but instead help raise the young of the breeding individuals. This behaviour is called co-operative breeding and is found in insects, birds and mammals.

> Individuals of many species form permanent groups and exhibit a variety of social behaviours, including co-operative hunting, defence and rearing of offspring. Social groups also generate competition over resources and opportunities for breeding.

Co-operative breeding in birds and mammals

Species of birds and mammals in which adult individuals regularly help rear the young of other individuals are called **co-operative breeders**. Helpers may provide the young with food or protection from predators. In some species, such as the Florida scrub jay, *Aphelocoma coerulescens*, or the silver-backed jackal, *Canis mesolmelas*, in Tanzania, helpers are young individuals that do not disperse from their parents' territory but instead help their parents raise the next brood of offspring. Although this is probably the most common system, other species have more complex social organisations. In some species, such as the Australian bell miner, *Manorina melanophrys*, groups are larger and comprise several 'nuclear' families that live together in a single large territory. Other co-operative breeders, such as the white-fronted bee-eater, *Merops bullockoides*, are highly gregarious and live in large colonies. Smaller extended family groups, including helpers, exist within the colony but, unlike miners, helpers attach themselves to a single nest. The social organisation is more complex in species in which the breeding individuals are not mono-gamous but males and females mate with several partners. This forms the basis of the social organisa-tion of some primates, such as saddle-backed tamarins, *Sanguinus fuscicollis*.

Co-operative breeding occurs in about 3% of all the species of birds in the world, but curiously over 22% of the Australian passerine (perching) birds breed co-operatively (Fig. 28.18). Why is co-operative breed-ing so common in the Australian bird fauna, compared with the rest of the world? The answer most likely lies in the biogeographical history of Australian birds (Chapter 41), rather than any peculiar feature of the Australian environment that might favour co-oper-ative breeding. Recent genetic evidence indicates that the Australian passerines have radiated from two groups: an old endemic radiation called the Corvi, and a more recent group of immigrants from Asia (Chapter 41). The Australian co-operative breeders

Fig. 28.18 The co-operatively breeding superb fairywren, *Malurus cyaneus*. This and several other co-operatively breeding Australian birds, including magpies, butcherbirds and some honey-eaters, are commonly seen in urban areas. This male has just fed the chicks in the nest

all belong to the Corvi, and co-operative breeding is not found in any of the comparatively recent in-vaders. Thus, the high frequency of co-operatively breeding birds in Australia is not the result of repeated evolution among several taxa, but rather of speciation from a common, co-operatively breeding ancestor.

The most extraordinary co-operatively breeding vertebrate is the naked mole-rat, *Heterocephalus glaber*, of Africa, whose social system was first reported in 1981. Naked mole-rats live underground in colonies of up to 300 individuals. The colony comprises one breeding female, who mates with several males. These males provide most of the direct care of the young. Other individuals in the colony, including the non-breeding females, defend and maintain the under-ground tunnel system. Younger individuals bring food to the breeding nests and generally keep the tunnels clean, while older individuals excavate new tunnels and defend the colony from predators, such as snakes. Why this lifestyle has evolved in the naked mole-rat is still a matter of contention, but this species is the closest mammalian analogue of a form of social organisation that is found among two groups of insects: the hymenoptera (bees, ants and wasps) and termites.

Insect societies

The social systems found among insects are among the most diverse and spectacular. The most simple social system involves groups of offspring that are cared for by their mother for extended periods, as in the female intertidal beetle, *Bledius spectabilis*, which remains in her mud-burrow with her young in order to prevent them from suffocating. At the other end of the continuum are the complex eusocial societies of many species of bees, wasps, ants and all termites. **Eusocial societies** are characterised by overlapping generations of individuals and reproductive division of labour, where some individuals reproduce and others do not. Such societies can be vast. A large ant or termite colony can contain as many as 10^5–10^7 individuals and occupy a complex home that is 10^9 times as large as any of the individuals within it (Fig. 28.19). Typically, these colonies are based on the reproductive output of a single *queen*. Her daughters, called *workers*, undertake duties, such as maintenance of the colony, obtaining food and caring for the queen's brood. Not all eusocial species have a single queen; many species have several queens in one colony. Colonies of some species of Australian *Iridomyrmex* ants can be very large, comprising several widely separated nests connected by trails. The queens can be found either in one central nest or distributed throughout the smaller satellite nests. Among ants, bees and wasps, workers are all female and males make no contribution to nest maintenance.

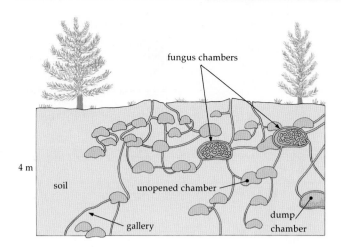

Fig. 28.19 The Texas leaf-cutter ant, *Atta texana*, constructs very large nests. One in the southern United States, excavated by a bulldozer, was over 4 metres deep and contained 110 chambers. The ants take leaf fragments below ground as a substrate for culturing fungus. The ants eat parts of the fungus and remove wastes to dump chambers

Typically, a colony is founded by a single mated female whose eggs hatch into workers that both forage and build the nest. The founding queen continues to produce workers and the colony increases in size. When the number of individuals in the colony reaches a certain size, which varies from species to species, the queen starts producing a few females that will eventually leave the nest as 'reproductives', or female *alates*. Males are also produced at this time, and the line is continued when these male and female alates leave the nest to mate and found new colonies.

In some species, individuals within the colony have different morphologies and perform different tasks. These different morphs are called *castes*, and the range and number of castes within a species can be quite spectacular (Fig. 28.20). Soldiers are usually larger than others in the nest and their large mandibles are effective in defending the nest from enemies. In some species, such as honey-pot ants (e.g. *Camponotus*), some individuals are essentially food containers, hanging from the roof of the nest with their abdomens distended with food. These ants live in arid zones where food availability is unpredictable. Foraging ants return to the nest and regurgitate their food to the 'honey-pots', which then effectively store it in their bodies. During times of food shortage, the members of the colony can obtain food from the 'honey-pots' through regurgitation (Fig. 28.21).

The organisation of colonies of eusocial insects, including the care of young, construction and maintenance of the nest, collection of food and provisioning of colony members, is mediated by a complex communication system. When the nest is threatened by enemies, some individuals release an *alarm pheromone*, and soldiers rush to the location of

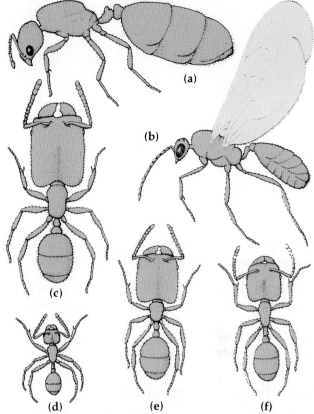

Fig. 28.20 The different castes of the myrmicine ant, *Pheidole tepicana*. The proportions of the body (and especially the head) change dramatically with size from the queen (**a**), male (**b**), and four kinds of workers (**c–f**)

Fig. 28.21 Honey-pot ants, *Camponotus*

the threat, while other workers will move the queen to the safest place in the nest. The location of food is indicated by chemical trails that workers leave on the substrate. Not all species rely entirely on chemical pheromones; honeybees communicate to nest mates the precise direction, distance and nature of food sources using a complex system of dances (Fig. 28.22).

Superficially, the nests of these eusocial insects may resemble a superorganism, in which the individuals are analogous to the cells of a single organism. However, while individuals within a nest are engaged in

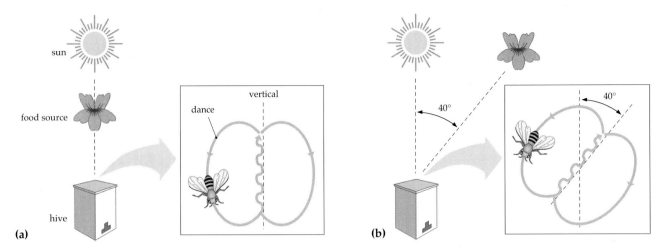

Fig. 28.22 The dance of the honeybee, *Apis mellifera*, a complex behaviour that conveys a variety of information about the source, location and quality of different food sources

extraordinarily co-operative relationships, the opportunity for behaving in more selfish ways, such as cannibalising eggs, occurs frequently in many eusocial species.

Evolution of co-operation

The evolution of co-operative breeding behaviour is puzzling because natural selection is not expected to favour behaviour that reduces an individual's reproductive success. Individuals that help others raise young are apparently forfeiting their own reproductive success while enhancing the reproductive success of others. The puzzle may be resolved for co-operatively breeding vertebrates by understanding why some individuals delay breeding; perhaps the younger helpers gain experience in raising offspring, which benefits them later in life. Alternatively, there may be insufficient breeding territories available, and the helpers must wait for a vacancy before breeding.

Explanations for the evolution of eusociality in insects are more complex because, with few exceptions, the workers remain workers and never become queens. Unlike co-operatively breeding birds or mammals, workers do not delay reproduction, they simply do not reproduce. We still do not understand fully the evolution of eusocial behaviour in insects, but three kinds of explanation are currently favoured. The first suggests there is a mutual benefit of helping to both the parents and workers. This idea is equivalent to the explanation of co-operative breeding in vertebrates, and only applies to those species in which workers are capable of producing eggs. The second explanation is that worker progeny are manipulated into helping by their parents. The mechanism by which parents are able to prevent workers from dispersing and reproducing but instead to remain in their natal nest as helpers is not fully understood.

The important point is that the workers' behaviour is not necessarily co-operative but rather the result of selection favouring parental self-interest.

The third explanation emphasises genetic considerations and is derived from the concepts of *inclusive fitness* and *kin selection* (see Box 28.4). These concepts reveal that co-operative behaviour between individuals that are close genetic relatives is likely to evolve because they share a higher proportion of genes, including those favouring this behaviour, than do individuals that are not closely related. Hence, the gene for co-operative behaviour can spread in the population. Eusociality has evolved independently several times in Hymenoptera and this may be due to their peculiar haplodiploid genetic system, in which females are diploid but males are unfertilised and therefore haploid. If the queen only mates with one male, then her daughters are more closely related to each other than they are to the queen. Thus, selection for helping raise sisters, at the expense of one's own offspring, may be more easily favoured.

These different kinds of explanations are not mutually exclusive, and the importance of each will depend upon the biology of each species. Nevertheless, the emerging consensus is that this extreme form of co-operative behaviour is not such an evolutionary puzzle. At best, the non-reproductive workers (or helpers at the nest) gain at least some level of reproductive success, while at worst their behaviour is the result of parental manipulation. The outcome is that among these apparently co-operating individuals there are considerable conflicts of interest, as is so frequently the case with other features of animal behaviour.

> The evolution of co-operative breeding is intriguing because natural selection is not expected to favour behaviour that reduces an individual's reproductive success. Several plausible explanations for this behaviour have been proposed.

BOX 28.4   INCLUSIVE FITNESS, KIN SELECTION AND CO-OPERATION

Inclusive fitness is a concept of evolutionary theory and particularly the evolution of co-operative behaviour. The logic behind this concept is that selection acts at the level of the gene rather than the individual; only genes are passed on from generation to generation. Individuals can be thought of simply as vehicles for genes to be represented in the next generation. For example, genes for parental care are likely to spread in the population because genes that ensure their 'bodies' are more effective at caring for their young will perpetuate themselves in the bodies of the cared-for young. However, while the only way that genes can be passed from generation to generation is by parents to offspring, care of one's own offspring is not the only behaviour that can ensure that genes for parental care will spread in the population. Genes for helping a relative raise offspring can also spread in the population. This is because close relatives share a higher proportion of genes than two distantly related individuals, and consequently have a higher chance of sharing the gene for this behaviour. Genes for helping sisters or brothers thus help the same genes carried by these relatives and can perpetuate themselves through the children of these relatives. Selection that takes account of relatives as well as direct descendants is termed *kin selection*.

Helping a relative reproduce will only be favoured if the benefit of helping is greater than the cost (i.e. reduction in reproductive success) to the helper. Helping a sister raise offspring is unlikely to evolve if the help does not result in an increase in the sister's reproductive output, and it also reduces the reproductive output of the helper. Clearly, the degree of relatedness between

individuals is important. Individuals that forfeit reproduction to help another unrelated individual raise its offspring are unlikely to be maintained in the population, because the gene for this behaviour is unlikely to be passed on to the next generation.

Balancing the costs and benefits of an individual's behaviour, and adjusting it according to the genetic relationship between the two individuals represents its inclusive fitness. If r is a measure of the relatedness between the individuals, and b and c are the benefit and cost respectively of the behaviour to the donor, then the donor is expected to help another when $rb - c > 0$. The benefit of the behaviour in terms of the recipient's reproductive output is adjusted by the degree of relatedness between the two, while the cost is the number of offspring the donor fails to produce as a result of performing the action. This simple mathematical inequality is known as Hamilton's rule, after W. D. Hamilton who developed the idea.

Biochemical techniques, such as DNA fingerprinting, can provide good estimates of the genetic relationships between different individuals. However, measuring the costs and benefits of co-operative behaviour is more difficult. For example, it is possible to measure the benefits of helping by comparing the number of offspring produced by parents with and without helpers, but it is not always possible to establish the number of offspring a helper might have produced had it not helped. Nevertheless, the ideas of kin selection and inclusive fitness have been extremely important in helping understand the evolution of co-operative behaviour and the social organisations that derive from this behaviour.

SUMMARY

- Ethology is the study of animal behaviour and its aims are to explain animal behaviour in terms of causation, development, adaptive value and evolutionary history.

- Behaviour, being an integral part of an animal's phenotype, has a genetic basis and is consequently subject to the influence of natural selection.

- An animal's behaviour may change during its lifetime; some changes develop with age while others occur as a result of learning.

- Animals obtain food in diverse ways, and for many species they must make choices about where, what and when to eat particular food items. These choices often reflect a trade-off between the benefits of particular foods and the costs of obtaining them.

- Numerous mechanisms have evolved in response to reducing the risk of predation, including camouflage, mimicry, living in groups, or even forming mutalistic associations with other species.

- Animals often compete for resources, but individuals can ensure exclusive use of these resources by defending territories from intruders. The size of a territory and its owner's tenure will depend on the balance between the benefits of the resource and the costs of its defence.

- Conflicts over access for resources can result in physical contests, however, they rarely result in serious injury to either contestant if the benefits from obtaining the resource are exceeded by the cost of injury.

- Males and females locate each other for mating using a variety of auditory, chemical and visual cues. Courtship behaviour subsequently takes place, which in many animals functions to allow species' identification and the female to choose her mating partner.

- The number of mating partners each sex has during its lifetime or during a mating season defines the species' mating system. There is a general pattern across species between which sex cares for the young and their mating system.

- Complex social behaviour is found in species that form essentially permanent groups, and individuals in these groups may co-operate in obtaining food, defence against predators, constructing places in which to live, and caring for offspring.

- Individuals of some species forfeit breeding in order to help raise the offspring of other individuals. The evolution of this behaviour is still not fully understood.

QUESTIONS

1. Define the four suggested answers to the question 'Why do male bowerbirds construct and decorate bowers?' (p. 618) as either ultimate or proximate explanations. Explain your reasoning.

2. What evidence is required to demonstrate that a particular behaviour has a genetic basis?

3. How can behaviour that has been learned be distinguished from that which has developed?

4. Compare the advantages of experimental approaches to understanding animal behaviour and purely observational studies.

5. What kinds of 'decisions' are made by a foraging animal? Illustrate your answer with examples.

6. What are the advantages of foraging in a group? Are all individuals in the group similarly advantaged?

7. Imagine supermarket shoppers queuing at two check-outs. What features of the shoppers and check-out attendants might the shoppers use if they were to queue at each check-out according to the ideal free distribution?

8. Explain why animal contests are usually resolved before serious injury occurs.

9. Why is it usually the case that the male courts the female, rather than the other way around?

10. Explain the difference between intrasexual and intersexual selection.

11. Why do helpers of co-operatively breeding birds remain with the parents rather than attempt to breed on their own?

12. Why is the naked mole-rat often referred to as a vertebrate analogue of termites?

COPING WITH ENVIRONMENTAL STRESS

Organisms are well suited to their usual environments as a result of evolutionary change by natural selection; that is, *adaptation* (Chapter 32). They have heritable anatomical and physiological characteristics that allow them to grow under particular environmental conditions. Some of these inherited characteristics also allow individual organisms to respond to environmental change in a way that increases their own chances of survival.

Organisms may have heritable characteristics that increase their chances of survival in environments unfavourable to the majority of organisms. Thus, a species of plant or animal may live in a harsh environment, not because it grows better there than elsewhere, but because it is one of the few that *can* survive there. The Australian plant old-man saltbush, *Atriplex numularia*, which is found in dry and saline habitats in western New South Wales, can grow more quickly in a well-watered garden in a milder climate.

Under optimal conditions, an organism should attain its full growth and reproductive potential as determined by its genotype. However, such optimal conditions may occur rarely in natural environments. Suboptimal conditions can occur on a daily basis, seasonally or at random. In a biological sense, **stress** refers to disturbances in the normal functioning of an organism that are induced either by environmental factors, such as other organisms, or chemical or physical factors, or by internal factors, such as disease.

> As a result of natural selection, organisms have heritable anatomical and physiological characteristics that allow them to grow under particular environmental conditions. They are adapted to their environments.

PHENOTYPIC PLASTICITY

Changes occurring in an individual organism during its development, or in response to an environmental factor, have been termed 'phenotypic plasticity', 'acclimatisation' or 'acclimation'. **Acclimatisation** refers to the response of an organism to changes in several aspects of its natural environment, such as occur seasonally or geographically, whereas **acclimation** refers to the response of an organism to change in a single environmental factor, usually within a laboratory. These changes in phenotype occur without any change in the genotype of the individual and take place within the limits set by the genotype. However, the *capacity* for such plasticity is genetically based and therefore subject to selection.

Interaction between the environment and the genotype of an organism determines the expression of some adaptive features of the phenotype. Phenotypic plasticity is best developed in plants, presumably because of their sedentary growth. For plants, acclimation is often called 'hardening'; for example, frost-hardening, where a period at low temperature, above freezing, can increase the tolerance of a plant to subsequent freezing temperatures.

There are many examples in animals of this type of response to environmental change. Young rats raised in large hypobaric (low pressure) chambers have been found, as adults, to have larger total alveolar surface areas available for gas exchange than do control animals raised at sea level barometric pressure. This change, termed 'alveolar proliferation', is also a response to low oxygen (O_2) levels at high altitude and can occur in all mammals if exposure to low O_2 levels occurs when an animal is young.

Adult mammals exposed to high altitude for even short periods of time show increases in the number of red blood cells and therefore the amount of haemoglobin available to transport O_2, as a response to the stress of low O_2 levels. Another example is seen in the wide range of animals that have the capacity to switch reversibly between enzymic systems with different temperature optima as the temperature of their environment changes seasonally.

> Responses of individual animals and plants to environmental stress may involve a change in phenotype, known as acclimatisation, acclimation or hardening.

For an animal, behavioural responses to environmental stress may also be important. For example, many animals dig shelters, such as elaborate burrows with a means of ventilation, which allows the animal to reduce the fluctuations in surface temperature and humidity to which it would otherwise be exposed. A striking example of a behavioural response to environmental stress is migration. Many large mammals of East Africa migrate to follow patterns of seasonal rainfall, and many whale species migrate seasonally before breeding. The muttonbird or short-tailed shearwater breeds in south-eastern Australia but migrates to the north Pacific each year.

Many insects undertake equally impressive migrations. The bogong moth, *Agrotis infusa*, occurs on the western slopes of the Great Dividing Range in New South Wales. In spring, most of the annual foods on which it feeds are no longer available and the moths migrate for hundreds of kilometres to congregate in caves over the summer (Fig. 29.1). In autumn, the moths move back to the breeding grounds to lay eggs.

Fig. 29.1 Bogong moths, *Agrotis infusa*, sheltering between rocks in summer in Canberra

Many organisms show a response to a particular *seasonal* stress. What stimulus could initiate such a response? A common characteristic of animals and plants is the ability to sense day length (photoperiod) and even the rate at which day length changes (Chapter 7). Biologists term photoperiod a 'proximate' factor, one that triggers an organism to initiate adaptive responses. In tropical areas, where day length is virtually constant throughout the year, or in cases where an environmental stress is unpredictable, the stress itself may provide the appropriate cues.

> The responses of an organism to changes in its environment can involve anatomical, physiological and behavioural changes within limits defined by its genotype.

In this chapter, we will look at selected examples of responses of organisms to environmental stress. These include animals living in Australian arid regions, at high altitude, and in intertidal and estuarine environments. For Australasian plants, lack of water, high salinity, high temperatures and high solar radiation are most important.

ANIMALS IN THE AUSTRALIAN ARID ZONE

The Australian arid zone includes about 70% of the continent and is characterised by high temperatures, long droughts and flooding rains (Chapter 41). Many desert animals, particularly small mammals, reptiles and invertebrates, avoid the extremes of desert temperatures by seeking shelter in burrows or cracks in the soil during the day and limiting their foraging to the cool evenings. Larger mammals, such as kangaroos, seek shelter in caves or shady areas to avoid the direct heat of the sun (Fig. 29.2). Thus, most animals in the desert *avoid* the direct impact of high temperatures.

Fig. 29.2 An eastern grey kangaroo, *Macropus giganteus*, sheltering during the heat of the day

The unpredictability in rainfall and therefore food availability is another critical stress for most Australian desert animals. In fact, it is this unpredictability that sets Australian deserts apart from other arid areas of the world. Few desert species of animals in arid Australia regularly require drinking water. CSIRO scientists have estimated that only about 4% of mammal species and 10% of bird species in arid Australia require drinking water. Of the 210 reptiles and thousands of insects that inhabit the arid zone, probably none require liquid water to drink.

Most desert animals avoid the direct impact of high temperatures and few animal species in arid Australia regularly require drinking water.

Termites in arid Australia

One group of animals that has adapted to living in the arid zone are termites. Because there are few resident herbivores in arid Australia, much of the plant production is immediately available to the abundant and diverse termite populations. These insects have symbiotic flagellates in their hindgut that have the capacity to hydrolyse this source of cellulose (Chapter 19).

Fluctuations in temperature inside termite nests are reduced by the insulative properties of the wood-pulp material of the walls (Fig. 29.3). 'Magnetic' termites, *Amitermes meridionalis*, which live in the Northern Territory, build their nest or mound with a north–south orientation, which exposes the maximum area to the sun when the air temperature is

Fig. 29.3 The termite nest is a good example of complex social behaviour resulting in a controlled microclimate that isolates the animals from an otherwise harsh climate

low but reduces the area exposed to the hot sun in the middle of the day. The air passages in these structures also aid air circulation. The relative humidity inside the nest is always around 95% and is maintained by water transported by the animals from the watertable and stored in the matrix of the nest walls.

Nests built by termites provide a controlled microclimate that isolates the animals from an otherwise harsh climate.

ANIMALS LIVING AT LOW TEMPERATURES

The upper altitudinal limit to forested areas, the 'timber line', reflects a marked change in habitat. Here the vegetation changes from trees to low shrubs and herbs, lichens and mosses. The annual mean temperature decreases, on average, 6.5°C for every 1000 m in altitude, and the reduction in dust and haze particles causes an increase in the amount of radiation reaching the ground. High altitudes are characterised by distinct *seasonal* changes in temperature but, even in summer, temperatures can change rapidly and marked local climatic differences can occur. In alpine environments, we might expect to find animals with a dependence on behavioural temperature regulation in summer and a tolerance of freezing in winter. Similar properties are found in animals of polar regions.

The ultimate cause of death in animals exposed to low temperatures is complex and not fully understood. Ice formation *within* cells causes cytoplasm to become more concentrated and pH may change. Ice may also form in the extracellular space, resulting in the osmotic withdrawal of water from the cell. The resultant dehydration of proteins is likely to be the chief cause of freezing damage.

Animals living at low temperatures are likely to be tolerant of freezing.

Do animals freeze?

The likelihood of survival for an over-wintering animal in alpine or polar regions ultimately depends on its ability to prevent intracellular freezing. However, many invertebrates, for example, some insects and intertidal molluscs, are able to tolerate freezing of at least a portion of their extracellular fluids. A number of these animals have nucleating agents that promote a slow rate of freezing and thus reduce the possibility of intracellular ice formation. A few species of frogs and reptiles are also able to tolerate freezing of extracellular fluid, especially in the limbs and in subcutaneous tissues.

Some animals prevent freezing by *supercooling,* a process that involves a reduction in the temperature of a fluid below its usual point of freezing. Ice will not form unless a nucleus (seed) for its formation is present. In animals employing this mechanism, the digestive system is emptied of food material that could act as sites of nucleation. In some cases, antifreeze compounds synthesised within the body inhibit ice formation. As a result of these mechanisms in insects and nematodes, whole body temperatures as low as −50°C have been reached experimentally. Much lesser degrees of supercooling (to −10°C) are found in benthic Arctic fishes, reptiles and one mammal, the Arctic ground squirrel.

A major disadvantage with supercooling is that contact with ice 'seeds' supercooled fluids, thus freezing the whole animal. Antarctic fishes avoid this problem by producing antifreeze proteins or glycoproteins in the blood that lower the freezing point as well as preventing ice formation. These molecules are rich in polar groups and interact with ice crystals, stopping them from growing. This is analogous to using antifreeze compounds in cooling systems of cars in the snow.

> Few animals tolerate freezing of tissues. Survival at very low temperatures usually involves supercooling, or biological compounds that lower the freezing point and prevent ice formation.

Insects at low temperatures

One of the most striking features among high-altitude insects is their small size when compared with related species living at lower altitudes. This reduction in size enables these animals to take advantage of sheltered microclimates, which would not be accessible to larger forms. Selection pressures associated with high winds may have also resulted in evolution of small size, especially in butterflies that forage in vegetation boundary layers. In the Himalayas above 6000 m, 60% of insects have reduced wings or are wingless (apterous), and species of flightless grasshoppers live in the Australian Alps. Insects from cold climates also tend to be darker in colour than those from lowlands. This colouration, 'thermal melanism', leads to increased body temperature due to an increased absorption of radiation. Various patterns of thermoregulatory basking behaviour have been described in butterflies, for example, the genus *Pieris,* which orient the dorsal thorax towards the sun and use their white wings to reflect radiation onto the body. Apart from such behaviours, alpine insects seek warm microclimates, including parabolic corollas of some alpine plants (Fig. 29.4).

> Insects from cold climates tend to be small, darker in colour and use various forms of basking behaviour.

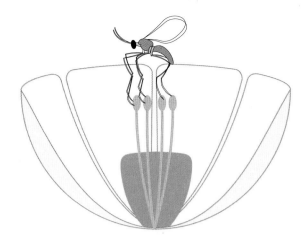

Fig. 29.4 The geometry of the flower of the Arctic plant *Dryas,* showing the parabolic inside cup that focuses the sun's rays

Hypothermia and the concept of torpor

Torpor is used to describe natural hypothermia involving a lowering of metabolic rate and a lessened responsiveness to external stimuli. Vertebrate ectotherms, fishes, amphibians and reptiles, may enter a period of torpor in which they are inactive or *dormant.* **Hibernation** is a specific term used to refer to long-term torpor in cold weather. Its use in describing the lethargic state of bears in winter, in which body temperature decreases by only a few degrees (Box 29.1), and the seasonal dormancy of reptiles and invertebrates, is misleading.

The major problems encountered during torpor are the dangers of freezing and desiccation, a lack of energy supply and, for animals that are underwater, a lack of oxygen. Like overwintering insects, animals in torpor avoid freezing by the use of cryoprotectants. Cryoprotectants are organic molecules, such as sugars, that protect membranes and enzymes against cold-induced damage. Others tolerate ice formation in extracellular fluids, including blood. Energy demands during this time are met by lipids, economically stored as fat bodies in muscle or liver; a major involvement of anaerobic pathways has been suggested. Novel anaerobic end products, such as ethanol in fishes, may also be used in addition to the more usual products of anaerobic metabolism, such as lactic acid.

Tolerance levels during hibernation or dormancy are impressive. Freshwater tortoises are dormant for up to six months of the year, some of them in water. Among fishes, the crucian carp, *Carassius carassius,* one of the few fish to occur in the ponds of northern Europe, is tolerant of anoxia for periods of months. Occasionally, however, there are problems for hibernating aquatic animals, and in Northern Hemisphere temperate lakes the phenomenon of 'winterkill' occurs. Frogs and fishes may die during a winterkill,

BOX 29.1   DO BEARS ENTER TORPOR?

Winter inactivity in the brown bear, *Ursus*, was once believed to involve low metabolic rates. However, for a large animal, this is not an energetically sensible thing to do. To increase body temperature from around 10°C to 38°C would require more than a day and consume a larger amount of energy than the animal would otherwise have consumed if it had simply remained at its normal body temperature. During hibernation, the bear reduces its body temperature only slightly, to between 31°C and 35°C, but does not drink, eat or void urine or faeces. This response may merely reflect an ability to withstand long periods of starvation. Surprisingly, females give birth and nurse their cubs during this period.

Biochemical studies of bears have shown just how extraordinary the process of mild hypothermia is in these animals. Nitrogenous end products, such as urea, do not accumulate during the time a bear is in its den but are hydrolysed and the nitrogen released is resynthesised to amino acids and then to protein.

Ursus arctos horibilis, the grizzly bear, a hibernating bear

resulting in thousands of dead animals in the spring thaw. Lakes in which this occurs are usually shallow and rich in organic matter. If snow on the frozen surface reduces light penetration so that photosynthesis ceases, oxygen levels are reduced and micro-organisms in sediments produce ammonia and hydrogen sulfide in sufficient quantities to be toxic.

Torpor is hypothermia involving a lowering of metabolic rate and reduced responsiveness to external stimuli. Long-term torpor in cold weather (hibernation) is associated with the dangers of freezing, dehydration and lack of energy supply.

Torpor in mammals and birds

Some small mammals, including rodents, marsupials and bats, and birds, such as hummingbirds and swifts, respond to seasonal changes in climate and food availability by reducing their rate of metabolism and body temperature and entering a period of torpor. In these animals, torpor is not an abandonment of body temperature regulation, but a controlled state that may involve an initial period of fattening, the preparation of a nest and several preliminary test drops in body temperature. In many species, temperature alarms are set off if body temperature goes below a certain point and arousal begins (Fig. 29.5). Bouts of torpor are often interrupted by days of arousal before the animal again drops its body temperature. Figure 29.6 shows measurements of metabolic rate in a small marsupial during entry into torpor and a subsequent period of arousal. This pattern is also typical of rodents that show daily torpor.

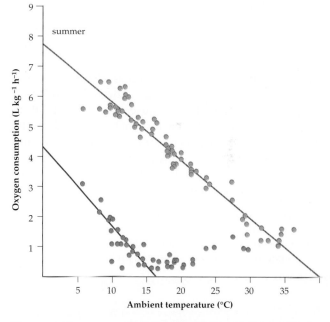

Fig. 29.5 The rate of O_2 consumption in a small marsupial, the fat-tailed dunnart, *Sminthopsis crassicaudata*, at different ambient temperatures. The open circles show the responses of animals that were not in torpor, the closed circles show data from torpid animals. At ambient temperatures below 15°C, the animals remain in torpor but their body temperatures show an increase to prevent dropping to dangerously low levels

There are dramatic circulatory changes that occur when mammals allow their body temperatures to drop as they enter hibernation. The lowering of body temperature is associated with a lowered metabolic rate and heart rate. The decrease in cardiac output

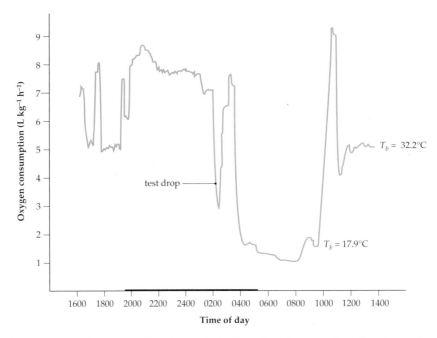

Fig. 29.6 The rate of O_2 consumption in a small marsupial, the fat-tailed dunnart, *Sminthopsis crassicaudata*, measured over a 21-hour period at a constant ambient temperature of 12°C. The dark bar represents the period of darkness; the body temperature is shown as T_b. The trace shows a 'test drop' in metabolism before the final decrease in metabolism during the 6-hour period the animal was in torpor

is associated with strong peripheral vasoconstriction, and blood flow is maintained to the brain, heart and lungs, rather similar to the diving response (Box 29.2). However, during arousal, blood is also sent to the brown fat and adrenal glands. Adrenaline released from the adrenal glands causes the brown fat to produce a great deal of heat, which is carried by the blood to the heart, increasing cardiac output and therefore blood flow to brown fat, and shivering of skeletal muscle. The cycle continues until the animal returns to normal body temperature.

> In mammals and birds, torpor is not an abandonment of body temperature regulation, but a controlled state that may involve an initial period of fattening, the preparation of a nest and several preliminary 'test drops' in body temperature.

ANIMALS IN HYPOXIC ENVIRONMENTS
Animal burrows

The proportion of O_2 in air is remarkably constant; however, many animals live in poorly ventilated environments, for example, in cavities in trees or in underground burrows. In these habitats, exchange of air with the outside is reduced and the rate of production of carbon dioxide (CO_2) and consumption of O_2 by an inhabitant may exceed the rate at which air is renewed. In animal burrows, most air exchange takes place through the walls of the burrow and diffusion through the opening is unimportant.

Plugging of burrow openings will therefore have little effect on the composition of gases in the burrow and many animals, such as the platypus, show this behaviour. However, anything that decreases diffusion through air spaces in the soil will affect O_2 and CO_2 levels in the burrow. The rate of gas diffusion in water is much slower than in air; for example, in the case of O_2, the factor is about 300 000. Therefore, if soil becomes waterlogged, a considerable barrier to gas exchange develops. The digging that North American gophers engage in after rain is a behaviour that increases burrow surface area to aid diffusion of air. In addition to water, the metabolism of bacteria and other micro-organisms within the soil can also decrease the effective replacement of burrow gases.

Oxygen concentrations as low as 15% (compared with 21% in atmospheric air) have been recorded within mammal burrows, with CO_2 levels above 5% (compared with 0.03% in atmospheric air). In nest cavities occupied by birds, changes from atmospheric gas levels are less severe, indicating a greater turnover with the atmosphere. Some species of marine fishes construct burrows in sand and mud. Recordings from these burrows suggest that their O_2 levels are among the lowest experienced by any vertebrate.

Adaptation to living in burrows is a combination of physiological and behavioural characteristics. Burrowing animals usually have low rates of O_2 consumption compared with others of the same body mass. This would appear to be adaptive in a closed burrow in that it reduces O_2 demand and CO_2 production and produces less metabolic heat. Haemoglobin in burrowing animals has a greater affinity for O_2 and is therefore fully saturated at lower O_2 concentrations (Chapter 20). Carbon dioxide is a powerful stimulus to breathing in most mammals but in burrowing forms, such as the echidna, this response seems to be reduced. Among behavioural adaptations to burrowing, fish move in and out of their burrows and this piston-like convective activity can increase burrow O_2 levels by up to 44%. Wombats are known to shuttle in and out of their burrows but whether this affects gas levels in the burrow is not known. Wombats are among the largest burrowing mammals and they have one of the lowest levels of O_2 consumption per unit body mass.

One of the best examples of a behavioural method of burrow ventilation occurs in the prairie dog, a North American member of the squirrel family. These animals construct burrows with two openings and, because the openings are at different heights, wind blowing across the ground induces a convective air current within the burrow (known as viscous entrainment; Fig. 29.7).

> Many animals occupy poorly ventilated environments, such as burrows, in which exchange of air with the outside is reduced, resulting in lower O_2 levels and higher CO_2 levels.

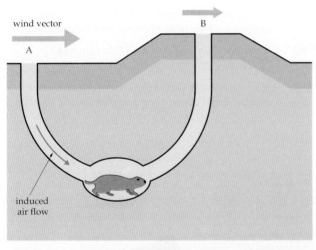

Fig. 29.7 A diagrammatic representation of an animal burrow in which an elevated mound at one end results in the corresponding opening being exposed to a greater wind speed as indicated by the length of the arrows. This results in the pressure at A exceeding that at B and air moves from A to B, thus ventilating the burrow

Responses to high altitude in humans

Investigations into altitude exposure probably date back to the Chinese, with accounts of the Great Headache Mountain in around 30 BC, and continue to the present, when all the peaks above 8000 m in altitude have been climbed without supplementary O_2 (Table 29.1). In between were descriptions of mountain sickness by the Spanish Conquistadors and Jesuit priests in South America, and the dramatic flights of early balloonists. Heroic experiments were conducted in the late 1800s by the Italian physiologist, Mosso, who placed his laboratory assistant in a chamber and exposed him for 33 minutes to an oxygen level equivalent to an altitude of 6500 m.

Low fertility among people living at high altitude is common, and the Spanish arriving in South America and Chinese immigrants to Tibet had high rates of *in utero* deaths of infants due to reduced O_2 transfer across the placenta. Every year large numbers of skiers and trekkers suffer acute mountain sickness and there are frequent deaths from altitude-induced pulmonary oedema and the comas that result from cerebral oedema. Such oedema, or swelling, is caused by changes in distribution of body fluid. Miners in Tibet and Chile work at altitudes of around 6000 m and for permanent high-altitude residents this seems close to a limit beyond which mental deterioration occurs. People born at sea level are not able to live permanently at altitudes above 4000 m.

As altitude increases, the atmospheric pressure, and therefore the partial pressure of O_2, decreases by about one-half every 5500 m. While the proportion of O_2 in air remains constant at 20.95%, the total amount available for exchange is less. The problem with exposure to high altitude is that the O_2 partial pressure gradient is reduced between different compartments of the total O_2 delivery system (Fig. 29.8).

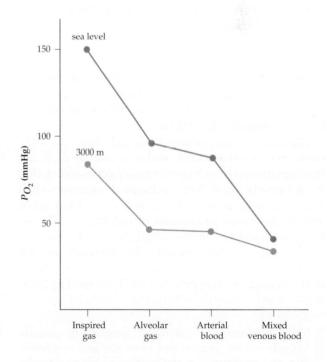

Fig. 29.8 Scheme of O_2 levels from inspired air to mixed venous blood taken from one group of people resident at sea level and a second group resident at an altitude of about 5000 m in the Andes

Table 29.1 Some milestones in the study of high-altitude physiology

Year	Event	Year	Event
c.30 BC	Reference to the Great Headache Mountain and Little Headache Mountain in the Tseen Han Shoo (classical Chinese history)	1913	T. H. Ravenhill publishes *Some experiences of mountain sickness in the Andes* describing in detail the symptoms of 'puna'
1590	Publication of the first edition (Spanish) of *Naturall and morall historie of the East and West Indies* by Joseph de Acosta with an account of mountain sickness	1920	Barcroft et al. publish the results of the experiment carried out in a glass chamber in which Barcroft lived in a hypoxic atmosphere for six days
1644	First description of mercury barometer by Torricelli	1921	A. M. Kellas publishes *Sur les possibilités de faire l' ascension du Mount Everest* (Congrés de l'Alpinisme, Monaco, 1920)
1648	Demonstration of fall in barometric pressure at high altitude in an experiment devised by Pascal	1924	E. F. Norton ascends to 8500 m on Mt Everest without supplementary oxygen
1783	Montgolfier brothers introduce balloon ascents	1946	Operation Everest I carried out by C. S. Houston and R. L. Riley
1786	First ascent of Mont Blanc by Balmat and Paccard	1948	C. Monge publishes *Acclimatization in the Andes* describing the permanent residents of the Peruvian Andes
1878	Publication of *La Pression Barométrique* by Paul Bert	1952	L. G. C. E. Pugh and colleagues carry out experiments on Cho Oyu near Mt Everest in preparation for the 1953 expedition
1890	Viault describes increase in red blood cells at high altitude		
1894	Angelo Mosso completes the high-altitude station, Capanna Regina Margherita, on a summit of Monte Rosa at 4560 m	1953	First ascent of Mt Everest by Hillary and Tensing (with supplementary oxygen)
1909	The Duke of Abruzzi reaches 7500 m in the Karakoram without supplementary oxygen	1978	First ascent of Everest without supplementary oxygen by Reinhold Messner and Peter Habeler
1910	Zuntz organises an international high-altitude expedition to Tenerife; members included C. G. Douglas and Joseph Barcroft	1981	American Medical Research Expedition to Everest, Scientific Leader J. B. West
1911	Anglo–American Pikes Peak expedition (4300 m)	1987	First ascent of Everest in winter (December) by Sherpa Ang Rita without supplementary oxygen

Partial pressure differences determine the rate of diffusion of a gas across membranes (Chapter 20); hence, exposure to high altitude ultimately reduces the O_2 availability to tissues.

The responses of humans to altitude hypoxia range from changes that occur within a couple of hours to adaptations, occurring over many generations, that have a genetic basis. Some individual responses, such as an acute increase in heart and respiration rates, and an increased production of red blood cells, are later reversed as other changes occur. Slower-developing changes include an increase in the number of capillaries in various tissues and a decrease in the ventilatory response to low O_2 levels (Fig. 29.9).

Initial responses of individual humans to altitude hypoxia include an increased heart rate and ventilation rate, and an increase in the number of red blood cells. Later changes include an increased number of capillaries supplying tissues and decreased ventilatory response to low O_2 levels.

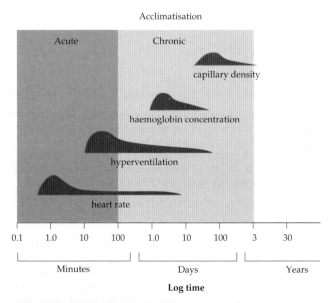

Fig. 29.9 The time courses of a number of acclimatisation and adaptive changes that occur in people during exposure to high altitude. The shape of the curve denotes the rate of change of each response

Birds at high altitude

Birds appear better adapted to high altitude than mammals. Geese fly over the Himalayas on migratory paths and there are documented cases of aircraft striking birds at altitudes above 8000 m. These birds may begin their journey near sea level and reach altitudes around 9000 m without long periods of acclimation. Experiments in which mice and sparrows were exposed to simulated altitudes of 6100 m resulted in the latter continuing to behave normally while the mice were unable to crawl. This difference in altitude tolerance may relate to the different respiratory systems in birds and mammals (Chapter 20), and to other changes in blood flow distribution that are not yet fully understood.

Low oxygen levels in aquatic systems

Swamps, tidepools and some coastal estuarine environments are often poorly mixed and large diurnal oscillations in respiratory gases occur. During the day photosynthetic organisms produce O_2 at a rate that exceeds the respiratory demands of all organisms in the community. At night, respiration continues but photosynthesis ceases. In tidepools, this may result in O_2 levels fluctuating from 0.3 kPa at night to 60 kPa during the day (Fig. 29.10). Adaptations to these stressful habitats are believed to centre on the remarkable plasticity seen in the properties of respiratory pigments (Chapter 20) and in the ability to use anaerobic metabolism.

Rockpools in the supralittoral fringe are one of the most stressful marine habitats. As well as low O_2 levels at night, reduction in salinity due to rain, increase in salinity due to evaporation, and build-up of decomposition products from drifting algae result in these rockpools being a demanding habitat for animals. As a result, isopods and molluscs characteristic of the littoral zone are more often found in sheltered moist crevices than in the pools.

RESPONSES TO OSMOTIC STRESS IN AQUATIC ANIMALS

Marine invertebrates have blood osmotic concentrations close to that of sea water (1000 mOsmol/l) and live in the open sea where they encounter no osmotic stress. Many are **stenohaline**, that is, they can tolerate little change in external salinity. However, invertebrates in intertidal zones and estuaries are exposed to periodic fluctuations in salinity. They show a range of behavioural and physiological adaptations that allow them to tolerate a wide range of salinities, that is, they are **euryhaline** (Chapter 22).

Behavioural adaptations of euryhaline invertebrates involve either burrowing in the substrate where ambient salinities are more stable, or seeking refuge in a closed shell. Burrowing strategies are adopted by annelid worms and crabs, and by this behaviour they also avoid the effects of currents. Hard-shelled species, such as mussels, close their valves as sea water concentrations fall and maintain salinities close to that of sea water within their mantle

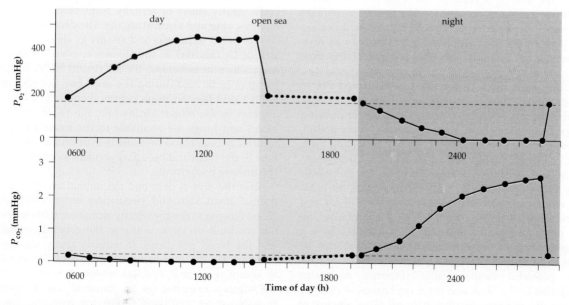

Fig. 29.10 The water in tidal rockpools shows great variation in dissolved O_2 and CO_2 levels when it is isolated from the sea at low tide. These values were recorded in a rockpool off the coast of France. Normal values of O_2 and CO_2 in the open sea are given by the dotted lines

cavity. Mussels have sodium and magnesium sensors on the tentacles associated with the incurrent siphon that control this response.

> Stenohaline animals tolerate little change in external salinity; euryhaline animals tolerate a wide range of salinities.

Osmotic conformers and intracellular osmotic regulation

In some aquatic animals, experiments have shown that exposure to a range of osmotic concentrations results in corresponding changes in body fluids. These animals have been termed 'osmotic conformers' (Chapter 22). If extracellular osmolarity decreases we would expect cells to take up water and consequently swell. With an increase in extracellular osmotic levels, water would move out of cells and they would decrease in volume. In the short term, such changes do occur in osmoconformers to some extent, however, they are followed by a slow recovery of cell volume to the initial level. This recovery phase is due to changes in the levels of low molecular weight organic molecules, such as non-essential amino acids, polyhydric alcohols and mono- and disaccharides. The control of this intracellular solute pool is responsible for reducing any osmotic gradient between the intracellular and extracellular fluid compartments, and appears to be the primary adaptation permitting osmotic conformers to inhabit areas of varying salinity. Recently it has been shown that a similar process occurs in mammalian cells, particularly in the regulation of brain cell volume. Although changes in plasma osmolarity of vertebrates do not show the degree of change seen in invertebrates, it is interesting that vertebrates may have retained this system from simpler animals.

> In some animals the composition of an intracellular solute pool may be varied to reduce the osmotic gradient across the cell membrane.

Migratory fishes

Most teleost fishes have a limited ability to move between the sea and freshwater. In freshwater, osmosis brings water into the body, which is then

---

BOX 29.2   DIVING

For mammals and birds, so dependent on high levels of O_2, diving is a considerable challenge. However, early studies showed that many can tolerate submersion for longer periods than would be expected on the basis of their O_2 stores and metabolic rates. More recent studies, centred on the diving behaviour of animals in their natural habitat, have shown that most voluntary dives are of a much shorter duration than the maximum of which the animal is capable. For example, the sperm whale is capable of remaining submerged for periods of 1 hour but sonar reports show that most dives are less than 10 minutes. Similarly, tufted ducks can dive for periods of 1 minute, but the preferred diving time is 20 seconds.

During most dives in the field, well-adapted divers take enough O_2 down with them to provide for the needs of their tissues. But there is always the danger of being trapped underwater for an unexpectedly long period, for example, while escaping a predator, dealing with food, or looking for a breathing hole in ice. Under these conditions, there may be insufficient O_2 for the whole body to remain aerobic. In birds and mammals, the heart and brain are particularly sensitive to lack of O_2 and permanent damage can occur in just a few minutes. So it is not surprising to find that these animals have a complex mechanism, the *diving response*, that can be invoked to reserve precious O_2 stores for these organs. This powerful cardiovascular response maintains circulation to the heart and brain (and lungs, of course), but greatly reduces circulation to all other organs of the body (Fig. a). The eminent physiologist, Per Scholander, called the mechanism 'the master switch of life'.

Although most vertebrates, including humans, show elements of a diving response, it is best developed in dedicated divers, such as marine mammals and diving birds. The primary reaction is an intense slowing of the heart, *bradycardia*, due to increased activity in the cardiac branch of the vagus nerve. In seals, heart rate can drop to less than a tenth of the normal rate, beating only five or six times a minute. At the same time, a powerful vasoconstriction of the major peripheral arteries of the body greatly slows or actually stops blood flow to the muscles, skin and visceral organs. The decrease in blood flow is called **ischaemia** and results in almost complete lack of O_2 (**anoxia**) in the tissues involved. Anaerobic metabolism in ischaemic tissues results in an accumulation of lactic acid during the dive; but, because there is no blood flow to these organs, very little lactic acid appears in the blood. Meanwhile, the O_2 stores of the blood and lungs are available to the heart and brain. Central arterial blood pressure is maintained, and blood flow to the brain and eyes may actually increase, despite the intense bradycardia.

After the dive is over and the animal breathes again, normal heart rate and circulation are re-established. Blood flowing to the previously ischaemic tissues flushes out the built-up lactic acid into the general circulation and the blood pH drops. It requires many minutes of recovery for the acid to be removed and the situation to return to normal. For this reason, it is obvious that the diving response is an unusual event; quickly repeated dives involving the diving response are not possible. In fact, recordings of heart rate and blood parameters in Antarctic Weddell seals show no diving response during repeated routine dives at sea, but the response can appear strongly if, for example, the animal is prevented from returning to its breathing hole in the ice.

The diving response is a pattern of reflexes under both conscious and unconscious control. It seems to be triggered when there is some doubt concerning the length of a dive. Animals forcibly submerged always show it, but when they are allowed to dive naturally they show it only if they are unfamiliar with the surroundings. In the experiment shown in Figure b, a freely diving duck decreases its heart rate only when it becomes aware that it cannot surface at the usual place.

Seals that have been trained to put their noses underwater on command show the response because they do not know how long the trainer wants them to stay under. However, if they are trained to dive for particular periods of time, they show a level of bradycardia from the outset of the dive that is appropriate for the length of the dive. It is interesting that even fishes show a 'diving response' when lifted from water (gills cannot take up sufficient O_2 from air). Apparently the response is a general, life-saving one, shown by vertebrates when removed from their normal respiratory environment.

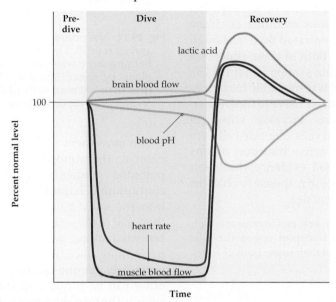

(a) The diving response. During the dive, heart rate decreases, but blood flow to the lungs and brain is maintained at the expense of flow to the skeletal muscles and gut. Lactic acid is produced by anaerobic metabolism in the ischaemic muscles and it is flushed out into the circulation, causing a decrease in blood pH after the dive

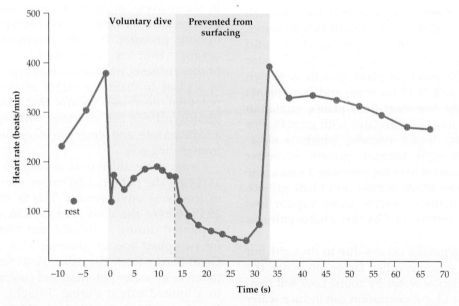

(b) The average heart rate during a dive in a tufted duck, *Aythya fuligula*. The dive was in two phases; a voluntary phase, and a second phase in which the duck apparently became aware that it was temporarily being prevented from surfacing and decreased its heart rate to a lower level

removed by the production of large volumes of dilute urine; solutes are lost by diffusion and are replaced by active transport across the gills and from food (Chapter 22). In sea water, fish are hypo-osmotic and the opposite water and solute gradients occur; water loss is replaced by drinking and production of more concentrated urine, and salts are actively excreted across the gills.

Some fishes, including lampreys, salmon and eel, migrate between the sea and freshwater streams as part of their normal breeding cycle. To compensate for this osmotic stress, the direction of salt transport across the gills can be reversed and the kidneys can change from producing copious amounts of dilute urine to produce a more concentrated fluid (although still hypo-osmotic to blood). Both of these processes are regulated by the endocrine system (Chapter 25). The steroid hormone cortisol is responsible for inducing the changes when the animals move to freshwater, and prolactin concentrations increase when in freshwater. Highly specialised cells, chloride cells, are primarily responsible for the active transport of ions across the gills. There is good evidence that their numbers and functions change in response to changes in the levels of these two hormones.

> Fishes that migrate between sea and freshwater can change the direction of salt transport across the gills and adjust the concentration of their urine. Both changes are under endocrine control.

WATER STRESS IN PLANTS

Water stress is the term used to describe the condition of impaired function in plants due to lack of water. Terrestrial plant biomass, productivity and distribution often correlate with supply of water (Chapter 44). The effect on plant growth is shown in Figures 29.11 and 29.12 for some arid zone plants in South Australia. For short-lived plants, rainfall in the previous 12 months correlates with growth. For bladder saltbush, *Atriplex vesicaria*, which is slow-growing and drought tolerant, growth is better correlated with rainfall over the previous 2 years. The close link between water supply and plant growth arises because of the inevitable water vapour loss associated with uptake of CO_2 for photosynthesis (Chapter 18).

Water stress generally occurs due to drought but can also occur if some other factor prevents the efficient absorption of water by roots. Low soil temperature and low O_2 concentration can induce water stress by increasing the resistance to water flow through roots, presumably by affecting membranes or the connections between cells (plasmodesmata, Chapter 6).

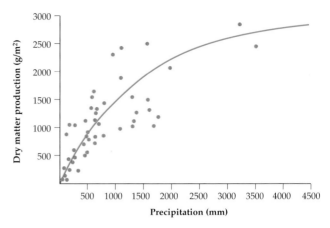

Fig. 29.11 Net primary productivity as a function of mean annual rainfall. The curve flattens at high rainfall because more water exits the sources available to plants by run-off and deep drainage. At low rainfall (less than 500 mm) each additional amount of rainfall results in a proportional increase in production

The movement of water from the soil through the plant to the atmosphere is governed by the water potential gradients in the soil–plant–atmosphere continuum (Chapter 18). If a plant is to extract water from the soil, it must generate a lower water potential than that of the soil. However, measurements of water potential alone will not indicate how a plant is performing since two plants may have identical water potentials yet one can be obviously wilted while the other can be turgid and still growing. Many plants, such as the grey mangrove, *Avicennia marina*, can lower their water potential to below −3.0 MPa and still carry out photosynthesis while more mesophytic plants wilt and cease photosynthesising at around −1.5 MPa. In mangroves, turgor pressure of leaf cells is above zero because cells are able to generate a high internal osmotic pressure. It is the maintenance of positive turgor, that is correlated with maintained photosynthesis, transpiration and growth. The ability of a plant to maintain turgor pressure as its water potential decreases means that transpiration and photosynthesis can continue, although probably at a reduced rate, and water can continue to be extracted from drying soil.

Plants show differences in their ability to function as their water potential decreases, even when growing in the same environment. This is shown in Figure 29.13, where the daily variation in water potential is plotted during a drought and after a wet period for two plant species growing on a coral cay in the Great Barrier Reef. The daisy, *Melanthera biflora*, is able to lower its water potential and continue to transpire to a limited extent during drought, while the tree, *Pisonia grandis*, is not able to do this.

> The extent to which a plant can continue to function during water stress depends on its ability to maintain a positive turgor pressure at low water potentials.

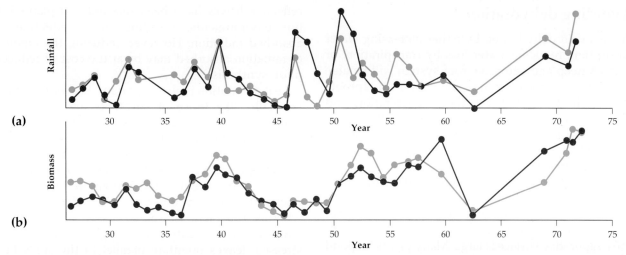

Fig. 29.12 Changes in relative biomass (red) and rainfall (blue) for arid zone vegetation at Koonamore in South Australia. **(a)** The biomass of herbs, annuals and dwarf shrubs is correlated with rainfall in the previous 12 months. **(b)** The biomass of bladder saltbush, *Atriplex vesicaria*, is better correlated with rainfall in the previous 48 months

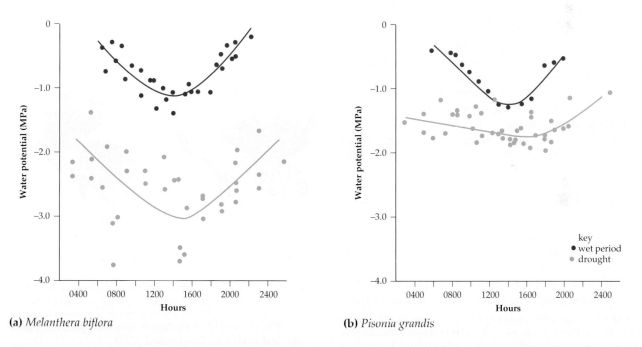

(a) *Melanthera biflora*   **(b)** *Pisonia grandis*

Fig. 29.13 Daily variation in water potentials for **(a)** *Melanthera biflora* (daisy) and **(b)** *Pisonia grandis* (a tree) growing on One Tree Island in droughts and wet periods. The daisy is able to lower its water potential below −3 MPa during droughts by its ability to attain high cell osmotic pressures

Ways that plants cope with drought

There are numerous ways in which plants live and survive in drought-prone environments. A plant may *escape* drought while in a dormant part of its life cycle, such as a seed. For example, when water becomes available in deserts, annual plants quickly grow and flower to produce more seeds. Sturt's desert pea grows quickly after rain, reaching seed set as quickly as 6 weeks from germination, and producing seed that is capable of withstanding dry periods and high temperatures.

In an arid zone, perennial plants are *drought-tolerant*. Dehydration is the main difficulty for these plants. Most drought-tolerant plants avoid dehydration, but a few, for example, some native grasses and desert mosses, can withstand total dehydration.

Avoiding dehydration

Plants avoid dehydration by either increasing water absorption, reducing water loss by transpiration, or both. A deep and extensive root system will be able to extract water from a large soil volume. If taproots can reach the watertable, then transpiration and photosynthesis can proceed normally. River red gums, *Eucalyptus camaldulensis*, growing along seemingly dry creek beds exploit underground water via deep taproots.

The proportion of root compared with shoot (root:shoot ratio) is often high for arid zone plants and plants subjected to water stress. A high root:shoot ratio means that the absorptive surface relative to the evaporative surface is large. Many plants respond to water stress by a change in growth pattern or by reducing the size of the shoot as a result of leaf shedding, as in blue bush, *Maireana sedifolia*. The size of the root system may be increased at the expense of the shoot by redirecting sugars to roots.

The soil under the canopy of vegetation is modified by the accumulation of leaf litter and by the shape of plants. As a result, water penetrates the soil more effectively and evaporation from the soil is reduced. Due to their shape, leaves and branches may intercept water during rainfall and channel it down the trunk to the roots, as in mulga, *Acacia aneura*, with its upward-pointing foliage.

The fleshy water-filled tissue of cacti and other succulent plants provides water storage. The cells composing this succulent tissue have large vacuoles and, by virtue of their flexible cell walls, are able to give up water to photosynthetic cells that are more dependent on maintaining a constant volume as water becomes scarce. In *Opuntia ficus-indica*, up to 82% of the water in water-storage tissue can be lost without irreversible tissue damage.

Control of water loss is closely related to control of leaf temperature. The shape of leaves enables them to capture light for photosynthesis but, in so doing, they also collect long-wavelength infra-red radiation. Most of the absorbed energy is reradiated but some must also be lost by other means if leaf temperature is to be kept within favourable limits. Transpiration from leaves is particularly important in dissipating such heat. If transpiration is reduced as stomata close during drought, leaf temperature increases and plants can suffer heat damage.

Leaf characteristics that reduce the absorption of radiation aid water conservation because less reliance needs to be placed on evaporative cooling. Leaf hairiness and increased reflectance (glaucousness), for example, due to surface waxes, reduce temperature and transpiration rate. Some saltbushes, *Atriplex*, change leaf reflectance during the course of leaf development such that higher reflectance coincides with summer. Varieties of wheat that have more reflective leaves have been selected and planted in hot environments, resulting in a reduction in absorbed radiation. However, reducing the amount of radiation absorbed may be at the cost of reduced photosynthesis.

Leaf orientation is also important. Leaves that hang vertically, like those of eucalypts and some salt-tolerant mangroves, avoid high radiation loads that are at a maximum in the middle of the day. Other plants are able to move their leaves to track the sun (heliotropic leaves). Siratro, *Macroptilium atropurpureum*, a pasture legume grown in Queensland, tracks the sun in different ways depending on whether the plant is water-stressed or not. If not stressed, leaves face the sun through the day; if water-stressed, leaves orientate parallel to the sun's rays (Fig. 29.14). This behaviour prevents damage from high light intensities to the photosynthetic apparatus when it is most sensitive under water stress.

Heat loss is greater for small leaves or highly dissected leaves than it is for large leaves because

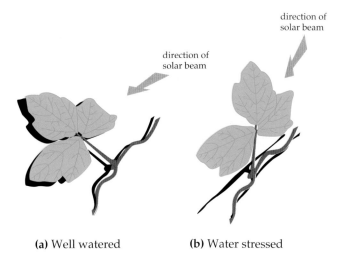

(a) Well watered   **(b)** Water stressed

Fig. 29.14 Changes in leaf orientation with respect to the direction of the sun's rays in response to water stress in Siratro, *Macroptilium atropurpureum*, a pasture legume used in Queensland: **(a)** leaf from a well-watered plant; **(b)** leaf from a water-stressed plant

small leaves are more efficient at losing heat to moving air. This is a physical effect related to the high perimeter:surface area relationship, and to the average thickness of the boundary layer of air around a leaf and how it varies with the velocity of air movement along the leaf (Fig. 29.15). It is not surprising that many arid zone plants have small leaves.

Plants able to cope with water stress may do so by entering a more tolerant stage of their life cycle (seeds) to escape dry conditions, or by either tolerating or avoiding dehydration.

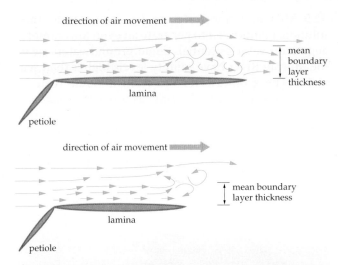

direction of air movement ▸

mean
boundary
layer
thickness

lamina

petiole

direction of air movement ▸

mean boundary
layer thickness

lamina

petiole

Fig. 29.15 Longitudinal section along leaves of different length showing air movement on the surface of the leaf and the average boundary layer thickness. The boundary layer is thinnest at the leading edge of the leaf and gets thicker towards the centre. At the trailing edge more turbulent flow occurs. The thickness of the boundary layer is greatly exaggerated and is normally less than a few millimetres. The longer the length of the leaf in the direction of air movement or the lower the velocity of air movement, the larger is the average boundary layer thickness. Heat loss from the leaf is rapid when the boundary layer is thin or the temperature difference between moving air and the leaf is large

Water-use efficiency

Minimising water loss under arid or water-stress conditions is only half the story. Plants that can minimise water loss yet simultaneously maximise the gain of CO_2 should perform well in dry conditions. We can assess this performance by measuring the efficiency of water use, that is, the carbon gained (e.g. dry weight) divided by the amount of water transpired in gaining the carbon:

$$\text{water-use efficiency} = \frac{\text{carbon gained}}{\text{water lost}}$$

Water-use efficiency varies widely among different species and within a species. Within-species variation can be due to different phenotypes resulting from different conditions during growth. Also, measurement of water-use efficiency for an individual crop plant grown in isolation may be very different from that measured for a whole crop on a farm where competition between plants occurs and the micro-climate is changed by the behaviour of the crop. A measure of water-use efficiency can be used to select for varieties of agricultural plants that may be more productive in Australia's arid climate.

In arid climates, perennial plants have high water-use efficiency, conserving water and maximising the amount of CO_2 gained.

Responsive stomata

The largest drop in water potential occurs at stomata, between the inside of the leaf and the atmosphere (Chapter 18). Therefore, to a large extent, the pattern of stomatal opening and closing over time controls the water-use efficiency of a plant. For example, if a leaf is water-stressed, it is better for stomata to close a little, thereby saving water but not greatly reducing the rate of photosynthesis as a result of lowered CO_2 uptake. Optimisation of carbon gain versus water loss accounts for the midday closure of stomata often seen in plants in summer (Fig. 29.16).

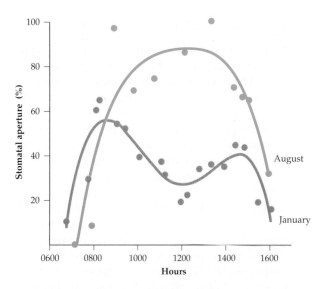

Fig. 29.16 Variation in stomatal aperture during the day for the same black box trees, *Eucalyptus largiflorens*, during January and August on the Chowilla floodplain near Renmark in South Australia. The stomatal aperture is relative to the maximum observed in the data times 100%. During summer (January) the stomata close in the middle of the day giving a double peak in the pattern. Despite the near vertical leaf orientation in these trees, the high temperature and low humidity in the middle of the day in summer results in a large evaporative demand. It is thought that the stomata close in response to low external humidity, thus preventing too much evaporation. Even if the stomata were to open to the values of winter, it is unlikely that any extra gain in photosynthesis (CO_2 uptake) would occur but there would be a large increase in transpiration

Stomatal control optimises water use and carbon gain on a daily basis for a particular set of conditions. To optimise use of soil water over a longer term, a plant may have to limit water loss, perhaps at the expense of carbon gain, so that it does not run out of water before it flowers and reproduces or before the next seasonal cycle of rainfall. There is good evidence that roots sense soil dryness and send a signal to the leaves, which affects stomatal aperture; for example, plants often close stomata as soil dries before there is any detectable change in the water potential of the plant. The signal may be the hormone abscisic acid (ABA; Chapter 24), or a compound that determines its synthesis, which is released by roots into the transpiration stream under these conditions. ABA in minute quantities closes stomata and is also produced by mesophyll cells in wilted leaves.

> Various controls are exerted on stomata in order to optimise the use of water. These involve an ability to sense water vapour flux and CO_2 concentrations, and the hormone ABA, which is released from roots and mesophyll cells in wilting leaves.

Photosynthesis and water stress

Evolutionary modifications to the basic C_3-type photosynthesis (Chapter 5) have enabled some plants to improve their water-use efficiencies considerably, particularly at high temperatures and high light intensities typical of arid environments. The C_4-modification allows photosynthesis to proceed at much lower internal CO_2 concentrations. By lowering the internal leaf CO_2 concentration, a larger gradient is established between the atmosphere and the inside of the leaf so that stomata need not open widely to allow a high rate of CO_2 diffusion into the leaf. With smaller stomatal aperture, less water escapes to the atmosphere and so water-use efficiency is increased.

The CAM mode of photosynthesis results in an even more spectacular water-use efficiency, but not high rates of photosynthesis (Chapter 5), because uptake of CO_2 occurs at night when evaporation is lowest. Some CAM plants have the requirement of low night-time temperatures for stomata to open, ensuring low evaporation rates. Some plants only use the CAM mode at certain times. These facultative CAM plants use C_3 photosynthesis when water is plentiful and, when water becomes limiting, CAM is invoked. The ice plant, *Mesembryanthemum crystallinum*, is such a facultative CAM plant. The switch to CAM is probably triggered by low turgor pressure in the leaf cells, a trigger that is also important in other plants as an emergency mechanism to close stomata under extreme water stress.

> C_4 and CAM modifications of photosynthesis allow plants to tolerate drought and high temperature by increasing water-use efficiency.

SALINITY AND MINERAL STRESS IN PLANTS

High salt concentrations in the root zone of a plant lower soil water potential and may lead to water stress (Chapter 18). Sea water has a water potential of about −2.5 MPa and, for water to be drawn from this solution, plants need to generate an even lower (more negative) water potential. Some groundwaters near the Murray River have salinity levels higher than sea water. The groundwater has risen in some regions due to locking on the Murray; as a result, black box trees, *Eucalyptus largiflorens*, are suffering (Fig. 29.17). Water stress may occur, in this case, because the volume of soil from which freshwater can be obtained has been reduced by the rising saline watertable.

Fig. 29.17 Salt-affected black box trees, *Eucalyptus largiflorens*, on the Chowilla floodplain near Renmark in South Australia. Two apparent varieties of box tree are observed based on leaf colour. The green variety (tree on the left) appears to be more tolerant of the conditions which have led to the demise of the grey variety on the right. This area has been subject to a rise in height of the saline groundwater to within a few metres of the surface

Salt (NaCl) can be toxic to most plants. Enzymes are inhibited by Na^+ at concentrations above about 0.1 M. This is so for all types of plants, halophytes included. Excess Na^+ taken into a cell must be excluded from the cytoplasm and it is generally sequestered in the vacuole. The osmotic pressure of the cytoplasm is balanced to that of the vacuole by the accumulation in the cytoplasm of solutes more compatible to metabolism, such as the amino acid, proline (Chapter 18). Proline is accumulated by a variety of plants and not just under salt stress. Compartmentalisation within the cell is, however, just one of the ways that may lead to resistance to salinity stress in plants.

Salt that enters the transpiration stream will be carried to leaves where it will become concentrated as water evaporates to the atmosphere. Even if only a very small fraction of the salt in the soil solution enters xylem, salt may build up to toxic levels in leaves, eventually killing them. Some halophytes, for example, white mangrove, *Avicennia marina*, have special salt glands in their leaves that excrete salt that is not excluded by roots. In contrast, most freshwater plants (glycophytes) cannot tolerate even small amounts of salt accumulating in their leaves. In these plants, there is a delicate balance between growth, which can dilute incoming salt, and the rate at which salt arrives in the leaves.

Since transpiration delivers salt to the shoot, efficient control of transpiration is one way to reduce the salt burden on leaves. For example, although mangroves are C_3 plants, they have very high water-use efficiencies, which would reduce the salt load to the shoot. Low transpiration also means that there is less evaporative cooling of leaves. Probably for this reason, some mangroves have small leaves vertically inclined to reduce their radiation load.

> High salt concentrations in soil can lead to water stress and Na^+ toxicity. Compartmentalisation of ions within cells and within the plant, the ability to exclude salt at roots and leaves, and the balance of transpiration and leaf growth with ion uptake all contribute to salt tolerance.

Another form of salt stress occurs from salt-laden air. Salt crystals can collect on leaves after evaporation of small water droplets. Some coastal plants have a mesh of cuticle over stomata, which prevents small water droplets from entering the leaf. Tiny amounts of detergent from sewage outfalls decrease the surface tension of salty droplets, allowing them to enter through the normally water-repellent mesh of cuticle over stomata. Norfolk Island pine has suffered salt damage as a result of this windborne detergent pollution (Fig. 29.18).

Other types of ions besides Na^+ and Cl^- that may be in excess in the soil can be toxic to plants. Soil acidity (excess hydrogen ions, H^+), either naturally occurring or induced by the use of ammonia- and amide-containing fertilisers, can release aluminium (Al) in a form that is toxic to plants. Aluminium interferes with the uptake of calcium by roots. Problems associated with aluminium toxicity in acid soils currently cost farmers in New South Wales and Victoria an estimated $100 million a year from lost crop and pasture. Adding lime to acid soils can eliminate aluminium toxicity in topsoil, but liming subsoil is economically unfeasible and technically difficult. Alternatively, aluminium-tolerant genotypes, for example, in wheat-breeding programs in Australia, would achieve better yields on acidic, aluminium-containing soils, with reduced input costs (eliminated or greatly reduced lime-related expenses).

Deficiencies of particular nutrients also produce stress symptoms in plants either directly or indirectly (Chapter 18). A deficiency of nitrogen affects photosynthesis since nearly 80% of the nitrogen in plants is used in the photosynthetic apparatus, particularly the enzyme ribulose bisphosphate carboxylase, which

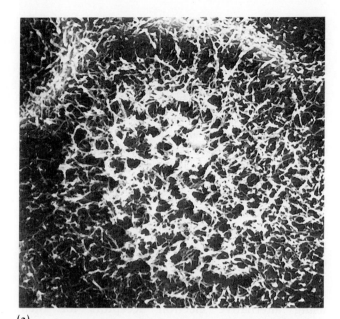

(a)

(b)

Fig. 29.18 Effect of detergent on the waxy fibrils that protect the stomata from the ingress of salt spray into leaves of Norfolk Island Pine, *Araucaria heterophylla*. **(a)** The normal condition. **(b)** Treatment with sea water spray containing detergent (30 ppm) causes the condition where the waxy fibrils are coalesced to form plates. The salt can traverse the stomata after such treatment

fixes CO_2 in all plants (Chapter 5). Mineral deficiencies may also lead to plants being more susceptible to disease, as is thought to be the case for a fungal disease in wheat where there is a strong correlation between zinc deficiency and presence of the disease.

> Ion toxicity can result from an increase in concentration of ions to toxic levels at low soil pH, or from the effect of pollutants that facilitate entry of toxic ions into a plant. Deficiencies of particular nutrients also cause stress in plants.

LACK OF OXYGEN AROUND PLANT ROOTS

Roots require O_2 for cellular respiration (Chapter 5), which generates ATP used by roots to actively accumulate particular ions, such as NO_3^-, or to excrete unwanted solutes, such as Na^+. When soil becomes flooded, air is displaced and, because of the slow diffusion of O_2 in water, the root environment quickly becomes anaerobic. ATP production in roots then occurs only by glycolysis, which yields much less ATP than does aerobic respiration and which leads to a rapid inefficient use of carbohydrate reserves. Thus, under waterlogged conditions, root growth, transpiration and translocation may decrease or even cease completely.

Toxic substances are also produced under anaerobic soil conditions due to micro-organisms using alternative electron acceptors for their respiration. These end products include hydrogen sulfide, methane and Fe^{2+} and Mn^{2+}. An immediate effect of anaerobiosis is an increase in the resistance to water flow across the root (an effect on the cell membranes), which can lead to water stress in shoots if the plant continues to transpire. There are also effects on membrane permeability and ion selectivity in uptake.

Many plants, such as mangroves, seagrasses and rice, are naturally adapted to anaerobic soils (Chapter 17). Adaptations include *anatomical* features, such as air canals in the roots (Fig. 29.19) or the production of lateral roots on the soil surface, which allow improved O_2 transport to root tissues; *biochemical* features, such as an increased ability to sustain anaerobic fermentation in roots and an increased resistance to toxic substances produced in anaerobic soil; and *hormonal signals* from roots to shoots, which slow growth, nutrient uptake and water use (Chapter 24). These adaptations allow roots in anaerobic conditions to sustain the aboveground parts.

> Lack of O_2 in waterlogged soil decreases cellular respiration by root cells. This impairs root growth, water uptake and nutrient uptake. Also, toxic substances are produced by anaerobic respiration of soil microbes.

TEMPERATURE STRESS IN PLANTS

Different parts of the leaf canopy and root environment may fluctuate in temperature so that each part of a plant is subjected to a different temperature regimen. Temperature differences between different parts of a plant can be as much as 30°C. Temperatures fluctuate most widely near the soil surface, which affects seedling establishment. As mentioned earlier, water stress, and leaf orientation and reflectance properties also influence leaf temperature.

Growth rate of plants shows a characteristic bell-shaped curve in response to temperature (Chapter 42). Temperature affects biochemical reactions in two quite different ways. It affects the rates of reactions due to its effect on the kinetic energy of reacting molecules, and it affects the tertiary structure of enzymes and membranes (Chapter 2). At low temperatures, membrane fluidity can abruptly change, resulting in a 'frozen membrane' and disruption of the bilayer structure. Many important enzymes in photosynthesis and respiration are embedded in membranes and require a fluid membrane for proper function. Damage to membranes due to chilling or freezing is often indicated by leakage of solutes and water from cells into intercellular spaces giving the tissue a water-soaked appearance.

Some plants survive freezing temperatures as low as −50°C by preventing intracellular ice formation. Ice formation usually occurs only in extracellular fluid because solute concentrations and lack of ice nucleation sites in cells tend to prevent intracellular freezing. Extracellular ice formation causes cells to dehydrate since water can easily be drawn out of the cell across the plasma membrane and into the forming ice crystal with its very low water potential. The problem then becomes one of tolerating dehydration and, when thawing occurs, the ability of a cell to be able to re-expand without damage to the cell membrane.

> Damage to plants at extremes of temperature is due to alteration in the tertiary structure of proteins, to changes in membrane fluidity or to disruption of biochemical processes. Dehydration can occur as a result of high temperature or freezing conditions.

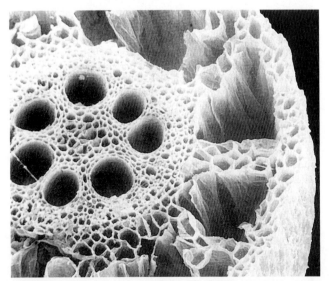

Fig. 29.19 Air canals produced in *Zea mays* as a result of several days' exposure of the roots to a solution of low oxygen (1%)

SUMMARY

- Responses of individual animals and plants to environmental stress may involve anatomical, physiological and behavioural changes in phenotype, within limits defined by their genotype.

- In animals, physiological adjustments to the environment are linked with behaviour that allows an animal to select microhabitats within its tolerance limits.

- Different environmental conditions, such as low temperature, low O_2 levels, drought and salinity, may lead to the same stress in an organism, for example, dehydration.

- Few organisms tolerate freezing of tissues. Survival of animals at very low temperatures usually involves supercooling, or biological compounds that lower the freezing point and prevent ice formation.

- Torpor is hypothermia involving a controlled lowering of metabolic rate and reduced responsiveness to external stimuli. Long-term torpor in cold weather (hibernation) is associated with the dangers of freezing, dehydration and lack of energy supply.

- Most desert animals avoid the direct impact of high temperatures and few animal species in arid Australia regularly require drinking water.

- Initial responses of humans to altitude hypoxia include an increased heart rate and ventilation rate, and an increase in the number of red blood cells. Later changes include an increased number of capillaries supplying tissues and decreased ventilatory response to low O_2 levels.

- Stenohaline animals tolerate little change in external salinity; euryhaline animals tolerate a wide range of salinities. Fishes that migrate between sea and freshwater can change the direction of salt transport across the gills and adjust the concentration of their urine. Both changes are under endocrine control.

- The extent to which plants can continue to function during water stress depends on their ability to maintain a positive turgor pressure at low water potentials. Plants able to cope with water stress do so by entering a more tolerant stage of their life cycle (seeds) to escape dry conditions, or by either tolerating or avoiding dehydration. In arid climates, perennial plants must conserve water and maximise the amount of CO_2 gained.

- Stomata are regulated to optimise water use. This involves the ability to sense water vapour flux and CO_2 concentrations, and the hormone ABA, which is released from roots and mesophyll cells in wilting leaves.

- C_4 and CAM modifications of photosynthesis allow plants to tolerate drought and high temperature by increasing water-use efficiency.

- Lack of O_2 in waterlogged soil decreases cellular respiration by root cells. This impairs root growth, water uptake and nutrient uptake. Also, toxic substances are produced by anaerobic respiration of soil microbes.

- High salt concentrations in soil can lead to water stress and ion toxicity. Compartmentalisation of ions within cells and within the plant, the ability to exclude salt at the roots and the balance of transpiration and leaf growth with ion uptake all contribute to salt tolerance.

- Ion toxicity can result from an increase in concentration of ions to toxic levels at low soil pH, or from the effect of pollutants that facilitate entry of toxic ions into a plant.

- Damage to plants at extremes of temperature is due to alteration in the tertiary structure of proteins, to changes in membrane fluidity or to disruption of biochemical processes. Dehydration can occur as a result of high or freezing temperatures.

QUESTIONS

1. What are biological antifreezes and how do they work?

2. If you took a saw, cut a termite mound in half, and rotated the severed end through 90°, describe how the microhabitat of the insects would change.

3. Describe the circulatory changes that occur when a diving mammal remains underwater for an unexpectedly long period, and explain why they are advantageous.

4. Explain why some athletes move to high altitude for some months to train before returning to sea level to compete. What are the short- and long-term changes that occur when they move to high altitude, and when they return to sea level?

5. Why is efficient water use in a terrestrial plant beneficial to salt tolerance?

6. Why is it unlikely that a single gene is responsible for either drought or salt tolerance in plants?

7. A leaf is detached from a plant and kept in its natural orientation. What could you conclude regarding its temperature regulation if the leaf's temperature remained constant and only slightly above ambient temperature?

8. In Figure 29.13, approximately how much would the osmotic pressure of the cell sap of the daisy have to increase in order for turgor pressure to be maintained constant during drought?

9. List the major differences in the ways that animals and plants cope with environmental stresses.

10. Lake Titicaca in the Andes of South America is at an elevation of 3812 m with a maximum water depth of 281 m. The temperature of the surface water averages 10°C, with less than a 4°C annual range. A frog, *Telmatobius*, lives in the lake. What would you expect to be its major adaptations to this habitat? Would you expect plants to occur around the edge of the lake? Explain your answers.

PART 6

EVOLUTION

AND

BIODIVERSITY

EVOLVING LIFE

During evolution new kinds of organisms arise and others disappear through competition, climatic changes or natural catastrophies. At a time when we realise that biological diversity is our most precious resource, species are being lost at a rate that is probably 1000 to 10 000 times the rate before human intervention. There is little precise information available regarding the number becoming extinct but, at the current rate, 20% may be gone in as few as 25 years.

It is this dramatic loss of species caused by human activity that is known as the **biodiversity crisis** (see also Chapter 45). Each species is a repository for an immense amount of genetic information, from 1000 to 400 000 genes, and each species is represented by many individuals, from less than a few thousand to many millions. Each time a species is lost it will never reappear. We do not know how many species a given ecosystem or the world as a whole can survive without. Nor have we explored the possible uses of species, let alone even recognised their existence.

Of the 6–10 million species estimated for the world, only about 1.4 million have names and have been documented. Information about biology, relationship (classification) and distribution is available for only a small proportion of species. Among the species that have been named, 750 000 are insects, 41 000 are vertebrates and about 200 000 are flowering plants. However, while most vertebrates and flowering plants have been documented, only a small proportion of invertebrates, lower plants, fungi, algae and bacteria have been named. The least-known floras and faunas are also among the most species-rich, including those in many areas of the tropics and subtropics.

At an international level biologists are working together to discover the world's biodiversity, so that it can be synthesised into a classification system that will reflect our knowledge of the **phylogeny** of organisms, that is, their history and evolutionary relationships.

BOX 30.1   AUSTRALIA'S BIODIVERSITY

Australia has a very rich flora and fauna—somewhere between 5% and 8% of the world's total species diversity. We do not know how many species occur in Australia, but there are well over 400 000, of which only about 130 000 species have been described to date (Fig. a). Some groups are relatively well known: there are nearly 25 000 species of vascular plants, 3600 fishes, 700 reptiles, 850 birds and 276 native mammals. It is estimated that Australia has a flora of nearly 250 000 species of fungi and nearly 200 000 species of insects and related groups. The insect order Coleoptera (beetles) alone comprises over 50% of Australia's named fauna (Fig. b).

Often the only knowledge we have of Australian insects, other invertebrates and lower plants is that associated with the specimens held in biological research collections, such as our state museums, herbaria or the National Insect Collection in Canberra. At the National Insect Collection, for example, CSIRO biologists have recently surveyed the insects of rainforests that occur in the McIlwraith Range on Cape York Peninsula. This

study is the first major attempt to sample insects in that region and, although far from complete, a very rich fauna has already been demonstrated. In the three-week survey during the beginning of the dry season, more than 2000 species of moths were found. Several species previously only recorded from New Guinea were found and several species of large moths were seen for the first time.

A very large proportion of all species occurring in Australia, perhaps over 80%, are endemic. For example, the moth family Oecophoridae, known as mallee moths, has approximately 6000 species in Australia. Many of these species occur in mallee and in dry sclerophyll *Eucalyptus* forest where their larvae, which feed on dead eucalypt leaves, play an important role in the breakdown of sclerophyllous leaf litter to humus. In comparison, only approximately 100 species of Oecophoridae occur in the whole of Europe and in America north of Mexico. Australian oecophorids probably diversified with the eucalypts. Another group of insects, the ancient moth family Lophocoronidae with six species, is endemic to the mallee and dry sclerophyll forests of southern Australia. This moth family, which probably had a wider distribution in the past (in the Cretaceous), is now known to be the closest relative of all the higher moths and

butterflies, nearly 250 000 species worldwide. The question of the value of species is a difficult one but it seems that the conservation of this unique, endemic family Lophocoronidae must have some priority.

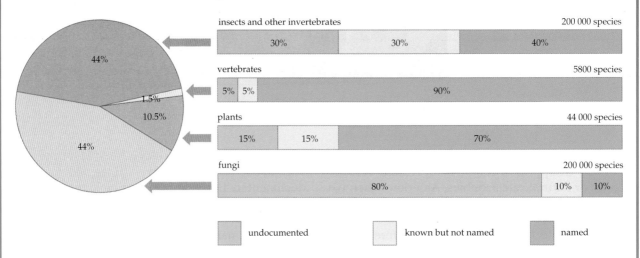

(a) Australia's biodiversity, showing the proportion of species that are named, known but as yet unnamed, and undocumented

(b) Diversity in size of beetles, order Coleoptera. The largest beetle shown here has a body length of 166 mm and the smallest of 0.5 mm

DISCOVERING PHYLOGENY

Evolutionary biology aims to discover the history of life, to answer the question 'How have living organisms come to be the way they are?'. One major task is to discover the evolutionary tree of life and another is to determine the mechanisms of evolution that lead to change. Before we look at how evolution happens (Chapter 32), we first need to understand the methods that biologists use to discover phylogeny.

Comparing morphology

Hypotheses about evolutionary relationships may be based on comparative morphology, the comparison of body form, including the study of embryology and the interpretation of fossils. Many people suppose that phylogeny can be discovered *directly* from the fossil record by studying a graded series of old to young fossils and by discovering ancestors, but this is not true. The fossil record supplies evidence of the geological ages of the forms of life, but not of their direct ancestor–descendant relationships. There is no way of knowing whether a fossil is a direct ancestor of a more recent species or represents a related line of descent (lineage) that simply became extinct. Phylogeny is discovered by comparing organisms, both living and fossil, with one another.

Comparative morphology as a scientific discipline has a long history and biologists of the nineteenth century showed it to be a key to reconstructing

phylogeny (see Box 30.2). For example, if we compare the early stages of development in vertebrates, we see that gill slits form in all embryos, even though they do not persist in adults other than fishes. The striking similarity of the embryos of fishes, frogs, lizards, birds and mammals (Fig. 30.1) is one reason why we hypothesise that these organisms had a single line of descent and why we classify them as vertebrates. The similarity of the early embryonic stages in all vertebrates indicates a fundamental step in their developmental programs. Modification of the later stages of development accounts for the great diversity of form within the vertebrates; for example, snakes do not develop legs, birds develop feathers, humans have a brain larger than that of other primates.

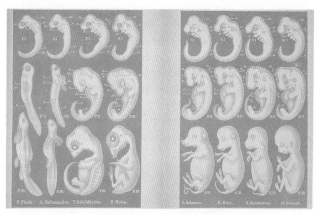

Fig. 30.1 There is a striking similarity between the embryos of vertebrates, from fishes to mammals

BOX 30.2 UNITY OF PLAN

That all of life is based on the same fundamental principle was an important idea of the French zoologist Etienne Geoffroy Saint-Hilaire (1772–1844). He tried to see the 'unity of plan' first through comparison of adult and then of embryonic structures of animals. He recognised corresponding parts of different animals with reference to their 'connections', namely by observing the position of the part relative to all other parts of the animal. He believed, for example, that the middle ear bones of mammals—malleus, incus and stapes—correspond to the opercular (cheek) bones of fishes. Today we use the term 'homologous', after the usage established by the English palaeontologist Richard Owen (1804–92), for parts with the same connections. Many of Geoffroy's comparisons were improved by later workers who found, for example, that the jaw and not the opercular bones of fishes—articular, quadrate and hyomandibular—are the correct homologues of the mammalian ear bones.

Geoffroy's students believed that homologies existed between adult parts of animals as diverse as molluscs, insects and fishes, a point of view opposed by Geoffroy's famous contemporary Georges Cuvier (1769–1832). Today, no biologist imagines that the chitinous

Etienne Geoffroy Sainte-Hilaire (1772–1844)

exoskeleton of insects and the bony skeleton of vertebrates are homologous. Geoffroy's 'unity of plan' led him eventually to believe that all life evolved through descent with modification, as first suggested in a scientific context by another of Geoffroy's famous contemporaries, Jean-Baptiste de Lamarck (1744–1829).

Biologists of today see a 'unity of plan' (or 'unity of composition') at the molecular level, because of the universal occurrence of DNA. They use Geoffroy's principle of connections to recognise homologous positions in base sequences of nucleic acids, and in amino acid sequences in proteins.

Divergence and convergence

Comparative morphology focuses on **divergence**, departure from the ancestral form. For example, study of bones reveals that the flipper of a porpoise has the same basic structure as the front leg of a frog or crocodile, and the wing of a bird or bat (Fig. 30.2). Structures that have the same basic plan but not necessarily the same function are **homologous**. Homologous structures indicate common inheritance and they allow us to recognise that frogs, crocodiles, birds, bats and porpoises are related as tetrapods (four-limbed vertebrates).

By **convergence**, organisms from different, distantly related lineages come to resemble one another. Plants of the families Cactaceae in the New World and Euphorbiaceae in the Old World look remarkably similar with their succulent, leafless stems (Fig. 30.3). Both groups of plants are adapted to living in arid environments. The Australian echidna, with its protective covering of spines, resembles the European hedgehog; the Australian lemuroid ringtail possum earned its common name because it is so monkey-like, leaping from tree to tree; porpoises and seals have become streamlined for swimming, as have fishes; social behaviour has evolved a number of times among bees and ants.

Convergent evolution results in features that are so similar that we could be misled in our search for

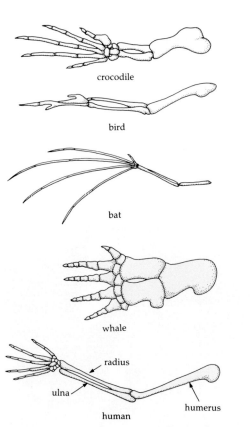

Fig. 30.2 The forelimb of a crocodile, bird, bat, whale and human have the same basic structure and are homologous

(a)

(b)

Fig. 30.3

(c)

Fig. 30.3 Convergence. Succulence occurs in different organs and different plant groups that inhabit arid environments throughout the world. Plants with leafless, succulent stems include (**a**) cacti (family Cactaceae) and (**b**) euphorbs (Euphorbiaceae). (**c**) Plants with succulent leaves include *Aloe*, a member of the monocot family Liliaceae

evolutionary relationships. For example, the branching pattern in the phylogenetic tree in Figure 30.4 shows the relationship of the platypus, koala and dingo to one another.

Comparison of the characteristics of these three species shows that they share a number of general features. All three have hair, internal fertilisation and suckle their young with milk from mammary glands. Both the koala and dingo have separate anal and urogenital openings as adults, mammary glands with teats and produce eggs that lack a shell. The platypus has only one opening, lacks teats and lays an egg with a thin papery shell. These characteristics of the platypus are more general, being shared with a larger group of organisms including snakes, lizards and birds. A more general character is considered ancestral or primitive, with an inheritance that is more remote. Characters that are less general, such as those shared by the koala and dingo but not the platypus, are derived or advanced, and are considered to have evolved more recently.

Of course, each species also has its own unique characters because each lineage, each branch in the tree, has undergone evolutionary change after divergence. An adult platypus lacks teeth, has a duck-like bill and webbed feet used for swimming; the koala is a highly specialised mammal with a gut adapted to digesting eucalypt leaves; the dingo is a carnivore specialised for hunting prey.

phylogeny if other features did not indicate that the organisms are unrelated. Thus, the different flowers of cacti and euphorbs indicate that these two plant groups are not closely related; porpoises and seals have the characteristics of other mammals and are clearly not fishes. Structures that have a similar function as a result of convergence, such as the wings of butterflies and birds, are **analogous**.

> Hypotheses about evolutionary relationships (phylogeny) are traditionally based on comparative morphology and the discovery of homologous structures. Homologous structures are indications of common inheritance. Analogous structures, which have similar functions, result from convergence, and are misleading in the search for phylogeny.

Finding a phylogenetic tree

The phylogeny of a group of organisms can be described by the pattern of branching depicting their

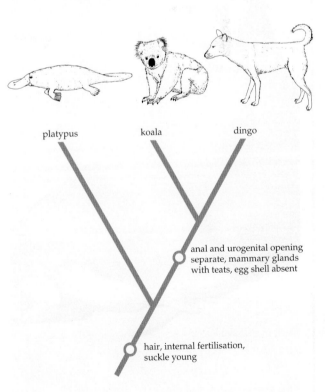

Fig. 30.4 A phylogenetic tree showing that the koala and dingo are related more closely than either is to the platypus

BOX 30.3   RIDDLE OF THE PINNIPEDS

Pinnipeds (seals, walrus and sea lions) have posed a persistent problem for evolutionary biologists. As early as last century, some biologists considered them to be two separate groups among the dog-like carnivores descended from different terrestrial ancestors, the sea lions and walrus being descended from bear-like ancestors and seals from otter-like ancestors. The argument for two groups was based on comparison of bones, principally cranial features. Features shared by pinnipeds were interpreted as convergent, relating to adaptations to an aquatic lifestyle. Other biologists considered that pinnipeds are a single group, descended from a common marine ancestor.

New and compelling evidence favours a single group and comes from molecular biology, biochemistry, genetics and parasitology. The eye lens protein, α-lens crystallin A, shows two amino acid substitutions (at positions 51 and 52), shared by sea lions and seals, that are unique among vertebrates. Seals, walruses and sea lions also share another amino acid substitution that occurs in a few other animals (whale and chicken)

outside the carnivores. Pinnipeds also share a unique bile acid (phocaecholic acid) found in no other animal yet sampled.

DNA–DNA hybridisation data also support the hypothesis of one group. DNA–DNA hybridisation is a technique where single strands of DNA from two organisms are combined to form hybrid double helices. The extent of hybridisation is directly related to the similarity of nucleotide sequences and gives an overall measure of the genetic similarity of a pair of organisms. Seals, walrus and sea lions are more similar to each other than they are to other dog-like carnivores. The karyotypes of the pinnipeds are similar; diploid numbers are 32, 34 or 36, and chromosome banding patterns appear to be homologous.

Finally, seals, walrus and sea lions share closely related ectoparasites. All three kinds of pinnipeds are hosts to a particular group of parasitic lice. These different data together with comparative anatomy support the phylogeny shown here.

(b) The Weddell seal from Antarctic waters keeps its hind limbs pointed posteriorly

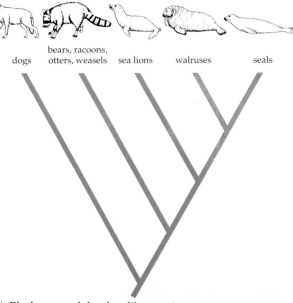
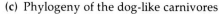

(c) Phylogeny of the dog-like carnivores

(a) A sea lion characteristically has external ears

Comparing molecules

Molecular biology, part of which is concerned with comparing DNA or its protein products from different species, is now making an important contribution to our knowledge of phylogeny. With the development of techniques for sequencing amino acids in proteins and nucleotides in DNA molecules, it is now possible to compare organisms at the most basic level. Furthermore, unlike morphology, molecules allow us to compare organisms that have no apparent morphological homologies, for example, a eucalypt tree and a kangaroo! Because all organisms, ranging from bacteria to fungi, algae, flowering plants, insects, birds and mammals, have nucleic acids, DNA sequencing promises to provide data for discovering the tree of all life.

Amino acid sequences

With the understanding of the role of DNA in protein synthesis (Chapter 11), it became clear that the sequence of amino acids in a polypeptide chain is a molecular record of the genetic information and evolutionary history of an organism. In the 1960s and 1970s, it was shown that humans and chimpanzees have identical amino acid sequences in the respiratory enzyme cytochrome *c* and in α- and β-haemoglobins, while they have one amino acid difference in myoglobin. However, not all point mutations in DNA cause amino acid substitutions, for the genetic code is degenerate. The same amino acid may be coded by different base triplets. For example, valine is specified by CAA, CAG, CAT or CAC, and any change in the third position of each of these codons is silent. For this reason, and because it is now technically easier and cheaper to sequence nucleic acids than proteins, amino acid sequencing is being replaced by DNA sequencing.

Nucleotide sequences

Nucleotide sequencing has been carried out on mitochondrial DNA, chloroplast DNA and nuclear DNA of various organisms. The amount of information is potentially enormous because every nucleotide base pair in a DNA sequence is a separate character. The following is a simple example to show how the sequence of bases in a piece of DNA can be used to discover the relationships of organisms to one another.

Table 30.1 shows three nucleotide positions (at positions 1156, 2311 and 3257) in the 28 S ribosomal gene of three vertebrates—mouse, cockatoo and frog—and an insect.

There are three possible phylogenetic trees for the three vertebrates (Fig. 30.5). Each tree seems to be supported by one base substitution. For example, mouse and frog share T at position 1156, suggesting that they are related (tree 1). On the other hand, cockatoo and frog share G at position 2311 (tree 2) and mouse and cockatoo share G at position 3257 (tree 3).

If we include the sequence for the insect *Drosophila*, which we assume is a branch early on the phylogenetic tree, we can determine which base positions are phylogenetically informative within the vertebrates. We assume that a base that is in both the *Drosophila* sequence and any of the vertebrate sequences (e.g. A at position 3257) must have been inherited from a more remote common ancestor. A different base (G) at the same position in the vertebrate sequences must have evolved within that group. Thus, in this case, G is a derived character (G substituted for A). G at position 3257 for mouse and cockatoo therefore tells us that tree 3 best fits

Table 30.1 A comparison of three nucleotide positions in the 28 S ribosomal gene of three vertebrates and an insect

	Base position		
	1156	2311	3257
Mouse, *Mus*	T	A	G
Cockatoo, *Cacatua*	C	G	G
Frog, *Xenopus*	T	G	A
Insect, *Drosophila*	**T**	**G**	**A**

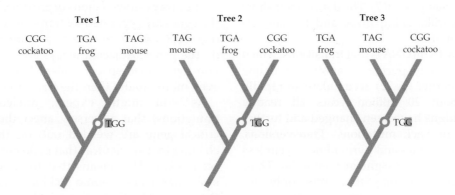

Fig. 30.5 Three possible phylogenetic trees for the frog, mouse and cockatoo

the data (Fig. 30.6). The bases at the other two positions are not informative of vertebrate relationship because only cockatoo has the derived character (C) at 1156, and only mouse (A) at 2311.

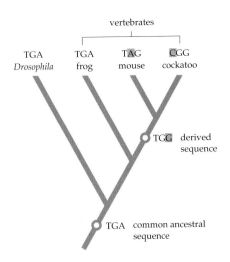

Fig. 30.6 The phylogenetic tree that best fits the DNA data when we include the insect *Drosophila*. The frog has the ancestral sequence (same as *Drosophila*). The mouse and cockatoo share a base substitution (G substituted for A) at the third position, indicating their relatedness. The mouse also has one unique substitution (A for G at the second position), as does the cockatoo (C for T) at the first position

This example uses only three base positions, but in reality a biologist would compare hundreds of positions (e.g. by sequencing a whole gene) to be confident of finding the correct phylogeny.

Sequencing amino acids in proteins and nucleotides in DNA is contributing to our knowledge of phylogeny.

Mitochondrial DNA

The early sequencing work was done on hominid mitochondrial DNA (mtDNA). Sequence data show that African apes (gorilla and chimpanzee) and humans share a common ancestor, and that the orang-utan diverged earlier (Fig. 30.7). The data do not show which two, of gorilla, chimpanzee and human, are most closely related.

One problem with mtDNA is that **transitions**, which are changes from one purine to another and one pyrimidine to another (p. 22) accumulate so rapidly that within about 20 million years all readily substituted positions have been changed and further changes result in back mutations. **Transversions**, changes between purines and pyrimidines, occur less frequently (by a factor of eight in hominids). Thus, mtDNA can give meaningful results only for organisms that have diverged from one another in a relatively short time (about 20 million years).

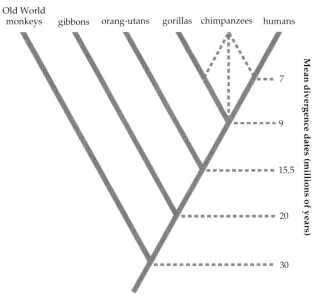

Fig. 30.7 Phylogeny of the great apes and humans. Nucleotide sequencing of mitochondrial DNA shows that African apes (gorilla and chimpanzee) and humans are closely related and that orang-utan is a lineage that diverged earlier. The position of chimpanzees in the tree is ambiguous. Amino acid sequencing and DNA–DNA hybridisation (see Box 30.3) suggest that chimpanzees are the closest relatives of humans. In contrast, two morphological characters, hind limb morphology associated with knuckle walking and enamel structure of teeth, link gorillas and chimpanzees as closest relatives

Nuclear-encoded RNA and molecular clocks

Many recent studies, such as in our example of the frog, mouse and cockatoo, have been made on nuclear DNA that encodes for ribosomal RNA (rRNA). There are three genes (in vertebrates they are 18 S, 5.8 S and 28 S rRNA genes) that form an array (unit) that is repeated many times along a chromosome. The three rRNA genes have evolved at a very slow rate and contain the most highly conserved sequences of DNA that occur in living organisms, and portions of these genes have been used to reconstruct phylogeny back to the origin of life nearly four billion years ago.

The arrays show regions of greater variation as well: the genes are separated by transcribed spacer regions (Fig. 30.8), which evolve at a more rapid rate than do the genes. Adjacent arrays are also separated by non-transcribed spacers (silent DNA), which evolve even more rapidly than the transcribed spacers.

As you might expect, nucleotide changes (mutations) that adversely affect the function of a critical gene are weeded out; on the other hand, changes that are neutral, that is, do not affect function, accumulate. This means that there are a variety of molecular clocks, some ticking slowly and some ticking quickly. The more variable spacers can be used to reconstruct the phylogeny of organisms back to

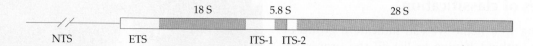

Fig. 30.8 The ribosomal DNA repeat unit of vertebrates consists of three genes (18 S, 5.8 S and 28 S) separated by internal spacer regions (ITS) and an external transcribed spacer (ETS). Each unit is separated by a non-transcribed spacer (NTS)

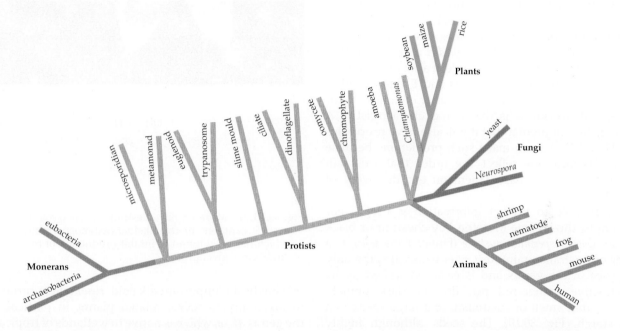

Fig. 30.9 A tree of major groups of life based on homologous parts of ribosomal RNA genes in the cytoplasm of prokaryotes and nucleus of eukaryotes. The base, or root, of the tree is not specified. The tree is interesting because it places oomycetes, previously classified as fungi, near photosynthetic organisms (chromophytes, see Chapter 35). Ciliates, traditionally considered animal-like, are placed in a more remote position, near the photosynthetic dinoflagellates. Microsporidians and metamonads are eukaryotes that evolved before the presumed endosymbiotic event that led to mitochondria (see Chapters 3 and 35)

1–50 million years ago, whereas rRNA genes can be used to reconstruct phylogenies back to four billion years ago. Thus, the various regions of ribosomal DNA (genes, transcribed spacers and non-transcribed spacers) can be used to construct phylogeny over the entire history of life (Fig. 30.9).

Different molecules evolve at different rates. Base substitutions accumulate rapidly in mitochondrial DNA, which can be used for studying the phylogeny of organisms back to about 20 million years; cytoplasmic ribosomal RNA genes are highly conserved and can be used to study organisms dating back to the origin of life, four billion years ago.

CLASSIFICATION

In discovering phylogeny, we recognise groups of organisms that have characters in common (homologies). On page 665, we grouped the platypus, koala and dingo together because they have hair, internal fertilisation and suckle their young. In fact, we can make the general statement that *all* mammals have these features. We can also make less general statements about subgroups within the mammals: koalas, wombats, kangaroos and possums all raise their young in a pouch and are classified as marsupials; dingos together with shrews, bats, primates, rodents, whales, cows and horses are 'placental' mammals. Thus, groups and subgroups of organisms can be defined, placed in a **classification** and given scientific names. What groups should be recognised and named are the primary concerns of classification.

Classification is an old notion dating back to the early Greeks. Centuries later the Swiss botanist de Candolle (1813) coined the term **taxonomy** for plant classification. Today, the word taxonomy is used in a broader way to refer to the methods and principles of classification of all organisms.

Purposes of classification

By referring to a name we can indicate a group of organisms without having to list all the properties known to occur in each of the organisms belonging to the group. Thus, given the vast diversity of organisms on earth (see p. 661), classification obviously performs an essential function in information storage and retrieval. Considerable information is subsumed into single words such as mammal or monocotyledon.

Classification, however, does more than store information; it serves also to *predict* information that we do not yet have. For example, if a set of organisms share characteristics A, B, C, D and E that no other organisms have, and we find another organism about which all we know is that it has characteristics A and B, we can predict that it will also have properties C, D and E. We can make such predictions because our classifications reflect the order that exists in nature—we have based them on our discovery of phylogenies.

Being able to predict information is useful. For example, the legume Moreton Bay chestnut or black bean, *Castanospermum australe* (family Fabaceae), is a tree of Australian rainforest river banks. It is the only species in the genus and is easily identified by its distinctive orange-red pea flowers, dark pinnate leaves that smell of cucumber, and large seeds rich in starch (Fig. 30.10). The seeds, although highly poisonous, were gathered by Aborigines for food; the toxins were removed by a process of grating, soaking and baking. One of the toxins is an alkaloid, castanospermine (Fig. 30.10), which has been found to control cancer cells in rats and inhibit production of the AIDS virus glycoprotein, an essential element in HIV infectivity. In a search for other sources of

(a)

(b)

Fig. 30.10 **(a)** Moreton Bay chestnut, *Castanospermum australe*, contains **(b)** the alkaloid castanospermine, which has been found to inhibit production of the AIDS virus glycoprotein

this medically important alkaloid, researchers turned to taxonomy to reveal related plants, in particular the genus *Alexa*, which is native to wetlands of tropical South America, including the Brazilian Amazon basin. *Alexa* and *Castanospermum* are known to have very similar and distinct pollen within the family Fabaceae, and their close relationship was confirmed by the finding of the alkaloid castanospermine in *Alexa*. This compound has not yet been found in any other plant genus.

BOX 30.4   BUSH MEDICINES

Bush medicines have always been important to Australian Aborigines and European colonists were not slow to try them. Native currants, *Leptomeria acida*, gave some relief from scurvy, oil in the leaves of peppermint trees, *Eucalyptus*, aided digestion, and chewing the leaves of native pepper, *Piper novae-hollandiae*, treated sore gums.

Bush medicines became important during World War II when normal supplies of drugs diminished. An alkaloid extracted from corkwood, *Duboisia myoporoides*, a tree known to Aborigines for its narcotic effects, was used to treat soldiers who suffered shell-shock and air sickness. After the war, 4000 species of Australian plants were surveyed in a major search for alkaloids that may have proved useful to medicine. Although 500 alkaloids were discovered, many in rainforest plants and many previously unknown to science, the survey did not directly lead to the commercial development of new drugs. However, in recent years there has been a renewed search for phytochemicals in rainforest species for contraceptives and drugs for the treatment of cancer and AIDS.

Family Solanaceae

Plant steroidal saponines, which closely resemble human hormones, can be converted into synthetic hormones and are used in the manufacture of the contraceptive pill. Eastern European manufacturers first used Mexican species of yams, *Dioscorea*, but with the decline in this resource due to habitat loss they turned, with success, to Australian plants in the family Solanaceae as a source of precursors for synthetic hormones. The kangaroo apple, *Solanum laciniatum* (Fig. a), which occurs naturally in Australia and New Zealand, contains the alkaloid salasodine, very similar to disogenin in the yams and easily converted to a steroid. Extensive plantations of kangaroo apple are now grown industrially in Eastern Europe.

The family Solanaceae includes food plants such as potatoes, tomatoes, eggplant and peppers, as well as tobacco and hallucinogenic plants, such as deadly nightshade and *Datura*. Plants in this family contain a vast range of phytochemicals and are a 'good bet' in the search for medicinal drugs. Harvesting corkwood,

also a member of the Solanaceae, has been a commercial success. Each year Australia exports large amounts of dried powdered leaves, representing half the world's market of the alkaloid hyoscine, used to treat motion sickness, stomach disorders and side-effects of cancer treatment.

In addition to the common corkwood, *Duboisia myoporoides*, which occurs on the margins of rainforests in Australia, New Guinea and New Caledonia, there is a rarer species, *D. leichhardtii*, restricted to vine thickets and scrub in southern Queensland. Both species produce hyoscine but the greatest yields are produced by hybrids of the two species (Fig. b), which are now cultivated in Queensland.

The potential use of other native species as 'bush medicines' is one argument for the conservation of Australia's remnant patches of rainforest.

(i)

(ii)

(a) The potato family Solanaceae includes the kangaroo apple (i) *Solanum laciniatum* and (ii) *S. aviculare*, which are similar shrubs with dark green leaves, purple flowers and orange or scarlet berries. *Solanum* is a source of steroids for the contraceptive pill

(b) The corkwood hybrid, *Duboisia leichhardtii myoporoides*, is grown commercially in southern Queensland for hyoscine, an alkaloid used to treat motion sickness and side-effects of cancer treatment

Classification: a hierarchy of taxa

In the biological system of classification, a group is called a taxon (pl. taxa) and is given a Latinised name; for example, Plantae (multicellular land plants), Magnoliophyta (flowering plants) and Liliopsida (monocotyledons: flowering plants with a single seed leaf in the embryo). Since the time of the Swedish naturalist Carl Linnaeus (1707–78), the biological system has been conceived as a hierarchy with specified levels or ranks. Thus, the taxon Plantae has the rank of kingdom, Magnoliophyta is a phylum, and Liliopsida a class.

The rank order of taxa commonly used today is:

> kingdom
>> phylum
>>> class
>>>> order
>>>>> family
>>>>>> genus
>>>>>>> species

Thus, species are grouped into genera, genera into families, families into orders, orders into classes, classes into phyla and phyla into kingdoms. Intermediate ranks can be designated by 'sub', such as subfamily or subclass. The platypus, together with two species of echidnas, are formally classified in subclass Prototheria (monotremes) within class Mammalia, while the koala and dingo and other related mammals are in subclass Theria (marsupials and 'placentals') within class Mammalia (Table 30.2).

The hierarchical classification of the platypus, koala and dingo expresses the branching pattern of the phylogeny of these three organisms that we constructed on page 7. Class Mammalia is the branch within the vertebrates that includes all three species (Fig. 30.11). Subclasses Prototheria and Theria are more recent branches within the mammals. Taxa recognised in this way are termed **monophyletic**, meaning an entire branch on a phylogenetic tree. However, many traditional classifications include taxa that have been found not to be monophyletic. For example, in the 1860s, Ernst Haeckel placed orang-utans, gorillas, and chimpanzees in one family, now called Pongidae, separate from the family for humans, now called Hominidae. Modern biologists agree that gorillas and chimpanzees are related more closely to humans than to orang-utans, as indicated by the branching pattern of the phylogeny in Figure 30.7. A classification based on phylogeny would group gorillas, chimpanzees and humans (one branch of the tree) in one monophyletic taxon (Fig. 30.12).

> In discovering phylogeny, we recognise groups of organisms about which we can make general statements. A classification results when we name these groups as taxa, at the rank of kingdom, phylum, class, order, family, genus and species. Essential functions of classification are information storage and retrieval, and prediction.

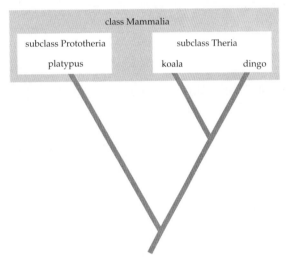

Fig. 30.11 A classification of the platypus, koala and dingo based on the branching pattern of their phylogeny. Taxa recognised in this way are termed monophyletic, that is, a taxon including all lineages that can be traced back to a common branch point

The binomial system

Naming taxa is governed by three international sets of rules called **codes of nomenclature**, one each for bacteria (and viruses), plants (and fungi), and animals. Having three sets of rules does lead to some confusion. There is no consistency between zoology and botany in the word ending of different taxonomic ranks; for example, botanical family names end in 'aceae', while zoological family names end in 'idae' (see Table 30.2). However, there are many similarities in the codes.

The ranks or categories of **genus** and **species** have a special significance because they form the basis of the **binomial system**, devised by Linnaeus. In this system, the name of each kind of organism consists of two parts. *Homo sapiens* (humans), *Pan troglodytes* (chimpanzee), *Gorilla gorilla* (gorilla) and *Acacia melanoxylon* (blackwood) are each an example of a binomial. Binomials, once created, do not necessarily remain constant. Although *Homo sapiens* is the only living species in the genus *Homo*, in the future the

Table 30.2 Biological classifications

Rank	Platypus	Koala	Dingo	Blue gum	Kangaroo paw
Kingdom	Animalia	Animalia	Animalia	Plantae	Plantae
Phylum	Chordata	Chordata	Chordata	Magnoliophyta	Magnoliophyta
Class	Mammalia	Mammalia	Mammalia	Magnoliopsida (dicots)	Liliopsida (monocots)
Subclass	Prototheria	Theria	Theria	Rosidae	Liliidae
Order	Monotremata	Marsupialia	Carnivora	Myrtales	Liliales
Family	Ornithorhynchidae	Phalangeridae	Canidae	Myrtaceae	Haemodoraceae
Genus	*Ornithorhynchus*	*Phascolarctos*	*Canis*	*Eucalyptus*	*Anigozanthos*
Species	*O. anatinus*	*P. cinereus*	*C. familiaris* subsp. *dingo*	*E. globulus*	*A. manglesii*

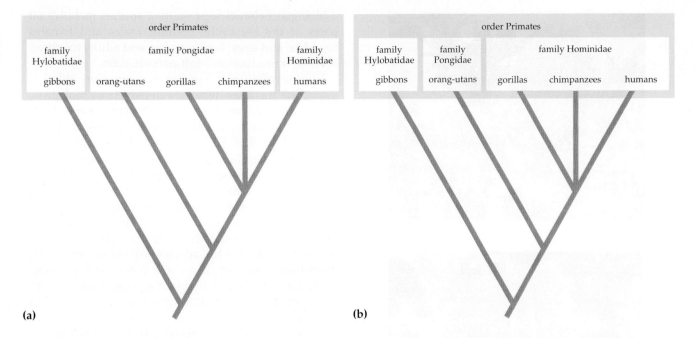

Fig. 30.12 (a) According to traditional classifications, great apes are grouped in the family Pongidae, separate from humans, family Hominidae, but this is not consistent with our knowledge of phylogeny (see Fig. 30.7). (b) An alternative classification would recognise the monophyletic taxon that includes gorilla, chimpanzees and humans (which share a common branch point), at the rank of family or even genus as originally proposed by Linnaeus

chimpanzee and gorilla could conceivably join humans as *Homo troglodytes* and *Homo gorilla*. It is interesting to note that, in 1758, Linnaeus placed the chimpanzee in *Homo*.

In a binomial, the genus name is always written first, as in *Eucalyptus viminalis* or *Eucalyptus globulus*. The second word (the specific epithet), *viminalis*, is meaningless alone because different species in different genera may have the same specific epithet, for example, *Eucalyptus viminalis* is the manna gum tree and *Callistemon viminalis* is a bottlebrush. Names given to taxa may be in honour of people. The genus *Banksia* was named after Sir Joseph Banks, a wealthy naturalist who voyaged to Australia with Captain Cook on the HMS *Endeavour* in 1770; the tropical plant genus *Aristolochia* was named after the celebrated philosopher Aristotle. Names may also be descriptive. *Banksia serrata* (saw-toothed banksia) has distinctive leaves with a serrated edge; *viminalis* refers to the willow-like habit of both the eucalypt and callistemon; *punctata* means spotted, as in *Sillaginodes punctata*, the southern Australian King George whiting, a fish easily distinguished by its spotted colour pattern.

Binomials are useful for unambiguous communication about organisms. Although common names exist for many taxa, they mean different things to different people. Thus, the Australian magpie, *Gymnorhina tibicen*, is not the same as the American magpie, *Pica pica*, and the prickly pear, the cactus *Opuntia stricta*, is very different from the pear that is cultivated for its edible fruit, *Pyrus*. The Australian

salmon, *Arripis trutta*, is not a salmon. The marsupial frog, *Assa darlingtoni*, of Australian rainforests is obviously not a marsupial, but an extraordinary frog in which the males form pouches on their flanks in which the young tadpoles are carried (Fig. 41.29).

In the binomial system, the name of each kind of organism consists of two parts based on the ranks of genus and species.

Species

As a taxonomic category, the species is the lowest rank in the Linnaean hierarchy to which all organisms must be classified. In practice, samples of organisms that a biologist can distinguish and tell others how to distinguish or diagnose are considered to represent different species. The specimens of three *Banksia* species illustrated in Figure 30.13 differ in the shape and size of leaves, flowers and fruits, characteristics which a plant taxonomist uses for identification. Some species, however, are morphologically very similar but differ in other ways; the fruitflies *Drosophila pseudoobscura* and *D. persimilis* were first recognised as different species by their chromosomes and ecological behaviour, and only later were subtle differences in the male genitalia observed.

Within many species, individuals are very different in form, such as males and females, and young and adults. Male birds are often more brightly coloured than females, and caterpillars are strikingly different from adult butterflies. The sample of organisms that

(a)

(b)

(c)

Fig. 30.13 These specimens are all members of the Australian genus *Banksia*, which is classified in the Southern Hemisphere family Proteaceae, together with *Protea, Macadamia, Persoonia* and *Grevillea*. (a) *B. oreophila*, (b) *B. coccinea* and (c) *B. ashbyi*. The genus is characterised by its flowers and fruits, which are woody follicles. Although all three specimens share characteristics typical of the genus, there are clear differences in the shape and size of their leaves, flowers and fruits by which botanists recognise them as different species

a biologist believes represents a species includes all parts of the life cycle, because males and females together, and eggs, larvae, pupae and adults together, involve reproduction—self-perpetuation.

In an evolutionary context, the question 'What is a species?' has been debated extensively by biologists. Numerous different species concepts have been proposed. One traditional concept, dating from the botanist John Ray (1627–1705), is of biological species, which the ornithologist Ernst Mayr described this century as 'groups of actually or potentially interbreeding natural populations which are reproductively isolated from other such groups'. This species definition is based on the idea that, if organisms of two species freely exchange genes, the difference between the species will break down. The definition, however, does not work well in some animal groups and many plant groups, such as ferns, eucalypts and oaks, that form fertile hybrids. Neither does it seem appropriate for organisms that reproduce asexually, such as bacteria and many unicellular eukaryotes. For evolutionary biologists, the debate centred on species' definitions is concerned with discovering the processes of evolution—the origin of species and higher taxa—which we shall return to in Chapter 32.

> As a taxon, the term species refers to the lowest rank in classifications. In an evolutionary context, the species concept has been debated extensively and there is no single definition that is agreed upon. The biological species concept emphasises reproductive isolation and the general lack of interbreeding between sexually reproductive species.

KINGDOMS OF LIFE

Linnaeus classified living organisms into two kingdoms—plants and animals—and Georges Cuvier classified all animals into just four phyla. However, with the development in this century of microscopy and biochemistry, our knowledge of the kinds of organisms on earth has increased remarkably. Cell biology has revealed fundamental differences between prokaryotes and eukaryotes. Bacterial classification is changing with the discovery of differences in ribosomal DNA between archaeo-bacteria (methanogens, halophiles and some sulfur-oxidising bacteria) and eubacteria (see Chapter 33). Observations of ultrastructure, cell wall biochemistry and photosynthetic pigments of algae show them to be a diverse assemblage of organisms. With the recognition that the oomycetes (water moulds, white rusts and downy mildews) have cellulose cell walls and flagellated zoospores, it appears that not all organisms traditionally called fungi are fungi!

How the major taxa are related to one another is still being discovered. Slime moulds have amoeboid, animal-like stages but their reproductive structures are like fungi—to what are they related? This uncertainty of relationships is reflected in continuing changes to classification and the expansion of the number of kingdoms (five or more) and phyla (nearly 100).

Most biology books follow a system that recognises five kingdoms: **Monera** (or Prokaryotae), **Protista**, **Fungi**, **Plantae** and **Animalia**. Monera include bacteria and cyanobacteria (blue-green algae), prokaryotic cells that lack a nuclear membrane, mitochondria and chloroplasts (see Chapter 3). The other four kingdoms are eukaryotes. Protista are mostly unicellular, including photosynthetic and non-photosynthetic forms. Fungi, plants and animals are multicellular organisms with different modes of nutrition: fungi absorb organic molecules from their surroundings, plants are producers of organic molecules, and animals are consumers that ingest other organisms for food. This classification, however, is not completely satisfactory, particularly with respect to the Protista, which are an assemblage of a number of different lineages and are not a monophyletic kingdom (see Fig. 30.9). Some of the green algae, for example, are closely related to land plants, yet this relationship is not apparent in the current classification, which groups green algae with other protists. We will explore such problems, and the biology and evolutionary history of living organisms, in more detail in Chapters 31 to 41. You should consult the Appendix, which gives a classification of each of the five kingdoms.

Most classification schemes recognise five kingdoms: Monera, Protista, Fungi, Plantae and Animalia. Monera are prokaryotes and the other four kingdoms are eukaryotes.

SUMMARY

- Hypotheses about evolutionary relationships (phylogeny) are traditionally based on comparative morphology and discovery of homologous structures.

- Homologous structures are indications of common inheritance and allow us to discover phylogeny. Analogous structures, which have similar functions and result from convergence, are misleading in the search for phylogeny.

- Sequencing of amino acids in proteins and nucleotides in DNA contributes to our knowledge of phylogeny. Different molecules evolve at different rates. Base substitutions (mutations) accumulate rapidly in mitochondrial DNA, which can be used for studying phylogeny of organisms back to about 20 million years ago. Ribosomal RNA genes are highly conserved and can be used to study organisms dating back to the origin of life, four billion years ago.

- In discovering phylogeny, we recognise groups of organisms, such as mammals or flowering plants, about which we can make general statements. Naming these groups (taxa) gives a classification. The science that deals with methods and principles of classification is called taxonomy. Essential functions of classifications are communication, information storage and retrieval and prediction.

- Classifications are hierarchical and include the ranks of kingdom, phylum, class, order, family, genus and species. Taxa that include all the lineages that can be traced back to a common branch point are termed monophyletic, but not all taxa in traditional classifications are monophyletic.

- Naming of taxa is governed by international codes of nomenclature. The ranks of genus and species form the basis of the binomial system. As a taxon, the term species refers to the lowest rank in the Linnaean hierarchy. In practice, if a biologist can distinguish between samples of organisms, including all stages in the life cycle, they are considered to represent different species. In an evolutionary context, the species concept has been debated extensively and there is no single definition that is agreed upon. The biological species concept emphasises reproductive isolation and the lack of interbreeding between sexually reproducing species.

- There are more than five million species classified in five kingdoms and nearly 100 phyla. The five kingdoms are Monera, Protista, Fungi, Plantae and Animalia.

QUESTIONS

1. What kinds of evidence support the theory that life evolves?

2. Distinguish between the terms 'homologous' and 'analogous', giving examples.

3. From your knowledge of the cellular characteristics of a bacterium, cyanobacterium (blue-green alga) and a green plant, construct a branching diagram to hypothesise the phylogeny of these organisms, as was done in the text for the platypus, koala and dingo (p. 7).

4. Given your knowledge of DNA, the genetic code and proteins, explain why some parts of DNA accumulate random mutations and evolve rapidly, while other parts are highly conserved and evolve slowly. How can this variation in the rate of base substitutions be useful in discovering phylogeny across a range of organisms?

5. Why is it useful to have a classification of organisms?

6. (a) What is meant by the terms 'taxon' and 'taxonomy'? (b) List in order from highest to lowest the ranks commonly used in biological classification.

7. (a) What are the major characteristics of each of the five kingdoms? (b) How is the kingdom Protista problematical? (c) If you were asked to classify all organisms into two kingdoms only, on what basis would you group them?

8. On what basis do traditional classifications place humans in a family (Hominidae) separate from the great apes (Pongidae)? Why do some taxonomists want to change the classification, and what do they mean when they recommend recognising only 'monophyletic taxa'?

9. Some biologists believe that species (e.g. the dog *Canis familiaris*) are real but that higher taxa (e.g. Mammalia) are unreal, being defined arbitrarily by taxonomists as convenient units of classification. Other biologists disagree and view all taxa (species, genera, families, etc.) as real in the sense that they all have evolved. What could be the arguments for each of these opposing views?

EVOLVING EARTH

How is the history of the earth relevant to the evolution of life? We saw in Chapter 5 that the environment of the early earth was conducive to the evolution of the first prokaryotes, cells that used ready-made organic molecules as a source of raw materials and energy. After the evolution of photosynthetic prokaryotes, which produce oxygen as a by-product of photosynthesis, the earth's atmosphere became oxygen-rich, destroying the conditions that first allowed life to evolve. Through geologic time the physical and chemical environment of the earth has influenced the evolution of life: most importantly, the movements of continents continue to alter the distribution and area of exposed land and sea and therefore the environment of organisms.

In this chapter, we follow the fossil record of marine and terrestrial organisms, from the first multicellular animals and plants to the appearance of humans in relation to the earth's history. The characteristics of modern, extant groups of organisms, particularly Australian flora and fauna, are described in more detail in Chapters 33–41.

GEOLOGIC TIME SCALE

Several centuries ago, before scientists were able to age rocks directly, geologists and palaeontologists devised a *relative* time scale for the evolution of the earth and life. This relative scale was based on the observation that sedimentary rocks, which contain characteristic fossils, are deposited one on top of another. It was assumed that the deepest rock strata and their fossils are the oldest, and that the uppermost strata and their fossils are the youngest. Thus, a geologic feature was established as older than, equal to, or younger than another feature.

Sometimes the same sediment is found in different geographic regions, providing a time reference point.

The best example of a sediment deposited simultaneously over large areas is a thin bed of volcanic ash blown over a large area from a single volcanic eruption. Sometimes the same stratigraphic sequence of fossils is found in different areas. Within north-western Europe, marine Jurassic strata are divided into numerous age zones on the basis of an orderly sequence of forms of ammonites, an extinct group of molluscs (see Chapter 39). The same sequence of ammonites occurs wherever these animals lived in that general region.

With the discovery of radioactivity at the end of the nineteenth century, absolute measurements of geologic age became possible (Box 31.1). The modern geologic time scale (Fig. 31.1) incorporates both actual age measurements (expressed in thousands or millions of years), as well as sequences of relative ages. One problem with measuring the age of ancient rocks is that an error of only 1% for rocks that are 500 million years old represents an error of 5 million years.

The geologic time scale is divided into eras and eras are subdivided into periods. Eras, from oldest to youngest, are the **Pre-Cambrian**, **Palaeozoic**, **Mesozoic** and **Cenozoic**. Boundaries between eras were initially recognised as abrupt changes in the fossil record, marking important biological events. The Pre-Cambrian extends from the origin of the earth through the formation of the earth's crust, oceans and atmosphere, the origin of life as prokaryotic cells, and the evolution of the first eukaryotes and multicellular organisms. The beginning of the Palaeozoic is marked by the appearance of a diversity of animals (the Cambrian explosion); its end is marked by the great Permian extinction. The Cretaceous extinction closed the Mesozoic.

The modern geologic time scale is based on relative ages of sequences of sedimentary rocks and fossils and quantitative measurements based on radiometric dating.

Eras of time	Periods of time	Epochs of time (Cenozoic era only)	Age (millions of years)	Major biological events
Cenozoic	Quaternary	Recent (Holocene)		Humans expand in range Major ice ages and extinction of large animals in Northern Hemisphere 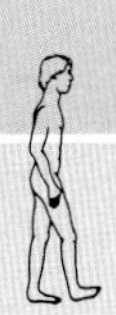
		Pleistocene		
			1.6	
	Tertiary	Pliocene		Extensive radiation of flowering plants and mammals Dominance of gastropods 
		Miocene	23	
		Oligocene		
		Eocene		
		Palaeocene		
			65	
Mesozoic	Cretaceous			First flowering plants Extinction of ammonites, marine and aerial reptiles 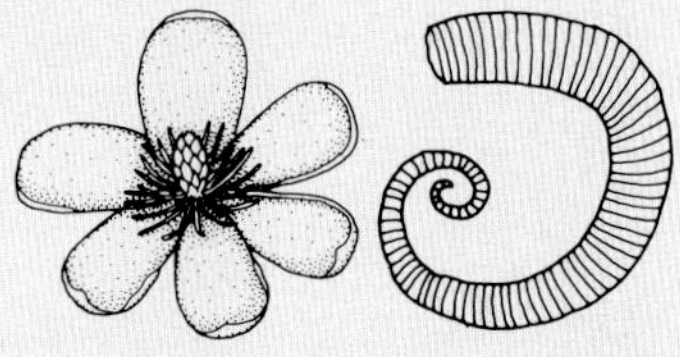
			146	
	Jurassic			Cycads, conifers, ginkgoes, dinosaurs dominant First birds, flying reptiles, marine reptiles 
			208	
	Triassic			First dinosaurs and mammals Dominance of mammal-like reptiles Dominance of ammonites 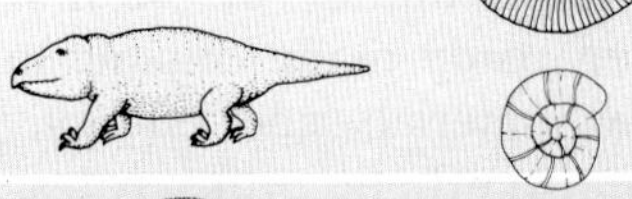
			245	
Palaeozoic	Permian			Extinction of trilobites and many invertebrates Reptiles more abundant as amphibians decline 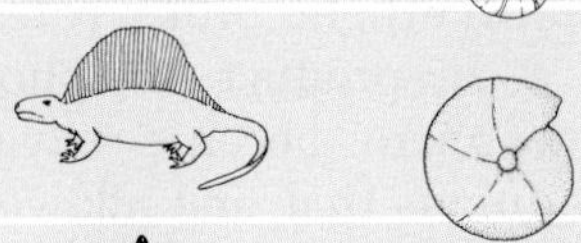
			290	
	Carboniferous			Coal swamp forests Amphibians on land First reptiles Algal-sponge reefs Echinoderms and bryozoans dominant 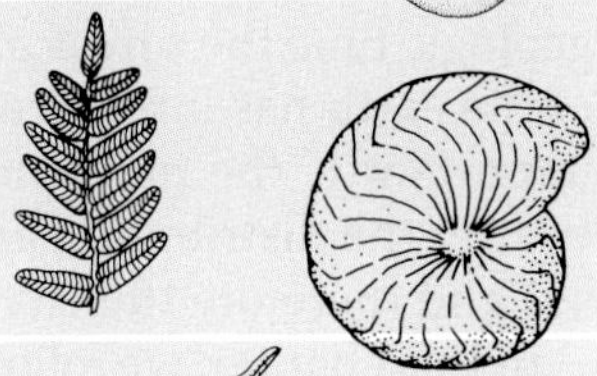
			362	
	Devonian			Oldest land vertebrates Radiation of land plants and fishes Corals, brachiopods and echinoderms 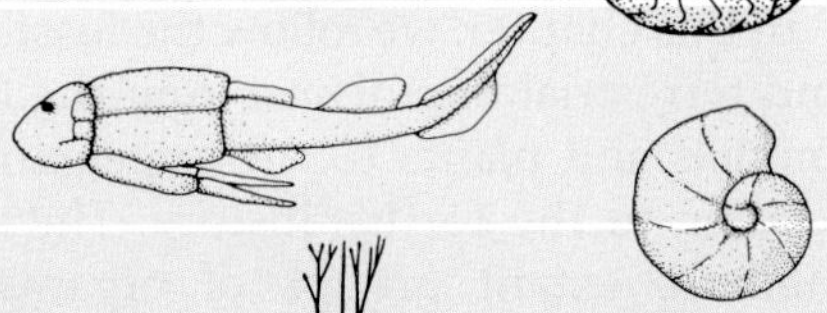
			408	
	Silurian			Oldest life on land: plants, scorpions First jawed fishes 
			439	
	Ordovician			First jawless fishes Diverse marine communities: brachiopods, bryozoans, corals, graptolites, nautiloids 
			510	
	Cambrian			Metazoan animals with skeletons Dominance of trilobites 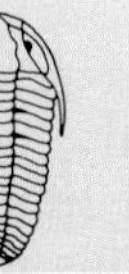
			570	
Pre-Cambrian				First metazoan animals Origin of eukaryotes Origin of prokaryotes
			4560	

Fig. 31.1 Modern geologic time scale

BOX 31.1 HOW OLD IS THAT ROCK?

Radiometric dating

Quantitative determination of geologic age is made possible by means of radioactivity. Atoms of radioactive elements emit radiation as they spontaneously decay. Each radioactive istope has its own constant rate of decay. This constant rate of decay allows definition of a geologic clock, although different methods give ages with different degrees of uncertainty.

Because the decay of individual radioactive atoms is random, the time required for all radioactive atoms of a given sample to decay is indefinite. For this reason, scientists calculate the time required for half of the atoms to decay, the *half-life*. Of the radioactive elements that occur naturally on earth, some are continuously being produced by neutron bombardment and have a relatively short half-life; others were inherited when the earth was formed and have a long half-life. Potassium-40, for example, has a half-life of 1.25 billion years, decaying to argon-40. By measuring the proportion of potassium-40 and argon-40 in a rock that has been undisturbed over geologic time we can estimate when the rock formed. If the rock contains the radioactive isotope and decay product in the ratio of 1:1, then it is 1.25 billion years old. If there is proportionately less argon-40, then the rock is younger; if more, then the rock is older.

Potassium-40, rubidium-87 (half-life of 50 billion years; decays to strontium-87), thorium-232 (half-life of 14 billion years; decays to lead-208) and uranium-235 (half-life of 4.5 billion years; decays to lead-207) all allow ancient rocks to be dated. The oldest rocks dated so far are 3.75 billion years old. Carbon-14 has a relatively short half-life of 5730 years and decays to nitrogen-14. Carbon-14 is produced in the atmosphere above an altitude of about 10 000 m. Some carbon-14 forms carbon dioxide, which is fixed by photosynthetic organisms and moves through food chains. Because of rapid decay, the ratio of carbon-14 to nitrogen-14 is only useful for dating fossil material less than 40–50 thousand years old.

Magnetic reversals

The earth's magnetic field occasionally reverses, being either in *normal* (present day) or *reversed polarity* (north becomes south and south becomes north) for a period

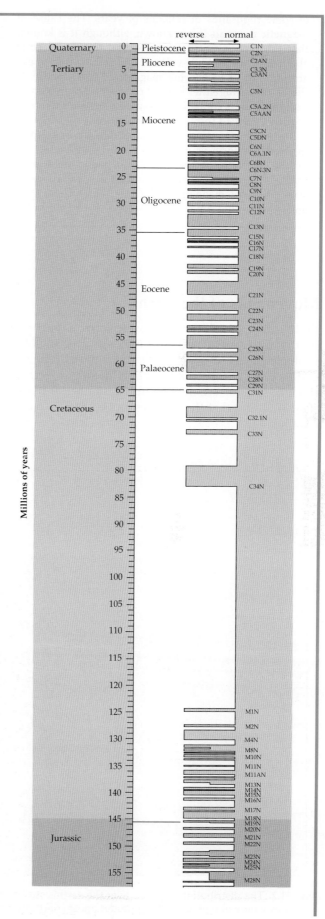

Reversals of the earth's magnetic field through the Cenozoic (C) and late Mesozoic (M). Normally magnetised intervals extend to the right of the polarity graph; reversely magnetised intervals extend to the left. The time between two successive reversals is a polarity interval, and one or more intervals, depending on length, is called a **chron**. Numerous chrons have been named for the last 160 million years of geologic history: for example C1N for the Pleistocene and C34N spanning more than 40 million years during the Cretaceous

of some thousands or millions of years. The cause of **magnetic reversals** is not known, although it is known that the time required for a reversal to take place is about 5000 years. While volcanic and sedimentary rocks are forming, they become magnetised in response to the earth's field, either normal or reversed. A record of past reversals is contained in the magnetism of the rock sequence formed throughout geologic time.

Rock magnetism may be measured directly for samples in the laboratory, or indirectly through measurement of the strength of the earth's magnetic field over extensive rock beds such as the sea floor. Variations in strength of the magnetic field, either more or less than an average value, are termed *anomalies*. *Positive anomalies* occur when the magnetism of rock beds is normal, locally enhancing the strength of the earth's magnetic field. *Negative anomalies* occur when magnetism is reversed, reducing the strength. Coherent patterns of magnetic anomalies are clearest in areas of the ocean. Symmetric patterns occur in areas to either side of a spreading ridge.

During the last 160 million years there have been two long periods of active reversal separated by a 40 million year magnetically quiet period (see figure). Where a sequence of four to six reversals is determined, the unique pattern of its reversals correlates with corresponding patterns elsewhere and identifies the time when the rock sequences were formed.

Movements of the earth

By the time the oldest rocks were formed, portions of the outer earth began to move. The modern geologic theory of *plate tectonics* recognises that the earth's crust and part of the upper mantle (together, the lithosphere) are divided into a number of plates (Fig. 31.2), and that these plates move relative to one another.

Plate tectonics

Lava upwells from part of the earth's mantle at **oceanic ridges**. New crust of sea-floor basaltic rocks moves outwards on either side of a ridge and, as the sea floor spreads, it carries with it continents made of lighter (sialic) rocks. Ocean ridges tend to be centrally located in ocean basins. For example, the Mid-Atlantic Ridge is centrally located in the Atlantic Ocean and

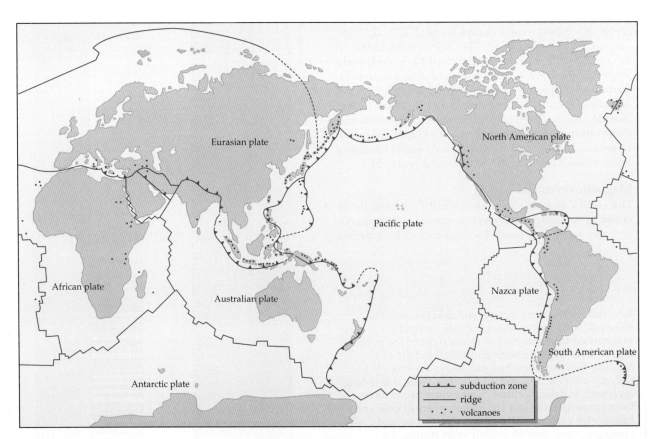

Fig. 31.2 The distribution of plates over the earth's surface. Ocean ridges where plates are moving apart, and subduction zones where one plate is sliding under another are shown. Volcanoes around the rim of the Pacific form the 'Ring of Fire'

is the site of sea-floor spreading, which separated the continents that now lie on either side of this ocean basin (Fig. 31.3).

If new crust forms and flows laterally, it must disappear somewhere else. **Deep-sea trenches** (Fig. 31.3) are the sites where sea floor descends back into the mantle in a process called **subduction**. Because continents are of lower density than the crust, they cannot descend into the mantle and remain on the surface. Most trenches of the world today encircle the Pacific Ocean, where the large Pacific plate is moving north-west relative to surrounding plates. Trenches mark sites of earthquakes and volcanoes, and explain the 'Ring of Fire' around the rim of the Pacific. Subduction explains why no modern ocean (Pacific, Indian and Atlantic) has any part of its floor older than about 200 million years (Jurassic); older parts have been destroyed through subduction at plate margins.

Immobile points at the surface of the earth's mantle where a column of hot upwelling asthenosphere rises are called **hot spots**. As a plate moves over a hot spot, a chain of volcanic islands may form, such as the Hawaiian Islands. A chain of successive volcanoes is a record of the movement of a plate over a fixed point, and allows rates of plate movement to be estimated. Most plates move about 5 cm per year!

Boundaries between two plates can be of three types. We have seen that they can be moving apart at mid-ocean ridges and moving under one another at trenches. Plates can also scrape past one another, deforming the earth's crust, a classic example being the San Andreas Fault in California. Mountains on continents have resulted from the collision of plates.

The earth's lithosphere consists of a number of plates, which arise at oceanic ridges, slide laterally and descend into the mantle again at deep-sea trenches. Plate movements deform the earth's crust, causing mountains to form and oceans to open and close. Continents, made of lighter rocks, ride on plates and shift position over geologic time.

Ancient positions of continents

Geologists have learnt about plate movements and the past positions of continents by studying rocks, hot spots and magnetic reversals (see Box 31.1). As we move back in time, the task of working out how land masses have shifted becomes harder because subduction and erosion destroy evidence of the past. Pre-Cambrian rocks, which represent the first 90% of the earth's history, form less than 20% of the rocks now exposed on the earth's surface.

In the early Palaeozoic (Upper Cambrian), the positions of the continents were different from today, with most land masses lying in the equatorial zone and none at the poles (Fig. 31.4a). Australia was part of the supercontinent **Gondwana**, and North America and Greenland were part of Laurentia. Europe was part of a separate land mass, Baltica. As the Palaeozoic progressed, the northern land masses united to form the supercontinent **Laurasia**. Gondwana remained intact and drifted southwards.

At the beginning of the Mesozoic, more than 200 million years ago, Laurasia and Gondwana joined to form a single supercontinent, **Pangaea** (Fig. 31.4b), uniting all northern and southern continents. Pangaea

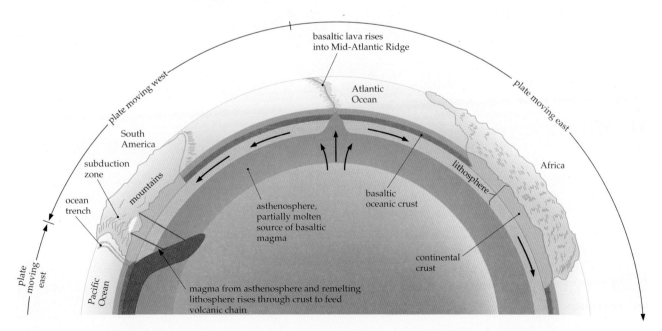

Fig. 31.3 Cross-section of the lithosphere and asthenosphere (partially molten mantle) in the Atlantic Ocean, showing the separation of South America (moving west) and Africa (moving east). South America meets the eastern part of the Pacific plate along a deep-sea trench

was such a large land mass, with inland areas a great distance from the sea, that much of it was arid. As the Mesozoic progressed, Pangaea separated into many fragments and, by the Jurassic, Gondwana was once again separated from northern land masses. Gondwana started to break up in the Cretaceous, and the southern continents gradually moved to their modern positions (Chapter 41).

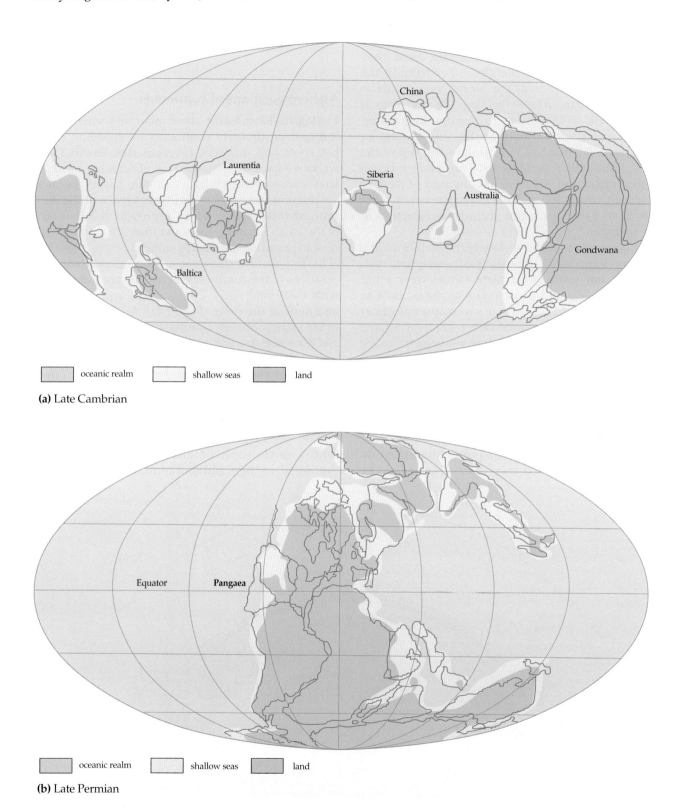

(a) Late Cambrian

(b) Late Permian

Fig. 31.4 Position of the world's land masses in the Palaeozoic era. **(a)** Early in the Palaeozoic era, in the late Cambrian. No continents were at the poles and many were inundated by shallow seas. **(b)** Towards the end of the era, in the late Permian, when the continents were forming one single supercontinent, Pangaea

EVOLVING LIFE
Fossils

Just as rocks tell us about the history of the earth, fossils in rocks tell us something about past forms of life on earth. How do fossils form and where do we find them? **Fossils** are the preserved remains of organisms (Fig. 31.5) or traces of them, such as dinosaur footprints, trilobite tracks or the tubes of soft-bodied worms—**trace fossils**—or even organic compounds produced by them—**chemical fossils**. Hard parts of organisms are most often fossilised, including teeth, bone and external shells of animals, or leaves and woody parts of plants. Soft parts are preserved rarely, with the Ediacaran fauna and Burgess Shale being two special exceptions (pp. 685, 687).

To be preserved, organisms must be buried by sediments of sand, silt or clay (Fig. 31.6) because burial prevents decay by bacteria. As sediments build up, much of the water they contain is squeezed out and chemical changes result in the formation of rocks— sandstone, shale, limestone and so on. In hard rocks, such as limestone, fossils may retain their three-dimensional shape but, in other rocks, such as shales that are highly compressed, fossils are flattened. Further changes can occur. Microscopic spaces in bone and wood can become filled in by minerals, making the fossil heavier than the original remains of the organism. Petrified wood is one example. Sections of such fossils can be cut to show the preservation of cell detail (Fig. 31.7). In a porous sandstone rock, the remains of the organism can be completely dissolved by ground water passing through. However, the fossil is not lost because the rock itself takes an impression of the organism, rather like a fingerprint. When such a rock is split open, one piece shows the organism preserved as a fossil cast, while the other piece shows its impression as a fossil mould. Sometimes the shell is replaced by another mineral, such as silica or opal; silicified fossils can be etched from limestone by application of acid, revealing intricate detail.

Fig. 31.6 The remains of a baby diprotodon that has fallen into a muddy pool will be preserved as fossil bones

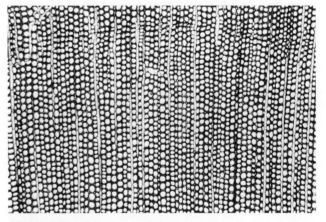

Fig. 31.7 Fossilised wood of a 'progymnosperm', *Callixylon trifilieri*, from the late Devonian

Fossils can be formed in other ways. Insects trapped in **amber**, the hardened resin of conifers, show wing venation, hairs and other fine details (Fig. 31.8). Black shales, rich in organic matter and deposited on the sea floor where oxygen is low, may have soft parts of organisms preserved as a thin carbon film on the

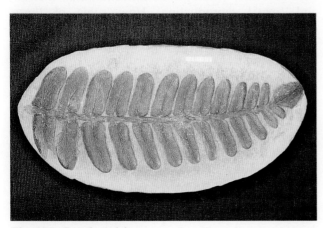

Fig. 31.5 Fossil seed fern

rock. Mobile organic compounds are lost during compaction, but carbon molecules are stable and remain behind, a process called **carbonisation**.

Fig. 31.8 Fossil insect trapped in amber

Fossils are the preserved remains of organisms, traces of them, such as footprints, or even chemicals produced by them.

Pre-Cambrian life

Before 1950, only a few fossils were known from Pre-Cambrian rocks. The mystery of early life (Fig. 31.9) unravelled as geologists and palaeontologists learned just what kinds of rocks to search in for fossils and how to recognise these fossils. Most Pre-Cambrian fossils have since been found in black chert, formed from gels of silica that precipitated on the surface of ancient sea floors. The gels trapped primitive organisms and hardened them into rock. The oldest *organisms* known from fossils are from cherts of

southern Africa dated at 3.2–3.3 billion years. These fossils resemble modern bacteria and cyanobacteria.

Other Pre-Cambrian fossils are **stromatolites**, which are concentrically layered rocks, the layers being formed by successive growth of thin mats of cyanobacteria, one on top of the other (see Box 31.2). Even in the absence of fossil remains of organisms, these rocks provide evidence of life. Stromatolites found in the Pilbara of Western Australia are dated at 3.5 billion years and are the oldest evidence of life on earth.

The oldest fossil organisms are from Pre-Cambrian rocks of southern Africa dated at 3.2–3.3 billion years. These fossils resemble modern bacteria and cyanobacteria. Stromatolites (rocks) are an indication of life 3.5 billion years old.

Pre-Cambrian organisms were prokaryotes and some at least were photosynthetic, as evidenced by the presence in black cherts of pristane and phytane, stable degradation products of chlorophyll. Banded iron rock, rich in iron oxides and ranging in age from 1.8 to 2.3 billion years, gives an indication of when the atmosphere became oxygen-rich through photosynthesis. Increase in oxygen in the atmosphere must have led to the extinction of many anaerobic forms of bacteria, although some lineages have persisted for 3.5 billion years to the present. These are the archaeobacteria (Chapter 33) that live in anaerobic environments.

Younger Pre-Cambrian rocks reveal a variety of small prokaryotes, some spheroidal cells and some filamentous strands of cells. Prokaryotes are the only organisms recorded for the next 2 billion years of the fossil record. The earliest eukaryotes are fossils from rocks found in Central Australia and dated at 0.9 billion years. These organisms resemble green algae.

The discovery of eukaryotes is significant because it marks the evolution of organisms with more complex cell structure (organised nucleus, chromosomes, chloroplasts, mitochondria), meiosis and sexual reproduction. How some of these characteristics evolved is unknown, but molecular data indicate that parts of the eukaryotic cell evolved by endosymbiosis (Chapters 3 and 35).

Ediacaran fauna

All phyla of shelly invertebrate animals, with the possible exception of the Bryozoa, are present as fossils in the Cambrian, the beginning of the Palaeozoic era, about 570 million years BP. The 'sudden' appearance of all the basic body plans of metazoans (animals) led palaeontologists to search for older fossils in Pre-Cambrian rocks.

Fig. 31.9 Artist's view of the earth 3.5 billion years ago

BOX 31.2 STROMATOLITES

In the calm waters of Shark Bay, Western Australia, living cyanobacteria build stromatolites as they have done for 3.5 billion years. Stromatolites are hard, dome-shaped rocks formed by a continuing process of sediment trapping, mineral precipitation and hardening. The cyanobacteria in Shark Bay form spongy mats on the surface of the stromatolites, trapping sediment (Fig. a). Motile, photosynthetic and responsive to light, they move through the fine sediment that accumulates around them and continue to live on top of the growing stromatolites. The cyanobacteria cause precipitation of the mineral calcite, producing sediment layers that harden. Stromatolites grow slowly, less than 1 mm per year, and reach a diameter of about 20 cm and height of 30 cm.

A variety of cyanobacteria build stromatolites (Fig. b). Coccoidal species build stromatolites with a coarse internal structure, and filamentous species in deeper water build stromatolites with a smooth structure. Micro-algae (eukaryotes) and other small organisms live with them, forming an intertidal community. Stromatolites generally form in shallow, saline water, such as the Persian Gulf, the Bahamas and Shark Bay, but they also form in hot springs, such as at Yellowstone National Park in the United States, and in freshwater lagoons rich in calcium carbonate, which provides the minerals for their structural framework.

The fossil record indicates that between 2.5 billion to 570 million years BP stromatolites were abundant and diverse in shape, being associated with a great variety of prokaryotes. Subsequent decline in importance of stromatolites in the fossil record coincides with the radiation of marine multicellular organisms.

(a) Living stromatolites at Shark Bay, Western Australia

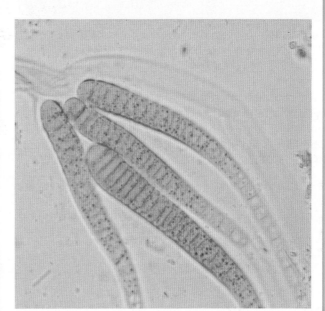

(b) *Dichothrix*, a cyanobacterium that forms benthic microbial communities that produce stromatolites in Lake Clifton, Western Australia

Fossil traces on sea floors of tracks and burrows, and impressions of soft-bodied animals (Fig. 31.10), possibly marine worms, jellyfishes and anemones, have been dated at 640–680 million years BP, although recent work suggests dates of 570–590 million years BP. Some are unlike any living animals and may represent extinct lineages. Collectively, these fossils are named the **Ediacaran fauna** after the beds in which they were found in the Ediacara Hills in South Australia, but the fauna is widespread, being found on all continents. The flora flourished for a relatively short time and disappeared before the appearance of the phyla of shelly invertebrates in the Cambrian. Ediacaran fauna are the best evidence that multicellular animals, with specialised tissues and organs allowing larger body size, had evolved in the Pre-Cambrian.

Fig. 31.10 A fossil animal, *Spriggina floundersi*, of the Pre-Cambrian Ediacaran fauna of South Australia

In contrast to animals, the fossil record of multi-cellular photosynthetic organisms—metaphytes—is poor and none is recorded for Pre-Cambrian rocks. The oldest metaphyte fossils are from the beginning of the Palaeozoic.

> The earliest eukaryotes, resembling green algae, are fossils from rocks found in central Australia and dated at 0.9 billion years. The Ediacaran fauna is evidence that multicellular animals had evolved in the Pre-Cambrian, perhaps by 640–680 million years BP.

Palaeozoic life

Palaeozoic means 'ancient life'. The **Palaeozoic** (570–245 million years BP) includes six periods: the Cambrian, Ordovician, Silurian, Devonian, Carboniferous and Permian (see Fig. 31.1).

Marine life

Ancient marine communities depended on phytoplankton—microscopic, photosynthetic organisms that float near the water surface. Phytoplankton fossils of cyanobacteria and green algae have been found from the Pre-Cambrian. In the Lower and Middle Palaeozoic (from the Cambrian to the Devonian) phytoplankton fossils are mainly **acritarchs**, tiny spherical cells with cellulose walls, some with small spines. Not until the beginning of the Mesozoic (in the Triassic) do we find fossil phytoplankton resembling modern taxa, such as diatoms and coccoliths (Chapter 35).

Marine microscopic zooplankton are not well represented as fossils because they lack hard components. One group of microscopic zooplankton, radiolarians, however, has an intricate skeleton of silica, and is recorded as far back as the Cambrian. Modern radiolarians occur today in deep, offshore waters. Larger zooplankton include **graptolites** (see Fig. 31.14), extinct colonial animals with skeletons of chitin.

Of the marine multicellular animals of the Lower Cambrian, most fossils are trilobites (Fig. 31.11). These are an extinct group of arthropods (Chapter 38), which have hard skeletal parts, large eyes, long antennae, and appendages for swimming, walking and feeding. Other fossils include brachiopods, with two shells, and rare specimens of echinoderms, living examples of which are starfishes.

Fig. 31.11 A trilobite, an extinct group of arthropods known from the early Cambrian

During much of the Ordovician, shallow seas were widespread on the continents. Marine communities at the beginning of this period were dominated by sessile, filter feeders, especially brachiopods, bryozoans and stalked echinoderms (Chapters 39, 40). Echinoderms were once a diverse group, including more than 20 classes, only five of which are alive today. The stalked **crinoids** were one of the most successful groups (Fig. 31.12).

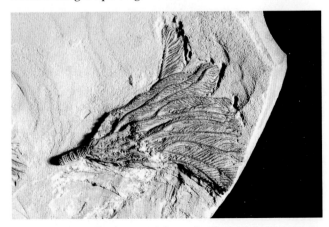

Fig. 31.12 A stalked crinoid from the Ordovician

BOX 31.3   THE BURGESS SHALE

One of the most famous fossil sites in the world is the **Burgess Shale**, in the Rocky Mountains of Canada. A crowd of marine animals was preserved in the Middle Cambrian in a deposit of black, organic-rich mud that lacked oxygen, preventing decay. Trilobites and brachiopods, with their hard shells, are well preserved, but more important is the great diversity of soft-bodied animals in the deposit (Fig. a). Unusual organisms include *Hallucigenia*, a segmented animal apparently with spikes for legs, an indefinite head, and projections from the back of each segment (Fig. b). Some animals were clearly predators, having a mouth with teeth.

The Burgess fossils suggest that the fossil record is generally incomplete with respect to soft-bodied organisms and that Cambrian marine communities were probably quite diverse.

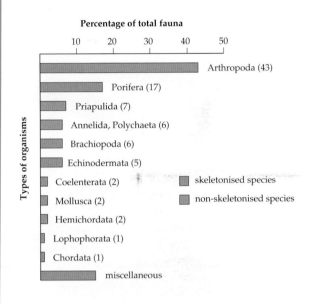

(a) Bar graph showing the major animal groups represented in the Burgess Shale

(b) A reconstruction of *Hallucigenia*, a segmented animal from the Cambrian Burgess Shale (Rocky Mountains, North America). The fossil animal was interpreted as having 14 stilt-like legs and seven tentacles, placing it in an extinct group unlike any modern phylum. However, scientists studying similar fossils from China interpret *Hallucigenia* the other way up, making it an animal with seven legs and many spines, and classifying it in the arthropod phylum

---

Two main groups of marine predators in the Ordovician were corals and cephalopods. The earliest cephalopods (molluscs, see Chapter 39) were nautiloids, related to the living *Nautilus*. Most nautiloids had a long, tapering, cone-shaped shell, which contained chambers. Another group of cephalopods, the ammonites, became dominant later, by the Devonian (Fig. 31.13).

Ordovician jawless fishes were the first vertebrate animals. The oldest fossils are small, only a few centimetres in length; the mouth was a simple hole at the front end of the gut and they were probably filter or detritus feeders. By the late Silurian and Devonian, larger armoured, jawed fishes were active predators (see Box 31.4). The Devonian, often called the 'Age of Fishes', was a time of rapid evolution of fishes, including sharks and bony fishes, comprising nearly all modern fishes. Among the bony fishes were fleshy-finned fishes (rhipidistians, coelocanths and lungfishes), which are related to the first four-legged vertebrates that colonised the land.

Fig. 31.13 A Jurassic ammonite found near Geraldton, Western Australia

The Palaeozoic era (570–245 million years BP) means 'ancient life' and includes six periods: the Cambrian, Ordovician, Silurian, Devonian, Carboniferous and Permian. Marine communities of the Palaeozoic show a sequence of organisms over time, including the first fishes in the Ordovician.

BOX 31.4 AUSTRALIA IN THE PALAEOZOIC

When the Palaeozoic fauna of Australia lived, Australia was further north, near the equator, at times closer to China; it later moved south towards Antarctica. The climate was warm and seas covered much of the eastern half of the continent, slicing it in half during the Ordovician (Fig. a). Seas then retreated and the climate cooled until late in the Palaeozoic when glaciation occurred.

There are many fossil fishes described from Australia. Armoured fishes, ray-finned fishes and lungfishes have been superbly preserved in Upper Devonian deposits from a reef complex that once fringed the Kimberleys in north-western Australia (the Gogo Formation). Fossil

fishes from an Upper Devonian highland stream in Victoria are illustrated in Figure b.

(b) Reconstruction of late Devonian fishes in a highland stream in Victoria, Australia. In the centre are two individuals of the armoured, jawed fish *Bothriolepis cullodenensis*; below, another armoured, jawed fish lies on the sandy stream bed; in the top to mid-water range are bony fishes

present day coastline

Antarctica

☐ land in the Ordovician

☐ sea in the Ordovician

(a) Much of Australia was covered by seas in the Ordovician period

Terrestrial life

The earliest fossil record in terrestrial rocks is of Silurian age. These fossils are invertebrates—scorpion-like arthropods. The earliest known land vertebrates are amphibians, found in late Devonian rocks of Greenland and Canada. By the Devonian, fishes lived not only in marine environments, but in freshwater ponds, streams (see Box 31.4) and deltas. As climate changed to alternating wet and dry periods and some aquatic habitats dried up in times of droughts, some fleshy finned fishes were able to move on land. The skull and skeleton of Devonian fleshy

finned fishes are almost identical to the earliest fossil amphibians except for the modification of fins into stubby limbs for locomotion on land.

The oldest land plant fossils are late Silurian in age, found mostly in Australia. The oldest terrestrial flora includes *Baragwanathia* (Fig. 31.14); younger fossils include *Psilophyta* and *Rhynia*. All of these fossils are small, vascular plants, having the main stem with a central core of xylem and phloem tissue, allowing transport of water and nutrients (Chapter 37). They lacked leaves and roots, and were anchored in marshy, wet bogs by rhizomes. Small non-vascular plants, liverworts and mosses, may be older than vascular plants, but there are no fossils of them in the middle Palaeozoic. Abundant vegetation provided an energy source for terrestrial fungi. Hyphae and spores of ascomycetes (Chapter 36) are recorded for the Silurian and Lower Devonian, and resting spores in the Rhynie chert in Aberdeenshire are suggestive of mycorrhizal associations between fungi and Devonian land plants.

Fig. 31.14 Early Devonian *Baragwanathia longifolia* fossil from Victoria, Australia, together with graptolites

By the end of the Devonian, every major group of land plants, except for flowering plants, is found as fossils. Forests of large tree-sized plants, some with woody stems had evolved, providing habitats for a variety of terrestrial animals. Carboniferous forests (Fig. 31.15) of low swampy areas were later buried by sediments of seas that moved in, forming extensive coal beds that are mined today.

Fig. 31.15 Reconstruction of a Carboniferous forest

The Permian marked the close of the Palaeozoic. It was a time when the continents coalesced as Pangaea, changing the distribution of seas and land, and affecting climate. It was a period of dwindling rainfall and extremes of temperature. Evolution of seeds was of great significance to the flora, allowing land plants to survive drier conditions. Reptiles, which first appeared in the Carboniferous, became more abundant in the Permian, and amphibians, more dependent on water, declined in dominance. In marine environments there was a progressive loss of shallow seas, which appears to have led to mass extinction of animal groups dominant for millions of years, including trilobites and some groups of brachiopods, bryozoans, ammonites and crinoids.

> Scorpion-like arthropods of the Silurian are the earliest fossil record of life on land. The earliest known land vertebrates are amphibians, found in Upper Devonian rocks. By the end of the Palaeozoic, every major group of land plant, except for flowering plants, is found as fossils.

Mesozoic life

The **Mesozoic era** ('middle life', 245–65 million years BP) includes the Triassic, Jurassic and Cretaceous periods, and is often referred to as the 'Age of Reptiles'.

During much of the Triassic the dominant vertebrates were the so-called mammal-like reptiles, a diverse and abundant group of bulky, large-headed animals with a sprawling posture. They did not look much like mammals, but it is believed that one group of them gave rise to mammals in the Upper Triassic. Living alongside these reptiles were forms that had limbs that were under the body, allowing more support of body weight and more mobility. Some were bipedal, with their forelimbs free for other functions, such as grasping prey or flying. Included among them were flying pterosaurs (Fig. 31.16), now extinct, crocodiles and dinosaurs. The major groups that evolved are shown in Figure 31.17.

Fig. 31.16 A marine turtle and a flying reptile (pterosaur), inhabitants of the great inland sea that covered much of Australia in early Cretaceous times. Flying reptiles are extremely rare in the Australian fossil record

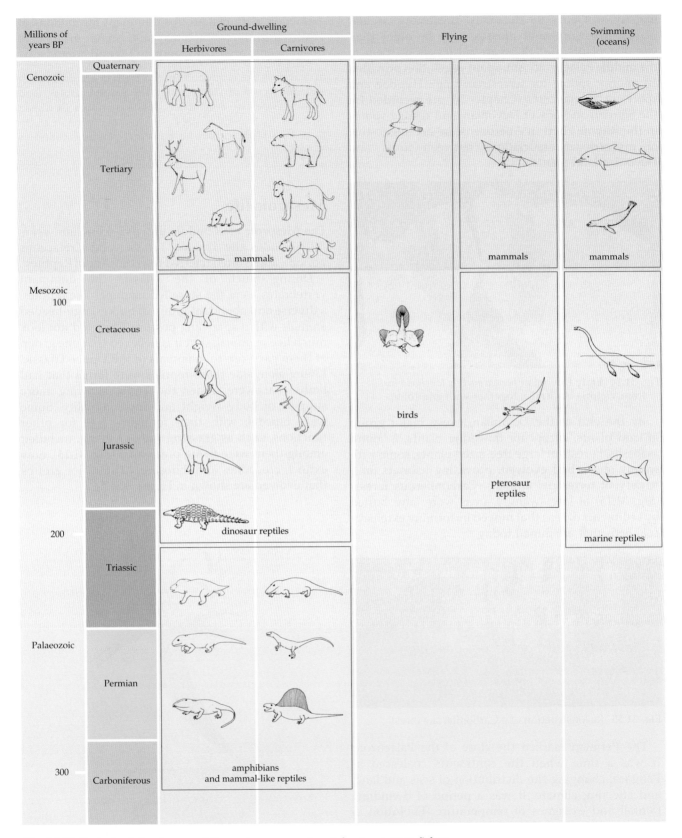

Fig. 31.17 Periods of dominance of the major groups of vertebrates, except fishes

Dinosaurs, including some of the largest animals ever to have lived on earth, reigned the land throughout the Jurassic and Cretaceous. They became extinct rather abruptly at the end of the Cretaceous. The reasons for this and other mass extinctions are uncertain. Some scientists point to geologic evidence of a large comet or asteroid that struck the earth 65 million years ago. They imagine that the impact of the comet sent enough debris into the atmosphere to block sunlight and plunge the earth into cold and dark, which led to extinctions. However, mass extinction of dinosaurs may be in part an illusion, because some of them at least evolved into birds by Jurassic times (Fig. 31.18).

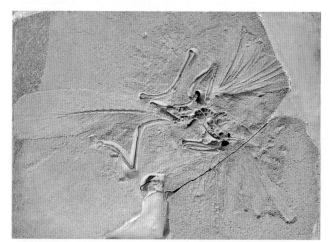

Fig. 31.18 The Jurassic bird, *Archeopteryx*, from southern Germany. A nearly complete fossil shows the impression of wing and tail feathers; the skeleton has teeth (unlike modern birds)

The Mesozoic era ('middle life', 245-65 million years BP) includes the Triassic, Jurassic and Cretaceous periods, and is often referred to as the 'Age of Reptiles'. Dinosaurs reigned but became extinct during the Cretaceous, marking the close of the era.

Whatever the fate of dinosaurs, land habitats became available for the expansion of mammals (Chapter 40). Although mammals first evolved in the Triassic, as evidenced by a few teeth and jaw fragments, they do not become abundant and diverse in the fossil record for almost another 100 million years. Jurassic forms were small, mostly shrew-like insectivores. Fossils recognisable as marsupials (the most common) and so-called placentals first appear in the Cretaceous.

In marine communities, large turtles and predatory dolphin-like ichthyosaurs (Fig. 31.19) evolved from terrestrial vertebrates that returned to an aquatic existence. Other marine reptiles were mosasaurs, large lizards that reached 8–9 m, and giant plesiosaurs with long necks. By the close of the Cretaceous, most modern groups of bony fishes appear in the fossil record. Among marine invertebrates, clams and other bivalves (inconspicuous components of Palaeozoic faunas) became common, attaching themselves to, or burrowing in, the sea bottom, or moving about with a muscular foot. Predatory gastropods that drilled holes in the shells of other animals and echinoderms that adapted to living buried in sediments became common in the fossil record. New forms of ammonites characterise different periods in the Mesozoic; and reef-building corals evolved by the Triassic.

Fig. 31.19 An ichthyosaur from the Australian Cretaceous period in shallow sea areas. An agile swimmer, this species, *Platypterygius australis*, grew to 6–7 m; it has caught a squid-like animal that became extinct in the Eocene period of the Cenozoic era

Terrestrial plant communities of the Triassic and Jurassic were dominated by ferns, seed ferns, cycads (with large palm-like leaves), ginkgoes and conifers (Chapter 37). The wood of these conifers indicates that they are relatives of extant araucarian pines, restricted today to the Southern Hemisphere, including New Guinea, Australia, New Zealand and South America (see Box 41.1).

One of the most significant events of the Mesozoic was the appearance and sudden expansion of flowering plants, which today are the dominant land plants. The oldest flowering plant fossils are rare and found in Lower Cretaceous rocks. By the late Cretaceous, angiosperm fossils are more abundant throughout the world. Herbaceous forms rapidly filled the understorey of plant communities, where ferns and mosses dominated for most of the Mesozoic. Flowering plants evolved a highly successful mechanism of fertilisation (Chapters 16 and 37) that involved insects rather than wind for the transport of pollen. Such features explain the success of flowering plants, which continued into the Cenozoic.

During the Mesozoic, land habitats became available for the expansion of mammals, which are first recorded as fossils in the Triassic. Also significant in the Cretaceous was the appearance and sudden expansion of flowering plants.

Cenozoic life

The **Cenozoic era** ('modern life', 65 million years BP to the present) includes the Tertiary and Quaternary periods, bringing us to today. Throughout the Cenozoic we can recognise more and more modern families and genera of flowering plants, and biogeographical patterns with groups characteristic of either the Northern or Southern Hemisphere.

By the Cenozoic mammals were abundant. The mammalian groups that flourished during the Palaeocene, Eocene and Oligocene in the early parts of the Tertiary are now largely extinct but, by the Miocene, modern groups of mammals had evolved, most of which are present today. Early primates (prosimians) were common in the early Cenozoic but decreased markedly in importance when true monkeys and apes evolved. Miocene and Pliocene deposits in Africa, Europe and Asia have yielded many fossil teeth and jaw fragments of apes, and fossil footprints are evidence of hominids 4 million years ago in Africa. *Homo* expanded in range rapidly in the last 2 million years, experiencing the Ice Age of the

Pleistocene. The history of modern humans is a story of increased use of tools, language, culture, and development of agriculture and domestication of animals (Chapter 40).

The Cenozoic era ('modern life, 65 million years BP to the present) includes the Tertiary and Quaternary periods. Mammals and flowering plants became abundant, with more and more modern groups becoming recognisable throughout the Tertiary. Hominid fossils date back 4 million years.

BIOGEOGRAPHIC REGIONS OF THE MODERN EARTH

The result of evolution, both of earth and life together, is the world we see today, with its great variety of organisms living in their great variety of habitats, in sum, the earth's present biodiversity. Since the eighteenth century, biogeographers (those who study the distribution and abundance of organisms) have attempted to characterise regions inhabited by unique forms of native life to understand the evolutionary history of life and earth (Box 31.5).

The main continental regions recognised today by botanists and zoologists are very similar (Table 31.1; Fig. 31.20). The names of the regions are based partly on the division into old (palaeo-) and new (neo-) worlds, with Australia treated as if it were a world apart.

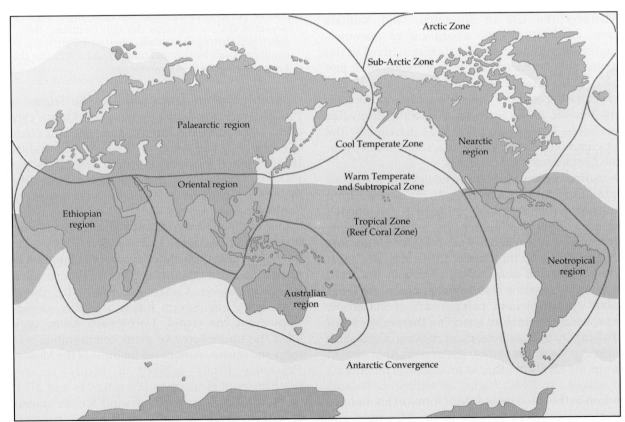

Fig. 31.20 Continental biogeographic regions and marine biogeographic zones

Table 31.1 Geographic regions recognised for plants and animals

Botanical regions	Zoological regions
Boreal	Holarctic 　　Palaearctic 　　Nearctic
Palaeotropical	Ethiopian Oriental
Neotropical	Neotropical
Australian	Australian

The **Boreal** region includes the northern fir forests, which are very similar in Europe, Asia and North America. Also known as the Holarctic region, it is sometimes divided into Nearctic (North America) and Palaearctic (Eurasian) regions. **Palaeotropical** plants occur in Africa (Ethiopian region) and in India and South-East Asia (Oriental region), in association with elephants and rhinoceroses among other animals. The **Neotropical** region includes all of South America and lower Central America to about the 200 km wide Isthmus of Tehuantepec in Mexico. Finally, the

Australian region includes both the mainland as well as islands on the continental shelf, such as Tasmania in the south and New Guinea and its own neighbouring islands in the north.

Marine organisms, too, are diverse in geographical distribution. Marine biogeographers commonly begin with a division between shallow water (shelf) and open ocean (pelagic) realms. Life of the **continental shelf** is most diverse in the tropics, particularly in coral reef communities (Fig. 31.20). Bounded by 20°C isotherms for the coldest months of the year, tropical marine regions include the Indo-West Pacific, East Pacific, West Atlantic, and East Atlantic. The Indo-West Pacific is the largest region, embracing the entire tropical Indian Ocean and the West and Central Pacific, including archipelagoes such as Hawaii and the Marquesas. Northern and Southern Temperate regions are commonly inhabited by organisms of related taxa, such as the commercially important pilchards (*Sardina*, *Sardinops*). Of six species of pilchards in the world, one lives in waters around southern Australia and New Zealand, and the other related species occur off the coast of southern Africa, Peru, California, Japan and northern Europe (Fig. 31.21).

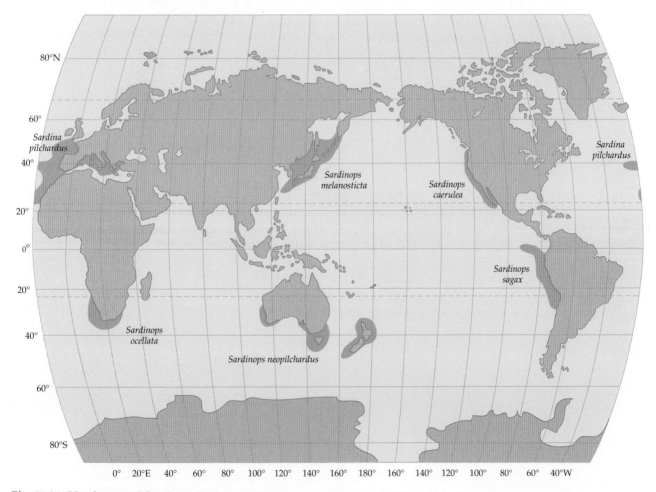

Fig. 31.21 Northern and Southern Temperate Regions are commonly inhabited by related organisms as illustrated by the distribution of six species of pilchards (*Sardina* and *Sardinops*)

Arctic and Antarctic seas have their own unique life forms but these, too, sometimes have their nearest relatives at the opposite poles. The phylum Priapulida, for example, consists of only eight species and six genera of cucumber-shaped animals, penis worms, that live buried in bottom sand or mud. The genus *Priapulus*, has a **bipolar distribution**, one species living in the Arctic and another in the Antarctic (Fig. 31.22). The phylum is believed to be an ancient group.

The **pelagic realm** is home to planktonic organisms typically confined to surface water (to 1000 m). The pelagic realm is governed by ocean currents, themselves a product of prevailing winds and ultimately of the earth's rotation. Major currents divide the marine environment into water masses, which tend to have their own physical properties of temperature and salinity. Water masses are bounded by major currents and their interactions.

Currents separate at a divergence, and come together at a convergence. A major feature of the Southern Ocean is the Antarctic Convergence (Fig. 31.20), which encircles the globe, where very cold Antarctic water sinks below warmer water to the north.

In the sea, as on continents, few species are cosmopolitan, that is, worldwide. Most species occupy only a relatively small area where they are native (endemic). Similar distributions of different organisms suggest common underlying causes, which in part relate to previous evolutionary history and in part to existing environmental conditions.

> The world is divided into a number of continental and marine biogeographic regions, characterised by endemic organisms. Similar distributions of taxa are explained by present-day environmental factors as well as previous evolutionary history.

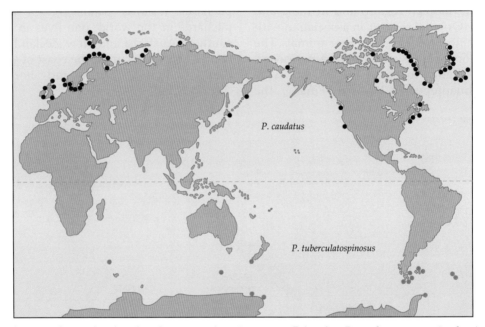

Fig. 31.22 Distribution of two closely related species of penis worms, *Priapulus*: *P. caudatus* occurs in the Arctic, and *P. tuberculatospinosus* in the Antarctic. This type of geographic distribution is termed 'bipolar'

BOX 31.5 WALLACE'S LINE

Alfred Russel Wallace (1823–1913) was the codiscoverer of the principle of evolution by means of natural selection (Chapter 33) with Charles Robert Darwin (1809–82). Wallace worked as a self-employed collector of museum specimens in the East Indies (1854–62) and travelled widely through the islands, often under primitive conditions of great hardship. He found that the Asian and Australian faunas divide along a line between Bali and Lombok in the south, continuing north between Borneo and Sulawesi (Celebes), and continuing east of the Philippines. **Wallace's Line** has been recognised ever since.

In his book *The Malay Archipelago*, Wallace wrote:

The great contrast between the two divisions of the Archipelago is nowhere so abruptly exhibited as on passing from the island of Bali to that of Lombok, where the two regions are in close proximity. In Bali we have barbets, fruit thrushes and woodpeckers; on passing over to Lombok these are seen no more, but we have abundance of cockatoos, honeysuckers and brushturkeys, which are equally unknown in Bali, or any island further west. The strait is here only fifteen miles wide, so that we may pass in two hours from one great division of the earth to another, differing as essentially in their animal life as Europe does from America.

Wallace noted that:

This division of the Archipelago into two regions, characterised by a striking diversity in their natural productions, does not in any way correspond to the main physical or climatal divisions of the surface. The great volcanic chain runs through both parts . . . Borneo closely resembles New Guinea, not only in its vast size and its freedom from volcanoes, but in the variety of geological structure, its uniformity of climate, and the general aspect of the forest vegetation that clothes its surface. The Moluccas are the counterpart of the Philippines in their volcanic structure, their extreme fertility, their luxuriant forests, and their frequent earthquakes; and Bali with the east end of Java has a climate almost as arid as that of Timor. Yet between these corresponding groups of islands constructed as it were after the same pattern, there exists the greatest possible contrasts when we compare their animal production. Nowhere does the ancient doctrine—that differences or similarities in the various forms of life that inhabit different countries are due to corresponding physical differences or similarities in the countries themselves—meet with so direct and palpable a contradiction.

Modern interpretation of Wallace's Line is based on the discovery that Australia, New Guinea and associated islands moved north during the last 60 million years. Wallace's Line approximates, but not exactly, the collision zone between Asian and Australian crustal plates.

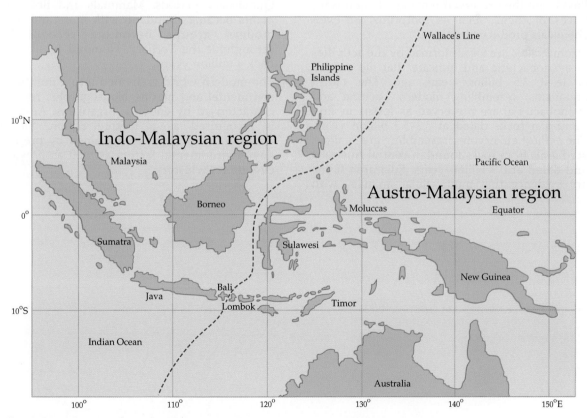

Wallace's Line showing the Indo-Malaysian region and the Austro-Malaysian region

SUMMARY

- The evolution of life has been influenced by the history of the earth: by the composition of the earth's early atmosphere and the movement of the earth's crust. The first organisms evolved in the absence of oxygen, in oceans and lakes. They used this store of molecules as a source of raw materials and energy. Photosynthetic organisms evolved later, enriching the atmosphere with oxygen.

- The modern geologic time scale is based on relative ages of sequences of sedimentary rocks and fossils, and absolute measurements based on radioactive dating. The main divisions of the geologic time scale are the Pre-Cambrian, Palaeozoic, Mesozoic and Cenozoic eras.

- Over geologic time, continents have shifted position. According to the theory of plate tectonics, the earth's lithosphere is divided into a number of plates, which arise at oceanic ridges, slide laterally, and descend into the mantle again at deep-sea trenches. Continents, made of lighter rocks, ride on plates but do not descend into the mantle. For hundreds of millions of years Australia was part of the supercontinent Gondwana, lying near the equator in the early Palaeozoic and to the south by the end of the era, when all of the continents coalesced to form one single supercontinent Pangaea.

- Fossils are the preserved remains of organisms, traces of them, such as footprints, or even chemicals produced by them.

- Stromatolites are fossils formed by the activities of cyanobacteria and indicate that life existed at least 3.5 billion years ago. The oldest organisms, resembling modern bacteria and cyanobacteria, are preserved as fossils in Pre-Cambrian rocks dated at 3.2–3.3 billion years. The earliest eukaryotes, resembling green algae, are fossils from rocks found in central Australia and dated at 0.9 billion years. Ediacaran fauna are evidence that multicellular animals, with specialised tissues and organs, allowing larger body size, had evolved in the Pre-Cambrian, perhaps by 640–680 million years ago.

- Palaeozoic means 'ancient life'. The Palaeozoic era (570–245 million years BP) includes six periods: the Cambrian, Ordovician, Silurian, Devonian, Carboniferous and Permian. Marine communities show a sequence of organisms over time, including the evolution of the first fishes in the Ordovician. The earliest fossil record of life on land is of Silurian age. These fossils are invertebrates: scorpion-like arthropods. The earliest known land vertebrates are amphibians, found in late Devonian rocks. By the end of this period, every major group of land plant, except for flowering plants, is found as fossils.

- The Mesozoic era ('middle life', 245–65 million years BP) includes the Triassic, Jurassic and Cretaceous periods, and is often referred to as the 'Age of Reptiles'. Dinosaurs reigned but became extinct during the Cretaceous, marking the close of the era. Land habitats became available for the expansion of mammals, which are first recorded as fossils in the Triassic. Also significant in the Cretaceous was the appearance and sudden expansion of flowering plants.

- The Cenozoic era ('modern life', 65 million years BP to the present) includes the Tertiary and Quaternary periods. Mammals and flowering plants became abundant, with more and more modern groups becoming recognisable throughout the Tertiary. Hominid fossils date back 4 million years.

- The modern world is divided into a number of continental and marine biogeographic regions, characterised by endemic organisms. Australia is recognised as a unique region. Similar distributions of taxa are explained by present-day environmental factors as well as previous evolutionary history.

QUESTIONS

1. What sort of data are used as a basis for the modern geologic time scale?

2. What geologic features help us to recognise past movements of continents?

3. Briefly explain how the earth's crustal plates move and what happens at their boundaries.

4. **(a)** What was the land mass that Australia was associated with in the Palaeozoic era? What was the position of this land mass in the early part and at the end of the era? **(b)** What was the single supercontinent with which all continents were associated that formed by the Upper Permian period?

5. How do fossils form?

6. What is a stromatolite? What does its structure indicate about the growth of its associated micro-organisms? Explain the historical significance of stromatolites.

7. **(a)** What significant biological events characterise each of the major geologic eras? **(b)** What are the approximate ages of the following fossils: oldest prokaryotes; first multicellular animals; land animals; land plants?

8. Briefly describe the Ediacaran fauna, where it is found and of what significance it is.

9. What are the major continental regions and marine zones of the modern world that are recognised by biogeographers? How can you explain the existence of such regions and zones.

10. What is 'Wallace's Line' and what is its significance?

MECHANISMS OF EVOLUTION

Evolution is the process of change and divergence in populations and taxa. As we have seen in Chapters 30 and 31, evidence for evolution is seen in the patterns of relationships among taxa, patterns of geographic distribution, homologous structures, the unity of life at the molecular level and the fossil record.

The most widely accepted modern theory of evolution is often described as *neo-Darwinian* because it is based on both Charles Darwin's idea of natural selection and new ('neo') knowledge of genetics discovered after Darwin. Darwin's observations and theorising relating to evolution by natural selection were constrained by the lack of a valid theory of inheritance at the time he published *On the Origin of Species by Means of Natural Selection* in 1859. Our understanding of heredity has improved since that time, beginning with the work of Gregor Mendel, published in 1866, on the transmission of simple traits in peas (Chapter 9). Developments in population and quantitative genetics in the early twentieth century established a theoretical foundation for modern evolutionary biology. In the mid-twentieth century, the recognition of DNA as the genetic material, and the elucidation of how genetic information is encoded, propelled us into a deeper understanding of evolutionary processes. These developments in knowledge continue to this day, aided by advances in technology that provide the means to obtain more information about evolution at the molecular level.

Before reading this chapter on evolutionary mechanism, you should review the genetics section (Chapters 9–12).

EVOLUTIONARY MODELS
Adaptive evolution

Information flowing from advancements in molecular technology has emphasised the importance of two types of evolution—adaptive and neutral evolution. Darwin was concerned with **adaptive evolution**, which refers to the acquisition of inherited physical

and physiological traits that benefit an organism by enabling it to compete better with other organisms in a particular environment.

Evolutionary change occurs in *populations*, a population being a group of interbreeding organisms (although some organisms reproduce asexually). The theory of adaptive evolution states that:

- within a population, the reproductive potential of organisms is always greater than the actual number of offspring that survive;

- individual organisms in a population vary, and much of this variability is inherited;

- because environmental resources, such as food or space, are limited, there is competition between organisms, so that some members of a population with characteristics that enable them to compete successfully in a particular environment are likely to leave more offspring than are other members with less favourable characteristics;

- as a result, the genetic composition of the population changes with time and diverges from that of other populations in different environments.

For example, increased aridity over the last 25 million years in Australia led to an expansion of open forests, woodlands and grasslands and a contraction of wetter forests (Chapter 41). These environmental changes are mirrored in the evolution of kangaroos (Box 32.1). At the beginning of this period, kangaroos were small and omnivorous, with unspecialised dentition, foot structure and diet. Today, the muskrat kangaroo, *Hypsiprimnodon moschatus*, which retains primitive features, is found only in rainforests of northern Queensland. Larger browsing, grazing and even carnivorous forms with specialised changes in dentition and foot structure evolved from primitive ancestral types as the physical environment and vegetation changed. Expansion of grassy forests and grasslands eventually favoured the grazing lineage, which dominates today.

Adaptive evolution is thought of as an innovative and progressive process, with selection leading to greater complexity. While greater complexity is attained in many cases, adaptive evolution also can

BOX 32.1   ADAPTIVE EVOLUTION OF KANGAROOS

During the Oligocene and early Miocene, Australia was bathed in rains that supported rainforest environments over much of the continent. As the Australian plate drifted north, aridity increased (Chapter 41). Evolutionary success and adaptations of kangaroos have been dictated by this increased aridity (Fig. a). Animals that survived more arid conditions, ate poor-quality grasses and escaped predators in more open environments, were favoured.

Change in diet and dentition

The most primitive living kangaroo is the muskrat kangaroo, *Hypsiprimnodon moschatus* (Fig. b). *Hypsiprimnodon* lives in rainforests and eats a variety of foods, including small invertebrates and fruits. It has specialised, blade-like premolars for slicing open the exoskeletons of insects, but the molars have somewhat generalised, relatively flat surfaces that break food items into more easily digested pieces. The muskrat's diet is relatively

Group	Carnivorous kangaroos	Muskrat kangaroo	Rat kangaroos	Sthenurine kangaroos	Macropodine kangaroos
Number of extant species	0	1	9	1	49
Group representative	*Propleopus*	*Hypsiprimnodon*	*Caloprymnus*	*Sthenurus*	*Macropus*
Diet	Flesh	Invertebrates, fruits	Roots, seeds, fungi, insects	Leaves, grasses	Grasses
Habitat	?Woodlands	Rainforests	Savanna, grasslands, forests	Open forests	Grasslands, savanna forests
Lower jaw structure					
Number of functional toes on hind foot	?	3	2	1	2

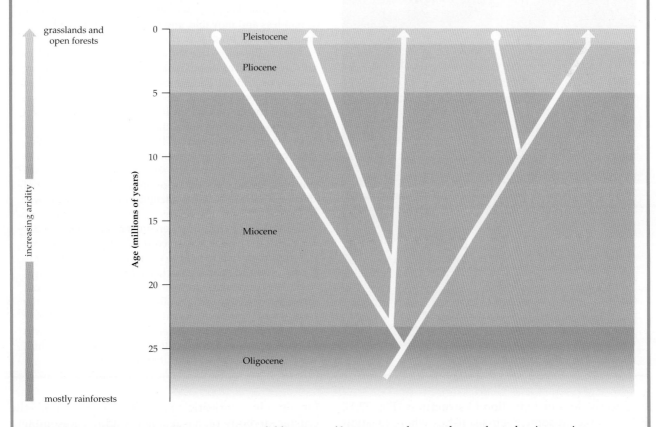

(a) Evolution of kangaroos. The most successful lineages of kangaroo today are those adapted to increasing aridity in Australia, in particular, the grazing macropods

(b) *Hypsiprimnodon moschatus*, the muskrat kangaroo, which lives in tropical rainforests in Queensland

(c) *Macropus rufus*, the red kangaroo

nutritious and food is abundant on the rainforest floor. It meets its nutritional needs easily and there is no need to grind food to extract all the nutritional value. Its stomach is a simple sac, not divided into several cavities and probably not capable of cellulose digestion.

Grasses are relatively poor quality foods and kangaroos that feed on them must be able to extract a high proportion of the nutrients available. Species of *Macropus*, such as the red kangaroo (Fig. c), have high-crested molar teeth that efficiently grind food into a paste, allowing nutrients to be extracted more efficiently in the gut, which is also adapted to harder, less digestible plant material. Grass-eaters have no need for a blade-like premolar and this tooth is reduced in size.

Escaping predators

On the floor of a rainforest, animals are easily concealed from predators by the undercover and speed is not essential for escape. In grasslands, animals are more conspicuous and speed is important for escaping predators. As with the large hoofed, grazing animals of other parts of the world, large kangaroos attain speeds of more than 80 km per hour over short distances.

Kangaroos have a hopping (saltatory) form of locomotion, and increased speed resulted from modification of the foot structure to a long lever, by reduction on the hind feet of the number of toes and increase in their length. *Macropus* and other true kangaroos (family Macropopidae) have two of their toes joined together, forming a single functional toe (with a long nail used for grooming). *Hypsiprimnodon* differs from all other kangaroos in retaining the first toe, which is nail-less and movable, and which can be spread away from the other toes (as in possums).

The phylogenetic tree shown in the figure summarises the evolution of kangaroos. The success of grazing kangaroos and rat kangaroos as rainforests contracted and drier open forests and grasslands expanded is reflected in the number of living species. Although rat kangaroos do not eat grass, they have other adaptations to dry, open environments. The losers were the browsing (leaf-cutting) or sthenurine kangaroos and the carnivorous kangaroos. The sthenurine kangaroos, with only one living species, were once more common but declined with reduced availability of low-level browse in forests.

Carnivorous kangaroos were never very common but their existence in the fossil record indicates another, very different path that kangaroo evolution may have taken. Carnivorous kangaroos such as *Propleopus* had large, knife-like premolars for slicing up flesh. Their foot structure is unknown. Unlike other kangaroos, the fortunes of *Propleopus* and its relatives were probably not dependent on increased aridity. Their extinction was likely the result of competition with other carnivores, notably the dingo (see also Chapter 39).

lead to the loss or reduction of structures (Fig. 32.1). For example, in some deep-sea fishes where location of mates is difficult, the male is reduced to little more than a 'parasitic gonad' permanently attached to the female. In parasitic barnacles (*Cirripedes*), the crustacean body plan is lost in the adults. Resembling a fungus, the body is reduced to a branching structure that penetrates the tissues of crab and lobster hosts so that nutrients can be extracted directly without the need for a complex alimentary system. The parasites reproduce asexually and have no structures for mate recognition and sexual reproduction. Many plants, too, have lost complex structures during evolution. Underground orchids of the genus *Rhizanthella* from Western and eastern Australia lack leaves and photosynthesis (Fig. 32.1). These remarkable orchids even flower underground. They obtain all nutrients from mycorrhizal fungi (Chapter 36), which are also associated with the roots of trees and shrubs.

(a)

(b)

Fig. 32.1 In addition to increased complexity, adaptive evolution can lead to the loss of structures.
(a) *Rhizanthella slateri* is an orchid that lives and flowers underground and lacks leaves. (b) Pygopodids, the legless lizards of Australia, live in cracks and crevices and some species even burrow through the soil. In these circumstances, an elongated body with reduced limbs is advantageous. Forelimbs are completely absent and the hind limbs are reduced to scale-like vestiges, as shown here

Adaptive evolution is evolutionary change that increases an organism's chance of survival and reproduction. Adaptive evolution sometimes results in greater complexity and sometimes in reduced complexity and loss of structures.

Neutral evolution

Neutral evolution does not confer any apparent advantage or disadvantage. If a change occurs in one member of a population, the number of descendants carrying it will be entirely a matter of chance. A neutral change will increase or decrease in frequency at random. In small populations, a random change may increase in frequency quite rapidly and even become universal, so that rapid divergence develops between populations. In large populations, there is less tendency for this to occur. The chance fixation of a change is termed **genetic drift** and is an evolutionary model different from natural selection.

Some evolutionary changes are neutral, having no apparent effect on the survival or reproduction of an organism. Neutral evolution is the result of changes fixed in small populations by genetic drift.

Adaptive and neutral evolution at the level of DNA

The difference between adaptive and neutral evolution is revealed at the sequence level of DNA. DNA base sequences are analogous to architectural plans. When we carefully compare two sets of building plans, we recognise differences that affect the building specifications; these are analogous to differences in the DNA of organisms specifying adaptive changes. If the plans differ only in the colour of the ink used for the drawings, then the specifications are unaffected. These differences in colour are analogous to neutral evolutionary changes, which have no functional impact. For example, an amino acid may be coded by a number of different triplets (Chapter 1) and a change from TAT to TAG does not affect the amino acid specified. The change from T to G in the third position is a neutral change. Furthermore, a protein nearly always occurs in a number of variant forms within a population as a result of single amino acid substitutions (e.g. there are many variants of human haemoglobin). It is difficult to imagine that all of these variants are adaptive.

The most convincing evidence for selection at the level of DNA is sequence conservation. Comparison of DNA sequences of the same gene in different species shows that non-coding and putatively non-functional regions (introns, Chapter 10) rapidly lose sequence similarity, whereas similarity is conserved in the coding (exon) regions. For example, comparison of the DNA sequence of the gene for scarlet eye colour in *Drosophila melanogaster*, the cosmopolitan vinegar fly, *D. buzzatii*, a species introduced into Australia on prickly pear, and *Lucilia cuprina*, the sheep blowfly of Australia, reveals that the length of the introns is not conserved and the sequence similarity is never greater than about 30% among these species. Thus, evolution of intron sequences is not subject to selective constraint and rapid neutral evolution occurs.

In contrast, the exons are highly conserved. There are severe functional constraints on the amino acid substitutions, which are consistent with function of the protein. In the nucleotide sequence of scarlet eye colour gene, *D. buzzatii* has 77% similarity with *D. melanogaster*, and 70% similarity with the more distantly related *L. cuprina*. At the amino acid level, the similarity is 81% in each case, further indication of the functional constraints on exon evolution. At this level, selection removes maladaptive mutations and is described as **stabilising selection**. However, a small proportion of sequence changes are favourable and permit adaptive evolution in response to changing environmental circumstances.

Lamarckism: inheritance of acquired characters

Jean-Baptiste Lamarck (1744–1829) was a French naturalist who published the first modern theory of evolution. His theory, known as **Lamarckism**, was that traits acquired during the lifetime of an organism are inherited by subsequent generations. This view was also shared by Darwin, who developed a theory of genetics, *pangenesis*, to explain inheritance of acquired characters. Acquired traits, such as increased muscular strength due to exercise, were believed to be transmissible to offspring. Experiments performed at the end of the nineteenth and early twentieth centuries, including one in which mice in 20 successive generations had their tails cut off to see whether the offspring were affected, demonstrated the Lamarckian theory to be incorrect.

Our current understanding of molecular genetics also suggests that inheritance of acquired traits is impossible because this would require a reversal of the flow of genetic information: from protein to RNA to DNA. The 'central dogma' of molecular genetics is that the flow of genetic information is from DNA to RNA to protein. Exceptions to this general rule are provided by some RNA viruses, which do not have DNA and thus the first step in this chain. More important exceptions are retroviruses (Chapter 34), which have added an extra step, reverse-transcribing their RNA genome into DNA before it is inserted into the genome of the host cell, where it behaves like host DNA. Retroviruses can reverse the transmission of information from RNA to DNA and sometimes they accidentally do this to host RNA sequences as well.

Recent attempts to revive Lamarck's theory postulate a role for retroviruses, suggesting that they permit movement along part of the path backwards from an organ to the genetic material (Fig. 32.2). However,

there is no example and no obvious mechanism of transmission of information backwards from protein to RNA or to the encoding DNA. Modifications of proteins by environmental factors during the lifetime of an organism are not known to be incorporated into the genetic blueprint for the protein.

> While the inheritance of traits gained during the lifetime of an organism was once proposed as a mechanism of evolutionary change—Lamarckism—it is now known that the mechanism of storage and expression of genetic information precludes the inheritance of acquired characters.

MUTATION AND GENETIC CHANGE

Variability in populations underlies evolutionary change and originates from *mutations*, change of the genotype (Chapter 9). Genetic information is stored in DNA sequences and a mutation is a change of this sequence. The effect of a mutation depends on what it is and where it is. For example, a change in, or even deletion of, one tandem repeat in non-coding, highly repeated DNA, may have no impact. On the other hand, even a single nucleotide change in a coding or regulatory region can have a dramatic effect on the survival or reproduction of an organism.

A change in the sequence of DNA is a **point mutation**. Changes causing breakage and rejoining of chromosomes are **chromosome mutations**.

Point mutations

A common type of point mutation is **substitution**, when a single base in DNA is replaced by another; for example, the substitution of G for T changes the triplet GAT to TAT. Two other types of mutations are **deletion** and **insertion** of one or more base (Fig. 32.3). These may have a significant effect by altering the reading frame of a sequence of bases. For example, deletion of G from the sequence GAT TCA CAT, converts the sequence to ATT CAC AT. Deletions are almost always unfavourable because they destroy gene function.

> Mutations provide variation, which underlies both adaptive and neutral evolution. Point mutations are small changes in DNA sequence and include base substitutions, deletions and insertions.

Chromosome mutations

Changes occur in the structure, size and number of chromosomes. These changes range from increase in the copy number, **duplication**, of tandemly repeated elements, which are visible at the chromosomal level,

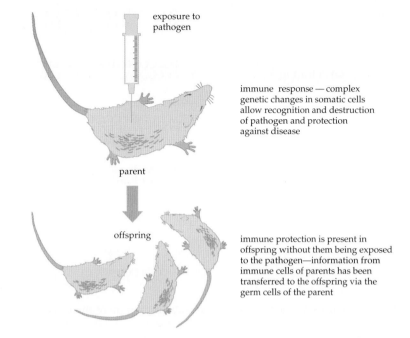

(a)

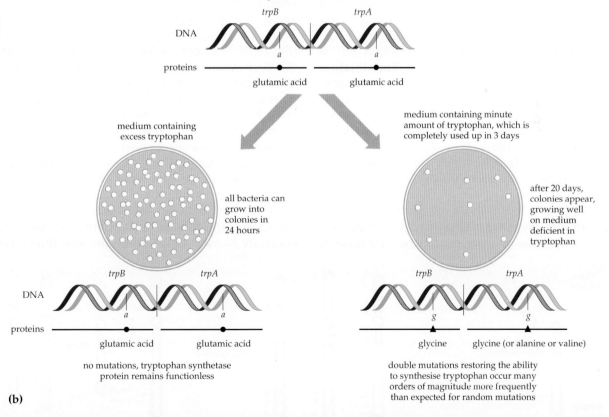

(b)

Fig. 32.2 Is Lamarckism still relevant? Two sets of experimental results from recent research support a modified version of Lamarckism. **(a)** The transmission of acquired disease resistance in mice. This has been suggested by results from one laboratory, although the experiments have not been repeatable in other laboratories. **(b)** Adaptive mutations. Several experiments with bacteria have suggested that mutations may occur in response to the need for them, rather than at random and independent of their potential usefulness. **(i)** With G (guanine) at the position indicated in both *trpA* and *trpB*, the amino acid glycine is incorporated into these proteins and a fully functional form of the enzyme is made; **(ii)** with a mutation (*m*) resulting in glutamic acid at these positions, the enzyme does not work at all. If bacteria are grown on a medium deficient in tryptophan, mutations may occur that restore the function of the enzyme

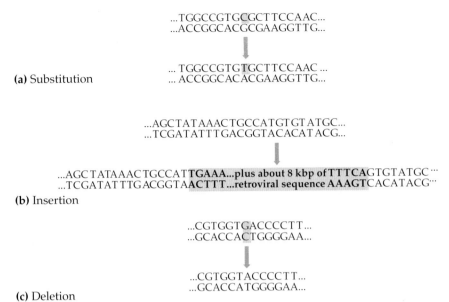

...TGGCCGTGCGCTTCCAAC...
...ACCGGCACGCGAAGGTTG...

↓

... TGGCCGTGTGCTTCCAAC ...
... ACCGGCACACGAAGGTTG...

(a) Substitution

...AGCTATAAACTGCCATGTGTATGC...
...TCGATATTTGACGGTACACATACG...

↓

...AGCTATAAACTGCCAT**TGAAA...plus about 8 kbp ofTTTCA**GTGTATGC···
...TCGATATTTGACGGTA**ACTTT...retroviral sequence AAAGT**CACATACG···

(b) Insertion

...CGTGGTGACCCCTT...
...GCACCACTGGGGAA...

↓

...CGTGGTACCCCTT...
...GCACCATGGGGAA...

(c) Deletion

Fig. 32.3 DNA sequences can be mutated by **(a)** substitution (here the upper sequence fragment is from the ryanodine receptor gene of a normal pig and the lower sequence is the equivalent region from a pig with porcine stress syndrome); **(b)** insertion (here the upper sequence is from the *hr* gene of a normal mouse and the lower sequence is from the mutant gene, which contains an 8000 base pair insert; the mouse sequence CCAT has been duplicated during the insertion); and **(c)** deletion (the upper sequence is from the glycosyltransferase genes responsible for the human A and B blood groups and the lower sequence is from the gene for the O blood group, where a single nucleotide deletion has abolished the function of the glycosyltransferase)

to a reshuffle of the genetic material by chromosomal rearrangements. Chromosomal rearrangements include **inversion** of segments within a chromosome: the segment remains in the same position but is turned around end to end (Fig. 32.4). **Translocation** occurs when segments are exchanged between non-homologous chromosomes. The number of chromosomes can decrease by **fusion**, when chromosomes join end to end, or increase by **fission** after chromosomes break.

Inversions and translocations are common types of chromosome mutations. Human chromosomes, for example, differ from those of chimpanzee by six inversions, and from those of gorilla by eight. Duplication of a segment of DNA may be an important way that new evolutionary possibilities are realised

rapidly. The variety of globins, including myoglobin and haemoglobins in vertebrates, is thought to have evolved after duplication of the primitive globin gene (Fig. 32.5). Different copies of the gene accumulated different mutations and evolved different functions. Chromosome fusions and fissions also seem to have played a part in evolution because related species often have different chromosome numbers (humans 46, chimpanzee and gorilla, 48).

Chromosomal analyses have revealed the existence of sibling species, which show no morphological divergence but considerable chromosomal divergence. Some chromosomal changes can prevent hybridisation between organisms with different chromosomal configurations. Chromosomal differences are barriers to interbreeding and thus hinder

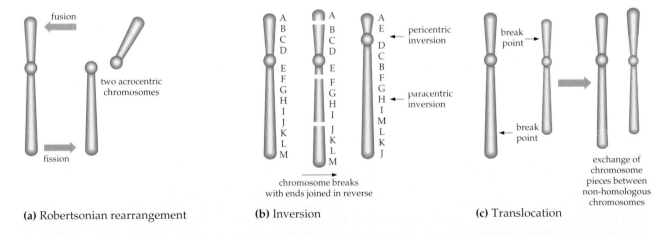

(a) Robertsonian rearrangement   **(b)** Inversion   **(c)** Translocation

Fig. 32.4 Major types of chromosomal mutations, causing variation in number, shape and size of chromosomes

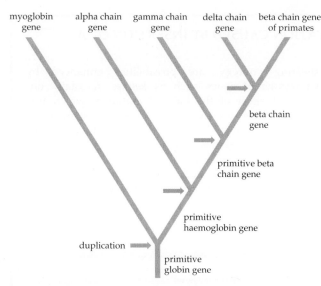

myoglobin gene   alpha chain gene   gamma chain gene   delta chain gene   beta chain gene of primates

beta chain gene

primitive beta chain gene

primitive haemoglobin gene

duplication

primitive globin gene

Fig. 32.5 The different globin genes in vertebrates evolved by gene duplication and divergence

or prevent genetic exchange between diverging populations (p. 711).

> Chromosome mutations involve changes in large segments of DNA or whole chromosomes during cell division. They include duplication, inversion and translocation of segments, and the fusion and fission of whole chromosomes.

Causes of mutation

Exogenous mutagens

Environmental agents of mutation, **exogenous mutagens**, are radiation and chemicals. Mutagenic radiation includes X-rays, ultraviolet radiation and gamma rays. The changes induced can be either point mutations or chromosome mutations.

Chemical mutagens are of three major types. Firstly, there are chemicals, such as mustard gas, that react with DNA and alter or remove bases from the DNA backbone, leading to misincorporation of bases in the next cycle of replication. They also cause more drastic changes, including breakage of the DNA backbone. Chromosomal rearrangements result if the repair process rejoins the incorrect ends. Secondly, there are chemicals, described as base analogues, that mimic some components of the structure of the four normal bases—A, C, G, T—but do not permit complementary pairing. These lead to mistakes during the next cycle of replication. An example is azidothymidine, better known as AZT, which is used as an anti-viral agent in the treatment of AIDS. Thirdly, some chemicals that bind with DNA interfere with the normal process of DNA replication. An example is ethidium bromide, a widely used fluorescent agent for staining DNA, which inserts neatly into double-stranded DNA.

Because of the high frequency of radiation and chemically induced damage to DNA, all organisms possess repair systems consisting of enzymes that recognise and remove damaged DNA. Provided that one strand of the DNA molecule is undamaged, the complementary strand can be used to direct repair of the damaged strand.

Transposable elements

Some sequences of DNA, **transposable elements**, are mobile and capable of moving to new locations, frequently taking other genes with them. Transposable elements, discovered first in maize by Barbara McClintock, occur both in prokaryotes and eukaryotes (Chapter 10). Although the probability of movement is very small for an individual element, the large number of elements means that substantial amounts of transposition occur over time.

Mutations due to transposition occur in several ways. Firstly, if a transposable element inserts itself into or near a coding or regulatory part of a gene, the sequence of the gene is disrupted, resulting in modification or total abolition of its expression. The precise consequences of an insertion are difficult to predict, but evidence from numerous organisms attests to their importance. In *Drosophila melanogaster*, the majority of spontaneous visible mutations affecting traits such as eye colour and wing shape are due to the insertion of a range of different transposable elements. One of these, the *P* element, is particularly interesting because it has invaded populations of *D. melanogaster* throughout the world in the past 40 or 50 years. Even the wrinkled seed trait of peas studied by Gregor Mendel (Chapter 9) is now known to be the result of a mutation caused by insertion of a transposable element into the gene coding for a starch branching enzyme.

The second way in which transposable elements cause mutations is by excision or movement away from their original position. Most transposable elements are capable of precise excision from the DNA of the host. Generally, precise excision will lead to reversion, as the sequence and function of the gene will be restored approximately to the original. However, on many occasions they excise imprecisely, taking flanking DNA sequences with them or sometimes even taking captive sequences to new locations in the genome. Deletion mutations, like those induced by radiation, can result.

Transposition can also cause chromosomal inversions and translocations. Transposable elements must be able to nick DNA to insert or excise. When multiple nicks are made in the DNA of the same or different chromosomes, the incorrect ends may rejoin.

> Mutations can be the result of radiation and chemical agents (exogenous mutagens) or the movement of transposable elements.

BOX 32.2   THE HAIRLESS GENE OF MICE: A MUTATION CAUSED BY INSERTION OF A RETROVIRUS

Hairless is a recessive mutation of a gene, situated on chromosome 14 in mice and involved in control of the hair cycle. Mutant mice grow the first (juvenile) coat of hair but are incapable of growing another coat. When hair of the first coat falls out, adult mice lack all hair (Fig. a), except for vibrissae (whiskers). The mutation is caused by insertion of a murine leukemia virus (Fig. b).

Genes that regulate hair growth are of particular interest to the Australian wool industry. Increasing wool growth, or temporarily halting it as part of a biological shearing strategy, are possibilities enhanced by knowledge of genes such as *hairless*. A totally unexpected effect of the hairless mutation, which was discovered because the mice are used as experimental models for human skin, is that it drastically increases susceptibility to skin cancer induced by ultraviolet light. Analysis of the gene and its expression and function may yield therapy and control strategies for the important problem of skin cancer in Australia.

(a) Hairless condition in mice

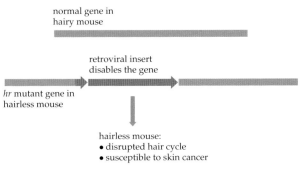

(b) *Hairless* in mice is a recessive mutation caused by insertion of a murine leukemia virus

FITNESS AND SELECTION

Mutations that are heritable, that is, those that occur in reproductive cells and can be passed on to the next generation, are the 'raw material' for evolution. Within a population, change in the frequency of variants created by mutation is the result of chance or selection. Change in gene frequency is completely random in the case of neutral evolution but is directed by selection in the case of adaptive evolution. Genetic variants can increase or decrease in frequency, or persist in, or be lost from a population. Over time, separate populations may develop different gene frequencies and diverge from one another.

As we saw in Chapter 9, in 1908, G.H. Hardy and W. Weinberg formulated a model for the behaviour of gene frequencies in a population not experiencing selection, large enough not to be influenced by genetic drift, and in which mating is random with respect to a particular gene. For a gene with two variants (alleles) A and a, with frequencies p and q respectively, the genotypes AA, Aa and aa can be predicted by the binomial expansion $(p + q)^2$. Thus, AA has frequency p^2, Aa has frequency $2pq$ and aa occurs with frequency q^2. For example, if the fraction of the population with allele A is 0.5, then the fraction with allele a is 0.5.

The frequency of the genotypes of the next generation are thus 0.25 (AA), 0.5 (Aa) and 0.25 (aa). The frequency of the phenotypes are 0.75 for the dominant ($AA + Aa$) and 0.25 for the recessive (aa). The gene frequencies remain constant from generation to generation and no evolution occurs, unless random or other evolutionary forces act.

With selection operating, the fate of a mutation depends on its fitness (Fig. 32.6). **Fitness**, in an evolutionary sense, refers to an individual organism's contribution of offspring to the next generation. Of particular concern is the effect of a mutation on the fitness of the individual carrying it. A mutation with a favourable effect on fitness will increase in frequency within a population (Box 32.3), whereas one with an unfavourable effect will decrease and eventually may be eliminated from the population. In our example above, suppose that 100% of organisms with the dominant A phenotype survive but only 10% with the recessive a phenotype survive. In the next generation, the frequency of the A allele will increase to 0.65 and the dominant A phenotype to 0.87. If some individual organisms contribute more fertile offspring to the next generation, selection is said to occur. Selection leads to adaptive evolution if the individuals

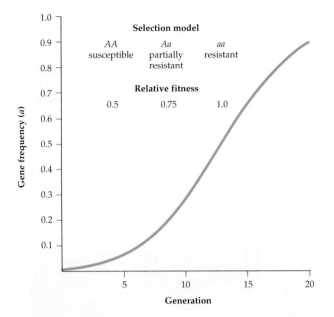

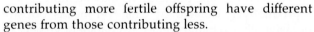

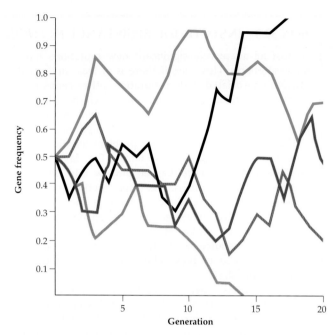

Fig. 32.6 Change in gene frequency in a large population due to selection. The plot shows the increase in gene frequency for an initially rare (1%) gene, such as an insecticide resistance gene, which becomes favourable because of the introduction of a new chemical. The selection model is shown with the relative fitness indicating the proportion of offspring transmitted to the next generation by the different genotypes

Fig. 32.7 Computer simulation of random drift in gene frequency in small populations. Five populations, each consisting of 10 breeding individuals, start out with an initial gene frequency of 50%. In the absence of selection, gene frequencies wander randomly and can increase to 100% or decrease to 0%. Differences between populations due to random drift constitute neutral evolution

contributing more fertile offspring have different genes from those contributing less.

Many mutations are neutral with no effect on fitness. Chance alone determines the fate of such neutral mutations within a population. As we saw earlier (p. 701), random genetic drift occurs in small populations, and in the absence of selection gene frequencies can increase to 1 or decrease to 0 (Fig. 32.7).

Between the extremes of genetic variants with no effect on fitness and those with a large effect on fitness, there is a complex interaction between fitness and population size. In small populations, genetic drift can swamp the influence of fitness, and a mutation with only a small favourable effect on fitness may be lost by chance. In large populations, the effects of differences in selection are more predictable and even selectively neutral variants have less tendency to drift in gene frequency.

MODES OF SPECIATION

Like Lamarck and Darwin, modern evolutionary biologists are interested in the origin of species. In nature, species occur usually as isolated populations. Members of a single population interbreed to varying extents, depending on the frequency of sexual reproduction. The opportunity for interbreeding

between members of different populations separated geographically may be much less. Furthermore, populations of a widespread species often occur in different environments. In different and changing environments, the effect of mutation on an organism's fitness can vary between populations. Natural selection, together with genetic drift, can thus lead to population divergence over time. The question is 'Do these processes lead to the formation of new species, **speciation**?'.

Speciation is an historical process and therefore hypotheses of speciation cannot be falsified by scientific experimentation. It is impossible to repeat a speciation event under controlled conditions except for some unusual modes of speciation not infrequently found in plants but very rarely in animals. Thus, indirect evidence from the fossil record, from comparative studies of morphology and development, DNA and proteins, from cytogenetics and from biogeography are used to test ideas and theories about speciation.

Understandably, there is still considerable contention about mechanisms of speciation. Creationists believe that species are created by God as described in the Bible, but since it is based on faith and belief, creation is outside the realm of science. Among biologists, debate about speciation mechanisms relates, in part, to arguments of species' definition.

BOX 32.3   INSECTICIDE RESISTANCE IN *LUCILIA CUPRINA*

Almost all adaptively significant new mutations have adverse effects. However, if there is at least one good attribute associated with a mutation, other genes that modify or mitigate the adverse effects will be selected. A good example is provided by insecticide resistance in the sheep blowfly, *Lucilia cuprina* (see figure).

The sheep blowfly is a major problem for the sheep industry in Australia because female flies lay eggs in wounds and wet fleece. Larvae burrow into the skin of sheep, causing distress, loss of production and even death if untreated. Chemicals, such as dieldrin and organophosphates, have been used extensively to treat and prevent fly strike. However, chance mutations conferring resistance to insecticides have arisen so that the effectiveness of these chemical control agents is greatly reduced.

Most newly arisen resistance genes are disadvantageous in the absence of insecticide and decrease to low frequency if the insecticide is withdrawn from use. Thus, rotation of use of different types of insecticides hinders the development of resistance. However, persistent use of an insecticide will cause the selection of modifier genes that reduce the disadvantageous effects of the resistance genes. A modifier gene, which removes the deleterious effects of the organophosphate resistance gene, has been recognised and mapped in Australian *L. cuprina*. This modifier gene enables the resistance gene to persist at high frequency after withdrawal of the insecticide, so that the insecticide can never be used effectively again.

Sheep blowfly *Lucilia cuprina*

What makes a species a species?

As we saw in Chapter 30, the question of 'What is a species?' has been a problem for biologists for many years and there is still no universal, generally agreed definition. Taxonomists use the word 'species' as the lowest taxon of formal classification, with the ranks of genus to kingdom indicating higher levels in the hierarchy of relationships of organisms. In this sense, species represent the terminal twigs on a phylogenetic tree (p. 672). They are recognised on the basis of character differences; for example, *Banksia serrata* has leaves with a serrated (saw-toothed) margin compared with *B. marginata*, which has an entire (smooth) margin.

Knowledge of genetics has led other biologists to seek a biological definition of species, which attempts to focus on mechanisms of speciation and the distinctiveness of the pool of genes possessed by different species. This search for a biological reality for species has problems. Most biological definitions of species apply only to sexually reproducing organisms. A biological species, according to Ernst Mayr's definition (p. 674), is a group of interbreeding natural populations reproductively isolated from other such groups. Some biologists define species as groups sharing a common, specific, mate-recognition system. Neither definition is relevant to organisms that normally reproduce asexually, because no mate recognition is involved and all individuals are reproductively isolated from all others so long as reproduction continues by asexual means.

An additional problem for biological definitions of species is that some populations may be more or less reproductively isolated, by whatever means, but still occasionally exchange genes. Exchange of genes can occur between closely related and very similar species or between more distantly related and dissimilar species. For example, *Eucalyptus obliqua* ('messmate', which is classified in the 'ash' group of eucalypts) hybridises with *E. baxteri* (a member of the 'stringybark' group) where the habitats of these two species overlap (Fig. 32.8). On morphological evidence, the two species are not each other's closest relative and are recognisably different. Their ability to interbreed is not used as a criterion for 'lumping' them together as a single species.

Other biologists define an 'evolutionary' species as a lineage of one or more populations bound by common ancestry and an ability to maintain integrity with respect to other evolutionary species. This definition does not concern itself with mechanisms of isolation or maintenance of integrity. However, the degree of integrity required is not explicitly spelled out and this definition is difficult to apply.

Despite differences in species' definition, all models of speciation emphasise restriction of interbreeding and thus gene flow between populations so that divergence may occur.

> The definition of a species is still an incompletely resolved problem. Biological definitions, based on the ability to interbreed, do not apply to the large number of organisms that reproduce asexually.

(a)

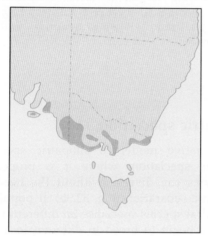

(i) *Eucalyptus baxteri*

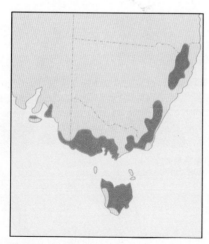

(ii) *Eucalyptus obliqua*

(b)

Fig. 32.8 **(a)** Hybrids (centre) occur in some mixed populations of *Eucalyptus obliqua* (right) and *E. baxteri* (left) and have fruits and leaves intermediate between the two parental species. **(b)** The species' geographic ranges in coastal Victoria and flowering time overlap

Allopatric speciation

The most widely accepted model of speciation is that of **allopatric speciation**. The allopatric model (Fig. 32.9a) assumes that populations of an ancestral

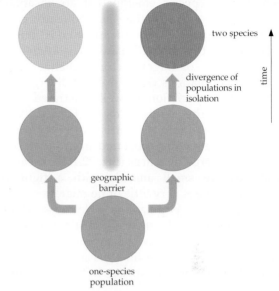

(a) Allopatric speciation

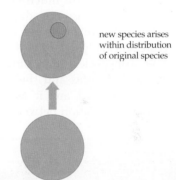

(b) Sympatric speciation

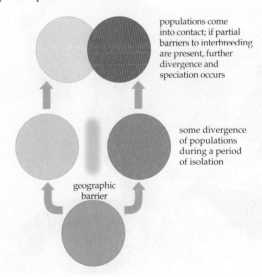

(c) Parapatric speciation

Fig. 32.9 Models of speciation: **(a)** allopatric; **(b)** sympatric; and **(c)** parapatric

species are separated by some geographic barrier, such as a desert or a mountain range, which completely inhibits migration. Over a period of time, the isolated populations diverge as a result of both adaptive and neutral evolution, so that eventually two distinct species emerge that are incapable of freely interbreeding if they subsequently come in contact. Complete geographic isolation is considered necessary in this model because it is argued that a small amount of migration or exchange of genes between populations would prevent divergence.

Allopatric speciation is well illustrated by rock wallabies of the genus *Petrogale* (Fig. 32.10a). The habitat of these animals consists of rocky outcrops, which are frequently separated from other suitable habitats by large distances (Fig. 32.10b). Because the wallabies are small animals with a tight social structure, there is very little or no migration between these disjunct populations. No less than 20 chromosomally and genetically distinct forms, 11 of which are classified as species, are recognisable within Australia. Many of the lineages are reproductively isolated, although some species hybridise where their ranges currently overlap.

Wallaroos and euros are much larger relatives of the rock wallabies and also are adapted to the rocky outcrop habitat. However, these animals disperse widely and are less likely to form isolated populations. Only a few subspecies, including the wallaroo *Macropus robustus robustus* and the euro *M. robustus erubescens* are recognised, and it seems that there is a continuous intergradation between them.

For species with good powers of dispersal, major geographic barriers are required for isolation and divergence. The eastern grey kangaroo, *Macropus giganteus*, is found on the east coast of Australia, whereas the related western grey, *M. fuliginosus*, is distributed in the south and south-west regions. Their distributions overlap in western Victoria and New South Wales but it seems likely that they evolved in isolation in the eastern and western parts of the continent respectively and subsequently expanded their range. Numerous other organisms have eastern and western counterparts, including bandicoots of the genus *Perameles*, frogs of the genus *Litoria*, whipbirds and plants such as banksias.

> Divergence of populations into different species is hindered by too much gene exchange through interbreeding. Geographic isolation (allopatric speciation model) provides an effective mechanism for preventing exchange, thus allowing adaptive and neutral divergence.

Sympatric speciation

An alternative model to allopatric speciation is **sympatric speciation**, where it is proposed that populations can diverge without the need for full geographic separation (Fig. 32.9b). If populations of an ancestral species specialise on different resources, they may begin to breed in different local areas or

(a)

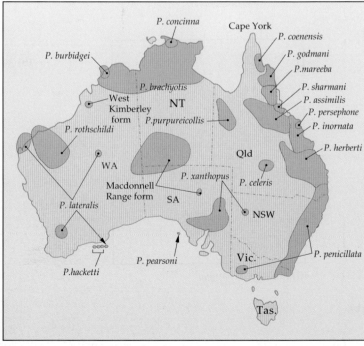

(b)

Fig. 32.10 (a) A typical rock wallaby of the genus *Petrogale* on a rock outcrop. **(b)** The distribution of rock wallabies in Australia, showing hybrid zones (H), suggests an allopatric model of speciation

at different times, so that gene flow is restricted. Thus, divergence may occur without the need for geographic isolation.

The most convincing evidence for sympatric speciation is perhaps provided by host races of the North American fruitfly, *Rhagoletis*, where very closely related species are confined to different host plant species, such as apples and hawthorn. These fruitfly species are believed to have arisen in the same geographical area. Divergence was possible because populations became established on new host plants with slight differences in time of flowering and fruit production. Populations of flies adjusted to the timing and life cycle of their host, and the possibility of migration was eliminated. Populations diverged and are now recognised as separate species.

Even in these cases, some biologists are not convinced by an explanation of sympatric speciation, claiming that the populations of fruitflies on different species of trees could have diverged during a period of geographic separation that was not observed. In other words, it is still possible to propose a plausible allopatric model of speciation.

> In the sympatric speciation model, populations specialising on different resources diverge without geographic isolation.

Parapatric speciation

Parapatric speciation is a model that assumes that some divergence between two populations occurs during a period of geographic isolation (Fig. 32.9c). If circumstances change so that the populations come into contact again, there are two possibilities. The populations may gradually merge by interbreeding and the distinction is lost. Alternatively, if partial barriers to interbreeding are in place, for example, chromosomal rearrangements that do not permit meiosis to occur normally in hybrids, there may be selection for avoidance of interbreeding between members of the two populations. Parapatric models assume that contact between diverging populations drives further divergence and speciation.

> The parapatric model of speciation assumes divergence between populations occurs initially in geographic isolation and is reinforced when the populations subsequently come in contact.

Rapid speciation

Allopatric speciation requiring geographic isolation followed by divergence is likely to be a slow process. There are many cases where related species differ by chromosomal rearrangements but are otherwise very similar genetically. Several examples are found among the rock wallabies, *Petrogale*. The low level of genetic divergence indicates that the species may have been isolated for a short time. Furthermore, the chromosome rearrangements that distinguish the species are often 'incompatible' and cannot persist in the same population for a long period of time. An individual with both forms of the chromosomes together will be less fit than an individual with only one of the forms. An Australian geneticist, the late M.J.D. White, proposed that chromosomal rearrangements have played an important role in establishing reproductive isolation between some species. A brief period of isolation of a small population may have permitted a new chromosomal mutation to become fixed, which has contributed to reproductive isolation when contact was re-established with the original population.

Instant chromosomal speciation

There is a mechanism of sudden or instant speciation not uncommon in plants but very infrequently found in fungi and animals, which involves doubling of the chromosome number. Sometimes this involves spontaneous doubling, from the diploid number ($2n$) to a tetraploid number ($4n$). This is **autopolyploidy** ('auto' since the chromosomes originate from the same population). Tetraploids may experience problems during meiosis as four chromosomes may associate during meiosis rather than the normal pair in the diploid, and thus fertility will be reduced. The problem is circumvented if the tetraploid can reproduce asexually.

In other cases of polyploidy, sets of chromosomes originate from different sources. For example, where two related species, *AA* and *BB*, coexist, *AB* hybrids may be produced frequently. If these hybrids are sterile due to problems of pairing of *A* chromosomes with *B* chromosomes, fertility will be restored if the chromosome number happens to double to *AABB*. *A* chromosomes will pair only with *A* chromosomes and *B* with *B*, and regular meiosis can occur so that normal sexual reproduction is possible. The **allopolyploid** *AABB* produced in this way cannot successfully reproduce with either the *AA* or the *BB* parent and therefore can be considered a new species.

Allopolyploid speciation has occurred in kangaroo grass, *Themeda triandra*, where diploid ($2n = 20$) and allopolyploid ($4n = 40$) forms are found (Fig. 32.11). Polyploidy is common in the grass family, where about 70% of species studied are believed to be polyploid. The most famous examples of allopolyploidy are found in wheat. Bread wheat, *Triticum aestivum*, is hexaploid ($6n$), with 42 chromosomes derived from three ancestors (Fig. 32.12). Two diploid ancestors, each with a diploid

Fig. 32.11 Kangaroo grass, *Themeda triandra*, in which there are diploid and allopolyploid forms

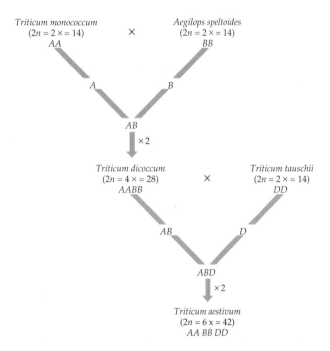

Fig. 32.12 Origin of hexaploid wheat, *Triticum aestivum*, by hybridisation and chromosome doubling. The *A* genome is derived from *T. monococcum*, the *B* genome from *Aegilops speltoides* and the *D* genome from *T. tauschii*

number of 14, initially hybridised to form a tetraploid species, which hybridised with another diploid to generate the hexaploid.

Instant parthenogenetic species

Species having a hybrid origin are not restricted to allopolyploid plants. In some animal species, *parthenogenesis*, which bypasses meiosis and problems of chromosome pairing (Chapter 13), provides an alternative method of overcoming the meiotic incompatibility of hybrids and may lead to instant speciation. The common Australian gecko, *Heteronotia binoei*, is a triploid (3*n*) all-female species derived from two chromosomally distinct sexual forms that hybridised to form a diploid hybrid occasionally producing diploid eggs (Fig. 32.13a). This hybrid then mated with one of the parental species to form a triploid hybrid that is self-perpetuating by parthenogenesis; triploid females produce triploid female offspring. The Australian morabine grasshopper, *Warramaba virgo*, arose in a similar fashion from two sexually reproducing ancestral species, but the parthenogenetic species in this case is diploid and one of the ancestral species has become extinct (Fig. 32.13b).

Parthenogenesis is considered to be an evolutionary dead end in the long term, although parthenogenetic species may be highly adapted and very successful under a particular set of environmental circumstances.

> In some cases, new species arise suddenly by increases in chromosome number (polyploidy) or non-sexual reproduction (parthenogenesis).

RATES OF EVOLUTION

Models of speciation consider the rate and uniformity at which evolutionary change occurs, and argument about evolutionary rate dates back to the time of Darwin. 'Gradualists' propose that evolutionary change and speciation occur slowly and gradually over time. Yet there are features of organisms that are difficult to imagine as evolving by gradual steps—what use is a half-developed eye or lung? A gradualist's explanation is that the feature has one function up to a point of evolution, beyond which it has another function (a half-developed lung providing buoyancy in a fish). 'Saltationists' propose that features arise suddenly (in addition to the case of polyploidy in plants) through mutations that have a dramatic effect. The recent idea of 'punctuated equilibrium' is that species remain relatively constant over long periods of time, with change resisted by the constraints of the developmental system or the availability of genetic variation. Perhaps in response to a changing environment or as a result of random genetic change due to small population size, sudden and dramatic change leads to rapid formation of new species. In other words, the usual situation of static equilibrium (stasis) is punctuated by a burst of change. As the sympatric model of speciation generally requires rapid development of reproductive isolation, the proponents of that theory have favoured the possibility of very rapid evolutionary change.

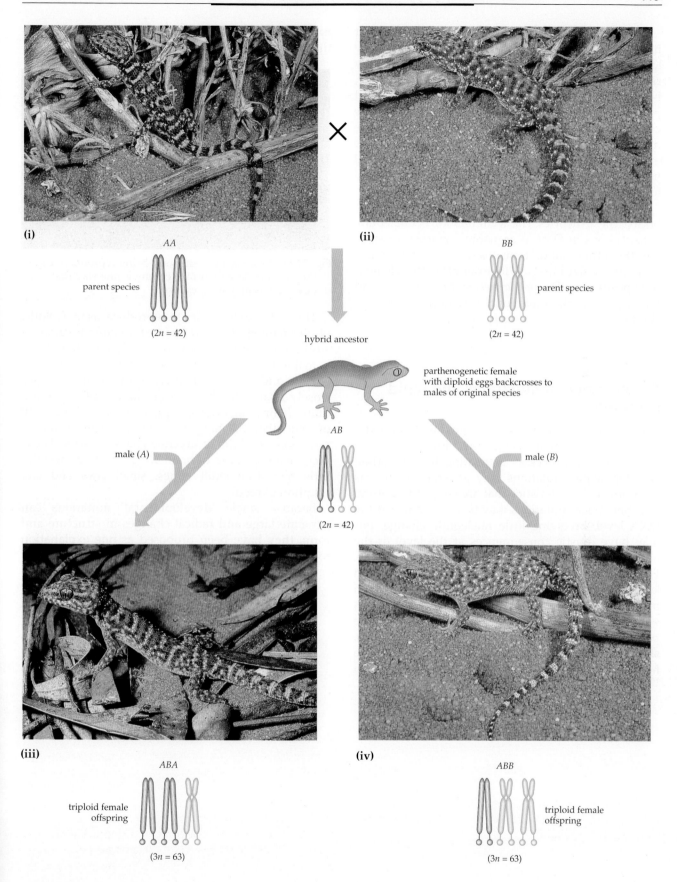

Fig. 32.13 Origin of the female parthenogenetic gecko, *Heteronotia binoei*. Two forms of gecko, with chromosomes *AA* and *BB*, crossed to form a hybrid, *AB*. At least some female hybrids produced eggs with diploid sets of chromosomes. When these females back-crossed with males of the original two forms, their diploid eggs, *AB*, were fertilised with haploid sperm (either *A* or *B*), resulting in triploid offspring (*ABA* and *ABB*), which were all female

The fossil record provides examples of taxa that have remained morphologically constant over very long periods of time, only to be apparently suddenly replaced by a variety of diverged forms. However, because the fossil record is inevitably incomplete and patchy, the gradualists can generally advance counter-arguments that the bursts of speciation are more apparent than real. Further, they can point to examples in the fossil record where gradual morphological change can be observed in series of fossils over a period of geologic time.

Undoubtedly there is some evidence supporting both of these theories. The accumulation of nucleotide substitutions in DNA is arguably a gradual process. On the other hand, transpositional mutation can occur in massive bursts, as for example with *P* element transposition in *Drosophila melanogaster* (p. 705), and might provide the mutational variation required for sudden bursts of evolutionary change.

Developmental mutations and rapid evolution

One way of creating evolutionary possibilities rapidly is by mutations that change the timing or rate of developmental processes in relation to each other. Developmental mutations may generate organisms with novel characteristics that are subject to natural selection. Such mutations may be quite minor at the DNA level, involving little nucleotide change, but may have drastic consequences at the level of the organism. Dramatic and sudden changes are possible. For example, axolotls reproduce in the aquatic, gilled larval stage, unlike most other salamanders, which moult and reproduce as terrestrial, adult amphibians with lungs (Fig. 32.14). Although axolotls retain the potential to develop into the terrestrial adult form, since moulting can be induced experimentally by the hormone thyroxin and adults sometimes occur in nature, natural selection appears to have favoured the aquatic form.

Fig. 32.14 Axolotls are neotenous, being reproductive in the juvenile (larval) form. Neoteny is one way that novel forms may have evolved

The ability of juveniles to reproduce, as in axolotls, or alternatively the retention of juvenile features in adults, is **neoteny**. Neoteny is considered to have been a major factor in human evolution. Although on genetic and chromosomal criteria humans are very closely related to chimpanzees and gorillas, we are quite distinct in external appearance. Practically all of our distinguishing features are the result of retention as adults of juvenile characteristics. These include a high ratio of brain weight to body weight, little hair, thin skull bones, small jaws and late eruption of teeth.

Because single 'developmental' mutations can generate large and radical changes in structure and form, they have been proposed as one explanation for **saltatory evolution**, jumps in the evolutionary record.

Gradualism is the idea that evolution occurs slowly by the accumulation of small changes over a long period of time. An alternative view—punctuated equilibrium or saltatory evolution—is that evolutionary change occurs suddenly followed by long periods of time in which there is little change (stasis).

SUMMARY

- Evolution is the process of change and divergence in populations and taxa. In modern biology, two models of evolution are recognised: adaptive and neutral evolution.

- Adaptive evolution is evolutionary change that benefits an organism in a particular environment by enabling it to leave more descendants than other organisms. Adaptive evolution is a process that sometimes results in greater complexity and sometimes in reduced complexity.

- Some evolutionary changes have no effect on the survival or reproduction of an organism. If there is no selection and population size is small, gene frequencies will change randomly (genetic drift) and evolution will be neutral.

- The inheritance of traits gained during the lifetime of an organism was once proposed as a mechanism of evolutionary change (Lamarckism). Current understanding of molecular genetics precludes the inheritance of acquired characters, although a role for retroviruses has been suggested.

- Mutations provide the variation that underlies both adaptive and neutral evolution. They are the result of radiation, chemical agents or the movement of transposable elements.

- Point mutations are changes in DNA sequence and include base substitutions, deletions and insertions. Chromosome mutations involve structural changes in segments of DNA

or whole chromosomes during cell division. They include duplication, inversion and translocation of segments, and the fusion and fission of whole chromosomes.

- Divergence of populations into different species is hindered by too much gene exchange or interbreeding. Geographic isolation provides an effective mechanism for preventing exchange, thus allowing adaptive and neutral divergence (allopatric speciation). In sympatric speciation, populations showing differences in behaviour and other traits diverge without geographic isolation. The parapatric model of speciation assumes that divergence between populations occurs initially in geographic isolation and is reinforced when the populations subsequently come in contact.

- In some cases, new species arise suddenly by an increase in chromosome number (polyploidy) or by hybridisation in organisms that have an ability for non-sexual reproduction (parthenogenesis).

- Evolution is generally thought to occur slowly by the accumulation of small changes over a long period of time (gradualism). An alternative view is that evolution involves periods of sudden changes (punctuated equilibrium) between which taxa remain constant for long periods of time.

QUESTIONS

1. Briefly explain how the application of molecular biological techniques, such as DNA sequencing, have enhanced our ability to study evolutionary change.

2. What did Charles Darwin mean by 'natural selection'?

3. Are changes in gene frequency always due to the action of selection? Briefly explain your answer.

4. If you had cloned the same gene from two species, explain how a comparison of the DNA sequences might permit you to recognise exons and introns. Explain adaptive and neutral evolution in relation to such DNA sequences.

5. Explain how the inheritance of acquired traits is incompatible with the mechanism of storage and retrieval of genetic information.

6. If a species lives in an area of high background radiation, do you think a more efficient DNA repair system would evolve or would the species evolve more rapidly?

7. Summarise and contrast the allopatric and sympatric models of speciation, giving an example of each.

8. Explain how (a) autopolyploidy and (b) allopolyploidy can lead to new species. Why do you think these processes occur more in plants than in animals such as vertebrates?

9. Discuss how developmental mutations can explain jumps or discontinuities in the evolutionary record.

10. A mutation occurs in a population of terrestrial salamanders so that the mutants never moult but reproduce exclusively at an aquatic 'larval' stage. If the aquatic form continues to coexist with the original form, which mates exclusively on land, at what stage would you be willing to recognise the aquatic (A) and terrestrial (T) forms as separate species? What would be the effect in an A animal of a reversion that allowed it to moult and mate with a T animal?

B A C T E R I A

The existence of bacteria has only been understood since Antonie van Leeuwenhoek, a Dutch microscopist, peering through his newly invented microscope about three hundred years ago, saw 'little animalcules'. But their invisibility to the naked eye is not the only reason for the mystery that they seemed. It took until the 1960s to discover that bacteria are not miniature animals at all, as van Leeuwenhoek supposed, nor, as many botanists suggested, tiny non-photosynthetic plants.

Bacteria (kingdom Monera) are micro-organisms and the smallest forms of *cellular* life on earth (Fig. 33.1). (As we shall see in Chapter 34, viruses are smaller but are not cellular organisms.) Bacteria exist generally as single cells with a thin cell wall and have simple shapes (rods, spheres, spirals, filaments; Fig. 33.2). Individually, they can be seen only with a microscope because they are only about 0.5–5 μm long. Their presence is often detected and their biochemical properties measured as a result of extensive 'colonial' growth, involving millions or billions of cells grown together in culture (i.e. in liquid media or on agar gels containing the necessary nutrients for that particular bacterium).

Bacteria may have beneficial, harmful or indifferent effects on other organisms, including humans. They are the continuing cause of many medically and agriculturally important diseases, and are the agents of many important industrial and agricultural processes. But their greatest significance is in the enormous variety of natural biochemical transformations that they carry out, transformations of material and energy that go on quietly in soil and water, and in and on animals and plants, without which other life on earth would grind to a halt (see Chapter 44). Some examples of the more obvious beneficial and harmful influences of bacteria on humans are given in Table 33.1, but these hardly describe the immensity and diversity of the impact that bacteria have as abundant members of the biosphere.

BACTERIA ARE PROKARYOTES

The application of modern techniques of electron microscopy was needed to discover the structure and diversity of bacteria. Examination of sections through bacterial cells shows the typical features of a *prokaryote* cell (Chapter 3), which has a less extensively developed internal structure than do eukaryotic cells (Figs 3.2, 3.3). Biochemists and molecular biologists are also uncovering important metabolic and genetic differences between bacteria (and from here on the

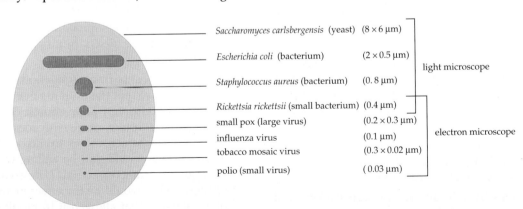

Fig. 33.1 Relative sizes of some micro-organisms and the effective range of microscopes. Bacteria are the smallest cells (viruses being non-cellular; 1 μm = 10⁻⁶ m)

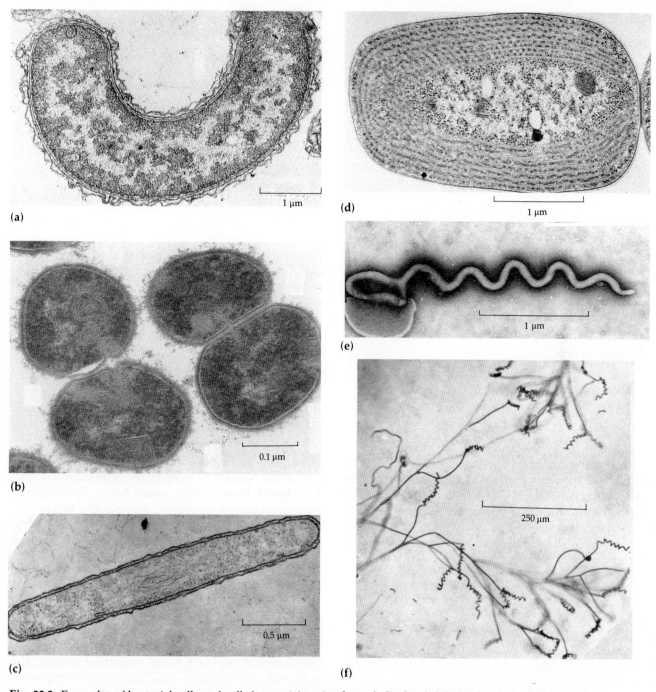

Fig. 33.2 Examples of bacterial cells and cell shapes: **(a)** a circular or helical rod, *Microcyclus major*; **(b)** a dividing coccus or sphere, *Gemella haemolysans*; **(c)** a long rod, *Nitrospina gracilis*; **(d)** a short rod, *Plectonema* sp., which is a cyanobacterium containing stacks of photosynthetic membranes; **(e)** a spiral bacterium, *Spiroplasma* sp.; and **(f)** a filamentous type, *Streptomyces* sp. **(a)**–**(d)** Electron micrographs of cut sections of cells lightly stained with an electron-dense substance, such as lead; **(e)** is an electron micrograph of an uncut and negatively stained cell (stained with a uranium salt to stain the immediate surroundings more intensely than the cell itself); and **(f)** is a light microscope image of an unstained live sample

term bacteria will be taken as synonymous with prokaryotes) and the rest of the cellular world. For example, ribosomes (the sites of protein synthesis) of bacteria are a different size and composition from those found in eukaryotes, and the process of protein synthesis occurring on these is also subtly different

(Chapter 11). An important practical consequence of this is that bacterial protein synthesis is susceptible to naturally occurring inhibitors—**antibiotics**. Antibiotics generally act through inhibiting a range of processes in prokaryotes at concentrations that are not toxic to eukaryotes (see also Box 11.3).

Table 33.1 Activities of bacteria that benefit or harm humans

Beneficial activities	Harmful activities
Food production and technology	
Cheese and yoghurt	Spoilage of fresh foods and beverages
Meat preservation	Food poisoning
Vinegar	
Medicine	
Synthesis of the majority of antibiotics (some are also produced by fungi)	Causation of: cholera, typhoid, dysentery, typhus, meningitis, gonorrhoea, syphilis, pneumonia, toxic shock, tetanus, botulism, tuberculosis, legionnaires' disease, wound infections, and many other important infectious diseases
Agriculture	
Biological insect control	Many animal diseases (mastitis, foot rot, fly strike,
Nitrogen fixation in soil	botulism, lumpy jaw, abscesses, wound infections)
Humus formation in soil (breakdown of dead plant and animal tissues)	Some plant diseases (wilt, damping off, galls)
Recycling and transformation of carbon, sulfur, phosphorus and nitrogen compounds in soil and water	Frost damage to plants (by ice nucleation)
Industry	
Sewage treatment and nutrient recycling from wastes	Steel and concrete pipe corrosion (acid-producing bacteria)
Ethanol production from carbohydrates	

Despite the important organisational and structural differences between prokaryotes and eukaryotes (as well as viruses), all life on earth is based on similar biochemical patterns of metabolism and synthesis, and the expression and replication of genetic information. This similarity is referred to as the *biochemical unity of life* (see Box 30.2).

> Bacteria are the smallest cellular life on earth. They are prokaryotes differing structurally, biochemically and metabolically from eukaryotes. Their protein synthesis is susceptible to inhibition by antibiotics.

ANCIENT ORIGINS OF BACTERIA

The formation of cellular structures that contained living cytoplasm was an early event in the evolution of life, probably because the cell unit provided the means to keep everything in one place and at a concentration high enough for the biochemical reactions of life to work efficiently.

The earliest cells evolved about 3.5 billion years ago, about one billion years after the earth itself formed (Chapter 31). At this time, parts of the earth's surface had cooled enough to permit biochemical reactions to take place in liquid water, that is, below 100°C, although temperatures were generally hotter and seas saltier than today. Microscopic cellular life

continued to evolve at the level of individual organisms, similar to some types of modern bacteria, for almost three billion years before multicellular macro-organisms appeared, leaving the recent (0.6 billion years old) heritage of fossilised life. It is interesting that, among modern bacteria, we still find species that thrive under hot and salty conditions (p. 724).

Early photosynthetic bacteria

Two other features of the early physical and chemical environment on earth were important. Firstly, the primeval environment was anaerobic (Chapter 5) with very little free oxygen in the atmosphere, although much was present as oxides and carbonates in rocks, and as carbon dioxide in the atmosphere. Secondly, the earth was bathed in a solar radiation far different from that which exists now. The atmosphere contained little ozone, which today forms a layer high in the earth's atmosphere, absorbing a large fraction of solar (UV) radiation. This protects life on earth by preventing the lethal and mutational damage of UV radiation on the genetic systems of living cells.

Since there was a limited supply of chemical energy in the form of organic carbon compounds ('fixed carbon'), the evolution of photosynthesis was an important part of the early history of life. Early photosynthesis was *anoxygenic* ('not generating oxygen');

in fact, oxygen would have been a very toxic substance to the early photosynthetic bacteria. Bacteria that carry out anoxygenic photosynthesis also exist today and retain these features of the early bacteria.

Also, because the amount of UV radiation arriving at the earth's surface was high, photosynthesising bacteria either must have been equipped in some way to prevent or overcome damage from UV radiation, or they adopted ecological strategies to avoid direct sunlight; for example, by floating deeply in aquatic habitats so that damaging UV radiation was filtered out but not the useful visible radiation needed for photosynthesis, or by occupying shaded habitats out of the direct rays of the sun so that light scattering removed most of the UV radiation but not the visible component.

The earth's surface has cooled considerably since those primeval days. Molecular oxygen has also appeared in abundance as a consequence of the development of *oxygenic photosynthesis* in bacteria about 2.5 billion years ago. This type of photosynthesis, in which oxygen is a product of the overall reaction to fix carbon dioxide as sugar (Chapter 5), is also characteristic of modern algae and plants because the chloroplasts in which this photosynthesis occurs are themselves the evolutionary remnants of bacteria (Chapters 3 and 35).

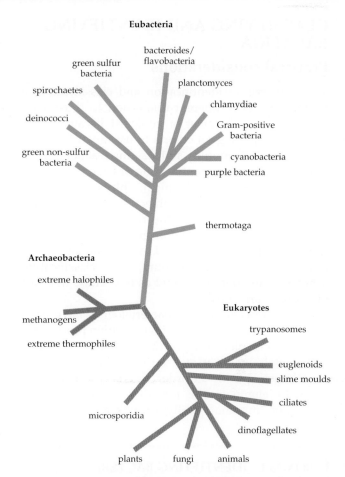

Fig. 33.3 The broad evolutionary relationships of bacteria showing two major groups—Archaeobacteria and Eubacteria. The root of the tree is unknown. The analysis was based on ribosomal RNA sequences

Two major groups

Except for colonial aggregations, such as those represented by stromatolites (Chapter 31), microorganisms rarely leave any fossil record because they have no hard structures, such as bones or teeth, and also because their size makes them difficult to distinguish among sedimentary rock grains and particles. Instead, their DNA and RNA molecules have been used to infer evolutionary relatedness. In Chapter 30 we saw that, on the basis of the nucleotide sequences in DNA or RNA, a phylogenetic tree can be constructed which summarises how closely related different groups (taxa) are to one another, and how long ago modern taxa diverged from each other's evolutionary paths.

The comparison of DNA sequences by molecular methods has caused a revolution in the way we now perceive the evolution of cellular life on this planet. Figure 33.3 summarises the view that is now generally held about the phylogenetic relationships between modern forms of bacteria and eukaryotes. This form of evolutionary tree is 'unrooted' and is like a tree seen from above. We do not know the root of the tree, that is, the position of the ancestral taxon that

gave rise to all cellular forms of living organisms now known to us (and in which, too, the common biochemical mechanisms of all living things on earth had their beginning).

A notable feature of this tree is that there are two major groups of bacteria, the **Eubacteria** ('good or true bacteria') and the **Archaeobacteria** ('very old bacteria'), which are considered separate kingdoms by some biologists. However, because our phylogenetic tree is unrooted, we do not know which of these two groups of bacteria is the nearest relative of the eukaryotes, although some biologists have suggested that the Archaeobacteria are probably more closely related to eukaryotes than to Eubacteria. A final important point is that the bacteria that formed symbiotic relationships with primitive eukaryotes, and ultimately became the organelles of modern eukaryotes (Chapter 35), were probably highly respiratory (mitochondria), photosynthetic (chloroplasts) or motile (cilia) Eubacteria.

Phylogenetic classifications based on comparisons of DNA sequences recognise two major evolutionary groups—Archaeobacteria ('old') and Eubacteria ('true').

CLASSIFYING AND IDENTIFYING BACTERIA

Practical considerations

The problems of identification and classification of bacteria are of a different dimension to those confronted by the botanist or zoologist because there are few morphological criteria to use, apart from cell shape (Fig. 33.2), motility and staining reactions. Most of the tests used in identification of bacterial species (Box 33.1) are biochemical, physiological or immunological, and are subject to a considerable degree of genetic variation, the mechanisms for which we will consider later. The propensity of bacteria to mix genes, to vary genes, and to gain and lose genes must be taken into account, and classification schemes must not be so rigid as to exclude this sort of natural variability.

From extensive testing, bacteriologists have been able to construct systematic identification procedures that are convenient and practical and therefore useful in hospitals where it may be necessary to identify a pathogen quickly in order to take the necessary steps to eliminate the infecting organism, either surgically, immunologically, or with antibiotics. For these purposes, a classification based on a phylogenetic tree (Chapter 30) is not essential. The important issue is to identify the organism quickly so that the seriousness of infection can be assessed and treatment begun. This also applies to identifying bacteria that inhabit a particular ecosystem, such as a city beach used for recreation, or a country river used to supply drinking water. Reliable tests that identify species or groups of bacteria are required to determine the extent to which a population of microbes represents a health threat, and how the threat might vary with place and time.

A detailed listing of bacterial types is given in Box 33.2. The sections listed are based on phenotypic and ecological characteristics. The phylogeny in Figure 33.3 is based on genetic criteria and thus reflects evolutionary relationships. However, the phylogenetic tree is made up of recent and incomplete molecular data and is therefore less comprehensive than the traditional classification. It will be interesting to see whether, as more phylogenetic data become available, the two schemes converge.

Bacteria are a diverse group and phenotypic classifications are used for practical purposes of identification.

BOX 33.1 IDENTIFYING BACTERIA

1. Isolation
Purify as visible colonies, containing a million or more cells grown from one original cell (which makes them visible on an agar medium containing nutrients [Fig. a]). Examine size, shape, texture and colour of colonies (Fig. b).

2. Staining and microscopy
Gram stain with crystal violet, Jensen's iodine, alcohol and carbol fuchsin (purple—Gram-positive bacteria; pink—Gram-negative bacteria; Fig. c). Examine shape (coccus, rod, spiral, other), cell grouping (chains, tetrads, pairs, clusters) and special structures (spores, capsules), and perform motility testing of live cells. More cell detail is revealed using electron microscopy.

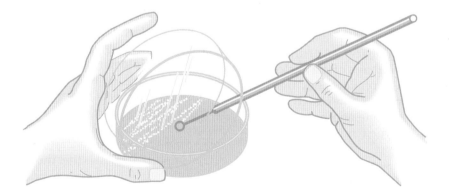

(a) Streaking bacteria to obtain single colonies. Material containing bacterial cells is collected onto a sterile wire loop, then streaked onto the agar medium in the petri dish so that individual cells are laid along the surface of the agar. Incubation at growth temperature results in many divisions of these individual cells to give rise to a colony (a million cells or more all derived from one individual cell originally at that spot). Each colony is thus a clone (a collection of identical individuals) of the original cell and can be used as starting material for the subsequent testing of that bacterium

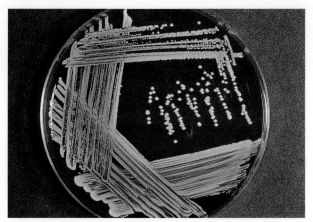

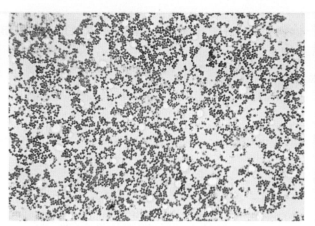

(b) Individual bacterial colonies of *Staphylococcus* sp. (and confluent areas, or unresolved colonies) generated by dilution streaking. This technique provides clonally pure material for subsequent identification testing. The purity of a culture is also checked by this means. Pure cultures of a bacterium will give rise to colonies of the same size, shape, texture, colour and contour, while mixed or impure cultures give rise to mixtures of colonial types

(c) Identification of *Staphylococcus aureus* in a stained sample of sputum from a case of pneumonia. Bacterial cells from these colonies yield Gram-positive (purple) cocci, clustered in bunches, typical of staphylococci. If necessary, scanning electron microscopy can be used to show cell shape in more detail. Further biochemical testing can be done (mannitol fermentation, growth in high concentrations of salt, serum coagulation, catalase activity) to confirm that the isolated bacterium was indeed *Staphylococcus aureus*

3. Physiological testing

Test for: growth on different carbon sources (e.g. sugars), nitrogen sources (e.g. protein, amino acids, ammonium or nitrate), sulfur compounds; need for oxygen; tolerance of salts or high osmolality; and optimum temperature for growth.

4. Biochemical and genetic testing

Test for specific enzymes, metabolic pathways, immunological reactions, resistance to antibiotics and other specific agents. Isolate DNA and test for proportion of guanine and cytosine (GC%). Hybridise DNA with DNAs from known organisms to estimate genetic similarity.

Eubacteria

In Box 33.2, all but sections 25 and 33 (Archaeobacteria, and other genera of uncertain type) are eubacterial groups. Eubacteria are an enormously diverse group of bacteria that share many environments with, or live on and in humans and other animals and plants, environments of moderate temperature that are moist, low in salt or other solutes, and where sunlight or organic compounds are plentiful. Oxygen is not so important since many of the Eubacteria have powerful biochemical systems of fermentation, or can replace molecular oxygen as a metabolic oxidant with oxidised anions such as nitrate and sulfate. Some are also capable of substituting reduced organic carbon compounds as a source of energy with reduced inorganic substances, such as sulfide or ferrous ions, or gaseous hydrogen. In fact, this group contains almost every variety and combination of biochemical energy extraction and carbon fixation that is thought to be feasible on the basis of the molecular composition of the biosphere (those few types not

represented here are found instead among the Archaeobacteria). Only a few plastics and organochlorine synthetics used as insecticides and herbicides have proven resistant to the capabilities of eubacterial metabolic versatility.

However, also in this group are some of the most fastidious and delicate cellular organisms known. These are often pathogens that have become highly adapted to special environments within warm-blooded hosts. The bacteria causing sexually transmitted diseases in humans (gonorrhoea, syphilis, chlamydial infections) are often difficult to grow in the laboratory but very easy to catch in bed! The rickettsias, which cause typhus, cannot be grown away from host cells, and *Legionella pneumophila*, which causes the respiratory infection known as legionnaires' disease, has a fastidious requirement for ferrous ions and organic sulfur compounds when grown alone, yet grows easily in mixed culture with certain free-living protists that occur in the soil or in water reservoirs and tanks.

BOX 33.2   THE BERGEY CLASSIFICATION OF BACTERIA

The Bergey classification of bacteria is a practical reference for bacterial taxonomists and provides a straightforward means of identifying bacteria. The nomenclature used is a standard binomial Linnaean scheme (Chapter 30). A bacterial species is defined as a group of bacteria with common phenotypic characteristics, occupying similar habitats, and whose DNA shows no major compositional differences (expressed as the percentage of the bases guanine plus cytosine). In turn, a genus is defined simply as a group of similar species. Beyond this, genera with similar physiological and ecological features are placed together in 33 'sections'. A brief description of these 33 sections, together with examples affecting humans, is given below.

1. **Spirochaetes**
   Spiral shaped cells; e.g. *Treponema*: cause of syphilis, yaws

spirals

2. **Aerobic motile helical Gram-negative bacteria**
   Helicobacter: diarrhoea in children, abortion in sheep

3. **Non-motile Gram-negative curved bacteria**
   A little-known bacterial group

4. **Gram-negative aerobic rods and cocci**
   Pseudomonas: soil and aquatic organisms, burns infections
   Rhizobium: legume nodulation and nitrogen fixation

5. **Facultatively anaerobic Gram-negative rods**
   Many enteric (gut) bacteria, including *Escherichia*, *Salmonella* (food poisoning, typhoid), *Shigella* (dysentery), and *Vibrio* (cholera)

6. **Anaerobic Gram-negative rods**
   Bacteroides and *Fusobacterium*, major members of the human gut and faecal microbiota

7. **Sulfate or sulfur-reducing bacteria**
   Soil and aquatic bacteria, which reduce sulfate to sulfur and hydrogen sulfide in rotting organic matter

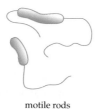

motile rods

8. **Anaerobic Gram-negative cocci**
   Abundant as part of the normal and harmless microbiota of the human gut and mouth

9. **Rickettsias and chlamydias**
   Bacteria that live only as parasites inside animal cells
   Rickettsia: typhus, scrub and spotted fevers
   Chlamydia: trachoma, venereal infections

10. **Mycoplasmas**
    Bacteria without a cell wall
    Mycoplasma: respiratory tract infections

11. **Endosymbionts**
    Bacteria that are symbiotically associated with and grow only inside the cells of host animals and fungi

12. **Gram-positive cocci**
    Staphylococcus and *Streptococcus* cause many important infections of humans and animals; industrially important species used in milk and meat processing

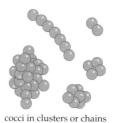

cocci in clusters or chains

13. **Endospore-forming Gram-positive rods and cocci**
    Bacterial spores are the most resistant and enduring form of life known—they resist heat, chemicals, drying and radiation
    Bacillus and *Clostridium* are abundant in soil, dust, and air, and cause food poisoning, tetanus, gangrene

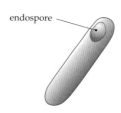

endospore

14. **Non-sporing Gram-positive rods**
    Lactobacillus: milk processing (yoghurt, buttermilk) and silage production

15. **Irregular non-sporing Gram-positive rods**
Corynebacterium: diphtheria in humans, and plant diseases
Eubacterium: a dominant type in the human gut
Actinomyces: filamentous, branching cells
Cellulomonas: degrades cellulose (rare but important activity in nature)

16. **Mycobacteria**
Mycobacterium: tuberculosis and leprosy

17. **Nocardioforms**
Partly filamentous cell growth common
Nocardia: includes producers of an important group of antibiotics, and hydrocarbon-utilising species

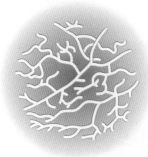

filaments

18. **Anoxygenic phototrophic bacteria**
Photosynthesising bacteria that do not produce oxygen as an end product; found in sunny but anaerobic environments (e.g. undisturbed ponds, water-saturated soils)

19. **Oxygenic photosynthetic bacteria**
Photosynthesising bacteria that (like algae and plants) produce oxygen as an end product; includes the cyanobacteria (e.g. *Anabaena*, *Spirulina*), which also fix atmospheric nitrogen; important in nitrogen and energy cycling in soils and lakes

20. **Aerobic chemotrophic bacteria**
Oxidise reduced inorganic compounds (e.g. H_2, NH_4^+, Fe^{2+}, H_2S) in place of organic substrates (e.g. sugars) to get energy for growth; important in the recycling and transformation of mineral nutrients in soil and water

21. **Budding and/or appendaged bacteria**
Morphologically unusual bacteria, usually found in soil and water

22. **Sheathed bacteria**
Appear filamentous but are short, rod-shaped cells packed within a common sheath or tube synthesised by the cells, and on which iron or manganese oxides may be deposited

23. **Non-photosynthetic, non-fruiting, gliding bacteria**
Bacteria whose cells move by gliding (a creeping motion across solid surfaces, whose mechanism is not well understood)

24. **Gliding fruiting bacteria**
Bacteria (e.g. *Myxococcus*) that move by gliding; cells mass together to form extensive, structured aggregates of cells (fruiting bodies) that contain resistant spore-like cells

25. **Archaeobacteria**
Bacteria that are phylogenetically remote from the Eubacteria (i.e. all other sections described here); include methane producers, and types that prefer high temperatures, or acidic, or salty environments

26. **Nocardioform Actinomycetes**

27. **Actinomycetes with multilocular sporangia**

28. **Actinoplanetes**

29. **Streptomyces and related genera**

30. **Maduromycetes**

31. **Thermomonospora and related genera**

32. **Thermoactinomycetes**
Sections 26–32 are mainly filamentous bacteria with some overlap with section 17; many produce 'spores', but these are not the highly resistant endospores produced by bacteria from section 13; *Streptomyces* includes many important antibiotic-producing species

33. **Other genera**
A small group with no known affinity with bacteria in the other 32 sections

Cyanobacteria

Cyanobacteria are the photosynthetic Eubacteria that resemble algae and plants in that they contain chlorophyll *a* and generate molecular oxygen during photosynthesis (Chapter 5). Many other photosynthetic bacteria do not do this. Modern cyanobacteria have evolved from the early bacteria that established the high concentrations of oxygen in the atmosphere. As you saw in Chapter 31, stromatolites, which formed from the activities of large aggregations of cyanobacteria, are the oldest fossil evidence of life.

Cyanobacteria occur as individual, often spheroidal, cells or as filamentous aggregates of many individual cells joined end to end. In addition to chlorophyll *a*, cyanobacteria contain water-soluble pigments, **phycobilins**; blue phycobilins present in most species give them a blue-green appearance. A slimy material may be present on the outside of the cell walls, forming a sheath.

Cyanobacteria are often more highly differentiated than other bacteria in having specialised cells such as akinetes and heterocysts (Fig. 33.4). An **akinete**

is a spore that develops from a cell that becomes enlarged and filled with food reserves. The spore can remain dormant and then germinate to produce a new filament. A **heterocyst** is relatively colourless, has a thick, transparent cell wall, may be involved in asexual reproduction, and (together with other cells) is a site of nitrogen fixation. The ability of some cyanobacteria to fix nitrogen (see Chapter 44) is made use of in rice cultivation, where the growth of such cyanobacteria is encouraged in rice paddies.

Cyanobacteria also often form dense *mats* in shallow marine or estuarine environments (Fig. 33.5), or ominous *blooms*, which are dense and extensive surface growths of cells in still lakes or slowly flowing rivers (Box 33.3).

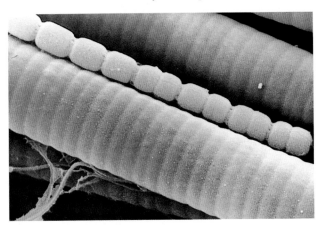

(a)

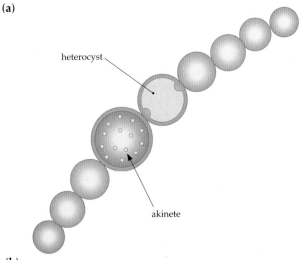

heterocyst

akinete

(b)

Fig. 33.4 (a) Two types of cyanobacteria. The large cylindrical form is *Microcoleus* and the smaller bead-like filament of cells is *Anabaena*, common in lakes and rivers. (b) Filament of *Anabaena* showing cell differentiation, with one akinete and one heterocyst

Fig. 33.5 Cyanobacterial mats formed in the intertidal zone of a mudflat in Spencer's Gulf, South Australia. In these mats, layers of cyanobacteria and other bacteria, which utilise sulfur compounds from the underlying mud, build up as flat strata, rather than as the mounds that form stromatolites

Archaeobacteria

Archaeobacteria (section 25, Box 33.2) are a more specialised group of bacteria and therefore restricted in the environments they inhabit. They might be thought of as primitive bacteria existing today but they are not, in fact, primitive—they have been evolving for as long and at similar rates as the Eubacteria and eukaryotes. However, because they

BOX 33.3 CYANOBACTERIA AND 'BLUE-GREEN ALGAL' BLOOMS

In the summer of 1991–2, extensive cyanobacterial blooms reportedly hundreds of kilometres long were reported on the Darling River. Towns and farms along the river had to be supplied with water from uncontaminated sources by tanker or had to install purification systems at considerable cost.

These blooms are usually referred to by the media as 'blue-green algae' or 'toxic algae', a consequence of earlier misidentification of these bacteria as algae. Species of cyanobacteria, such as *Microcystis aeruginosa* and *Anabaena flosaquae*, which form freshwater blooms of this sort, also produce toxins that are capable of causing illness and even death to animals and humans that drink water containing the bacteria. These toxins are small peptides (5–10 amino acids chemically linked together) and have their primary effects on liver metabolism; swimming in toxin-contaminated water may also cause serious skin and eye reactions.

High concentrations of nutrients, such as phosphate, in water and the slow movement of the water (so that cells remain in the surface layers capturing maximum light) promote blooms of this sort. The disposal into lakes, dams and slow-flowing streams of sewage and other wastes, or surface runoff from over-fertilised pastures and crops, sets the scene for serious cyanobacterial blooms.

Cyanobacterial bloom in Lake Burley Griffin, Canberra. Blooms of *Microcystis* often form in inland lakes, dams and rivers in Australia, and can be a problem for towns and farms that rely on them for drinking or recreational water. Lake Burley Griffin catches street-water runoff from Canberra, rich in organic material and nutrients from animal droppings and from gardens and lawns. On calm sunny days conditions are then ripe for development of a cyanobacterial (or 'algal') bloom in the lake. Natural agitation of the lake surface by wind or rain is usually sufficient to disturb the conditions needed for the organism's growth, so these blooms usually do not persist for long periods. In dryer, calmer areas inland, a bloom may last for weeks

---

have stayed within specialised environments (e.g. anaerobic, hot places) they have retained their 'primitive' features and not developed the biochemical diversity of the Eubacteria or the structural variability and multicellular style of the eukaryotes.

The archaeobacterial groups, as shown in Figure 33.3, include halophiles and acidophiles (salt- and acid-loving types), and thermophiles (preferring high temperatures and unable to grow at normal temperatures). All tend to be found in extreme environments not colonised by Eubacteria (or any other organisms): hot thermal springs, where the water may be salty or acidic, or near volcanic vents on the deep-sea floor (Fig. 33.6). Although temperatures in the latter environment may exceed 250°C, the water is present as super-heated liquid rather than as steam because of the high pressures that prevail at the depths concerned. Indeed, it is

likely that bacteria have evolved to grow in the hottest environments on earth, provided that appropriate energy sources, nutrients and water in liquid form are available. Thus, on the earth's surface, bacteria will grow at temperatures up to about 110°C.

Another important group within the Archaeobacteria are the **methanogens**. These do not occupy extreme environments but are strict anaerobes with a unique metabolism. Methanogens can utilise hydrogen gas and carbon dioxide to generate energy and make sugars. An important end product of this metabolism is the gas methane, which can be thought of as a product of the reaction:

$$4H_2 + CO_2 \longrightarrow CH_4 + 2H_2O$$

The biochemistry of this apparently simple reaction is complex and involves multiple enzymes and

(a)

(b)

Fig. 33.6 **(a)** This thermal valley near Rotorua, New Zealand, is a habitat for heat-tolerant cyanobacteria, seen here as zones of green. (The white areas are silica dioxide and the brown areas are iron oxides and hydroxides.) **(b)** Close up of mats of bacteria

chemical cofactors not known in other living organisms. Methane is produced in huge amounts in the biosphere, reflecting the abundance of methanogenic bacteria. Any anaerobic environment rich in organic material, which supports the fermentative production of H_2 and CO_2 by Eubacteria and other organisms, will contain methanogens: in the rumen, in the lower gut of all mammals, in peat bogs, marshes and water-logged soils, in the muddy bottoms of deep lakes and reservoirs, and in the garbage tips of urban life. Methanogens thus live symbiotically with other bacteria and eukaryotes, utilising the apparently final products of the fermentation of organic material by other organisms. Bacteria have devised remarkable ways to extract energy and nutrients from what seem to be the most 'wasted' of biological end products.

Archaeobacteria live in extreme conditions, such as saline, acidic or hot environments. Methanogens utilise hydrogen gas and carbon dioxide to generate energy and to make sugars, producing the gas methane as an end product.

Abundance of bacteria in, on and around humans

The lives of bacteria are more clearly understood when they are contrasted with macroscopic eukaryotes. Apart from the diversity in metabolism and ecology referred to earlier, two other aspects show remarkable contrast: the size of typical bacterial populations, a feature often shared with other, eukaryotic micro-organisms; and the ease with which genetic information varies and mixes within large, heterogeneous bacterial populations. An important outcome of this latter aspect is that new genes and gene combinations appear and spread rapidly in response to environmental change and challenge to bacteria; they are remarkably adaptive, a fact that humans have learnt at great practical cost in medicine, agriculture and technology.

The density and dimensions of bacterial populations in favourable habitats is astonishing. Our own personal resident populations provide good examples. On the skin of the average clean-living citizen there are about 100 000 (10^5) bacterial cells (of a great range of species) per square centimetre. The total skin surface of the average citizen is about 2 m² or 2×10^4 cm². Thus, the skin microbial population of a single human amounts to about $2 \times 10^4 \times 10^5$, or 2×10^9 cells, or approximately equal to one-third the present world population of humans! These microbes are not evenly distributed, being more abundant in the moister and hairier realms of the body, and the mix and total numbers tend to vary between different persons.

The human gut is even more astounding. In the large intestine or bowel, bacterial populations reach 10^{11} cells per gram of contents; assuming a lower bowel content of approximately 1 kg, the total number of bacteria in this part of the gut alone is about 10^{14} cells (one hundred billion). On a weight basis, this amounts to about one-third of the mass of faecal material. Higher up the gut the numbers of bacteria are fewer, and at the top of the digestive tract, the stomach, there are very few bacteria (a few million) or any other micro-organisms present, largely because of the acidity of the stomach (pH 1–2). In the gut generally there is an enormous diversity of bacterial species, many of which have yet to be described; like the skin, the composition of the gut microbial population is characteristic of each human individually, and in turn varies according to diet, age, general health, transient infections, and so on.

And so it goes for the many environments in which bacteria flourish—soil, lakes and streams, hair and skin of animals, plant surfaces, rotting materials, spoiled foods, and foods in which controlled bacterial growth is encouraged (cheese, yoghurt, salami, sauerkraut, silage, to name a few).

GENETIC SYSTEMS OF BACTERIA

How do bacteria achieve the genetic flexibility that permits populations of them to respond so quickly and appropriately to change of environment? If they did not vary and evolve, then bacteria would not have lasted on this ever-changing planet for so long.

In eukaryotes, meiosis allows genetic recombination, but meiosis does not happen in bacteria (Chapter 8). Mostly, bacterial multiplication occurs as a result of simple mitotic or vegetative division of cells, in which the progeny cells contain chromosomes that are identical to those in the original single parental cell. How then do bacteria become genetically variable?

Firstly, genetic variation due to mutation (Chapter 32) is much more important in bacteria, if only for the reason that bacteria are haploid and therefore mutations in genes will usually be expressed in the cell in which they occur: mutated alleles are not obscured by the presence of non-mutated alleles as they may be in a diploid organism.

Secondly, bacteria have other mechanisms of genetic variation that are different from eukaryotic sexual reproduction but with the same outcome: the introduction into a cell of new and sometimes very different DNA, its integration or recombination into the DNA of the recipient cell, and its genetic expression as a change of phenotype. This phenotypic variation provides the population with individuals able to take advantage of new or changing environments.

Three major mechanisms operate as follows.

Transformation: the role of free DNA

Many bacteria are able to take into their cells molecules of DNA that occur free (released from cells that have died and disintegrated) in their environment. Once inside the cell, such DNA may be degraded by restriction enzymes if these recognise the DNA as 'foreign' to the cell. Alternatively, if the genetic information encoded by the DNA is similar to that which is already present in the bacterium, then the incoming DNA may either recombine with the chromosomal or plasmid DNA (a small, circular molecule of DNA not part of the bacterial chromosome; Chapter 10) or, if it has the necessary information to direct its own replication, establish itself as a plasmid without recombining with existing DNA in the cell. Whether recombination or establishment as a plasmid occurs, new genetic information has been introduced into the cell, and therefore phenotypic variation is a possible outcome. Selection of the new genetic variant will follow if the new phenotype is more adapted to the challenge and competition operating in that particular environment. **Transformation** is thus the process of taking up and

utilising free DNA present in the bacterial cell's environment (Fig. 33.7). Information present in this DNA may cause genetic variation.

Bacterial transformation is probably not a common process in nature because, in the environments in which bacteria abound, free DNA is rare; free DNA is highly susceptible to enzymes that many micro-organisms secrete in their never-ending search for nutrients. However, it is believed to be an important process for some pathogenic bacteria, such as members of the genus *Neisseria*, which includes the cause of gonorrhoea (*N. gonorrhoeae*) and meningitis (*N. meningitidis*), and it was first observed in pneumonia in mice caused by *Streptococcus pneumoniae*. Transformation is also a process used extensively in recombinant DNA manipulation because in the laboratory it is a very convenient method for inserting recombinant DNA molecules into a range of bacteria (Chapter 12).

Conjugation: the role of plasmids

The second means of incorporation of new DNA into bacterial cells is by **conjugation**, which is the process whereby bacteria transfer DNA directly from one cell to another (Chapter 13). This process requires contact between cells and the presence of particular types of plasmids in the cell from which transfer is to take place. The spread of plasmids (and the genes they carry) in a population of bacteria can be quite rapid; in this sense they are sometimes referred to as 'infectious'.

Plasmids either transfer themselves, or they act as **vectors** to transfer other DNA picked up from the chromosome or from other plasmids in the cell they occupy. By a process we don't understand yet, these plasmids transfer a copy of themselves (together with any hitch-hiking DNA) from the donor cell across the contact junction into the recipient cell (Fig. 33.7). There, the DNA may recombine or integrate into the chromosome or a resident plasmid, or may establish itself as a resident plasmid in the recipient cell. Whichever of these alternatives, new DNA is introduced into a bacterial cell and the prospect of phenotypic variation and selection occurs once again.

Transduction: the role of bacteriophages

The third general method of bacterial DNA transfer is **transduction** (Fig. 33.7). Bacteria, though the smallest cellular organisms known, are themselves parasitised by smaller living organisms—viruses.

Viruses that multiply in bacterial cells are called **bacteriophages** (meaning bacteria-eating), commonly abbreviated to **phages**. Some phages are temperate (Chapter 34) and their presence in a bacterial cell does not necessarily result in cell death. Many temperate phages that have DNA as their genetic

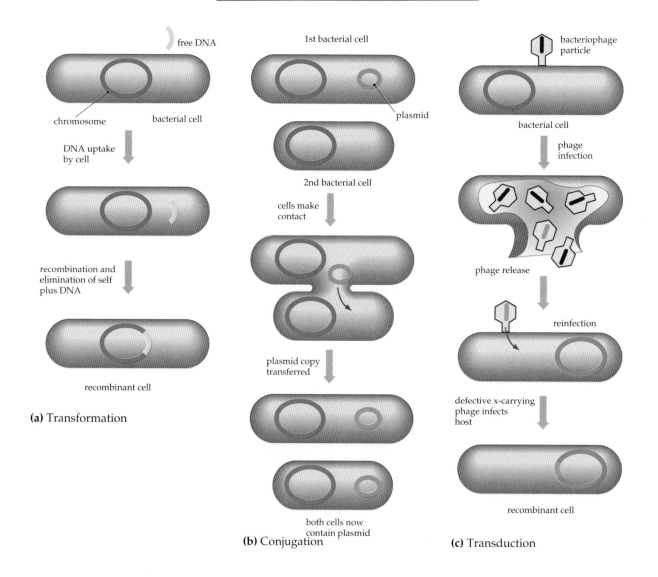

Fig. 33.7 Methods of gene transfer. (**a**) Transformation involves the uptake of free DNA by living cells. The incoming DNA recombines into the cell's chromosome or becomes a plasmid within the cell. Whichever path is taken, new genetic information is brought into the cell so that it becomes a genetic variant of the original cell (a recombinant). The new genetic information may be important in the adaption of the recombinant to new environments in which the original cell type could not exist. (**b**) In conjugation, pairs or groups of cells make contact and a plasmid copy is passed from one to another. Once a plasmid is established in the new host, it then excludes the transfer of another copy of the same plasmid, or a similar plasmid. However, a copy of this newly received plasmid can be transferred in turn to cells without plasmids with which contact is made. Some plasmids can also integrate themselves into the chromosome, and they then make parts of the surrounding chromosome behave as a plasmid. Alternatively, genes on the surrounding chromosome may become part of a new, recombinant plasmid when it reverses out of the chromosome and becomes a plasmid again. Transfer of the recombinant plasmid will result in transfer of genes from one bacterium to another, so that a recombinant cell is generated; this results in genetic variation in the population. (**c**) In transduction, DNA from one cell is moved to another as a result of infection of the first by a virus (bacteriophage), and the subsequent infection of the second cell by virus particles released after infection of the first. During the infection cycle of the first cell, a small number of newly forming virus particles will pick up host cell genes in error. These particles then transmit these genes to the second cell and because the virus particles contain host cell genes in place of virus genes, they are defective in that they cannot complete the virus life cycle and thus do not kill the second cell. This second cell is again a recombinant, having received genes from another cell. Once again, genetic variation occurs in the bacterial population due to a gene transfer system

material (some phages use RNA, Chapter 34) integrate their DNA into the host cell's chromosome and exist in this state indefinitely. However, with appropriate external stimuli, usually threatening to the life of the host cell, many temperate phages become **virulent** and in this state will multiply within the host cell, kill and lyse (break down) the host cells, with the resultant release of a hundred or more phage particles, **virions**. These virions, in turn, infect any susceptible new host cell with which they come into contact, and once again a temperate or virulent relationship is established with a host cell.

During the formation of new virions within the host cell in the virulent phase of infection, the developing virions may, in error and quite rarely, pick up pieces of host cell DNA rather than their own. The virion carrying host DNA is released during lysis along with many normal ones. Infection of a new host by a virion carrying bacterial rather than viral DNA will result in delivery of the original host's DNA into the new host cell. Recombination or integration of the DNA into the host chromosome takes place and a change of phenotype again may result. The analogy with plasmid vectors of DNA in conjugation is a close one, except that in most cases host cell death is required for release of the transducing phage virion, whereas the transfer of a copy of a conjugative plasmid does not destroy the donor cell.

Gene transfer and gene mixing in bacteria by each of the three processes described above usually involve small numbers of genes, largely because the dimensions of the DNA molecules transferred are small compared with the bacterial chromosome (10–1000 times larger). In this respect, these processes differ fundamentally from sexual reproduction in eukaryotes, where whole chromosomes and sets of chromosomes from each parent interact and precisely exchange complementary parts to generate entire recombinant and reassorted chromosomes.

> Bacteria achieve genetic variation, not by sexual reproduction as in eukaryotes, but by other genetic mechanisms:
> - transformations (taking up free DNA in the environment);
> - conjugation (transfer of DNA from cell to cell by plasmids);
> - transduction (DNA transfer by viruses—bacteriophages).

IMPORTANCE OF PLASMIDS AND PHAGES TO BACTERIAL POPULATIONS

Both conjugation and transduction are thought to be widespread in natural populations of bacteria because plasmids and temperate phages abound in wild populations. The circulation and mixing of genes by these vectors can readily be demonstrated in both natural environments and the laboratory, and a wide range of genes of bacterial origin are found as part of phage and plasmid DNA. These genes are particularly obvious in medical practice because they often cause the host bacterium to become resistant to antibiotics and antiseptics, or they provide additional genes (e.g. for toxins), which make the host bacterium a more virulent pathogen. Examples of important genetic characteristics that are found on moveable genetic elements in bacterial populations are shown in Table 33.2.

Plasmids and phages are often referred to as **accessory genetic elements** since they are not vital to the host cell's normal capacity to exist and can be lost from the cell relatively easily. Their presence in a population of bacteria, even in a very small proportion of cells, provides additional functions or capabilities that enable such bacterial populations to expand their range of habitat, for example, into the wounds of a person who is receiving heavy antibiotic therapy.

The host range of plasmids and phages is quite variable. Many will exploit different bacterial genera as hosts, while others enter only members of the same species or even strain of bacterium. There is much yet to be learnt about host range and compatibility between plasmids and phages, and their host bacteria. Nevertheless, the abundance of these mobile accessory genetic elements and the wide host range they may have mean that the circulation of genes and groups of genes in the bacterial world occurs relatively easily.

Table 33.2 New properties given to cells as a result of carriage of plasmids or phages

Bacteria	Plasmid or phage	Property
Many species that cause human and animal disease	Plasmids or phages	Antibiotic and disinfectant resistance; toxin production; attachment to animal cells; production of tissue-damaging enzymes
Pseudomonas spp.	Plasmid	Degrade and metabolise complex synthetic chemicals (e.g. DDT), which would otherwise accumulate in the environment
Rhizobium spp.	Plasmid	Nodulate roots of legume plants and fix nitrogen
Agrobacterium spp.	Plasmid	Plant galls
Streptomyces spp.	Plasmid	Antibiotic synthesis
Bacillus thuringiensis	Plasmid	Insecticidal toxin synthesis
Many species	Phage	Resistance to attack by other virulent phages

It has been said that all the genes in the Eubacteria and Archaeobacteria are theoretically accessible (although perhaps not directly) to any species of bacterium; if this is so, then bacterial genes are part of a common pool that any bacterial species might sample from. In support of this, it is known that genes in a number of distantly related Eubacteria are closely related in the sequence of bases that make up their DNA. Moreover, it is a straightforward procedure to isolate genes from one bacterium and clone them in a distantly related bacterium, and vice versa, using recombinant DNA methods (Chapter 12). Hybrid plasmids can also be constructed, for example, with one part that will multiply in and transfer genes between one group of bacteria (e.g. the Gram-negative enteric bacteria) and the other part that will permit replication and transfer among Gram-positive bacteria, such as streptococci and staphylococci. There is nothing 'unnatural' about these molecules or the procedures by which they are made, and there is no reason to believe that the processes by which we produce them in the laboratory do not occur, although perhaps with less purpose, in mixed wild populations.

Thus, the bacterial or prokaryotic world, although lacking the powerful processes of meiotic recombination and gamete fusion which ensure abundant phenotypic variation in eukaryotes, obtains its variation through processes of genetic transfer and genetic 'promiscuity', which have the same end result.

All modern bacteria are the result of genetic variation, adaptation and evolution in response to changes that have occurred as the earth and other living organisms have evolved. Exploitation of highly competitive niches (e.g. the gut or soil), or chemically and physically extreme niches (e.g. hot or salty water) is simpler for prokaryotes than for eukaryotes, probably because of the simpler cellular structures and greater genetic flexibility of prokaryotes.

The abundance of plasmids and phages in bacterial host cells means that genetic material can be readily transferred between different bacteria. Such transfer of genetic material has been important in the evolution and adaptation of bacteria.

SUMMARY

■ Bacteria (prokaryotes) are microscopic, unicellular organisms, with particular types of cell structure, biochemistry and genetic systems that distinguish them from eukaryotic cells.

■ There are two major phylogenetic groups of bacteria: the Eubacteria and the Archaeobacteria. Classification within these two groups is currently based largely on practical needs of identifying species.

■ Bacteria are enormously diverse ecologically, occupying not only all the habitats that eukaryotes do, but also extreme environments. Archaeobacteria live in environments that are hotter, saltier and more acid than is tolerated by any other life forms on earth.

■ Bacteria are important in the cycling of nutrients, such as carbon, nitrogen, sulfur and phosphorus, in the biosphere, converting forms of these that are unusable by plants and animals into useful nutrients. They often do this by way of biochemical processes that are unique to bacteria. Methanogens are Archaeobacteria, for example, that utilise hydrogen gas and carbon dioxide as an energy source and produce the gas methane as an end product.

■ Some bacteria are photosynthetic, and among Eubacteria, cyanobacteria produce oxygen during photosynthesis and were responsible for the earth's atmosphere becoming oxygen-rich two billion years ago. Some modern cyanobacteria are of nuisance value, causing toxic blooms in aquatic habitats.

■ Bacteria are major agents of disease in animals and plants, and through their transmission by, or excessive growth in, water and food may make these unusable or hazardous to other organisms, such as humans.

■ Bacteria also carry out important industrial and technological processes, and their natural abundance in the gut and on the skin of humans, for example, is an important element in protection of the human body from invasion by pathogenic bacteria and other micro-organisms. Bacteria are thus agents both of harm and of benefit for humans. Without them the biosphere would lose much of its biological and chemical diversity, processes of decay and renewal would slow or cease, and life as we know it would change for the worse in many ways.

■ Bacteria achieve genetic variation, not by sexual reproduction as in eukaryotes, but by transformation (taking up free DNA from the environment), conjugation (transfer of DNA from cell to cell by plasmids), and transduction (DNA transfer by viruses—bacteriophages).

QUESTIONS

1. References to bacteria in the media often depict them as the 'enemies' of humans. Respond on behalf of the bacterial world illustrating how, through their natural activities or their technological exploitation, many of them are really 'friends' of humans.

2. Describe the internal structural features that distinguish the prokaryotic from the eukaryotic cell.

3. Outline two important ways in which bacterial classification and identification differ from that for plants and animals.

4. If you were asked to identify a bacterium isolated from a person, what techniques would you apply in a preliminary identification? What characteristics would you consider to determine the major group (section) it belongs to?

5. What are (a) Archaeobacteria and (b) Eubacteria? To which group do methanogens belong and what are their characteristic features?

6. In the summer of 1992 there were a number of reports in the media of 'blue-green algal blooms'. What organisms cause these blooms and what are their characteristic features? Why are these blooms significant to humans and what are their likely causes?

7. If a bacterium such as *Staphylococcus aureus* (which has a doubling or generation time of 30 minutes) produces sufficient toxin to cause food poisoning when it is present at 1000 cells per gram of food, how long, to the nearest hour, would it take a single cell introduced into food to become a hazard to consumers of the food?

8. How does the production of genetic variation in bacteria differ from that which occurs in eukaryotes?

9. In a hospital outbreak of bacterial infections that do not respond as expected to a new antibiotic, it is noticed that resistance to the antibiotic is spreading into bacterial species different from the one in which it was first identified. Describe in genetic terms how this spread might happen.

10. Mobile genetic elements such as plasmids and bacteriophages are important in transmitting genes between cells in bacterial populations. Give examples of the sorts of genes that are spread in this way, and describe how they could provide evolutionary benefits to bacterial populations.

V I R U S E S

Viruses are subcellular genetic parasites that reproduce only in the cells of susceptible hosts. Viruses have been isolated from animals, plants, fungi, protists and bacteria, indeed from almost all cellular organisms that have been properly tested. There are even some viruses, called satellite viruses, that only reproduce in cells already infected with another virus. Viruses probably have origins as ancient as life itself and it is likely that they originated more than once, that is, they are **polyphyletic** in origin. Perhaps the best known viruses are those that infect human beings and cause the common cold, influenza, AIDS and various childhood diseases such as mumps, measles and chickenpox; but others cause diseases as diverse as foot-and-mouth disease of cattle and myxomatosis of rabbits (Fig. 34.1a), lysis of bacteria, La France disease of mushrooms and various yellowing and mosaic diseases of plants (Fig. 34.1b).

THE DISCOVERY OF VIRUSES

There are many historical records of viruses. There are scars, probably of smallpox (Box 34.1), on the faces of Egyptian mummies 3500 years old; a person with a withered leg, typical of poliomyelitis, is part of a bas relief on an Eygptian tomb 3000 years old; and flowers with viral symptoms are common in paintings from the sixteenth century.

It was not until the end of last century that the unusual nature of viruses was realised. At that time many diseases of humans, domesticated animals and plants had been shown to be caused by bacteria or fungi, most of which could be grown in nutrient media, and all of which had cells that could be seen in light microscopes. However, there remained several diseases that seemed to be caused by invisible pathogens. No characteristic particles could be seen in material from diseased individuals, and no pathogens could be cultured from them. One of these diseases was tobacco mosaic (Fig. 34.2), which could be transmitted from plant to plant by pricking sap from a diseased plant into the stem of a healthy one.

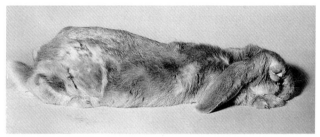

(a)

(b)

Fig. 34.1 **(a)** European rabbit suffering from myxomatosis caused by myxoma poxvirus. The natural hosts of this virus are the cottontail rabbits of the Americas. The virus is transported by mosquitoes and only causes warts on the cottontail's ears and head where mosquitoes feed. When European rabbits were taken to South America in the 1890s for laboratory use, some became infected with the virus and developed myxomatosis. The potential of using myxoma virus for the biological control of rabbit plagues was realised but it was not until the 1950s that the idea was tested on those in Australia. The first releases in Victoria were over-effective as the virus killed all the rabbits infected, but in the following wet year with large populations of mosquitoes, slightly less virulent variants of the virus spread rapidly along the Murray–Darling river system and effectively controlled rabbit populations. Rabbit strains that are more resistant to myxoma have appeared in areas, such as the Mallee of Victoria, where myxoma epidemics occur every year. **(b)** White clover leaf infected with white clover mosaic potexvirus. This virus is common wherever white clover grows

BOX 34.1 SMALLPOX

Smallpox virus, with its ability to kill up to a third of the people it infects, has been a scourge of humanity for many centuries. Human communities free from smallpox, like those in the Americas before Europeans invaded in the sixteenth century, were particularly susceptible. When smallpox was inadvertently introduced into Mexico in 1520 by the Spanish invaders led by Cortés, millions of the indigenous people of the Americas were killed.

Even in Roman times it was known that those who recovered from smallpox rarely became infected again, and so children were deliberately infected with mild strains of the virus, a risky technique called variolation. Safety was greatly improved when Edward Jenner, a British physician, reported in 1798 that deliberate infection with cowpox (vaccinia) gave stable immunity to smallpox. This technique became known as *vaccination*. It was used in a campaign organised by the World Health Organization and led to the worldwide eradication of the virus infection by 1979. This unique operation was successful because:

- no *animal reservoir* of the virus was available;
- a *stable freeze-dried vaccine* was available that could be administered simply by scratching, giving *lifelong immunity*;
- the virus spread obviously but relatively slowly, often only to members of a family, so that 'contacts' of infected people could be vaccinated and protected from infection, thus *breaking the infection cycle*.

Recently, ways have been found to transfer selected genes into the genome of vaccinia virus, so that they are expressed along with the genes of the virus. Vaccinia virus clones carrying the genes encoding the immunising antigens of a range of other microbes are being tested as multivalent vaccines for controlling diseases of humans and other animals, including rinderpest of cattle and even rabies in wild foxes.

Fig. 34.2 Capsicum leaf infected with a tobamovirus, probably pepper mild mottle virus, in Jamaica

Experiments by several plant pathologists in the 1890s found that the tobacco mosaic pathogen had rather unusual properties. It diffused from sap into a block of gelatin jelly, could be precipitated with ethanol and remained infective and, most importantly, could pass through a porcelain filter with pores small enough to retain all known bacteria. These and other tests suggested that the pathogen was very small, and W. M. Beijerinck, a Dutch pathologist, concluded that tobacco mosaic was caused by an unusual pathogen, which he described as a '*contagium vivum fluidum*', namely, a contagious living fluid.

The filtration test was also used at about the same time by Loeffler and Frosch in Germany to test the saliva of cattle suffering from foot-and-mouth disease. They showed that healthy cattle became infected when saliva from infected animals was sprayed into their nostrils, even if it had been filtered. Soon, many other atypical pathogens of animals, plants and even bacteria were found to behave similarly, and became known as 'ultra-filterable viruses', from the Latin for poison. This name was later contracted to 'viruses', although those that infect bacteria (Chapter 33) have retained the name of bacteriophage (i.e. bacteria eater), or just phage.

The nature of viruses, however, remained a mystery until the 1930s, when they were studied by various new biochemical techniques developed in the branch of science now called molecular biology. In fact, most of the important advances of molecular biology have involved work with viruses.

In the 1930s, serological tests showed that the sap of tobacco plants with mosaic disease contained a novel antigen. The serum of rabbits immunised with sap of infected plants reacted specifically with the sap of infected plants but not with sap from healthy plants. It was also shown that sap of infected plants, like some shampoos, showed streaming birefringence in that, when the sap was clarified and stirred, the plane of polarisation of light passing through it was changed. This indicated that the sap contained particles that were either disc- or rod-shaped. No particles could be seen in this sap using a light microscope, indicating that the particles were less than about 0.3–0.5 µm in diameter, as this is the limit of resolution of a light microscope. However, minute particles had been seen in fluids from chickens infected with fowlpox, indicating that the particles of that virus might be larger than those of tobacco mosaic virus (TMV).

In the late 1930s, the electron microscope was invented and, with its greatly increased resolution, the sap of tobacco plants with mosaic disease was always found to contain characteristic rod-shaped particles, the structure of which was soon determined using the new technique of X-ray diffractometry. In the 1950s and 1960s the characteristic particles, or **virions**, of a great range of viruses were purified and their beauty and complexity revealed (Fig. 34.3). Virions were shown to contain nucleic acids, proteins and, in some, lipids. It was experiments with T2 phage and turnip yellow mosaic virus that first showed that nucleic acids, not proteins, contained the genetic information.

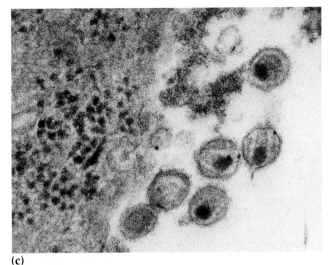

(c)

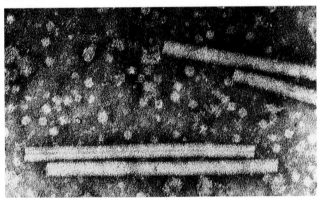

(a)

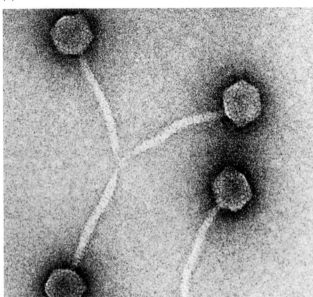

(d)

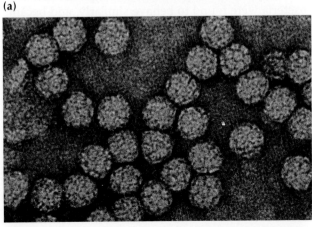

(b)

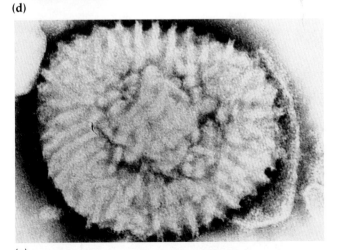

(e)

Fig. 34.3 Virions of: **(a)** tobacco mosaic tobamovirus (300 nm long); **(b)** turnip yellow mosaic tymovirus (28 nm in diameter); **(c)** human immunodeficiency lentivirus (about 100 nm in diameter); **(d)** lambda phage (head diameter about 60 nm, tail about 150 nm long); **(e)** vaccinia poxvirus (about 300 nm in diameter)

VIRUSES: SUBCELLULAR PARASITES

Features distinguishing viruses from cellular organisms

The fact that virions could be purified like chemicals, and some even crystallised, led to much speculation about whether viruses were living or dead. Biochemical work showed that virions are metabolically inert and merely consist of the genome of the virus in a protective and infective package ready to be transported to another susceptible cell; they are not equivalent to the cells of cellular organisms. They are able to be crystallised merely because they are of uniform shape and size.

When a virion enters the cell of a suitable host, it disassembles, the genome it contains usurps the host's own genes, and it diverts the metabolism of the host into virus reproduction. Thus, viruses differ from cellular organisms in having two clearly defined phases in their biochemical life cycle. Firstly, there are the characteristic but metabolically inert virions, which are the transmission phase of the virus. These alternate with the reproductive phase, in which the virus consists of metabolically active viral genes. The viral genes use the metabolic systems of the host to replicate themselves and, using host ribosomes, produce a range of viral proteins; the progeny genomes assemble with virion proteins to form virions. By contrast, cellular organisms in all stages of their life cycles consist of cells that are bounded by cellular membranes, and contain complete and largely independent metabolic systems that include mitochondria and ribosomes (Chapter 3).

> Viruses are subcellular genetic parasites that infect all types of cellular organisms and cause many diseases.

Virions: the transmission phase

The simplest virions are regular geometric structures built from the viral genome together with many copies of a virus-encoded protein, usually called the virion, or coat, protein.

The virions, for example, of TMV (Fig. 34.3a) are rod-shaped and consist of a tube made of about 2100 helically arranged copies of a single type of coat protein (Fig. 34.4a). Where the protein subunits make contact, a groove is formed, and wound into this is the genome of the virus, which is a single molecule of ribonucleic acid about 6400 nucleotides long; three nucleotides are tucked into and between each protein subunit. Many other viruses, including turnip yellow mosaic virus (Fig. 34.3b), have isometric virions with the genome centrally folded within a protein shell made from 180 protein subunits arranged as an icosahedron. An **icosahedron** has a surface with 20 triangular facets (Fig. 34.4b). These two basic structures (rod and icosahedron) recur, in various forms, in the virions of most viruses. In viruses of animals, virions often have an outer lipid envelope (Fig. 34.4c). This contains viral proteins, and is acquired as the virion buds from the surface of the infected cell.

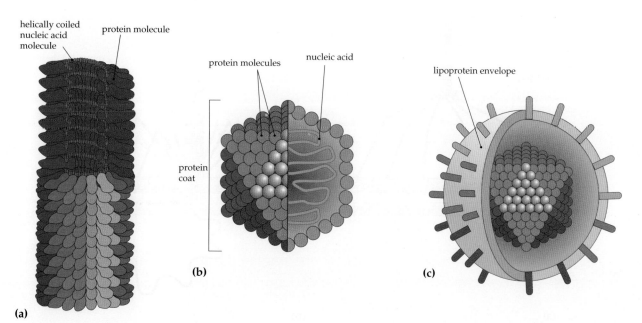

Fig. 34.4 Virions have two basic structures. **(a)** In rod-shaped virions, such as those of tobamovirus, clusters of proteins form a coat around a helical coil of nucleic acid. **(b)** In an icosahedral virion, the coiled nucleic acid core is surrounded by a protein coat that forms a surface of 20 triangular facets. **(c)** Virions of some viruses are enclosed in a lipoprotein envelope containing viral proteins

The largest virions, those of poxviruses (Fig. 34.3e), are brick-shaped and, at 250–300 nm in size, are just visible in a light microscope. They contain over 100 different protein species, and are as genetically complex as the simplest bacteria.

Bacteriophages

The most obviously complex virions are those of the T2 and T4 phages. Each virion has a large rounded 'head', which contains the genome, and a complex tail with which it attaches specifically to a bacterial host cell by terminal fibres (Fig. 34.5). The outer layer of the tail then shortens, and the core of the tail penetrates the host wall, like a hypodermic syringe, so that the viral genome can enter the host cell. Other phages, such as lambda phage (Fig. 34.3d), have particles of similar size but their tails are not contractile.

Many of the simplest virions, such as those of TMV, can be disassembled into their constituent parts and reassembled in vitro; however, the larger virions, such as those of T2 phage, cannot as their assembly involves several irreversible steps.

> As virions, viruses are metabolically inert, contain the genome in a protective package, and are infective. When in a metabolically active phase, they consist of viral nucleic acids and proteins that replicate in a host cell.

Viral genomes

The **genome** of a virus is the minimum set of genes required to cause infection. In most viruses the genome is packaged in a single infectious virion, but some viruses with divided genomes package the parts separately. Viral genomes are biochemically much more varied than those of cellular organisms. Viral genomes, although smaller than those of cellular organisms, differ greatly in size and composition. The genomes of some viruses are RNA, while others are DNA. Some, of both types, are single-stranded molecules, others are double-stranded. Many of the smallest viruses, such as TMV, have genomes that are single molecules of about 6000 nucleotides of single-stranded RNA (ssRNA). The largest, such as those of the poxviruses of vertebrates, are single molecules of about 130–260 thousand base pairs of double-stranded DNA (dsDNA); those of the poxviruses of insects may be even larger. Some of the smaller ssDNA genomes are circular, and some of the larger dsDNAs have the ends of the complementary strands covalently linked so that they too are, in essence, circular. Many of the ssRNA and all the dsRNA genomes are divided into several parts.

> Genomes of viruses vary in size and may be DNA, RNA, single- or double-stranded.

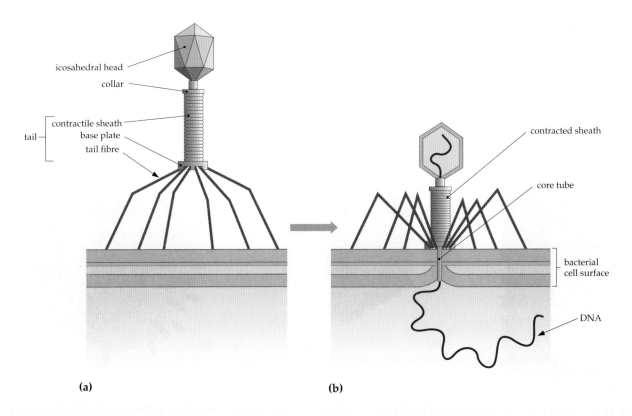

(a)   (b)

Fig. 34.5 (a) Many bacteriophages have an icosahedral head and a contractile tail. (b) When infecting a bacterial host cell, the tail sheath contracts and forces an internal protein tube through the cell wall, and the nucleic acid in the head passes into the bacterium

BOX 34.2 GENOMES OF BACTERIOPHAGES

Bacteriophages are diverse. Many have been involved in pioneering experiments of molecular biology and are used in many of the standard techniques of biotechnology. T2 and lambda phages have large dsDNA genomes: that of lambda phage is of 48 500 base pairs, and that of T4 is four times larger. Phages, together with plasmids, are an integral part of the gene pool of bacteria. Phages of this sort have been isolated from Archaeobacteria as well as Eubacteria (Chapter 33). Some, such as T2 phage, infect, replicate in and lyse a series of host cells, whereas others, such as lambda phage, sometimes also integrate their own genome into that of the bacterial

host and several cell cycles later they lyse their host and become virulent. The virions of the tailed phages are constructed from subassembled components involving large numbers of genes (see figure).

There are also other types of phage. M13 phage infects male bacteria by attaching to their sex pili. It has a circular ssDNA genome of only 6407 nucleotides, and variants of it are used for cloning and sequencing genes. There are also other male-specific phages, such as MS2 and Qb phages, that have isometric virions and ssRNA genomes of only about 3500 nucleotides.

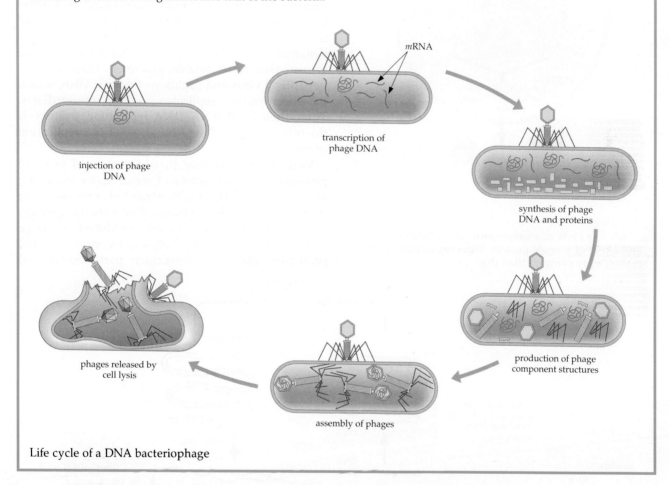

*m*RNA

transcription of phage DNA

injection of phage DNA

synthesis of phage DNA and proteins

production of phage component structures

assembly of phages

phages released by cell lysis

Life cycle of a DNA bacteriophage

Replication

The replication strategies of viruses are as varied as their genomes. They reproduce asexually (Fig. 34.6), but in almost all groups of viruses there is clear evidence of genetic recombination, or of reassortment of the parts of divided genomes.

Large double-stranded genomes replicate using biochemical pathways closely similar to those used by the genomes of cellular organisms. The single-stranded genomes replicate via a complementary strand. For example, the ssRNA genomes of TMV and many other viruses, including foot-and-mouth

disease virus and poliovirus, are transcribed into complementary strands by *replicase complexes*, which include viral and host proteins. These complementary strands then are transcribed repeatedly to produce progeny genomes. Some genomes, called plus-stranded genomes, are translated directly, although some of their genes may be translated from subgenomic mRNAs, and these genomes can infect cells even after being chemically deproteinised. By contrast, the genomes of other viruses, such as influenza and rabies, are negative-stranded RNA. They are not infectious after being chemically

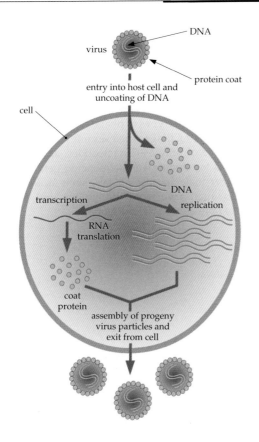

Fig. 34.6 Life cycle of a simple virus consisting of dsDNA and a coat of a single protein. Virus replication is, in reality, more complex than this

extracted from virions because, before their genes can be translated, they must be transcribed into a complementary plus strand. This is done by a viral replicase that is carried in infectious virions.

The most bizarre replication strategy is that of the **retroviruses**, such as human immunodeficiency virus (HIV). Retroviruses have ssRNA genomes which, upon infection, are transcribed into dsDNA and incorporated into the chromosomes of their host (Fig. 34.7). They are then transcribed into progeny ssRNA genomes. It is easy to understand how such a process could disrupt control systems in the host genome and thus it is no surprise that many retroviruses cause cancers. Sequencing studies have found that some retroviral enzymes are distantly related to those of transposons. Even more unexpected, retrovirus genomes show clear sequence similarities to some viruses that have dsDNA genomes. These include hepatitis B virus and cauliflower mosaic virus, which have been found to replicate via ssRNA intermediates. Thus, their DNA–RNA–DNA replication strategy is the mirror image of the RNA–DNA–RNA replication of retroviruses.

Viruses also translate their genomes to produce proteins in a great variety of ways. Many, including all of those with double-stranded genomes, are translated from mRNAs transcribed from the genome, but some viral proteins are produced as large *polyproteins* and then hydrolysed, by virus-encoded proteases, into their constituent parts. Some viral

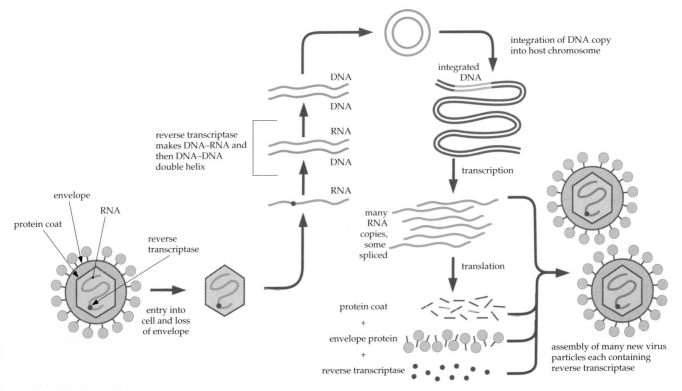

Fig. 34.7 Life cycle of a retrovirus, showing the enzyme *reverse transcriptase* that makes a DNA copy of the viral RNA, then a second DNA strand to make dsDNA. The DNA is then integrated into a host chromosome for the manufacture of new virus particles

genomes also have overlapping genes, which are read from a single stretch of genome but using different reading frames (i.e. they start translating from nucleotides that are not multiples of three nucleotides apart). Others, both dsDNA and ssRNA genomes, are ambisense in that different parts of one genome, or genome segment, are translated in opposite directions.

> Viruses reproduce asexually but there is evidence of genetic recombination. They use a variety of mechanisms for genome replication and protein production.

Satellite viruses

In addition to the conventional types of viruses discussed so far, there are **satellite viruses**, which are only able to replicate in cells infected with a specific helper virus. The first one described was the satellite virus (STNV) of tobacco necrosis virus (TNV). STNV has a genome of only 1200 nucleotides and just one protein, the virion protein. It relies for all its other functions on TNV and the host plant. Simpler still are the **satellite nucleic acids**. These rely on a helper virus for all their proteins and are transmitted in the virions of the helper.

Most satellites depress the replication of their helper viruses. Many of the RNA satellites of helper viruses with ssRNA genomes are around 350 nucleotides in length and are ssRNA circles. They reproduce by transcribing the complementary genomic strand into a long ssRNA consisting of many copies of the genome. This is then cut into unit satellite genomes by a folded part of each genome. This fold is called a **ribozyme** (see Box 11.2), and it attaches, by hybridisation, to several nucleotides on either side of those where it hydrolyses; thus, ribozymes are very specific and there is much interest in designing ribozymes for specific functions. For example, DNA encoding a ribozyme to cut the genome of a virus might protect the host of that virus against infection.

Viroids and prions: virus-like infectious agents

Finally, there are some unusual infectious agents that are structurally different from viruses. **Viroids** are virus-like but lack a protein coat. They resemble satellite RNAs in structure but are able to replicate alone in a suitable host. They cause important plant diseases, especially of tropical trees, and are possibly spread by insects.

Most peculiar of all pathogens are **prions** (proteinaceous infectious particles). They have some similarity to viruses in that they reproduce in host cells, but differ in structure. They appear to lack nucleic acid and consist only of protein. In mammals, they cause the nervous system to accumulate myeloid proteins and to degenerate slowly, and are the cause of kuru and Jakob–Creutzfeldt diseases of humans as well as mad cow disease (bovine spongiform encephalopathy) and scrapie of sheep. The proteins that accumulate are encoded by the host's genome but are altered in some way. These altered proteins seem to be able to alter the same protein in healthy animals. Thus the alteration is 'infectious'.

ECOLOGY OF VIRUSES

In addition to their biochemical life cycle, all viruses have an ecological life cycle, and this is similarly diverse. Many animal viruses rely on the active lifestyle of their hosts for dispersal. They spread directly: coughing and sneezing produce infective droplets containing influenza and common cold viruses; faecal contamination of food and water spreads poliovirus and many others; Epstein–Barr virus, the cause of glandular fever, is spread during kissing; the transfer of body fluids during sexual intercourse spreads HIV; and some viruses are spread congenitally, or in milk, from mother to offspring. Few plant viruses (Box 34.3) are transmitted directly by contact, but some of these, such as TMV, are important in crop and glasshouse crops. More plant viruses are transmitted by pollen and by seed, but most, together with a significant proportion of the viruses of mammals and birds, are transmitted by **vector** organisms, which are usually mobile pests or parasites of the viral host.

Vectors of viruses of vertebrates, such as mammals or birds, include mosquitoes, ticks and midges, whereas those of plants are mostly aphids, beetles, leaf-hoppers, mites, thrips, whitefly and other arthropods above ground, and nematodes and fungi below. Some vectors act merely as mobile needles but others are infected by the virus they are carrying, and hence the distinction between host and vector is blurred. Some viruses of sap-feeding aphids and leaf-hoppers spread between individuals through their plant host without replicating in it; thus the plant is the vector!

Symptoms of virus infection

Virus infections cause a great variety of symptoms. These are not merely a sign of illness but are often a specific feature of the ecology of each virus and are important in its dispersal. For example, the coughing and sneezing caused by the common cold and influenza viruses are essential for their spread. Mosquito-borne viruses often cause fevers in their vertebrate hosts; the resultant increased body

BOX 34.3   PLANT VIRUSES

Tobacco mosaic tobamovirus (TMV), as described earlier (Fig. 34.2), has been the subject of many pioneering viral studies. It and other tobamoviruses are transmitted when plants touch, and some, such as tomato mosaic tobamovirus (ToMV), are transmitted in seeds. Nowadays, TMV and ToMV are controlled in crops by using varieties into which natural resistance genes from wild relatives have been transferred.

Viruses continue to be important crop pathogens despite success with controlling some. Most problems come from **potyviruses**, named after the type species, potato virus Y, and the **luteoviruses**, which cause yellowing symptoms. Potyviruses are the commonest plant viruses and over 300 have been described. They have filamentous virions that contain the ssRNA genome, which is translated as a single polyprotein, like that of the picornaviruses. Interestingly, some potyvirus and picornavirus proteins have similar sequences. Potyviruses cause leaf mosaics, and some cause attractive colour changes in flowers that may be prized; in this way, for example, tulip flower-breaking potyvirus clearly benefits the tulips it infects (see figure).

Luteoviruses cause many important plant diseases, especially of temperate cereals. They have isometric virions and an ssRNA genome. Like the potyviruses, they are transmitted by aphids but neither replicate in those vectors. However, whereas potyviruses are spread

Tulip flower infected with virus

by probing aphids flitting from plant to plant in search of a suitable host and are rapidly lost by aphids, luteoviruses only infect phloem cells and are thus only spread by aphids after long feeding periods and may survive in aphids for their lifetime. This difference in *vector relations* affects the pattern of spread of these viruses: luteoviruses can be transmitted over long distances but may be controlled by insecticides, whereas potyviruses only spread over short distances and are not controlled by many persistent insecticides.

temperature and production of carbon dioxide both attract mosquitoes, which ensures continued spread. Similarly, viruses of plants that are transmitted by aerial vectors, such as aphids, whitefly and leafhoppers, often cause bright yellowing symptoms which, it is known, attract these flying vectors.

All viruses have an ecological life cycle. Many animal viruses rely on the active lifestyle of their hosts for dispersal. Most plant viruses are spread by vectors.

CLASSIFICATION AND RELATIONSHIPS OF VIRUSES

Viruses, like other organisms, fall into more or less well-defined species. These then fall into higher groups or genera, and some genera form even higher groupings (Box 34.4). However, there is no universal phylogenetic tree joining all viruses, indicating that they are probably polyphyletic in origin.

The virus isolates that form a single species are closely related in all their features so they usually cause similar symptoms in the same host species and are transmitted in the same way. Their genomes may differ in nucleotide sequence but by no more than

a few percentage points, and so the proteins they encode are closely related in serological tests.

So what is it, in the absence of sexual reproduction, that holds a viral species together? The answer came first from experiments with an isolate of Qβ bacteriophage, which had been passaged many times without change in its genomic sequence. The Qβ genome is ssRNA, which has a much larger replication error frequency than DNA. It was found that although new variants of Qβ phage were constantly arising in the apparently stable population, one genome sequence, the 'master copy', was favoured during competition between variants. Thus **stabilising selection** (Chapter 32), sometimes called 'purifying selection', operating on variants of a genome inherited from a common ancestor maintains well-defined viral species.

The several viral species that form each genus or group usually share all the major features of their biochemical and ecological life cycles, such as the structure and composition of their virions, their replication strategy and mode of spread. They differ most often in host specificities; for example, tobacco and tomato are the preferred natural hosts of the closely related tobacco and tomato mosaic viruses, and herpes simplex viruses I and II are similarly close but infect the lips and genital regions respectively.

Viruses have been isolated from almost all cellular organisms that have been carefully examined and, from some, such as humans and *Escherichia coli*, several dozen have been isolated. Many viruses seem to be host-species specific, and thus there are probably more species of viruses than of cellular organisms. However, at present only about 2000 of the 10 000 named viruses are well-studied enough to be recognised by the International Committee on Taxonomy of Viruses, and these fall into about 80 genera, some of which are classified into higher taxa.

Recent work on the nucleotide sequences of viral genomes and the amino acid sequences of viral proteins has started to reveal the complex interrelationships of viral genera and their origins. It has been found that many viral genes, in particular the genes that encode the essential 'housekeeping' proteins, such as viral enzymes involved in genome replication, viral proteases and virion proteins, are members of gene families (Table 34.1). These gene families span viruses of plants, animals and bacteria. Some are encoded in DNA genomes and others in RNA genomes, and some are clearly related to host genes. Thus they are clearly very ancient. By contrast, some genes are only found in particular virus groups or individual viruses, and it is likely that many of these have arisen recently. These genes include the overlapping gene of tymoviruses and the regulatory overlapping genes of some retroviruses.

The relationships indicated by one gene frequently do not coincide with those indicated by others, so it is clear that modern viral genera have arisen by genetic recombination between genes from diverse sources. Thus, genetic recombination has had a much greater influence on the evolution of viruses, at all taxonomic levels, than on the evolution of cellular organisms, where it is mostly confined to the individuals of single species. This is not surprising given the fact that one host cell may have the genomes of several different viruses mingling within it, whereas the genomes of cellular organisms are almost always separated from one another by cell membranes.

Although viruses are probably polyphyletic in origin, species closely related to one another are classified into genera and higher taxa. Stabilising selection operating on variants of a genome inherited from a common ancestor maintains well-defined viral species.

Table 34.1 Amino acid sequences of part of the replicase proteins of various viruses. The triplet GDD (gly-asp-asp) sequence, aligned here, is shared by almost all viral replicases (a)

Tobacco mosaic tobamovirus	QRKSGDVTTFIGNTVIIAACLASMLPMEKIIKGAFC**GDD**SLLYFPKGCEFPDVQHSA
Cucumber mosaic cucumovirus	QRRTGDAFTYFGNTIVTMAEFAWCYDTDQFDRLLFS**GDD**SLAFSKLPPVGDPSKFTT
Sindbis alphavirus	KSGMFLTLFVNTVLNVVIASRVLEERLKTSRCAAFI**GDD**NIIHGVVSDKEMAERCAT
Poliovirus	SGTSIFNSMINNLIIRTLLLKTYKGIDLDHLKMIAY**GDD**VIASYPHEVDASLLAQSG
Potato Y potyvirus	TVVDNSLMVVLAMHYALIKECVEFEEIDSTCVFFVN**GDD**LLIAVNPEKESILDRMSQ
Qβ levivirus (phage)	SMGNGYTFELESLIFASLARSVCEILDLDSSEVTVY**GDD**IILPSCAVPALREVFKYV

(a) Note that the sequences are grouped by their similarities, but that the top group is of two plant viruses and one animal virus, the second is of one of each and the third is a phage.

BOX 34.4 VIRUSES INFECTING HUMANS

Viruses that infect humans and other animals are classified into a number of genera and families, the names of which may describe the group's characteristics. For example, picornaviruses are small ('pico' meaning small) and poxviruses cause smallpox among other diseases (see figures). In humans, viruses may cause death (AIDS), permanent disability (polio), or be relatively harmless (warts) depending on the type of virus and the cells it infects. The following are a few examples.

'Flu virus

Influenza orthomyxovirus (from Italian meaning 'the influence') infects the lungs and respiratory tract of human beings and causes a disease that has been known for several centuries. The disease is spread by droplets produced during coughing and is commonest in winter when people congregate indoors. The number of people infected and the severity of the disease vary from year to year; in the 1918–19 world epidemic, millions died. Influenza virions are of irregular shape (Fig. c) and have a lipid outer membrane containing two types of surface protein: haemagglutinin (HA) and neuraminidase (NA). Infected individuals develop antibodies to HA and NA and, as a result, resist infection; however, HA and NA vary antigenically from year to year and so people may be reinfected.

The major structural elements of the virions of viruses of different groups that are directly important to humans

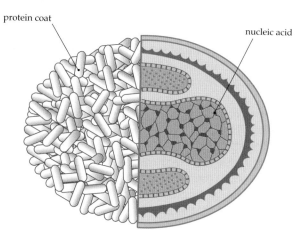

(a) Poxviruses: smallpox, vaccinia and myxoma viruses

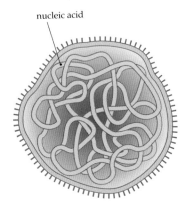

(b) Paramyxoviruses: mumps and measles viruses (a paramyxovirus also causes dog distemper)

(c) Orthomyxoviruses: influenza viruses

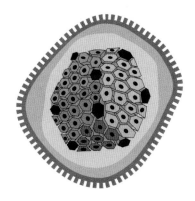

(d) Herpesviruses: herpes I and II, varicella zoster (chickenpox/shingles), Epstein–Barr (glandular fever), cytomegalovirus

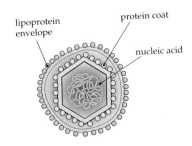

(e) Retroviruses: human immunodeficiency I and II, maedi-visna and various leukosis viruses

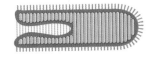

(f) Rhabdoviruses: rabies and vesicular stomatitis viruses, and, smallest of all, picornaviruses: common cold and poliomyelitis (a picornavirus also causes foot-and-mouth disease in cattle)

There are three types of influenza viruses, the most important of which is the influenza A virus. Their major source of antigenically different HAs and NAs is the many strains found in wild water birds, especially ducks, in which the virus usually causes a symptomless gut infection. The genome of influenza A is in eight parts, which reassort in mixed infections and produce influenza isolates with new combinations of HA and NA genes. This seems often to occur between bird and human influenzas in South-East Asia, where farmers live close to their stock, especially ducks and pigs. This is the likely source of the viruses that produced the great Hong Kong and Asian influenza A epidemics of 1957 and 1968.

AIDS

Human immunodeficiency lentivirus (HIV) is the cause of the acquired immunodeficiency syndrome (**AIDS**), and perhaps the best known member of the retrovirus family (Fig. e). (Remember that retroviruses are so-called because, during replication, they transcribe their ssRNA genome into dsDNA.) HIV is a species of the genus *Lentivirus*, which includes many chronically infecting viruses of monkeys and other mammals. HIV is spread during sexual intercourse, by blood products, and from mother to fetus. It is not surprising that there has been little immediate success in designing a vaccine to control HIV as its major long-term effect is to destroy parts of the immune system. It has only recently spread widely among humans, probably from monkeys in Africa.

Murray Valley encephalitis

Murray Valley encephalitis flavivirus (MVEV) is an **arbovirus** (arthropod-borne virus). There are several unrelated genera of arboviruses, all with an ecological life cycle that involves alternately infecting and replicating in blood-feeding arthropods, usually mosquitoes or ticks, and vertebrates, usually birds or mammals. MVEV is a flavivirus, named after the type species, yellow fever virus. It is a virus of river systems and the tropical north of Australia, and normally infects birds and mosquitoes without causing obvious symptoms. When human beings are infected, most develop a fever, but some develop a severe encephalitis that may be lethal. **Ross River alphavirus** is also an arbovirus, but from another viral genus. It is widespread throughout Australia and infects mosquitoes and mammals, probably rodents and flying foxes. In humans it causes fever and an arthritis-like swelling of the joints.

Chickenpox and shingles

Varicella zoster herpesvirus (VZV) causes the diseases chickenpox and shingles, being named after the medical names of the two diseases it causes. Like other herpesviruses (Fig. d), VZV has complex isometric virions with a loose envelope, and a dsDNA genome of 125 000 nucleotide pairs. Chickenpox typically occurs in children, and shingles in adults. When children recover from chickenpox, the virus remains in one or more of the dorsal root ganglia of their spinal cords. In later life, stress activates one of these *latent infections* and the virus spreads causing a painful rash of the skin area supplied with nerves from that particular ganglion. The virus spreads to children from adults with shingles. Thus, the long period of latency in a ganglion, safe from the immune system of the host, enables the ecological life cycles of VZV and its host to be matched in time despite them having biochemical life cycles measured in minutes and years respectively.

SUMMARY

- Viruses are subcellular genetic parasites and have been found infecting all sorts of cellular organisms. Viruses are probably polyphyletic in origin, not so much a family but more a way of life that is probably as ancient as life itself.

- Viruses have a characteristic two-phase life cycle. One phase consists of particles or virions, which are metabolically inert and contain the genome in a protective and infective package. The alternate phase consists of metabolically active viral nucleic acids and proteins replicating in the cell of a susceptible host.

- Viruses have genomes of diverse size and composition, some being single-stranded, others double-stranded, some being DNA, others RNA. Genomes range from 1000 nucleotides to over 300 000 base pairs in size.

- Viruses spread from one host individual to another in various ways, some directly, others are vector borne and, for some, vectors are alternate hosts.

- Many viruses cause important diseases of humans and domesticated animals, plants, bacteria and fungi. Many animal viruses rely on the active life cycle of their hosts for dispersal. Other plant and animal viruses are transported by vectors, particularly insects. Viruses are often best controlled by ecological intervention.

QUESTIONS

1. (a) What features distinguish viruses from cellular organisms? (b) What features do viruses and cellular organisms share? (c) Make a case for and against viruses being considered living organisms.

2. Viruses are thought to be polyphyletic in origin. What does this mean?

3. What is a bacteriophage? Describe the structure of a phage virion.

4. Distinguish between (a) a virus, (b) a satellite virus, (c) a viroid and (d) a prion.

5. What do the acronyms HIV and AIDS mean? What is the name of the virus genus to which HIV belongs? HIV is a retrovirus. Describe how retroviruses replicate.

6. Describe two viruses that are beneficial to humans.

7. What might be the importance, to a virion, of having a protein coat?

8. How do plant viruses spread from one host individual to another? How are viruses of plants controlled?

9. What measures can be used to control viruses of mammals? Why are antibiotics ineffective in the control of virus infections?

10. Why was it possible to eradicate smallpox virus, and why is it unlikely that influenza virus and HIV will also be eradicated?

THE PROTISTS

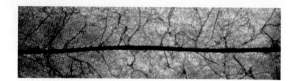

Kingdom Protista includes a weird and wonderful potpourri of eukaryotic organisms that few people ever see. Most protists are unicellular and live in aquatic habitats. There are at least 100 000 species and new ones are being discovered daily. Photosynthetic protists are a major source of primary productivity in lakes, rivers and oceans, producing at least 30% of the planet's oxygen. Herbivorous protists are the link in food chains between algal primary producers and larger animal consumers, such as fishes and invertebrates. Parasitic protists are responsible for serious illnesses in humans, such as malaria, sleeping sickness and certain types of dysentry. Protists also parasitise animals and plants causing agricultural losses.

Protists are diverse. Comparing two protistan phyla is analogous to comparing elephants with moths, or eels with tomatoes. In the past, phyla were grouped together based on their form of nutrition—whether they were autotrophic (able to produce food by photosynthesis) or heterotrophic (consumers of other organisms). Photosynthetic protists were known as algae, protists that ate smaller organisms were known as protozoa (simple animals), and some protists that absorbed small food molecules from the environment were considered to be fungi. It is now obvious that this system is far too simplistic. Numerous photosynthetic protists, for example, swim about like animals and even capture smaller cells and eat them. These organisms are both animal-like and plant-like, and cannot be classified on the basis of nutrition. A more natural classification based on morphological, biochemical and molecular features is now emerging. Some natural groups include organisms with various modes of nutrition. Alveolates, for example, have photosynthetic, parasitic and predatory members, but all are close relatives based on ultrastructure and DNA sequence data.

From the evolutionary tree in Chapter 30 (Fig. 30.9) you can see that protists are not a monophyletic group. The kingdom is essentially a default taxon, containing all the eukaryotic organisms that are not plants, animals or fungi. Kingdom Protista is thus artificial and many members are now known to be more closely related to other kingdoms than to each other. Green algae, for example, are the closest relatives of land plants (Chapter 37), chytrids are close relatives of fungi, and collar cells probably have the same ancestors as sponges and other animals. So why do we put them all together in one chapter as though they were one evolutionary lineage? The answer is partly historical and partly practical. There are still groups of unicellular eukaryotes of unknown evolutionary relationships, some not even named. It is for convenience that these organisms are temporarily collected together under the banner of protists—a vernacular term that serves to describe them as well as any.

> Protists may be photosynthetic, parasitic, predatory or absorb small food molecules from the environment. Relationships within protists are still unclear but they are a diverse range of eukaryotic cell types.

ORIGIN OF EUKARYOTIC CELLS

The oldest fossils of eukaryotic organisms do not appear until about 1.5 billion years ago. Since fossils of prokaryotes are older, it is generally thought that eukaryotes evolved from prokaryotic organisms.

As we have seen in earlier chapters, prokaryotic and eukaryotic cells share many cellular processes but the internal layout of their cells is different. Prokaryotic cells are essentially one single compartment, whereas eukaryotic cells contain several membrane-bound subcompartments. So how did these subcompartments originate? The answers turn out in some instances to be quite a surprise.

Origin of the nucleus

The eukaryotic nucleus differs from the prokaryotic nucleoid in numerous respects. Two major distinctions are the nuclear envelope and multiple linear chromosomes of eukaryotes. Prokaryotes lack a nuclear envelope and usually have a single circular

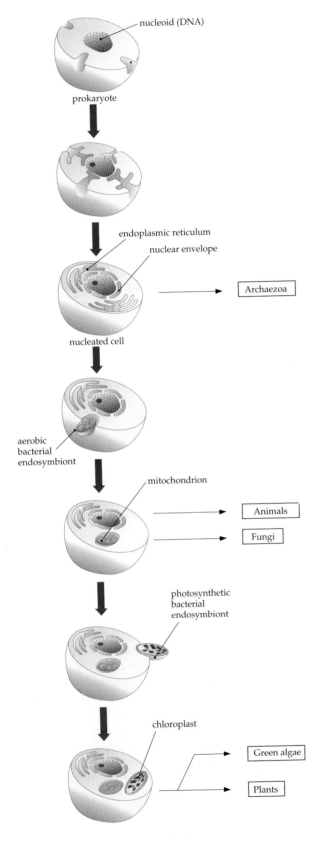

nucleoid (DNA)

prokaryote

endoplasmic reticulum

nuclear envelope

Archaezoa

nucleated cell

aerobic
bacterial
endosymbiont

mitochondrion

Animals

Fungi

photosynthetic
bacterial
endosymbiont

chloroplast

Green algae

Plants

Fig. 35.1 Evolution of eukaryotic cells. Origin of the nucleus and endomembrane system of a eukaryote that lacked mitochondria and chloroplasts (Archaezoa). Mitochondria originated from a bacterial endosymbiont. Chloroplasts originated from a photosynthetic endosymbiont

chromosome. Transformation from a circular chromosome to linear chromosomes might have arisen from a break in the circle and duplication of the linearised chromosome to give multiple copies.

Origin of the nuclear envelope can be explained by accumulation of vesicles resulting from cell membrane invaginations around the prokaryotic nucleoid. If the vesicles flatten around the nucleoid, as shown in Figure 35.1, then they form a rudimentary double envelope complete with gaps or nuclear pores. Such accumulations of membrane vesicles around the nucleoid are known to occur in certain cyanobacteria (Chapter 33).

Origin of the endomembrane system

The endomembrane system forms a conduit from the nuclear envelope to various subcellular compartments and also to the exterior of the cell via the plasma membrane. It probably evolved as a means of sorting and transporting proteins and glycoproteins in large eukaryotic cells. The endoplasmic reticulum probably developed from protrusions of the nuclear envelope, to which it still remains attached (Fig. 35.1). These protrusions could then have become elaborated into the Golgi apparatus and other components of the endomembrane network.

The nuclear membrane and endomembrane system probably evolved from invaginations of the bacterial cell membrane that enveloped the nucleoid.

Origin of cilia or flagella

Cilia or flagella occur in most eukaryotic organisms. Although they go under two names (cilia in animals and animal-like cells, and flagella in plants, sperm, algae and flagellates), the two organelles are homologous, derived from a common ancestral structure (Fig. 35.2). Bacterial and eukaryotic flagella, however, are fundamentally different in both chemical composition and structure (Chapter 3), and appear not to be homologous. They are a case of convergent evolution: two similar solutions to the one problem—how to get around in a liquid medium.

So where did eukaryotic cilia and flagella come from? This is presently one of the most contentious questions in evolutionary cell biology. One school of biologists suggests that cilia or flagella arose as extensions of the cytoskeleton. A second school suggests that flagella or cilia are derived by **endosymbiosis**, one organism living inside another, in this case a spirochaete bacterium living within a eukaryotic cell. Some controversial experimental work suggests that cilia and flagella contain DNA, supporting the notion that they were originally organisms in their own right.

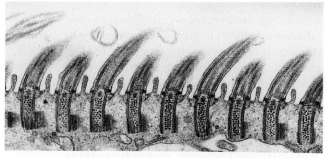

Fig. 35.2 Opalinid flagella. With the invention of the electron microscope it was discovered that cilia and flagella are essentially identical and differ only in length

Origin of mitochondria and chloroplasts: endosymbiotic theory

While the idea that flagella or cilia are derived from a foreign organism living inside a eukaryotic cell is still speculative, an endosymbiotic origin of two other organelles seems almost certain. Chloroplasts and mitochondria have long been recognised as having a degree of autonomy within the cell. They divide before the rest of the cell by fission, like bacteria. This led nineteenth-century microscopists to remark that chloroplasts were reminiscent of cyanobacterial cells living inside plant cells. The organelles also have membranes separating them from the main cell compartment. The discovery of DNA in chloroplasts and mitochondria in the 1960s revived speculation that these structures were derived from prokaryotes. When it was found that the DNA in chloroplasts and mitochondria was a circular chromosome (Chapter 10) and that many of the genes were typically prokaryotic, the endosymbiotic theory of the origin of these organelles gained almost universal acceptance.

In fact, the more we look at chloroplasts and mitochondria, the more convincing is the argument. Chloroplasts and mitochondria have 70 S ribosomes that contain rRNAs with nucleotide sequences similar to bacteria. Like bacterial ribosomes, ribosomes of chloroplasts and mitochondria are sensitive to the antibiotic chloramphenicol but insensitive to cycloheximide, which stops translation in eukaryotic, cytoplasmic ribosomes. Phylogenetic trees based on nucleotide sequences of rRNAs actually group mitochondria and chloroplasts with bacteria, not with eukaryotes. Chloroplasts relate to cyanobacteria and mitochondria to α-purple bacteria.

The circular chromosomes of chloroplasts and mitochondria are considerably smaller than those of their bacterial counterparts. They are so small that their DNA can only encode a fraction of the proteins needed in the organelle. The remaining proteins are encoded by nuclear genes. Messenger RNAs (mRNAs) from these nuclear genes are translated on 80 S ribosomes in the cytoplasm, then translocated into the chloroplast or mitochondrion. This was initially rather puzzling but it is now believed that many of the endosymbiont's genes moved from the organelle's chromosome into the nucleus of the host. Exactly why this should have occurred remains unknown but it certainly serves to 'hobble' the endosymbiont by making it absolutely dependent on the host for its survival.

One feature of chloroplasts and mitochondria is the presence of a double membrane. The two membranes most probably derive from the two membranes that surround Gram-negative bacteria. The host plasma membrane that surrounded the endosymbiont during engulfment has apparently been lost (Fig. 35.3).

The evolutionary tree in Figure 35.3 shows some of the major lines of descent. An endosymbiotic origin of eukaryotic organelles means that the tree actually has two grafts joining the prokaryotic line of descent to the eukaryotic line in at least two places: one for the mitochondrion of all eukaryotes and a second for the chloroplast of plants.

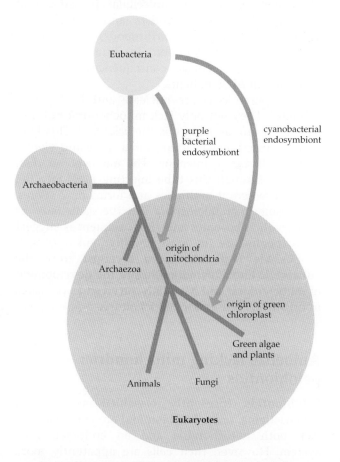

Fig. 35.3 Evolutionary tree showing descent of Eubacteria and Archaeobacteria, Archaezoa, animals, fungi and plants. Grafts joining lines of descent are formed by eukaryotic cells engulfing eubacteria (see Fig. 35.1), once for the origin of mitochondria, a second time for the origin of chloroplasts. Animal and fungal cells are chimaeras (derived from cells of two different organisms) of two evolutionary lineages and plant cells are chimaeras of three lineages

> Chloroplasts and mitochondria are almost certainly derived from endosymbiotic bacteria that have become organelles in eukaryotic cells.

EARLIEST EUKARYOTES: THE ARCHAEZOA

We begin our survey of the protists by looking at three groups of archaezoans, primitive forms that have some features typical of eukaryotes but retain prokaryotic features. While each of these primitive protists has a nucleus, they lack typical eukaryotic organelles. All lack mitochondria and some lack endomembrane systems. Nucleotide sequences of archaezoan rRNAs are very similar to prokaryotic sequences.

Simplest eukaryotes: phylum Microspora

Microspora are obligate unicellular parasites that infect most types of protists and animals, but particularly fishes and arthropods. Microsporan infection is also common in humans and is increasingly prevalent as a consequence of weakened immunity in AIDS patients.

Microsporan cells are nucleate and 1–20 μm in diameter. They not only lack mitochondria but also flagella. A unique feature is the *polar tube*. This long narrow structure is everted from a spore and pierces a host cell during infection. The microsporan then squeezes through this tube injecting itself into the host cell where it multiplies and forms new spores.

Ribosomes of microspora are the same size as prokaryotic ribosomes and rRNA sequences suggest microsporans are the most primitive eukaryotes. It is possible that microspora diverged from the eukaryotic line of evolution before the development of mitochondria and have only survived as parasites in organisms possessing mitochondria.

Amoebae lacking mitochondria: pelobiontids

Pelobiontids (presently classified as phylum Zoomastigina, order Pelobiontida) are amoebae that lack both mitochondria and an endomembrane system. However, their cells are apparently more complex than microspora because they have rudimentary flagella. The DNA is surrounded by a nuclear envelope but it is not known whether the nucleus divides mitotically with a microtubular spindle, or whether it simply pinches in two, such as a prokaryotic cell. The best known amitochondriate amoeba is *Pelomyxa palustris*, a giant, multinucleate, free-living, herbivorous cell. *Pelomyxa palustris* may compensate for its lack of mitochondria by harbouring numerous endosymbiotic bacteria that apparently perform oxidative phosphorylation for the host. In any case, it seems to live only at the bottom of ponds where oxygen is fairly scarce.

Flagellates lacking mitochondria: diplomonads

Diplomonads (phylum Zoomastigina, class Diplomonadida) are unicellular, heterotrophic flagellates. The name refers to the presence of two nuclei, each of which is associated with a pair of flagella. Diplomonads inhabit the gut of various animals. They lack mitochondria and are restricted to an anaerobic environment.

Giardia, an intestinal parasite causing severe dysentry, is the best known diplomonad (Fig. 35.4). It is one of the first protists on record, accurately described by Leeuwenhoek in 1681 from his own diarrhoeic stools. Nucleotide sequence data from *Giardia* show it to be a primitive eukaryote with rRNAs most similar to prokaryotes. Diplomonads thus diverged from the main eukaryotic evolutionary lineage at a similar time to microspora, before the evolution of mitochondria and elaborate endomembrane systems. These organisms are perhaps relicts, survivors of ancient protists that evolved before the accumulation of oxygen in the atmosphere (Chapter 31). Anaerobic environments such as the alimentary canal are the only places in which they survive.

> Archaezoan protists have a nucleus but all lack mitochondria and some lack endomembrane systems. These organisms are believed to be descendants of the first eukaryotes.

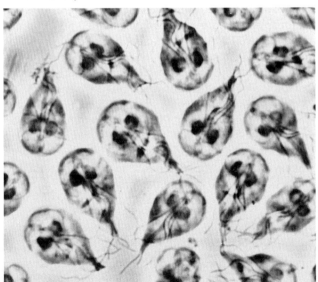

Fig. 35.4 (a)

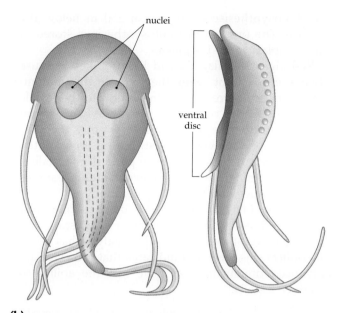

(b)

Fig. 35.4 **(a)** *Giardia* is a primitive eukaryote (a diplomonad) that parasitises humans and other animals. **(b)** Cells have two nuclei (*n*), each of which is associated with a set of flagella. On the ventral side of the cell is a disc (*vd*) through which the cell attaches to the host's gut lining. Infection is spread by cysts excreted in faeces. The cysts, which remain viable in water for several months, can infect the gut of animals drinking from the contaminated water source. *Giardia* is not restricted to polluted waters and can occur in metropolitan water supplies or even in wilderness streams. The most effective means of purification is to boil the water; cysts are resistant to iodine and chlorine

PHOTOSYNTHETIC PROTISTS

It is not yet clear whether all chloroplast-containing protists are related. One possibility is that chloroplasts were obtained by eukaryotes on more than one occasion, that is, different host organisms engulfed a different prokaryote and kept it as an endosymbiont. In this way, chloroplasts could be present in different eukaryotic lineages. To further complicate things, it is now known that some eukaryotes with chloroplasts stole them from other eukaryotes. This means that heterotrophic eukaryotes can convert to autotrophy by taking the photosynthetic organelle from a distant relative. From this you can see that it is not valid to unite all chloroplast-containing protists into one group, traditionally labelled algae.

The phyla discussed here have chloroplasts and are primarily photosynthetic. In some organisms the chloroplasts lack photosynthetic pigment and are called **leucoplasts**. Even though species with leucoplasts obtain their nutrition by means other than photosynthesis, they still retain bleached chloroplasts, suggesting that the organelle provides something to the cell in addition to food.

Flagellates with a photosynthetic endosymbiont: phylum Glaucocystophyta

Glaucocystophyta seem to be living examples of an intermediate stage in the evolution of a chloroplast from a photosynthetic prokaryotic endosymbiont. Chloroplasts of glaucophytes are **cyanelles**, which have a peptidoglycan wall, as do bacteria. The presence of the wall is evidence that the cyanelle was once a bacterium before it took up residence in the host cell. Cyanelles contain chlorophyll *a* and phycobilin pigments identical to cyanobacteria. Like a cyanobacterium, a cyanelle has a circular chromosome, but it is no longer fully autonomous, having lost genes to the nucleus during the endosymbiotic relationship. Some genes for producing peptidoglycan have been found on the cyanelle chromosome, which is otherwise the same as a chloroplast chromosome. Cyanelles are thus partially dependent on the host cell and cannot survive independently. Host cells are typically flagellates with two laterally inserted smooth flagella.

> Glaucocystophytes are photosynthetic flagellates with chloroplasts, cyanelles, that have a peptidoglycan wall, as do bacteria.

Red algae: phylum Rhodophyta

Red algae are common seaweeds on rocky seashores around the world. There are some 4000 species, many of which are endemic to Australia (Chapter 41). Red seaweeds are of commercial importance in the production of agar for microbiology and molecular biology, and as food in the Orient, North America and Ireland. Sushi is prepared with dried *Porphyra*, Japanese *nori*. About 60 000 hectares of *nori* are grown by mariculture around the Japanese coast. Carrageenan from red algae is used also as a stabilising agent in confectionery, ice cream, cosmetics and pet foods. The red seaweed industry is worth about 1 billion dollars per annum worldwide.

Most red algae are multicellular, adjacent cells often being attached by **pit plugs** (Fig. 35.5), and a few are unicellular. They form a thallus with branches and blades plus extensions attaching the plant to the substrate. Red algae have complex life histories with alternating stages that are different in morphology. Some red algae are calcified and are known as coralline red algae because they were mistakenly thought to be coral animals.

Chloroplasts contain chlorophyll *a* and phycobilin pigments—phycocyanin and phycoerythrin (the latter producing the typical red colouration). Red algae absorb long wavelength blue and green light that penetrates deepest into the ocean, allowing them

(a)

(b)

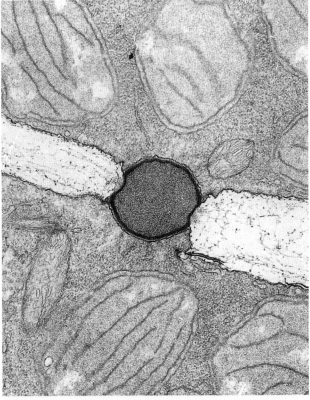

(c)

Fig. 35.5 Red algae range from **(a)** fine feathery structures to **(b)** crusty calcified plants resembling corals. **(c)** Adjacent cells are often attached by pit plugs

to photosynthesise at depths of 250 m below the surface. The product of photosynthesis is stored in the cytoplasm as α-1, 4 glucan.

Red algae lack flagella and basal bodies. Because their sperm cannot swim, they rely on the randomness of ocean currents to bring sperm to the female part of the thallus containing the egg. When a sperm does contact an egg to form a zygote, the alga capitalises on the event by distributing copies of the diploid nucleus to other female parts of the thallus. Thus, from a single fertilisation event, multiple spores can be produced for the next generation.

The lack of flagella and basal bodies was originally interpreted as a primitive character suggesting that red algae are ancient. Molecular analysis has failed to confirm this view and indicates that red algae are advanced organisms that have lost the ability to produce flagella.

> Red algae (phylum Rhodophyta) are familiar seaweeds. Most are multicellular and macroscopic, and they lack flagella. They contain chlorophyll *a* and phycobilin pigments.

Green algae: phylum Chlorophyta

Green algae are a large group (about 16 000 species) including unicellular, colonial and multicellular forms. Their chloroplasts are grass green and contain the same pigments as land plant chloroplasts—chlorophylls *a* and *b*, β-carotene and other carotenoid derivatives. Like land plants, the product of photosynthesis of green algae is stored as starch (an α-1, 4 glucan) within the chloroplast, and cell walls are primarily cellulose (β-1, 4 glucan). These and other similarities leave us in no doubt that green algae are closely related to land plants (Chapter 37).

Green algae are common in most habitats and fix an estimated 1 billion tonnes of carbon from the atmosphere per annum. They are used as food (*Spirogyra* as vitamin supplement tablets) and are being tested in biotechnological applications (Box 35.1).

Green algae were classified traditionally on the basis of their form—unicellular, colonial, filamentous, coenocytic (technically unicellular but multinucleate and greatly enlarged to form a macroscopic thallus), and multicellular three-dimensional forms. Closer investigation with the electron microscope shows these categories to be somewhat artificial, with several cases of convergent evolution. Studies of mitosis, for example, have shown that two species from the filamentous genus *Klebsormidium* actually belong in different classes. Although superficially similar, the two species of *Klebsormidium* have different types of mitosis (the phragmoplast and phycoplast types described in Chapter 37) and fundamentally different motile cells. A filamentous thallus therefore seems

BOX 35.1   GREEN ALGAE AND BIOTECHNOLOGY

Not only do green algae grow in a wide range of habitats, such as freshwater, oceans, salt lakes and snow, but they also show a great diversity in their chemistry. It is this chemical diversity, combined with the ability of some species to grow in extreme environments, that makes green algae attractive to biotechnologists.

Early studies in the 1970s focused on using algae, such as *Chlorella*, *Scenedesmus* and *Oocystis*, as sources of protein. Unfortunately, the algal protein proved to be expensive and was not always palatable. Since the early 1980s, the focus of algal biotechnology has shifted to commercial production of high value chemicals, such as carotenoids, lipids, fatty acids and pharmaceuticals.

An important alga is *Dunaliella salina*. When grown at high salinity (about 10 times the concentration of sea water) and with high light intensity, *D. salina* accumulates large amounts of an orange-red carotenoid, β,β-carotene. This pigment compound is used to colour products, such as margarine, noodles and soft drinks, and as a vitamin supplement because it is readily converted to vitamin A. There is also evidence that β,β-carotene may help prevent lung cancer. Pure β,β-carotene is worth more than $A600 per kilogram. Production of β,β-carotene from *D. salina* means growing and harvesting vast quantities of algae in 'farms'. The world's largest algal farms are at Hutt Lagoon in Western Australia (see figure) and Whyalla in South Australia.

Another alga under study is the freshwater chlorophyte *Haematococcus pluvialis*, which is the best natural source of the carotenoid, astaxanthin. Astaxanthin is used in aquaculture as a fish food additive to ensure trout and salmon flesh is the natural pink colour. Fish food currently contains synthetic carotenoids and astaxanthin is a desirable natural alternative.

Green algae may also be a future source of alternative fuels. *Botryococcus braunii* produces long-chain hydrocarbons similar to crude oils, and these can be cracked in a refinery to produce petrol and other useful fractions. *Tetraselmis* species accumulate fats and oils and, once extracted, the lipids can be used as a diesel fuel substitute.

The scope for algal use in producing pharmaceuticals, antibiotics, fuels and foods, and as an adjunct in waste treatment is enormous. Manipulation of strains by genetic engineering will contribute to the production of useful natural substances.

With its wide flat spaces and intense sunshine, Australia is the perfect place for algal farms producing food, fuel and pharmaceuticals. The ponds of *Dunaliella salina* at Hutt Lagoon, Western Australia, range in colour from green to brick red depending on how much of the valuable β, β-carotene cells have accumulated

---

to have evolved more than once in the green algae. Current classification recognises five classes: Prasinophyceae, Chlorophyceae, Ulvophyceae, Charophyceae and Conjugatophyceae.

Primitive green algae: class Prasinophyceae

Prasinophytes are the most primitive green algae. Most are unicellular and have no cell wall. Instead they are covered with layers of delicate scales that have elaborate shapes and ornamentation. Fossils of prasinophyte cysts found in Tasmania show that prasinophytes are at least 600 million years old.

Chlamydomonas and relatives: class Chlorophyceae

The flagellate *Chlamydomonas* and green algae that produce motile cells similar to *Chlamydomonas* are included in this class. Several levels of organisation, including unicellular, filamentous, colonial and parenchymatous, occur among the Chlorophyceae.

A unique form of crystalline glycoprotein cell wall has evolved in this class.

Common forms include *Dunaliella*, a unicellular flagellate, and *Volvox*, a spherical, colonial form composed of *Chlamydomonas*-like cells. *Chlamydomonas* is a model organism for cell biology. It is readily grown in the laboratory, reproduces sexually, and produces a range of mutants able to be mapped by classical and molecular genetic techniques.

Sea lettuce and miso soup: class Ulvophyceae

The sea lettuce *Ulva* (Fig. 35.6a) is the best known member of this class, which includes most of the green seaweeds occurring along coastlines around the world. Most Ulvophyceae are multicellular or coenocytic, up to 3 m in size, although typically much smaller. Motile cells are produced during sexual and asexual reproduction. Common forms include the sea cactus, *Caulerpa* (Fig. 35.6b), and *Codium*. *Ulva* is used as a garnish for Japanese *miso* soup. The Ulvophyceae diverged from the other green algae early on.

(a)

(b)

Fig. 35.6 Green algae. **(a)** The sea lettuce *Ulva lactuca* is used as a food but **(b)** its close relative *Caulerpa* can be poisonous. Both are common members of the class Ulvophyceae found on rocky shores around the south-eastern coast of Australia

Stoneworts: class Charophyceae

Charophytes are essentially restricted to freshwater habitats. They are delicate and typically small (2–30 cm in length) with some, the **stoneworts**, encrusted with $CaCO_3$ (calcite). Gametes are asymmetrical and mitosis involves a phragmoplast—characteristics that identify charophytes as the closest relatives of the land plants (Chapter 37).

Desmids and relatives: class Conjugatophyceae

These green algae have no cells with flagella. They reproduce by **conjugation**, a process involving formation of an interconnecting tube between two strains of the one species through which an amoeboid gamete crawls to fuse with its opposite gamete. Most conjugatophytes live in freshwater and are either unicellular (Fig. 35.7) or filamentous (Fig. 35.8). The unicells are **desmids**, which have beautiful symmetrical shapes comprising two semicells connected by an isthmus of cytoplasm passing through a central constriction. There are more than 10 000 species of desmids (Fig. 35.8).

> Green algae (phylum Chloropophyta) are unicellular, colonial or multicellular, and one group is the closest relative of land plants. Chloroplasts contain chlorophyll *a* and *b*, the product of photosynthesis is stored as starch, and cell walls are cellulosic.

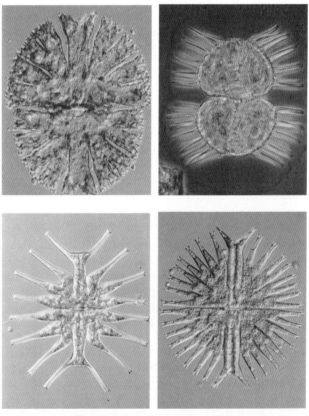

Fig. 35.7 In addition to bird life, lilies and crocodiles, the water holes of Kakadu National Park in Northern Territory contain this splendid selection of desmids (phylum Chlorophyta)

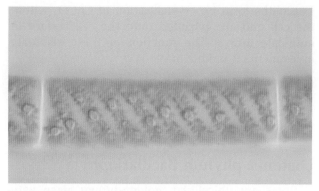

Fig. 35.8 The green filamentous alga *Spirogyra* (class Conjugatophyceae) is named for the spiral chloroplast that winds its way around the periphery of the elongate cells

HETEROKONT PROTISTS

Heterokonts are characterised by flagellar architecture. They have one smooth flagellum directed posteriorly and one hairy flagellum directed anteriorly. The hairy flagellum has numerous thin, tubular appendages that alter the direction of thrust produced by the flagellar beat. The beat of the hairy flagellum thus drags the cell through the water. If the cell happens to be fixed in place, the flagellar beat draws the water down and over the cell. Heterokonts include chrysophytes, haptophytes, diatoms and brown algae, which are photosynthetic, and oomycetes, which are non-photosynthetic (heterotrophs).

Golden flagellates: phylum Chrysophyta

Chrysophyta are golden-brown flagellates of marine and freshwater habitats. Cells are unicellular or colonial (Fig. 35.9) and have heterokont flagellation.

Fig. 35.9 *Synura* is a colonial chrysophyte common in freshwater

Chloroplasts contain chlorophylls *a* and *c* plus **fucoxanthin**, an accessory pigment giving the golden colour. Numerous heterotrophic forms have a colourless chloroplast or no chloroplast whatsoever, and even coloured photosynthetic forms can ingest food particles. Chrysolaminarin (β-1, 3 glucan) is stored in a vacuole. Various cell coverings, including spines and scales composed of silica or a **lorica** (external vase-shaped shell) made of either cellulose or chitin, adorn the cells. Silicoflagellates contain spectacular, star-shaped silica skeletons (Fig. 35.10).

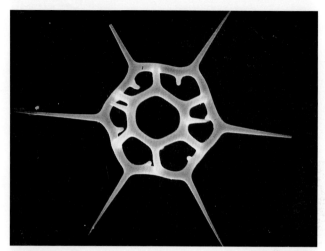

Fig. 35.10 Chrysophytes have various cell coverings. The beautiful star-shaped skeleton of a silicoflagellate is made from silica

Chrysophytes (golden-brown flagellates) are heterokonts, cells with one smooth and one hairy flagellum.

Haptophytes: phylum Prymnesiophyta

Haptophytes are extremely abundant in oceans. *Emiliana huxleyi* (named after T. H. Huxley) occurs in massive blooms visible in satellite photographs. A global correlation between these satellite photographs and water samples taken from oceanographic vessels at the same time indicates that *E. huxleyi* may have the largest biomass of any single species on earth. Enormous chalk deposits, such as the white cliffs of Dover, were formed from haptophytes and other protist skeletons. Several haptophytes are toxic to fish and shellfish, and blooms of these algae can result in total decimation of marine life over great areas.

Haptophytes are thought to be close relatives of chrysophytes because they have similar chloroplasts and mitochondria. Their flagella are, however, quite different. The two flagella of haptophytes are both smooth and lack hairs, which means that haptophytes are not true heterokonts. The name haptophytes refers to the curious **haptonema**, a thread-like (filiform) extension situated between the two flagella.

The haptonema can move, either bending or coiling, and can capture prey, drawing them down to a 'mouth' on the posterior of the cell for ingestion.

A major group of haptophytes are **coccolithophorids** (Fig. 35.11), which are covered with intricately sculptured calcite plates, **coccoliths.**

Coccoliths form by crystallisation of $CaCO_3$ within the cell and are extruded onto the cell surface in imbricate arrays. The function of these elaborate investments is unknown.

> Haptophytes are unicellular and have chloroplasts similar to chrysophytes. Although classified as heterokonts, they have two identical, smooth flagella between which extends a haptomena for capturing prey.

Diatoms: phylum Bacillariophyta

Diatoms are unicellular, golden-brown algae with siliceous walls (Fig. 35.12). They are ubiquitous in aquatic environments and are important producers. Chloroplasts and storage products of diatoms are the same as their close relatives, the chrysophytes.

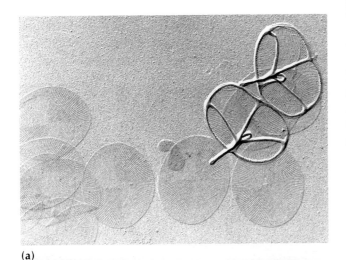

(a)

(b)

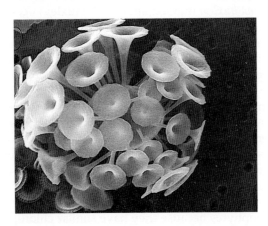

(c)

Fig. 35.11 Haptophytes. **(a)** The filmy scales of *Chrysocromulina* are only visible by high resolution electron microscopy. The calcium carbonate armour plating of coccolithophorids can vary in shape from **(b)** flat discs, as in *Pontosphaera*, to **(c)** the elaborate trumpet-shaped structures of *Discosphaera tubifera*

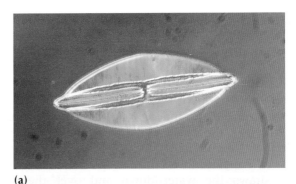

(a)

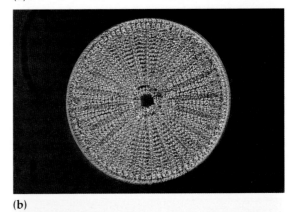

(b)

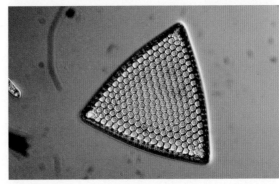

(c)

Fig. 35.12 Diatoms are typically either **(a)** pennate such as *Navicula lyra* or radially symmetrical, such as **(b)** *Arachnoidiscus* and **(c)** *Triceratium*. The silica valves have an opalescent appearance in the light microscope

Diatoms have a unique silica cell wall. Each cell has two silica dishes, valves or **frustules**, interconnected by silica hoops, girdle bands. The valves are highly ornamented with pores and spines creating some remarkable patterns (Fig. 35.13). The valves and bands are perhaps derived from silica scales of an ancestor resembling modern-day chrysophytes. The silica valves form some of the best preserved fossils of any protists but, in older deposits, they have been converted to formless chert, destroying early diatom fossils. Massive deposits of diatom valves (diatomaceous earth) exist in recent strata and are mined for use as a very fine, high-grade filtration material or as an abrasive in toothpaste and metal polishes.

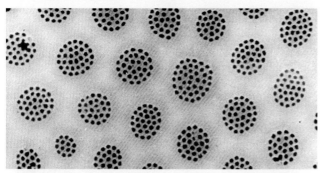

Fig. 35.13 Seen in detail under the scanning electron microscope, the markings observed on diatom valves by light microscopy are revealed to be small, regularly shaped pores in the silica. The pores allow transfer of materials through the cell's otherwise impervious, glass-like case

Diatoms are classified into two groups—*centrics*, radially symmetrical, and *pennates*, bilaterally symmetrical (Fig. 35.12). Many pennate forms have a longitudinal slit, a **raphe**, in the valve, which enables them to move by crawling along the substrate. Wall-less, motile sperm with a single flagellum are released during sexual reproduction.

> Diatoms (phylum Bacillariophyta) are unicellular golden-brown algae with a unique silica wall that forms two valves.

Brown algae: phylum Phaeophyta

There are about 900 species of brown algae, nearly all of which are marine and multicellular. They include the giant **kelps**, *Macrocystis pyrifera*, growing off the coast of California, which are as long as a blue whale and as tall as the biggest mountain ash trees in south-west Tasmania. Kelps form underwater forests that are home to a variety of temperate ocean marine life (see Box 41.2). They are also a source of alginic acid, a gelling agent used in foods, adhesives, paint and explosives. The large thallus of kelp is differentiated into a *holdfast*, which attaches to the substrate, a *stipe*

and *blades* (Fig. 35.14). This organisation parallels that of terrestrial plants, and kelps were once regarded as 'underwater trees' that were the marine ancestors of land plants.

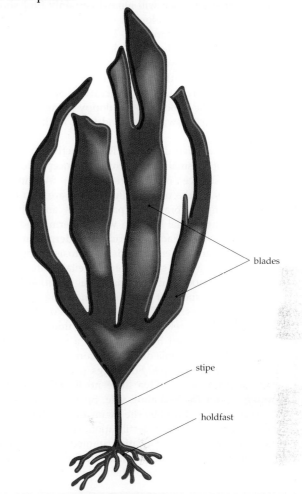

(a)

(b)

Fig. 35.14 Phaeophytes (brown algae). **(a)** Diagram of the thallus of a kelp. The stipe contains a vascular system that translocates material down from the photosynthetic blades to the holdfast, which may be many metres below the surface. **(b)** Bull kelp, *Durvillea potatorum*, occurs on southern Australian rocky shores subject to high surge action. The disc-shaped holdfast adheres tenaciously to rocks, preventing the thallus from being ripped away by waves

BOX 35.2   NEPTUNE'S NECKLACE *HORMOSIRA BANKSII*

If you poke around in the tide pools on the eastern coast of Australia, you will almost certainly find short strings of drab, olive-coloured beads splayed over the rocks. These beads are the brown alga Neptune's necklace, *Hormosira banksii*. Like other intertidal life forms, *H. banksii* must withstand exposure to the air twice daily, and the leathery, fluid-filled beads, termed **receptacles**, are resistant to desiccation. Supported by sea water on the flood tide, the floppy strings of beads fan up and out to sway back and forth in the surging waves.

Hormosira banksii is dioecious. Reproductive structures are found within small warty growths, **conceptacles**, that stud the surface of the receptacles. Within the conceptacles on the male thallus (plant) are two types of hairs: long, unbranched paraphyses and shorter, branching antheridial hairs on which sperm-producing antheridia develop. Each antheridium undergoes meiosis and several subsequent rounds of mitosis to produce 64 sperm cells. Motile sperm are biflagellate heterokonts (having one smooth and one hairy flagellum) and bear an orange eyespot. At low tide, an orange ooze of antheridia exudes from the conceptacles on the male thallus. Sperm are released on the flood tide.

Eggs are produced by oogonia on a female thallus. Like antheridia, oogonia develop in conceptacles, which also contain paraphyses. Four eggs (ova) are released from each oogonium. Ova have no flagella and drift motionless on the incoming tide. Sperm are attracted to a secretion produced by the ovum and cluster around the ovum until one effects fertilisation. The zygote settles and, if it finds a suitable location, immediately develops into a new, diploid, male or female thallus. The gametes are the only haploid stage of the life cycle.

Hormosira banksii

While kelps are large and highly visible, many other brown algae are small inconspicuous tufts or simple filaments barely visible to the naked eye. Even some of the larger kelps have a microscopic filamentous life form as one of their alternating generations.

Brown algae have chloroplasts with the same pigments as chrysophytes. The storage product *laminarin*, a β-1, 3 glucan, is similar to chrysolaminarin. The heterokont motile cells released as gametes or zoospores closely resemble chrysophyte flagellates and it is probable that multicellular brown algae evolved from unicellular chrysophytes.

> Brown algae (phylum Phaeophyta) include the largest protists with a differentiated, multicellular thallus. Pigments and the storage product, laminarin, are similar to chrysophytes.

Water moulds and downy mildews: phylum Oomycota

Water moulds and downy mildews, **oomycetes**, have a superficial resemblance to fungi (Chapter 36) since they produce a network of filaments (hyphae) that permeate their food substrate. The hyphae are coenocytic, having no septa (cross-walls). Oomycetes are different from fungi, however, in that cell walls are cellulosic rather than chitinous. Oomycetes are so-named for their distinctive **oogonium**, the female reproductive structure containing oogonia. Male gametes are produced in nearby antheridia and non-motile 'sperm' are brought to the oogonium through a fertilisation tube. Fusion of gametes (syngamy) produces a diploid oospore within which meiosis usually occurs to produce zoospores with heterokont flagella. These zoospores are remarkably similar to chrysophytes (golden algae) and comparison of rRNA sequences from oomycetes and chrysophytes indicates that they are related.

BOX 35.3 DIEBACK DISEASE

In the 1920s there were a number of reports of mysterious deaths of jarrah trees, *Eucalyptus marginata*, in Western Australian forests (Fig. a). Tree deaths appeared to follow bush tracks and logging sites, and were at first attributed to soil disturbance. When sand and gravel from these cleared areas was transported to other regions, trees at these sites also died.

It was not until the late 1960s that the cause of the forest dieback was identified as the oomycete, *Phytophthora cinnamomi*. This pathogen attacks the roots of susceptible plants, causing problems in water uptake and translocation. Infected trees show symptoms of water stress, with leaf yellowing and dieback of upper branches. Spread of the disease occurs underground by movement of flagellated zoospores, which are able to

swim in moist soil. Zoospores seek a host rootlet, attach themselves and produce hyphae that invade the plant's root system (Fig. b). This mechanism of disease transfer explains how transport of contaminated soil or flushing of floodwater spreads the disease.

Phytophthora cinnamomi is thought to originate from cinnamon trees in Sumatra, and was probably introduced to Australia by European colonists. Many endemic plants have no apparent resistance to dieback and some highly susceptible *Banksia* species are threatened with extinction. The massive scale of the problem prevents the use of fungicide, and outbreaks of the disease must usually run their course before natural antagonistic soil microbes bring the epidemic under control.

 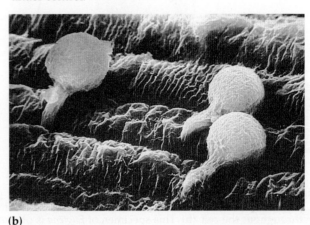

(a)   (b)

(a) Dieback of jarrah trees in Western Australia caused by the oomycete *Phytophthora cinnamomi*. (b) Cysts of *P. cinnamomi* germinating on a plant rootlet. *P. cinnamomi* zoospores swim through soil water and encyst when they contact a plant root. The cyst then germinates to produce hyphae, which penetrate the root and invade the vascular system of the host, eventually causing dieback

Oomycetes are of considerable commercial and environmental importance. *Phytophthora infestans*, which causes late blight of potatoes, destroyed potato crops in the 1840s in Ireland. Potatoes, which were introduced from South America, had become the staple food of workers in Europe. The average Irish farm worker ate 5 kg of potatoes—boiled, mashed, roasted or fried—every day. However, due to cool, damp summer weather, the potatoes became infected with *P. infestans* and all rotted. During the resultant famine, 1 million people perished, prompting many Irish to seek a new life in the United States and Australia. Also last century, another oomycete, *Plasmopara viticola*, attacked French grapevines and almost obliterated the French wine industry in a single season.

Oomycetes have coenocytic hyphae with cellulosic walls. The gametes are non-motile and sperm are brought to the female reproductive organ, an oogonium, through a fertilisation tube. Oomycetes are related to chrysophytes.

EUGLENOIDS AND KINETOPLASTS

This group includes flagellated unicells that are, photosynthetic or heterotrophic, some being parasitic, others free-living. They are currently classified in different phyla but are related. They all have an anterior depression, *gullet*, from which the flagella emerge. Some of the heterotrophic forms ingest food particles through this anterior gullet.

Euglenoid flagellates: phylum Euglenophyta

Euglenoids (Fig. 35.15) are flagellates of both marine and freshwater habitats. There are about 800 species, a third of which are photosynthetic. The other species lack chloroplasts and are heterotrophic. Even some of the chloroplast-containing forms are facultative heterotrophs and when kept in darkness their chloroplasts shrivel and they revert to heterotrophy, engulfing prey through the gullet.

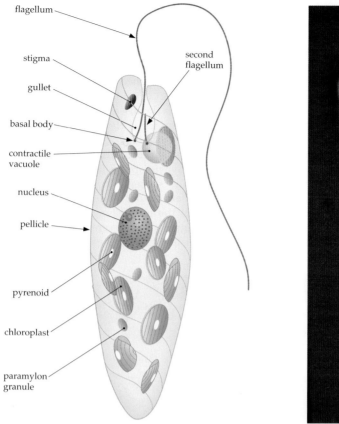

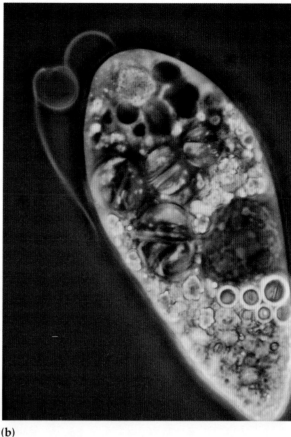

(a)   **(b)**

Fig. 35.15 **(a)** Diagram of a euglenoid. These flagellates can possess chloroplasts but are often heterotrophic. The helical proteinaceous strip forms a spiral pellicle giving the cell its shape. Several paramylon granules are distributed throughout the cell. **(b)** This specimen of *Euglena* is photosynthetic but can also live heterotrophically

The euglenoid chloroplast is similar to that of green algae and land plants in that it contains chlorophylls *a* and *b* and β-carotene. However, the euglenoid chloroplast is bounded by three membranes and the organisation of genes on the chloroplast chromosome is unique. Unlike green algae, euglenoids do not store any starch in the chloroplast. It is not clear how photosynthetic euglenoids came by their chloroplasts. Some biologists think euglenoids stole their chloroplasts from green algae but others think that they acquired them at an early stage of evolution.

Products of photosynthesis are stored as *paramylon*, a β-1, 3 glucan, that forms solid granules in the cytoplasm. At the anterior end of the cell is a red eyespot able to detect light. Euglenoids usually swim with one long flagellum (a second short flagellum does not usually emerge from the gullet) and many species perform a sinuous gyration or crawling motion known as 'metaboly'. Euglenoids are technically naked, having no cell wall or ornamentation outside the plasma membrane, but many species have an elaborate proteinaceous *pellicle* comprising overlocking helical strips that interslide as the cell moves. Reproduction is principally by asexual division.

Flagellate parasites: kinetoplasts

Flagellate parasites (phylum Zoomastigina, class Kinetoplastida) include trypanosomes and leishmanias, which are disease-causing organisms of major medical and veterinary significance. Species of *Phytomonas* infect plants and are a major problem in coconut palms, oil palms, coffee trees and various fruit crops in Latin America.

Kinetoplasts are unicellular biflagellates with an apical depression into which the flagella are inserted. Kinetoplasts are named for the large mass of DNA, the kinetoplast, present in the single mitochondrion at the base of the flagella. A kinetoplast is composed of thousands of catenated DNA mini-circles (linked together like a chain), often forming an elongated rod-shaped structure in the mitochondrion (Fig. 35.16). The kinetoplast DNA also contains normal circular mitochondrial chromosomes.

Trypanosomes cause African sleeping sickness and nagana. These parasitic flagellates are free-swimming in the blood of humans and other vertebrates. Infection is usually transmitted by the blood-sucking tsetse fly. Occasional cross-infection occurs through bites from vampire bats. The trypanosome that causes

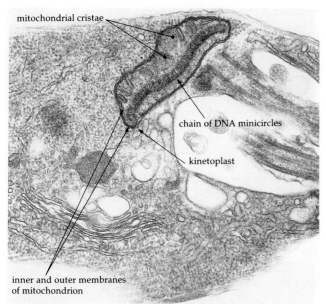

Fig. 35.16 Kinetoplasts are parasitic flagellates. The name kinetoplast refers to a specialised mitochondrion containing thousands of DNA mini-circles, which are visible here as a tangled mass of threads forming an elongate body. The kinetoplast lies at the base of the flagellum

Chagas' disease in South and Central America (Fig. 35.17) invades the heart and other muscles. About 10–12 million people are infected. Again, the disease is transmitted by blood-sucking insects.

Leishmaniasis, for example, is an infection of macrophage cells (white blood cells that normally ingest foreign particles in the bloodstream) caused by the kinetoplast parasite *Leishmania*. Disease transmission is by sandflies, and the parasite occurs in South and Central America, Africa, the Middle East, the Mediterranean and Asia. Relatively benign forms cause cutaneous lesions (Fig. 35.18), but visceral leishmaniasis attacks macrophages of the liver, spleen and bone marrow, often resulting in fatal anaemia.

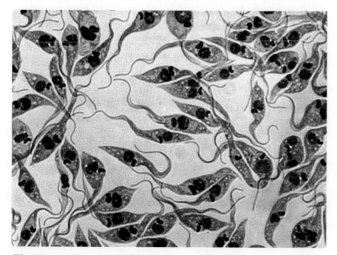

Fig. 35.17 *Trypanosoma cruzi*, a kinetoplast from Central and South America, causes Chagas' disease

Fig. 35.18 In Costa Rica, the small scars produced by cutaneous leishmaniasis, caused by the kinetoplast parasite *Leishmania*, are known as the 'seal of the forest'. More severe forms of leishmaniasis often result in death

A remarkable feature of trypanosomes, such as *Leishmania*, is their ability to survive in the host's bloodstream and avoid elimination by the immune system. Trypanosomes do this by constantly changing the glycoproteins on their surface. Thus, no sooner does the host mount an immune response (Chapter 23) to the invader, than the trypanosomes slip into another 'jacket' that the immune system cannot yet 'see'. The parasite has up to 1000 different versions of surface molecules that it produces by sequentially rearranging the genes that code for surface glycoproteins. In this way, the parasite can stay one step ahead of the host's immune system.

Morphologically, kinetoplasts are rather similar to euglenoids and studies of rRNA confirm that these groups are related. Euglenoids differ in that they are free-living, can have chloroplasts, and never have kinetoplast DNA.

Euglenoids and kinetoplasts are related, flagellated cells. Euglenoids are free-living, some of which have chloroplasts and some of which engulf prey through an anterior gullet. Kinetoplasts are parasitic with a unique mitochondrion.

ALVEOLATES

Members of this group all have distinctive vesicles, **cortical alveoli**, just beneath the plasma membrane. In some species, the cortical alveoli are involved in the formation of the cell's covering, such as plates and scales. Although they are a diverse group, including photosynthetic, parasitic and predatory organisms, DNA sequence data confirm that they are monophyletic.

Dinoflagellates: phylum Dinophyta

Dinoflagellates are an extremely diverse phylum of unicells. About half the species are photosynthetic and major primary producers in tropical seas (Fig. 35.19). Some cause red tides, which may be toxic (Box 35.4). Their name refers to the characteristic spinning motion of the cells as they swim through the water. By protist standards, dinoflagellates are quite vigorous swimmers and can swim at speeds of 1 m per hour. Cells have one posteriorly directed flagellum that steers the cell, plus a unique *transverse flagellum* positioned in a *girdle* encircling the cell (Fig. 35.19c). This transverse flagellum is corkscrew-shaped, and its beat causes the cell to spin and aids forward movement.

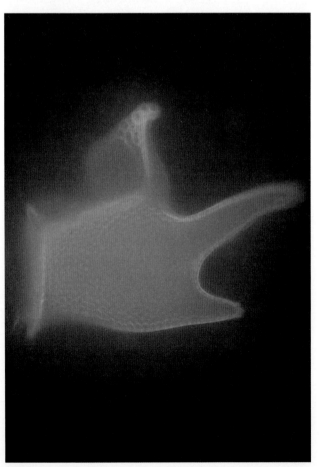

(a)

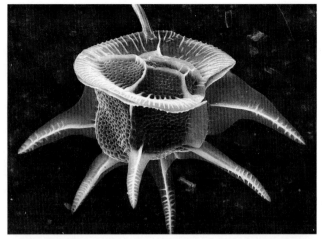

(b)

(c)

Fig. 35.19 Dinoflagellates. **(a)** Stained with a fluorescent dye, the cellulose armour plating of dinoflagellates glows an eerie blue when viewed with an ultraviolet microscope. **(b)** This dinoflagellate from the Coral Sea has wing-like extensions of its plates that are believed to act like sails and catch water currents, moving the cell through the ocean. **(c)** The distinctive girdle formed by a constriction in the mid-region of dinoflagellates is where the spiral transverse flagellum is normally located. In this cell, prepared for scanning electron microscopy, the delicate flagellum is lost

Chloroplasts of photosynthetic dinoflagellates contain chlorophylls *a* and *c*, plus a xanthophyll, **peridinin**. Three or more membranes, rather than the usual two, surround the chloroplast. Starch is stored in the cytoplasm. Some dinoflagellates are naked, some have scales, and some are armoured with cellulosic plates. Dinoflagellates such as *Noctiluca* (night-light) are bioluminescent and congregate in the surf, creating phosphorescence. The luminescence is perhaps a mechanism to startle would-be predators.

Dinoflagellates known as **zooxanthellae** are endosymbionts in the tissues of corals, sea anemones and molluscs, supplying the host animal with nutrition in return for protection and a supply of nitrogen from the animal's excretory products

BOX 35.4 TOXIC DINOFLAGELLATES

Red tides occur when the concentration of dino-flagellates in sea water becomes so high that they discolour the surface of the sea (see figure). The explosive burst of growth results in millions of cells per litre and is induced by a particular set of environmental conditions, such as high temperatures, excess nutrients, and a stratified, stable water column. The majority of red tides, such as those caused by the biolumines-cent dinoflagellate *Noctiluca scintillans*, appear to be harmless events. However, under exceptional conditions, blooms of dinoflagellates can cause severe problems. Sometimes the algae become so densely concentrated that they generate anoxic conditions, suffocating fish and invertebrates in sheltered bays. Other dinoflagellates such as *Gymnodinium mikimotoi* cause serious damage to fish in intensive aquaculture systems, either by the production of mucus, which causes mechanical damage to fish gills, or by the production of haemolytic substances that destroy red blood cells in gill tissues.

About 30 species of dinoflagellates produce potent toxins that move through food chains, via fish or shellfish to humans. Dinoflagellate toxins are so potent that a pinhead-size quantity (about 500 µg), an amount easily accumulated in just one 100 g serving of shellfish, could be fatal to humans. The toxins involved rarely affect the nervous systems of fish or shellfish but they evoke a variety of gastrointestinal and neurological symptoms in humans. The resulting illnesses are known as paralytic shellfish poisoning (PSP), diarrhoetic shellfish poisoning (DSP) and ciguatera fish food poisoning.

Algal blooms seem to be increasing in frequency and geographic spread. On a global scale, close to 2000 cases of human poisoning through dinoflagellate toxins occur each year. While toxic plankton blooms appear regularly on a seasonal basis in temperate waters of Europe, North America and Japan, until recently they were unknown in Australian waters. Tasmania was the first Australian state to suffer problems with toxic dinoflagellates contaminating the shellfish industry. In 1986, dense blooms of the chain-forming species *Gymnodinium catenatum*, a species causing PSP, resulted in temporary closure of 15 Tasmanian shellfish farms. In 1988, the dinoflagellate *Alexandrium catenella*, which causes PSP, caused limited toxicity in wild mussels from Port Phillip Bay, but fortunately no commercial shellfish farms were affected. Since 1986, blooms of *Alexandrium minutum* have occurred annually in the Port River area near metropolitan Adelaide. Ciguatera poisoning caused by the coral reef dinoflagellate *Gambierdiscus toxicus* poses an increasing danger in the Great Barrier Reef region.

Four possible explanations for this apparent increase in red tide problems have been suggested: increased scientific awareness of toxic species; increased utilisation of coastal waters for aquaculture; stimulation of plankton blooms by coastal eutrophication; and accidental introduction of toxic dinoflagellates to new areas. Recent discovery of resistant resting cysts in ships' ballast water suggests that international shipping could be distributing toxic dinoflagellates around the world. Introduction of toxic species may also be associated with the movement of shellfish stocks from one area to another.

Red tides occur when explosive plankton growth produces so many algal cells that they discolour the water. This bloom of the harmless dinoflagellate *Noctiluca scintillans* occurred in Lake Macquarie, New South Wales. Blooms of toxic algae have recently caused alarm in Port Phillip Bay and the Gippsland Lakes, Victoria

---

(Chapter 43). Dinoflagellate species lacking chloroplasts are predatory, capturing smaller cells and ingesting them. Several predatory species have feeding tentacles that pierce prey and suck out the contents. An extraordinary feature of certain dinoflagellates is their 'eye'. The eye-like structure has a refractile lens that changes shape, seeming to focus images onto a light-sensitive retinoid. Dinoflagellates may be able to 'see' their prey.

Several characteristics distinguish dinoflagellates as a protist phylum. Dinoflagellate DNA appears to be permanently condensed (like a prophase nucleus) and is complexed with proteins that are different from typical eukaryotic histones. Originally thought to be a primitive feature described as *mesokaryotic* (intermediate between prokaryotic and eukaryotic), it is now thought that dinoflagellates lost their histones secondarily. Molecular studies of rRNA sequences suggest that dinoflagellates are closely related to ciliates and apicomplexans (see p. 762).

Dinoflagellates are alveolates with two flagella, one of which encircles the cell. Many are photosynthetic, containing chlorophylls *a* and *c*, and some are predatory.

Small but deadly: phylum Apicomplexa

There are at least 5000 species of **Apicomplexa**, all of which are endoparasites of animals. Previously classified as Sporozoa, apicomplexa are now named for their **apical complex**, a structure involved in penetration of host cells. The apical complex is a conical arrangement of microtubules and secretory structures that release a lytic product thought to break down the host cell. The parasite then enters the host through this opening.

Apicomplexa include **gregarines**, **coccidians** and **haematozoa**. Gregarines, which only parasitise invertebrates, are probably the most primitive apicomplexans. Coccidia, which cause coccidiosis and toxoplasmosis in humans, can infect both invertebrates and vertebrates. Some coccidia alternate between a vertebrate host and an invertebrate host. Coccidia leave one host as spores in the faeces and remain in the open environment until they can infect the second host. Humans can contract toxoplasmosis by ingesting spores of *Toxoplasma* present on the fur of cats carrying the infection. Haematozoa, the most advanced apicomplexans, invade blood cells of vertebrates where they feed on haemoglobin. The most notorious haematozoan is *Plasmodium*, the causal agent of malaria (Box 35.5). Like coccidia, haematozoa also alternate between vertebrate and invertebrate hosts, and have efficient ways of effecting cross-transfer between host species.

> Apicomplexa are endoparasites of animals. They parasitise two hosts and cause diseases such as malaria. An apical complex is used to penetrate host cells.

BOX 35.5 MALARIA

An estimated 200–400 million people suffer from malaria, and each year the disease kills about 1 million people. The life cycles of species of *Plasmodium* involve two different hosts, a vertebrate and a blood-sucking insect. The parasite is transferred from one host to the other when insects suck blood from vertebrates. Humans become infected with *Plasmodium* by the mosquito *Anopheles*.

As **sporozoites** (the stage that has the apical complex), *Plasmodium* cells pass into the human bloodstream from the salivary glands of the mosquito. The sporozoites quickly move to the liver and undergo asexual reproduction to produce numerous **merozoites**, which feed in red blood cells. Merozoites divide synchronously, and every 48 or 72 hours they induce lysis of red blood cells, causing release of toxins and the cycles of fever

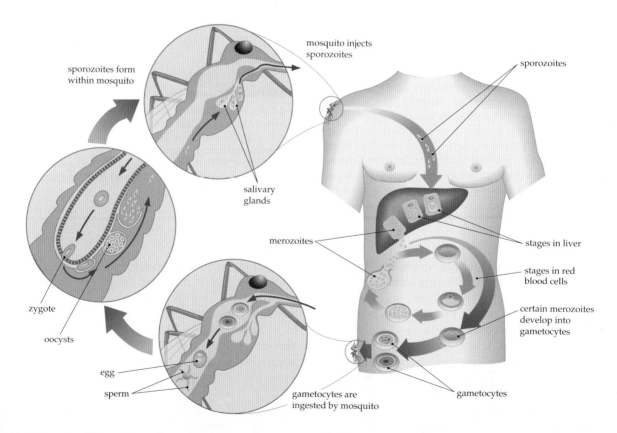

(a) Life cycle of *Plasmodium*, the apicomplexan that causes malaria

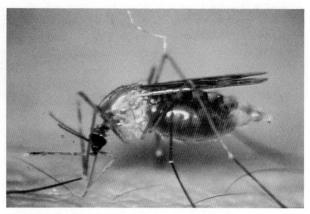

(b) Blood-sucking insects, like this *Anopheles* mosquito, are both host and vector for parasitic haematozoans

and chills characteristic of malaria. Merozoites produce the next cell type, gametocytes. **Gametocytes** are ingested by mosquitoes sucking blood from a malaria sufferer. They pass into the mosquito's gut where they develop into sperm and eggs, which fuse to form a zygote. The zygote then undergoes meiosis to produce new haploid sporozoites, which move into the mosquito's salivary glands to complete the cycle.

The transfer from one host to the other can be risky, and it is here that haematozoa excel. When a female mosquito takes her meal of blood from a vertebrate, she infects the vertebrate with haematozoa. The parasites multiply in the vertebrate and are then available to back-infect the next generation of mosquitoes. By exploiting the relationship between blood-sucking insects and vertebrates, haematozoa ensure their own reproduction and distribution. They also reduce the defence strategies available to hosts since only a part of the life cycle occurs in each host.

Ciliates: phylum Ciliophora

Ciliates are unicellular organisms, ranging from 10 μm to 3 mm in length, with numerous cilia on the surface (Fig. 35.20). Cilia are often arranged in clusters, **cirri**, that beat synchronously and work as paddles or feet for the cell. Sometimes cilia are arranged evenly over the surface and beat in rhythmic waves to propel cells. Cilia are all embedded in a *cortex* of protein fibres that cross-link the network. Elements of the cortex are contractile and, like miniature muscles, they can modify shape in certain species.

There are about 7500 species of ciliates and most are predatory, bacteria being the favoured prey. Prey are driven into an invagination, the *buccal cavity*, through which they are ingested. Undigested material is excreted through the *cytoproct* (Fig. 35.21).

Ciliates have two types of nuclei, a *micronucleus* and a *macronucleus*. Cells usually have one of each type but some cells have several. The micronucleus is diploid, contains normal chromosomes with all the

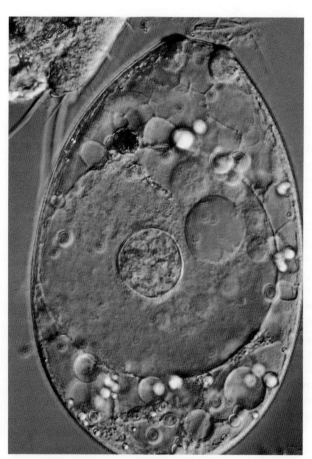

(b)

Fig. 35.20 Ciliates. **(a)** Scanning electron micrograph of a ciliate from the Pacific Ocean. The cilia are distributed uniformly across the surface. **(b)** A ciliate is a unicell that is capable of a variety of activities. It moves about, senses the environment, responds to stimuli and captures prey. All these functions must be managed by the individual cell, not by numerous complex organs as occurs in multicellular animals

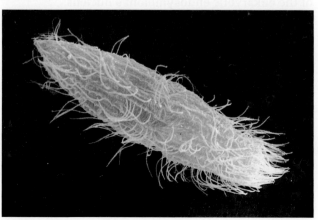

(a)

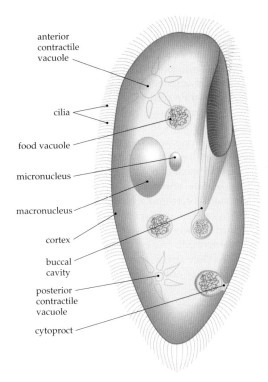

anterior
contractile
vacuole

cilia

food vacuole

micronucleus

macronucleus

cortex

buccal
cavity

posterior
contractile
vacuole

cytoproct

Fig. 35.21 Diagram of a ciliate. Food particles are wafted into the buccal cavity by the surface cilia and ingested through the cytostome into food vacuoles for digestion. Undigested material is ejected via the cytoproct

ciliate's genes, and divides mitotically. The macronucleus develops from the micronucleus and contains multiple copies of genes on relatively short pieces of DNA, some of which are circular. Macronuclei divide amitotically, simply by pinching into two approximately equal halves. Because the macronucleus contains multiple copies of the ciliate's genes (sometimes several thousand copies) it is not essential that the genetic material be accurately apportioned, as occurs in mitosis. Messenger RNAs are transcribed from the multiple gene copies in the macronucleus while the master genes in the micronucleus are not transcribed. The macronucleus seems to be the working copy of the DNA blueprint, while the micronucleus is retained as a master copy to be passed on during the sexual process. This is analogous to making several working copies of a computer program for daily use, while keeping the original program safely stored away.

Ciliates reproduce asexually by binary fission along the short axis of the cell. Sexual reproduction is by conjugation. Two cells of opposite mating type become attached and over a period of several hours they exchange haploid versions of their micronuclei produced by meiosis. The haploid micronuclei fuse to regenerate diploid micronuclei. After conjugation, the macronuclei degenerate and new macronuclei are produced from the recombinant micronuclei. Conjugation in ciliates is essentially a form of sex

in which haploid nuclei from two individuals are brought together without ever needing to form gamete cells.

> Ciliates are predatory unicells with numerous cilia and two types of nuclei. They reproduce sexually by conjugation.

PROTISTAN PIRATES

Recent work has demonstrated that two groups of protists have stolen the ability to photosynthesise from chloroplast-bearing cells. By cannibalising parts from photosynthetic prey, they acquired chloroplasts and adopted an autotrophic way of life.

Flagellates with second-hand chloroplasts: phylum Cryptophyta

Cryptomonads have a small anterior invagination (the 'crypt') into which their two flagella are inserted. They are unicellular and usually reproduce asexually. All genera, except *Goniomonas*, which is heterotrophic, possess a chloroplast. *Chilomonas* is also heterotrophic but contains a leucoplast. Chloroplasts have chlorophylls a and c_2 plus a phycobilin pigment, either phycocyanin or phycoerythrin. The product of photosynthesis is stored outside the chloroplast as starch.

Cryptomonads have a second small nucleus associated with the chloroplast. They apparently obtained their chloroplasts by endosymbiosis, like other photosynthesising protists. However, the endosymbiont was not a prokaryote but a photosynthetic eukaryote. Cryptomonads appear to have acquired the capacity to photosynthesise second-hand by cannibalising a eukaryote that had already formed a permanent association with a prokaryote (Fig. 35.22). The second nucleus associated with the cryptomonad chloroplast is the remnants of the eukaryotic endosymbiont's nucleus.

Amoebae with second-hand chloroplasts: phylum Chlorarachnida

This phylum includes only one genus, *Chlorarachnion* (Fig. 35.23), which exists principally as an amoeboid **plasmodium**, individual cells being linked by a network of cytoplasmic strands, **reticulopodia**. The plasmodial network captures small prey, which are ingested. When starved, the plasmodium separates and forms individual walled cysts that release uniflagellate swarmers. These swarmers regenerate a plasmodium.

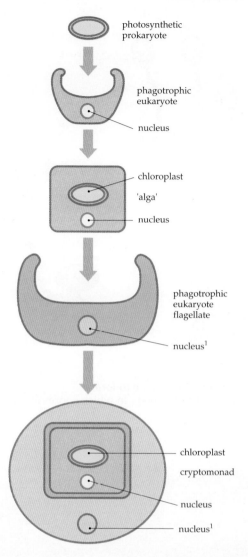

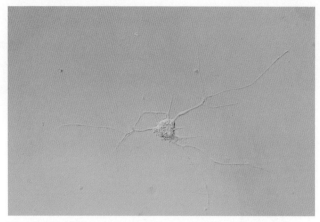

Fig. 35.23 With its amoeboid movement, *Chlorarachnion reptans* (phylum Chlorarachnida) is more like an animal than a plant, but the cells have grass-green chloroplasts with which they are able to photosynthesise

Not all eukaryotes with chloroplasts are close relatives. Photosynthetic cryptomonads and *Chlorarachnion* stole chloroplasts from protist prey cells.

SLIME MOULDS

Slime moulds are amoeboid protists that produce fruiting bodies, **sorocarps**, as part of their life history. They were often classified as fungi because they absorb nutrients directly from the environment, but this is the only similarity to fungi. The term **slime mould** refers to the habit of the most conspicuous part of the life cycle, which is a small slimy mass.

Cellular slime moulds

You could perhaps mistake a cellular slime mould for a minute slug if you found one creeping across the forest floor. The 'slug' or **pseudoplasmodium** is a mass of amoebae that have aggregated to form a single perambulatory colony. The amoebae, which are normally free-living individuals that prey on bacteria, congregate when their food supply runs short and move off collectively as a 'slug'. Having found a suitable location, the slug differentiates into a fruiting body that produces numerous spores (Fig. 35.24). Spores are released and eventually produce amoebae, completing the life cycle.

Cellular slime moulds (phylum Acrasea, phylum Dictyostelida and the informal group protostelids) inhabit damp places in forests and can usually be found on rotting plant material or animal dung. The amoebae are often referred to as **myxamoebae** (slime amoebae) to distinguish them from normal amoebae, but the two may be closely related. Most cellular slime moulds do not have flagella.

Fig. 35.22 Cryptomonads are animal-like flagellates that acquired the capacity to photosynthesise by enslaving an algal cell, which now exists as a permanent endosymbiont within the cryptomonad host. Double endosymbiosis explains the origin of cryptomonad chloroplasts. The primary endosymbiosis involved a eukaryote engulfing a prokaryote to create a eukaryotic alga. The second endosymbiosis resulted in the alga being engulfed by another eukaryotic phagotroph. The second nucleus in the cryptomonad cell is apparently the remnants of the algal endosymbiont's nucleus

Chlorarachnion is also photosynthetic and each amoeba has several grass-green chloroplasts containing chlorophylls *a* and *b*, like green algae. *Chlorarachnion* chloroplasts are different, however, in that they are bounded by four membranes rather than two, and no starch is stored within the chloroplast. Associated with the *Chlorarachnion* chloroplast is a nucleus-like structure, suggesting that *Chlorarachnion* acquired its chloroplast by engulfing a photosynthetic eukaryote, presumably a green alga.

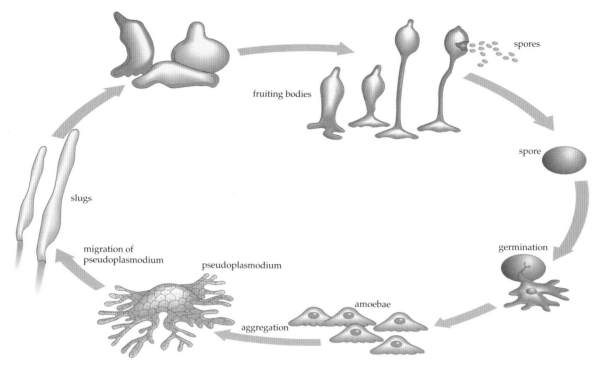

Fig. 35.24 Life cycle of the cellular slime mould *Dictyostelium discoideum*. Amoebae aggregate to form a pseudoplasmodium. They move off together as a slug, eventually forming a fruiting body in which spores are produced

Acellular slime moulds: phylum Myxomycota

Whereas the pseudoplasmodium of cellular slime moulds consists of numerous individual cells aggregated together, the **plasmodium** of a myxomycete is one large multinucleate cell. The plasmodium resembles a slimy scum, sometimes vivid yellow or orange in colour (Fig. 35.25), and is the major feeding stage, absorbing organic matter and ingesting bacteria and other micro-organisms. Should the plasmodium encounter a nutrient-poor region or other adverse environmental conditions, it differentiates into a fruiting body (Fig. 35.26), with cells dividing by meiosis to produce haploid spores. Spores germinate to produce haploid amoebae, which are the gamete stage. In the presence of sufficient water, they convert to biflagellate forms. Two amoebae (or two biflagellates) fuse to form a zygote. The diploid nucleus of the zygote divides mitotically but no cell membranes separate the daughter nuclei, resulting in a multinucleate plasmodium.

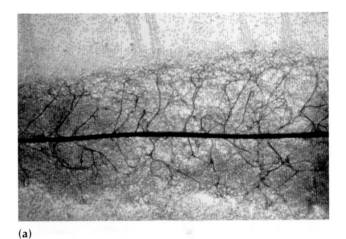

(a)

(b)

Fig. 35.25 Acellular slime moulds. **(a)** Plasmodium of the acellular slime mould *Stemonitis fusca*. **(b)** The brilliant yellow plasmodium of *Physarum polycephalum* coats plant leaves

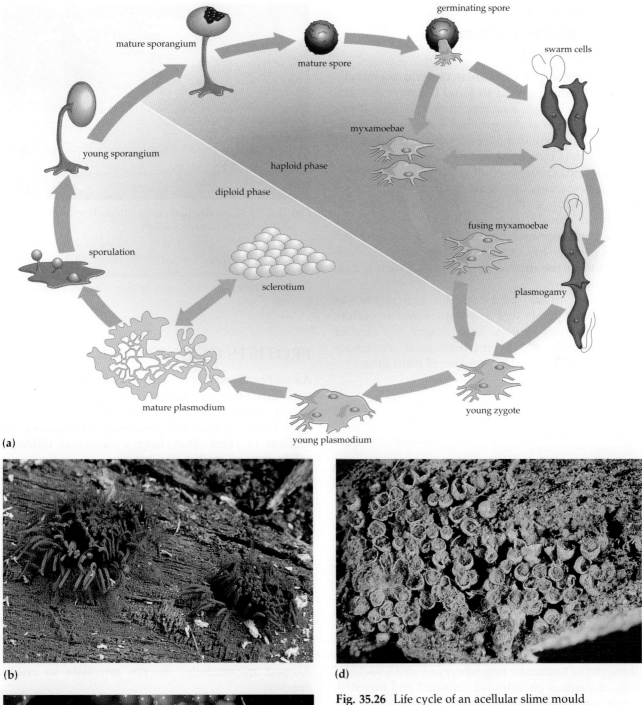

(a)

(b)

(c)

(d)

Fig. 35.26 Life cycle of an acellular slime mould (myxomycete). **(a)** A plasmodium is multinucleate and, in response to lack of food, forms a fruiting body (sporangium). Cells divide by meiosis to form haploid spores, which germinate as amoebae. They convert to biflagellate cells, which fuse to form a diploid zygote. **(b)** The sporangia of *Stemonitis fusca* take the form of tufts of brown threads on a log of wood. **(c)** Fruiting bodies of *Arcyria* are brilliant orange and of **(d)** *Trichia* look like little cups

Slime moulds are amoeboid protists, which aggregate to form colonies, either cellular or acellular, with fruiting bodies that produce spores.

FUNGAL-LIKE PROTISTS

Fungal cousins: phylum Chytridiomycota

Chytrids are either unicellular or form coenocytic threads (hyphae) similar to fungi (Chapter 36). The cell wall of chytrids contains chitin, also like fungi. Chytrids and fungi also produce lysine by the same pathway (the aminoadipic pathway), and molecular sequence data indicate that they are related.

Motile cells (zoospores) of chytrids have a single, smooth, posteriorly directed flagellum. A second basal body, without a flagellum, is located near the functional basal body. Although hyphae are coenocytic, a septum forms to isolate the sporangium. Spores produce haploid uniflagellate gametes, which fuse to form the diploid zygote from which a new thallus develops.

Chytrids cause problems in agriculture as pathogens of plants such as potatoes, and as vectors for several plant viruses. *Neocallimastix* is a chytrid living anaerobically in the rumen of vertebrate herbivores, where it aids digestion of plant fibre.

Chytrids are fungal-like protists, with cell walls containing chitin. They include pathogens of plants and symbionts of herbivorous vertebrates (digesting cellulose in the gut).

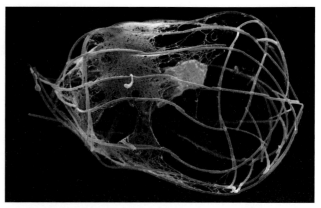

Fig. 35.27 Choanoflagellate cells (collar flagellates) live within a basket of silica strips

Choanoflagellates are marine protistans that eat bacteria and detrital particles. They resemble sponge collar cells and choanoflagellates and sponges are close relatives.

PROTISTS OF UNKNOWN AFFINITY

A number of protist groups are not known to be allied with any other eukaryotes. They may represent a number of early, divergent eukaryotic lineages that have been evolving independently for hundreds of millions of years. They have a variety of different cell forms and modes of nutrition.

SPONGE-LIKE PROTISTS

'Collar' flagellates: choanoflagellates

Choanoflagellates are free-living, usually unicellular heterotrophs found in marine, brackish and freshwater environments. The cell has a single flagellum that is surrounded by a ring of tentacles. If the choanoflagellate is sessile (attached to a substrate by a stalk), the flagellar beat draws water through the tentacular ring where any small bacterial cells or detritus particles are captured and ingested. Some choanoflagellates swim freely using the flagellum to push them through the water. Cells are small (less than 10 μm) but they are surrounded by a basket-shaped enclosure, the *lorica*. The choanoflagellate lorica is composed of several silica strips cemented together and surrounded by a membranous web (Fig. 35.27). Reproduction is asexual and the parent cell releases a smaller juvenile cell. In some forms, the juvenile cell inherits some silica strips from the parent lorica and uses them to commence construction of its own lorica.

Collar cells (choanocytes) of sponges (Chapter 38) bear a striking resemblance to choanoflagellates, and sequence data show that they are close relatives.

Radially symmetrical unicells: phylum Actinopoda

Actinopoda are characterised by **axopods**, long slender radial projections. Axopods contain a thin layer of cytoplasm bounded by plasma membrane, and are reinforced with a highly ordered bundle of microtubules. Axopod microtubules collectively form an *axoneme*, which should not be confused with the microtubules of flagella and cilia given the same name. Axopod microtubules do not interslide to create bending.

The main function of axopods is prey capture. Food particles stick to their surface and are transported to the cell for ingestion. In one group, *Sticholonche*, axopods are modified to function as oars and 'row' the cell through the water. The axoneme microtubules of these oar-like axopods are attached to the nucleus by ball-and-socket articulations, and the axopod is moved by co-ordinated contraction/relaxation of non-actin fibres that interconnect the axopods.

The cells of actinopods are highly variable in organisation and are often partitioned into inner and outer zones. The outer zone can harbour zooxanthellae. Some actinopods are amoeboid and others produce flagellate swarmers. Skeletons can be composed of organic material, accreted sand particles and diatom valves, celestite (strontium sulfate) or silica with traces of magnesium, copper and calcium, depending on the class of actinopod. Skeletons form fossils and huge deposits of 'radiolarian ooze', a sludge found on the ocean floor. Like diatom valves, actinopod skeletons are metamorphosed to chert with time and no extremely old fossils are known. The best known actinopods are radiolarians (Fig. 35.28), called sun animalcules because they resemble a minuscule sun with radiating rays.

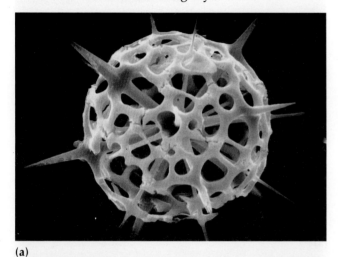

(a)

(b)

Fig. 35.28 Actinopods include radiolarians, such as (a) *Dictyacantha* and (b) *Trizona*, which produce spectacular siliceous skeletons that accumulate on the sea floor forming 'radiolarian ooze'

Actinopods have radial skeletons and projections known as axopods with which they capture food.

Amoebae: subphylum Sarcodina

Sarcodines are **amoebae** that are able to produce transiently extensions of the cell surface, **pseudopodia** ('false feet'), involved in locomotion or feeding (Fig. 35.29). One of the first amoebae to be named was *Amoeba proteus* after the sea god Proteus of Greek mythology who could change his shape at will. Many sarcodines are naked but some produce internal or external skeletons. Most species are unicellular and uninucleate. Sarcodines are ubiquitous in aquatic habitats where they prey on bacteria and other protists. The parasitic form *Entamoeba* lives in the human alimentary canal where it causes amoebic dysentry. Another form, *Acanthamoeba*, was recently shown to infect the human brain, causing encephalitis, and the eye, causing keratitis. Amoebae in soil can play host to pathogenic bacteria such as *Legionella*.

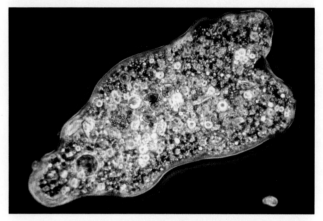

Fig. 35.29 A characteristic trait of sarcodines (amoebae) is their ability to transiently alter cell shape to produce pseudopodia. *Amoeba villosa* has several pseudopodia ('false feet') projecting from the cell in different directions

There are two main groups of sarcodines: **rhizopods** and **foraminifera**, which have different pseudopodia. Foraminiferan pseudopodia are reticulated (one pseudopod connects to others) and contain small granules.

Foraminifera are mostly marine heterotrophs that produce calcareous ($CaCO_3$) or arenaceous shells (*tests*). Tests can be quite elaborate and have multiple chambers for flotation. The chambers are often arranged in a spiral. Some foraminifera are particularly large (up to 12 cm in diameter in the case of *Nummulites*) and contain many symbiotic algae. The shells are mini-greenhouses with algal endo-symbionts housed in thin-windowed chambers around the surface to collect light. The spiny foram-inifera *Globigerina* (Fig. 35.30) acts as 'shepherd' to a 'flock' of dinoflagellate symbionts. At night, algae are harboured safely inside the foraminifera's shell, but each morning they venture out along the spines into the sunlight to photosynthesise.

About 40 000 species of foraminiferans have been described but 90% are extinct and known only from fossil shells up to 600 million years old. Foraminiferans were once so numerous that deposition of their skeletons produced large chalk deposits. When

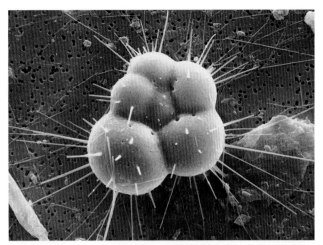

Fig. 35.30 The foraminiferan *Globigerina* has a calcareous shell through which pseudopods extend

building the great pyramids, Egyptian engineers noticed that the limestone blocks contained numerous nummulites (Fig. 35.31), fossil remnants of the large foraminifera *Nummulites gizehensis*. Foraminifera fossils are indicators of geological strata and are used extensively by the petroleum industry to characterise sediments in the search for fossil fuels. They occur in great abundance both in plankton and on the sea bed down to 10 000 m deep.

Fig. 35.31 Nummulites are coin-shaped foraminiferans. These examples with holes in the middle are from Great Keppel Island beach and measure approximately 1 cm in diameter, but much larger forms are known

Sarcodines (amoebae) can transiently alter their shape. Most are heterotrophs but some have algal endosymbionts. Foraminiferans are marine amoebae with shells, often housing photosynthetic symbionts.

Living as commensals: opalinids and proteromonads

Opalinids (phylum Zoomastigina, class Opalinata) are large, unicellular and multinucleate, and have numerous flagella. They are shaped like cigars or the fin of a surfboard (scalene triangle), and their surface is covered by rows of short flagella (Fig. 35.32). Between the flagellar rows are surface folds supported by ribbons of microtubules. First found in frog faeces by Leeuwenhoek, opalinids live mostly in the rectum of ectothermic vertebrates. The relationship between host and opalinid is commensal, meaning that the opalinid uses the host as a place to live without apparently causing harm (Chapter 43). Opalinids feed by pinocytosis.

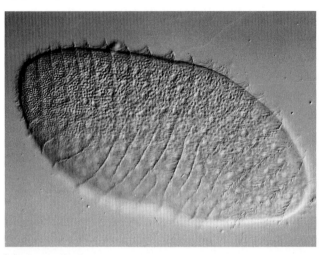

(a)

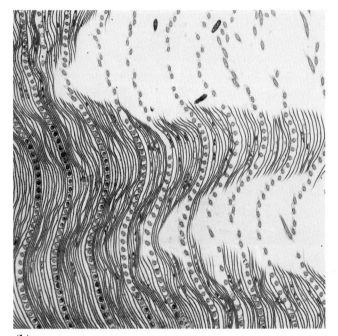

(b)

Fig. 35.32 Opalinids are covered with rows of flagella, between which are numerous flaps of the plasma membrane. Opalinids superficially resemble ciliates but do not have a cortex interconnecting the flagella

Proteromonads (class Proteromonadida) have either two or four flagella and are uninucleate, spindle-shaped cells. They occur as commensals in the rectum of amphibians, reptiles and rodents. The surface of the cell has microtubule-supported folds like opalinids. On the cell surface are hairs similar to the flagellar hairs of heterokonts and opalinids.

Symbionts *par excellence:* parabasalids

Parabasalids (phylum Zoomastigina, class Parabasalia) are uninucleate flagellates involved in commensal or parasitic relationships with animals. They typically have a parabasal body, a large Golgi-type complex beside the basal body. An **axostyle**, a stiff rod-like bunch of microtubules, runs the length of the cell. *Trichomonas vaginalis* is a parabasalid that infects the human genital tract. A relatively benign sexually transmitted disease, *T. vaginalis* is estimated to infect 3.5% of the world's population. Two types of parabasalids (*Trichonympha* and *Mixotricha*) are symbionts in termite guts where they are responsible for digestion of wood. *Trichonympha* has several thousand flagella. *Mixotricha* has only four eukaryotic flagella but also has thousands of spirochaete bacteria attached to its surface that create propulsion (Fig. 35.33).

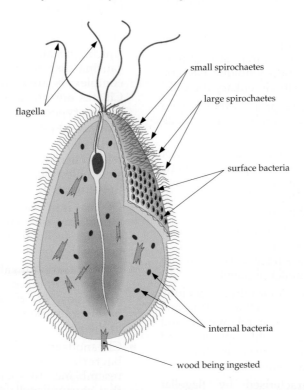

Fig. 35.33 The parabasalid *Mixotricha paradoxa* is a symbiont *par excellence*. The cell is actually a co-operative involving as many as 500 000 individual organisms. The host cell is a quadriflagellate eukaryote. On the surface are two forms of spirochaete bacteria that propel the cell. The spirochaetes attach to the cell surface via anchor bacteria embedded in the host cell membrane. Numerous internal bacteria within the host cell aid metabolism. The entire collection is an endosymbiont within the gut of Australian termites and is responsible for the digestion of wood

SUMMARY

- Protists are a diverse group of eukaryotic organisms. Unicellular forms are the most prevalent, but many groups include colonial or multicellular forms. The majority are aquatic and have flagella or cilia. Protists have a wide variety of ways of gaining nutrition, including photosynthesis, parasitism, predation and absorption.

- The first eukaryotic organisms were probably similar to modern-day protists. The nuclear membrane and endomembrane system probably evolved from invaginations of the bacterial cell membrane that enveloped the nucleoid. Chloroplasts and mitochondria are almost certainly derived from endosymbiotic bacteria that have become organelles in eukaryotic cells.

- Archaezoans are primitive protists that have a nucleus but lack mitochondria; some lack endomembrane systems.

- Photosynthetic protists are diverse and are not all related. Glaucophytes are photosynthetic flagellates with apparently primitive chloroplasts (cyanelles) that have a peptidoglycan wall, like bacteria.

- Red algae (phylum Rhodophyta) are familiar seaweeds. Most are multicellular and macroscopic, and they lack flagella. They contain chlorophyll *a* and phycobilin pigments.

- Green algae (phylum Chlorophyta) include unicellular, colonial and multicellular forms, one group of which is the closest relative of land plants. Chloroplasts contain chlorophylls *a* and *b*, the product of photosynthesis is stored as starch, and cell walls are cellulosic.

- Heterokonts are characterised by flagellar architecture. They have one smooth flagellum directed posteriorly, and one hairy flagellum directed anteriorly. They include a number of phyla: Chrysophyta (golden flagellates), Prymnesiophyta (such as coccoliths), Bacillariophyta (diatoms), Phaeophyta (brown algae, the largest protistans) and Oomycota (water moulds and downy mildews). Most are photosynthetic, with chlorophylls *a* and *c*, but oomycetes absorb food through filamentous hyphae.

- Euglenoids and kinetoplasts are related, flagellated cells. Euglenoids are free-living, some of which have chloroplasts and some of which engulf prey through an anterior gullet. Kinetoplasts are parasitic with unique mitochondria.

- Alveolates are unicells with distinctive vesicles, cortical alveoli, beneath the cell membrane. They include dinoflagellates, apicomplexans and ciliates. Dinoflagellates have two flagella, one of which encircles the cell. Many are photosynthetic, containing chlorophylls *a* and *c*, and some are predatory. Apicomplexans are endoparasites of animals, causing diseases such as malaria. An apical complex is used to penetrate host cells. Ciliates are predatory unicells characterised by two types of nuclei and many cilia.

- Photosynthetic cryptomonads (biflagellates) and chlorarachniophytes (amoebae) are not related to other photosynthetic protists. They stole chloroplasts from eukaryotic cells that they ingested.

- Of the non-photosynthetic protists, slime moulds are amoeboid and aggregate to form colonies (cellular or acellular) with fruiting bodies that produce spores.

- Chytrids are fungal-like protists, with cell walls containing chitin. They include pathogens of plants and symbionts of herbivorous vertebrates (digesting cellulose in the gut).

- Choanoflagellates are marine protists that eat bacteria and detrital particles. Their resemblance to sponge collar cells suggests that choanoflagellates and sponges are close relatives.

- Protists of unknown affinity include actinopods, sarcodines, opalinids and parabasalids. Actinopods are cells with radial skeletons and projections (axopods) with which they capture food. Sarcodines are amoebae, most of which are heterotrophs but some, such as foraminiferans, have algal endosymbionts. Opalinids and parabasalids are commensals and parasites.

QUESTIONS

1. What major technological advances have allowed biologists to discover great diversity and previously unrecognised evolutionary relationships in the protists? Why is the kingdom Protista considered to be an artificial taxon?

2. What is the significance of protists that lack mitochondria and endomembrane systems? Give one example.

3. Make a table comparing the main distinguishing features of red, green and brown algae. Why are brown algae called heterokonts and what are their closest relatives? What organisms are green algae most closely related to? What useful substances are obtained from red, green and brown algae?

4. What features would enable you to identify a protist as (a) a euglenoid or (b) a dinoflagellate? Describe how each of these protists moves and how they gain their nutrition.

5. What are red tides? Why have they become a problem in Australia?

6. Which protist causes Jarrah dieback? What features of this protist explain the extensive spread of the disease through Australian forests? How would you control the spread of disease in a logging area or national park?

7. What is unusual about the genetic material in ciliates? How do ciliates reproduce sexually?

8. What organism causes malaria? Describe the life cycle of this protist and suggest why parasitising two different hosts may have been a selective advantage?

9. What type of protist is a trypanosome? How do these parasites avoid being eliminated by the immune system when they are in the bloodstream of their host?

10. In this chapter you have seen that 'flagellate' and 'amoeboid' cell types occur in photosynthetic, fungal-like and animal-like categories of protists. This suggests that these cell types are either primitive or that they evolved more than once. Discuss the role of endosymbiosis in the origin of chloroplasts and mitochondria and in the diversification of protists.

F U N G I

Have you ever returned from holidays to find that your fruit bowl or refrigerator has turned into some kind of living science experiment? Have you wondered about those bright-coloured mushrooms that grow on rotting logs in the forest? Have you eaten bread and wondered why it is full of air bubbles? How do bubbles get into beer and champagne? What produces alcohol in beer and wine? Was Joan of Arc's religious experience intensified by eating bread made from flour containing hallucinogenic drugs? The answers to all of these questions involve members of the kingdom Fungi.

WHAT IS A FUNGUS?

Fungi are eukaryotic heterotrophs that digest their food externally. Because of external digestion, fungi are secretory organisms—they secrete enzymes onto their substrate for digestion as well as a variety of other substances. These secretions may increase their competitive ability in environments usually teeming with other organisms. Many of these secretions have immense importance to humans as antibiotics and toxins.

The body of a fungus, **mycelium**, generally grows as filamentous **hyphae** (s.: hypha), which are microscopic tubes of cytoplasm bounded by tough, waterproof cell walls (Fig. 36.1). Hyphae extend apically, branch and, theoretically, have an unlimited life. Fungi reproduce by spores (Fig. 36.2). During their life cycles, fungi may undergo several changes in ploidy and morphology. Their vegetative cells usually contain more than one haploid nucleus, and the diploid state is typically very brief. During mitosis the nuclear envelope remains intact.

Fungi occupy all habitats in the biosphere, including land, sea, water and air. They tolerate a wide range of environmental extremes, which means that they

can be found in any environment where life is possible. They may be saprophytes, which play a crucial role in litter decomposition and nutrient recycling, or parasites, including many important plant and animal pathogens.

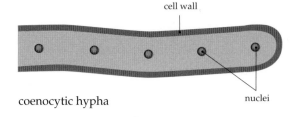

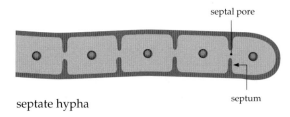

Fig. 36.1 Coenocytic and septate fungal hyphae

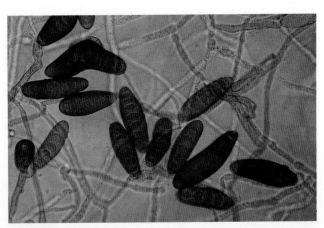

Fig. 36.2 (a)

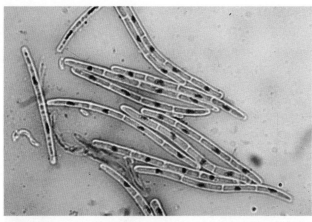

(b)

(c)

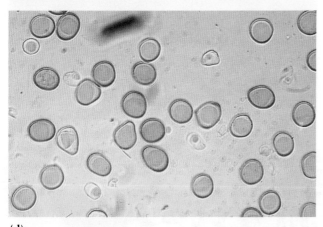

(d)

Fig. 36.2 Fungal spores come in many shapes:
(a) *Bipolaris sorokiniana* (Deuteromycota) conidia;
(b) *Septoria tritici* (Deuteromycota) conidia stained to
reveal nuclei; (c) *Fusarium graminearum*
(Deuteromycota) conidia; and (d) *Tilletia foetida*
(Basidiomycota) urediniospores

EVOLUTION OF FUNGI

There are at least 69 000 named species of fungi,
although this probably represents only 5% of the real
number, estimated at about 1.5 million species. Fungi
are classified into five groups: three natural groups

recognised by their different sexual life cycles—the
phyla Zygomycota, Ascomycota and Basidiomycota;
one group of anamorphs that has no sexual stage—
the Deuteromycota; and one group, the lichens, which
only exist in a mutualistic symbiosis—the
Mycophycota.

Recent evidence from nucleic acid sequencing and
developmental studies has shown close affinities
between chytrids (kingdom Protista, phylum Chytridiomycota, Chapter 35) and fungi. It seems likely that
fungi and chytrids evolved from a common ancestor.
A previous suggestion, based on similarities in morphology and life cycles, that fungi evolved from a
primitive rhodophyte (red alga) has not been supported by ribosomal DNA sequence analysis. Much
more detailed comparative molecular, biochemical
and morphological work needs to be completed
before a definitive answer emerges.

The most widely accepted view is that the kingdom
Fungi represents a distinct evolutionary lineage (Fig.
30.9). Morphological similarities with other organisms,
such as the Rhodophyta and Oomycota (Chapter 35),
demonstrate convergent evolution toward the fungal
way of life.

> Fungi are eukaryotic heterotrophs that digest their food
> externally. They are probably related to chytrids among
> the protists.

FUNGAL GROWTH

Vegetative growth

Fungal cells are bound by a cell wall that encloses
the cell membrane, nucleus, organelles and cytoplasm
(Fig. 36.3). They contain all the usual eukaryote cell
machinery except chloroplasts. Except in some
specialised structures, hyphae grow as strands one
cell thick.

Fungi grow and explore their substrates by
extension from the hyphal tip. As a hypha absorbs
nutrients and swells, it expands at the tip, where the
cell wall is incomplete and elastic. Branches also
develop at the growing tip of hyphae. As a consequence of this type of tip growth, fungal colonies
grow uniformly outwards by radial extension.

A mass of branched and apparently tangled hyphae
is a mycelium. Macrofungi, such as lichens, mushrooms, puffballs and bracket fungi, produce a partially
differentiated **pseudoparenchyma**. A few fungi, the
yeasts, grow as single cells rather than as hyphae.
These grow by budding.

Fungal cell walls are produced from prefabricated
subunits that are synthesised in the Golgi apparatus.
These subunits are enclosed in membrane-bound
vesicles, which bud off from the Golgi apparatus and
migrate to the hyphal tip. The vesicles and their
contents fuse with the existing membrane and wall

at the tip, and the cell elongates. The walls are made of layers of **chitin** (*N*-acetylglucosamine) microfibrils embedded in a matrix of non-glucose polysaccharides (mannans, galactans and glucomannans).

Chitin is also used by arthropods, such as insects and crustaceans, to form their tough external skeletons, and has a number of industrial uses (e.g. speaker cones) because of its strength and lightness. Fungal cell walls are similar in structure to plant cell walls but differ in that the microfibrils in plant cell walls are cellulose.

Septa (s.: septum) are cross-walls that form after mitosis. Septa grow centripetally inwards from the cell wall, leaving a pore at the centre. These pores allow for nuclear migration and cytoplasmic continuity (Fig. 36.3). In this sense, fungi are not truly multicellular like plants. Complete septa form only to separate spores, sexual reproductive structures or old hyphae from the actively growing mycelium.

> A mycelium grows as filamentous hyphae, which extend at the tip. Fungal cell walls are composed of chitin microfibrils embedded in a matrix of polysaccharides.

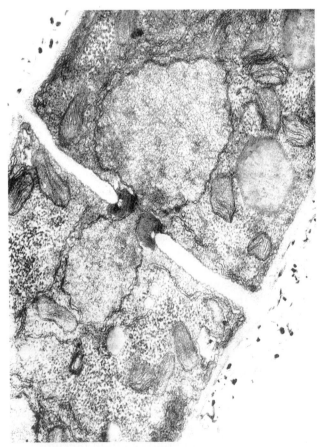

Fig. 36.3 Electron micrograph showing some ultrastructural features of *Neurospora crassa* (phylum Ascomycota). A nucleus is migrating through the simple pore in the septum. In the cytoplasm you can see endoplasmic reticulum, ribosomes, mitochondria, fat globules and crystalline inclusions

Hyphal aggregates

Nearly all fungi produce microscopic spores, sometimes on very beautiful and elaborate fruiting bodies, such as mushrooms and toadstools (Fig. 36.4), or on a mat of hyphae called a **stroma**. These fruiting bodies are made from a tight mesh of hyphal strands, constructed like fibreglass, forming pseudoparenchyma tissue.

(a)

(b)

Fig. 36.4 (c)

(d)

Fig. 36.4 Examples of macroscopic fungal fruiting bodies, all from the phylum Basidiomycota. **(a)** *Amanita muscaria*, the fly agaric, an ectomycorrhizal fungus on many forest trees. **(b)** *Mycena pullata*, a saprophytic species growing in temperate forests. **(c)** *Coprinus disseminatus* growing on the forest floor. **(d)** A giant puffball, *Calvatia gigantea*, from open eucalypt woodland

Some soil-inhabiting fungi produce hard resting bodies, composed of masses of tightly compacted mycelium, which are able to survive dormant in soil for several years. These bodies are **sclerotia** if they are composed only of fungal tissue, or **pseudosclerotia** if they contain host tissue as well. When they germinate they may form a mycelium, or they may bear sexual fruiting bodies that produce and disperse many small spores (Fig. 36.5).

Armillaria luteobubalina (phylum Basidiomycota), which attacks the roots of forest trees, is known as the bootlace fungus because it produces thick, dark **rhizomorphs** made of many hyphae growing together.

Fig. 36.5 The dark resting structure, called a sclerotium, of *Sclerotinia sclerotiorum* (phylum Ascomycota), which is embedded in a fine web of mycelium, has germinated to produce fruiting bodies called apothecia. These apothecia bear the asci and sexual ascospores

These grow faster than individual hyphae, as much as a few centimetres a day, and have tough, melanised walls that are resistant to drying. They allow the fungus to grow from tree to tree very quickly, and penetrate the bark of roots (Fig. 36.6).

(a)

(b)

Fig. 36.6 The bootlace fungus, *Armillaria luteobubalina* (phylum Basidiomycota), is a serious root-rotting pathogen of forest trees. **(a)** It grows between trees as coarse, dark hyphal aggregates called rhizomorphs, which are the bootlace-like threads growing on the surface of this pine trunk. **(b)** It produces fruiting bodies, or basidiocarps, that bear millions of spores

Specialised hyphae

Delicate membrane-bound feeding structures, **haustoria**, are produced by many parasitic fungi. These penetrate the cell walls of host plants, in direct contact with but without damaging the host cell membrane (Fig. 36.7). This allows the parasite to feed off the living host cell without killing it, an important ability for obligate parasites.

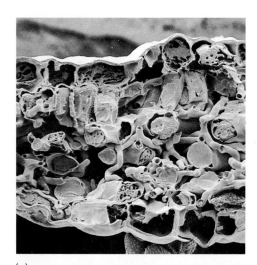

(a)

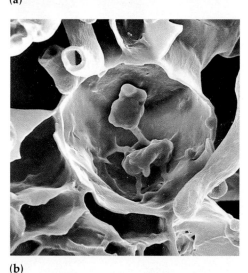

(b)

Fig. 36.7 **(a)** Hyphae of the rust fungus, *Hemileia vastatrix* (phylum Basidiomycota), seen under the scanning electron microscope, growing through a coffee leaf. **(b)** Like many parasitic fungi, this species uses fine membrane-bound structures, haustoria (seen here as club-like intrusions in the mesophyll cells), to penetrate the cells of its host and absorb nutrients without causing severe disruption

Many plant parasitic fungi form **appressoria**, swellings at the tip of hyphae that adhere to the surface of the host, facilitating penetration. Some saprophytic fungi form **rhizoids**, short, thin, root-like hyphal branches that anchor them to their substrate. Fine individual stalks that bear spores, **conidiophores**, are also examples of specialised hyphae (Fig. 36.8).

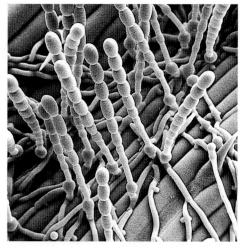

Fig. 36.8 Electron micrograph of conidiophores of the cereal powdery mildew fungus, *Erisyphe graminis* (phylum Ascomycota) bearing conidia

> Fungi produce a wide variety of specialised structures, including spores, sclerotia, rhizomorphs and haustoria, for survival, spread or parasitism.

Fungal nuclei

A number of unusual nuclear conditions exist in the Fungi. Vegetative hyphae normally contain haploid nuclei. A **monokaryon** contains one nucleus per cell. A **dikaryon** contains two haploid nuclei per cell, one from each parent, and is usually formed sexually after cytoplasmic fusion (*plasmogamy*) and before nuclear fusion (*karyogamy*). In basidiomycetes, karyogamy is delayed until just before meiosis, so that the dikaryon is the normal nuclear state of vegetative mycelium.

Multinucleate vegetative cells are called **homokaryons** if the nuclei are the same, and **heterokaryons** if the nuclei are different (Figs 36.9, 36.10).

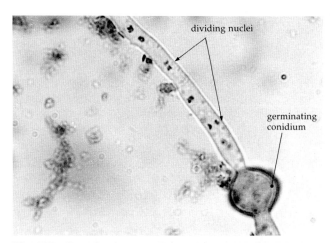

dividing nuclei

germinating conidium

Fig. 36.9 Germinating conidium of *Botrytis cinerea* (phylum Deuteromycota). The cell has several nuclei, in different stages of mitosis. Chromosomes do not all line up along a metaphase plate but separate asynchronously, seen here as a 'double track'

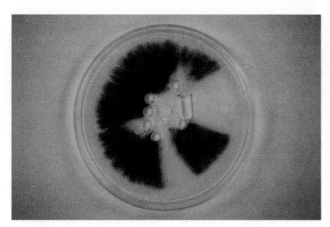

Fig. 36.10 A colony of *Verticillium dahliae* (phylum Basidiomycota) growing in artificial culture. The different coloured sectors are due to hyphal branches that developed from different nuclei in the same heterokaryotic mother cell

Heterokaryons form by two mechanisms. In the first, hyphal fusions, **anastomoses**, are followed by migration of the nucleus from one cell into the other. Such hyphal fusions are common in many fungi. In the second mechanism, a mutation in one of the nuclei in a homokaryon is propagated by mitotic divisions.

Fungal mitosis

Mitosis (Chapter 8) is hard to study in fungi because of their small nuclei and clandestine chromosomes, which are difficult to see under the light microscope. For these reasons, details of the mitotic processes in fungi are not yet completely understood.

In ascomycetes and zygomycetes, the nuclear envelope remains intact and the nucleus remains visible during mitosis. The nucleolus also remains visible throughout most of mitosis. In basidiomycetes, the nucleus sometimes disappears when viewed under the light microscope, suggesting that the nuclear envelope may be disrupted at some stage of mitosis.

In fungi, chromosomes do not line up in a usual metaphase plate, but rather seem to separate asynchronously along the mitotic spindle, giving a characteristic 'double track' appearance at metaphase (Fig. 36.9). Dense **spindle pole bodies**, attached to the nuclear membrane at the spindle poles, are connected to the kinetophores of each chromosome by microtubules. These spindle pole bodies replace the centrioles found in other eukaryotes.

FUNGAL NUTRITION

Nearly all fungi, whether saprophytes or parasites, can use glucose as a carbon source. Some species are also able to use more complex carbon molecules,

requiring the induction of special enzymes. Because fungi can use different carbon sources, there are often complex successions of different fungal species on natural substrates, such as dung. The first fungi to appear are usually fast-growing zygomycetes that reproduce rapidly. Slower-growing species, usually ascomycetes or deuteromycetes, which use cellulose as well as simple sugars, reproduce next. Finally, the slowest-growing species, usually basidiomycetes, which use simple sugars, cellulose and lignin, are able to grow. Once established, these slower-growing species inhibit the growth of competitors.

Basidiomycetes and some ascomycetes are practically the only organisms able to digest lignin, which means they are very important in the decomposition of plant material. They are recorded in the fossil record associated with early land plants during the early Devonian period (Chapter 31). *Phanerochaete* (phylum Basidiomycota), the white rot fungus of wood, produces lignin-degrading enzymes when exposed to nutritional stress, for example, after all the easily obtained carbon and nitrogen sources are exhausted by other fungi. Even then, lignin is only used by this fungus if traces of another carbon source, such as cellulose, are available. It may be that lignin is only degraded to allow the fungus access to cellulose.

Amorphotheca resinae (phylum Ascomycota) can utilise kerosene jet fuel as a carbon source, and was once responsible for grounding the Indonesian Air Force when it contaminated the fuel tanks of jet aircraft. Related species are becoming very important as biodegrading organisms that are able to detoxify chemical pollutants, such as chlorinated hydrocarbons. Other species attack computer floppy discs, or even glass lenses (Box 36.1).

Trichophyton (phylum Deuteromycota) produces keratinases, enzymes that degrade the tough protein, keratin, in hair and skin. Most species are saprophytes that live off dead tissue or tissue products, but some species cause skin, nail and hair diseases.

Apart from carbon, nitrogen is the most common limiting nutrient for fungi. Fungi that decompose leaf litter grow fastest when the carbon:nitrogen ratio is between 10:1 and 20:1. In dry leaf litter this ratio is about 30:1, so nitrogen is often a limiting factor for these fungi. This is why a handful of nitrogenous fertiliser or manure often speeds up decomposition in compost heaps. Fungi also require potassium, phosphorus, magnesium, sulfur, calcium and trace amounts of iron, copper, manganese, zinc and molybdenum, and generally prefer slightly acidic substrates. They are able to synthesise their own vitamins.

Parasitic fungi are often fastidious, and many have never been cultured outside their hosts. Those that have often require specific amino acids or vitamins, and even then do not sporulate properly outside their host. This indicates the degree of dependence resulting from coevolution of host and parasite.

BOX 36.1   FUNGI AND WORLD WAR II

Fungi play an unwanted role in biodeterioration, which causes mould on damp walls and timber rots. Nearly any organic, and even inorganic, material in the tropics may be attacked by moulds. Even glass lenses on cameras and other optical equipment are affected by tropical moulds that produce corrosive organic acids.

Tropical heat and humidity, encountered by Australian soldiers during the Pacific war, provided ideal conditions for the rapid growth of many fungi. Lenses of optical equipment, such as bombsights, field glasses and cameras, were attacked by fungi that grew on dust particles that settled on these lenses. The mycelium not only blurred the image, but the organic acids they

produced also etched the lenses and destroyed them. Working at The University of Melbourne, Professor J. S. Turner and his team developed effective antifungal coatings for polished glass lenses, which were eventually adopted by all Allied units.

Also at The University of Melbourne, Drs Ethel McLennan and Sophie Ducker maintained and added to an important collection of antibiotic-producing fungi, mainly from the genus *Penicillium*. This collection had been started in Britain by Sir Howard Florey and Sir Ian Fleming, but was moved to Melbourne in 1944 because of a fear that the collection may be lost due to invasion or the Blitz.

J. S. Turner, Ethel McLennan and Sophie Ducker, three pioneers of mycology in Australia

Fungi, like animals and unlike plants, store their energy reserves as glycogen rather than as starch. Thay also store unusual sugars—trehalose and mannitol—glycerol, fats and oils.

> Nearly all fungi use glucose as a carbon source, but specially adapted species, with novel enzymes, digest other carbon molecules. Most fungi are able to use inorganic sources of other nutrients and can synthesise their own amino acids and vitamins. They store energy reserves as glycogen, trehalose, mannitol, glycerol, fats and oils.

Environmental conditions

Fungi tolerate a wide range of environmental extremes. They can be found wherever other organisms are in air, soil, freshwater and marine environments.

Water

Xerotolerant fungi are able to grow at extremely low water potentials compared with plants, animals or

bacteria. **Xerotolerance** is achieved through the interconversion of storage compounds, releasing water

$$\text{mannitol} \rightleftharpoons \text{glycerol} + \text{water}$$

or through the use of water released from the hydrolysis of cellulose by dry-rot fungi

$$\text{cellulose} \longrightarrow \text{glucose} + \text{water}$$

Temperature

Most species of fungi are **mesophiles** that grow best between 10°C and 30°C, where most life is found. Some species, **thermophiles**, grow best between 30°C and 50°C. These fungi are found in hot mineral springs, in soil and in compost heaps, while some are animal pathogens. They are able to adjust the lipid composition of their membranes to maintain a constant membrane fluidity at high temperatures. Some fungi are able to tolerate temperatures down to −5°C, although they grow best at about 5°C. These species are **psychrophiles** and are found in refrigerators, cold rooms or in other cold environments (Box 36.2).

BOX 36.2   DEADLY DEREK AND A FUNGUS COMBINE TO DEFEAT AUSTRALIA

Microdochium nivale (phylum Deuteromycota) attacks grasses during very cold weather. In Northern Hemisphere wheat crops, which can be covered by snow over winter, this fungus causes snow mould, a serious disease that becomes obvious after the spring thaw. This fungus is also notorious for its role in deciding the fourth cricket test at Headingley in 1972. The wicket had been covered for several cold, wet days before the match and, when the covers were removed, snow mould had killed the turf, leaving a soft, sticky, substandard surface. The bowling of Derek Underwood, the talented English spin bowler, was almost unplayable. He took 10 wickets and Australia lost by nine wickets.

Temperature may also be important in breaking the dormancy of spores. For example, the teleutospores of the rust fungus, *Puccinia graminis* (phylum Basidiomycota), require a cold shock to germinate, while the ascospores of many ascomycetes require a heat shock.

Light

Fungi do not require light, as do photosynthetic organisms, and most fungi thrive when kept in the dark. However, ultraviolet (UV) and visible light may be necessary to induce the formation, release and germination of spores. Some fungi have light sensors that help them to release their spores into the air, where they can be picked up and scattered by wind currents.

Gravity

Mushrooms and all macrofungi have gravity-sensing mechanisms, which enable them to orient their fruiting bodies vertically.

Topography

Some fungal leaf pathogens invade their hosts through stomata (Fig. 36.11). They may locate stomata by chance, although some fungi are able to locate stomata on a leaf surface. One mechanism used is a chemical attraction to photosynthetic by-products released through the stomata; another involves the ability to sense the topography of the leaf.

> Fungi tolerate a wide range of environmental extremes. They can be found wherever other organisms are in air, soil, freshwater and marine environments.

FUNGAL REPRODUCTION

Spores

Spores are protectively packaged genomes with the two primary functions of survival and dispersal. There is evidence that the asexual **urediniospores** of *Puccinia graminis* (phylum Basidiomycota), the wheat stem rust fungus, released from rusted wheat leaves in Africa, have been blown on the wind to Australia. This journey involves a climb of about 10 km in thermal updraughts to the jet stream, and a trip across the Indian Ocean, about 13 000 km. These spores have melanised cell walls, which protect cell contents from UV radiation, are dehydrated to resist desiccation and freezing, and are produced in millions on each rust-infected plant. The probability that a spore would travel so far, survive the extreme conditions faced during its journey and find a susceptible wheat plant on which to land and germinate—all matters of chance—is infinitesimal, yet it happens.

The form of a spore reflects its particular function. Spores may look the same as other cells, or they may be single or multicelled, have thick or thin walls, be pigmented, or they may be dry or 'wet' (enveloped in a mucilage; Fig. 36.2). Because their morphology and method of formation is so much more distinctive than hyphae, spores and spore-forming structures are useful characters for fungal classification and identification.

Physiologically, sporulation is induced by a switch from vegetative to reproductive growth. This may be triggered by nutrient depletion or staling products as the fungus exhausts its substrate, or an environmental stimulus. Hormones are known to mediate the switch in at least some species. For example, the common mould, *Phycomyces blakesleeanus* (phylum

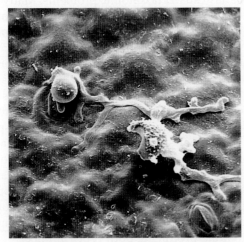

Fig. 36.11 Hyphae from germinating fungal spores often follow the topography of the surface of leaves in their search for stomata

Zygomycota), has two sexually compatible strains, which must meet to form sexual spores (Fig. 36.12). The hormone, trisporic acid, is released by these compatible hyphae and stimulates the formation of gametangium (where gametes form). In ascomycetes, there is a family of hormones that induces the change from vegetative growth to reproductive growth. However, very little is known about hormones in other fungi. There is a great potential application for fungal hormones as 'designer' fungicides, which could be used to suppress sporulation selectively in pathogenic species.

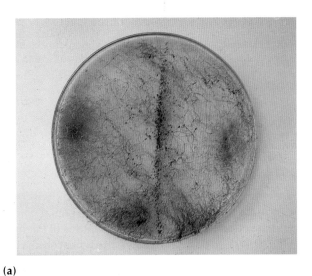

(a)

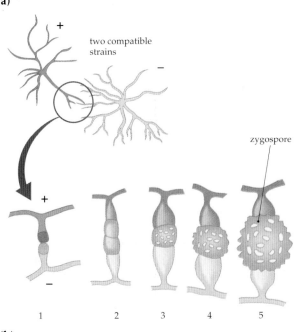

two compatible strains

zygospore

1   2   3   4   5

(b)

Fig. 36.12 Sexual reproduction in *Phycomyces blakesleanus*, a heterothallic zygomycete. **(a)** Two compatible strains, + and −, grow together to form a distinct line of zygospores where they meet. **(b)** Zygospores in different stages of development (1–5), from the initial contact bridge (1) to the spiny, mature zygospore (5). Compare these stages to the sequence shown in Figure 36.14

Many fungi produce spores both asexually and sexually during their life cycles. Only one type of sexual spore is produced by any one species, but some fungi produce up to four different types of asexual spores or **conidiospores** (conidia). Sexual spores commonly function as survival spores, **memnospores**, which assist the fungus in persisting over unfavourable seasons. Asexual spores are usually produced in great numbers and assist in dispersing the fungus rapidly during periods of favourable environmental conditions. This type of spore is a **xenospore**. There are many exceptions to this generalisation resulting from the diversity of fungal lifestyles.

Life without sex

Production of asexual spores by fungi is more common than sexual spores. Many fungi (e.g. deuteromycetes) do not produce sexual spores, and a few, classified as the **mycelia sterilia**, produce no spores at all. Sexual reproduction provides genetic diversity through recombination during meiosis. Without this source of diversity, an individual can only reproduce clones of itself, which are less able to respond to changes in the environment (Chapter 14). However, those fungi that have dispensed with sexual reproduction are able to generate sufficient genetic diversity by the release of incredibly large numbers of asexual spores. Mutations and rare mitotic recombinations can result in enough variant propagules that their chances of survival are real. Heterokaryons are also able to 'carry' mutations that are only expressed under unusual selection pressures when they may be more 'fit' than the wild-type nucleus. A gene for fungicide resistance, for example, may reduce the competitiveness of the organism if expressed, but can be hidden and 'kept in reserve' until the fungus is exposed to a fungicide.

The rarity of sexual reproduction in some species of fungi has caused taxonomic problems. Many fungi have two names—one for the asexual, or **anamorphic**, form, and another for the sexual, or **teleomorphic**, form. The asexual state, which is usually more common in nature, was often described before the sexual state was discovered. The asexual state is often called the imperfect state because botanists considered that an organism that does not reproduce sexually must be missing something. Correspondingly, the sexual state is called the perfect state. In many instances, it was some time before the two were recognised as different forms of the same organism. After all, if all you can see are microscopic spores, two fungi that have different spores would reasonably be expected to be different species, unless you saw an individual producing both types at once.

One of many examples is *Talaromyces flavus* (Klöcker) Stolk & Samson (phylum Ascomycota). It is the perfect (sexual) state of *Penicillium dangeardii*

Pitt (phylum Deuteromycota). The entire fungus, consisting of both the anamorphic and teleomorphic states as described in the literature, is the **holomorph** and, under article 59 of the Botanical Code, takes the name of the teleomorph. In practice, unless the sexual spores are observed, it is common to use the name of the anamorph.

> Fungi reproduce by spores, produced asexually and sexually. Spores are protectively packaged genomes with the two primary functions of survival and dispersal.

FUNGAL PHYLA

Of the five groups of Fungi, three—Zygomycota, Ascomycota and Basidiomycota (Table 36.1)—are 'natural' or monophyletic taxa (Chapter 30). They are characterised by their type of sexual reproduction. The Deuteromycota is an informal grouping of anamorphic fungi, organisms that do not reproduce sexually but have obvious affinities with the three natural phyla. The final group is the lichenised fungi, Mycophycota, which are 'dual organisms', where a fungus and an alga or cyanabacterium live together closely in a mutualistic association. The fungal partner is usually an ascomycete, rarely a basidiomycete, and is unable to live outside the relationship: it is an **obligate symbiont**.

Phylum Zygomycota

The phylum **Zygomycota** is probably the oldest line of fungi to evolve, with the Ascomycota and Basidiomycota diverging later. The Zygomycota is distinguished from the other fungal phyla by: coenocytic hyphae, which have no septa, except to seal off reproductive structures; sexual zygospores; and the fact that asexual spores are formed inside a **sporangium**, or spore-forming cell. There is great diversity within the 600 named species. Zygomycetes include fast-growing saprophytic moulds that are the first visible colonisers of decomposing organic matter (Fig. 36.13), insect pathogens, some of which are potential biological control agents (Chapter 42), and the fungal partners of arbuscular mycorrhizae (p. 790).

Fig. 36.13 *Rhizopus nigricans* (phylum Zygomycota), a common spoilage fungus, growing on a tomato

Zygomycetes may be **homothallic** (sexually self-compatible) or **heterothallic** (self-incompatible). When compatible strains meet each other, they produce short hyphal branches that develop single-celled gametangia, which fuse at their tips (Fig. 36.14). Nuclear exchange takes place, followed soon after by nuclear fusion. A thick wall develops around the zygote, resulting in the formation of a dormant **zygospore**. Upon germination, meiosis occurs, forming haploid, multinucleate mycelium.

The cytoplasm of specialised hyphal branches that develop as sporangia undergoes a series of mitotic divisions to produce haploid **sporangiospores** within the sporangial wall. The sporangium is separated from the vegetative mycelium by a simple septum and is often swollen at its tip. Eventually, the sporangial wall ruptures to release thousands of these asexual spores. Often these spores are released in a wet mucilage and depend on insects or water for dispersal. A few species, such as the mould *Rhizopus nigricans*, produce dry spores that are dispersed by wind. There is an evolutionary trend within the phylum towards a reduced number of spores in each sporangium.

> The hyphae of zygomycetes are coenocytic—tubes of cytoplasm with many haploid nuclei. Zygomycetes produce both sexual spores (zygospores) and asexual spores (sporangiospores). Unlike other fungi, hyphal fusions are rare and complex macroscopic structures are not formed.

Table 36.1 Three natural fungal phyla

Characteristics	Phylum		
	Zygomycota	Ascomycota	Basidiomycota
Sexual spore	zygospore $(2n)$	ascospore (n)	basidiospore (n)
Asexual spores	sporangiospores (common)	conidia (common)	conidia (uncommon)
Septa	absent	simple pores	complex pores and clamp connections
Vegetative hyphae	multinucleate $(n + n \ldots + n)$	heterokaryon $(n_a + n_b \ldots n_j)$	dikaryon $(n_a + n_b)$
Size	microfungi	micro- and macrofungi	micro- and macrofungi

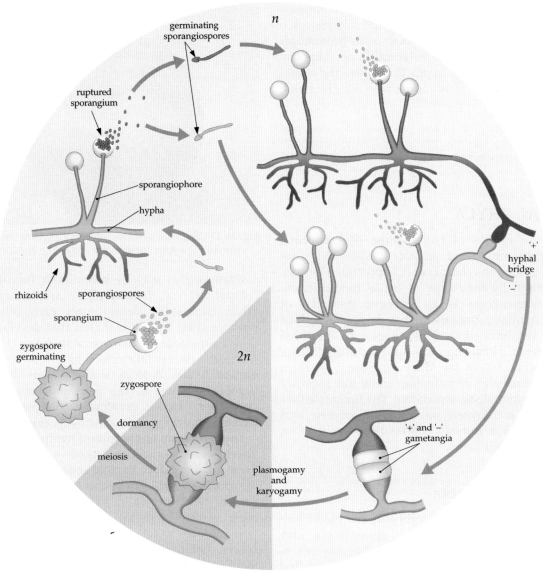

Fig. 36.14 Life cycle typical of Zygomycota

Phylum Ascomycota

The **Ascomycota** are the largest group of fungi (about 30 000 described species). They are the most important cellulose degraders in ecosystems and are also important as plant and animal pathogens. We use yeast in food and beverage preparation, and savour the exquisite flavours of morels and truffles.

Ascomycetes have septate hyphae, produce sexual **ascospores** inside a sac-like cell, the **ascus** (pl. asci), and produce abundant asexual conidia. Ascospores create genetic diversity and often function as memnospores, or survival spores, while conidia function as xenospores, which allow rapid dispersal.

The female gametangium, the **ascogonium**, forms a short bridge, which fuses with the male gametangium, the **antheridium** (Fig. 36.15). Haploid nuclei migrate into the ascogonium but do not fuse

immediately. Dikaryotic, heterokaryotic ascogenous hyphae, containing several nuclei from each parent, curl to form **croziers**. Mitosis occurs and asci form as short branches from the crozier. Each ascus contains two haploid nuclei, one from each parent. Many asci form from each ascogonium. Karyogamy within the ascus produces a zygote, but the diploid nucleus divides immediately by meiosis to form four haploid daughter nuclei. Each usually undergoes another mitotic division and a cell forms around each nucleus, producing eight ascospores within each ascus (Fig. 36.16). In this way, ascomycetes are able to maximise the amount of genetic variation possible from each fertilisation, as each fertilised ascogonium zygote produces many asci, each containing eight offspring. This gives ascomycetes a tremendous ability to adapt to altered environments.

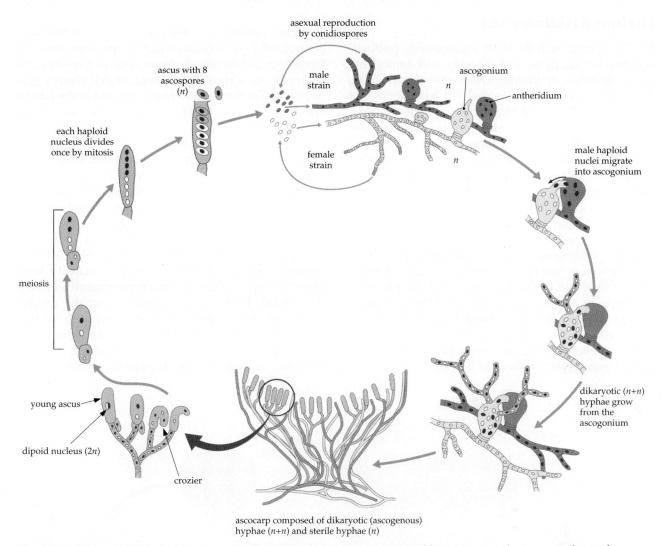

asexual reproduction
by conidiospores

ascus with 8
ascospores
(n)

male
strain

ascogonium

antheridium

n

female
strain

n

male haploid
nuclei migrate
into ascogonium

each haploid
nucleus divides
once by mitosis

meiosis

dikaryotic (n+n)
hyphae grow
from the
ascogonium

young ascus

dipoid nucleus (2n)

crozier

ascocarp composed of dikaryotic (ascogenous)
hyphae (n+n) and sterile hyphae (n)

Fig. 36.15 Life cycle typical of Ascomycota. The type of ascocarp represented here is an apothecium, similar to the ones in Figure 36.5

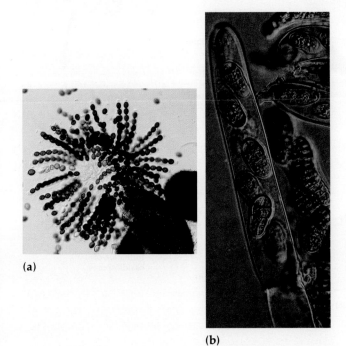

(a)

(b)

Vegetative hyphae around the ascogonium may form a fruiting body, the **ascocarp**. Classification of ascomycetes is based on the presence and type of ascocarp produced, and on the structure of the ascus.

The tip of the ascus wall often ruptures violently, releasing ascospores explosively, sometimes ejecting them 30 cm into the air. They can then be picked up by wind currents and scattered widely. Since fungi are non-motile, their ecological success depends on the production of vast numbers of spores, which, even when widely dispersed, have a very small chance of landing on a suitable substrate.

> The sexual spore of ascomycetes is an ascospore, formed inside an ascus. Asexual conidiospores are borne directly on hyphae. The vegetative mycelium is haploid, multinucleate and heterokaryotic, and has septa with simple pores. Hyphal fusions are common and some species form macroscopic structures.

Fig. 36.16 Mature asci, each containing eight ascospores of **(a)** *Sordaria macrospora* and **(b)** *Pleospora herbarum*

Phylum Basidiomycota

Basidiomycota include large mushrooms, puffballs and bracket fungi, as well as rusts and smuts that cause diseases of plants. There are about 25 000 described species. The life cycle of basidiomycetes (Fig. 36.17) differs from that of ascomycetes because, while nuclear fusion is delayed in both groups, dikaryotic cells in the ascomycetes are only found in the ascogenous hyphae. Normal vegetative hyphae of basidiomycetes are dikaryotic. **Basidiospores** are borne externally on a club-like cell, the **basidium**, rather than inside a sac.

Nuclear fusion takes place inside the basidium, which is formed from the dikaryotic mycelium, and may be surrounded by a large **basidiocarp**, such as a mushroom or bracket. Basidiomycetes are classified by the structure of the basidium, and the presence and type of basidiocarp produced. Meiosis rapidly follows nuclear fusion, forming four basidiospores from each basidium. Basidiospores are produced in great profusion; for example, every gill in a mushroom cap is covered with millions of basidia.

Microscopically, two features distinguish the basidiomycetes from ascomycetes. **Clamp connections** are formed in mitotically dividing cells to preserve the dikaryon. As a result, the daughter cell always gets one of each of the haploid nuclei present in the parent cell (Fig. 36.18a). Many basidiomycetes also have complex **dolipore septa**, rather than the simple pores of ascomycetes (Fig. 36.18b).

Asexual reproduction is less common than in the Ascomycota, but when it occurs it can be part of a very complex life cycle. *Puccinia graminis tritici*, the stem rust pathogen of wheat (Box 36.3), produces five morphologically different spore types on two plant hosts, wheat and barberry, between which it alternates. One spore type will infect only barberry, another only wheat. The others germinate to form mycelium upon which another type of spore is formed (Fig. 36.19). Think for a moment about the precise control of gene expression that is required for this organism to complete its life cycle. The fungus never completes its sexual life cycle in Australia because barberry, introduced by European settlers as an

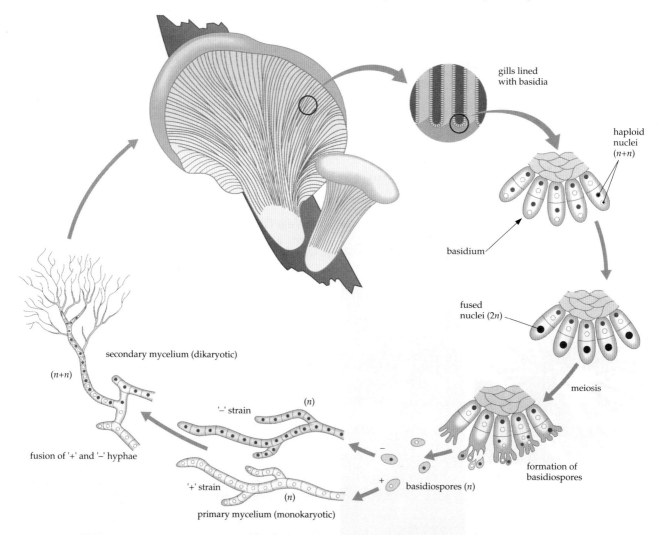

Fig. 36.17 Life cycle typical of Basidiomycota

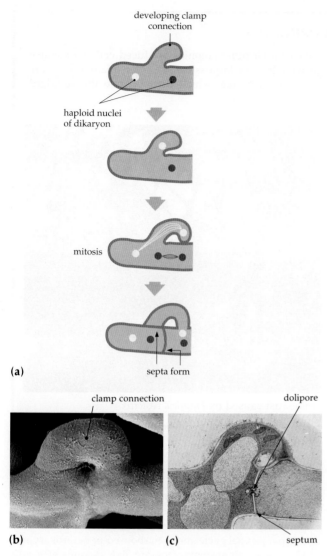

(a)

(a)

clamp connection

dolipore

(b)   **(c)**

septum

Fig. 36.18 Two microscopic features used by
basidiomycetes to maintain the dikaryon during
vegetative growth: clamp connections ensure the
distribution of one of each of the parental nuclei to
each daughter cell; the complex dolipore septum
restricts nuclear migration. (a) Mitosis and clamp
connection formation. (b) SEM of a mature clamp
connection. (c) Section through a clamp connection
showing new septa and dolipore

(a)

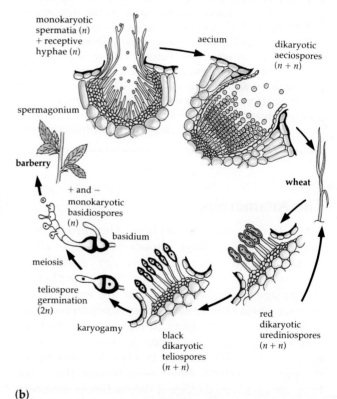

(b)

ornamental plant, never became widespread, and
because our winters are not cold enough to break
the dormancy of the spore that produces the
basidium. Nevertheless, stem rust remains a dangerous pathogen because the *disease cycle* in Australia
involves only one asexual spore type, the urediniospore, which continually infects wheat.

> Basidiomycetes produce basidiospores from a basidium,
> often borne on a mushroom or toadstool (a basidiocarp),
> and conidiospores. The vegetative mycelium is
> dikaryotic and divided by complex septa. Hyphal fusions
> are common. Some basidiomycetes have complex life
> cycles with more than one type of asexual spore and
> host.

Fig. 36.19 (a) Stem rust of wheat, *Puccinia graminis tritici*,
causes large red pustules to form on wheat stems and
grains to shrivel up before they mature. The red
pustules are the asexual fruiting bodies that release
millions of urediniospores, conidia, which can infect
healthy plants. (b) The full life cycle involves two hosts
and several spore types. In Australia, the disease cycle
involves continuous infection of wheat by only one
spore type, the urediniospore. Stem rust remains the
most serious potential disease of wheat in Australia,
and it is only controlled by a concerted ongoing
program by scientists to breed disease-resistant wheat
varieties. In the past, many varieties retained their
resistance for only a few years, because the genetic
composition of the fungal population can change
rapidly due to selection of random mutants or arrival of
new forms with virulence against each new variety

BOX 36.3   RUST FUNGI AND WHY THE ENGLISH DRINK TEA

Rust diseases get their name because the pathogens, all basidiomycetes, produce masses of red spores on their hosts. There are about 7000 species of rusts and they are all plant pathogens, infecting leaves and stems. Hyphae grow between host cells but push membrane-bound haustoria into cells. These haustoria become intimately associated with the host plasma membrane, forming a surface across which nutrients are transferred to the fungus. Nutrients are diverted away from seeds, fruits and other growing parts of the plant, and eventually leaves drop off. The fungus sporulates heavily producing red, brown or black pustules.

Rusts have played an interesting role in human affairs throughout history. Before the nineteenth century, the most common drink in polite English society was coffee. Coffee is produced from the berries of two species of *Coffea* trees, *C. arabica*, which is native to the area around Ethiopia, and *C. canephora*, a native of west and central Africa. English coffee came mostly from extensive plantations in what was then the colony of Ceylon. In 1869, coffee rust, caused by the basidiomycete *Hemileia vastatrix*, reached Ceylon from Africa, probably on infected trees. This disease defoliates and eventually kills coffee trees, and by the 1880s coffee growing in Ceylon was abandoned. The problem for the British was that

none of their other colonies produced coffee. As coffee growing became impossible in Ceylon, plantations were turned over to tea, which then became the standard English beverage.

Coffee rust caused by *Hemileia vastatrix*

The Anamorphs

Anamorphic fungi, phylum **Deuteromycota**, sometimes called **fungi imperfecti**, include about 25 000 named species, many of which are economically important. They are a polyphyletic assemblage containing fungi that are more related to other phyla than to each other. Anamorphic fungi have either lost the ability to reproduce sexually in nature, or have a sexual spore that has not yet been discovered and thus cannot be classified into any of the phyla described above, even though they may have obvious morphological similarities to members of one phylum. Sometimes sexual reproduction may be induced, and the teleomorph, or perfect stage, is usually an ascomycete. Less commonly it is a basidiomycete or zygomycete. Classification of anamorphs is based on the type of asexual spores. However, members of the subgroup Mycelia Sterilia produce no spores at all, making the taxonomist's job very difficult.

Because some fungi named as Deuteromycota were later discovered to produce sexual spores, some species have two names (p. 783). Some Mycelia Sterilia, such as the common soil fungus, *Rhizoctonia*, have been found to have teleomorphs that form basidia (*Tulasnella* or *Thanetephorus*) or asci (*Trichophaea*). This has always caused confusion among mycologists.

> Classification of deuteromycetes is based on conidiospore formation because these organisms do not produce sexual spores. Genetic diversity is created by the parasexual cycle and is harboured in heterokaryons. Otherwise, deuteromycetes are like ascomycetes and basidiomycetes.

Lichens

The lichens, phylum **Mycophycota**, are not so much a taxonomically distinct group of fungi as an ecological grouping of fungi that are totally dependent on a mutualistic relationship. There are about 18 000 species of **lichenised fungi**, and all but about 20 are ascomycetes, the rest being basidiomycetes. The fungal partner provides most of the substance of the lichen; the photosynthetic partner, either a chlorophyte, xanthophyte or cyanobacterium, is embedded in the mass of mycelium. The photosynthetic partner captures light energy to produce sugars, while the fungus absorbs minerals and water. Lichens containing cyanobacteria, such as *Nostoc*, are able to fix atmospheric nitrogen. The fungus controls the metabolism of its partner using biochemical signals, making it produce foods that only the fungus can use. The photosynthetic partner is sometimes able to live separately from the fungus, but the fungus is always, to some extent, dependent on the food provided by its photosynthetic companion.

Different morphological forms of lichens include: **crustose** ('crusty') lichens, which adhere very closely to their substrate; **foliose** ('leafy') lichens, which have more loosely attached leaf-like structures; and **fruticose** ('bushy') lichens (Fig. 36.20).

Lichen reproduction may simply be a matter of fragmentation, or they may produce 'spores', **soredia**, made of an algal cell embedded in hyphae. Each partner can also reproduce independently, the fungus usually producing ascospores sexually and the photosynthetic partner usually reproducing asexually. The two partners come together again under favourable conditions to form a new lichen.

Because of their structure, lichens are among the hardiest of organisms, and are, for example, found on rocks in Australian deserts, in the Snowy Mountains, in Antarctica and the New Zealand Alps, as well as in the rainforests of eastern Australia. They are also extremely slow growers. However, because of their efficiency in absorbing water and minerals they are important in nutrient cycling. This efficiency also makes them sensitive to pollution, and lichens almost disappeared from forests around cities after the Industrial Revolution, although as a result of improved pollution control they are now beginning to return. Reindeer in Northern Europe accumulated very high doses of radioactivity after the explosion at the Chernobyl nuclear reactor in the Ukraine in 1986 because their winter diet consists exclusively of lichens, which absorbed very high levels of radio-activity.

> Lichens are mutualistic partnerships between an ascomycete (or some basidiomycetes) and a chlorophyte, xanthophyte or cyanobacterium. They are classified on the basis of the fungal partner and their appearance, which may be crustose, foliose or fruticose. Lichens reproduce by simple fragmentation or as spores— soredia—an algal cell surrounded by hyphae.

ECOLOGICAL IMPORTANCE OF FUNGI

Decomposers

Fungi play a number of important ecological roles. They are the primary decomposers of organic matter, which not only releases minerals to the various nutrient cycles (Chapter 44), but saves us from being buried under a pile of dead animals and plants. Without decomposing fungi and bacteria, the entire land surface of the earth would soon be covered by metres of dead plant and animal bodies.

(a)

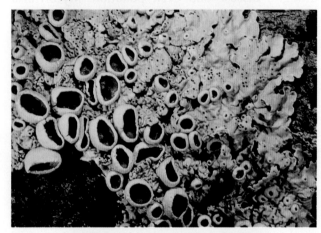

(b)

(c)

Fig. 36.20 Morphological forms of lichens: (a) crustose; (b) foliose; and (c) fruticose

Mycorrhizae

Some fungi have evolved the ability to use both saprophytic and parasitic modes of nutrition by forming a 'nutrient bridge' between the soil and plant roots. About 90% of land plants are infected by **mycorrhizal fungi**.

A mycorrhizal fungus absorbs, and transfers to the plant, mineral nutrients such as phosphorus, zinc and copper. Because of the efficiency with which fungi absorb mineral nutrients, they enable plants to grow on soils that would otherwise be deficient in these nutrients (Chapter 18). One estimate is that mycorrhizal fungi increase the efficiency of phosphorus uptake in wheat plants growing on phosphate-deficient Australian soils by a factor of 10 000. In return for these nutrients, the plant supplies the fungus with organic carbon. Plants infected with mycorrhizal fungi are also much more tolerant to environmental extremes, such as high temperature, drought, disease and even pollution. Mycorrhizal associations are probably almost ubiquitous because of the tremendous advantages to both partners.

Mycorrhizae are well represented in the fossil record and there is evidence that the successful colonisation of land by plants was assisted by the ability of mycorrhizae to extract nutrients from the almost totally inorganic soils of the time (see Chapter 31).

There are several types of mycorrhizae: **arbuscular mycorrhizae, orchid mycorrhizae, epacrid mycorrhizae** and **ectomycorrhizae**. The most widespread are the arbuscular mycorrhizae, involving over 100 species of the phylum Zygomycota, which infect the cortical cells of the roots of at least 200 000 species of plants, about 85% of land plants. The association is characterised by a network of fine branched hyphae, **arbuscules**, within cortical cells of the plant root. Arbuscules, with their large surface area, function as the organ across which nutrient exchange with the plant cell occurs. A sparse network of very fine hyphae extends from the root into the soil, but otherwise root morphology is not radically changed. Up to 3 m of hyphae may grow out of each centimetre of root, exploring the soil beyond the nutrient-depleted zone around the root. Hyphal swellings or spores, vesicles, may also be formed inside and between cortical cells, and outside the root. These function as storage organs.

Epacrid mycorrhizae allow sclerophyllous heaths to grow on acid, nutrient-depleted soils in Australia (Chapters 18, 41). Orchid mycorrhizae are essential for the germination and growth of many native orchids, some of which have no chlorophyll and depend totally on their fungal partner.

Ectomycorrhizae, or sheathing mycorrhizae, usually involve basidiomycetes, but sometimes ascomycetes or zygomycetes. The fungi form much more selective symbioses, mainly with temperate forest trees, such as eucalypts and pines, and are responsible for many of the mushrooms and toadstools found growing in forests. The morphology of ectomycorrhizal roots is very different from that of uninfected roots. The fungus forms a thick sheath or **mantle** of mycelium over the root, replacing the epidermis and root hairs. The fungus forms a **Hartig net** of mycelium, which grows between the cortical cells, facilitating nutrient transfer (Fig. 36.21). Around 15–20% of Australian plants form both arbuscular and ectomycorrhizal associations.

(a)

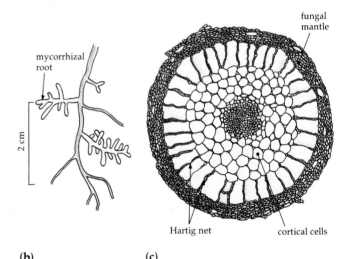

(b)   (c)

Fig. 36.21 Mycorrhizal roots. (**a**) A network of ectomycorrhizal hyphae under the leaf litter on a forest floor. (**b**) An ectomycorrhizal root. (**c**) Transverse section of a mycorrhizal root showing the Hartig net

Pathogens and predators

Fungi are also important pathogens (disease-causing organisms) of soil-inhabiting organisms, including plants, nematodes, insects and other fungi. Fungi cause plant diseases such as 'damping-off' of seedlings, root rots and wilt diseases. Fungi are also important pathogens of organisms in aquatic and aerial environments.

Some soil fungi trap nematodes using sticky branches or rings of hyphae that act like lassoes. Once trapped, the unfortunate nematode is invaded by hyphae, and then digested by enzymes secreted by the killer fungus (Fig. 36.22).

Because of these pathogenic and predatory abilities, some fungi are used in biological control programs against insects and weeds (Chapter 43).

> Fungi are important in nutrient cycling as decomposers and form mycorrhizal associations with most plant roots. Some fungi cause plant or animal diseases.

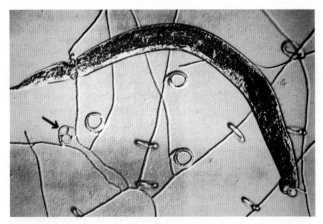

Fig. 36.22 One method used by fungi to trap nematodes is lassoe-like loops

Fig. 36.23 A number of fungi are eaten directly as food or used in the manufacture of food products

FUNGI AND HUMANS

Eating and drinking fungi

One of the most common ways in which fungi are eaten is as mushrooms, the fruiting bodies of a number of species of basidiomycetes and ascomycetes (Fig. 36.23). Their nutritive value is limited because of their high water content, but they are a useful source of easily digestible proteins and vitamins and minerals.

Some examples of commonly eaten fungi include the cultivated mushroom, *Agaricus bisporus* (Basidiomycota); padi straw mushroom, *Volvariella volvacea* (Basidiomycota); oyster mushroom, *Pleurotis ostreatus* (Basidiomycota); shiitake mushroom, *Lentinus edodes* (Basidiomycota); truffles, *Tuber* spp. (Ascomycota); and morels, *Morchella* spp. (Ascomycota).

Fungi can be eaten directly, like mushrooms, or they can be used to increase the nutritive value or palatability of foods. Soybeans, although high in protein, are almost totally indigestible for humans. However, when they are fermented by *Aspergillus oryzae* (Ascomycota) to make soy sauce and miso, or by *Rhizopus oligosporus* (Zygomycota) to make bean curd or tempeh, soybeans become quite palatable.

The basic process of cheese-making involves bacteria, but fungi are used during ripening to impart the delicate, or not so delicate, flavours to fancy cheeses. Camembert and brie are ripened by two aerobic fungi, *Endomyces geotrichum* (Ascomycota), which removes the tart lactic acid from the curd, and *Penicillium camembertii* (Deuteromycota), which partially hydrolyses the proteins, creating a smooth, creamy texture. Because they are both aerobes, these cheeses are made flat to allow air and fungal enzymes to penetrate and allow ripening to proceed into the cheese. Blue cheeses, such as roquefort, stilton and gorgonzola, are large, crumbly cheeses with blue veins of spores of *Penicillium roquefortii* (Deuteromycota), a species that tolerates low oxygen conditions. The raw cheese is inoculated with metal spikes that are covered in fungal spores. The holes made by the spikes allow enough air to enter, and the spores germinate to produce the characteristic blue veins. The flavour is due to the action of fungal lipase enzymes acting on fatty acids in the cheese.

Certain types of fungi, the yeasts, *Saccharomyces cerevisiae* (Ascomycota), ferment sugars to alcohol and carbon dioxide. In bread-making, the carbon dioxide released causes the dough to rise, and the alcohol is evaporated during baking. In traditional beer brewing, the carbon dioxide produced is contained under pressure in a sealed bottle, making bubbles, and the alcohol is produced by anaerobic fermentation (Chapter 5). The yeast cell can be seen as a sediment in home-brewed and some commercial beers. In most beers, the yeast is filtered out before bottling, and the residual yeast is used in making Vegemite. Every brewery has its own closely guarded, prized culture of yeast, which is the result of years of research.

In wine-making, the carbon dioxide is allowed to escape, except during the secondary fermentation of sparkling wine. Traditionally, natural yeasts living on the grapes were used, but today more highly controlled fermentations using added yeasts and cool temperatures are responsible for a more consistent wine. Bacterial contamination of wines causes them to undergo another fermentation, which converts the ethanol to acetic acid, making vinegar.

Poisonous fungi

Some fungi are not so good to eat. Many toadstools contain poisonous alkaloids, which affect the human nervous system, sometimes with fatal results. Throughout history, toadstools have been associated with witches and magic because of their hallucinogenic properties. For example, *Amanita muscaria* (Basidiomycota; Fig. 36.4), the fly agaric, has been used by shamans and witchdoctors in religious ceremonies in many cultures. In small doses it is a powerful hallucinogen, which the shamans used to assist them in their spiritual work and healing. In slightly higher doses it can be fatal. It is called fly agaric because in Europe it was mashed and mixed with milk and left to stand in dishes. It attracted flies, which were killed by the poison. Other hallucinogenic mushrooms include the common genus *Psiloscybe* (Basidiomycota), the 'blue meanies', gold tops and so on (Fig. 36.24). These are rarely fatal to humans, but may trigger psychotic reactions and should therefore be treated with caution.

Fig. 36.24 The hallucinogenic mushroom, *Psilocybe subaeruginosa* (Basidiomycota)

Claviceps purpurea (Ascomycota) infects the flowers of wheat, rye and other grasses. The fungus replaces the seed with its own sclerotia, or resting bodies, which are known as ergots. These ergots contain alkaloids very similar to the hallucinogen, LSD (lysergic acid diethylamide). Because they are the same size as cereal grains, the ergots may contaminate flour made from infected crops. Epidemics of ergot poisoning are frequent throughout history and have been linked to unusual phenomena (such as Joan of Arc's vision, St Anthony's fire and outbreaks of witchcraft). Epidemics still occur, for example, in France in the 1970s, but livestock are more common victims than humans nowadays. In a recent case in New South Wales, over 300 cattle and sheep were killed by eating ergot-infested ryegrass. Other alkaloids from the ergot fungus cause veins and arteries to constrict, and are used in medicine to assist in childbirth and to treat migraines caused by hypotension.

Allergens and mycoses

Allergens are agents that provoke an over-reaction of the immune system, which can cause the respiratory diseases hay fever and asthma (Chapter 16). Fungal spores may trigger allergic responses in sensitised people, or in people exposed to unusually large doses of spores, such as farmers handling bales of mouldy hay. The saprophytic fungi *Alternaria* and *Cladosporium* (Deuteromycota) grow on fallen leaves and dead grass, and release millions of spores into the air, particularly in late summer and autumn. When inhaled, these spores trigger severe allergic reactions in sensitive individuals.

There are a number of important human and animal diseases caused by pathogenic fungi. Most are cosmetic diseases of the skin, for example, ringworm and tinea, that are caused by keratin-digesting fungi such as *Trichophyton* and *Microsporum* spp. (Deuteromycota; Fig. 36.25). These are not usually life-threatening because the body's immune system is very effective against fungi, and fungi generally do not tolerate low-oxygen conditions. A few species are able to cause systemic infections of bone tissue, and some anaerobic fungi that infect lungs are fatal. Fungal pathogens that normally are mild, such as *Candida albicans* (Deuteromycota), the cause of thrush, may become lethal if the immune system is suppressed, for example by HIV, the cause of AIDS (Chapter 34), or by immunosuppressants, which are given to organ transplant patients. Immunosuppressants that are used in medicine are themselves fungal products, **cyclosporins**, produced by *Tolypocladium inflatum* (Deuteromycota).

BOX 36.4   FUNGICIDES OR MYCOTOXINS IN FOOD

Aspergillus flavus (Deuteromycota) is a poisonous and common mould on stale peanuts. It produces **aflatoxins**, which are converted in animal livers to extremely potent carcinogens. As a result, the permitted levels of these toxins in peanuts is measured in parts per billion. In the early 1960s, grain-fed turkeys in Britain were fed mouldy peanut meal from South America. Most died and there was a critical shortage of Christmas turkeys that year.

Many other fungi produce mycotoxins that affect both humans and animals. New Zealand sheep are regularly affected by outbreaks of diarrhoea, liver damage, facial eczema and photophobia caused by *Pithomyces chartarum* (Deuteromycota), a saprophyte that grows on dead leaves of pasture grasses and produces the mycotoxin, sporodesmin. Residues of chlorophyll, normally excreted by the liver, accumulate, causing eczema and photophobia.

Most fruit and vegetables that we eat are treated with one or more fungicides to suppress the growth of fungi before and after harvest. The use of agricultural chemicals is coming under increasing scrutiny because of their potential toxicity to the environment and to humans. In Australia, the only chemicals used have to pass strict safety tests for toxic and chronic effects on human health. Although most of us would prefer to eat pesticide-free food, the judicious use of pesticides may actually make food safer. Many fungi that grow on our food produce potent toxins and carcinogens. There is no doubt that there are also many fungal secretions of unknown toxicity in much of our food.

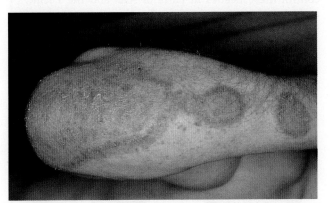

Fig. 36.25 A skin infection caused by ringworm, *Trichophyton rubrum* (Deuteromycota)

Agriculture

Fungi are the most important causes of plant diseases (see Box 36.3), which account directly for losses of about 10% to the world's food production, plus an unknown loss due to storage decay.

Fungal **epiphytotics** (plant disease epidemics) in agriculture result from the way we grow our crops as extensive monocultures of uniform genotypes. Once a virulent pathogen gets established on these monocultures, it is very difficult to stop. Monocultures are unusual in nature, where plants grow in mixed, genetically diverse communities (Chapter 43), usually preventing the build-up and rapid spread of a virulent pathogen. Selection pressure is for mild virulence, resulting in host and pathogen populations being in equilibrium, and sudden changes being buffered. Occasionally, epiphytotics develop in natural plant communities, but even these can usually be traced to human intervention.

Humans have also played an important part in the intercontinental spread of plant disease by inadvertently introducing new encounters between crops and pathogenic fungi, many of which have been disastrous. Two examples of this are the spread of Dutch elm disease and chestnut blight in forests of the United States. Melbourne has some of the world's healthiest remaining elm trees because Australia has, until now, been quarantined from the Dutch elm pathogen, *Ophiostoma ulmi* (Ascomycota) and the bark-eating beetle that carries the fungus. Inauspiciously, the beetle has recently been found near Melbourne.

Eucalypts are attacked by many fungi, mostly ascomycetes that cause leaf spots. Although leaf spots sometimes cause a lot of damage, they are rarely destructive. *Armillaria luteobubalina* (Basidiomycota) causes dieback of primary branches, crown thinning, wilting and epicormic shoot growth of eucalypts, melaleucas, grevilleas and other forest trees (Fig. 36.6). Death may follow quickly. A white mat of fungal mycelium grows under the bark of infected roots, causing a white rot of the sapwood. Sometimes thick, dark rhizomorphs occur under the bark, on roots, stumps and dead trees. The fungus uses these to spread rapidly from tree to tree. In autumn and early winter, dense clusters of yellow-brown mushrooms, 4–15 cm in diameter, appear at the base of infected trees. Clouds of basidiospores are released and carried by the wind to infect other trees. This type of dieback is scattered throughout temperate, subtropical and tropical Australian forests, woodlands, orchards, gardens and parks, and is often most severe when the topsoil has been compacted by roads or carparks.

Biological control

Fungi are also used in the biological control of pests and weeds (Box 36.5). Blackberries, skeleton weed, Paterson's curse and heliotrope are all important

BOX 36.5   COCONUTS, RHINOCEROS BEETLES AND THE GREEN MUSCARDINE FUNGUS

The coconut tree is so important in Asia and the Pacific that it is known as 'the tree of life'. Rhinoceros beetles, *Oryctes rhinoceros*, are one of the most destructive pests of coconut trees because they cut the leaves and eat the growing bud and inflorescences, sometimes killing a tree. Insecticide sprays are expensive and environmentally dangerous, and not very effective because the beetles are often hidden in the foliage.

However, one feature of their life cycle makes them vulnerable to biological control using the green muscardine fungus, *Metarhizium anisopliae* (Deuteromycota). Rhinoceros beetles congregate in dead tree trunks or piles of sawdust to breed. One or two infected larvae, or spores of the green muscardine fungus, released into a breeding log are sufficient to start an epidemic, which decimates the beetle population.

This technique is successfully used by farmers to control these beetles in parts of Asia and the Pacific, and also shows potential for limiting the severity of locust plagues.

(a) Coconut palm damaged by rhinoceros beetle

(b) Rhinoceros beetle larvae (*Oryctes rhinoceros*). The larva on the right is clearly infected with the green muscardine fungus (the spores of which are green). Scale bar = mm

(b)

---

weeds introduced to Australia from Mediterranean Europe. The CSIRO runs a program based in Montpellier, France, that searches for pathogens of these weeds in their native habitats, and screens them for their selectivity and virulence against these weeds. All four weeds are now being attacked by specific rust fungi selected from this program. Although these fungi do not kill their host plants, they reduce their vigour and thus aggressiveness as weeds.

Fungi and biotechnology

Fungi are model eukaryotes for biotechnological research and application. Because they are eukaryotes, they have the potential to be 'engineered' to produce large amounts of useful biochemicals.

Because many fungi are easily cultured on a range of substrates, and because they naturally secrete enzymes, antibiotics and toxins, they have the ability to convert low-value wastes into useful compounds, such as vitamins, hormones and antibiotics. The use of fungi for this purpose is 'industrial mycology' and, in fact, the use of fungal cultures in large industrial vats in this way has pioneered the development of biotechnology. Some examples have already been given: the production of Vegemite, penicillin and so on. Organic acids, such as oxalic and citric acids, which are used in the food industry, are produced by strains of *Aspergillus niger* (Deuteromycota). Penicillin is produced by strains of *Penicillium chrysogenum* (Deuteromycota).

The efficiency and potency of fungal strains used in industrial mycology is improved by selection and mutation, and more recently by direct gene transfer.

Once a eukaryotic gene has been identified and cloned, it is a relatively simple matter to incorporate it into the haploid genome of a fungus, and have it expressed in large quantities. For example, the human insulin gene has been spliced into the genome of baker's yeast, *Saccharomyces cerevisiae* (Ascomycota), and insulin, for use by diabetics, is now produced from bulk cultures of this fungus.

Fungi are also becoming important as biodegradation agents, being used to clean oil spills and toxic organochlorine dumps.

Fungi are directly important to humans as foods, as causes of food spoilage, plant disease and famine, as producers of mycotoxins and medicines, and as causes of human diseases and allergies.

SUMMARY

- Fungi are filamentous eukaryotes that form non-motile spores both sexually and asexually. A fungal mycelium grows as hyphae, the cell walls of which are composed of microfibrils of chitin embedded in a matrix of complex carbohydrates. Fungi probably evolved from a conjugating protist-like ancestor, possibly shared with the chytrids.

- Fungi are food absorbers, secreting enzymes that digest their food externally. They are saprophytes (feeding on dead organisms), parasites (feeding on living host cells, some of which are pathogens causing disease) or partners in mutualistic associations (lichens and mycorrhizae).

- The vegetative mycelium is haploid and often dikaryotic. Sexual reproduction involves conjugation of hyphae. Nuclear fusion after conjugation is delayed and the life cycle of many fungi includes a prolonged dikaryotic stage. When it finally occurs, nuclear fusion is usually followed immediately by meiosis.

- Within the kingdom Fungi there are five phyla. Three phyla—Zygomycota, Ascomycota and Basidiomycota—are natural groups, which differ in their hyphal structure, sexual and asexual spores, and size.

- Zygomycetes have coenocytic hyphae and produce diploid zygospores (sexual) and sporangiospores (asexual). Hyphae of ascomycetes are multinucleate and heterokaryotic and have septa with simple pores; their sexual spore is an ascospore, formed inside an ascus. Basidiomycetes have dikaryotic hyphae that are divided by complex septa; sexual basidiospores form from a basidium often borne on a basidiocarp, such as a mushroom or toadstool. Both ascomycetes and basidiomycetes form asexual conidiospores and include macroscopic forms.

- Within the fungal kingdom there is an evolutionary trend towards reliance on asexual reproduction. Anamorphic fungi (phylum Deuteromycota) are classified by their conidiospores because they do not produce sexual spores. Genetic diversity results from a parasexual cycle and is harboured in heterokaryons. Otherwise, deuteromycetes are like ascomycetes and basidiomycetes.

- The phylum Mycophycota includes lichens, which are mutualistic partnerships between an ascomycete and a chlorophyte, xanthophyte or cyanobacterium. They are classified on the basis of the fungal partner and their appearance, which may be crustose, foliose or fruticose.

- Fungi are important to humans as foods, decomposers, plant and animal pathogens, and in biological control and biotechnology.

QUESTIONS

1. What features distinguish fungi from other organisms?

2. List both the common and characteristic features of: (a) the fungal-like Protists, (b) the Rhodophyta and (c) the Fungi. What features justify separating these groups into different kingdoms?

3. Draw life cycles of a 'typical' ascomycete and a 'typical' basidiomycete. List the major differences between the two phyla.

4. Why are ascomycetes and basidiomycetes thought to be more closely related to each other than they are to the zygomycetes?

5. Describe some unusual features of nuclear behaviour in fungi.

6. What features allow fungi to be found in all environments where life is possible? Why is this important?

7. What is a 'fungal succession'? Give an example.

8. How do imperfect fungi attain variability and diversity without sexual reproduction?

9. Describe (a) lichens and (b) mycorrhizae. Why are fungi particularly well suited to their roles in these partnerships?

10. You are an industrial mycologist working for the Environment Protection Authority. An old factory site is to be redeveloped as a park, but toxic levels of chemical pollutants have been detected in the soil. Your job is to select, from a range of potential biodegrading fungi, the most suitable isolate. List the important criteria you would use in selecting the best isolate.

11. In what ways have plant diseases affected the way we live? Think of present and historical examples.

12. What are the risks and benefits of using fungi in biological control programs?

PLANTS

This chapter summarises the characteristics of the major groups of land plants, emphasising their evolutionary history and characteristics. Before reading this chapter, you should first be familiar with the material in Chapters 16–18, which detail plant reproduction, architecture and structure. Remember also from Chapter 31, that over geologic time, the earth and organisms have changed together and continue to do so. The movement of continents affects climate and terrestrial environments, and has played a part in the rise and fall of plant groups through time.

EVOLUTION OF PLANTS

Plants, kingdom **Plantae**, are those photosynthetic organisms that have adapted to life on land (Fig. 37.1), and include liverworts, mosses, ferns, conifers and flowering plants among the major modern groups (Table 37.1). They are multicellular and have cells specialised to form tissues and organs.

Plants first colonised the land about 410 million years ago, towards the end of the Silurian period (see Chapter 31). At this time, the land surface was mostly bare and the earliest plants, which were small and inconspicuous, were restricted to the margins of wetlands and waterways.

Fig. 37.1 Australian heathlands are rich in plant species

The first characteristics important in the transition of plants from water to land were a cuticular covering of the plant body and a protective layer around the sex organs, both of which reduced loss of water by evaporation and prevented desiccation. The evolution of an internal transport system of xylem and phloem, specialised organs, such as leaves and roots, and woody tissue allowed vascular plants to attain greater body size than smaller non-vascular plants, which rely on simple diffusion for transport of water and nutrients. Transport of male gametes, in pollen, to the female reproductive organ by wind or animals rather than by water finally freed plants from their dependence on a moist environment.

Although the earliest vascular plants were relatively small and probably confined to swampy areas, the Devonian was a time of rapid diversification of plants. By the end of the period, large tree-sized plants formed complex forests. These Palaeozoic forests (330 million years BP) would have looked very different from the forests we know today. Many of the early plant groups are extinct or represented by only a few living taxa today.

Green algae and the origin of land plants

Green algae (phylum Chlorophyta, Chapter 35), which grow in marine habitats, freshwater or on soil, include the closest relatives of land plants. Modern green algae and land plants have similar pigments (chlorophylls *a* and *b*), chloroplast structure and cell wall chemistry (cellulose) and both have starch as a storage material. However, there has been considerable disagreement over which particular green algal group is most closely related to land plants. Electron microscopy has resolved this debate.

From the study of algal cell divisions in a diversity of green algae, it was found that their dividing nucleus is enclosed within an intact nuclear envelope and, during cytokinesis, their spindle collapses and is replaced by a system of fibres (microtubules) that

Table 37.1 Major groups of land plants living today (a)

Phylum	First appearance in fossil record	Evolution of important characteristics for living on land
NON-VASCULAR PLANTS		
Hepatophyta: liverworts **Anthocerotophyta:** hornworts **Bryophyta:** mosses	Late Devonian (360 million years BP) first unequivocal liverwort fossil; late Silurian fossils of unidentified thalloid plants	Cuticular covering; resistant spores; small body size; stomata on sporophyte of hornworts and some mosses
VASCULAR PLANTS	Late Silurian (410 million years BP)	Sporophyte dominant, vascular tissue
Psilophyta: *Psilotum* and *Tmesipteris*	No Palaeozoic fossils recognisable as these extant forms	Stems and underground rhizomes with simple vascular cylinder of phloem and xylem; lignin; stomata
Lycophyta: clubmosses and quillworts	Late Silurian fossils are non-woody forms (e.g. *Baragwanathia*); early Carboniferous (350–290 million years BP) tree forms	Roots, stems and leaves (small—microphylls); heterospory in some; reproductive structures in strobili (cones)
Sphenophyta: horsetails (*Equisetum*)	Late Devonian; many fossils; only one living genus	Roots, stems, small leaves
Filicophyta: ferns	Early–middle Devonian with increased diversity by late Carboniferous	Leaves large—megaphylls
SEED PLANTS	Late Devonian—fossil ovules	Ovules and seeds with dormancy; pollen and pollination
Cycadophyta: cycads	Early Permian (290–245 million years BP)	Seeds borne on scales in cones; modern forms insect pollinated; some secondary growth
Ginkgophyta: *Ginkgo*	Late Carboniferous to early Permian	Secondary growth
Coniferophyta: conifers	Late Carboniferous primitive forms; Triassic (245–208 million years BP—modern conifers	Secondary growth; pollen tube deposits non-flagellate sperm
Gnetophyta	Triassic to early Cretaceous (130 million years BP)	Vessels with circular perforations; flower-like cones; modern forms insect pollinated
Magnoliophyta: angiosperms or flowering plants	Early Cretaceous (130 million years BP) monosulcate pollen (single germination aperture) suggestive of magnoliid dicots or monocots	Vessels with ladder-like perforations; flowers; double fertilisation; seeds in fruits; seeds with nutritive endosperm; diversity of animal pollinators

(a) See Chapter 31 for geologic time scale.

are all oriented in the plane of cell division. The latter system is called the **phycoplast**. In comparison, the mitotic spindle of land plants is 'open' (i.e. the nuclear envelope disappears) and the new cross wall is formed in the **phragmoplast**, the structure containing the remnants of spindle fibres oriented at right angles to the new cross wall (Chapter 8; Fig. 37.2).

The significance of these differences soon became obvious. A few green algae, including *Coleochaete* and the order Charales (commonly called charophytes), were found to have a phragmoplast and open spindle. *Coleochaete* (Fig. 37.3) is an odd alga, with additional characteristics that are typical of land plants. For example, it has a covering of protective cells around

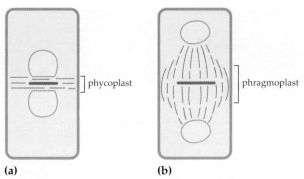

(a)   **(b)**

Fig. 37.2 **(a)** Cell division of many green algae involves a system of microtubules called a phycoplast. **(b)** Those green algae thought to be the closest relatives of the land plants have a different cell division, involving a phragmoplast

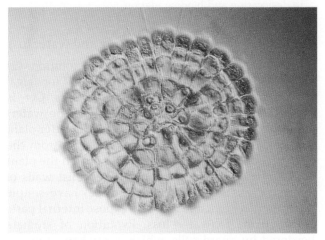

Fig. 37.3 *Coleochaete* is a small green alga that grows on other aquatic plants. It is a disc generally one cell thick

its delicate reproductive structures. *Chara* (Fig. 37.4) is a common example of the Charales. It grows on the bottom of shallow, freshwater lakes and dams. Some species become encrusted with lime so that when the plant dies a calcareous 'skeleton' remains intact, hence the common name stonewort.

These algae, as well as liverworts, mosses, ferns, cycads and even the tree *Ginkgo* (see p. 818), all have similar asymmetric, motile flagellated cells. In contrast, all those green algae with a phycoplast produce symmetric motile cells during vegetative or sexual reproduction. Charophytes and land plants both have the enzyme glycolate oxidase, equivalent to, but different from, the enzyme glycolate dehydrogenase found in other green algae. More recently, molecular biology has shown that charophytes and land plants share similarities in their 5 S rRNA (see Chapter 30). It is now apparent that of the several lines of evolution in the green algae, *Coleochaete* and Charales are two of the closest relatives of the land plants (Fig. 37.5). The different groups of green algae must have diverged a long time ago, well before the colonisation of the land.

Plants (kingdom Plantae) are multicellular photosynthetic organisms adapted to living on land. Their closest relatives are green algae. The oldest plant fossils date back 410 million years.

Fig. 37.4 *Chara* is a common member of the Charales, found in freshwater

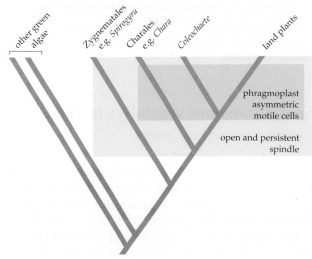

Fig. 37.5 Green algae include a number of evolutionary lines, with *Coleochaete* and Charales being the closest relatives of the land plants. Of the other major lines of evolution in the green algae shown here, the large order Zygnematales, which includes the familiar *Spirogyra*, have persistent spindles like those of *Coleochaete* and *Chara*. In *Spirogyra*, the persistent spindle even develops into a rudimentary phragmoplast. The Zygnematales have completely lost the flagellum, which is unusual for organisms living in water. They undergo sexual reproduction by conjugation, a process analogous to that seen in terrestrial fungi (see Chapter 36). Perhaps these algae adapted to terrestrial existence to some extent, and then moved back into water—the algal equivalent of whales, penguins, seals and sea grasses!

LIVING ON LAND

The increase in complexity and specialisation of vegetative and reproductive features associated with colonisation of the land surface is the key to understanding the remarkable diversity of plant life on earth. The differences between living in water and living on land are enormous, particularly with respect to two critical aspects of plant metabolic function—water balance and gas exchange. Many of the evolutionary modifications of plants relate to these factors.

Algae are surrounded by water. Passive diffusion across cell walls keeps them in equilibrium with the surrounding environment; there is no need for special water-absorbing or water-conducting structures, and desiccation is not a problem except when seaweeds are exposed on rock platforms at low tide. In contrast, land plants inhabit an environment where water is limiting and usually confined to the soil in which they grow. They require specialised organs and complex tissues for several functions:

- roots for extracting water and dissolved nutrients from soil;
- vascular tissue for transporting water and nutrients to above-ground parts of the plant;
- a protective, water-resistant coating to minimise water loss to the atmosphere;
- a system of support in an aerial environment.

The degree to which these features are developed largely determines growth form and ecological tolerance of plants.

Specialisation of vegetative features

Cuticle: a waterproof barrier

Plant **cuticle**, composed of insoluble lipid polymers and waxes, is deposited over the entire above-ground surface of plants, and was essential for land colonisation. It is completely impermeable and acts as a waterproof barrier to prevent diffusion of water from turgid plant cells to the drier atmosphere. Cuticle also filters a substantial component of ultraviolet (UV) radiation from sunlight.

Sporopollenin: protecting spores

Sporopollenin is a polymer, tougher even than lignin but with similar properties, composed chiefly of carotenoids. Some green algal cells and spores have a sheath of, or are impregnated with, sporopollenin. This protects them from attack by grazing organisms and micro-organisms. It also makes spores of land plants (Fig. 37.6) tough and flexible and resistant to biodegradation; for example, it is found in the wall of pollen grains of flowering plants, and is the reason that pollen grains preserve well as fossils.

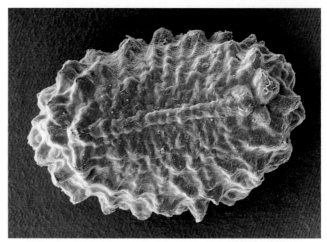

Fig. 37.6 Spores of land plants, for example, *Polypodium formosanum* (grub fern or caterpillar fern), are resistant to desiccation and biodegradation

Stomata: controlling gas exchange and water loss

In aquatic environments, CO_2 is in solution (as HCO_3^-) and diffuses across cell walls for algal photosynthesis. In terrestrial environments, CO_2 is a gaseous component of air. Because of the water-resistant nature of the cuticle covering the outer plant surfaces, CO_2 cannot be absorbed directly from the air. Instead, it enters through small pores in the plant surface, where it is absorbed by the moist walls of internal cells. Some liverworts (p. 802) have simple pores for gas exchange that also expose internal parts of the plant to water loss. Evolution of stomata bordered by guard cells, which control the opening and closure of pores (Chapter 18), must have allowed plants to expand their range and inhabit drier environments. Stomata are present in modern hornworts, mosses and vascular plants, as well as simple, ancient fossil plants such as *Rhynia* (see Box 37.2).

Vascular supply and lignin: transport and support

Vascular supply of ferns, fern allies, and seed plants consists of phloem for the transport of sugar, and xylem for the transport of water and mineral ions throughout the plant (Chapters 17, 18). The cell walls of xylem tracheids, vessels and fibres are conspicuously reinforced by a rigid layer of lignin that is deposited between the cellulose microfibrils.

Biochemical conversion of primary metabolic products, such as sugars and amino acids, to secondary compounds, such as lignin (together with alkaloids and tannins), would have protected early land plants from UV radiation and pathogenic fungi. The ability to form lignin proved to be of great importance because it gave strength to upright terrestrial plants. Lignin prevents xylem cells from collapsing under hydrostatic pressure that is developed around them. Greater structural support, together with a conducting system, allowed plant size to increase. The largest land plants, trees, produce large amounts of wood through secondary growth.

Stems, roots and leaves: division of labour

Early vascular land plants (e.g. *Rhynia*, Box 37.2) were small, rootless and leafless. They had simple upright, photosynthetic stems, and simple underground stems, **rhizomes**, with extensions of epidermal cells, **rhizoids**, for anchorage to the substrate and absorption of water and dissolved nutrients. True roots with a vascular system and aerial leaves for photosynthesis developed later (Fig. 37.7).

> At various stages in their evolution, land plants developed cuticles, resistant spores, vascular tissue, lignin, stomata, stems, roots and leaves, which overcame problems of water balance, gas exchange and physical support.

Alternating generations and sexual reproduction

Remember from Chapter 14 that there are two alternating generations in the life cycle of plants. The *gametophyte* generation, which is *haploid*, produces male and female gametes by mitosis. Male and female gametes fuse and the zygote grows into the *diploid sporophyte* generation. The diploid sporophyte produces spores by meiosis, which are dispersed and germinate into new haploid gametophytes.

The successful transition from water to land was associated with the sperm-producing and egg-producing cells being protected by a layer of sterile jacket cells, forming the sex organs **antheridia** (male) and **archegonia** (female). Unlike their green-algal relatives, land plants also retain the zygote on the gametophyte, where the developing embryo is protected and nourished.

In land plants, the haploid and diploid generations are markedly *heteromorphic*, that is, gametophyte and sporophyte differ in size and shape. There is a distinct evolutionary trend regarding the contribution of each generation to the life cycle. In bryophytes, the most primitive plants, the gametophyte is the dominant, photosynthetic generation. The sporophyte is simple in structure, short-lived and totally dependent on the gametophyte for nutrition. In seed plants, the opposite is true. The conspicuous plant is the sporophyte generation. The gametophyte is very reduced, commonly just a few cells. Ferns have a dominant sporophyte but a gametophyte that is not as reduced as those of seed plants.

Several anatomical features are correlated with the diploid generation. They provide insight into the functional reasons for the evolutionary change from the gametophyte stage being dominant, as in liverworts and mosses, to the sporophyte being dominant, as in vascular plants. Stomata and lignified vascular tissue only occur in sporophytes. The rare examples of tracheids in gametophytes are polyploid ferns and fern allies, that is, the gametophytes are not strictly haploid. If diploidy is essential for these features, then complex growth forms could only have occurred in the sporophyte. Also, the diploid provides for greater genetic variability and expression of alleles.

Evolution of pollen and seeds

In the life cycle of some ferns, fern allies and seed plants (see Table 37.1), two types of spore are produced by the sporophyte: **megaspores** in **megasporangia** and **microspores** in **microsporangia**. Megaspores germinate into female gametophytes, which bear egg cells in archegonia; microspores develop into male gametophytes, which produce sperm in antheridia. Having two sorts of spores is referred to as **heterospory** (having one type of spore is homospory). Heterospory, giving rise to *separate* male and female gametophytes, allowed the evolution of pollen, enabling wind and animal dispersal of male gametes, and the ovule, which houses an egg cell and, after fertilisation, develops into a seed.

> In land plants, haploid and diploid phases are markedly different. Haploid gametophytes produce sex organs, antheridia and archegonia, while diploid sporophytes produce spores by meiosis. Heterospory (development of two types of spores) allowed the evolution of pollen and ovules.

BRYOPHYTES: LIVERWORTS, HORNWORTS AND MOSSES

Non-vascular plants, commonly called bryophytes, include **liverworts** (phylum Hepatophyta), **hornworts** (phylum Anthocerotophyta) and **mosses** (phylum Bryophyta). The similarity of these plants lies in their lack of true vascular tissue. They have a photosynthetic, free-living gametophyte, which is the dominant generation compared with their simple and largely 'parasitic' sporophyte. These features suggest bryophytes are early lineages of land plants (Fig. 37.7). However, fossils of non-vascular land plants, clearly recognisable as bryophytes (Fig. 37.8), are not as old as are those of vascular plants. Bryophytes first appear in the late Devonian, about 360 million years ago. Because non-vascular plants are not easily preserved, they are rare and older fossils may yet be found.

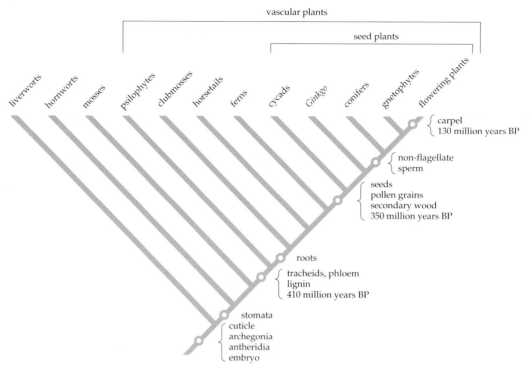

Fig. 37.7 Phylogeny of living groups of land plants, showing the evolution of important features and approximate age of oldest known fossils for some groups

Fig. 37.8 Early Cretaceous fossil (about 120 million years old) from south-west Victoria clearly recognisable as a thalloid liverwort

Gametophytes

Gametophytes are the conspicuous generation of the life cycle (Fig. 37.9) but rarely grow larger than several centimetres in height. They are usually attached to the substrate by elongated cells or filaments of cells, rhizoids. Male gametes (antherozoids or spermatozoids) are released from sac-like antheridia (Fig. 37.10a) and swim using their flagella to the archegonia (Fig. 37.10b), entering through the neck and fertilising the non-motile egg cell *in situ*. Sexual reproduction involving motile sperm requires the presence of free water (rain, dew, etc.) and, together

with the absence of vascular tissue, explains why the larger bryophytes are mainly confined to moist environments.

Liverworts have either thalloid or leafy gametophytes (Fig. 37.11). The common garden and pot-plant weeds *Marchantia* (Fig. 37.11a) and *Lunularia* both have a relatively large and conspicuous thallus that is anchored to the soil by a mat of rhizoids. They are easily recognised by the circular or crescent-shaped cups on their surface that contain small, asexual, vegetative propagules, **gemmae**. In favourable conditions, they produce sexual reproductive organs on peculiar umbrella-shaped structures.

Marchantia and its relatives are among the most complex of the liverworts. The thallus is differentiated into a thin upper spongy layer of photosynthetic cells. These enclose airspaces that open by pores to the external environment, allowing CO_2 uptake. Below the photosynthetic tissue, there is a thick lower layer of more closely packed storage cells (Fig. 37.11b). However, the majority of liverworts are smaller, with a delicate leafy habit. Although they are divided into stem and simple leaves, they are far less complex with respect to tissue differentiation and sex organ production than *Marchantia*.

The half dozen genera of hornworts resemble thalloid liverworts (Fig. 37.12). Mosses, on the other hand, have leafy gametophytes, but the leaves are extremely diverse in relation to their function and the environment.

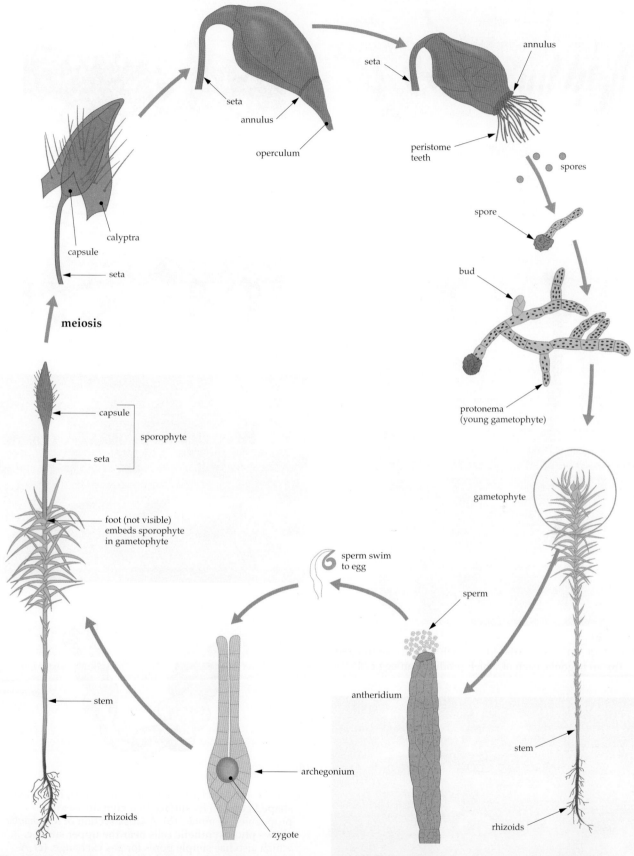

Fig. 37.9 Life cycle of a moss (phylum Bryophyta). Sperm and egg cells are produced in antheridia and archegonia respectively on the gametophyte (haploid) plant. Sperm swim to the neck of an archegonium where one fertilises the single egg cell. The zygote divides and develops into the sporophyte (diploid) plant, which consists of a seta and capsule, and is parasitic on the gametophyte. Haploid spores form by meiosis, are released from the capsule, and germinate into filamentous protonemas, from which leafy gametophytes grow

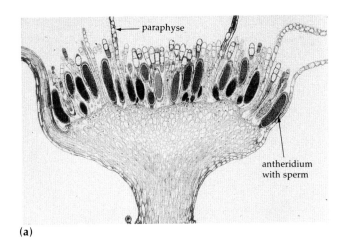

(a)

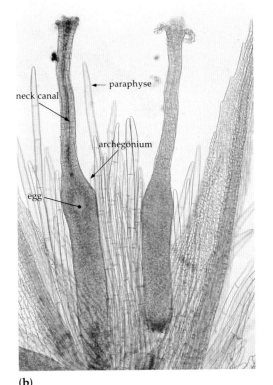

(b)

Fig. 37.10 (a) Moss antheridia, which produce sperm and (b) archegonia, each of which produces an egg cell

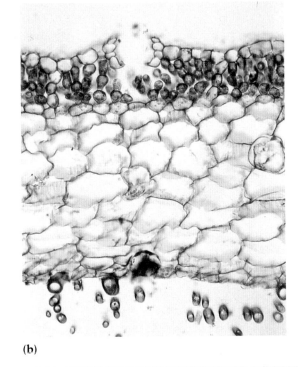

(b)

(c)

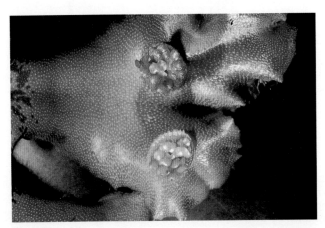

Fig. 37.11 (a)

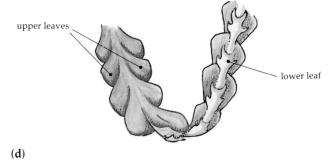

(d)

Fig. 37.11 (a) *Marchantia*, a thalloid liverwort, has circular-shaped cups on its surface that contain asexual propagules, gemmae. (b) A cross-section of *Marchantia* shows photosynthetic cells near the upper surface, which also has simple pores for gas exchange. (c) A leafy liverwort. Leaves are simple, consisting of a single layer of undifferentiated cells, with no midrib unlike moss leaves. (d) The leaves are generally arranged in two rows, with a third row of reduced leaves on the undersurface

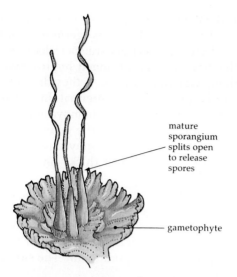

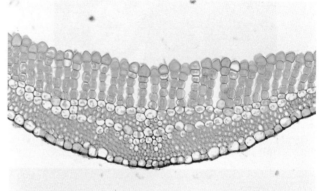

Leaf cells of peat moss, *Sphagnum*, are of two types: a network of small, photosynthetic cells entwine large, perforated cells that are devoid of contents and act in a water-holding capacity, much like a sponge (Fig. 37.13). This water-holding capacity, which is an important ecological feature of peat bogs in the highlands of Australia, also makes *Sphagnum* a favoured component of horticultural potting mix. The upper leaf surface of *Polytrichum* and its relatives is ornamented with parallel, longitudinal files of cells several cells high, which increase the photosynthetic area of the leaf (Fig. 37.14).

Fig. 37.12 The hornwort *Anthoceros*. Cavities within the thallus are filled with mucilage and house cyanobacteria (*Nostoc*), which fix atmospheric nitrogen

Fig. 37.14 Transverse section of a leaf of the moss *Polytrichum*, showing the upper leaf surface with longitudinal files of cells that are thought to increase the photosynthetic area

(a)

Although mosses lack specialised strengthening and supporting tissue and are generally small in stature, some species are quite stiff and robust. The largest moss, *Dawsonia superba* (Fig. 37.15a) grows up to about 40 cm in height in moist eucalypt forests of south-eastern Australia. In *Dawsonia* and most other mosses, antheridia and archegonia are borne in separate clusters at the tips of gametophytes (Fig. 37.15b). Antheridia form in leafy rosettes, and are interspersed with sterile hairs, **paraphyses**, that presumably function in trapping moisture and preventing desiccation of the sex organs (see Fig. 37.10).

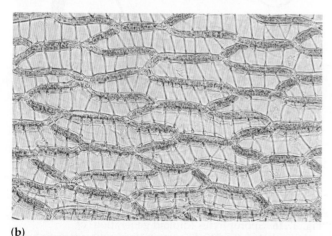

(b)

Fig. 37.13 (a) Peat moss, *Sphagnum*, has leaves with (b) two types of cells: small and photosynthetic or large and lacking cytoplasm. These large cells have an enormous water-holding capacity, which is why peat bogs in the highlands of south-eastern Australia are an important store of water

Sporophytes

Bryophytes retain the egg cell within the archegonium on the gametophyte (Fig. 37.9). Only the sperm are released to rely on free water (rain, dew, etc.) to swim to the archegonia down the neck canal to the egg cell, attracted by chemical substances produced during the breakdown of the archegonial neck canal cells. After fertilisation, the sporophyte is retained and develops on the gametophyte. The sporophyte remains dependent upon the gametophyte for nutritional requirements.

(a)

(b)

Fig. 37.15 (a) The largest moss, *Dawsonia superba*, grows in moist mountain ash (*Eucalyptus regnans*) forest.
(b) Antheridia (male), shown here, and archegonia (female) are usually clustered at the tips of gametophytes, in leafy rosettes (moss heads)

The sporophyte of bryophytes is simple in structure, consisting of a single sporangium, **capsule**, usually borne terminally on a slender stalk, **seta**. It is embedded in the gametophyte by a **foot**, which contains specialised transfer cells for nutrient exchange with the gametophyte. Hornwort and moss sporophytes have some capacity for photosynthesis—for maintenance but not enough for growth.

Liverwort sporophytes usually develop a seta that is white and delicate, elongating only when the spore capsule is mature and the spores are ready to be shed (Fig. 37.16a). The brown, spherical capsule either disintegrates or dehisces along one or more, usually

longitudinal, lines of weakness, releasing spores and elaters. The haploid spores germinate into new gametophytes. The **elaters** are specialised, elongated cells that have helically arranged, hygroscopic (water-absorbing) wall thickenings. As they dry out, stresses build up because of the differential thickening of the cell walls, causing the elaters to move and flick spores from the capsule. However, the genus *Riccia* has no elaters or seta and the sporophyte is no more than a temporary wall, one cell thick, embedded in the thallus.

Hornwort sporophytes (Fig. 37.12) lack a seta. The sporophyte consists of a foot and a long green sporangium. A meristem between the foot and the base of the sporangium is responsible for the growth and elongation of the sporangium. The surface of the

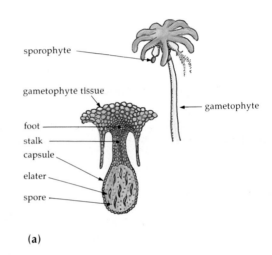

(a)

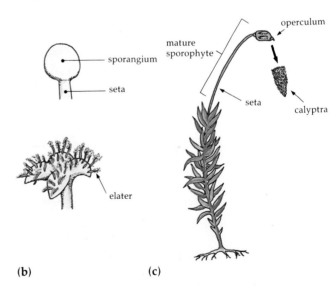

(b)   **(c)**

Fig. 37.16 Sporophytes (diploid generation) of bryophytes. **(a)** The liverwort *Marchantia*, consisting of a sporangium, here emerging from an archegonium, elevated on a stalk (archegoniophore). **(b)** The liverwort *Frullania* splits open to release spores. **(c)** Many species of mosses have a stiff seta, with a sporangium (capsule) crowned by a calyptra (part of the old gametophyte archegonial neck), which is shed

sporangium is covered with stomata, making hornworts the simplest living land plants to possess these structures. They shed their spores and elaters through one or more vertical slits that extend down from the apex.

Some mosses, such as *Sphagnum* (peat moss), also lack a seta. However, all mosses lack elaters, and advanced mosses possess a capsule that is borne on a stiff, wiry seta that elongates before sporangium maturation. The capsule is generally more elaborate in mosses than in liverworts. It is often crowned by a **calyptra**, which is modified tissue of the old gametophytic archegonial neck (Fig. 37.16c). The apical portion of the capsule is a cap-shaped **operculum**. It is shed when the spores are mature to reveal one or two specialised rows of tooth-like structures around the top of the sporangium known as the **peristome**. The peristome shelters the spores when they are not being released. In mosses with a double peristome, the spores must also filter through the inner row of teeth (Fig. 37.17). The peristome teeth are impregnated with a lignin-like compound. Changes in humidity cause the teeth to reflex, opening and closing the aperture of the capsule, thus controlling spore discharge. Moss spores germinate into a **protonema**, a mass of branched, single-celled filaments (see Fig. 37.9). Upright leafy gametophytes develop as buds along the protonema.

> Bryophyte plants are small, photosynthetic, free-living gametophytes that lack vascular tissue. Sexual reproduction requires water for sperm to swim from an antheridium to an egg in an archegonium. The sporophyte grows on and is dependent on the gametophyte for nutrition.

VASCULAR PLANTS

Vascular plants, characterised by the presence of xylem and phloem, include fern allies, ferns and seed plants (Fig. 37.7). Their tracheids and vessels are resistant to decay and have been detected in fossils of early land plants (Box 37.2). In contrast to bryophytes, the sporophyte of vascular plants is the dominant generation of the life cycle, existing as a free-living plant. The gametophyte is usually short-lived, and degenerates once the sporophyte is established.

> The dominant generation of living vascular plants is the sporophyte, which is characterised by phloem and lignified xylem.

FERN ALLIES

The term **'fern allies'** refers to a loose assemblage of living, spore-producing vascular plants that are classified in three phyla (Table 37.1): **Psilophyta** (*Psilotum* and *Tmesipteris*), **Lycophyta** (*Lycopodium, Selaginella* and *Isoetes*) and **Sphenophyta** (*Equisetum*).

Psilophytes: 'living fossils'

Two closely related genera, *Psilotum* and *Tmesipteris* (sometimes called whisk ferns), are the most primitive living vascular plants (Fig. 37.7). The aerial shoots of the sporophyte branch simply and dichotomously.

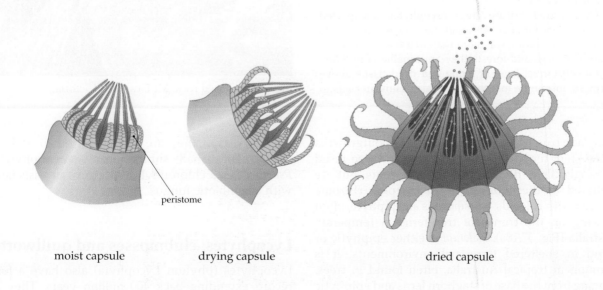

moist capsule   drying capsule   dried capsule

peristome

Fig. 37.17 The peristome at the top of the sporangium (capsule) controls the release of spores in most mosses; in some mosses it consists of two rows of interlocking teeth. As the capsule dries out, the teeth pull away to allow spore dispersal by wind

BOX 37.1   BRYOPHYTES IN AUSTRALIAN ENVIRONMENTS

Although bryophytes occur in virtually every habitat where vascular plants are found, except the sea, their anatomical and physiological properties make them ecologically very different.

Although they lack lignified vascular tissue, many have central strands of elongated cells that are capable of conducting water. However, mostly they are **ectohydric**, conducting water up the outside of the plant by capillarity among the small spaces between leaves and stem, like a wick. Fine hair-like rhizoids, which anchor them to the ground and often assist this capillary conduction by coating parts of the stem like cotton-wool, are generally not responsible for absorbing and conducting nutrients from the substrate. That function is performed by the whole plant, which soaks up, like a sponge, the mineral nutrients in rainwater or dissolved from dust blown onto the plant surface.

Apart from rainforest species, the majority of Australian bryophytes are **poikilohydric**, that is, they are resurrection plants capable of withstanding desiccation to dryness, sometimes for months on end, but rehydrate whenever rain returns. They assume full metabolism within minutes, or at most a few hours, only to dry up and suspend activity again when dry conditions resume. This opportunistic physiology is the secret of their success in intermittently dry habitats, such as rock surfaces, tree trunks and even city buildings, where 'rain tracks' are generally colonised by bryophytes (see figure). In the Mallee and similar areas of Australia, this characteristic permits mosses and liverworts, together with lichens, cyanobacteria and the roots of dead flowering plants, to form tenacious soil crusts that hold the soil in place even in times of drought. The importance of this role has long been underestimated, and it was the disruption of the soil crust by clearing and ploughing that allowed the notorious Mallee dust storms to occur.

One further feature gives bryophytes unexpected ecological importance. Although they are surface plants seldom penetrating soil more than to a few centimetres depth, in bogs and forests they are capable of building up a substantial layer of living and dead tissue above ground, forming a mat through which rainwater has to filter before reaching plant roots. These mats have cation-exchange properties, preferentially retaining some ions while releasing others, and thus modifying the nutrition of the entire ecosystem.

Perhaps the most remarkable of all bryophyte properties is their totipotency, that is, their ability to regenerate new plants from casual fragments even as small as one or two cells (Chapter 16). There are many bryophytes for which sexual reproduction and spore formation are unknown and may never occur; these bryophytes, and even most others, rely on vegetative reproduction from detached fragments blown or washed away to new habitats where they can establish. This extraordinary power of asexual reproduction, which can be shown easily by air-drying a handful of mosses, grinding them to a powder, and sowing on moist earth or peat, probably explains the prevalence of mosses and liverworts throughout the world today. Furthermore, in the case of prostrate species, especially thallose liverworts, but also many other mosses and liverworts, new plant material is produced at the tip, as the product of the apical cell and its divisions. As tissue ages and dies at one end of the plant, new growth extends at the other, so that the plant slowly creeps forwards over the ground, subdividing as it goes.

Moss on exposed rock, Mt Baw Baw, Victoria

They lack true leaves but produce small outgrowths of the stem (enations) that bear two or three fused sporangia. They do not produce true roots, but are anchored to the substrate by a creeping rhizome covered with rhizoids. *Tmesipteris* is an epiphyte, often growing on the trunks of tree ferns in temperate Australia (Fig. 37.18a). *Psilotum* is either epiphytic or found in sheltered terrestrial environments. It is common in tropical Australia, often found in trees, hanging from the base of staghorn ferns and epiphytic orchids (Fig. 37.18b). The growth form of both genera, but particularly *Psilotum*, resembles closely the earliest vascular plant fossils, such as *Rhynia* (Box 37.2), and for this reason they are often referred to as living fossils. The small subterranean gametophyte of *Psilotum* lacks chlorophyll and grows in association with a symbiotic fungus.

Lycophytes: clubmosses and quillworts

Lycophytes (phylum Lycophyta) also have a fossil record extending back 400 million years. They are the first plants to have evolved true roots (Fig. 37.7). Roots are initiated deep within the tissue of the stem, often growing downward through it for some distance

BOX 37.2 EARLY VASCULAR PLANTS

The oldest vascular plant fossils are found in late Silurian sediments around 410 million years old, although much of our knowledge of the structure and diversity of early land plants comes from slightly younger fossils, from the early Devonian (about 400 million years BP). Current research on plant fossils of these ages is revealing that early land floras in different parts of the world were relatively uniform and shared common taxa. Plants were generally quite small, ranging from several centimetres to about a metre in height. The larger specimens may have grown semisubmerged in swampy environments and relied on buoyancy for support. They were all relatively simple in construction, consisting of leafless, dichotomously branched stems with a small central strand of primary vascular tissue. They reproduced by spores.

These extinct plants are classified into three phyla of which the phylum Rhyniophyta is the most primitive. Given the critical part these plants play in our understanding of early land plant evolution and the often fragmentary nature of their fossil record, it is not surprising that aspects of their structure are under some dispute.

Rhyniophyta

Rhyniophytes were small plants with creeping rhizomes that produced erect, leafless, dichotomous stems with terminal sporangia. They are the simplest of the early land plants, characterised by the two most widely known examples, *Rhynia* (Fig. a) and *Cooksonia* (Fig. b). Two species of *Rhynia* were described from exquisitely preserved silicified peat beds near the Scottish village of Rhynie. Several features of these plants attest to their existence in a terrestrial habitat. They are covered by a layer of cuticle, they have stomata, their meiotic spores possess a resistant sporopollenin wall, and the stems have a darker zone of cells forming a central strand where vascular tissue would be expected. *Cooksonia*, named after the eminent Australian palaeobotanist Isabel Cookson, was a similar but smaller plant.

Close examination of the central strand cells of *Rhynia* and *Cooksonia* has yielded some interesting observations. Firstly, central strand cells of *Cooksonia* and one species of *Rhynia* (*Rhynia major*, now classified as *Aglaophyton*) have smooth, unthickened cell walls. These cannot really be defined as tracheids as we know them in living vascular plants. Central strand cells of the second species of *Rhynia* (*R. gwynne-vaughanii*) possess wall ingrowths that form rings or spirals, but these are very different in size and structure to tracheids of modern plants. Consequently, the status of rhyniophytes as true vascular plants is not proven to everyone's satisfaction. There has also been some debate as to whether the *Rhynia* plant was sporophytic or gametophytic. However, their spores have a clearly defined trilete (spore tetrad) mark, which is evidence that they resulted from a meiotic division. They have stomata, which, in living plants, occur only on sporophytes. This combined evidence suggests that *Rhynia* was a sporophyte. Several unusual fossils that may be gametophytes are known from the Rhynie chert locality.

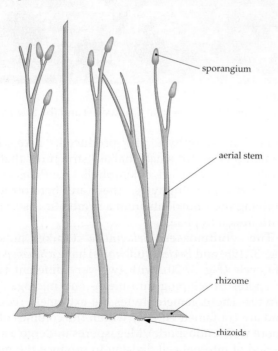

(a) *Rhynia* is one of the simplest of the early extinct vascular plants. *Rhynia* fossils are of early Devonian age (about 400 million years old). The plant had an aerial branch system about 20 cm high, with a vascular cylinder of tracheids and phloem-like tissue, stomata and cuticle. Spores were produced in sporangia on the ends of aerial branches. There were no roots, but a prostrate rhizome system with rhizoids

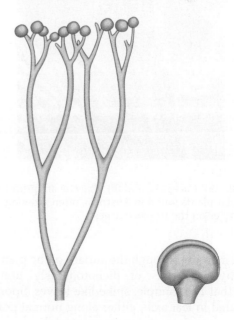

(b) An early land plant fossil, the rhyniophyte *Cooksonia*

(a)

(b)

Fig. 37.18 **(a)** *Tmesipteris* and **(b)** *Psilotum* are primitive vascular plants found in Australia, often growing as epiphytes on the trunks of trees

(a)

(b)

Fig. 37.19 Clubmosses: **(a)** *Lycopodium* and **(b)** *Selaginella*

before emerging through the surface of the plant. The sporophyte consists of dichotomously branched shoots that bear simple, spike-like leaves. Sporangia are located in leaf axils, either along normal portions of stem or condensed into cones, **strobili**.

The clubmoss *Lycopodium* (Fig. 37.19a) exhibits the *homosporous* life cycle, thought to be the primitive condition for plants. Only one type of spore, and

hence gametophyte, is produced. *Lycopodium* gametophytes are small thalloid structures that are subterranean and lack chlorophyll. Like *Psilotum* and *Tmesipteris* gametophytes, they are heterotrophic, deriving their nutrients from a symbiotic association with fungal hyphae.

The clubmosses *Selaginella* (spike mosses, Fig. 37.19b) and *Isoetes* (quillworts) have a *heterosporous* life cycle (Fig. 37.20), with two very different types of spores developing into mega- and microgametophytes. These gametophytes are extremely reduced and are contained entirely within the confines of the spore wall (endosporic). Megaspores undergo a short period of internal cell division to produce the megagametophyte (female), which develops archegonia with eggs. The archegonia are exposed when the megaspore wall splits along the trilete (spore tetrad) mark. Microspores divide internally to produce the microgametophyte (male). The microgametophyte is essentially a single antheridium. The antheridium produces many biflagellated sperm, which are released when the microspore splits open.

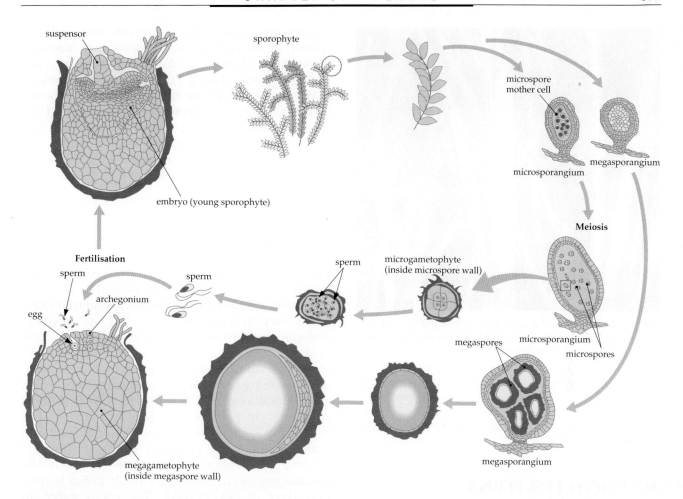

Fig. 37.20 Heterosporous life cycle of *Selaginella* (phylum Lycophyta). The sporophyte (diploid) is the dominant generation. On the sporophyte, two kinds of haploid spores (megaspores and microspores) develop in sporangia. The male gametophyte (microgametophyte) develops within the microspore and produces biflagellate sperm, which swim to the egg cell. The female gametophyte (megagametophyte), which produces archegonia, each with an egg cell, develops within the megaspore. As the megagametophyte grows, the megaspore wall splits and the gametophyte protrudes to the outside. After fertilisation, the young sporophyte develops within the megagametophyte tissue for a time

Sphenophytes: horsetails

Horsetails (phylum Sphenophyta) have their origins in the late Devonian (360 million years BP). Today the phylum consists of a single genus, the herbaceous, often weedy *Equisetum* (Fig. 37.21). It mostly occurs in the Northern Hemisphere, but horsetails were once diverse and formed an important component of the earth's vegetation. In the Carboniferous (360–290 million years BP) giant horsetails (*Calamites*) were large trees. Together with tree-like lycophytes, these giant horsetails dominated forests in massive coal-forming swamps across what is now North America and Europe (Chapter 31).

Equisetum consists of a creeping, underground rhizome that produces upright stems. Small and insignificant leaves are borne in sheath-like whorls at nodes in the stem, and the stem surface itself is green and photosynthetic. The stem epidermis is impregnated with crystals of silicate, which give it

a coarse, abrasive texture. During the Middle Ages, *Equisetum* was extensively collected and used as an abrasive, hence its more antiquated common name, 'scouring rush'.

Sporangia are produced in small **strobili** (cones) borne terminally on either normal vegetative or specialised reproductive shoots. *Equisetum* spores are unusual. The outer layer of the spore wall unravels to form four sporopollenin threads that remain attached to the spore. These threads are hygroscopically sensitive, and their movement assists in spore dispersal. They are termed elaters but they have a different structural origin to the cellular elaters of either liverworts or hornworts. Spore dispersal by movement of structures that are differentially thickened with cellulose and lignin or sporopollenin is an interesting feature that has evolved independently in liverwort and hornwort elaters, moss peristome teeth, *Equisetum* spores and, as we will see in the following, the sporangia of true ferns.

Fig. 37.21 *Equisetum*, the only living representative of the phylum Sphenophyta (horsetails), some of which were large trees in Carboniferous forests

'Fern allies' are a loose assemblage of spore-producing vascular plants. Lycophytes were the first plants to evolve true roots, and tree forms of extinct lycophytes and sphenophytes (horsetails) formed Carboniferous forests.

FILICOPHYTES: FERNS

With over 12 000 living species, ferns, phylum **Filicophyta**, are the most diverse and conspicuous of the spore-producing vascular plants. Although a large proportion of species are tropical, ferns are well represented in cool temperate climates. In Australia and New Zealand, they range in size from small, delicate filmy ferns and floating aquatics (see Box 37.3) to tree ferns exceeding 20 m in height (Fig. 37.22). Epiphytes, climbers and xerophytes add to a remarkable variety of ecological preferences and life strategies.

Ferns are an assemblage of plants that probably consist of several distinct evolutionary lineages, representing stages of evolution between fern allies and seed plants (Fig. 37.7). They have their origins in the late Devonian. By far the greatest proportion of living species are **leptosporangiate ferns**. They possess small, delicate sporangia (usually with less than 64 spores) that develop from a single cell, compared with the **eusporangiate ferns**, including fewer species, that possess massive sporangia (with 256 or more spores).

Fern sporophytes

The fern sporophyte, like that of the fern allies, is the free-living, dominant generation (Fig. 37.23). It consists of a stem bearing true leaves and true roots. Stems vary in form, but often they are creeping, underground rhizomes, such as in bracken, *Pteridium esculentum*. Only in a few species are stems erect and trunk-like as in tree ferns. Many fern stems have a dense covering, indumentum, of scales or hairs, which are modifications of epidermal cells. Leaves (**fronds**) are prominent parts of most fern sporophytes. The leaf lamina is often deeply dissected or divided into leaflets (**pinnae** and **pinnules**), giving the frond a feathery appearance. A distinctive feature is the unusual manner in which fern fronds unfurl in their final stage of development. The young rolled-up leaves are commonly called 'fiddle-heads' (see Fig. 37.22) because of their resemblance to the end of a violin. Although many ferns contain toxic substances, 'fiddle-heads' of certain species are considered a culinary delicacy in some countries.

(a)

(b)

Fig. 37.22 **(a)** Australian tree ferns. **(b)** The young rolled-up leaves of ferns are called 'fiddle heads'

BOX 37.3   WEIRD, WATERY AND WEEDY FERNS

There are three families of unusual aquatic ferns and each has representatives in Australia. They grow in water or mud and form two types of spores, microspores and megaspores. Spore production occurs in special nut-like structures, sporocarps. Gametophytes are reduced to the development of sperm within a single antheridium within the microspore, and an archegonium and egg cell within the megaspore.

Marsilea

Marsilea (family Marsileaceae) includes more than 50 species, many of which occur in tropical Africa and Australia. Australian species are commonly called nardoo. They do not look like other ferns, having leaves with two pairs of leaflets, rather like four-leaf clover (Fig. a). The leaves of some forms often have 'sleep movements', folding or unfolding according to the light intensity. The leaves arise from a long creeping rhizome with adventitious roots at the nodes. The plant is usually rooted in mud in shallow water, with the leaves floating on the surface.

The sporocarps occur near the base of the leaves. They are resistant to desiccation and can survive for up to 20 years when ponds dry out. Common nardoo, *M. drummondii*, is widespread in mainland Australia, especially in drier inland areas. Aborigines ground up the nutritious sporocarps into a starchy paste to bake as cakes. *Marsilea salvatrix* provided some food for the men on the fateful Burke expedition in central Australia, which is how the species came to be named.

Azolla

Azolla (family Azollaceae) is a cosmopolitan genus, two species of which occur in Australia. *Azolla* is a small, free-floating aquatic fern (Fig. b). It has tiny overlapping leaves, green to red in colour, that cover a branching stem. This gives individual plants a triangular shape. Because cyanobacteria live in a cavity at the base of each leaf and fix atmospheric nitrogen, the fern is grown in rice fields of China and Vietnam as a 'green' fertiliser. *Azolla* multiplies easily by fragmentation and grows in dense populations that can cover extensive areas of still water. It can smother other organisms, such as troublesome mosquito larvae.

Salvinia

Salvinia (family Salviniaceae) is also free-floating and is related to *Azolla*. These plants have leaves in whorls of three—two floating and one submerged. The submerged leaf is highly dissected resembling a root system, but there are no true roots. The upper surface of the floating, boat-shaped leaves are covered with silvery, stiff hairs, which make them waterproof.

Salvinia occurs in tropical regions of the world, and *S. molesta* (Fig. b) was introduced into Australia for use in dams, ponds and fish tanks as an ornamental aquatic plant and refuge for fishes. It is sterile and believed to be of hybrid origin. Its effective vegetative reproduction and very fast growth have allowed it to become a major weed species of aquatic ecosystems in many tropical countries.

(a)

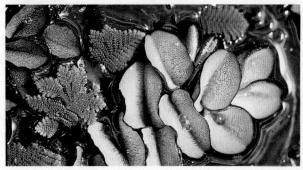

(b)

Water ferns: **(a)** *Marsilea*; and **(b)** *Azolla* (red)

Sporangia are usually clustered in **sori**. These are found on the margins or undersurfaces of either normal or specialised reproductive fronds. In many ferns, they are covered by a protective scale-like membrane or **indusium** (Fig. 37.24). Sporangia of leptosporangiate ferns are small delicate structures consisting of a spore sac, with walls that are only one cell thick, and a short, slender stalk. Embedded in the sporangium wall is an encircling ring of enlarged cells, the **annulus**, with lignin-thickened walls. When the sporangium is ripe, stress develops in the annulus, causing the sporangium to rupture

in a non-thickened zone, the **stomium**, explosively releasing spores into the environment (Fig. 37.23). This hygroscopic sensitivity is similar to that in liverwort elaters and moss peristomes.

Fern gametophytes are small, free-living, heart-shaped thalloid structures that bear rhizoids and sex organs on their ventral (lower) surface. Globular antheridia release motile, multiflagellated sperm that swim to the protruding necks of embedded archegonia containing the eggs. Like bryophytes and fern allies, ferns rely on free water for sexual reproduction. The young sporophyte develops in

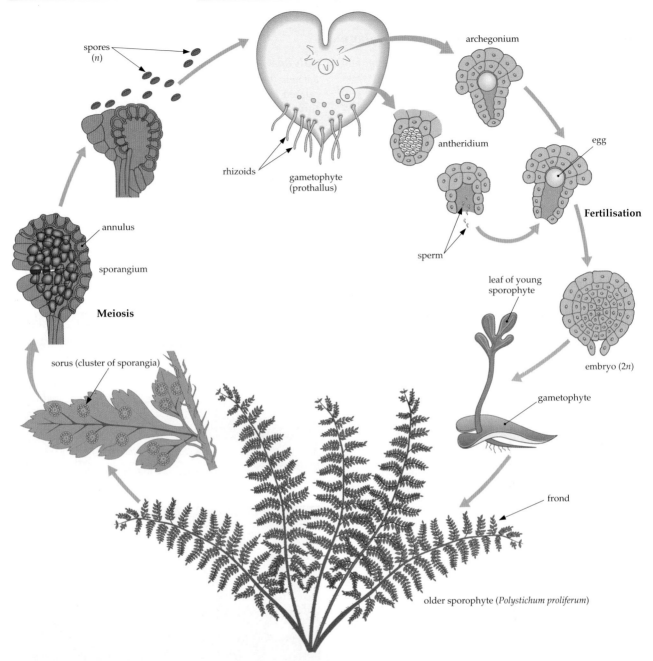

Fig. 37.23 Life cycle of an homosporous fern, such as *Polystichum proliferum*. Spores are produced by meiosis in sporangia located on the sporophyte (the conspicuous fern plant). The sporangium dehisces by the action of the annulus. Haploid spores germinate and develop into small, delicate, heart-shaped gametophytes (prothallus), which attach to the substrate by rhizoids. Each gametophyte forms antheridia and archegonia on the undersurface. Spirally coiled, multiflagellated sperm swim from an antheridium to the neck of an archegonium to fertilise the egg cell. The zygote divides immediately and a new sporophyte plant begins development, gaining nutrition from the gametophyte for a time

place on the gametophyte, which dies and decays once the sporophyte is established. Most ferns have a homosporous life cycle, but a group of peculiar water ferns are heterosporous (see Box 37.3).

> The sporophyte of homosporous ferns is photosynthetic, free-living, and the dominant phase. Like bryophytes, ferns have flagellated sperm and rely on free water for sexual reproduction. Fern antheridia and archegonia are borne on a small, free-living heart-shaped gametophyte (prothallus).

SEED PLANTS

There are five groups of living seed plants. Flowering plants, phylum **Magnoliophyta**, are by far the most numerous and well known of these, with about one-quarter of a million species, although conifers, phylum **Coniferophyta** (with about 600 species), and cycads, phylum **Cycadophyta** (about 150 species), are familiar to most people. The remaining two phyla are the bizarre **Gnetophyta** with three living genera, and

(a)

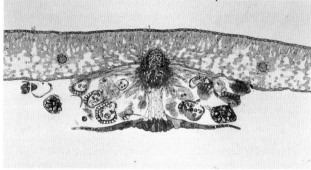

(b)

Fig. 37.24 **(a)** Filmy fern, *Hymenophyllum*, with sporangia clustered in sori partly covered by an indusium. **(b)** Transverse section of a sorus with indusium

Ginkgophyta with just a single living species, *Ginkgo biloba*. Cycads, *Ginkgo*, conifers and gnetophytes are commonly called 'gymnosperms', although they are not a single evolutionary lineage.

Seed plants have a number of derived features that are widely considered to represent 'major events' in land plant evolution (Fig. 37.7). These features include:

■ enclosure of the female gametophyte and embryo within an ovule, which becomes the seed;

■ microspores transported directly to the ovule as pollen (pollination);

■ secondary growth (vascular cambium).

Seeds from ovules

Evolution of the seed is one of the key reasons for the success of higher vascular plants. This depended on a heterosporous life cycle. As we saw on page 801, in heterospory there are two different types of spores, and the larger megaspores grow into megagametophytes, which produce egg cells. In seed plants, the megagametophyte is retained within the megasporangium, which is further surrounded and protected by one or more layers of cells, the integuments. Megasporangia of other vascular plants are not so protected. This whole structure (integuments enclosing megasporangium with developing megagametophyte) is the **ovule** (see Fig. 37.25). After fertilisation and development of the embryo, the ovule matures into a **seed** (Chapter 16).

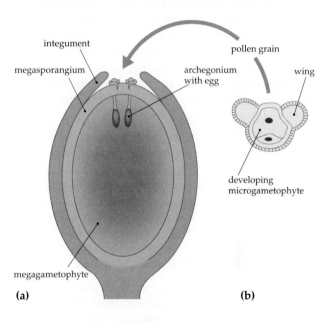

(a) **(b)**

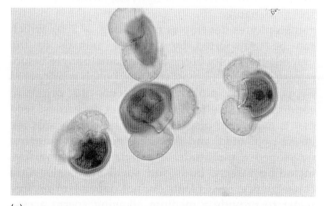

(c)

Fig. 37.25 Seed plants are characterised by ovules and pollen. **(a)** An ovule includes the developing (female) megagametophyte, with developing archegonia and egg cells, surrounded by protective integuments. An ovule becomes a seed after fertilisation. **(b)** Pollen grains encompass the (male) microgametophyte, in which sperm develop. **(c)** In pines, a pollen grain has two bladder-like wings for wind transport to the ovule

Within the megasporangium, four megaspores are produced by meiosis. Three of these degenerate, leaving only one to grow into the female gametophyte. In cycads, *Ginkgo* and conifers, the mature female gametophyte contains as many as several thousand sterile cells and several archegonia; after fertilisation of the egg cell, the embryo utilises the gametophytic remains (nucellus) as a nutritive source. In flowering plants, the megagametophyte is less elaborate, consisting of only eight cells (the embryo sac) of which one is the egg cell (Chapter 16).

The seed is significant because it is a protective structure encasing the embryo of the next sporophytic generation until conditions become optimal for its germination and growth. The embryo lies dormant, embedded in nutritive tissue and surrounded by a seed coat or testa derived from the one or two integuments which thicken and harden. Protected in this way and with a supply of stored food, the dormant embryo can be shed from the parent plant and dispersed by wind or animals.

Seed plants represent the other extreme to mosses and liverworts, in which the gametophyte is photosynthetic and free-living and the sporophyte grows *in situ* and is nutritionally dependent. In seed plants the opposite is true; the megagametophyte is reduced and retained within the dominant sporophyte.

Pollen transport of male gametes

A second significant feature of seed plants is their pollen. In heterospory, the smaller microspores develop into microgametophytes, which produce sperm (Chapter 16). In seed plants, these microgametophytes develop as pollen (Fig. 37.25b), which can be transported by wind or animals to the immediate vicinity of the female gametophyte before release of sperm. In most conifers, pollen is dispersed by wind. This probably represents the primitive condition, although it is possible that the earliest seed plants had water-dispersed microspores. Cycads are insect pollinated, and flowering plants have further developed a multitude of pollination mechanisms, including insects, birds and small mammals. As a consequence, seed plants are not dependent on free water to provide a medium allowing sperm access to eggs. This has allowed seed plants to develop a far greater ecological tolerance to dry environments, and is an important factor allowing their widespread dominance. Although some ferns, fern allies and bryophytes do inhabit quite xerophytic habitats, their opportunities for sexual reproduction are intermittent and limited.

Secondary growth

The production of woody tissue by a vascular cambium is a characteristic feature of seed plants and a few extinct but closely related plants (progymnosperms). Vascular cambium produces large quantities of secondary xylem (wood) to the inside of a stem or root and smaller quantities of secondary phloem (bark) to the outside (see Chapter 17).

Vascular cambium adds girth to stems and roots (the apical meristems generate length), and has a dual function. The tracheids and sieve cells produced by the cambium continually add new capacity for water and food transport within the plant. Older water-conducting elements become non-functional and filled with waste products. These cells, together with any lignified fibres, constitute a rigid support characteristic of tree trunks and branches. It is these two functions that have allowed woody plants to grow to massive sizes (Fig. 37.26).

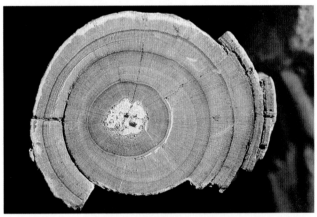

Fig. 37.26 Fossil conifer wood (210 million years old) from Antarctica. Secondary growth allowed seed plants to grow to a large tree size

> Important features of seed plants are: the enclosure of the female gametophyte and embryo within an ovule which becomes the seed; microspore transported as pollen (pollination); and secondary growth (vascular cambium).

Cycadophytes: ancient seed plants

Cycads (phylum Cycadophyta) are the most ancient of living seed plants (Fig. 37.7), and have a fossil record extending back early in the Permian (290–245 million years BP). Twenty-three extant species in four genera (*Cycas*, *Bowenia*, *Lepidozamia* and *Macrozamia*) occur in Australia.

The most common growth form of cycads consists of a thick, unbranched or sparsely branched trunk and a crown of pinnate leaves (Fig. 37.27), although stems of some species are subterranean. Below

(a)

(b)

Fig. 37.27 Australian cycad, *Macrozamia moorei*
(a) growing in Queensland and (b) a female seed cone

the axil of leaves) are completely lacking. In vegetative regions of the stem, these dichotomies are even, resulting in a branch point from which two stems arise. In reproductive regions, dichotomies are uneven, each resulting in a large woody cone and a dormant meristem that will later continue with stem growth. Ovules and microsporangia are borne in these prominent cones, on separate female and male plants (Fig. 37.27).

Cycads were long thought to be wind pollinated, but it is now clear that many, if not all, cycads are pollinated by beetles, one of the oldest insect groups. Although the boat-shaped pollen grains are transported directly to the ovules and free water is not required for gamete transfer, cycad sperm are multiflagellated, a condition retained from their spore-producing ancestors. Pollen grains germinate inside the micropyle to produce a microgametophyte of about five cells with a structure known as a 'feeder' pollen tube (haustorium) that penetrates nucellar tissue of the ovule and absorbs nutrients needed for its growth (Fig. 37.28). Two flagellated sperm cells are produced by each microgametophyte, and are liberated into the micropylar chamber close to the megagametophyte. The megagametophyte is a mass of starch-filled cells with several archegonia differentiated at the micropylar end of the ovule. Sexual fusion occurs when sperm contact an egg within this aqueous environment.

Like all seed plants, the megagametophyte remains enclosed by the ovule integument and retained on the parent sporophyte. The seed that develops has a fleshy outer coat, coloured red, orange, yellow, brown or bluish, and a thick inner coat. Seeds of *Cycas media*, although toxic, were a staple food for Aborigines in Arnhem Land and Cape York Peninsula. They were cracked open and their kernel pounded into flour, which was then leached of toxins and made into dough for baking. Seeds of *Macrozamia* were processed in a similar manner in coastal New South Wales and Queensland.

ground, cycads have a tap root that produces highly specialised lateral roots, coralloid roots, in which nitrogen-fixing cyanobacteria live.

Superficially, many cycads resemble small palm trees, but palms are flowering plants and the two groups are not related. Cycad stems have a vascular cambium that produces a limited amount of secondary growth. However, most of the tissue contributing to the thickness of the trunk is derived from primary thickening by a special meristem that develops at the base of the most recently formed leaves. A meristem somewhat like this is present in palms (see Box 17.1), which retain leaf bases on the trunk, hence the similarity in appearance.

Branching in cycads is dichotomous, that is, the shoot apex divides in two. Axillary buds (buds in

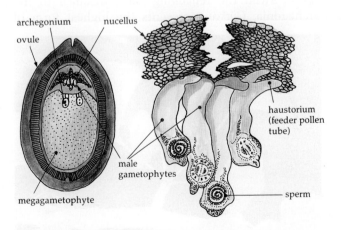

Fig. 37.28 The pollen tube of cycads forms a haustorium, which absorbs nutrients from the nucellus

Ginkgophytes

Ginkgophytes are represented in the world's modern flora by a single species, *Gingko biloba* (Fig. 37.29), which is endemic to a restricted region of China. It is the sole survivor of a lineage of plants that were prominent in the past. For example, *Ginkgo* fossils are common in mid-Mesozoic (100–130 million years BP) sediments of Europe, North and South America, and Australia. *Ginkgo biloba* is a commonly cultivated tree, valued for its shape, delicate foliage and autumn colours.

The growth habit of *Ginkgo* is arborescent (a tree), the result of axillary branching and a well-developed vascular cambium. In this respect it is similar to conifers and woody flowering plants. The leaves are very distinctively wedge-shaped with an apical notch, giving *Ginkgo* its common names—maiden hair tree, because of its resemblance to the fern fronds of the same name, and the Chinese name duck's foot tree. Branches consist of two different kinds of shoots, fast-growing long shoots and slow-growing short shoots. Most leaves are borne on short shoots.

Micro- and megasporangiate structures are also borne on short shoots, on separate male and female trees. Loose cones shed large quantities of wind-dispersed pollen. Ovules are borne in pairs at the end of an elongated stalk, and produce a **pollination droplet** that traps pollen grains and draws them through the micropyle into the ovule. Gametophyte growth and fertilisation is very similar to that of cycads. The megagametophyte derives its nutritional requirements from the nucellus of the ovule by way of specialised feeding structures. During this period the megagametophyte matures into a multicellular mass of tissue with several archegonia. When both micro- and megagametophytes are mature, which may be up to six months after pollination, the microgametophyte ruptures to deposit a pair of flagellated sperm into a pool of fluid that covers the archegonia, essentially the same as for cycads. The sperm swim the remaining short distance to the archegonia. *Ginkgo* represents the highest form of

Fig. 37.29 *Ginkgo biloba*

vascular plant life to retain motile, multiflagellated sperm. The flagella retain the 9 + 2 structure of microtubules described for algae, and may even be branched.

> Cycads are the most primitive of the living seed plants. They produce ovules and pollen in cones. Cycads and *Ginkgo biloba* rely on insects or wind for transport of microspores (pollination). They have multiflagellated sperm and rely on water for the final stage of transport to the egg.

Coniferophytes

Conifers (phylum Coniferophyta) are the most diverse and widespread of the non-flowering seed plants. They range from prostrate shrubs, such as the mountain plum pine, *Podocarpus alpina*, of Australian alpine regions (see Chapter 41), to the largest trees in the world, *Sequoia gigantea* of south-western United States. Their fossil record dates from the Permian (290–245 million years BP), but they were dominant during the Mesozoic era (245–65 million years BP). Even today conifers are the most important elements in some vegetation types, for example, the extensive pine forest communities of the Northern Hemisphere, and *Callitris* woodlands of Australia (see Chapter 41).

Conifer wood is a valuable commercial resource worldwide, and in Australia it is the basis for a diverse range of products. Extensive plantations of Monterey pine, *Pinus radiata*, are the basis of the paper industry and provide wood for furniture and building materials. Several native conifers (Box 37.4), such as white cypress pine or Murray pine, *Callitris columellaris*, and Huon pine, *Lagerostrobos franklinii*, are highly valued as specialty timbers. Because conifer wood is composed largely of tracheids, as opposed to flowering plant wood, which has large vessels embedded in a matrix of fibres, it is soft and easily crafted, and hence commonly called 'softwood'. Many conifers are extremely slow-growing, adding small amounts of new wood in annual increments. This results in the very fine-grained appearance sought after by woodcarvers and decorators.

> Conifers are the most diverse and widespread of the living non-flowering seed plants. They are all woody plants, many of which are important softwood timber trees.

Most conifers bear their micro- and megasporangiate structures in cones (Fig. 37.30). However, a few conifers have such a highly modified ovule-bearing structure that their conelike nature can only be seen very early in development. In the Southern Hemisphere, genus *Podocarpus* (family Podocarpaceae) and its relatives bear a single ovule on a fleshy receptacle that represents a swollen cone

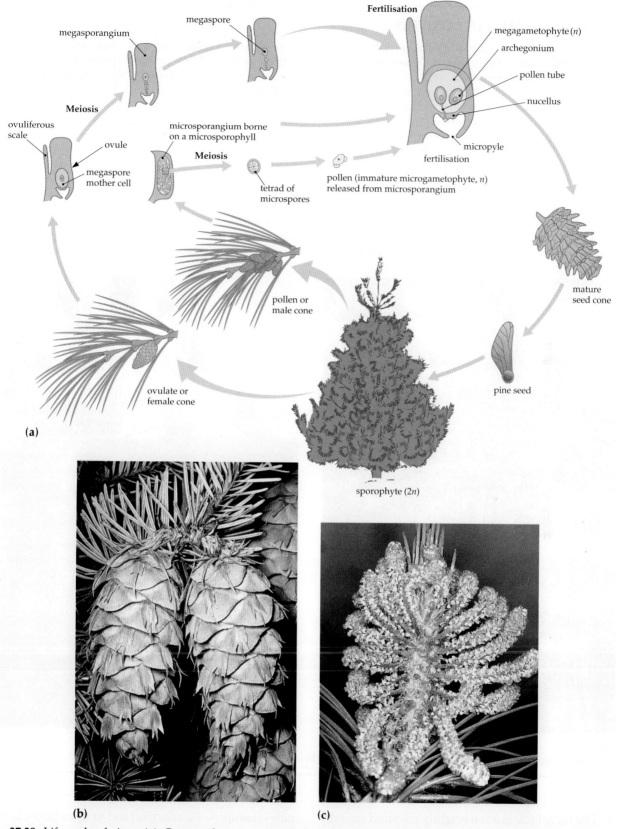

Fig. 37.30 Life cycle of pines. **(a)** Gametophytes are very reduced and enclosed within sporophyte tissue. The ovule, produced in female cones, encloses the megagametophyte. Pollen grains develop in male cones and are transported by wind to the micropyle of the ovule. The tube of the germinating pollen grain (immature microgametophyte) slowly digests its way toward the female gametophyte. After many months, the pollen tube reaches the egg cell, and the now-mature microgametophyte discharges its two non-flagellate sperm. The fertilised ovule develops into a seed. **(b)** A cluster of mature pollen cones of Monterey pine, *Pinus radiata*. **(c)** Female seed cones of Douglas fir, *Pseudotsuga menziesii*

BOX 37.4   AUSTRALIAN CONIFERS

In terms of number of species, gymnosperms are few in Australia. Only two phyla, Cycadophyta and Coniferophyta, are represented.

Within the conifers (see figure) there are four families in Australia. Kauri pines and araucarias (family Araucariaceae) are tall evergreen trees found predominantly in tropical and warm temperate regions of the Southern Hemisphere. Queensland kauri, *Agathis robusta*, hoop pine, *Araucaria cunninghamii*, and bunya pine, *A. bidwillii*, are prized for their timber, and many populations have been lost. Podocarpaceae are also largely a Southern Hemisphere family, once dominant, together with Araucariaceae, in ancient Gondwanan forests (see Chapter 41). *Phyllocladus*, *Podocarpus* and *Lagerostrobos* are cultivated as ornamentals or for timber.

The family Taxodiaceae was once more diverse, particularly in the Northern Hemisphere. There is only one genus in Australia, *Athrotaxis*, including two species, pencil pine, *A. cupressoides*, and King Billy pine, *A. selaginoides*, which are endemic in western Tasmania.

Cypress pines, of the worldwide family Cupressaceae, are recognised by their small scale-like leaves. Of the Australian genera, *Callitris* is widespread, often on sandy soils in dry areas. All species are killed by fire but regenerate from seed. The wood of *Callitris columellaris* was used by Aborigines for canoe paddles and spear throwers, and European settlers used the wood for houses and fences. Two other Australian genera are restricted to the south-east of Western Australia (*Actinostrobus*) and south-west Tasmania (*Diselma*).

(a)

(b)

Australian conifers: **(a)** *Callitris glavea* in dry regions and **(b)** *Araucaria muelleri* in tropical rainforests

axis. The receptacle is often brightly coloured (usually red), and attracts birds that eat the fleshy part and disperse the seeds (Fig. 37.31).

Microsporangia (containing pollen) are borne on **microsporophylls** (fertile leaves) that are aggregated into pollen cones. Pollen grains of conifers are wind dispersed. Entry of pollen into the ovule is effected in one of several ways. In kauri and Norfolk Island

pines (family Araucariaceae) and fir trees (*Abies*, family Pinaceae), pollen lands on the surface of the cone adjacent to the ovule and the pollen tube grows over the surface of, or sometimes through, the cone scales into the micropyle. More commonly, as in the cypress pines (*Callitris* and *Cupressus*, family Cupressaceae), the ovule secretes a sticky drop of fluid that protrudes from the micropyle. Pollen blowing through the cone

Fig. 37.31 In the conifer *Podocarpus*, an ovule is borne on a fleshy, coloured receptacle, which attracts birds

archegonium are induced by pollen germination. Gametophyte development is slow and in many conifers not complete until the following year. Fertilisation of the egg cell by the sperm cell may occur a year or more after the pollen grains are actually deposited in the micropyle. Development of the embryo to the stage where the seed is ready to be shed can take another year, making the time elapsed from pollination to seed maturity in excess of two years in some cases.

> Conifers bear their pollen and seeds in separate cones. Pollen is wind dispersed and, on germination, produces a pollen tube that delivers non-flagellated sperm cells directly to archegonia.

is trapped on the surface of the micropylar drop. As the droplet evaporates it shrinks back through the micropyle, drawing the pollen into the ovule.

Pollen of many conifers has two (sometimes one or three) air bladders or saccae developed from the pollen wall that are thought to aid in both wind dispersal and adhesion to the micropylar droplet. Once in the micropyle of the ovules, pollen grains undergo several cell divisions within the confines of the pollen grain wall to form the microgametophyte. A pollen tube germinates through an aperture in the pollen grain wall, grows through the nucellus (megasporangium) and provides a direct route to the archegonia and egg cell. The nucellus acts as a screen for foreign pollen, with the pollen tubes of other species being arrested within this tissue. Such events result in incompatibility between species of pines, a phenomenon first studied in detail in Canberra by Dr James McWilliam. In this way, the female cone is able to screen out unwanted potential male partners, selecting only those of its own species.

In conifers, sperm lack flagella and they are transported passively within the tip of a pollen tube. When the tube arrives at the egg, two sperm are released, and one fuses with it. The other sperm aborts. In some cypresses, however, many sperm are produced within a single pollen tube, an adaptation to ensure fertilisation of multiple archegonia.

Female cones of conifers are compound structures of several orders of branching. The structures of the cone that bear ovules, **ovuliferous scales** (Fig. 37.30), although flattened and leaf-like, are thought to be very reduced shoots. Male cones, on the other hand, are simpler in structure. In a comparative sense, a female cone is equivalent to a group of male cones.

One unusual feature of conifers (and cycads and *Ginkgo*) is the timing of events involved in reproduction. Ovules are pollinated when they are very small and female gametophyte development is not triggered until after pollination. The megaspore undergoes meiosis and then becomes dormant. The post-meiotic changes to produce the egg within an

Gnetophytes: relatives of flowering plants

Gnetophytes (phylum Gnetophyta) are a bizarre group of plants consisting of just three extant genera that are markedly different from each other in appearance. Switch bushes, *Ephedra*, are shrubs of arid areas; species of *Gnetum* are lianes or small trees of the tropics; and the single species of *Welwitschia* is a xerophyte of the Namibian desert that was described by one botanist as resembling a giant octopus stranded on the beach (Fig. 37.32). Gnetophytes are of only minor importance in the present day vegetation of the earth but are botanically very important because they are the closest relatives of the flowering plants (Fig. 37.7), thus providing some clues to the origin and evolution of that group.

Although at first glance the three genera of gnetophytes look very different, their close relationship is evidenced by similar reproductive structures. The micro- and megasporangia are clustered into small flower-like structures that are borne in a cone-like inflorescence (Fig. 37.32c). The droplet produced from the micropyle of the ovule has a high content of sugar and nitrogen, and acts as an attractant for insects. In 'flowers' that produce only pollen, a central sterile ovule produces the nectar-like pollen drop necessary to attract pollinators.

The fertilisation process and the structure of xylem cells in gnetophytes have attracted great interest, as both features have marked similarities to flowering plants. In *Ephedra*, there is some evidence for a kind of double fertilisation. Here, one sperm nucleus fuses with the egg nucleus, while the other may occasionally fuse with the sterile neck canal nucleus of the archegonium. This accessory fusion does not appear to play any further role in seed production.

Water-conducting cells in the xylem of gnetophytes are open-ended, forming a continuous vessel network throughout the plant which is far more efficient than the discrete tracheids of other vascular plants. Almost identical xylem vessels are a characteristic of flowering plants.

(a)

(b)

(c)

Fig. 37.32 (a) *Ephedra viridis* in California, (b) *Welwitchia* from the Namibian desert and (c) *Gnetum* from the tropics. These three genera are the only living gnetophytes in the world

Gnetophytes include only three living genera. They are the closest relatives of flowering plants.

Magnoliophytes: flowering plants

The flowering plants, or angiosperms (phylum Magnoliophyta), are dominant vascular plants of modern floras. They are major producers in terrestrial ecosystems, and provide food, hardwood timber, natural fibres, spices and medicinal drugs that sustain the human population.

Flowering plants are adapted to a great range of habitats, from the tropics to polar regions, including fresh and salt water. The smallest forms are free-floating aquatic duckweeds (Fig. 37.33a), which are highly reduced and not even differentiated into leaf and stem. The tallest flowering plant is the mountain ash tree, *Eucalyptus regnans*, of south-eastern Australia, which can reach heights of more than 100 m (Fig. 37.33b).

(a)

(b)

Fig. 37.33 (a) The duckweed *Wolffia*, one of the smallest angiosperms. (b) The Australian mountain ash, *Eucalyptus regnans*, is the tallest flowering plant in the world

Flowering plants are classified into two major groups (Fig. 37.34): the monocotyledons (class Liliopsida) and the dicotyledons (class Magnoliopsida). **Monocotyledons** are generally non-woody plants, with flower parts in threes, parallel leaf veins, scattered vascular bundles in the stem, and one cotyledon (seed leaf) in the embryo. They include grasses (e.g. cereals), lilies, grass trees, kangaroo paws (Fig. 37.35a), orchids, palms and duckweeds. **Dicotyledons** are either woody or herbaceous, with flower parts in fours or fives (Fig. 37.35b), net-like venation, vascular bundles arranged as a cylinder in the stem, and two cotyledons. Dicots include magnolias, buttercups, daisies, roses, peas, acacias, banksias and the mountain ash eucalypt.

(a)

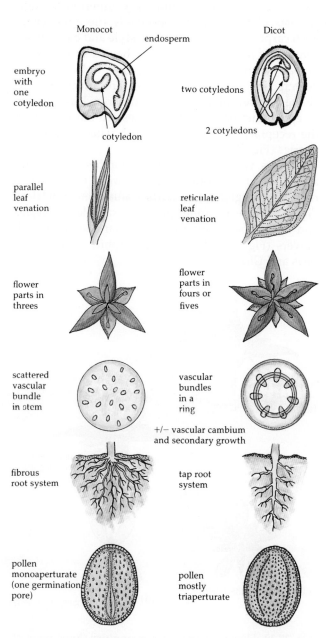

Fig. 37.34 Major characteristics of the two major groups of angiosperms: monocotyledons and dicotyledons

(b)
Fig. 37.35 **(a)** Australian kangaroo paw, *Anigozanthos manglessii*, is a monocot. **(b)** Australian *Thryptomene maissonneuvei* is a dicot

The term **angiosperm** (literally, a vessel seed) describes one of the most important characteristics of flowering plants—the enclosure of the ovules within a hollow **carpel** (Fig. 37.36). As the ovules develop into seeds, the carpel becomes the **fruit**. This is in contrast to other seed plants, in which seeds are borne 'naked' on the surface of the megasporophylls of the female cone.

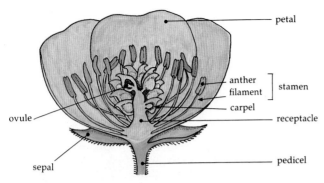

Fig. 37.36 Buttercup flower (family Ranunculaceae) has numerous stamens and free carpels. When mature, each carpel becomes a fruit with seeds developing from ovules

In addition to carpels, angiosperms are characterised by a number of vegetative features. Phloem sieve tubes lacking nuclei at maturity are associated with nucleated companion cells (Chapter 18). Xylem consists of vessels as well as tracheids. Vessels also occur in *Selaginella*, *Equisetum*, the ferns *Pteridium* and *Marsilea*, as well as gnetophytes, but these seem to have evolved independently. Angiosperm vessels are different in having a range of structural patterns, including ladder-like (scalariform) perforations (Fig. 37.37).

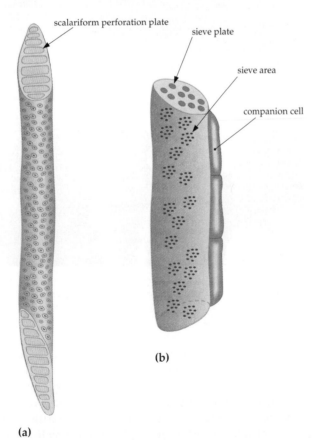

(a)

Fig. 37.37 (a) Xylem vessels with scalariform perforations and (b) phloem sieve elements with companion cells are characteristics of angiosperms

Flowers

Chapter 16 describes the process of sexual reproduction in angiosperms. We can now compare angiosperms with other seed plants to discover the homologies of the different floral structures and the evolution of the flower.

You will remember that flowers typically consist of four whorls or circles of similar parts attached to a central axis (Fig. 37.36). In summary, the outermost whorl is the calyx, consisting of sepals, usually leaf-like in appearance and protective of the inner parts when the flower is in bud. Next is the corolla, consisting of petals, often colourful and adapted to attract animal pollinators. The next whorl consists of the stamens with their anthers (with pollen sacs), collectively the androecium. The innermost whorl consists of the carpels, collectively the gynoecium. Each carpel consists of a stigma, style and basal ovary containing ovules. Flowers may have a single carpel or many, either free or fused together. In **hypogynous** flowers (Fig. 37.38a), the ovaries are attached to or above the receptacle and are described as *superior*. In **epigynous** flowers, the ovaries are buried within the receptacle, below the perianth. These are *inferior* ovaries (Fig. 37.38b).

Flowers vary in size, colour, number, shape and arrangement of floral parts. Flowers that have a regular arrangement of parts (radially symmetrical, Fig. 37.38c) are **actinomorphic**, for example, *Eucalyptus* and *Leptospermum* (tea tree). Irregular (asymmetrical) flowers (Fig. 37.38d) are **zygomorphic**, for example, peas and *Grevillea*.

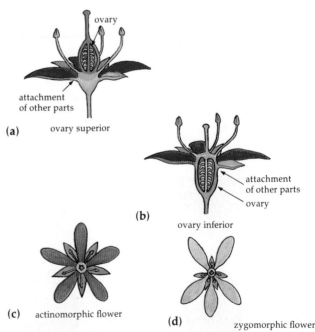

Fig. 37.38 Flowers with (a) superior ovary (hypogynous), (b) inferior ovary (epigynous), (c) regular (actinomorphic) and (d) irregular (zygomorphic) arrangement of parts

Fertilisation: a 'double' event

Microspores develop into microgametophytes within the pollen sacs of the anther, and are shed as pollen grains, as in other seed plants. The microgametophytes of angiosperms have three cells when mature—two sperm, which lack flagella, and one cell that plays a role in the growth of the pollen tube (see Chapter 16) when pollen reaches the female stigma. The pollen tube emerges through one of several thin areas in the pollen grain wall (apertures). Monocots have only one aperture (monosulcate pollen), and most dicots have three (a useful diagnostic character when identifying pollen).

Within the ovule, one of the four haploid products of meiosis, a functional megaspore, develops into the megagametophyte, which is the embryo sac described in Chapter 16. In most flowering plants, the embryo sac consists of only eight nuclei enclosed within seven cells (a result of three mitotic divisions and cell wall formation). At one end of the embryo sac is the egg cell, flanked by two cells; in the middle is one large cell containing two polar nuclei; and at the other end are three cells (Fig. 16.3). The embryo sac is thus a very reduced megagametophyte and, unlike in cycads, *Gingko* and conifers, eggs do not form in archegonia.

Double fertilisation is a unique event in angiosperms. Of the two sperm delivered to the ovule by the growth of the pollen tube, one fuses with the egg to form the zygote and the other fuses with the two polar nuclei to form the triploid endosperm nucleus, which divides by mitosis to produce endosperm tissue (Fig. 37.34). Endosperm only occurs in angiosperms and supplies the developing embryo with nutrition (Chapter 16).

> In flowering plants, the megagametophyte, which develops in the ovule, is reduced to an embryo sac. The microgametophyte, shed as pollen, develops in anthers. Double fertilisation and endosperm formation are unique to flowering plants.

Origin of the flower

Flowers are determinate shoots, which means that the shoot apical meristem stops growing once the last floral organs are formed: sepals being produced first, followed by petals, stamens and finally carpels. The classical theory of the origin of the flower therefore interprets the various floral parts as modified leaves borne on a shoot that has a limited capacity for elongation. Under this interpretation, stamens and carpels are merely specialised leaf-like appendages (sporophylls) that bear sporangia.

These classical ideas receive some support from the observation that sepals resemble leaves in position, venation, colour, form and early development, and also from recent work in molecular biology, which shows that genes that affect the development of leaves also affect the development of sepals. However, the evolution of other floral structures from leaves is much less clear. Petals, for example, may have arisen in a number of ways. Some petals may be leaves that have been modified directly while others may be modified stamens that have become sterile and function to attract pollinators rather than produce pollen. It has also been suggested that stamens represent a modified, reduced branch system with terminal microsporangia rather than a leaf-like organ. Controversy also surrounds the interpretation of carpels. In many respects they resemble leaves, and can be influenced by the same developmental genes that are expressed in sepals, but in some flowering plants the early stages of carpel development are very distinctive and unlike that of leaves or any other floral organ.

Whatever their evolutionary origin, carpels are the most distinctive reproductive structures of angiosperms, and the enclosure of ovules by these structures undoubtedly led to a series of important modifications of the reproductive process. Most obviously, carpels provide additional protection for developing ovules and, after fertilisation, they may also be modified to form a variety of specialised fruit types for animal, wind and water dispersal (see p. 828). A more subtle consequence of the origin of the carpel is that pollen is no longer trapped directly by the ovule, as in most other seed plants, but instead becomes trapped on the stigmatic surface. Pollen tubes must then grow through some of the carpel tissues in order to deliver the male gametes to the embryo sac, thus introducing new possibilities for 'selection' of male gametes by the female plant (see Chapter 16).

Traditionally, flowers such as those of the living members of the magnolia family have been considered the most 'primitive' flower among angiosperms (Fig. 37.39a). Such flowers are bisexual, with numerous, often spirally arranged, perianth parts, stamens and carpels; they are often pollinated by beetles. It has frequently been hypothesised that all other flowers evolved from this type by reduction in number and fusion of parts to form more complex structures, such as corolla tubes. In support of this idea, there is evidence that magnolia-like flowers existed in the mid-Cretaceous (about 95 million years BP) at an early stage in the angiosperm fossil record.

Although the magnolia theory has been widely accepted for almost a century, recent studies have begun to challenge some of its basic tenets. Firstly, there is a great variety of floral types among living flowering plants, which range from very small flowers with only one carpel, one stamen and no petals to the large multiparted magnolia-like forms. Secondly, recent studies of angiosperm relationships using both morphological and molecular data favour the view that the magnolia group may be an early but perhaps

(a)

(b)

Fig. 37.39 **(a)** Flowers of the tulip tree, *Liriodendron tulipifera* (family Magnoliaceae) are considered one of the primitive types of flowers. **(b)** Mid-Cretaceous fossil flowers, however, include small separate male flowers, shown here (*Spanomera mauldinensis*), and female flowers

rather specialised offshoot in angiosperm evolution. Thirdly, the fossil record shows that a great variety of other angiosperm flowers (Fig. 37.39b) were already present by the time magnolia-like forms are first recorded (approximately 100 million years ago). As a result of these studies, recent ideas of floral evolution suggest that the earliest, least specialised flowers may have been small, with relatively few parts and were probably organised on a three-part ground plan as occurs in some primitive monocots. Although the earliest flowers were almost certainly bisexual and insect pollinated, it is also quite probable that unisexual, perhaps wind-pollinated flowers also developed at a very early stage in angiosperm evolution.

> The enclosure of the ovules within a hollow carpel led to important modifications of the reproductive process of angiosperms. Carpels protect ovules and, when mature as fruits, aid seed dispersal. Pollen can also be trapped directly on the stigma of the carpel.

Pollination

The agents of pollination in modern flowering plants are wind, water, insects, birds, bats and other small mammals. Pollen is usually transported between flowers of different individual plants, sometimes separated by considerable distance, promoting outcrossing. Among the flowers that are wind pollinated are grasses and sedges (Fig. 37.40a). They have inconspicuous flowers that are odourless, greenish in colour, and with petals reduced or absent. They produce abundant pollen, which lands on stigmas by chance. Flowers that attract animals are more effective in ensuring the transfer of pollen. This is of considerable advantage since a one-to-one relationship between a plant and animal species reduces wastage of pollen by ensuring that it is deposited on the 'right' flower.

Animals that act as pollinators search in flowers for a meal of nectar, a sugar-rich liquid secreted by floral nectaries, or pollen. The Australian honey possum is one of the few mammals that specialises in eating flower nectar (Fig. 37.40b). In the process of robbing the plant for food, the pollinator brushes against the sticky stigma, depositing a load of pollen.

Attracting pollinators

Individual plants that produce only a single flower, such as a tulip, are not common. Most plants produce many flowers, usually in discrete clusters or **inflorescences** (Fig. 37.41). A good example is wattle, where each golden head is actually an inflorescence containing a few to many flowers (see Fig. 41.24). Inflorescences provide a mass display of flowers to attract pollinators.

Once in the vicinity of an inflorescence, an insect may be attracted to a flower by its scent. Heavy, sweet-smelling scents are produced by flowers that are pollinated by night-flying moths or bats that do not rely on vision. Some flowers produce foul-smelling scents that mimic rotting meat and attract flies.

Colouration and markings are important attractants in other flowers. Birds do not perceive scents but are often attracted to red flowers, such as the conspicuous flowers of the New South Wales waratahs (Fig. 37.40c). Waratah flowers are long, tubular and slightly curved; their rate of nectar production is relatively high and they are commonly visited by nectar-feeding honeyeaters.

Red is not a distinct colour to insects. Bees, which are especially important as pollinators, are generally attracted to yellow or blue flowers. Petals that look uniformly yellow to our eyes can appear quite

(a)

(b)

Fig. 37.41 Flowers, such as *Grevillea*, are usually clustered as inflorescences, which make them more conspicuous for pollinators

(c)

Fig. 37.40 (a) Grass flowers, such as *Lolium perenne*, are small, clustered in inflorescences, and wind pollinated. (b) The Australian honey possum, *Tarsipes rostratus*, feeding on and pollinating *Dryandra quercifolia*. (c) The red colour of the New South Wales waratah, *Telopea speciosissima*, attracts bird pollinators

different to a bee. In addition to reflecting yellow light, the petals can have markings that reflect ultraviolet light, giving a mixture of colours called *bee's purple*. Markings on petals, like lights on an airport runway, help the bee to locate nectaries (Fig. 37.42).

Flower shapes are also varied to ensure pollination. Petals may be fused to form a trumpet shape, with nectaries placed deep in the throat of the flower. This arrangement forces a pollinator to push into the flower to reach its reward. Anthers may be positioned in a way that maximises transfer of pollen to the pollinator. Some insects can also rob flowers without effecting pollination by cutting a hole in the petals and getting at the food source more directly.

Pollination by deceit

There are some orchids whose flowers mimic the shape and colouring of female insects. The mimics are so realistic that male insects will attempt to copulate with the flower, thereby pollinating them. Some of these orchids do not even present a nectar reward for the insect. In the genus *Drakaea*, the hammer orchids of Western Australia, the labellum (central petal) of the flower resembles a wingless insect, with shiny eyes, hairy thorax and fat body

(a)

(b)

Fig. 37.42 **(a)** *Rhododendron lochae*, a Queensland species, has bright red tubular flowers with no scent or visible markings to guide pollinators. **(b)** However, under ultraviolet light, the outer parts of the petals are reflecting and the inner region of the flower absorbing. The dark central region probably attracts and guides insects and birds

(a)

(b)

Fig. 37.43 **(a)** Flying male thynnine wasp, *Megalothynnus klugii*, approaching a wingless female, which lives underground, emerging only to attract a mate. **(b)** A hammer orchid, *Drakaea glyptodon*, mimics the female wasp, whereby the male is deceived and mates with the orchid flower, effecting pollination

(Fig. 37.43). The flower is held outwards by a hinged arm. When triggered by an insect, the hinged arm moves towards the column of the flower, where the stigma and pollinia (pollen masses) are located. Female thynnine wasps are wingless and spend much of their time underground, emerging to mate. Above ground, they climb to the top of a small shrub or grass stalk and attract a male wasp by releasing a chemical scent (pheromone) and posing in a characteristic manner—mimicked by the orchid! The male wasp swoops on the female, carrying her off to copulate in the air or on a nearby plant. However, if the male attempts to copulate with a hammer orchid, the hinged labellum of the flower flings him against the column, effecting pollination.

Pollination of flowers by animals is an example of how interactions between organisms play a role in the evolution of each group. The topic of species interactions is dealt with further in Chapter 43.

Pollination by animals is an advantage to flowering plants, being less wasteful and more effective than wind pollination. Flower scent, colour, markings, shape and nectaries are important in attracting animals.

Fruits and seed dispersal

After successful pollination and fertilisation, the fruit develops as the product of growth in the persistent parts of a flower. Fruits come in all colours and forms, including peanuts and their shells, corn kernels, pea pods, tomatoes, cucumbers, apples, oranges and eucalypt gumnuts (capsules). They are usually classified as simple, aggregate and multiple.

Simple fruits

A **simple fruit** is derived from the ovary of a single flower with one or more carpels, together with, in some species, adhering sepals. Simple fruits are moist and fleshy, or dry, when mature.

BOX 37.5 POLLINATION IN AQUATIC ANGIOSPERMS

In the pond weed *Vallisneria*, male flower buds abscise and rise to the surface. There, they float across the water surface, opening their anthers and eventually colliding with their stigma targets. Female flowers with receptive stigmas are raised to the water surface by remarkable elongation of the flower stalk. A high probability of pollen contacting a stigma is achieved by releasing the male flowers to float across the surface of the water rather than individual pollen grains, which have a much smaller diameter. Surface tension near the female plant causes the pollen mass to be drawn into the flower.

Pollination of the aquatic plants *Ruppia* (family Ruppiaceae) and *Lepilaena* (family Zannichelliaceae) is achieved in a different way. *Ruppia* pollen is released from the anther underwater and floats to the surface in a gas bubble (see figure). In *Lepilaena*, the boat-like anthers are released underwater and float to the surface, where they then release their pollen. As in *Vallisneria*, the female flowers of these two aquatics have an elongated stalk, so that they also float on the surface.

Angiosperms that have adapted to aquatic habitats have novel mechanisms to ensure pollen reaches the female stigma. In *Ruppia*, large pollen grains stick together as rafts, which float on the surface of the water

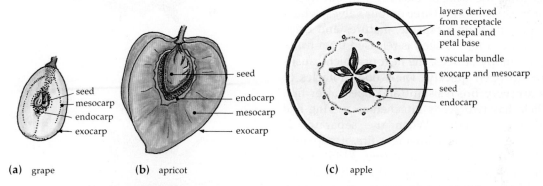

(a) grape   (b) apricot   (c) apple

Fig. 37.44 Simple fleshy fruits include **(a)** berries—grape, *Vitis*; **(b)** drupes—apricots and plums, *Prunus*; and **(c)** pomes—apple, *Pyrus malus*. The fruit wall (pericarp) consists of three layers: exocarp, mesocarp and endocarp

In fleshy fruits, the wall of the ovary increases greatly after fertilisation. The fruit wall, **pericarp**, becomes differentiated into three layers of tissue—the **exocarp**, **mesocarp** and **endocarp**. The exocarp is the outermost layer, usually a single layer of epidermal cells; in peaches, it is the skin. The mesocarp is the middle layer and varies in thickness; in peaches, it is the thick edible layer. The endocarp, the innermost layer, is also variable in thickness and structure; in peaches, it is the stone.

Cucumber, grape, orange and tomato are all **berries** (Fig. 37.44). These are fleshy fruits in which the seeds, usually many, are enveloped within a fleshy mesocarp and a well-defined outer skin-like exocarp. In an orange, the segments correspond to the carpels. The juicy flesh comprises filaments of cells that arise from the carpel walls.

A **drupe** is a fleshy fruit, such as a plum, containing a single seed enclosed in a hard stony endocarp (Fig. 37.44). The coconut is a drupe, the shell being the hard endocarp; the mesocarp is fibrous rather than fleshy.

A **pome** is a fruit derived from an ovary of an epigynous flower. The fleshy, edible part of the fruit includes non-ovarian tissues. A good example is the apple (Fig. 37.44). The true fruit arises from an inferior ovary to form only the core, while the bulk of the fruit derives from the receptacle together with the sepal and petal bases of the flower.

Fleshy fruits, particularly those coloured red or orange, may be eaten by birds, who unwittingly transport the seeds to new locations, depositing them by defecation. (Some seeds are covered with material that acts like a laxative!) The fleshy layers of fruits sometimes provide additional nourishment for developing embryos. Aborigines have always relied on native fruits as 'bush tucker', knowing which ones are edible and which ones are poisonous.

Of the dry simple fruits (Fig. 37.45), some are dehiscent, opening and shedding their seeds at

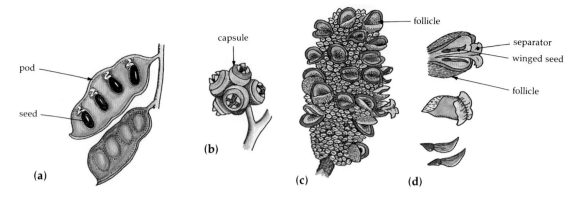

Fig. 37.45 Simple dry fruits include **(a)** pods or legumes, *Acacia*; **(b)** capsules, *Eucalyptus*; **(c)** follicles, *Banksia*. **(d)** In the *Banksia* follicle, a separator, when wet, pulls the two seeds out of the fruit

maturity. A **follicle**, which opens on the lower side, is formed from a single carpel, such as in *Grevillea*. **Pods** or **legumes** open along two sides, as in beans, peas and *Acacia*. The dry fruits of many Australian native plants have thick, woody walls that protect the seeds from the heat of bushfires. In eucalypts, the top of the ovary forms three or four valves, which split open as the fruit, a **capsule**, matures and dries out. After a fire, a rain of seed is released from fruits held in the canopy of trees, the heat of which causes the valves to open quickly. The woody follicles of banksias are very fire resistant and open after fire. Each follicle has two seeds attached to a wing-like structure, the separator. When the separator is sufficiently wet by rain following a fire, it expands, pulling the seeds out of the fruit (Fig. 37.45). Seeds will otherwise remain protected in the fruit until rain falls, providing suitable conditions for germination.

Aggregate and multiple fruits

An **aggregate fruit** arises from a cluster of separate carpels from *one* flower. Raspberries and blackberries are familiar examples. The strawberry is an aggregate of numerous seed-like fruits (achenes) from a single flower that develop on a sweet, fleshy receptacle (stalk), which is the edible part (Fig. 37.46).

A **multiple fruit**, such as pineapple and figs, is a cluster of many carpels produced from *several* flowers in an inflorescence. In pineapple (Fig. 37.46), the inflorescence axis and various floral organs contribute to its structure.

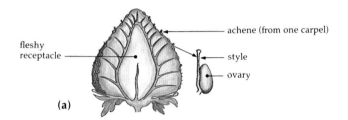

Fig. 37.46 **(a)** A strawberry is an aggregate fruit, consisting of numerous small achenes, from separate carpels, on a fleshy receptacle derived from one flower. **(b)** A pineapple is a multiple fruit derived from a number of flowers in an inflorescence

SUMMARY

- Plants are those photosynthetic organisms that have adapted to life on land, and include liverworts, mosses, ferns, conifers and flowering plants among the major modern groups. They are multicellular, have cells specialised to form tissues and organs, have an outer cover of cuticle, and have sex organs with an outer layer of vegetative cells to protect gametes and, within the female reproductive organ, the developing zygote.

- In plant life cycles, the haploid gametophyte generation forms male antheridia, which produce sperm, and female archegonia, each of which produces a single egg. Male and female gametes fuse to form a zygote, which grows into the diploid sporophyte generation. Meiosis occurs on the sporophyte, resulting in the formation of haploid spores for dispersal. In some ferns, fern allies and seed plants, two types of spores are produced. These plants are heterosporous.

- Bryophytes include liverworts (phylum Hepatophyta), hornworts (Anthocerotophyta) and mosses (Bryophyta). Bryophyte plants are small, photosynthetic, free-living gametophytes, which lack vascular tissue. Sexual reproduction requires water for sperm to swim to an egg in an archegonium. The sporophyte grows on and is dependent on the gametophyte for nutrition.

- 'Fern allies' refers to a loose assemblage of living, spore-producing, vascular plants that are classified in three phyla: Psilophyta (*Psilotum* and *Tmesipteris*), Lycophyta (*Lycopodium*, *Selaginella* and *Isoetes*) and Sphenophyta (*Equisetum*).

- The sporophyte of homosporous ferns, like that of fern allies, is photosynthetic, free-living, and the dominant generation. Like bryophytes, ferns have flagellated sperm and rely on free water for sexual reproduction. Fern antheridia and archegonia are borne on a small, free-living heart-shaped gametophyte (prothallus).

- Cycads (phylum Cycadophyta) are the most primitive of the living seed plants. They are palm-like in appearance but produce ovules and pollen in cones. Cycads and *Ginkgo biloba*, the only living representative of the phylum Ginkgophyta, rely on insects or wind for transport of microspores (pollination) but have flagellated sperm.

- Conifers (phylum Coniferophyta) are the most diverse and widespread of the living non-flowering seed plants; they are all woody plants, many of which are important softwood timber trees. They bear their pollen and seeds in cones. Pollen is wind dispersed and on germination produces a pollen tube that delivers non-flagellated sperm cells directly to archegonia. Reproduction may take up to several years.

- Phylum Gnetophyta includes only three living genera. They are the closest relatives of flowering plants. Their reproductive structures are surrounded by leaf-like structures, which make them resemble flowers. Their xylem contains vessels in addition to tracheids.

- Phylum Magnoliophyta, the flowering plants or angiosperms, are classified into two major groups: the monocotyledons (class Liliopsida) and dicotyledons (class Magnoliopsida). Monocots are generally non-woody plants, with flower parts in threes, parallel leaf veins, scattered vascular bundles in the stem, and one cotyledon (seed leaf) in the embryo. Dicots are either woody or herbaceous, with flower parts in fours or fives, net-like venation, vascular bundles arranged as a cylinder in the stem, and two cotyledons.

- In angiosperms, the male (micro-) and female (mega-) sporangia are located in the flower. The microgametophyte is shed from anthers as pollen. The megagametophyte is the embryo sac, which often has eight nuclei within seven cells.

- Double fertilisation is unique to angiosperms. One sperm fertilises the egg, the other sperm fuses with the two polar nuclei of the embryo sac, resulting in a zygote surrounded by triploid, nutritious endosperm.

- The enclosure of the ovules, containing the megagametophytes, within a hollow carpel led to important modifications of the reproductive process of angiosperms. Carpels provide protection for ovules and, when mature as fruits, they aid seed dispersal. Pollen can also be trapped directly on the stigma of the carpel.

- Angiosperms have the advantage of attracting animals for pollination and fruit and seed dispersal.

QUESTIONS

1. What evidence is there that green algae such as *Coleochaete* (class Charaphyceae) are the closest relatives of land plants?

2. What are the major problems of living on land compared with living in water? What adaptations to living on land are associated with (a) the gametophyte generation and (b) the sporophyte generation in the life cycle of plants? What phyla of living plants have vascular tissue?

3. Plant life cycles show an alternation of generations. On what stage do sporangia develop? Does spore production involve mitosis or meiosis? What is the main function of spores? What are antheridia and archegonia and on what stage do they develop?

4. What is the age of the oldest land plant fossils? Briefly describe the form of *Rhynia*.

5. Mosses not only occur in damp situations, but some species live in inland Australia or Antarctica. How are such mosses able to live in an arid environment?

6. Compare the transport of sperm to the egg cell in a moss, fern, cycad and conifer.

7. What is meant by 'homosporous' and 'heterosporous' plant life cycles? Is a pine tree homosporous or heterosporous? What is a megaspore and a microspore, and what do they develop into?

8. What is a seed and from what structure(s) does it develop? Name the major groups of plants that develop seeds and state whether their seeds are borne in cones or fruits. In what ways is the seed an important advantage to plants living on land?

9. Name the parts of a flower and state a function for each. Given that a flower is a determinate shoot, discuss the possible origin of the flower and its various parts.

10. Of the vascular plants, flowering plants dominate modern floras. What features of flowering plants explain their success? How has their evolution been affected by animals?

SIMPLE ANIMALS: SPONGES TO NEMATODES

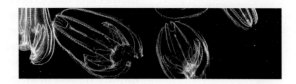

All members of Kingdom Animalia, also known as **Metazoa**, are *multicellular*. Multicellular organisation of animals, as in plants (Chapter 37), allows division of labour between individual cells or groups of cells. For instance, some animal cells function in locomotion, while others function in nutrition or reproduction. Cells differentiated into distinct types aggregate to form *tissues* and, in most animals, except sponges, tissues are organised into structural and functional organs. Metazoan cells recognise other cells that belong to the same animal, and thus, if cells are separated in tissue culture, they will reaggregate. This 'self-recognition' depends on the presence of glycoproteins on the cell surface (Chapter 1).

Metazoans undergo *embryonic development* (ontogeny), a multicellular individual developing from a single-celled zygote (Chapter 15). Metazoans produce gametes in multicellular organs (gonads) or in aggregations surrounded by non-sexual (somatic) body cells. Metazoans are usually diploid, with meiosis occurring only during production of gametes. Metazoans are *heterotrophs*, feeding on other organisms, and are motile during some part of their life cycle.

The animal kingdom includes an amazing variety of species classified in some 30 phyla. Each phylum has characteristic features, reflecting a common body plan. Phyla differ in pattern of development, body symmetry, body segmentation, body cavity, external covering, internal skeleton, appendages, and in other ways that we will study in Chapters 38–40.

This chapter describes simple animals, including sponges (simplest in structure), jellyfish, anemones, corals and comb jellies (radially symmetrical), flatworms (which lack a body cavity or coelom) and nematodes (roundworms with a pseudocoelom). Chapter 39 describes coelomate animals (with a mesodermal body cavity), including segmented annelid worms, arthropods characterised by an external skeleton and jointed appendages, and molluscs (usually with a shell). Chapter 40 deals with animals

as different as echinoderms, such as starfish, and chordates, which include vertebrates from fishes to humans. Echinoderms and chordates show a similar pattern of embryonic development.

> Metazoa are multicellular animals that undergo embryonic development from a zygote and have cells and tissues differentiated for different functions.

ORIGIN OF METAZOA

How metazoans evolved from protistan-like organisms and how phyla are related to one another are not fully understood. Various theories have been proposed for the origin of multicellularity. One theory is that a multicellular organism arose by partitioning of a ciliate protist, producing a flatworm (Fig. 38.1a). Another idea is that metazoans evolved from a colony of flagellate protists (Fig. 38.1b). A third theory suggests that the earliest multicellular animal was a two-layered organism that crept along the sea floor, rather like *Trichoplax*, a marine animal living today (Fig. 38.1c). *Trichoplax* is the least differentiated of all known metazoans and has the smallest amount of DNA per nucleus of any multicellular animal.

Fossil faunas

Fossil evidence suggests that the first multicellular animals originated in Pre-Cambrian seas. Fossils representing the transition from protists to metazoans are not known. The earliest known metazoan fossils, the Ediacaran fauna, are more than 600 million years old and were present and already well diversified in the Pre-Cambrian. They are known from several continents but are best represented in Australia, in the Ediacara Hills of South Australia (Chapter 31).

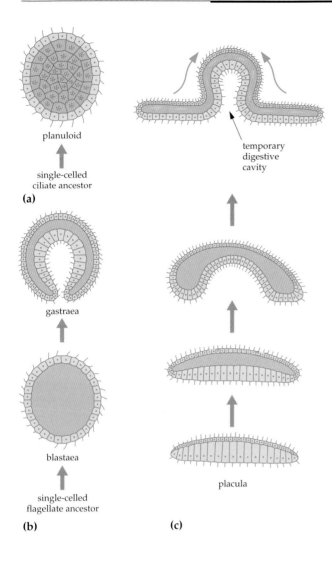

Fig. 38.2 *Dickinsonia costata*, of the Pre-Cambrian Ediacaran fauna

Most of them resemble the medusa stage of modern jellyfish. Some, reaching a length of 40–60 cm, have been likened to sea pens, in the Cnidaria (see p. 836), while others, such as *Dickinsonia*, almost a metre long but only 3 mm thick (Fig. 38.2), resemble annelid worms.

By the Cambrian period a vast diversity of metazoans had evolved. The most remarkable record of this radiation, especially of arthropods, occurs in shales from the Burgess Pass in the Canadian Rocky Mountains. The Burgess Shales (Chapter 31) were derived from sediments of fine mud which allowed minute details of animals to be preserved.

New data, particularly DNA and RNA sequences, are providing some of the answers to the history of the Metazoa. Figure 38.3 summarises what we know of the phylogeny of living representatives based on data from both morphology and molecules.

SPONGES: PHYLUM PORIFERA

Sponges, phylum Porifera, are aquatic animals (Fig. 38.4) that are similar to higher multicellular animals in having three layers of cells. However, in sponges these layers are not organised into tissues or organs. Sponges have no mouth or gut, no circulatory system and no nervous system, and they represent a simple level of organisation.

There are about 5000 species of sponges, which are mostly marine and many of which are colonial. Adults are sessile on a substrate such as rocks, but larvae are free-swimming.

In the simplest sponges (Fig. 38.5a), the body is tubular and its wall has three functional layers. The inner layer, lining an internal cavity, consists of flagellated cells, **choanocytes** (collar cells), which have a distinctive collar of microvilli surrounding the flagellum (Fig. 38.6). The outer surface of the sponge consists of a layer of flattened cells, **pinacocytes**,

Fig. 38.1 Various theories explain how multicellular animals originated. (a) The syncytial theory is that multicellularity arose by partitioning of a ciliate protist to produce a flatworm (planuloid). In this theory it is suggested that the cell membranes of the first metazoan were incomplete, hence the word syncytial, meaning continuous cells. (b) An alternative to the syncytial theory, the colonial theory, is that Metazoa arose from a colony of flagellate protists. In the nineteenth century, Ernst Haeckel proposed that the first multicellular stage was a hollow spherical organism, a blastaea, consisting of a single layer of cells resembling the blastula typically seen in the embryonic development of metazoans. Haeckel envisaged that the multilayered condition was acquired in evolution by invagination (infolding) of the blastaea to give a gastraea, resembling the gastrula stage of metazoan development. (c) Another theory proposed last century is that the earliest multicellular animal was a placula—a two-layered organism that developed an internal space, became invaginated, and crept along the sea floor. *Trichoplax adhaerans* is a marine organism considered a living example of a placula. *Trichoplax* has a lower layer of endoderm, which has resorptive and glandular cells, and the animal raises its body to form a temporary digestive cavity. This hypothesis offers an explanation for why the endodermis of modern animals is specialised for nutrition

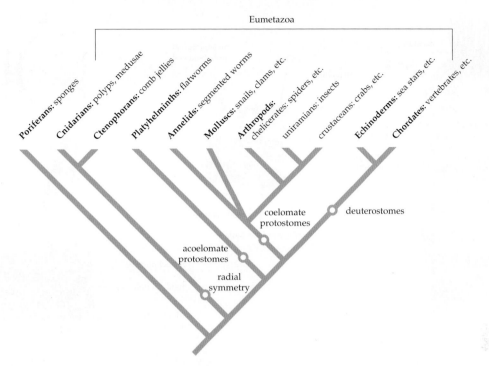

Eumetazoa

coelomate protostomes

deuterostomes

acoelomate protostomes

radial symmetry

Fig. 38.3 Phylogeny of major groups of living metazoans based on both morphological and molecular data. The relationships of some animal groups are not fully known

Fig. 38.4 A cluster of small calcareous sponges of the genus *Callispongia*, at Heron Island

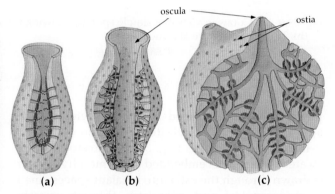

oscula

ostia

(a)   (b)   (c)

Fig. 38.5 Structure of (a) simple, (b) intermediate and (c) advanced forms of sponges

making up the **pinacoderm**, analogous to the epithelium of other animals. Between these two layers is the **mesohyl**, which consists of a gelatinous protein matrix containing amoeboid cells, **amoebocytes**, dispersed collagen fibres (a protein common to all animals), and skeletal elements. Some amoeboid cells are large and phagocytic. They are important in digestion but are also capable of giving rise to any other type of cell; that is, they are totipotent.

The skeleton is composed of calcium carbonate or silica **spicules**, or fibres of a coarse collagenous proteinaceous material termed **spongin**, or a combination of siliceous spicules and spongin fibres. Collagen fibres are secreted by amoebocytes as well as by fixed cells. The components of the skeleton, particularly

the structure of the spicules, are used in classification of sponges. Skeletons are sold for bath sponges and other uses!

Sponges have a unique method of filter feeding (Fig. 38.6). Water is drawn in through pores, the **ostia** (sing. ostium), in the external wall of the sponge into a central cavity, the **atrium** or **spongocoel**. This feeding and respiratory current then leaves through one or more large openings, the **oscula** (sing. osculum). The current is propelled by beating of the flagella of the choanocytes lining the passages.

Sponges feed on extremely fine, mostly submicroscopic, particulate matter drawn in by the water current. Larger particles are phagocytosed by pinacocytes lining the inhalant canals but particles of bacterial size and below (less than 1 μm) are trapped on the exterior of the choanocyte collar and ingested in food vacuoles at its base. The structure of a choanocyte, as revealed by electron microscopy, is shown in Figure 38.6. Both choanocytes and

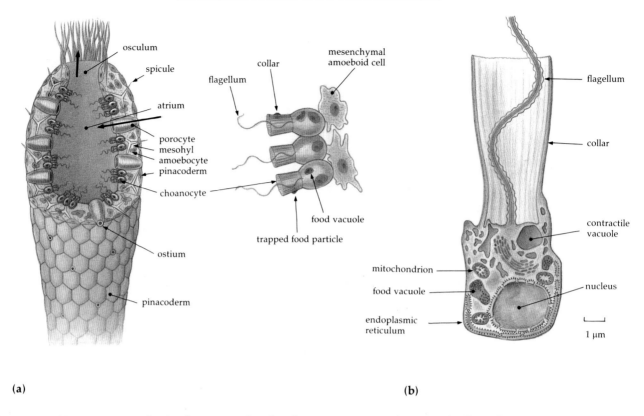

(a)   (b)

Fig. 38.6 (a) Cutaway view of a simple sponge, showing the arrangement and action of collar cells (choanocytes). (b) The ultrastructure of a choanocyte

pinacocytes transfer food to amoebocytes, which appear to be the principal sites of digestion.

In advanced sponges (Fig. 38.5c), choanocytes are located in chambers embedded in the mesohyl. Water is drawn through the ostia into inhalant spaces, lined by pinacocytes, and percolates through narrow passages into the choanocyte-lined chambers. From these, exhalant canals allow water to escape through the oscula. Evolution of this body form allows attainment of a much larger size and diversity of shape than is possible in simpler sponges. Intermediate conditions occur (Fig. 38.5b).

> Sponges (phylum Porifera) are simple aquatic filter-feeders, with three layers of cells but no tissues and organs, mouth or nervous system.

EUMETAZOA

Multicellular animals more complex than sponges are grouped together as the **Eumetazoa**. Eumetazoans differ from sponges in having a mouth, nervous system and the germ layers—ectoderm and endoderm—with or without intervening mesoderm. They are classified into approximately 29 phyla.

> Multicellular animals with germ layers, tissues and organs are the Eumetazoa.

RADIALLY SYMMETRICAL ANIMALS

Polyps and medusae: phylum Cnidaria

The phylum **Cnidaria** includes jellyfish, sea wasps, hydrozoans, sea anemones and corals. Cnidarians are marine or freshwater animals with radial symmetry (Box 38.1). Some, a few anemones, are radiobilateral; that is, they are primarily radially symmetrical but the development of one or two ciliated grooves at the edges of the mouth means that the animal can only be divided into two equal halves in one or two planes.

The life cycles of cnidarians include one or two life stages—polyp and medusa—each with a terminal mouth fringed with tentacles and lacking an anus. The **polyp** is an attached tubular form with its mouth upwards; the **medusa** is free-floating (pelagic) and bell-shaped, with its mouth pointing downwards (Fig. 38.7). The mouth opens into a cavity that is lined by endoderm, gastrodermis, and termed the **gastrovascular cavity** or **coelenteron**, which is homologous with the alimentary canal of other metazoans. The mouth is borne on a projection, termed the **hypostome** in the polyp and the **manubrium** in the medusa.

BOX 38.1 RADIAL AND BILATERAL SYMMETRY

Two living metazoan phyla, Cnidaria and Ctenophora, have **radial symmetry**, and any longitudinal plane bisects the animal into similar halves (Fig. a). Radial symmetry is characteristic of animals that are attached to a substrate (e.g. polyps) or float freely in water (e.g. medusae or jellyfish). Such animals receive equal stimuli and opportunity to obtain food from all sides. Radial symmetry is seen also in echinoderms (Chapter 40), for example, starfish, but here it is secondary and is acquired as a result of metamorphosis from a bilaterally symmetrical ciliated larva.

In animals with **bilateral symmetry**, only one plane divides the animal into equal halves (Fig. b). Bilateral symmetry is typical of animals that move along a substrate. Locomotory mechanisms are more effective if one part of the animal, the anterior or head end on which sensory receptors can be concentrated, leads the way. Bilaterally symmetrical animals often have a lower ventral surface specialised for movement; the upper dorsal surface may be protective.

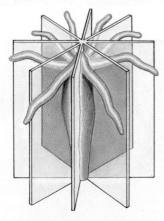

(a)

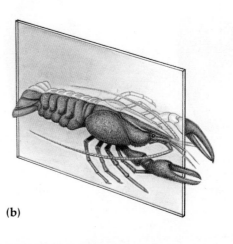

(b)

(a) Radial and (b) bilateral symmetry

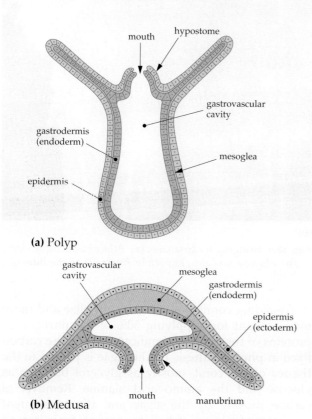

(a) Polyp

(b) Medusa

Fig. 38.7 The basic structure of **(a)** a polyp and **(b)** a medusa are the same. A medusa is like an upside-down flattened polyp

Cnidarians are characterised by stinging organelles, **nematocysts (cnidae)**, which function in defence and capture of prey. Each nematocyst is produced in a cell, the **cnidocyte**. There are many types of nematocysts, but all basically consist of a capsule and an enclosed, usually hollow, thread that is eversible and often armed (Fig. 38.8). In penetrant nematocysts, a proteinaceous toxin is injected from the capsule through the thread into the prey or predator.

In Australia, people may encounter stings from cnidarians. The commonest encounters are those with the medusae of the dangerous sea wasp or box jellyfish, *Chironex fleckeri* (Box 38.2), the hydroid polyp, *Aglaophaenia cupressina* (Fig. 38.9a), which is common on the Great Barrier Reef, and the blue bottle or Portuguese man o'war, *Physalia utriculus* (Fig. 38.9b).

Cnidarians are described as **diploblastic** because the body wall has two cellular layers, an ectoderm and endoderm. Between them is an intermediate, largely gelatinous, layer, the **mesoglea**, which differs from true mesoderm (Chapter 14). The mesoglea forms the bulk of the body mass in the medusoid phase.

Cnidarians are at the tissue level of organisation. In the absence of mesoderm, organ formation is minimal although tentacles, reproductive structures, and sensory structures concerned with balance and light detection can be considered organs.

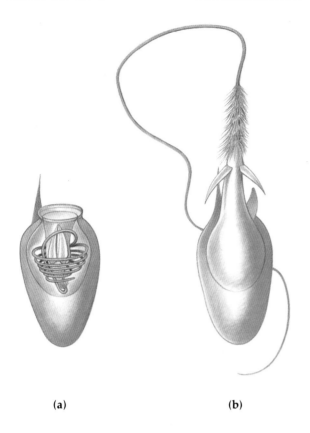

(a)   (b)

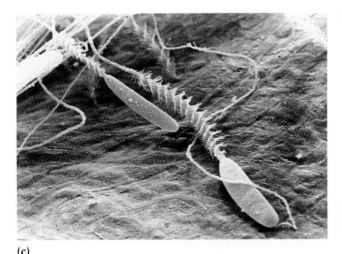

(c)

Fig. 38.8 Structure of a penetrant nematocyst:
(a) undischarged nematocyst; (b) nematocyst
discharged after contact with the prey. (c) Discharged
nematocyst photographed under a scanning electron
microscope

(a)

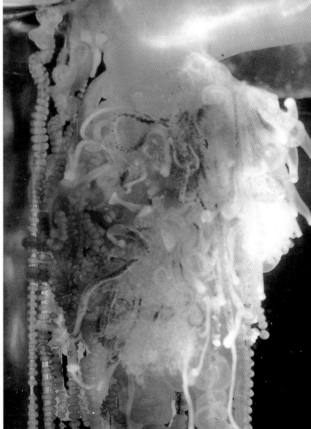

(b)

Fig. 38.9 Stinging hydrozoans: (a) *Aglaophaenia cupressina;*
(b) *Physalia utriculus,* known in Australia as the blue
bottle

All cnidarians are carnivorous but some members
of all three classes, particularly corals, harbor symbiotic intracellular algae (zooxanthellae or zoochlo-
rellae, or both), usually in the endoderm. These algae
enable corals to build reefs, of which the Great Barrier
Reef of eastern Australia is the most extensive
example. The cnidarian derives much of its nutrition
from the algae which, in return, are provided by the
animal with a favourable environment. Virtually all
reef-building corals possess zooxanthellae and these
may account for supplying 50% of the nitrogen in
proteins of the coral. A significant part of the carbon
fixed in photosynthesis by the algae is passed to the
tissues of the coral, mostly as glycerol but also as
glucose and the amino acid alanine. Removal of
carbon dioxide by the algae and the consequent
formation of calcium carbonate accounts for increased
growth of the coral skeleton in the presence of algae;
growth is slowed in corals deprived of algae (e.g. if

kept in the dark). Prey caught by the coral provides nitrogen and phosphorus to both the coral and the algae.

> Cnidarians (polyps and medusae) are characterised by stinging cells, nematocysts (cnidae). They have a body wall consisting of two cellular layers and an intermediate layer, also containing cells, the mesoglea. Their body cavity is a coelenteron.

Jellyfish: class Scyphozoa

Scyphozoans are exclusively marine, the majority living in coastal waters. The conspicuous phase of their life cycle is the jellyfish or medusa (Fig. 38.10), while the polyp is very small or sometimes absent. Scyphozoan medusae include the largest individual cnidarians. Although most are free-swimming, some rest on or attach to the ocean floor. Typically, the medusa consists of a flat or domed bell (Fig. 38.11). The mouth is borne on the manubrium, a trunk-like structure in the centre of the subumbrellar surface. In *Pelagia* (Fig. 38.12), a luminescent jellyfish of Australian waters, and *Aurelia*, the margins of the manubrium commonly extend as four oral arms (Fig. 38.11). The mouth leads into a central stomach from which radiate four radial canals. The radial canals connect with a ring canal, which runs around the margin of the bell. The muscular system is ectodermal

in origin. The margin bears tentacles armed with nematocysts and is usually scalloped. In the indentations there are often sense organs, the **rhopalia**, which contain balance structures, **statoliths**, and sometimes eyespots.

Adult medusae, usually of separate sexes, produce eggs and spermatozoa in gonads situated on the septa dividing the stomach or on the walls of the four gastric

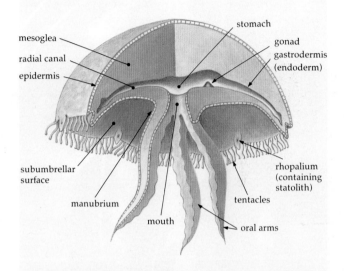

Fig. 38.11 Structure of the jellyfish, *Aurelia*. One-quarter of the body has been cut away to show internal features

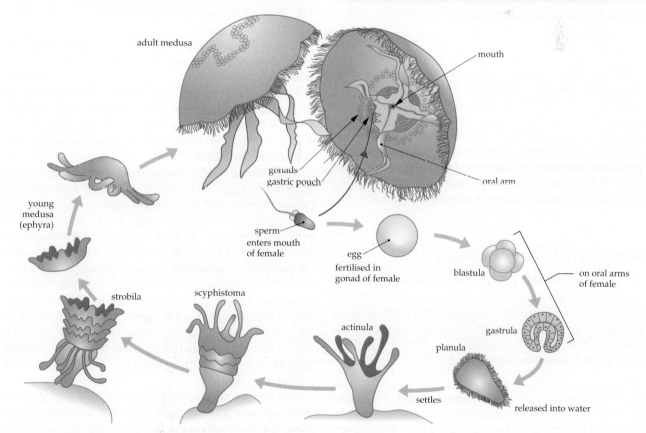

Fig. 38.10 Life cycle of the jellyfish, *Aurelia*

Fig. 38.12 The medusa of the schyphozoan *Pelagia*

Hydras: Class Hydrozoa

In the class Hydrozoa, the life cycle typically includes a dominant, although often very small, attached polyp stage, which reproduces asexually, and a microscopic pelagic medusa stage, which reproduces sexually. In some species, however, either the polyp or the medusa may be reduced or absent. The freshwater genus *Hydra* exemplifies cnidarian structure (Fig. 38.13),

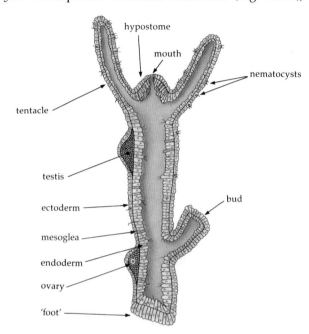

Fig. 38.13 Structure of *Hydra*

pouches. When mature, eggs and sperm break into the gastrovascular cavity and pass out the mouth. In some scyphozoans, such as *Aurelia*, eggs lodge and zygotes develop on the oral (mouth) arms. A ciliated **planula** larva is released and later attaches to the substrate to become a polyp. The polyp produces medusae asexually by horizontal splitting of the body (Fig. 38.10). Attached polyps may persist for years, outliving the individual medusae they produce.

Box jellyfish: class Cubozoa

In box jellyfish, the bell is square in cross-section and the margin bears four extensions from which hang one or more hollow tentacles. The medusae are strong swimmers and their sting is often fatal to humans. They include the sea wasp, *Chironex fleckeri*, in tropical Australian waters (Box 38.2).

> Scyphozoans (jellyfish) are marine animals with the conspicuous life cycle stage being the jellyfish or medusa. Cubozoans (box jellyfish), are also marine, with a cube-shaped bell and tentacles in groups of four. Their stings are often fatal to humans.

although its mesoglea is very thin in comparison with hydrozoan medusae where it forms the bulk of the body.

Hydroid polyps are commonly colonial, with an interconnecting gastrovascular cavity. They usually have a chitinous, sometimes calcified, exoskeleton, *Hydra* being one exception. An individual in the colony is a **hydranth**.

In contrast with jellyfish (scyphozoans), the hydroid medusa lacks radial partitions and rhopalia (sense organs), and the gametes ripen in the ectoderm. In the life cycle of *Obelia* (Fig. 38.14), the colony of polyps grows by asexual budding. Two types of polyps are produced, feeding hydranths and reproductive **gonangia** (sing. gonangium). Free-swimming medusae of both sexes bud off from the gonangia (Fig. 38.15a) and give rise to eggs or sperm, which are released into the water. Fertilisation is external and the zygote divides to form a blastula, which becomes a ciliated planula larva. The planula rapidly attaches to the substrate and forms a new colony by budding.

The medusoid stage may be reduced and permanently attached to the hydroid colony as a gonophore, which produces gametes. In *Tubularia*, polyps are born from the gonophore within the tentacle crown of the

BOX 38.2 SEA WASPS IN AUSTRALIAN WATERS

One of the greatest threats to people in coastal waters of tropical Australia is the sea wasp, *Chironex fleckeri* (see Fig. 27.4). Stings from its tentacles cause near-sudden death of, on average, one Australian per year. *Chironex fleckeri* is chiefly restricted to Australia, although it may occur in Indonesian and Malaysian waters to the north. It is an inshore species, not normally occurring in offshore waters, such as those surrounding the Great Barrier Reef.

Populations are sighted in shallow water off the coast of Queensland from late November to March each year, after which they disappear. Young are carried in the bell of adult female medusae and are released in late November when adults move into shallow water.

Like all box jellyfish (class Cubozoa), *C. fleckeri* medusae can be recognised by their cube-shaped bell and tentacles in groups of four. Eyespots are present marginally and stinging cells (nematocysts) occur in bands on the tentacles. A large specimen can have a bell as large as a human head with trailing tentacles reaching a total length of 90 m. The nematocysts discharge a toxin into the skin of a victim. The toxin elicits sustained contraction of all muscle types. There is variation in the stinging ability of individual sea wasps, but when a victim is severely stung over a large area of the body, death usually results within minutes.

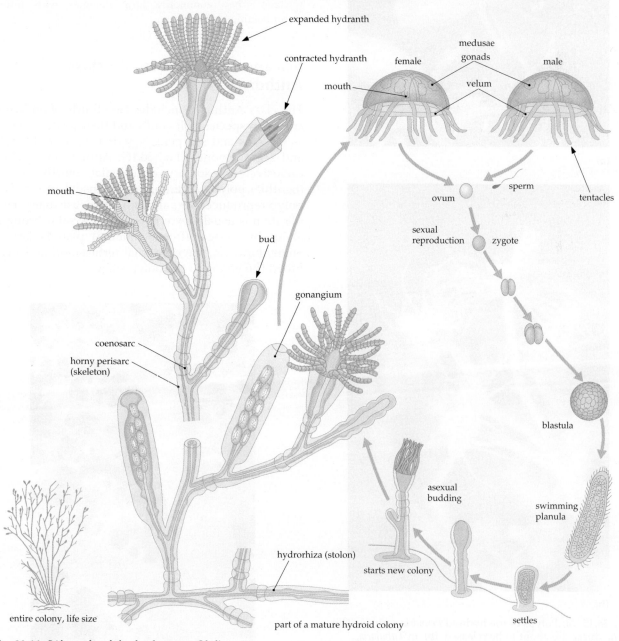

Fig. 38.14 Life cycle of the hydrozoan *Obelia*

(a)

(b)

Fig. 38.15 (a) Polyps of the hydroid *Pennaria australis*. The medusae are about to be released. (b) In *Tubularia*, medusae are permanently attached to gonophores, which release young polyps

hydroid (Fig. 38.15b). In *Hydra* and some marine genera, the medusa is totally absent. In a few species, the polyp is reduced or absent and the medusa is conspicuous. The thick, colourless transparent bells, several centimetres in diameter, which are washed up on Australian shores, are the medusae of the luminescent hydrozoan *Aequoria*.

The polyp-like individuals in colonial hydrozoans may assume several forms in the same colony. In the blue bottle, *Physalia* (Fig. 38.9b), there is a large gas-filled float. By budding, the initial polyp produces both feeding forms with long, fishing, nematocyst-laden tentacles, and elongate female forms (gonophores). At the base of these female gonophores there are grape-like clusters of male gonophores.

> Hydrozoans (hydras) usually have a dominant polyp stage. They commonly form colonies with interconnecting coelenterons.

Sea anemones and corals: class Anthozoa

The class Anthozoa includes two-thirds of all living cnidarian species: soft corals and their relatives, such as sea fans and sea pens, together with hard corals and sea anemones (Fig. 38.16). Anthozoans consist exclusively of solitary or colonial, mostly sessile (benthic) polyps. There is no medusoid phase. The polyp reproduces sexually and, often, asexually. Fertilisation is usually external and dispersal is brought about by pelagic (free-swimming) planula larvae. Some, however, have internal fertilisation and may brood the young to the polyp stage.

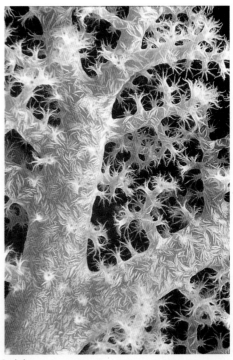

Fig. 38.16 (a)

(b)

(c)

Fig. 38.16 Anthozoans: **(a)** soft coral; **(b)** anemone, *Phlyctenanthus australis*; and **(c)** hard corals, forming part of the Great Barrier Reef

The anthozoan polyp is usually a hollow cylinder. The polyp is topped by an **oral disc** fringed by hollow tentacles (Fig. 38.17), the cavity of which is continuous with the gastrovascular cavity, the coelenteron. Unlike those of hydrozoans, gonads are endodermal and are located on radial septa (mesenteries) that divide the coelenteron. Muscle bands, which retract the polyp, run longitudinally on one surface of each mesentery.

Polyps such as anemones are 'naked' or may have a chitinous cuticle but, in hard corals, polyps are embedded in a skeleton of calcium carbonate secreted by and external to the ectoderm. Soft corals have spicules in the mesoglea.

During the day, coral polyps withdraw into spaces, corallites, within the skeleton of calcium carbonate that they secrete. When feeding, polyps evert and extend their tentacles, which bear nematocysts and catch zooplankton. The tentacles pass the prey through the polyp's mouth and into the coelenteron, where digestion occurs. Passing through connections between individual polyps, the food becomes part of the nutrition of the whole colony.

Coral colonies, such as those that form the Great Barrier Reef, occur in a great variety of colours, shapes and sizes (Fig. 38.16). Their colours come from the pigments of symbiotic zooxanthellae in their tissues.

oral disc   tentacle   polyp

Fig. 38.17 In hard corals, the polyps, here retracted, are embedded in a skeleton of calcium carbonate

Their shape depends on the way individual polyps build their skeletons and how they bud. In brain corals, polyps divide without forming complete walls so that long lines of polyps have a common wall around them. In staghorn corals, long branches arise from single polyps.

> Anthozoans (e.g. soft corals, hard corals and sea anemones) are exclusively polyps. Hard corals secrete a skeleton of calcium carbonate, forming reefs such as the Great Barrier Reef.

COMB JELLIES: PHYLUM CTENOPHORA

Ctenophores (comb jellies) are pelagic marine animals, numbering about 100 species, which somewhat resemble cnidarian medusae in their transparent, sac-like bodies and jelly-like mesoglea (Fig. 38.18). The fact that one species, *Euchlora rubra*, has its own nematocysts (some borrow them!) suggests that ctenophores are related to the Cnidaria.

Ctenophores are distinguished by eight longitudinal rows of ciliary plates, **comb rows**, formed by fused cilia. These beat in unison and propel the animal, mouth forward. The combs cause light interference patterns, which make the rows iridescent while the remainder of the body is transparent (Fig. 38.18). A line of ciliated cells connects each comb row to a complex **apical organ**, which lies at the aboral pole, the point on the upper surface that is diametrically opposite the mouth, which is on the lower surface. The most conspicuous structure in the apical organ is a balance organ, the **statocyst**.

Ctenophores have an endodermal digestive system equivalent to the coelenteron of cnidarians. The mouth leads into an elongate tube, the walls of which are greatly folded and secrete enzymes that rapidly break down food. Digested food enters the dilated stomach and is circulated, by the action of cilia, along longitudinal branches, like a canal system, to the comb

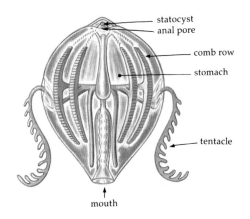

Fig. 38.18 Fragile comb jellies or sea gooseberries move by beating rows of ciliated plates

rows. A notable difference from cnidarians is the presence of two posterior openings to the digestive system, the **anal pores**.

Ctenophores prey on other animals, which they catch with two branched, retractile, adhesive tentacles or, in species where these are reduced, directly with the mouth. They are hermaphroditic and even in species in which the adult body is depressed or elongated, the larva is spherical.

> Ctenophores (comb jellies) resemble cnidarian medusae but, except for one species, lack nematocysts. They have eight rows of ciliary plates (comb rows), the cilia of which beat in unison and propel the animal.

PROTOSTOMES AND DEUTEROSTOMES: TWO MODES OF DEVELOPMENT

During animal development the embryonic cavity, the archenteron, develops into the digestive tract (Chapter 15). With the notable exception of cnidarians and nematodes (p. 851), Eumetazoa can be assigned to one or other of two groups chiefly on the basis of whether the opening to the archenteron, the blastopore, develops into the mouth or the anus.

Protostomes are animals in which the blastopore becomes the mouth. Cleavage (cell division) of the embryo is spiral, which means that at the eight-cell stage the four upper cells (blastomeres) of the embryo are rotated 45° relative to the four lower blastomeres (Fig. 38.19). The fates of various parts of the embryo are determined or fixed at a very early stage in cleavage, or even in the egg, and development is *determinate*. The mesoderm, for example, derives from a single cell. If a coelom (body cavity) develops, it is schizocoelic, produced by splits in the mesoderm. Major protostome phyla are Platyhelminthes, Annelida, Mollusca and Arthropoda.

Deuterostomes are animals in which the anus forms at the site of the blastopore, the mouth forming at a secondary opening. Cleavage is radial, so that the upper blastomeres of the embryo at the eight-cell stage are directly above the lower blastomeres (Fig. 38.19). The fate of individual cells is not fixed until a stage in cleavage much later than in protostomes, and development is *indeterminate*. The mesoderm and enclosed coelom arise by outpocketing from the embryonic precursor of the gut (except in vertebrates, which are schizocoelic). The two chief deuterostome phyla are Echinodermata and Chordata.

Division into 'protostomes' and 'deuterostomes' has been criticised as artificial, and some phyla (e.g. lophophorates, Chapter 39) have characteristics of both. However, the groupings have proved useful and receive support from molecular studies, particularly ribosomal RNA sequence analysis (see Fig. 38.2).

> Protostomes are animals in which the blastopore of the embryonic gut (archenteron) becomes the mouth in the adult. In deuterostomes, the anus forms at the site of the blastopore.

ACOELOMATE PROTOSTOMES

Flatworms: phylum Platyhelminthes

Platyhelminths are dorsoventrally flattened, bilaterally symmetrical animals ('flatworms') termed **acoelomate** because the mesoderm surrounding the endodermal gut contains no large body cavity (no perivisceral cavity). Absence of a coelom in flatworms is generally regarded as primitive.

Flatworms live in marine, freshwater and moist terrestrial habitats. They are abundant as free-living scavengers and carnivores but many are parasites. There are about 12 000 species classified in four major classes, including free-living turbellarians, and parasitic monogeneans (skin or gill flukes), trematodes (endoparasitic flukes) and cestodes (tapeworms).

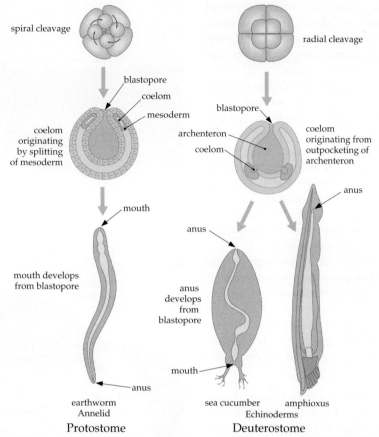

Fig. 38.19 Protostomes and deuterostomes differ in development. In protostomes, the blastopore becomes the mouth, and in deuterostomes it becomes the site of the anus. If a coelom develops in a protostome, it is schizocoelic, produced by splitting of the mesoderm. In deuterostomes, such as echinoderms and amphioxus, the coelom arises from outpockets of the embryonic gut (archenteron)

Platyhelminths share with cnidarians a number of characters, all probably primitive: no coelom; no anus; no circulatory system; and no respiratory system. Flatworms appear more advanced than cnidarians in having a differentiated excretory system of **protonephridia**, the basic unit of which is the **flame cell** (Fig. 38.20). They also differ from cnidarians in possessing a third 'germ layer', the mesoderm, between ectoderm and endoderm. Mesoderm differs

from the intermediate layer (mesoglea) of cnidarians in appearing early in embryonic development, and is related to the endoderm. Possession of a true mesoderm is associated with organ formation on a scale impossible in cnidarians and ctenophores.

In primitive flatworms, the nervous system consists of a simple network or plexus with some anterior concentration of neurons. In the more advanced turbellarians it is a ladder-like system, with two or

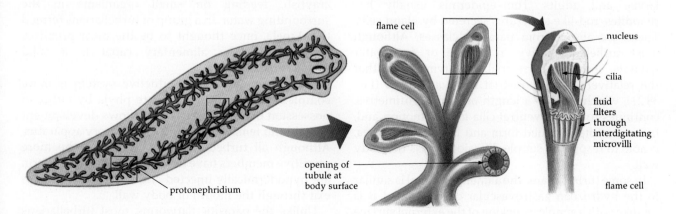

Fig. 38.20 Excretory system of a platyhelminth, showing the structure of a flame cell

more longitudinal nerve cords and anterior cerebral ganglia. The simplest eyes are merely pigment spots with associated nerve endings. In most free-living turbellarians and larval trematodes, the eyes are pigment cups enclosing photoreceptors.

Most flatworms are hermaphroditic (monoecious), having both sexes in each individual, but the blood fluke, *Schistosoma*, provides a rare example of separate sexes (dioecious condition). In hermaphroditic species, self-fertilisation is frequent but cross-fertilisation usually occurs. Parthenogenesis (reproduction from unfertilised eggs) is also known. Cross-fertilisation, and the genetic variability it allows (Chapter 13), is perhaps less necessary in parasites, which are highly adapted to particular and constant conditions of a host as opposed to free-living animals.

Some turbellarians and tapeworms use hypodermic impregnation, in which the penis is injected through the body wall of the partner. In most cases, however, sperm are passed directly into receptacles in the female reproductive system. Most species are oviparous (releasing unfertilised eggs), but ovoviviparity and viviparity (eggs developing within the female) also occur.

> Platyhelminthes (flatworms) are acoelomate, dorsoventrally flattened worms, including free-living and parasitic forms. Their digestive cavity (not always present) has a single opening and lacks an anus. Their excretory system consists of protonephridia, the basic unit of which is the flame cell.

Free-living flatworms: class Turbellaria

There is convincing evidence that the class Turbellaria is not a single evolutionary lineage, but we will use the term turbellarians as a convenient collective name for groups of flatworms.

Turbellarians are mostly free-living, a common example being *Dugesia*, a freshwater planarian. The other three classes of flatworms (trematodes, monogeneans and cestodes) are exclusively parasitic.

Turbellarians have a ciliated epidermis in both larvae and adults. The epidermis usually has **rhabdites**, rod-like structures secreted by gland cells. These are absent from parasitic classes. Although most turbellarians live in marine or freshwater habitats, some are terrestrial. It is remarkable that the relatively large terrestrial *Bipalium kewense* (Fig. 38.21), which reaches a length of many centimetres, continues to use its ventral cilia for locomotion and, because of its flattened form and large surface area, is able to respire by simple diffusion across the body wall.

In some turbellarians, the alimentary canal is similar to the hydrozoan gastrovascular cavity in that it is sac-like with a simple intucking of the external surface forming a simple pharynx. Some turbellarians have a more complex tubular (Fig. 38.22) or bulbous

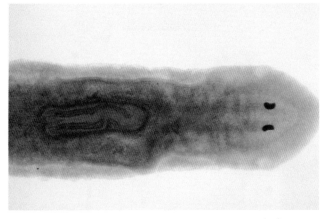

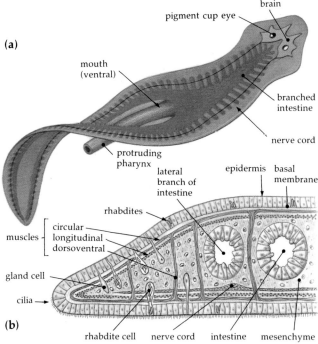

Fig. 38.22 (a) Dorsal view of a flatworm, showing digestive and nervous systems, with (b) a cross-section to show the anatomy. Note the lack of a coelom (mesodermal body cavity)

pharynx, which can be protruded from the mouth. *Craspedella* (Fig. 38.23) and *Temnocephala* have a bulbous pharynx. They live on the surface of Australian crayfish, feeding on small organisms in the surrounding water. In a group of turbellarians termed the **Acoela**, once thought to be the most primitive eumetazoans, the alimentary canal is a solid endodermal core.

The platyhelminth reproductive system is more complex than any in the radiate phyla by virtue of possession of mesoderm, which allows development of male and female ducts and a copulatory apparatus. Although all turbellarians have a penis, the more primitive members have no genital ducts; their sperm are hypodermically injected and their fertilised eggs exit through the mouth or body wall.

Unlike the parasitic flatworms, most turbellarians have no larval stages, although there are exceptions that have a distinctive free-swimming, ciliated larva.

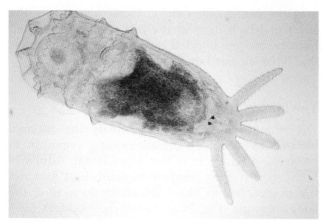

Fig. 38.23 The flatworms *Temnocephala* and *Craspedella spenceri* (shown here) live on the surface of crayfish. They have anterior tentacles used in locomotion and in catching prey, a bulbous pharynx and sac-like intestine, lacking an anus

Of the four classes of Platyhelminthes, turbellarians are free-living flatworms, and trematodes, monogeneans and cestodes are all parasites.

Ectoparasitic flukes: class Monogenea

Monogeneans are ectoparasites and rarely endoparasites, chiefly of aquatic vertebrates. One exception lives in the eye socket of the hippopotamus. *Gyrodactylus* attaches to the skin and gills of fishes and can be a serious pest in fish hatcheries. *Polystoma* is unusual in occurring in the urinary (cloacal) bladder of frogs and toads. Reproduction in *Polystoma* is triggered by the reproductive hormones of the frog, with eggs of the parasite being shed in the urine as the frog enters water to breed. This ensures that the larvae of the parasite, which attach to the gills of tadpoles, find a new host.

Monogeneans have a posterior holdfast device that attaches the adult parasite to the host. The holdfast bears anchors, suckers or pincers or is fringed with hooks. Although a larval stage is typical, monogeneans do not have the series of larval stages seen in trematodes.

Endoparasitic flukes: class Trematoda

Trematodes are endoparasitic flukes, often having complicated life cycles involving a number of different hosts. One group of about 7000 species, the digeneans, becomes sexually mature in a vertebrate host, especially fishes. Several larval stages occur in a snail host. A few digeneans are precociously sexually mature in the snail and may dispense with the vertebrate host, for example, the Sydney species, *Parahemiurus bennettae* (Fig. 38.24) in the estuarine snail *Salinator fragilis*. Most flukes of vertebrates are gut parasites but some, for example, *Fasciola hepatica*, the liver fluke of sheep, live in the bile ducts. A few,

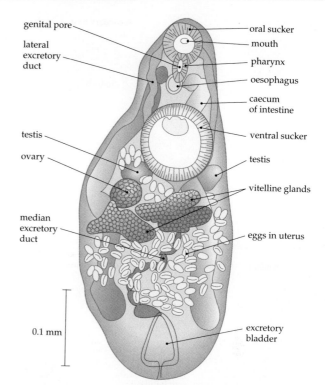

Fig. 38.24 The trematode fluke, *Parahemiurus bennettae*, has dispensed with a vertebrate host and matures in the snail host

including *Schistosoma mansoni*, the human blood fluke, occur in blood vessels, and others occur in lungs or other sites.

Adult flukes attach themselves to their vertebrate hosts by suckers. There is usually one oral sucker, around the mouth, and one ventral sucker. Their body has an outer covering, **tegument**, which is highly resistant to the host's enzymes but capable of absorbing some food materials. Their digestive system consists of a mouth, pharynx, very short oesophagus and two intestinal caeca (rarely one caecum), which are usually unbranched but may bear numerous lateral branches, as in *Fasciola hepatica*. Very rarely there may be one or two anal apertures, which evolved independently of the anus found in other protostomes. Almost all are hermaphrodites, except for the family Schistosomatidae. In *Schistosoma*, the male has a ventral groove in which it holds the female.

Liverflukes

The life cycle of *Fasciola hepatica*, the liver fluke of sheep (Fig. 38.25), includes a free-swimming ciliated **miracidium**, which hatches from the fertilised egg. The miracidium penetrates an aquatic snail, losing its ciliated epidermis on entry, and transforms into a sac-like **sporocyst**. Sporocysts give rise to numerous **rediae**, which, with or without the production of secondary rediae, produce free-swimming **cercariae**. A cercaria encysts on vegetation and becomes a **metacercaria**. Grazing sheep become infected with liver flukes by ingesting metacercariae.

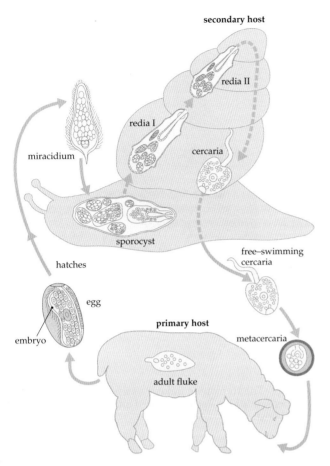

secondary host

redia II

redia I

cercaria

miracidium

sporocyst

free–swimming
cercaria

hatches

egg

embryo

primary host

metacercaria

adult fluke

Fig. 38.25 Life history of the liver fluke of sheep, *Fasciola hepatica* (an endoparasite)

In the human liver fluke, *Clonorchis sinensis*, the adult of which lives in the bile ducts of the liver, humans become infected by ingesting metacercariae that have encysted under the scales of fishes.

The blood fluke, *Schistosoma*

The blood fluke, *Schistosoma*, rivals the malarial parasite as a serious disease-causing organism of humans, although it is less often lethal. Adults of *Schistosoma* species, common in South-East Asia and the tropics, live in blood vessels in the wall of the large intestine (Fig. 38.26) (or of the bladder). Eggs break through the wall of the intestine into the lumen and pass out with the host's faeces. If the faeces enter water, the eggs undergo further development. The miracidium hatches and penetrates a freshwater snail. Two generations of sporocysts occur in the snail before release of fork-tailed cercariae. The cercaria loses its tail during penetration of the human skin, and becomes a juvenile fluke or schistosomulum. Schistosomula travel via the heart and lungs to the liver. The female is unable to make the journey unaided down the hepatic portal vein to its tributaries on the wall of the intestine and is carried, against the portal blood flow, in the ventral groove of the male.

Schistosoma causes incalculable human suffering. The four major species parasitic in humans infect some 200 million people. Disease symptoms are

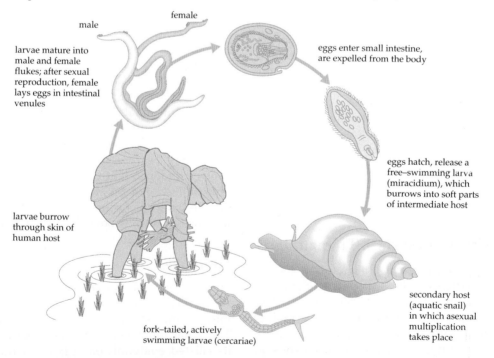

female

male

larvae mature into
male and female
flukes; after sexual
reproduction, female
lays eggs in intestinal
venules

eggs enter small intestine,
are expelled from the body

eggs hatch, release a
free–swimming larva
(miracidium), which
burrows into soft parts
of intermediate host

larvae burrow
through skin of
human host

secondary host
(aquatic snail)
in which asexual
multiplication
takes place

fork–tailed, actively
swimming larvae (cercariae)

Fig. 38.26 Life cycle of the human blood fluke, *Schistosoma japonicum*, which occurs in South-East Asia. Larvae (cercariae) in water burrow into the skin of the human host. Larvae mature into adults and females lay eggs in the host's intestinal venules. Blood vessels rupture and eggs make their way into the gut, from where they are expelled from the host's anus. The eggs hatch in water and the ciliated, free-swimming larvae (miracidia) burrow into their secondary host, a snail. The parasite reproduces asexually (as sporocysts and cercariae) in the snail, and the life cycle is complete when cercariae escape into the water to reinfect and mature in the human host

caused by eggs traversing the wall of the intestine, resulting in blood and mucus in the stools or, in the case of *S. haematobium*, by eggs penetrating the bladder. The most severe symptoms arise because of trapping of eggs, many of which are swept along with the hepatic portal flow and pass in the vascular system to all parts of the body. Severe stunting of the human host and swelling of the spleen, liver and abdomen commonly occur.

> Monogeneans are ectoparasitic flukes infecting one host, usually an aquatic vertebrate. Digeneans are endoparasitic flukes (trematodes) that have complicated life cycles with two or more hosts, often a snail and a vertebrate. Digeneans include important parasites of humans.

Tapeworms: class Cestoda

Trematodes, although showing important adaptations to parasitism, are not greatly different from free-living flatworms, many of which have suckers. Tapeworms (class Cestoda), in contrast, are much more specialised. As adults, most are parasites of the vertebrate gut. There is usually a vertebrate or invertebrate intermediate host.

In general structure, tapeworms retain basic platyhelminth features but they show several striking adaptations. They have a special anterior holdfast, the **scolex**, with suckers and often hooks (Fig. 38.27a). Just behind the scolex, there is a growth zone that continuously buds off segment-like body units, **proglottids** (see Fig. 38.27b), in a process termed **strobilisation**. The external covering of the body, the tegument, is highly permeable to food materials, including glucose, amino acids, fatty acids and vitamins, in the gut of the vertebrate host and is the sole absorptive surface for food. Tapeworms have lost the mouth and alimentary canal. Strobilisation produces many proglottids, each with at least one hermaphroditic set of reproductive organs.

Pork tapeworm

Taenia solium, the pork tapeworm, produces many proglottids and lacks asexual reproduction. The primary host, in which sexual reproduction occurs, is human and the intermediate host is the pig. The scolex (Fig. 38.27a) attaches to the wall of the human intestine by four suckers and an apical protrusion, the **rostellum**, armed at its base with 32 sickle-shaped hooks. Strobilisation produces a tape, **strobila**, 3–5 m long (Fig. 38.27c). Younger, more anterior proglottids cross-fertilise with older, more posterior ones. The uterus becomes filled with eggs and expands to occupy most of the proglottid, which is then said to be **gravid**. Gravid proglottids detach from the strobila and pass out in human faeces. If eggs or

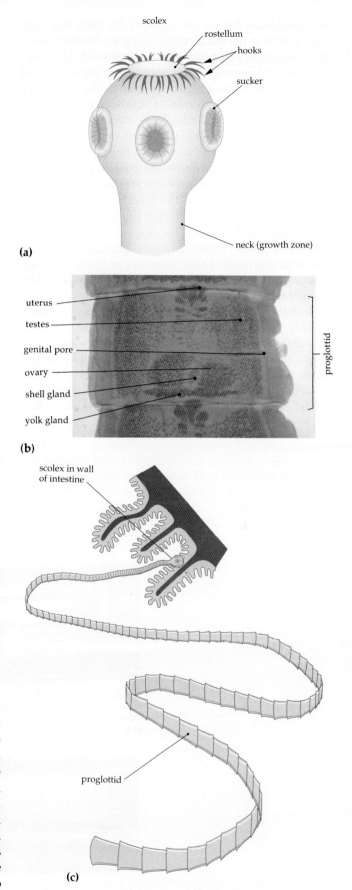

Fig. 38.27 *Taenia* species. **(a)** Scolex, **(b)** mature proglottid and **(c)** scolex and strobila attached to the intestinal wall

proglottids in faeces are ingested by pigs, the eggs may hatch in the duodenum, and a six-hooked larva, the **hexacanth** or **onchosphere**, is released. This uses its hooks to penetrate the wall of the pig's gut and passes through the vascular system to the voluntary muscles. There it forms a **cysticercus**, a larva consisting of a bladder-like structure and an invaginated scolex. The pea-sized cysticerci in the muscle produce what is termed 'measly' pork. If a cysticercus is infected and undercooked pork is eaten by a human, the scolex evaginates and attaches to the wall of the intestine, the bladder is cast off, and strobilisation occurs, producing the tape.

If eggs of *T. solium* are ingested by humans, cysticerci can be produced in the muscles, including those of the heart. Large numbers of cysticerci may prove fatal to the infected person and thus to the parasite.

Taemiarhynchus saginatus, the beef tapeworm has a similar life cycle, but the cysticerci occur in cattle. It differs in that eggs do not produce cysticerci in humans and it lacks a rostellum and hooks in the adult worm.

Echinococcus granulosus is unusual in tapeworms in undergoing larval asexual multiplication (see Box 38.3).

> As parasites of vertebrates, tapeworms (cestodes) have specialised features, including an anterior holdfast, the scolex, and many body segments, proglottids, produced by strobilisation. They include species that infect and harm humans.

Proboscis worms: phylum Nemertinea

Nemertinea or Rhynchocoela contain about 900 species of elongate, sometimes flattened, often highly coloured ribbon or proboscis worms, so-named because of their remarkable eversible anterior **proboscis**, which is used in capturing prey. There are many marine representatives (Fig. 38.28a), one freshwater genus, and one terrestrial genus (*Geonemertes*) that occurs in Australian rainforests. A few species are parasitic in marine invertebrates, for example, crabs.

(a)

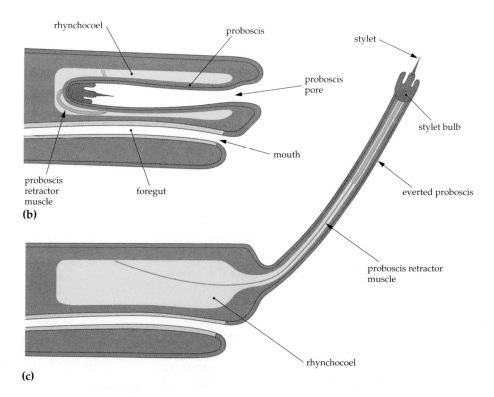

Fig. 38.28 (a) A marine nemertine worm, *Gorgonorhynchus*, with everted proboscis. (b) An undischarged and
  (c) discharged proboscis of a nemertine

BOX 38.3   HYDATID CYST TAPEWORM

Taenia solium is very rare in Australia but the hydatid cyst tapeworm, *Echinococcus granulosus*, is alarmingly common. Here the primary host is the dog and the intermediate host is the sheep, a kangaroo, or human. It is the smallest tapeworm, with only four proglottids, only one of which is gravid at any one time.

The scolex closely resembles that of *T. solium*. *Echinococcus granulosus* compensates for its relatively poor powers of sexual reproduction by producing asexually in the tissues of the intermediate host—it produces a cyst with many invaginated scolices (protoscolices) instead of the single cysticercus of *Taenia*. The wall of the cyst buds off numerous brood capsules, each with protoscolices on its walls (see figure). This hydatid cyst is usually about the size of a golf ball, with

hundreds of thousands of protoscolices; it has been known, in one Australian woman, to reach a volume of 8 L, with an estimated two million protoscolices.

Cysts commonly cause pathological symptoms. There are about 100 surgical cases a year in Australia, although surgical removal is difficult because rupture of the cyst releases proteins, which may cause fatal anaphylactic shock, and freed protoscolices may establish cysts elsewhere in the body.

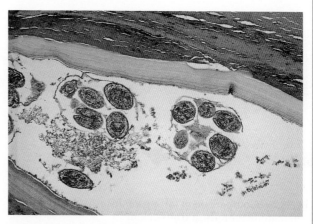

(b) A portion of a hydatid cyst of *Echinococcus granulosus* from a sheep, showing a few brood capsules and their protoscolices

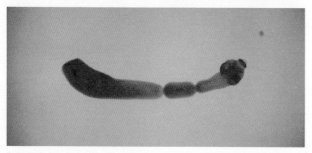

(a) Adult of the hydatid cyst tapeworm, *Echinococcus granulosus*

Nemertines probably share a common ancestry with platyhelminths. Characters that they have in common include: protostome type of development, lack of coelom (acoelomate) and often flattened body form with parenchyma between the body wall and intestine, flame cell excretory system (protonephridia), and ciliated epidermis. The presence in some of mucus-producing epidermal rodlets, rhabdites, also links nemertines and flatworms.

Advances in nemertines beyond flatworm organisation are the presence of a limited closed circulatory system, a one-way digestive system with mouth and anus, and the proboscis. The proboscis (Fig. 38.28b, c) is not connected to the gut but is continuous with an ectodermal invagination and is housed in a body cavity, the **rhyncocoel**. This body cavity is located dorsal to the intestine and opens anteriorly to it, and is considered by some to be the homologue of a coelom. The proboscis is often armed with a spike for penetrating prey, into which toxins may be injected.

The nervous system is like that of higher turbellarians. The brain has a dorsal and a lateral ganglion on each side interconnected to form a ring around the rhynchocoel. A pair of lateral ganglionated nerve cords with, sometimes, additional longitudinal cords, give an arrangement reminiscent of that in some flatworms.

Most nemertines are dioecious, but many freshwater and terrestrial species are hermaphroditic. Fertilisation is external or internal. In the more primitive members, there is a ciliated free-swimming larva, which resembles the larva of some flatworms (p. 846).

> Nemertinea (proboscis worms) are slender worms with a characteristic, eversible anterior proboscis used in defence and capturing prey.

PSEUDOCOELOMATES: PHYLUM NEMATODA

Some groups of animals are artificially classified together as 'pseudocoelomates' because they possess a body cavity, a **pseudocoel**, which is not a true coelom (Fig. 38.29). A true coelom is lined on all sides by mesoderm. The inner wall of a pseudocoel is the endoderm of the gut and the outer wall consists of mesoderm, hence the term pseudocoel (false coelom).

An important pseudocoelomate phylum is the Nematoda, including 12 000 known species, many of which are parasites of plants and other animals. Nematodes (roundworms and eelworms) are found, often in huge numbers, in virtually every habitat, including soil and water. Almost every species of plant

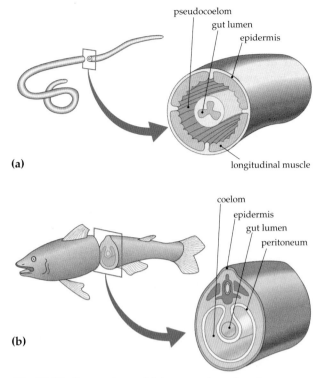

(a)

pseudocoelom
gut lumen
epidermis
longitudinal muscle

coelom
epidermis
gut lumen
peritoneum

(b)

Fig. 38.29 Comparison of **(a)** a pseudocoelom in a nematode with **(b)** a true coelom

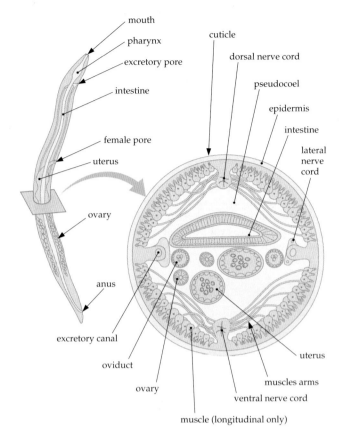

mouth
pharynx
excretory pore
intestine
female pore
uterus
ovary
anus
excretory canal
oviduct
ovary
cuticle
dorsal nerve cord
pseudocoel
epidermis
intestine
lateral nerve cord
uterus
muscles arms
ventral nerve cord
muscle (longitudinal only)

Fig. 38.30 Structure of a roundworm (nematode), which is a pseudocoelomate animal. The pig roundworm, *Ascaris suum*, is one of the largest nematodes

or animal that has been studied has been found to have at least one parasitic species of nematode living in it. Nematode parasites of agricultural crops cause millions of dollars of damage annually.

Nematodes have bilateral symmetry with elongate, cylindrical bodies tapering at both ends. They are never segmented and are covered by a proteinaceous cuticle secreted by epidermal cells. In parasitic species, the body may be worm-like, pear-shaped, lemon-shaped or sac-like. They lack circular muscle in their body wall and most move by vertical undulations.

Nematodes are herbivorous, carnivorous, or saprophagous (sucking up liquid food). Many species are microscopic but *Ascaris suum* (Fig. 38.30), one of the largest roundworms, reaches a length of approximately 20 cm. It is a parasite of the pig intestine.

In nematodes, the mouth is anterior and often surrounded by specialised lips. The anterior end of the body bears seta-like or papilla-like, probably tactile, sense organs and chemoreceptors. The mouth leads into the pharynx or stoma (often muscular and glandular), intestine and rectum, which opens through a subterminal, ventral anus. The excretory system opens through a single, anterior ventromedial pore.

The nervous system consists of a ganglionated ring surrounding the oesophagus. Ventral, dorsal and lateral nerves extend posteriorly from the brain and run within longitudinal cords of epidermis. Longitudinal epidermal cords divide the body wall musculature into two main blocks and the opposing

contraction of these blocks (upper half and lower half) accounts for the characteristic locomotory undulations of nematodes. The musculature, which is mesodermal, forms the outer lining of the spacious pseudocoel, the inner wall of which, unlike a coelom, is formed by the endoderm of the gut. Cilia are absent from nematodes, although structures resembling the bases of cilia occur in some sense organs. The excretory system consists either of small gland cells or a single, large H-shaped cell, the arms of which run, as lateral excretory canals, in the longitudinal epidermal cords. In *Ascaris*, the female of which may be half a metre long, the lateral canals of this single excretory cell reach a correspondingly great length. The sexes are usually separate.

Nematodes are unusual in their development with a fixed number of mitotic divisions, which take place before hatching. Mitotic divisions do not occur after hatching, and growth occurs by increases in cell size. This pattern makes nematodes useful experimental animals in developmental studies.

Vertebrate nematode parasites include the human hookworms, *Ancylostoma duodenale* and *Necator americanus*, which infect more than 50 million people in Asia alone. Hookworms are armed with teeth for gripping the wall of the small intestine, from which blood is sucked continuously. *Enterobius vermicularis*, the pinworm or seatworm, is very common in children. *Ascaris lumbricoides* is a roundworm that also infects humans. This produces enzyme inhibitors that protect the worm from the host's digestive system. Related to *Ascaris* is *Toxocara canis*, a roundworm of the gut of puppies, which can cause serious larval infection of human tissues, especially the nervous system. *Wuchereria bancrofti*, a widespread tropical parasite of humans, is the cause of **elephantiasis**, causing symptoms of grotesque swellings of lymphatic tissues, including genital organs and mammary glands (Fig. 38.31). Another tropical nematode is *Dracunculus medinensis*, a subdermal parasite of humans. *Dracunculus* is credited with being the biblical 'fiery serpent', a testimony to the pain it causes when it comes to the surface of the skin to discharge its larvae into water.

Nematodes (roundworms and eelworms) are pseudocoelomate animals, many of which are parasites of plants and animals. They are round in cross-section, covered with a cuticle and are never segmented.

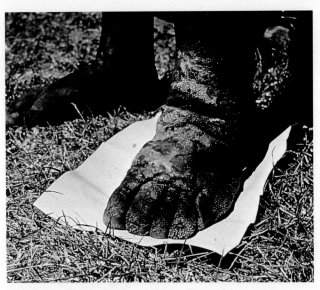

Fig. 38.31 Elephantiasis is caused by a roundworm, *Wuchereria bancrofti*, which blocks lymphatic tissue, causing grotesque swelling

SUMMARY

- All members of Kingdom Animalia, also known as Metazoa, are *multicellular*. Metazoans undergo embryonic development from a zygote and have cells differentiated for different functions.

- Sponges (phylum Porifera) are simple aquatic animals and an early lineage in the Metazoa. They consist of three layers of cells, which differ from germ layers, and they lack tissues and organs characteristic of other metazoan phyla. They are filter-feeders, drawing water through external pores by the action of flagellated cells, choanocytes, that line internal cavities.

- Metazoans with germ layers, tissues and organs are the Eumetazoa. Among eumetazoan phyla, Cnidaria (polyps and medusae) are characterised by radial symmetry, and stinging cells, nematocysts (cnidae). They have a body wall consisting of two cellular layers and an intermediate layer, the mesoglea. Their body cavity is a coelenteron.

- Cnidarians include: class Scyphozoa (jellyfish), which are marine animals with jellyfish or medusa as the conspicuous life cycle stage; class Cubozoa (box jellyfish), also marine, which have a cube-shaped bell with tentacles in groups of four, and stings that are often fatal to humans; class Hydrozoa (hydras and siphonophores), which usually have a dominant polyp stage, and commonly form colonies with interconnecting coelenterons; and class Anthozoa (e.g. soft corals, hard corals and sea anemones), which are exclusively polyps and often colonial, hard corals secreting a skeleton of calcium carbonate, forming reefs such as the Great Barrier Reef.

- Phylum Ctenophora (comb jellies) resemble cnidarian medusae but, except for one species, lack nematocysts. They have eight rows of ciliary plates (comb rows), the cilia of which beat in unison and propel the animal.

- Other eumetazoans have bilateral symmetry. Most can be assigned to one or other of two groups based on differences in embryological development. Protostomes are animals in which the blastopore of the embryonic gut (archenteron) becomes the mouth in the adult. In deuterostomes, the blastopore is the site of the anus in the adult.

- Phylum Platyhelminthes (flatworms) are acoelomate, dorsoventrally flattened worms, including free-living and parasitic forms. Their digestive cavity (not always present) has a single opening and lacks an anus. Their excretory system (sometimes lost) consists of protonephridia, the basic unit of which is the flame cell.

- Phylum Nemertinea (proboscis worms) are long slender worms with a characteristic, eversible anterior proboscis used in defence and capturing prey.

- Phylum Nematoda (roundworms and eelworms) are pseudocoelomate animals, many of which are parasites of plants and animals. The inner wall of a pseudocoelom is the endoderm of the gut; it is not homologous to a true coelom, which is lined by mesoderm. Nematodes are round in cross-section, covered with a cuticle and are never segmented.

QUESTIONS

1. List the main features that characterise the animal kingdom compared with the other eukaryote kingdoms. For animals, what are the advantages of multicellularity?

2. Which groups of metazoans have radial symmetry and which have bilateral symmetry? What are the advantages of each body form relative to an animal's particular lifestyle?

3. How do sponges feed and digest food? How do sponges respire? What evidence is there for the suggestion that sponges are related to flagellate protists (see also Chapter 35)?

4. What phylum is characterised by nematocysts? Describe the structure and function of these cells. Name one Australian marine animal that is dangerous to humans because of its nematocysts.

5. In what phylum is *Hydra* classified? Describe the body symmetry and structure of *Hydra*. How does *Hydra* feed?

6. Explain the difference between the terms acoelomate, pseudocoelomate and coelomate. Give an example of an animal for each term.

7. What are the differences between a protostome and a deuterostome? Of the following animals, which are protostomes and which are deuterostomes: starfish, nemertine, flatworm, human?

8. Explain the structure and function of a flatworm flame cell.

9. Phylum Platyhelminthes includes a variety of parasitic animals. Which classes in this phylum are parasites? Describe the life cycle of one example that infects humans.

10. List the characteristics that would enable you to identify an animal as a nematode. What is the ecological importance of nematodes?

ANNELIDS, ARTHROPODS AND MOLLUSCS

Annelids (segmented worms), arthropods (spiders, crustaceans, insects and their relatives), and molluscs (snails, clams, octopuses and their relatives) are coelomate protostomes. In structure, they are more complex than the animals described in Chapter 38, and each group has been highly successful in terms of abundance of animals and species diversity.

The phylogeny in Figure 38.2, based on both molecular and morphological data, shows the relationship of the three phyla, Annelida, Arthropoda and Mollusca. Annelids and arthropods share, for example, many structural similarities in early development. A glance at an earthworm reveals a body made of a series of 'rings' or segments. Typically, each segment contains key structures, such as nephridia (for excretion), ganglia (clusters of nerve cells), muscles and blood vessels. Such a body plan, in which there is a linear repetition of functional units, is called **metameric segmentation**. It is evident also in arthropods and is most easily seen in millipedes (Fig. 39.1). Segmentation in annelids and arthropods is not, however, homologous to the kind of repetition (strobilisation) of tapeworm proglottids. Unlike tapeworms, in the metameric segmentation of annelids and arthropods, new segments are added at the posterior end of the series and there is a fixed number of segments in the adult.

Repetition of functional units probably allowed primitive segmented animals to become larger, probably because it was associated with development of a vascular system. More complex and diverse body structures evolved as particular segments became specialised for certain tasks (e.g. head end with concentration of sensory structures). Differentiation of segments is best developed in arthropods, the most species-rich phylum in the animal kingdom.

The only sign of segmentation in modern molluscs is in *Neopilina*, a limpet-like animal that was dredged from deep water in the Pacific Ocean, off the coast of Costa Rica in 1952. It is the only living example of a class (Monoplacophora) that includes early fossil molluscs. *Neopilina* has repeated series of some structures, including five pairs of gills, six pairs of

(a)

(b)

Fig. 39.1 (a) The external appearance of a megascolecid earthworm, (annelid) and (b) a millipede (arthropod) reveal body segmentation

nephridia and ten sets of lateral nerve connections, which may indicate segmentation, linking molluscs to annelids and arthropods.

SEGMENTED WORMS: PHYLUM ANNELIDA

Annelids are soft-bodied segmented worms of soils and marine and freshwater habitats. They include marine bristle worms, earthworms, leeches and 'jawed' worms that live on crayfish.

Annelids are protostomes characterised by:

- segmentation;
- a spacious, schizocoelic coelom (a coelom that develops by splitting of the mesoderm);
- a closed vascular system;
- a through-gut with an anus;
- a solid ventral ganglionated nerve cord;
- epidermal setae borne in four bundles per segment;
- no exoskeleton.

The annelid body has three parts (Fig. 39.2). At the anterior end is the presegmental **prostomium**, which houses the brain. Behind this is the segmented **soma**. At the posterior end is the postsegmental **pygidium**. Immediately anterior to the pygidium there is a **growth zone**, which is where embryonic growth occurs by addition of new segments at the anterior edge. The first apparent segment, surrounding the mouth, is the **peristomium**.

Advantages of segmentation and a coelom

Locomotion

The longitudinal division of a worm into a series of segments, each with muscles and nerves that control it, and the hydrostatic pressure provided by the coelom are important in the mechanics of locomotion. It is probable that evolution of the coelom in the first annelids was an adaptation for burrowing in an animal moving by alternate constriction and widening of the body. This process of **peristaltic locomotion** requires a deformable, fluid-filled body cavity, the coelom. Development of transverse septa (Fig. 39.2), typical of the annelid coelom, allows different hydrostatic pressures in various parts of the body. The differences cause local changes in shape along the worm's body, allowing fast and effective crawling, burrowing and response to danger from predators.

Respiration and circulation

Because septa interrupt the coelom, septation may have been accompanied by evolution of a vascular system for circulating body fluid for respiration and nutrition. A vascular system augments simple diffusion for transporting oxygen, carbon dioxide, food materials, hormones, and excretory substances (Chapter 22). Consequently, there is potential for greater size and complexity and increased levels of activity. It has been estimated that, without a circulatory system, the diameter of the annelid body could not exceed 1.5 mm. Some Australian giant earthworms reach 25 mm in diameter.

The respiratory pigment is commonly erythrocruorin (haemoglobin), which is not contained in blood cells and has a molecular weight of approximately 4 million Daltons. Alternative pigments, such as chlorocruorin (also iron-containing but green), may occur. In the annelid circulatory system (Fig. 39.2), blood passes anteriorly in the dorsal vessel but anteriorly and posteriorly in the ventral vessel. Oxygenation of blood occurs at or near the epidermis.

Excretion

Excretory products are released into the coelom. Tubes, **nephridia**, are repeated in successive segments and transport the waste products from the coelom to the exterior. Protonephridia end as blind tubes with terminal flagellated cells and occur in the embryos of all classes and in some adult polychaetes. Open and funnelled **metanephridia** (Fig. 39.2) occur in many polychaetes, all oligochaetes and leeches.

Digestion

A coelom allowed internal organs to be free from attachment to the body wall. Annelids have a through-gut, which allows one-way flow of food, and specialisation of consecutive regions for successive stages in food digestion.

> Annelids are coelomate protostomes. They are segmented worms with a digestive system, closed circulatory system, solid ventral ganglionated nervous system and nephridia for excretion. Their coelom and segmentation allow for peristaltic locomotion.

Marine bristle worms: class Polychaeta

Most polychaetes are marine but a few are freshwater or terrestrial. They are either free-moving (errant) or sedentary. Many sedentary species construct and live in burrows or tubes (Fig. 39.3). Polychaetes usually have lateral paired appendages, **parapodia**, on each body segment. A parapodium is a lateral fleshy lobe of the body that functions for both respiration and locomotion (Fig. 39.4). The name 'Polychaeta' means many setae and refers to the chitinous bristles, **setae**, associated with the parapodia. Bundles of setae tend to spread out like a fan in a variety of patterns. Some tropical polychaetes living in coral have tubular, poisonous setae that are used for defence. Polychaetes

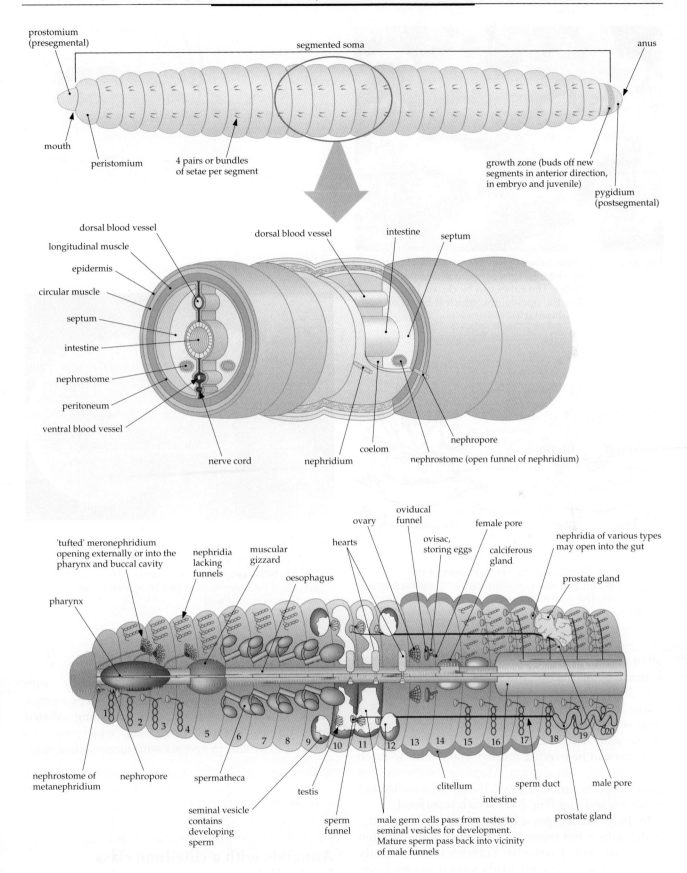

Fig. 39.2 Structure of an earthworm of the Australian family Megascolecidae

Fig. 39.3 Calcareous tubes of the polychaete, *Galeolaria caespitosa*, mark the mid-tidal level on rocky shores in New South Wales. The tubes can be closed by a calcareous plug, the operculum, which is borne on one of the prostomial tentacles

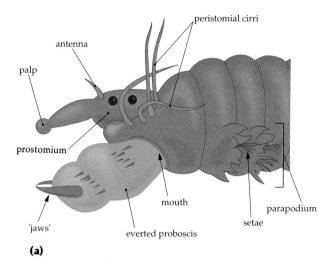

(a)

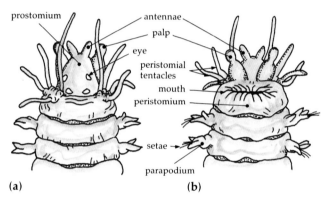

(a)   **(b)**

Fig. 39.4 (a) Dorsal and (b) ventral views of the prostomium and anterior segments of the polychaete *Perinereis*. The parapodia on the body segments bear setae

(b)

Fig. 39.5 (a) Head of a feeding nereid polychaete showing the pharynx everted through the mouth. (b) *Marphysa sanguinea*, the common 'bloodworm', favoured by fishermen as bait in Moreton Bay, southern Queensland

often have sensory structures such as dorsal antennae, eyes and ventral palps on the prostomium (Fig. 39.4).

The mode of feeding in different polychaetes (Box 39.1) is reflected in the structure of the anterior end of the alimentary canal, the pharynx. In some, the pharynx can be everted through the mouth to catch prey. In the bloodworm *Marphysa sanguinea*, the pharynx bears cuticular 'jaws', the dorsal maxillae and ventral mandibles (Fig. 39.5), which seize food.

In polychaetes, gas exchange occurs across the body surface, but many polychaetes have gills, often associated with parapodia. Polychaetes are mostly dioecious but some, particularly minute worms living buried in sand, are hermaphroditic. They usually have long series of gonads, which are simply swellings containing masses of developing gametes that are shed into the coelom. Gametes escape into the

surrounding sea water through special ducts, nephridia or sometimes through rupture of the body wall. Some polychaetes copulate, either with hypodermic impregnation or insemination by a penis. Many polychaetes have a free-swimming ciliated larva, the trochophore (Fig. 39.6), which may be reduced or absent in species with large-yolked eggs.

> Polychaetes are marine segmented worms characterised by parapodia—paired appendages that function in locomotion and respiration.

Annelids with a clitellum: class Euclitellata

Earthworms, leeches and their near relatives are annelids that possess a **clitellum**, a thickening of the epidermis, which secretes the cocoon in which eggs

BOX 39.1   FEEDING IN POLYCHAETES

Polychaete worms show a variety of feeding methods. *Chaetopterus*, for example, is a marine polychaete that lives in a parchment-like tube (Fig. a). Each tube is composed of cuticle elevated above the epidermis of the animal. Except for its ends, the tube is buried in mud. Water is drawn into the tube by the beating of fan-like parapodia of three segments. A mucus net filters out particles from the water and forms a bag, which is gathered by a ciliated cup. Periodically, a food bolus is detached from the bag and placed on a dorsal groove, which takes the bolus forward to the mouth for ingestion.

Other polychaetes feed by means of ciliated, pinnate tentacles that arise from the prostomium and filter particles from water. An example is the sedentary polychaete *Galeolaria caespitosa*, the white calcareous tubes of which form a conspicuous mid-tidal band on rocks and pier pylons in New South Wales and Victoria (Fig. 39.3).

Many polychaetes feed actively by means of a well-developed cylindrical pharynx, which may or may not be armed with cuticular jaws (in some species poisonous). In carnivorous species, the pharynx can be everted rapidly to catch prey. Herbivorous or scavenging polychaetes with jaws use them to tear off pieces of food, such as algae. In *Sabellastarte indica* (Fig. b), a filter-feeder, bipinnate prostomial tentacles form a crown which draws in organic particles to the base for sorting.

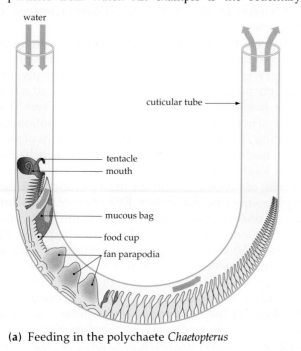

(a) Feeding in the polychaete *Chaetopterus*

(b) *Sabellastarte indica*, a polychaete of shores from Queensland to Western Australia, feeds by filtering organic particles with its crown of prostomial tentacles at the base of which they are sorted

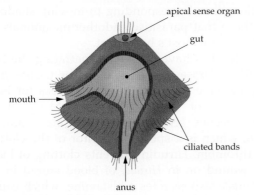

Fig. 39.6 Many polychaetes have a trochophore larva, typical of protostomes

are deposited. The clitellum is located around and/or behind the female genital pores. Unlike polychaetes, euclitellates lack parapodia. Setae, although they may be numerous, are never jointed and are absent from advanced leeches. There is no trochophore or other larval stage.

> Earthworms, leeches and their relatives are annelids that have a clitellum, a thickening of the epidermis, which forms a cocoon in which eggs are deposited. Unlike polychaetes, they lack parapodia.

Earthworms and relatives: subclass Oligochaeta

The name 'oligochaetes' means 'few setae', and there are usually only four pairs per segment, although some oligochaetes have many more. Oligochaetes have a reduced prostomium and most are burrowing animals. They feed on vegetation or on organic matter in soil, and a few are carnivorous. Although the best known oligochaetes are earthworms and related freshwater worms, increasing numbers of taxa are being found in marine habitats. It has been estimated that there are 20 000 worms per cubic metre in sublittoral sand of Heron Island on the Great Barrier Reef. There may prove to be more oligochaete species in marine than in other habitats. There are some 25 families of oligochaetes, of which six are native to Australia. Five of these have species restricted to this continent but no family is endemic. The largest Australian oligochaete is *Megascolides australis* from Gippsland, which may exceed 3 m when extended; *Digaster longmani*, from Queensland and New South Wales, is also gigantic (Fig. 39.7).

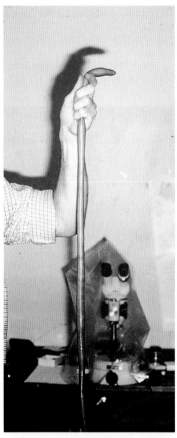

Fig. 39.7 A giant Australian earthworm, *Digaster longmani*

In contrast to most polychaetes, oligochaetes have gonads in only a few segments, usually from two to four pairs. They are hermaphroditic and in earthworms there are usually two pairs of testes, in segments 10 and 11, followed by one pair of ovaries, in segment 13. Reproduction (Fig. 39.8) involves reception of sperm from a partner, almost always into small sacs, **spermathecae**. Sperm are then ejected from spermathecae into a cocoon secreted on the outside of the body by the clitellum. The cocoon receives eggs from the female genital pore(s), and fertilised eggs develop into juvenile worms in the cocoon after it has been shed from the anterior end of the body. The African night-crawler (family Eudrilidae) has internal fertilisation.

Asexual reproduction is common in aquatic oligochaetes; the parent worm simply divides into two or more worms.

> Oligochaetes (earthworms) are euclitellates with typically few setae on each segment. They are hermaphroditic, with gonads in only a few segments.

Leeches: subclass Hirudinea

Hirudinea are related to oligochaetes and all have a clitellum. Leeches, however, are more specialised than oligochaetes. Although a few are carnivores, many are external parasites feeding on host blood. Terrestrial leeches that live in moist Australian forests attach to leaves to await passing prey. A few species are endoparasitic in the respiratory passages of mammals.

The medicinal leech, *Hirudo*, used from antiquity for 'bleeding' human patients and still used in plastic surgery, is ectoparasitic on amphibians when juvenile and on mammals when mature. Leeches attach to their hosts by posterior and anterior suckers (Fig. 39.9). They use their suckers for locomotion by 'looping', in contrast to the lateral undulations of most polychaetes or the peristaltic creeping of oligochaetes.

In leeches, the coelom is reduced and forms a system of tubes, which in the most advanced leeches replaces the vascular system. Reduction of the coelom is correlated with loss of peristaltic creeping. Correspondingly, septa are lost and a submuscular tissue is developed. Internal segmentation is evident only in the nervous and excretory systems. External segmentation is obscured by the development of secondary rings. The number of segments is fixed at 34 (six in the head region and anterior sucker, 21 body segments, and seven in the posterior sucker). Most leeches have lost setae, probably associated with locomotion by looping.

In leeches, single-celled photoreceptors are clustered as **ocelli**. Some leeches find a host, for example, a fish, by responding to moving shadows, while those that parasitise endothermic animals are attracted by the host's body heat.

Many leeches have jaws for penetrating the host (Fig. 39.9b). Those that suck blood or tissues from their prey have crop and intestinal caeca for long-term storage of ingested material. Many blood-sucking leeches secrete an anticoagulant protein, **hirudin**, which is a specific inhibitor of the clotting factor thrombin. Hirudin prevents clotting of blood at the wound or, in *Hirudo*, of blood stored in the crop. *Hirudo* also secretes a histamine, which causes dilatation of blood vessels of the host and augments bleeding. There are no proteolytic enzymes in the crop of *Hirudo* and digestion of blood depends upon symbiotic bacteria (*Pseudomonas*), the presence of which appears to prevent putrefaction of the blood.

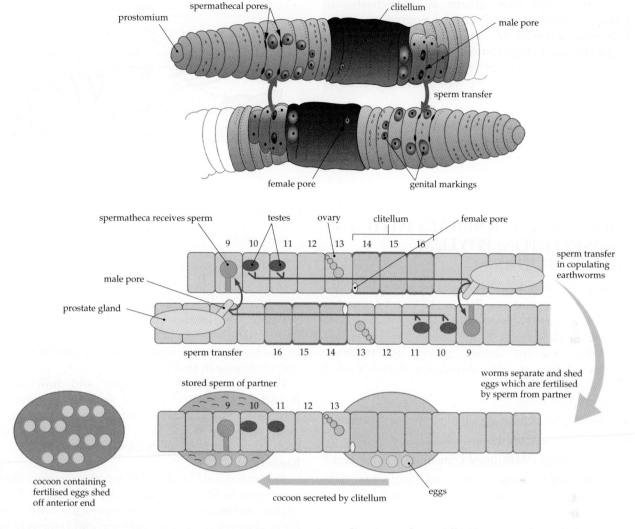

Fig. 39.8 Copulation in earthworms, here exemplified by the Australian genus *Spenceriella*. The worms lie with their anterior regions in contact, facing in opposite directions. The male pores deposit sperm in spermathecae of the partner. The partners separate. Eggs pass out of the female genital pore into a cocoon, which is secreted by the clitellum. The worm wriggles backwards out of the cocoon, which effectively moves forwards and receives stored sperm from the spermathecae. Eggs are fertilised, and the cocoon is shed off the anterior end of the body

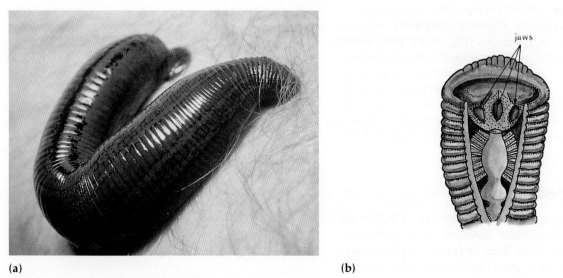

(a)   (b)

Fig. 39.9 (a) A leech attached to a human host by its suckers. (b) Jaws of a leech

Leeches differ from almost all oligochaetes in having internal fertilisation. Sperm may be transferred directly by a penis or by hypodermic impregnation. In hypodermic impregnation, a bundle of sperm (a spermatophore) is expelled from one worm and penetrates the body wall of another (Chapter 15).

> Leeches (Hirudinea) have an anterior and posterior sucker, generally lack setae, and have reduced segmentation and a reduced but specialised coelom. Most leeches are specialised as external parasites.

JOINT-LIMBED ANIMALS WITH AN EXOSKELETON: PHYLUM ARTHROPODA

Arthropods, including crabs, insects and spiders, are the most abundant of all animals in total number of species, more than one million species having been described. They occur in the sea, in freshwater and on land, and the largest group, insects, have conquered the air. This abundance depends on a combination of features.

Arthropods have:

- a hard, chitinous external skeleton—**exoskeleton**;
- metameric segmentation;
- jointed limbs ('Arthropoda' meaning 'jointed feet');
- **tagmatisation**—organisation of segments into functional groups, **tagmata** (sing. tagma).

Advantages of an exoskeleton

The exoskeleton protects soft parts of the body, allows development of internal projections to which muscles attach and, in insects, has resulted in moveable wings. Limbs must be jointed because the exoskeleton is rigid, but this rigidity allows a complexity of form and function of appendages (Fig. 39.10) that is not possible in the thin-walled parapodia of polychaetes. For example, the chela (claw) of a crab is able to grasp and crush, and is structurally different from a walking limb. However, a rigid exoskeleton restricts growth and is periodically shed as an arthropod grows; moulting is under hormonal control (see Chapter 25).

The arthropod body

Arthropods include three subphyla: **Chelicerata** (spiders, scorpions, ticks, mites and their relatives); **Crustacea** (from water fleas to crabs); and **Uniramia** (chiefly myriapods, onychophorans and insects). Uniramians are so-called because their appendages are unbranched. Appendages of crustaceans and

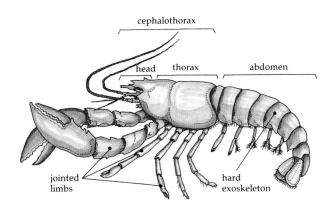

Fig. 39.10 Arthropods have a hard exoskeleton, jointed limbs and body segments grouped into tagmata (e.g. the thorax)

primitive chelicerates are made up of several branches. Uniramians chew with the tips of whole limbs (mandibles). In other arthropods biting is done by the *bases* of limbs, termed mandibles in crustaceans and chelicerae in chelicerates.

The coelom (well developed and functionally important as an internal fluid skeleton in annelids) is greatly reduced and virtually absent in adult arthropods. There are, however, large spaces in the body that are filled with blood and which form a **haemocoel**.

The arthropod blood vascular system is open. Blood is pumped forward into arteries by an elongate dorsal vessel, the heart. Blood then enters the haemocoel, bathes tissues and organs directly, and returns to the posterior end of the heart (Fig. 39.11).

The nervous system, like that of annelids, consists of a dorsal anterior brain followed by a basically paired, solid, ventral nerve cord with segmental ganglia.

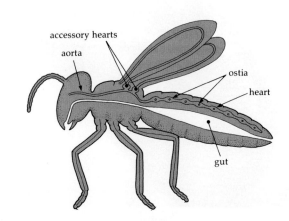

Fig. 39.11 An arthropod's circulatory system is an open one (haemocoel)

Evolution of arthropods

There has been a long debate whether arthropods are a monophyletic or a polyphyletic taxon. Those supporting a polyphyletic origin propose that Chelicerata, Crustacea and Uniramia have separate origins. They consider that chelicerates, which alone in the arthropods have spiral cleavage, have no close affinities with insects and related forms. Supporters of arthropod monophyly argue that the chitinous exoskeleton and jointed limbs are unlikely to have evolved more than once. Recent phylogenies based on DNA sequences do not support the polyphyletic theory and suggest that arthropods are monophyletic (Fig. 38.2). Arthropod fossils have been found in early Cambrian rocks, and thus the group is at least this old.

> Arthropoda, the largest animal phylum, have body segments grouped as tagmata, a hard exoskeleton for protection and attachment of muscles, jointed limbs, and a haemocoel.

Fig. 39.12 Horseshoe crabs, *Limulus polyphemus*, coming ashore

Fig. 39.13 A pycnogonid—a sea spider

Spiders, scorpions and relatives: subphylum Chelicerata

Chelicerates lack antennae, and only some primitive forms (horseshoe crabs) have compound eyes, typical of crustaceans and uniramians. The body is basically divided into two or three tagmata: an anterior cephalothorax, **prosoma**; an abdomen, **opisthosoma**, with at most 12 metameres; and, in primitive members, a postsegmental part, the **telson**. Typically, the first appendages behind the mouth are a pair of food-handling chelicerae followed by a pair of sensory, prey-catching (and often copulatory) pedipalps and four pairs of legs.

Marine chelicerates

Fossil evidence suggests that chelicerates had a marine origin, beginning in the early Palaeozoic. Now only horseshoe crabs (class **Merostomata**) and sea spiders (class **Pycnogonida**) are marine.

Horseshoe crabs, of which there are only five living species (Fig. 39.12), have five or six pairs of abdominal appendages modified as gills, **book gills**, and a long spike-like telson at the posterior end of the body. The prosoma is covered by a large **carapace**.

The relationship of sea spiders to other chelicerates is uncertain. Their often narrow body (Fig. 39.13) bears an anterior proboscis and four, five or six pairs of usually long legs. Organs for gas exchange and excretion are absent, possibly as a correlate of the pycnogonids' large surface area for diffusion and their aquatic existence. The male carries the eggs on a pair of legs located in front of the first walking legs.

Terrestrial chelicerates: the arachnids

The class **Arachnida** includes spiders, scorpions, ticks and mites and their relatives. The first arachnids, including the earliest scorpions, were aquatic, but most modern forms are terrestrial. A waxy epicuticle contributes significantly to the success of arachnids in terrestrial habitats.

The arachnid body

In arachnids, the prosoma is covered by a carapace fused onto the body. The ventral surface of the prosoma has one or more sternal plates or is covered by the coxae (basal joints) of the appendages. In scorpions, the long and segmented abdomen is divided into a pre- and post-abdomen. In primitive spiders the abdomen may be segmented, but in mites, primary segmentation is lost and the abdomen is fused with the prosoma.

Gas exchange

Large arachnids (scorpions and some spiders, Fig. 39.14) have **book lungs** and small forms (pseudo-scorpions, some spiders, and mites) have **tracheae** (respiratory tubes). The blood contains the copper-containing respiratory pigment haemocyanin.

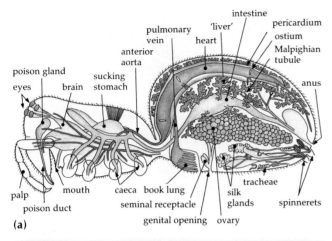

(b)

Fig. 39.14 **(a)** Anatomy of a spider in longitudinal section. **(b)** Spiders produce silk, often used to build webs to catch prey

Excretion

Excretion is by coxal glands and **Malpighian tubules**. Water conservation is greatly aided by production of guanine as the nitrogenous excretory substance, a material requiring very little water for its excretion.

Catching prey

Most arachnids are carnivorous, and poison is often used to immobilise prey. The way in which arachnids deliver poison varies and mechanisms have evolved independently in different groups. In spiders, poison is delivered by the chelicerae, in scorpions by the terminal abdominal barb, and in pseudoscorpions through fingers of the pedipalps. Many spiders use silk, produced in spinning organs, **spinnerets**, at the end of the abdomen, to build webs to capture prey (Fig. 39.14b). Other arachnids, such as many harvestmen, are scavengers, and many mites are parasites.

Sense organs

Most arachnids have three types of sense organs: sensory hairs, eyes and slit sense organs. Hairs on the body and appendages are sensitive to slight air vibrations and are probably the most important sense organs. Slit sense organs, composed of slits in the cuticle covered by a diaphragm in contact with a basal sensory seta, detect sound vibrations. Many arachnids possess simple eyes, but only a few, such as hunting spiders, see detailed images.

Reproduction

In many arachnids, the male transfers sperm in spermatophores. Spermatophores are picked up by the female, or transferred to her by the male, as in scorpions, pseudoscorpions and some mites (Chapter 14). Often the male shows courtship behaviour to attract a female to the spermatophore. In spiders, sperm are exuded by the pedipalps onto a web and then transferred by the pedipalps to the female genital opening. In harvestmen and many mites, however, sperm transfer is direct by insemination during copulation.

> Chelicerate arthropods have anterior appendages (chelicerae) that function as fangs or pincers. The body is divided into a cephalothorax, an abdomen and, in primitive forms, a telson, and they lack antennae.

Prawns, crabs and relatives: subphylum Crustacea

With approximately 40 000 species, crustaceans (Fig. 39.15) are the only major group of aquatic arthropods and relatively few species, such as wood lice and slaters, are terrestrial. They live at most depths of the sea, in freshwater up to elevations of 3.6 km, in waters from 0°C to 55°C, and in leaf litter of forests. Most are omnivores, often scavengers; others are filter-feeders or carnivores.

The crustacean body

In crustaceans, metameres (segments) are organised into three more or less distinctly demarcated tagmata: head, thorax and abdomen. Crustaceans are unique among arthropods in having two pairs of antennae (antennules and antennae; Fig. 39.16). These are on their first two head segments. Antennae are typically followed by one pair of mandibles, with grinding and biting surfaces, and two pairs of accessory feeding appendages, maxillae. Trunk specialisations and appendages vary greatly but a carapace covering all or part of the body is common. Crustacean appendages are typically made of two parts (*biramous*)—endopodite and exopodite (Fig. 39.16)—a major

(a)

(b)

Fig. 39.15 Crustaceans: **(a)** soldier crab, *Mictyris longicarpus*, and **(b)** goose barnacle, *Lepas fascicularis*

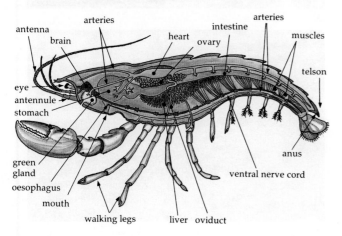

Fig. 39.16 Anatomy and appendages of a crustacean

difference from insects and other uniramians. In filter-feeders, closely placed setae on appendages function as filters.

Circulation and gas exchange

Crustaceans have a heart with perforations (ostia) through which venous blood enters from the pericardial sinus. The blood of most crustaceans is blue owing to the presence of haemocyanin but a few species have red, haemoglobin-containing blood. Delicate processes of the limbs function as gills in many species.

Excretion

Excretion is by a pair of blind sacs in the head that open onto the bases of the second pair of antennae (antennal glands) or the second pair of maxillae (maxillary glands).

Nervous system

Typically, the nervous system is a double, solid ventral ganglionated nerve cord (Fig. 39.16). Crustacean eyes are of two main types: a pair of compound eyes and a small median dorsal eye, which consists of three or four closely placed simple eyes, **ocelli**.

Reproduction

Sexes are usually separate but hermaphroditism is the rule in sessile barnacles, in some primitive crustaceans and in some parasitic forms. Parthenogenesis (Chapter 14) is frequent in water fleas such as *Daphnia* and other members of the class Branchiopoda. Copulation is usual and egg brooding is common. In many crustaceans, sperm lack a flagellum.

The earliest hatching stage, when it is not suppressed, is a larva or **nauplius** (Fig. 39.17). The nauplius has only three pairs of appendages, namely antennules, antennae and mandibles. A single eye, the nauplius eye, is at the front of the head. The nauplius is free-living in many lower crustaceans but

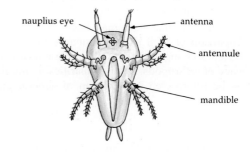

Fig. 39.17 A nauplius larva of a crustacean

in higher forms the egg hatches as a more advanced larva. After the nauplius stage in some groups, there are remarkable larvae, such as the free-swimming cypris of barnacles; the phyllosome of spiny lobsters; and the zoea and megalopa of crabs.

> Crustaceans have two pairs of antennae, a pair of mandibles and two pairs of maxillae. Their appendages are typically biramous. Segments are organised into head, thorax and abdomen.

Insects and relatives: subphylum Uniramia

Uniramia are arthropods with one pair of antennae. The name refers to the unbranched limbs compared with the biramous appendages of crustaceans. However, some fossil 'uniramians' have branched limbs. Uniramians include millipedes, centipedes and insects. The velvet worms, onychophorans, are also often classified as uniramians, but their relationships have been debated (Box 39.2). Uniramians are worldwide but there are few marine representatives.

> Uniramians (millipedes, centipedes and insects) have one pair of antennae, and unbranched limbs.

Millipedes and centipedes: classes Diplopoda and Chilopoda

Millipedes and centipedes have a head and a segmented body (Fig. 39.18). Millipedes appear to have two pairs of legs per segment but each 'segment' is a tagma of two fused segments. Centipedes have one pair of legs per segment. Living in leaf litter and soil, millipedes are mostly herbivores and centipedes are carnivores. In millipedes, each segment contains a pair of glands, which secrete an offensive chemical as a defence against predators.

Fig. 39.18 Centipedes (class Chilopoda) are uniramian arthropods

---

BOX 39.2   ONYCHOPHORANS: ARTHROPODS OR ANNELIDS?

Onychophora are velvet worms, such as the Australian *Peripatoides* and the remarkable, recently discovered, *Cephalofovea* (see figures). They look rather like slugs with legs, and most live in moist habitats, such as tropical rainforests of the world, under logs and litter. Their body is covered by tubercles and has a velvety and iridescent appearance.

There has been much debate about the relationships of onychophorans because in some respects they are similar to annelids and in others to arthropods. Like annelids, their body is soft and covered by a flexible, permeable cuticle, not a rigid exoskeleton as in arthropods. Their body plan also is more like that of an annelid, and their legs are not jointed like those of arthropods. However, like arthropods they have mandibles, their body cavity is a haemocoel, and their gas exchange organs are tracheae. Their embryonic development is similar to uniramians, and they have one pair of antennae.

Recent molecular data may have solved the riddle. Comparison of 12 S rRNA sequences for a variety of invertebrates suggests that onychophorans are an early lineage of arthropods. Their similarities to annelids can thus be interpreted as retention of general primitive characters.

(a)

(b)

Two examples of onychophorans: **(a)** a velvet worm, *Peripatoides*; and **(b)** the recently discovered *Cephalofovea*

Insects: class Insecta

Insects are the largest group of animals. About three-quarters of a million species have been described and insects are among the most numerous inhabitants of terrestrial environments. Their success relates to a number of features, but especially their ability to fly and to exploit an aerial environment. Moths, butterflies, bees, wasps and flies are important pollinators of flowering plants, and the evolutionary histories of the angiosperms and insects are closely interlinked (Chapter 37).

Body structure

Insects have the uniramian features of unbranched limbs and a single pair of antennae but differ from other classes in having a tagma of three segments that forms a thorax, which is distinct from the head and abdomen (Fig. 39.19).

The thorax bears three pairs of walking legs and, in all but primitive insects, two pairs of wings on the two posterior segments. In flies (order Diptera), the last pair of wings is modified as a pair of mobile knobbed rods, the halteres, involved with balance during flight. The insect abdomen generally lacks appendages apart from a pair of cerci on the last segment in males. Short structures homologous to limbs are present on the abdomen in bristle tails.

Mouth parts and feeding

The mouth parts consist of three kinds of appendages: a pair of mandibles, a pair of maxillae and a labium derived from the fused pair of second maxillae. Each maxilla bears a lateral maxillary palp, and the labium has a pair of labial palps. An anterior plate of the exoskeleton of the head, the labrum, covers these mouth parts and a tongue-like hypopharynx projects behind the mouth.

Primitively, the mouth parts are used for chewing, for example, in grasshoppers and many beetles. In other insects, the mouth parts are modified for specialised feeding. In flies, the labium is a sponge-like pad, used to take up liquid food. In mosquitoes, the mouthparts are elongated as a proboscis. The proboscis consists of a fleshy, grooved labium, over-arched by the labrum to enclose the elongate mandibles and maxillae, which form sharp stylets. Stylets penetrate skin for blood-sucking in female mosquitoes (Fig. 39.20), or plant vascular tissue for sap feeding in the males. As food is sucked through a channel in the stylet, saliva can also pass down through the slender hypopharynx.

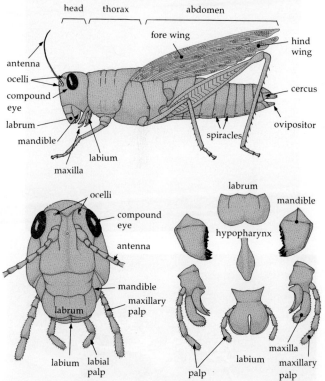

Fig. 39.19 External features of a grasshopper, a typical insect. Details of the head and mouthparts are shown. The head and biting mouthparts of a grasshopper are adapted for cutting and then crushing food between the mandibles. The mandibles of carnivorous insects are sharp and pointed blades

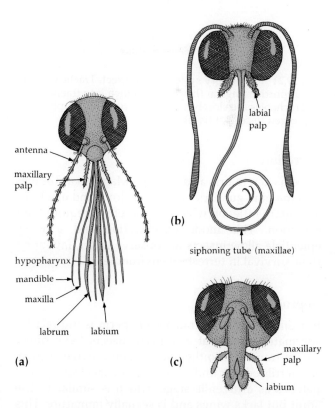

Fig. 39.20 Three types of insect mouthparts. (a) The piercing and sucking mouthparts of a mosquito. (b) The long coiled sucking proboscis of a butterfly is a highly modified maxilla. The mandibles are absent. At rest the tube is coiled. (c) The lobes at the end of the mouthparts in the housefly have grooves on the lower surface that direct liquid towards a central channel. They sponge or lap up surface liquids

Sensory organs

Insects have a pair of lateral compound eyes, three dorsal ocelli and a pair of antennae (Fig. 39.19). A variety of sensory organs bearing olfactory receptors and mechanoreceptors, including ones modified for hearing, is distributed over the body, particularly on the antennae and legs.

Gas exchange

Gas exchange (Chapter 20) is by a system of air-filled tracheae, the openings of which are **spiracles** or stigmata, located along the sides of the thorax and abdomen (Fig. 39.19). The finest branches of the tracheae, **tracheoles**, are fluid-filled and exchange oxygen and carbon dioxide directly with tissue cells. Tracheoles are enclosed within a tracheal end-cell (Fig. 39.21).

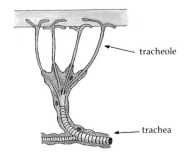

Fig. 39.21 Respiratory system of an insect. Tracheae, with supportive rings in their walls, branch into tracheoles. Tracheoles are closely associated with and supply oxygen to cells of tissues

Excretion

Insects excrete uric acid through Malpighian tubules into the hind gut. Uric acid is non-toxic and insoluble in water, little of which is therefore required for excretion. This mode of excretion and a waxy epicuticle allow insects to conserve water, favouring their survival in terrestrial environments.

Reproduction

In primitive insects, transfer of sperm is indirect by spermatophores, but advanced insects have direct sperm transfer into the female reproductive system. The fertilised egg undergoes superficial cleavage and hatches as a juvenile stage, which is similar to the adult but lacks wings and is sexually immature. This is termed a **nymph** and undergoes a series of moults giving a succession of stages, **instars** (Chapter 13).

Insect diversity

The subclass **Apterygota** includes bristletails (orders Protura and Thysanura) and springtails (Collembola), insects that lack wings. More advanced insects with wings constitute the subclass **Pterygota**. Australian examples of some orders in these two subclasses are shown in Figure 39.22.

In the lower pterygote orders, successive instars have increasingly developed wing buds. They have gradual or incomplete metamorphosis and are termed **exopterygotes** because the developing wings are externally visible in nymphs. Exopterygote insects include, among others, mayflies (order Ephemeroptera), dragonflies and damselflies (Odonata), grasshoppers and crickets (Orthoptera), cockroaches (Blattodea), mantids (Mantodea) and stick and leaf insects (Phasmatodea), termites (Isoptera), stoneflies (Plecoptera), earwigs (Dermaptera), biting and sucking lice (Anisoptera), bugs (Hemiptera), and lacewings (Neuroptera).

In higher insects, the juvenile or **larva** is often radically different in appearance and mode of feeding

(a)

Fig. 39.22 (b)

(c)

(d)

(e)

(f)

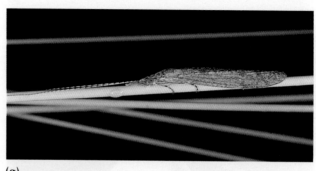

(g)

Fig. 39.22 Australian examples of some orders of insects: (a) violet stick insect, *Didymuria violescens*, (order Phasmatodea); (b) grey-back cane beetle, *Dermolepida albohirtum* (order Hemiptera); (c) native cockroach, *Polyzotteria mitchelli* (order Blattodea); (d) northern Australian locust, *Chortoicetes terminifera* (order Orthoptera); (e) hover fly, *Syrphus damastor* (order Diptera); (f) dung beetle (order Coleoptera); and (g) caddis fly (order Trichoptera)

from the adult (contrast a caterpillar with a butterfly, Fig. 39.23). A profound metamorphosis to the adult stage, **imago**, occurs in a **pupa**. Such insects have complete metamorphosis and, because wing development is concealed in the pupa, are termed **endopterygotes**. They include, among others, beetles (order Coleoptera), scorpion flies (Mecoptera), caddis flies (Trichoptera), moths and butterflies (Lepidoptera), true two-winged flies (Diptera), ants, bees and wasps (Hymenoptera) and fleas (Siphonaptera).

Insects and humans

Many insects are pests of humans, and of their domestic animals and crops, and some are major carriers (vectors) of human diseases (Fig. 39.24). The mosquito, *Anopheles*, is the intermediate host for the malarial parasite, *Plasmodium*. A moth fly, *Phlebotomus*, transmits typhus. Fleas, including the genus *Pulex*, transmit several diseases, including bubonic plague and tapeworms. *Simulium*, the buffalo gnat, causes blindness by transmitting the nematode *Onchocerca*. The sand fly, *Culicoides*, and the horse fly, *Chrysops*, also carry nematode parasites. The tsetse fly, *Glossina*,

(a)

(b)

(c)

Fig. 39.23　Three stages of the wanderer butterfly life cycle: (a) caterpillar, (b) pupa and (c) imago

transmits the protist *Trypanosoma*, the causative agent of sleeping sickness. The louse *Pediculus* transmits *Rickettsias*, although another louse, *Pthirus*, seems innocent of carrying disease.

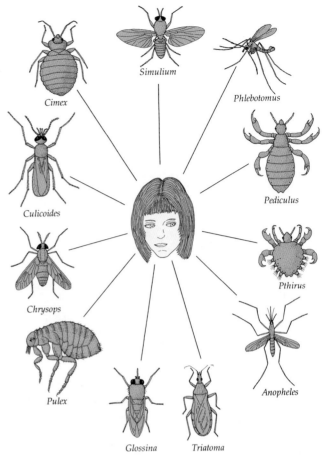

Fig. 39.24　Winged insects that parasitise humans and may carry diseases

Insects are the largest group of animals and are well adapted to terrestrial environments. They respire through tracheae, excrete uric acid, have a waxy epicuticle, and have evolved the ability of flight.

SNAILS, CLAMS AND RELATIVES: PHYLUM MOLLUSCA

After arthropods, molluscs constitute the second largest animal phylum. There are more than 100 000 living species and 35 000 fossil species dating back to the Cambrian. They are aquatic and terrestrial animals, including chitons, gastropods (snails and nudibranchs), bivalves (mussels, oysters and cockles), tusk shells and cephalopods (squids and octopuses).

Molluscan body plan

Morphologically, molluscs are highly diverse. Their basic structure can best be described in terms of a hypothetical primitive mollusc, the morphology of which has been deduced by careful consideration of the anatomy of all molluscan classes. However, there is no organism that conforms exactly to this plan.

The primitive mollusc (Fig. 39.25) is envisaged as a dorsoventrally flattened, unsegmented, almost worm-like animal with a head, a ventral gliding surface (the foot) and a visceral mass containing the body organs. Covering the visceral mass was a dorsal **mantle**, a fold of the body wall with a cavity beneath it opening backwards. The mantle secreted a shell, which was simply a chitinous cuticle invested with more or less overlapping scales consisting of a calcareous mineral (aragonite). The mantle cavity contained a pair, possibly several pairs, of respiratory gills and received the excretory, alimentary and reproductive openings. The alimentary canal was straight and possessed a mid-gut with ventrolateral, serial outpockets, which alternated with dorsoventral muscles.

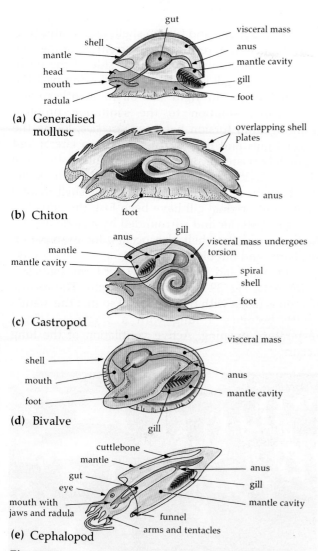

(a) Generalised mollusc

(b) Chiton

(c) Gastropod

(d) Bivalve

(e) Cephalopod

Fig. 39.25 **(a)** Generalised structure of a mollusc with head, visceral mass and flattened gliding foot. Modifications of this basic structure in the evolution of **(b)** chitons, **(c)** gastropods, **(d)** bivalves and **(e)** cephalopods

One of the most characteristic features was a broad, tongue-like structure, the **radula** (Fig. 39.26), rows of rasping teeth borne on a cuticular strip present in the floor of the foregut. All modern molluscs, except bivalves, have retained the radula.

The nervous system consisted of paired, dorsal cerebral ganglia, a nerve ring around the oesophagus and two pairs of longitudinal nerve cords innervating the gills and a possibly paired sensory organ associated with the mantle cavity. The circulatory system was open and consisted of a dorsoposterior sac, the **pericardium**, containing a ventricle and a pair of lateral auricles. The animals are thought to have been dioecious with paired gonads discharging into the mantle cavity. Whether the pericardium and the cavity in which the gonads lie are homologous with the coelom of other protostomes is debatable.

Fertilisation was external, cleavage spiral, and development included a pelagic, ciliated trochophore larva. The larva developed into a second stage, the **veliger**, which develops a foot, mantle and shell. The adult animal was marine, living on sea floors and shores, and probably fed by scraping micro-organisms off the substrate by means of the protrusible, rasping radula.

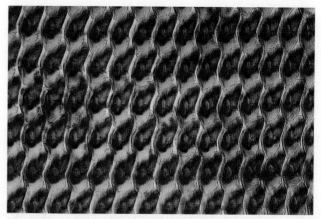

Fig. 39.26 Teeth of the radula of a gastropod

> The basic molluscan body has a head, visceral mass, and a ventral gliding surface, the foot. All modern molluscs, except bivalves, have a rasping tongue, the radula.

Chitons: class Polyplacophora

Polyplacophorans (chitons or coat-of-mail shells) are exclusively marine, bilaterally symmetrical molluscs (Fig. 39.27). They are elongated and dorsoventrally flattened with eight overlapping calcareous shell plates. The shell plates are held in place and sometimes entirely covered by a girdle. The epithelium of the girdle produces a cuticle in which are embedded calcareous or chitinous spicules. The broad foot is surrounded by a groove, the mantle cavity, in which gills (six to 88 pairs) are arranged. The mouth is

(a)

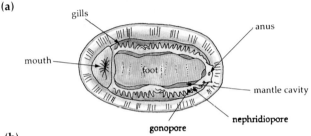

(b)

Fig. 39.27 (a) Dorsal view of a chiton. (b) Ventral view of a chiton

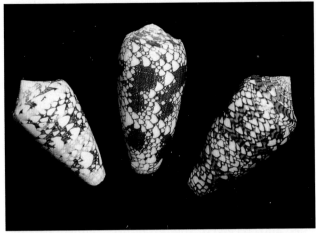

Fig. 39.28 *Conus*, the cone shell, is a venomous gastropod. The barbed end of a single radula tooth is hurled into prey, which is quickly immobilised by a neurotoxin that enters the wound through a hollow cavity in the tooth. Some South Pacific species of cone shells are highly toxic to humans

anterior but there is no distinct head. Sexes are separate. Although there are usually pelagic larvae, some chitons brood their young in the mantle cavity until the young reach the creeping stage.

Torsion and the class Gastropoda

Gastropods (Fig. 39.28) have undergone three major changes relative to the hypothetical primitive mollusc (Fig. 39.25). They show greater development of a head, **cephalisation**. The shell has become spiral. The upper part of the body, the visceral mass, has undergone **torsion** through 180° relative to the foot so that the originally posterior mantle cavity, with its two gills and anus, is now anterior. The evolutionary significance of torsion is uncertain. One theory suggests that torsion is specifically a larval adaptation, perhaps allowing protective withdrawal of the head into the mantle cavity. A further view is that it increases the ventilating current over the gills during forward locomotion.

A disadvantage of torsion was that the anus then discharged directly over the head and mouth! Much of the diversity of gastropods relates to different ways of overcoming this sanitation problem. In the most primitive gastropods, slit-limpets and abalone, a slit, or a row of holes in the shell (Fig. 39.29a), allows a flow-through ventilating current to sweep wastes out. True limpets (Fig. 39.29b) retain a posteriorly directed ventilating current but lose one or both gills with or without replacement by lateral gills in the

mantle cavity. Top and turban shells and nerites (Fig. 39.29c) have the respiratory current entering the anterior mantle cavity on the left side of the head, where a single gill is retained, then flowing past the anus to exit on the right side of the mantle cavity.

A further solution to the sanitation problem resulting from torsion has been *detorsion*, losing the original gills and moving the anus to a lateral and ultimately posterior position. In side-gilled slugs, the mantle cavity and anus are lateral. In some nudibranchs or sea slugs (Fig. 39.30), the shell, mantle cavity and original gill have been lost; the anus has become posterior and surrounded by a circle of gills. Reduction of the shell also occurs, for example, in sea hares and bubble shells.

Land snails and related freshwater forms (subclass Pulmonata, Fig. 39.31) have lost all gills. The mantle margin is fused with the body wall so that the mantle cavity becomes a lung with a single, restricted respiratory opening. Active ventilation of the lung occurs.

Fig. 39.29 (a)

(b)

(c)

Fig. 39.29 Gastropods. **(a)** A solution to the problem of sanitation in gastropods is seen in the shell of the abalone *Haliotis*, which has a series of holes for a ventilation current. **(b)** The limpet *Cellana*. **(c)** *Nerita*

Fig. 39.30 An Australian nudibranch, *Chromodoris*

Fig. 39.31 (a)

(b)

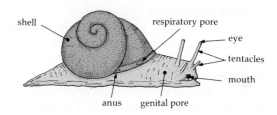

(c)

Fig. 39.31 Land snails. **(a)** *Sphaerospira*, a large land snail of Queensland rainforests. **(b)** *Triboniophorus*, a large slug. **(c)** Structure of the garden snail, *Helix*

Filter feeding and the class Bivalvia

Bivalvia, as their name suggests, have two shell plates that are bilateral (Fig. 39.32). They have no distinct head nor radula. The foot, like the body, is laterally compressed and in advanced forms is shaped like a hatchet. The general features of the class appear to have arisen in relation to burrowing in soft substrates.

Fig. 39.32 A bivalve, showing the exhalent and inhalent siphons and the flattened foot

Most bivalves are filter-feeders, with gills having both the function of food collection and gas exchange. As their descriptive name—lamellibranchs—suggests, most bivalves have gills greatly expanded as sheets or lamellae, extending from the anterior to the posterior end of the long mantle cavity on either side of the foot (Fig. 39.33). The ventilating current, which supplies both gases and food, enters posteriorly, passing forwards then upwards through the gills, before exiting posteriorly. Each of the two gills is folded on itself in a V-shape. Cilia on the surface of the gills and mantle cavity create the ventilating current. Specialised cilia on the gills sweep food particles filtered from sea water downwards to a food groove, running along the ventral edge (the bottom edge of the 'V') of the gill. Cilia in the groove convey the particles in an anterior direction to labial palps for sorting and transfer to the mouth.

Evolution of filter feeding led to radiation of lamellibranchs. Although many bivalves live on soft substrates, others live on hard substrates. These anchor one valve of the shell to the substrate by horny (byssal) threads, as in the mussel, *Mytilus*, or by fusion to the substrate, as in the oyster, *Crassostrea*. Others bore into rock, coral and even wood. A few bivalves,

such as scallops (*Pecten* and *Chlamys*), are unattached and can swim vigorously when disturbed by clapping the shell valves together

Most bivalves are dioecious. In protobranchs, gametes exit through nephridia but in other bivalves through special gonoducts. Fertilisation is usually external, although it may occur in the mantle cavity. A trochophore and a bivalved veliger larva typically develop. In some freshwater bivalves, such as *Alathyria*, the Australian freshwater mussel, a modified veliger termed the glochidium is released from the parent and grips onto fishes, which it parasitises until it becomes a cyst. Eventually, an immature mussel falls from the cyst to the bottom and undergoes development to the free-living adult.

'Brainy' molluscs: class Cephalopoda

Cephalopoda (Fig. 39.34) include the largest living invertebrate, the giant squid *Architeuthis*, known from South Australian waters. It attains a body length of 16 m, including the tentacles. Cephalopods are bilaterally symmetrical molluscs in which the shell, although primitively external, as in *Nautilus* and fossil ammonites (see Chapter 31), is usually internal or even absent. In *Nautilus* (Fig. 39.34a), the most

Fig. 39.33 The gills of bivalves are greatly expanded, forming sheets or lamellae. They function in both gas exchange and filter feeding. **(a)** Position of the gills on either side of the foot—visceral mass shown in cross-section. **(b)** Detail showing folding of gill and position of food groove

(a) (i)

Fig. 39.34 (a) (ii)

(a) (iii)

(b)

(c)

Fig. 39.34 Cephalopods include **(a) (i)** squid, **(ii)** paper nautilus and **(iii)** octopus. **(b)** The internal shell of the cephalopod *Spirula* is often washed up on our beaches with attached goose barnacles. **(c)** The common Australian octopus, *Octopus cyaneus*, showing the highly developed eye and respiratory and locomotory siphon. Colour changes occur during courtship and in defensive and offensive situations

primitive living cephalopod, the animal occupies the largest, last formed chamber of the shell. In the cuttlefish *Sepia*, the shell is shield-shaped and in *Spirula* it is coiled (Fig. 39.34b); it may be reduced to a thin plate, as in the squid *Loligo* (Fig. 39.34a), and occasionally it has been lost, as in *Octopus* (Fig. 39.34c).

Buoyancy and locomotion

In *Sepia* and *Spirula*, as in *Nautilus*, gas-filled chambers of the shell, separated by transverse septa, give buoyancy. Regulation of the amount of gas and of buoyancy is brought about by a cord of body tissue, the **siphuncle**, which runs through all the chambers. Water flows from the mantle cavity through a specialised funnel, the **siphon**, which can produce a jet of water for propulsion. Octopuses, which are benthic and crawl on the sea bottom, use jet propulsion only intermittently, often for escape. In *Nautilus*, which is a slow swimmer, force is generated only by contraction of the funnel, but in most living cephalopods the mantle provides the contractile force that permits very rapid movement. Some cephalopods are even capable of leaping out of the water to glide through the air. It has been suggested that development of rapid swimming, with streamlining of the body by internalisation of the shell, was a selective response to competition from bony fishes, which had replaced nautiloids and ammonites as the dominant pelagic animals by the end of the Mesozoic (Chapter 31).

The cephalopod head

The name cephalopod means 'head–foot', because the head is surrounded by a ring of arms and tentacles (usually equipped with suckers or hooks), which represent a modified foot. *Nautilus* has 38 tentacles, without suckers, while most cephalopods have eight arms and two extensible, prehensile tentacles, as in cuttlefish, squids and vampire squids. Octopuses have lost the two tentacles.

In cephalopods the mouth leads to a strong beak, shaped like that of a parrot, with dorsal and ventral mandibles accompanied by a radula. Salivary glands are usually present and their secretion is sometimes poisonous, as in the blue-ringed octopus, *Hapalochlaena maculosa* of southern Australian shores.

Nervous system, sense organs and behaviour

Cephalopods have a well-developed brain, which provides advanced capabilities in vision, touch, memory and recognition of shapes, and swimming. While the eyes of *Nautilus* lack lenses, those of squids, cuttlefish and octopuses have a lens and camera-like organisation that in many ways resembles the eyes of vertebrates. Cephalopods have colour vision and are perhaps unique in invertebrates in displaying signals and emotions by changes in the colour and pattern of the skin. Pigment granules are located in **chromatophores**, branching cells that expand under the action of tiny muscles attached to their periphery. The muscles are under nervous and hormonal control. Colour changes occur during courtship and defensive responses, merging an animal with its background, or in offensive situations, as when an animal is alarmed. In cephalopods other than *Nautilus* and some deep-water species, an ink-sac opening just inside the anus can release a cloud of alkaloid-containing inky fluid into the water when the animal is alarmed. This acts as a distraction to predators, and is possibly objectionable or even anaesthetic to chemoreceptors or other sense organs.

Gas exchange and circulation

The mantle encloses a ventral cavity, which houses one or two pairs of gills and receives the openings of the alimentary canal, and the excretory and reproductive systems. Cephalopods are the only molluscs with a closed vascular system and are exceptional in the invertebrates in having a true endothelium lining the blood vessels as in the vertebrates. Blood passes through a muscular heart before entering the gills. Accessory pumping structures, **branchial hearts**, may be present at the bases of the gills. The closed, high pressure blood system, presence of the respiratory pigment haemocyanin, absence of gill cilia, highly developed eyes, complex nervous system and behaviour, presence of chromatophores and production of 'ink', are features that relate to the active life and relatively high metabolic rate of cephalopods.

Reproduction

The sexes are separate in cephalopods. There is a single gonad. In the female, the gonad usually has a pair of ducts. In the male, one duct is retained while the other forms a sac that contains sperm packets (spermatophores). Copulation occurs 'head on'. Males often have the lower left arm specialised for transferring spermatophores. After fertilisation in the female oviduct, yolky eggs are usually deposited or shed into the sea water. A few cephalopods brood their eggs. Development is direct, with no trochophore or veliger larvae. Female octopuses die after laying eggs.

> Cephalopod features, including a closed vascular system, complex nervous system and behaviour, chromatophores, ink gland, jet propulsion, and highly developed eyes, relate to their active, carnivorous life.

BRYOZOANS AND RELATIVES: SUPERPHYLUM LOPHOPHORATA

In structure and development, three phyla of animals straddle the borderline between protostomes and deuterostomes. These three phyla are **Phoronida**, **Bryozoa** and **Brachiopoda**, constituting the superphylum **Lophophorata**. The **lophophore**, which gives them their name, is a feeding structure of ciliated tentacles, which contain extensions of the coelom.

Lophorates can be considered to consist of only three segments. They have three coelomic cavities in longitudinal sequence, the protocoel, mesocoel and metacoel, corresponding with three body divisions, the protosome, mesosome and metasome. The metacoel forms the main body cavity.

Bryozoans, moss animals, form the largest group and are the most common lophophorates, with about 4000 living marine and freshwater species (Fig. 39.35). They are sessile and colonial, and have a chitinous covering with or without calcareous layers, with a pore for protrusion of the lophophore. The coelom is spacious and the digestive tract is U-shaped. Colonies may form a thin film-like lacework on seaweed or on the underside of boulders and look like small, branching plants. They are a major marine fouling organism on boats and other structures.

Fig. 39.35 Bryozoans, the largest and most common group of lophophorate animals. Bryozoans are sessile and colonial, looking like lacework or seaweed

Phoronids are worm-like animals generally regarded as the most primitive lophophorates and there are only about 10, exclusively marine, species. Brachiopods (lamp shells) are today represented by some 280 species in contrast with the 30 000 or so known fossil species of the Palaeozoic and Mesozoic eras.

It is likely that deuterostomes, which we discuss in Chapter 40, arose from ancestors shared with lophophorates.

> The superphylum Lophophorata includes three phyla of animals that have few body segments. They are characterised by a lophophore, a feeding structure of ciliated tentacles, which contain extensions of the coelom.

SUMMARY

- Annelids, arthropods and molluscs are coelomate protostomes. They are relatively complex in structure and each group has been highly successful in terms of abundance of animals and species diversity.

- The body of annelids and arthropods shows metameric segmentation, where there is a linear repetition of functional units. Typically, each segment contains key structures such as nephridia (for excretion), ganglia (clusters of nerve cells), muscles and blood vessels. The only suggestion of segmentation in molluscs is in the class Monoplacophora, which includes early fossil molluscs and the extant *Neopilina*.

- Phylum Annelida contains segmented worms, including bristle worms (Polychaeta) and clitellum-bearing worms (Euclitellata). They have a digestive system, closed circulatory system, nervous system, and nephridia for excretion. A coelom and metameric segmentation allow peristaltic locomotion.

- Polychaetes are marine worms characterised by parapodia, paired appendages that function in locomotion and respiration.

- Euclitellate worms lack parapodia. They are characterised by a clitellum, which secretes a cocoon in which eggs are deposited. Euclitellate worms include oligochaetes, which have few setae on each segment, and leeches, which have none. Most leeches are specialised as external parasites and attach to hosts by posterior and anterior suckers. The coelom is reduced and leeches have lost peristaltic locomotion.

- Arthropoda are the most abundant and diverse phylum of animals. Arthropods include subphyla Chelicerata (spiders, scorpions and ticks), Crustacea (from water fleas to crabs) and Uniramia (insects and their relatives). Arthropods have a hard exoskeleton for protection and attachment of muscles, subdivision of the body into groups of segments (tagmata), and jointed limbs, necessary because of the rigid nature of the skeleton. The coelom, as seen in annelids, is reduced and virtually absent. They have an open blood vascular system (the haemocoel).

- Chelicerate arthropods have anterior appendages (chelicerae) that function as fangs or pincers and they lack antennae. The body is divided into a cephalothorax, an abdomen and, in primitive forms, a telson.

- Crustaceans are unique among arthropods in having two pairs of antennae. They have a pair of mandibles and two pairs of maxillae. Segments are organised into head, thorax and abdomen, and appendages are biramous (with two branches).

- Uniramians have one pair of antennae and appendages that are unbranched compared with crustaceans. Insects, the largest class of uniramians, have a well-defined head, thorax and abdomen, three pairs of legs attached to the thorax, specialised mouthparts, and usually compound eyes as well as ocelli. They respire by tracheae, excrete uric acid through Malpighian tubules, and are well adapted to terrestrial life. Evolution of flight has enabled them to exploit aerial environments.

- Phylum Mollusca is second in size only to the arthropods. Molluscs are a diverse group of aquatic and terrestrial animals including chitons, snails and nudibranchs, bivalves, tusk shells and squids and octopuses. The basic molluscan body has a head, visceral mass, and a ventral gliding surface, the foot. All except bivalves have a rasping tongue, the radula.

- In gastropods, the visceral mass undergoes torsion so that the mantle cavity, with its two gills and anus, is anterior, allowing the head to be withdrawn for protection into the mantle cavity.

- Bivalve molluscs have two bilateral shell plates. They have no distinct head, no radula, and a laterally compressed foot for burrowing. Most bivalves are filter-feeders, with gills functioning in both gas exchange and food collection.

- Cephalopod features, including closed vascular system, complex nervous system and behaviour, chromatophores, ink gland, jet propulsion, mobile arms, and highly developed eyes, relate to their active, carnivorous life.

- The superphylum Lophophorata includes three phyla of animals that have few body segments. They are characterised by a lophophore consisting of ciliated feeding tentacles that contain extensions of the coelom. Structurally and developmentally they have features similar to both protostomes and deuterostomes, and their phylogenetic relationships are unclear. It is likely that true deuterostomes arose from ancestors shared with lophophorates.

QUESTIONS

1. By reference to an annelid, such as an earthworm, and an arthropod, such as a centipede, explain what is meant by metameric segmentation.

2. Which animal phyla are coelomate protostomes? Explain what this term means.

3. You are collecting in a marine habitat and find an elongated animal with numerous bristles on body segments. Each segment also bears a pair of lateral fleshy lobed appendages. Name the phylum and the class to which this animal probably belongs. Name the type of appendage described and state its function(s).

4. Earthworms move by peristaltic creeping. Describe this movement and explain the role of the coelom.

5. Discuss the key body features that relate to the success of arthropods.

6. Prepare a table and list the features that distinguish chelicerates, crustaceans and uniramians from one another.

7. Describe some of the ways that arachnids locate (sense) and catch prey.

8. Briefly describe the basic mouthparts of an insect and how they are modified in specialist feeders such as a house fly, and a female mosquito.

9. Describe the basic body structure of a mollusc. Compare and contrast the method of feeding in a gastropod and a bivalve.

10. In this chapter, cephalopods are described as the 'brainy' molluscs. Explain what is meant by this, and highlight the features of cephalopods that relate to their active way of life.

ECHINODERMS AND CHORDATES

In Chapter 38 we saw that, with the exception of cnidarians and nematodes and some borderline cases in the lophophorates, eumetazoan animals are either protostomes or deuterostomes, depending on their pattern of embryonic development (Chapter 38). Of the coelomate animals, annelids, arthropods and molluscs (Chapter 39) are protostomes. Despite the remarkable differences between an adult sea star and a vertebrate, such as a trout, they are both deuterostomes, in which the anus of the adult develops at the site of the blastopore of the embryonic precursor of the gut (archenteron). Sea stars and their relatives are classified in the phylum Echinodermata, while vertebrates and their distant relatives, acorn worms, tunicates and lancelets, are classified in the phylum Chordata. The very different body organisations of these coelomate deuterostomes reflect ancient divergence of the two phyla (see Fig. 38.2), probably more than 600 million years ago.

ECHINODERMS

Phylum **Echinodermata**, totalling some 6000 marine species, includes six classes: **Crinoidea**, feather stars and sea lilies; **Asteroidea**, sea stars or star fishes; **Ophiuroidea**, brittle stars and basket stars; **Echinoidea**, sea urchins, sand dollars and heart urchins; **Holothuroidea**, sea cucumbers; and the little known **Concentricycloidea**, which are related to sea stars. Echinoderms are unsegmented coelomate animals. As larvae they are bilaterally symmetrical but develop into radially symmetrical adults, with a basic five-rayed (pentameric) symmetry and no head or brain. Echinoderms are never colonial and are distinguished from all other animals by structural peculiarities of the skeleton and coelom.

The echinoderm skeleton

The skeleton of echinoderms is internal in origin and mesodermal. It is composed of **spicules** or plates, **ossicles**, of protective armour just beneath the skin.

Each ossicle represents a single crystal of calcium carbonate (calcite) with small amounts of magnesium carbonate. The crystal, like a spicule, is first formed within a cell and enlarges into an ossicle. Spicules are usually embedded beneath the skin, as in sea cucumbers. Ossicles may articulate with each other, as in sea stars and brittle stars (Fig. 40.1), or may be sutured together to form a rigid skeleton (test), as in sea urchins and sand dollars. Typically, the body surface has a warty or spiny appearance because of projecting spines or tubercles, which give the group the name echinoderm, 'spiny skin'.

The water vascular system

Unique to echinoderms is a system of coelomic canals, the **water vascular system** (Fig. 40.2). This includes a circular 'water ring', around the mouth, from which radial canals run to the arms in sea stars and brittle stars or their equivalent areas in other echinoderms. The water ring connects via a single vertical stone canal, so-called because of calcareous deposits in its walls, to a porous disc, the **madreporite**. Water is drawn in through the madreporite by cilia, which form tracts on its surface and line the water vascular system. The radial canals bear rows of **tube feet** (Fig. 40.1). Tube feet function in attachment to the substrate, locomotion, gas exchange and, in predatory forms, such as sea stars, manipulation of prey. Many short side branches from the radial canals of the water vascular system lead to the tube feet. A tube foot is extended when watery fluid of the water vascular system is forced into it by contraction of a sac, the **ampulla**, at its base (Fig. 40.2). A valve prevents the fluid from being forced back into the lateral canal. The flattened tip of the tube foot is sucker-like. When it comes into contact with the substrate, the centre of the sucker is withdrawn with consequent production of a partial vacuum. Adhesion is aided by a sticky secretion.

The tube feet, in addition to other thin areas of the body, also function in removal of nitrogenous wastes by diffusion. Echinoderms are usually

(a)

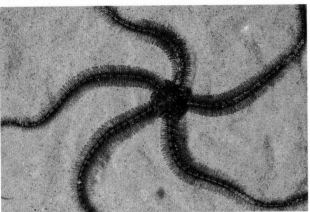

(b)

(c)

Fig. 40.1 Echinoderms include **(a)** sea stars, here with
tube feet extended, **(b)** brittle stars and **(c)** sea
cucumbers

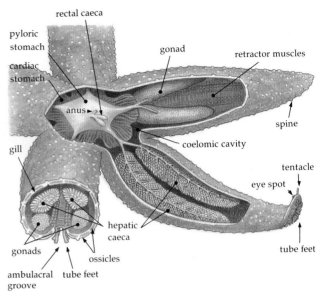

(a)

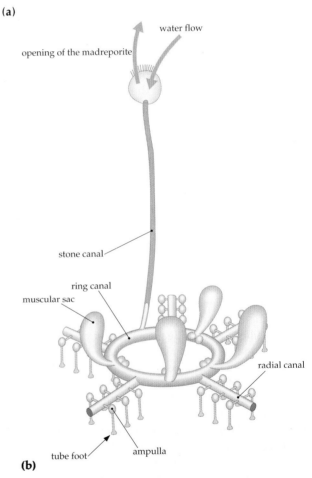

(b)

Fig. 40.2 (a) Structure of a sea star, showing **(b)** the
water vascular system. The water vascular system
operates as a hydraulic system. The ampulla contracts
and forces water into the tube foot, which becomes
extended. When the flattened, sucker-like end of the
tube foot comes into contact with a substrate, the
centre of the sucker is withdrawn, creating a vacuum,
which results in adhesion. The other parts of the water
vascular system (the madreporite, stone canal, ring and
radial canals) appear to control the water pressure
needed for the ampullae to function

considered to lack excretory organs and the ability
of osmoregulation. They are isosmotic, conforming
to their sea-water environment, and thus are not
found in brackish water. However, cells termed
podocytes, lining part of the water vascular system,
have recently been said to provide evidence of a type
of metanephridium.

Reproduction, development and regeneration

In echinoderms, sexes are usually separate and fertilisation is external after release of gametes directly into the sea. During development the embryo transforms from a single-layered blastula to a two-layered gastrula. In approximately half of all echinoderm species, the gastrula becomes a free-swimming larva, dipleurula (Fig. 40.3) with winding ciliated bands, different from the trochophore larva of protostomes (Chapter 38). A later larval stage usually develops projecting arms, the arrangement of which distinguishes the various classes. These arms do not survive into the adult and are not homologous with arms seen in adult sea stars and brittle stars. The larva then undergoes a complete metamorphosis. It settles on its left side, which becomes the *oral* surface of the adult, the anus being at the other, *aboral*, pole.

Sea stars and brittle stars have remarkable powers of regeneration and some reproduce asexually by splitting in half and regenerating the missing half. Some holothurians regularly regenerate their entire gut, which they discard as a distraction to predators.

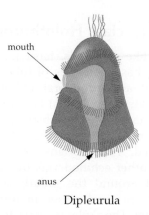

Dipleurula

Fig. 40.3 An echinoderm larva (a dipleurula) is bilaterally symmetrical, with winding bands of cilia. Compare this larval type to the trochophore of many protostomes (Fig. 39.6)

Sea lilies and feather stars: class Crinoidea

Crinoids are the only living echinoderms with the oral surface normally directed upward (Fig. 40.4a), the extended arms being used for suspension feeding on microplankton. They are the only survivors of a Palaeozoic group well represented by fossils (see Fig. 31.12). Sea lilies are sessile, with a five-armed body attached to the substrate by a stalk. Feather stars, which are free-swimming, detach from their stalk shortly after metamorphosis of the ciliated larva. Crinoids show a great range of often bright

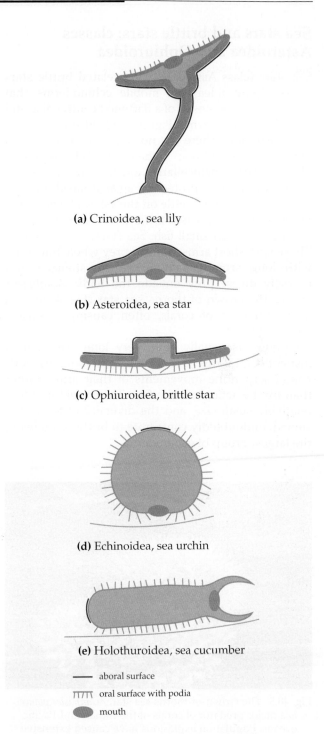

(a) Crinoidea, sea lily

(b) Asteroidea, sea star

(c) Ophiuroidea, brittle star

(d) Echinoidea, sea urchin

(e) Holothuroidea, sea cucumber

—— aboral surface

⊓⊓⊓ oral surface with podia

⬤ mouth

Fig. 40.4 Comparison of the basic body structures of the different classes of echinoderms: **(a)** sea lily; **(b)** sea star; **(c)** brittle star; **(d)** sea urchin; and **(e)** sea cucumber

colouration, which advertises the fact that they are toxic. They therefore have few predators. They do, however, have numerous parasites and commensals (Chapter 43).

Sea stars and brittle stars: classes Asteroidea and Ophiuroidea

Sea stars (class Asteroidea) and related brittle stars (class Ophiuroidea) are mobile echinoderms that have a body composed of a flattened central disc and radially arranged arms (Figures 40.1, 40.4b).

In asteroids there is no distinct demarcation between arms and disc. Some asteroids have pincer-like structures, **pedicellariae**, on their body surface. These are used in protection against small animals or larvae that might settle on the asteroid. *Stylasterias forreri* of the north-east Pacific Ocean uses the pedicellariae to catch small fish. Sea stars are carnivores. Those with short arms swallow prey whole but those with long arms usually evert the stomach and partially digest prey outside the body. *Acanthaster plancei*, the crown-of-thorns sea star (Fig. 40.5), is a major predator of corals, often causing extensive damage to reefs.

Ophiuroids usually have very long arms and a distinct central disc (Fig. 40.4c). They move by horizontal serpentine movements of their arms, rather than by the action of tube feet. Their rapid motility, relatively small size, and the diversity of food consumed, undoubtedly contribute to brittle stars being the largest group of echinoderms.

Fig. 40.5 The crown-of-thorns sea star, *Acanthaster plancei*, is a major predator of corals of the Indian and Pacific oceans. Population explosions have caused extensive damage to parts of the Great Barrier Reef. An adult sea star everts its stomach over coral and digests the soft bodied polyps, leaving behind the dead skeleton of the coral. Juvenile sea stars graze on algae. The crown-of-thorn's large size, armour of spines and nocturnal habit enable it to avoid being eaten by most predators

Sea urchins, sand dollars and heart urchins: class Echinoidea

Echinoids lack arms and the body, with five bands of tube feet, is spherical or flattened along the oral–aboral axis (Fig. 40.4d). Skeletal ossicles form a test and the animal is typically covered with movable spines (Fig. 40.6). Although sea urchins are radially symmetrical, many soft-bottom dwellers (sand dollars and heart urchins) have varying degrees of secondary bilateral symmetry. Sea urchins feed by rasping algae or other materials from hard substrates using a complex organ of bony appearance, '**Aristotle's lantern**', which has five large jaws (Fig. 40.6).

Fig. 40.6 Most sea urchins feed by rasping algae from rocks using a complex organ, 'Aristotle's lantern', which has five jaws

Sea cucumbers: class Holothuroidea

Sea cucumbers, like echinoids, have no arms (Fig. 40.1). Their body has five bands of tube feet and is greatly elongated along the oral–aboral axis (Fig. 40.4e). With this form the animal lies with one side of its body in contact with the substrate. As a result, the tube feet in the bands away from the substrate are often reduced. Holothurians are further distinguished from other echinoderms by modification of the tube feet around the mouth as branched tentacles, which are used either in deposit or in suspension feeding. The skeleton consists of spicules (microscopic ossicles) embedded in the body wall.

> Phylum Echinodermata includes deuterostome marine animals that have a basic five-rayed symmetry and no head. A unique feature is their water vascular system connected to tube feet, which function in locomotion, attachment, gas exchange and, in some groups, manipulation of prey.

CHORDATES

Phylum **Chordata** includes subphyla: **Hemichordata**, acorn worms; **Urochordata**, tunicates or sea squirts; **Cephalochordata**, lancelets, including *Branchiostoma*, commonly called amphioxus; and **Vertebrata**, fishes, amphibians, reptiles, birds and mammals.

Chordates are so-named because of the presence of a **notochord**, a dorsal, cylindrical rod situated below the nerve cord and above the gut (Chapter 15). The notochord gives the chordate's body longitudinal support and lateral (side to side) flexibility. The notochord, however, cannot with certainty be said to occur in hemichordates, which are sometimes classified in a separate phylum. Chordates have **pharyngeal slits**, paired openings in the pharynx, at some stage during development. They are often called gill slits, but the earliest pharyngeal slits were for feeding, acting as exits for a water current entering through the mouth and supplying food. During evolution, pharyngeal slits have become modified for hearing, vocalisation or gas exchange (as true gills) in different chordates. All chordates have a hollow **dorsal nerve cord** that controls body movement.

> Chordates have a dorsal notochord, pharyngeal slits and a dorsal hollow nerve cord.

Relationship between echinoderms and chordates

A group of hemichordates termed pterobranch appear to be the nearest relatives of echinoderms. It has been suggested that the water vascular system of echinoderms originated from a lophophore (Chapter 39), which the pterobranchs, alone among the chordates, retain. It is possible therefore that chordates and echinoderms are descended from a pterobranch-like ancestor. Recently, a living pterobranch has been discovered, which can be classified in the Palaeozoic group graptolites. Acorn worms may have lost the lophophore in acquiring a worm-like burrowing body, and tunicates lost the lophophore in favour of filter feeding through gill slits. Larvae of acorn worms are strikingly similar to those of echinoderms (Fig. 40.3), lending further support to an echinoderm–hemichordate relationship.

Acorn worms and pterobranchs: hemichordates

Subphylum Hemichordata is a small group of worm-like marine animals with pharyngeal slits. The 'notochord' or stomochord of hemichordates may not be homologous with a true notochord and, on this basis, the subphylum is often excluded from the Chordata and raised to the rank of an independent although unquestionably deuterostomatous phylum.

Acorn worms, the most common hemichordates, live mostly in burrows in shallow marine environments. They have three body divisions—a **proboscis, collar** and **trunk** (Fig. 40.7). Pharyngeal slits, located in the anterior part of the trunk, filter off excess water from material, often sand and mud, collected by mucus secreted by the proboscis and conveyed by cilia to the mouth. The mouth lies in a groove between the proboscis and collar. The dorsal tubular

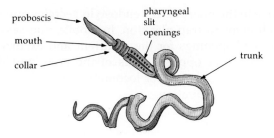

Fig. 40.7 Acorn worms are marine hemichordates, animals possessing gill-slit openings

nerve cord, characteristic of chordates, is situated in the collar. Elsewhere the nervous system resembles that of echinoderms in consisting of a sheet of nerve fibres and cells lying under the epidermis over the entire body. A short structure beneath the dorsal nerve cord at the junction of the proboscis and trunk, and forming a dorsal diverticulum of the pharynx, constitutes the **stomochord** and is associated with a skeletal plate.

Blood flows forward in the dorsal blood vessel and backward in the ventral vessel. The dorsal vessel expands into a heart-like sinus, surrounded by a muscular pericardium. The front of the sinus forms a series of glomeruli which are covered by a region of the proboscis coelom that is specialised for excretion. The entire body surface is ciliated.

Tunicates: urochordates

The relationship between tunicates (subphylum Urochordata) and other chordates is supported by impressive evidence, including the identical mode of development of pharyngeal gill slits. The tadpole-like larva of tunicates (Fig. 40.8a) resembles other chordates, although adult tunicates are very different. Larvae have a hollow dorsal nerve cord and a notochord in the tail, absent in adults.

Most tunicates, totalling about 1300 species, are sessile but others are pelagic. They are marine and worldwide, found in a range of water depths, from intertidal to great depths. They are filter-feeders in which the adult body (Fig. 40.8b) is typically covered by a complex **tunic** or supportive and protective 'coat' secreted by ectoderm. The tunic is unique among animals in containing a type of cellulose (tunicin) and, although it lies outside the epidermis, in some animals the tunic includes amoeboid cells and blood vessels. Strangely, tunicates lack a true coelom.

The pharynx, perforated by numerous pharyngeal slits, typically forms a large basket. Water taken in at the incurrent siphon passes through the pharyngeal slits into a space, the atrium, between the pharynx and the body wall before being ejected through an excurrent siphon (Fig. 40.8b). In this way, tunicates can filter an enormous volume of water to extract plankton as food. In addition, a ciliated ventral groove in the pharynx, the **endostyle**, secretes mucus, which helps to trap plankton. In mobile, non-sessile tunicates, openings for the incurrent and excurrent siphons are at opposite ends of the body and the water current is used for locomotion in addition to feeding and respiration.

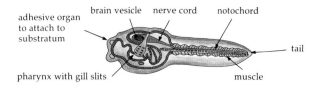

(a)

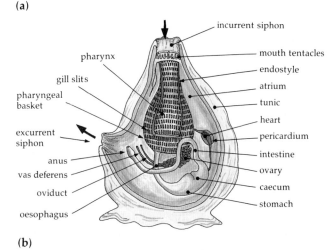

(b)

(c)

Fig. 40.8 **(a)** The tadpole-like larva of some tunicates resembles chordates and is evidence of their relationship. The larva does not feed, and swims for only a short time before undergoing metamorphosis. The tail retracts, parts of the body degrade, the pharynx develops and the pharyngeal slits become functional. **(b)** Structure of an adult tunicate, which is a sessile filter-feeder. **(c)** Several species of sea squirts (tunicates) are present here

The heart, a short U-shaped tube lying in a pericardium, periodically reverses the direction in which it pumps blood through vessels. Most tunicates are hermaphrodites, fertilisation occurring externally or within the atrium.

Lancelets: cephalochordates

Subphylum Cephalochordata includes small marine, fish-like animals (up to 8 cm long) that occur in all tropical oceans (Fig. 40.9). They show typical chordate features: a notochord, pharyngeal slits and hollow dorsal nerve cord. The notochord extends to the tip of the head, where there is a cerebral vesicle but no true brain. Unlike hemichordates and urochordates, lancelets have blocks of muscles, **myotomes**, on each side of the body (Fig. 40.9), as in vertebrate chordates. Although a lancelet swims only sporadically, alternate contraction of the myotomes of the two sides of the body, in a wave from head to tail, propels the animal forward.

The alimentary canal has a liver diverticulum and a pharyngeal endostyle. Numerous pharyngeal slits open from the pharynx into a chamber, the atrium, which opens to the exterior posteriorly at the atriopore and is covered by lateral folds of the body wall, the metapleural folds (Fig. 40.9). A lancelet lies buried in the substrate with buccal cirri exposed, and feeds by straining minute organisms from the water. The feeding current is maintained by cilia of a 'wheel organ' in the oral hood, and by cilia of the gill bars and endostyle. Sensory tentacles are present within the hood. The anus is well anterior to the hind end of the body, leaving a definite postanal tail as is typical of chordate metamerism.

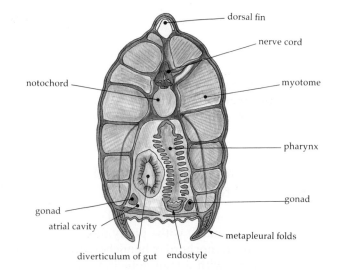

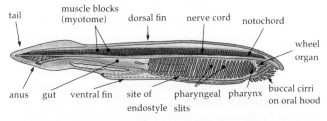

Fig. 40.9 Lancelets, cephalochordates, are small fish-like marine animals. The structure of *Branchiostoma* is shown here, illustrating true chordate features: notochord, pharyngeal slits, hollow dorsal nerve cord and muscle blocks (myotomes)

VERTEBRATES

Vertebrates have a distinct head with a skull (cranium) protecting the brain, eyes and cranial nerves, and are sometimes called the craniate chordates. Most vertebrates, as the name implies, have a backbone composed of **vertebrae**. Vertebrae develop around and replace the notochord during embryonic development (Fig. 40.10). Dorsal projections from the vertebrae form a neural arch that protects the hollow nerve cord. Ventral projections may form a haemal arch, which protects the aorta, the major artery carrying blood posteriorly from the gills and heart. The backbone forms part of the endoskeleton, but most vertebrates have, in addition, an exoskeleton of **dermal bones**, which develop in the skin. Dermal bones include tissues such as dentine and enamel, which often form teeth or tooth-like structures (denticles). Many features unique to vertebrates develop from an embryonic tissue, the **neural crest**, composed of cells associated with the embryonic nerve cord. Neural crest cells eventually develop into sensory organs and some nerve ganglia, branchial cartilages, dentine of teeth and denticles, and other structures such as the adrenal medulla (source of arenaline, Chapter 25).

Different groups within the subphylum Vertebrata, including fishes, amphibians, turtles, snakes, lizards, crocodiles, birds and mammals, have characteristic

internal organ systems, which are described in Chapters 19 to 22. The following is a comparative overview of the major groups of vertebrates, commencing with the most primitive—jawless fishes.

> All vertebrates have a distinct head with a skull, and most have a backbone composed of vertebrae. A key feature is the embryonic neural crest, the cells of which develop into many of the structures characteristic of the subphylum.

Jawless vertebrates

Jawless vertebrates, modern lampreys (Fig. 40.11) and hagfishes, have a notochord, not a vertebral column. They lack bone, although ancient fossil jawless fishes, 400–500 million years old, had plates of dermal bone. Modern and fossil forms are classified together in the superclass **Agnatha** solely on the basis of the absence of jaws, and the group is an artificial taxon comprising a number of separate evolutionary lineages, lampreys and hagfishes being the only surviving ones.

Fossil agnathans were filter-feeders, but some lampreys are specialised as parasites on other fishes, and hagfishes are scavengers. Lampreys are eel-like in shape and have a mouth like a suction cup. Parasitic forms attach to a host and rasp through the skin with their tongue, sucking blood. Adults of parasitic species are marine, but during spring they ascend freshwater streams where they build nests of stone on the stream bottom in which they spawn. After spawning, adults die. Free-living larvae remain in freshwater streams for some years before they undergo metamorphosis.

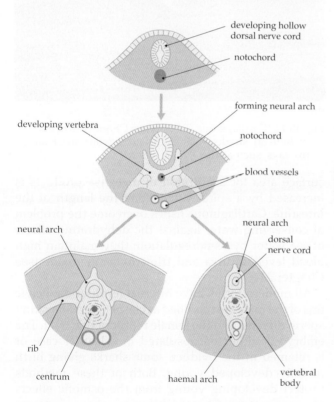

Fig. 40.10 Embryonic development of vertebrae. A vertebral column develops around and replaces the notochord. Paired projections of vertebrae protect the developing dorsal hollow nerve cord (neural arch; see Chapter 14)

Fig. 40.11 A female pouched lamprey, *Geotria australis*, a jawless vertebrate, with a mouth like a suction cap. Males of the species are unusual in having a pouch in the throat

Jawed vertebrates

Jawed vertebrates (superclass **Gnathostomata**) have jaws derived from anterior branchial arches of the gill system (Fig. 40.12). Evolution of jaws enabled vertebrates to become predators and to attain a wide range of body sizes. Jawed vertebrates include fishes and **tetrapods**—vertebrates with four limbs with separate fingers and toes. Within jawed fishes there are two classes—cartilaginous fishes and bony fishes.

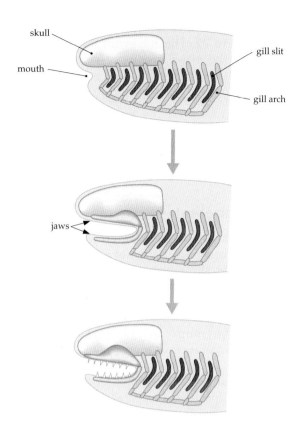

Fig. 40.12 Evolution of jaws by modification of the anterior gill arches of jawless vertebrates

> Among vertebrates, agnathans (hagfishes and lampreys) lack jaws. All other vertebrates (gnathostomes) have jaws derived from gill arches.

Cartilaginous fishes

Cartilaginous fishes (class **Chondrichthyes**), include two groups, chimaeras (subclass Holocephali) and sharks, skates and rays (subclass Elasmobranchii). Two pairs of fins, pectoral (anterior) and pelvic (posterior), together with unpaired dorsal and tail fins make sharks agile swimmers. The pectoral fins of skates and rays are extended as wings (Fig. 40.13) and the tail is whip-like, sometimes armed with a poisonous spine. Pectoral and pelvic fins are supported by cartilaginous girdles to which muscles attach. These fin girdles are homologous to the shoulder and hip bones of tetrapods.

The cartilaginous skeleton is often calcified but it never has true bone. This light skeleton helps chondrichthyans maintain buoyancy despite the lack of a swim bladder or lung. The skin of chondrichthyans is covered with small denticles (p. 885) and is rough, like sandpaper. Teeth (large denticles) are not fused to the jaws and are successively replaced during life. Nasal openings are usually ventral in position, unlike the double nasal openings on each side of bony fishes. The intestine is short, but the

(a)

(b)

Fig. 40.13 Cartilaginous fishes (class Chondrichthyes) include **(a)** sharks, such as this great white shark and **(b)** rays, such as this Pacific manta ray

surface area for absorption of digestive products is increased by a spiral valve along the length of the intestine. Cartilaginous fishes overcome the problem of conserving water against the dehydrating effects of sea water by osmoregulation: they maintain high blood levels of urea and trimethylamine oxide (see Chapter 22).

All modern forms have internal fertilisation. Pelvic fins of the male are modified as claspers for depositing sperm directly into the female reproductive tract. The embryo is either encapsulated in a leather case or is retained in the oviduct, some sharks giving birth to fully developed young. Both of these methods protect developing young from the osmotic effects of sea water. Chondrichthyans show extreme sensitivity to electric fields, and skates and rays have electric organs that produce either weak or strong discharges.

Bony fishes

All remaining fishes are members of the class **Osteichthyes**, classified into two subgroups: ray-finned fishes, which include most of the world's fishes (subclass Actinopterygii) and fleshy finned or lobe-finned fishes (subclass Sarcopterygii).

The skeleton of bony fishes, as their name Osteichthyes indicates, is made of true bone, at least in part. The skull has sutures; teeth are usually fused to the jaws; fin rays are bony; nasal openings on each side are usually double and lateral in position; and the biting edge of the jaws are usually formed by dermal bones, the maxilla and premaxilla of the upper jaw and the dentary of the lower jaw. The success of bony fishes is attributed in part to the **swim bladder** or lung, a gas-filled sac, which forms as an outgrowth of the pharynx and allows the fish to regulate buoyancy at any water depth and, in some cases, to breathe air.

Internal fertilisation is rare and has apparently been acquired secondarily. Even when fertilisation is internal, claspers are absent, although the anal fin is sometimes modified as a gonopodium. Embryos are not encapsulated in a case.

Ray-finned fishes

Ray-finned fishes (subclass **Actinopterygii**) have dermal, ray-like supports within the fins (Fig. 40.14). They include primitive bichirs of African freshwaters, sturgeons of the Northern Hemisphere, paddlefishes of freshwater in North America and China, and the worldwide group, the teleosts.

Primitive ray-finned fishes (bichirs and gars) have unique scales with an outer layer of enamel, termed **ganoine**, and a lower layer of dentine (normally present in teeth). The structure of the pectoral fin is unique, including an extensive surface for articulation with the shoulder girdle of the endoskeleton. A further feature is the general presence of a single dorsal fin. The internal skeleton does not extend into the fins at the fin base.

Teleosts, with 20 000 species, are by far the most diverse group of the ray-finned fishes (Box 40.1). They include well over half of all vertebrates. Teleosts are distinguished by fusion of vertebrae in the caudal fin, and loss of dentine and enamel from the scales.

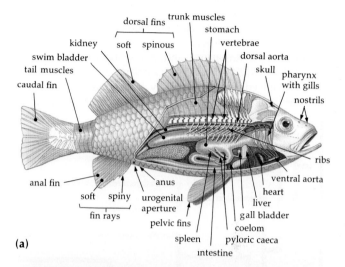

(a)

(b)

Fig. 40.14 **(a)** Structure of a ray-finned bony fish (class Osteichthyes). **(b)** An Australian teleost fish, the Murray cod, *Maccullochella peeli*

The term 'fishes' covers a number of vertebrate groups. Fishes with a cartilaginous skeleton include chimaeras, sharks and rays. Other fishes have a bony skeleton, and a swim bladder used in regulating buoyancy.

BOX 40.1 TELEOST DIVERSITY

Teleost fishes occur wherever there is permanent water, and even as 'annual fishes' in some tropical regions of seasonal drought, where eggs left in drying pools hatch with the advent of the rainy season. Teleosts are classified in about 40 orders, 400 families, and 4000 genera.

Primitive teleosts typically lack fin spines, and have large cycloid or smooth scales. The pelvic fins, located beneath the dorsal fin in the middle of the body, have more than six rays and the caudal fin has 17 branched rays. There are more than 25 vertebrae, and the upper

jaw has two tooth-bearing bones (premaxilla and maxilla) on each side, the premaxilla firmly united to the skull.

One large group of primitive teleosts is the Ostariophysi (see figure), so-named because of the bones derived from anterior vertebrae that interconnect the swim bladder and internal ear. Ostariophysans are predominantly freshwater, abundant in all continents except Australia, represented there only by tandan catfishes. They include more than 5000 species of minnows (carps), loaches, electric eels, characins and catfishes. Ostariophysans are best known as the 'tropical fish' of the international aquarium trade. Certain species, such as the goldfish, native to the Far East, have been kept and bred for centuries and are represented by numerous domestic varieties.

Other major groups of primitive teleosts are freshwater bony tongues, including the saratogas of tropical Australia, marine tarpons and true eels; herrings, sprats and anchovies; and salmons, trouts and smelts, including the smelts, *Retropinna*, and native trouts, *Galaxias*, of southern Australia and New Zealand.

Advanced teleosts typically have ctenoid or prickly scales, spines in all fins except the pectorals and caudal;

pelvic fins with one spine and five soft rays located anteriorly in the body beneath the pectoral fins; a caudal fin with 15 branched rays; 25 vertebrae; and the upper jaw with a single tooth-bearing bone (premaxilla), freely movable from the skull.

The major group of advanced teleosts is the Percomorpha, including numerous subgroups and more than 10 000 species. The group is named after the genus *Perca*, freshwater perch, of which there are only two closely related species—the yellow perch of eastern North America and the redfin perch of Europe, introduced into southern Australia.

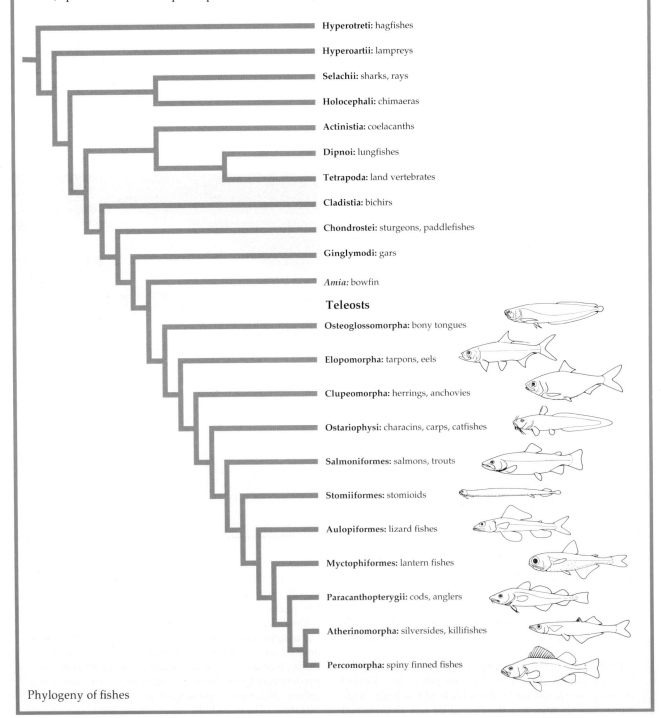

Hyperotreti: hagfishes

Hyperoartii: lampreys

Selachii: sharks, rays

Holocephali: chimaeras

Actinistia: coelacanths

Dipnoi: lungfishes

Tetrapoda: land vertebrates

Cladistia: bichirs

Chondrostei: sturgeons, paddlefishes

Ginglymodi: gars

Amia: bowfin

Teleosts

Osteoglossomorpha: bony tongues

Elopomorpha: tarpons, eels

Clupeomorpha: herrings, anchovies

Ostariophysi: characins, carps, catfishes

Salmoniformes: salmons, trouts

Stomiiformes: stomioids

Aulopiformes: lizard fishes

Myctophiformes: lantern fishes

Paracanthopterygii: cods, anglers

Atherinomorpha: silversides, killifishes

Percomorpha: spiny finned fishes

Phylogeny of fishes

Coelacanths and lungfishes

Fleshy finned fishes (subclass **Sarcopterygii**), many of which are only known as fossils, include coelacanths and lungfishes. Coelacanths were known only from Devonian and Cretaceous fossils until a specimen, later named *Latimeria chalumnae* (Fig. 40.15a), was caught off the South African port of East London in 1938. Since then, more specimens have been caught in what appears to be their normal habitat, deep water off the Comoros, islands north of Madagascar. Recently, a deep submersible (deep-sea vessel) has allowed this already endangered species to be observed in its natural habitat. *Latimeria* reaches 1.8 m in length and fertilisation of the huge, 9 cm diameter eggs is internal, with birth of well-developed young.

The relationships of *Latimeria* and extinct coelacanths to other vertebrates have been the subject of controversy. Some workers place coelacanths near sharks. However, morphological evidence, and new data from DNA and sperm ultrastructure, now seem to unequivocally place *Latimeria* with lungfishes and tetrapods (Fig. 40.16).

(a)

(b)

Fig. 40.15 One subclass of bony fishes is the fleshy finned or lobe-finned fishes, including coelacanths and lungfishes. (**a**) *Latimeria chalumnae*, from waters off the Comoros, islands near Madagascar, is the only living example, here shown at a depth of about 200 m off Grande Comore Island. (**b**) There are only three living genera of lungfishes in the world. One occurs only in Australia, the lungfish, *Neoceratodus forsteri*

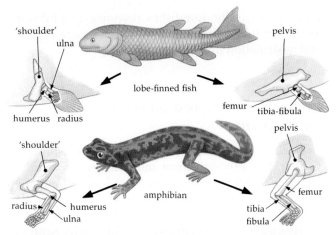

Fig. 40.16 Comparison of the limbs of a fleshy or lobe-finned fish and a tetrapod showing the homology of structure

There are only three extant genera of lungfishes, one in Africa, one in South America and one, including a single species (*Neoceratodus forsteri*), endemic to Australia (Fig. 40.15b). African lungfishes are freshwater fishes that live in shallow rivers and lakes, which dry up periodically. As the water level recedes, the fish burrows into the mud and makes a protective cocoon of mucus and mud. The single or paired swim bladder, which has become a functional lung, allows the fish to have sufficient gas exchange to survive. The fleshy lobed fins also allow the fish to 'walk' along the lake bottom.

> Of the bony fishes, fleshy finned or lobe-finned fishes are the closest relatives of four-limbed vertebrates, the tetrapods.

TETRAPODS

Tetrapods, as their name suggests, have four limbs with separate digits instead of the two pairs of fins of fishes. Tetrapods are conventionally divided into the classes **Amphibia**, **Reptilia**, **Aves** (birds) and **Mammalia**. However, the term 'reptile' covers a range of organisms that are not a single evolutionary group. Some reptiles are related to birds and others are related to mammals.

A large number of advanced characters is shared between lungfishes and tetrapods, suggesting that they are related. These characters include the presence in lungfishes of a pair of internal nostrils (nares), although in lungfishes the nares are not used for breathing as they are in tetrapods. Other shared characters include: type of gill arch muscles; ciliation of the larvae; structure of the pelvic girdle; structure of the pelvic and pectoral appendage; dermal bone pattern covering the braincase; and numerous features of the soft anatomy, such as the partially divided conus arteriosus (the anterior continuation of the heart), and division of the atrium of the heart by a septum.

Embryologically, lungfishes resemble amphibians and differ from other fishes in that cleavage is total and gastrulation produces a yolk plug.

Living in water and on land: amphibians

The class Amphibia includes the oldest land vertebrates, although most modern species require water in which to reproduce. Most have an aquatic larval stage. Modern amphibians, the **Lissamphibia** (Fig. 40.17), include salamanders, mud puppies and newts (order Urodela), frogs and toads (order Anura), and worm-like, legless forms of the tropics (order Gymnophiona or caecilians). They are basically quadrupedal, tailed vertebrates that have two occipital condyles, articulating the skull with the neck vertebrae, and a single sacral (pelvic) vertebra. The skin is glandular and lacks epidermal structures such as scales, feathers and hair, which are typical of the other vertebrates. It is moist and used for gas

(a)

(b)

Fig. 40.17 The oldest tetrapods are amphibians, (a) including modern frogs such as *Litoria moorei* and (b) the barred tiger salamander

exchange (Chapter 20). They are the first vertebrates to have a true tongue. Amphibians have the antidiuretic hormone vasopressin, which decreases the amount of water excreted (Chapter 22). Eggs are laid in freshwater and fertilisation is usually external, although some amphibians have internal fertilisation.

In extant amphibians, the number of bones in the skull is greatly reduced relative to fossil amphibians and there are a number of unique characters. Typically, modern amphibians have a maximum of four digits on the forefeet. In some salamanders, hind limbs and pelvic girdles have been lost. Caecilians have lost both sets of limbs and girdles (pectoral and pelvic). Frogs are characterised by elongation and modification of the hind limbs for jumping. Lungs are lost in some salamanders and frogs, and the left lung is reduced in caecilians.

> Amphibians are land vertebrates, but most require water for reproduction. Their skin is glandular and moist and used, usually with lungs, in gas exchange.

An embryo with a pond of its own: amniotes

All tetrapods other than amphibians are amniotes. In the Amphibia, the whole of the fertilised egg, the zygote, contributes to the structures of the embryo. In amniotes, large regions of the early embryo give rise to a number of extra-embryonic membranes including one, the **amnion**, which encloses the embryo in a fluid-filled sac, providing the embryo with an aquatic environment—a pond of its own (Chapter 14). This is a major adaptation to life on land. Correspondingly, fertilisation is always internal. Many amniotes lay eggs, but in others live young are born.

Amniotes have a set of distinctive features in addition to the amnion. The **allantois**, an outgrowth of the embryonic hindgut, is used for excretion during development. The amniote skin is thick and horny, and epidermal scales (or feathers or hair) provide a water-resistant barrier suitable for a terrestrial environment. Unless the animal is live-bearing, eggs have shells and are generally rich in yolk. Cleavage of the fertilised egg is not total, only part of the egg dividing into cells (**meroblastic cleavage**; Chapter 14). Development from egg to juvenile and adult is direct. In the skeleton, **intervertebral discs** are formed and the first two neck vertebrae are differentiated into an anterior **atlas**, supporting the skull, and a posterior **axis**. Usually the atlas is capable of rotation, allowing movement of the head. There are at least two sacral (tail) vertebrae, contrasting with only one in Amphibia.

A vexing question of classification and phylogeny of vertebrates is the position of the so-called reptiles within the Amniota. The term reptile signifies to most

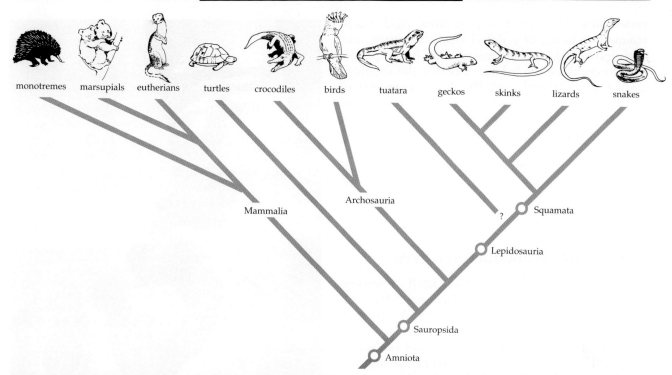

Fig. 40.18 One hypothesis of the relationships of amniotes. Note that 'reptiles' do not form a single evolutionary group (i.e. they are not monophyletic). The name Sauropsida is used for the group that includes turtles, squamates (lizards, etc.), crocodiles and birds

people 'cold-blooded' amniotes with scales rather than 'warm-blooded' amniotes with feathers or hair, but it is improbable that reptiles are a single group. Mammals are commonly regarded as separate from but related to all other amniotes, giving two groups, the Mammalia and the Sauropsida (Fig. 40.18). Sauropsids include turtles, the New Zealand tuatara, squamates (geckos, lizards and snakes), crocodiles and birds. The name **Sauropsida** is used instead of Reptilia since the birds are within this group (Fig. 40.18). The term 'reptile' should perhaps be restricted to common usage and will here be used in a descriptive sense.

Several other hypotheses of phylogenetic relationship exist and have yet to be disproved. Most workers agree that crocodiles and birds are related, with their extinct precursors, dinosaurs, forming the Archosauria. It is less certain the tuatara and squamates comprise a group.

> All tetrapods other than amphibians are amniotes. Extra-embryonic membranes, including the amnion, surround and protect the embryo and are a major adaptation to life on land.

Turtles and relatives

Turtles (Fig. 40.19) and their relatives (order **Chelonia**) have all the basic amniote features mentioned above. Some of them live in water, but others are successful on land. On land they have a sprawling gait, presumed typical of the first four-legged land

Fig. 40.19 Green turtles remember their terrestrial origin by coming ashore to mate

vertebrates. They have a protective, rigid trunk armour consisting of a dorsal shield, **carapace**, and a ventral shield, **plastron**, covered by horny plates. A feature unique to chelonians is that the shoulder girdle is within the rib cage. The skull is **anapsid** (Fig. 40.21a), lacking the one or two pairs of openings in the temples that accommodate the jaw muscles of other reptiles. The jaws are toothless, being covered by a horny sheath.

The relationship of turtles to other groups of amniotes is a subject of much debate but they are probably at the base of non-mammalian amniotes (Fig. 40.18).

Snakes, lizards and their relatives

Snakes, lizards, goannas and their relatives (**Lepidosauria**, Figs 40.18, 40.20) have teeth that are firmly fused to the edges of the jaws. A common feature of lizards is the ability to shed the tail at preformed points of fracture in several tail vertebrae, a device that is often effective in diverting a predator's attention to the writhing tail, allowing the escape of the tailless individual. In snakes, external limbs have been lost, probably as an adaptation to a subterranean existence. Many snakes are poisonous, manufacturing venom in their salivary glands. After immobilising its prey, a snake can open its jaws wide and swallow the prey whole because its jaws can be disarticulated.

The tuatara, *Sphenodon* (Fig. 40.20), is a 'living fossil', the only surviving member of its order, found today on a few offshore islands of New Zealand. It differs from lizards and snakes (squamates) in having two well-defined temporal openings, the **diapsid** condition (Fig. 40.21b), in its sperm ultrastructure, and in having a rigidly fixed upper jaw with a beak-like process. Its pineal or 'third eye' is better developed than in most other amniotes.

(b)

(c)

Fig. 40.20 **(a)** The Australian western blue-tongued lizard and **(b)** the children's python are two Australian examples of squamates. **(c)** Distantly related to squamates is the New Zealand tuatara, *Sphenodon*, the only surviving member of its order. It has the diapsid condition of two well-defined temporal openings in the skull

Fig. 40.20 **(a)**

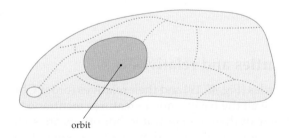

orbit

Fig. 40.21 **(a)** Anapsid

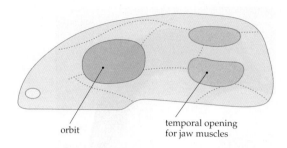

orbit

temporal opening
for jaw muscles

(b) Diapsid

Fig. 40.21 (a) An anapsid skull is characteristic of
chelonians (turtles and their relatives). This is in
contrast to the skulls of other sauropsids (lizards,
crocodiles and birds), such as **(b)** the diapsid skull,
with one or two pairs of openings in the temples to
accommodate jaw muscles

Crocodiles and birds

Crocodiles and birds (**Archosauria**) share a large
number of features. Extinct fossil birds and crocodiles
have teeth inserted into deep pits. Teeth, so
conspicuous in living crocodiles, are present only in
embryos of living birds. A gap (foramen) in the skull
over the pineal is closed over. In addition to two
temporal openings typical of diapsids there is a
further opening in front of the eyes (preorbital
opening).

Crocodiles (Fig. 40.22), caimans and alligators are
large carnivores of the tropics and subtropics. In
Australia, there are two species of crocodile, the
freshwater crocodile, *Crocodylus johnstoni*, and the
larger, more dangerous, saltwater crocodile, *Crocodylus
porosus*. They spend most of their time in water, and
venture on land to catch prey and to lay their eggs.
Many peculiar features can be attributed to their
semiaquatic mode of life. As examples, the nostrils
are united to form one opening at the tip of the snout;
the tail is compressed laterally as the main swimming
organ; and the pupils are vertically elongated.

Birds (class Aves) are a highly evolved group and
are phylogenetically younger than mammals. The
oldest bird fossil, *Archaeopteryx lithographica*, was found
in limestone dated at 150 million years ago, in the
Jurassic (Chapter 31). *Archaeopteryx* had clawed
forelimbs (seen on the wings of one modern bird from
South America), teeth, and a long tail containing
vertebrae. Little more than its feathers distinguish
it from those dinosaurs (a name that means 'terrible
lizards') called theropods.

Most of the peculiar features of birds are connected
with the acquisition of flight. Birds, although basically
quadrupedal animals, have forelimbs modified as
wings and skin covered by **feathers**, which are
modified scales (Fig. 40.23). In recent birds there are
specialised flight feathers on the wings, and
specialised tail feathers are attached to a short, plate-
like structure formed by fused caudal vertebrae, the
pygostyle, unlike the long tail of *Archaeopteryx*. There
is a large, bony keeled **sternum** for insertion of the
larger, pectoral flight muscles. Bones tend to be
hollow, making the skeleton light.

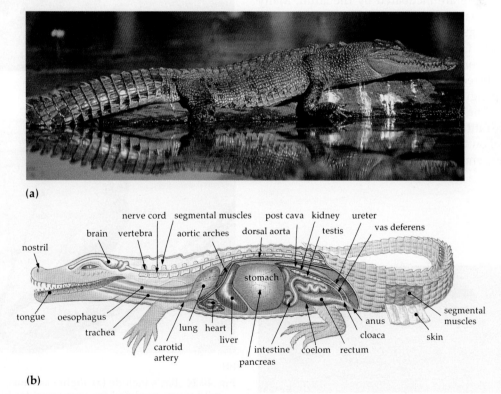

(a)

(b)

Fig. 40.22 (a) Australian crocodile, *Crocodylus porosus*. Crocodiles and alligators lay eggs on land and their skin has horny
scales. They share a number of features with birds, to which they are related. **(b)** Anatomy of a crocodile

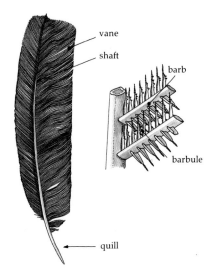

Fig. 40.23 Structure of a feather. The vanes or secondary branches are linked together, allowing the feather to function in flight

(b)

The jaws of modern birds are covered by a horny **bill**. Like turtles, crocodiles, lizards and snakes, they have only one knob on the back of the skull (the occipital condyle) for articulation with the vertebral column, in contrast with two in amphibians and mammals. All birds are endothermic with a four-chambered heart and right aortic arch, compared with the left retained in mammals. The brain is highly developed, and birds have excellent vision, hearing and balance. They lay eggs with a large yolk and a shell, and these are incubated by the adult. Many birds are social and have elaborate courtship involving songs, bright plumage and displays (Chapter 28). Colour vision, a basic feature of sauropsids, is correspondingly highly developed.

There are three groups of extant birds (Fig. 40.24). One is the penguins, which have wings modified for swimming. The second group is the Southern Hemisphere ratites, mostly flightless birds all of which lack the keel on the sternum, including the tinamou, ostrich, rhea, emu and cassowary (Chapter 45). The

(c)

(d)

Fig. 40.24 Birds include (a) flightless emus, (b) penguins, with wings modified for swimming, (c) black swans and (d) songbirds, such as this brown songlark

(a)

third group includes the vast majority of birds, more than 9000 species. They have a keeled sternum, like penguins, but unlike both other groups they have **contour feathers** arranged in a regular fashion.

> Within amniotes, birds are a highly evolved group with unique features associated with flight. They have feathers and a sternum for insertion of flight muscles. Their closest living relatives are crocodiles.

Mammals

Mammals (class Mammalia) appear from the fossil record to have originated from mammal-like reptiles, therapsids. The oldest fossil mammals date from the Triassic period, 190 million years ago. They coexisted with dinosaurs throughout the Mesozoic period, but with the extinction of dinosaurs in the Cretaceous, mammals underwent a great radiation.

Mammals are amniotes in which epidermal **hair** is present at some stage of development, although absent from the adult in some species, and young are nourished by secretions of the **mammary glands** of the female. They are endothermic, maintaining a high internal body temperature. Mammals, like crocodiles and birds, have a four-chambered heart and double circulation (Chapter 21). As in birds, the heart is fully divided into two auricles and two ventricles, but it is the left aortic arch that empties the left ventricle, the right arch being absent from adults. An efficient circulatory system is able to deliver sufficient oxygen to tissues to maintain body heat. The ability to maintain a high body temperature, together with the insulating effect of hair, probably allowed mammals to survive in a range of habitats.

> Mammals are amniotes that have hair at some stage, and mammary glands for feeding young. Mammals appear to have evolved from mammal-like 'reptiles' called therapsids.

There are two living subclasses: **Prototheria**, with one order, the Monotremata (echidnas and platypus), and **Theria**, consisting of the Metatheria (marsupials) and Eutheria (so-called placental mammals). Protheria are the only egg-laying mammals. Theria are mammals that produce live young with the embryo being nourished from the wall of the uterus by means of a specialised organ, the placenta (Chapter 14).

Monotremes

Monotremes are now represented by the web-footed duck-billed platypus, *Ornithorhynchus anatinus* (Fig. 40.25), and the echidnas, *Tachyglossus* and *Zaglossus*, which are restricted to Australia, New Guinea and some neighbouring islands. Survival of these monotremes is probably accounted for by their individual peculiarities. Thus, the platypus has an aquatic way of life in an environment where there are few competitors or predators. Echidnas typically have strong protective spines, similar to those of the eutherian hedgehog in the Northern Hemisphere, and are remarkable in their highly adaptive torpor (Chapter 29). In both echidnas and the platypus, there is a large horny venomous spur on the ankle of the male, evidence that monotremes are a single evolutionary lineage.

Monotremes are the most primitive living mammals and retain several basic amniote features. These include, among others, the presence of 'reptilian' bones in the pectoral girdles; shell glands in the female reproductive tract; and a cloaca. They have two completely separate uteri and, correlated with this, the penis is bifurcate. There is no scrotum as the testes are retained in the abdomen. Eggs of monotremes are incubated by the parent. Although the young feed on a milky secretion from mammary glands, there are no nipples.

Fig. 40.25 The platypus, *Ornithorhynchus anatinus*, of Australia and New Guinea, and the echidna, are egg-laying, prototherian mammals (monotremes)

Marsupials

Metatheria, marsupials (Fig. 40.26), occur mostly in the Australian region (see Chapter 31) but also in South America. The young of marsupials are nourished from the uterine wall via a yolk-sac placenta. However, a simple chorioallantoic placenta typical of eutherians, although lacking villi, occurs in bandicoots. As in monotremes, the two uteri are completely separate, there are two distinct lateral vaginal canals, and the penis is bifurcate.

Marsupials are characterised by birth of young at a stage that appears equivalent to an early fetal stage in eutherians. The young then actively find their way, unaided, to the nipples. There is usually a pouch, the **marsupium**, for concealment and carriage of the young (Fig. 40.27). Early parturition has contributed to the idea that marsupials are more primitive than eutherians, but they are highly adapted to their environment.

(a)

(b)

Fig. 40.26 Marsupials include: (a) the Eastern quolls and (b) the bilby

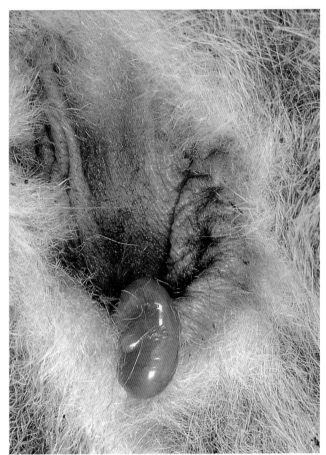

Fig. 40.27 Marsupials are characterised by a pouch. Here, the young of a kangaroo is entering the pouch

Eutherian mammals

Eutheria include a diversity of forms, totalling about 4500 species. Eutherians are viviparous mammals with a chorioallantoic placenta, formed from the chorion and the allantois, both extra-embryonic membranes of the developing young, and a portion of the uterine wall. We will avoid using the common name 'placental mammals' because of the presence of a placenta, albeit of different structure, in marsupials.

Typically in eutherians, and unlike marsupials, finger-like villi project from the embryonic portion of the placenta into the portion formed by the wall of the uterus. The two uteri may be separate to completely fused but there is only a single terminal tube, the vagina. In males, the penis is not bifurcated, a scrotum may be present and, if so, is situated posterior to the penis in all eutherians except rabbits and hares (order Lagomorpha), in which it is anterior, as in marsupials.

From fossil evidence it appears that eutherians and marsupials diverged from a common ancestor about 80 to 100 million years ago. The existing diversity of orders of the Eutheria (Fig. 40.28) evolved during the late Cretaceous and early Tertiary periods, about 70 to 45 million years ago from a common stock of small insectivorous mammals. In Australia, marsupials became the predominant mammalian fauna. Eutherians appear to be relatively recent immigrants from the north (Chapter 41) but recently a single eutherian tooth has been found in Cretaceous deposits in Queensland.

> Echidnas and platypuses (monotremes) lay eggs and are the only living members of the subclass Prototheria. The subclass Theria, including marsupials and eutherians, produce live young, with the embryo being nourished by a placenta.

Evolution of primates

The order Primates (Fig. 40.29) was named by Linnaeus, who included in it not only the primates as we define them but also the bats (Chiroptera) and flying lemurs (Dermoptera). In 1781, Thomas Pennant rejected the grouping primates, solely 'because my

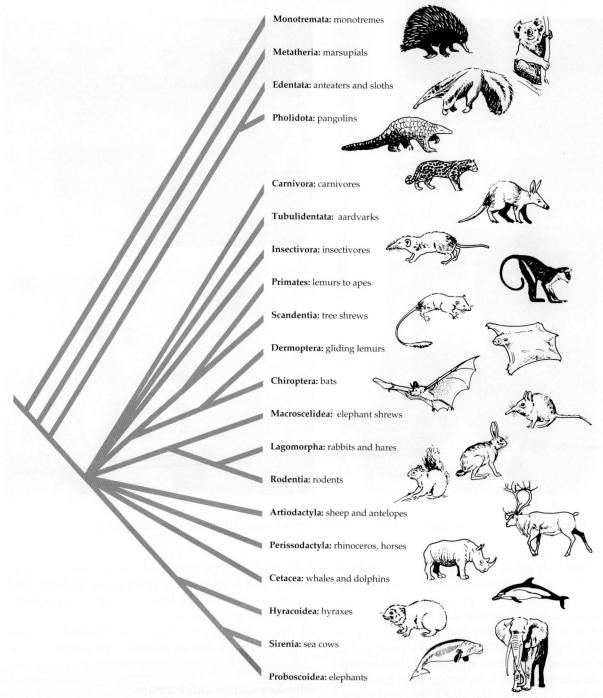

Fig. 40.28 Relationships of the orders of eutherian mammals, from anteaters and sloths to elephants. The monotremes (prototherians) and marsupials (metatherians) are shown at the base of the tree

vanity will not suffer me to rank mankind with Apes, Monkeys, Maucacos and Bats, the companions Linnaeus has alloted us'. The order continued to be recognised, however, although without inclusion of bats.

Early primates were probably small insectivore-like arboreal animals. Whether the living tree shrew, *Tupaia*, is representative of this stock is the subject of continuing debate. A fossil genus, *Purgatorius*, from the Cretaceous–Tertiary boundary of North America, about 65 million years ago, is thought by some to be the oldest known primate. However, its fossils are

no more than teeth and jaw fragments, and its primate affinities are questionable. The earliest undoubted fossil primates are lemur-like animals from the Eocene of Europe and North America, 55 million years ago. In a recent study, the origin of primates has been estimated to have been as early as 90 to 100 million years ago.

Premolar teeth of primates are *bicuspid*, that is, with two cusps, while molars have three to five cusps. The nose is moderate to very short, and eyes are well developed, with *binocular vision*. The brain is usually large, exceptionally so in apes, including

(a)

(b)

(c)

Fig. 40.29 Primates include: (**a**) this siamang gibbon, hanging from a branch; (**b**) this male gorilla; and (**c**) these black-faced monkeys

humans. Limbs are *plantigrade*—the whole foot, or hind foot in apes, touches the ground. Digits are *prehensile* (grasping) and thumbs are apposable to the second digit. Fingers are very mobile and sensitive. Claws are often replaced with *nails* and there is always a nail on the big toe. A well-developed *clavicle* (collar bone) is present and allows rotation of the forelimb on the shoulder girdle for swinging from branches, aided by lack of fusion of the long bones of the arm, the radius and ulna. A long tail is a primitive feature of primates such as lemurs. A tail is lost in apes and some others. The uterus is usually *bicornuate* (two-horned) and the penis is pendent. Births are usually single and there is long parental care and a progressive deferment of maturity, particularly in humans.

Prosimians

Primates are classified into two main groups, the **Strepsirhini** and the **Haplorhini**, named for features of the nostrils (Figs 40.30, 40.31).

Strepsirhini are 'prosimians', meaning precursors of monkeys. A naked nose pad (rhinarium) surrounds both nostrils. The nose pad has a marked median groove and is attached to the gums anteriorly. Each nostril has a crescentic lateral slit, hence the group's name. Examples are lemurs and the aye-aye, from Madagascar, African galagos, and South-East Asian lorises and their African relatives, the pottos.

Monkeys, apes and humans

Haplorhini do not have the bilateral slit at the nostrils. They include the anthropoids or human-like animals (**Anthropoidea**). Whether or not tarsiers should also be included in the haplorhines is debatable.

There are two groups of anthropoids. One is the New World monkeys, the **platyrrhines** (meaning flat-nosed), which evolved in South America. They have nostrils far apart and earholes (not a simple tube). Many of these monkeys are arboreal (tree-dwelling) and many have a prehensile tail, a feature never seen in Old World monkeys. Any monkey you see hanging from its tail would be from the New World!

The second group of anthropoids is the **catarrhines** (downward-nosed). They have nostrils close together and the earhole leads to a tube, as in humans. They

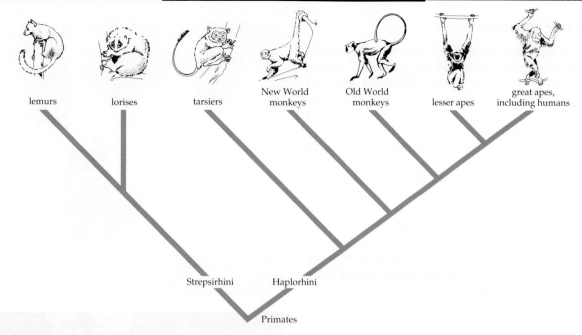

lemurs   lorises   tarsiers   New World
monkeys   Old World
monkeys   lesser apes   great apes,
including humans

Strepsirhini   Haplorhini

Primates

Fig. 40.30 Phylogenetic trees of primates have been constructed from estimates of the genetic distances between the different groups by cross-hybridisation of the total genome of single copy DNAs, and from DNA sequences of β-type globin. The various trees agree closely with one another and with those based on morphological characters

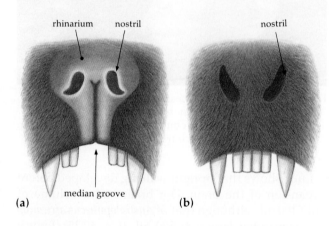

rhinarium   nostril   nostril

median groove

(a)   (b)

Fig. 40.31 Primates are classified into two groups, Strepsirhini and Haplorhini, on the basis of features of the nostrils. (a) In the strepsirhine condition in lemurs and lorises, the nostrils are surrounded by the rhinarium, which has a marked median groove. (b) In haplorhines, hairy maxillary processes have completely obliterated the rhinarium

are the Old World monkeys, apes and humans, some of which are arboreal. All of these forms occur in the Oriental region and only gibbons are absent from the Ethiopian region (Africa, south of the Sahara). Monkeys extend into the Australian geographical region on Sulawesi (Celebes).

The great apes have been traditionally set apart from humans as a separate family, **Pongidae**, but it is apparent from mitochondrial DNA (Chapter 30), that chimpanzees and the gorilla are more closely related to humans than they are to other apes and monkeys. It is clearly inconsistent with the evidence to place humans in a different group from great apes.

In a new and somewhat radical taxonomic system (Table 40.1), therefore, humans (*Homo*) are grouped with chimpanzees (*Pan*) and gorilla (*Gorilla*) in the tribe Homini and, with the orang-utan (*Pongo*) and gibbons (*Hylobates*), in the family **Hominidae** (previously restricted to humans).

Table 40.1 A new classification of the great apes?

> **Family Hominidae**
> Gibbons (*Hylobates*)
> Orang-utans (*Pongo*)
> **Tribe Homini**
> Gorillas (*Gorilla*)
> Chimpanzees (*Pan*)
> Humans (*Homo*)

From DNA and morphological studies it is deduced that apes descend from a catarrhine ancestor shared with Old World monkeys. The earliest known fossil apes are *Catopithecus* and *Proteopithecus*, described from Egypt in 1990 and of Eocene age (36–55 million years ago). They are believed to have been precursors of early apes that spread from Africa into Eurasia in the Miocene, from about 25 million years ago. From this stock arose the modern apes. Of these, the orang-utans and gibbons remained arboreal, but the ancestor of the gorilla–chimpanzee–human tribe presumably was more terrestrial, like its three living descendant genera.

> Primates have bicuspid teeth, short noses, well-developed eyes and brains, prehensile digits, and usually long parental care of offspring.

Human evolution

We have seen that the evidence seems indisputable that humans share a common ancestry with chimpanzees and gorillas. The genetic difference between the chimpanzee and humans is, in fact, extremely small and is less than would be expected from morphological differences between the two genera. It appears that in these apes small genetic shifts may give quite large morphological, and behavioural, changes (Chapter 32). To state that humans are the closest *living* relatives of chimpanzees does not, however, imply that, after divergence of the ancestor of the present-day chimpanzees, the ancestors of modern humans, *Homo sapiens*, may not have given rise to other human-like species that have become extinct. In fact, the fossil evidence suggests that there have been such sidelines in human evolution.

The earliest known fossils of forms that walked upright and had hands and teeth closely similar to ours are those of *Australopithecus africanus*, first discovered in 1924 in South Africa. *Australopithecus* (meaning southern ape) lived in the African savanna from about three million to one million years ago.

In 1974 a further species, *A. afarensis*, was described from a remarkably complete female skeleton, only about a metre tall, found in the Afar region of Ethiopia. The age of Lucy (Fig. 40.32), as she was nicknamed (after a Beatles song that was being played on a tape at the time of her discovery), has been the subject of controversy. If, as first suggested, it was 3.5 million years old, it might have represented a stock that gave rise to *A. africanus* and to *Homo*. If, however, a more recent age is correct, it might be part of an australopithecine branch that had already diverged from the ancestor that also gave rise to *Homo*. Richard Leakey considers that numerous individuals discovered at the same site in 1975 were indeed specimens of *Australopithecus* and *Homo*. *Australopithecus* nevertheless gives us important clues as to the probable appearance of early members of the branch that ultimately gave rise to *Homo*.

The erect posture of *Australopithecus* may have partly been an adaptation for allowing a creature with few defences (it did not even make tools) to detect predators in long savanna grass, but probably had its origins in using prehensile hands. Functions of hands include grasping food (Fig. 40.33), holding young and possibly nest building, which is seen in the gorilla. Apes, like monkeys, have retained the prehensile abilities of the feet, being able to appose the big toe to the other toes just as the thumb can touch the tips of the other fingers; however, the human foot has lost this ability. The big toe projects forwards and is longer than the other toes, providing thrust in the striding bipedalism that distinguishes human locomotion. Even the gorilla and chimpanzee, the most bipedal of the apes, normally progress when on the ground by 'knuckle-walking' (Fig. 40.34).

Fig. 40.32 Reconstruction of the fossil 'Lucy', possibly 3.5 million years old

Fig. 40.33 Why did our ancestors take to upright walking? Bipedalism may have allowed our ancestors to grasp objects, as this chimpanzee is doing and hence carry food away, or reach fruit in trees

Lucy shows that upright walking developed before expansion of the brain. Her brain had a volume of only 400 mL, although that of *Australopithecus africanus* was somewhat larger, at 500 mL (Fig. 40.35). Lucy's brain size was no greater than that of *Ramapithecus*, a 'large ape' with teeth like those of hominids, which occurred in Pakistan over a period of 8–15 million years ago. By the later part of australopithecine time, about two million years ago, the brain had enlarged to about 650 mL. With this advance, about a million years after hominids first 'stood up', came the making of simple stone tools, a criterion that persuades us to admit the maker into our genus as *Homo habilis* ('handy man'). Recently, however, chimpanzees, especially those of Asia, have been shown to fashion a variety of simple tools.

Homo

Some workers prefer to regard *H. habilis* as a tool-making australopithecine but *H. habilis* is the only fossil yet known that could have been in the direct line to modern humans. It coexisted for nearly a million years with the smaller-brained *Australopithecus*.

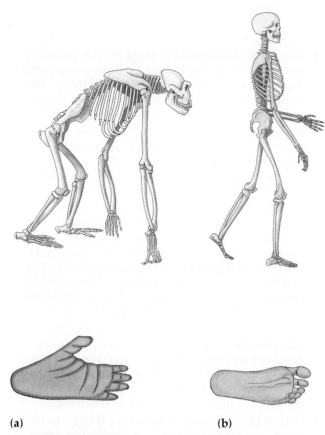

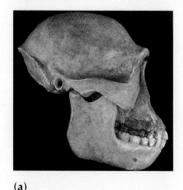

(a)   (b)

Fig. 40.34 Comparison of **(a)** a gorilla and **(b)** a human, illustrating skeletal changes associated with the evolution of bipedalism. Note the shortening of the pelvis, the S-shape of the vertebral column, and the position of the head in the human. In the human, the big toe is the longest digit and not divergent as in the gorilla

From what we know of human propensities, the extinction of *Australopithecus* may well have been brought about by *H. habilis*. Expansion in the brain case was accompanied by gradual reduction in the size of the jaws relative to the overall size of the skull.

The first species of *Homo* known outside Africa, in Europe and Asia, was *Homo erectus*, referring to the bipedal gait that had been acquired, we have seen,

much earlier. Early finds of this species were named Java Man (in Indonesia) and Peking Man (in China). Fossils date from 1.5 million years to about 300 000 years ago. Enlargement of the brain continued over this period and it finally exceeded 1000 mL (Fig. 40.35). The intelligence conferred by the larger brain size of *H. erectus* equipped it to make more sophisticated tools than those of *H. habilis* and its wide distribution indicates its success.

Anatomically modern forms of *Homo sapiens* occur first in the fossil record in Africa, about 100 000–140 000 years ago. Palaeontologists use this observation to argue that *H. sapiens* originated in Africa and migrated to other parts of the world. Early fossils in West Asia, at Zuttiyeh (75 000–150 000 years ago) and Qafzeh (50 000–70 000 years ago) may be records of this first migration out of Africa by *H. sapiens*. African populations were making blade tools 80 000–90 000 years ago, long before they replaced the simpler flake tools in Asia or Europe. Humans have been in Australia for at least 40 000 years (Chapter 41).

Homo sapiens has also been recorded as **Neanderthal man**, found originally in the Neander Valley in Germany. Neanderthals are first recorded from about 130 000 years ago. They had slightly larger brains than ourselves and made elaborate tools. The fact that they left evidence of rituals and burials suggests that they were capable of abstract thought. They had heavier brow ridges and less developed chins than modern humans. Some consider Neanderthals to be a distinct subspecies or even separate species from *H. sapiens*.

The modern West Asian forms of *H. sapiens* overlapped Neanderthals for many millenia but Neanderthals disappeared some 30 000 years ago, after which all human fossils are wholly modern. It has been suggested that Neanderthals interbred with humans with modern features.

Finally, characteristic of humans is language. It has recently been proposed that, because of the complex involvement of the brain and anatomy of the larynx and face in speech, language is at least two to three million years old. It is likely that even *Australopithecus* had a form of speech.

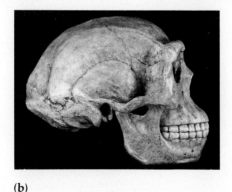

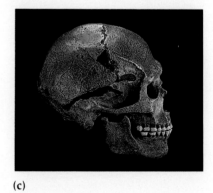

(a)   (b)   (c)

Fig. 40.35 African hominid skulls: **(a)** *Australopithecus africanus;* **(b)** *Homo erectus;* **(c)** *Homo sapiens.* Note the progressive expansion in size of the brain case and gradual reduction in size of the jaw, relative to the overall size of the skull

SUMMARY

- Despite their obvious differences in body organisation and their divergence nearly 600 million years ago, echinoderms and chordates are related. Both are coelomate deuterostomes, in which the anus of the adult develops at the site of the blastopore of the embryonic gut.

- Phylum Echinodermata includes marine animals that have a basic five-rayed symmetry with no head. A unique feature is their water vascular system connected to tube feet that function in locomotion, attachment, gas exchange and in some groups, manipulation of prey.

- Phylum Chordata includes animals characterised by a dorsal notochord, pharyngeal slits and a dorsal hollow nerve cord. They include hemichordates (acorn worms), urochordates (tunicates), cephalochordates (lancelets) and vertebrates, although there is some doubt as to whether hemichordates develop a true notochord.

- A key feature of vertebrates is the embryonic neural crest, the cells of which develop into many of the structures characteristic of the subphylum. All vertebrates have a distinct head with a skull, and most have a backbone composed of vertebrae. Among the vertebrates, agnathans (hagfishes and lampreys) lack jaws. All other vertebrates (gnathostomes) have jaws derived from gill arches.

- Of the bony fishes, fleshy finned or lobe-finned fishes are the closest relatives of four-limbed vertebrates, the tetrapods. Amphibians are land vertebrates, but most require water for reproduction. Their skin is glandular and moist and used, usually with lungs, in gas exchange.

- All tetrapods other than amphibians are amniotes. Extra-embryonic membranes, including the amnion, surround and protect the embryo and are a major adaptation to life on land.

- The name Sauropsida is used for the group of amniotes that includes turtles, squamates (lizards, etc.), crocodiles and birds. The traditional term 'reptile' is not used because so-called reptiles are not a monophyletic group.

- Within the sauropsids, birds are a highly evolved group with unique features associated with flight. They have feathers (modified scales) and a sternum for insertion of flight muscles. Their closest living relatives are crocodiles.

- Mammals are amniotes that have hair at some stage, and mammary glands for feeding young. Mammals appear to have evolved from mammal-like 'reptiles' called therapsids. Echidnas and platypuses (monotremes) lay eggs and are the only living members of the subclass Prototheria. The subclass Theria includes marsupials and eutherian mammals.

- Early primates were probably small insectivore-like animals. Modern forms are shrews, lemurs, monkeys and apes, including humans. Primates are characterised by bicuspid teeth, short nose, well-developed eyes and brain, prehensile digits, and usually long parental care of offspring.

- Within the primates, humans are related to the great apes (chimpanzees and gorillas) of Africa. The evolutionary lineage that led to modern humans is characterised by bipedalism. The earliest hominid fossils, *Australopithecus*, dating from three to one million years ago, stood upright but had skull features like other apes. *Homo habilis*, the oldest fossils known for our genus, had a larger brain size and used stone tools. *Homo erectus* had more advanced tools and an even larger brain size. *Homo sapiens*, recorded as Neanderthals, date from 130 000 years ago. Neanderthals disappear from the fossil record about 30 000 years ago, after which all humans are of a modern form.

QUESTIONS

1. What embryonic features do chordates share with echinoderms?

2. Briefly describe the structure and function of the water vascular system of a sea star. Which echinoderms do *not* use tube feet for locomotion?

3. What animals are characterised by pharyngeal gill slits? What functions do gill slits perform?

4. Adult chordates have either a notochord or vertebral column but not both. Explain why not.

5. What is the importance of the embryonic neural crest of vertebrates (see also Chapter 15)?

6. Bony fishes have a swim bladder, believed to have been important in their success. What is the swim bladder or its equivalent and what is its function in (**a**) a typical teleost fish, (**b**) a lungfish and (**c**) a tetrapod?

7. In this book, we did not classify turtles, lizards, snakes, and crocodiles together as Reptilia, as in more traditional classifications. With reference to the relationships of the amniotes, explain why use of 'Reptilia' is problematic.

8. Describe the features of birds that are associated with flight.

9. What features are unique and characteristic of mammals? Name the three main groups of mammals and explain how you would identify a representative of each.

10. How does the skeleton of australopithecines and other hominids differ from that of the gorilla? Discuss the differences in brain size, and the implication for behaviour, between *Homo habilis*, *Homo erectus* and *Homo sapiens*.

AUSTRALIAN ENVIRONMENTS AND BIOTA

Chapters 30–40 describe the diversity and evolution of organisms in general. This chapter focuses on Australian plants and animals. The Australian biota is the product of millions of years of evolution influenced by isolation and climatic changes resulting from alteration in the positions of continents and seas. The history of Australia's present-day flora and fauna is directly linked by geologic processes to the history of other southern continents.

AUSTRALIAN BIOTA: SOUTHERN CONNECTIONS

It has long been recognised that Australia and other lands in the Southern Hemisphere share many plant and animal taxa. The English botanist Joseph Hooker was one of the first, in 1853, to point out that many Indian plant genera occur in similar monsoonal environments in northern Australia, that Malaysian rainforest genera occur in tropical eastern Australia, and that cool temperate and alpine genera are shared between southern mainland Australia, Tasmania and New Zealand. Southern beech trees, *Nothofagus*, have a modern-day and fossil geographic distribution in Australia, New Guinea, New Caledonia, New Zealand, Antarctica and South America (Fig. 41.1). Bony-tongue fishes (family Osteoglossidae) occur today in tropical regions of Australia, New Guinea, South-East Asia, South America and Africa (Fig. 41.2). Flightless birds, the ratites, include cassowaries (Australia, New Guinea), emus (Australia), extinct moas (New Zealand), kiwis (New Zealand), rheas (South America), extinct elephant birds (Madagascar) and ostriches (Africa), all restricted to the Southern Hemisphere. Other largely Southern Hemisphere distributions of animals include marsupials, parrots, chelid turtles, tree frogs, galaxiid fishes and insects (Fig. 41.3).

In the past, these similarities in biotas were explained by recent long-distance dispersal—over waterways or across ancient land bridges that connected stationary continents. With acceptance of the theories of continental drift, sea floor spreading and plate tectonics (see Chapter 31), the distributions of *Nothofagus*, bony-tongue fishes, ratites and many other animal and plant groups are now best explained by old union of the southern continents and their moving apart to their present positions on the earth's surface.

Rainforests of present-day eastern Australia are probably relics of forests similar to those that once covered much of Australia and Antarctica during the early Tertiary dating back at least 40 million years ago. Native marsupials, birds, reptiles, frogs, fishes and insects and the vegetation that now characterises the Australian landscape—eucalypts, wattles, banksias, casuarinas, heaths, grass-trees and spinifex—have expanded their range in response to increased aridity.

AUSTRALIA IN GONDWANA

The last major tectonic–climatic cycle of earth's history started 320 million years ago, with the coalescence of all continents into the supercontinent Pangaea—a coalescence that was completed by about 230 million years BP (Fig. 31.4). Within Pangaea, the southern land masses of Australia, New Guinea, South America, Africa, Madagascar, Antarctica, India, New Zealand and maybe parts of South-East Asia were in close proximity and together formed Gondwana.

With the start of sea-floor spreading in the Atlantic and Indian Oceans, Pangaea began to break up in the mid-Jurassic (160 million years BP). India separated from Australia and Antarctica, the Indian Ocean began to widen, and southern Africa separated from

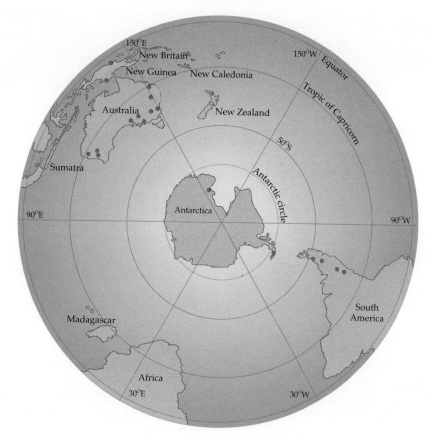

Fig. 41.1 Distribution of fossil and living *Nothofagus*, the southern beech trees

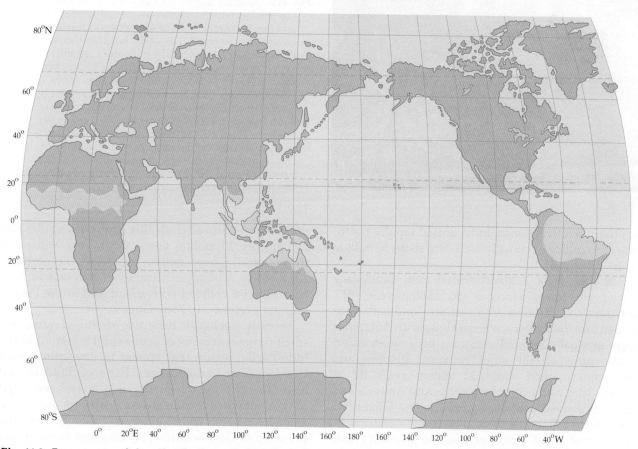

Fig. 41.2 Bony-tongue fishes (family Osteoglossidae) occur in tropical parts of the Southern Hemisphere

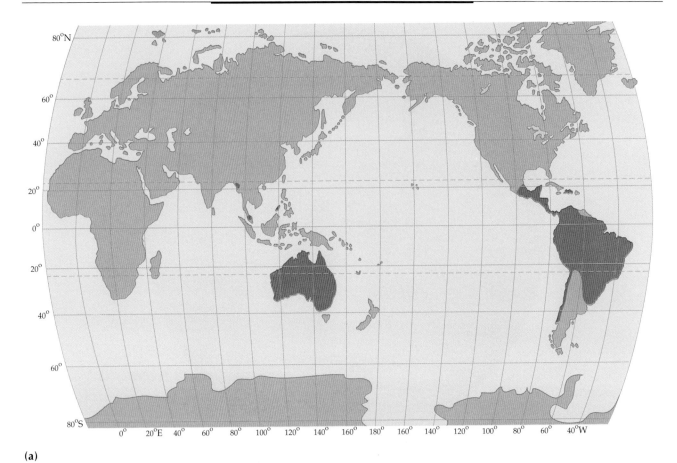

(a)

(b)

Fig. 41.3 *Synemon* moths are classified in the family Castniidae, which includes about 120 species in Central and South America, about 40 species in Australia, and two species in South-East Asia. **(a)** The distribution of *Synemon* and related moths of family Castniidae. This suggests a Gondwanan origin for the group. Many species live in rainforests, but Australian forms live in seasonally dry habitats, feeding on sedges and grasses. The greatest diversity of species in Australia is in south-western Western Australia. **(b)** Female of *Synemon plana*, an Australian species whose larvae live in soil and feed on the roots of native wallaby grass (*Danthonia*). With the loss of many areas of native grasslands and the invasion of introduced plants, this moth is now an endangered species

southern South America. India drifted rapidly northwards to collide eventually with Asia about 50 million years BP (Fig. 41.4).

Although New Zealand drifted from Gondwana during the late Cretaceous (80 million years BP), Australia's connection with eastern Gondwana continued into the Tertiary, when its links with Antarctica were gradually severed. Australia became an island continent when the strait between Australia and Antarctica opened about 30 million years BP. A narrow land connection between South America and Antarctica was maintained until the late Oligocene.

From the Oligocene, Australia drifted northwards, at a rate of about 7 cm per year, on a journey that took it from high latitudes to its present position in mid-to-lower latitudes. (If this continues at this rate for another 20 million years, Cape York will reach the equator!) Fifteen million years ago, the northern edge of the Australian plate, what is now southern New Guinea, collided with island arcs in the Pacific, forming present-day New Guinea.

Recently, geologists have also identified numerous small terranes (slivers of continental material) in the Indo-Pacific region that appear to have rifted from Gondwana at various times. Parts of Sumatra and Kalimantan (Borneo), for example, may have rifted from Australia in the Cretaceous, carrying plants and animals north. These terranes account for the presence of Gondwanan-type flora and fauna in the Indo-Pacific region.

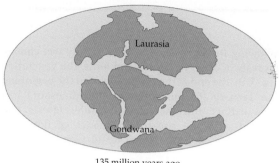

135 million years ago
(early Cretaceous)

65 million years ago
(late Cretaceous)

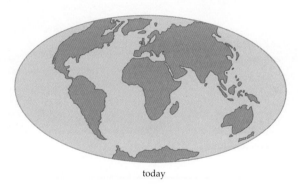

today

Fig. 41.4 With the start of sea-floor spreading, Pangaea began to break up in the mid-Jurassic (160 million years BP). By the Cretaceous, Gondwana was breaking up

Australia was part of the southern supercontinent Gondwana during the Cretaceous and early Tertiary. Australia severed its link with Antarctica and became an island continent by 30 million years ago. Environmental changes and isolation moulded the evolution of the modern Australian biota.

EVOLUTION OF AUSTRALIAN ENVIRONMENTS

Changing climate

During the early Tertiary (Palaeocene to Eocene, 65–40 million years BP), the climate of Australia and adjacent Antarctica was humid temperate, and rainforest covered much of the land. Separation of land masses created the Southern Ocean and its current began to circle Antarctica (Fig. 41.5). This current, known as the circum Antarctic current, has a profound effect on wind patterns and climate. It does not mix with warmer tropical water, and thus surface water temperature and air temperature are low. With development of the circum Antarctic current, ice began to form on Antarctica, and by the late Miocene (7 million years BP) it had reached its present-day extent. These changes reduced air temperature and rainfall over parts of Australia, heralding the onset of aridity and contraction of rainforests dominated by *Nothofagus* and conifers.

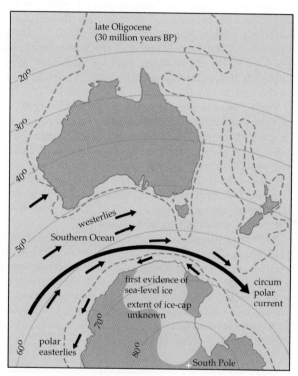

Fig. 41.5 Australia's connection with Antarctica was severed during the late Oligocene. This severence created the Southern Ocean and the circum Antarctic current, which resulted in declining temperatures and ice formation on the Antarctic continent. The arrows show atmospheric circulation patterns

Changing landforms

Australia is often described as one of the oldest landmasses because of the great age of rocks in regions such as the Pilbara (more than 3.8 billion years old). Landscapes from the Kimberley and Arnhem Land plateaus through the desert regions across to the eastern highlands, largely undisturbed by mountain building and volcanoes, became extensively weathered. Exposed rocks wore down to low hills and ranges. Soils were deeply weathered during warm wet periods, with nutrients, such as phosphorus and nitrogen, removed by millions of years of leaching.

BOX 41.1   ANCIENT FORESTS

In the mountains and dry valleys of Antarctica, fossils embedded in rocks and coal seams reveal plants and animals of the past. Fossil evidence of ancient Palaeozoic forests dominated by *Glossopteris* seed ferns was collected by Edward Wilson on Captain Robert Scott's last and fatal expedition to the South Pole. These important collections of rocks and fossils were found with their frozen bodies. Preserved leaves and stems of *Glossopteris* are the same as those found in deposits of the same age (250 million years BP) in India, South America, South Africa and Australia.

Glossopteris leaves are tongue-shaped with net-like venation and pronounced mid-rib (see figure). Glossopterids were a diverse group of trees and shrubs of swampy habitats, which favoured the formation of coal, particularly in the eastern part of Australia. Glossopterids had gymnospermous wood and seeds borne in clusters on leaves.

The *Glossopteris* flora dominated the Permian (286–248 million years BP) and replaced earlier forests composed of giant clubmosses and horsetails (see Chapter 37). A variety of insects inhabited the *Glossopteris* forests together with amphibians and reptiles. By the early Triassic (about 230 million years BP), glossopterids disappeared from the fossil record, and a new flora developed, characterised by forked-frond seed ferns, *Dicroidium*, early conifers and cycads.

From the Jurassic through to the early Cretaceous (213–100 million years BP), forests were dominated by conifers of the modern-day families Araucariaceae, which includes the kauri pine *Agathis*, and Podocarpaceae, which includes *Podocarpus*. *Ginkgo*, now restricted to China, was also widely distributed during this period. Dinosaurs, too, became established in the Australian part of Gondwana (see Box 41.6).

Following the Jurassic with its uniformly wet climate and forests of conifers, cycads and ferns, the vegetation changed. By the end of the Cretaceous and the beginning of the Tertiary (65 million years BP), dinosaurs had become extinct and the ancient conifer forests had contracted with the rise to dominance of flowering plants. By this time, the break-up of Gondwana was well underway.

The earliest record of flowering plants in Australia is fossil pollen, of uncertain modern affinity, from Cretaceous sediments. Pollen identifiable as *Nothofagus* and the family Proteaceae is up to 80 million years old. Similar pollen is recorded from New Zealand. As flowering plants diversified, so did the early marsupials, probably feeding on flowers, fruits and insects attracted to flowers.

With the increasing isolation of Australia from Antarctica in the Tertiary, the stage was set for the evolution of a unique Australian flora and fauna.

Glossopteris leaf fossil

By the middle Tertiary, weathered land surfaces were mainly **laterite**, still evident today where ironstone gravel or boulders occur at the surface or subsurface overlying a layer of kaolinitic (bleached white) clay (Fig. 41.6). Where weathering of laterite subsequently occurred, sediments of low fertility were blown over the continent, adding to sandy soils derived from rocks, such as sandstones and coarse granites.

In Miocene times, there were huge inland lakes where deserts now occur. Lake Eyre and Lake Frome basins once contained freshwater and provided environments for crocodiles, flamingoes and plants typical of warm and humid conditions. When these areas became arid, winds eroded and transported sediments to form extensive desert dune systems, and the aquatic life became extinct. Evaporation of lakes left saline mudflats. Seas, which had invaded low-lying areas across southern Australia, retreated, exposing extensive deposits of limestone formed from

Fig. 41.6 Laterite is an ancient, weathered land surface, evident in this outcrop, eroded on the edges and surrounded by younger sand plains, in Western Australia

shelled invertebrates. The Nullarbor Plain and Murray Basin, previously submerged, were exposed as dry land.

Climatic changes through the Tertiary altered landforms and soil, which in turn influenced the history of Australia's plants and animals.

> During the early Tertiary (65–40 million years BP), the climate of Australia and adjacent Antarctica was humid temperate, and rainforest covered much of the land. Separation of land masses created a circum Antarctic current, causing rainfall to decrease over parts of Australia, air temperatures to decrease, and conditions to become more arid.

Fire

Fire is an integral part of the present Australian environment (Fig. 41.7). Its history is recorded by charcoal particles preserved in sedimentary deposits. A combination of preserved charcoal and pollen reveals past vegetation–fire relationships.

Fig. 41.7 Fire is an important part of the present and past Australian environment

There is evidence of fire throughout the period of evolution of the Australian biota, but its frequency and effects on vegetation have varied. In the early part of the Tertiary, when rainfall was high, occasional fires were caused by volcanic activity and lightning during dry periods. With the dominance of rainforest and rarity of fire-promoting plants, such as eucalypts, fire did not cause large-scale destruction of vegetation. However, with the development of a drier and more variable climate towards the end of the Tertiary period, fires became more frequent and rainforest was replaced by more fire-tolerant and arid-adapted open forests.

Ice ages

About 2.5 million years ago (around the Pliocene–Pleistocene boundary), Australia, for the first time, came under the influence of westerly winds with wet winters and hot dry summers. This influence also coincided with the beginning of climatic fluctuations that were being experienced worldwide. Throughout Australia during the Quaternary period cool 'glacial' periods were associated with aridity, and warmer 'interglacial' periods, such as that existing today (the Holocene), were associated with higher rainfall. Wetter conditions than today occurred at times of slightly lower temperatures when evaporation was reduced.

Australia did not experience major expansions of ice, covering large areas of continents at high latitudes, as in the Northern Hemisphere, but some glaciation did occur in the Snowy Mountains and Central Highlands of Tasmania. Characteristic of the glacial periods in Australia, dry and windy conditions led to the movement of sand. A lack of continuous vegetation cover facilitated sand movement, and sand dune systems developed well beyond the present arid centre. During certain wet phases, large lakes formed over what are now small salt pans, and river discharge and erosion were much higher.

Variability in climate and associated landscape has prevented the development of continuous sedimentary sequences, which provide palaeobiologists with a full record of environmental changes; however, a good chronology for the Quaternary is preserved in ocean sediments. An excellent example showing the number, amplitude and increasing magnitude of climatic cycles through this period is illustrated in Figure 41.8.

The climatic fluctuations of the Quaternary eventually restricted rainforest to patches not much larger than those surviving today. During interglacial periods, forests were dominated by she-oak (*Casuarina*) and during glacial periods, herbaceous plants such as daisies (composites) were conspicuous.

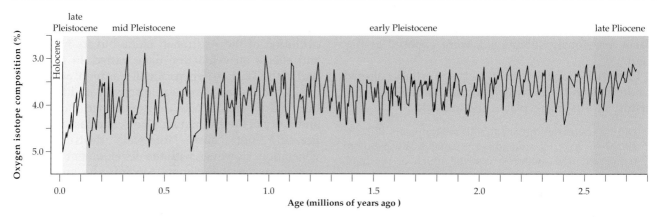

Fig. 41.8 Changes in the oxygen isotope composition of shells preserved in a sediment core from the north Atlantic Ocean, illustrating fluctuating environmental conditions during the last 2.8 million years. High values of oxygen-18 reflect low temperatures and extensive ice-cap development. Fluctuations increase in magnitude close to the Pliocene–Pleistocene boundary and increase further around the early to mid Pleistocene boundary, although there is a decrease in amplitude here

This general pattern persisted until the last glacial–interglacial cycle, when it is probable that increased burning due to the activities of Aboriginal people caused a substantial replacement of casuarinas and composites by eucalypts and grasses, as well as reducing rainforest to localised fire shadows. The present vegetation, dominated by eucalypts, achieved its present expansion within the last 100 000 years.

> With the development of a drier climate, fires became more frequent towards the end of the Tertiary. During the last 2.5 million years (the Quaternary) cool, arid glacial periods alternated with warm, wetter interglacial periods.

Arrival of humans

There is a great deal of debate over the first humans in Australia, their time of arrival and the impact that they had on the environment. Examination of fossil skulls shows that there were a number of different immigrant groups, most likely derived from both Indonesia and China. Some ocean travel would have been required, although passage to Australia would have been easier from Indonesia when sea levels were much lower during the last glacial period. At this time, New Guinea was connected to Australia. It is generally accepted that the first colonisation occurred at least 40 000 years ago, but some archaeologists believe that it may have been much earlier. A major problem in determining the time of arrival of humans has been that of obtaining absolute dates on archaeological sites beyond 35 000–40 000 years ago—the effective limit of radiocarbon dating. However, recent studies in northern Australia, using an alternative dating method, thermoluminescence, suggest that humans arrived more than 50 000 years ago.

Some indirect evidence for the long residence time of humans is provided by increased levels of charcoal within pollen records, which may be attributed to burning by humans. These increases date from about 38 000 years ago at Lynch's Crater in north-east Queensland and perhaps as early as 128 000 years ago at Lake George near Canberra.

Changes in the pollen record measure the way in which, and degree to which, burning by humans altered the landscape. Around Lynch's Crater, for example (Fig. 41.9), rainforest, which dominated during wetter interglacial periods, was generally replaced by drier rainforest with some fire-tolerant vegetation during glacial periods. Drier rainforest, associated with increase in charcoal during the last glacial period, was totally replaced by fire-tolerant vegetation. Drier rainforest exists now only as small isolated patches. Some rainforest plants became extinct in this region, including the conifer *Lagerostrobus*.

The extent of environmental change is best illustrated by the number of large animals that became extinct during the period in which people have been in Australia (see also Box 41.7). Most extinctions occurred between 35 000 and 15 000 years ago, at a time of driest conditions during the last glacial period. Although climate has been considered to be the main cause of these extinctions, there is no evidence to suggest that climatic conditions at this time were more extreme than during the previous glacial phases. However, increased environmental instability resulting from human burning and associated vegetation changes may have sufficiently altered stream flow and lake levels to produce a more drought-prone environment.

A generalised picture summarising the development of the present vegetation and environment in south-eastern Australia is shown in Figure 41.10.

> Humans colonised Australia at least 40 000 years ago, and caused increased burning.

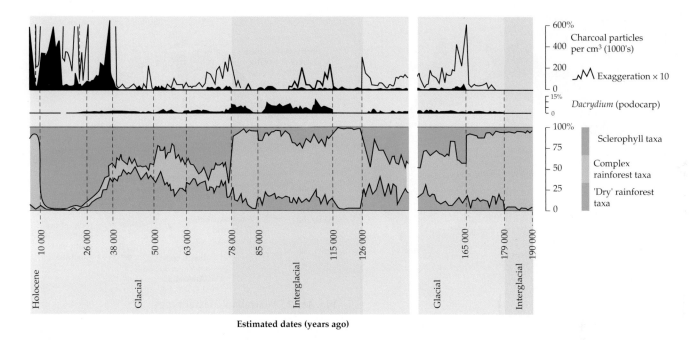

Fig. 41.9 Changes in abundance of pollen and charcoal from Lynch's Crater in Queensland illustrate vegetation responses to climatic fluctuations in the late Quaternary and changes associated with increased burning

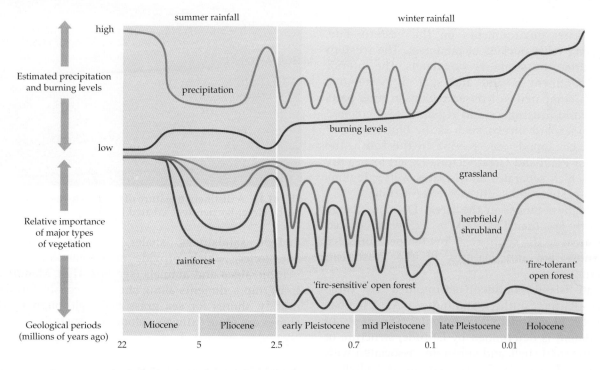

Fig. 41.10 Summary of vegetation and environmental changes through the last 14 million years in south-eastern Australia

MODERN AUSTRALIAN ENVIRONMENTS

Terrestrial environments of modern Australia are varied. The continent spans the latitudinal belt 10°–44°S, and climate varies with latitude from tropical (in the north), subtropical, warm temperate to cool temperate (in the south). The seasonal distribution and effectiveness of rainfall determine the type of plant and animal communities; the monsoonal climate of the north, with summer maximum rainfall (the 'wet' season), contrasts with the Mediterranean climate of the south, with winter maximum rainfall (Fig. 41.11).

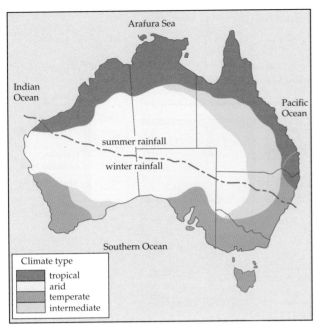

Fig. 41.11 Australian climatic regions and major landforms

Fig. 41.12 Australian coastal types

Fig. 41.13 A distinctive feature of the north-east coast is the Great Barrier Reef, created by living corals

The Great Dividing Range slices Australia into two unequal parts from east to west. It is a watershed, with moister environments on the eastern side, supporting relict pockets of rainforest. The western side is drier and supports woodlands. The interior of the continent is arid and sparsely vegetated, although, compared with desert regions in other parts of the world, Australia's arid region is well covered with grasses and shrubs, such as the blue bush and saltbush chenopods (see p. 916). In the south, the Nullarbor Plain and Great Victoria Desert form a dry region, which separates and isolates wetter regions of south-eastern and south-western Australia.

The highest ranges in the Great Dividing Range are mostly less than 1500–1800 m. Except in isolated alpine areas in the south-east, low temperatures, ice and snow are not features of the Australian environment.

As an island continent surrounded by three oceans—Pacific, Southern and Indian—Australia has a variety of marine environments. The type of environment is determined by latitude, structure of the continental shelf, and substrates associated with the sea bed (Fig. 41.12). The continental shelf is continuous around Australia, varying in width from 15 km to 400 km, with its outer limit at a depth of about 150 m. The adjacent continental slope has an incline up to 40° and plunges to abyssal depths greater than 4 km. The Great Barrier Reef, created by living corals and coralline algae, is a distinctive feature of the north-east coast and extends over 2000 km (Fig. 41.13). With many lagoons and cays (small islands), the reef has a range of different environments for marine organisms. Rocky shores in the south and east include sandstones, shales, volcanic rocks and limestones, which provide anchorage for marine benthic algae and animals (see Box 41.2). Mangrove trees trap sediments along sheltered coasts, forming muddy shores for sea grass beds, although sandy shores are more extensive along the Australian coast.

AUSTRALIAN FLORA

The modern Australian terrestrial flora is strikingly different from that of other continents and islands. Numerous species, genera and families are endemic to Australia, while others have their centre of diversity in Australia but are also represented in adjacent areas. The distinctive appearance of most Australian vegetation is due to the dominance of *Eucalyptus* and *Acacia* trees and shrubs over about 70% of the continent. Gymnospermous trees such as conifers are

BOX 41.2 MARINE FORESTS

The impact of Australia's geologic history on the diversity of its biota is usually identified with larger, more obvious land-based plants and animals. However, the less conspicuous seaweeds and sea grasses that form complex intertidal and submarine communities along the region's coasts are equally novel and diverse, and constitute a marine flora that is one of the richest in the world.

For example, the red seaweeds (phylum Rhodophyta) from southern Australia include more than 1100 recorded species, with approximately 70% being endemic. Because large areas of the Australian coastline are still unexplored, particularly in the north and north-west of the continent and in Tasmania, the diversity and endemism of Australian red seaweeds will probably prove to be greater than in any other area of similar size in the world.

In addition to the Rhodophyta, there are two other major phyla of benthic seaweeds that populate most marine habitats. The largest of all the algae, and most abundant in terms of biomass, are the brown algae (phylum Phaeophyta). Included within this group are the true kelps *Ecklonia* and *Macrocystis* (Fig. a), the bull kelp *Durvillaea*, and the large fucoids, such as *Hormosira*, *Xiphora* and *Cystophora*. All genera are conspicuous on rocky, temperate coastlines in southern Australia, Tasmania and New Zealand. Harvested commercially in a number of locations around the world, including King Island (in Bass Strait) and New Zealand, brown algae provide a range of products, such as gelling agents, fertiliser and food for both livestock and human consumption.

The highly diverse green algae (phylum Chlorophyta) are abundant in most marine habitats, but are particularly obvious in tropical regions, including the Great Barrier Reef, where *Chlorodesmis* (turtle grass),

Caulerpa and the calcified genus *Halimeda* dominate. Although *Caulerpa*, *Codium*, *Ulva* and other genera are used for food in Asian countries and some Pacific islands, green seaweeds are of relatively little economic use and are not commercially exploited in Australia.

Red algae are generally less conspicuous in size and overall biomass than other seaweeds. However, like brown algae, the gelling and emulsifying properties of the 'slimes' (polysaccharides) extracted from their cell walls provide products of considerable commercial value. Economically important genera, such as *Gelidium*, *Pterocladia*, *Gracilaria* and *Gigartina*, often attach to rocks in the intertidal zone. Red algae have generally been harvested by hand, thereby restricting this activity to countries such as Chile, the Philippines and Indonesia where labour costs are low. However, due to a recent increase in both demand and price of red algal agars and carrageens, these organisms are now intensively studied by biotechnology companies interested in generating genetically improved strains for mass culture.

Seaweed populations are integral parts of the coastal environment and, as primary producers, help to form the basis of the marine food web (Chapter 44). The importance of marine algae to total primary productivity is significant, with estimates suggesting that they account for up to 25–30% of global production.

In addition to algae, marine flowering plants, which grow submerged on sheltered and soft-bottomed coasts, contribute to primary productivity. Marine flowering plants are monocotyledons, commonly called sea grasses (Fig. b) because of their ribbon or strap-like leaves arising from nodes on a creeping rhizome. There are 31 species in Australasian waters, including five species of eelgrasses (*Zostera*) and eight species of strapweeds (*Posidonia*).

(a) The kelp, *Macrocystis*

(b) Australian sea grass beds are important primary producers, such as for this dugong

far less common, although there are a number of distinctive endemic Australian taxa, such as *Athrotaxis* (King Billy pine) in Tasmania (see Chapter 37). The she-oak family (Casuarinaceae, Fig. 41.14) is ubiquitous in Australia, with representatives also in

tropical regions of Malaysia. The 'grass-trees', *Kingia* (Fig. 41.15) and *Xanthorrhoea* (family Xanthorrhoeaceae, related to lilies) are unlike any other monocotyledons; of 75 known species in the Australasian region, 70 occur in Australia. Endemic

Fig. 41.14 The she-oaks, *Casuarina* and *Allocasuarina*, have leaves reduced to scales or toothed sheaths, needle-like photosynthetic stems and small flowers clustered in cone-like heads

Fig. 41.15 The grass tree *Kingia* from Western Australia has a woody trunk, linear leaves, and flowers clustered in globular heads. The related *Xanthorrhoea* was used by Aborigines as a source of resin, used as an adhesive and to make torches to set fire to vegetation to catch game

Fig. 41.16 The pungent-leaved hummock grasses *Triodia* and *Plectrachne* are endemic to Australia and have probably undergone fairly recent speciation, occurring in vast areas of the arid and semiarid zones

hummock grasses, such as *Triodia* and *Plectrachne* (Fig. 41.16), form a characteristic ground cover over vast areas of arid and semiarid Australia.

The major components of the Australian flora have a Gondwanan origin. Today, a relict component is represented by plants confined to moist rainforest habitats. However, the major component has diversified since the middle of the Tertiary, predominantly in temperate and more arid environments.

Dominance of sclerophylls

Why do eucalypts, wattles, banksias, casuarinas, grass trees, heaths, native peas, hummock grasses and other familiar 'bush flowers' dominate the modern Australian flora? These plants possess features that allow them to survive low soil nutrients, water stress and fire. All are **sclerophyllous**, characterised by rigid, often small leaves, short internodes, and often small plant size.

In 1916, E. C. Andrews, and later N. C. W. Beadle in 1954, developed the idea that sclerophylly was primarily an adaptation to infertile soils, even where rainfall was relatively high, since sandy soils dry out and can cause water stress in summer. Sclerophylly is well developed in banksias (family Proteaceae) and the southern heaths (family Epacridaceae), which were present in the early Tertiary when rainfall was high and fires infrequent. Beadle and other workers suggested that sclerophyll plants formed wet heathlands on the margins of rainforest where soil fertility was low. Heathlands occur today on the most extremely infertile soils, in both wet and dry climates.

Beadle suggested that low levels of soil phosphorus were particularly important in limiting the growth of rainforest species and favouring sclerophylls. He argued that low phosphorus levels affect physiological processes and lead to reduction in the number of cells formed and length of stem internodes. Such changes result in smaller leaf sizes and smaller plants, which would have been able to survive low and variable rainfall as aridity increased, and also droughts of cool montane environments during the onset of the ice ages.

Sclerophylls of the modern flora are characterised by a number of features in addition to small leaves and short internodes (Fig. 41.17): thick cuticle on the leaf epidermis, which may over-arch guard cells; stomates sunken in grooves or pits, or protected by a dense covering of hairs; high proportion of lignified cells such as fibres; and thick-walled hypodermal cells between the epidermis and photosynthetic mesophyll.

(a)

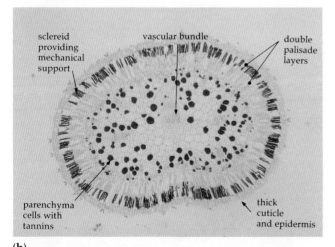

(b)

Fig. 41.17 **(a)** *Hakea* (Proteaceae) has sclerophyll leaves, one of which is **(b)** shown in cross-section. Note the thick cuticle, sunken stomates, distribution of photosynthetic tissue and cells with thick walls

With increased frequency of fire towards the end of the Tertiary (p. 909), sclerophylls became even more widespread. Many of their characteristic features promoted their fire resistance.

Modern species (e.g. eucalypts) survive fire by having thick, insulating bark, at least on the lower half of the main trunk. If the shoot of a tree or shrub is damaged, resprouting after fire may take place from dormant buds on the trunk, or from a mass of dormant buds that form swellings, **lignotubers**, at the base of the stem (Fig. 41.18). Lignotubers are well developed in mallee eucalypts, but resprouting ability is absent from species of wet forests such as mountain ash, *Eucalyptus regnans*. Grass trees, *Xanthorrhoea*, are stimulated by fire to flower and produce seed.

Fig. 41.18 Many woody Australian sclerophylls, such as this eucalypt, resprout from dormant buds beneath the bark or in lignotubers

Some plants are killed by fire but reproduce from seed that is protected in hard, woody cones (cypress pines) or fruits, such as follicles (hakeas, banksias) and capsules (eucalypts, tea-trees). Seeds of peas and wattles have a hard seed coat, which is cracked by the heat of fire; fast-growing wattles are often among the first woody plants to regenerate after fire.

Many of these characteristics that favour plant survival after fire may have evolved for other reasons. The ability to resprout, for example, may have evolved in response to grazing or drought, while it has been suggested that hard seed coats and woody fruits were a response to low nutrients.

> Sclerophylly was primarily an adaptation to soils of low fertility, but features of sclerophyll plants allowed them to survive water stress and fire.

Succulent survivors

The Australian arid flora also includes a number of **succulents**, which like sclerophylls are **xerophytes** (plants that live in dry conditions in contrast to **mesophytes**, which live where water is in adequate supply). Succulents survive harsh conditions by having fleshy leaves or stems and a highly mucilaginous cell sap. The outer epidermis is often heavily cutinised and stomates may be sunken, but plants are not lignified. Examples of Australian succulents are stoneflowers and pigface (family Aizoaceae), parakelias (family Portulacaceae), and saltbushes (Fig. 41.19) and samphires (family Chenopodiaceae), some of which are C_4 plants (Chapter 5). Succulents occur in dry or salty areas, either inland or on the coast, although they are rarer in Australia than in other parts of the world (e.g. North and South America, where the cactus family occurs).

Fig. 41.19 *Halosarcia* is a succulent plant that grows near Oodnadatta in South Australia

Characteristics of Australian flowering plants

Myrtaceae: the eucalypt family

Australian eucalypts (*Angophora* and *Eucalyptus*), tea-trees (*Leptospermum*), paperbarks (*Melaleuca*), bottle-brushes (*Callistemon*) and lilly pilly (*Acmena*) are five examples of the family Myrtaceae. The distribution of the family is primarily in the Southern Hemisphere (Fig. 41.20a), with nearly 50% of genera occurring naturally in Australia.

Myrtaceae have leaves with oil glands, flowers with four or five perianth parts, and numerous stamens borne above the inferior ovary (Fig. 41.21). Dry fruited

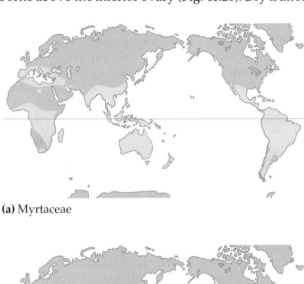

(a) Myrtaceae

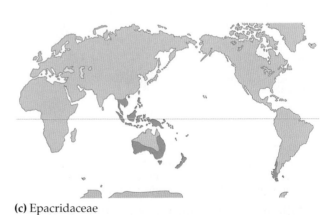

(b) Proteaceae

(c) Epacridaceae

Fig. 41.20 Distributions of the families **(a)** Myrtaceae, **(b)** Proteaceae and **(c)** Epacridaceae

forms are most numerous and widespread in drier parts of Australia. Fleshy fruited forms occur more in wetter tropical habitats, such as rainforests. Many species are used in ornamental horticulture: brush box (*Lophostemon confertus*) is a commonly planted street tree; and cloves (*Syzygium aromaticum*) and allspice (*Pimenta dioica*) provide spices. The most familiar Australian group, the eucalypts, are fast-growing trees planted in many parts of the world for timber, paper pulp, firewood and aromatic oils (Box 41.3).

Fig. 41.21 *Leptospermum laevigatum* is typical of the family Myrtaceae

BOX 41.3   EUCALYPTS: NATIVE AUSTRALIANS

To overseas visitors, eucalypts are as 'Australian' as koalas and kangaroos. Eucalypts dominate Australian landscapes, except rainforests and the arid interior. They form tall forests in high rainfall areas (mountain ash, *Eucalyptus regnans*, in the south-east and jarrah, *E. marginata*, in the south-west), low forests and woodlands in drier regions (dominated by gums, boxes, peppermints, stringybarks or ironbarks), mallee shrublands in semiarid areas, and subalpine forest and woodlands (snow gums, *E. pauciflora* and *E. coccifera*) at higher altitudes, up to 2000 m. A few species occur on islands to the north of Australia.

Despite the present-day dominance of eucalypts, little is known of their evolutionary history from the fossil record. Fossil eucalypt pollen is recorded for the Oligocene, and there are younger records of leaves and fruits. There is also evidence of fossil eucalypt pollen from New Zealand, although no extant species occurs there naturally. It is thought, however, that *Eucalyptus* is an ancient Australian group that has increased in dominance relatively recently as the climate developed a marked dry season, and fire became more frequent.

Eucalyptus L'Hér. is a large genus of about 700 species. It was named in 1788 by Charles Louis L'Héritier de Brutelle, a French botanist working in London at the time. L'Héritier published a description of *E. obliqua* (messmate), based on specimens collected in 1777 from Bruny Island, Tasmania, on Cook's third voyage to Australia.

Eucalyptus flowers lack showy petals and sepals, but develop either one or two opercula—caps that enclose the stamens and style of the flower (Fig. a). These opercula are homologous to sepals and petals that have become continuous during floral development. In contrast, the closely related genus *Angophora*, which grows in eastern Australia, has flowers with creamy free petals and green free sepals. Variation in the development and number of opercula, together with variation in ovules, seeds, fruits, leaf hairs and other characters, has led botanists to recognise a number of subgroups (usually treated as subgenera) within *Eucalyptus*.

One feature of eucalypts is their development of different types of foliage as they grow and mature (Fig. b). Young plants have characteristic juvenile foliage, with leaves often held horizontally, sessile and in opposite pairs; saplings develop intermediate leaves, and mature plants have petiolate leaves, which hang in the vertical plane, reducing heating and water loss.

(a) *Eucalyptus pulverulenta* has two operculum; the outer one is the small brown cap being pushed off the flower bud by growth of the inner operculum

(b) A sapling of *Eucalyptus nitens* (shining gum) showing different forms of leaves, from juvenile to adult

Proteaceae: the banksia family

Characteristic of Australian vegetation, the family Proteaceae, particularly subfamily Grevilleoideae, includes many genera endemic to Australia. Well-known Australian genera are *Grevillea*, *Telopea* (waratahs), *Macadamia* and, perhaps the most distinctive of all, *Banksia* (see Fig. 30.13). There are 92 species of *Banksia*, with 61 in south-western Western Australia. Proteaceae are a Gondwanan group, occurring in South Africa, India, South-East Asia and South America, as well as Australia (Fig. 41.20b). The family has a long fossil record in the Southern Hemisphere, including Antarctica. Pollen and macrofossils of *Banksia* have been found, for example, in Victorian and South Australian Oligocene deposits. The greatest number of Australian species occurs in heaths and sclerophyll forests, with primitive genera in rainforests of north-eastern Queensland.

Proteaceae are characterised by flowers with a four-lobed perianth (tepals), four stamens often attached to the tepals, and a one- or two-celled ovary (Fig. 41.22). They are sometimes spectacular when flowering because their flowers cluster in brightly coloured inflorescences and attract bird pollinators (Chapter 37). *Banksia* flowers cluster in dense cylindrical or globular heads.

Epacridaceae: the southern heath family

Epacridaceae are commonly called southern or Australian 'heaths' because of their similarity to Northern Hemisphere heaths and heathers of the related family Ericaceae. Epacridaceae are largely Australian, but with species in New Guinea, Indo-China, Malaysia, Indonesia, the Philippines, New Zealand, New Caledonia, Pacific islands and South America (Fig. 41.20c). Epacrids are shrubs or small trees with stiff, hard leaves, distinctive leaf venation (palmate or sub-parallel), and tube-like flowers (Fig. 41.23). Epacrids are common on infertile soils and, like the Myrtaceae, they typically have root mycorrhizae (Chapter 36).

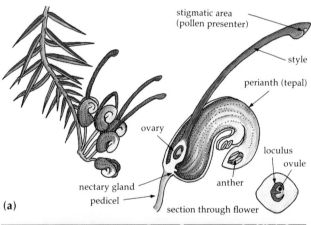

(a)

(labels: stigmatic area (pollen presenter), style, perianth (tepal), ovary, loculus, ovule, nectary gland, pedicel, anther, section through flower)

(b)

Fig. 41.22 (a) Typical flower of family Proteaceae.
  (b) *Grevillea longistyla*

Fig. 41.23 *Epacris impressa*, typical of the southern heath family Epacridaceae, and floral emblem of Victoria

Mimosaceae: the wattle family

Mimosaceae (one of three families of legumes) are best known in Australia for *Acacia*, the wattles. There are about 835 Australian species, making *Acacia* the largest genus represented on the continent. *Acacia* is also well represented in Africa and tropical America. It is known in the Australian fossil record only from the early Miocene, but it is believed to have had a longer Tertiary history.

Wattles are trees or shrubs with foliage of either compound bipinnate leaves or **phyllodes**. Bipinnate ('twice-divided') leaves consist of small leaflets (pinnules) arising from pinnae arranged along a central axis, the rachis. Phyllodes are laterally compressed or flattened petioles and rachis (Fig. 41.24a). Usually longitudinal veins (nerves) are visible on their surface, and glands may occur along the edges. Phyllodinous acacias (Fig. 41.24b) are almost entirely Australian (and are treated as a separate genus by some taxonomists). They are particularly dominant in the woody vegetation of arid central Australia, for example, mulga, *A. aneura*, although only a small number (118) of *Acacia* species occur there. Phyllodes are a scleromorphic feature, and in arid-zone species they are often small in size, sometimes reduced to spines, with stomates sunken in pits.

Acacia flowers are small, clustered in globular heads or spikes. The familiar golden-yellow colour of the flower heads is due to the stamens, which are often numerous. The fruit is a leathery or woody legume (a pod).

Many wattles are grown as ornamentals and some, such as blackwood, *Acacia melanoxylon*, are prized for timber. Seeds and seed pods were roasted and eaten by Aborigines, who also used the bark of some species as a poison to catch fish. Wattles are ecologically important because they increase the nitrogen content of soil through the activity of symbiotic species of *Rhizobium*, which live in nodules on *Acacia* roots (Box 44.5).

Fabaceae: the pea family

The pea family is found throughout the world, but 1100 species in a large number of genera occur in Australia, especially in the tropics. They include trees, shrubs (e.g. the familiar 'eggs and bacon' plants, Fig. 41.25), herbs and climbers. Their leaves are often compound, commonly with three leaflets, as in the common clover. Sclerophyllous forms have leaves reduced to spines or small scales. Their butterfly-shaped flowers make peas easily recognisable. The flower has five petals, three of which are free and two of which are united. The standard is the posterior and usually largest petal, the wings are the two lateral petals, and the keel consists of the two anterior petals united along their lower margin to enclose 10 stamens and the ovary (Fig. 41.25). Pea seeds, like those of acacias, have hard seed coats, which are cracked by the heat of bush fires. Peas also have root nodules and fix nitrogen.

Asteraceae: the daisy family

The daisy family (Asteraceae, old name Compositae), is one of the largest angiosperm families in the world with more than 25 000 species. Some taxa in the family are confined to the Southern Hemisphere and, within Australia today, native species occur in habitats ranging from alpine herbfields, where silvery carpets of snow daisies of the genus *Celmisia* are conspicuous, to semiarid and arid regions, where there is a large number of species. The abundance of daisies relates to their high reproductive rate and 'plasticity' of form; many species are widely distributed weeds, such as the common ragwort, dandelions, thistles, capeweed and boneseed.

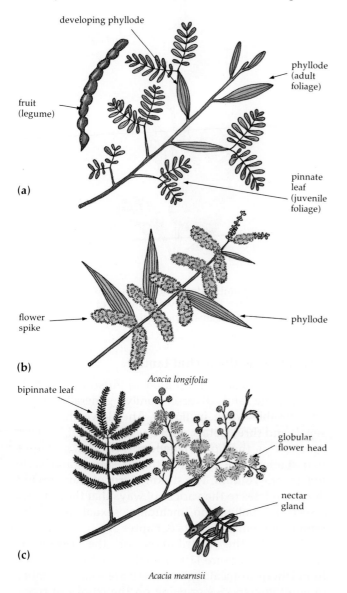

Fig. 41.24 **(a)** Phyllodinous acacias first develop bipinnate, compound leaves and then phyllodes. **(b)** *Acacia longifolia* is a phyllodinous acacia. Note that the flowers are typically clustered in spikes. **(c)** *Acacia mearnsii* is one Australian species that retains bipinnate foliage when mature. Note that its flowers are clustered in globular heads

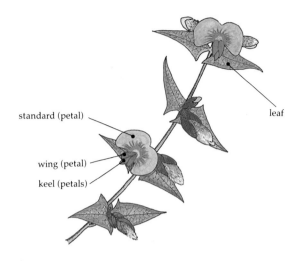

standard (petal)

wing (petal)

keel (petals)

leaf

(a)

(b)

(c)

Fig. 41.25 **(a)** Pea flowers have distinctive butterfly-shaped flowers, as shown for the showy parrot pea, *Dillwynia sericea*. **(b)** Australian 'eggs and bacon' flowers. **(c)** Sturt's desert pea, *Clianthus formosus*, floral emblem of South Australia

Daisies are easily recognised by their flowers, which are clustered together in an inflorescence called a head (or capitulum). Individual flowers are, in fact, small florets, which are of two types. **Ligulate** (ray) florets on the outside of the head have five united petals strongly developed to one side, forming the showy rays. These flowers are often unisexual. **Tubular** (disc) florets form towards the centre of the daisy head (Fig. 41.26). They usually have both fertile stamens and an ovary. The inflorescence is surrounded by one or more rows of **involucral bracts**, the arrangement, shape and texture of which are important characters for identification.

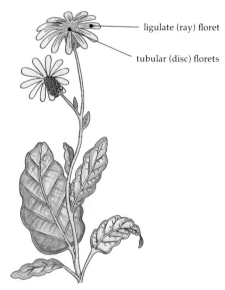

ligulate (ray) floret

tubular (disc) florets

Fig. 41.26 The familiar daisy 'flower' is a cluster of small flowers (florets), the showy outer ones of which often function to attract insect pollinators

Orchidaceae: the orchid family

Of the monocots represented in Australia, orchids are a large and diverse family, generally easily recognisable when in flower. The flower has three sepals and three petals, with one of the petals usually modified as the lip or **labellum** (Fig. 41.27). One or two stamens are united with the stigma and style to form the **column**. The striking form of many orchid flowers relates to the variety of ways that they attract insects for pollination, including 'sexual deceit' by mimicking female insects (Chapter 37).

Although the distribution of the family is worldwide, there are 1100 orchid species in Australia. In northern tropical regions there are a large number of epiphytic species growing on the trunks of trees; one example is the cooktown orchid, *Dendrobium phalaenopsis*, which is the floral emblem of Queensland. Epiphytes and **lithophytes**, which grow among rocks, are xeromorphic, often having stems that are swollen into fleshy pseudobulbs, which appear to function as water-storage organs; some are also C_4 plants (Chapter 5). In the southern more

(a)   **(b)**   **(c)**

Fig. 41.27 **(a)** The orchid flower is recognised by one of the petals, the labellum, which is usually very different in shape, size and markings, and the column (fusion of stamens, style and stigma). Orchid flowers usually twist as they develop to present the labellum petal as a landing platform for insects. **(b)** The Cooktown orchid, *Dendrobium phalaenopsis*, the floral emblem of Queensland, and **(c)** the terrestrial greenhood, *Pterostylis revoluta*, are two Australian examples

temperate regions of Australia, many native orchids are **terrestrials**, the majority of which are of Gondwanan origin and related to African taxa. Terrestrials include greenhoods, *Pterostylis* (Fig. 41.27c), sun orchids, *Thelymitra*, and donkey orchids, *Diuris*; some have root–stem tuberoids, and survive dry periods by retreating beneath the ground.

Today, the relict Gondwanan component of the Australian flora is represented by plants confined to moist rainforest habitats. The major component has diversified predominantly in temperate and more arid environments and includes, among the flowering plants, *Eucalyptus*, *Acacia*, *Banksia*, heaths, peas, daisies and orchids.

BOX 41.4 NORTHERN PLANTS

Some plant groups represented in Australia have distribution patterns that suggest origins elsewhere than in Australia. For example, about 120 tropical and subtropical rainforest genera of dicotyledons are each represented in Australia by a single species and in Asia by several to many. There are 250 species of *Cinnamomum* (family Lauraceae) in Asia and New Guinea and only four in Australia. There are 850 species of *Rhododendron* (family Ericaceae), many of which occur in the Himalayas, South-East Asia and Malaysia; 155 species are endemic in New Guinea, but only one occurs in Australia. Such patterns are sometimes interpreted as plant migration into Australia from the north, an explanation requiring dispersal of seeds across water barriers. Alternatively, some plants of New Guinea and Australia may have dispersed northwards as the Australian plate made contact with South-East Asia. This is one explanation for the occurrence of a single species

of eucalypt, *Eucalyptus deglupta*, on Mindanao in the Philippines.

Some plants are cosmopolitan. For example, there are many species of *Epilobium* (fuschia family, Onagraceae) in western North America, on arctic and tropical mountains, as well as in Australia. *Epilobium* seeds have a tuft of hairs, making them buoyant in air and easily dispersed; some species have spread as weeds after habitat disturbance this century. Of the cosmopolitan grass family Poaceae, the swamp reed, *Phragmites australis*, is probably the most widespread species of flowering plant in the world. In fact, many widespread species are freshwater aquatics.

Humans are responsible for recent introductions of a variety of plants into Australia. Familiar examples are *Pinus radiata*, *Lantana* and *Opuntia* (prickly pear), deliberately introduced for cultivation, as ornamentals or hedges.

AUSTRALIAN FAUNA

Like Australian plants, much of the modern terrestrial fauna is markedly different from that found elsewhere in the world. Also, it includes many animal groups that show Gondwanan patterns of relationships (Box 41.5) as well as groups with relationships to Asian and Northern Hemisphere groups.

Biogeographic patterns of terrestrial vertebrates

For terrestrial animals, major waterways provide a substantial barrier to dispersal, so that islands such as Australia and New Zealand often have many unique or endemic species, which commonly include surviving groups that have long since become extinct in the rest of the world. The tuatara (Chapter 40) of New Zealand is an example of island survival of an ancient reptilian group for perhaps as much as 160 million years. During the break-up of Gondwana many such 'islands' of differing ages were created and the biota of these continental fragments often provide important insights into past biotic assemblages. For example, the native vertebrate fauna of New Zealand includes only three species of very primitive frogs, some skinks and geckos as well as the tuatara, no snakes, no terrestrial mammals (except for two species of bat) and a limited bird fauna, although the birds include an ancient group of ratites—the flightless kiwis (Fig. 41.28)—and the recently extinct moas.

BOX 41.5   INSECTS WITH SOUTHERN CONNECTIONS

Insects have a long evolutionary history, being known in the fossil record for around 350 million years. The fossil record in Australia stretches from the upper Carboniferous (300 million years ago) and includes over 400 different species. The modern insect fauna of Australia is represented by over 660 families and around 86 000 species. General patterns of relationships suggest that the primitive members of many groups are descendants of the Gondwanan insect fauna, while the more modern highly evolved members of each group have affinities with Asia. Orders of Australian insects that have clear Gondwanan elements include mayflies (Ephemeroptera, Fig. a); dragonflies and damselflies (Odonata); stoneflies (Plecoptera); termites (Isoptera); bugs (Hemiptera); beetles (Coleoptera); scorpion flies (Mecoptera); flies (Diptera); caddis flies (Trichoptera); moths and butterflies (Lepidoptera); and wasps, bees, ants and sawflies (Hymenoptera).

Many of the ancient groups maintain associations with early plants, which presumably represent coevolved relationships (see Chapter 43) that have existed throughout most of the evolutionary history of herbivore and plant. Examples of such associations in Australia include beetles of the subfamily Paracucujinae, which feed as adults and larvae entirely on pollen of cycads (Cycadaceae); larvae of the primitive moth family Agathiphagidae, which feed on seeds of *Agathis* (Araucariaceae); cynipoid wasps (subfamily Austrocynipinae), which are associated with seeds of *Araucaria*; and primitive weevils (family Nemonychidae), which feed on pollen of Araucariaceae and Podocarpaceae. Larvae of an as-yet-unnamed genus of Australian leaf miners feed on *Banksia* (Fig. b). Relatives of these insects in Africa feed on proteas, relatives of banksias in the family Proteaceae.

Murphyella
South America

Coloburiscoides
Australia

Coloburiscus
New Zealand

(a) The evolutionary relationships of three genera of mayflies show a geographic pattern that is consistent with our knowledge of the historical sequence of the break-up of Gondwana. The South American genus is more closely related to the Australian genus than it is to the older New Zealand taxon. This pattern is seen in other groups of mayflies as well as other groups of animals and plants

(b) Australian leaf miners are host-specific, feeding only on banksias

Fig. 41.28 The flightless kiwi, endemic to New Zealand

Three of the more important island land masses throughout much of the Tertiary were Antarctica, Australia and South America. Decreasing temperatures and development of the southern polar ice sheet made Antarctica an unsuitable place for all but a few animal groups. In South America and Australia, however, highly regionalised faunas developed, essentially in isolation from the rest of the world. For South America, this isolation came to an end with the formation of the Panamanian isthmus some six million years ago. The first effect of this land connection was a mixing of northern and southern faunas and a marked faunal enrichment in both North and South America. Inevitably, the interaction between many ecologically similar species led to large-scale extinctions. Now, many of the animal groups we associate with South America, such as those represented by llamas, jaguars and peccaries, are really recent immigrants from North America.

Australia, since rifting from Antarctica around 50 million years ago, has maintained much of its insularity together with much of its unique fauna, although new groups have reached the Australian plate since it has come in contact with South-East Asia.

While much of the fauna shows similar biogeographic patterns of relationships (see, for example, Box 41.5), these patterns are best illustrated by two groups of vertebrates: the amphibians (order Anura) and mammals (class Mammalia), both of which evolved in the early to middle Mesozoic, around the time Gondwana began to break up. In general, these animals have relatively poor powers of dispersal over water, so that any global patterns of biogeographic relationships are less likely to be confounded by major dispersal events.

The modern Australian fauna is diverse and includes many unique and endemic groups, some of which evolved around the time of the break-up of Gondwana.

Adaptive radiation of frogs

Modern amphibians are represented by three orders, but only one, the Anura, which includes the frogs and toads, is found in Australia. The Anura includes two ancient families, two families derived more recently from Asian stocks and one family recently introduced by Europeans (the introduced cane toad, *Bufo marinus*, now posing one of the most significant threats to tropical Australian ecosystems).

Most of the fauna comprises leptodactylids (family Leptodactylidae) and hylids (family Hylidae), two families with Gondwanan affinities. Except for one genus, *Heleophryne*, in southern Africa, modern leptodactylids are restricted to South and Central America and Australia. *Indobatrachus*, a fossil frog from India, is thought to be related to living leptodactylids. Within Australia, leptodactylids include a variety of often bizarre species (Fig. 41.29) classified into 20 different genera.

Fig. 41.29 A leptodactyl frog, *Assa darlingtoni*, with tadpoles crawling to the male pouch

Adaptive radiation, that is, the rapid evolution of a large number of closely related species, within the leptodactylids is manifest in morphological, developmental, behavioural and ecological specialisations. For example, female frogs of the genus *Rheobatrachus* from southern Queensland brood their larvae in the stomach, which functions as a uterus during incubation of the young. When tadpoles complete their larval development froglets are 'born' through the mouth of the mother (Fig. 13.26). Another example of specialisation of breeding behaviour is shown by *Assa darlingtoni*, a frog of high cool rainforests of northern New South Wales and southern Queensland. In this species, the male has specialised lateral 'pockets' to protect developing tadpoles. The male sits among the egg mass that has been laid under the forest leaf litter and, when the eggs hatch, the tadpoles find their way into the male's pouches, where they complete their larval development. These examples are two of the more spectacular leptodactylid life-history patterns, but a trend towards increasing independence of aquatic environments for larval development is seen throughout the family.

The family Hylidae, like the leptodactylids, has two centres of diversity—Australia and South America—although the family does have a worldwide distribution. Within Australia, hylids are represented by three genera: *Cyclorana*, *Litoria* and *Nyctimystes*. Except for some specialisations for living in and around torrential streams, life-history patterns throughout the family all follow the primitive pattern of aquatic development from eggs through tadpole stages followed by metamorphosis to a terrestrial frog. Nevertheless, the family has been just as successful as the leptodactylids in utilising most available habitats within Australia.

The highest diversity of frogs is found in mesic and warmer areas of the continent but, given their dependence on the availability of water for survival and breeding, there is a surprising diversity of frogs in the two-thirds of Australia described as arid or semiarid (see Fig. 41.11). In these generally dry environments where rainfall is unpredictable, frogs avoid extreme conditions by burrowing into the soil (35% of Australian frog fauna burrow) or by seeking shelter under rocks or in cracks in the ground. Many species reduce water loss through the skin. Species of *Cyclorana*, the so-called water-holding frogs, develop a cocoon from shed skin, which, when hardened, becomes relatively impermeable to water and within which the frog survives underground for many months (Fig. 41.30). Most arid-zone species reduce the time spent as eggs and tadpoles, the most vulnerable stages of the life cycle. Whereas, some species from mesic environments spend more than a year as tadpoles, the successful species in dry areas commonly complete the aquatic stages of their life in less than one month.

The remaining two families of amphibians in Australia, the Microhylidae and Ranidae, are of Asian

origin, having entered Australia through New Guinea after collision between the Australian and Oriental crustal plates. Throughout the rest of the world, the Ranidae are a successful and diverse family but in Australia they are represented by a single species, *Rana daemeli*, restricted to north-east Queensland and Arnhem Land. Even though microhylids are more diverse, represented by two genera, *Cophixalus* and *Sphenophryne*, and 18 species, their distribution is still restricted to the tropical north of Australia, particularly north-east Queensland.

> The largest component of Australia's frog fauna are leptodactylids and hylids, two Gondwanan families that radiated on the continent and adapted to dry environments. Two other families of amphibians in northern Australia are of Asian origin.

Adaptive radiation in mammals

Australia is the only continent where all three subclasses of the Mammalia are represented: Prototheria—egg-laying monotremes; Metatheria—marsupials; and Eutheria—so-called placental mammals. The mammal fauna of modern Australia is extremely diverse and includes five orders of native terrestrial mammals as well as marine mammals (e.g. sea lions, seals, porpoises and whales) and a large variety of introduced species. Nearly all of the introduced mammals, many of which, like the rabbit, fox and house mouse, have become serious pest species, were deliberately or accidentally brought to Australia by Europeans. Even so, the first introduced species, the dingo, came with Aboriginal colonisers, and records of its presence in southern Australia date from around 7000 years ago.

Introduced mammals, including humans, have had a profound impact on Australian ecosystems and have led to the extinction of many native species. This issue is discussed further in Chapter 45, but one example is the interaction between the introduced dingo, *Canis familiaris dingo*, and its ecological equivalent, the Tasmanian tiger or marsupial wolf, *Thylacinus cynocephalus*.

Thylacinus cynocephalus (Fig. 41.31) was among the largest of the dasyuromorphs and thus the largest mammalian predator in Australia. This dog-like marsupial was the last descendant of a once quite diverse assemblage of dog-like predators (Fig. 41.32). Until the introduction of the dingo, *Thylacinus* was widely distributed across Australia (Fig. 41.31b), but subsequently it became extinct except on Tasmania, an area the dingo did not reach before the island became isolated from the mainland at the close of the last glacial period of the Pleistocene (around 13 000 years ago). It was not until Europeans colonised Australia

Fig. 41.30 *Cyclorana platycephalus*, the water-holding frog, is well adapted to withstanding long periods of drought. It burrows deep into sand and makes a cocoon from its cast-off skin, which it sheds after rain

(a)

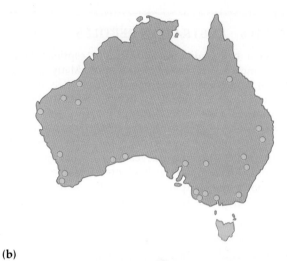

(b)

Fig. 41.31 **(a)** *Thylacinus cynocephalus*, the marsupial wolf. **(b)** Despite numerous reports of 'sightings' of thylacines on the mainland, the known range of *T. cynocephalus* at the time of European settlement was restricted to Tasmania. However, the localities on the map show sites where late Quaternary and Recent fossils, mummified remains or Aboriginal rock art, have been recorded. These indicate that the marsupial wolf was once widely distributed on mainland Australia, as well as New Guinea

Fig. 41.32 Australian scientists have discovered a wealth of fossils at Riversleigh in north-western Queensland. In these 20 million year old deposits, five types of fossil thylacines have been recognised, ranging from a species the size of a domestic cat to one as large as a medium-sized dog. Once widespread in Australia, this predatory lineage was nevertheless in decline. Deposits 15 million years in age reveal only two fossil species and by eight million years ago only one species, *T. potens*. By the early Pliocene, this species is replaced in the fossil record by *T. cynocephalus*, which survived until recent times. It was driven to extinction in its final stronghold, Tasmania, by an active program of extermination undertaken because thylacines were seen as serious pests for sheep farmers

that *Thylacinus* was finally hunted to extinction. The last known individual died in the Hobart Zoo in 1936.

The Australian fauna includes species introduced by humans, such as the dingo brought by Aborigines about 7000 years ago, and other species brought in by Europeans in the last 200 years. Introduced species have led to the extinction of many native species.

Prototheria: the platypus and echidnas

This remarkable group of mammals is endemic to the Australian plate where it has had a long history. The oldest known fossil of an Australian mammal is a 100 million year old platypus-like jaw (Fig. 41.33).

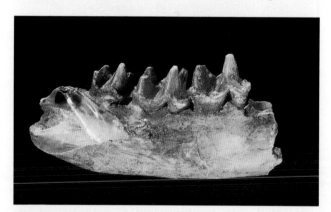

Fig. 41.33 Oldest known Australian mammal fossil: a platypus jaw 120 million years old

Prototherians display many primitive features, such as egg laying, secreting milk from glands that do not form nipples (Fig. 41.34), a cloaca, and markedly reptilian features in the bones of the limbs and girdles. Nevertheless, the two extant species in Australia, the platypus, *Ornithorhynchus anatinus*, and the echidna, *Tachyglossus aculeatus*, show some remarkable morphological and ecological specialisations, for example, the recently reported electrosensory system. The platypus, which is aquatic and widespread in most unpolluted streams in mesic eastern Australia, uses this *sixth* sense to locate its prey. Specialised sensory pits or pores on the bill of the platypus (see Fig. 7.11)

BOX 41.6 AUSTRALIAN REPTILES

Australia had a diverse and ancient reptilian fauna, which included a variety of dinosaurs. Many of those known from early Cretaceous deposits were small to medium sized (human sized) hypsilophodonts, bipedal plant-eaters. One of the largest dinosaurs was *Rhoetosaurus brownei*, a giant plant-eater, which lived in Queensland about 190 million years ago. Some dinosaurs from southern Victoria had large eyes and enlarged optic lobes in the brain, seemingly well adapted to the polar light conditions they experienced in the southern high latitudes (Fig. a). These reptilian groups all became extinct, leaving no derivative modern descendants. Even so, on some small parts of Gondwana that are part of New Zealand, the tuatara (*Sphenodon punctatus*), now an endangered species, is the last living representative of an ancient group of reptiles that was once widespread in Triassic and Jurassic times.

There are many arguments concerning the origins and relationships of the modern reptilian fauna of Australia. The side-necked tortoises (family Chelidae) are a Gondwanan group as are some of the geckos and their relatives (subfamily Diplodactylinae) but most modern forms appear to have been derived from Asian groups and colonised Australia after the Australian plate drifted north.

Even so, Australian dragon lizards (family Agamidae), which traditionally are believed to have been derived from Asia, may have Gondwanan affinities. Recent molecular data suggest that one group of dragons, including the bearded dragon (Fig. b), may be more closely related to African rather than Asian agamids.

(a) Victorian dinosaur with large eyes adapted to the longer polar night that the animal experienced when Australia was part of Gondwana

(b) Australian bearded dragon

allow it to detect extremely weak electric fields generated by neuromuscular activity of its invertebrate prey. Underwater, with its eyes, ears and nostrils (nares) closed, it has been suggested that the field generated by a tail flick of a freshwater shrimp can be detected by a platypus as far away as 10 cm. More recently, the system has been found in a less well-developed form in the snout of the echidna.

Such specialisations enabled these ancient mammals to survive in the face of increasing competition from other animal groups, particularly other types of mammals. If a diverse monotreme fauna existed in Mesozoic Australia, it was largely displaced by the marsupials.

Prototherians, the platypus and echidnas, are endemic to the Australian plate. They have primitive features and are known from fossils as old as 100 million years.

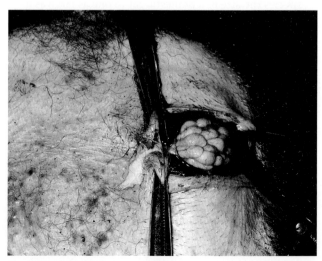

Fig. 41.34 Echidna secreting milk

Metatheria: marsupials

Like hylid and leptodactylid frogs, marsupials have two centres of diversity, South America and Australia, although the oldest marsupial fossil comes from North America, indicating that the group had a widespread distribution during its early history. Even so, the past distribution of various mammalian groups is unclear, mainly because few fossils of Mesozoic age (200–65 million years ago) are available for many places, especially Australia and Antarctica. The earliest South American mammal faunas included both marsupial and placental groups, but placental mammals failed to become established in the original mammal fauna of Australia. It is difficult to explain this anomalous pattern. Placentals may never have lived in the eastern, Australian, part of Gondwana. If this were true, then marsupials must either have arisen in Gondwana, or dispersed across a barrier, from west to east Gondwana, that placentals were unable to cross. Alternatively, placentals may have become extinct in east Gondwana, although this is unlikely given their universal success in the rest of the world. However, this idea has been given some recent support with the discovery of a presumed placental tooth in early Tertiary deposits (55 million years old) at Tingamarra, Queensland.

The present marsupial fauna of Australia includes two orders: the **Polyprotodonta** and the **Diprotodonta**. The Polyprotodonta include four suborders: South American marsupials (Didelphimorphia); quolls, dunnarts, planigales, Tasmanian devil (Fig. 41.35), numbat and marsupial wolf (Dasyuromorphia); the bizarre marsupial mole, *Notoryctes typhlops* (Notoryctemorphia); and bandicoots and bilbies (Peramelomorphia). As the name Polyprotodonta suggests, members of this group all have more than one pair of incisor teeth in the lower jaw, but in general these animals are characterised by having a large number of sharp, relatively unspecialised teeth (Fig. 41.36). All are carnivores, although a trend towards a more omnivorous diet is seen in the bandicoots and bilbies.

Fig. 41.36 Skull and teeth of a Tasmanian devil, typical of the Polyprotodonta

Diprotodonta (Fig. 41.37) include those marsupials that most people recognise as typical Australian animals. There are four distinctive superfamilies: koalas and wombats (Vombatoidea); ringtail, brushtail and pygmy possums, cuscuses and gliders (Phalangeroidea); the honey possum, *Tarsipes rostratus* (Tarsipedoidea); and a large group that includes rat kangaroos, bettongs, potoroos, kangaroos and wallabies (Macropodoidea). Although these four superfamilies include a variety of strikingly different types of animals, all of them are related and form a monophyletic group, characterised by several derived characteristics. The most obvious features of Diprotodonta are a general reduction in the number of teeth, with only one pair of incisors in the lower jaw, together with a tendency to greater specialisation of tooth structure, especially in development of molars with grinding and crushing surfaces used in browsing and grazing (Fig. 41.38). While the majority of the Diprotodonta are herbivores, fossil material shows that a reversion to a carnivorous lifestyle did occur, although all of these predators, like the carnivorous kangaroo (*Propleopus*), are now extinct. The group also included many extinct species that were part of Australia's megafauna (Box 41.7).

> The present marsupial fauna of Australia includes two orders: the Polyprotodonta (carnivores, with more than one pair of incisor teeth in the lower jaw) and the Diprotodonta (mostly herbivores, with only one pair of incisors in the lower jaw).

Eutheria: placentals

While Australia is thought of as being the 'land of marsupials', the native, terrestrial mammal fauna includes a remarkable diversity of placental mammals. Even so, this large number of endemic genera and species belong to only two orders: bats (Chiroptera) and rodents (Rodentia).

Fig. 41.35 Tasmanian devil (Polyprotodonta)

(a)

(b)

(c)

Fig. 41.37 Examples of Diprotodonta: **(a)** the common wombat, **(b)** yellow-bellied glider and **(c)** the red-necked (Bennett's) wallaby

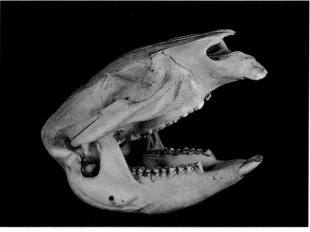

Fig. 41.38 Skull and teeth typical of the Diprotodonta

Bats (Fig. 41.39) have a long history in Australia, being known as fossils from the late Oligocene (about 23 million years ago). Because they are flying mammals, water is less of a barrier to dispersal and the early arrival of bats from Asia to the north is perhaps not so surprising. The order Chiroptera includes two suborders and both are well represented in the Australian fauna. The familiar fruit and blossom bats and the flying foxes (suborder Megachiroptera, see Chapter 40) are generally large herbivorous bats.

Fig. 41.39 A microchiropteran bat, *Pipistrellus tasmaniensis*, with large ears for echo location

The small predatory bats (suborder Microchiroptera) hunt with a sophisticated mechanism of echo location of prey. In Australia, all microchiropterans are insectivorous except the ghost bat, *Macroderma gigas*, which, although taking some large insects and other invertebrates, preys on a variety of small vertebrates, including other bats.

Even though Australia has a large and diverse fauna of bats, the native rodents (Fig. 41.40) provide an even

BOX 41.7 MEGAFAUNA

Unlike the rest of the world where mammals are commonly the conspicuous large predators, the largest carnivorous vertebrates in the terrestrial fauna of Australia were reptiles: a giant lizard, *Megalania*, a large land-dwelling crocodile, *Pallimnarchus*, and a giant snake, *Wonambi* (Fig. a). These predators were part of the Australian megafauna, which included also a great many other large and impressive organisms (Fig. b), two of the most spectacular being *Diprotodon optatum*, a lumbering diprotodontid that reached the size of a modern hippopotamus, and the giant mihirung birds (family Dromornithidae), which were flightless emu-like animals that included the largest bird that has ever lived. During the late Pleistocene, the megafauna became extinct; the last *Diprotodon* and mihirung birds died out around 20 000 years ago.

These large-scale extinctions of megafauna over the past one to two million years were a worldwide phenomenon and causal explanations are generally of two kinds. The first implicates human hunting, and there is evidence that humans did hunt and kill elements of the megafauna. The second suggests that environmental deterioration accompanying increasing aridity brought about the megafaunal extinctions. Clearly both processes could have been acting in concert. In Australia, humans coexisted with the megafauna for at least 20 000 years so that direct hunting, along with human-induced environmental changes, may have driven some of these large animals to extinction. At the same time, increasing aridity, which reached its most extreme about 17 000 years ago, led to marked environmental change. For example, Lake Callabonna, to the north-east of the Flinders Ranges in South Australia, is now one of a series of dry salt pans. In the recent past, this area supported large populations of *Diprotodon*—in fact, it is one of the best localities for finding fossils of this animal.

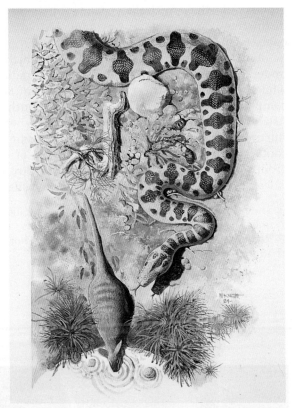

(a) 'Wonambi', a giant python of Australia's extinct megafauna, stalking a young wallaby

(b) A large and a small *Palorchestes*, one of Australia's most unusual extinct mammals. It was the size of a large bull and may have inspired the legend of the bunyip

more remarkable example of adaptive radiation. All native rodents belong to only one family, Muridae. Because there are so many different types of animals among the 50 or more species of native rodents (ranging from mouse and rat-like forms to large aquatic animals), it was once thought that this group had had a very long history in Australia. It is now clear that this diverse fauna has arisen over a relatively short period and from as few as two episodes of colonisation.

Fig. 41.40 A native rodent, *Hydromys chrysogaster*, a water rat of eastern Australia

Rodents first make their appearance in the Australian fossil record in the Pliocene, so native rats and mice have evolved over the past few million years, certainly within the last 15 million years. Most of the rodent fauna (subfamily Hydromyinae) derive from this initial colonisation. The remainder (subfamily Murinae) is made up of a small number of native species of the cosmopolitan genus *Rattus* (excluding the introduced *R. rattus*, *R. norvegicus* and the Melanesian rat *R. exulans*; together with the house mouse *Mus musculus*), that derive from a much later colonisation event that occurred within the last one million years. Both groups of rodents colonised the Australian plate from Asia. Ancestors of the Hydromyinae may have rafted across a decreasing water gap between South-East Asia and Australia or they may have been part of an existing rodent fauna on Asian islands that were incorporated into the north coast of New Guinea during the collision between the crustal plates. The more recent colonisation of

Rattus undoubtedly involved chance dispersal across a water gap, although the distances involved may have been quite small.

During the last million years the world has been subject to times of very cold conditions (see p. 909). The importance of glacial periods for dispersal of terrestrial animals is that much of the world's water is bound up as ice and the resulting lower sea levels both connect continental islands to the mainland as well as reduce the extent of seaways between neighbouring land masses. During the last glacial period, continental islands such as Tasmania, New Guinea and Kangaroo Island were part of mainland Australia. Importantly, during glacial periods the Australian plate was separated from South-East Asia in some places by water gaps only a few kilometres wide. It was across such barriers that ancestors of Australia's native rodents dispersed.

In summary, the biogeographic patterns illustrated by the frogs and mammals are repeated in many groups of the Australian fauna. Usually two distinctive components are evident: an old faunal component, which has affinities with similar groups in other fragments of Gondwana, and which has undergone long and extensive adaptive radiation while isolated in Australia; and a more recent faunal component, which has affinities with the South-East Asian fauna and which entered Australia when the Australian plate drifted nearer to and came into contact with the Asian plate. The history of other groups of animals, such as the song birds (Box 41.8), is less well known, but being actively investigated.

> The placental fauna of Australia is diverse but includes only bats (order Chiroptera) and rodents (order Rodentia). Rodents radiated over a relatively short time after dispersal from the north.

BOX 41.8   AUSTRALIAN SONG BIRDS

Australia has a large and diverse bird fauna, including many taxa that have had a long evolutionary history on the continent. Some, such as the emus and cassowaries, have relationships with other Gondwanan ratites (p. 904), while in other groups, such as the song birds (Fig. a), the evolutionary relationships are not so easily detected. Using modern techniques to study the structure of DNA and various proteins, systematic zoologists are able to determine the remarkable pattern of adaptive radiation and relationships within and between various groups. Even though early European settlers gave familiar names to many native song birds (e.g. robins, wrens and warblers), the similarity with 'Old World' species is only superficial, the similar phenotypes being the result of *convergent evolution* through adaptation

(a) The lyrebird, an endemic Australian song bird

to exploit similar ecological conditions. A simplified phylogeny of song birds (Fig. b), shows the long history of many of these groups in Australia. The information is derived from comparing the differences in protein structure between the groups together with a knowledge of the rates of change in bird DNA.

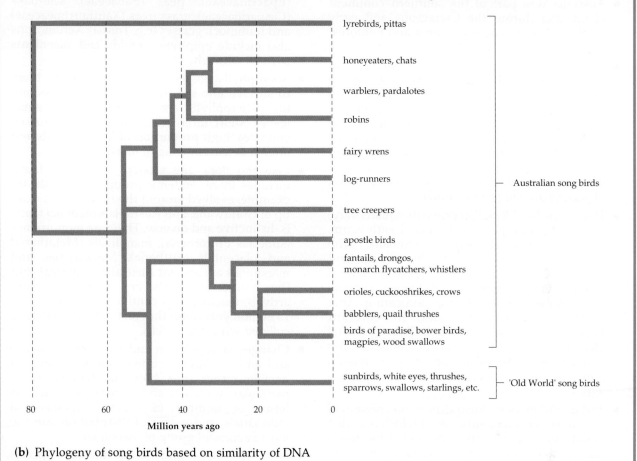

(b) Phylogeny of song birds based on similarity of DNA

SUMMARY

- Australia was part of the southern continent Gondwana during the Cretaceous and early Tertiary. The climate was humid and rainforest covered much of the land.

- Australia severed its links with Antarctica and became an island continent about 30 million years ago. Separation of land masses created a circum Antarctic current, causing rainfall and temperatures to decrease, and climate to become more arid. Climate changes altered landforms and soils, and fires became more frequent towards the end of the Tertiary. Together with isolation, these changes influenced the evolution of Australian plants and animals.

- During the last 2.5 million years (the Quaternary) cool, arid glacial periods alternated with warm, wetter interglacial periods. Humans, who colonised Australia at least 40 000 years ago (and possibly much earlier), caused increased burning. These environmental changes correlate with pollen assemblages, indicating that rainforest contracted and fire-tolerant vegetation became dominant. Environmental changes since the arrival of humans also correlate with the extinction of large animals, which occurred between 35 000 and 15 000 years ago.

- Today, the modern Australian biota consists of a number of components, one of which is a relict Gondwanan component represented by plants and animals confined to moist rainforest habitats. The major component has evolved from Gondwanan stock and diversified since the middle of the Tertiary, predominantly in temperate and more arid environments.

- Native plants include xeromorphs, such as the sclerophylls *Eucalyptus* (family Myrtaceae), *Acacia* (Mimosaceae), *Banksia* (Proteaceae), heaths (Epacridaceae), peas (Fabaceae), she-oaks (Casuarinaceae), grass trees (Xanthorrhoeaceae) and hummock grasses (e.g. *Triodia*). Xeromorphs also include epiphytic orchids and succulents such as saltbush.

- Sclerophylls possess features that allow them to survive low soil nutrients, water stress and fire. Sclerophylls are characterised by small leaves, short internodes, thick cuticle, sunken stomates, high proportion of lignified cells and often a dense covering of hairs.

- Like the flora, the modern Australian fauna includes many endemic groups. Mammals, for example, evolved around the time of the break-up of Gondwana, and Australia's mammal fauna is distinctive and diverse. There are monotremes (subclass Prototheria), marsupials (Metatheria) and placentals (Eutheria). Monotremes and marsupials are ancient Australian mammals but the native placentals are relatively recent arrivals, reaching this continent within the last 15 million years after the Australian crustal plate collided with that of Asia.

- Changes in vegetation and range expansion of arid and semiarid flora led to a dominance of grazing rather than browsing herbivores, in particular, wallabies and kangaroos, which, in terms of number of species, numbers of individuals, and extent of distribution, are the most successful group of marsupials.

- Both the flora and fauna include species introduced by humans, such as the dingo brought by Aborigines about 7000 years ago, and other species brought by Europeans in the last 200 years. Introduced species have had an impact on the Australian environment, leading to extinction of many native species.

QUESTIONS

1. On what land masses are **(a)** living representatives and **(b)** fossils of *Nothofagus* found? Name two other plant or animal groups that have a Southern Hemisphere distribution. What is the accepted geologic explanation for these biogeographic patterns?

2. Briefly explain why the climate of Australia changed from humid-temperate to more arid during the Tertiary period. How did this affect the dominant type of vegetation?

3. List a characteristic feature of plants in each of the following angiosperm families: **(a)** Myrtaceae; **(b)** Proteaceae; **(c)** Orchidaceae; and **(d)** Mimosaceae. What feature characterises most Australian species of *Acacia*?

4. What are some of the features of Australian sclerophyll plants, such as wattles, banksias and eucalypts, that enable them to grow on soils of low nutrient level (refer to Chapter 18)?

5. Briefly discuss the various adaptations that reduce water loss in Australian frogs.

6. Prototherians are only found in Australia and have been here for at least 100 million years. The group displays many features that reflect the mammal's reptilian heritage. Describe four of these primitive characteristics found in prototherians.

7. Give an example of an animal in each of the following superfamilies of diprotodontid marsupials: **(a)** Vombatoidea; **(b)** Phalangeroidea;

(c) Tarsipedoidea; and (d) Macropodoidea. What features characterise species of the Diprotodonta?

8. In contrast to other continents, Australia's largest mammalian predators were dog-sized thylacines. Within the recently extinct megafauna of Australia, which animals filled the role of large predators, ecologically similar to tigers, bears and lions?

9. Australia is known as the 'land of the marsupials', yet many of our native mammals are placentals. Describe in general terms the composition of the native placental fauna and discuss its evolutionary and biogeographical relationships.

10. What impact did the arrival of humans at least 40 000 years ago have on Australian environments and biota?

11. There is increased attention being focused by Governments and the public on Australia's 'biodiversity'. What is meant by 'biodiversity' (see Chapter 30)? Why should people be concerned with documenting and understanding the history and diversity of the world's biotas?

PART 7

ECOLOGY

POPULATION ECOLOGY

Until now this book has been concerned primarily with biological processes, such as the functioning of cells, organs and individual organisms, and the evolution and diversity of species. This section deals with the **ecology** of organisms—the ways in which organisms interact with one another and their environment. We can study the ecology of organisms at three levels of organisation: at the level of **populations** of single species; at the level of **communities**, which consist of groups of species living together; and at the level of **ecosystems**, which are communities together with their non-living surroundings. This chapter examines the first of these.

The primary concern of population ecologists is to understand the dynamics of the distribution and abundance of organisms. This understanding is essential to applied biology, including conservation biology, pest control and the management of natural populations that are harvested to provide resources, such as food, for humans. Population ecology provides the essential building blocks of community and ecosystem ecology. It is important to remember that humans are a species of animal, and many of the concepts and phenomena described in this section apply equally well to us.

> Population ecologists try to understand the dynamics of the distribution and abundance of organisms, knowledge that is essential to applied biology and the management of natural populations.

WHAT IS A POPULATION?

A **population** is a number of organisms of the same species which inhabit a defined geographic area, for example, noisy scrub-birds, *Atrichornis clamosus*, at Two People's Bay, Western Australia (Fig. 42.1), or mountain plum pines, *Podocarpus lawrencia*, on the Errinundra Plateau, Victoria. Members of species usually occur in a number of local populations. This subdivision often depends on geographic features that tend to dissect an otherwise continuous population into fragments. The definition of a

population is particularly important with regard to evolution, since populations, and not individuals, evolve (see Chapter 32). In this section, however, we will concentrate on non-evolutionary processes influencing populations, although remember that a population's evolution and its dynamics are inextricably linked.

(a)

(b)

Fig. 42.1 The noisy scrub-bird, *Atrichornis clamosus*, was known from six areas in Western Australia before 1889 but was not sighted again until 1961, when a small population was discovered at Two People's Bay. Since then, the population has expanded and new populations are being formed by translocating individuals into other suitable localities, thus avoiding the possibility that a single catastrophe, such as fire, extinguishes the species

Populations have many properties. Some of the most important properties are: number of individuals in the population; area occupied by the population; age structure of the population; and sex ratio of the population. For species with complex life cycles, understanding the ecology of the entire species is more complicated. For example, plants usually exist in a particular locality, not only as growing individuals, but also as dormant seeds; barnacles occur as both sessile adults adhering to a rocky shore and as larvae in the ocean. The distribution and abundance of organisms at each phase of the life cycle may be affected by different processes.

Two questions immediately spring to mind: 'Why is a species only present in particular places?' and 'What determines the number of individuals of a population in a particular place?'. In other words, 'What determines **distribution** and **abundance** of populations?'. Ecologists studying, for example, the effects of pollutants on a river, would monitor the distribution and abundance of a number of species and compare these with a similar but unpolluted river.

BOX 42.1   MEASURING THE SIZE OF ANIMAL POPULATIONS USING THE MARK–RELEASE–RECAPTURE METHOD

It is usually not possible to count all the individuals in a population, especially when organisms are mobile. An alternative method is to estimate the total population from a sample. Stationary organisms, such as plants, can be counted in a small sample area—a quadrat—marked out on the ground. Soil organisms can be sampled in a core of soil, and aquatic organisms in a volume of water. Another approach used to estimate the abundance of mobile animal populations in the wild is the *mark–release–recapture method*.

Consider a population of sleepy lizards, *Trachydosaurus rugosus* (see Fig. a), that lives in a grazed mallee woodland in South Australia. During a single day, two population ecologists systematically search a square kilometre marking any sleepy lizards they find with pink dye. Assume that they marked M_1 individuals on the first day. The next day they mark lizards in the same area with blue dye. Some of these individuals will have been marked with pink dye already and some will not. Let M_2 be the number of individuals marked with blue dye that were already marked with pink dye and let N represent the total number of individuals marked with blue dye on the second day. (The number of individuals marked on the second visit that were not marked on the first visit is $N - M_2$). The proportion of individuals marked on the second day that were already marked should equal the proportion of individuals in the population that are marked:

$$\frac{M_1}{\text{Total population size}} = \frac{M_2}{N}$$

Hence, an estimate of the number of lizards in the mallee woodland, subject to statistical error, is:

$$\text{Total population size} = N\left(\frac{M_1}{M_2}\right)$$

For example, if 52 animals were marked on the first day, and 64 on the second (of which 16 had already been marked), then the estimate of the total population size is 208. This estimate relies on many assumptions that can be difficult to satisfy:

1. The population is assumed to be closed. If individuals die or move into and out of the area between the first and second marking days, the estimate of population size will be inaccurate.
2. All individuals are equally likely to be marked. In practice, some individuals may be more likely to be found than others.

3. Marked individuals do not lose their mark (see Fig. b). Despite these problems this method can be a useful means of assessing the population size of a species as it changes from year to year and place to place.

(a) The sleepy lizard, *Trachydosaurus rugosus*, in threat posture

(b) Ecologists have used natural marks on whales, such as this humpback whale, and dolphins to track animals off the coast of Queensland when estimating population sizes

DISTRIBUTION AND ABUNDANCE OF POPULATIONS

The environment of an organism can be divided into two parts: the **biotic** (other organisms) and the **abiotic** (the physical inorganic components). Both abiotic and biotic factors influence the distribution and abundance of populations.

Plants spend their entire life in a single locality. For a plant population to persist in a particular locality it must be able to survive and to reproduce successfully. Successful reproduction of flowering plants may depend on the availability of a specific animal pollinator, a biotic factor. However, the survival of many plants appears to be closely correlated with abiotic factors, particularly light, temperature, rainfall and soil. The way in which these abiotic factors limit the distribution of a species can often be related to physiological processes that directly affect a plant.

To assimilate carbon for growth and reproduction, plant stomata must be open to allow the entry of carbon dioxide. This inevitably entails a loss of water, and if a plant loses too much water it may wilt and die. The distribution of many species is limited by the availability of water, which, in turn, is determined by a number of environmental parameters, including temperature, rainfall, soil type, humidity and altitude. In this case, available water can be regarded as a direct environmental factor that is controlled by a number of indirect environmental factors.

By recording the presence or absence of eucalypt species in over 5000 sites in south-eastern New South Wales, and knowing the mean annual temperature at each of these sites, M. Austin from the CSIRO was able to plot the probability of occurrence of different species as a function of temperature (Fig. 42.2). Plots of species' occurrence against an environmental gradient, such as temperature, typically show a peak in the probability of occurrence. Such curves define the realised distribution of a species in environmental (not geographic) space. The limits to the distribution of the eucalypt species in the environmental space is often determined by a plant's ability to tolerate extreme physiological stress (e.g. very cold nights), although it may also be determined by biotic factors such as herbivory, pathogens or competition with other plant species.

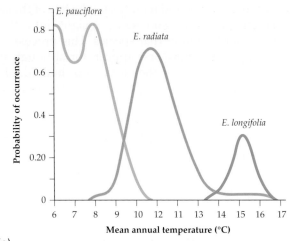

(a)

(b)

Fig. 42.2 **(a)** Plots of the probability of occurrence of different eucalypt species in south-eastern New South Wales and Victoria as a function of temperature. Some species show typical bell-shaped curves, such as *Eucalyptus longifolia*, while the response of other species is skewed, such as *E. radiata*. By mapping environmental space onto a geographical map it is possible to use curves like these to predict the probability of occurrence of a species at any locality in a region. **(b)** *E. pauciflora*, snow gum, is tolerant of low temperatures and snow. It forms subalpine woodlands at high altitudes in south-eastern Australia

Potential and realised distributions

The **potential distribution** of a species is the range over which individuals could theoretically survive and reproduce. The **realised distribution** of a species is the range in which it does live and reproduce. The two are rarely the same.

One way of determining the potential distribution of a species is to conduct transplant experiments. By introducing a population of a plant or animal into a number of localities and monitoring the fate of those populations, we can gain some insight into the potential distribution of that species. This process is not as simple as it may first appear, for although it may be easy to monitor the survival of individuals in new localities, the long-term persistence of a population relies on successful reproduction and survival over several generations. Conducting such experiments, especially with long-lived organisms, may be impractical.

> A knowledge of the response of a species to environmental gradients can be used to make predictions about the potential distribution and abundance of species.

One indirect method of estimating the potential distribution of a species is to determine the abiotic and biotic properties of the places in which it currently lives, and extrapolate these to other areas to determine its potential range. Although this procedure has numerous pitfalls, it may be a useful predictive tool for determining the effect of global warming and climatic change (p. 1013) on the distribution of species. This is particularly relevant to pests that damage crops. For those pest species with distributions and population growth rates limited by low temperatures, an idea of a species' **environmental niche** (the total range of conditions under which the members of a species live and reproduce) helps us to identify those areas in cooler, southern regions of Australia that the pest will invade as climate changes. For example, since its introduction to Queensland at the beginning of this century, the cane toad, *Bufo marinus*, has increased its distribution. Its potential distribution has been modelled on the basis of temperature tolerance of the tadpoles. It continues to spread west to the Northern Territory but appears to have reached its temperature limit in the south. With climatic change, the southern limit is likely to shift.

The realised distribution is often a surprisingly small subset of the potential distribution. There are a variety of reasons for this. One is that a species may not occur in an environmentally suitable locality simply because it has not been able to colonise that place. This simple concept often explains the limited distribution of sessile organisms, such as Antarctic beech trees, which have poor dispersal abilities (Fig. 42.3). Historical factors, such as past climates and fires, play a key role in restricting species to subsets of their potential range.

It seems reasonable to expect that all the individuals of a species would attempt to live in places that maximise their chances of survival and reproduction, that is, their 'evolutionary fitness'. This idea is the basis of **habitat selection theory**. Very mobile species, such as migratory animals, can choose between different localities from year to year and season to

(a)

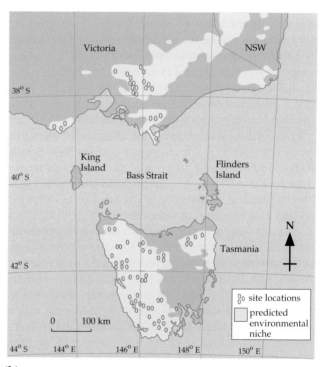

(b)

Fig. 42.3 (a) The Antarctic beech tree, *Nothofagus cunninghamii*, occurs in cool temperate rainforests in south-eastern Australia and Tasmania. **(b)** Its distribution is of considerable interest to biologists because of its Gondwanan origin (Chapter 32). By means of data on the current distribution of Antarctic beech, its environmental niche based on rainfall, temperature and altitude, has been determined. In general, the species occurred in all locations in which the environment was suitable except for areas in the north-eastern highlands of Victoria and southern New South Wales. It is suggested that the species is slowly migrating north-east along a corridor of suitable habitat

season. Their absence from a particular place in any one year may reflect the unsuitability of that habitat in that year. Also, in years when a species is abundant, the most favourable habitat may fill up with the territories of individuals of the same species, so some are forced to occupy less favourable habitats. To determine the potential distribution of species, we need to know the way populations shift in space and change in size over time.

Biotic factors, such as competition between species and predation, also limit distributions, but we will examine these in the next chapter.

> Understanding the distribution of a species requires a knowledge of the way in which populations fluctuate in space and time.

DENSITY-INDEPENDENT POPULATION DYNAMICS

The size of a population can only change (increase or decrease) as a consequence of four processes: **birth**, **death**, **immigration** and **emigration**. Adults reproducing within the population and the immigration of individuals into a population from other populations increase the size of a population. The death of individuals in a population and the emigration of individuals out of a population decrease population size. Using these principles, it is possible to formulate an equation that forms the basis for studies of population dynamics:

$$N_{t+1} = N_t + B_t - D_t + I_t - E_t$$

where N_{t+1} is the number of individuals in the population at time $t + 1$, N_t is the number of individuals at time t, B_t is the number of births between time t and time $t + 1$, D_t is the number of deaths between times t and $t + 1$, I_t is the number of immigrants that arrive between times t and $t + 1$, and E_t is the number of individuals leaving the population between times t and $t + 1$. Usually, population size is recorded each year, so t is measured in units of years.

When the per capita birth and death rates of a population are independent of the size of that population, we say that the population dynamics are **density independent**. Conversely, when the birth and death rates depend on population size or density, then the population dynamics of the species are **density dependent**.

Population ecologists distinguish between two types of population: *closed* and *open*. There is no immigration into or emigration out of closed populations, so that I_t and E_t are zero for all time. A closed population with density-independent birth and death rates will experience an exponential increase or decrease in size or remain constant.

Exponential population growth in discrete time

Imagine a closed population living in a place where the environment is constant. Each year each pair of individuals produces a constant number of offspring, say four, and then dies. In other words, the population size changes over a discrete period of time. If the initial number of individuals in the population is N_0, then after one year the number of individuals in the population will have doubled:

$$N_1 = 2N_0$$

After two years the number will have doubled again and, using this concept inductively, we can see that the number of individuals in year t will be:

$$N_t = 2^t N_0$$

This is known as **exponential growth**. Under these circumstances the size of the population increases rapidly. An initial population of only 10 pairs will result in more than 10 million pairs after 20 years. Clearly there is potential for enormous growth in any population. Before we consider some of the factors that limit this potential for increase let us examine the mathematics of exponential growth in more detail.

Modelling density-independent growth

Although populations are often made up of two sexes, it is convenient to consider only the females in the population so, in the following model, the population size is equivalent to the number of females in the population. The number of females in a closed population at time $t + 1$, N_{t+1}, is related to the number at the previous time, N_t, according to the following equation:

$$N_{t+1} = N_t b_t + N_t(1 - d_t)$$

where b_t is the average number of female offspring that one female gives birth to each year and that survive to be counted the following year, and d_t is the proportion of adult females that die each year. Rewriting this equation we find that:

$$N_{t+1} = N_t(1 + b_t - d_t)$$
$$= R_t N_t$$

where R_t is the growth rate of the population in year t. If R_t is constant, then, as in our example before (when the population doubled each year, i.e. $R = 2$):

$$N_t = R^t N_0$$

If R is greater than one, then the population will increase in size indefinitely. If R is less than one, then the population size will approach zero asymptotically and the species will become extinct (Fig. 42.4).

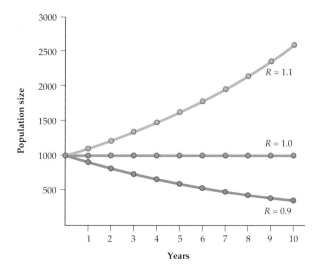

Fig. 42.4 The dynamics of a population with different growth rates: 0.9, 1 and 1.1. If the growth rate is less than one, the population declines towards zero. If the growth rate is greater than one, the population will increase geometrically and reach an enormous size surprisingly quickly. If the growth rate is exactly one, the population size remains constant

Environmental variability and exponential growth

In a single population, birth and death rates will vary from year to year, so the growth rate will be different every year. If we know the value of R for each year, then we can predict the population size at time t from the initial population size, N_0, by applying the above equation repeatedly

$$N_t = R_{t-1} \; R_{t-2} \; R_{t-3} \dots R_0 N_0$$

where R_{t-1} is the growth rate in the year before year t and so forth.

Because we cannot predict changes in the environment that affect birth and death rates, we are unlikely to know the growth rates for each year. However, we can estimate the expected population size at some future date if we know the geometric mean of R from a record of observations of R made in past years.

$$N_t = R^t N_0$$

where R is the geometric mean of the growth rates

$$R = \sqrt[t]{R_{t-1} \; R_{t-2} \dots R_0}$$

This is a simple example of a **stochastic model**, that is, in any year the growth rate is a random variable with a mean and variance. The final prediction of the population size is only an expectation, not a precise prediction.

It is important to remember that long-term population dynamics are determined by the geometric mean of the rate of population increase and not the arithmetic mean (i.e. $[R_{t-1} + R_{t-2} + \dots + R_0]/t$). This fact is illustrated by a simple example.

Consider a population that experiences good and bad years with equal probability. In a good year the growth rate is $R = 2$, while in a bad year $R = \frac{1}{8}$. The arithmetic average of the growth rate is 1.0625 (>1) but the geometric average is $\frac{1}{2}$ (<1). In Table 42.1, the population size is tabulated for a number of years when the initial population size is 480. The geometric mean, not the arithmetic average, correctly predicts the decrease in population size.

Table 42.1 The geometric mean of the rate of population increase (R) correctly predicts population size

Year	Growth rate (R)	Population size
0	—	480
1	2	960
2	$\frac{1}{8}$	120
3	2	240
4	2	480
5	$\frac{1}{8}$	60
6	2	120
7	$\frac{1}{8}$	15
8	$\frac{1}{8}$	<2
Geometric mean	$\frac{1}{2}$	

A closed population with density-independent death and birth rates will ultimately either increase to an infinite size or decline to extinction depending on whether the geometric mean of its growth rate (R) is greater than or less than one. Theoretically, a population will remain constant if R exactly equals one.

Exponential population growth in continuous time

So far we have only considered population growth over discrete time intervals where the population size is 'updated' from year to year. The model we developed is especially appropriate for populations in which births occur within a short space of time. Species that live in environments with strong seasonal conditions behave in this way. For example, mammals, birds and plants living in temperate and sub-Antarctic conditions breed almost exclusively in spring (Fig. 42.5). In contrast, some populations, in particular the human population, change continuously as births and deaths occur throughout the year. The dynamics of these populations are more accurately described using differential equations.

Fig. 42.5 Gentoo penguins at Heard Island. Birds, mammals and plants of sub-Antarctic islands breed almost exclusively in spring, that is, births occur in a short space of time

If the per capita birth rate, that is births per unit time per individual, is b, and the per capita death rate is d, then the rate of change in the population size $N(t)$ is $(b - d)N(t)$. Written as a differential equation this is:

$$\frac{dN(t)}{dt} = (b - d)N(t) = rN(t)$$

where the intrinsic rate of increase of the population is r. This equation can be solved using elementary calculus to generate an expression for the population size as a function of time:

$$N(t) = N(0)\exp(rt)$$

In the continuous time model of population growth, the intrinsic rate of increase determines the fate of the population. If $r > 0$, then the population grows exponentially; if $r < 0$, the population declines towards zero; and in the special case $r = 0$, the population remains constant.

After the introduction, or colonisation, of a species to a new area, the population may grow exponentially and often shows a rapid expansion in range. One intriguing quality of this expansion in range, especially in the early stages of an invasion, is that the limit of the species' distribution spreads at a constant rate. A simple model has been produced for this phenomenon, showing that the rate of spread of the population is proportional to the growth rate of the population. This phenomenon has been observed for many plant and animal species. Figure 42.6 illustrates the spread of a species of mangrove in New Zealand. A mathematical model was used to predict successfully the spread of the population using a population growth rate of 1.2 based on life-history data.

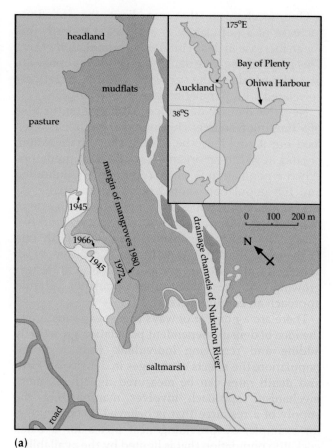

(a)

(b)

Fig. 42.6 **(a)** The expansion in the size of a population of *Avicennia marina* mangroves in Ohiwa harbour, New Zealand, from 1945 to 1980. At this locality the species is at the southern limit of its range. **(b)** *Avicennia resinifera* estuary in the Bay of Islands, North Island, New Zealand

DENSITY-DEPENDENT POPULATION DYNAMICS

Given that all populations have a capacity for exponential growth, and the observation that none, including human populations, increases indefinitely in size, the obvious question is, 'What controls population increase?'.

One mechanism that can limit the indefinite increase of a population is a decrease in birth rate (or an increase in death rate) as the population grows. This is called negative density-dependent population growth because the growth rate is negatively correlated with population size.

Negative density-dependence is based on the idea of a **limiting resource**. This limiting resource is often food but it could also be nesting sites, space, water, or sites in which an organism is safe from predators. When a population is small, resources are unlimited and the population should grow exponentially. As the population becomes large, a deficiency in one or more of the resources critical to a species' survival causes the death rate to increase or the birth rate to decrease until the growth rate is reduced to an average of one, so that the population is no longer growing. At this point the population should approach an equilibrium state.

There are a number of ways to test experimentally the idea of density-dependent population growth. By adding and removing individuals from a closed population, the effects of changes in density on birth and death rates can be measured directly. A more mechanistic approach involves manipulating the supply of a resource that is believed to be limiting, for example, by providing nest boxes for a black cockatoo population that is limited by the availability of tree hollows. Figure 42.7 illustrates the effect of increasing food supply on a population of small marsupial carnivores.

When there is density-dependent population growth, an increase in population size causes a decrease in growth rate. Eventually, an equilibrium may be reached when the number of births and deaths balances. This is the carrying capacity of a

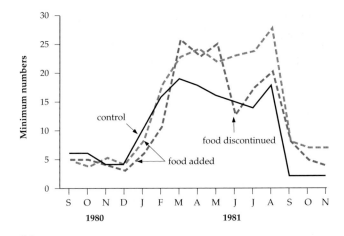

(b)

Fig. 42.7 (a) Supplementary food was provided to a marked population of an Australian marsupial *Antechinus stuartii*. **(b)** The population increased in numbers, mean body weight and survival, and showed smaller overlaps in home ranges when contrasted to a control population. These responses in *A. stuartii* indicate that the density of the populations studied was limited by food supply

population. The logistic equation is a means of representing the population dynamics of a resource-limited species:

$$\frac{dN}{dt} = rN \left(1 - \frac{N}{K} \right)$$

Now, r is equal to the rate of increase per head of population only when the population size is very low and density-dependent effects have not decreased the rate at which the population grows. The carrying capacity is represented by the parameter K. When the population size reaches K, it ceases to grow (Fig. 42.8).

Fig. 42.7(a)

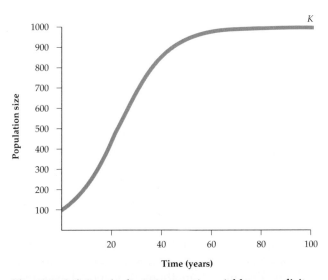

Fig. 42.8 Solving the logistic equation yields an explicit expression for the dynamics of a population in a constant environment. This is an example of logistic growth when the initial population size is 100

The logistic growth model is a naive representation of the population dynamics of a species limited by resources. Although it has conceptual value, we would not expect the abundance of wild populations to obey such a simple equation. In Australia, the variability of the environment means that the parameters in the model vary from year to year and there is no such thing as a fixed carrying capacity except under special circumstances (Fig. 42.9). Catastrophic events such as drought or severe predation may keep population sizes well below their carrying capacity most of the time. Under these circumstances populations will usually exhibit density-independent dynamics.

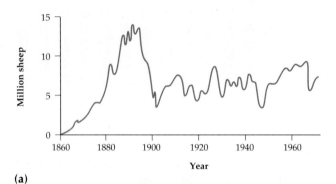

(a)

(b)

Fig. 42.9 The population dynamics of sheep in the Western Divisions of New South Wales between 1860 and 1972. The initial increase in population size was approximately exponential. Numbers appeared to initially overshoot the carrying capacity of about 8 million sheep. The large fluctuations in population size from year to year may reflect environmental changes, but are equally likely to be a consequence of changes in the economy

Whether or not population growth is density-dependent has been a subject of controversy for decades. A. J. Nicholson, an Australian biologist, used populations of blowflies to test experimentally a simple model of density-dependent population growth. The importance of density-dependence in controlling the dynamics of species was challenged strongly by H. G. Andrewartha and L. C. Birch in their classic text *The Distribution and Abundance of Animals*. They believed fluctuations in environmental conditions to be the primary influence on population size, that competition for resources was over-emphasised, and that simple mathematical models of population dynamics were of little value. Other scientists studying plant populations have emphasised the importance of density-dependent population growth (see Box 42.2).

Recent controversy about the importance of density-dependent growth can often be traced to differences in the kinds of organisms that population ecologists study. Vertebrates, especially those high in the food chain, such as birds, often appear to maintain populations that only fluctuate slightly about a fairly constant carrying capacity. Many invertebrates, such as locusts, however, display enormous fluctuations in abundance, completely disappearing from parts of their range for several years and then reappearing in plague proportions. **Density-vague population dynamics** is a new term that has been used to describe the dynamics of species whose population dynamics usually appear to be independent of population size. In these species, resource limitation may be rare, however, that does not mean it is unimportant.

> A finite resource, such as food, may constrain the exponential growth of natural populations; however, environmental fluctuations in space and time often mean that resources are rarely limiting for many species.

Space-limited populations

For some species the resource that limits population growth is not food but space. This is particularly true for sessile organisms, such as plants and adult barnacles. Many of the inhabitants of the rocky intertidal zone live on particles of food that they filter from water. For these species food may not be limiting, however, space for attachment can be in short supply. Two individuals cannot occupy the same place and the amount of available space puts an upper limit on the number of individuals that can inhabit a particular area. For example, if the average size of a barnacle is 1 cm², then a square metre can, at most, contain 10 000 individuals (Fig. 42.10).

The same may be true for plants. Although their canopies often overlap, competition for space may also be critical. For plants, it is usually competition for light that limits the biomass of plant material that a particular area can support.

Two features characterise space-limited sessile populations: they live in habitats that experience disturbance and they have widely dispersing

Fig. 42.10 These barnacles, packed onto the surface of rocks in the intertidal zone of Australia and New Zealand, are a graphic illustration of competition for space—a limiting resource

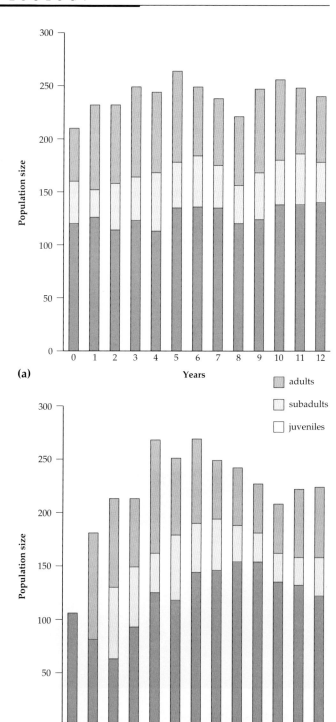

Fig. 42.11 An example of age-structured population dynamics. The graphs show the age structure of a population with three age classes: two juvenile age classes and adults. In both graphs, the birth and death rates are identical. In (**a**), the initial population has a balanced age structure. The population size and the age structure (proportion of individuals in each age class) remain relatively constant. In (**b**), the initial population is 100 adults. Although the parameters of the population are identical, this imbalanced initial age structure generates population cycles in both the age structure and the total number of individuals

propagules. For populations limited by the availability of space, disturbance is invariably necessary for recruitment of new individuals. This disturbance may be the death of an individual tree, creating a gap in the canopy, a starfish eating a patch of mussels, or a fire that destroys thousands of hectares of forest. After a disturbance, space is liberated and new individuals can find a place to grow. In marine systems, new individuals often come from far away and the population in a small area is considered open.

Age- and size-structured population dynamics

So far we have assumed that all the individuals in a population are the same age and size with equal probabilities of dying and giving birth. Clearly this is never true. Most detailed population studies have shown that the age and/or size of an individual has a significant effect on its **fecundity** (probability of giving birth) and probability of survival. For example, in human populations, the probability of a female giving birth before the age of 10 or after the age of 50 is negligible; for most plant species the fecundity of an individual (often measured as the number of seeds set) is positively correlated with its size. Figure 42.11 gives an example of how the age structure of a population affects population growth.

BOX 42.2   SELF-THINNING AND THE LAW OF CONSTANT YIELD

A combination of density-dependence and plasticity in plant growth produces regularities in plant population dynamics that are rarely seen in animals, except for clonal invertebrates, such as corals.

Assume that several plots are sown with the seed of a particular plant species at different densities. The average weight per seedling is measured for each plot a number of times after sowing. For each plot, the logarithm of the average weight of a plant is graphed as a function of the density of plants on the plot (see figure).

At the beginning the seedlings are very small and do not compete with each other for resources. Consequently, the per seedling weight will initially increase at a rate that is independent of the initial sowing density. As time passes, the seedlings in the denser plots begin to compete and the weight per seedling decreases,

whereas the seedlings in the sparsely sown plots grow unabated.

If we harvest the denser plots after a fixed time we find that the yield—plant number times average plant biomass—is constant. This is the **rule of constant yield** and it can be expressed mathematically:

$$Y = WN$$

where Y is the total yield and is constant for all plots, W is the per individual weight and N is the density of plants. The equation can be rewritten to show that plant weight is inversely correlated with plant density.

The law of constant yield can be used to determine the density at which crop plants are sown. A low sowing density will yield a small number of large plants, while a high sowing density will yield a large number of small plants. Depending on the relationship between plant size and value to the farmer, and the cost of seeds and sowing, the rule of constant yield helps us to choose the optimal sowing density.

As time passes, individuals on the denser plots begin to die, in addition to growing more slowly. For a particular plot, the average weight per plant becomes related to the density of survivors according to the following law:

$$W = k/N^{-3/2}$$

where k is a constant. This is referred to as the $-\frac{3}{2}$ **self-thinning rule** because plant weight is inversely proportional to the density of plants raised to the power $-\frac{3}{2}$.

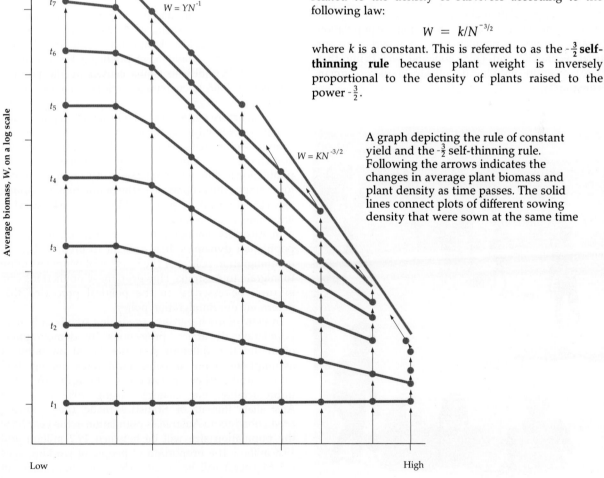

A graph depicting the rule of constant yield and the $-\frac{3}{2}$ self-thinning rule. Following the arrows indicates the changes in average plant biomass and plant density as time passes. The solid lines connect plots of different sowing density that were sown at the same time

Age and size affect the fecundity and survival of individuals of most species. Treating all members of a population as identical is only a crude representation of reality.

A useful way of describing the effect of age on survival and reproduction is to construct a life table (Table 42.2). A **life table** consists of age-specific probabilities of survival and expected reproduction (that is, the average number of young successfully reared by a female of that age). The age-specific survival probabilities can be used to determine a survivorship probability, that is, the probability of surviving to a particular age. The expected number of births per female during an entire lifetime can be estimated by combining the age-specific fecundities and survivals.

Table 42.2 Life table for the Himalayan thar, *Hemitragus jemlahicus* (Fig. 42.12), in New Zealand (females only)

Age class	Fecundity (a)	Probability of dying	Survivorship
0	0.000	0.533	1.000
1	0.005	0.006	0.467
2	0.135	0.028	0.461
3	0.440	0.046	0.433
4	0.420	0.056	0.387
5	0.465	0.062	0.331
6	0.425	0.060	0.269
7	0.460	0.054	0.209
8	0.485	0.046	0.155
9	0.500	0.036	0.109
10	0.500	0.026	0.073
11	0.470	0.018	0.047
>12	<0.470		

(a) The fecundity of a thar of a particular age is the average number of live births per female per year.

BOX 42.3   PREDICTING AUSTRALIA'S HUMAN POPULATION DYNAMICS

One animal population about which we know a great deal is the human population. Predicting the population dynamics of human populations is so important to our social, economic and environmental well-being that there is an entire discipline devoted to its study—**human demography**.

Predicting the population dynamics of human populations is the discipline of human demography

Predicting population trends in humans is notoriously difficult and riddled with pitfalls. Most of these pitfalls arise because of rapid and unexpected changes in human behaviour. Who would have predicted the enormous decline in fecundity in China caused by the political decision, in 1968, to encourage families to have only one child?

Age structure plays a central role in human population dynamics. Not only does the age structure of populations change, but the age-specific fecundity and survival rates also change over time. Predicting the proportion of the population in different age classes is of economic interest; for example, the proportion of the population of working age (15–64 years) or pensionable age (>64 years) is important to the economy.

In Australia, immigration has a large impact on our population dynamics. It is difficult to predict changes in immigration policy, or the number of people wanting to migrate to Australia. However, these predictions can, in turn, be essential to the political processes that determine our immigration policy.

In view of the hazards of predicting trends in human population dynamics, projections by demographers often describe different scenarios based on different assumptions about possible birth rates, immigration rates, and other parameters thought to vary unpredictably. For example, using 1987 data, the Australian Bureau of Statistics made the following predictions about Australia's population in the year 2000: the population size will be between 19.2 million and 19.8 million; the proportion of people of working ages (15–64 years) will be about 63.5%; and the number of people aged 65 is expected to increase greatly. The trend towards an ageing population will become more evident as the next century progresses.

Fig. 42.12 Himalayan thar, *Hemitragus jemlahicus*

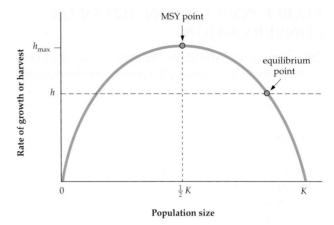

Fig. 42.14 Assume that a biological population grows logistically in the absence of harvesting, and that the harvesting rate is *h*. The rate at which the population is growing can be plotted as a function of the current population size. When the population size is zero, or at the carrying capacity (*K*), the population ceases to grow. When the population is half its normal carrying capacity (½*K*), then the growth rate reaches a maximum. If we harvest the population at this rate, while maintaining the population at half its maximum, then we achieve maximum sustainable yield, MSY. If the population is harvested at a greater rate than the MSY rate, then the population will become extinct. If the population is harvested at less than the MSY rate, then the population size will reach an equilibrium at greater than half its carrying capacity, if and only if its initial size is large enough

BIOECONOMICS: MANAGING EXPLOITED POPULATIONS

Bioeconomics, or biological resource economics, is the study of the interaction of economic and biological systems. Bioeconomic theory addresses the problem of harvesting a natural resource, often a population (e.g. trout), for maximum economic and social benefit (e.g. recreational fishing) (Fig. 42.13). One of the major applications of population ecology is in bioeconomics.

One of the paradigms of bioeconomic theory is **maximum sustainable yield** (MSY). Harvesting populations at MSY maximises the long-term supply of resources, although environmental fluctuations and uncertainty about the biology of a species often mean that it is more prudent to harvest at rates below the MSY rate (Fig. 42.14).

Unfortunately, the value of a resource is discounted through time, so maximising long-term harvesting rates does not always make sense economically. If maximising profit in the short term is the basis for a natural resource industry, then it may be more profitable to over-exploit a resource and drive the species in question close to extinction. This has occurred with many species of whale, starting with the largest species, the blue whale, and ending with the over-exploitation of populations of even small whales, such as the minke whale.

Despite over-exploitation, no species of great whale has yet become extinct. As each species was exploited, its catchability, and hence profitability, was reduced to a level such that it was more profitable to switch to a smaller species. This phenomenon has allowed many highly mobile organisms to avoid extinction. For species that are concentrated in small areas (seabirds), or for sessile species, such as trees, reduced population size does not necessarily decrease the profitability of harvesting the species. This situation is illustrated by the continual harvest of old-growth forest around the world, despite an enormous reduction in the area of forest. Government intervention is often necessary to ensure that natural resources are harvested prudently, particularly regarding fisheries and forests.

> A bioeconomic strategy that maximises short-term profits invariably leads to the over-exploitation of natural resources.

Fig. 42.13 Commercial fishing of school and gummy sharks in Bass Strait

VIABLE POPULATION SIZES FOR CONSERVATION

One aim of conservation biology is to maintain biodiversity by minimising the possibility that species become extinct. A fundamental question relevant to the maintenance of biodiversity is, 'What is the population size above which a species needs to be maintained to ensure its long-term survival?'. This population size is termed the **minimum viable population size** (MVP) of a species. The precise definition of what is 'viable', together with the characteristics of the ecology and behaviour of a species, have an enormous impact on a species' MVP.

What is an acceptable probability of extinction? There is no population size that can guarantee the persistence of a species indefinitely. All species will eventually become extinct! Consequently, the definition of a minimum viable population size must be accompanied by an acceptable probability of extinction over a fixed time interval. For example, an acceptable extinction probability over a period of 1000 years may be 2%, while an acceptable extinction probability over a period of only 100 years may be 1%. These figures will depend on the value society places on the species concerned and the cost of management practices that increase the chance the species will persist. In general, the probability that a species becomes extinct is a function of its distribution and abundance.

> There is no single minimum viable population size that applies to all species under all circumstances.

Before discussing causes of extinction and the contribution they make to the MVP of a species, we need to discuss the related concepts of chance, uncertainty and stochasticity. It is tautological to say that a species becomes extinct when the last individual of that species dies. However, it is worth noting that this extinction event is, by its very nature, a chance event. Precisely when an individual dies is largely a matter of chance. Chance or uncertainty are characteristics of the various processes that may cause extinction. Scientists call processes that include elements of chance and uncertainty stochastic processes.

Four processes of chance (stochasticity) may cause, alone or in conjunction, the extinction of a species: genetic stochasticity, demographic stochasticity, environmental stochasticity, and catastrophes. These processes interact in complex ways with attributes of the population, such as its age structure and distribution in space, to determine the likelihood that a species becomes extinct over a given period of time. For example, it is estimated that there is an 86% chance that the Sumatran rhinoceros, *Dicerorhinus sumatrensis*, will become extinct within the next 30 years (Fig. 42.15).

Fig. 42.15 The Sumatran rhinoceros, *Dicerorhinus sumatrensis*, is threatened with extinction

The degree of genetic heterozygosity of a population is known to affect the average fecundity and survivorship of natural populations. Small populations lose genetic variability rapidly as a consequence of genetic drift (Chapter 32).

Demographic stochasticity is a process that describes the random nature of births and deaths. The birth and death of individuals are instantaneous events so population sizes do not change smoothly. For example, given two individuals, each with a probability $\frac{1}{2}$ of dying in one year, and each with a probability $\frac{1}{2}$ of giving birth in one year (before death), the probability of the number of individuals next year will be: 4 individuals, $\frac{1}{16}$; 3 individuals, $\frac{4}{16}$; 2 individuals, $\frac{6}{16}$; 1 individual, $\frac{4}{16}$; and 0 individuals, $\frac{1}{16}$. On average, this population should remain constant; however, because of demographic stochasticity, there is a significant chance that the population may become extinct after only one year; indeed this population is very unlikely to persist for more than 10 years.

The variable nature of the environment and the possibility of catastrophes (e.g. cyclones, epidemics, fire) impose further levels of stochasticity on the dynamics of populations. In the eucalypt forests of Australia, fires may wipe out entire populations of arboreal marsupials (Fig. 42.16). Conversely, maintaining an appropriate fire frequency, and thereby a series of vegetation successions (Chapter

Fig. 42.16 Fire is one factor that causes variability in an environment and imposes a level of stochasticity on the dynamics of populations

43), is essential to the survival of species, such as the heath-dwelling ground parrot, *Pezoporus wallicus*.

Incorporating all the processes of extinction into a model with details of age structure, life history and population spatial pattern has not yet been achieved for any Australian animal; however, a detailed population viability analysis has been carried out for the northern spotted owl *Strix occidentalis caurina* in an attempt to assist the management of the old-growth forests in the western United States. In general, conservation biologists estimate that the MVP of a species is of the order of several thousand individuals under normal conditions. This sort of population size ensures not only a very good chance of persistence in the long term but also gives the population some chance of adapting to environmental change through natural selection.

SUMMARY

- Population ecology is devoted to a mechanistic understanding of the dynamics of the distribution and abundance of species.

- Experiments identify biotic and abiotic factors that affect the four processes of population change: birth, death, immigration and emigration. Mathematical theory is used to formalise these experimental results and make predictions about population change.

- Theoretically, a closed population with density-independent death and birth rates ultimately either will increase to an infinite size or will decline to extinction depending on whether the geometric mean of its growth rate is greater or less than one.

- In the continuous time model of population growth, the intrinsic rate of increase (r) determines the fate of the population. If $r > 0$, the

population grows exponentially; if $r < 0$, the population declines towards zero; and if $r = 0$, the population remains constant.

- A finite resource, such as food, may constrain the exponential growth of natural populations; however, environmental fluctuations in space and time often mean that resources are rarely limiting for many species.

- Age and size affects the fecundity and survival of individuals of most species. Thus, the age structure of a population affects population growth.

- Mathematical models are used to manage natural resources, control pest species, and reduce the probability that species become extinct. There is no single minimum viable population size that applies to all species under all circumstances.

QUESTIONS

1. The Australian National Parks and Wildlife Service is interested in determining the number of saltwater crocodiles, *Crocodylus johnstoni*, in a 10 kilometre stretch of the Adelaide River. On the first day they catch and mark 40 crocodiles with a blue tag. On the second day they catch and mark 50 crocodiles with a red tag, 20 of which were already marked with a blue tag. What would be your estimate of the size of the population? During a third visit to the area they catch 40 more crocodiles, 10 have both tags, 10 have no tags, 10 have only a blue tag and 10 have only a red tag. Use this result to generate two more estimates of population size. Discuss problems associated with the census method.

2. An ecologist transplants 60 snow gums, *Eucalyptus pauciflora*, into three sites in the Australian Alps, 20 plants above the current treeline, 20 at the treeline and 20 below the treeline. After 10 years of careful monitoring the ecologist finds that the snow gums survive equally well at all three sites but that the individuals above the treeline have not grown as fast as those at or below the treeline. Discuss the implications of these results for the environmental distribution of *Eucalyptus pauciflora*.

3. Soil moisture, temperature and nutrients are environmental gradients that are said to affect

plant distribution and abundance directly. Altitude, rainfall, and soil type are said to affect plant distribution and abundance indirectly. Discuss.

4. In a study of frog competition, a zoologist manipulates the numbers of frogs in a number of isolated farm dams. Fifteen initially empty, almost identical, dams are stocked with tadpoles at five different densities (three dams for each density) in the early spring. The number of adult frogs in each dam is determined by sampling on three successive years. The mean adult abundance for each treatment is tabulated below.

Initial stocking density	1992	1993	1994
10	8	0	0
20	30	30	80
40	60	40	100
80	60	80	100
160	20	30	60

Discuss these results in the context of intraspecific competition and a variable environment. What other experiments could be conducted to explore the possibility that resources are limiting?

5. A butterfly population in the mallee is believed to respond directly to summer rainfall. When the summer rainfall is below 100 mm, each adult female produces, on average, 0.1 adult females; when between 100 mm and 200 mm the annual fertility is 1; when above 200 mm, annual fertility is 5. If the probability of summer rainfall being one of these three rainfall classes is equal, will the population increase or decrease in the long term? Do you think that this is a permanent or ephemeral population? If you are familiar with programming, write a computer program to simulate the dynamics of the species and estimate the probability that the population is extinct after 10 years given an initial population size of 100.

6. Consider a biennial herb that sets seed twice in its life. In its first year the average plant produces 100 seeds and in its second year the average plant sets 200 seeds (and then dies), but there is a 25% chance that the herb dies between the first and second seed set. If the chance of a seed germinating and reaching the age at which it first sets seed is 0.7%, on average, how many mature plants does each mature plant produce?

7. A butterfly species lives in two regions. In one region, the species exists as only one population that inhabits a large area capable of supporting 1000 individuals. In the other region the species can inhabit three areas each capable of supporting 100 individuals. Discuss the effects of catastrophes and migration between subpopulations on the viability of the species in the two regions.

8. Years of study suggest that a closed fish population displays dynamics similar to that predicted by the logistic equation

$$\frac{dN\,(t)}{dt} = rN\left(1 - \frac{N}{K}\right)$$

with $r = 1$ and $K = 20\,000$. Predict the change in population size when the initial population size is 1000 fish and the harvesting rate is (a) 1000 fish per annum, (b) 500 fish per annum and (c) no fish per annum.

9. After a cyclone, an ecologist monitors the recovery of trees in a tropical rainforest. The number of trees of different size classes is counted in a 1 ha plot and the results tabulated below. Discuss the data in the context of self-thinning, the law of constant yield and plant growth.

Size (m³)	Year			
	1950	1960	1970	1980
0.0–0.5	1000	200	100	50
0.5–1.0	10	180	75	40
1.0–2.0	20	5	70	30
2.0–4.0	10	5	2	40
4.0–8.0	6	1	0	0

10. Consider a common species of barnacle that dominates part of the rocky intertidal zone. A starfish predates on the barnacle, selectively choosing patches of about 10 cm² and consuming all individuals in the patch. Discuss the consequences of this predation on patterns in the distribution of adult barnacles at different spatial scales.

LIVING IN COMMUNITIES

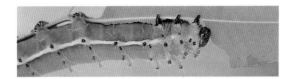

In Chapter 42, we examined the ecology of populations. Populations, however, usually occur in natural assemblages, **communities**, and the neighbours of any one individual may be individuals of other species as well as those of its own. Individuals of different species interact in a variety of ways: animals eat and fight, plants shade one another, and animals pollinate plants. These interactions include symbiosis, predation and competition, all of which affect the distribution and abundance of species. In the long term, such interactions may lead to evolutionary change. For example, in rainforests in northern Queensland, there is synchronisation of flowering and insect activity: peaks in the flowering of tree species coincide with peaks in the number of insects in the same area. Such interactions are important in determining the structure and species' composition of communities, the characteristics of which are dealt with in the second part of this chapter.

SYMBIOSIS

Symbiosis is a term used for interactions in which two organisms (symbionts) live together in a close relationship that is beneficial to at least one of them. A lichen is a symbiotic relationship between an alga and a fungus. A mycorrhiza is an association between a fungus and the root of a higher plant (Chapter 36).

Where the characteristics of interacting species evolve in concert, resulting in a one-to-one relation, the relationship is regarded as an example of **coevolution**. An orchid flower that mimics the shape of a female wasp, tricking males into copulating with it, thus resulting in pollination (Chapter 37), is one such case.

Symbiotic relationships are the rule rather than the exception in most ecological communities, especially in complex communities, such as tropical rainforests and coral reefs. Symbiotic interactions are of three

kinds: one species benefits and the other is unaffected—**commensalism**; both species in the relationship benefit from the association—**mutualism**; or one species benefits and the other is harmed—**parasitism**. In some associations, one organism is physiologically dependent on another; in others, organisms can just as easily live apart. Symbiosis usually involves providing protection, food, cleaning or transportation.

> Symbiosis is where two organisms (symbionts) live together in a close relationship that benefits at least one of them. It includes mutualism (where both species benefit), commensalism (one benefits, one is unaffected) and parasitism (one benefits and one is harmed).

Commensalism

In moist forests, most tree trunks are covered with mosses, small ferns and orchids. These plants are **epiphytes**: they benefit from living on the trunk of a host tree, which provides a substrate, a catchment for rainwater in which mineral nutrients are dissolved, and a position closer to the light. Epiphytes do not appear to affect the host tree adversely. In marine communities, epiphytic algae attach to crabs and molluscs, thus having somewhere to live and gaining free transportation. In some cases, there may be a benefit to the host through camouflage but the host does not depend on the algae.

Interactions between species can change over time. In tropical forests of South-East Asia and Australia, the strangler fig commences its life as an epiphyte. A seed delivered in the droppings of a bird germinates on the host tree and the young fig commences to grow (Fig. 43.1a). In the early stages of the association, the fig benefits and the host is not affected. However, when the fig grows, it extends its roots down to the soil and envelops the host, eventually crushing it (Fig. 43.1b). The relationship changes from commensalism to one of aggressive competition for space (see p. 962).

(a)

(b)

Fig. 43.1 (a) This strangling fig in a Queensland rainforest starts off life as an epiphyte (commensalism), germinating from a seed that has landed on the host tree. (b) An older fig has enveloped its host tree and is now an aggressive competitor

Mutualism

The symbiotic relationships of lichens and mycorrhizae (Chapter 36) are examples of mutualism, since both partners in each association benefit. Mutualistic relationships are particularly common in coral reef communities.

All reef-building corals (Chapter 38) have abundant symbiotic algae within their tissues—up to 30 000 algal cells per cubic millimetre of coral tissue (Fig. 43.2). The algae are dinoflagellates (Chapter 35), called zooxanthellae, which have yellow–brown pigments that give the coral its colour. Without this mutualistic relationship, the formation of coral reefs such as the Great Barrier Reef would be impossible. Within the cells of coral polyps, zooxanthellae live, reproduce, photosynthesise and utilise the waste products of the animal host (carbon dioxide, phosphorus and nitrogen). In turn, the coral uses oxygen and food produced by the algae during photosynthesis to grow, reproduce and form its hard skeleton, which is the basis of the reef. Corals regulate the number of zooxanthellae they contain, picking up new ones as required and 'spitting out' the excess. Since zooxanthellae need light for photosynthesis, reef-building corals grow only in clear waters less than 100 m deep; disturbance that makes water murky threatens the life of reefs.

Fig. 43.2 Reef-building corals have a mutualistic relationship with algal cells, zooxanthellae, living in their tissues. Close up view of branching coral, *Acropora sarmentosa*, branch shows dark colouration of tentacles due to populations of zooxanthellae

Among other coral-reef animals, sea anemones have a number of mutualistic relationships. The most well-known example is the association between the sea anemone and the clownfish (Fig. 43.3). The clownfish is one fish that is neither stung nor eaten by the anemone. The clownfish acclimatises itself to the sting of the anemone by repeatedly brushing itself against the tentacles until its own mucous coating inhibits the anemone's sting. The fish is thus protected from predators and feeds on scraps of the anemone's food. The anemone benefits in turn when the clownfish cleans its host and lures other animals into the anemone's tentacles.

Some species can have more than one symbiotic association. The 'ant plant' *Myrmecodia* is an epiphyte that also has a mutualistic relationship with a species of ant. The plant has a swollen base or tuber, in which there are specialised chambers. The ants form large colonies within these chambers, and they carry their excreta and the corpses of their prey to parts of the chambers (a cemetery) where the plant is able to absorb the waste nutrients. This relationship is called **myrmecotrophy** (Box 43.1).

Fig. 43.3 The clownfish is unharmed by the tentacles of its symbiotic partner the anemone

BOX 43.1 ANTS FEEDING PLANTS

The flowering plant genera *Myrmecodia* and *Hydnophytum* (family Rubiaceae) from South-East Asia and north-eastern Australia have spectacular swollen tuber-like organs that are inhabited by ants of the genus *Iridomyrmex*. The ants live in special chambers called **domatia** (see Fig. a). They pack adjacent tunnels with their waste material. Radioactive labelling of prey or honey water given to foraging ants from these colonies shows that nutrient ions are absorbed from the colony waste, translocated and incorporated into the plant's tissue. Bacteria and fungi in the chambers probably facilitate decomposition of the wastes. The mineral nutrients of the plant are thus augmented and the ant colony functions as a second root system.

In South American bromeliads (plants of the pineapple family) that also have myrmecotrophic associations, absorption of such waste nutrients appears to be in areas of plant tissue that lack cuticle or where there are specially absorptive trichomes (hairs etc.). In other plants, where the walls of domatia are thick and apparently impenetrable to nutrients, these ants tend homopteran insects that also live in the domatia; the ants chew through to the vascular tissue of the host plant so that the homopterans have access to sap. The ants use the insects rather like cattle, cleaning them, feeding them and collecting the honey dew that the insects secrete.

There are about 350 known plant species (in more than 70 families), some of which occur in Australia, that have domatia, that is, structures that accommodate ants. These species occur in tropical and subtropical environments. Many structures function as domatia.

(a) The epiphytic 'ant plant', *Myrmecodia tuberosa*, photographed in a forest in Borneo, has specialised chambers in its swollen pseudobulbs, in which ants live in a mutualistic relationship

Those that are lived in more or less permanently include tubers, swollen thorns (e.g. found in *Acacia*) and hollow stems and twigs (see Fig. b). Leaf pouches and swollen petioles are less permanent (ephemeral) homes.

Benefits of ant–plant associations

Analysis of the costs and benefits of myrmecotrophy is not simple. There is a cost to the plant in producing domatia or having its sap tapped for homopterans to feed. There is the benefit to the plant of increased nutrients from the ants' wastes, especially for a plant that is an epiphyte living in a nutrient-poor environment, such as on the trunk of a tree. Many of the ant species implicated in myrmecotrophy also perform other services for the plant: they defend and protect the plant from herbivores and seed-eaters. In turn, some host plants offer food rewards such as nectar secreted in extrafloral nectaries and other food bodies.

We assume that these relationships are mutualistic and that they are the result of coevolution—the plants are certainly structured to accommodate ants. However, we are less sure that the ants' traits are special adaptations because the traits also occur in ant species not involved in myrmecotrophy.

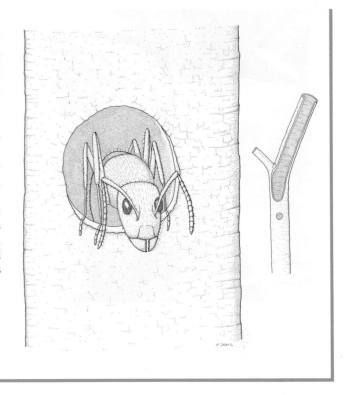

(b) A domatium formed from the hollow stem of *Endospermum myrmecophilum*. The entrance hole has been gnawed through the twig by the ant *Camponotus quadriceps*

Parasitism

Parasites live in an obligatory association with one or more host species for most of their lives. They gain sustenance from their host, who is harmed but not often killed. Parasites are usually smaller than their host. They may live on the surface of their host (**ectoparasites**, e.g. lampreys that attach to other fishes) or internally (**endoparasites**, e.g. tapeworms, stem-rotting fungi). **Parasitoids** are insects that are free-living in the adult stage but are parasites as larvae (Fig. 43.4). Adults lay their eggs in a host, usually the larvae of another insect species. Larvae of the

parasitoid develop in the host doing no harm for a while, but they eventually consume and kill the host before pupation.

We have looked at the life cycles and biology of some parasites in Chapters 35 and 38, but how do parasites affect the abundance of their hosts? Parasitism is a special type of predation, and the relationship between host and parasite abundance is similar in some ways to that of predators and their prey.

Fig. 43.4 A female ichneumonid wasp, *Diadromus collaris* (Gravenhorst), ovipositing into the pupa of the cabbage moth, *Plutella xylostella* (Linnaeus), its natural enemy. The wasp was introduced into Australia as a biological control agent

PREDATION

Predation in a broad sense includes one kind of animal, a true **predator**, capturing and eating another as well as an animal, a **herbivore**, consuming a plant or part of a plant. Some plants even consume animals (Fig. 43.5).

Predators and herbivores differ from parasites in not usually having a close obligatory association with a host, but all three can affect the abundance of their animal or plant food. Herbivorous insects that feed on flowers or seeds, for example, can reduce significantly the number of viable seeds produced by a plant. Larvae of the moth *Agathiphaga queenslandensis*, which is a highly specific feeder, destroy large quantities of the seed of kauri pines (*Agathis*) in rainforests of northern Australia.

Fig. 43.5 The pitcher plant traps and digests insects in highly modified leaves

Abundance of predators and prey

The abundance of a predator and its prey can oscillate through time, with the predator numbers tracking those of the prey. When there are large numbers of prey available, the predator population increases in size. As prey are consumed, their numbers decline, leading to a shortage of food for the predators, whose numbers also decline (Fig. 43.6). Such oscillations are termed **predator–prey cycles**.

Biologists have known for centuries that the numbers of individuals in many natural populations fluctuate in a predictable, regular fashion, and that such cycles can occur in both predators and their prey. In the early part of the 20th century, ecologists began to develop mathematical models to predict these changes. Currently in the literature, there are numerous models that predict fluctuations, ranging from simple damped oscillations to extraordinarily complex, even chaotic, dynamics. Predator–prey cycles are not, however, universal. Even when cycles

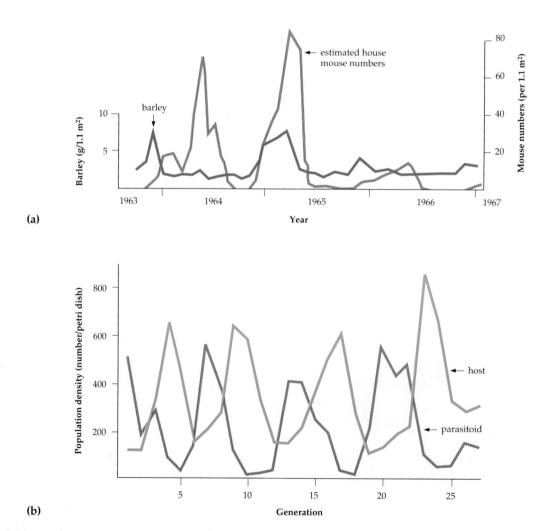

Fig. 43.6 (a) The relationship of predator and prey abundance: the house mouse, *Mus musculus*, and barley, *Hordeum vulgare*, in South Australia 1963–7. **(b)** The number of parasitoid wasps, *Heterospilus prosopidis*, oscillates in relation to its host the bean weevil, *Callosobruchus chinensis*

are observed, it can be difficult to distinguish between fluctuations that are due to interactions between predators and their prey, and correlated fluctuations that are due to random variations in the environment.

One reason that cycles are not universal is that predators and their prey do not normally exist in simple two-species' communities isolated from other organisms. Once a food source becomes limiting, a predator may switch to another food source, or the predator itself may be kept in check by its own predators. The relationship between two species may also depend on their stage of development. For example, insect herbivores switch food sources as they develop from larvae to adults; caterpillars often need young leaves as a suitable nutrient source to complete development, while adult moths may feed on flower nectar (Fig. 43.7).

Another factor that complicates the relationship between predator and prey is the *quality* of food, that is, the prey itself and not just its abundance. Despite being abundant, plant food may be of such low quality that herbivores barely have enough energy for maintenance, let alone growth and reproduction (see also Chapter 19). Herbivores are affected particularly by the nitrogen content of their food (Fig. 43.8). Peaks in insect populations, including outbreaks of pest species such as psyllids in eucalypt forests, have been related to peaks in the quality of available food measured as soluble nitrogen in phloem sap (see Box 44.4).

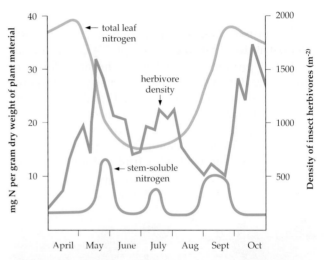

Fig. 43.8 Seasonal changes in the number of insects on the grass *Holcus mollis* in relation to changes in food quality, measured as nitrogen in leaves and stems

The abundance of a predator or herbivore and its prey may show regular oscillations—predator–prey cycles.

Predation and biological control

Predators and herbivores can be useful for biological control of pest species. Control is best where a predator that is monophagous (feeds on a single prey type) virtually exterminates its prey and then becomes scarce itself but does not become extinct. The survival in small numbers of both species makes it possible for a small population of predator to control the prey indefinitely.

The most famous example of successful biological control is the use of the moth *Cactoblastis*, which was introduced to control vast outbreaks of prickly pear cactus, *Opuntia*, in Queensland and New South Wales. The larvae of the moth fed on the cactus and reduced the *Opuntia* population to a tiny fraction of its original size. Other Australian examples are water hyacinth and *Salvinia* (Chapter 37), which are introduced aquatic weeds that grow unchecked by local herbivores. One of the organisms used to control

(a)

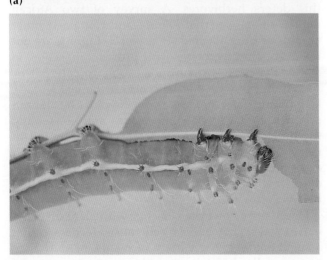

(b)

Fig. 43.7 **(a)** Grapevine hawk moth feeding on nectar of a grevillea flower. **(b)** Larvae, such as this caterpillar of the gum emperor moth, utilise a different food source from adult moths

water hyacinth, *Eichhornia crassipes*, is the South American weevil *Neochetina eichhorniae*, the larvae of which tunnel into the plant's tissues, chiefly into petioles. The Brazilian weevil *Cyrtobagous singularis* is being used as part of a program for biological control of *Salvinia* in Australia.

Attempts at biological control can, of course, also be disasterous. The introduction of the cane toad, *Bufo marinus*, to Australia to control pests of sugar cane led to a population explosion of the toad, which is now a major pest species.

Fruit eaters and seed dispersal

Some herbivores specialise as fruit eaters, **frugivores**. In a northern Queensland rainforest, a CSIRO scientist noted that about 84% of 774 plant species examined produce fleshy fruits. The abundance of fruit in a rainforest means that a diversity of frugivores can exist, including many mammals and birds (Table 43.1). The plants also benefit because these frugivores ensure the dispersal of seeds, which usually pass unharmed through the animal's digestive tract. They digest only the soft wall of the fruit and not the seeds. Fruit pigeons (Fig. 43.9) and the cassowary have a thin-walled gizzard with hardened lumps that

Fig. 43.9 The rose-crowned fruit pigeon, *Ptilinopus regina*, is a frugivore and important to a number of plant species for dispersal of seeds

Table 43.1 Major bird frugivores in Australian tropical rainforests (*a*)

Common names of frugivores	Number of plants	
	Species	Families
Southern cassowary	115	46
Superb fruit-dove	54	27
Rose-crowned fruit-dove	16	11
Wompoo fruit-dove	61	30
Torresian imperial pigeon	37	20
Topknot pigeon	20	10
White-headed pigeon	14	9
Brown cuckoo-dove	32	23

(*a*) Note the large number of species and families of fruits recorded in their diets.

massage the soft parts away from the seed and do not grind up the seed, as happens in the gizzards of other birds.

> Frugivores, animals that eat fruit, are important for seed dispersal.

Plant defences

Plants have many defence mechanisms to avoid being eaten. Plants often have *physical* defences, such as tough leaves, spines and hairs, which generally decrease a plant's susceptibility to insect attack (Fig. 43.10). The volatile oil and hairs of many young eucalypt seedlings makes them less palatable to herbivores (see Chapter 6).

Fig. 43.10 Spines on plants, such as *Solanum cardleyi* in the Northern Territory, help to deter herbivores

A second mechanism of defence is *chemical*. Many plants are unpalatable or toxic to some extent, and this deters herbivores. A survey of two African rainforests showed that 90% of plants contained either tannins or toxic alkaloids (Fig. 43.11). Tannins are not toxic but they combine with proteins in leaves in a way that makes them indigestible to caterpillars. As tannin-filled vacuoles are nearer the leaf surface, leaf-mining insect larvae can avoid the tannin problem by burrowing into leaves and consuming the inner tissue. Chemical defence is also found in marine organisms. Many marine red, brown and green algae produce **terpenoids** and other compounds that deter herbivorous fishes (as well as inhibiting bacterial growth).

Herbivores, however, can cope with plant defences in two ways. They may be **generalised feeders**, eating a small amount of a variety of plants, or they may be **specialised feeders**, with a mechanism for detoxifying their preferred plant food. Most herbivores possess an array of detoxifying enzymes, which enable generalised feeders to eat small quantities of each toxic species. Colobine monkeys,

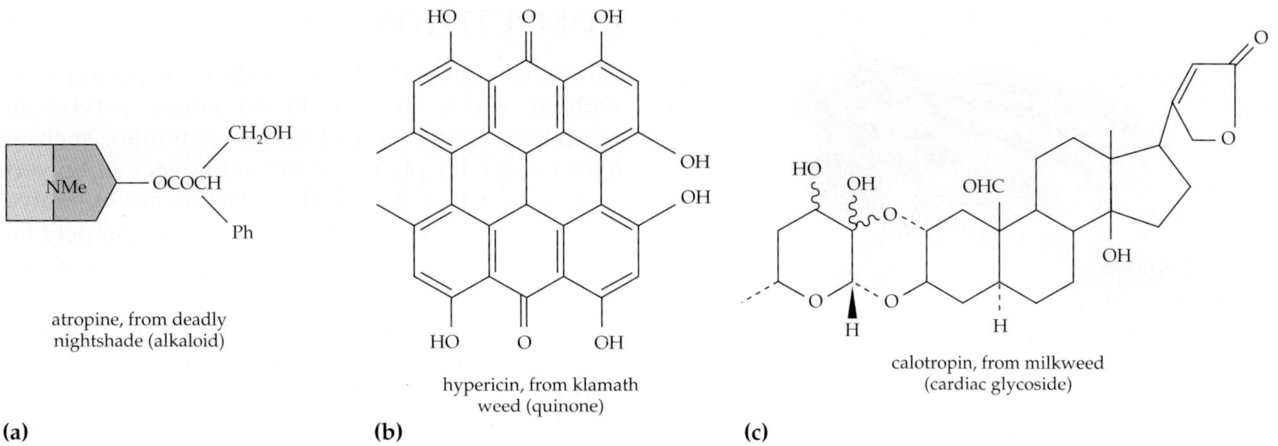

atropine, from deadly
nightshade (alkaloid)

(a)

hypericin, from klamath
weed (quinone)

(b)

calotropin, from milkweed
(cardiac glycoside)

(c)

Fig. 43.11 Plant secondary compounds, that is, compounds that do not play a role in primary metabolism, are important chemical defences against herbivores. There is a variety of compounds, many of which are useful medicinally. **(a)** The potato family (Solanaceae) is rich in alkaloids, one of which is atropine, the principal toxin of deadly nightshade, *Atropa belladona*. **(b)** Hypericin is a quinone found in the klamath weed, *Hypericum perforatum*. **(c)** Calotropin is the major cardiac glycoside found in milkweeds (*Asclepias*), which monarch butterflies feed on

for example, have microflora in the forestomach which can detoxify at least some alkaloids. The grey kangaroo (*Macropus*) in Western Australia can tolerate poisons present in the leaves of many members of the pea family (Fabaceae). One of the compounds, fluoroacetate, is similar to the highly toxic rabbit poison known as '10/80'. The natural levels of these poisons can be lethal to introduced animals of similar body size, such as sheep and cattle.

> Plants are protected from herbivores by physical or chemical defences, although many animals have evolved ways to cope with plant toxins.

Animal defences

Animal defences against predators include speed and agility, large size, camouflage, toxic chemicals, physical barriers such as thick shells, physical defences such as horns and antlers, and behavioural adaptations such as living in social groups (Chapter 28). Modification of shape, colour and form for camouflage can be extreme in insects. Species of *Phyllium* (family Phasmatidae, 'stick insects') mimic leaves or twigs; the insects have large, leaf-like expansions of the legs and abdomen and flattened green wings. A catydid insect mimics the shape of an *Acacia* phyllode (Fig. 43.12). Black and white larvae of some papilionid butterflies resemble bird droppings. Disruptive black and white colour patterns afford camouflage in the dappled light beneath a forest canopy.

Many animals release toxic or irritating compounds that kill, damage or at least ward off predators; the secretions of some caterpillars, the stings of bees, and the bite of the blue-ringed octopus are examples. Animals that use chemicals for defence often advertise

Fig. 43.12 A small, unnamed catydid insect from Central Australia on mulga, *Acacia aneura*. As an adult, the catydid mimics the phyllodes of the wattle on which it spends the daylight hours. After dark, the insect becomes active, feeding on the phyllodes and perhaps the flowers of the host. This species is, as yet, unnamed even to the genus level (it is known in the Australian National Insect Collection as Genus 20, Species 1). It is not known to occur on any species other than *A. aneura*

the fact that they are toxic by **warning colouration** (Fig. 43.13). Bold patterns of red, orange or yellow, often with black stripes, are common colour patterns, and it is thought that predators learn to avoid warning colouration. Red-back spiders, *Latrodectus basseltii*, have a warning mark on their abdomen, which they expose when they hang upside down from their webs.

The monarch butterfly, *Danaus plexippus* (Fig. 43.14), is unpalatable because, as a larva, it accumulates toxic compounds, cardiac glycosides, from its diet of plants in the milkweed family (Asclepidaceae). It is a specialist feeder that is able to sequester the toxin and store it. The toxin, which passes through to the adult butterfly stage (and even the eggs), will cause a predatory bird to vomit and regurgitate its prey.

Fig. 43.13 Warning colours are often red, orange or yellow, striped with black. The Australian frog *Pseudophryne corroboree* produces toxic alkaloids in the skin and is distinctively coloured gold with black stripes

Fig. 43.14 Insects have not only become resistant to the chemical defences of plants but some, such as the monarch butterfly, *Danaus plexippus*, use plant food as a source of chemicals for their own defence against predators. Monarch butterflies advertise their unpalatability by warning colouration

The bird quickly learns to leave other individuals alone, especially because monarch butterflies, both as caterpillars and adults, have bold warning colouration patterns.

Other species of butterflies that are palatable have exploited the unpalatability of the monarch by **mimicry**: their colouration pattern is so similar to the monarch's that they too are afforded protection from predators. Bees, wasps and ants, which have stings, are also models for mimics. One family of flies mimics bees and wasps. Some spiders mimic ants by holding their first pair of walking legs in the air to look like antennae. When harmless species mimic toxic models it is called **Batesian mimicry** (see also Chapter 28).

> Animal defences against predators include speed, agility, size, camouflage, toxic chemicals, physical barriers and behavioural adaptations. Toxic animals often advertise their unpalatability by warning colouration. Other species mimic unpalatable forms so that they too are protected from predators.

COMPETITION

All animals and plants have a number of requirements without which they could no longer survive or reproduce. It is easy to think of examples, such as light energy for plants, or the abundance of flowers and fruit for fruit bats. If these factors are in limited supply, individuals within a species will compete for them. We measure the relative success with which individuals cope with such limitations in terms of the number of offspring they produce compared with their competitors. Competition between members of a species is **intraspecific competition**.

Different species living in the same geographic space also compete for limited resources. This is **interspecific competition**. It happens when, for example, different species of lizards compete for a limited supply of insects, or when different species of rainforest trees germinating in a newly created gap in the canopy compete for sunlight. Usually, interspecific competition is less intense than intraspecific competition.

Just as for competition among individuals within a species, some species are, on average, better competitors than others. In an experiment in the 1950s, Charles Birch observed the population sizes of two species of grain beetles living in wheat. In isolation, populations of both species could persist indefinitely. However, when the species were sharing the same environment, one species was always driven to very low numbers or became extinct (Fig. 43.15). Individuals of the less successful species were out-competed for food by individuals of the species that eventually replaced it.

Such graphic experiments have the ability to colour the thinking of biologists for decades. The phenomenon demonstrated in such experiments became known as the **competitive exclusion principle**: if two species have the same mode of life and use the same kinds of resources, they will be unable to coexist.

Of course, these experiments were based on populations of very simple organisms living in a very simple environment, namely an experimental tank filled with wheat. The real world is much more complex and we should not expect that, for any two species that compete for a resource, one will always replace the other. We only need to look around to observe that similar species coexist. Many species of leaf-eating insect exist on any rainforest tree. Numerous native mammal species roam the Australian semiarid plains, competing for limited supplies of water and edible plants.

> Interspecific competition is competition between members of different species for limited resources. Competitive exclusion occurs when one species outcompetes another for a limited resource, resulting in its local extinction. Usually, interspecific competition is less intense than intraspecific competition.

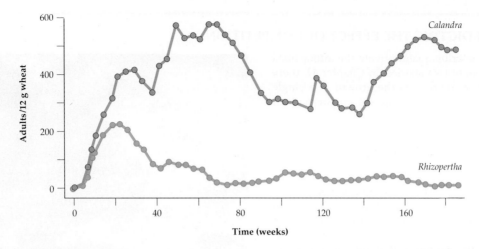

Fig. 43.15 Population abundances of two species of grain beetle, *Calandra* and *Rhizopertha*, living together. Birch found he could reverse the winner and loser of the competition by adjusting the temperature of the environment
From Charles J. Krebs, *Ecology: The Experimental Analysis of Distribution and Abundance*, Harper & Row, New York, 1985, with permission

COEXISTENCE
Ecological niche

In the late 1950s, G.E. Hutchinson defined the concept of the **fundamental niche** as that region of the environment within which a species can persist indefinitely. It is defined by all the biotic and abiotic factors that impinge on the survival and reproduction of the species. Hutchinson saw the environment as composed of a number of independent dimensions, each one representing a factor that could limit the distribution and abundance of a species. The range of tolerance of a species to each of these factors determines a multidimensional niche.

The giant kelp *Macrocystis pyrifera* lives in the ocean near the south-eastern coast of Australia and on several Pacific coastlines. Three things, among others, that limit where this species lives on any particular coastline are:

1. Light. The species can tolerate water depths down to about 20 m. Gametes settle to the bottom, where fertilisation takes place, but at depths below about 20 m there is usually insufficient light for zygotes to develop. The maximum depth the species can tolerate is less when the turbidity of the water column is increased.

2. Substrate. The species is only found where there is rocky substrate. The holdfasts require it to keep a grip on the sea floor and the gametophytes may become buried in sand or mud.

3. Nutrients. The species is restricted to relatively cool waters. Cool ocean waters are derived from upwellings that bring nutrients from the ocean floor. Warm surface waters often are depleted of nutrients, especially nitrogen, and the species is intolerant of nutrient-poor conditions.

Keeping Hutchinson's model of a multidimensional niche in mind, we can visualise each independent factor as a resource axis. Thus, for *M. pyrifera* we can define three independent axes, one each for light, substrate and nutrients (Fig. 43.16). It is possible that

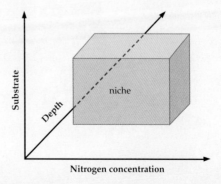

Fig. 43.16 An idealised three-dimensional niche space for *Macrocystis pyrifera*. Each axis represents a range of environmental conditions for three factors that limit the growth of the species. The edges of the cube represent the limits of each factor that the species can tolerate, and the cube represents the fundamental niche for the species

BOX 43.2 PREDICTING THE EFFECT OF COMPETITION

It is possible to describe quantitatively the interactions between members of two species. In Chapter 42, there is an equation that represents the growth of a single population in an environment that will support a given number of individuals:

$$\frac{dN}{dt} = rN\left(\frac{K - N}{K}\right) \qquad [43.1]$$

where N is the population size within a confined region, K is the carrying capacity of the environment, and r is the maximum rate at which the population can increase when resources are not limiting. If we use this equation to represent a population, we are saying that one or more of the resources that the individuals in the population require are limited, so that no more than K individuals can ever be supported.

We can represent the competitive interactions between members of two species in an equation analogous to 43.1. We let α represent the amount of a resource taken up by a member of species 2, relative to the amount taken up by a member of species 1. Then the growth of the population of species 1 is given by

$$\frac{dN_1}{dt} = rN_1\left(\frac{K - N_1 - \alpha N_2}{K}\right) \qquad [43.2]$$

(b) Two species in the intertidal zone of a Victorian seashore involved in aggressive competition for space. The red animal is a Victorian species of bryozoan, *Mucropetraliella ellerii*. The colonial ascidian, *Distaplia viridus*, is enveloping the bryozoan and will eventually kill it and overgrow it. Such interactions together with differential resistance to desiccation may lead to patterns of zonation on rocky seashores

This equation says that the growth rate of species 1 is determined by the number of individuals of species 1 (N_1), and by the number of individuals of species 2 (N_2), multiplied by a term, α, that represents the strength of the competitive effect of species 2 on species 1. If the effect is small, α will be small and species 2 will not reduce the growth rate of species 1 very much unless it is present in very large numbers.

Usually, α is less than 1, which means that adding another member of species 1 is felt more by species 1 than adding another member of species 2. Most species use the environment in different ways. Even if two species compete for a single limited resource, such as a set of food items, there will be other resources for which they do not compete. For example, there will be food palatable to one species but not the other.

When α is greater than 1, either interference or exploitation competition is implied. In a sense, species 2 is aggressively interfering with the growth of the population of species 1. For example, the ecologist J. Connell, researching intertidal communities in Scotland, showed that when two species of barnacle compete for limited space on rocks, the faster growing species (*Balanus balanoides*) excludes the other (*Chthamalus stellatus*) from the lower reaches of the intertidal zone by smothering, undercutting or crushing young individuals before they become established. *B. balanoides* is sensitive to desiccation and is unable to colonise the upper zone where *C. stellatus* persists.

(a) Coral on the Great Barrier Reef. The patterns of zonation are due in part to interspecific competition for light

species compete on one resource axis but not on another. For example, another species of kelp may compete with *M. pyrifera* for light. However, if the second species has different metabolic requirements, there may be no competition between the species for nutrients.

Competition with other species limits the ability of a species to take full advantage of the environment. Competition on the light axis will reduce the ability of *M. pyrifera* to live in places where it otherwise may have survived and reproduced. This is especially true in those places where *M. pyrifera* is at the edge of its tolerance. Any reduction in available light through competition will make the place uninhabitable. The **realised niche** is that part of the environment where a species *does* persist (Chapter 42).

> The fundamental niche of a species is that region of the environment, that is, of the multidimensional niche, within which a species can survive and reproduce.

Niche breadth

Some species can make use of a broader range of an axis than other species. Two species of small marsupials, the yellow-footed antechinus, *Antechinus flavipes*, and the brown antechinus, *A. stuartii*, coexist in open forest north of Canberra. Both species are nocturnal, partly arboreal, and are similar in many aspects of behaviour. Both species eat insects and we can view the available insect population as a resource axis, ranging from very small insects at one end to very large ones at the other (Fig. 43.17). One

way to measure the use the species make of the insect population is to count how many of each type of insect they eat. By comparing their diets, we may see the ways in which these two species use the insect resource axis in slightly different ways. One species might only eat large insects, the other only eat small ones.

The range and abundance of insects of different sizes eaten by a species will represent its **niche breadth** on this axis. Figure 43.17 shows that the yellow-footed antechinus concentrates on eating insects mostly in the size class range 5–7.4 mm (class mark 6.25), whereas the brown antechinus takes insects more evenly from the entire range of insect sizes. As a result, the brown antechinus has a greater niche breadth than that of the other species.

Niche overlap and character displacement

In the late 1950s, Robert MacArthur and G. E. Hutchinson noted that species competing for a limited resource appear sometimes to exploit the resource axis in different ways. This is termed **resource partitioning**. In 1972, MacArthur raised the question of how *different* species need to be to coexist in the same habitat. A series of studies followed in which various people tried to measure the amount of **niche overlap** between species. Many of these studies took a single axis on which it was considered two or more species were competing. The amount of overlap in the use of the resource by the different species was measured, just as in the example of antechinus above.

These studies showed that when two species compete for a limiting resource, they sometimes adapt to minimise competition. The adaptations are sometimes reflected in the morphology of competing populations, a phenomenon that became known as **character displacement**. The phenomenon has been inferred from comparative studies in which the form of a species is different when it is in competition with a second species than in places where no competition exists, and the changes are directly attributable to some limiting resource, usually food.

For example, when two species of mud snails (*Hydrobia*) found in Limfjorden, Denmark, live apart, **allopatry**, their sizes are the same. In places where they coexist, **sympatry**, there is a marked difference in their sizes (Fig. 43.18). The size differences are related to the size of food eaten by the two species and appear to be the result of selection for resource partitioning leading to character displacement.

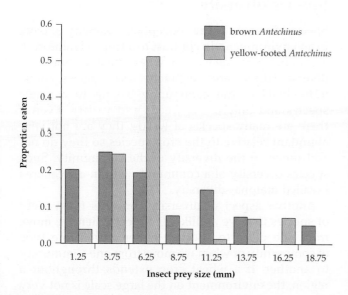

Fig. 43.17 The proportion of insects of different sizes eaten by two species of *Antechinus*. The values for each category of insect prey sizes are the class marks for the classes, each with a range of 2.5 mm

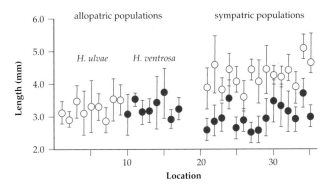

Fig. 43.18 Character displacement in mud snails, *Hydrobia ulvae* and *H. ventrosa*. In allopatric populations there are no important differences in size but, when the species are sympatric, there are significant differences

Niche breadth is the range of a resource that a species uses. Niche overlap and competition for a resource sometimes lead to selection for resource partitioning, reflected in character displacement in coexisting populations.

Character displacement is just one of the mechanisms by which competing species can coexist in the same habitat. Spatial variability can also make it easier for species to share the same habitat. Groups of closely related species that use a habitat in the same fashion, and between which we may expect competition, are called **guilds**. Members of the same guild may exist in separate habitat patches but within the same general environmental niche. The species that colonises a vacant patch first has a competitive advantage over new arrivals.

Variation in time may be just as important as variation in space. The environment changes from one year to the next and temporal fluctuations provide some species with a competitive advantage in some years, allowing them to persist if they are at a disadvantage in other years, a kind of temporal refuge. Environmental variation can ensure the persistence of several competing species in the same habitat even when, in a constant environment, a competitively superior species would have forced the other species to extinction.

Variability in the environment in space and time allows species that naturally compete to exist in the same habitat.

COMMUNITY STRUCTURE
Species richness

We have seen above that one of the important components of any community is the number of species within it, a characteristic of a community that is termed **species richness**. There are about one and a half million species of all kinds recorded by scientists, but there are probably more than 10 million species in existence. The total number is uncertain because estimates are based on extrapolations from samples of species richness taken from only a few environments. Some habitats support a large number of species per unit area, while others support very few. The majority of species occur in tropical forests, while the land mass of Antarctica supports little more than lichens, mosses and a few hardy species of penguins, strong flying birds and seagoing mammals. The land mass of Australia supports hundreds of thousands and probably millions of different species (Fig. 43.19).

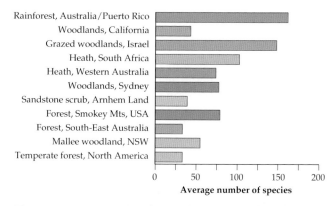

Fig. 43.19 A conventional way of measuring the richness of botanical communities is to count the number of vascular plant species in a 1000 m² (0.1 ha) area. Various people have sampled 1000 m² plots in a range of vegetation types throughout the world, and the ranges of species richness vary greatly as shown here. The data provide a guide to patterns of species richness in Australia and throughout the world

Species diversity

Species diversity is the concept of variability of biota in an environment. In contrast to a tropical rainforest, a monoculture such as a wheat field is not very diverse; there are relatively few species, most individuals in the community belong to a single species and only a few weeds add variety. Even if there are many species of weeds, they are not very abundant relative to the crop species so they do not add much to the diversity of the community. Such species diversity of a community within a local area is called α(alpha)-**diversity**.

Another aspect of diversity concerns the variety of species that we are likely to meet when we move from one place to another, either within one type of community or when we move from one community to another. If a wheat crop extends throughout a region, the environment on the large scale is not very diverse. In contrast, as one moves from place to place in a tropical rainforest, one will find many new species at each place. The diversity *among* different communities in different habitats in spatially separate areas is called β(beta)-**diversity**.

Species richness is the number of species within a habitat or a defined geographic region. Species diversity accounts for the number of species and their relative abundances within a community in a local area (α-diversity) and between different communities separated spatially (β-diversity).

Structural diversity

So far we have viewed diversity from the perspective of the species composition of a community. Another aspect of community diversity is structural diversity, which focuses on variations in the size and shape of plants, irrespective of the species. In a pine plantation, the structural diversity may be low, not because all of the individual trees are of the same species, but because they are very often of the same age.

One of the most important characteristics of community structure is *projective foliage cover* of the plants. This is the percentage of the ground surface above which there is foliage. If we know the percentage cover and at what height the foliage is found, it tells us a great deal about the biomass and productivity of the community (see Chapter 44).

Tropical rainforests are tall and have very high values for projective foliage cover (Fig. 43.20). More than 90% of the surface of a site may be covered by foliage that is more than 30 m above the ground. Temperate eucalypt forests may be as tall or taller than their tropical counterparts, but projective foliage cover of the tallest stratum never reaches 90%. Values around 30% are more common (Fig. 43.20). Arid zone shrublands dominated by saltbush often have a projective foliage cover of less than 10%, including all height classes.

Information on the size and shape of plants, their life forms and growth habits, is also used in classification of communities. In 1970, the Australian ecologist R.L. Specht designed a table for Australian

(a)

(b)

Fig. 43.20 Overhead cover of **(a)** an Australian rainforest compared with **(b)** an open eucalypt forest

vegetation. Specht's table (Table 43.2) is based on the projective foliage cover and growth form of plants in the tallest stratum and has been used, in slightly varying forms, as a basis for vegetation maps of all parts of Australia.

BOX 43.3   MEASURING DIVERSITY

α-diversity

Diversity within communities, α-diversity, has two components: the number of species and their relative abundances. There are many ways that α-diversity can be expressed quantitatively. One way is Simpson's Index,

$$D = 1 - \sum_{i=1}^{s} \left(\frac{n_i}{N} \right)^2 \qquad [43.3]$$

where n_i is the number of individuals in species i, N is the total number of individuals in the community, and S is the number of species present. This index is

equivalent to calculating the probability of picking at random two individuals that are different species. It ranges from 0 in communities made up of single species, to almost 1 in communities made up of many species, each present in equal numbers.

The data in Table 1 represent the relative abundances of eight small mammal species in three vegetation types in an area within Kakadu National Park. Sandstone forest and mature open forest have three species each, and the riparian woodland contains five species. Total abundance, on the other hand, is highest in mature open forest and lowest in riparian woodland.

Table 1 Abundances (numbers) of small mammals in three vegetation types at Jabiluka in Kakadu National Park (a)

Mammal species	Sandstone forest	Mature open forest	Riparian woodland
Antechinus bilarni (sandstone antechinus)	6.3		
Dasyurus hallucatus (northern quoll)	8.6	16.7	1.2
Zyzomys woodwardi (large rock rat)	5.5		
Sminthopsis virginiae (red-cheeked dunnart)		1.1	0.4
Trichosurus arnhemensis (northern brushtail possum)		33.3	
Pseudomys nanus (western chestnut mouse)			3.7
Pseudomys delicatulus (delicate mouse)			0.6
Melomys burtoni (grassland melomys)			6.5
Species richness	3	3	5
Total abundance	20.4	51.1	12.4
Simpson's α-diversity	0.65	0.47	0.62

(a) Values are based on trapping success. Data are used to calculate α-diversity.

Even though both sandstone forest and mature open forest have the same number of species, and total abundance is greater in mature open forest, species diversity is greater in sandstone forest. In other words, there is a greater chance of picking up two individuals that are *different* in the sandstone forest. Species diversity is higher in sandstone forest because the relative abundances of the three species are more even than in mature open forest, in which most individuals are northern brushtail possums. For the same reason, the species diversity values in sandstone forest and riparian woodland are much the same, even though riparian woodland supports more species.

β-diversity

β-diversity is the change in community composition we meet as we move through the landscape. If the *turnover* of species between sites is high, communities will be relatively different.

There are numerous indices for measuring the similarity between communities in different locations. One common coefficient of similarity is **Jaccard's coefficient (S_J)**. It is employed when the only information available about the communities is lists of species. The coefficient is used to compare pairs of sites. Lists can be arranged in a table in which the rows (or columns) are sites, and the columns (or rows) are species (Table **2**).

Jaccard's coefficient is calculated for pairs of sites from the equation:

$$S_J = \frac{a}{t} \qquad [43.4]$$

where a is the number of species shared by the two sites and t is the total number of species, including those species that are unique to either of the sites.

The coefficient may be calculated to represent the degree of similarity between any pair of sites such as those in Table **2**. The coefficient between sites 1 and 2 is 1/7, while between sites 2 and 3 it is 3/4. This suggests that the community at site 2 is more similar to the community at site 3 than it is to the community at site 1 because sites 2 and 3 have more species in common.

Table 2 A site by species table for grass, sedge and orchid species growing in three different communities on low granite rocks in the semiarid mallee zone of southern Western Australia (a)

Species	Sites 1	2	3
Neurachne alopecuroides (grass)	1		
Gahnia drummondii (sedge)		1	1
Lepidosperma drummondii (sedge)	1		
Lomandra mucronata (needle lomandra)	1		
Stypandra imbricata	1		1
Borya nitida (resurrection plant)		1	1
Diurus longifolia (orchid)	1	1	1

(a) The 1's indicate the presence of a species in a site.

In such a small data set as that in Table **2**, the coefficient is hardly necessary. One can see by inspection that the communities at sites 2 and 3 are relatively similar. The index becomes useful in larger studies where there are many rows and columns of information. The results can then be used to make decisions about which areas to include in a conservation park to ensure that the park includes representative samples of the variety of communities.

Table 43.2 Specht's structural classification of Australian vegetation

Growth form of the tallest stratum	Foliage cover of the tallest stratum			
	>70%	30–70%	10–30%	<10%
Tall trees (>30 m)	Tall closed forest	Tall open forest	Tall woodland	
Medium trees (10–30 m)	Closed forest	Open forest	Woodland	Open woodland
Low trees (<10 m)	Low closed forest	Low open forest	Low woodland	Low open woodland
Tall shrubs (>2 m)	Closed scrub	Open scrub	Tall shrubland	Tall open shrubland
Low shrubs (<2 m)	Closed heath	Open heath	Low shrubland	Low open shrubland
Hummock grasses			Hummock grassland	
Tufted/tussock grasses	Closed tussock grassland	Tussock grassland	Open tussock grassland	Sparse open grassland
Graminoids	Closed sedgeland	Sedgeland	Open sedgeland	
Other herbaceous spp.	Dense sown pasture	Sown pasture	Open herbfield	Sparse open herbfield

> Structure in plant communities is recorded usually as percentage projective foliage cover within height classes.

Biomes

Regardless of the particular species in a community, the vegetation structure found in different parts of the world is similar in similar environments. On a global scale, communities with the same structure are called **biomes** (Box 43.4). For example, forests inhabit cool temperate regions of most parts of the world. Regions at high latitudes, close to the poles, and at high altitudes, support tundra, herbfields and low shrubs. Biomes are a result of evolutionary convergence (Chapter 32).

Gradients and ecotones

Gradients represent continuous, usually gradual changes in environmental variables that occur with geographic distance. Dive in the ocean and you will notice that available light is a simple function of depth. Drive from the coast inland to the top of a mountain range and there will be a more or less gradual change in temperature. Such gradients are the axes in Hutchinson's multidimensional niche space (Fig. 43.15). They are factors that determine the distribution and abundance of species which change in a more or less continuous fashion from place to place in the landscape.

Apparent abrupt changes in the environment from one place to another are a matter of scale. In fact, it is possible to argue that all changes in environmental conditions are gradual; it is only a matter of how closely you look. Near the treeline on alpine slopes, trees gradually become more and more reduced in height, vigour and density with increasing altitude.

Because individuals respond to environmental conditions in their immediate vicinity, and because these conditions change gradually in space, the interfaces between different habitats and different communities are usually blurred. The boundaries between two different communities are termed **ecotones**. Often, ecotones are species rich because they support individuals from both community types. Some species that persist there are at the edge of their niche, either because of competition or because they barely tolerate the combination of physical conditions.

For example, mangrove trees and shrubs are distributed differentially along salinity gradients such that banded zonation patterns form along salinity gradients (Fig. 43.21). Despite differences in distribution, most mangrove species grow best under low salinity conditions but differ in the range of salinities that they can tolerate. In general, the greater the salt tolerance of a species, the lower its growth rate under optimal conditions. It appears that the attributes that enable species to tolerate highly saline conditions involve trade-offs that reduce growth and competitive ability under low salinity conditions. Thus, species that dominate in optimal conditions are limited by their susceptibility to highly saline conditions. Other species that can tolerate salt are limited to highly saline environments in which conditions for growth are suboptimal but in which there is limited competition from less tolerant, faster growing species. In this way, interspecific differences in salt tolerance may influence the structure of mangrove forests along natural salinity gradients.

Gradients are gradual changes in environmental parameters that occur with geographic distance.

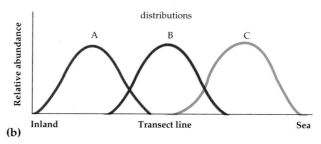

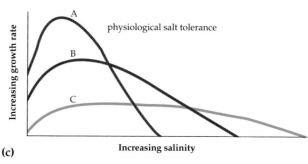

(a)

Fig. 43.21 (a) Mangroves in New Guinea.
(b) Distribution of three mangrove species, A, B and C, along a line transect from the sea to inland. The species closest to the sea, C, is the most tolerant of saline conditions but is the slowest growing under **(c)** a range of salinities

BOX 43.4 WORLD BIOMES REPRESENTED IN AUSTRALIA

Biomes are ecological communities delineated by climate. They occur widely across the earth, and each has a characteristic structure. Some important biomes that occur in Australia (see Figs a and b) are described here.

Biomes are broad categories and each includes several structurally different classes of vegetation. Specht's vegetation classification is shown in Table 43.2 and Fig. b.

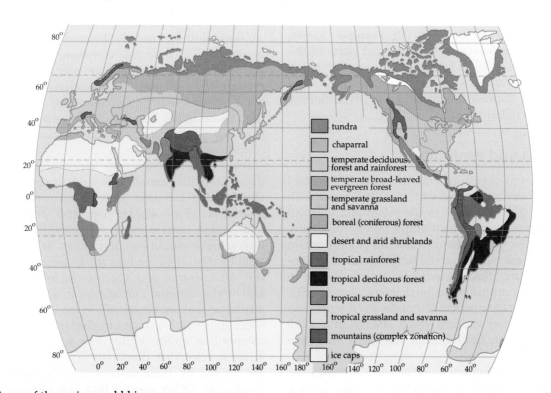

(a) Some of the major world biomes

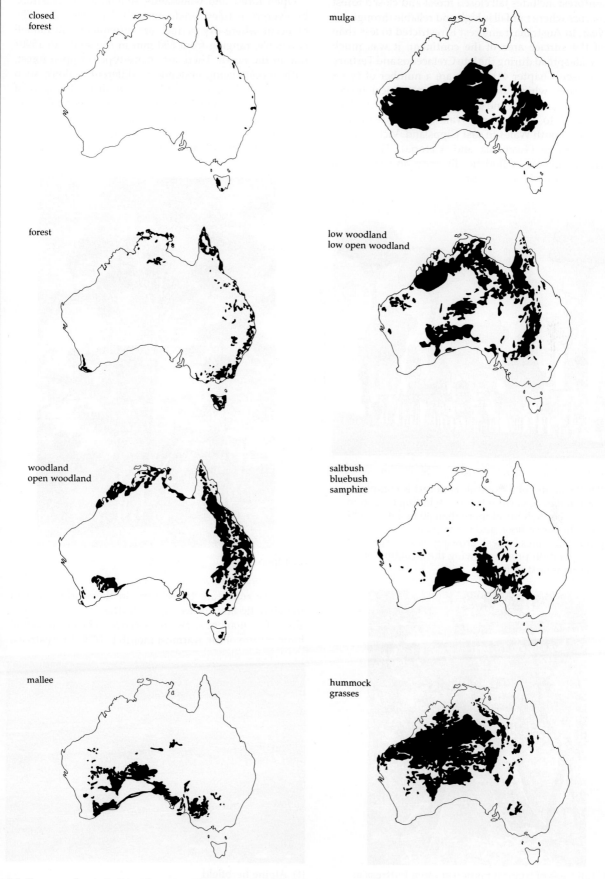

(b) Structural vegetation classes within biomes represented in Australia

Rainforest includes tall closed forest and closed forest and occurs where rainfall is high and reliable throughout the year. In Australia, rainforest is restricted to less than 1% of the surface area of the continent; it was much more widespread during the late Cretaceous and Tertiary periods (see Chapter 41). There are a number of types of rainforest, which are described as tropical (north Queensland, see Figs c and d), subtropical (from MacKay, Queensland to New South Wales), warm temperate (from New South Wales to east Gippsland, Victoria) and cool temperate (Tasmania and Victoria). There is a general reduction in floristic diversity and structural complexity from north to south.

Open forest and woodlands, dominated in Australia by evergreen trees, predominantly *Eucalyptus* (see Fig. e), occur where dry periods are infrequent and rainfall is reliable, ranging from 650 mm in the south to 1500+ mm in the north. There are many types of open forest, with species being endemic to different regions, such as Cape York, Queensland, the south-west region of Western Australia or Tasmania. Tree heights range from 30 m or more to 10 m, and overhead canopy cover varies from 70% to 30%. In drier climates, open forests are replaced by woodlands, where trees are more sparse.

(c) The structure of a tropical rainforest is complex with a number of vertical layers, including emergent trees reaching heights of more than 30 m. Little light reaches the forest floor, and many plants are epiphytes or lianes (vines), growing on the trunks of trees. Vegetation profile showing the structure of a tropical rainforest in Brunei

(e) Open forest

Alpine zones are those areas in the world between the climatic limit of tree growth (the treeline) and the zone of permanent snow and ice cover. The mean annual temperature of the warmest month is 10°C. In Australia,

(d) Tall trees of tropical rainforest often buttress at the base

(f) Alpine herbfield

the alpine biome is restricted to the south-eastern highlands, such as Mt Kosciusko, where the treeline occurs at about 1830 m altitude, and the 'snow country' of Victoria and Tasmania, a combined area of 0.15% of Australia. The biome includes a number of plant communities such as herbfields (see Fig. f), heaths, fens and bogs. Many of the plants have their renewal buds close to the ground, protected from the cold, or they are annuals with a short life cycle. Snowgum woodlands (*Eucalyptus pauciflora* or *E. coccifera*) grow at lower altitudes in the subalpine zone.

Savanna is a kind of low open woodland with hummock grasses (see Fig. g). In northern Australia it occurs in a tropical, semihumid climate with seasonal (summer) rainfall. Trees are scattered through the region, including *Brachychiton* (bottle tree), *Callitris* (native pine) and *Eucalyptus* (such as *E. alba*, white gum), and tall grasses form an understorey. Some tree species are deciduous in the dry season. Savanna in the south-east of Australia experiences a semiarid, subtropical to temperate climate, with some winter rainfall. Different species of eucalypts and acacias, such as brigalow, *Acacia harpophylla*, occur here. In treeless areas within the savanna biome, grasses dominate to form semiarid **hummock grassland** (*Triodia* and/or *Plectrachne*) or **tussock grassland** (*Astrebla*, Mitchell grass). The *Triodia* grasslands are sometimes called spinifex.

Mallee is a tall shrubland dominated by mallee-form (multistemmed) eucalypts that are usually less than 10 m tall. It predominates in extensive areas of the southern part of Australia where the climate is semiarid or Mediterranean. The mallee communities grade into **heath** and **shrublands** in which sparse low mallees sometimes persist. Mallee also intergrades with **saltbush** and **bluebush** low shrublands (succulent chenopods; see Chapter 41). The type of community within the biome depends on factors such as rainfall and soil. Typically, heath communities are found on nutrient-deficient sands, whereas mallee communities are found on sandy clays with higher nutrient status, and chenopod

(g) Savanna

shrublands (shrub steppe) dominate on dry and saline or calcareous soils such as the Nullabor Plain.

Desert occurs in arid areas where rainfall is low (less than 350 mm per year) and where temperatures can be extremely high. Australian deserts are in the interior of the continent and consist of sandhills, sand plains, salt lakes and gibber plains (stony deserts). Deserts are sparsely vegetated, with grasses such as *Triodia* dominating. Parts are treeless but tree species form a tall shrubland or open woodland bordering the desert called **mulga**. Mulga is dominated by *Acacia*, mainly *A. aneura* (also called mulga), as well as species of *Casuarina* and *Eucalyptus*.

Succession

Communities change with time. Individuals are born and others die. Species become locally extinct and others colonise. Even within a single year, the composition and structure of any community will change because of the behavioural and physiological characteristics of individuals that enable them to adapt to changing environmental conditions.

When communities are observed over long periods of time, say tens of years, the relative abundances of different species change. In areas of competition, the process of replacement of one species by another within an area of habitat may be a gradual process, especially if the mechanism of competition occurs among the juvenile stages. If one plant species is a

better competitor for light than another, but only as a seedling, it will have to wait for the adults of its competitor to die before it gets its chance.

Succession is a theory about how communities change, for example, how species invade and colonise empty patches of habitat. Species interact by eating each other, providing protection and resources for each other, competing with each other for limited resources and so on. The assemblage of species found at any given place is the result of which species were able to colonise the habitat since it was vacated, the tolerance these species have for the local conditions, and the interactions that have taken place between these species. **Primary succession** takes place in a habitat that was previously uninhabited, and **secondary succession** takes place in a habitat that has been modified by other species.

The process of replacement of one community by another sometimes proceeds through a series of assemblages or **seres**, each with an identifiable character. If one clears a patch of forest, and then leaves it alone, species will colonise from surrounding vegetation (Fig. 43.22). Among the species to arrive first are those that disperse quickly in large numbers and over relatively large distances. Often, individuals of these species are relatively small, produce large numbers of offspring, and have relatively short life spans. They are adapted to an opportunistic lifestyle. Examples of these so-called 'disturbance opportunists' include many members of the flowering plant family Asteraceae (daisies and everlastings).

Fig. 43.22 Succession: species colonise a cleared patch in the Victorian highlands. In the foreground, forest was cleared for timber harvesting. In the background, the standing stems were killed by a wildfire in 1983

Eventually, the community (the 'climax' community) will be dominated by individuals of species that are longer lived and slower growing. These individuals are often larger, and are able to tolerate the rigours of limited resources better than members of the opportunistic species, but are not as well adapted to dispersal or rapid population increase. Early colonists may be lost entirely if they are unable to tolerate the competition for limited resources. This concept of succession is known as the tolerance model. Examples of tolerant species include eucalypts, which eventually dominate Australian landscapes after disturbance.

One can make a qualitative distinction between organisms that employ different strategies to compete with other species. Species have been called '*r*' and '*K*' strategists, in reference to the two parameters of the logistic equation (equation 43.1). **r strategists** are the opportunists, those species that disperse rapidly, have large numbers of offspring (and hence high population growth rates, r) and short life cycles, for example, many plant weeds and insects that periodically reach plague numbers. **K strategists** are better long-term competitors, slower growing but more likely eventually to encounter intraspecific competition when they approach the carrying capacity of the environment (hence the K). The stages in the development of communities through time are a result of the different strategies adopted by different species. The species making up the so-called climax community are simply the best long-term competitors.

It is questionable whether many communities ever reach a 'climax' state because disturbances and changes in environmental conditions, such as fire, cyclones and floods, are commonplace. In Australia, fire is a particularly important factor affecting community change.

> Succession is the process of replacement of one community by another. Competition between species with different life-history strategies may lead to a climax community dominated by the best long-term competitors.

Fire

Many of the plant and animal species that inhabit the Australian landscape are adapted to periodic fires (see Chapter 41). For example, several species of *Banksia* are killed by fire and their seeds will only germinate in the ash beds left after a fire. Similarly, the seeds of numerous *Acacia* and other species, stored in the soil, depend on the heat generated by fire to germinate.

Fire may enhance species richness and diversity through its interactions with community processes. Vegetative regrowth, flowering and seedling survival after fire in south-western Australian woodlands (Fig. 43.23) are facilitated by the post-fire environment, which includes changes such as improved soil nutrient and water availability, and reduced predation pressure from insect grazers.

Fig. 43.23 This mallee vegetation north of Esperance in Western Australia is regenerating after a wildfire. Repeated fires at short intervals in this region may change open woodland to low open shrubland

Communities in the subtropical savanna of northern Australia, for example, are structured by fire. The long fire season, from April to December, provides a range of burning conditions, which in turn provide a resource that is used in different ways by numerous species of birds (Fig. 43.24) and many other animals. The behaviour and feeding preferences of different species determine when they make intensive use of the vegetation. Brown and black falcons and whistling kites, for example, hunt grasshoppers, lizards and small mammals that escape before the fire front. The hot ash phase after fire is exploited by a second group of birds, including butcherbirds and kingfishers, who forage on vulnerable vertebrates and invertebrates. Corellas, rosellas and galahs, which are granivorous species, do not exploit the habitat heavily until regeneration commences. In general, the abundance and quality of food is enhanced for a large number of bird species for the few months after a fire.

The interesting thing about fire is that it usually burns in a mosaic pattern, unless environmental conditions result in very severe fires. The mosaic of fires provides a variety of habitats in space while the dynamic processes after fire, such as succession, provide a variety of habitats through time.

Fire causes variability in habitats in both time and space and may promote species richness and diversity.

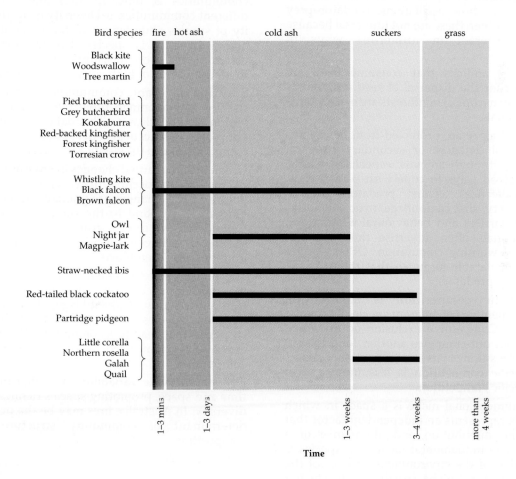

Fig. 43.24 Generalised diagram of the use of fire and the vegetation stages immediately following it by eight groups of birds in Kakadu National Park. The heavy horizontal bars represent periods of intensive use

SUMMARY

- Communities are natural assemblages of species that interact with one another. Species interactions include symbiosis, predation and competition.

- Coevolution is where interactions between organisms lead to reciprocal evolutionary change in the characteristics of species.

- Symbiosis is where two organisms live together in a close relationship and have coevolved. It includes mutualism (where both species benefit), commensalism (one benefits, one is unaffected) and parasitism (one benefits and one is harmed).

- The abundance of a predator or herbivore and its prey can show oscillations: predator–prey cycles. However, these are not universal because species interact in numerous ways in communities.

- Frugivores, animals that consume fruit, are important for the dispersal of seed. Frugivores in Australian tropical rainforests are mainly birds and mammals.

- Plants and algae are protected from herbivores by physical defences or secondary chemical defences, although animals, especially insects, have evolved ways to cope with plant toxins. Animal defences against predators include speed, agility, size, camouflage, toxic chemicals, physical barriers and behavioural adaptations. Toxic animals often advertise their unpalatability by warning colouration. Other species mimic unpalatable forms so that they too are avoided by predators.

- Interspecific competition is competition between members of different species for limited resources. Competitive exclusion occurs when one species out-competes another for a limited resource, resulting in its local extinction. Usually, interspecific competition is less intense than intraspecific competition.

- A multidimensional niche is a space in which each axis represents an independent factor that limits the distribution and abundance of a species. The fundamental niche of a species is that region of the environment (that is, of the multidimensional niche space) within which a species can survive and reproduce.

- Niche breadth is the range of a resource a species uses. Niche overlap is the length of a resource axis over which the use by different species overlaps. Competition for a resource sometimes leads to selection for resource partitioning, reflected in character displacement in sympatric populations.

- Variability in the environment in space and time promotes coexistence of species with similar resource requirements.

- Species richness is the number of species within a habitat or a defined geographic region. Species diversity accounts for the number of species and their relative abundances within sites, within communities at different sites and between different communities. α-Diversity is the diversity of species abundances within communities and β-diversity is species diversity between communities. These measures can be quantified to identify areas of high conservation status.

- Structure in plant communities is recorded usually as percentage projective foliage cover within height classes. The data are used to compare structural diversity between communities and to classify and map vegetation.

- Gradients are gradual changes in environmental parameters that occur with geographic distance. Because these changes are gradual and because individuals respond to the conditions in their immediate vicinity, zones of transition (ecotones) between species, or between communities, are often blurred.

- The process of replacement of one community by another, succession, sometimes proceeds through a series of assemblages. Competition between species with different life-history strategies may lead to a community (the climax community) dominated by the best long-term competitors.

- Fire may provide variability in habitat in both time and space, promoting species richness and diversity. In Australia, fires may be the primary determinant of community structure and composition.

QUESTIONS

1. (a) What is meant by 'symbiosis'? (b) Distinguish between commensalism and mutualism. (c) Distinguish between parasitism and predation.

2. What is the significance of 'warning colouration'? Give an example.

3. Why are mimics that are themselves non-toxic usually present in low numbers relative to the distasteful species that they resemble?

4. How could species diversity in a community be determined by the activities of predators?

5. In the tropics, leaf-cutter ants remove pieces of leaves and take them to underground nests. There they chew and innoculate them with fungal spores. The fungi grow on the chewed leaves and are cultivated in this way. The fungi, not the leaves directly, are the main food of the ants. What sort of interaction, between the fungus and ant, do you think this is an example of? Explain your reasoning. How would you test your hypothesis?

6. Two species of similar-sized insectivorous lizards coexist in rocky outcrops scattered in dry sclerophyll forest in a region of eastern Australia. You hypothesise that the two species are competing for food. Discuss observations that could be made and manipulative experiments that could be carried out to test your hypothesis.

Discuss the relative merits of different approaches.

7. Describe the multidimensional niche space of a species in isolation from other species (in allopatry) compared with its volume in the same niche space when in competition with another species for food. How will character displacement affect the shape of the niche space?

8. Suggest experiments that would test the hypothesis that a physical difference between two closely related species has been caused by selection resulting from interspecific competition for a limited resource.

9. How might you expect measures of species richness and α-diversity to change as you move along an environmental gradient and across an ecotone between forests dominated by different species?

10. What role does fire play in determining species richness and diversity in the Australian biota? Given circumstances that require fire control to protect human life and property, how would you go about establishing the most appropriate fire control regimen to manage the flora and fauna?

11. How do the ideas underlying the theory of succession relate to the concepts of environmental gradients and ecotones?

E C O S Y S T E M S

Individual animals or plants aggregate into populations, populations assemble into communities, and assemblages of communities interacting with their physical environments constitute ecosystems. **Ecosystems** are the basic functional units of ecology. They are living, changing systems encompassing organisms, their biotic and abiotic environments, and exchanges within and between each of these. Ecosystems are difficult to define more precisely than this because their boundaries are seldom fixed or precise. An ecosystem is neither static nor closed; instead, an ecosystem is a changing, self-modifying ecological system.

> Ecosystems are changing, self-modifying ecological systems consisting of interacting biotic and abiotic components.

ECOLOGICAL PYRAMIDS

Many ecological interactions take the form of exchanges and flows of energy and matter, and the nature of the interactions is often associated with the ways in which the component organisms obtain their food. As a result, the structure and function of ecosystems are often described in terms of their feeding or trophic relationships. Autotrophs synthesise complex organic compounds from inorganic precursors, using external energy sources. Green plants and algae are the largest group by far, manufacturing all their required organic molecules from simple inorganic molecules, using sunlight as the energy source for photosynthesis (Chapter 5). Chemosynthetic bacteria also manufacture their own organic molecules (Chapter 5), but their contribution to the volume of energy flowing through ecosystems is far outweighed by that of plants and algae. Ultimately, autotrophs are the producers of food for all other organisms.

All other organisms are heterotrophs because they cannot synthesise organic matter but can only reorganise it (Chapter 19). They must consume other organisms in order to gain the organic molecules they need for life. Heterotrophs include primary consumers (mostly herbivores), which feed directly on producers, secondary consumers (mostly carnivores and parasites), which feed on herbivores, and tertiary and higher order consumers (also carnivores and parasites), which feed on other consumers. Among heterotrophs there are also organisms that feed on dead organisms and organic wastes from several trophic levels. These are degraders, which include scavengers (animals that eat dead organisms) and detritivores (animals that eat organic litter, or detritus) and decomposers (fungi and bacteria that cause chemical decay of organic matter from all trophic levels) (Fig. 44.1).

Of course, this view is simplified. Omnivores (e.g. many ants, humans, many fishes) overlap more than one trophic level, as do degraders and detritivores. Some organisms change trophic levels during their life cycles. Some insects, for example, are carnivorous as larvae but herbivorous as adults, and many species of fishes move up the trophic scale as they grow. The simple trophic relationships of Figure 44.1 serve to demonstrate some useful general characteristics of ecosystems.

> Trophic interactions are probably the most important relationships determining ecosystem structure and functioning. All organisms are either autotrophs (producers) or heterotrophs (consumers, degraders and decomposers).

Pyramids of numbers

Why are there few large carnivores? Why are there billions of plankton, millions of krill, but only hundreds of whales in Antarctic waters (Fig. 44.1)? Why can a patch of eucalypt woodland support thousands of leaf-eating insects but only hundreds of insect-eating birds and only one or two bird-eating foxes or cats? Or, as the ecologist Paul Colinvaux has put it, why are big fierce animals rare? An early British ecologist, Charles Elton, was one of the first

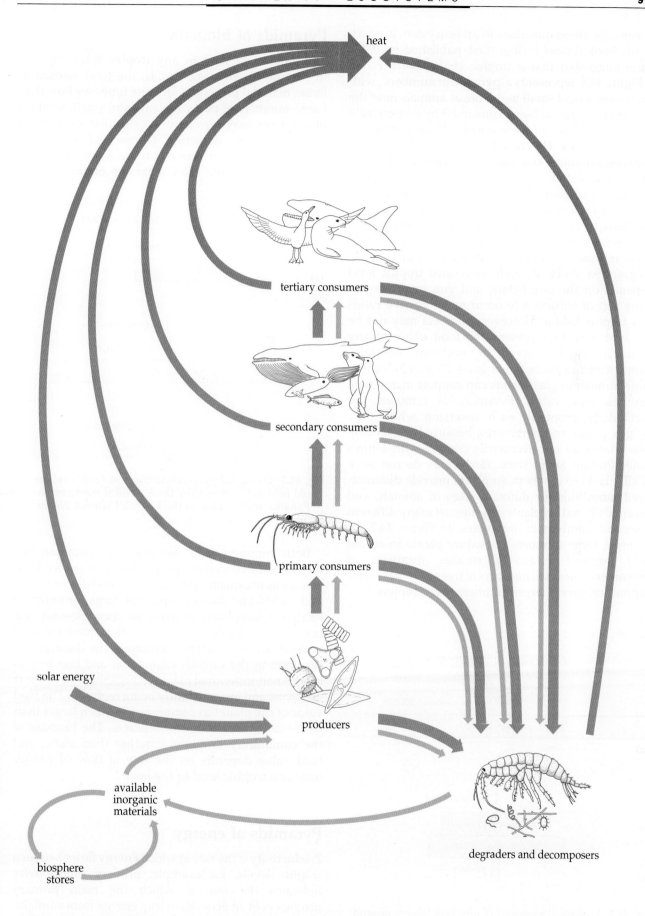

Fig. 44.1 A simplified diagram of the flow of energy (red) and materials (blue) through an ecosystem. Note that although there is a net loss of energy from the system, materials may be recycled

to consider these questions in an ecosystem context. In his book *Animal Ecology* (first published in 1927), Elton suggested that a trophic chain, like the one in Figure 44.1, represents a 'pyramid of numbers', with large numbers of small herbivorous animals near the base of the pyramid being consumed by successively smaller numbers of increasingly large carnivores toward its apex. He thought that the relationships between organism size and the mechanics of eating and being eaten were the key to this pyramid structure. For example, insectivorous birds have to be big enough to kill and eat insects, and foxes and cats have to be big enough to kill and eat birds.

It seems sensible that trophic interactions should result in some sort of pyramidal relationship that can be graphed. After all, each successive trophic level depends on the one below, and you might expect some sort of reduction to occur up successive rungs of a trophic ladder. However, numbers may not be the best way to represent the food value of any particular trophic level. In northern Australia, termites and cattle both eat grass (Box 44.3) but the same number of grass plants can support many more termites than cattle. Pyramids of numbers are particularly prone to such inversion when they include plants and herbivores, because their feeding interactions do not necessarily comply with Elton's generalisation about sizes. Herbivores do not have to kill plants to eat them, so plants provide different-sized mouthfuls for different sizes of animals, and pyramids based on plants can support many different sizes and numbers of herbivores. In Figure 44.2, for example, large numbers of pasture plants in an old field support relatively smaller numbers of consumers, but small numbers of trees in an oak forest support relatively larger numbers of consumers.

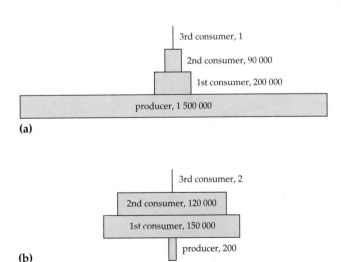

(a)

(b)

Fig. 44.2 Ecological pyramids of numbers (above-ground individuals per 0.1 ha). **(a)** An old field in Michigan, USA. **(b)** An oak forest near Oxford, UK

Pyramids of biomass

The amount of food in any trophic level depends in part on its **biomass**, that is, the total amount of living material present at any one time (see Box 44.1). Large numbers of pasture plants and small numbers of oak trees may represent very similar amounts of food. Diagrams depicting biomass of successive trophic levels in an ecosystem are more frequently pyramidal in shape than diagrams of numbers of organisms (Fig. 44.3).

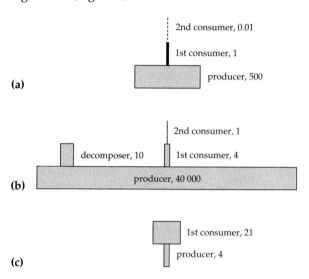

Fig. 44.3 Ecological pyramids of biomass (g/m^2). **(a)** An old field in Georgia, USA. **(b)** A tropical rainforest in Panama. **(c)** Plankton in the English Channel, North Sea

Both numbers and biomass are instantaneous measures of ecosystem properties, and are liable to change as organisms are born, grow and die. In Figure 44.3, fields and forests represent large amounts of relatively long-lived biomass, so their biomass is a reasonably reliable indication of their food value to consumers. In contrast, although the biomass of plankton in the English Channel at any one time is quite small, individual plankton organisms have short life spans and are constantly being replaced. The food value of the plankton community is much larger than its instantaneous biomass indicates. The biomass of the community is dynamic rather than static, and food value depends on the rate of flow of energy from one trophic level to the next.

Pyramids of energy

Productivity is the rate at which energy flows between trophic levels, for example, primary productivity indicates the rate at which the main primary producers of an ecosystem trap energy from sunlight. The energy flowing into one trophic level sets an upper limit to the food available to the next. Not all the energy and material taken in by one trophic group is passed onto the next, because not all organisms at one trophic level are consumed by the next; there are also losses associated with cellular respiration and elimination of wastes. Thus, there is a net loss of energy and materials as the trophic scale is ascended. Energy flow indicates the food value of trophic levels more accurately than either numbers or biomass and, for most ecosystems, a pyramid of energy flow conforms most closely to Elton's classical ascending pyramid (Fig. 44.4).

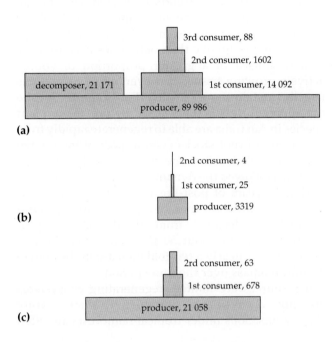

(a)

(b)

(c)

Fig. 44.4 Ecological pyramids of energy flow (kJ/m²/year). (a) A freshwater spring in Florida, USA. (b) A desert grassland in New Mexico, USA. (c) A prairie in Oklahoma, USA

Graphs of the energy flowing into successive trophic levels represent ascending ecological pyramids, but pyramids based on numbers or biomass may sometimes be inverted.

PRODUCTIVITY OF DIFFERENT ECOSYSTEMS

How much does the productivity (energy flow) of ecosystems vary, and what does this tell us about functioning of ecosystems? If the productivity of an ecosystem changes little over time, it suggests that either the ecosystem's environment is unchanging, or that its organisms are compensating for changes that are occurring. If productivity changes dramatically, it could mean that an important environmental change is occurring, or that there has been an important change in the interactions of organisms within the system.

Because plants and algae are at the base of most energy flow/productivity pyramids, the amount of energy passing through ecosystems is determined by its net primary productivity, which is usually measured as the rate at which the main primary producers of an ecosystem synthesise energy equivalents (indicated by biomass) from sunlight (see Box 44.1). Since productivity is limited by the availability of all resources, material as well as energy, the average net primary productivity of different terrestrial ecosystems varies in ways that we might have guessed (Fig. 44.5): tropical forests are generally more productive than grasslands, which are generally more productive than deserts. In other words, communities of plants growing in sunny, warm, wet, fertile

---

BOX 44.1 BIOMASS, ENERGY AND PRODUCTIVITY

Biomass ('living mass') is usually expressed as a weight (e.g. grams). Each unit of biomass can also be considered as a parcel of energy, usually measured as joules (J) or calories. The energy represented by a unit of biomass at any particular time may be stored, used in chemical reactions, converted to movement or heat, or consumed by other organisms. The energy content of different types of biomass varies but not by very much. For example, plants contain about 19 kJ of energy per gram of dry weight, algae contain about 22 kJ, insects contain about 23 kJ and vertebrates contain about 24 kJ per gram of dry weight. Thus, biomass can also be expressed as energy equivalents and vice versa.

Of particular importance to ecosystem function is **net primary productivity**. This is the portion of total (or gross) primary productivity that remains after the respiratory losses of primary producers have been accounted for; it is this portion that is available for harvest by the

ecosystem's consumers or decomposers. Since energy can be expressed in biomass equivalents, net productivity of an individual plant is usually measured as the biomass (usually dry weight) that it synthesises and accumulates in tissues per unit time. Net primary productivity of an ecosystem is most commonly measured as dry weight of biomass synthesised per unit area of the earth's surface per unit time, and is expressed as grams per square metre per year (g/m²/year).

Some part of the net production of a plant or community may be eaten, shed or may die, and this must be taken into account when measuring productivity. Hence, the growth of a tree through a year may appear as a 10 kg increase in its weight but the net production by the tree includes this growth plus net production expended in leaves, fruits, flowers, branches and roots that were lost during the year because of herbivory, abscission or senescence.

environments are more productive than those growing in environments in which the availability of sunlight, heat, water or nutrients is sometimes limiting.

What about marine ecosystems? Rather surprisingly, most marine ecosystems are relatively unproductive. This is largely because ocean nutrients (particularly phosphorus and nitrogen) accumulate at great depths where they are unavailable to oceanic algae (mostly phytoplankton), which are restricted to growing in the top 100 m or so of ocean that sunlight can penetrate. The major oceanic fishing grounds are upwelling zones such as the North Sea, the Newfoundland Banks and the Antarctic Ocean. These are more productive because they are regions in which deep sea currents (and their nutrients) are forced upward and thereby continuously replenish the nutrient supply for phytoplankton in the surface layers. Coastal waters are also relatively productive because nutrients are continuously provided by freshwater rivers and continuously circulated by coastal currents. However, the total primary productivity of oceans (including the upwelling and coastal zones) is still less than half that of terrestrial ecosystems, even though oceans cover 70% of the world's surface.

For the systems shown in Figure 44.5, productivity and biomass rankings are very similar. However, as we saw when considering ecological pyramids, biomass and productivity are not always directly related. Productivity relates to the rate of turnover of biomass, not just the total amount present at any one time, so biomass can be misleading as an indicator of productivity. The duration and season of measurement of productivity are not accounted for. Daily rates of productivity in tropical rainforests are not very different from daily rates during the growing season in most other forests. Annual rates, however, are much greater in rainforests (Fig. 44.5) because the growing season in the humid tropics lasts for most of the year instead of the six months (or less) that is more usual in other regions.

As well as seasonal influences on primary productivity there are trends associated with the age of an ecosystem: for the same amount of available resources, young or regenerating ecosystems are often more productive than older ones because they usually consist of a greater proportion of young, actively growing tissue. However, older ecosystems usually contain more biomass. For example, despite the apparent destruction caused by fire, many plant species in Australia are able to regenerate rapidly from underground root stocks even if most of their aerial parts are killed. After one fire, scientists in South Australia followed the dynamics of a mallee woodland during 12 years of post-fire recovery (Fig. 44.6). As the woodland regenerated, the above-ground productivity declined from about 180 g/m²/year before the fire to about 50 g/m²/year after the fire, but there was nearly a ten-fold increase in the above-ground biomass over the same period.

If resources are limited, regenerating ecosystems are not always more productive than mature ecosystems. Soils under tropical rainforests are often

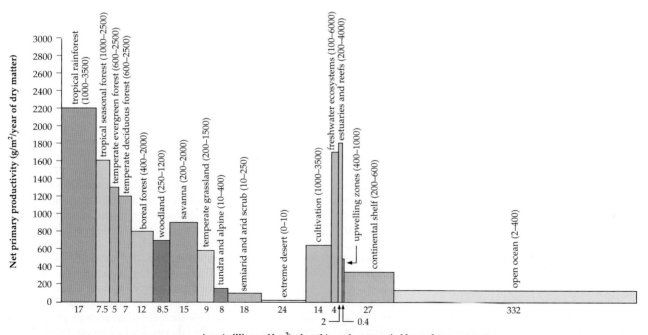

Fig. 44.5 Net primary productivity of world ecosystems

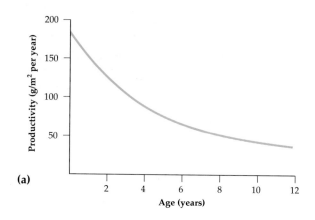

(a)

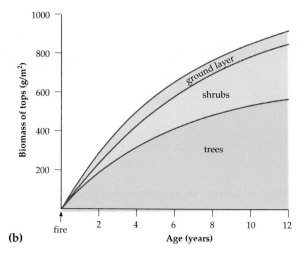

(b)

Fig. 44.6 Productivity (**a**) and biomass (**b**) of a mallee woodland recovering from fire

Fig. 44.7 Grasslands in the lower Jimi Valley of Papua New Guinea. Repeated prehistoric clearing and burning of rainforest has led to depletion of soil nutrients and replacement of rainforest by grassland

Fig. 44.8 In the mulga, *Acacia aneura*, rangelands of semiarid Australia, prolonged drought and heavy grazing by sheep, cattle or kangaroos, leads to perennial grasses being replaced by herbs, with the ecosystem becoming less productive

nutrient-poor because of the intensity of tropical weathering processes. Clearing of mature rainforest may lead to sufficient loss of nutrients to cause a shift in species composition in the regenerating ecosystem, and may result in a change from rainforest to less productive open grassland (Fig. 44.7). In the mulga, *Acacia aneura*, rangelands of semiarid Australia (Fig. 44.8), prolonged drought can lead to death of most of the perennial grasses that are utilised as dry-season food by grazing animals, such as sheep, cattle and kangaroos. Under conservative grazing management, these perennial grasses will regenerate when drought-breaking rains fall. However, if ecosystem resources are depleted by heavy grazing, regenerating pastures may come to be dominated by ephemeral herbs rather than perennial grasses, leading to marked reductions in long-term ecosystem productivity.

> The amount of energy flowing through any ecosystem is determined by its net primary productivity. The productivity of forests and freshwater swamps is generally higher than that of grasslands and streams, which exceed that of oceans and deserts.

FOOD WEBS

Primary productivity tells us something about the energy base of ecosystems, but to what extent is ecosystem structure limited by energy flow? Primary productivity is limited at least as much by the availability of material resources, such as nutrients and water, as it is by energy. And, while energy reduction up the trophic scale helps explain ecological pyramids, how much more does it tell us about the detail of trophic relationships within ecosystems? A useful way to analyse this question is to consider the flow of energy up the trophic scale as a simple sequence of organisms eating and being eaten; this can be diagrammatically represented as a food chain (e.g. Fig. 44.9) or, more realistically (since food chains are not isolated in ecosystems) as food webs, which depict patterns of interlocking food chains (e.g. Fig. 44.10).

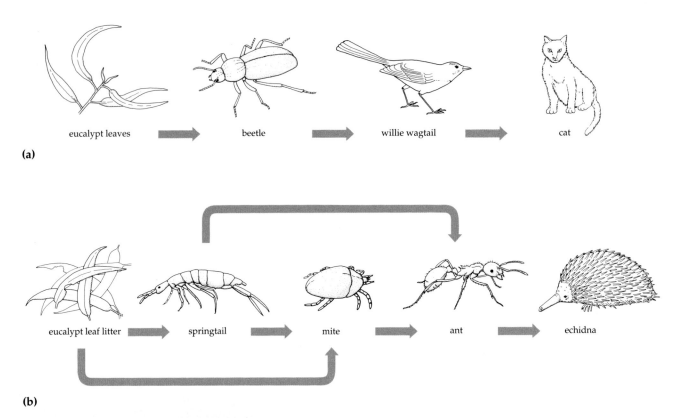

(a)

(b)

Fig. 44.9 Examples of simplified food chains that might occur in eucalypt woodland: **(a)** a grazing food chain; **(b)** a detritus food chain

Although green plants and algae are the ultimate base of most food chains, it is sometimes convenient to separate those food chains that are directly dependent on green plants (grazing food chains; e.g. Fig. 44.9a) from those whose primary food base is the mixed debris (or detritus) that is shed from organisms and their remains (detritus food chains; e.g. Fig. 44.9b). The relative importance of these food chains varies among ecosystems. In grasslands such as mulga rangelands, grazing food chains are prominent, commencing with pastures and shrubs and proceeding through herbivores such as red and grey kangaroos, sheep, cattle and goats, to consumers such as people, dingoes, foxes and cats. In forested ecosystems, detritus food chains are often more prominent, with much of the primary production proceeding via leaf litter through detritus food chains based on soil arthropods, such as springtails and mites (Fig. 44.9b). However, the distinction is not always clear. For example, in the rangelands, termites are also important primary consumers, so much so that their biomass may exceed that of domestic livestock (Box 44.3), but they eat both living and dead plant material. In this case, the distinction between grazing and detritus food chains is not very meaningful.

Despite their apparent complexity, food webs from different ecosystems show basic patterns:

- Food chains are short, with typically three or four trophic levels.

- Omnivores, defined here as organisms that feed on more than one trophic level, are usually scarce. Typically, food webs have one omnivore per top predator. For example, in Figure 44.10 currawongs are omnivores but echidnas are not.

- Omnivores usually feed on species in adjacent trophic levels. For example, in Figure 44.10, currawongs feed on willy wagtails and beetles but not on eucalypt leaves.

- Insect- and detritivore-dominated food webs are often exceptions to these patterns. For example, in Figure 44.10, the detritus food chain that proceeds from litter through springtails to cats has six levels, and two omnivores—ants and mites.

- Trophic interactions are confined to within a habitat but, within that habitat, further sub-groupings (compartments) are uncommon. For example, because skinks live on the ground they are unlikely to feed on leaf-eating beetles, but within their ground habitat they may eat any potential prey they encounter, regardless of whether it depends on a grazing or detritus food chain. (If the pupae of the leaf-eating beetles shelter in litter, then they may be eaten by skinks.)

- Food webs are not too complex; there are a limited number of relationships in which any one species may be involved.

- There are a number of commonsense patterns in food webs: predators do not exist in the absence of prey; a predator cannot feed on all the species it encounters; and loops (e.g. a species of bird able to eat and be eaten by a species of insect) do not generally occur.

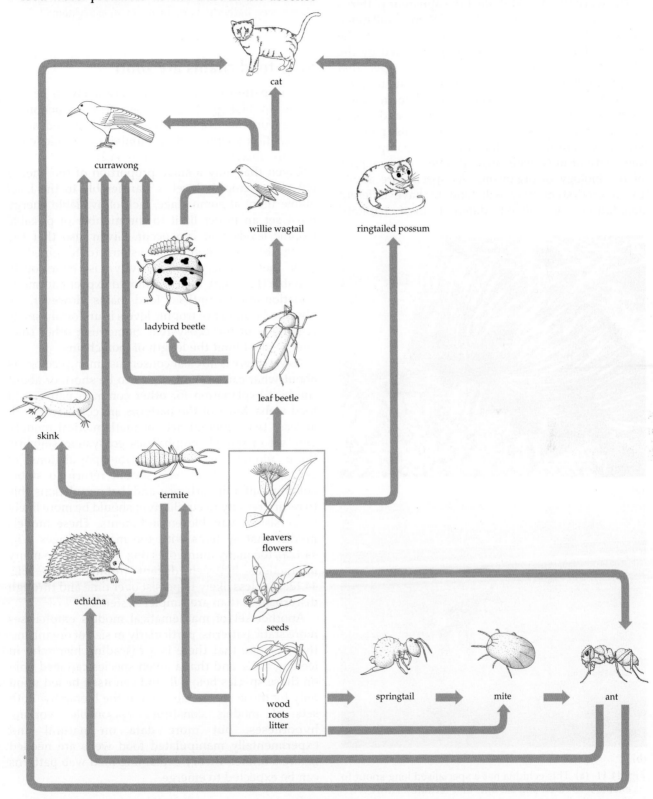

Fig. 44.10 A simplified food web, illustrating how the food chains in Figure 44.9 might interconnect with some other food chains in eucalypt woodland

Some of these patterns have been known for a long time; by 1927 Elton had recognised that the length of food chains was usually limited to four or five links. More recently, mathematical ecologists have identified other apparent regularities.

The work that has identified the apparent patterns shown above is still exploratory and there is still much debate about the ecological significance of the patterns. Food webs are obviously restricted by the characteristics of organisms. Some specialist adaptations (e.g. an efficient snout for eating termites and ants; Fig. 44.11a) are likely to preclude an animal from some other activities (e.g. catching and killing large animals; Fig. 44.11b). The commonsense patterns are largely a result of such logistics. The comparative simplicity of most webs may also be a consequence of the biology of organisms. As species are added to an ecosystem they will tend to have different adaptations or inhabit different local habitats.

(a)

(b)

Fig. 44.11 **(a)** This echidna has a specialised long snout to enable it to catch and eat ants and termites. However, this precludes it from eating large prey compared with **(b)** the dingo, seen here catching an eastern grey kangaroo

Consequently the number of species a predator may exploit will be limited and will not continue to increase linearly with the number of species in the food web.

> Food webs are diagrams depicting trophic interactions between individual organisms in an ecosystem.

Why food chains are short

Not all patterns in ecosystems are likely to be due solely to biological restrictions. The reduction of energy up the trophic scale is often quoted as an explanation for the limited number of trophic levels that are observed in nature (Box 44.2).

Given that only a small proportion of the energy entering a trophic level is transferable to the level above then, at some stage, lack of available energy must set an upper limit to the number of possible trophic levels that can occur. Given also that the energy base of different ecosystems (determined by their net primary productivity) is enormously variable (Fig. 44.5), then we should expect enormous variation in the lengths of food chains. However, the observed number of trophic levels is uniformly small, regardless of the ecosystem. Something other than energy must limit the length of food chains.

As yet, there is no clear consensus among ecologists about what causes food chains to be short, or about the best explanation for other common patterns in food webs. Most of the patterns are consistent with at least two different sets of mathematical models. One set of models emphasises ecosystem stability: these models assume that the populations of organisms in a food web should return to stable equilibria after disturbance, and that populations able to return rapidly to equilibrium should be more likely to persist in variable environments. These models predict that systems with too many linkages (Fig. 44.12a), too many omnivores (Fig. 44.12b) or too many interactions between distant trophic levels (Fig. 44.12c) are less likely to persist over time and through disturbances than are simpler systems.

Another set of mathematical models emphasises more static patterns, particularly in size of organisms: they assume that there is a cascading hierarchy in feeding links; and that a given species can feed only on those species below it, and can itself be fed upon only by those species above it in the hierarchy. Both sets of models constitute reasonable working hypotheses, but more data on natural and experimentally manipulated food webs are needed before a unified theory explaining food web patterns can be expected to emerge.

> Food chains are usually short and many food webs share other patterns that may be related to logistic constraints, body size and/or ecosystem stability.

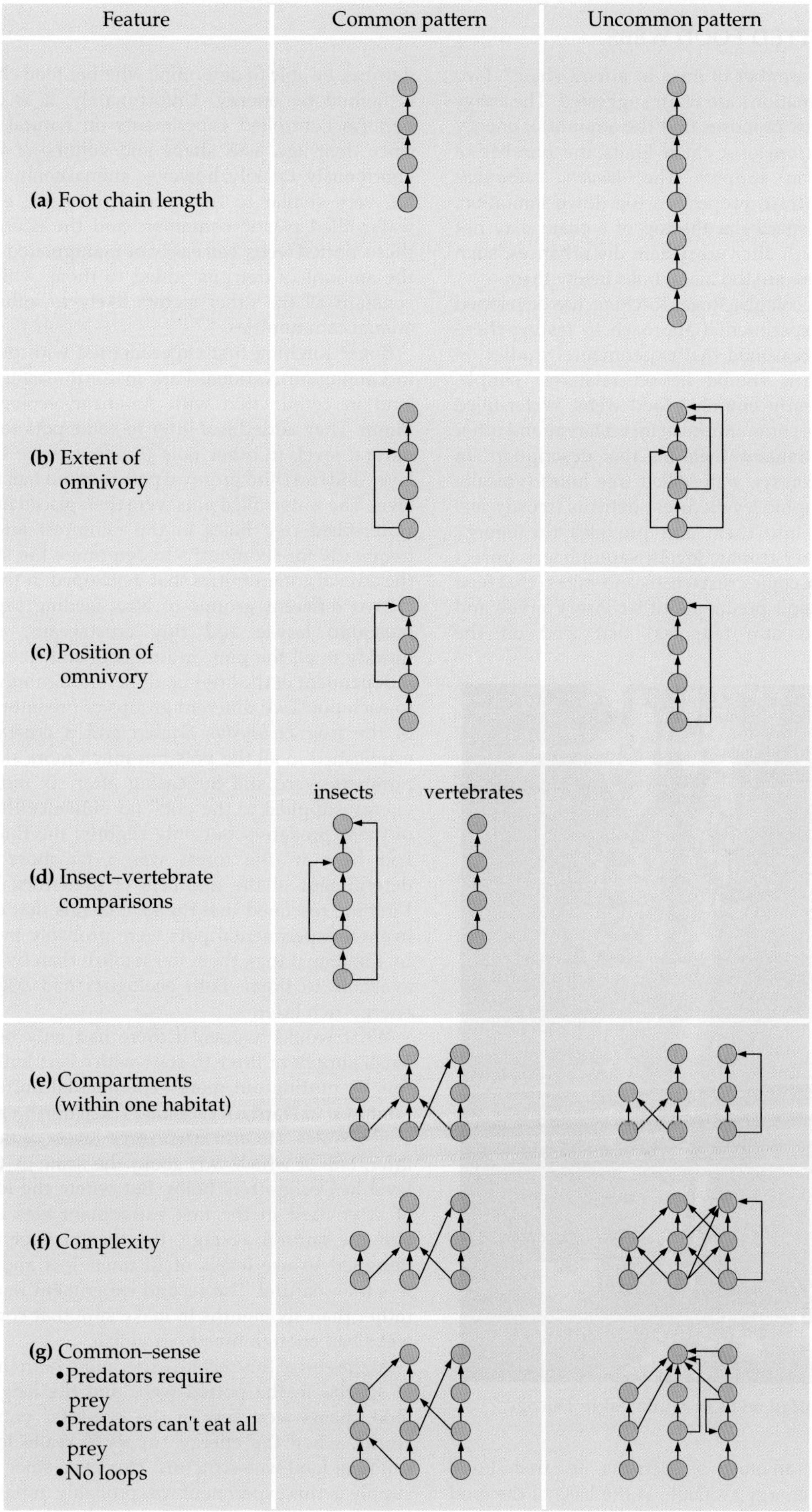

Feature	Common pattern	Uncommon pattern
(a) Foot chain length		
(b) Extent of omnivory		
(c) Position of omnivory		
(d) Insect–vertebrate comparisons	insects vertebrates	
(e) Compartments (within one habitat)		
(f) Complexity		
(g) Common–sense •Predators require prey •Predators can't eat all prey •No loops		

Fig. 44.12 Models of common and uncommon patterns in food webs

BOX 44.2   POTTED FOOD WEBS

What limits the number of links in a food chain? Two alternative explanations are often suggested. The *energy limitation hypothesis* proposes that the amount of energy entering the bottom of a chain limits the number of consumers it can support. The *dynamic constraints hypothesis*, in contrast, proposes a top-down limitation; it suggests that species at the top of a chain may not be able to establish after ecosystem disturbances, such as famines, if there are too many links below them.

An Australian ecologist, Roger Kitching, has developed an interesting experimental approach to testing these hypotheses. He reasoned that experimental studies on food chain length should be on relatively simple, common and clearly bounded food webs. Water-filled tree holes and the communities of insect larvae and other organisms that inhabit them fit this description. In subtropical rainforests, water-filled tree holes typically contain three trophic levels: forest detritus (mostly leaf litter) that falls into them and provides the energy available to higher trophic levels; saprophages (insect larvae and microscopic crustaceans and mites) that feed on the detritus; and predators (other insect larvae and crustaceans, and also tadpoles) that feed on the saprophages.

Experimental potted webs in rainforest in Dorrigo National Park

Because the amount of detritus in tree holes determines the energy available at the base of the food chains, we should, by measuring the lengths of food chains in tree holes containing varying amounts of detritus, be able to determine whether food chain length is limited by energy. Unfortunately, it is difficult to perform controlled experiments on natural tree holes since their age, size, shape and volume of water vary enormously. Luckily, however, animal communities that are very similar to the natural ones will establish in water-filled plastic containers and the energy base of these 'potted webs' can easily be manipulated by varying the amount of detritus added to them, while keeping constant all the other factors likely to influence their animal communities.

Roger Kitching first experimented with potted webs in Lamington National Park in south-eastern Queensland in conjunction with American ecologist, Stuart Pimm. They added leaf litter to some pots to about the natural level, to other pots to about twice the natural level, and to a third group of pots to about half the natural level. The water-filled pots were then placed near natural water-filled tree holes in the rainforest and sampled frequently for six months to determine the structure of the animal communities that developed in them.

Two different groups of litter-feeding saprophages, mosquito larvae and tiny crustaceans, established quickly in all the pots, in numbers and sizes that were independent of the litter (and therefore energy) supplied to each pot. Two different groups of predators, tadpoles of the frog *Lechriodus fletcheri* and a crustacean, also established in all the pots but much more slowly (their numbers were still increasing after six months). The energy supplied to the pots did influence the numbers of these predators but only slightly; the time the pots had been in the forest was a far more important determinant of the numbers of predators. Pimm and Kitching reasoned that the food chains that established in their experimental pots were probably more limited by the time it took them to establish than by the energy available to them. Both ecologists had used a lot of energy-rich litter.

What would happen if there had only been a very small supply of litter to start with? Bert Jenkins tested this by putting out more experimental potted webs in rainforest in Dorrigo National Park in north-eastern New South Wales. He also used three levels of input of leaf litter, one of which was about the same as the natural level in Dorrigo tree holes. But where the lowest level of litter used in the first experiment was only about half the natural average, the second experiment was designed to use levels of 10 times less and 100 times less than natural. The second experiment ran for a year rather than six months to make sure that complex food webs had enough time to establish.

At the end of this second experiment both the numbers of species in the potted webs and the length of their food chains were less in the low-litter pots; in other words, when the energy supply is really low, energy can limit food web structure. However, since the energy supply in this experiment was probably unnaturally low, it is also unlikely that energy is a limiting factor at the base of tree hole food webs in the real world.

BIOGEOCHEMICAL CYCLES

Energy flows through the biosphere. It is continuously radiated by the sun and, despite the temporary divergence of a small fraction of solar energy into ecosystems, most is eventually dissipated and lost as heat (Fig. 44.1). The inorganic materials necessary for life, however, are not continuously produced. Instead, the elements from which they are synthesised are withdrawn from world atmospheric and geological stores to which they are eventually returned. As we have already learnt, the productivity of ecosystems is limited at least as much by the supply of materials as of energy. Thus, the efficiency of cycling of materials is a major determinant of ecosystem structure and function.

For materials actively cycled through the atmosphere (e.g. water, carbon, nitrogen and sulfur) it is often useful to think of **global cycles** operating on a world scale because atmospheric circulation patterns result in effective transfers between ecosystems. Disturbances of these cycles can have drastic effects far removed from the source of the disturbance.

Other minerals, such as phosphorus, potassium, calcium and magnesium, generally undergo far less mixing; their movements are often considered as local cycles because they operate mostly at the local, rather than the global, scale. However, because minerals are transferred between local ecosystems, the availability and efficiency of recycling of minerals is highly variable within any one ecosystem. Accumulation of minerals in the biotic part of an ecosystem often occurs during regeneration (e.g. after a forest fire) or succession (Chapter 43), when productivity is high and biomass is accumulating. A small depletion of minerals (negative balance between inputs and outputs) may result from normal weathering processes in a mature ecosystem. A large deficit in minerals indicates a disruption of the usual ecosystem processes.

> Energy flows through the biosphere but materials are recycled within it. Materials cycled through the atmosphere circulate in global cycles. Other materials cycle mostly in local ecosystems.

Water

We saw at the beginning of this book that water is an essential and major component of all life. The bulk of water on earth (about 97%) resides in oceans, where it is directly accessible to marine life. Water used by terrestrial life (less than 0.5% of the total) circulates between the oceans and atmosphere through processes of evaporation and precipitation. Within the atmosphere it circulates by convection, resulting in some transfer to land surfaces as precipitation. Eventually most of the water precipitated over land is transferred back to the oceans through processes of runoff and stream flow (Fig. 44.13; Table 44.1).

Within this larger cycle, subcycles also operate. Some water is returned from the land surface directly to the atmosphere in the processes of transpiration and respiration. Some resides in non-oceanic pools as either soil moisture or freshwater bodies; it is from these stores that terrestrial plants and animals derive the water that they require. Most of the world's water circulates between active pools that are theoretically available to life, but a small fraction (less than 3%) is in relatively inaccessible storage pools, such as icecaps, glaciers and deep ground water.

> Water cycles globally, with the ocean as an enormous active pool. Within local ecosystems its movement is effectively a one-way flow that is replenished only by new input.

Table 44.1 Active and storage pools within the global water cycle

Pools	Pool capacity (10^{12} moles)
Active	
Atmosphere	720 000
Soil moisture	3 700 000
Stream channels	69 000
Fresh water lakes	6 900 000
Saline lakes	5 800 000
Oceans	74 000 000 000
Total active pools	74 017 189 000
Storage	
Icecaps and glaciers	1 600 000 000
Ground water	460 000 000
Total storage pools	2 060 000 000

Water is not equally available in all ecosystems. At a local ecosystem level, water behaves more like energy than most other materials in that it effectively flows through systems rather than being recycled within them. The amount of water recycled from plants and animals back to soil and local water bodies is negligible and relatively little evaporated or transpired water is recycled locally. Most of it is lost from the local ecosystem to the atmosphere to be precipitated in a distant ecosystem. Thus, for local ecosystems water is essentially a non-cyclable, periodically exhaustible resource, replenished only by new input.

Australia is particularly dry by world standards. Two-thirds of mainland Australia is desert (Fig. 44.14) and in many regions water is the material most limiting for ecosystem processes. The *variability* of rainfall in Australia (except for a narrow zone along the southern coast) is also particularly high by world standards. This has major implications for the ways

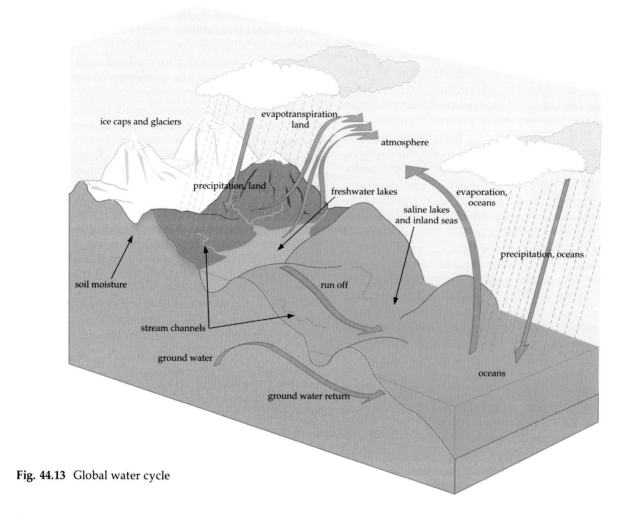

Fig. 44.13 Global water cycle

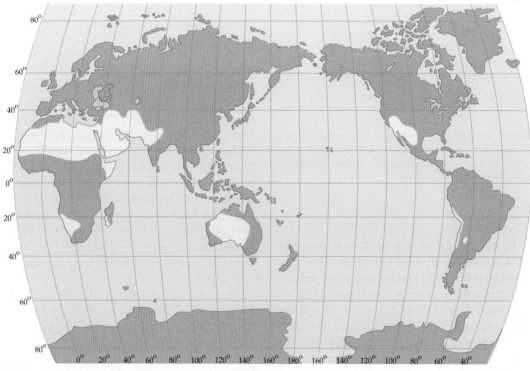

Fig. 44.14 Desert areas of the world, showing the high proportion of Australia that is desert

in which many Australian ecosystems function. The flow of energy into primary producers occurs via photosynthesis, with water loss an inevitable consequence. Thus, primary productivity in water-controlled ecosystems is closely coupled with water availability. This means that desert primary producers effectively operate as converters of water flow to energy flow.

The seasonal rate of energy flow can usually be regarded as fairly constant for most ecosystems. However, because desert rainfall is infrequent and unpredictable, the driving input into desert ecosystems, and therefore their productivity, is pulsed. Thus, rather than considering desert water cycles as systems of pools and flows, it may be more appropriate to consider them as systems of 'pulses-and-reserves' (Fig. 44.15). Consumers in these eco-systems must either adopt a pulse-and-reserve pattern themselves (e.g. insects such as grass-hoppers), utilise the reserves of other organisms (e.g. seed or wood eaters) or adopt extremely flexible feeding habits, using whatever pulse or reserve is available (Box 44.3).

> Water moving through desert ecosystems does not usually flow through active pools in a steady state, but instead moves episodically, as a series of pulses-and-reserves.

Carbon

Most of the world's carbon, unlike its water, exists in relatively inaccessible storage pools; the amount circulating between active pools is only about one five hundredth of this store (Fig. 44.16; Table 44.2). The vast bulk occurs as carbonate in rocks (e.g. chalk, limestone, marble) and as fossil fuels (e.g. coal, oil, natural gas) formed from carbon-rich sediments, with the amount in rocks far outweighing that in fossil fuels (Table 44.2). Although the amount of carbon in the atmosphere is tiny (the concentration of carbon dioxide is only about 0.03% of total atmospheric gases), the atmospheric pool is the most active. Carbon dioxide is withdrawn from the atmosphere during photosynthesis (at a rate determined by the primary productivity of ecosystems) and returned by cellular respiration, mostly by decomposers. Because it is readily soluble, atmospheric carbon dioxide also dissolves in the oceans, where the largest active pool resides and where it is available in solution to marine plants. Oceanic and atmospheric pools are generally in equilibrium and rates of vaporisation tend to balance rates of solution.

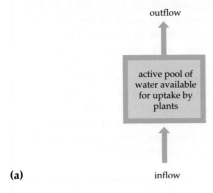

(a)

outflow

inflow

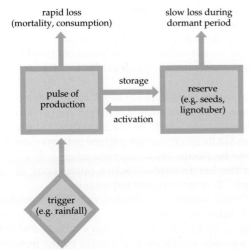

(b)

Fig. 44.15 Different ways of representing water flows through a desert ecosystem. (a) Level-and-pool model. The level of a resource is assumed to be controlled by uniform rates of inflow and outflow. Thus, there is always an active pool of resource available for uptake and use. (b) Pulse-and-reserve model. A rainfall 'trigger' sets off a pulse of production. Much of this pulse is rapidly lost to consumption and mortality but some is diverted back into storage reserves such as seeds or lignotubers. The reserve compartment diminishes only slowly during the no-growth period and from it the next pulse is activated

Table 44.2 Active and storage pools within the global carbon cycle

Pools	Pool capacity (10^{12} moles)
Active	
As CO_2 or CO_3 in solution	
Atmosphere	58 000
Ocean surface	43 000
Ocean depths	2 900 000
As organic carbon	
Land organisms	38 000
Decaying land organic matter	58 000
Marine organisms	830
Decaying marine organic matter	250 000
Total active pools	3 347 830
Storage	
Carbonate sediments	1 700 000 000
Fossil fuels	830 000
Total storage pools	1 700 830 000

BOX 44.3   TERMITES IN TROPICAL AUSTRALIAN ECOSYSTEMS

Although termites are commonly called 'white ants' they are not ants at all. They belong to the insect order Isoptera and are more closely related to cockroaches than to ants. Termites are social insects and live in nests. Termite nests are often quite complex structures largely isolated from their external environment. In size, they range from small and diffuse to vast interconnecting networks of chambers, galleries and covered walkways. Some are subterranean or within trees, tree stumps and timber; above-ground earth mounds are often of great architectural complexity (see figure). There are nearly 2000 species of termites, with the majority of them living in tropical and subtropical regions.

Some termites gain a lot of nutritional benefit from wood although they do not eat it themselves. Few organisms, and termites are no exception, can digest the cellulose of wood. However, the guts of termites contain a variety of micro-organisms that can break down cellulose, and this symbiotic interaction allows termites to utilise a range of cellulose-rich foods unavailable to most other herbivores. Wood is only one such food and wood-feeding termites are only one of several termite trophic groups. Grass, herbs and plant litter are the main diet of another major group; harvester termites cut down standing grass and herbs and scavenger termites collect various types of freshly fallen

Termite mounds in subtropical woodland, northern Queensland

and decaying plant litter and dung. Humivores (humus-feeding termites) eat organically rich soil. Not all termite species fall neatly into just one of these feeding categories; many are polyphagous and consume an enormous variety of cellulose-rich materials, including wood, plants, litter, dung and timber products.

In Australia, the largest numbers and majority of species of termites are found in subtropical and tropical sclerophyll forests, woodlands and savannas. In these habitats their numbers may be extraordinarily high. For example, in lightly timbered pastoral country in Queensland, where cattle are grazed at about one beast to 25 ha (about 24 kg of cattle/ha), CSIRO scientists calculated that the biomass of termites may be as high as 40–120 kg/ha.

Although much of inland Australia is both infertile and dry, plant biomass is often surprisingly high. This is because perennial grasses are able to establish during infrequent but heavy rains. For much of the time this biomass is unavailable to most herbivores because it has insufficient water and nutrients for their requirements. It is, however, an abundant food resource for termites, which also have the advantage in arid regions of the insulating properties of their nests, their co-operative social structure and their ability to store food harvested during pulses of plant production for use in leaner times. In this they mirror the pulse-and-reserve strategy followed by many species of arid zone plants.

Some species of termites are undoubtedly commercial pests, particularly of timber products. Grass-eating termites are sometimes so numerous that they compete with domestic livestock and occasionally harvest so much grass and litter that large areas become denuded. While their destructive potential is usually obvious it is probably far outweighed by their contribution toward ecosystem functioning, particularly in dry tropical regions.

Decomposer organisms such as soil bacteria, fungi and earthworms, play a critical role in the cycling of nutrients through ecosystems, both in the breakdown of organic matter and in its incorporation into soil in forms available for uptake by plants. What happens when soils are too dry and too hot to support active populations of soil organisms? Termites become important!

The CSIRO scientists estimated that decomposition of plant litter by termites may be responsible for recycling up to 20% of the ecosystem's carbon, and that termite mounds, which are kept humid by the termites, are also important centres of activity for moisture-requiring soil micro-organisms. Organic matter that termites bring to their mounds is relatively rich in most nutrients and, as a result, termite mounds contain much more nitrogen, phosphorus and other valuable nutrients than do the surrounding soils. Although it may take as long as 30 years, this nutrient-rich mound material is eventually redistributed to surface soils, at rates equivalent to adding 300–400 kg of fertiliser per hectare every year.

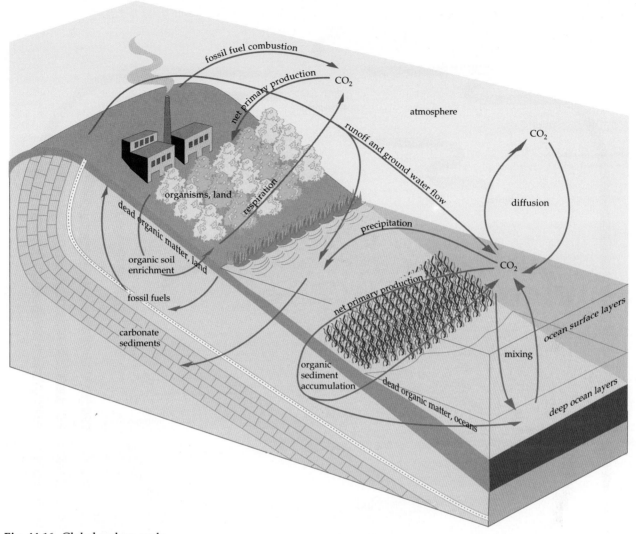

Fig. 44.16 Global carbon cycle

Exchanges do occur between active and storage carbon pools but, until recently, only at rates measurable on geological time scales. Much of the carbonate in freshwater and oceanic sediments eventually becomes lithified into carbonate rocks and fossil fuels. Carbon returns from storage very slowly during the weathering of carbonate rocks, but much more rapidly during the combustion of fossil fuels. Since the Industrial Revolution, fossil fuels have been mined and burned at unprecedented rates, with the result that carbon dioxide is currently being returned to the atmosphere faster than it can be cycled, resulting in a net increase in the concentration of atmospheric carbon dioxide. Some of the implications of this for world climates are discussed in Chapter 45.

Carbon cycles globally through a small but active atmospheric pool from which it is withdrawn by photosynthesising organisms. Most of the world's carbon resides in storage pools in rocks and sediments.

Nitrogen

Although nitrogen is the most abundant gas in the atmosphere (78%), the functioning of many ecosystems is limited by the availability of the element. This is because most plants cannot absorb atmospheric nitrogen but obtain their nitrogen as ammonium or nitrate ions from symbiotic soil micro-organisms or from soil solution (Chapter 18). Since most animals obtain their nitrogen requirements by eating plants (or animals that have eaten plants), most of the nitrogen requirements for the biotic components of ecosystems therefore depend on active pools of nitrogen cycling in soil (Table 44.3).

The concentration and composition of soil nitrogen pools are determined mainly by the activities of soil micro-organisms (Box 44.4; Fig. 44.17). Some atmospheric nitrogen is also fixed during thunderstorms when nitrogen is converted to oxides by lightning and washed from the atmosphere by rain. An

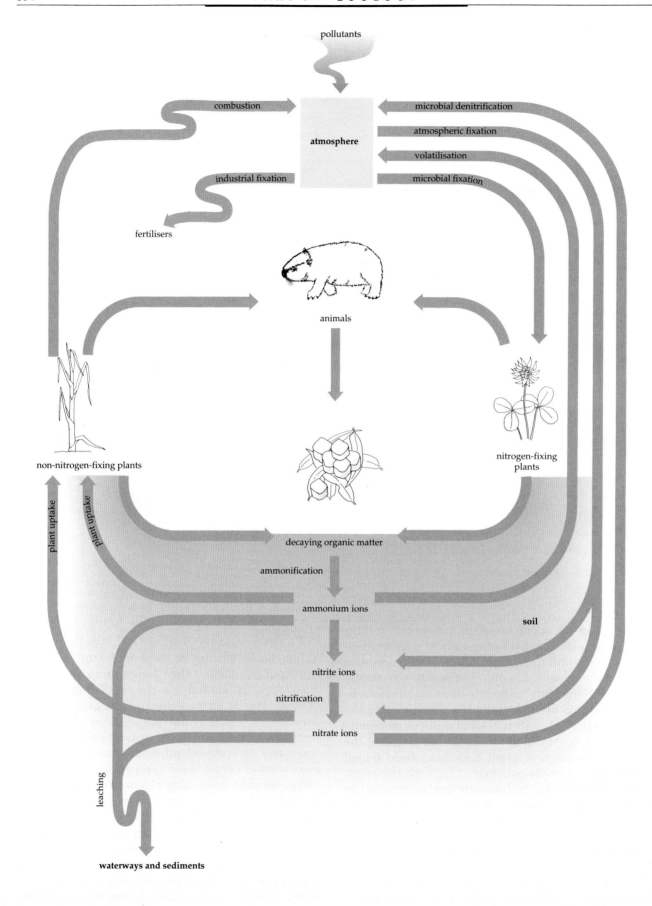

Fig. 44.17 Global/local cycle of nitrogen in a terrestrial ecosystem. Wavy lines indicate distant sources and sinks

Table 44.3 Active and storage pools within the global nitrogen cycle

Pools	Pool capacity (10^{12} moles)
Active	
As organic nitrogen	
Land organisms	870
Marine organisms	69
Decaying organic matter	120 000
As inorganic nitrogen	
Soils	10 000
Marine sediments	7 100
Total active pools	138 039
Storage	
As N$_2$	
Atmosphere	270 000 000
Oceans	1 400 000
As rock-forming minerals	
Sedimentary rocks	290 000 000
Crustal rocks	1 000 000 000
Total storage pools	1 561 400 000

increasing amount is fixed industrially, in high-temperature reactions that convert atmospheric nitrogen to chemical fertilisers. A variable amount is lost from local ecosystems by volatilisation of organic wastes. This amount can become major during bushfires and from stockpiles of animal excreta accumulated in intensive livestock operations. Some volatilised and ashed compounds may be deposited later, often in distant ecosystems, as atmospheric fall-out, but much is lost to the atmospheric store. Some industrial processes, particularly automobile use, also contribute volatile nitrogen compounds (particularly oxides) to the atmosphere, which may be precipitated elsewhere as fall-out.

Because the total soil nitrogen pool is so highly dependent on soil micro-organisms, nitrogen is most likely to be limiting in environments in which microbial activity is inhibited (e.g. in cold or dry regions) or in soils with low organic contents (e.g. sandy soils). Any ecosystem disturbance large enough to modify soil microclimates or destroy organic matter will have drastic effects on the availability of nitrogen in the ecosystem. For example, cutting and burning of tropical forests completely changes the living microbial community that fixes and conserves soil nitrogen. Ammonium, which would be taken up rapidly by the trees in an unlogged forest, is converted to nitrate; crop species use only a small fraction of this nitrate and the rest is lost through leaching or denitrification.

> Soil micro-organisms are the main agents of fixation of atmospheric nitrogen, mineralisation of nitrogen in organic matter and denitrification.

Phosphorus

Phosphorus is essential for life because it is used by all organisms in the form of ATP, but its supply often constrains ecosystem functioning because it is not a common element in the earth's crust. The pathways and storage of elements that are cycled less actively, such as phosphorus, operate through essentially similar processes, although the size of storage compartments and rates of flow vary between elements and between ecosystems (Fig. 44.18).

The rate of phosphorus input from the atmosphere is very low compared with most other nutrient

---

BOX 44.4   MICRO-ORGANISMS INVOLVED IN THE NITROGEN CYCLE

Micro-organisms causing **nitrogen fixation**, the conversion of unavailable atmospheric nitrogen to forms available for plant uptake, include:

- symbiotic nodule-forming bacteria (mostly *Rhizobium* spp.) on roots of agricultural legumes (e.g. clover, lucerne, beans) and non-agricultural legumes (e.g. *Acacia, Cassia*);

- symbiotic nodule-forming bacteria (mostly *Frankia* spp.) on roots of non-leguminous plants (many genera, including she-oaks, *Casuarina* and *Allocasuarina*, and alders, *Alnus*);

- symbiotic cyanobacteria, including those in lichens, liverworts, some ferns, and root associations on cycads (*Macrozamia*) and a few other genera;

- free-living bacteria and cyanobacteria (many genera; low fixation rates but total contribution may be substantial).

Micro-organisms causing **nitrogen mineralisation**, the conversion of organic nitrogen compounds in humus to simple ions in soil solution, include:

- ammonifying bacteria and fungi, which release ammonium ions by breaking down decaying organic matter in soil;

- nitrifying bacteria, chemautotrophs that convert soil ammonium to nitrite (*Nitrosomonas*) and nitrite to nitrate (*Nitrobacter*).

Micro-organisms causing **denitrification**, the conversion of nitrate ions to gaseous nitrogen compounds released to the atmosphere, include:

- denitrifying bacteria, anaerobic organisms that are most active in soils with poor aeration.

BOX 44.5   TOO MUCH OF A GOOD THING: NITROGEN AND TREE DIEBACK

Trees 'dieback' whenever they are subjected to long or repeated periods of sublethal stress. Some trees recover from dieback but, if widespread stress continues for long enough or becomes severe enough, many trees may start to die. Sometimes the cause of stress is well understood, although none-the-less damaging and difficult to reverse. For example, jarrah dieback, which affects *Eucalyptus marginata* forests in Western Australia, is primarily caused by the destructive introduced soil-borne pathogen, *Phytophthora cinnamomi* (Chapter 35).

Sometimes, dieback is the result of a number of interacting factors. This is the case for rural dieback, which refers to premature and relatively rapid decline and death of native trees (usually eucalypts) in rural Australia. Rural dieback is connected with changes that agriculture and pastoralism have wrought but the specific factors involved vary in importance from region to region. In the New England and Southern Tablelands of New South Wales, leaf-feeding insects are an important part of the dieback equation (Fig. a) but are not the whole story, since the same species of insects also feed on trees growing in natural woodlands, where rural dieback does not occur.

Insect damage is probably just one link in a long chain of events that are still not well understood. Nitrogen cycling may also be a crucial factor. Grazing is intensively managed on the Tablelands. Since the 1950s, farmers have been steadily improving the fertility of their pastures by planting nitrogen-fixing legumes and spreading fertilisers, such as superphosphate. This has enabled pastures to carry more and better sheep and has increased wool production. Carrying capacity of improved pastures is higher, not just because there is more pasture available, but also because its quality, particularly its protein content, is enhanced. Rates of growth and reproduction of sheep depend on an intake of dietary protein, which depends on the supply of nitrogen to food plants.

Sheep are not the only herbivores that benefit from pasture improvement. White curl grubs, which are the larvae of beetles belonging to the scarab family, live in soil and feed on grass roots and other organic matter. Therefore, improving the quantity and quality of pasture available to sheep may also unintentionally benefit these unseen underground herbivores. In some improved pastures, the weight of scarab larvae grazing on pasture roots may actually equal the weight of sheep grazing the tops. This is bad news for the farmer because feeding by scarabs damages and sometimes kills pasture. It is even worse for eucalypt trees growing among or adjacent to pasture. If the weather is favourable during December, swarms of large golden 'Christmas beetles' emerge from pupae of scarab larvae and fly to the nearest eucalypt trees where they feed voraciously on foliage, sometimes defoliating whole trees. By late summer all the adult Christmas beetles mate, lay their eggs in the soil under the trees and die, giving the trees a chance to produce new foliage to carry them through the winter.

(a) Rural dieback of eucalypts in the New England Tablelands of New South Wales

The trees' problems are not restricted to Christmas beetles. Increased soil fertility, as well as improving the protein content of pasture plants, also enhances the protein content of the foliage of trees growing among the pasture. If sheep congregate under trees, nutrient enhancement of foliage may become very marked as concentrated nutrients are applied to the soil in the form of urine and dung. This means that all the insect herbivores of trees (beetles, caterpillars, leaf hoppers, stick insects and many others) will benefit from feeding on fertilised foliage in the same way that sheep and scarab larvae benefit from fertilised pasture. This is probably why trees under which sheep graze may carry as much as ten times the weight of insects as trees in more natural woodlands, and why much nitrogen creates unexpected problems for eucalypts in rural Australia (Fig. b).

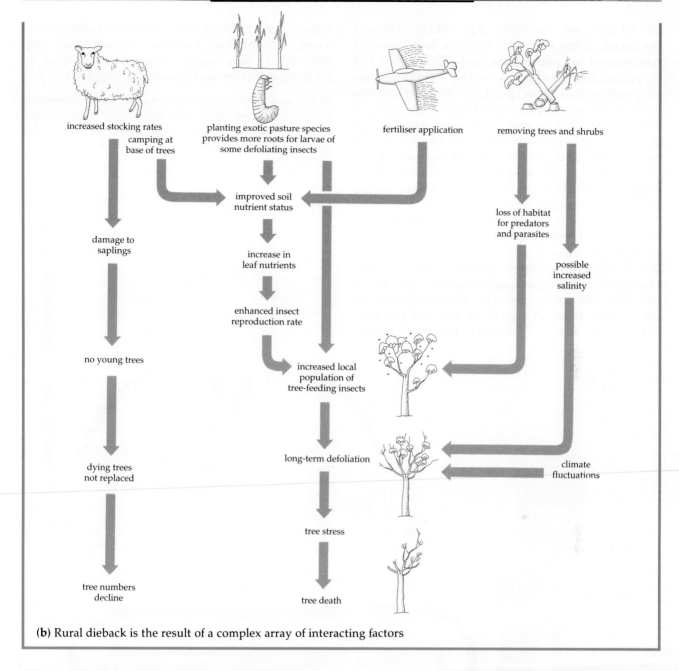

(b) Rural dieback is the result of a complex array of interacting factors

elements, so the supply in soil is precious. Thus, it is important that phosphorus is far less mobile in soils than most other nutrients. Soluble phosphate ions are readily available to plants. Sparingly soluble phosphates (believed to be ions bound to the surfaces of iron, aluminium and calcium salts) are released slowly into the soil solution, and phosphates incorporated into clay minerals or concretions of iron or aluminium are immobilised and largely unavailable for plant growth. During the weathering of soil minerals, an increasingly large proportion of phosphorus becomes immobilised. Phosphorus immobilisation is most extreme in soils of the tropics that experience intense weathering. Because most Australian soils have been subjected to weathering for a long time (Chapter 41), their content of available

phosphorus is extremely low by comparison with soils of other continents. This is why most non-native crops in Australia require the addition of large quantities of superphosphate fertiliser to achieve adequate yields. Native plants, however, are adapted to make the most of the naturally available phosphorus levels in Australian ecosystems (Chapter 18).

Associations between plant roots and mycorrhizal fungi are very common in plants on infertile soils. In addition to enhancing the rate of absorption of nutrients, mycorrhizas probably also play an important role in nutrient cycling. The fungal hyphae readily penetrate litter and compete with other decomposers for the utilisation of phosphorus and other nutrients, thereby short-cutting the labile pool of soluble soil nutrients (Fig. 44.18). Efficient extraction from soil and litter, a low absolute requirement for phosphorus, and efficient use and recycling of phosphorus within biomass, are adaptations that allow plants of low phosphorus ecosystems to achieve remarkably high rates of productivity. For example, on sand islands near Brisbane, highly productive eucalypt forests grow on some of the most phosphorus-deficient soils in the world. Most of the phosphorus requirements for primary production are stored and recycled within the living vegetation.

Because the productivity of low phosphorus ecosystems usually depends on biomass stores of phosphorus, these ecosystems are easily disrupted. When tropical rainforests growing on extremely infertile soils are cleared, agriculture is possible for the first few years because of mobilisation of phosphorus stores in organic matter and ash from the original forest. Eventually, however, surface

organic matter decomposes, soil nutrients are leached away, phosphorus becomes immobilised, crop productivity fails, and the rainforest is replaced by far less productive open grasslands (Fig. 44.7).

The main inputs of phosphorus into local ecosystems are from sparingly soluble soil storage pools. In low phosphorus systems, vegetation constitutes the largest active pool of phosphorus and is efficient at internal recycling.

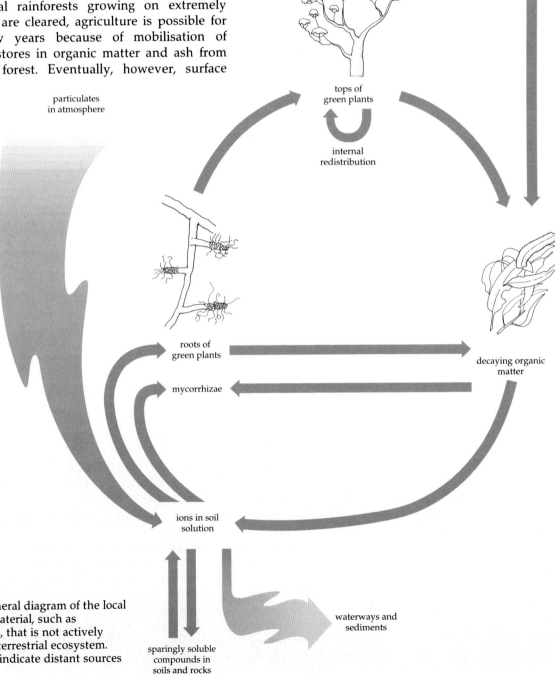

Fig. 44.18 General diagram of the local cycle of a material, such as phosphorus, that is not actively cycled in a terrestrial ecosystem. Wavy lines indicate distant sources and sinks

SUMMARY

- Ecosystems are changing, self-modifying ecological systems consisting of interacting biotic and abiotic components.

- Trophic interactions are probably the most important relationships determining ecosystem structure and functioning. All organisms can be considered as either autotrophs (producers) or heterotrophs (consumers, degraders and decomposers). Graphs of the energy flowing into successive trophic levels represent ascending ecological pyramids, but pyramids based on numbers or biomass may sometimes be inverted.

- The amount of energy flowing through any ecosystem is determined by its net primary productivity, which is also limited by availability of materials. The productivity of forests and freshwater swamps is generally higher than that of grasslands and streams, which exceed that of oceans and deserts, but there are many exceptions to these trends. Net productivity also depends on the duration and season of measurement, and on the age of the ecosystem.

- Food webs are diagrams depicting trophic interactions between individual organisms in an ecosystem. Although food webs appear complex, their component food chains are usually short. Many food webs share other patterns that may be related to logistic constraints, body size and/or ecosystem stability.

- Water cycles globally, with the ocean as an enormous active pool. Within local ecosystems its movement is effectively a one-way flow that is replenished only by new input. Much of Australia is desert, characterised by low and highly variable rainfall. Water moving through desert ecosystems does not usually flow through active pools in a steady state but moves episodically, as a series of pulses and reserves.

- Carbon cycles globally through a small but active atmospheric pool from which it is withdrawn by photosynthesising plants. Most of the world's carbon resides in storage pools in rocks and sediments; post-Industrial Revolution combustion of the fossil fuel fraction of these pools has resulted in a global increase in the atmospheric pool of carbon.

- The nitrogen cycle is dependent on local soil and microbial pools. Soil micro-organisms are the main agents of fixation of atmospheric nitrogen, mineralisation of nitrogen in organic matter and denitrification, by which soil nitrogen is returned to the atmosphere. External inputs of nitrogen in plant-available forms can cause major ecosystem disturbance.

- The main inputs of phosphorus into local ecosystems are from sparingly soluble soil storage pools. In low phosphorus systems, such as most tropical rainforests and most Australian ecosystems, vegetation constitutes the largest active pool of phosphorus and is very efficient in its uptake, use and internal recycling.

QUESTIONS

1. Under which circumstances would you expect a trophic diagram based on biomass to be pyramidal in shape? Give an example of an ecosystem in which this might occur.

2. Explain the difference between the ecological meaning of the terms biomass, energy and productivity.

3. Which ecosystem would you expect to have the highest annual productivity: a coastal rainforest north of Cairns in Queensland or a monsoonal rainforest in Kakadu National Park in the Northern Territory? Why? What ecosystem properties would you need to know to test your prediction?

4. Construct a food web showing possible feeding interactions between cattle, grass, trees, wood-feeding termites, harvester termites, scavenger termites, legless lizards, kookaburras and people.

What are the least, most and average number of links in the food chains you have drawn?

5. Design an experiment to test whether the amount of leaf litter that accumulates in water-filled tree holes depends on their depth. If your experiment showed that very shallow tree holes accumulated little leaf litter, what consequences might this have for the development of food webs in very shallow tree holes?

6. What are some advantages harvester termites might have over kangaroos, as primary consumers of semiarid rangelands? What are some possible disadvantages? Which represents more energy: 100 g of termites or 100 g of kangaroo?

7. In what way(s) have industrialised societies affected the global carbon cycle?

8. Which processes in the nitrogen cycle are controlled primarily by soil micro-organisms? Under what circumstances might you expect there to be limited soil nitrogen available for plant uptake?

9. Which fertiliser is most commonly used by Australian farmers? Why do Australian farmers use relatively more of this fertiliser than farmers in Europe and North America? What sorts of problems could result from excessive reliance on fertilisers?

10. Why does rainforest in many tropical regions fail to regenerate after widespread clearing?

HUMAN IMPACTS

Understanding the principles of ecology that we reviewed in Chapters 42–44 is vital to the future of life on earth. The human population continues to grow rapidly and the rate of destruction of habitat, with the consequential loss of species, is greater than in any other period of history. In this chapter, we have chosen the important issues of loss of biodiversity, impact of introduced species and degradation of the physical environment to highlight the impact of the human species on ecosystems, and to consider positive solutions through the application of our knowledge of biology.

HUMANS IN AUSTRALIA

As we discussed in Chapter 41, the history of the Australian continent has dictated, to a large extent, the composition of the Australian biota. Isolated as an island continent after its separation from Gondwana, Australia evolved unique animals and plants. As the climate became more arid during the Tertiary period, the typically Australian element of the biota evolved under the new environmental conditions. However, once the continent was close enough to the South-East Asian plate it became possible for some organisms, for example, bats and rodents, to be exchanged between the island continent and the island archipelago to the north.

With proximity to islands and other land masses, it was also possible for people to enter northern Australia. The first humans arrived in Australia possibly as early as 100 000 years ago (see Chapter 41). Europeans arrived more recently. Portuguese sailors drew maps in the 1500s that bear a striking likeness to the east coast of Australia, suggesting that they were the first Europeans to sight Australia. William Dampier, a Dutchman, is credited with the first landing in Western Australia during the late 1600s.

Sydney Cove was the site of the first permanent European settlement (Fig. 45.1) with the arrival of five small sailing ships carrying about 1500 people

from Great Britain in 1788. The small fleet called at Cape Town, South Africa, for provisions and water. There the settlers obtained plant cuttings and seeds to plant in their new home. The fleet became a veritable Noah's Ark as they took on board sheep, goats, cattle, horses, pigs and poultry.

As well as alien animals and plants, the European colonisers brought with them European attitudes. Australia was seen as a wild and inhospitable place, to be conquered and 'civilised'. In 1855, the explorer William Howitt said of Victoria:

> the choked-up valleys, dense with scrub and rank grass and weeds, and the equally rank vegetation of swamps, cannot tend to health. All these evils, the axe and the plough, and the fire of settlers, will gradually and eventually remove; and when it is done here, I do not believe that there will be a more healthy country on the globe.

The impact of Europeans on the Australian environment was rapid and extensive. Open grasslands and grassy woodlands were regarded as productive country and quickly turned over to pasture for domestic stock. Heavily wooded country was extensively cleared for agriculture with little regard

Fig. 45.1 The first European settlement at Sydney Cove

for the impact on the soil or native wildlife. Cities were built with little planning for effluent disposal and raw sewage was discharged directly into streams.

In Australia today, ecosystems most under threat from human activities are those of agricultural value. Extinction rates and declines in abundance and range of native flora and fauna have been highest in regions where settlement first occurred (Box 45.1), attributable to land use, habitat loss and fragmentation, over-exploitation, and the spread of exotic herbivores, predators and disease.

> Human impact on the Australian environment has greatly accelerated over the last 200 years due to extensive forest and woodland clearing and the introduction of exotic animals and plants.

CHANGING AUSTRALIAN ENVIRONMENTS
Losing biodiversity

Biodiversity is the number, relative abundance and genetic diversity of species found in an area. We learnt earlier that species are becoming globally extinct at a rate of 1000 to 10 000 times the natural rate before human intervention and that it is likely that 20% of all species will become extinct within the next 30 years. Australian biologists have recently reviewed the changes in the distribution and abundance of the vertebrate fauna of Australia over the last 200 years. More than 90% of the 1600 terrestrial vertebrate species are endemic. Of these, nearly 300 are considered endangered. Altogether, 17 mammals, three birds and one lizard are known to have become extinct (Fig. 45.2). Many other species have undergone dramatic reductions in range and abundance. For example, eight mammal species that were once present on the mainland persist only on small offshore islands.

There is no reason to believe that any equilibrium has been reached between agricultural development and species conservation. In fact, accelerated rates of extinction are expected, particularly among birds, which have so far fared better than mammals. The paradise parrot is, to date, the only species of Australian bird thought to have become extinct (Fig. 45.3).

(a)

Fig. 45.2 The pig-footed bandicoot, *Chaeropus ecuadatus*, is now an extinct Australian mammal

(b)

Fig. 45.3 **(a)** The paradise parrot, *Psephotus pulcherrimus*, is, to date, the only species of Australian bird thought to have become extinct. **(b)** Australia has a number of geographic areas (as shown here) that are recognised centres of diversity on a world scale. Accelerated rates of extinction are expected for birds

BOX 45.1   THE USE OF AUSTRALIAN FORESTS

At the time of European settlement, Australia was covered with about 80 million ha of forest, about 10% of the surface area of the continent. Since then, about half of the forests, approximately 40 million ha, have been cleared, largely for agriculture.

In each Australian state, European impacts on forest vegetation and fauna followed very similar patterns, with initial settlement being followed by almost a century of largely uncontrolled clearing. The first export commodity of the new colony of New South Wales was red cedar, *Toona australis* (see Fig. a), discovered on the banks of the Hawkesbury River in 1790. In Victoria, the discovery of gold near Clunes in 1851 led to a rapid population increase. Forests were heavily exploited to provide timber and fuel for mines and towns.

The Conservator of Forests in Victoria reviewed the condition of the forests in 1896 and emphasised the 'forest vandalism' practised by miners, timber splitters and sleeper hewers who produced more waste than timber. Forest Commissions were established in most Australian states to halt the over-use and degradation of forests, and formal and relatively effective control of forest exploitation was in place by the early 1900s. The first timber reserves in New South Wales were set aside in 1871 to protect forest resources from uncontrolled exploitation. By 1920, most of the forest estate in New South Wales was established. In Tasmania, there was no effective legislative control over forest use until 1920.

The Australian Association for the Advancement of Science lobbied successfully for the creation of a national park in South Australia in 1893. In 1894, they succeeded in forming a large 65 000 ha park in the centre of the jarrah forest in Western Australia but failed to have the park properly vested. The purpose of the park was reclassified to 'timber—government requirements' in 1911 following pressure from the timber industry. Other smaller parks were created around the same time and were more permanent. Most of the northern jarrah forest was dedicated as state forest in the late 1920s to protect it against uncontrolled clearing and unmanaged timber exploitation.

The current situation

Of the 40 million ha of forest remaining in Australia, 26 million ha are publicly owned native forest. Of this, more than 10 million ha (about 40%) is in national parks and flora and fauna reserves. More than 7 million ha of forest is managed for commercial forestry. In 1989–90, Australia exported about $350 million worth of woodchips and over $10 million of other forest products. Over the same period, Australia imported about $1000 million worth of pulp and paper and $500 million worth of sawn timber.

More than 22% of all endangered plant species occur in forest environments. The current rate of clearing is 0.22–0.27% of the total forest area per year and, if it continues, almost all unreserved and unmanaged forests will have been lost within 170–250 years (see Fig. b). Forest clearing continues relatively unabated on private land.

(a) Red cedar, *Toona australis*, grows in the rainforests of New South Wales and Queensland. Because it provides a fine furniture timber, it has been heavily harvested and is now a rare species in its native habitat

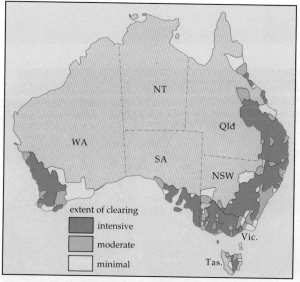

(b) Australian forest and woodland that has been significantly modified in the 200 years since European settlement

Habitat fragmentation

Habitat fragmentation (the breaking up of large habitats into many smaller habitats) is a major problem (Fig. 45.4). Many habitat remnants are too small to support viable populations of resident organisms (see Chapter 42). Consequently, the probability of organisms becoming extinct is greater in smaller patches, although the effects of fragmentation depend on the dispersal ability of organisms and the variability of the environment.

Scientists from the CSIRO Division of Wildlife and Ecology in Western Australia have found that bird species are rapidly lost from forest remnants. As each population becomes extinct, the chance of the forest remnant being recolonised by the same species is diminished. The effective size of fragments can be increased by linking up adjacent fragments with *corridors* of similar vegetation, allowing organisms to communicate between them.

Fragment shape is also important. The edge of a remnant fragment in an agricultural landscape experiences different environmental conditions from the interior of the fragment. Edges are typically windier and receive more light and inputs of nutrients from surrounding pastures. Certain species of plant and animal prefer edges to interiors of remnant fragments. For example, noisy miners are birds usually associated with edges, whereas yellow robins prefer the interior of a remnant. A long thin remnant patch has a greater edge to area ratio than a circular patch. Corridors that consist entirely of edge habitat are thus unsuitable for the movement of interior species. However, edges may be managed to enhance species diversity because edges typically support more species than are found in adjacent habitats.

> Biodiversity is the diversity of species in an area and is under threat from habitat destruction and fragmentation.

Why conserve biodiversity?

Does it matter whether we lose species? There are a number of reasons why the answer is 'yes'. The first reason is an *ecological* one. Organisms play roles in ecological systems that are essential for ecosystem survival. Our failure to discern the role of a species within an ecosystem does not mean the species is unimportant. Ecosystem processes, including species interactions (Chapters 43, 44), are complex and loss of a key species may have a substantial impact. For example, in Chapter 43 (p. 960) we discussed the role of frugivores, such as cassowaries (Fig. 45.5), that eat the large fleshy fruit of some rainforest trees in northern Queensland. The seeds of these trees are still viable after passing through a bird's digestive tract. Because of this, cassowaries play an important role in the dispersal of these seeds and therefore the ecology of the trees. Cassowaries are thought to be the only species that are capable of eating many of these fruit, but the birds are threatened by rainforest

(a)

(b)

Fig. 45.4 Habitat fragmentation is a major threat to biodiversity. **(a)** Narrow corridors of native vegetation, shown in the wheat belt area of Western Australia, are the only remaining habitat for the survival of native species, such as **(b)** *Banksia cuneata*

Fig. 45.5 It is essential to conserve organisms that play a pivotal role in ecosystems. The cassowary, *Casuarius casuarius*, is a keystone species in Queensland rainforests

clearing and feral animals. The loss of cassowaries from rainforests could have a substantial effect on the ability of many tree species to disperse and regenerate.

The second reason is a *practical* one: we use organisms for our own benefit (as bioresources) for food, fibre, medicines, beverages and so on. We need a diversity of species to maintain our quality of life. Since so much biodiversity remains undiscovered there must be equivalent quantities of bioresources still to be discovered. The potential of these undiscovered bioresources is being recognised and efforts are underway to identify uses for them. For example, it has been discovered recently that ants possess specialised glands for producing antibiotics to reduce disease in their colonies. This type of antibiotic is totally new to medicine and holds great potential for therapeutic use.

The third reason for conserving biodiversity is one of *aesthetics*. We enjoy the beauty of plants and animals, their complexity, their appearance of wildness. There is even uncertainty about whether humans can actually survive psychologically, let alone physically, in a world with reduced biodiversity.

The fourth reason is a matter of *philosophy*. Should we allow a species to become extinct if we can save

it? Humans are, after all, just another species. From an ethical point of view, other species have as much right to exist as we do.

The fifth reason is *custodial*. Countries conserve their natural and cultural heritage, which is passed down ('inherited') as a birthright.

On a world scale, about 1.4 million species have been described so far out of an estimated total of 6–10 million species. We know little about the genetic diversity of these species, their relationships or their biology, and there will never be sufficient time to describe all species. Consequently, in addition to efforts to describe individual species, classifications of communities and ecosystems are being developed based on the *functional* attributes of organisms. By dividing up the animals according to their feeding habits it is possible to work out their ecological relationships without naming all the organisms. For example, rainforest ecologists often use this approach because of the complexity of rainforest ecosystems. Keystone, indicator and vulnerable species can be identified and action taken to conserve them.

> Human survival depends on conserving biodiversity in order to maintain ecosystem processes, provide food and resources, and maintain the beauty of the natural environment.

BOX 45.2   CLEARING TROPICAL RAINFOREST

Tropical rainforests of the world (see Fig. a) are shrinking faster than any other ecological biome on earth. In 1992, around 100 000 km² of tropical rainforest (1% of their total area) was cleared and burnt for agriculture. This is an area slightly larger than Ireland! Each year up to 10 000 species are estimated to become extinct.

(a) Tropical rainforests are being lost at a rate faster than any other world biome. This forest in Borneo is home to the orang-utan

The diversity of species in tropical rainforests is unparalleled in any other ecosystem. Tropical rainforests cover only 7% of the earth's land surface but they contain at least 50% of all species. Ecologists have recorded 700 tree species in ten 1 ha plots in a Borneo rainforest—equal to the entire diversity of trees in North America. Forty-three species of ant belonging to 26 genera were collected from a single tree in a Peruvian rainforest—about equal to the entire ant fauna of the British Isles.

Disturbance is a natural process in rainforests, as it is in most forests. The most common natural disturbances in Amazonian rainforests are tree falls and forest fires. Gaps created by disturbance provide increased light, water and nutrients, and the forest rapidly recovers through a process of succession, the progressive replacement of one group of species by another (Chapter 43). Rainforests are dynamic ecosystems made up of a mosaic of vegetation patches at various stages of succession.

Rainforest ecosystems are also characterised by rapid nutrient cycling (Chapter 44). Organic material that falls to the forest floor is rapidly decomposed by fungi, invertebrates and bacteria in the litter and top layer of soil. Nutrients, which are released by the decomposition process, are mopped up by a network of plant roots that form a mat at the top of the soil. In this way, nutrients are returned to the living biomass without entering the mineral soil and are not subjected to leaching by abundant rainfall. On average, the above-ground biomass in rainforest contains approximately one-half of the total nutrients in the system. This contrasts with most other terrestrial vegetation types where the soil is the major nutrient store, and has important consequences in relation to human disturbance, which is typically more intense and more prolonged than natural disturbance.

The traditional agriculture practised by Indians in Amazonian rainforests is **slash-and-burn agriculture** (Fig.b). This involves clearing a small area of forest (usually about 1 ha) and burning the fallen timber. This releases abundant nutrients from the ashed plant material, enough to support an agricultural crop for two to three years. However, once this supply is exhausted, the soils are too poor to support further agriculture and the farmers move on to another patch of forest.

The re-establishment of rainforest trees on slash-and-burn sites is dependent on dispersal of seed into the clearing from surrounding forest. Some 'pioneer' species, that is, species that colonise disturbed sites, would naturally germinate from the soil 'seed bank' where they may have been dormant for many years. However, the constant weeding of farm plots severely depletes the soil seed bank. The seeds of many rainforest trees stay viable only for a short period, as little as six weeks, so they do not accumulate in the soil. These species must be introduced into clearings from surrounding vegetation. However, the clearings produced for slash-and-burn agriculture are large compared with natural gaps produced by tree falls. Many species are dispersed by animals that do not frequent large clearings, further

(b) Disturbance by slash-and-burn clearing for small-scale agriculture in an Amazonian rainforest

slowing the process of rainforest regeneration. It may take 200 years or more for rainforest to establish on an abandoned farm plot in the Amazon Basin.

Much Amazonian rainforest is now being cleared on a larger scale for cattle pasture. As with slash-and-burn agriculture, pastures on cleared rainforest sites are only productive for a short period of time. The ground is compacted by cattle and seed dispersal must occur over large distances. Natural re-establishment of rainforest on these sites is extremely slow, if it happens at all.

Introducing new species

Some of the greatest impacts that Europeans have had on the Australian environment have been caused by the introduction of new species (Fig. 45.6). Many species have been unintentionally introduced. In Chapter 35 we mentioned cinnamon fungus, *Phytophthora cinnamomi*, which was introduced into Australia last century, probably accompanying plant material brought in for agriculture. Cinnamon fungus (an oomycete) causes dieback disease (see Box 35.3) and results in great changes in the species' composition of a forest, leaving behind a much less diverse community. Susceptible plant families (such as Proteaceae) are common components of the diverse and highly endemic Western Australian flora. At least 10% of the remaining 21 000 km² of jarrah forest in south-west Western Australia is infected with cinnamon fungus.

Great efforts are now being taken to prevent the introduction of foreign diseases into Australia. For example, for an animal to be legally introduced into Australia it must be kept in quarantine, out of contact with other animals and people, until it is confirmed

Fig. 45.6 Native heathland on French Island, Victoria, invaded by introduced cluster pine, *Pinus pinaster*

that it is not carrying any foreign disease. Some animals must be kept in quarantine for years before they can be declared disease-free!

Impact of introduced animals

The introduction of many species has been quite deliberate. All our important crop plants and farm animals have been introduced from other countries. In 1861 the Victorian Acclimatisation Society was formed with the principal objective: 'the introduction, acclimatisation and domestication of all innoxious animals, birds, fishes, insects and vegetables whether useful or ornamental'. The key word in this statement is 'innoxious', meaning not harmful, but this is a matter of perception. These well-intentioned people, who wanted to make the Australian environment more like 'home' in Britain, were responsible for introducing two of Australia's most serious pest animals: European rabbits, *Oryctolagus cuniculus* (Fig. 45.7), and European foxes, *Vulpes vulpes*.

(a)

(b)

Fig. 45.7 (a) One of Australia's most serious pest species is the introduced European rabbit, *Oryctolagus cuniculus*. (b) Native *Callitris* pines in the mallee ecosystem of Australia are killed by fire. Normally they would grow again from seed but the rabbit finds young pine seedlings palatable, preventing regeneration

Australia has many feral animals, which are domestic animals that have gone wild. Apart from feral cats and horses (brumbies), there are feral dogs, donkeys, sheep, cattle, goats, pigs, water buffalo, camels and ostriches (Fig. 45.8). Feral animals may damage vegetation and soils, foul water or compete with native animals for habitat and food. In total, about 30 species of mammals, 20 species of birds, five amphibians and 20 species of freshwater fishes make up the feral vertebrate populations of Australia.

Fig. 45.8 Feral water buffaloes, *Bubalus bubalis*, cause environmental damage in the 'top-end' of Australia

Most animals do not successfully establish and become invasive when introduced into a foreign environment. The success of establishment depends on the conditions of the ecosystem and the characteristics of the invading organism. Ecosystems susceptible to invasion are often disturbed and free of competitors and predators of the invading organism. Common characteristics of successful invaders are tolerance of a wide range of environmental conditions, high mobility, high fecundity (either producing many progeny in each reproductive cycle or reproducing frequently) and a broad diet.

> Ecosystems susceptible to invasion by introduced organisms are often disturbed and free of competitors and predators of the invading species.

Impact of weeds

The eminent botanist Baron Ferdinand von Mueller went to great lengths to ensure that the European blackberry, *Rubus fruticosus*, was spread throughout Victoria so that a hungry traveller could always obtain a meal of berries. Blackberry is now the most severe weed of streamsides in Victoria, choking native vegetation and providing a haven for rabbits and foxes.

A weed is a plant that is growing where it is not wanted (see Box 45.3). Thus, what constitutes a weed is largely a matter of perception. A weed to one person may be dinner to another! Dandelion, *Taraxacum officionale*, is a common urban weed in Australia, the leaves and tubers of which are a popular food in Europe. Many weeds are of agricultural importance since they may compete with crops for light, water and nutrients. Others interfere with wool production by lodging in the fleece of sheep and decreasing its value. Devil's claw, *Proboscidea louisianica*, is a weed originally from the Americas. It is common in eastern Australian grazing country. It produces a large, hooked seed capsule (Fig. 45.9), which can lodge in the knee joint or around the nostrils of sheep, greatly distressing and debilitating them.

Weeds are often opportunistic organisms. They usually produce many seeds that germinate under a range of conditions, are quick growing, flower early, are self-compatible, can reproduce vegetatively and are good competitors. Weeds invade environments that have been disturbed as a result of natural events, such as fire or wind, or human activities, such as forest clearing, fertiliser application, grazing or mining.

Weeds pose a serious threat to the conservation of natural plant communities. Bitou bush, *Chrysanthemoides monilifera* subsp. *monilifera* (Fig. 45.9b), is a serious weed of coastal New South Wales. Introduced from South Africa to help stabilise sand dunes, it now forms impenetrable thickets on foreshores and chokes native coastal plant species. Bitou bush out-competes coast wattle, *Acacia sophorae*, because it has a more efficient leaf arrangement to harvest sunlight and a deeper root system to tap soil water. In South Africa, bitou bush is not considered to be a pest since it has natural insect and fungal predators that keep it in check.

Many popular ornamental plants can be serious pests if they are allowed to spread into native bush. Some of the worst weeds are plants with fleshy fruit such as sweet pittosporum, *Pittosporum undulatum*, as well as bitou bush. Sweet pittosporum is a native plant from eastern Victoria and New South Wales but also a popular garden plant, which has become a pest species outside its natural range. The seeds of these plants are dispersed over large distances by birds, which eat the fruit and defecate the seeds into native eucalypt forest. The introduced European blackbird, *Turdus merula* (Fig. 45.9c), is one of the main dispersers of berry-fruited weeds in bushland adjacent to urban areas.

Weeds threaten the conservation value of native communities by competing with native plant species as well as degrading animal habitats.

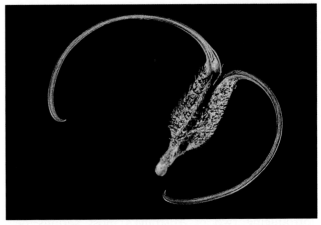

(a)

(b)

(c)

Fig. 45.9 **(a)** The hooked fruit of the weed devil's claw, *Proboscidea louisianica*, causes damage to sheep.
(b) Bitou bush, *Chrysanthemoides monilifera* subsp. *monilifera*, is a weed that invades coastal vegetation.
(c) In Victoria, the European blackbird, *Turdus merula*, is responsible for the spread of sweet pittosporum, *Pittosporum undulatum*, a native species turned weed

BOX 45.3 *MIMOSA PIGRA*: PEST PLANT OF THE TROPICS

The giant sensitive plant, *Mimosa pigra* (see figure), gets its name from its ability to quickly close up its leaves when touched. It is native to central America but has spread to many parts of the world including Africa, South-East Asia and northern Australia. It is a plant of the tropics, thriving in hot humid conditions and forming dense thickets that exclude almost all other species.

This plant has been grown in Darwin for almost a century, causing few problems. However, since the 1950s its population has 'exploded' in the Northern Territory and impenetrable thickets now cover more than 45 000 ha of river floodplains.

River floodplains in the Northern Territory were heavily grazed by another pest, water buffalo, introduced from Indonesia in the early 1800s. *M. pigra* is thorny and unpalatable to water buffalo, so it thrived while many of the palatable native species suffered. *M. pigra* also has other traits that give it an edge over native species. These include floating seeds, which are widely dispersed by flood waters, drought tolerance, and a rapid growth rate of up to 1 cm per day under good conditions.

The giant sensitive plant, *Mimosa pigra*, is a pest plant of tropical Australia

M. pigra is in the family Mimosaceae and is related to wattles. It produces large quantities of hard seeds that remain viable for long periods. Ecologists have estimated that soil under thickets contains up to 12 000 *M. pigra* seeds per square metre. They also discovered that the longevity of seed is determined by the depth to which they are buried in the soil. At 1 cm depth, 50% of seeds are viable after nine weeks, whereas at 10 cm depth, 50% of seeds are still viable after 99 weeks. This means that deeply buried seeds may be viable for many years, so that if adult trees are removed there will be numerous seeds to germinate and replace them.

Why should we worry about an invasive, introduced plant such as *M. pigra*? The dense thickets of this plant stop much light from reaching the ground. This causes the diversity of plant species under the thicket to decline. Many reptiles also find *M. pigra* thickets inhospitable, probably because of the reduced radiation from the sun, essential for the maintenance of their body temperature. Birds such as ducks, magpie geese, brolga and jabiru, whose open sedge and grassland habitat is under threat from *M. pigra*, are also affected. However, some species, in particular small mammals, benefit from *M. pigra* since the thickets provide cover from predatory animals, such as hawks.

Although it is presently limited to a region within 200–300 km of Darwin, *M. pigra* has the potential to spread across the tropical north of Australia. Search for a biological control is centred on two species of moth from the plant's natural habitat in Mexico. After stringent tests to ensure that the moths do not attack other native species of plant, they have been released into infested areas. The larvae of these moths bore into stems causing plants to die. One of the moth species has successfully established, but it is not yet known whether it has reduced plant vigour or seed production. In the meantime, other biological controls are being tested in the hope of stopping the spread of this highly invasive weed.

Degrading the physical environment

In addition to the effects of introducing alien species, human activities have caused degradation of the physical environment. In recent years, for example, the world has witnessed ecologically disastrous oil spills (Fig. 45.10): the *Exxon Valdez* spill that fouled the Alaskan coast (1989); the Iranian tanker spill that poured tens of thousands of tonnes of oil into the sea off the Moroccan coast (1990); and, during the Gulf War, millions of barrels of oil were spilled into the Arabian Gulf (1991), representing the biggest oil spill on record. An oil spill on the Victorian Ninety-Mile Beach, in February 1990, pales into insignificance in comparison with these spills, but reminds us of the vulnerability of our local coastal ecosystems.

Fig. 45.10 This fish was a victim of a heavy oil spill in the Yarra River, Melbourne, in October 1980

Holding onto soil

Erosion, a loss of soil by the action of either wind or water, is a natural process that shapes the surface of the earth. However, the clearing of land for agriculture has, in many cases, greatly increased the rate of erosion, stripping away nutrient-rich topsoil and silting up river systems.

Soil in forested ecosystems is stabilised by a dense mat of plant roots. The soil surface is also covered by a layer of litter (dead leaves and twigs), which protects the soil surface from the erosive action of wind and water. In a typical rainfall, water is quickly absorbed into the top layers of soil. This is because soil is naturally porous due to the adhesion of soil particles into larger crumbs, producing a matrix with many air spaces into which water can flow. Earthworms and other soil invertebrates assist this process. They burrow through the soil, breaking down organic material and creating a network of tiny channels.

Substantial amounts of soil erosion would usually only occur in forested ecosystems following intense fire. Fire removes stabilising vegetation cover and the heat of fire can cause the soil surface to become water-repellent. Heavy rainfall after fire can therefore wash away large amounts of soil and associated nutrients.

Agricultural soils, in contrast to forest soils, have little protection from the action of wind and rain, particularly during dry periods when pasture dies back. Grazing by hard-hoofed stock, such as sheep and cattle, compacts the ground and reduces soil porosity. This slows the rate of rainwater absorption leading to increased run-off and water erosion. The reduced vegetation cover also results in greatly increased wind velocities at the surface of the soil, leading to loss of soil through wind erosion. It has been estimated that more than half of Australian agricultural land requires treatment to repair erosion damage (Fig. 45.11).

Wind and water erosion can be greatly reduced by sensible land management. Agricultural methods to improve soil conservation are becoming more widespread as the seriousness of soil loss is realised. Ceasing to cultivate erodible soils can reduce erosion significantly; this is called the 'no-till' system of agriculture. In cropping systems, leaving the stubble (the stalks of the grain) behind after harvest rather than ploughing or burning as is traditional, is one of the best ways to reduce soil erosion. The standing stubble reduces wind velocities and the organic material helps to cement soil particles together, improving soil structure.

Soil salinity

More than 20 000 km² of land in Australia is currently salt-affected and this figure is expected to double by the year 2000. The worst salt-affected lands are in the drier parts of the continent where extensive forest clearing has occurred for grazing and cropping. Native forest cover has been reduced to less than 10% in most of these regions. Many parts of Australia are salt-affected, with the wheat belt of Western Australia, southern South Australia and the Mallee and Wimmera of Victoria particularly affected (Fig. 45.12).

(a)

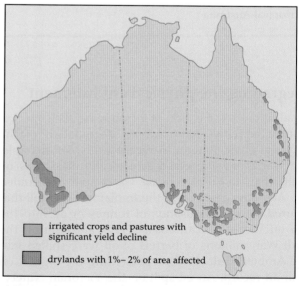
(b)

Fig. 45.12 **(a)** Dryland salinity in mallee of wheat belt of Western Australia. **(b)** Salt-affected areas in Australia

irrigated crops and pastures with significant yield decline

drylands with 1%– 2% of area affected

Fig. 45.11 Land degradation caused by wind erosion is a serious problem in Australian agricultural ecosystems

Salt is a natural part of the environment. With every shower of rain, salt falls out of the atmosphere, having been carried as salt spray great distances on the wind from the surface of the ocean. This salt has accumulated in the soil and ground water over millions of years. In parts of south-west Western Australia, natural soil salt levels of 50–100 kg/m² are common. In wetter regions, large amounts of rain wash the salt out of the soil. Thus, high levels of soil salinity are mainly found in drier regions.

Under normal circumstances, salt in the soil is harmless since it remains below the root zone of the vegetation. However, soils become salinised when the hydrological system is altered by clearing vegetation (Fig. 45.13).

Trees transpire large amounts of water through open stomata when photosynthesising (Chapter 18). The combined process of transpiration from plants and evaporation of water directly from the soil surface is known as **evapotranspiration**. The position of the watertable, the level to which water rises in the soil (Fig. 45.13), is largely determined by the balance between rainfall and evapotranspiration. When land is cleared of vegetation, the rate of evapotranspiration greatly decreases. More rainwater reaches the watertable and the watertable rises. Regions where water enters the watertable are recharge zones, typically found on hillsides where porous sedimentary rocks outcrop or are close to the surface.

As the watertable rises, it becomes increasingly salty because it collects the salt that has accumulated

in the subsoil. Once the watertable is within about 2 m of the surface, salty water is carried to the surface by capillary action. This leads to an increase in saltiness of the soil at the surface where water evaporates, leaving salt behind. The areas where the salt reaches the surface are discharge zones. This process leads to **dryland salinity**.

An important feature of dryland salinity is that the recharge and discharge zones may be many kilometres apart. Thus, farmers in one area may be suffering salinisation from over-clearing elsewhere. Salinity must therefore be managed co-operatively on a catchment basis, rather than at the level of the individual farm.

Irrigation salinity occurs in much the same way as dryland salinity, but the rise in the watertable is largely due to the increased water applied to the soil surface for irrigation.

The first sign of increasing soil salinity is usually death of the most deeply rooted plants. Salt-sensitive species, such as clovers and rye-grasses, are replaced by species of increasing salt tolerance, such as buckshorn plantain, *Plantago coronopus*, and sea barley-grass, *Hordeum leporinum*, until eventually no plants can tolerate the salt, and a bare 'salt scald' results. At this stage, a white crust of salt crystals is usually visible on the soil surface during dry periods. During wet periods the salt scald may become a lake.

The increased amount of salt in the ground water eventually finds its way into streams. Streams become increasingly saline downstream as more and more

water, taken up from soil, is transpired by plants

trees removed; transpiration reduced and more water filters through to ground water

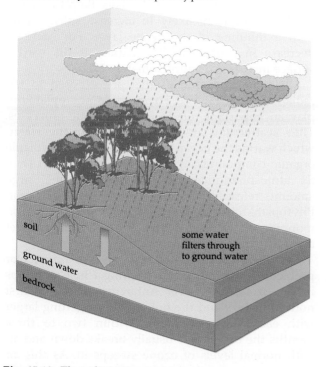
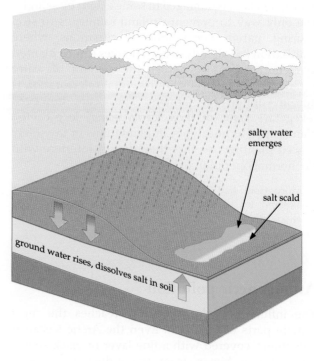

Fig. 45.13 The salinisation process

saltwater flows in. The Murray River is a good example of this. It drains more than 1 million km² of Queensland, New South Wales, Victoria and South Australia (Fig. 45.14). All the salt that is washed out of the soils in the Murray River catchment passes out the river mouth at Lake Alexandrina—an average of 5.5 tonnes per minute!

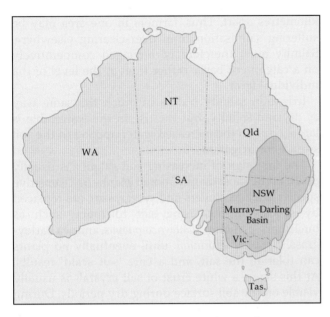

Fig. 45.14 The Murray–Darling Basin drains a large area of eastern Australia, and the rivers accumulate salt that is washed out of the soils in the catchment

Most of the nations of the world suffer, to some degree, from dryland or irrigation salinity; almost 25% of irrigated land is damaged to some extent by salinity. The only way to prevent dryland salinity is to stop clearing native vegetation. In situations where salinisation has already occurred, trees can be planted in recharge areas to reduce the amount of water entering the aquifer. Of course, total reforestation is not an option for farmers. They can, however, revegetate areas of their farms using deep-rooted plants, such as eucalypts, in plantations or in widely spaced plantings so that they do not interfere with the normal operation of the farm.

> More than half the agricultural land in Australia is affected by land degradation, with salinity and soil erosion being major problems. These problems can be managed by revegetation.

Atmospheric pollution

The influence of humans now reaches the most remote parts of the earth. Even the Arctic icecap is sometimes covered with a fine layer of black soot—air pollution blown from the industrial cities of Europe. Such global effects of pollution are best exemplified by atmospheric changes, such as the ozone 'hole' and global warming.

The ozone layer

The diffuse layer of ozone gas (O_3) in the stratosphere, between 13 and 20 km above the earth, plays an essential role for life on earth, shielding it from destructive ultraviolet (UV) radiation. Ozone gas is constantly created by the effect of sunlight on oxygen in the upper atmosphere. It is also constantly destroyed by the impact of various atmospheric constituents, both natural and artificial. However, in recent years the rate of ozone destruction has exceeded its formation and the ozone layer has been thinning, allowing more UV radiation to reach the earth.

Some of the main causes for the destruction of the ozone layer are chlorofluorocarbons (CFCs). CFCs are used as aerosol propellants (Freon-11) and in refrigerators, air-conditioners and freezers (Freon-12). The chlorine in CFCs reacts with ozone, breaking it down and thereby destroying its UV screening effect.

The damaging effects of UV radiation depend on its wavelength (Fig. 45.15). UV-C radiation has the shortest wavelength and is the most harmful. Fortunately, almost no UV-C radiation reaches the earth's surface since most of it is screened out by the ozone layer. UV-B radiation is the next most damaging form and it also is largely screened out. It is primarily UV-A radiation that causes sunburn and substantial amounts of UV-A radiation reach the earth's surface, even on cloudy days. Both UV-A and UV-B radiation are likely to increase substantially with a thinning of the ozone layer. For each 1% decrease in ozone, UV-B radiation at the earth's surface increases by approximately 2%.

All plants and animals that are exposed to UV radiation are threatened. For example, phytoplankton play an important role in many climatic processes. They scatter and absorb light entering the water, which warms the ocean surface. They release volatile organic compounds, which help clouds form. They fix atmospheric carbon dioxide, including large amounts released by industrial processes each year. Phytoplankton therefore play an important role in mopping up elevated levels of carbon dioxide, yet are under threat from increased UV radiation.

The thinning of the ozone layer is also a particular problem for Australians. Each spring a 'hole' develops in the ozone layer over Antarctica. This was first noticed in 1979 and the hole has been getting larger with each successive year. Within two to three months the ozone hole usually breaks down and air with normal levels of ozone sweeps in. As this air sweeps in, ozone-depleted air from the hole drifts

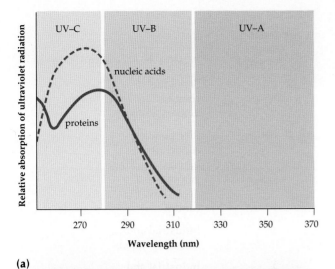

(a)

(b)

Fig. 45.15 (a) UV-B and UV-C bands of ultraviolet radiation are absorbed by proteins and nucleic acids and therefore are the most damaging to organisms. (b) The bronze-coloured leaves on the left have been exposed to UV-B radiation for 20 days relative to control leaves on the right

up over southern Australia, at times increasing UV-B radiation reaching the earth's surface by 20% or more. Australia has the highest rate of skin cancer of any country in the world, so any increase in cancer-causing UV radiation is of serious concern. Coincidentally, Australia releases a very high level of CFCs into the atmosphere per head of population.

Since ozone depletion is an international problem, it must be tackled at an international level. In 1987, a large group of countries met in Montreal and agreed to have no further increases in CFC production immediately and to halve production by 2000. This agreement is the Montreal Protocol. However, evidence is mounting that an even greater reduction is required. It should be possible to phase out CFC

use entirely without a substantial impact on any country's economy and with obvious benefits to global ecosystems.

> Thinning of the ozone layer by CFCs is allowing more UV radiation to reach the earth's surface, causing problems such as increased occurrence of skin cancer.

The greenhouse effect and global warming

The greenhouse effect (Fig. 45.16) is the natural warming of the earth by heat trapped due to the presence of certain heat-absorbing atmospheric gases. On reaching the earth's surface, sunlight, largely short wavelength radiation, is converted into longer wavelength heat energy. Much of this heat is reradiated from the earth but is absorbed by atmospheric gases, the greenhouse gases (CO_2, N_2O, CH_4) and water vapour, warming the lower atmosphere. The heat trapped by atmospheric gases is essential for providing suitable conditions for life on earth. Without greenhouse gases to trap heat the temperature at the surface of the earth would average 33°C less than at present.

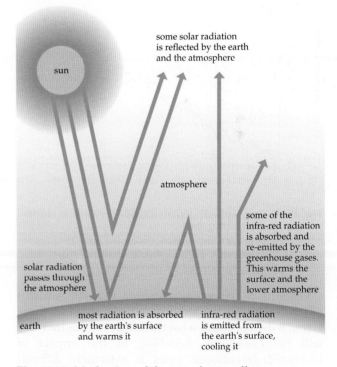

Fig. 45.16 Mechanism of the greenhouse effect

Since industrialisation in the mid 1800s, the concentrations of several atmospheric gases have been steadily increasing (Table 45.1): CO_2 has increased by about 25% (Fig. 45.17); N_2O by 19% and CH_4 by 100%. Ozone in the lower atmosphere and CFCs also act as greenhouse gases.

Table 45.1 Greenhouse gases

Greenhouse gas	Concentration		Present trend (% per year)	Possible sources of the increases
	1850	1989		
CO_2	275 ppmv	350 ppmv	0.4	Fossil fuel combustion, deforestation
CH_4	750 ppbv	1700 ppbv	0.8	Rice paddies, ruminants, biomass burning, gas and coal, fields, land fills, tundra
CFC-11	nil	250 pptv	4	Industrial and consumer goods
CFC-12	nil	450 pptv	4	Industrial and consumer goods
N_2O	285 ppbv	310 ppbv	0.3	Biomass burning, agriculture, fossil fuel combustion
O_3 (tropospheric)	15–20 ppbv	20–30 ppbv	0.5	Urban and industrial pollution

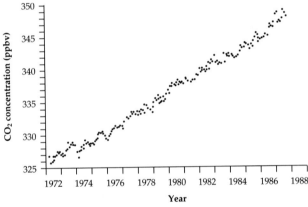

Fig. 45.17 Atmospheric carbon dioxide levels over south-east Australia, 1972–88
ABS, *Australia's Environment Issues and Facts 1992*, AGPS, Canberra, 1992, Commonwealth of Australia copyright reproduced by permission

While it is commonly accepted that the amount of these gases in the atmosphere is increasing, there is much debate about what the impact of this increase will be. Computer modelling of the impact of increased greenhouse gases is extremely difficult since many factors need to be considered and little is known about how global climate functions. Most scientists agree that a significant warming of the earth through enhancement of the greenhouse effect is very likely but at present this is only a hypothesis. Global mean temperatures appear to have risen by 0.5°C over this century. This warming appears to be confined to the lower layers of the atmosphere, with the upper layers cooling over the same period. These changes support the predictions of greenhouse climatic change but they cannot be taken as evidence since they may be part of natural temperature fluctuations.

How the hydrological cycle will respond to climate change is particularly important. Most models predict more clouds, which have the capacity to enhance the greenhouse effect by trapping more heat or, alternatively, to reduce it by reflecting solar radiation back into space, thereby cooling the earth. The response of the oceans will also be of great importance since ocean currents redistribute energy across the globe and play an important role in determining rainfall patterns.

One of the greatest concerns is that a warmer climate will lead to higher sea levels due to heating and expansion of seas, and partial melting of icecaps and glaciers. Estimates of sea level rise range from 50 to 200 cm by the year 2100 depending upon assumptions about the rate of climate change and changes in ice and water temperature and density. If all the world's ice sheets melted, sea levels could rise by as much as 70 m!

The impact of a rise in sea level would be greatest in developing countries since they have the most extensive regions of highly productive, densely populated, low-lying river deltas. Many large cities, including Bangkok, Calcutta, Shanghai and Hanoi, are on low-lying river banks and are under threat of inundation by rising seas unless major engineering works, such as dykes, are constructed to hold back the rising waters. Engineers in coastal cities are taking the threat of rising sea levels seriously and are making allowances in plans for new coastal developments.

The US Environmental Protection Agency has estimated that carbon emissions need to be reduced by 50–80% worldwide to stabilise atmospheric CO_2 concentrations at current levels. Currently, fossil fuel burning accounts for the majority of net CO_2 production, so it is here that the greatest reductions can be made.

Fossil fuels presently provide 78% of energy requirements. The amount of CO_2 produced for each unit of energy varies according to the fuel. Natural gas is the cleanest, while oil produces 45% more CO_2 per unit of energy and coal 75% more. Thus, changing to 'cleaner' fossil fuels may provide some benefits.

The greatest gains in CO_2 reduction are immediately available through improved energy efficiency. More efficient lighting, economical motor vehicles and energy-efficient industry could greatly reduce the amount of CO_2 released each year. Clean or non-carbon based forms of energy, such as solar thermal (trapping the sun's heat; Fig. 45.18), photovoltaic (direct conversion of sunlight to electricity), wind, geothermal and tidal energy hold the greatest potential for long-term reductions in carbon emissions but their implementation on a large scale to substantially replace fossil fuels is still several decades off.

Fig. 45.18 Use of clean (non-carbon) forms of energy such as solar power has the potential to reduce carbon dioxide emissions and the greenhouse effect

The clearing of native forests also contributes substantially to rising atmospheric CO_2 levels. Many developing countries add far more carbon to the atmosphere through deforestation than through burning fossil fuels. Brazil currently contributes six times as much CO_2 through deforestation as it does through fossil fuels, making it the fourth largest carbon emitter in the world. Of course, the impacts of deforestation go far beyond the carbon that is released. Loss of biodiversity, soil erosion and soil salinity all add up to make conservation of forests a top priority.

> The greenhouse effect is the warming of the earth's lower atmosphere due to heat-absorbing 'greenhouse' gases, such as CO_2, CH_4, N_2O and CFCs and water vapour. The most effective ways to reduce emission of greenhouse gases is by reduced fossil fuel use, improved energy conservation and reduction in deforestation.

RESTORING THE BALANCE?

The ecological crises that we face are a result of human interference with natural processes. In order to conserve ecosystems and their component organisms, we need legislative control to save species and habitats, and application of ecological knowledge to restore degraded systems.

Legislative control

Each Australian state or territory has legislative power to protect native species. For example, the *Native Flora and Fauna Guarantee Act* 1987 protects all species of wattles, banksias, grevilleas, orchids and many other species of plant and animal (Fig. 45.19). However, conservation is best achieved through protection of whole habitats. Although some areas have to be used to grow food and timber, and to mine resources, other areas can be protected through legislation as nature reserves and national parks. Australia's first national park, for example, was established near Sydney in 1879 and formal control of forest exploitation was in place by the 1900s (see Box 45.1).

Many Australian state governments have introduced planning regulations that require land-owners to obtain a permit before clearing native vegetation. This enables control of clearing and conservation of habitats of value. Unfortunately, there are no controls over the effects of grazing of remnant vegetation, such as along roadsides. Thus, even though a forest or woodland may survive as a remnant patch, its value as habitat may be destroyed through grazing.

Fig. 45.19 The eastern pygmy possum is a rare species of marsupial protected in Victoria under the *Flora and Fauna Guarantee Act* 1987

CITES

Protection of species that are traded between countries requires international legislative control. For example, the flourishing worldwide trade in exotic wildlife includes many endangered species, and the rarer a species, the more 'collectable' it is (Fig. 45.20). Extremely rare giant pitcher plants from Borneo can sell for $1300 each on the black market. Rare Peruvian butterflies may fetch $4000 each. In Japan, South-East Asian clouded leopard skin coats sell for as much as $165 000. Each year, some 40 000 primates, at least 1 million orchids, 4 million live birds, 10 million reptile skins, 15 million pelts from wild furred animals, over 350 million tropical fish, tusk ivory from tens of thousands of African elephants and other products as diverse as ginseng roots, black bear gall bladders and fur seal penises are traded. One year's wildlife trade is worth at least $6 billion.

Wildlife trade is controlled through the Convention on International Trade in Endangered Species of Fauna and Flora (CITES) to which many countries, including Australia, are signatories. It is designed to promote conservation of endangered species while allowing the commerce of species that can withstand the pressures of trade.

Fig. 45.20 Worldwide trade in exotic wildlife includes endangered species such as leopards an tigers

Restoration ecology: resetting the clock

Restoration is the process of replacing a community of organisms so that it resembles its former condition and functions much as it would have naturally. It can be likened to resetting the biological clock so that natural processes, such as succession and evolution, continue as they did before human interference.

Restoration is usually applied to barren, degraded land, such as that left behind by open-cut mining or poor farming. By restoring the biotic diversity and ecosystem processes, degraded land may again function as a valuable component of the environment.

Most successful restoration mimics the natural process of succession (Fig. 45.21). Degraded sites are initially inhospitable for most plants and animals. The

Fig. 45.21 Restoration of mining sites often mimics stages of ecological succession as shown here in this aerial view of Fraser Island, where the mined area was seeded with the dominant tree and shrub species three years ago to re-establish the plant communities

soil may be compacted, saline or leached of nutrients. The microclimate will be different to the natural community, with lower humidity and more wind. Pioneer species of plant are introduced first to stabilise the soil, increase soil nutrients and organic matter, and create suitable conditions for species of later successional stages. Animals can be introduced when suitable habitat becomes available or else they may migrate from surrounding areas. The biodiversity of the site is gradually increased until, ideally, it resembles the original community.

Managing natural communities for conservation

Even more important than restoration is the *management* of natural communities to preserve biodiversity and *prevent* degradation. National parks, world heritage areas and flora and fauna reserves, which are covered by legislation to protect significant natural resources, must be managed carefully.

The Anangu people are traditional owners of Uluru (Ayers Rock) and Kata Tjuta (The Olgas), which are today a part of Uluru National Park (Fig. 45.22). Before the arrival of Europeans, Anangu practised 'fire stick' farming to manage the environment so that a regular supply of green feed was available for the animals that they hunted (see Chapter 41). The Anangu burnt the country in a patchwork fashion, providing a diversity of habitat patches and impeding the passage of large fires.

When Europeans arrived with cattle, sheep and fences, traditional fire management was abandoned with dire consequences for native plant and animal communities. Instead of regular and patchy small fires, irregular and extensive wildfires occurred. Extensive areas of vegetation of the same age resulted, limiting the available habitat for animals and helping to drive some species to extinction.

Fig. 45.22 Uluru National Park

In 1985, freehold title to Uluru National Park was given to the Anangu. They leased the park back to the Australian National Parks and Wildlife Service who set out to manage it, together with Anangu people, using traditional ideas of stewardship. An integral part of this was reintroducing patch-burning to minimise fuel for wildfires and to maintain a diversity of plant successions and therefore animal habitat.

Management of fire is an important aspect of conservation management, whether it be excluding fire from tropical rainforest or encouraging regular patchwork fires to maintain diversity in coastal heath. Other impacts must also be managed: feral animals and weeds, and disturbance by off-road vehicles, which damage vegetation and cause soil erosion.

> Land reserved for conservation is managed using knowledge of the ecology of natural communities, including the role of fire, and the effects of feral animals, weeds and human impacts.

Looking to the future

In this chapter we have considered many human impacts on the environment, most of them negative. In the future, we have the opportunity to greatly increase the positive impacts that we have on our environment through the application of biological knowledge.

Ecologically sustainable development

A sustainable society is one in which all human activity takes place within the limitations set by the environment. This includes its capacity to assimilate waste, provide food and supply other resources such as minerals and energy.

Ecologically sustainable development (ESD) means using, conserving and enhancing the communities' resources so that ecological processes, on which life depends, are maintained, and the total quality of life can be increased. Historically, a rising standard of living has been linked to increasing energy and resource usage, environmental degradation and pollution. During the 1980s and 1990s this has been reversed, with a trend towards decreasing intensity of resource and energy usage. Today, we can do more with much less than we could a decade ago because technology has allowed industries to become more efficient.

To achieve ecologically sustainable development, a number of objectives must be met.

1. *Intergenerational equity.* The present generation should ensure that the next generation is left an environment that is, at least, as healthy, diverse and productive as the one we presently enjoy.

2. *Conservation of biological diversity and ecological integrity.* This should be a fundamental consideration in all economic activity.

3. *Constant natural capital.* Natural capital, such as biological diversity, healthy environments and fresh water supplies, cannot be substituted by human-made capital and should be maintained from one generation to the next.

4. *Anticipatory and precautionary policy approach.* Policies should err on the side of caution. The onus of proof should be placed on developers to demonstrate that their activities are ecologically sustainable.

5. *Social equity.* Equity, or fairness, in social and economic policies is an essential part of a sustainable society.

6. *Limits on natural resource use.* The limits of the environment to absorb wastes and provide resources must be recognised and accommodated.

7. *Qualitative development.* The key objective in using natural resources is the improvement of human welfare rather than quantitative growth.

8. *Pricing environmental values and natural resources.* Wherever possible, prices for natural resources should be set to recover the full social and environmental costs. Many environmental values cannot be expressed in monetary terms, hence pricing policies must form a part of a broader framework for decision making.

9. *Global and regional perspectives.* These are required so that environmental problems are not just moved from country to country.

10. *Efficiency.* Efficient use of resources should be a major objective in economic policy.

11. *Resilience.* Economic policies should be developed that are resilient to external ecological and economic changes.

12. *Community participation.* Strong community involvement is an essential prerequisite for a smooth transition to an ecologically sustainable society.

One of the most important changes that must occur is decoupling environmental degradation from economic development. Increased recycling of materials can make a substantial contribution towards this since it decreases the amount of resources and energy required to produce goods. 'Think global, act local' was a catchcry of the 1970s and early 1980s. It essentially meant 'think about the global implications of what you do; lots of small changes in people's behaviour can add up to change on a global scale'. If one person uses recycled lecture pads, this will not save many trees, but if every student does the same, this might mean a whole forest still standing.

Changes in lifestyle will also be important. Motor cars cause more pollution than any other single source. A greatly improved public transport system, together with better urban planning, could substantially decrease pollution from transport. Likewise, a shift from non-renewable, highly polluting sources of energy, such as oil, coal and gas, to renewable sources of energy, such as wind, water and solar radiation, can greatly reduce the environmental impact of economic growth. Developing a global society along ESD principles is ultimately the only possible solution.

SUMMARY

- The impact of humans on the Australian environment has greatly accelerated over the last 200 years since European settlement. Extensive land clearing has resulted in loss and fragmentation of habitat, the single major cause of loss of biodiversity. Human survival depends both physically and psychologically on the maintenance of biodiversity.

- Introduced animals, plants and micro-organisms, which have become pests or weeds, have also been detrimental to the environment and affected the conservation value of native biological communities.

- More than half the agricultural land in Australia is affected by land degradation, with salinity and soil erosion being major problems. Salinity can be managed by using vegetation to increase evapotranspiration and lower watertables.

Erosion can be reduced by minimising cultivation in susceptible areas and by maintaining vegetation cover.

- Human impacts are being felt on a global scale. Although the greenhouse effect is a natural process, increased warming of the earth's lower atmosphere is predicted because of the emission of greenhouse gases CO_2, CH_4, N_2O and CFCs. Thinning of the ozone layer, which normally absorbs harmful UV radiation, is of concern in Australia.

- Application of biological knowledge is important for the future of the human species. A sustainable society is one in which all human activity takes place within the limitations set by the environment. Ecologically sustainable development means using, conserving and enhancing resources so that ecological processes, on which life depends, are maintained.

QUESTIONS

1. Biologists and conservationists are concerned about the 'biodiversity crisis'. What is biodiversity and why is its conservation considered important?

2. What is meant by CITES? Name three Australian species that are listed under CITES.

3. What are feral animals? Name six feral animals found in Australia and research the impact of one of these species.

4. What is a weed? What are some of the characteristic features of weedy plants? Explain why a plant that is not a problem in its native country can become a serious weed when introduced into another country.

5. Discuss why agricultural soils are more susceptible to erosion than forested soils.

6. What is dryland salinity and what causes it? How can the planting of native vegetation help to reduce salting?

7. Explain what is meant by 'fire stick' farming. Why has it been reintroduced at Uluru National Park?

8. Briefly explain the 'greenhouse effect', naming the major causes. Discuss the positive and negative results of the greenhouse effect.

9. The little pigmy possum is found in mallee scrubs and heaths of north-eastern Victoria and south-eastern South Australia. The species lives within certain environmental limits: 33% of its current range is within conservation areas. However, ecologists concerned with the effects of global warming have predicted that, with a rise in temperature of 1°C, the animal's range would shift southwards into areas outside nature reserves. If the potential range of a species can change because of temperature increase, we need to take this into account when designing conservation areas. What sort of design features should we consider to ensure saving species affected by climate change? What, for example, are habitat corridors and what function do they serve for nature conservation?

10. Discuss what is meant by ecologically sustainable development.

CLASSIFICATION OF CELLULAR ORGANISMS

The following classification is of living cellular organisms and excludes many fossil groups as well as viruses. Some groups of organisms, particularly within the protists, have not yet been formally named by biologists and are here listed as informal groups. Overall, the order of the taxa reflects our current knowledge of the hierarchy of life.

PROKARYOTES

Kingdom Monera: the bacteria

Archaeobacteria: methanogens, thermophiles and halophiles
Eubacteria: spirochaetes, Gram negative, Gram positive, purple and sulfur
  bacteria, rickettsias, chlamydias, mycoplasmas and cyanobacteria

EUKARYOTES

Kingdom Protista

Archaezoa

Protists that have a nucleus but lack some other eukaryotic organelles
Phylum Microspora: unicellular parasites, pelobiontids (presently classified in
  the phylum Zoomastigina, order Pelobiontida), diplomonads (phylum
  Zoomastigina, class Diplomonadida, e.g. *Giardia*)

Photosynthetic protists

Phylum Glaucocystophyta: flagellates with a photosynthetic endosymbiont
Phylum Rhodophyta: red algae
Phylum Chlorophyta: green algae (e.g. *Chlamydomonas*, *Spirogyra*, *Ulva* and
  Chara)

Heterokonts

Phylum Chrysophyta: golden flagellates
Phylum Prymnesiophyta: haptophytes (e.g. coccolithophorids)
Phylum Bacillariophyta: diatoms
Phylum Phaeophyta: brown algae (e.g. *Durvillea*, *Hormosira*)
Phylum Oomycota: water moulds, downy mildews, *Phytophthora*

Euglenoids and kinetoplasts

Phylum Euglenophyta: flagellates (e.g. *Euglena*); kinetoplasts: flagellate
  parasites (phylum Zoomastigina, class Kinetoplastida) including
  trypanosomes and *Leishmania*

Alveolates

Phylum Dinophyta: dinoflagellates (e.g. zooxanthellae of corals)
Phylum Apicomplexa: animal parasites (also called sporozoans), including gregarines, coccidians, haematozoa (e.g. *Plasmodium*, which causes malaria)
Phylum Ciliophora: ciliates (e.g. *Paramecium*)

Protists with second-hand chloroplasts

Phylum Cryptophyta: cryptomonads (flagellates)
Phylum Chlorarachnida: *Chlorarachnion*, an amoeboid plasmodium

Slime moulds

Cellular slime moulds: Phylum Acrasea; Phylum Dictyostelida, protostelids
Acellular slime moulds: Phylum Myxomycota

Fungal-like protists

Phylum Chytridiomycota: chytrids

Sponge-like protists

Phylum Zoomastigina
　　Class Choanomastigotes: choanoflagellates

Protists of unknown affinity

Phylum Actinopoda: actinopods (e.g. radiolarians)
　Subphylum Sarcodina: amoebae, including rhizopods and foraminiferans
　　Phylum Zoomastigina
　　　　Class Opalinata: opalinids—large, commensal flagellates
　　　　Class Proteromonadida: proteromonads—commensals
　　　　Class Parabasalia: parabasilids—commensals and parasites

Kingdom Fungi

Phylum Zygomycota: moulds (e.g. *Rhizopus*) and other microfungi
Phylum Ascomycota: sac fungi—cup fungi, yeast, morels, truffles, *Penicillium* and *Neurospora*
Phylum Basidiomycota: club fungi—rusts, smuts, mushrooms, puffballs and bracket fungi
Phylum Deuteromycota: anamorphs or Fungi Imperfecta
Phylum Mycophycota: lichens

Kingdom Plantae

Non-vascular land plants

Phylum Hepatophyta: liverworts
Phylum Anthocerotophyta: hornworts
Phylum Bryophyta: mosses

Vascular land plants

Phylum Psilophyta: *Psilotum* and *Tmesipteris*
Phylum Lycophyta: clubmosses and quillworts
Phylum Sphenophyta: horsetails
Phylum Filicophyta: ferns

Seed plants

Phylum Cycadophyta: cycads
Phylum Ginkgophyta: *Ginkgo*
Phylum Coniferophyta: conifers
Phylum Gnetophyta: gnetophytes
Phylum Magnoliophyta: flowering plants (angiosperms)
    Class Liliopsida: monocotyledons
    Class Magnoliopsida: dicotyledons

Kingdom Animalia: the Metazoa

Phylum Porifera: sponges

Radiate phyla

Phylum Cnidaria
    Class Scyphozoa: jellyfish
    Class Cubozoa: sea wasps
    Class Hydrozoa: hydras and siphonophores
    Class Anthozoa: sea anemones and corals
Phylum Ctenophora: comb jellies

Acoelomate protostomes

Phylum Platyhelminthes: flatworms
    Class Turbellaria: free-living flatworms
    Class Monogenea: ectoparasitic flukes
    Class Trematoda: endoparasitic flukes
    Class Cestoda: tapeworms
Phylum Nemertinea: proboscis worms

Pseudocoelomates

Phylum Nematoda

Coelomate protostomes

Phylum Annelida
    Class Polychaeta: marine bristle worms
    Class Euclitellata: clitellum-bearing worms
      Subclass Oligochaeta: earthworms and aquatic relatives
      Subclass Hirudinea: leeches
      Subclass Branchiobdellida: 'jawed' worms
Phylum Arthropoda
  Subphylum Chelicerata
    Class Merostomata: horseshoe crabs
    Class Arachnida: spiders, scorpions, ticks and mites
    Class Pycnogonida: sea spiders
  Subphylum Crustacea
    Class Branchiopoda: water fleas
    Class Copepoda: water fleas
    Class Ostracoda: mussel or seed shrimps
    Class Cirripedia: barnacles
    Class Branchiura: fish lice
    Class Pentastomida: tongue worms
    Class Peracarida: slaters and woodlice
    Class Malacostraca: shrimps, crabs, and so on

Subphylum Uniramia
    Class Diplopoda: millipedes
    Class Chilopoda: centipedes
    Class Insecta: insects
      Subclass Apterygota: wingless insects
      Subclass Pterygota: winged insects
    Class Onychopora: velvet worms (e.g. *Peripatus*, *Peripatoides*)
Phylum Mollusca
    Class Aplacophora: primitive, worm-like molluscs
    Class Monoplacophora: *Neopilina*
    Class Polyplacophora: chitons
    Class Gastropoda: snails, limpets, and so on
    Class Bivalvia: mussels, oysters, scallops, and so on
    Class Scaphopoda: tusk shells
    Class Cephalopoda: *Nautilus*, cuttlefish, squids, octopuses

Lophophorates

Phylum Phoronida: primitive lophophorates
Phylum Brachiopoda: lamp shells
Phylum Bryozoa: bryozoans or moss animals

Deuterostomes

Phylum Echinodermata
    Class Crinoidea: feather stars
    Class Asteroidea: sea stars
    Class Concentricycloidea: relatives of sea stars
    Class Ophiuroidea: brittle stars and basket stars
    Class Echinoidea: sea urchins, sand dollars and heart urchins
    Class Holothuroidea: sea cucumbers
Phylum Chordata
    Subphylum Hemichordata: acorn worms
    Subphylum Urochordata: tunicates or sea squirts
    Subphylum Cephalochordata: lancelets
    Subphylum Vertebrata
     Superclass Agnatha: jawless fishes (lampreys and hagfishes)
     Superclass Gnathostomata: jawed vertebrates
      Class Chondrichthyes: cartilaginous fishes (chimaeras, sharks, skates
        and rays)
      Class Osteichthyes: bony fishes
        Subclass Actinopterygii: ray-finned fishes including teleosts
        Subclass Sarcopterygii: fleshy-finned fishes (coelocanths and lungfishes)
     'Tetrapods'
      Class Amphibia: salamanders, mud puppies, newts, frogs, toads and
        caecilians
      'Sauropsida': including traditional classes Reptilia (turtles, snakes, lizards,
        crocodiles) and Aves (birds)
      Class Mammalia
        Subclass Prototheria: platypus and echidnas
        Subclass Theria: Metatheria (marsupials) and Eutheria (so-called
          placental mammals, including the order Primates)

GLOSSARY

abiotic Pertaining to physical and inorganic components.

abscisic acid A growth-inhibitory hormone controlling responses of plants to stress (such as drought), frost tolerance and seed dormancy; synthesised in the carotenoid pigment pathway.

abscission zone A zone of tissue at the base of a petiole that regulates abscission or shedding of the leaf.

absorption spectrum Pattern of absorption of photons at different wavelengths of light.

acclimation An adaptive change in an organism in response to a change in a single feature of the environment, for example, acclimation to temperature.

acclimatisation An adaptive change in an organism in response to broadly defined changes in weather conditions.

acetyl CoA A 2-carbon compound that is the substrate for the citric acid cycle; produced in mitochondria during the final stage of glycolysis when O_2 is available, and as a product of β-oxidation.

acid Substance that is a proton donor, releasing hydrogen ions (H^+) into solution; possessing a pH in solution below 7.

acoelomate Containing no coelom (body cavity within mesoderm).

acrosome Secretory vesicle in the head of the sperm containing hydrolytic enzymes, which are released during fertilisation.

actin A globular protein; main structural component of microfilaments.

actin filaments Thin myofilaments composed of two actin molecules (in turn consisting of a chain of globular actin molecules), tropomyosin and troponin molecules.

actinomorphic Describes a flower with parts arranged in a regular way (radially symmetric).

action potential Electrical event that is conducted the full length of an axon without loss of amplitude because it regenerates itself at successive points; triggered by depolarisation that reaches the threshold potential for the membrane; involves a rapid non-linear opening of voltage-dependent sodium channels, followed by opening of voltage-dependent potassium channels.

action spectrum Absorption spectrum of light that activates photosynthesis.

activation Increase in metabolism of an egg following penetration by a sperm cell.

activation energy Energy required to initiate a reaction; is more than the minimal level to break existing bonds at the moment molecules collide.

activators Regulatory proteins of eukaryotes, which bind to promotor elements upstream of the coding region in order for it to be optimally transcribed.

active response Amplification of a local depolarisation of a neuron that dies away with distance from the point of initiation; triggered by depolarisation that reaches the threshold potential for the membrane.

active site Specialised region of an enzyme into which substrate molecules fit; pocket or groove formed by folding of the polypeptide chains of the enzyme (quaternary structure).

active transport Carrier-mediated process requiring energy derived from hydrolysis of ATP; can move substances against a concentration gradient.

activin A protein believed to be responsible for the induction of mesoderm in the amphibian *Xenopus*.

adaptation Characteristic of an organism that fits it for a particular environment (*see also* adaptive evolution).

adapter molecule Transfer RNA; delivers amino acids to nucleotides during protein synthesis.

adaptive evolution The aquisition of inherited anatomical, physiological and behavioural traits that benefit an individual organism by enabling it to survive and reproduce better than other individuals in a particular environment.

adaptive radiation Rapid evolution and divergence of members of one lineage into different niches.

adduction Movement (e.g. of limb) towards the mid-line of the body.

adenosine triphosphate (ATP) An adenine-containing compound that releases free energy when its terminal phosphate bond is hydrolysed; this energy is used to drive energy-requiring reactions in cells.

adherant junction Junction that forms cross-links between contractile bundles of actin filaments in the cortical cytoplasm of neighbouring cells; are important in developmental processes.

adhesion belt A band of microfilaments often found in the apical cytoplasm of epithelial cells and associated with adherant junctions.

adrenal cortex Endocrine gland that secretes steroid hormones—mineralocorticoids (such as aldosterone) and glucocorticoids (such as corticosterone).

adrenal gland A gland located anterior to the kidneys; composed of the adrenal medulla (neurosecretory cells) surrounded by the adrenal cortex (non-neural endocrine cells).

adrenal medulla Neurosecretory gland that secretes catecholamines—adrenaline, noradrenaline and dopamine.

adventitious root Root that arises from deep within the stem of certain plants such as at the nodes of grasses and palms.

aerenchyma Type of parenchyma found in aquatic plants; spongy in appearance with large gas-filled intercellular spaces.

aerobic metabolism Metabolic processes that require molecular oxygen.

aerobic organism Organism that relies on oxidation of organic substrates for energy.

afferent Leading to.

aleurone layer The outer layer(s) of the endosperm of cereal grains that produces enzymes required for endosperm breakdown.

alga (pl. algae) Any photosynthetic, aquatic protist, including unicellular and multicellular forms, such as seaweeds.

allantois In amniotes; outgrowth of the embryonic hindgut used for excretion during development.

allele One of two or more forms of a gene located in the same position on homologous chromosomes.

allele frequency The proportion of a particular allele in a population.

allergen Agent that provokes an over-reaction of the immune system.

allergy Immune response that is excessive; may cause tissue damage or be potentially life-threatening.

allopatric speciation Geographic separation of populations leading to divergent evolution and formation of new species.

allopatry Living in different areas.

allopolyploid A polyploid organism with sets of chromosomes originating from different species.

alveolus (pl. alveoli) A cup-shaped chamber of extremely small diameter located after repeated branching of the airways in mammals; forms the gas-exchange surface of the lungs.

amino acid An organic molecule with an amino group ($-NH_2$, except for proline, which has an imino group, $-NH-$), an acidic carboxyl group ($-COOH$), a hydrogen atom and a unique side chain (R-group) all bonded to a central carbon atom; structural unit of proteins.

aminoacyl-tRNA Transfer RNA charged with the amino acid corresponding to its anticodon.

aminoacylation Attachment of an amino acid to its specific tRNA molecule catalysed by specific synthetase enzymes.

ammonotelic Pattern of excretion of nitrogen wastes in the form of ammonia.

amnion Extra-embryonic membrane enclosing an amniote embryo in a fluid-filled sac.

amniote All tetrapods, other than amphibians; having an amnion.

amoeboid Able to move by means of pseudopodia (temporary cytoplasmic extensions of a cell).

amphipathic molecule Molecule in which there is a difference in water solubility between one end and the other, such as a phospholipid, which has a phosphate head (hydrophilic) and a fatty acid tail (hydrophobic).

amyloplast A kind of plastid containing large starch granules and very few, if any, membranes within the stroma.

anabolism Metabolic reactions involving the building or synthesis of molecules.

anaemia Low concentration of haemoglobin in the blood.

anaerobic Not requiring oxygen.

anaphase A phase of mitosis in which the two kinetochores of the centromeres separate and sister chromatids move apart forming two groups.

anapsid Describing a vertebrate skull that lacks one or two pairs of openings in the temples.

anastomose Hyphal fusion in fungi allowing migration of nuclei from one hyphal cell to another.

anatropous Type of ovule in flowering plants in which the micropyle is located at the base, adjacent to the funicle (stalk).

anchoring junction Junction linking neighbouring cells and providing mechanical support; includes desmosomes, hemidesmosomes and adherant junctions.

androgen Steroid hormone including testosterone and dihydrotestosterone; secreted by the Leydig cells of the testes.

angiosperm Flowering plant.

angle of attack The orientation (angle) of an aerofoil or hydrofoil in relation to the direction of travel. A disc travelling through air edge-on has an angle of attack of $0°$ and travelling flat surface first of $90°$.

animal Multicellular heterotrophic eukaryote; member of kingdom Animalia.

animal pole The pole of the egg containing relatively low concentrations of yolk.

anisogamy Having gametes that differ in appearance, the male being small and motile, the female large and non-motile.

annulus Ring of enlarged cells with lignified wall found in sporangia of leptosporangiate ferns.

anomorphic Refers to the asexual form of a fungus, which may have been given a taxonomic name in addition to the sexual stage, often discovered later.

anoxygenic Photosynthetic bacteria that do not generate oxygen during photosynthesis.

antagonist muscle A muscle that can oppose the action of another.

antenna (pl. antennae) Paired appendages, located on the head of an arthropod and bearing sensory receptors.

antennal gland The paired metanephridial excretory organs of crustaceans, which have excretory openings at the base of the antennae.

anterior determinant A developmental factor that specifies both the head and thoracic segments of *Drosophila*; released from the anterior pole of the egg.

anther In a flower, part of the stamen that houses developing male reproductive cells.

antheridium Male gametangium (sex organ) producing sperm (or male haploid nuclei); antheridia are unicellular in algae and fungi and multicellular in plants (bryophytes and lower vascular plants).

antibiotic Naturally occurring inhibitor of bacterial protein synthesis and thus bacterial growth.

antibody Protein molecule produced by B cells in response to antigen and which reacts specifically with that antigen.

anticodon Trinucleotide sequence occurring at the end of a loop in transfer RNA molecules that is complementary to a specific mRNA codon.

antidiuretic hormone Hormone that increases the permeability of the renal collecting duct to water, and thus osmoconcentrates urine.

antigens Any molecule that can be recognised by one of the specific molecules (antibodies or T-cell receptors) of the immune system.

antipodal cells Nutritive cells of the embryo sac of flowering plants; lie at the end opposite the micropyle.

aorta Main artery that carries oxygenated blood from the heart to the body in higher vertebrates.

aortic arches Main arteries in fishes that stem from the ventral aorta, flow past the gills and join the dorsal aorta. Some aortic arches persist in vertebrates without gills, while others have been lost.

apical cell Single meristematic cell in the shoot and root apices of mosses and ferns.

apical complex A structure of endoparasitic protists (sporozoans) involved in penetration of host cells.

apical meristem Specialised growth region at the tip of shoots and roots; cells divide continually to produce the primary tissues and organs of the plant.

apoplastic pathway Pathway of least resistance of water from soil into the plant; through cell walls and intercellular spaces.

appressoria Swelling at the tip of a hypha of a parasitic fungus that penetrates the cell wall of a host plant.

apterous Without wings; especially in insects.

Archaeobacteria 'Very old bacteria'; one of two major groups of bacteria that includes methanogens, halophiles and thermoacidophiles (*see also* Eubacteria).

Archaezoan Primitive form of eukaryote (protist) with a nucleus but lacking typical eukaryote organelles such as mitochondria.

archegonium Multicellular, female gametangium (sex organ) producing egg cells; in all plants except flowering plants.

archenteron Embryonic cavity in animals that becomes the gut.

Archimedes' principle A body immersed in a fluid is subject to an upward force equal in magnitude to the weight of fluid it displaces.

Aristotle's lantern In sea urchins; complex, bony organ with five teeth and used for feeding.

arterial Of an artery; usually referring to blood flowing in arteries.

arterialised Referring to blood that has been oxygenated in a gas-exchange organ. Thus, arterialised blood flows in the pulmonary vein.

arteriole Small muscular arteries leading to the capillaries.

artery Large blood vessel carrying blood from the heart.

arthrology The study of joints, from the Greek *arthron*, joint and *logos*, study.

ascocarp Fruiting body formed from vegetative hyphae of an ascomycete fungus; encloses asci with ascospores.

ascogonium Female gametangium enclosing female gametes or female haploid nuclei (e.g. in ascomycete fungi).

ascospore Haploid, sexual spore produced in an ascus of an ascomycete fungus.

ascus (pl. asci) Sac-like cell that produces ascospores (ascomycete fungi).

asexual reproduction Reproduction in which offspring are clones of the parent organism.

aspiration Generation of negative pressure that results in air being sucked in.

assortative mating Mating between organisms with similar phenotypes and genotypes.

aster An organelle associated with nuclear division; comprises bundles of microtubules produced from a central centrosome.

atom The smallest part of an element that can exist and retain the properties of that element; comprising a central nucleus made of protons (positively charged) and neutrons (neutral charge) surrounded by one or more orbiting electrons (negatively charged).

atomic number Number of protons in the nucleus of an atom; characteristic for each element.

atrioventricular bundle (Bundle of His) A group of rapidly conducting cells that leads from the atrioventricular node to the Purkinje fibres; together, they produce a co-ordinated ventricular contraction.

atrioventricular node A patch of modified muscle cells lying between the right atrium and ventricle in higher vertebrates; slows the conduction of excitation between atria and ventricles.

atrium (pl. atria) Heart chamber that receives blood from veins or the sinus venosus and delivers it to the ventricle.

Australian region Biogeographic region including Australian mainland and islands on the continental shelf, such as Tasmania and New Guinea.

auto- From the same or a genetically identical individual; as in autoimmune, autoantigen, autograft.

autocrine hormones Hormones that interact with receptors on the surface of the cell that releases them.

autoimmune disease Disease resulting from development of an immune response to an individual's own antigens; usually chronic diseases, such as diabetes and arthritis.

autoregulation Control of blood flow to a tissue caused by direct effects of metabolites on smooth muscle of arterioles and precapillary sphincters.

autosomes Chromosomes that are the same in appearance and number in males and females of a species.

autotroph Organism able to synthesise its own food by photosynthesis or chemosynthesis.

auxin A type of plant hormone controlling stem elongation; synthesised in the growing shoot and root tips of plants; synthetic auxins used as selective herbicides.

axon A long neuronal process, which carries the output of the neuron to the next cell.

axon hillock The first part of an axon, the membrane properties of which allow action potentials to be generated.

axoneme A precise array of microtubules covered by the plasma membrane; structural basis of cilia or flagella.

axopod Long, slender radial projection of cytoplasm bounded by plasma membrane; characteristic of unicellular, actinopod protists.

B cell (B lymphocyte) Lymphocyte that makes antibodies; produced in the bone marrow, spleen or gut lymphoid tissue.

β-oxidation Process involving the release of chemical energy stored as C—C bonds in lipids; results in long-chain fatty acids being degraded by two carbon atoms at a time to form acetyl-CoA.

back-cross In plant breeding, a cross between F_1 (heterozygous) individuals and either of their pure-breeding parents.

bacteriophage Virus that infects and multiplies in bacteria; commonly called phage.

bacterium (pl. bacteria) The smallest cellular life form on earth (*see* prokaryote).

balanced polymorphism Genetic polymorphism that is stable and can be maintained in balance in terms of Hardy–Weinberg principle; occurs if heterozygotes for particular alleles are fitter than either homozygote.

baroreceptor (pressoreceptor) Nerve endings in the walls of blood vessels that sense blood pressure.

base Substance that can accept hydrogen ions (H^+) causing a decrease in their number in solution; possessing a pH above 7.

basement lamina Type of extracellular matrix that underlies epithelial cell layers.

basidiocarp Fungal fruiting body such as a toadstool or mushroom (basidiomycete fungi).

basidiospore Spore of a basidiomycete fungus borne externally on a basidium.

basidium A club-shaped fungal cell that bears basidiospores on its surface (basidiomycete fungi).

bilateral symmetry Symmetry of an organism where only one plane divides the organism into two similar halves.

binary fission Process of cell division involving cleavage to create two equal-sized cells each containing one copy of the genetic information and approximately half the cytoplasm.

binomial system The system devised by Linnaeus whereby the name of each kind of organism (each species) consists of two words: genus name and specific epithet.

biodiversity Number, relative abundance and genetic diversity of organisms on earth (*see also* species diversity).

bioeconomics The study of the interaction of economic and biological systems.

biogeochemical cycle The movement of material through an ecosystem, from atmospheric and geologic stores through food webs and back again.

biological control Control of a pest species by biological means, for example, introducing herbivore to consume and control a plant pest.

biomass 'Living mass'; amount, usually expressed as weight, of organisms in a particular area at a particular time.

biome On a global scale, ecological communities with the same structure and delineated by climate (e.g. grasslands of the world).

biomechanics The field of study involving the application of engineering principles to biology.

biota Fauna and flora of a given habitat or region.

biotic Pertaining to organisms.

biramous Referring to appendages consisting of two parts (e.g. in crustaceans).

blastocoel First cavity of the embryo; appears during cleavage.

blastocyst Mammalian blastula.

blastodisc Layer of cells of the avian embryo, forming a disc on the uncleaved yolk mass.

blastomeres Cells of an embryo during cleavage.

blastopore Depression on the surface of the gastrula marking the site at which inward cell movement occurs.

blastula Embryo at the end of cleavage.

Bohr effect A decrease in the oxygen affinity of a respiratory pigment when the partial pressure of CO_2 and/or proton (H^+) concentration increases.

book gill In some chelicerates; abdominal appendage, modified as a gill that has many leaf-like folds (lamellae); for gas exchange.

book lung Gas-exchange organ of spiders and scorpions; similar in structure to a book gill but internal on the ventral side of the abdomen.

bordered pits Pores with an overarching lip; connecting water-conducting tracheids of vascular plants.

Boreal region From or belonging to the north; biogeographical region of the world extending from the Polar Sea southwards and including North American and Eurasian regions (also called Holarctic).

boundary layer A barrier to diffusion that is established in the layer of fluid next to a surface across which a diffusion gradient exists.

bradycardia Decrease in heart rate.

buccal chamber Mouth cavity.

buccal force pump Mechanism that forces gas in the buccal chamber into the lungs under positive pressure.

budding A form of asexual reproduction involving the development of a new individual from outgrowths of the body wall of the parent.

buffer A substance that minimises changes in the pH of a solution by taking up or releasing H^+ ions when extraneous acids or bases are added to the solution.

bundle sheath Layer of cells surrounding veins of leaves of some plants; provide a link between photosynthetic mesophyll cells and vascular tissue.

bursicon (tanning hormone) Brain neurosecretory hormone that produces hardening and darkening of the adult cuticle in insects.

C phase Phase of the cell cycle; cytokinesis (division of the whole cell).

C_3 photosynthesis The process of carbon fixation in most plants in which the 3-carbon compound phosphoglyceric acid is the first stable product.

C_4 photosynthesis The process of carbon fixation in which 4-carbon compounds (e.g. malate) are the first stable product; found in tropical and subtropical grasses and cereals.

callus cells Cells produced at a cut surface of a plant; wound tissue; widely used in *in vitro* culture.

calyptra Tissue of the old gametophyte neck that persists on the top of a moss spore capsule.

cambium A secondary (sheet) meristem in vascular plants; vascular cambiun increases the girth of stems and roots.

canine teeth Conical teeth used for stabbing and gripping.

capillary Smallest blood vessel where exchange of substances between blood and extravascular fluid occurs.

capillary action Combined effect of cohesion and adhesion; means by which water rises up a fine capillary tube.

capsule A dry simple fruit that opens by valves on the top; fruit typical of eucalypts.

carapace Dorsal, protective covering over the thorax or anterior trunk segments of crustaceans; dorsal, protective shield in turtles.

carbohydrates Most abundant organic compounds in nature, composed of carbon, hydrogen and oxygen; basic unit a sugar molecule.

carbon fixation The capture of atmospheric carbon dioxide and its conversion into carbohydrates; occurs in the stroma of chloroplasts in eukaryotes.

carbonic anhydrase An enzyme that speeds up the hydration of CO_2 by a factor of 10^9.

cardiac centre A centre in the brain that controls the rate and strength of the heart beat.

cardiac muscle Branching network of individual striated muscle cells linked by many communicating junctions; found in the heart, hence its name.

cardiac output Rate of blood flow (mL/min) from one ventricle of the heart.

carnassial teeth Large molar teeth for shearing flesh and/or bone; found in carnivores.

carnivore Animal that catches other animals for food.

carotid body A peripheral chemoreceptor found in mammals and birds near the bifurcation of the common carotid; responds to changes in the partial pressure of oxygen.

carpel Female reproductive organ of a flowering plant; encloses ovules; ripens to become a fruit.

carrier proteins Membrane proteins involved in the transport of molecules across cellular membranes.

Casparian strip Strip of suberin thickening on the radial walls of the endodermis or exodermis of plant roots; regulates uptake of water and solutes.

caste A set of individuals within a colony of ants that are morphologically distinct and behaviourally specialised; includes queens, workers, soldiers and alates.

catabolism Metabolic reactions involving the breakdown of molecules.

catalase Enzyme involved in breakdown of hydrogen peroxidein; present in large amounts in microbodies.

catalysis Process by which the activation energy of a reaction is lowered; affects only the rate of the reaction.

cavitation Breaking of the water column in xylem vessels under water stress.

cDNA library cDNA clones spliced into vectors and transformed into eukaryotic host cells.

cell-adhesion molecules Molecules involved in the adhesion between cells, or between a cell, and the extracellular matrix.

cell cycle   The continuous cycle of growth and division in cells.

cell lineage   Series of cell divisions that precedes the formation of a particular cell.

cell plate   Region of new cell wall that forms during cytokinesis in eukaryotic, walled cells.

cellular immunity   Active destruction of foreign cells or of the body's own virally-infected cells by T cells.

cellular respiration   Oxidation of fuel molecules, that is, removal of electrons, coupled to synthesis of ATP.

cellulases   A group of enzymes able to hydrolyse cellulose molecules; produced by some micro-organisms and a few invertebrates.

cellulose   A structural polysaccharide present in cell walls of plants and some protists; composed of long chain of glucose molecules.

Cenozoic   The youngest era in the geologic time scale, from 65 million years to present day.

centimorgan   A map unit in chromosome mapping, expressing the frequency of the number of recombinant progeny in relation to total progeny.

central nervous system   The brain and spinal cord of vertebrates.

centriole   An organelle near the nucleus of animal and protist cells; associated with the spindle during mitosis and meiosis.

centromere   A constricted region of a chromosome at which the two chromatids are held together.

centrosome   Organelle containing two centrioles at right angles at which microtubules are assembled; the microtubular organising centre.

cephalisation   *See* encephalisation.

cephalothorax   United head and thorax in crustaceans and spiders.

cercaria (pl. cercariae)   The fluke larva that develops from a redia.

CFCs   Chlorofluorocarbons; used as aerosol propellants and in refrigerators, air conditioners and freezers; chlorine in CFCs reacts with and breaks down atmospheric ozone.

chela   Claw of a chelicerate (e.g. crab).

chemical digestion   Breakdown of food molecules by hydrolytic enzymes into smaller molecules.

chemoreceptor   A type of receptor that binds to a particular signal molecule, a ligand.

chiasma (pl. chiasmata)   Attachment point between chromosomes where crossing over occurs.

chlorenchyma   Photosynthetic parenchyma cell.

chloride shift   Maintenance of electrical balance within an erythrocyte by the exchange of Cl^- for HCO_3^- across the membrane.

chlorophyll   A light-absorbing, green pigment involved in photosynthesis.

chloroplast   An organelle (plastid) containing membrane-bound light-absorbing pigments; functions in photosynthesis.

choanocyte   Collar cell; flagellated cell lining the internal cavity of a sponge.

chorioallantois   A vascular extraembryonic membrane consisting of the fusion of the chorion and allantois; underlies the eggshell and functions for respiration and osmotic and ionic regulation in avian and reptilian embryos.

chorion   Outermost embryonic membrane in amniotes.

chromatid   A single chromosomal strand.

chromatophore   Cell that contains pigment granules and expands and contracts under muscular control, allowing body colour change.

chromophore   Light-absorbing region of a protein photoreceptor; absorbs light of a particular wavelength.

chromoplast   Plastid containing carotenoid pigments; responsible for red, orange or yellow colours of some plant organs.

chromosomal inversion   Chromosomal mutation where a segment within a chromosome is turned around end to end.

chromosomal translocation   Chromosomal mutation where segments are exchanged between non-homologous chromosomes.

chromosome   Structure containing a single DNA molecule; in prokaryotic cells and in the nucleus, mitochondria and chloroplasts of eukaryotic cells; nuclear chromosomes are visible during cell division.

citric acid cycle   Also known as the Krebs cycle; cyclic series of reactions involving oxidation of fuel molecules; occurs in mitochondria in eukaryotes.

cladode   A photosynthetic stem in plants with leaves reduced or absent.

clamp connection   Feature of basidiomycete fungal hyphae that ensures that the two nuclei of the dikaryon remain together following mitosis.

classification   A hierarchy of groups and subgroups of organisms reflecting their phylogenetic relationships. Each group (taxon) is given a name and rank: kingdom, phylum, class, order, family, genus, species (*see also* taxonomy).

cleavage   Series of mitotic cell divisions that take place in the egg after fertilisation and that result in a progressive decrease in cell size.

cleistogamy   In flowering plants, a process that ensures self-fertilisation; anthers open and self-pollination occurs within unopened flowers.

clitellum   Thickened region of the epidermis of a euclitellate worm (e.g. earthworm) that secretes the cocoon in which eggs are deposited.

clot (thrombus)   An aggregation of blood platelets, fibrin and trapped blood cells, usually at the site of a wound where it prevents blood loss.

cnidocyte   Cell in a cnidarian that contains a nematocyst.

cocurrent flow   Flow of two media in the same direction.

codominance   The full expression of two alternative alleles in a heterozygote.

codon   Triplet of nucleotide bases in DNA and RNA; codes for a particular amino acid.

coelenteron   In cnidarians; gastrovascular cavity lined with endoderm.

coelom   Body cavity of an animal, lined on all sides by mesoderm.

coelomoduct   A tubular excretory organ that has a ciliated, funnel-like opening in the coelomic cavity to draw coelomic fluid into the tubule; develops from the interior of an animal towards the outside, unlike nephridia.

coenocytic   A term used to describe a cell or non-septate hypha containing numerous nuclei.

cofactor   An additional chemical component, such as a metal ion or organic molecule, required by certain enzymes in order to function.

cohesion   The attraction between similar polar molecules, such as hydrogen bonding between water molecules.

colchicine   Drug derived from the bulbs of the autumn crocus (*Colchicum*); causes disassembly of spindle microtubules, preventing completion of cell division.

collagen   A structural protein of the extracellular matrix that is the most abundant protein of mammals; collagens associate into a strong sheet-like meshwork in basment lamina and form fibrils in interstitial matrices.

collecting duct   The terminal portion of the vertebrate nephron, which conveys fluid from the distal convoluted tubule into the renal pelvis.

collenchyma Living plant cells strengthened with primary thickening either at the corners or on the tangential walls; have a support function.

colloid osmotic pressure (oncotic pressure) Osmotic pressure due to large proteins, chiefly albumin, in the blood; involved in the balance between filtration and reabsorption of fluid in tissues.

commensalism Symbiotic interaction between two species where one benefits and the other is unaffected; usually one organism living with another for shelter or support (e.g. epiphyte).

communicating junctions (gap junctions) Junctions specialised for chemical and electrical communication between cells.

community In an ecological sense, an assemblage of populations of different species, interacting with one another, living in a particular area (e.g. pond or forest).

companion cell A cell type of phloem; transfer cell involved in the loading of sucrose into the sieve cells.

competition Individuals of either one species (intraspecific competition) or different species (interspecific competition) striving for the same resource, which is in limited supply.

competitive exclusion When one species outcompetes another for a limited resource, resulting in its local extinction.

complement system A series of about 20 serum proteins that activate sequentially in a cascade of reactions. Triggering of the complement cascade leads to activation of non-specific defensive cells, facilitation of phagocytosis and lysis of cells.

complementary DNA (cDNA) Complementary DNA copy of mRNA transcript active in a particular cell type or tissue.

compliance Relationship between change in volume and change in pressure.

compound A molecule composed of more than one type of atom.

compound leaves Leaves divided into leaflets, each with its own stalk.

condensation reactions Reactions involving removal of water molecules in the assembly of complex molecules from simpler ones.

conformation (of a protein) Three-dimensional shape of a protein molecule.

conidiophore Specialised fungal hypha that forms a stalk and bears spores (conidia).

conidiospore (conidium) Asexual fungal spore formed on a conidiophore; for dispersal and spread of the fungus.

conjugated protein A protein molecule consisting of amino acids and other organic or inorganic components, such as glyco-, lipo-, nucleo- and phosphoproteins.

conjugation Process by which organisms, such as bacteria, make direct contact and transfer DNA via plasmids from one cell to another, leading to genetic variation.

connective tissue Tissue that provides structural, metabolic and defensive support for other tissues; for example, blood, bone and cartilage; extracellular matrix usually more abundant than cells.

constitutive secretion Secretion that continues throughout the life of a cell.

consumer An organism that derives its energy by consuming other organisms.

contractile vacuole An organelle of cells that excretes fluid by a pulsating action, first filling the vacuole with fluid then ejecting the fluid from the cell.

convection Mass movement of a fluid; transfer of heat or substances along with a moving fluid, usually air or water.

convection requirement Amount of medium passed over a gas-exchange surface in order to extract a given amount of oxygen.

convergent evolution Evolution whereby organisms from different, distantly related lineages come to resemble one another.

co-operative breeding Occurs when adults in a group do not themselves reproduce, but instead help raise the young of the breeding individuals.

coprophagy (caecotrophy) Reingestion of special faecal pellets (the contents of the caecum, which are very rich in microbes) released at night, allowing utilisation of this high protein source.

copulation Mating between sexes associated with internal fertilisation.

coralloid roots Coral-like upward growth of roots of certain plants (alders, cycads, she-oaks) following root hair infection by the bacterium *Frankia* or cyanobacteria.

cork cambium Meristem responsible for producing bark at the surface of stems and roots of plants.

coronary artery Artery that supplies blood to heart muscle.

corpora allata Non-neural endocrine gland in insects, which secretes juvenile hormone.

corpus In the shoot apex of flowering plants, the inner layers of the apical dome of cells that contribute to stem formation.

corpus luteum Ovulated follicle, which secretes the steroid hormone progesterone.

corpuscle A blood cell.

cortex (adj. cortical) Outer region of structure.

cotyledon Leaf of a plant embryo functioning in food storage and digesting; first leaf to emerge following seed germination.

countercurrent flow Flow of two media in opposite directions.

countercurrent exchange Transfer of heat or a substance from a fluid flowing in one direction to fluid flowing in the opposite direction in an adjacent vessel.

courtship behaviour Interactions between members of the opposite sex that take place before mating.

covalent bond Bond formed between atoms due to sharing of electrons in their outermost orbitals.

Crassulacean acid metabolism (CAM) A variation of the C_4 pathway of photosynthesis, in which C_4 and Calvin–Benson cycle reactions occur in the same cells but at different times. CAM plants fix CO_2 at night and convert it to carbohydrate during the day.

cretinism Impairment of growth and development, particularly of the nervous system, as a result of lack of thyroid hormones in children.

cristae Folds of the inner membrane of mitochondria.

cross-current flow Flow of one medium perpendicular to the flow of another.

crossing over Exchange of DNA between chromatids of homologous chromosomes; involves cutting and rejoining strands of DNA.

crozier Curled branch of a dikaryotic, heterokaryotic hypha where mitosis occurs and acsi form in ascomycete fungi.

crustose A flat, 'crusty' growth form (e.g. of a lichen).

cuticle Outer water-resistant layer secreted by epidermis.

cyanobacterium Photosynthetic eubacterium that has chlorophyll *a* and produces oxygen as a byproduct of photosynthesis.

cyclosis Cytoplasmic streaming. Circulation of cytoplasm inside cells.

cyclosporine Immunosuppressant used in medicine and derived from a fungus (*Tolypocladium inflatum*).

cysticercus Tapeworm larva consisting of a bladder-like structure and inverted scolex.

cytochalasins Anti-actin agents derived from certain fungi, which act by specifically disrupting actin microfilament-based cytoskeletal systems.

cytokinesis Division of the cytoplasm of a cell following mitosis or meiosis.

cytokinins Plant hormones promoting cell division.

cytoplasm The cytosol and organelles of eukaryotic cells, excluding the nucleus.

cytoplasmic determinants Substances present in the egg cytoplasm that are believed to determine embryonic cell fate.

cytoskeleton Network of microtubules and microfilaments in eukaryotic cells; involved in functions such as maintenance and change in cell shape, movement of organelles within the cytoplasm and cell movement.

cytosol An aqueous solution of molecules with a gel-like consistency within the cytoplasm of eukaryotic cells.

cytotoxic cells (T_C cells, killer T cells) T cells that, when stimulated by antigen and lymphokines produced by T_H cells, directly lyse or kill target cells recognised by T_C cells on the basis of their particular antigen.

Darwinian fitness Estimates fitness (adaptive value) of individual genotypes of a species in terms of relative reproductive success.

dead space Volume of air in the conducting airways that takes no part in gas exchange.

decomposer Organism, such as some fungi and bacteria, that consume and break down organic matter for energy, releasing inorganic nutrients.

dendritic cell Cell of the immune system that has long branching processes and is able to break down foreign molecules and present them to lymphocytes.

density-dependent In ecology, when the per capita birth and death rates of a population depend on the size of the population.

density-independent In ecology, when the per capita birth and death rates of a population are independent of the size of the population; also termed density-vague.

dentrites Branching processes of neurons that are generally short and receive information from other cells.

deoxyribonucleic acid (DNA) A nucleic acid that is the hereditary material of an organism, stored as a coded sequence of nitrogenous bases; comprising two complementary double helical strands of nucleotides made up of a pentose sugar, phosphate group and nitrogenous base.

depolarisation Decreased voltage difference across a membrane; brings membrane potential closer to threshold potential and therefore is excitatory.

dermal bone Bone that develops in the skin without going through a cartilaginous phase.

desmosomes Provides structural support by cross-linking between cytoskeletons of neighbouring cells.

determination Process by which the developmental fate of a cell is fixed.

detritus Debris and dead remains of organisms in an ecosystem.

deuterostome Animal in which, during development, the anus forms at the site of the blastopore and the mouth forms as a secondary opening.

development Series of events leading to the formation of an adult organism from a zygote.

diabetes mellitus Insulin-dependent diabetes caused by the destruction of beta cells of pancreatic islets, apparently due to an autoimmune response.

diapause A period of dormancy during development in which growth and differentiation virtually cease.

diaphragm Mammalian structure composed of muscle and tendons; separates the thoracic cavity from the abdominal cavity; a major source of inspiratory force.

diapsid Describes a vertebrate skull with two, well-defined temporal openings.

diastole Phase of the cardiac cycle involving muscle relaxation and filling of a heart chamber.

dichotomous branching Dividing into two equal branches.

differentiation Process leading to changes in structure and function of cells during development often leading to specialisation; in animal cells such changes are often irreversible.

diffusion Net passive movement of molecules from a region where they are in high concentration to one where they are in low concentration; due to random thermal motion of molecules; passive movement of molecules along their electrochemical gradient.

dihybrid cross Cross involving individuals that are heterozygous at two different loci; involving the segregation of alleles of two genes.

dikaryon Fungal cell containing two haploid nuclei, one from each parent; usually formed after sexual fusion of parent hyphae.

dilate To increase in diameter, as in blood vessels or pupil of the eye.

dioecious Organism in which sperm and eggs are produced by separate individuals.

diploblastic In an animal, having two cell layers.

diploid Having two sets of chromosomes ($2n$).

direct development Development in which an animal is born with the general form of the adult.

disaccharide Two monosaccharide molecules joined by a glycosidic bond.

disulfide bond Covalent bond that can cross-link polypeptide chains; formed between the sulfur atoms of two cysteine residues.

divergent evolution Evolution that leads to descendants becoming different in form from their common ancestor.

diving response A co-ordinated set of cardiovascular and metabolic changes that occurs strongly in diving, air-breathing vertebrates.

DNA polymerases Enzymes that catalyse the replication (template-dependent synthesis) and repair of DNA.

dolipore septum Septum with complex pore separating adjacent cells in hyphae of basidiomycete fungi.

dominance hierarchy Physical domination of one individual over another; usually established by aggressive behaviour and once established the relationship remains stable without subsequent high levels of aggression.

dorsal lip of the blastopore Region of the blastopore which, in amphibian embryos, initiates gastrulation and induces formation of a dorsal axis.

drag Backward component of force acting on a moving body produced by a fluid resistance.

drupe Fleshy fruit, such as a plum, containing a single seed enclosed in a hard stony layer (endocarp).

ductus arteriosus An embryonic blood vessel that connects the pulmonary artery with the aorta so that much of the blood from the placenta bypasses the non-functional lungs of the fetus.

ecdysial gland Non-neural endocrine gland in the head of some insects; secretes ecdysone.

ecdysone Insect hormone, secreted by the ecdysial glands, which stimulates moulting, growth and differentiation of adult tissues.

eclosion   Hormonally induced change from pupal state to emergence of the adult in holometabolous organisms.

ecological gradient   Continuous or gradual change in environmental variables that occurs over geographic distance.

ecological pyramid   Diagram showing the change in energy, biomass or numbers of organisms at successive trophic levels in an ecosystem.

ecological sustainable development (ESD)   Using, conserving and enhancing resources so that ecological processes in ecosystems are maintained.

ecosystem   An ecological community together with the physical environment with which its members interact.

ecotone   Boundary between two different ecological communities.

ecotoxicology   Field of biology that deals with the effects of pollutants on organisms and ecosystems.

ectoderm   Outermost germ layer of animal embryos, giving rise to the outer body covering and associated structures.

ectotherm   Animal whose body temperature is more or less determined by the temperature of its surrounding environment.

Ediacaran fauna   Fossil traces of a collection of soft-bodied animals dated at 640–680 million years BP; best evidence that animals had evolved in the Pre-Cambrian; found on all continents.

efferent   Leading away from.

egg   Female gamete.

elastin   Structural protein of the extracellular matrix that is unusual because it remains in an unfolded, random coil configuration.

electrocardiogram (ECG)   Electrical activity from the heart that can be measured with electrodes on the body surface. The sequence of cardiac events can be interpreted from changes in electrical potential between the electrodes.

electron transport system   A group of membrane-bound enzymes and cofactors, which operate sequentially in a highly organised manner.

element   A substance made up of only one type of atom with the same atomic number.

elephantiasis   Grotesque swelling of lymphatic tissue caused by the tropical, nematode parasite *Wuchereria bancrofti*.

elimination   Loss of undigested and unabsorbed food from the digestive tract (not to be confused with excretion).

embryo   A developing organism.

embryo sac   Female gametophyte of flowering plants typically containing seven cells and eight haploid nuclei.

embryogenesis   Series of events between fertilisation and hatching or birth.

embryonic induction   The determination of embryonic cell fate as a result of influences from a neighbouring cell or tissue.

encephalisation   Over the course of evolution, the progressive aggregation of nerve cells in groups at the anterior end of animals.

endarch xylem   Pattern of primary xylem development in which new xylem is added to the outside of the protoxylem in stems of plants.

endergonic   Reaction in which the change in free energy is positive; energy is needed for the reaction to proceed.

endocrine glands   In animals, glands of internal secretion; usually secrete into the circulatory system.

endocytosis   Process of invagination of the plasma membrane to form a vesicle containing extracellular material that is transported into the cell.

endoderm   In animals, innermost germ layer; lines the archenteron and gives rise to the epithelial mucosa of the gut and associated structures, such as glands, and the lining of the lungs.

endodermis   In plants, layer of cells immediately outside the pericycle of a root; regulates uptake of water and solutes into the central vascular cylinder by means of the Casparian strip.

endoplasmic reticulum (ER)   Network of membranous sacs (cisternae) extending thoughout the cytoplasm of a eukaryotic cell; usually flat and sheet-like but can be linked by tubular cisternae.

endosperm   Triploid nutritive tissue in seeds of angiosperms.

endostyle   Ciliated ventral groove in the pharynx of tunicates that secretes mucus to trap planktonic food.

endosymbiosis   *See* symbiosis.

endothecium   Secondarily thickened layer of cells that lies within the epidermis of an anther of a flower, and following loss of water through the filament, causes the anther to open, releasing pollen.

endotherm   Animal with a relatively constant body temperature that is usually higher than the temperature of the surrounding environment; maintained by regulation of internal heat production.

energy   Capacity to do work; exists in a number of forms, including chemical, heat, sound, electricity and light.

enhancer   Activator-binding site, which may be far-removed from the gene (upstream or downstream) but on the same chromosome.

entropy   Measure of disorder (randomness) in a system; energy becomes lost as heat in every energy conversion, resulting in increased entropy.

environmental niche   Total range of conditions under which members of a species live and reproduce.

enzyme   Biological catalyst, usually a protein, which increases the rate of a reaction.

epiboly   Overgrowth of one cell layer by another layer during gastrulation.

epicotyl   Growing meristem or shoot of a germinating seed that lies above the cotyledons.

epidermis   Outer cellular layer of a multicellular organism.

epigynous   Describes a flower with an inferior ovary, that is buried within the receptacle below the perianth.

epiphyte   Plant that grows on another plant for support, but is not parasitic.

epistasis   Gene interaction that occurs when the phenotype of one gene masks that of the other.

epithelium   Tissue that forms a continuous layer covering internal or external surfaces of most multicellular organisms.

epitoky   Morphological changes in some invertebrates that lead to sexual maturity, from the atoke (non-reproductive) to the epitoke (reproductive) form.

epitope   Portion of an antigenic molecule that is recognised by a receptor. A large protein may have hundreds of different epitopes.

equilibrium (of a chemical reaction)   When a chemical reaction is in this state, there is no net change in either the concentration of reactants or products.

erythrocyte   Mature, anucleate red blood cell; contains haemoglobin.

erythropoiesis   Sequence of synthesis of erythrocytes, mainly in bone marrow.

erythropoietin   Glycoprotein hormone produced by the kidney in response to low oxygenation of the blood; stimulates erythropoiesis.

essential amino acid   Amino acid that cannot be manufactured by an animal and so must come from a dietary source.

ethology   The study of animal behaviour.

ethylene (C_2H_2) Gas of low molecular mass, influencing a wide range of processes in plant development, including promoting fruit ripening, shoot growth and flower senescence.

etioplast Plastid that develops in darkness which, on exposure to light, develops into a chloroplast.

Eubacteria 'Good or true bacteria'; one of two major groups of bacteria, with features typical of most bacteria (*see also* Archaeobacteria).

euchromatin Lightly staining regions in an interphase nucleus; consisting of dispersed strands of chromatin that are sites of active gene transcription.

eukaryotic cell Cell with a nucleus and other membrane-bound organelles (*compare* prokaryotic cell).

euryhaline Able to tolerate a broad range of salinities.

eusocial Social group in which individuals co-operate in raising young; essentially sterile workers care for the young of reproductively active individuals (reproductive division of labour).

eusporangiate Describes a fern with large sporangia, containing numerous spores.

evolution Process of change and divergence in populations and taxa.

evolutionarily stable strategy Strategy (or set of behaviours) that, if adopted by most members of a population of interacting individuals, cannot be bettered (in terms of reproductive success) by another.

exarch xylem Pattern of xylem development in roots in which the xylem forms from the outside, filling the centre of the root.

excretion Loss of ions, solutes, metabolic waste products or water from body fluids (not to be confused with elimination).

exergonic Term used to describe a reaction when the change in free energy is negative; energy is released in the reaction.

exine Outer patterned layer of pollen grains.

exocrine glands Glands of external secretion.

exocytosis Fusion of a vesicle with the plasma membrane, expelling its contents from the cell.

exodermis Layer of suberised cells at the junction of epidermis and cortex of certain roots; regulates uptake of water, solutes and ions into the cortex.

exogenous mutagen Environmental agent causing mutation (e.g. radiation and chemicals).

exon Individual coding sequence of a eukaryotic gene that is transcribed into mRNA.

exoskeleton External hard body covering of some animals.

exponential growth Population growth in which the size of the population regularly doubles; population size thus increases rapidly (geometrically).

expressivity Degree to which an allele is expressed in an individual.

extracellular environment The physical and chemical environment that exists outside cells; important aspects of the extracellular environment are ion concentrations, organic solute concentrations and total osmotic concentration.

extracellular matrix Forms the extracellular environment of animal cells; a fluid matrix containing an extensive network of proteins and polysaccharides linked together by covalent and non-covalent bonds that fills the spaces between cells.

fecundity Probability of giving birth.

fermentation Anaerobic production of alcohol, lactic acid or similar molecules from carbohydrates by the glycolytic pathway.

fertilisation Specific interaction between an egg and sperm leading to formation of a zygote.

fibre (in diet) Cellulose and pectins of plant cell walls, which are not easily digested and which form bulk in the diet.

fibronectin An adhesive protein of the extracellular matrix occurring in interstitial matrices; has a high relative molecular mass (about 460 kD) and two polypeptide chains.

filopodia Long, thin projections seen in many migrating cells and in the growing tips of axons of nerve cells.

filtration Loss of fluid through holes in capillary walls (or membranes) due to hydrostatic pressure. Water and small dissolved substances move through but large proteins and blood cells remain.

First law of thermodynamics *Energy can be neither created nor destroyed*; energy can be transformed from one form to another but the total energy of the universe remains constant.

first polar body One of the products of the first meiotic division of the oocyte, a small cell that eventually degenerates.

fitness Biological success as measured by an individual organism's contribution of offspring to the next generation.

flame cell Cell of excretory organs, protonephridia, found in flatworms and annelids.

flower Sexual reproductive structure of angiosperms; comprises four whorls or layers: sepals, petals, stamens (male organs) and carpels (female organs).

fluid mosaic Describes cell membranes; fluidity referring to the lateral movment of lipid molecules, and mosaic referring to the irregular arrangement of proteins.

foliose Leaf-like, describing some types of lichen.

follicle A dry simple fruit from one carpel and which opens on the lower side (e.g. banksias and grevilleas).

follicle cells Somatic cells surrounding the maturing oocyte and serving a protective and nutritive function.

food chain A sequence of organisms from producer to consumers along which energy flows in an ecosystem; usually with three or four trophic levels.

food web A number of interacting food chains in an ecosystem.

foramen ovale A valved opening between the right and left atria of embryonic mammals that allows blood from the placenta to bypass the non-functional lungs of the fetus.

foregut fermentation Digestion of cellulose by symbiotic micro-organisms located anterior to the true stomach.

fossil Preserved remains of an organism or traces of it, such as footprints (trace fossil), or chemical compounds produced by it (chemical fossil).

frameshift mutation Removal or addition of a single nucleotide base in a DNA sequence, resulting in the reading frame for all subsequent triplets being changed.

free energy (*G*) Usable energy in a chemical system.

frond Fern leaf.

frugivore Fruit-eating animal.

fruit Mature ovary of a flowering plant; contains seeds; may be dry or fleshy; simple (from one carpel), aggregate (from a cluster of separate carpels on one flower) or multiple (from a cluster of many carpels from different flowers).

fruiting body Specialised, spore-producing structure of a fungus.

frustule Valve or silica dish, two of which make up the cell wall of a diatom.

fruticose Shrub-like, describing some types of lichen.

fucoxanthin An accessory photosynthetic pigment found in the chloroplasts of chrysophytes (golden flagellates).

fuel molecules Molecules such as carbohydrates and fats with energy-rich chemical bonds that are broken down to give energy.

fulcrum Pivot point of a lever.

fundamental niche That region of the environment within which a species can persist indefinitely; defined by all the abiotic and biotic factors that impinge on the survival and reproduction of the species.

fungus (pl. fungi) Eukaryote with cell walls, lacks chlorophyll and absorbs its food (e.g. moulds, yeasts, mushrooms etc.).

fusion Combination of egg and sperm pronuclei to form the zygote nucleus during fertilisation.

G-proteins (guanosine triphosphate-binding regulatory proteins) Intermediate molecules in many cellular signalling pathways, which can couple to receptors altering the activity of an ion channel or intracellular enzyme.

G1 phase Gap 1, usually first and longest phase of synthesis and growth in the cell cycle.

G2 phase Gap 2 phase, the main period of synthesis of cellular molecules that occurs following S phase in the cell cycle.

gait A characteristic pattern of locomotion.

gametes Mature male and female germ cells that fuse to form the zygote.

gametogenesis Formation of the gametes.

gametophyte Haploid stage of a plant life cycle that produces gametes.

ganglion (pl. ganglia) Organised group of neurons.

gastrula Embryonic stage following cleavage; characterised by reorganisation of cells to form the three germ layers and development of bilateral symmetry.

gastrulation Bulk movement of cells leading to the formation of the gastrula.

gene family Genes of related function that have arisen by duplication and have been modified during the course of evolution.

generative cell Male reproductive cell of a pollen grain, formed by asymmetric cell division of the microspore and lies entirely within the vegetative cell; the progenitor of the sperm cells.

genetic drift Evolutionary change in small populations due to chance not natural selection.

genetic code Relationship between codons of DNA and RNA and specific amino acids.

genetic drift Random change in allele frequencies in small populations of organisms.

genetic engineering Techniques making use of recombinant DNA to produce transgenic (transformed) organisms for applications in pharmaceutical, agricultural, horticultural and veterinary industries.

genetic polymorphism The presence in a population of more than one allelic form of a gene, or two or more genotypes at higher frequencies than would be expected on the basis of mutation.

genome In a prokaryotic cell, the genome is the single circular DNA molecule (chromosome); in a eukaryotic cell, the nuclear genome is the chromosomal DNA in the nucleus; chloroplast and mitochondrial genomes are circular chromosomes.

genomic equivalence Cells of the same organism having the same genes, although they may be differentially expressed in different tissues.

germ cells The line of cells that gives rise to gametes.

germ layers The three basic tissue layers formed during gastrulation—endoderm, mesoderm and ectoderm.

gibberellin A type of plant hormone promoting stem elongation and seed germination; composed of small

molecules each containing 19 or 20 carbon atoms; synthesised in the shoot and germinating seeds.

gill Outgrowth of the body surface used in gas exchange.

gill chamber Chamber that houses gills.

glial cells Supporting cells of the nervous system; provide insulation, and mechanical and nutritional support for neurons, and guide their development and repair.

glomerular filtrate Fluid filtered from glomerular capillaries into the renal capsule; primary filtrate.

glomerulus A spherical tuft of capillaries associated with the vertebrate nephron; filtration of fluid from the glomerular capillaries forms the primary filtrate.

glucagon Hormone released by alpha pancreatic islet cells; causes increased blood glucose levels due to breakdown of glycogen and synthesis of glucose from amino acids.

gluconeogenesis Synthesis of glucose from non-carbohydrate sources such as amino acids.

glycocalyx Carbohydrate chains on the outer surface of the plasma.

glycogen A polysaccharide that serves as the principal storage form of carbohydrate in animals.

glycolipid Lipid with a short chain of sugar residues; in membranes, occurs on the non-cytosolic side.

glycolysis Anaerobic catabolism of glucose to pyruvic acid, producing two molecules of ATP.

glycoprotein Chain of sugar molecules attached to protein; occurs on the non-cytosolic side of plasma membranes.

glycosaminoglycans Large polysaccharide molecules composed of repeating dissaccharide units, usually linked to a protein core; are a major component of the extracellular matrix and are responsible for gel hydration.

glycosidic bond Bond linking two monosaccharides in which the first carbon atom of one sugar molecule reacts with a hydroxyl group of another sugar molecule, with loss of a water molecule.

goitre Enlargement of the thyroid gland caused by lack of iodine in the diet.

Golgi apparatus Stacks of four to 10 disc-shaped cisternae functioning in storage and modification of secretory products.

gonadotrophic hormone Hormone that stimulates gonads to produce gametes.

gonad Testis or ovary.

gonangium (pl. gonangia) A reproductive hydranth polyp.

Gondwana Past supercontinent uniting all southern land masses.

grana Stacks of thylakoids that form part of the internal membrane system of chloroplasts.

granulocyte A leucocyte produced in bone marrow that migrates to sites of infection where it engulfs and kills foreign organisms.

gravid Filled with eggs.

greenhouse effect Natural warming of the earth by heat trapped due to the presence of certain heat-absorbing gases in the atmosphere.

guanine ($C_5H_5ON_5$) A nitrogen waste product excreted by spiders; formed by nucleic acid metabolism, and from ammonia.

guanotelic Pattern of excretion of nitrogen wastes in the form of guanine.

guard cells Pair of kidney-shaped cells regulatoring stomata.

habitat selection theory The idea that all individuals of a species attempt to live in places that maximise their chances of survival and reproduction (their evolutionary fitness).

haematocrit Volume fraction of whole blood occupied by the blood cells.

haemocoel Large spaces in the body that are filled with blood.

Haldane effect Decrease in the affinity of respiratory pigments for CO_2 with increased oxygenation of the pigment.

halophyte Plant adapted to saline environments.

haploid A cell possessing only one set of chromosomes (*n*), as in egg or sperm.

Hardy–Weinberg principle In an infinitely large, inter-breeding population, in which there is random mating but no migration, mutation or selection, frequencies of genes and genotypes will be constant in each generation.

hartig net Mycorrhizal fungus mycelium that grows between root cortical cells facilitating nutrient transfer.

heartwood Mature secondary xylem, in which the rays have degenerated, and the vessels and tracheids are filled with secondary organic compounds that make the wood hard and durable.

helper cells (T_H cells) Regulatory T cells that produce and secrete lymphokines.

hemidesmosome Junction formed by cross-linking between the cytoskeleton of a cell and the extracellular matrix; provides structural support by anchoring cells to the matrix.

hemimetabolous development Development in which newly hatched young resemble adults in all except size and sexual maturity.

hemitropous Type of ovule in flowering plants in which the micropyle is located at the side.

herbivore Animal that consumes algae or plants as food.

hermaphrodite Type of animal in which both male and female reproductive organs occur within the same individual (also called monoecious). Type of plant in which the flowers contain both male and female organs.

heterochromatin Densely staining regions in an interphase nucleus; consists of aggregated strands of chromatin that are inactive in gene transcription.

heterokaryon Multinucleate vegetative cell of a fungus where the nuclei are genetically different.

heterokont Protist characterised by one smooth flagellum directed posteriorly and one hairy flagellum directed anteriorly.

heterospory Having two types of spore; heterosporous plants develop separate male and female gametophytes.

heterotroph Organism that consumes other organisms as food; unable to synthesise organic molecules from inorganic compounds.

heterozygote advantage Occurs when the Darwinian fitness of heterozygotes for particular alleles is greater than either homozygote.

heterozygote An individual carrying different alleles of a gene.

hindgut fermentation Digestion of cellulose by symbiotic micro-organisms located posterior to the true stomach.

holometabolous development Development in which the young have the form of worm-like larvae bearing little resemblance to adults. Larvae enter a quiescent stage during which they undergo complete structural reorganisation (metamorphosis) into adult form.

homeobox Characteristic DNA sequence of about 180 nucleotides that occurs in homeotic genes.

homeostasis Maintenance of a relatively constant internal environment.

homeotic gene Gene that regulates the activity of other genes during development.

homokaryon Multinucleate vegetative cell in fungi where the nuclei are genetically all the same.

homologous chromosomes In diploid cells, one chromosome of a pair that associate at the first stage of meiosis; carry equivalent genes.

homology Similarity indicating common ancestry. Homologous structures in different organisms have the same basic plan but not necessarily the same function.

homozygote An individual carrying two copies of the same allele of a gene.

hormone Chemical messenger secreted by cells of an organism in response to specific stimuli. Hormones modify the activity of cells as a result of interaction with specific receptors.

hot spot Immobile point at the surface of the earth's mantle where a column of hot, upwelling asthenosphere rises, which may form islands.

humoral immunity Immunity mediated by soluble antibody molecules secreted by B cells in the serum or other body fluids.

hybrid Offspring of two different varieties or species.

hybrid vigour The basis of hybrid seed production in crop plants, occurring when two or more pure lines (homozygous) are crossed, and the resulting progeny show increased yield and vigour (heterozygous).

hydathode Pore-like structure on the tips or margins of leaves of rainforest plants that permit water to be extruded when required by high root pressure.

hydranth An individual in a colony of hydroid polyps.

hydrofoil Structure that generates a lifting force when moving through a fluid.

hydrogen bond Relatively weak bond formed between hydrogen and another polar atom such as oxygen.

hydrolysis Reaction involving the addition of water in the breakdown of a complex molecule to simpler ones such as a protein to amino acids.

hydrophilic Substances such as polar molecules that dissolve readily in water because they can readily form hydrogen bonds with water molecules.

hydrophobic Substances such as non-polar molecules that are insoluble in water because the hydrogen bonds between water molecules tend to exclude non-polar molecules.

hydrostatic pressure Pressure exerted by a liquid, such as blood. A misnomer because the pressure can be exerted by liquids other than water, and the liquids can be moving.

hydroxyapitite Main mineral of bone and teeth; complex form of calcium phosphate.

hyperosmotic solution Solution with a higher osmotic concentration than another.

hyperpolarisation Increased voltage difference across a membrane; moves membrane potential further from the threshold potential and therefore is inhibitory.

hypersaline Salt solutions that have a higher salt concentration than sea water.

hypha (pl. hyphae) Microscopic tube of cytoplasm bounded by a tough, waterproof cell wall; form fungal mycelia.

hypocotyl The part of the axis of a germinating seed below the point of attachment of the cotyledons.

hypogynous Describes a flower with a superior ovary, that is, attached to or above the receptacle.

hypo-osmotic solution Solution with a lower osmotic concentration than another.

hypostome Projection on which the mouth of a polyp is borne.

hypothalamus In vertebrates, region in the midbrain surrounding the third ventricle; receives information regarding the well-being of an animal and provides central neural and hormonal control of many functions.

hysteresis The failure of a system to follow identical paths of response upon application of and withdrawal of a forcing agent.

ideal free distribution The distribution of animals between two resource sites; individuals are *free* to choose between the sites, and the distribution *ideal* because each individual goes to the place that provides the highest returns.

imago Sexually developed adult stage of an insect life cycle.

immunodeficiency Absence of T cells as a result of absence of the thymus in which T cells mature, may be due to certain genetic abnormalities or removal of the thymus early in development.

immunoglobulin (Ig) Another term for antibody.

immunological memory Retention of stimulated B cells, called memory cells; allows a rapid immunological response to a subsequent interaction with the same antigen (secondary immune response).

imprinting Occurs when a newborn animal recognises the first moving object it sees (usually its mother) and follows it for the next few weeks; occurs usually during a limited period, the sensitive period.

immunogen Antigen that can stimulate an immune response.

incisors Teeth used to grasp, hold and cut food.

incomplete dominance The phenotype of a heterozygote shows incomplete dominance when it is intermediate between that of each homozygote parent.

indirect development Development that involves a larval stage followed by metamorphosis into the adult stage.

indole-3-acetic acid (IAA) Plant hormone; main auxin occurring naturally in plants.

indusium Protective leaf-like structure covering sori of some ferns.

inflorescence A cluster of flowers.

initiation complex Complex formed when a ribosome binds to mRNA and the first specified aminoacyl-tRNA; initiates protein synthesis.

inner cell mass A group of cells at one end of the mammalian blastocyst, part of which gives rise to the embryo proper.

inorganic compound Substance that does not contain carbon atoms; usually associated with non-living sources.

inspiration Bulk movement of air from outside the body into the lungs.

instars Juvenile stages between moults during development of arthropods.

insulin Hormone released by alpha cells of the islets of Langerhans. Binds to membrane receptors and increases the membrane permeability to glucose and amino acids. Leads to increased storage of glucose and increased fat production.

integument Layer of cells surrounding megasporangium in seed plants (*see also* ovule).

intercellular Between cells.

interferon Group of proteins secreted by some virus-infected cells that assist uninfected cells to resist infection by that virus.

intermediate filaments Filaments of the cytoskeleton of eukaryotic cells that are intermediate in size between microtubules and microfilaments, 8–10 nm in diameter, and provide mechanical support for the cell.

interneuron Neuron that transmits information from one neuron to another.

internode Portion of stem between successive nodes (site of leaf attachment).

interstitial fluid Liquids in the spaces between cells.

interstitial matrix Type of extracellular matrix prominent in connective tissues.

intine The inner wall layer of microspores and pollen grains of flowering plants that is pectocellulosic in nature, and forms the pollen tube.

intracellular environment Physical and chemical environment that exists within cells; important aspects of the intracellular environment are ion concentrations, organic solute concentrations and total osmotic concentration.

intracellular Inside cells.

intron Intervening non-coding sequence present in the coding region of a eukaryotic gene that is transcribed but excised before translation.

invagination Local buckling of an epithelial cell layer resulting in a depression opening to the outside.

involution A process whereby groups of cells roll under their neighbours and move inwards, for example, during gastrulation of frog embryos.

ion An atom that loses or gains electrons, becoming positively or negatively charged.

ion channels Fastest enzymes known; permit passive transport through cellular membranes; highly selective for particular ions; opened by a change in voltage across a membrane, or by binding with specific signal molecules.

ionic bond Bond formed when ions of opposite charge are attracted to each other.

ionisation The process in which a substance dissociates in solution to form ions.

ionoconform To have the same ionic concentrations in the body fluids as occur in the external medium.

ionoregulate To maintain body fluid ionic concentrations different from those of the external medium.

ischaemia Severe reduction in blood flow to a tissue.

islets of Langerhans Endocrine cells of the pancreas that secrete glucagon (alpha cells), insulin (beta cells) and somatostatin (gamma cells).

isogamy Situation where gametes are similar in appearance but differ in mating type.

isotope One or more kinds of an element whose atoms have a similar number of protons and electrons but differ in number of neutrons.

juvenile hormone Insect hormone secreted by the corpora allata; stimulates development of nymphal structures.

juxtamedullary nephron Nephron with its glomerulus located near the junction of the renal cortex and medulla; has a long loop of Henle.

kidney Excretory organ of vertebrates; the nephron is the functional excretory unit of the kidney.

kinetic energy Energy of movement, as in running water.

kinetochore The two protein discs of a centromere, into which microtubules are inserted.

kinetoplast Large mass of DNA, composed of thousands of catenated DNA mini-circles (linked as in a chain) present in the mitochondrion of certain flagellate parasites (e.g. trypanosomes).

labellum Petal modified as the lip of an orchid flower.

lactic acid In animals; three-carbon molecule derived from pyruvic acid as a product of anaerobic respiration, especially in muscle.

Lamarckism Theory of evolution by Jean-Baptiste Lamarck that traits acquired during the lifetime of an organism are inherited by subsequent generations.

lamella (pl. lamellae) Thin layer or plate-like structure.

lamellar bodies Intracellular storage form of surfactant.

lamellipodia Veil-like extensions of the plasma membrane of migrating cells or the growing tip of nerve cells.

laminin An adhesive protein of the extracellular matrix, occurring in basement lamina; has a very high relative molecular mass (about 850 kD) and three polypeptide chains arranged in form of a cross.

larva Juvenile stage in the life cycle of many animals.

laterite Weathered land surface with ironstone gravel occurring at the surface or subsurface overlying a layer of bleached, white clay.

Laurasia Past supercontinent uniting all northern land masses.

learning Any change in an individual's behaviour that is due to its experience.

lenticel Special site for gas exchange in the periderm (outer layer of bark) of woody plants; raised area of cells with extensive intercellular spaces.

leptosporangiate Fern with small, delicate sporangia.

leucocyte A class of nucleated white blood cells that protect the body against invasion and collect cellular debris.

lichen Mutualistic relationship between a fungus and a green alga or cyanobacterium.

life cycle The sequence of stages in the growth and development of organisms from zygote to reproduction; in sexually reproducing organisms, these show an alternation between diploid and haploid stages.

lift The component of force acting at right-angles to the direction of motion of a hydrofoil.

ligament A type of connective tissue linking two bones in a joint.

ligand Signal molecule which is capable of interacting with a receptor.

lignin Main component of secondary walls and wood of plants; composed of phenylpropanoid units, which provide a rigid matrix for cellulose fibres.

limiting resource Environmental requirement of an organism that is in limited supply (e.g. food, nest site etc.).

lipid Biological compound that functions in membranes, energy storage and transport, and insulation; insoluble in water as a result of the non-polar (hydrophobic) nature of their numerous C—H bonds; composed principally of carbon, hydrogen and oxygen together with phosphorus and nitrogen.

lipid bilayer Double layer of lipid molecules that forms the basic structure of cell membranes.

lithophyte Plant that grows among rocks.

loop of Henle Part of the mammalian nephron (and some avian nephrons) that lies between the proximal and distal convoluted tubules; enables the osmoconcentration of urine.

lophophore Feeding structure of ciliated tentacles containing extensions of the coelom in lophophorates.

lorica External vase-shaped shell of some chrysophyte protists.

lung Invaginated gas-exchange surface connected to air outside the body via narrow tubes.

lymph A transparent fluid formed by filtration of liquid from capillaries into the interstitial space; collected by primary lymphatic vessels and returned to the blood; contains white blood cells that attack invading organisms; transports proteins and fats into the blood.

lymphatic system A system of interconnected vessels, nodes and spaces through which lymph is transported throughout the body.

lymphocyte (lymphoid cell) Mononuclear cells that are the predominant cells in immune organs; responsible for the immune response; two principal classes are T cells and B cells.

lymphokines Molecules secreted by helper T cells that control the development and function of other T and B cells, as well as of accessory cells such as macrophages.

lysosome Cellular organelle; vesicle bounded by membrane; contains hydrolytic enzymes involved in the breakdown and recycling of many types of molecules.

M phase Phase of nuclear division (mitosis) in the cell cycle.

macronutrients Nutrients that are required in large amounts.

macrophage Phagocytic white blood cell.

madreporite Porous disc, opening of the water vascular system of echinoderms.

magnetic reversal Reversal of the earth's magnetic field, from normal (present day) to reversed polarity (north becomes south and south becomes north).

major histocompatiblitity complex (MHC) Presents (shows) antigen to T cells, without which T cells will not respond to antigen; fundamental to identification of 'self' and in graft rejection.

male germ unit In flowering plants; single transmitting unit for transfer of male gamete to female gamete; consisting of generative or sperm cells and tube nucleus.

Malpighian tubule Blind-ended excretory tubule of arthropods; urine is formed by active K^+ secretion into the tubule and passive solute and water influx. The urine is emptied into the hindgut.

mandibles First pair of insect appendages for feeding; modified for grinding and chewing (jaw-like) or piercing and sucking (stylets).

mantle (of molluscs) Dorsal fold of the body wall with a cavity beneath it; secretes the shell.

mantle (mycorrhizal) Thick sheath of a mycorrhizal fungal mycelium surrounding a root, replacing epidermis and root hairs.

manubrium Projection in a medusa on which the mouth is borne.

mass number Combined number of protons and neutrons in a nucleus.

mate-guarding Male behaviour that ensures the female he mates with does not mate with another male.

maternal inheritance Situation in which the genotype of the female parent alone determines a particular phenotype of the individual.

mating system Defined according to the number of partners each sex may have during its lifetime or during the mating season.

maxilla (pl. maxillae) Second pair of appendages; mouthparts in insects.

maximum sustainable yield (MSY) Harvesting a population at a rate that allows the population size to be maintained at half its maximum.

mechanoreceptor A type of receptor that detects stimuli such as mechanical pressure or stretch.

medusa Free-floating (pelagic), bell-shaped form of a cnidarian, with its mouth pointing downwards; jellyfish.

megaspore Haploid spore of ferns, fern allies and seed plants that germinates into a female gametophyte, which bears egg cells.

megasporangium Female sporangium of plants in which megaspores develop.

meiosis A type of nuclear division that takes place in germ cells and that results in halving of the number of chromosomes; each daughter cell (gamete) receives one set of chromosomes.

memnospore Sexual spore of a fungus; allows survival during harsh conditions.

memory cell Long-lasting B cell formed after binding of antigen to the specific receptor of a B cell in the presence of helper T cells. Is the basis of immunological memory.

mesenchyme A class of mesodermal cells, characterised by a tendency to migrate as individual cells.

mesoderm One of the three germ layers, lying between the ectoderm and the endoderm; gives rise to many of the internal organs and internal epithelia.

mesoglea Intermediate gelatinous layer, between ectoderm and endoderm, in a cnidarian (e.g. jellyfish).

mesohyl Middle layer of a sponge consisting of a gelatinous protein matrix containing amoeboid cells, collagen fibres and skeletal elements.

mesophile Organism that grows best between 10° and 30° C.

mesophyte Plant that lives where water is in adequate supply (*compare* xerophyte).

Mesozoic Geologic era from 245 to 65 million years BP.

messenger RNA (mRNA) An RNA molecule with a sequence complementary to a DNA coding sequence; carries the information specifying the amino acid sequence of a given polypeptide.

metabolic rate The rate of metabolism of an organism; usually related to level of activity. Basal metabolic rate refers to an inactive endotherm in thermoneutral environment, standard metabolic rate refers to an inactive ectotherm at a particular temperature.

metabolism All the chemical processes occurring within the cells of a living organism.

metameric segmentation A body plan in animals in which there is linear repetition of functional units, which are added at the posterior end (e.g. in annelids).

metanephridium (pl. metanephridia) Tubular excretory organ with a ciliated, funnel-like opening in the coelomic cavity that draws coelomic fluid into the tubule.

metaphase A phase of mitosis, in which chromosomes become arranged equatorially on the mitotic spindle.

metaxylem Xylem tissue that forms outside the protoxylem in stems of plants; has larger and thicker-walled cells than protoxylem, with reticulate secondary thickening.

metazoan An animal.

micelle Spherical structure formed when phospholipids are added to water; forms because fatty acid tails of phospholipids are hydrophobic.

microbody Organelle in eukaryotic cells involved in removal of compounds generated within a cell; spherical in shape and surrounded by a single membrane; often contain crystalline inclusions.

microcirculation Circulation and exchange in arterioles, capillaries and venules.

microfilaments Fine fibres composed of filamentous chains of actin molecules; part of the cytoskeleton.

micronutrients (trace elements) Nutrients that are required in small amounts.

microsporangium Male sporangium of plants in which microspores develop.

microspore Haploid spore of ferns, fern allies and seed plants that germinates into a male gametophyte, which produces sperm.

microsporocyte Diploid male spore-forming cell of plants that undergoes meiosis to form microspores and pollen grains.

microsporophyll Fertile leaf that bears microsporangia; aggregated into male (pollen) cones.

microtubules Major tubular scaffolding components of the cytoskeleton; form hollow cylinders 25 nm in diameter; associated with plasma membranes and forming spindle fibres during eukaryotic cell division.

midbody During cytokinesis in animal cells, a contractile ring of actin between the two daughter cells causes cleavage and the fibres running between the poles aggregate into a tight rod, the midbody.

middle lamella A thin layer between the primary wall of a plant cell; rich in pectins.

mineral elements Inorganic elements required by organisms.

minimum viable population size (MVP) Population size above which a species needs to be maintained to ensure its long-term survival.

miracidium Free-swimming ciliated stage of parasitic flukes.

mitochondrion (pl. mitochondria) DNA-containing organelle of eukaryotic cells; surrounded by a highly permeable double membrane; contain circular DNA molecules, RNA and small ribosomes; site of cellular respiration.

mitosis Nuclear division in eukaryotes in which the daughter nuclei are identical to the parent.

mitotic spindle An elaborate cytoskeletal structure that causes chromosomes to move towards the equator at metaphase of mitosis, and the chromatids to separate and move toward the poles at anaphase.

molar concentration Number of moles of solute disolved in a litre of solution.

molar teeth Cheek teeth involved in mechanical processing of foods, such as cutting, grinding and chewing.

molecule Stable association of two or more atoms due to sharing of electrons in their outer orbitals.

momentum The mass of a body multiplied by its velocity.

moneran Member of the prokaryote kingdom Monera (*see* bacterium).

monocistronic transcript Primary transcript from a typical eukaryotic messenger RNA transcript; contains one functional reading frame and codes for one polypeptide.

monocyte A motile leucocyte capable of engulfing old erythrocytes and infectious organisms.

monoecious Type of animal in which both male and female reproductive organs occur within the same individual (also called hermaphrodite). Type of plant in which male and female organs are present in different flowers on the same plant.

monohybrid cross Cross involving organisms that are heterozygous at one locus such as $Aa \times Aa$.

monokaryon Fungal cell containing one nucleus.

monophyletic Referring to a group or taxon of organisms that includes all of the lineages descended from a common ancestor (i.e. an entire branch on a phylogenetic tree).

monosaccharide Sweet-tasting simple sugar that cannot be broken down into smaller sugar molecules; most common are 5-carbon pentoses and 6-carbon hexoses.

monosome Functional ribosome composed of a large and a small subunit.

morphogenesis Generation of new shape during development.

morphology The study of body form, including embryology, anatomy and the study of both fossil and living organisms.

mosaic development Development in which the fate of many blastomeres is predetermined in the absence of interactions with surrounding cells.

mutation Change of the genotype. A *point mutation* is a change in the base sequence of DNA (base substitution, insertion or deletion). A *chromosome mutation* is a change causing breakage and rejoining of chromosomes.

mutualism Symbiotic interaction between two species where both benefit from the association (e.g. lichen).

mycelium The body of a fungus, generally growing as filamentous hyphae.

mycorrhiza (pl. mycorrhizae) A mutualistic association between certain types of fungi and the roots of plants; enhances nutrient uptake by the plant; includes arbuscular, orchid, epacrid and ectomycorrhizae.

myelin sheath Insulation layer around some axons in

vertebrates; formed by wrapping the axon in many layers of glial cell membrane.

myocardium The three-layered muscle composing the heart wall.

myofibril Rod-like bundle of myofilaments found in muscle cells.

myogenic Initiation of the heart beat within cardiac muscle itself, for example, in vertebrate hearts.

myoglobin Oxygen-binding pigment in muscle; has a greater affinity for oxygen than does haemoglobin.

myosin Protein that commonly interacts with actin filaments to generate cytoplasmic movements or changes in cell shape; organised into thick filaments in muscle cells.

myotomes Blocks of muscle on each side of the body.

myrmecotrophy Mutualistic relationship where ants live within special chambers formed in certain plants.

myxoedema Condition associated with accumulation of water as a result of lack of thyroid hormones in adult humans.

natural selection Term proposed by Darwin for 'survival of the fittest': some members of a population with characteristics that enable them to compete successfully in a particular environment are likely to survive and leave more offspring than are members with less favourable characteristics.

nauplius First, free-swimming, planktonic larva of most crustaceans.

negative feedback control Control system where the response produced to a particular stimulus reduces the size of the original disturbance; leads to homeostasis.

nematocyst Stinging organelle of cnidarians (e.g. jellyfish), which functions in defence and capture of prey; nematocysts are also called cnidae.

neotropical region Biogeographical region of the world including South America and lower Central America.

nephridium (pl. nephridia) Tubular excretory organ of invertebrates that develops from the body surface into the coelomic cavity; it is either a protonephridium or a metanephridium, depending on the structure of the coelomic end.

nephron Tubular excretory unit of the vertebrate kidney; derived from a coelomoduct but lacks a ciliated, funnel-like opening.

net primary productivity Portion of total (gross) primary productivity that remains after the respiratory losses of primary producers are accounted for.

neural crest In vertebrates, a group of cells formed as the neural tube detaches from the overlying ectoderm. These cells migrate individually throughout the embryo and ultimately differentiate into many different cell types. Neural crest tissues are important in the evolution of many vertebrate features.

neural plate Region of dorsal ectoderm in vertebrate embryos that forms the neural tube.

neural tube Structure in vertebrate embryos from which the spinal cord and brain form.

neurogenic Initiation of the heart beat by nerves leading to the heart muscle, for example, in many invertebrate hearts.

neurohaemal organ Aggregation of neurosecretory cells into a discrete, highly vascularised organ.

neuron Cell specialised for receiving, conducting and transmitting information to other cells; basic unit of the nervous system.

neurosecretion Secretion of hormones by nerve cells.

neurosecretory cells Specialised nerve cells that secrete hormones.

neurotoxins Chemicals that cause nervous systems to malfunction, produced as offensive or defensive weapons in both plants and animals.

neurotransmitter Water-soluble signal molecule released from nerve endings at a synapse with an effector cell; acts on receptors located on other nerve cells, muscle cells or glands.

neurula Embryonic stage in vertebrates during which the neural tube forms.

neurulation The process of neural tube formation.

neutral evolution Evolutionary change having no apparent effect on the survival or reproduction of an organism; result of changes fixed in small populations by genetic drift.

Newton's Laws of Motion (i) A body remains at rest or moves at constant velocity unless acted on by external forces. (ii) An unbalanced force F acting on a body of mass m gives the body an acceleration a in the direction of the force: $F = m \cdot a$. (iii) For every action (force) there will be an equal and opposite reaction (force).

niche *See* fundamental niche.

niche breadth For any abiotic or biotic factor, the range of tolerance of the species.

nitrogen fixation Conversion of gaseous, atmospheric nitrogen by certain bacteria to ammonia, nitrites and nitrates.

node Site on stem at which leaves are attached.

nodes of Ranvier Small bare regions of axon between Schwann cells that form myelin sheaths; allow saltatory conduction of action potentials from node to node.

nuclear envelope Double membrane surrounding the nucleus in eukaryotic cells.

nuclear pore Channel in the nuclear envelope that allows movement of certain molecules between nucleus and cytoplasm.

nucleolus A spherical fibrillar and granular structure within the nucleus of eukaryotic cells; composed chiefly of rRNA in the process of being transcribed from multiple copies of rRNA genes.

nucleosome core particle A complex of eight histone proteins that form a core around which the double helix of DNA is coiled.

nucleosome Particle about 10 nm in diameter comprising nucleosome core particle (histone proteins) and associated DNA; found in large numbers in chromatin.

nucleotide Five-carbon sugar, a phosphate group and a nitrogenous base; nucleotides are linked together by phosphodiester bonds between the sugar and phosphate groups to form nucleic acids.

nucleus The principal membrane-bound compartment of the eukaryotic cell; control centre of the cell; contains chromosomal DNA.

nurse cells Found in many invertebrates; a class of cells that provides nutrients and other materials to a maturing oocyte. In insects, nurse cells are cytoplasmically connected to the oocyte.

nutrient Particular substance required by an organism that must be obtained from its environment; includes organic compounds (carbohydrates, amino acids and fats), vitamins and minerals.

occluding junction (tight junction) Junction where the plasma membranes of adjacent cells are fused tightly, preventing passage through the extracellular space.

oceanic ridge Site where lava upwells from part of the earth's mantle.

ocellus (pl. ocelli) Simple eye of adult arthropods.

oedema Build-up of fluid in tissues when filtration exceeds reabsorption and lymph flow. Usually associated with vascular disease.

oestrogens Steroid hormones including oestradiol, oestrone and oestriol; produced by ovarian follicles.

omnivore An animal that feeds on a variety of organisms, for example, plants and animals.

onchosphere Six-hooked larva that hatches from the egg of a tapeworm.

oncogene Gene whose products are able to change (transform) eukaryotic cells so that they grow in a way similar to tumour cells.

oocyte Female germ cell undergoing meiosis within the ovary.

oogenesis Process by which eggs form from primordial germ cells.

oogonia Diploid female germ cells undergoing mitosis; in animals within the ovary; in protists within oogonia (female reproductive structure).

open-flow ventilation Unidirectional flow of respiratory medium past a gas-exchange surface.

operculum Cap-shaped covering, for example, apical portion of a moss spore capsule or covering formed from fused sepals or petals of a eucalypt flower bud.

operons In bacteria, transcription units that cover more than one polypeptide coding region.

opisthosoma In arthropods, abdomen, posterior region of the body behind the prosoma.

organ Combination of tissues forming a functional unit.

organelle Specialised part of a cell, such as nucleus or ribosome.

organic compound Molecule that contains one or more carbon atoms; largely produced by living organisms.

organogenesis Process of organ formation.

origin Start sequence of DNA replication; there is a single origin in the circular DNA molecules of prokaryotes, but multiple origins in eukaryote chromosomes.

osculum (pl. oscula) Opening in the wall of a sponge through which water leaves.

osmoconcentration Physiological process whereby water is resorbed from the urine by excretory tubules making the urine more osmotically concentrated than body fluids.

osmoconform To have the same osmotic concentration of body fluids and the external medium.

osmoregulate To maintain an osmotic concentration of the body fluids different from that of the external medium.

osmosis The movement of water from a region of low osmotic concentration (high water concentration, high water potential) to one of high osmotic concentration (low water concentration, low water potential), as a result of the random thermal motion of water molecules through a selectively permeable membrane.

osmotic adjustment A plant response to drought, involving an increase in amount of vacuolar solutes; leads to decreased osmotic potential and reduced cell water potential without adversely affecting cell turgor, allowing growth and photosynthesis to continue in drier conditions.

osmotic concentration The total concentration of all solutes dissolved in solution, expressed as moles of solutes per litre of solution, or osmols per litre.

ossicle Crystal of calcium carbonate; first formed within a cell and enlarging to a plate beneath the skin; forming the skeleton of echinoderms.

osteoporosis Depletion of bone calcium in women as a result of decreasing oestrogen levels after menopause.

ostium (pl. ostia) Pore in the wall of a sponge, through which water enters during filter-feeding.

ovulation The release of a mature egg cell from the ovary.

ovule Megagametophyte retained within the megasporangium, which is further surrounded and protected by one or more layers of cells, the integuments; following fertilisation the whole structure (the ovule) develops into a seed.

ovuliferous scale Leaf-like structure (thought to be a reduced shoot) that bears ovules; aggregated into female cones.

oxidation Reactions involving the breakdown of molecules by removal of electrons, especially those in which oxygen combines with, or hydrogen is removed from, a compound.

oxygen affinity The ease with which oxygen binds to a respiratory pigment.

oxygen equilibrium curve Relationship between the percentage of oxygen saturation of a respiratory pigment and the partial pressure of oxygen.

oxytocin Posterior pituitary hormone that influences reproductive functions.

ozone Triatomic form of oxygen, O_3; highly reactive molecule that increases in concentration in the atmosphere as a result of air pollution.

P_{50} Po_2 at which a respiratory pigment is 50% saturated with oxygen; is a measure of the oxygen affinity of the pigment.

palaeotropical region Biogeographical region of the world including Africa (Ethiopian region) and India and South-East Asia (Oriental region).

Palaeozoic 'Ancient life'; geologic era 570 to 245 million years BP.

Pangaea The supercontinent uniting all northern (Laurasia) and southern (Gondwana) landmasses that formed, and then fragmented, during the Mesozoic.

parabronchi The smallest airways in the gas-exchange region of the bird lung in which flow is unidirectional; give rise to the air capillaries.

paracrine hormones Animal hormones that usually act over very short distances, travelling by diffusion through extracellular fluid.

parapatric speciation Divergence in populations, initially in geographic isolation and subsequently when the populations again come in contact; leads to formation of new species.

paraphysis (pl. paraphyses) Sterile hair (e.g. associated with reproductive structures in mosses).

parapodium (pl. parapodia) Lateral, paired appendage on the body segments of polychaetes; functions in gas exchange and locomotion.

parasite An organism that lives and feeds on or in another organism, the host, which is usually larger than the parasite.

parasitoid Insect that is free-living as an adult but parasitic as a larva.

parastism Symbiotic interaction between two species where one benefits and the other is harmed.

parathyroid gland Endocrine gland closely associated with the thyroid gland; secretes parathormone, which is involved in increasing blood calcium levels.

parenchyma Large living cells that form the ground tissue of plants, comprising large, thin-walled cells, large central vacuole, and a peripheral nucleus and cytoplasm.

parthenogenesis A mode of reproduction, found in some sexually reproducing organisms, in which the male plays no role.

partial pressure of a gas Total pressure in a mixture of gases is the sum of the partial pressures of each gas.

partitioning sequence DNA sequence in cells of prokaryotes that attaches to the plasma membrane in the region where the cell is to divide; similar function to the centromere of eukaryotic cells.

pathogen An organism capable of causing disease.

pedicellaria (pl. pedicellariae) Pincer-like structure on the body surface of some echinoderms.

pedipalps Second pair of appendages of the cephalothorax in arachnids; modified for various functions; corresponding to insect mandibles.

pelagic realm Marine biogeographic region including the surface water of open oceans and planktonic organisms.

penetrance The proportion of individuals in a population carrying a particular allele and in which the allele has a phenotypic effect.

penis Intromittent organ of males used for transferring sperm during internal fertilisation in many animals.

pepsinogen Inactive precursor (zymogen) of the protease, pepsin; released into the stomach.

peptide bond Bond formed when the acidic carboxyl group (−COOH) of an amino acid attaches to the amino group (−NH$_2$) of another, with the release of a molecule of water.

perfusion Convective movement of an internal fluid.

pericardial Of the fluid-filled space and membranes surrounding the heart.

pericycle Layer of cells in a root that surrounds the vascular cylinder; site of initiation of lateral roots.

periderm A protective outer tissue that replaces the epidermis in secondary stems and roots of woody plants; corky tissue.

peridinin A xanthophyll; an accessory photosynthetic pigment found in dinoflagellates.

peripheral resistance Resistance to blood flow due mainly to friction of the blood in small diameter arterioles.

peristaltic locomotion Movement of an animal by alternate constriction and widening of the body; depends on a fluid-filled body cavity, as in annelids.

peristome Specialised rows of teeth-like structures around the top of a moss sporangium; shelters spores.

peristomium First segment of an annelid.

permeability coefficient An estimate of the ease by which a molecule can pass through a cellular membrane; dependent especially on the lipid solubility of the molecule.

peroxisome Type of microbody in eukaryotic cells that contains numerous enzymes; involved in the production and degradation of peroxides and oxidition of amino acids and uric acid.

peudosclerotium (pl. pseudosclerotia) Sclerotium composed of both fungal mycelium and host tissue (*see* sclerotium).

pH The concentration of hydrogen ions (H$^+$) in solution; measured on a logarithmic scale ranging from 0 to 14.

phage *See* bacteriophage.

phagocyte White blood cell able to engulf micro-oganisms or damaged tissues.

pharyngeal slits Paired openings appearing in the pharynx of chordates at some stage of development.

phenocopy Phenotype that resembles a mutant phenotype, even though it is genetically wild type; due to severe environmental effects.

pheromone Hormone released into the external environment for chemical communication between individuals.

phloem Transport tissue of vascular plants comprising several cell types, including sucrose-transporting sieve cells, companion cells and sclerenchyma fibres.

phosphorylation Addition of a phosphate group to a compound; often results in the formation of a high energy bond (e.g. ATP from ADP).

photon Elementary particle of electromagnetic radiation (light).

photoperiodism Plant response (e.g. flowering) to the length of light and dark periods in a 24-hour cycle.

photoreceptor A type of receptor that detects light by absorbing it at a particular wavelength.

photorespiration A light-activated type of respiration found in the chloroplasts of plants in which Rubisco uses O$_2$ as a substrate to oxygenate RuBP and produce CO$_2$ as a product of oxygenation.

photosynthesis The process by which solar energy is harvested and used to convert CO$_2$ and water into carbohydrates; involves: absorption of energy from sunlight by means of pigments, reactivation of pigments, and carbon fixation to produce sucrose in 'dark reactions'.

phragmoplast Structure containing the remnants of spindle fibres orientated at right angles to the new cross wall forming during cell division in all plant cells and some related green algae (charophytes).

phycoplast System of microtubules orientated in the plane of cell division following the collapse of the spindle during cytokinesis; feature of many green algae.

phyllode Laterally compressed petiole and rachis; foliage that replaces true leaves in most Australian acacias.

phyllotaxy Geometric pattern of leaf arrangement on a stem of plants (e.g. spiral).

phylogeny Evolutionary relationships of organisms, usually depicted as a branching tree diagram (phylogenetic tree).

physical digestion Breakdown of food into small particles, such as by grinding or chewing.

phytochrome A plant pigment that absorbs light; exists in two interconvertible forms, P$_r$ (inactive form) and P$_{fr}$ (active form); involved in the timing of a number of processes (e.g. flowering, dormancy and seed germination).

pigment Molecule that appears coloured because it absorbs photons with particular energy levels and reflects others.

pinacoderm Outer surface of a sponge, consisting of a layer of flattened cells (pinacocytes).

pineal gland In vertebrates, an outgrowth in the midline of the roof of the third ventricle, which is used to measure photoperiod. In fishes, amphibians and some reptiles may contain photoreceptor cells (the 'third eye'). In mammals and birds, is a neurosecretory organ releasing melatonin at night.

pinna (pl. pinnae) Leaflet of a compound leaf.

pistil In a flower, a female reproductive organ, comprising a stigma, style and ovary.

pith Parenchyma cells that lie centrally within the vascular tissue of stems and some roots.

pituitary gland (hypophysis) Gland at the base of, and largely controlled by, the hypothalamus; composed of the median eminence, the anterior pituitary gland (adenohypophysis—an endocrine gland) and the posterior pituitary gland (neurohypophysis—neurosecretory gland).

placenta An organ that enables exchange between an embryo and the maternal circulation; present in marsupial and placental mammals.

plant Multicellular, photosynthetic eukaryote (kingdom Plantae) that lives on land.

plantigrade Whole foot touches the ground, as in primates.

planula A ciliated type of animal larva.

plasma cell Non-dividing, antibody-secreting B cell formed after binding of antigen to the specific receptor of a B cell in the presence of helper T cells; produces the humoral response.

plasma membrane The boundary of living cells separating a cell from its environment; formed from a phospholipid bilayer.

plasma The transparent, slightly yellowish fluid component of blood without the cellular components; obtained by centrifugation.

plasmid Small unattached circular DNA molecule occurring in bacterial cells, often carrying genes conferring antibiotic resistance.

plasmodesma (pl. plasmodesmata) Special channel through primary cell walls, connecting plasma membranes and cytosols of adjacent plant cells.

plasmodium A multinucleate mass of protoplasm surrounded by a membrane; stage in the life cycle of acellular slime moulds (myxomycetes).

plasmolysis Shrinkage of cytoplasm due to loss of water by osmosis, drawing the plasma membrane away from the wall.

plastid Organelle of plant and algal cells that is surrounded by a double membrane; generally larger in size than mitochondria and contain circular DNA molecules, RNA and small ribosomes; function in processes such as photosynthesis, starch storage and geotrophism.

plastron In turtles and tortoises; a ventral, protective shield covered by horny plates.

plate tectonics Modern geological theory that recognises that the earth's crust and part of the upper mantle (together the lithosphere) are divided into a number of plates that move relative to one another.

platelet Anucleate disc-shaped cell fragments derived from megakaryocytes in bone marrow; involved in several aspects of blood clotting.

pleural cavity Coelomic space surrounding the lungs in mammals; separated from the rest of the visceral coelom by the diaphragm.

pluripotent Stem cells that produce a whole range of cell types, for example, blood-forming stem cells.

pneumatophore Upright aerial root of mangroves; exposed at low tide and function in gas exchange.

pod A fruit (legume) that opens along two sides, as in beans and peas.

poikilohydric Capable of withstanding desiccation, as in resurrection plants.

polar nuclei The two nuclei present in the central cell of the embryo sac of flowering plants.

polarisation Imbalance of electrical charge across the plasma membrane due to uneven distribution of ions, with the inside of the cell usually negative with respect to the outside.

pollen A collective term for pollen grains; microgametophytes of seed plants, which develop sperm and can be transported by wind or animals.

pollination The process in which pollen of flowering plants is transferred by animals, air or water currents to the stigma for fertilisation.

pollution The addition of materials to air, soil or water that adversely affects the environment.

polyarch xylem Pattern of xylem development in roots of monocotyledons among flowering plants, in which the xylem does not fill the centre, but is divided into many ridge-like projections (archs).

polycistronic transcript Primary transcript from operons that contains more than one functional reading frame and codes for more than one polypeptide.

polygamy The acquisition of more than one mate during a lifetime or mating season; polygyny refers to more than one female per male; polyandry is the reverse.

polygene A set of genes that together control a quantitative character such as height or mass of an organism.

polymerase chain reaction (PCR) Selective and repeated replication of segments of DNA *in vitro*.

polyp An attached tubular form of a cnidarian with its mouth upwards.

polypeptide A molecule composed of a chain of amino acid residues.

polyphyletic Referring to a group or taxon of organisms that has had multiple origins (e.g. viruses).

polyploid An organism with more than two sets of chromosomes.

polysaccharide A carbohydrate composed of many monosaccharides joined in long linear or branched chains (polymers).

polysome A number of ribosomes attached to the same mRNA strand, each at different stages of protein synthesis; may be free in the cytosol or attached to endoplasmic reticulum.

pome A fleshy fruit from an ovary of an epigynous flower.

population A group of organisms of the same species living in a defined geographic area.

portal system System of blood vessels carrying blood between two capillary networks.

potential distribution Of a species; the range over which individuals could theoretically survive and reproduce.

potential energy Stored energy, such as chemical energy stored in the bonds of atoms and molecules.

Pre-Cambrian The oldest era in the geologic time scale; before 570 million years ago.

precapillary sphincter Band of smooth muscle at the entrance of a true capillary; opens or closes the vessel in response to local influences.

predation One kind of animal (a predator) eating another; an animal (herbivore) eating a plant.

predator–prey cycle Oscillations in population size where predator numbers follow those of the prey.

preferential channel Main capillary route for blood flow through a tissue when metabolic demands are minimal.

prehensile Able to grasp, as with a tail or fingers.

primary cell wall First wall of a plant cell, composed of cellulose, pectins and non-cellulosic polysaccharides.

primary response Initial immune response to an antigen, usually results in immunological memory, which causes a later immune response to the same antigen to be larger and more rapid.

primary transcript Messenger RNA product of transcription units of DNA.

primer Short sequence of bases that pairs with a complementary sequence in a strand of DNA and provides a free hydroxyl end for DNA polymerase to commence synthesis of a nucleotide chain.

primordial germ cell Embryonic germ cell prior to its proliferation (mitotic division) in the ovary.

Principle of independent assortment Alternative forms of a gene controlling one trait assort into gametes independently of alternative forms of another gene controlling a different trait.

Principle of segregation Alternative forms of an inherited trait, controlled by alternative forms of a gene (alleles), segregate into gametes after meiosis with equal probability.

prion Infectious agent that is virus-like but appears to lack nucleic acid and consists only of protein.

producer An organism that converts the energy of sunlight (by photosynthesis) or inorganic chemicals (chemosynthesis) to chemical energy in the form of carbohydrates and other molecules.

productivity Rate at which biomass accumulates; primary

productivity refers to productivity of producer organisms (e.g. plants).

progesterone Hormone, released by the corpus luteum, which prepares the reproductive tract for embryo implantation or stimulates secretion of coatings for eggs.

proglottid Segment-like body unit of a tapeworm; new segments are added at the anterior end.

prokaryotic cell Cell with a simple structure, lacking a nucleus and other internal membrane-bound organelles; bacterial cell.

prometaphase A phase of mitosis in which the nuclear envelope breaks down, allowing the mitotic spindle to interact with and move chromosomes.

promotor Specific sequence of DNA that binds to RNA polymerase, promoting initiation of transcription of the coding region by the enzyme.

pronucleus The male or female nucleus in the zygote, prior to fusion.

prophase The initial phase of mitosis in which dispersed chromatin in the cell condenses into chromosomes composed of paired chromatids, and the mitotic spindle is formed.

proplastid Precursor organelle of all types of plastids.

prosoma In arthropods, anterior part of the body, the head or cephalothorax.

prostomium Anterior, presegmental part of an annelid (segmented worm), which houses the brain.

protandry In animals, situation in which an individual starts life as a male changing to a female at some later stage. In plants, condition where a flower first opens in the male phase (anthers dehisce) and later becomes female (stigmas receptive).

proteoglycans O-linked glycoproteins containing many sugar chains; produced in the Golgi apparatus from proteins containing serine, threonine or hydroxyproline residues; mature to form surface slimes and mucus.

prothoracic gland Non-neural endocrine gland in the thorax of some insects that secretes ecdysone.

protist A eukaryote including unicellular, multicellular, photosynthetic and non-photosynthetic organisms. Member of the kingdom Protista, which is a diverse array of organisms not a single evolutionary group.

protogyny In animals, situation in which an individual starts life as a famale, changing to a male at some later stage. In plants, condition where a flower firsts opens in the female phase (stigma receptive) and later becomes male (anthers dehisce).

protonephridium (pl. protonephridia) Tubular excretory organ of animals that has a flame-cell to filter coelomic fluid into the tubule for excretion.

protoplasm The cytoplasm and nucleus of eukaryotic cells.

protostome Animal in which the blastopore becomes the mouth; primary mouth.

protoxylem The first formed and earliest maturing xylem tissue that forms towards the centre of the stem in plants.

pseudocoel Body cavity of an animal, such as a nematode, that is not lined on all sides by mesoderm.

pseudoparenchyma Mass of branched hyphal mycelium of macrofungi resembling tissue.

pseudoplasmodium Mass of amoebae of a cellular slime mould that aggregate to form a single mobile colony ('slug').

pseudopodia Transient extensions of a cell surface ('false feet' of amoebae).

psychophile Organism, such as a fungus, that can live in cold conditions.

pulmonary artery Main artery that carries deoxygenated blood from the heart to the lungs in higher vertebrates.

pulmonary Referring to the circulation of blood in the lungs.

pulvinus A motility organ at the base of the leaf petiole in certain plants, such as legumes, that controls the position of the leaf.

pupa Developmental stage in some insect life cycles between the larva and adult; non-feeding, immobile and sometimes encapsulated or in a cocoon.

pure-breeding Term used to describe a group of interbreeding organisms in which the same phenotype is observed across many generations; the genotype determining the phenotype is homozygous.

pygidium Posterior, postsegmental part of an annelid.

pyrimidine Single ring molecule; forms three of the bases of nucleic acids—cytosine, thymine and uracil.

pyruvate ($C_3H_3O_3^-$) End product of glycolysis.

qualitative character Inherited character that is distinctively different from others and can be used to group organisms into distinct phenotypic classes.

quantitative character Character of an organism, such as height or weight, that shows continuous variation and is controlled by both environment and polygenes.

radial symmetry Symmetry of an organism such that any plane passing through the central axis bisects the organism into equal halves (e.g. jellyfish).

radicle Root axis of a germinating seed.

radula Tongue-like structure, with rows of rasping teeth; present in the floor of the foregut of molluscs, except bivalves.

ram ventilation Ventilation of gills that results from the forward movement of the body.

raphe Longitudinal slit in the valve of a pennate diatom.

ray initial A type of meristematic cell produced by the vascular cambium differentiating into wood rays.

reabsorption Physiological process in animals whereby solutes are actively transported from the urine or the gut contents back into body fluids.

reaction centre Specialised chlorophyll complex functioning as a photosynthetic unit capable of channelling energised electrons to an acceptor molecule; located on the thylakoid membranes of chloroplasts.

reading frame Pattern by which nucleotide sequences of DNA are decoded as a sequence of triplets (codons) from the 5' end to the 3' end of messenger RNA.

realised distribution Of a species; the range over which individuals live and reproduce.

receptor A molecule, usually a protein, in the plasma membrane or within a cell, that undergoes a change as a result of a specific interaction with a signal, leading to a particular response.

recombinant Individual or cell that carries a new genotype, not present in either parent, typically generated by genetic recombination during meiosis. Also said of genes that have been isolated, cloned and expressed in bacteria by recombinant DNA technology.

recombinant DNA Technology involving cutting DNA into fragments using restriction enzymes, isolating the fragments, identifying genes and multiplying them in bacteria, where the gene products can be expressed.

redia (pl. rediae) Larval stage in the life cycle of parasitic flukes that develops within sporocysts and that forms cercariae.

reduction Reactions involving the breakdown of molecules by addition of electrons, especially those in which oxygen is lost and hydrogen accepted.

regeneration A form of asexual reproduction involving the production and differentiation of new tissues of an organism.

regulated secretion Secretion that only occurs in response to a specific signal.

regulative development Development in which the fate of many blastomeres becomes determined as a result of interactions with surrounding cells.

repetitive DNA Small sequences of DNA in the genome of eukaryotes that are expressed many times and have no known function.

replication Process of DNA duplication; involves unwinding of the double strand of DNA and synthesis of new complementary strands using the parent strands as templates. Replication is semiconservative, meaning that each of the parental strands pairs with a newly synthesised strand.

replicon Unit region of replication, between origins, in chromosomes of eukaryotes.

reproductive success The number of surviving offspring produced by an individual.

resource partitioning Where species exploit a limited resource in different ways, reducing direct competition and allowing species coexistence.

respiratory pigments Metalloproteins (hence the colour) that bind oxygen reversibly to the metal; primarily responsible for increasing the amount of oxygen carried in blood.

restriction fragment length polymorphisms (RFLPs) Variation in length of fragments of DNA generated by presence or absence of restriction sites; used to distinguish different individuals; detected by separating fragments by electrophoresis.

restriction (in immune system) Inability of T cell receptors to recognise antigen unless presented by MHC molecules.

restriction enzyme Site-specific nuclease enzyme (endodeoxyribonuclease) that recognises short sequences of unmethylated DNA and cleaves the double strands either at specific sites (the recognition sequence) or nearby.

restriction mapping Technique for mapping (or physically locating) sites at which restriction enzymes cut cloned DNA.

rete Network of interdigitated arteries and veins that functions for efficient countercurrent exchange.

rhabdite Rod-like structure secreted by epidermal gland cells in free-living flatworms.

rhizoid Short root-like structure that anchors bryophytes and some fungi to their substrate.

rhizome Underground stem.

rhizosphere Soil zone immediately surrounding the root hairs in plant roots where interactions occur between plant, soil and micro-organisms.

rhodopsin An important photoreceptor of animals, found in rod cells of the retina of the vertebrate eye, associated with the chromophore retinal, which undergoes a chemical rearrangement after absorbing light.

rhopalia Sense organ of a jellyfish that contains a statolith or sometimes an eyespot.

rhyncocoel Coelomic body cavity that houses the proboscis of a proboscis worm (nemertine).

ribonucleic acid (RNA) Single-stranded nucleic acid characterised by a ribose sugar in each nucleotide and the bases adenine, cytosine, thyamine and uracil; see also mRNA, rRNA and tRNA; involved in transcribing and translating coded information of DNA during the production of proteins.

ribosomal RNA (rRNA) RNA that forms the major part of ribosomes.

ribosome Cytoplasmic organelle where protein synthesis occurs; formed from two rRNA subunits in association with an mRNA molecule.

ribulose bisphosphate (RuBP) The 5-carbon sugar that binds to CO_2 in the first step of carbon fixation in photosynthesis.

ribulose bisphosphate carboxylase-oxygenase (Rubisco) Enzyme that catalyses the first step of carbon fixation (see RuBP); constitutes 50% of protein in chloroplasts.

RNA polymerase Enzyme that catalyses transcription of a DNA template to form mRNA.

root cap Parenchyma tissue covering a root tip; secretes a mucigel to aid the root penetration through the soil; usually contain numerous starch granules that may act as sensors of gravity.

root cluster (proteoid roots) Group of rootlets that form dense mats at the soil surface in family Proteaceae and certain legumes.

root nodule Outgrowth of roots in a wide range of plants, especially legumes, in which symbiotic nitrogen-fixing bacteria (*Rhizobium* or *Frankia*) occur.

root pressure Positive pressure generated by roots that is responsible for the exudation of sap from tapped or cut stems in spring.

rough endoplasmic reticulum (RER) Endoplasmic reticulum with attached ribosomes; involved in synthesis of proteins usually destined for export from a cell.

ruminant Foregut fermenter that ruminates, that is, regurgitates food from the first part of the stomach and rechews it to reduce further the particle size.

S phase Phase of the cell cycle in which DNA replication occurs.

salt gland Gland that can secrete a salt solution that is more concentrated than body fluids (e.g. reptiles, birds and mangroves).

saltatory conduction Conduction of action potentials along myelinated axons involving action potentials skipping from node to node.

saltatory evolution Evolutionary change that occurs suddenly followed by long periods of time in which there is little change.

saprophyte An organism that lives on dead organic matter.

sapwood The outer region of wood (secondary xylem) in a tree trunk containing living rays; of lesser strength than heartwood.

sarcolemma Electrically excitable membrane that surrounds muscle cells.

sarcomere Region of myofibril between adjacent Z-lines.

satellite virus A type of virus only able to replicate in cells infected with a specific helper virus.

satellite DNA Small sequence of repetitive DNA in the genome of eukaryotes that is associated with the centromere.

sclerenchyma Plant tissue that has a support role; cells with thickened lignified secondary walls that impart rigidity as well as strength; includes sclereids, branched or more-or-less even-shaped stone cells that form the hard tissue of fruits and seed coats and fibres; elongate cells.

sclerophyll Plant characterised by rigid, often small leaves, and short internodes; able to survive low soil nutrients, water stress and fire.

sclerotium (pl. sclerotia) Hard, resistant resting body, composed of masses of tightly compacted mycelium, formed by some soil-inhabiting fungi.

scolex Anterior attachment organ, with suckers and often hooks, of a tapeworm.

scutellum The single cotyledon of monocotyledonous flowering plants; forms an interface tissue between the embryo and endosperm.

Second law of thermodynamics *The entropy of the universe* *is increasing*; thus an input of energy is needed to maintain the ordered state of the universe.

second messengers Activation of first messengers, G-protein-linked receptors by extracellular signals leads to changes in the concentration of one or more small intracellular signalling molecules, second messengers, which then produce a cellular response; examples of second messengers are calcium ions and cyclic AMP.

secondary response Larger, more rapid immune response to a particular antigen as a result of immunological memory due to an earlier primary immune response.

secondary walls During development, plants cells strengthen their walls and make them more rigid by differentiating secondary walls, whose main component is lignin.

seed Structure from which a new plant develops; produced from a fertilised ovule, containing an embryo, a food source (cotyledons or endosperm) and usually a hard outer seed coat (testa).

selective permeability The ability of membranes to allow passage of some molecules and not others.

self-incompatibility Genetically controlled process preventing self-fertilisation; in flowering plants recognition of self-pollen leads to a rejection response usually in the style, while compatible pollen from another individual of the same species is accepted.

semiconservative replication *See* replication.

septum (pl. septa) Wall that divides, or partially divides a structure or cavity; cross-wall in fungal hypha; septum dividing the ventricle of the heart.

serum The yellowish fluid isolated from clotted blood; plasma without some of the constituents bound to the clot.

sessile Fixed in one position, immobile.

seta (pl. setae) Chitinous bristle.

sex chromosomes Chromosomes that differ in appearance or number between males and females and carry genes involved in sex determination.

sex-linked genes Genes that are located on sex chromosomes.

sexual reproduction Formation of offspring by the fusion of haploid gametes from two different organisms.

sexual selection The differential ability of individuals to acquire mates, involves contests between males or choice by females; leads to the selection of morphological or behavioural traits relating to attracting mates.

sieve cells Phloem cells that transport sucrose; long and tubular in shape, bounded by a plasma membrane and containing mitochondria and ribosomes, with perforated sieve areas on their end walls.

signal A stimulus, chemical (e.g. food or hormones) or physical (e.g. light or heat), that can be detected by cells leading to a particular response.

simple leaf Leaf with a single lamina.

sinoatrial node The heart's 'pacemaker'. A small group of non-contractile muscle cells in the right atrium of higher vertebrates that initiates the cardiac cycle; evolutionarily derived from the sinus venosus of fishes.

sinus A fluid- or gas-filled space within the body.

sinus gland In crustacea, highly vascular reservoir for neurosecretions of the x-organ (eye stalk hormones).

sinus venosus The first chamber of the fish heart that collects blood from the major veins and leads to the atrium.

siphon Specialised funnel in cephalopods that can produce a jet of water for propulsion.

siphuncle Structure that regulates the amount of gas and buoyancy in some cephalopods.

skeletal muscle Cylindrical fibres of multinucleate striated muscle, with nuclei located peripherally. Sometimes referred to as voluntary muscle, reflecting its voluntary control.

smooth endoplasmic reticulum (SER) Endoplasmic reticulum that lacks attached ribosomes.

smooth muscle Spindle shaped contractile cells with a central nucleus and less regular arrangement of myofilaments than striated muscle cells; lines the walls of internal organs and arteries and veins and is under involuntary control.

solutes Ions or organic molecules dissolved in a liquid.

soma The cell body of a neuron.

soma Middle, segmented part of an annelid.

somatic cells All cells in the body of a multicellular organism other than germ cells.

somites Blocks of mesoderm that form adjacent to the notochord in the vertebrate embryo.

soredium (pl. soredia) Structure of a lichen analagous to a spore; consisting of an algal cell embedded in fungal hyphae.

sorocarp Fruiting body produced during the life cycle of slime moulds.

sorus (pl. sori) Cluster of sporangia on the margins or undersurface of a fern frond.

specialisation Occurs during differentiation of cells when certain functions develop at the expense of others.

speciation Formation of new species (*see* allopatric, sympatric and parapatric speciation).

species As a taxonomic category, the species is the lowest rank in a taxonomic classification (*see also* binomial system). There is no single definition that is agreed upon by evolutionary biologists. The biological species concept is a group of actually or potentially interbreeding natural populations that are reproductively isolated from other such groups.

species diversity Variability (species richness and abundance) of biota in an area; *alpha diversity*, within a local area; *beta diversity*, diversity among different communities in different areas.

species richness Number of species within a community.

specific heat The amount of heat required to raise the temperature of a substance by 1°C.

sperm The male gamete.

spermatheca (pl. spermathecae) A small sac containing sperm.

spermatocyte Male germ cell in the process of meiosis within the testis.

spermatogenesis Process by which sperm form from primordial germ cells.

spermatogonium Diploid male germ cell in the process of mitosis in the testis.

spermatophore A structure enclosing many sperm (*see also* spermatheca).

spicule Skeletal component in sponges, composed of calcium carbonate or silica.

spindle pole body Dense region of microtubules attached to the nuclear membrane at each spindle pole during mitosis in fungi; analogous to the centriole of other eukaryotes.

spinneret Spinning organ that produces silk in spiders.

spiracle Small external opening of the air-filled gas-exchange system (tracheae) of spiders and insects.

spongin Course collagenous proteinaceous material forming skeletal fibres in sponges.

spongocoel Internal cavity of a sponge; also called atrium.

sporangiospore Haploid asexual spore that develops in a sporangium.

sporangium A sac-like cell or multicellular structure in which asexual spores form.

spore A cell capable of producing a new individual; often a dormant resistant structure or functioning in dispersal.

sporocyst Sac-like structure in the life cycle of parasitic flukes that produces rediae.

sporophyte Diploid stage of a plant life cycle that produces spores.

sporopollenin A polymer, tougher than lignin but with similar properties, composed chiefly of carotenoids; makes spores and pollen grains of plants resistant to biodegradation.

stamen In a flower, the male reproductive organ, comprising a bi-lobed anther on an elongated filament.

starch Insoluble polymer of glucose; composed of amylose (long chains of glucose units) and amylopectin (short branched chains of glucose); chief storage polysaccharide of green algae and plants; formed in chloroplasts and amyloplasts.

statolith Small calcareous body found in sensory organs for balance.

stele The central vascular cylinder of stems and roots.

stem Main part of the aerial shoot of plants, usually bears leaves, lateral branches and reproductive organs.

stem cells Relatively undifferentiated cells that can divide continuously throughout the lifetime of an animal; daughter cells can either remain as stem cells or differentiate fully.

stenohaline Tolerant of a narrow salinity range only.

stigma The terminal cells of the pistil of a flower that receive and recognise pollen grains during interactions that may lead to fertilisation.

stolon A horizontally growing stem or runner.

stoma (pl. stomata) Specialised pore in epidermis of leaves and stems that allows uptake of CO_2 from the atmosphere for photosynthesis; pore through which transpiration occurs.

stomium Zone of thin-walled cells where sporangia or anthers rupture to release spores or pollen grains.

stomochord Short structure beneath the dorsal nerve cord at the junction of the proboscis and trunk of a hemichordate, forming a dorsal diverticulum of the pharynx.

storage parenchyma Tissue of parenchyma cells containing storage reserves such as starch granules, lipid droplets or protein storage organelles.

striated muscle Muscle with a highly organised array of actin and myosin filaments giving the appearance of cross-striations when viewed under the light microscope; includes skeletal and cardiac muscle.

strobilisation Process of adding new proglottids in the growth of a tapeworm; occurs at the anterior end.

strobilus (pl. strobili) In plants, a cone, which is a collection of sporangia; body of a tapeworm.

stroma (of chloroplast) Matrix enclosed within inner membrane.

stroma (fungal) Mat of fungal hyphae bearing spores.

stromatolite Concentrically layered rock, the layers being formed by successive growth of thin mats of cyanobacteria; fossil and present-day.

style The pathway for pollen tubes between stigma and ovary in the pistil of flowering plants; may comprise solid transmitting tissue or a canal.

subduction Descent of sea floor back into the earth's mantle (at deep-sea trenches).

succession Process of replacement over time of one ecological community by another.

succulent Xerophytic plant with fleshy leaves or stems and highly mucilaginous cell sap.

surface tension Surface tension at an air–liquid interface is the result of intramolecular attractive forces in the liquid, providing the potential energy that draws molecules from the surface and therefore shrinks the interfacial area.

surfactant A substance predominantly composed of phospholipid that forms a film on air–liquid interfaces and decreases surface tension.

suspensor A filament of cells below an embryo of flowering plants that connects it to the ovule.

swim bladder A gas-filled sac that forms as an outgrowth of the pharynx, allowing fishes to regulate buoyancy; homologous with lungs of land vertebrates.

symbiosis Interactions in which two organisms (symbionts) live together in a close relationship that is beneficial to at least one of them.

sympatric speciation Populations specialising on different resources diverge and form new species without geographic isolation.

sympatry Living in the same area.

symplastic pathway Pathway in plants for uptake of water, solutes and ions from cell to cell via the cytosol; only pathway for crossing the endodermis of roots.

synapse Small area of close contact between an axon terminal and a post-synaptic cell across which information is transmitted, usually by chemical neurotransmitters. May be excitatory or inhibitory.

synapsis Pairing of homologous chromosomes during prophase of meiosis I.

synaptonemal complex The molecular scaffold on which crossing over occurs between paired chromosomes at prophase of meiosis.

synergids In flowering plants, pair of cells adjacent to the egg at the micropylar end of an embryo sac, one of which acts to receive the pollen tube.

synonymous codons Different triplets of messenger RNA (codons) that encode the same amino acid.

synovial fluid Slippery fluid found inside joints, produced by the synovial membrane; lubricates joint surfaces and nourishes articular cartilage.

systemic Referring to the circulation of blood to the body organs except the lungs.

systems In animals, several organs may combine to form systems, which carry out entire processes such as digestion or gas exchange.

systole The phase of the cardiac cycle involving muscle contraction and ejection of blood from a heart chamber.

T cell (T lymphocyte) Lymphocyte that matures in the thymus and recognises antigen by means of the T-cell receptor; functions independently to kill micro-organisms and controls B-cell responses.

T-cell receptor (TCR) The glycoprotein molecule on the surface of T cells that recognises an antigen.

tachycardia Increase in heart rate.

tagmatisation Organisation in arthropods of body segments into functional groups; tagmata (sing. tagma).

tapetum The inner layer of the anther wall of a flower, comprising cells dedicated to nutrition of the developing microspores.

target cells Cells that respond specifically to a particular hormone by means of specialised receptor molecules located on the surface or inside the cell.

taxonomy Methods and principles of classification of organisms.

tegument Outer, resistant body coat of parasitic animals such as flukes and tapeworms.

telomere DNA sequences occurring at the ends of chromosomes of eukaryotes.

telomorphic Refers to the sexual form of a fungus.

telophase The final phase of mitosis, in which new nuclear envelopes form, surrounding each group of chromosomes.

telson Posterior tagma (tail-end) of arthropods, which bears the anus.

terminus   The sequence of DNA that terminates replication.

territory   An area defended and occupied exclusively by one (or sometimes more) individuals.

test-cross   In plant breeding, a cross between F_1 (heterozygous) individuals and the parent with the recessive phenotype.

testa   The hard outer coat of a seed, formed from the integuments of an ovule.

tetrapod   Vertebrate with four limbs with separate fingers and toes.

thallus   Body of an alga, fungus or plant that lacks special tissue systems or organs.

thermophile   Organism that grows best in hot conditions, between 30° and 50°C.

thermoreceptor   A type of receptor that detects heat or cold.

threshold potential   Potential difference across a membrane at which certain voltage-dependent channels (usually sodium, occasionally calcium) increase their permeability.

thrombin   An enzyme that converts soluble fibrinogen into insoluble filaments of fibrin during the formation of a blood clot.

thylakoid   Flattened disc-like sac that forms part of the internal membrane system of chloroplasts; site of location of photosynthetic pigments.

thyroid hormones   Thyroxine and triiodothyonine; lipid soluble hormones that bind to intracellular receptors where they regulate the production of enzymes important in normal growth and maturation of cells. They stimulate metabolic rate of cells, and growth and metamorphosis in some immature vertebrates.

thyrotoxicosis   A condition of elevated metabolic rate caused by excessive thyroid hormone activity.

tidal ventilation   Movement of air into and out of an invaginated gas-exchange surface through the same opening; results in some rebreathing.

tissues   Groups of similar, differentiated cells and associated extracellular matrix, which together carry out a particular function.

tolerance   Failure of the immune systems to respond to a potentially immunogenic antigen. Occurs early in development, when lymphocytes become tolerant to self-antigens.

torpor   Physiological state in animals in which metabolism decreases and body temperature is maintained at a lower level than normal.

totipotent   The ability of plant cells to regenerate a new plantlet from a single cell.

trachea (pl. tracheae)   Tiny spirally ringed tube that carries air internally to cells in insects and some other arthropods; in mammals, initial portion of the airway before it branches dichotomously into the bronchi leading to the lungs.

tracheid   Elongated thick-walled cell of xylem of vascular plants; conducts water.

tracheole   Finest tube arising from an arthropod trachea, where gas exchange occurs directly with body tissues.

transcription   The process involving the copying of information stored in the sequence of one strand of DNA, the template, to produce a precise complementary strand of mRNA.

transcription unit   The transcribed region of DNA in a gene between promotor and termination sequences.

transfer cell   Type of parenchyma cell of plants characterised by primary wall ingrowths, resulting in massive increase in surface area of plasma membrane, allowing rapid transfer of molecules to adjacent cells, especially the vascular system.

transfer RNA (tRNA)   RNA that carries specific amino acids to ribosomes to add to growing polypeptide chains according to the base sequence in mRNA.

transformation   Process in bacteria of taking up free DNA from the cell's environment; may lead to genetic variation in bacteria. Process made use of in biotechnology.

translation   The process in which the information stored in the sequence of mRNA specifies the amino acid sequence of a polypeptide.

translocation   Transport of assimilates (sugar) in the phloem of vascular plants, from the site of production in leaves (source) to other parts of the plant (sink).

transpiration   Loss of water from a plant by evaporation through stomata in leaves; requiring energy from incoming solar radiation to vaporise water.

transport proteins   Enzymes that catalyse the movement of specific solutes across cellular membranes.

transposable element   Sequence of DNA that can move either directly to a new location within chromosomes or can replicate during transposition inserting the new copy at another site, resulting in mutation.

triglyceride   A fat molecule, having a backbone of glycerol with three carbon atoms, to each of which is attached a fatty acid chain.

trophic level   Position of an organism in a food chain (e.g. primary producer or first-order consumer).

trophoblast   Outer layer of epithelial cells in the mammalian blastocyst, gives rise to the chorion and forms the barrier between fetal and maternal tissues.

true capillary   Capillary vessels across which exchange occurs; open in response to high metabolic demands.

trypsinogen   Inactive precursor (zymogen) of the protease, trypsin, released into the duodenum.

tube foot   Short, tubular, external projection of the body wall of echinoderms containing an extension of the radial canals of the water vascular system; functioning in gas exchange, attachment, locomotion and catching prey.

tubulin   Protein that forms the major cytoskeletal scaffolding elements, microtubules, composed of equal amounts of two forms: α- and β-tubulin.

tunic   Supportive and protective 'coat' of tunicates secreted by ectoderm.

tunica   In the shoot apex of flowering plants, the outer one to three layers of the apical dome of cells that contributes to leaf and flower formation.

turgor pressure   Hydrostatic pressure within a cell that has a cell wall.

ultimobranchial bodies   Endocrine glands that secrete calcitonin, which is involved in decreasing blood calcium levels.

unipotent   Stem cells that produce only one type of cell, for example, skin stem cells.

urea (CON_2H_4)   A common nitrogen waste product of terrestrial animals that is formed from ammonia by the urea cycle.

urediniospore   Asexual spore of a rust fungus.

ureotelic   Pattern of excretion of nitrogen wastes in the form of urea.

uric acid ($C_5H_4O_3N_4$)   Nitrogen waste product of insects, reptiles and birds; formed by purine metabolism and from ammonia by uricogenesis.

uricotelic   Pattern of excretion of nitrogen wastes in the form of uric acid.

urine   Fluid formed and modified by the excretory system to excrete ions, organic solutes or water.

vacuole   Large membrane-bound organelle, prominent in plant cells.

vaporisation   The change from a liquid to a gaseous state, such as water to steam.

vascular bed The blood vessels in a region of the body, usually an organ, that are supplied from a common artery. Blood flow rates to individual vascular beds are usually independent of one another.

vascular bundle Prominent structural feature of primary growth of shoots, roots and leaves, comprising the transport system; consists of xylem and phloem.

vascular cambium Meristem responsible for producing wood (secondary xylem and secondary phloem).

vascular tone Tension in the walls of blood vessels which partly controls blood pressure and vascular resistance. Venous tone is important in controlling the rate of blood flow into the heart and arteriolar tone is largely responsible for determining the pattern of blood flow to vascular beds.

vasomotor centre Centre in the brain that controls the pressure and distribution of blood by affecting the contraction of smooth muscle in arterioles.

vasopressin Posterior pituitary hormone that influences blood pressure and water balance.

vector An agent, such as an insect, able to transfer a pathogen from one organism to another.

vegetal pole The pole of the egg containing high concentrations of yolk.

vegetative cell The largest cell of a pollen grain that contains the generative cell, and in some cases the sperm cells, and produces the pollen tube at pollen germination.

vein In animals, large blood vessel channelling blood towards the heart. In vascular plants, network of vascular bundles in the leaf lamina.

velamen Thick, multilayered epidermis of aerial roots of epiphytic orchids; covers all but the absorptive tip of the root.

veliger Second stage in the life cycle of molluscs; develops from the larva and forms a foot, mantle and shell.

venous Of veins. Usually referring to blood flowing in veins, or deoxygenated blood flowing in arteries or veins.

ventilation Movement of the external medium for purposes of gas exchange.

ventricle A heart chamber with strong muscular walls that develops most of the force necessary to pump blood through the circulatory system.

venule Small blood vessels that collect blood from the capillaries.

vernalisation Induction of flowering in certain plants by exposure to low temperatures.

vertebra (pl. vertebrae) A segment of backbone.

vessel element A cell type of the xylem of flowering plants; typically long and tubular in shape, lignified in reticulate, spiral or other patterns; end walls are broken down to allow unimpeded transport of water.

virion A complete mature virus particle, which is metabolically inert and is the transmission (infective) phase.

viroid Infectious agent that is virus-like but lacks a protein coat.

virus Subcellular genetic parasite that reproduces only in the cells of a susceptible host and may cause disease.

viscosity Physical property of a fluid that determines its tendency to flow.

vitamin Organic compound required by animals in small amounts for normal growth and maintenance.

vitelline membrane A tough, clear, elastic envelope found in many eggs and embryos, lying adjacent to the cell membrane.

vitellogenesis Incorporation of lipids, proteins and carbohydrates into developing oocytes.

vitellogenic hormones Hormones that stimulate vitellogenesis in egg-laying animals.

water vascular system System of coelomic canals, lined with ciliated cells and filled with fluid; includes a circular water canal and radial canals leading to tube feet; unique to echinoderms (*see also* tube foot).

weed Plant growing where it is not wanted; often an introduced species.

whole blood Blood containing both plasma and blood cells, as drawn directly from a blood vessel.

x-organ In crustacea, neurosecretory cells located in the largest eyestalk ganglion.

xenospore Asexual fungal spore for dispersal and spread of the fungus (*see* conidiospore).

xerophyte Plant tolerant of dry conditions.

xylem A major tissue of vascular plants comprising several cell types, including water-conducting vessels and tracheids, sclerenchyma fibres for support and parenchyma.

xylem sap The liquid contained as a continuous column in the tracheids and vessels of xylem; a dilute solution of inorganic ions and some organic nitrogen compounds.

y-organ In crustacea, non-neural endocrine glands located in the head, usually at the base of the first antennae, which secretes a moult-stimulating hormone.

zona pellucida The vitelline membrane of the mammalian egg and embryo.

zooxanthella (pl. zooxanthellae) Dinoflagellate endo-symbiont found in the tissues of corals, sea anemones and molluscs.

zygomorphic Describes a flower with parts arranged asymmetrically.

zygospore Dormant spore with thick cell wall enclosing a zygote.

zygote Diploid cell resulting from the fusion of the male and female gametes.

zymogen An inactive precursor of a digestive enzyme released into the gut by cells to prevent damage to the secretory cells. Zymogens are activated by conditions in the gut.

A C K N O W L E D G M E N T S

Every care has been taken to trace and acknowledge copyright. The publisher apologises for any accidental infringement where copyright has proved untraceable and would be pleased to come to a suitable arrangement with the rightful owner in each case.

Part 1 Cell biology and energetics

Chapter 1
1.1a: Brett Gregory, Auscape
1.1b: Kev Deacon, Auscape
1.1c: Jean-Paul Ferrero, Auscape
1.10a: C. A. Henley, Auscape
1.10b: Colin Monteath, Auscape
Box 1.1a: Jean-Paul Ferrero, Auscape
Box 1.1b: Kathie Atkinson, Auscape
Box 1.1c: Andrew N. Drinnan
1.18d: Norman Nicholls
1.18e: G. R. Roberts, Documentary Photographs
1.18f: The Photo Library—Sydney/Science Photo Library
1.19: Australian Cotton Foundation
1.23: Reg Morrison, Auscape
1.31: Ferrero/Labat, Auscape

Chapter 2
2.2a: N. N. Birks, Auscape
2.2b: J. M. Labat, Auscape
2.10: Jean-Paul Ferrero, Auscape

Chapter 3
3.1a, 3.6c, 3.9a, 3.13b: A. Hardham
3.1b: P. A. Janssens
3.1c: L. Marotte
3.5a: Electron Microscope Unit, The University of Sydney
3.9b, 3.12c, d: M. Fitzgerald
3.10a: The Photo Library—Sydney/Science Photo Library
3.12a, b, 3.15a, 3.16b: B. Gunning & M. Steer, *Plant Cell Biology: An Ultrastructural Approach*, Edward Arnold Publishers, 1975
3.15b: D. Barkla
3.18a: Nancy Kedersha/The Photo Library—Sydney
3.18d: R. E. Williamson & G. O. Wasteneys, *Fortschritte der Zoologie*, 34:17, Fig. 2A, Stuttgart, 1987
3.19a: Francis Leroy/The Photo Library—Sydney
3.19b: Dr Gerald Schatten/The Photo Library—Sydney
3.20a: Dr Peter Dawson/The Photo Library—Sydney
3.21a: K. A. Pickerd, EM Unit, CSIRO Division of Entomology
3.21b: M. Abbey/The Photo Library—Sydney
3.22b: Chris Bjornberg/The Photo Library—Sydney

Chapter 4
4.11: G. R. Roberts, Documentary Photographs
4.13: S. Tyerman
4.19b: R. Marginson

Chapter 5
5.19: Queensland Sugar Corporation
5.23b: K. Winter & C. B. Osmond

Chapter 6
6.1: Ian Lovegreen/*Gardens & Backyards Magazine*
6.2: J. Chamley-Campbell
6.3a: Alberts et al, *Molecular Biology of the Cell*, Garland Press, New York, 1989, p. 819

6.5: Joan Clarke
6.10, 6.11, 6.12: Anatomy Department, The University of Melbourne
6.13: H. M. Young
6.14a: A. Hardham
6.14b: M. C. McCann, B. Wells, K. Roberts, *Journal of Cell Science* 96, 1990, The Company of Biologists Ltd
6.16: Robyn Overall
6.17, 6.22d: Ripon Microslides, USA
6.18b: J. Carpenter
6.20: John Considine, Department of Agriculture, University of Western Australia

Chapter 7
7.1: Ron Oldfield & Keith Williams, Biological Sciences, Macquarie University
7.9: P. Roach/NPIAW
7.10: Bill & Peter Boyle, Auscape
7.11: C. Marcroft

Chapter 8
8.6: Jeremy Pickett-Heaps

Part 2 Genetics and molecular biology

Chapter 9
9.1a: Archiv für Kunst und Geschichte, Berlin
9.1b: David Smyth
9.5, Box 9.2: Stirling Macoboy
9.8b: Reproduced from *Ishihara's Tests for Colour Blindness* published by Kanehara & Co., Ltd, Tokyo
9.10b: © Owen Newman, Oxford Scientific Films
9.14: Jean-Paul Ferrero, Auscape
9.16: Murdoch Institute
9.19a: G. D. Anderson, ANT Photo Library
9.19b: L. H. Schmitt, *Evolution* (1978) 32, 1–14, Fig. 2
9.20a: Breck P. Kent
Box 9.4: Mourant, Kopec & Domaniewska-Sobczak, *The Distribution of the Human Blood Groups*, 2nd edn, OUP, London, 1976, Fig. 16

Chapter 10
10.1: Margaret Lawrence, CSIRO Division of Horticulture
10.7: P. Sloof, A. Maagdellin & A. Boswinkel, *J. Mol. Biol.*, 1983, 163, p. 278
10.13: R. G. Wake, *J. Mol. Biol.*, 1973, 77, p. 571
10.15: H. J. Kriegstein & D. S. Hogness, *Proc. Nat. Acad. Sci.*, 71(1), Jan 1974
10.17: © Scientific American Inc., George V. Kelvin
Box 10.4: L. Chasin & G. Urlaub
10.18: D. Riley & H. Weintraub, *Proc. Nat. Acad. Sci.*, 76(1), Jan 1979

Chapter 11
11.9a: Jeremy Pickett-Heaps

Chapter 12
Box 12.1: Reprinted courtesy *The Boston Globe*
12.6b, c: J. D. Watson & J. Tooze, *The DNA Story*, 1981, © 1981 by James D. Watson & John Tooze, reprinted with permission of W. H. Freeman & Co.
12.21a: L. Stryer, *Molecular Design of Life*, 1989, © 1989 by Lubert Stryer, reprinted with permission of W. H. Freeman & Co.
12.21b: R. L. Brinster, University of Pennsylvania
Box 12.6: Robin Lovell-Badge, *Nature*, 351(6322), p. 119, © 1991, Macmillan Magazines Ltd

Part 3 Reproduction and development

Chapter 13
13.2, 13.21b: Otto Ridge, ANT Photo Library
13.3a: G. R. Roberts, Documentary Photographs
13.4, 13.14: L. Newman & A. Flowers, Auscape
13.5a: John Cooke, Oxford Scientific Films
13.6: NHPA, ANT Photo Library
13.9b: Jeremy Pickett-Heaps
13.9c: N. Yanagishima in G. P. Chapman et al, *Eukaryote Cell Recognition: Concepts and Model Systems*, Cambridge University Press, Cambridge, 1988, p. 52, Fig. 4.3
13.10a, d: M. Renfree
13.10b: J. Anthony, *Proceedings of The Royal Society of London* 1980, p. 208, 349–67, plate 2
13.12, 13.19a, 13.22: D. Parer & E. Parer-Cook, Auscape
13.16: R. J. Shiel
13.17a: G. I. Bernard, Oxford Scientific Films
13.17b: C. A. Henley, Auscape
Box 13.2a: Jamie Oliver/GBRMPA
Box 13.2b, c, d: Dr Peter Harrison
Box 13.2e: R. Babcock
13.18a: D. Rentz, Australian National Insect Collection, CSIRO Division of Entomology
13.18b: G. I. Bernard, Oxford Scientific Films
13.19b: Kathie Atkinson, Auscape
13.20: Jan Aldenhoven, Australia, Auscape
13.21a: J. M. Labat, Auscape
Box 13.4b: Geoff Shaw
13.23, 13.27: Hans & Judy Beste, Auscape
13.24: John Cancalosi, Auscape
13.25: C. Andrew Henley, Auscape
13.26: M. J. Tyler, ANT Photo Library
13.31: Arancha Arbeloa & Maria Herrero, Diputacion General de Aragon, Zaragoza, Spain
13.32a: Steve Read

Chapter 14
14.3: Yuh-Nung Jan, University of California, San Francisco
14.6b: Barrie Jamieson
Box 14.1: Lynne Selwood
Box 14.2: D. Newgreen
14.21: S. J. Smith, 'Neuronal Cytomechanics: the Actin-based Motility of Growth Cones,' *Science*, 1988, 242, pp. 708–15, Fig. 14.2a © AAAS
14.25a: Alberts et al, *Molecular Biology of the Cell*, Garland Press, New York, 1989, p. 915, Fig. 16.44b
14.25b: *Scientific American*, 239(October), p. 125
14.28: J. R. Whittaker in S. F. Gilbert, *Developmental Biology*, 3rd edn, Sinauer Press, frontispiece photograph

Chapter 15
15.2: David Beardsell & Tony Slater
15.9a, b: I. Lambiris & A. Drinnan
15.9c, d: W. Verkerke & F. Bouman, University of Amsterdam
15.19a, b, 15.20: S. Y. Zee, Department of Botany, University of Hong Kong
15.21: C. O'Brien
15.22b, 15.23: E. G. Williams
15.25: L. Wolpert

Chapter 16
16.5b: Dr Jon Martin

Part 4 Regulation of the internal environment

Chapter 17
17.6b, c, 17.8, 17.9, Box 17.1b, 17.24: Andrew Drinnan
17.7: G. R. Roberts, Documentary Photographs
17.16b: A. Ashford
17.19: L. A. J. Thomson, *ECOS*, 67, 1991, p. 21
17.30a, b: D. Albrecht, National Herbarium of Victoria

Chapter 18
Box 18.1: A. Drinnan
Box 18.3: Stirling Macoboy
18.8, 18.18a, b: I. Ridge, (ed.), *Plant Physiology*, Edward Arnold
18.14a, c: J. A. Milburn, *Water Flow In Plants*, Longman, London, p. 101, Fig. 5.9
18.21: Bill Bachman, ANT Photo Library

Chapter 19
19.1a: Peter Atkinson, ANT Photo Library
19.1b: C. & S. Pollitt, ANT Photo Library
19.1c: V. Tunnicliffe
19.4: D. A. Denton, *The Hunger For Salt*, Springer-Verlag, 1982, Fig. 3.18
Box 19.3a: T. Bauchop, *AEM*, 37, 1979, p. 1219

Box 19.3c, d: W. J. Foley
19.9b: Anatomy Department, The University of Melbourne
19.9c: Megan Klemin
19.14a, b: R. Ortner
19.20b: Cyril Webster, ANT Photo Library
19.21: K. F. Liem, *Journal of Morphology*, 158, Fig. 7, © 1978, reprinted by permission of John Wiley & Sons, Inc.
19.22: Australian Picture Library
19.25, 19.26, 19.27: G. Samson
19.28a: Stephen Burnell
19.28b: B. Willis

Chapter 20
20.5a: Kelvin Aitken, ANT Photo Library
20.10: Peter Greenaway
20.14b: C. B. Daniels
20.15b: Dr Jill Hallam
20.18: H. R. Duncker
Box 20.3ci, ii: CSIRO Entomology, *The Insects of Australia*, 2nd edn, vol. 1, Melbourne University Press, Melbourne, 1991

Chapter 21
Box 21.1c: P. J. Wynne, in H. Rahn, A. Ar & C. V. Paganelli, *How Bird Eggs Breathe*, © 1979 by Scientific American, Inc., all rights reserved
Box 21.2c: Australian Picture Library

Chapter 22
22.21: P. Withers

Chapter 23
23.9a, b: Dr P. J. Holt, WAICH
23.25: L. E. Hood, I. L. Weissman & W. B. Wood, *Immunology*, 2nd edn, The Benjamin/Cummings Publishing Company, Menlo Park, CA, 1984, p. 354, Fig. 4.43

Part 5 Responsiveness and co-ordination

Chapter 24
24.6: P. F. Wareing & I. D. J. Phillips (eds), *The Control of Growth and Differentiation in Plants*, 3rd edn, Pergamon Press, Sydney, 1981, p. 144
24.7: Courtesy of Prof. A. Sievers & Dr K. Schroter, 'Versuch einer Kausalanalyse der geotropischen Reaktionskette im Chara-Rhizoid', *Planta* 96, 339–53 (1971)
24.9: Brian & Hemming, *Plant Physiology*, 8: 669–81, 1955
24.17c: Dr T. L. Setter, International Rice Research Institute, The Philippines
24.23: Stirling Macoboy

Chapter 25
25.16a, b: Ken Griffiths, ANT Photo Library
25.17: C. Donnell Turner & J. T. Bagnara, *General Endocrinology*, 6th edn, © 1976 by Saunders College Publishing, reproduced by permission of the publisher
25.19a: Drs S. K. Loo & B. Freeman
25.19b: Professor A. W. J. Lykke, The University of New South Wales
25.24b: Dr B. Tuch, Prince of Wales Hospital

Chapter 26
26.3b: D. Parer & E. Parer-Cook, Auscape
Box 26.4a: Darran Leal, Auscape
Box 26.4b: J. C. Wombey, Auscape
Box 26.4c: R. Hill/Botanic Gardens of Adelaide
Box 26.6a: Bruce G. Thomson, ANT Photo Library
Box 26.6b: Cyril Webster, ANT Photo Library

Chapter 27
27.1a: Mike Tinsley, Auscape
27.3: Alby Ziebell, Auscape
27.4: Ron & Valerie Taylor, ANT Photo Library
27.5a: CSIRO Division of Fisheries
27.6a, bi, ii, iii: Alexander R. McNeill, *Dynamics of Dinosaurs and Other Extinct Giants*, 1989, © Columbia University Press, New York, Fig. 9.5b, p. 130, reprinted by permission of the publisher
27.11a: Jean-Paul Ferrero, Auscape
27.14b: J. O'Neil, ANT Photo Library

Chapter 28
28.1, 28.5a, 28.5c, 28.18: Kathie Atkinson
28.2: Sybille Kalas & Klaus Kalas
28.4: R. J. Allingham, ANT Photo Library
28.5b: Pictor/Austral
28.6, 28.7, 28.8: Densey Cline/Mantis Wildlife Films
28.9: N. N. Birks, Auscape
28.11: Jean-Paul Ferrero, Auscape
28.12: D. Parer & E. Parer-Cook, Auscape
28.13: Glen Threlfo, Auscape

28.15: Rudie Kuiter, ANT Photo Library
28.16: Robert Hernandez/The Photo Library—Sydney
28.17: K. Sheridan, Australian Antarctic Division
28.19, 28.20a–f: J. R. Krebs & N. B. Davies, *Behavioural Ecology*, 3rd edn, Blackwell Scientific, Melbourne, 1978, Fig. 11.1, p. 339
28.21: Mike Gillam, Auscape
Box 28.3a: The Photo Library—Sydney
Box 28.3b: Jan Taylor, ANT Photo Library
Box 28.3c: Silvestris/ANT Photo Library

Chapter 29
29.1: Kerrie Ruth, Auscape
29.2: Kathie Atkinson
29.3: J. Burt, ANT Photo Library
Box 29.1: Australian Picture Library/ZEFA

Part 6 Evolution and biodiversity

Chapter 30
Box 30.1b: Australian National Insect Collection, CSIRO Division of Entomology
30.1: E. Haeckel, *The Evolution of Man*, vol. 1, D. Appleton & Co., New York, 1896
Box 30.2: American Philosophical Society
30.3: D. Stevenson, New York Botanic Garden
Box 30.3a: Kelvin Aitken/ANT Photo Library
Box 30.3b: Tony Howard/ANT Photo Library
Box 30.4ai: I. R. McCann/ANT Photo Library
30.10a, Box 30.4aii, Box 30.4b: Tim Low
30.13: K. Thiele

Chapter 31
Box 31.1: W. Harland et al, *Geologic Timescale*, Cambridge University Press, Cambridge, 1989, p. 164
31.5, 31.7, 31.14: A. Drinnan
31.6, Box 31.4b, 31.16, 31.19: Reproduced with permission of P. V. Rich, G. F. van Tets & F. Knight, *Kadimakara, Extinct Vertebrates of Australia*, Pioneer Design Studio, Melbourne, 1985
31.8, 31.11, 31.12, 31.15: Courtesy Field Museum of Natural History, Chicago
Box 31.2b: Linda S. Moore
31.9: Artist: Peter Sawyer, Smithsonian Institute
31.10: N. Pledge, South Australian Museum
Box 31.3b: D. E. G. Briggs, Department of Geology, University of Bristol
31.13: Kristine Brimmell, WA Museum
31.18: The Natural History Museum, London

Chapter 32
Box 32.1b: Martin Harvey/ANT Photo Library
Box 32.1c: D. Whitford/ANT Photo Library
32.1a: Mark A. Clements, Australian National Botanic Gardens
32.1b: Peter Robertson
Box 32.2a, 32.7: Chris Moran
Box 32.2b: Q. Lang
32.10a: B. G. Thomson/ANT Photo Library
32.11: R. H. Groves, CSIRO Division of Plant Industry
32.13: Adapted from *ANH*, 23 (8), 1991, p. 647 by permission of I. Faulkner
32.14: N. H. P. A./ANT Photo Library

Chapter 33
33.2: Peter Stewart
33.4a: Dr Tony Brain/The Photo Library—Sydney
Box 33.1b, c: Microbiology Department, The University of Melbourne
33.6a,b: G. R. Roberts, Documentary Photographs

Chapter 34
34.1a, b, Box 34.3, 34.3a–e: Adrian Gibbs
34.2: David Guest

Chapter 35
35.2, 35.32: David Patterson
35.4: Karen Kabnick, Harvard School of Public Health, USA
35.5, 35.6, 35.12, 35.14, 35.23, 35.31: Geoff McFadden
35.7: Tim Entwistle, State Herbarium, Melbourne
35.8: Tim Spurck
Box 35.2: Michael Borowitzka
35.9: Jeremy Pickett-Heaps
35.10, 35.11b, c, 35.13, 35.19, Box 35.4, 35.20, 35.28, 35.30: Gustaaf Hallegraeff
35.11a: Peter Beech, Yale University, USA
Box 35.2: Paddy Ryan, ANT Photo Library
Box 35.3: Gretna Weste
35.15, 35.20a, 35.26c, d, 35.29: Biophoto Associates, Bristol
35.16: Rosey Van Driel
35.17: Jeanetta Brewer
35.18, Box 35.5: D. Reifsynder
35.19a: David Hill

35.25: Ian Staff
35.27b: Phillip Keane
35.27: Harvey Marchant, Antarctic Division, Hobart

Chapter 36
36.2, 36.12, 36.13, 36.16: G. Bompeix, Laboratoire Biochimie et Pathologie Vegetale, University of Pierre et Marie Curie, Paris
36.3: M. Bourne
36.4, 36.6: H. Swart
36.5: C. O'Brien
36.7, 36.8, 36.11: R. Guggenheim, REM-Labor, Universitaet Basel-Geowissenschaffen, Basel
36.9, 36.10: D. Guest
36.18: D. Orlovich
36.23: Norman Nicholls
36.24: B. Fuhrer
36.25: R. StC. Barnetson, The University of Sydney
Box 36.1: The University of Melbourne Archives
Box 36.3: P. Keane

Chapter 37
37.3: J. Pickett-Heaps
37.4, 37.8, 37.10, 37.11, 37.13, 37.14, 37.15, 37.18, 37.19a, 37.21, 37.25, 37.26, 37.27, 37.30, 37.42, 37.44: A. Drinnan
37.6: J. Carpenter
37.24, 37.28: M. Regan
Box 37.3b: G. Sainty
37.34a: D. Stevenson
37.34b: T. Chambers
37.35a: M. Duretto
37.35b: D. Ashton
37.37: K. Thiele
37.40b: Reg Morrison, Auscape
37.40c: Jean-Paul Ferrero/Auscape
37.43a, b: Babs & Bert Wells

Chapter 38
38.2: N. Pledge, South Australian Museum
38.4: L. Newman & A. Flowers, Auscape
38.9, 38.12, 38.15, 38.16b, 38.21, 38.23, 38.27: Barrie Jamieson
38.16a: Becca Saunders, Auscape
38.17, 38.18, 38.28a: Kathie Atkinson
38.31: R. StC. Barnetson, The University of Sydney

Chapter 39
39.1a: Kathie Atkinson, Auscape
39.1b, 39.22c: C. A. Henley, Auscape
39.3, Box 39.1b, 39.26, 39.27a, 39.28, 39.29b, c, 39.31b, 39.32: Barrie Jamieson
39.5b: Kathie Atkinson
39.9a: Martin Dohrn/The Photo Library—Sydney
39.12, 39.15a, 39.15b, Box 39.2b, 39.23, 39.29a, b, c, 39.34b, c: Kathie Atkinson
39.13: Kev Deacon/Ocean Earth Images
39.14b: R. Raven, Queensland Museum
39.18, 39.22e, 39.22g: Otto Ridge, ANT Photo Library
Box 39.2a: R. W. Taylor, Australian National Insect Collection
39.22a: Jean-Paul Ferrero, Auscape
39.22b: Klaus Uhlenhut, ANT Photo Library
39.22d: John McCammon, Auscape
39.29a, 39.35: L. Newman & A. Flowers, Auscape
39.30, 39.31a: R. Willan
39.34a: Jan Aldenhoven, D. Parer & E. Parer-Cook, Auscape

Chapter 40
40.1a, 40.26: D. Parer & E. Parer-Cook, Auscape
40.1b, 40.6: Kathie Atkinson, Auscape
40.1c, 40.19: Kathie Atkinson
40.5: C. Moran
40.8c: Michael Aw/The Photo Library—Sydney
40.11: Babs & Bert Wells
40.13a: Kev Deacon, Auscape
40.13b: Francois Gohier, Auscape
40.14b: Gunther Schmida
40.15: Copyright Schauer/Fricke
40.16a, 40.21b, 40.26, 40.27: Jean-Paul Ferrero, Auscape
40.17b: Zig Leszczynski/The Photo Library—Sydney
40.21a: D. Rosen
40.21c: G. R. Roberts, Documentary Photographs
40.22a: Darran Leal, Auscape
40.24a: D. Cantrill
40.24b, 40.25: Australian Tourist Commission
40.24d: Roger Brown, Auscape
40.24b, 40.25, 40.26: Australian Tourist Commission
40.29a: I. R. Van Nostrand/The Photo Library—Sydney
40.29b: Brett Allatt/The Photo Library—Sydney
40.29c: David Messent/The Photo Library—Sydney

40.32: neg. no. 338315. Courtesy Department of Library Services, American Museum of Natural History

40.33: George Holton/The Photo Library—Sydney

40.35a: neg. no. 316188, photo Julius Kirschner, courtesy Department of Library Services, The American Museum of Natural History

40.35b: neg. no. 315446, courtesy Department of Library Services, The American Museum of Natural History

40.35c: Natural History Museum, London

Chapter 41

41.7: M. Gill, CSIRO

41.13: L. Newman & A. Flowers, Auscape

Box 41.2b, 41.30: D. Parer & E. Parer-Cook, Auscape

41.14: Stirling Macoboy

41.16, 41.19: K. Thiele

41.17a: R. Gleadow, The Botanic Gardens of Adelaide

Box 41.3a, 41.25b: A. Drinnan

41.23: J. Bruhl

41.28: G. R. Roberts, Documentary Photographs

41.29: H. Ehmann/NPIAW

41.31a: Vincent Serventy/The Photo Library—Sydney

Box 41.5b: E. Nielsen, CSIRO Division of Entomology

Box 41.6a: © Peter Schouten

41.32: Michael Archer, University of New South Wales

41.33: John Fields, Australian Museum

41.34: Mervyn Griffiths

41.35, 41.37b: Jean-Paul Ferrero, Auscape

41.36, 41.38: D. Paul

41.37a, Box 41.8a: Hans & Judy Beste, Auscape

41.37c: John Cancalosi, Auscape

41.39: L. Lumsdon

41.40: Esther Beaton, Auscape

Part 7 Ecology

Chapter 42

42.1a: Babs & Bert Wells, NPIAW

Box 42.1a: C. A. Henley, Auscape

Box 42.1b: Ross Isaacs, Auscape

42.6a: B. R. Burns & J. Ogden, *Australian Journal of Ecology*, 10, 1985, Fig. 1

42.6b: G. R. Roberts, Documentary Photographs

42.7a: C. A. Henley, Auscape

42.7b: G. R. Dickman, *Australian Journal of Ecology*, 11, 1986, Fig. 2

42.8b: J. R. Busby, *Australian Journal of Ecology*, 11, 1986, Fig. 1

42.9b, Box 42.3: Australian Tourist Commission

42.10: Paddy Ryan, ANT Photo Library

42.12: Sylvian Cordier, JACANA-Auscape

42.13: Dave Watts, ANT Photo Library

42.15: Jean-Paul Ferrero, Auscape

Chapter 43

43.2: Dr Peter Harrison

43.3: Michele Hall, Oxford Scientific Films

Box 43.1a, 43.5: D. Albrecht, National Herbarium of Victoria

Box 43.1b: A. Beattie & C. Turnbull

43.4: Australian National Insect Collection, CSIRO Division of Entomology

43.6a, 43.8: M. Begon & M. Mortimer, *Population Ecology*, Blackwell Scientific Publications, Oxford, © 1981, pp. 90, 103

43.7a: A. P. Smith, ANT Photo Library

43.7b, 43.14: G. R. Roberts, Documentary Photographs

43.9: Glen Threlfo, Auscape

43.10: K. Thiele

43.12: D. Rentz, CSIRO Division of Entomology

43.13: C. Myers, American Museum of Natural History

43.15: Fig. 13.9 from *Ecology: The Experimental Analysis of Distribution and Abundance*, 3rd edn, by Charles J. Krebs, © 1985 by Harper & Row Publishers, Inc., reprinted by permission of HarperCollins Publishers, Inc.

Box 43.2: M. Keogh

43.22: C. Weston

43.23: M. Burgman

Chapter 44

Box 44.2: Bert Jenkins

Box 44.3: G. Chapman, CSIRO Division of Wildlife and Ecology

44.11a: Kathie Atkinson, Oxford Scientific Films

44.11b: Jean-Paul Ferrero, Auscape

Chapter 45

45.1: W. Bradley, Sydney Cove, Port Jackson, 1788, watercolour, Mitchell Library, State Library of New South Wales

45.2: Karen Wynn-Moylan, NPIAW

45.3a: A. D. & M. C. Traunson, NPIAW

45.3b: International Council for Bird Preservation, *Putting Biodiversity on the Map: Priority Areas for Global Conservation*, BirdLife International, Cambridge, p. 35

45.4a, b: Byron Lamont, Western Australian Herbarium

45.5: A. Gillison, CSIRO Division of Wildlife and Ecology

Box 45.1a: State Forests of NSW

Box 45.2a: D. Albrecht, National Herbarium of Victoria

Box 45.2b: Pavel German, ANT Photo Library

Box 45.3: D. Curl, CSIRO Division of Plant Industries

45.6: M. Vanderzee

45.7a: Jean-Michel Labat, Auscape

45.8: Paddy Ryan, ANT Photo Library

45.9b: Ted Hutchison, ANT Photo Library

45.9c: Jean-Michel Labat, JACANA/Auscape

45.10: Ralph & Daphne Keller, ANT Photo Library

45.12a: Bill Bachman, ANT Photo Library

45.15b: Dr Jan Anderson, CSIRO Division of Plant Industry

45.17: Australian Bureau of Statistics, *Australia's environment issues and facts* 1992, p. 100, ABS cat. no. 4140.0, Commonwealth of Australia copyright, reproduced by permission

45.18, 45.22: Australian Tourist Commission

45.19: Kathie Atkinson, Oxford Scientific Films

45.20: Belinda Wright, Oxford Scientific Films

45.21: CRA Corporate Services

INDEX

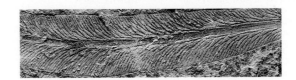